PHYSICAL GEOLOGY
& the Environment

SECOND CANADIAN EDITION

Charles C. Plummer
California State University at Sacramento

Diane H. Carlson
California State University at Sacramento

David McGeary
Emeritus of California State University of Sacramento

Carolyn Eyles
McMaster University

Nick Eyles
University of Toronto

Toronto Montréal Boston Burr Ridge, IL Dubuque, IA Madison, WI New York San Francisco
St. Louis Bangkok Bogotá Caracas Kuala Lumpur Lisbon London Madrid Mexico City
Milan New Delhi Santiago Seoul Singapore Sydney Taipei

Physical Geology and the Environment
Second Canadian Edition

ISBN-13: 978-0-07-095633-9
ISBN-10: 0-07-095633-2

1 2 3 4 5 6 7 8 9 10 QPV 0 9 8 7

Printed and bound in the United States

EDITORIAL DIRECTOR: Joanna Cotton
PUBLISHER: Lynn Fisher
SENIOR SPONSORING EDITOR: Leanna MacLean
MARKETING MANAGER: Claire Morrison
DEVELOPMENTAL EDITOR: Jennifer Bastarache
ASSOCIATE DEVELOPMENTAL EDITOR: Alison Derry
PHOTO RESEARCHER: Christina Beamish
SUPERVISING EDITOR: Graeme Powell
COPY EDITOR: Kelli Howey
SENIOR PRODUCTION COORDINATOR: Paula Brown
COVER DESIGN: Valid
COVER IMAGE: The Image Bank/Mary Liz Austin Canada, Banff National Park, Mount Rundle and Sulphur Mountain, dusk Vermillion Lakes
INTERIOR DESIGN: ArtPlus Limited
PAGE LAYOUT: ArtPlus Limited
PRINTER: Quebecor Printing Versailles

Library and Archives Canada Cataloguing in Publication

Physical geology and the environment / Charles C. Plummer ... [et al.]. — 2nd ed.

Includes index.
ISBN-13: 978-0-07-095633-9
ISBN-10: 0-07-095633-2

1. Physical geology—Textbooks. I. Plummer, Charles C., 1937–
QE28.2.P49 2007 550 C2007-900158-0

CONTENTS

1 Introduction to Physical Geology and the Environment *1*

2 Plate Tectonics *21*

3 Earthquakes *65*

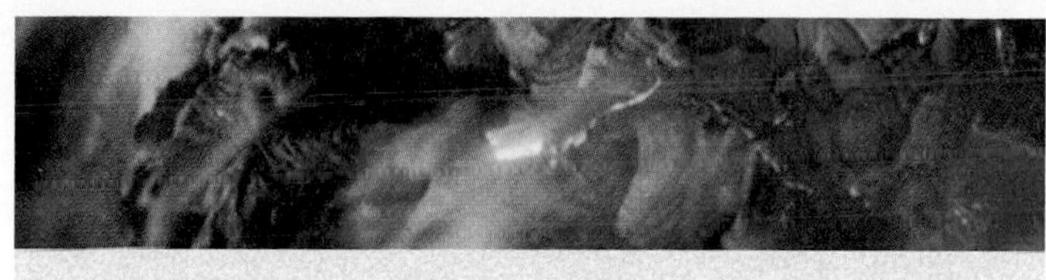

7 Volcanism and Extrusive Rocks *174*

8 Weathering and Soil *202*

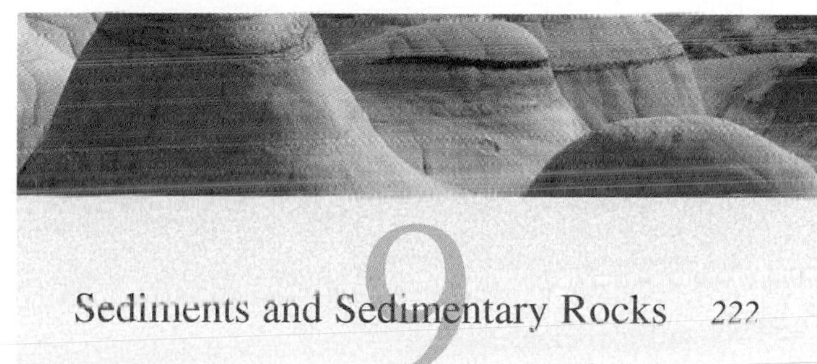

9 Sediments and Sedimentary Rocks *222*

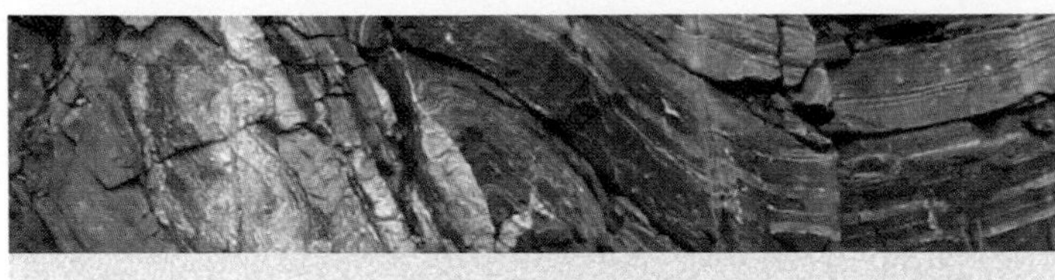

10 Metamorphism, Metamorphic Rocks, and Hydrothermal Rocks 252

11 Geologic Structures 276

12 Geologic Resources 298

13 Mass Wasting 330

14 Streams and Floods 355

15 Groundwater 390

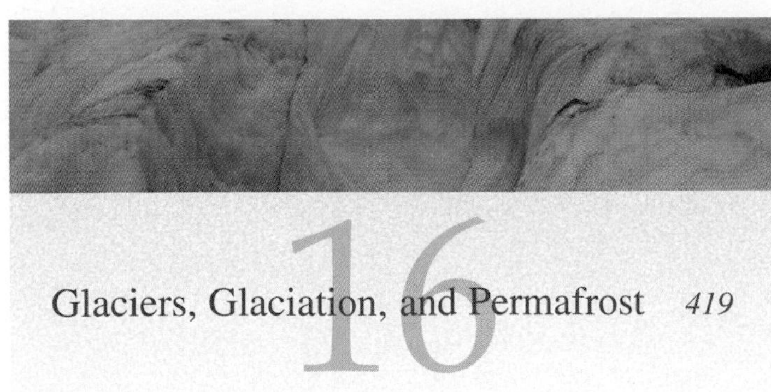

16 Glaciers, Glaciation, and Permafrost 419

17 Deserts and Wind Action 455

18 Coasts and Oceans 479

PREFACE

PREFACE TO THE SECOND CANADIAN EDITION

The Canadian edition of ***Physical Geology and the Environment*** was created in response to a common concern that most introductory geoscience textbooks lacked sufficient Canadian content to make them relevant or attractive to professors and students working in Canada. The second Canadian edition includes more Canadian content, more environmental applications, and updated examples of "globally significant" events triggered by geological phenomena, such as the 2004 Indonesian earthquakes and tsunamis and the flooding caused by Hurricane Katrina in 2005. In addition, this edition introduces students to the range of skills required by the modern professional geoscientist, including analysis of various forms of digital data and visualization of these data in three dimensions (a developing field known as geomatics). A new chapter (Chapter 21: The Earth in Space) has also been added to the Online Learning Centre.

Much of the content of any introductory geoscience textbook is quite naturally concerned with descriptions of geological processes and products. It is possible, for example, to generalize about the workings of volcanoes, rivers, or glaciers worldwide. Nonetheless, the discipline of geoscience holds vital importance in the economic, social, and political realms of Canada, and it is for this reason that detailed references to Canadian examples have been updated for this new edition. Geology and geological processes have created the country's diverse resources and landscapes and have helped shape the varied human response to them. The huge expanse of the Canadian Shield, the barrier of the Rocky Mountains, the Great Lakes, the Arctic barrens, and the cold coasts have all created unique environments. In turn, the discovery of natural resources and the ensuing gold rushes, oil booms, and access to deep prairie soils have all played their part in populating and developing the country. Currently, we are witnessing a modern-day diamond rush in the Northwest Territories that is stimulating geological exploration elsewhere. The search for oil and gas continues in the Arctic and Atlantic provinces. As Canadians, we are economically dependent on our natural resources. Our future well-being depends on how well we can identify, protect, and efficiently extract additional new resources.

In a society that is increasingly concerned with environmental issues and geological hazards, understanding and managing these issues requires an appropriate knowledge of our own geological history. Because we live on the surface of a dynamic planet where no place is fixed but always moving, Canadians must learn to live with hazards of landslides and earthquakes; as more and more Canadians live in cities, the risk from these events increases. More than 80 percent of Canadians now live in urban areas. As such, we face the challenge of keeping our lakes and rivers clean, safely disposing of wastes (ranging from household garbage to high-level nuclear wastes), and minimizing the impacts of a wide range of contaminants released into the air, water, and soil by industrial, urban, and agricultural activities. Much of our water moves in the subsurface as groundwater. Mapping and tracking its movement along with any contaminants that are present requires a geoscientist's skills and an understanding of both the bedrock and glacial geologic history of the area. This textbook is designed as a first step in the education of a new generation of geoscientists.

APPROACH

The organization of the content of the book reflects a systematic journey from the large-scale tectonic processes that shape the body of our planet to those that affect and change its surface characteristics. The initial chapters (Chapters 1 through 4) describe how the Earth formed and the operation and significance of fundamental processes such as plate tectonics, earthquakes, and the Earth's internal heat engine. Chapters 5 through 12 examine the materials that make up the Earth, including minerals, rocks (igneous, metamorphic, and sedimentary), soils, and the processes that create and deform these materials. Chapters 13 through 18 look at the many dynamic processes that shape the surface of the Earth, including the actions of gravity, ice, water, and wind. The final section of the book (Chapters 19 and 20) examines the concept of geologic time and provides a unique overview of the geological history of Canada. Where possible, we have included examples of environmental geological applications to demonstrate the importance of geoscience to modern society. We have also emphasized the many and varied professional careers open to students with a background in geoscience. The addition of Chapter 21 (Earth in Space) to the Online Learning Centre provides students access to information about other planets and objects in our solar system and facilitates comparisons with characteristics and features of the Earth.

Perhaps one of the most important features of the text that has been retained in the second Canadian edition is the chapter on the geological history of Canada (Chapter 20). This chapter is significant for many reasons, not least the fact that Canada still depends on natural resources for much of its economic well-being. The Canadian Shield, formed between 3,500 and 1,000 million years ago, is the largest and oldest shield in the world, and hosts a wealth of geological resources such as gold, zinc, and nickel. Much that can be learned about the geology of Canada can be applied globally. This is well illustrated by the Lithoprobe project, which aims to understand the origin and mineral wealth of the shield and adjacent mountain belts (Chapter 4). Canadian expertise in finding minerals deep in the shield is being employed worldwide.

Carolyn Eyles
Nick Eyles

KEY FEATURES

- Inquiry-based learning — A series of questions are provided at the beginning of each chapter. These questions are appropriate to the main topics covered in the chapter, and are repeated at the beginning of relevant sections, stimulating students to create their own questions.

CHAPTER 20

Geological History of Canada

What are the main geological "building blocks" of North America?
How did the North American continent evolve?
How were the Atlantic provinces added to Canada?
Where did British Columbia come from?
Where can we find dinosaurs in Canada?
How did the Canadian Rockies form?
When did the ice sheets develop?
Why is it important to understand the geological history of Canada?

- New "Geomatics" boxes show one of the fastest growing areas of the geosciences in action:

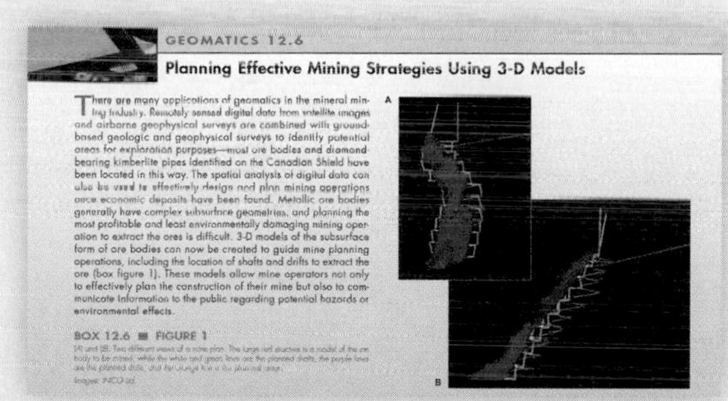

GEOMATICS 12.6

Planning Effective Mining Strategies Using 3-D Models

BOX 12.6 ■ FIGURE 1

- "Environmental Geology" boxes show how material pertaining to physical geology relates to environmental concerns:

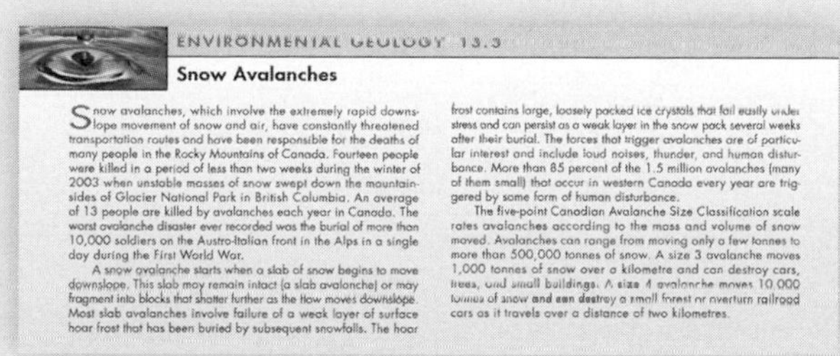

ENVIRONMENTAL GEOLOGY 13.3

Snow Avalanches

- "Astrogeology" boxes relate topics discussed in the text to what has been discovered on other planets in our solar system:

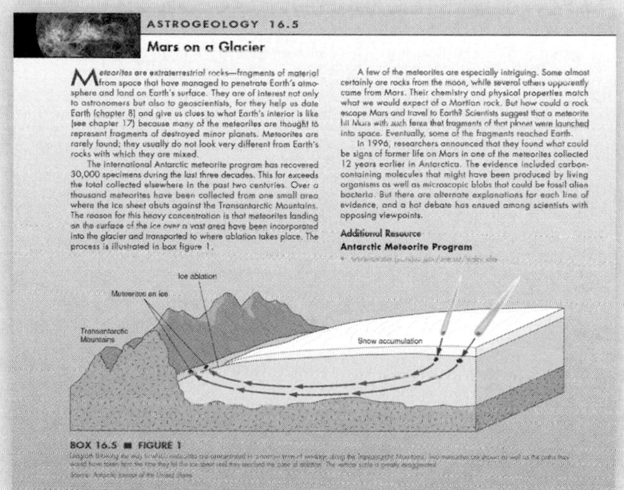

ASTROGEOLOGY 16.5

Mars on a Glacier

BOX 16.5 ■ FIGURE 1

- "In Greater Depth" boxes focus on interesting topics that are not usually an essential part of an introductory geology course:

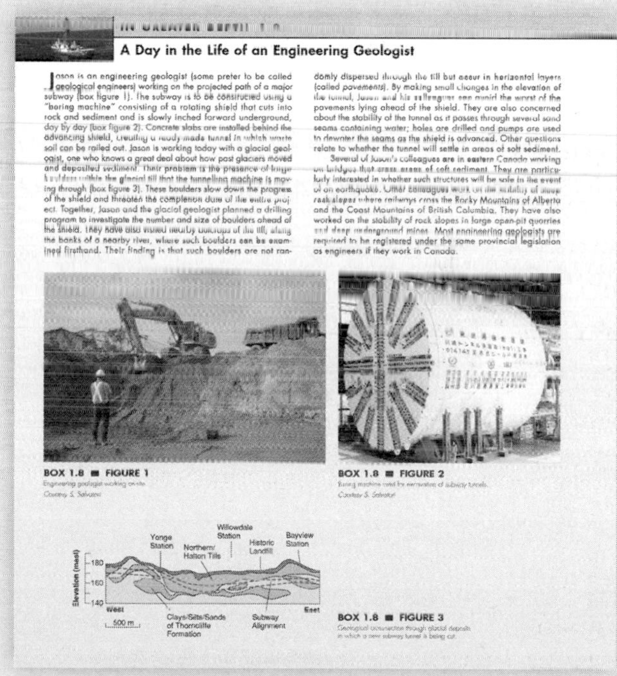

IN GREATER DEPTH 1.8

A Day in the Life of an Engineering Geologist

BOX 1.8 ■ FIGURE 1

BOX 1.8 ■ FIGURE 2

BOX 1.8 ■ FIGURE 3

- Study aids are found at the end of each chapter and include the following:

  Summaries that bring together and summarize the major concepts of the chapter.

  Terms to Remember that include all the boldfaced terms covered in the chapter so that students can verify their understanding of the concepts behind each term.

  Testing Your Knowledge Quizzes that stimulate a student's critical thinking by asking questions with answers not found in the textbook.

  Exploring Web Resources that describe some of the best sites on the Web relating to the chapter.

NEW TO THE SECOND CANADIAN EDITION

Improved Art Program

Geology is a visually oriented science, and one of the best ways a student can learn is by studying illustrations and photographs. This new edition includes an updated art program that not only will aid understanding, but also will engage a student's interest.

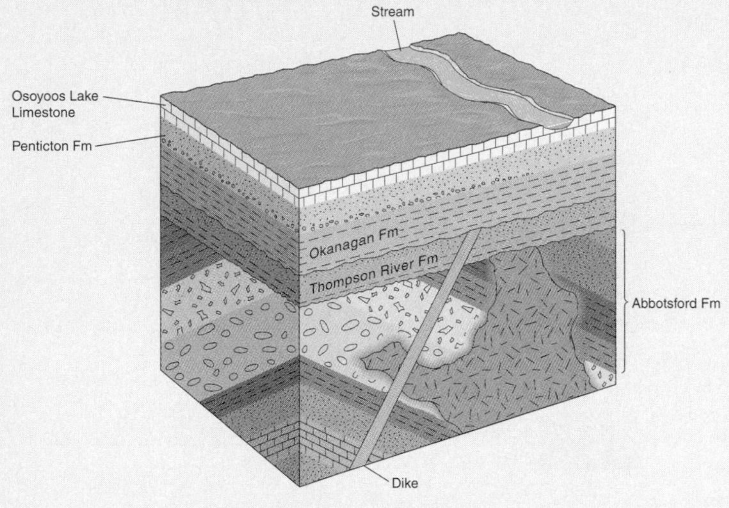

New and updated photos have replaced many images from the previous edition, with an emphasis on a Canadian perspective and enhanced visuals to illustrate the concepts and topics that are covered. Chapter openers now include an interesting and eye-catching photo that is unique to each chapter:

Supplementary Chapter: "The Earth in Space" Available at www.mcgrawhill.ca/olc/plummer

A focus on astronomy and the solar system is being featured not only as an online supplementary chapter, but also with some coverage within the text on the planets, moon, and terrestrial planets.

Animations

McGraw-Hill is proud to bring you an assortment of animations like no other. These include new and retained animations from previous editions, and are located on the Online Learning Centre. A special animations icon has been placed beside every figure in the text that has a corresponding animation. These animations offer students a fresh, dynamic method of learning about geology concepts such as the dynamics of groundwater movement, isostacy, plate tectonics, and more.

SUPPLEMENTS FOR STUDENTS

Student Online Learning Centre

The robust Student Edition of the Online Learning Centre at www.mcgrawhill.ca/olc/plummer features quizzes for study and review, interactive exercises and animation, as well as additional boxed readings, searchable glossary, suggestions for further reading, and much more!

SUPPLEMENTS FOR INSTRUCTORS

Digitized Videos

This exciting DVD offers short (3–5 minute) videos on topics ranging from conservation to volcanoes. Begin your class with a quick peek at science in action.

Visual Resource Library

Available on the Instructor's CD-ROM, this resource contains illustrations, photographs, and tables from the text, as well as animations, lecture outlines, and additional photos.

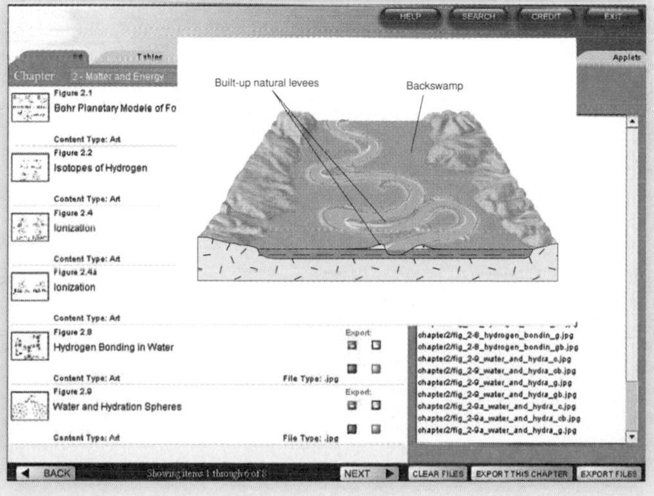

Instructor's Online Learning Centre

The OLC at www.mcgrawhill.ca/olc/plummer includes a password-protected Web site for instructors. The site offers downloadable supplements and access to PageOut, the McGraw-Hill Ryerson Web site development centre.

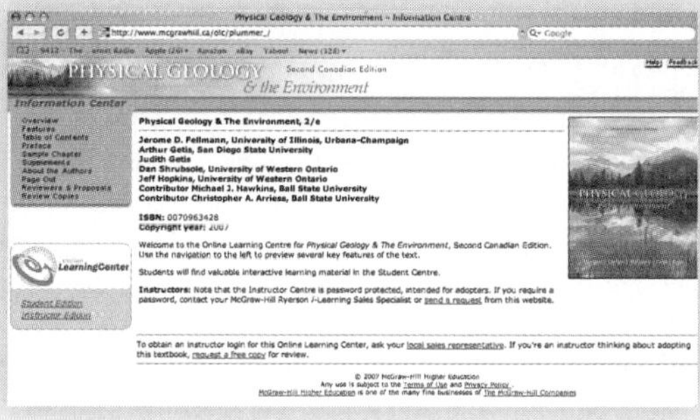

Instructor's Manual

The Instructor's Manual contains a chapter overview, list of changes per chapter, chapter learning objectives, further information on boxed features, short and long discussion/essay questions, and details on additional readings and resources outside the textbook.

Test Bank

Available on the Instructor's CD-ROM, this resource contains a bank of approximately 60 multiple-choice and true/false questions for each chapter.

ACKNOWLEDGEMENTS

We would like to thank the many undergraduate and graduate students at McMaster University who have provided feedback on the text and helped us research, design, and prepare material included in this second Canadian edition of *Physical Geology and the Environment*. Special thanks to John Maclachlan, Mark Radomski, Aislyn Trendell, Sean Fletcher, Sean Dickie, Marianne Stoesser, Luke Hunt, Katie Shea, Katherine Card, Viktor Terlaky, Kate Dekeyser, and Kelsey MacCormack.

We would also like to gratefully acknowledge our many colleagues who supplied photographs, illustrations, advice, and input.

We are also very grateful to the following reviewers of the second Canadian edition for their careful evaluation and useful suggestions:

Kevin Ansdell, *University of Saskatchewan*
Ingrid Bajewsky, *Nipissing University*
Sandra M. Barr, *Acadia University*
Paul Batson, *Nova Scotia Community College*
Venetia Bodycomb, *Concordia University*
Bruce E. Broster, *University of New Brunswick*
Philippe Erdmer, *University of Alberta*
Wayne Haglund, *Mount Royal College*
Tark Hamilton, *Camoson College*
K. Maggie McColl, *Malaspina University-College*
Mungandi Nasitwitwi, *Douglas College*
Terrence Neufeldt, *Trinity Western University*
Tim Patterson, *Carleton University*
Leslie Reid, *University of Calgary*
Mark Smith, *Langara College*
Allan Turnock, *University of Manitoba*

ABOUT THE AUTHORS

Charles Plummer

Professor Charles "Carlos" Plummer grew up in the shadows of volcanoes in Mexico City. There, he developed a love for mountains and mountaineering that eventually led him into geology. He received his B.A. from Dartmouth College. After graduation, he served in the U.S. Army as an artillery officer. He resumed his geological education at the University of Washington, where he received his M.S. and Ph.D. degrees. His geologic work has been in mountainous and polar regions, notably Antarctica (where a glacier is named in his honour). He taught at Olympic Community College in Washington before joining the faculty at California State University, Sacramento. At CSUS he taught optical mineralogy, metamorphic petrology, and field courses before his semi-retirement. He continues to teach introductory courses. He flies airplanes, skis, and recently became a certified open-water SCUBA diver.

Diane Carlson

Professor Diane Carlson grew up on the glaciated Precambrian shield of northern Wisconsin and received an A.A. degree at Nicolet College in Rhinelander and her B.S. in geology at the University of Wisconsin at Eau Claire. She continued her studies at the University of Minnesota, Duluth, where she studied the structural complexities of high-grade metamorphic rocks along the margin of the Idaho batholith for her master's thesis. The lure of the West and an opportunity to work with the U.S. Geological Survey to map the Colville batholith in northeastern Washington led her to Washington State University for her Ph.D. Dr. Carlson accepted a position at California State University, Sacramento after her Ph.D. and teaches physical geology, structural geology, environmental geology, and field geology. Professor Carlson is a recipient of the Outstanding Teacher Award from the CSUS School of Arts and Sciences. She is also actively engaged in researching the structural and tectonic evolution of part of the Foothill Fault System in the northern Sierra Nevada in California.

David McGeary

Dave McGeary died in December 2002. He was born in 1940 and grew up in the town of State College, Pennsylvania. He received his B.A., majoring in geology, from Williams College in 1962. He earned an M.S. degree from University of Illinois and a PhD in marine geology at Scripps Institution in La Jolla, California. While at Scripps, he taught SCUBA diving. He began his college teaching career at Sacramento State College (later to become California State University) in 1969. Dave and Elly, his wife, had two sons born during this time. Dave was known as a demanding, but brilliant, teacher. He developed and taught a broad range of courses, most of them outside his specialty of marine geology. He loved teaching in the field. His weeklong field trips to classic geology locales (e.g. Yellowstone, Grand Canyon) were legendary and he organized and taught field method courses in the Mojave Desert. He retired from CSUS in 1992 and from coauthoring this book in 1995. After his retirement he indulged in his love of the theatre. He played leading roles in various community productions and travelled with Elly to see performances in New York and London.

Carolyn Eyles

Professor Carolyn Eyles holds a B.Sc. in environmental sciences from the University of East Anglia, a postgraduate Certificate of Education from the University of Newcastle upon Tyne in Great Britain, and an M.Sc. and Ph.D. in geology from the University of Toronto. She is a professor in the School of Geography and Geology at McMaster University in Hamilton, Ontario. Her research interests lie in the fields of glacial sedimentology and environmental geology, and she has worked extensively in Alaska, Australia, Norway, Great Britain, and Canada. She teaches courses in Earth and the environment, Earth history, sedimentology, and glacial sedimentology, and has won several teaching awards including an OCUFA Teaching Award and the McMaster University President's Award for Excellence in Instruction.

Nick Eyles

Professor Nick Eyles holds a Ph.D. and D.Sc. and is a professor of geology at the University of Toronto. He has worked at the universities of Leicester, Newcastle upon Tyne, and East Anglia in Great Britain and at Memorial University in Newfoundland, and has been at Toronto since 1981. He has authored more than 150 publications in leading scientific journals on glacial geology and environmental geology in urban areas and has conducted geological fieldwork from the Arctic to the Antarctic, including work on the ocean drillship *Resolution*. He has written several books, including *Toronto Rocks*, which won an Award of Merit from the Toronto Heritage Board in 1999, and the best-selling *Ontario Rocks*, published in 2002.

CHAPTER

Introduction to Physical Geology and the Environment

What is geology?
What do geoscientists do?
What is the scientific method?
How did the Earth form?
What was the early Earth like?
What is the "Earth system"?

This chapter provides an introduction to the discipline of geology. It begins with a discussion of the early history of the discipline and why it developed, and looks at some of its early pioneers. Geoscientists are now involved in a wide range of professional occupations, and these are examined with an emphasis on areas of geology that pose new and exciting challenges, particularly those in the field of environmental geoscience. An introduction to geology is not complete without a brief overview of the Earth's beginnings, its early history, and its characteristics. The chapter ends with a discussion of the Earth as a complex system of interacting spheres and presents the rock cycle as one of the many cyclic processes operating on our planet.

Mount Robson, 3,954 metres above sea level, is the highest peak in the Canadian Rocky Mountains. *Photo © J. A. Kraulis/Masterfile*

WHAT IS GEOLOGY?

The word *geology* is derived from the Greek words *geo* and *logos* and means the "the study of the Earth."

Many past civilizations and societies were highly skilled in finding rocks and minerals, and the first known geological map was created by ancient Egyptians over 3,000 years ago. However, the scientific discipline of **geology** as we understand it today came into being only in the late eighteenth century. At that time, the Industrial Revolution in northern Europe created a growing demand for energy and minerals such as coal, limestone, iron, and water. Finding and exploiting these resources on the scale required forced new ways of investigating planet Earth by a new professional called a *geologist*. As urban centres rapidly expanded, the construction of mines, railways, tunnels, and canals brought to light details of how rocks are arranged below the ground, how they vary over distance, and how they can be matched from one place to another using distinguishing features such as fossils. One of the earliest detailed geologic maps was published in England in 1815 by William Smith; it is for good reason that he is referred to as the "Father of English Geology" (see box 1.1). In North America, geological mapping began in the mid–nineteenth century and, again, was driven by the need to locate resources for an expanding population. Sir William Logan, who in 1842 became the founding director of the Geological Survey of Canada, was the first to systematically describe the geology of Canada (see box 1.2).

Through the next century and a half, geologists have gathered and interpreted many data from the continents and ocean floors and have built a fairly detailed picture of the Earth's structures and processes; they have also gained understanding of the vast age of the planet. In the late seventeenth century it was widely believed that the Earth was only about 6,000 years old and had been made essentially as we see it today. Now, the age of the Earth has been determined to be at least 4,500 million years; in that time, the life forms and physical geography of the planet are known to have changed dramatically. Indeed, the ever-changing nature of physical environments on planet Earth (such as the positioning of the continents), together with the role of extra-terrestrial processes (such as meteorite impacts), are seen as having largely controlled the evolution of life forms.

Moving Continents

The movement of continents on the Earth's surface was suggested early in the twentieth century by the German meteorologist Alfred Wegener (box 1.3), who wrote in 1912 of *continental drift*. Wegener recognized that today's continents had previously been clustered together in a large land mass but had subsequently moved apart; he called the land mass **Pangea** (Greek for "all the lands"; figure 1.1). Wegener collected a wealth of evidence to support his ideas of continental movement but he couldn't convincingly explain how it had happened, which led many geologists to reject the entire theory. It took many years of gathering geophysical and geological data from oceans and the margins of continents for geoscientists to demonstrate that continents do, in fact, move. This ultimately allowed development of the plate tectonics theory.

IN GREATER DEPTH 1.1

William Smith (1769–1839)

William Smith (box figure 1) spent his early years in the English village of Churchill in Oxfordshire. He had an inquiring mind and a keen interest in geology, which led him to take notes on what he observed during his travels through the country.

In 1793, while conducting surveys for the excavation of a canal in Bath, Smith noted the regular succession of the rock layers he encountered. Later, as he toured other areas of England to investigate similar canals being dug, Smith compared the rocks he saw with some of the successions he had observed near Bath. Several years later, in 1799, he outlined and coloured the distribution of geological features on a map of the local area around Bath. He produced the first comprehensive geological map of England in 1815 (box figure 2).

On the basis of these two geological maps, along with a document he wrote on the rock layers and fossils found at Bath, William Smith is widely recognized as one of the founders of the science of geology—and is the subject of a popular book by Simon Winchester entitled *The Map That Changed the World*.

BOX 1.1 ■ FIGURE 1
William Smith.

BOX 1.1 ■ FIGURE 2
William Smith created the first geologic map of England and Wales, published in 1815.

IN GREATER DEPTH 1.2

Sir William Logan (1798–1875)

Born in Montreal, William Logan (box figure 1) was widely recognized as Canada's premier scientist during his time. University-educated in Scotland, Logan went on to manage a coal mining and copper smelting venture in Swansea, Wales during the late 1830s. Realizing that maps indicating coal-seam locations would benefit the coal industry, Logan developed a professional interest in geology. His geological maps of South Wales were revolutionary in that they introduced the **cross-section**, a hypothetical vertical slice through the land, and were subsequently published by the British Geologic Survey.

After becoming the first director of the Geological Survey of Canada in 1842, Logan made it his task to ascertain the country's mineral resources. He travelled on foot across Canada (at that time consisting of Ontario and Quebec), generating full scientific descriptions of rocks, soils, and minerals. By using maps, diagrams, and drawings, he created the first systematic layout of the geology of Canada. Logan discovered several ore bodies north of Lakes Huron and Superior that represent a large portion of Canada's mineral wealth (see chapter 12). He also worked in Nova Scotia and New Brunswick and is credited with the initial suggestion that coal is an organic deposit. Canada's highest mountain, Mount Logan (5,959 m) in the Yukon Territory, is named after him.

BOX 1.2 ■ FIGURE 1

Sir William Logan.

Web image of Sir William Logan from: http://gsc.nrcan.gc.ca/hist/150_e.php Reproduced with permission of the Minister of Public Works and Government Services Canada, 2005 and Courtesy of Natural Resources Canada, Geological Survey of Canada.

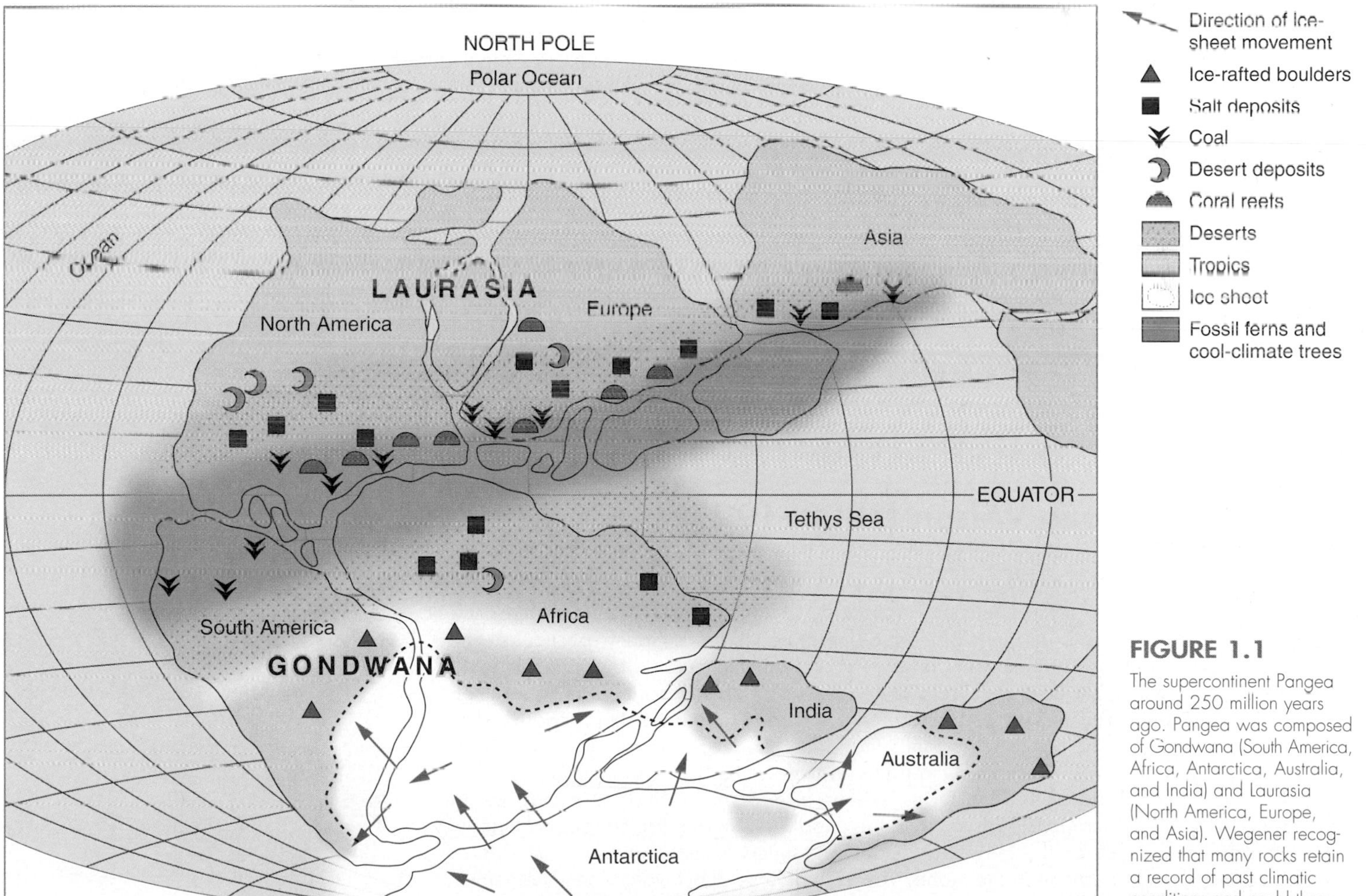

FIGURE 1.1

The supercontinent Pangea around 250 million years ago. Pangea was composed of Gondwana (South America, Africa, Antarctica, Australia, and India) and Laurasia (North America, Europe, and Asia). Wegener recognized that many rocks retain a record of past climatic conditions and could therefore be used to reconstruct climate belts on the ancient supercontinent.

IN GREATER DEPTH 1.3

Alfred Wegener (1880–1930)

Alfred Wegener (box figure 1) was a German meteorologist who first suggested the theory of continental drift in 1912. In his book *The Origin of Continents and Oceans*, he proposed that a supercontinent called Pangea, an assemblage of all the continents, existed more than 250 million years ago. He also argued that Pangea broke up into smaller continents that subsequently drifted apart to their present locations. Wegener offered a variety of evidence in support of his theory of continental drift, including the geographic "fit" of South America and Africa and similarities of rock types, structures, and fossils on now widely separated continents. Wegener's ideas were discredited by most geologists and failed to gain any support until the 1960s, when the identification of mid-ocean ridges and paleomagnetic data provided a mechanism for the movement of continents on the Earth's surface.

BOX 1.3 ■ FIGURE 1
Alfred Wegener.
Reinke-Kunze, 1994. Alfred Wegener: Polarforscher und Entdecker der Kontinentaldrift. Birkhaüser Verlag, Boston, 188 pp.

Development of the plate tectonics theory and the nature of processes and geologic events that occur at plate boundaries are discussed fully in chapter 2. However, it is important to note that it was a Canadian geophysicist, J.Tuzo Wilson (box 1.4), who in the early 1970s was responsible for bringing together several of the key elements of what we now know as plate tectonics theory.

Time and Geology

Great amounts of time are required for most geological processes. As humans, we think in units of time related to personal experience—seconds, hours, years, a human lifetime. It stretches our imagination to contemplate ancient history that involves 1,000 or 2,000 years. Geology involves vastly greater amounts of time, often referred to as *deep time.*

To be sure, some geological processes occur quickly, such as a great landslide or a volcanic eruption. These events occur when stored energy is suddenly released. Most geological processes, however, are slow but relentless, reflecting the pace at which the Earth's processes work. It is unlikely that a hill will visibly change in shape or height during your lifetime (unless through human activity). However, in a geologic time frame, the hill probably is eroding away quite rapidly. "Rapidly" to a geologist may mean that within a few million years the hill will be reduced nearly to a plain.

The rate of plate motion is relatively fast. If new magma erupts and solidifies along a mid-oceanic ridge (a giant mountain range that lies under the ocean), we can easily calculate how long it will take that igneous rock to move 1,000 kilometres away from the crest of the ridge. At the rate of 1 centimetre per year, it will take 100 million years for the currently forming part of the crust to travel the 1,000 kilometres.

IN GREATER DEPTH 1.4

J. Tuzo Wilson (1908–1993)

Born in Ottawa, Jack Tuzo Wilson (box figure 1) was the first Canadian to complete a degree in geophysics at the University of Toronto (1930). Wilson is remembered primarily for his contributions to the development of the plate tectonics theory, particularly his ideas on transform faults and hot spots. These ideas were first stimulated by his work on the Canadian Shield, which showed the shield to be composed of a mosaic of "terranes" separated by long linear features that he interpreted as ancient fault lines. He recognized on modern ocean floors similar large-scale faults that offset the crust laterally but neither created nor destroyed material. He named the faults "transform faults," and they are now recognized as a major plate boundary type. Wilson was also the first to recognize the significance of young volcanic islands stuck in the middle of oceans. He suggested that volcanic island chains, such as the Hawaiian Islands, resulted from a moving plate drifting over a stationary magma plume in the mantle. He called these active volcanoes "hot spots." Wilson's ideas were initially written off as being too radical, but are now accepted as important components of the plate tectonics theory and helped convince many people that the Earth's surface is in constant motion.

BOX 1.4 ■ FIGURE 1
J. Tuzo Wilson.
Family photo given to Nick Eyles

Although we will discuss geologic time in detail in chapter 19, table 1.1 shows some reference points to keep in mind. The Earth is estimated to be at least 4.55 billion years old (4,550,000,000 years). Fossils in rocks indicate that complex forms of animal life have existed in abundance on the Earth for about the past 545 million years. Reptiles became abundant about 230 million years ago. Dinosaurs evolved from reptiles and became extinct about 65 million years ago. Humans have been here only about the last 3 million years. The eras and periods shown in table 1.1 comprise a kind of calendar for geologists into which geologic events are placed (as explained in the chapter on geologic time).

TABLE 1.1 Some Important Events in the Development of Life on Earth

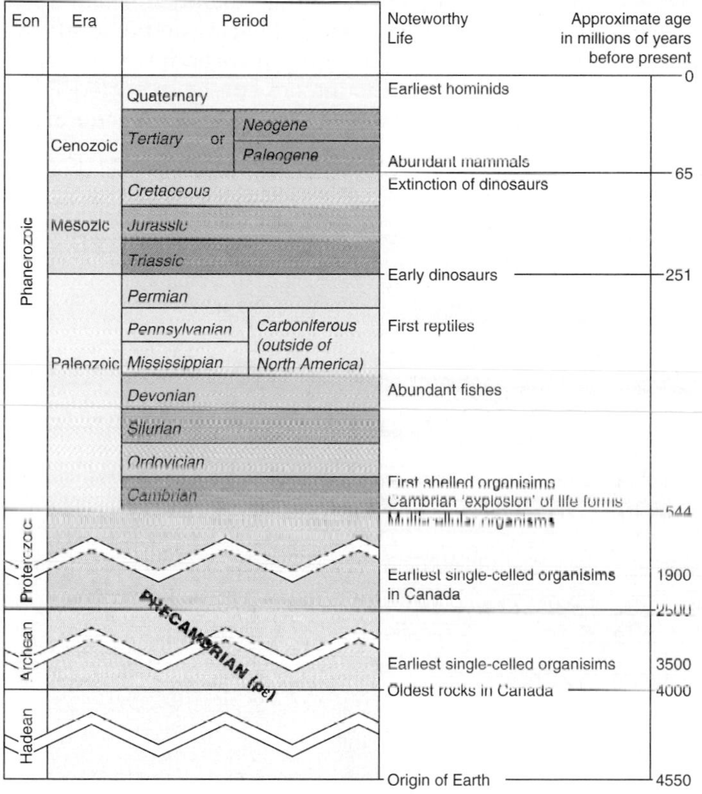

Modified from Okulitch, A.V., *Geological Time Chart*, 2003

Not only are the immense spans of geologic time difficult to comprehend, but very slow processes are impossible to duplicate. A geologist who wants to study a certain process cannot repeat in a few hours a sluggish chemical reaction that takes a million years to occur in nature. As Mark Twain wrote in *Life on the Mississippi*, "Nothing hurries geology."

WHAT DO GEOSCIENTISTS DO?

Geoscientists come in many different shapes and sizes and perform a variety of different tasks. In the past, traditional geologists spent most of their time in the field looking (*prospecting*) for tell-tale signs of minerals; to be a prospector, a geologist had to love the outdoors and be able to shrug off such hardships as biting insects and bad weather (figure 1.2).

FIGURE 1.2
Early geologists at camp on the banks of the Red River.
National Archives of Canada, #e000009486

Prospectors still rough it in the bush, but now they are called **exploration geologists** and their ranks include both men and women (see box 1.5). These geoscientists may work for an exploration company looking for gold, silver, and other metals—or, more recently, for diamonds. They get around in four-wheel-drive trucks and helicopters and are in constant touch by e-mail with financiers, market analysts, and other business professionals. After all, the whole point of prospecting is to find something and then get it to market before anyone else does.

The discipline of geology has broadened its scope over the past several decades. The location of geological resources is more important than ever, but at the same time the skills of the geoscientist are required to address additional issues such as mitigation of the effects of natural hazards and other environmental concerns. As the global population becomes increasingly concentrated in large urban "supercities," the risk to public safety from hazards such as volcanism, earthquakes, and severe storms is greatly increased. In addition, there is increased urgency to find and protect water resources and dispose of waste materials appropriately. Today, no one can claim to be educated and aware of environmental issues without some knowledge of the discipline of geology. Expanding the scope and responsibilities of the modern geologist into additional scientific fields means that these professionals are now more correctly referred to as **geoscientists**; some of their tasks and responsibilities are examined below.

Modern geoscientists can specialize in a number of areas within the discipline. *Geochemists* are comfortable working in the ordered environment of the laboratory and use high-technology equipment to analyze the chemistry of rocks or minerals. They may consult with *mineralogists*, who study minerals, or *petrologists*, who study the makeup of rocks and how they form (more often than not by volcanic activity and by being altered by heat and pressure either at the Earth's surface or deep below). Other geoscientists employ high-tech equipment in the field, using boats, ships, planes, or satellites to learn more of

IN GREATER DEPTH 1.5

A Day in the Life of an Exploration and Mining Geoscientist

Dave works in exploration geology for a mining company and spends many weeks of the year, often in winter, working and collecting geological data from the remote northern reaches of Canada (box figure 1). His aim is to find gold in the ancient rocks of the Canadian Shield. Gold can be found in volcanic rocks, in quartz veins, and where rocks have been stretched along faults (box figure 2). Some of these rocks may be exposed on the Earth's surface, but many are hidden by glacial sediments deposited during the several ice ages that affected Canada in the geologically recent past. These glacial deposits often contain gold and other minerals eroded from the bedrock by overriding glaciers. Exploration geoscientists looking for gold in previously glaciated areas have to reconstruct past ice-flow directions to find the highly mineralized source of disseminated gold particles found in the glacial deposits. Drilling is often necessary to sample mineralized rocks deeply buried beneath the Earth's surface (box figure 3).

Dave often works in the winter; because the ground is frozen, heavy drilling machinery and equipment needed to keep his team working and living in the harsh environment can be brought in without damaging the ground surface. Dave works with a large team including geophysicists and a support crew of drillers, cooks, and camp managers. Drill core has to be described and sampled, geophysical data pored over for clues, and the next round of drilling planned. Work continues in shifts, 12 hours long, for 24 hours a day. Mineralogical data collected from the glacial sediments overlying bedrock are very promising and indicate the nearby presence of gold-bearing rocks. Dave will be back in the field next winter, after spending several months in the head office with his colleagues working on similar projects elsewhere. They will work with economic geologists, who make the final determination as to whether a "prospect" such as the gold vein can be developed at a profit. If it can, Dave will be working with other engineering geologists to design the mine so the ore can be extracted safely and economically and to Canadian environmental standards.

BOX 1.5 ■ FIGURE 1
Collecting field data from natural exposures.
Photo courtesy of Westmin Exploration

BOX 1.5 ■ FIGURE 2
Exploration geoscientist looking for gold in quartz-rich rocks.
Photo courtesy of Westmin Exploration

BOX 1.5 ■ FIGURE 3
Examination of core drilled at a potential gold mine site.
Courtesy of Prospectors and Developers Association of Canada

IN GREATER DEPTH 1.6

A Day in the Life of a Petroleum Geoscientist

Matt is a petroleum geoscientist with a mid-sized oil company (Petrox Inc.) based in Calgary. He has an earth sciences degree and is part of a large team that includes other geologists, geophysicists, petroleum engineers, and financial land accountants (a.k.a. "land men").

Petrox Inc. holds the land rights to several oil fields in the Western Canada sedimentary basin, which underlies much of Alberta. Here, oil and gas are produced from deeply buried shallow marine sandstones, carbonates, and bioclastic deposits. New sources of methane gas found in association with coal beds (coal bed methane) are currently being explored and developed in the basin. Matt spends much of his time working at the Alberta Energy and Utilities Board Core Research Centre, located in northwest Calgary, describing drill core extracted from wells (box figure 1). He uses his core descriptions to interpret the depositional environments in which the rocks formed and to begin to piece together the geological history of the rocks within and around the reservoir. Frequent field trips to study outcrops exposed in the foothills of the nearby Rocky Mountains provide useful comparisons with the sedimentary rocks he sees in core. Matt combines his analysis of rock characteristics with geophysical data such as seismic reflection and downhole geophysical well logs to build up a picture of the subsurface extent of porous reservoir rocks (box figure 2). When combined with his knowledge of sedimentary basins and how sediments accumulate, this information helps Matt reconstruct the three-dimensional architecture of reservoirs using subsurface modelling programs.

Most oil fields in Canada, whether found on land or offshore (such as at Hibernia in eastern Canada or in the Beaufort Sea in northern Canada), are complex and oil is often trapped by geological structures such as anticlines, pinch-outs, and large faults that developed as the sedimentary basins on the North American craton were deformed by tectonic stresses (see chapter 20). A detailed knowledge of Canada's geologic history and the various tectonic events that have created sedimentary basins is a key requirement for oil and gas exploration anywhere in Canada. Matt's dream is to one day form his own junior oil company and to use his knowledge to find new productive oil fields in the Western Canada Sedimentary Basin.

BOX 1.6 ■ FIGURE 1

Drill core laid out for inspection at the Alberta Energy and Utilities Board Core Research Centre, Calgary.

Photo courtesy of M. Radomski.

BOX 1.6 ■ FIGURE 2

Geoscientist examining downhole geophysical well logs.

Photo courtesy of M. Radomski.

the nature of the physical conditions on or under the Earth's surface; these are called *geophysicists*. Similar techniques may be used by *petroleum geologists* in their search for oil and gas (see box 1.6) or by *coal geologists*. *Seismologists* study how to measure and mitigate earthquake activity. A *paleontologist* is a specialist who studies the fossilized remains of ancient organisms, whether gigantic dinosaurs or the remains of organisms too small to see without a powerful microscope (*micropaleontologists*). In Canada, much of the landscape is made of landforms and sediments left behind from when ice sheets covered the northern part of North America. These landforms and sediments are studied by *glacial geologists*. Such sediments contain and transmit water, an increasingly important mineral that is studied and protected by *hydrogeologists*.

Today's geoscientist has the advantage of working in many different areas, both by geography and by topic. They may teach in a high school, college, or university, work in all levels of government, especially those concerned with mapping resources and/or hazards, or be involved in planning and environmental issues such as where to locate disposal sites for wastes from households or nuclear generating plants.

Regardless of their chosen field within geology, all geoscientists are professionals and have the qualifications to join a professional association such as the Association of Professional Geoscientists of Ontario (APGO) or the Association of Professional Engineers, Geologists and Geophysicists of Alberta (APEGGA). Geoscientists often deal with information that is sensitive and/or financially significant (e.g., a parcel of land has been extensively contaminated by chemicals, or a mineral deposit has been discovered). Because geoscientists are registered professionals, the public has the assurance that certain skills have been learned and will be correctly applied, just as in the case of a dentist or a medical doctor. A university degree in geology (or geoscience or Earth science, as it is now being called) is a first step in a rewarding career—and this book is designed to help you get there.

Environmental Geology: New Challenges for Geoscientists

One might think that there would be a decreasing demand for geoscientists as Canada's economy moves from its long dependence on extracting natural resources to a service- and information-based economy focused principally in urban areas. This is not the case. Basic commodities such as metals, oil, and gas still need to be found and exploited (and Canadian mining and oil companies are world leaders, working around the globe). There is, nonetheless, a strong and growing demand for a new breed of geoscientist working with new tools on very different tasks. These new challenges reflect the fact that Canada is now one of the most urbanized countries in the world, with more than 75 percent of its population living in cities and towns. Urban populations create large amounts of waste, consume vast quantities of water, and create many environmental problems.

The new challenges for today's geoscientist relate to the finding and managing of drinking water (mostly groundwater), and in dealing with a wide variety of wastes ranging from radioactive waste to household (municipal) waste. These environmental problems are dealt with by *environmental geoscientists* (see box 1.7). Rapid urban development places great demands on our knowledge of where sufficient groundwater is located and how it can be protected. Real estate and insurance businesses also employ environmental geoscientists to provide key information regarding past use of lands and buildings as part of the "due diligence" process. Questions such as "Have past land uses released contaminants into the ground—and if so, where are they?" must be answered. In order to do this, environmental geoscientists increasingly need to "see" underground using geophysical techniques, geochemical data, and flow models to create a 3-D picture of what is below our feet. The 3-D arrangement of strata, and the type of strata themselves, controls the movement of groundwaters (and also contaminants). These are key considerations in finding clean drinking water, determining safe locations for storing wastes, and identifying the environmental impact of past waste-disposal activity.

Other geoscientists work with the challenges of engineering structures that form part of the human landscape. *Engineering geologists* work closely with civil engineers and are concerned with providing information about the substrates on which buildings, roads, bridges, and other infrastructure are to be built (see box 1.8). They need to know what soils or rocks might create problems for foundations. The majority of geological work in British Columbia is related to assessment and mitigation of landscape stability problems.

Creation of digital maps and 3-D geological models for resource exploration and environmental and engineering applications is one of the roles of a new type of geoscientist called a **geomatician**. Geomaticians collect, organize, analyze, and create images from any spatial and geographic data available in digital form (see box 1.9). They usually work with extremely large databases such as remotely sensed satellite data (e.g., RADARSAT, LANDSAT, ASTER, Hyperion), deep probing ground geophysical surveys (e.g., seismic, gravity, and magnetic), or compilations of geological data (e.g., provincial water well databases) and require high-speed computers and sophisticated imaging software to carry out their work. Geomaticians are involved in all types of geological and environmental applications, and in each chapter of this book we will outline these types of projects.

Common Themes

While the day-to-day work of the geoscientists described in the case examples above is varied, several common themes emerge. Today's geoscientist usually works as part of a team with biologists, lawyers, engineers, planners, and policy makers. He or she commonly works for an environmental consulting company or, increasingly, municipal or provincial government. Modern society needs geoscientists because they appreciate and understand geological processes and they are aware of all the different rock and sediment materials that arise from the billions of years of Earth's history. Geoscientists are used to thinking in 3-D and far back in time, using a wide range of geological, physical, and chemical data—along with their imagination—to reconstruct the distribution of subsurface layers to show how an area has evolved through geologic time. In short, they use the scientific method outlined in the next section.

IN GREATER DEPTH 1.7

A Day in the Life of an Environmental Geoscientist

Cheryl is 27 years old. She graduated with a geology degree five years ago and now works for an environmental consulting company that has offices in all the major cities of Canada. A degree, and some practical job experience working after graduation, enabled her to register as a professional geoscientist in her province. She is currently working on an old factory site that used to manufacture small engines (box figure 1). The site lies within a residential area close to a major river. Making engines requires extensive use of solvents, which are employed to degrease engine parts prior to assembly. Many years of manufacturing and the continued leakage of solvents into the ground now poses an environmental threat in the form of a "plume" of contaminants that is being carried by groundwater away from the site, toward the river (box figure 2). The geology is complex, consisting of glacial sediments resting on bedrock below. Understanding the three-dimensional variation in the geological layering is critical, as this layering controls the migration and shape of the contaminant plume. Cheryl works closely with a drilling company to locate drill holes and describe and sample the cores, and with a geophysicist who has carried out a seismic investigation. By working together, a picture of the subsurface layers emerges. However, the problem is rendered even more complicated because some solvents in the plume are heavier than water and sink through groundwater to rest on bedrock (DNAPLs); others (LNAPLs) are lighter and float on the water table.

Identifying the size and shape of the contaminant plume is only the first step for Cheryl; the next is to identify what action will be taken to control and deal with it. A wide variety of possible techniques are open to her, and to decide among them she requires knowledge of the water chemistry and geology below the site (box figure 3). Can the contaminated sediment simply be dug out and treated on-site, or do *in situ* techniques such as injecting steam into the ground and recovering contaminants have to be used? Are these techniques economically feasible? Local residents and politicians need to be informed of how the work is progressing, and Cheryl works closely with both. She also works with the provincial ministry of environment to ensure that cleanup will meet provincial water quality standards.

BOX 1.7 ■ FIGURE 1

Measuring water levels in test wells at a contaminated industrial site.

Photo courtesy of S. Salvatori

Some of Cheryl's colleagues are working with old garbage dumps that are leaking contaminants into surrounding areas and are affecting local groundwater supplies. Another city is running out of space for garbage at its disposal sites, and other colleagues are trying to identify new sites using the latest engineering techniques to make them safe. These include the use of plastic "liners" at the base of the dump to hold in contaminated water (leachate) moving through the waste pile.

The skills and techniques Cheryl uses are in wide national and international demand to clean up ground and surface waters impacted by mining and urban development. She will soon be working on a project involving the cleanup of mine wastes at one of Canada's largest mining districts in Northern Ontario.

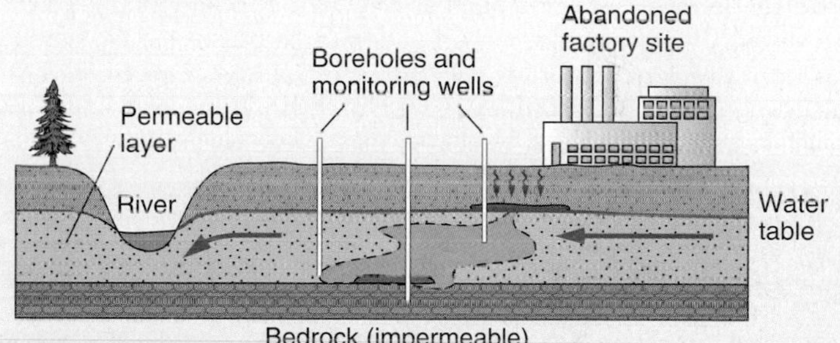

BOX 1.7 ■ FIGURE 2

Diagram of contaminant plume below an abandoned industrial site.

BOX 1.7 ■ FIGURE 3

Installing a monitoring probe within a well to record changes in groundwater chemistry.

Photo by Nick Eyles

IN GREATER DEPTH 1.8

A Day in the Life of an Engineering Geologist

Jason is an engineering geologist (some prefer to be called geological engineers) working on the projected path of a major subway (box figure 1). The subway is to be constructed using a "boring machine" consisting of a rotating shield that cuts into rock and sediment and is slowly inched forward underground, day by day (box figure 2). Concrete slabs are installed behind the advancing shield, creating a ready-made tunnel in which waste soil can be railed out. Jason is working today with a glacial geologist, one who knows a great deal about how past glaciers moved and deposited sediment. Their problem is the presence of large boulders within the glacial till that the tunnelling machine is moving through (box figure 3). These boulders slow down the progress of the shield and threaten the completion date of the entire project. Together, Jason and the glacial geologist planned a drilling program to investigate the number and size of boulders ahead of the shield. They have also visited nearby outcrops of the till, along the banks of a nearby river, where such boulders can be examined firsthand. Their finding is that such boulders are not randomly dispersed through the till but occur in horizontal layers (called *pavements*). By making small changes in the elevation of the tunnel, Jason and his colleagues can avoid the worst of the pavements lying ahead of the shield. They are also concerned about the stability of the tunnel as it passes through several sand seams containing water; holes are drilled and pumps are used to dewater the seams as the shield is advanced. Other questions relate to whether the tunnel will settle in areas of soft sediment.

Several of Jason's colleagues are in eastern Canada working on bridges that cross areas of soft sediment. They are particularly interested in whether such structures will be safe in the event of an earthquake. Other colleagues work on the stability of steep rock slopes where railways cross the Rocky Mountains of Alberta and the Coast Mountains of British Columbia. They have also worked on the stability of rock slopes in large open-pit quarries and deep underground mines. Most engineering geologists are required to be registered under the same provincial legislation as engineers if they work in Canada.

BOX 1.8 ■ FIGURE 1
Engineering geologist working on-site.
Photo courtesy of S. Salvatori

BOX 1.8 ■ FIGURE 2
Boring machine used for excavation of subway tunnels.
Photo courtesy of S. Salvatori

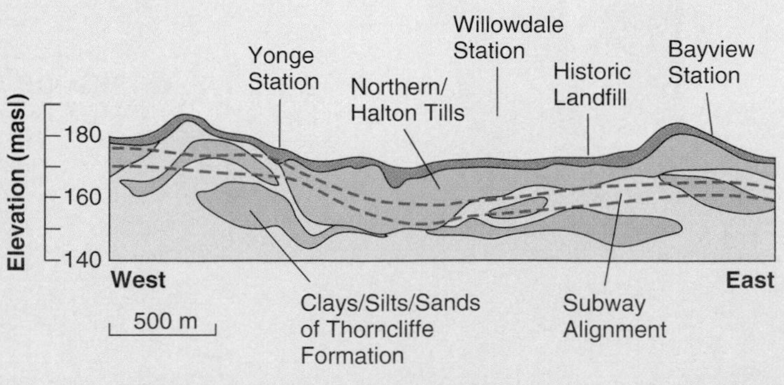

BOX 1.8 ■ FIGURE 3
Geological cross-section through glacial deposits in which a new subway tunnel is being cut.

IN GREATER DEPTH 1.9

A Day in the Life of a Geomatician

Patty is a geomatician and works for Natural Resources Canada on a variety of projects involving the production and interpretation of digital maps for resource exploration, coastal protection, hazard management, and environmental impact assessment in Arctic Canada. Geomatics is one of the fastest-growing areas of the geosciences and involves the collection, manipulation, presentation, and analysis of spatial and geographic data in digital form. Canada is considered to be a world leader in the provision of software, hardware, and support services for geomatics applications. The project Patty is working on at present is assessing the potential impacts of drilling and pipeline engineering activities in the Beaufort Sea—it requires examination of coastal and near-shore stability, artificial island stability (see box figure 1), ice scouring (see box figure 2), sub-sea and coastal permafrost, sea-floor sediment characteristics, shallow gas hazards (see box figure 3), and habitat/ecosystem sensitivity. She is part of a team of geoscientists who are reviewing existing data and carrying out new surveys to generate digital maps, databases, and interpretive reports. Patty's work essentially involves the integration of a number of sources of digital data obtained from satellite images, aerial photography, bathymetric surveys, multibeam and side scan sonar, sub-bottom seismic profiles, and sea-floor photography. She manipulates and analyzes these data using GIS and 3-D modelling software and produces a series of maps and 3-D images that may be used to identify potential hazards and can be used to make informed decisions regarding the location and types of development that are suitable in Canada's Arctic regions.

Geomaticians are most likely to be found working in the areas of natural resources management, infrastructure development and maintenance, and environmental management. In the future Patty may work on projects for offshore companies positioning oil and gas rigs or monitoring pipelines, or she may map environmentally sensitive areas for local municipalities. She could also work for utility and transport companies, supermarkets, or any other industry that requires the spatial analysis of digital data. She has a B.Sc. in earth and environmental sciences and a graduate certificate in GIS analysis, and her skills are in demand in both national and international employment markets.

BOX 1.9 ■ FIGURE 1

3-D multibeam sonar image of the artificial island Issungnak built on the Beaufort Shelf, Canadian Arctic, in approximately 18 m of water.

Image courtesy of S. Blasco, Geological Survey of Canada.

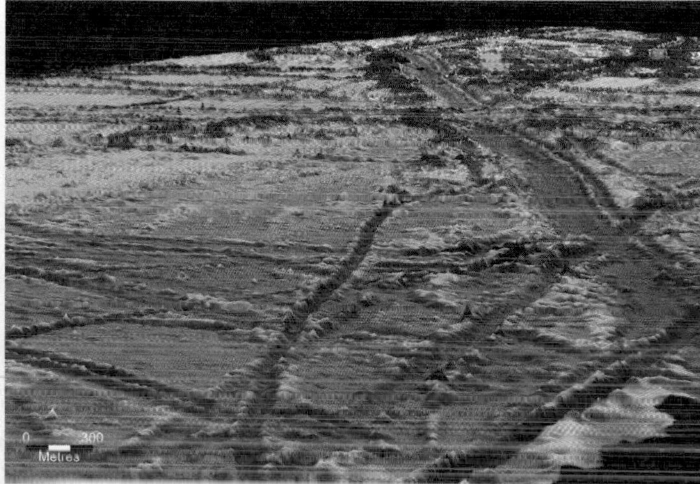

BOX 1.9 ■ FIGURE 2

3-D multibeam sonar image of the Canadian Beaufort Sea seabed (12 m water depth). The seabed is covered with linear scours created by pressure ridge keels within the sea ice.

Image courtesy of S. Blasco, Geological Survey of Canada.

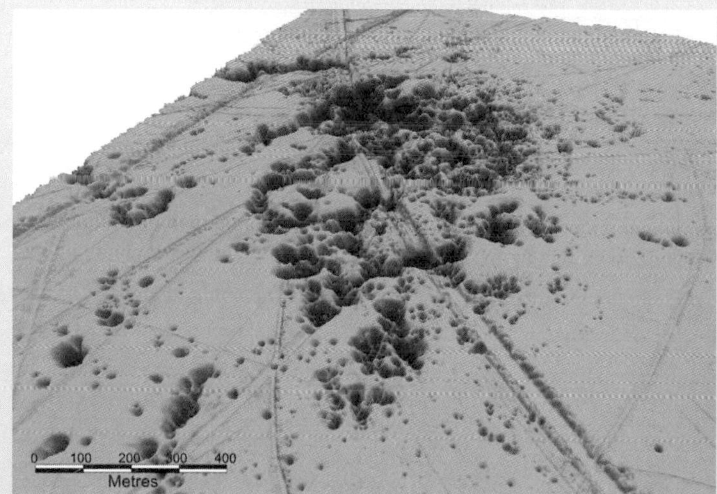

BOX 1.9 ■ FIGURE 3

3-D multibeam sonar image of a series of gas vents on the seabed of the Canadian Beaufort Shelf.

Image courtesy of S. Blasco, Geological Survey of Canada.

WHAT IS THE SCIENTIFIC METHOD?

Geologists are scientists; they routinely employ the **scientific method** (figure 1.3). This is the process by which scientists first identify a *problem* (e.g., finding diamonds in Nunavut or oil in Newfoundland, or explaining why earthquakes happen more often in British Columbia than in Ontario). They then select a *methodology* to collect data in order to help solve the problem (e.g., using a drill rig to drill holes and collect rock samples). Having collected the data they then *analyze* and *interpret* the information, possibly using maps or computer graphics as tools, and come up with a solution. Of course, many data are not useful—because the information was not collected at the right time, or from the right place, or at the right scale, or because the instrument used was faulty. Using data that are appropriate to solve the problem, the geologist creates a *hypothesis*. A **hypothesis** is simply a theoretical explanation where the geologist is essentially saying "Well, I don't know the whole story but I think this is a good explanation." In any investigation, there may not be just one answer to any problem. Good science invariably turns up more questions to be investigated than concrete answers; in these cases, the scientists may use *multiple hypotheses* (e.g., I think *this* or *that* might explain the greater frequency of earthquakes in British Columbia). Many hypotheses will prove to be incorrect and will be discarded. A hypothesis that passes thorough and repeated testing will become a *theory*, which has an excellent chance of being true.

Geoscientists make these types of problem-solving decisions when they carry out their scientific investigations. Such investigations are often highly complex (and thus require the help of different specialists) because the Earth and its geology reflect a very long history. This history is reviewed briefly below.

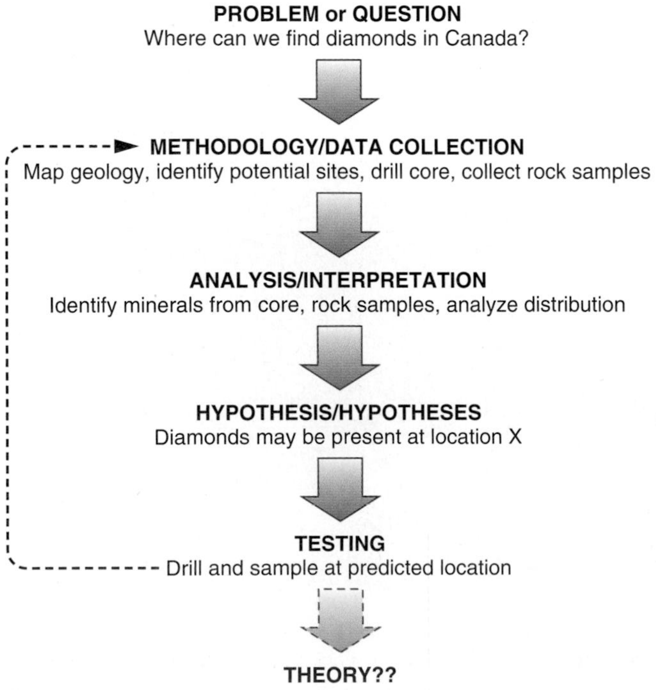

FIGURE 1.3
The scientific method.

HOW DID THE EARTH FORM?

You will learn the meaning of many new geological terms in this book. To gain a complete understanding of these terms, however, we must go beyond vocabulary and become acquainted with fundamental geological *concepts*. First of all, the Earth is very old and is unique within the solar system in having not only a solid body, but also oceans, an atmosphere, and life. These individual components continuously interact to form the complex and dynamic system we refer to as the *Earth system*. In order to understand the nature of our planet and the complex and ever-changing Earth system, we must first briefly examine how the Earth formed, and the characteristics of its early atmosphere and life forms. More detailed information regarding the formation of the solar system and the other planets is presented in chapter 21 on the Online Learning Centre.

The universe was formed by the clumping together of gas and debris in the aftermath of the Big Bang that is thought to have occurred some 15 billion years ago. There are billions of galaxies in the universe; one of these, the Milky Way, contains our own solar system and planet Earth (figure 1.4). The solar system consists of the sun and the nine planets and space debris that orbit the sun. It was created from a cloud of gas and dust particles called a **nebula** (figure 1.5). This cloud of gas and dust began to rotate and contract, creating a bulbous *core* surrounded by a flattened *disc*. The core progressively collapsed to the point where nuclear fusion began, and our sun was formed sometime fewer than five billion years ago. Dust in the outer disc condensed to form rocks and metals that combined to form large rounded *planets* and much smaller, irregularly shaped **planetismals**. This process of building large bodies of matter through collisions and gravitational attraction is called **accretion**.

The differing densities of the planets of the solar system (table 1.2) show that the planets differ in composition. Those planets that formed close to the sun—Mercury, Venus, Earth, and Mars—are small, dense, and rocky and are called the **terrestrial planets**. The giant planets of Jupiter, Saturn, Uranus, and Neptune have low densities and are called the **Jovian planets** (figure 1.6).

WHAT WAS THE EARLY EARTH LIKE?

Soon after Earth had formed about 4.5 billion years ago, it collided with a planetismal—the Earth's moon was created from the debris that was flung off into space (see chapter 21 on the Online Learning Centre for more on this phenomenon). During its final stages of formation, until about 3.9 billion years ago, the Earth swept up chunks of space debris and in the process was bombarded by huge meteorites. Few impact craters produced by these ancient meteorite collisions can be identified on the Earth's surface today

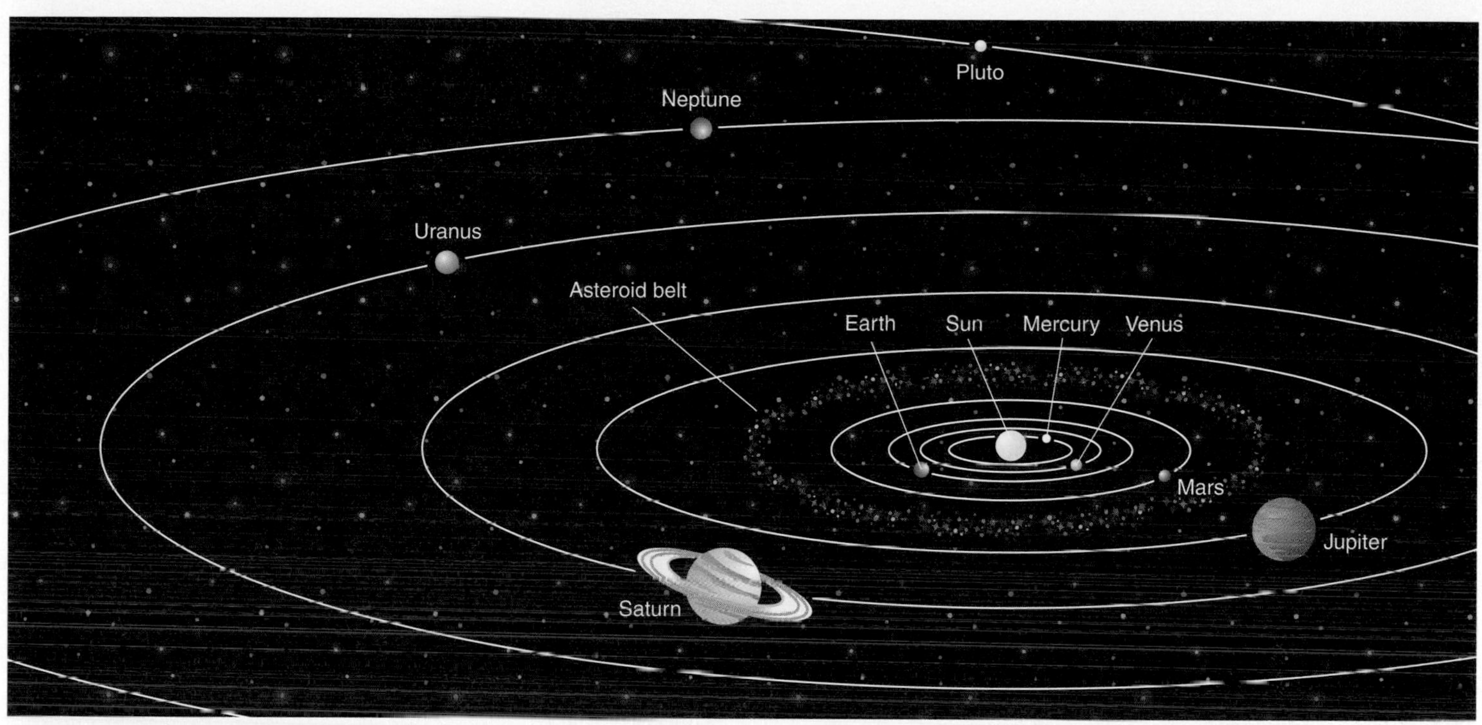

FIGURE 1.4

Orbits of the planets in the solar system.

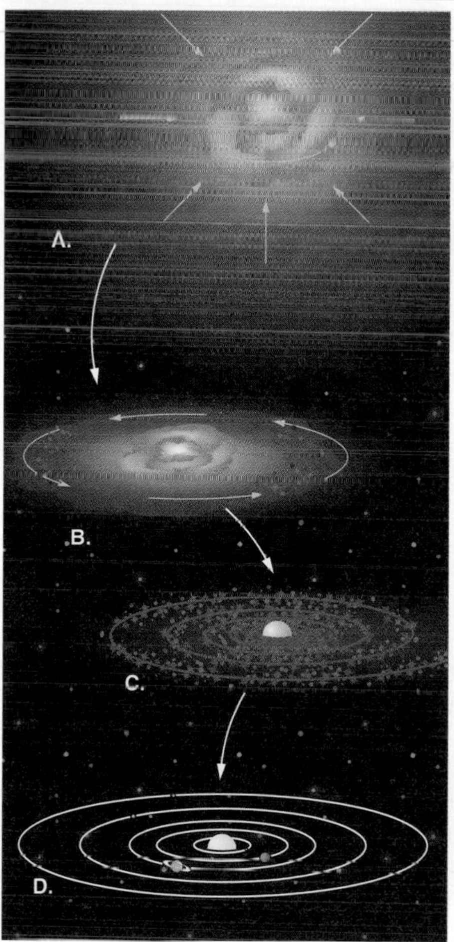

FIGURE 1.5

Nebular hypothesis for origin of the solar system. (*A*) A huge rotating cloud of dust and gases (nebula) begins to contract. (*B*) Most of the material is gravitationally pulled toward the centre, producing the sun. However, because of rotational motion, some dust and gases remain, orbiting the central body as a flattened disc. (*C*) The planets begin to accrete from the material that is orbiting within the flattened disc. (*D*) In time, most of the remaining debris collects into the nine planets and their moons.

TABLE 1.2 Planetary Data

Planet	Distance from sun (millions of kms)	Diameter (kms)	Average density (g/cm^3)
Mercury	58	4,878	5.4
Venus	108	12,104	5.2
Earth	150	12,756	5.5
Mars	228	6,794	3.9
Jupiter	778	143,884	1.3
Saturn	1,427	120,536	0.7
Uranus	2,870	51,118	1.3
Neptune	4,497	50,530	1.6
Pluto	5,900	~2,300	2.1

as they have either been buried by subsequent deposits or removed by tectonic movements or erosion. However, more geologically recent impact craters can be recognized and provide evidence of a violent past (figure 1.7). Many more crater forms can be seen on the less geologically active planet Mercury, and on our moon.

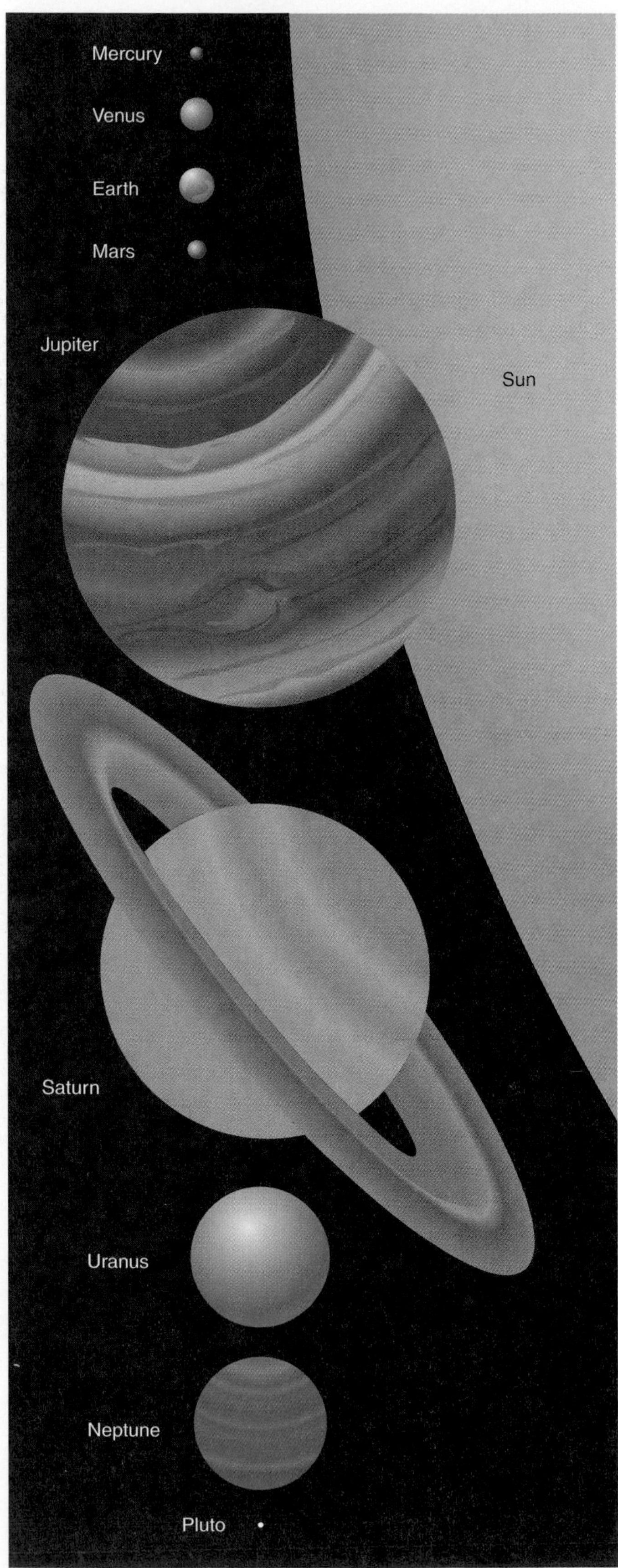

FIGURE 1.6
The planets drawn to scale.

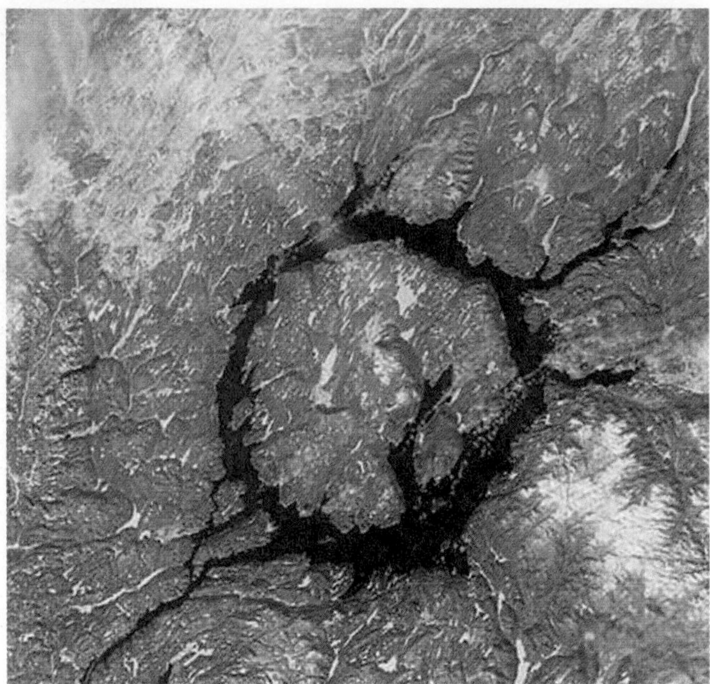

FIGURE 1.7
Meteorite craters on Earth. Manicouagan Crater, Quebec, at approximately 100 km in diameter, is one of the largest known impact craters. It was produced by impact of a 5-km-wide meteorite around 214 million years ago.
Photo courtesy of NASA.

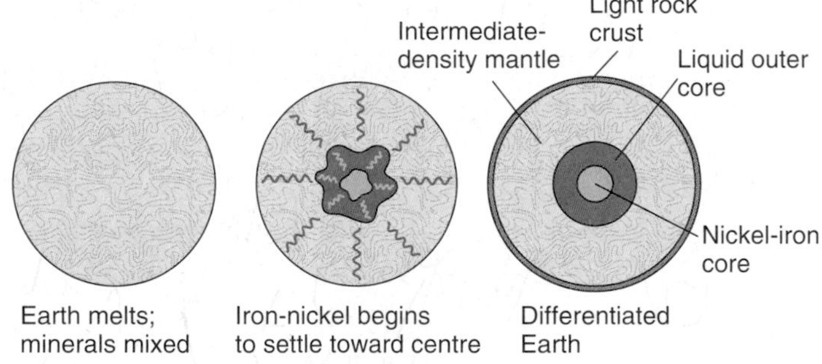

FIGURE 1.8
Differentiation of the Earth.

Heat generated by accretion of debris and gravitational compression of the growing Earth and from decay of radioactive isotopes allowed the planet to melt. Under such conditions, the heavier materials such as iron and nickel settled toward the planet's centre, while lighter materials such as silica and oxygen rose toward the Earth's surface (figure 1.8). This process of zonation of different materials within a planet is termed **differentiation**. Meteorites that fall to Earth today provide geoscientists with valuable information about the composition of planets in the solar system. Most meteorites have the same isotopic age of around 4.6 billion years, and probably represent materials that were present in the initial solar nebula. (For more on meteorites, see box 1.10.)

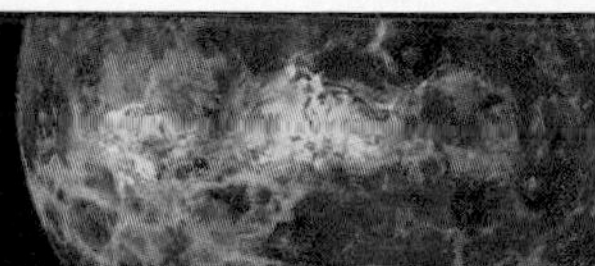

PLANETARY GEOLOGY 1.10

Meteorites

Small solid particles of rock, metal, and/or ice orbiting the Sun are called meteoroids. When these particles enter Earth's atmosphere, they are heated to incandesence by friction; these glowing particles are called meteors (or "shooting stars" or "falling stars"). Most meteors are small and burn up while still in the atmosphere, but about 150 per year are large enough to strike Earth's surface. Those that do are called *meteorites* (box figure 1). The largest fragment of a meteorite found (in South Africa) weighs 50 tons; much larger meteorites have hit Earth in the past.

Three basic types of meteorite are iron, stony-iron, and stony meteorites. Stony meteorites are by far the most common, but they look like Earth rocks, so they are hard to find. Iron meteorites are rare, but look so unique that they are commonly found; most museum meteorites are of the iron type.

Iron meteorites are mostly iron alloyed (mixed) with a small percentage of nickel. Small amounts of other metals or minerals may be present. Iron-nickel meteorites give an important source of information regarding the composition of Earth's core.

Stony iron meteorites are made of iron-nickel alloy and silicate minerals in about equal parts.

Stony meteorites are made of silicate minerals such as plagioclase, olivine, and pyroxene; they may contain a small amount of iron-nickel alloy. *Chondrites* comprise about 90 percent of stony meteorites and contain round silicate grains called *chondrules*. The other 10 percent of stony meteorites are *achondrites*, which lack chondrules.

Chondrules consist mostly of olivine and pyroxene, and range from distinct spheres to large bodies with fuzzy outlines. The composition of chondrite meteorites resembles the ultramafic rock peridotite, but peridotite lacks the chondritic texture and iron-nickel content of the meteorites.

One kind of chondrite is composed mostly of serpentine or pyroxene and contains up to 5 percent organic materials, including carbon, hydrocarbon compounds, and amino acids. These meteorites are called *carbonaceous chondrites*. All available evidence indicates that the organic compounds were in fact produced by inorganic processes. Carbonaceous chondrites are of particular interest to scientists because they are believed to have the same composition as the original material from which the solar system was formed.

Achondrites are generally similar to terrestrial rocks in composition and texture. In composition they are most similar to basalts. Some have textures like ordinary igneous rocks, and others are breccias with fragments of different compositions and textures.

The origin of meteorites is controversial. Many meteorites have a coarse-grained texture, probably formed by slow cooling within a larger body, such as a planet. The similarity in iron-nickel composition among iron meteorites also suggests that they are fragments from a single, large body. The larger body may have differentiated into a heavy, iron-rich core and a lighter, rocky mantle before it fragmented into meteoroids. Isotopic dating shows that most meteorites have the same age, 4.6 billion years old. No terrestrial rocks have ages greater than 3.8–3.9 billion years; therefore, meteorites provide the best clue as to the age of the solar system and the formation of the planets.

BOX 1.10 ■ FIGURE 1

The largest meteorite found in Canada (156 kg) was found near Madoc, Ontario in 1855 and is now on display at the Geological Survey of Canada geological museum in Ottawa. The front part of the meteorite has been cut away for sampling.
Photo by C. H. Eyles

Early heating of planetary interiors also generated large quantities of magma that rose to the surface along fractures produced either by meteorite impacts or by some other tectonic processes. Vast areas of the moon, Mercury, and Mars are covered with volcanic rocks, most formed early in the history of the solar system. Similar early extrusive activity must have occurred on Venus and on Earth. Volcanic activity on a much smaller scale still occurs on Venus and on Earth.

Internal Structure of the Earth

The Earth is said to be *differentiated*, because intense heat and pressure within the body of the planet gives rise to "onion-like" layers of different chemical composition and physical behaviour (figure 1.9). No well has yet been drilled that penetrates the crust, so our understanding of the Earth's interior is based on indirect evidence. This evidence suggests the layers form an innermost **core**, composed of iron alloy (iron with nickel and silicon); a **mantle**, composed of Fe-Mg silicates (forming a rock called **peridotite**); and an outer **crust**, composed of "lighter" rocks such as basalt and granite. Volumetrically, peridotite is the most common rock on Earth. Much of what we know of the Earth's internal structure is derived by the study of how earthquake energy travels through the body of the planet (see chapter 4); volcanic eruptions and violent plate tectonics collisions bring samples of deeper materials to the surface where we can study them.

The hot rock of the Earth's mantle behaves much like wax and can slowly flow and change shape as it is moved around in

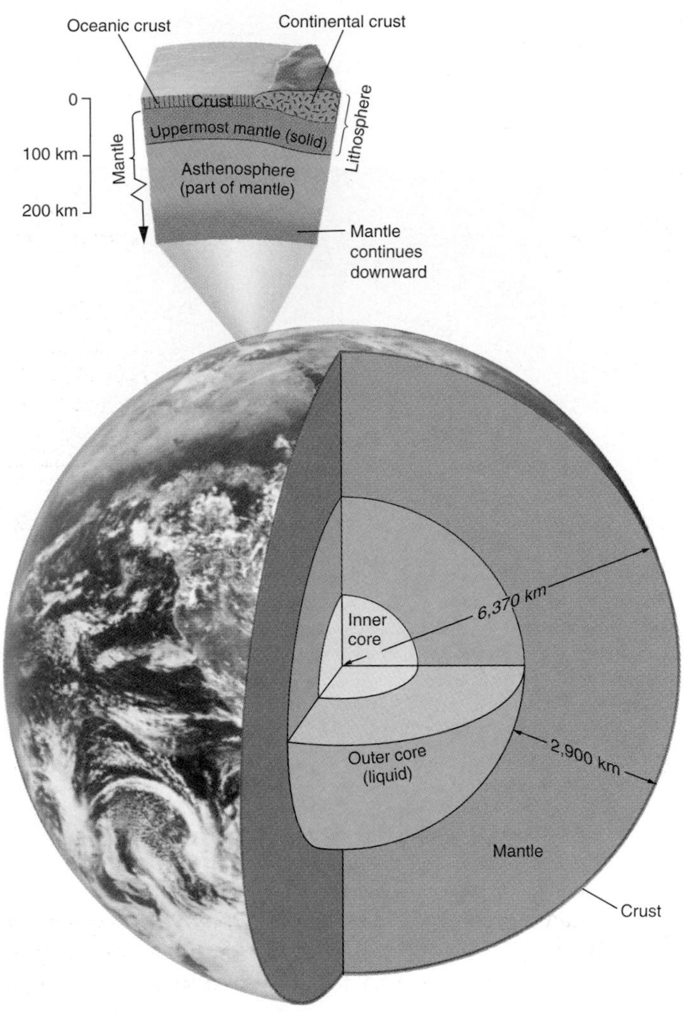

FIGURE 1.9

Cross-section through the Earth. Expanded section shows the relationship between the two types of crust, the lithosphere and the asthenosphere, and the mantle. The crust ranges from 5 to 75 km thick.

Photo by NASA

enormous convection cells (just as in a lava lamp). Movement in the mantle creates stresses in the overlying layer (the crust), which is thin and rigid much like an eggshell. Mantle convection breaks the crust and uppermost rigid mantle into large pieces, which geoscientists call **lithospheric plates**. These pieces are pushed around the surface of the planet over a weak layer called the **asthenosphere** (figure 1.9). Movement of rigid lithospheric plates over the more mobile asthenosphere is the fundamental process involved in plate tectonics (see chapter 2).

Formation of the Early Atmosphere

Earth's early atmosphere was derived predominantly from water and gaseous elements released during volcanic eruptions in a process called **outgassing**. Some water and gases would have been present in the original materials that aggregated to form the growing planet, but large amounts were also supplied by meteorites and comets. Most outgassing of water probably occurred within the first billion years of Earth's history, forming extensive oceans and lakes on the Earth's surface and allowing sediments to be eroded and deposited. The oldest known sedimentary rocks are around 3.8 billion years old. Earth's early atmosphere was rich in carbon dioxide released from volcanoes and created rainwater and ocean waters that were highly acidic. Evolution of the earliest life forms occurred in these acidic waters, and the oldest fossil life forms are 3.5 billion years old. The oxygen-rich atmosphere in which we live today was itself created by the evolution of photosynthetic life forms that could separate carbon dioxide into carbon and free oxygen.

Early Life Forms

The earliest life forms preserved in the geologic record are microorganisms called **prokaryotes**, which were similar to modern cyanobacteria. These single-celled organisms were able to trap sediment and grow organic structures called **stromatolites** (figure 1.10). The oldest known fossils on Earth are 3.5-billion-year-old stromatolites found in the rocks of Australia's Pilbara Shield.

The so-called Cambrian "explosion" of life forms about 600 million years ago saw the appearance of more complex organisms with backbones and hard shells. One of the richest repositories of Cambrian-age fossils is the Burgess Shale, which outcrops in the Canadian Rockies and has allowed paleontologists to gain a remarkable view of the strange creatures that lived at this time (see box 1.11). All of the major animal groups evolved during the period between 530 and 520 million years ago, and this may have been a response to dramatic changes in geography and climate created by the breakup of the supercontinent Rodinia.

WHAT IS THE "EARTH SYSTEM"?

The **Earth system** is a small part of the larger solar system but also has its own component parts, or subsystems. These subsystems, or *spheres*, include the atmosphere (the gases that envelop Earth), the hydrosphere (water on or near the Earth's surface), the biosphere (living or once living materials), and the geosphere (rock or other inorganic Earth materials). This book concentrates on the geosphere, but this component cannot be fully understood without consideration of the many interactions between it and the other subsystems.

The entire Earth system is fuelled by two major sources of energy that continually interact to create complex and ever-changing landscapes and geographies. The Earth's *external energy source* is the sun, which drives atmospheric and hydrologic processes and controls weather, climate, weathering, ocean circulation, erosion, and deposition. Geothermal heat, remaining from our planet's formation and generated by radioactive decay of minerals within the Earth, creates an *internal energy source* that drives plate movements, volcanic eruptions, and earthquakes. Interactions between these externally and internally driven forces are responsible for the many dynamic processes operating on our planet and the continual cycling of materials and chemicals within the Earth system.

FIGURE 1.10

(*A*) 1,300-million-year-old stromatolite from Northern Ontario. (*B*) Modern stromatolite mounds (thrombolites) built by blue-green algae in hypersaline lake environments north of Perth, Western Australia.

Photo *A* from NASA. Photo *B* by Nick Eyles

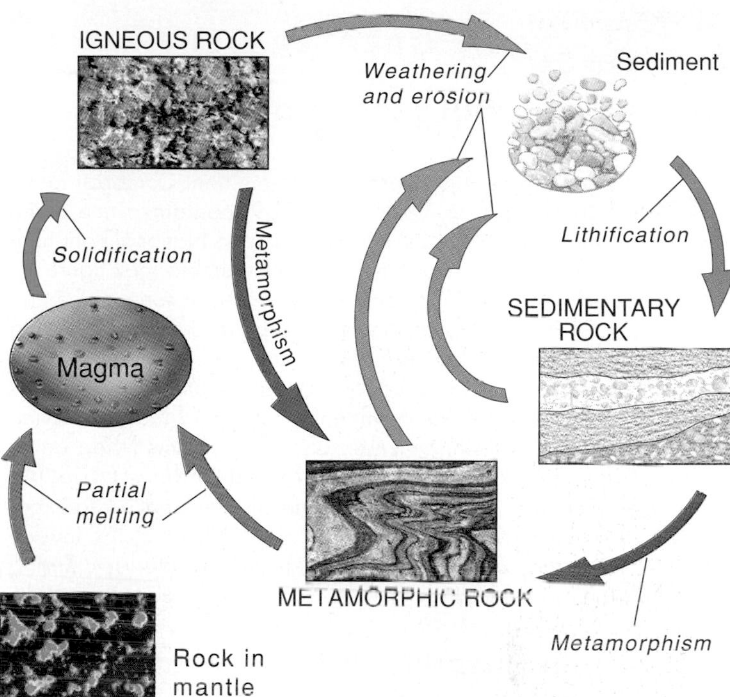

FIGURE 1.11

The rock cycle.

For example, if the Earth's internal energy source were no longer functioning (and tectonic forces had therefore stopped operating), the external engine plus gravity would long ago have levelled the continents, and the resulting sediment would have been deposited on the sea floor. Everything would be at rest. Nothing would be changing. That is to say, everything would be in *equilibrium* (and geology would be a dull subject). But this is not the case. The internal and external forces continue to interact, forcing substances out of equilibrium. Therefore, the Earth has a highly varied and ever-changing surface. Minerals and rocks change as well.

A useful aid in visualizing these relationships is the **rock cycle** shown in figure 1.11. This is a conceptual model that links rock-forming processes that operate in the Earth's crust. The three major rock types—igneous, metamorphic, and sedimentary—are shown. As you see, each may form at the expense of another if it is forced out of equilibrium with its physical or climatic environment by either internal or surficial forces.

Magma is molten rock. **Igneous rocks** form when magma solidifies. If the magma is brought to the surface by a volcanic eruption, it may solidify into an *extrusive* igneous rock. Magma may also solidify very slowly beneath the surface. The resulting *intrusive* igneous rock may be exposed later after uplift and erosion remove the overlying rock. The igneous rock, being out of equilibrium, may then undergo *weathering* and *erosion,* and the debris produced is transported and ultimately deposited (usually on a sea floor) as *sediment.* If the unconsolidated sediment becomes *lithified* (cemented or otherwise consolidated

IN GREATER DEPTH 1.11

Burgess Shale

The Burgess Shale, discovered by the eminent Cambrian paleontologist Charles D. Walcott in 1909, contains some of the world's most important fossils. Exposed in Yoho National Park high in the Rocky Mountains near Field, British Columbia (box figure 1), the Burgess Shale contains a remarkable record of some of the earliest forms of animal life as they first appeared on Earth during the Cambrian "explosion" (box figure 2).

The Burgess Shale was deposited underwater at the base of a large reef located on the continental margin of North America approximately 540 million years ago. Organisms living on or around the reef were periodically buried under mud, dying instantly in the sulphur-enriched environment, which allowed excellent preservation of their body parts. Through time, the muddy sediments changed into shale and were thrust onto land during the building of the Rockies around 175 million years ago.

The Burgess Shale contains a remarkable assemblage of soft-bodied organisms that are only rarely preserved in rocks. Fossils of arthropods, sponges, onycophorans (velvet worms), crinoids, molluscs, worms, corals, chordates, and many other bizarre species—different from any known today—are found in the Burgess Shale (box figure 3). The great diversity and excellent preservation of these fossil organisms has allowed scientists to learn a great deal about the evolutionary complexity of early life forms on our planet and has earned the Burgess Shale its status as a UNESCO World Heritage Site.

BOX 1.11 ■ FIGURE 1
The Burgess Shale exposed in Yoho National Park near Field, British Columbia.
Photo by Nick Eyles

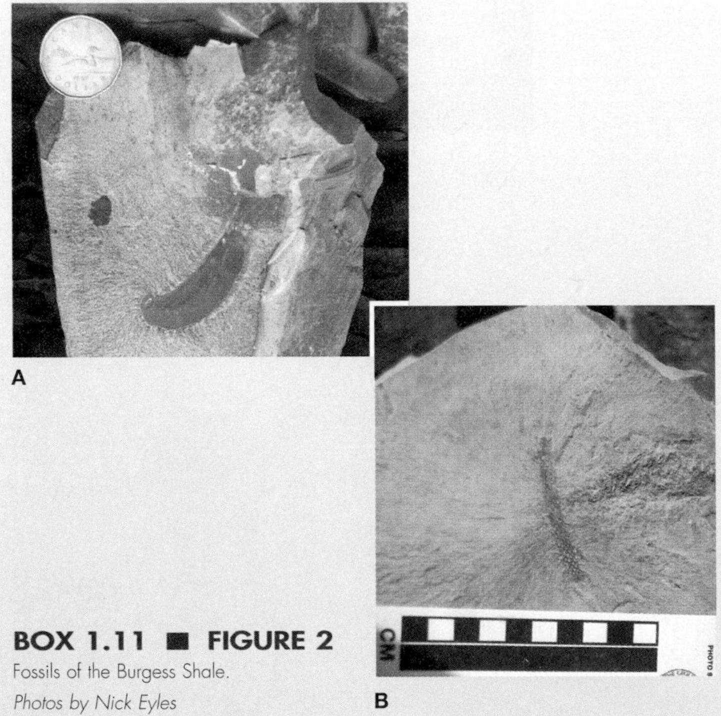

BOX 1.11 ■ FIGURE 2
Fossils of the Burgess Shale.
Photos by Nick Eyles

into a rock), it becomes a **sedimentary rock**. As the rock is buried by additional layers of sediment and sedimentary rock, heat and pressure increase. Tectonic forces may also increase the temperature and pressure. If the temperature and pressure become high enough, usually at depths greater than several kilometres below the surface, the original sedimentary rock is no longer in equilibrium and recrystallizes. The new rock that forms is called a **metamorphic rock**. If the temperature gets very high, the rock partially melts producing magma, completing the cycle.

The cycle can be repeated, as implied by the arrows in figure 1.11. However, there is no reason to expect all rocks to go through each step in the cycle. For instance, sedimentary rocks might be uplifted and exposed to weathering, creating new sediment.

The rock cycle diagram reappears on the opening pages of chapters 6 through 10. The highlighted portion of the diagram will indicate where the material covered in each chapter fits into the rock cycle. In chapter 2 we will begin by examining the most important concept in geology today—that of plate tectonics.

SUMMARY

Geology is the scientific study of the Earth. Early geologists used their knowledge of rocks and minerals to find and exploit resources required by newly industrialized nations. Today, geoscientists are involved in a wide range of activities including resource exploration, environmental geology, geomatics, and engineering geology, and knowledge of geological processes is required by individuals working in both wilderness and urban settings.

One of the most important theories in geology today is that of *plate tectonics*, which proposes that the surface of the Earth is broken into plates that move relative to each other and allow continents to move over time. Most geological features and processes operating on the Earth can be related to plate tectonics.

The Earth is estimated to be at least 4.5 billion years old and is thought to have formed at the same time as other planets in our solar system. During formation of the Earth, intense heat caused it to melt and allowed the most dense materials to settle toward the centre of the planet, forming a *core*. The lightest materials "floated" to the surface to form a *crust*. This process of *differentiation* created the zoned internal structure of the Earth consisting of a dense core, thick *mantle*, and relatively thin crust.

The earliest life forms on Earth were microorganisms (similar to modern cyanobacteria) that formed mound-like structures called *stromatolites*. More complex life forms emerged after 600 million years ago, and by 100 million years ago dinosaurs wandered across our continental interiors.

Erosion and transport of rocks exposed on the continents creates sediments that may form *sedimentary rocks* over time. If these are deeply buried and heated they may change into *metamorphic rocks* or may melt completely and form *magma*. Solidification of magma, either underground or on the Earth's surface, produces *igneous rocks*. The "recycling" of Earth materials as different rock types is conceptualized as the "*rock cycle*," and illustrates some of the many interactive processes operating within the Earth system.

Terms to Remember

accretion 12
asthenosphere 16
core 15
cross-section 3
crust 15
differentiation 14
Earth system 16
exploration geologist 5
geology 2
geomatician 8
geoscientist 5
hypothesis 12
igneous rock 17
Jovian planets 12
lithospheric plates 16
magma 17
mantle 15
metamorphic rock 18
nebula 12
outgassing 16
Pangea 2
peridotite 15
planetismals 12
prokaryotes 16
rock cycle 17
scientific method 12
sedimentary rock 18
stromatolite 16
terrestrial planets 12

Testing Your Knowledge

Use the questions below to prepare for exams based on this chapter.

1. What historical events/changes caused the rapid expansion of our geologic knowledge? What were early geologists looking for?
2. How has the profession of geoscience changed in the past 100 years? Which geoscience specialization deals with the formation and makeup of rocks?
3. What is the scientific method? Explain how it works and how you use it in your everyday life.
4. What are the main issues in environmental geoscience in your part of the country?
5. How did the Earth accumulate enough heat to melt and allow differentiation to occur? Explain how differentiation shaped the internal structure of the Earth.
6. What are the relationships among the mantle, the crust, the asthenosphere, and the lithosphere?
7. The largest zone of Earth's interior by volume is the
   a. crust
   b. mantle
   c. outer core
   d. inner core
8. The lithosphere is
   a. the same as the crust
   b. the layer beneath the crust
   c. the crust and uppermost mantle
   d. only part of the mantle
9. How old is the universe as we know it today? Compare this to the age of the solar system and to the age of the Earth.
10. What is the rock cycle? Explain why it is a simplification of the real processes at work.

CHAPTER

2

Plate Tectonics

What is plate tectonics?
How did the plate tectonics theory evolve?
What is sea-floor spreading?
What are plates and how do they move?
How do we know that plates move?
How fast do plates move?
What happens at plate boundaries?
How do mountain ranges form?
How do plates change over time?
What causes plate motions?
How are mantle plumes and hot spots related?
Why is it important to understand plate tectonics?

Earth is a dynamic planet. Many of the geological processes that allow the Earth to constantly change are elegantly explained by the plate tectonics theory. This theory suggests that the surface of the Earth is divided into several large plates that change in position and size. Intense geologic activity, such as volcanic eruptions and earthquakes, occurs at plate boundaries.

The history of the concept of plate tectonics is a good example of how scientists think and work and how a hypothesis can be proposed, discarded, modified, and then reborn. In the first part of this chapter we trace the evolution of an idea—how the earlier hypothesis of moving continents (continental drift) and a moving sea floor (sea-floor spreading) were combined to form the theory of plate tectonics. The second part of this chapter examines the processes that operate at plate boundaries and the economic deposits and environmental hazards that may be associated with them.

Satellite image of the Red Sea and Arabia. Plate motion has torn the Arabian Peninsula (centre) away from Africa (lower left) to form the Red Sea (left centre). *Photo provided by the SeaWiFS Project, NASA/Goddard Space Flight Center, and ORBIMAGE*

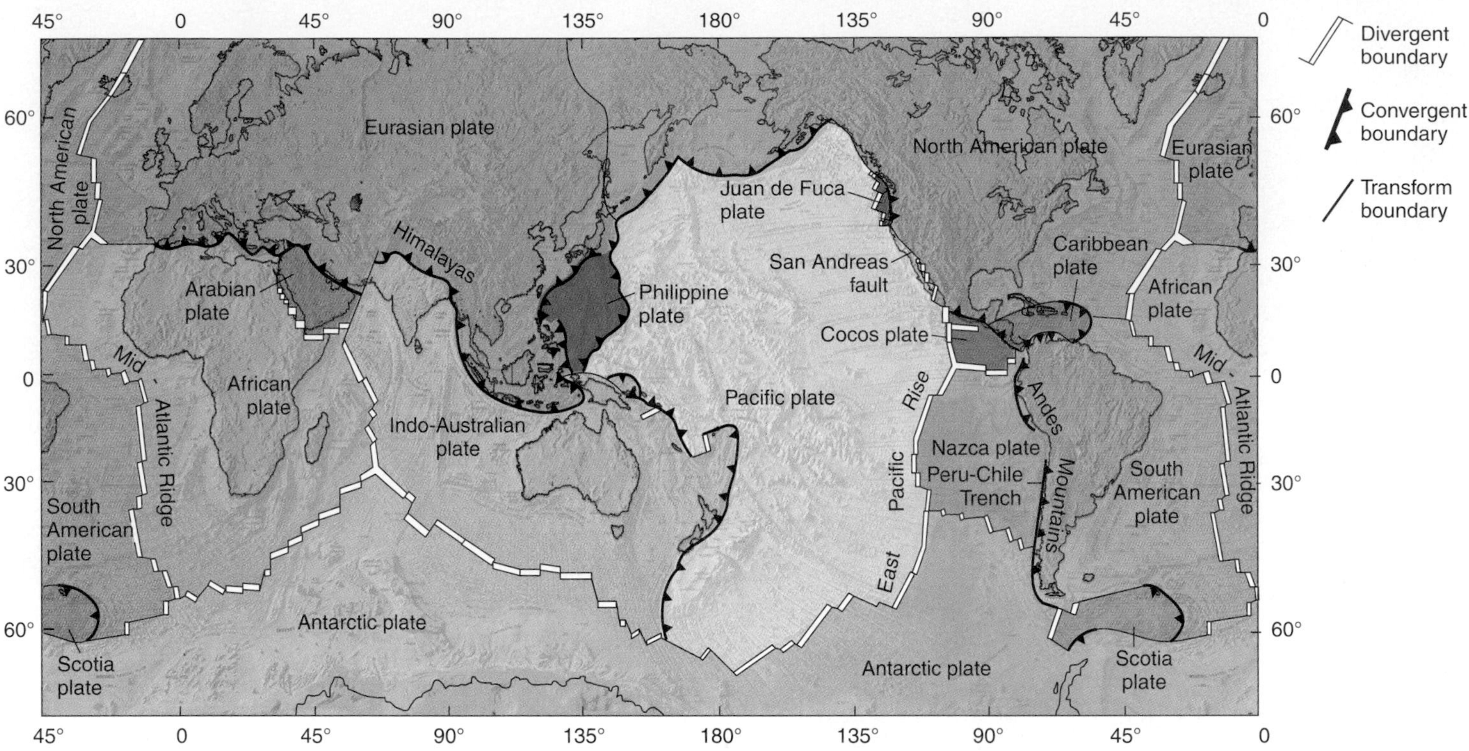

FIGURE 2.1

The major plates of the world. The western edge of the map repeats the eastern edge so that all plates can be shown unbroken. Double lines indicate spreading axes on divergent plate boundaries. Single lines show transform boundaries. Heavy lines with triangles show convergent boundaries, with triangles pointing down subduction zones. *Modified from W. Hamilton, U.S. Geological Survey*

WHAT IS PLATE TECTONICS?

Tectonics is the study of the origin and arrangement of the broad structural features of the Earth's surface, including not only folds and faults, but also mountain belts, continents, and earthquake belts. Tectonic models such as an expanding Earth or a contracting Earth have been used in the past to explain *some* of the surface features of the Earth. Plate tectonics has come to dominate geologic thought today because it can explain so *many* features. The basic idea of **plate tectonics** is that the Earth's surface is divided into a few large, thick plates that move slowly and change in size. Intense geologic activity occurs at *plate boundaries* where plates move away from one another, past one another, or toward one another. The eight large plates shown in figure 2.1, plus a few dozen smaller plates, make up the outer shell of the Earth (the crust and upper part of the mantle).

Plate tectonics has become a unifying theory of geology because it can explain so many diverse features of Earth. Earthquake distribution, the origin of mountain belts, the origin of sea-floor topography, the distribution and composition of volcanoes, and many other features can all be related to plate tectonics. It is a convenient framework that unifies geologic thought, associating features that were once studied separately and relating them to a single cause: plate interactions at plate boundaries.

The concept of plate tectonics was developed in the late 1960s by combining two pre-existing ideas—continental drift and sea-floor spreading. **Continental drift** is the idea that continents move freely over the Earth's surface, changing their positions relative to one another. **Sea-floor spreading** is a hypothesis that the sea floor forms at the crest of mid-oceanic ridges, then moves horizontally away from the ridge crest toward an oceanic trench. The two sides of the ridge are moving in opposite directions like slow conveyor belts.

Before we take a close look at plates we will examine the earlier ideas of moving continents and a moving sea floor, because these two ideas embody the theory of plate tectonics.

HOW DID THE PLATE TECTONICS THEORY EVOLVE?

The Early Case for Continental Drift

Continents can be made to fit together like pieces of a picture puzzle. The similarity of the Atlantic coastlines of Africa and South America has long been recognized. The idea that continents were once joined together, and have split and moved apart from one another, has been around for more than 130 years (figure 2.2).

In the early 1900s Alfred Wegener, a German meteorologist (refer to box 1.3), made a strong case for continental drift. He noted that South America, Africa, India, Antarctica, and Australia had almost identical late Paleozoic rocks and fossils.

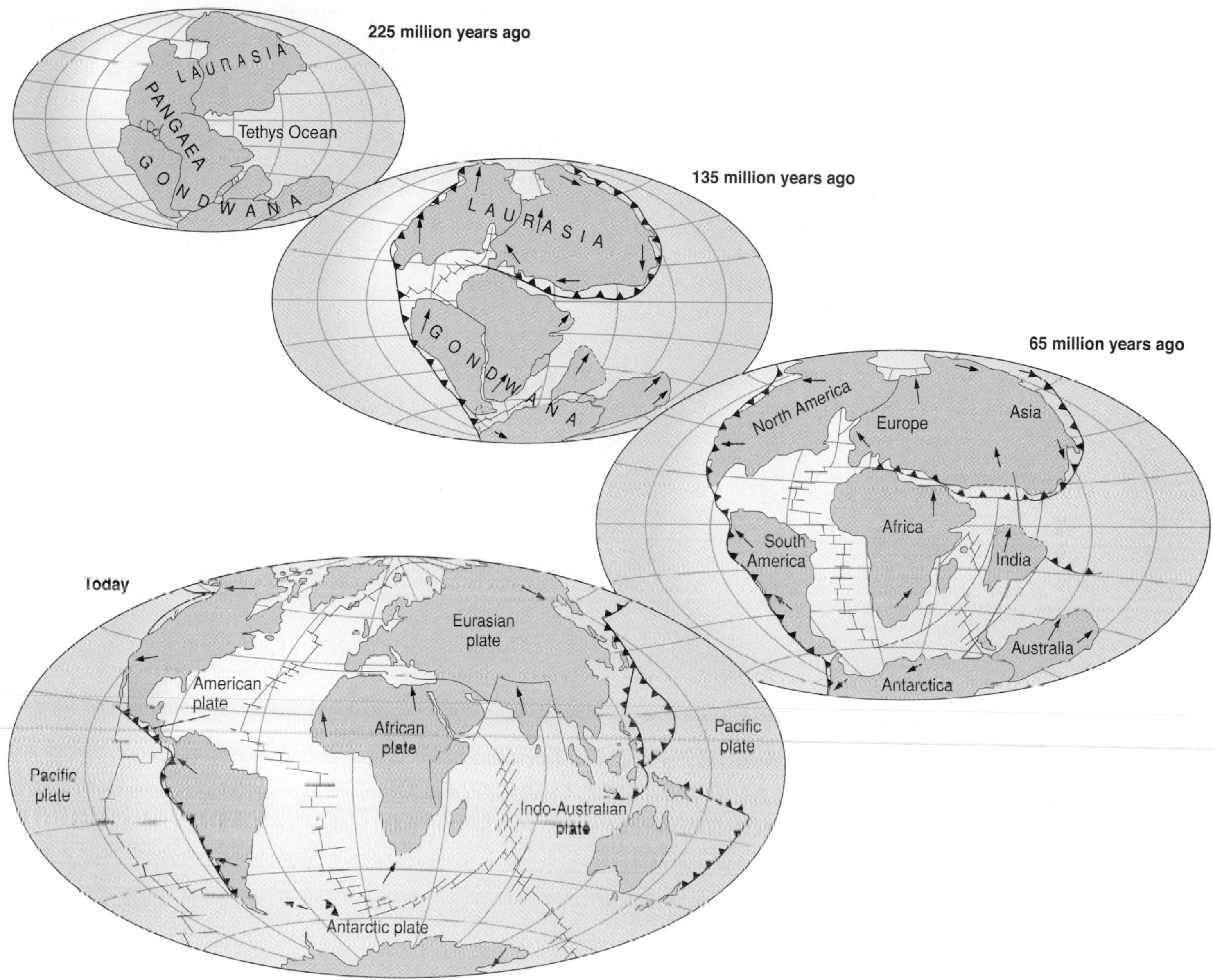

FIGURE 2.2

Pangea breakup and continental drift.
After C. R. Scotese (www.scotese.com)

The plant *Glossopteris* is found in Pennsylvanian and Permian-age rock on all five continents, and fossil remains of *Mesosaurus,* a freshwater reptile, is found in Permian-age rocks only in Brazil and South Africa (figure 2.3). In addition, fossil remains of land-dwelling reptiles *Lystrosaurus* and *Cynognathus* are found in Triassic-age rocks on all five continents.

Wegener reassembled the continents to form a giant supercontinent, called *Pangea* (also spelled Pangaea). Wegener thought that the similar rocks and fossils were easier to explain if the continents were joined together, rather than in their present, widely scattered positions.

Pangea initially separated into two parts. *Laurasia* was the northern supercontinent, containing what is now North America and Eurasia (excluding India). *Gondwanaland* was the southern supercontinent, composed of all the present-day southern-hemisphere continents and India (which has drifted north).

The distribution of Late Paleozoic glaciation strongly supports the idea of Pangea (figure 2.4). The Gondwanaland continents (the southern-hemisphere continents and India) all have glacial deposits of Late Paleozoic age. If these continents were spread over the Earth in Paleozoic time as they are today, a climate cold enough to produce extensive glaciation would have had to prevail over almost the whole world. Yet no evidence has been found of widespread Paleozoic glaciation in the northern hemisphere. In fact, the late Paleozoic coal beds of North America and Europe were being laid down at that time in swampy, probably warm environments.

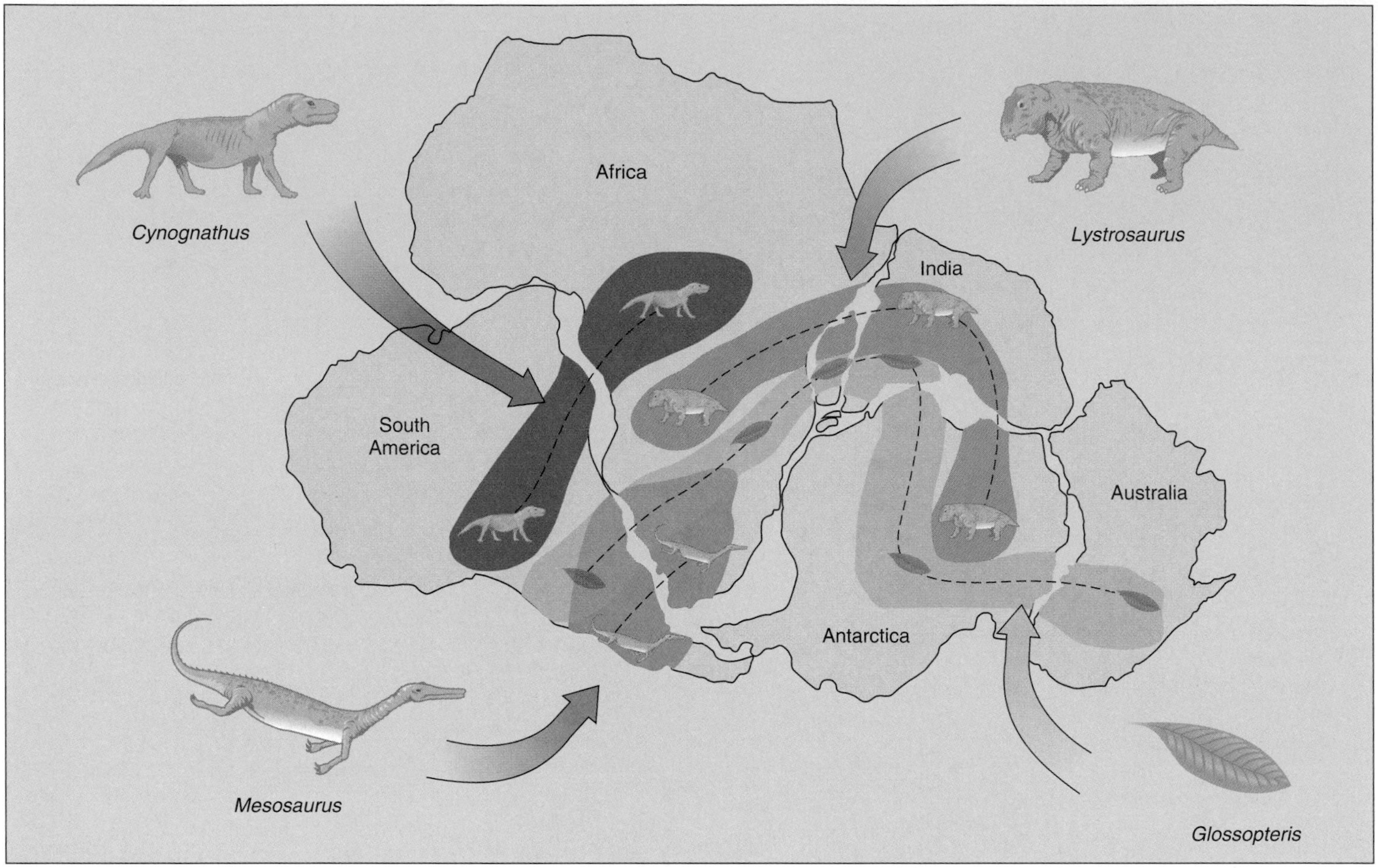

FIGURE 2.3

Distribution of plant and animal fossils that are found on the continents of South America, Africa, Antarctica, India, and Australia give evidence for the southern supercontinent of Gondwana. *Glossopteris* and other fernlike plants are found in Permian- and Pennsylvanian-age rocks on all five continents. *Cynognathus* and *Lystrosaurus* were sheep-sized land reptiles that lived during the Early Triassic Period. Fossils of the freshwater reptile *Mesosaurus* are found in Permian-age rocks on the southern tip of Africa and South America.

If the continents are arranged according to Wegener's Pangea reconstruction, then glaciation in the southern hemisphere is confined to a much smaller area (figure 2.4), and the absence of widespread glaciation in the northern hemisphere becomes easier to explain. Also, the present arrangement of the continents would require that late Paleozoic ice sheets flowed from the oceans toward the continents, which is impossible.

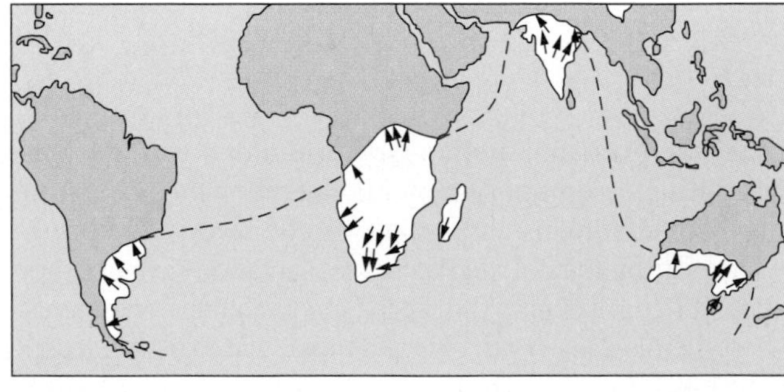

A

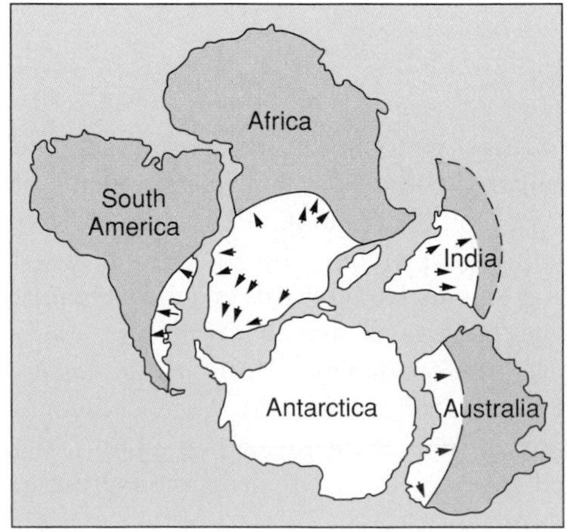

B

FIGURE 2.4

Distribution of late Paleozoic glaciation; arrows show direction of ice flow. (*A*) Continents in present positions show wide distribution of glaciation (white land areas with flow arrows). (*B*) Continents reassembled into Pangea. Glaciated region becomes much smaller.

From Arthur Holmes, 1965, Principles of Physical Geology, *2d ed., Ronald Press*

Wegener also reconstructed old climate zones (the study of ancient climates is called *paleoclimatology*). Glacial features indicate a cold climate near the North or South Pole. Coral reefs indicate warm water near the equator. Crossbedded sandstones can indicate ancient deserts near 30 degrees North and 30 degrees South latitude. If ancient climates had the same distribution on Earth that modern climates have, then sedimentary rocks can show where the ancient poles and equator were located. When Wegener examined ancient sedimentary rocks he discovered that paleoclimatic reconstructions suggested polar positions very different to those at present. This could be interpreted as evidence for changes in the position of the poles over time. Wegener, however, hypothesized that this indicated the continents had moved and gave strong support to his concept of continental drift.

Skepticism about Continental Drift

Although Wegener presented the best case possible in the early 1900s for continental drift, much of his evidence was not clear-cut and he had no good mechanism to account for continental movement.

Wegener proposed that continents ploughed through the oceanic crust (figure 2.5), perhaps crumpling up mountain ranges on the leading edges of the continents where they pushed against the sea floor. Most geologists thought that this idea violated what was known about the strength of rocks at the time. The driving mechanism proposed by Wegener for continental drift was a combination of centrifugal force from the Earth's rotation and the gravitational forces that cause tides. Careful calculations of these forces showed them to be too small to move continents. Because of these objections, Wegener's ideas received little support in North America or much of the northern hemisphere (where the great majority of geologists live) in the first half of the twentieth century. The few geologists in the southern hemisphere, however, where Wegener's matches of fossils and rocks between continents were more evident, were more impressed with the concept of continental drift.

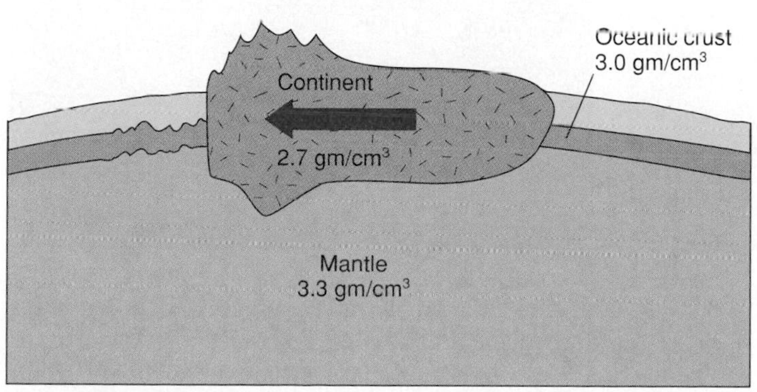

FIGURE 2.5

Wegener's concept of continental drift implied that the less dense continents drifted *through* oceanic crust, crumpling up mountain ranges on their leading edges as they pushed against oceanic crust.

Renewed Interest in Continental Drift

Much work in the 1940s and 1950s set the stage for the revival of the idea of continental drift and its later incorporation, along with new ideas about sea-floor spreading, into the concept of plate tectonics. The new investigations were in two areas: (1) study of the sea floor and (2) geophysical research, especially in relation to rock magnetism.

Study of the Sea Floor

Oceans cover more than 70 percent of the Earth's surface and make the study of sea-floor rocks a difficult and challenging task. However, new methods of investigating sea-floor rocks, sediment, and topography that were developed during the mid–twentieth century have provided most of the information that led to the concept of plate tectonics.

Samples of rock and sediments can be taken from the sea floor in several ways. Rocks can be broken from the sea floor by a *rock dredge,* which is an open steel container dragged over the ocean bottom at the end of a cable. Sediments can be sampled with a *corer,* a weighted steel pipe dropped vertically into the mud and sand of the ocean floor. Both rocks and sediments can be sampled by means of *sea floor drilling.* Offshore oil platforms drill holes in the relatively shallow sea floor near shore. A ship with a drilling derrick on its deck can drill a hole in the deep sea floor far from land (figure 2.6*A*). The drill cuts long, rodlike rock cores from the ocean floor (figure 2.6*B*). Thousands of such holes have been drilled in the sea floor, and the rock and sediment cores recovered from these holes have revolutionized the field of marine geology. In the 1950s more was known about the moon's surface than about the floor of the sea. Sea-floor drilling has been instrumental in expanding our knowledge of sea floor features and history. Small research submarines, more correctly called *submersibles,* can take geologists to many parts of the sea floor to observe, photograph, and sample rock and sediment (figure 2.6*C*).

A basic tool for indirectly studying the sea floor is the *single-beam echo sounder,* which measures water depth and draws profiles of submarine topography. A sound signal sent downward from a ship bounces off the sea floor and returns to the ship (figure 2.7*A*). The water depth is determined from the time it takes the sound to make the round trip. Multibeam sonar uses a variety of sound sources to produce detailed shaded relief images of the sea-floor topography (figure 2.7*B*). **Sidescan sonar** measures the intensity of sound reflected back to the tow vehicle from the ocean floor and provides detailed images of the sea floor and information about sediments and bedforms (figure 2.7*A*). A *seismic reflection profiler* works on essentially the same principles as the echo sounders but uses a louder noise at lower frequency. This sound penetrates the sea floor and reflects from layers within the underlying sediment and rock. The seismic profiler records water depth and reveals the internal structure of the rocks and sediments of the sea floor, such as bedding planes, folds and faults, and unconformities. *Magnetic, gravity,* and *seismic refraction* surveys (see chapter 4) also can be made at sea. *Deep-sea cameras* can be lowered to the bottom to photograph the rock and sediment.

A

B

C

FIGURE 2.6

(A) The *Joides Resolution* is a ship specially built for sampling both sediment and rock from the deep ocean floor. (B) Sampling core on board *Joides Resolution*. (C) The small research submersible ALVIN of Woods Hole Oceanographic Institution in Massachusetts; it is capable of taking three oceanographers to a depth of about 4,000 metres.

Photos A, B by Nick Eyles, photo C by Woods Hole Oceanographic Institution

A

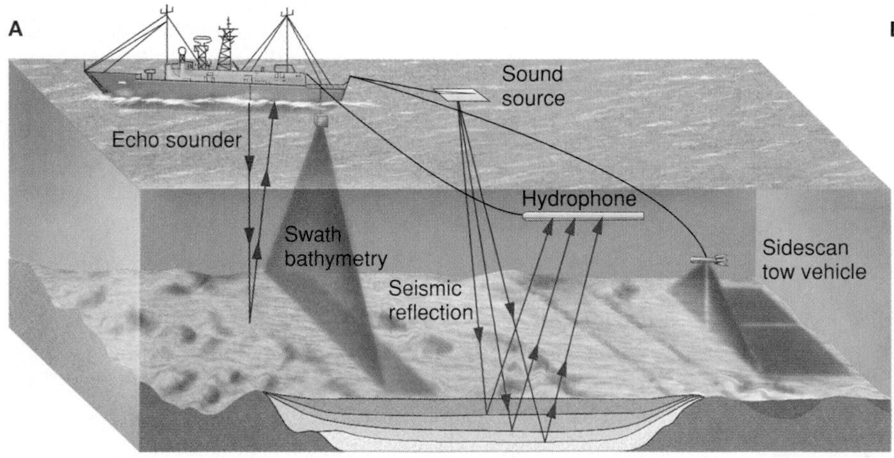

B

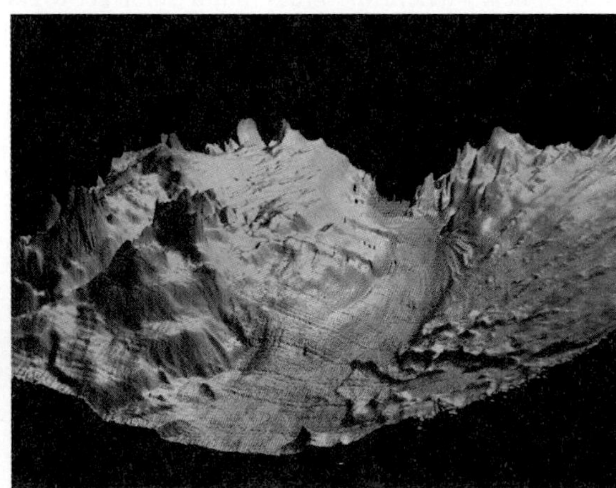

FIGURE 2.7

(A) Diagram showing how echo sounding, seismic reflection and sidescan sonar are used to study the sea floor. Modified from US Geological Survey Fact Sheet 039-02. (B) Colour-coded shaded relief image of the sea floor along the eastern side of Placentia Bay, Newfoundland produced from multibeam bathymetric data. This image shows a channel in the centre with bedrock outcrops to the west (left) side.

Reproduced with the permission of the Minister of Public Works and Government Services Canada 2006 and courtesy of Natural Resources Canada, Geological Survey of Canada.

Using new data collected by these methods, detailed maps of the ocean floors were created in the early 1960s. These maps showed extensive, linked submarine mountain chains (mid-ocean ridges), deep trenches (ocean trenches), fracture and fault zones, and conical seamounts (figure 2.8). All of these features indicated that the sea floors were sites of intense geologic activity and supported the notion of a dynamic Earth surface.

FIGURE 2.8

The floor of the ocean. Note continental margins, mid-oceanic ridge, fracture zones, and oceanic trenches (compare with figures 2.21 and 2.27). Conical mountains are volcanic seamounts; aligned seamounts are aseismic ridges.

From World Ocean Floor by Bruce C. Heezen and Marie Tharp, 1977. Copyright © Marie Tharp 1977. Reproduced by permission of Marie Tharp, 1 Washington Ave., South Nyack, NY 10960

Geophysical Research

Convincing new evidence about **polar wandering** came from the study of rock magnetism. Wegener's work dealt with the wandering of Earth's *geographic* poles of rotation. The *magnetic* poles are located close to the geographic poles. Historical measurements show that the position of the magnetic poles moves from year to year, but that the magnetic poles stay close to the geographic poles as they move. As we discuss magnetic evidence for polar wandering, we are referring to an apparent motion of the magnetic poles. Because the magnetic and geographic poles are close together, our discussion will refer to apparent motion of the geographic poles as well.

As we will discuss in chapter 4, many rocks record the strength and direction of the Earth's magnetic field at the time the rocks formed. Magnetite in a cooling basaltic lava flow acts like a tiny compass needle, preserving a record of Earth's magnetic field when the lava cools below the *Curie point.* Sedimentary rocks such as red shale contain other iron oxides and can also record Earth's magnetism. The magnetism of old rocks can be measured to determine the direction and strength of the magnetic field in the past. The study of ancient magnetic fields is called *paleomagnetism.*

Because magnetic lines of force are inclined more steeply as the north magnetic pole is approached, the inclination (dip) of the magnetic alignment preserved in the magnetite minerals in the lava flows can be used to determine the paleolatitude at which the flow formed (figure 2.9). Old rocks reveal very different magnetic pole positions to those at present and it was once thought that this indicated movement of the poles ("polar wandering"). It is now recognized that the rocks (and their paleomagnetic records) have moved as part of migrating tectonic plates. Apparent polar wandering paths are now used to reconstruct continental movement over time (figure 2.10).

Permian rocks in North America point to a pole position in eastern Asia, but Permian rocks in *Europe* point to a different position (closer to Japan), as shown in figure 2.10. Does this mean there were *two* north magnetic poles in the Permian period? In fact, every continent shows a different position for the Permian pole. A different magnetic pole for each continent seems highly unlikely. A better explanation is that a single pole stood still while continents split apart and rotated as they moved.

Note the polar wandering paths for North America and Europe in figure 2.10. The paths are of similar shape, but the path for European poles is to the east of the North American path. If we mentally push North America back toward Europe, closing the Atlantic Ocean, and then consider the paths of polar wandering, we find that the path for North America lies exactly on the path for Europe. This strongly suggests that the continents were joined together.

Recent Evidence for Continental Drift

As paleomagnetic evidence revived interest in continental drift, new work was done on fitting continents together. By defining the edge of a continent as the middle of the continental slope, rather than the present (constantly changing) shoreline, a much more precise fit has been found between continents (figure 2.11).

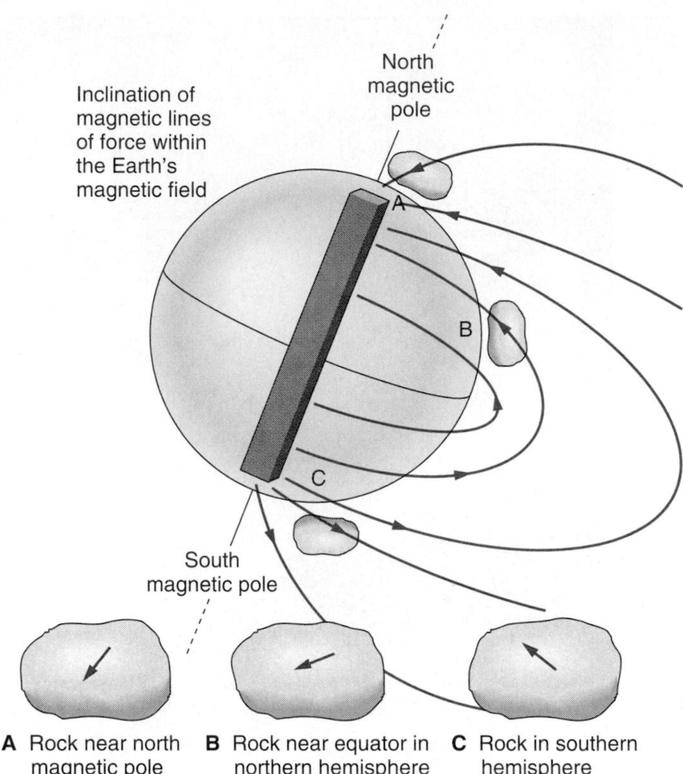

FIGURE 2.9

Magnetic inclination is steep at the south magnetic pole, decreases steadily toward the horizontal at the equator and then increases toward magnetic north. Rocks in bottom part of figure are small samples viewed horizontally at locations A, B, and C on the globe. Magnetic inclination can therefore be used to determine the distance from a rock to the north magnetic pole.

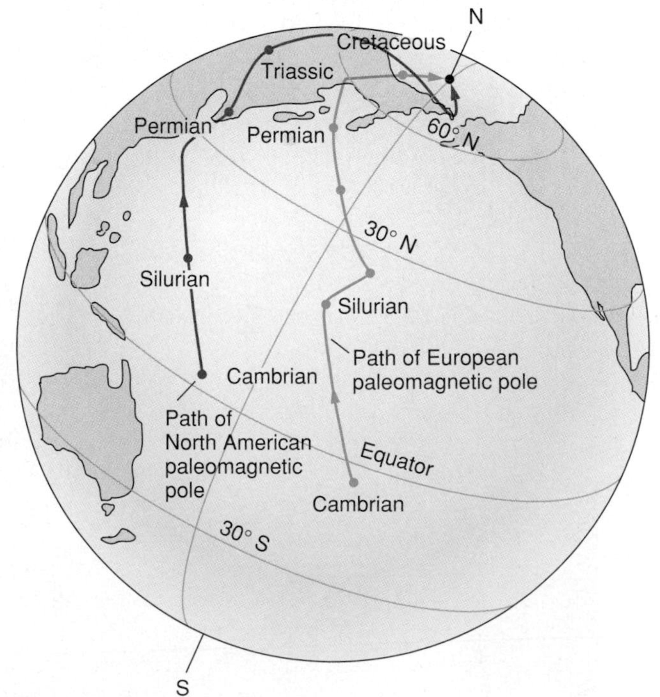

FIGURE 2.10

Apparent polar wandering of the north magnetic pole for the past 520 million years as determined from measurements of rocks from North America (red) and Europe (green).

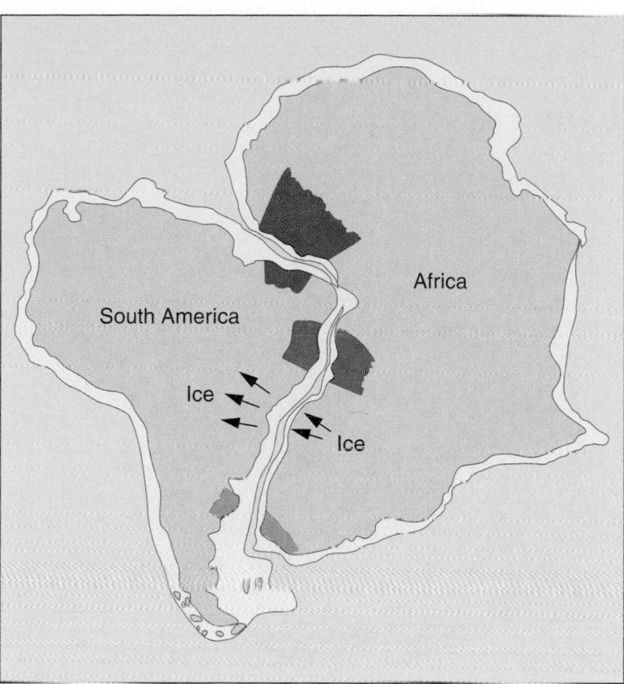

FIGURE 2.11

Jigsaw puzzle fit and matching rock types between South America and Africa. Light-blue areas around continents are continental shelves (part of continents). Coloured areas within continents are broad belts of rock that correlate in type and age from one continent to another. Arrows show direction of glacier movement as determined from striations on bedrock surfaces.

The most convincing evidence for continental drift came from greatly refined rock matches between now-separated continents. If continents are fitted together like pieces of a jigsaw puzzle, geological features should match from piece to piece.

The matches between South America and Africa are particularly striking. Some distinctive rock contacts extend out to sea along the shore of Africa. If the two continents are fitted together, the identical contacts are found in precisely the right position on the shore of South America (figure 2.11). Isotopic ages of rocks also match between these continents.

Glacial striations show that during the late Paleozoic Era continental glaciers moved from Africa toward the present Atlantic Ocean, while similar glaciers seemingly moved *from* the Atlantic Ocean *onto* South America (figure 2.11). Continental glaciers, however, cannot move from sea onto land. If the two continents had been joined together, the ice that moved off Africa could have been the ice that moved onto South America. This hypothesis has now been confirmed; from their lithology, many of the boulders in South American glacial deposits have been traced to a source that is now in Africa.

Some of the most detailed matches have been made between rocks in Brazil and rocks in the African country of Gabon. These rocks are similar in type, structure, sequence, fossils, ages, and degree of metamorphism. Such detailed matches are convincing evidence that continental drift does, in fact, take place. There is now an abundance of satellite geodetic data from the Global Positioning System (GPS), that allow us to watch the continents move in real time (see box 2.1).

History of Continental Positions

Rock matches show when continents were together; once the continents split, the new rocks formed are dissimilar. Paleomagnetic evidence indicates the direction and rate of drift, allowing maps of old continental positions, such as figure 2.2, to be drawn.

Although Pangea split up 200 million years ago to form our present continents, the continents were moving much earlier. Pangea was formed by the collision of many small continents long before it split up. Recent work shows that continents have been in motion for at least the past 2 billion years (some geologists say 4 billion years), well back into Precambrian time. For more than half of Earth's history, the continents appear to have collided, welded together, then split and drifted apart, only to collide again, over and over, in an endless, slow dance (refer to box 2.4 later in this chapter).

WHAT IS SEA-FLOOR SPREADING?

At the same time that many geologists were becoming interested again in the idea of moving *continents,* Harry Hess, a geologist at Princeton University, proposed that the *sea floor* might be moving, too. This proposal contrasted sharply with the earlier ideas of Wegener, who thought that the ocean floor remained stationary as the continents ploughed through it (figure 2.5). Hess's 1962 proposal was quickly named *sea-floor spreading*, for it suggests that the sea floor moves away from the mid-oceanic ridge as a result of mantle convection (figure 2.12).

According to the initial concept of sea-floor spreading, the sea floor is moving like a conveyor belt away from the crest of the mid-oceanic ridge, down the flanks of the ridge, and across the deep-ocean basin, finally to disappear by plunging beneath a continent or island arc (figure 2.12). The ridge crest, with sea floor moving away from it on either side, has been called a *spreading axis* (or *spreading centre*). The sliding of the sea floor beneath a continent or island arc is termed **subduction.** Hess's original hypothesis was that sea-floor spreading is driven by deep mantle convection. **Convection** is a circulation pattern driven by the rising of hot material and/or the sinking of cold material. Hot material has a lower density, so it rises; cold material has a higher density, and sinks. A slow convective circulation of a few centimetres per year is set up by temperature differences within the mantle, and convection can explain many sea-floor features as well as the young age of the sea-floor rocks.

If convection drives sea-floor spreading, then hot mantle rock must be rising under *mid-oceanic ridges*. Hess showed how the existence of ridges and their *high heat flow* are caused by the rise of this hot mantle rock. The *basalt eruptions* on ridge crests are also related to this rising rock, for here the mantle rock is hotter than normal and begins to undergo partial melting.

As hot rock continues to rise beneath ridge crests, the circulation pattern splits and diverges near the surface. Mantle rock moves horizontally away from ridge crests on each side of the ridge. This movement creates tension at the ridge crest, cracking open the oceanic crust to form *rift valleys* and associated *shallow-focus earthquakes.*

GEOMATICS 2.1

Measuring Plate Movement in Real Time

Satellite and Global Positioning System (GPS) data can now be used to measure plate movement directly and on a global scale. During the late 1970s there was rapid development of *space geodesy*, a space-based technique for taking very precise measurements of points on the Earth's surface. The three most commonly used space-geodetic techniques are very long baseline interferometry (VLBI), which uses pairs of radio telescopes pointed toward a common quasar; satellite laser ranging (SLR); and the Global Positioning System (GPS) (box figure 1).

GPS has proved to be the most useful of the three techniques for studying movements of the Earth's crust and is routinely used to measure the relative motion between plates. Twenty-four satellites currently orbit the Earth as part of the NavStar system and continuously transmit radio signals back to Earth. A series of GPS ground stations have been established to receive these signals and record the precise time and location of each satellite when the signal was received. Using these data, geoscientists are able to measure changing distances between GPS sites and determine if there has been active movement along faults or between plates (box figure 2). Plate motions are now recorded on a yearly basis throughout the world. Rates and directions of plate movement measured from space geodetic data and averaged over periods of several years are remarkably similar to rates and directions calculated on the basis of geological data and averaged over millions of years.

BOX 2.1 ■ FIGURE 1

Schematic diagram showing the 24 satellites that form the global GPS space-based system. These satellites are constantly moving, making two complete orbits in less than 24 hours, and travel at speeds of more than 11,000 km/h.

Courtesy of Garmin International

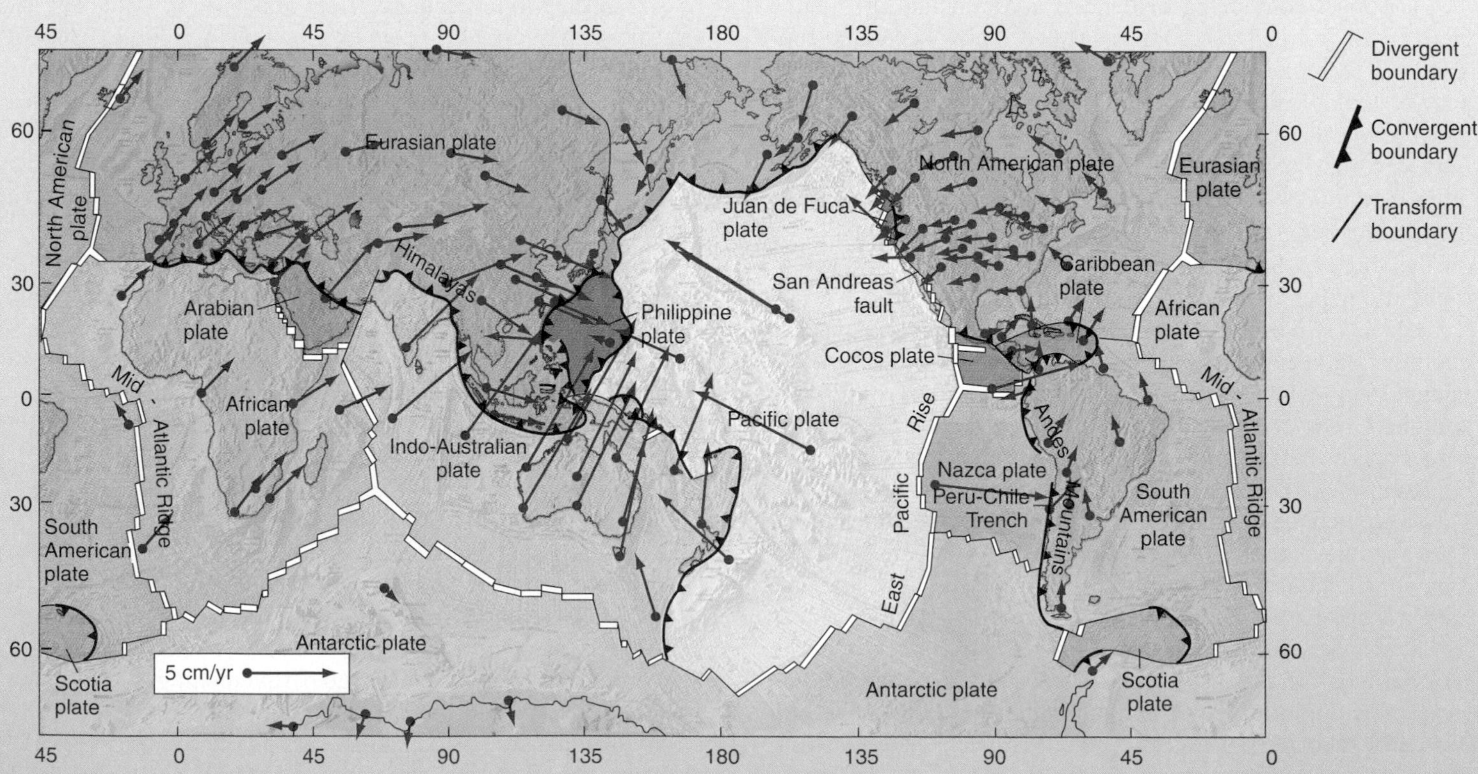

BOX 2.1 ■ FIGURE 2

Yearly plate motions measured from GPS stations around the world.

From NASA (http://slideshow.jpl.nasa.gov/mbh/series.html)

Measurement of relative plate movement also provides valuable data regarding potential earthquake activity along active plate margins. GPS data are currently being used to monitor interactions between the Pacific plate and surrounding continental plates in the hope of learning more about earthquake and volcanic activity in the Pacific Ring of Fire (see box 2.5). The Western Canada Deformation Array (WCDA) is a permanent network of GPS stations established by the Geological Survey of Canada to investigate crustal movements in southwestern British Columbia (box figure 3). Recent data show that GPS sites on southern Vancouver Island occasionally reverse their motion and move seaward for periods of about two weeks before resuming their landward track. One of the most recent interpretations of these data suggests that slow-slip events are occurring on the subduction fault between sections of the Juan de Fuca and North American plates. This has implications for prediction of the timing and severity of future earthquakes as the increased stress caused by these slow-slip events may be sufficient to trigger a great earthquake. Monitoring the slow but variable movements along plate boundaries using GPS data will significantly enhance understanding of plate behaviour.

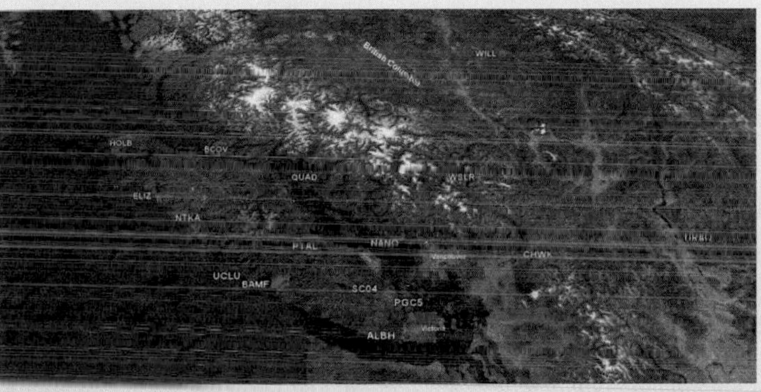

A

B

BOX 2.1 ■ FIGURE 3

A. Location of GPS stations that form the Western Canada Deformation Array (WCDA).

B. GPS antenna on concrete pier at Nootka Island, B.C. This station forms part of the WCDA. Photo. M. Schmidt, Geological Survey of Canada, NRCan.

From http://gsc.nrcan.gc.ca/geodyn/wcda/info_e.php M. Schmidt, Geological Survey of Canada, NRCan. Reproduced with the permission of the Minister of Pulbic Works and Government Services Canada, 2006 and Courtesy of Natural Resources Canada, Geological Survey of Canada.

As the mantle rock moves horizontally away from ridge crests, it carries the sea floor (the basaltic oceanic crust) piggyback along with it. As the hot rock moves sideways, it cools and becomes denser, sinking deeper beneath the ocean surface. Hess thought it would become cold and dense enough to sink back into the mantle. This downward plunge of cold rock accounts for the existence of the *oceanic trenches* as well as for their *low heat flow* values. It also explains the large *negative gravity anomalies* associated with trenches, for the sinking of the cold rock provides a force that holds trenches out of isostatic equilibrium (see chapter 4).

As the sea floor moves downward into the mantle along a subduction zone, it interacts with the stationary rock above it. This interaction between the moving sea-floor rock and the stationary rock can cause the *Benioff zones of earthquakes* associated with trenches. It can also produce *andesitic volcanism,* which forms volcanoes either on the edge of a continent or along an island arc (figure 2.12).

Hess's ideas have stood up remarkably well over more than 40 years. We now think of plates moving instead of sea floor riding piggyback on convecting mantle, and we think that several mechanisms cause plate motion, but Hess's explanation of sea-floor topography, earthquakes, and age remains valid today.

How Old Is the Sea Floor?

As marine geologists began to determine the age of sea-floor rocks (by isotopic dating) and sediments (by fossils), an astonishing fact was discovered. All the rocks and sediments of the sea floor proved to be younger than 200 million years old. This was true only for rocks and sediments from the *deep* sea floor, not those from the continental margins. The rocks and sediments currently found on the deep sea floor formed during the Mesozoic and Cenozoic eras, but not earlier.

By contrast, as discussed in chapter 1, the Earth is estimated to be 4.5 billion years old. Every continent contains some rocks

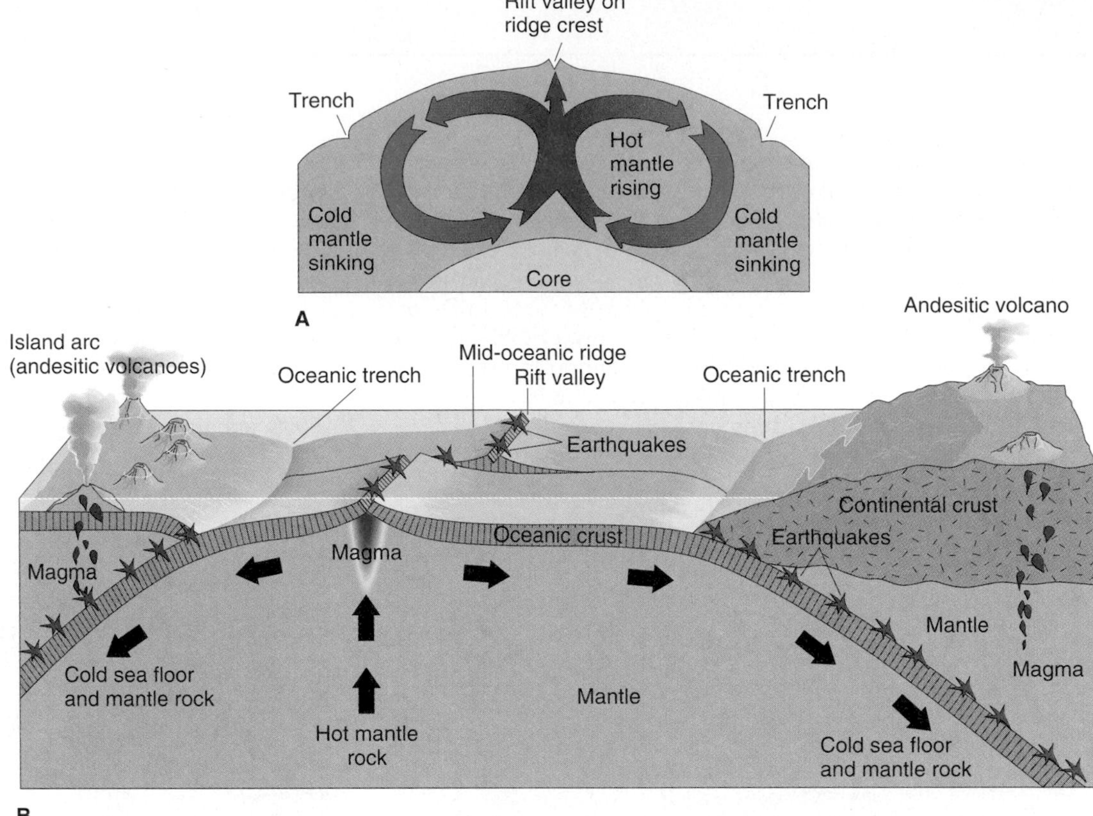

FIGURE 2.12

Sea-floor spreading hypothesis of Harry Hess. (*A*) Hess proposed that convection extended throughout the mantle. (Scale of ridge and trenches is exaggerated.) (*B*) Hot mantle rock rising beneath the mid-oceanic ridge (a spreading axis) causes basaltic volcanism and high heat flow. Divergence of sea floor splits open the rift valley and causes shallow-focus earthquakes (stars on ridge). Sinking of cold rock causes subduction of older sea floor at trenches, producing Benioff zones of earthquakes and andesitic magma.

formed during the Paleozoic Era and the Precambrian. Some of the Precambrian rocks on continents are more than 3 billion years old, and a few are almost 4 billion years old. Continents, therefore, preserve rocks from most of the Earth's history. In sharp contrast to the continents is the deep sea floor, which covers more than half of Earth's surface but preserves less than one-twentieth of its history in its rocks and sediment.

The *young age of sea-floor rocks* is neatly explained by Hess's sea-floor spreading. New, young sea floor is continually being formed by basalt eruptions at the ridge crest. This basalt is then carried sideways by convection and is subducted into the mantle at an oceanic trench. Thus, old sea floor is continually being destroyed at trenches, while new sea floor is being formed at the ridge crest. (This is also the reason for the puzzling lack of pelagic sediment at the ridge crest. Young sea floor at the ridge crest has little sediment because the basalt is newly formed. Older sea floor farther from the ridge crest has been moving under a constant rain of pelagic sediment, building up a progressively thicker layer as it goes.)

Note that sea-floor spreading implies that the youngest sea floor should be at the ridge crest, with the age of the sea floor becoming progressively older toward a trench. This increase in age away from the ridge crest was not known to exist at the time of Hess's proposal but was an important prediction of his hypothesis. This prediction has been successfully tested, as you shall see later in this chapter when we discuss marine magnetic anomalies.

WHAT ARE PLATES AND HOW DO THEY MOVE?

By the mid-1960s the twin ideas of moving continents and a moving sea floor were causing great excitement and emotional debate among geologists. By the late 1960s, these ideas had been combined into a single theory that revolutionized geology by providing a unifying framework for Earth science—the theory of plate tectonics.

A **plate** is a large, mobile slab of rock that is part of Earth's surface (figure 2.1). The surface of a plate may be made up entirely of sea floor (as is the Nazca plate), or it may be made up of both continental and oceanic rock (as is the North American plate). Some of the smaller plates are entirely continental, but all the large plates contain some sea floor.

The plates are part of a relatively rigid outer shell of Earth called the **lithosphere.** The lithosphere includes the rocks of the crust and uppermost mantle (figure 2.13).

The lithosphere beneath oceans increases in both age and thickness with distance from the crest of the mid-oceanic ridge. Young lithosphere near the ridge crest may be only 10 km thick, while very old lithosphere far from the ridge crest may be as much as 100 km thick (figure 2.13).

Continental lithosphere is thicker, varying from perhaps 125 km thick to as much as 200 to 250 km thick beneath the oldest, coldest, and most inactive parts of the continents.

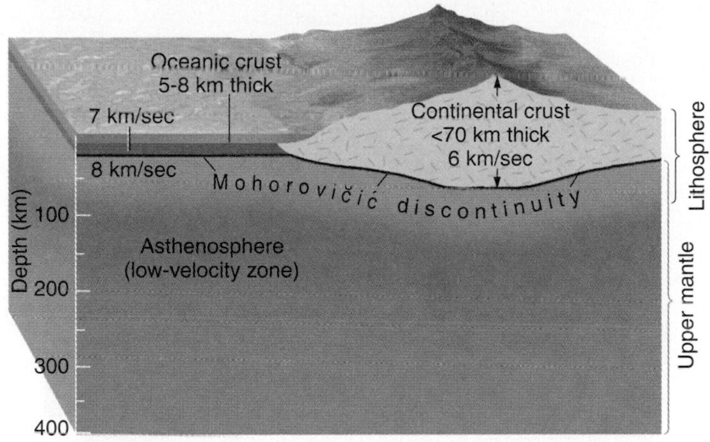

FIGURE 2.13

The rigid lithosphere includes the crust and uppermost mantle; it forms the plates. The ductile asthenosphere acts as a lubricating layer beneath the lithosphere. Oceanic lithosphere averages 70 kilometres thick; continental lithosphere varies from 125 to 250 kilometres thick. Asthenosphere may not be present under continents.

Below the rigid lithosphere is the **asthenosphere,** a zone of low-seismic-wave velocity that behaves plastically because of increased temperature and pressure. The plastic asthenosphere acts as a lubricating layer under the lithosphere, allowing the plates to move. The asthenosphere, made up of upper-mantle rock, is the low-velocity zone and will be described in chapter 4. It may extend from a depth of 70 to 200 km beneath oceans; its thickness, depth, and even existence under continents is vigorously debated. Below the asthenosphere is more rigid mantle rock.

The idea that plates move is widely accepted by geologists, although the reasons for this movement are debated. Plates move away from the mid-oceanic ridge crest. Some plates move toward oceanic trenches. If the plate is made up mostly of sea floor (as are the Nazca and Pacific plates), the plate can be subducted down into the mantle, forming an oceanic trench and its associated features. If the leading edge of the plate is made up of continental rock (as is the South American plate), that plate will not subduct. Continental rock, being less dense (specific gravity 2.7) than oceanic rock (specific gravity 3.0), is too light to be subducted.

To a first approximation, a plate may be viewed as a rigid slab of rock that moves as a unit. As a result, the interior of a plate is relatively inactive tectonically. Plate interiors generally lack earthquakes, volcanoes, young mountain belts, and other signs of geologic activity. According to plate-tectonic theory, these features are caused by plate interactions at plate boundaries.

Plate boundaries are of three general types, based on whether the plates move away from each other, move toward each other, or move past each other. A **divergent plate boundary** is a boundary between plates that are moving apart. A **convergent plate boundary** lies between plates that are moving toward each other. A **transform plate boundary** is one at which two plates move horizontally past each other.

HOW DO WE KNOW THAT PLATES MOVE?

The proposal that the Earth's surface is divided into moving plates was an exciting, revolutionary hypothesis, but it required testing to win acceptance among geologists. You have seen how the study of paleomagnetism supports the idea of moving continents. In the 1960s two critical tests were made of the idea of a moving sea floor. These tests involved marine magnetic anomalies and the seismicity of fracture zones. These two successful tests convinced most geologists that plates do indeed move.

Paleomagnetic Evidence

The Earth's magnetic field has periodically reversed its polarity in the past. At such times, the north magnetic pole and the south magnetic pole exchange positions (this is discussed further in chapter 4). These changes in the polarity of the magnetic field are called **magnetic reversals**. During a period of **normal polarity** (such as the present), magnetic lines of force flow from the south pole to the north pole and our compass needles point to the north. During times of **reversed polarity** the lines of magnetic force run the other way and our compass needles would point toward the south.

Many rocks behave like compasses and record both the strength and direction of the Earth's magnetic field at the time they are formed. Study of ancient rocks can therefore tell us about changes in the Earth's magnetic field in the past. Study of ancient magnetic fields recorded in rocks is called **paleomagnetism**.

Lava flows contain abundant magnetic minerals and can be isotopically dated. Stacked continental lava flows have been used to construct a **magnetic polarity time scale** (figure 2.14) that records the pattern of magnetic reversals over time. It appears that the Earth's magnetic field reverses about every 500,000 years and it takes about 10,000 years for a reversal to develop.

The strength of the Earth's magnetic field has also changed over time. Any deviation of magnetic strength from average readings is called an **anomaly**. A **magnetometer** is an instrument that measures the strength of the Earth's magnetic field; it may be carried over the land surface or flown over land or sea.

Marine Magnetic Anomalies

In the mid-1960s, magnetometer surveys of the sea floor disclosed some intriguing characteristics of marine magnetic anomalies. Most magnetic anomalies on the sea floor are arranged in bands that lie parallel to the rift valley of the mid-oceanic ridge. Alternating positive and negative anomalies (chapter 4) form a stripelike pattern parallel to the ridge crest (figure 2.15).

The Morley-Vine-Matthews Hypothesis

Working independently, Canadian geophysicist Lawrence Morley and British geologists Fred Vine and Drummond Matthews made several important observations about these anomalies.

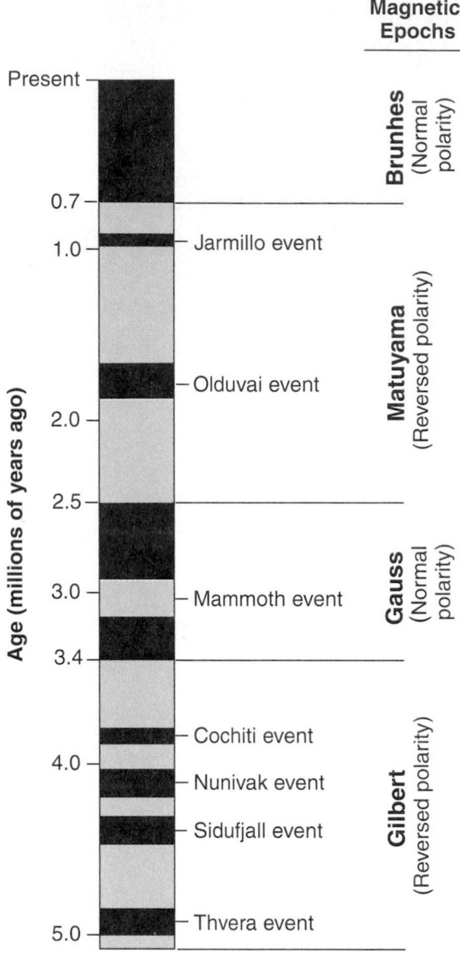

FIGURE 2.14

Magnetic reversals during the past 5 million years determined from lava flows that have been radiometrically dated. Black represents normal magnetism; tan represents reverse magnetism.

After Mankinen, E. A. and Dalrymple, G. B., 1979. Revised geomagnetic polarity time scale for the interval 0–5 m.y. B.P. Journal of Geophysical Research, *v. 84, p. 615–626.*

They recognized that the pattern of magnetic anomalies was symmetrical about the ridge crest. That is, the pattern of magnetic anomalies on one side of the mid-oceanic ridge was a mirror image of the pattern on the other side (figure 2.15). Morley, Vine, and Matthews also noticed that the same pattern of magnetic anomalies exists over different parts of the mid-oceanic ridge. The pattern of anomalies over the ridge in the northern Atlantic Ocean is the same as the pattern over the ridge in the southern Pacific Ocean.

The most important observation that Morley, Vine, and Matthews made was that the pattern of magnetic *anomalies* at sea matches the pattern of magnetic *reversals* already known from studies of lava flows on the continents (figure 2.14 and chapter 4). This correlation can be seen by comparing the pattern of coloured bands in figure 2.14 (reversals) with the pattern in figure 2.15 (anomalies).

Putting these observations together with Hess's concept of sea-floor spreading, which had just been published, Morley, Vine, and Matthews proposed an explanation for magnetic anomalies.

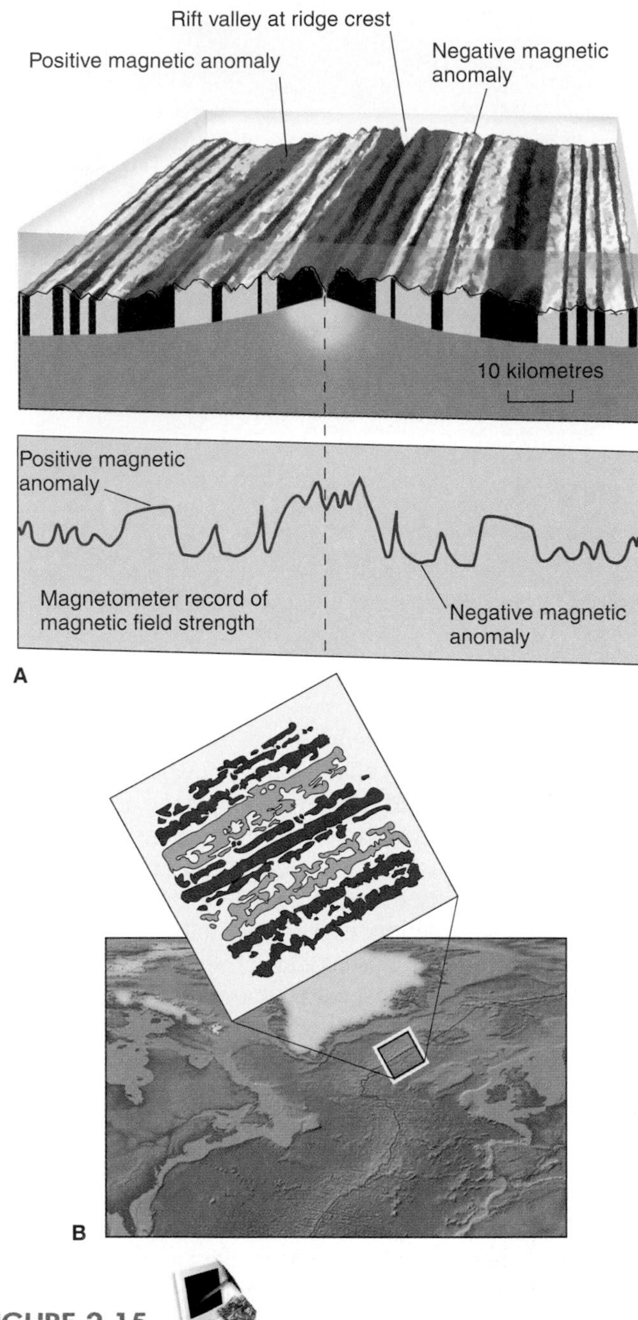

FIGURE 2.15

Marine magnetic anomalies. (*A*) The red line shows positive and negative magnetic anomalies as recorded by a magnetometer towed behind a ship. Positive anomalies are shown in black and negative anomalies are shown in tan. Notice how magnetic anomalies are parallel to the rift valley and symmetric about the ridge crest. (*B*) Symmetric magnetic anomalies ("stripes") from the mid-Atlantic ridge south of Iceland.

They suggested that there is continual opening of tensional cracks within the rift valley on the mid-oceanic ridge crest. These cracks on the ridge crest are filled by basaltic magma from below, which cools to form dikes. Cooling magma in the dikes records Earth's magnetism at the time the magnetic minerals crystallize. The process is shown in figure 2.16.

When Earth's magnetic field has a *normal polarity* (the present orientation), cooling dikes are normally magnetized. Dikes that cool when the field is reversed (figure 2.16) are reversely

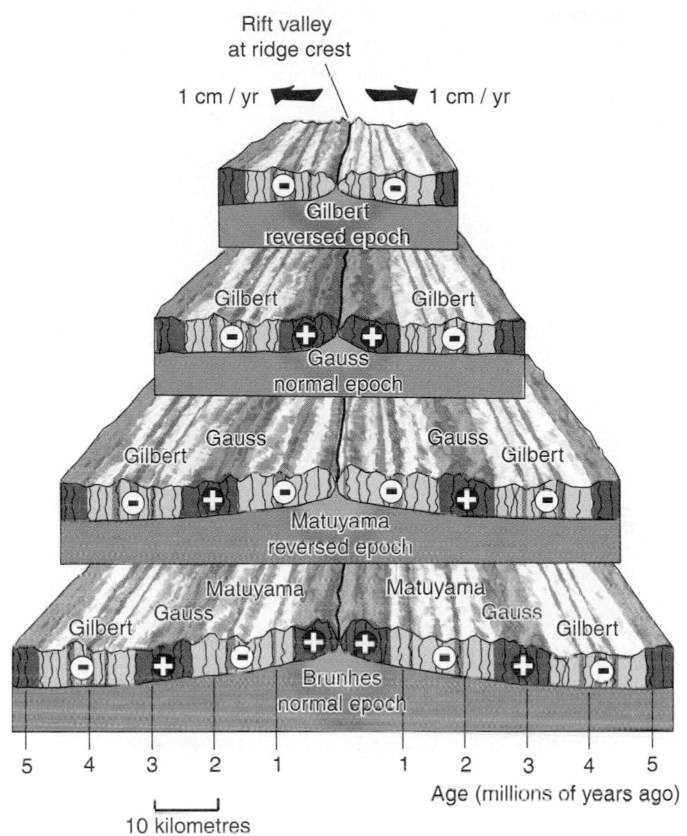

FIGURE 2.16

The origin of magnetic anomalies. During a time of reversed magnetism (Gilbert reversed epoch), a series of basaltic dikes intrudes the ridge crest, becoming reversely magnetized. The dike zone is torn in half and moved sideways, as a new group of normally magnetized dikes forms at the ridge crest. A new series of reversely magnetized dikes forms at the ridge crest. The dike pattern becomes symmetric about the ridge crest. Correlating the magnetic anomalies with magnetic reversals allows anomalies to be dated. Magnetic anomalies can therefore be used to predict the age of the sea floor and to measure the rate of seafloor spreading (plate motion).

magnetized. So each dike preserves a record of the polarity that prevailed during the time the magma cooled. Extension produced by the moving sea floor then cracks a dike in two, and the two halves are carried away in opposite directions down the flanks of the ridge. New magma eventually intrudes the newly opened fracture. It cools, is magnetized, and forms a new dike, which in turn is split by continued extension. In this way a system of reversely magnetized and normally magnetized dikes forms parallel to the rift valley. These dikes, in the Morley-Vine-Matthews hypothesis, are the cause of the anomalies.

HOW FAST DO PLATES MOVE?

There are two important points about the Morley-Vine-Matthews hypothesis of magnetic anomaly origin. The first is that it allows us to measure the *rate of sea-floor motion* (which is the same as plate motion, since continents and the sea floor move together as plates).

Because magnetic reversals have already been dated from lava flows on land (figure 2.14), the anomalies caused by these reversals are also dated and can be used to discover how fast the sea floor has moved (figure 2.16). For instance, a piece of the sea floor representing the reversal that occurred 4.5 million years ago may be found 45 km away from the rift valley of the ridge crest. The piece of sea floor, then, has travelled 45 km since it formed 4.5 million years ago. Dividing the distance the sea floor has moved by its age gives 10 km/million years, or 1 cm/year for the rate of sea-floor motion here. In other words, on each side of the ridge, the sea floor is moving away from the ridge crest at a rate of 1 cm per year. Such measured rates generally range from 1 to 10 cm per year.

Predicting Sea-Floor Age

The other important point of the Morley-Vine-Matthews hypothesis is that it *predicts the age of the sea floor* (figure 2.17). Magnetic reversals are now known to have occurred back into Precambrian time. Sea floor of *all* ages is therefore characterized by parallel bands of magnetic anomalies. In chapter 4 we will discuss the pattern of marine magnetic anomalies (and the reversals that caused them) during the past 160 million years. The distinctive pattern of these anomalies through time allows them to be identified by age, a process similar to dating by tree rings.

Now, even before they sample the sea floor, marine geologists can predict the age of the igneous rock of the sea floor by measuring the magnetic anomalies at the sea surface. Most sections of the sea floor have magnetic anomalies. By matching the measured anomaly pattern with the known pattern of anomalies, the age of the sea floor in the region can be predicted, as shown on the map in figure 2.17.

This is a very powerful test of the hypothesis that the sea floor moves. Suppose, for example, that the sea floor in a particular spot is predicted to be 70 million years old from a study of its magnetic anomalies. If the hypothesis of sea-floor motion and the Morley-Vine-Matthews hypothesis of magnetic anomaly origin are correct, a sample of igneous rock from that spot *must* be 70 million years old. If the rock proves to be 10 million years old or 200 million years old or 1.2 billion years old, or any other age except 70 million years, then both these hypotheses are wrong. But if the rock proves to be 70 million years old, as predicted, then both hypotheses have been successfully tested.

Hundreds of rock and sediment cores recovered from holes drilled in the sea floor were used to test these hypotheses. Close correspondence has generally been found between the predicted age and the measured age of the sea floor. (The sea-floor age is usually measured by fossil dating of sediment in the cores rather than by isotopic dating of igneous rock.) This evidence from deep-sea drilling has been widely accepted by geologists as verification of the hypotheses of plate motion and magnetic anomaly origin. Most geologists think that these concepts are no longer hypotheses but can now be called theories. (A *theory*, as illustrated in figure 1.3 in connection with the scientific method, is a concept with a much higher degree of certainty than a hypothesis.)

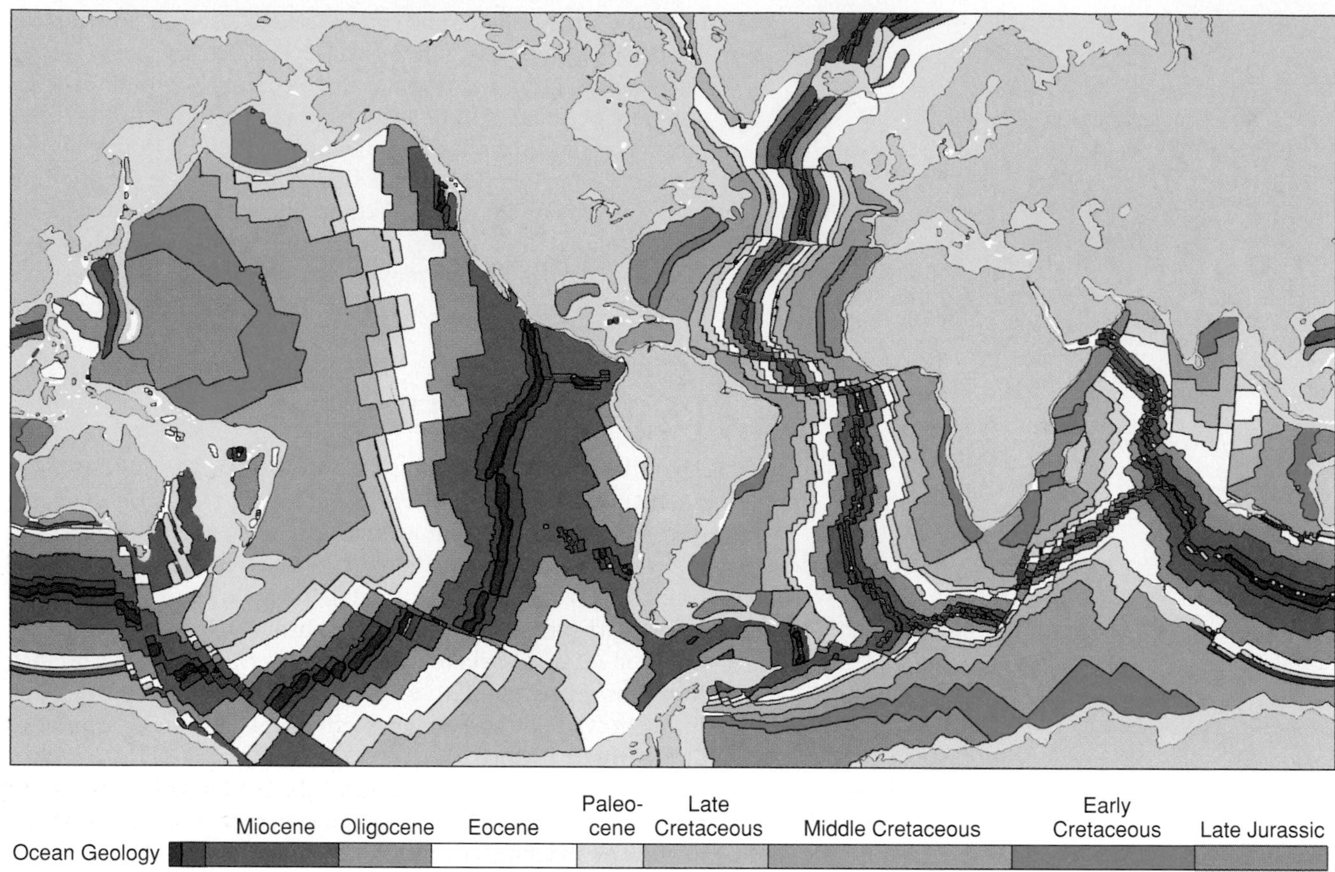

FIGURE 2.17

The age of the sea floor as determined from magnetic anomalies.

After The Bedrock Geology of the World *by R. L. Larson, W. C. Pitman, III, et al., W. H. Freeman*

Plate movement can now be measured directly using satellites, radar, lasers, and the Global Positioning System (GPS) (refer to box 2.1).

Another Test: Fracture Zones and Transform Faults

Cores from deep-sea drilling tested plate motion by allowing us to compare the actual age of the sea floor with the age predicted from magnetic anomalies. Another rigorous test of plate motion has been made by studying the seismicity of fracture zones.

Mid-oceanic ridges are offset along fracture zones (figure 2.8). Conceivably, mid-oceanic ridges were once continuous across fracture zones but have been offset by strike-slip motion along the fracture zone (figure 2.18*A*). If such motion is occurring along a fracture zone, we would expect to find two things: (1) earthquakes should be distributed along the entire length of the fracture zone, and (2) the motion of the rocks on either side of the fracture zone should be in the direction shown by the arrows in figure 2.18*A*.

In fact, these things are not true about fracture zones. Earthquakes do occur along fracture zones, but only in those segments between offset sections of ridge crest. In addition, first-motion studies of earthquakes (see chapter 3) along fracture zones show that the motion of the rocks on either side of the fracture zone during an earthquake is exactly opposite to the motion shown in figure 2.18*A*. The actual motion of the rocks as determined from first-motion studies is shown in figure 2.25*B*. The portion of a fracture zone between two offset portions of ridge crest is called a **transform fault**, a term first introduced by the Canadian geophysicist J. Tuzo Wilson (refer to box 1.4).

The motion of rocks on either side of a transform fault was predicted by the hypothesis of a moving sea floor. Note that sea floor moves away from the two segments of ridge crest (figure 2.18*B*). Looking along the length of the fracture zone, you can see that blocks of rock move in opposite directions only on that section of the fracture zone between the two segments of ridge crest. Earthquakes, therefore, occur only on this section of the fracture zone, the transform fault. The direction of motion of rock on either side of the transform fault is exactly predicted by the assumption that rock is moving

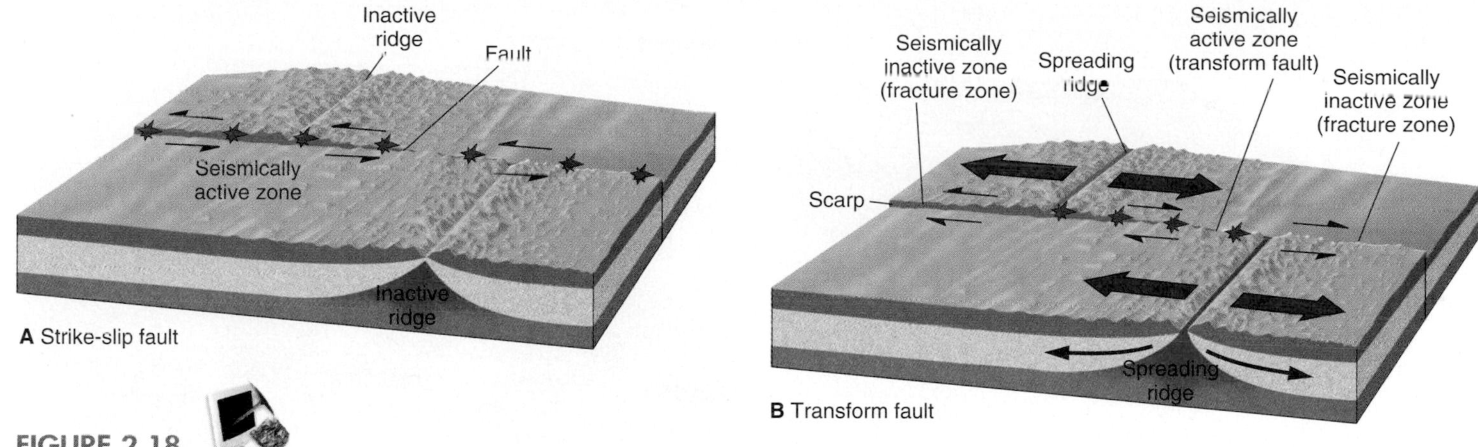

FIGURE 2.18

Two possible explanations for the relationship between fracture zones and the mid-oceanic ridge. (*A*) The expected rock motions and earthquake distribution assuming that the ridge was once continuous across the fracture zone. (*B*) The expected rock motions and earthquake distribution assuming that the two ridge segments were never joined together and that the sea floor moves away from the rift valley segments. Only explanation (*B*) fits the data. The portion of the fracture zone between the ridge segments is a transform fault.

away from the ridge crests. Verification by first-motion studies of this predicted motion along fracture zones was another successful test of plate motion.

WHAT HAPPENS AT PLATE BOUNDARIES?

Divergent Plate Boundaries

Divergent plate boundaries, where plates move away from each other, can occur in the middle of the ocean or in the middle of a continent. The result of divergence at plate boundaries is to create, or to open, new ocean basins. This dynamic process has occurred throughout the geologic past.

When a supercontinent such as Pangea breaks up, a divergent boundary can be found in the middle of a continent. The divergent boundary is marked by rifting, basaltic volcanism, and uplift. During rifting, the continental crust is stretched and thinned. This extension produces shallow-focus earthquakes on normal faults, and a rift valley forms as a central *graben* (a down-dropped fault block). The faults act as pathways for basaltic magma, which rises from the mantle to erupt on the surface as cinder cones and basalt flows. Uplift at a divergent boundary is usually caused by the upwelling of hot mantle beneath the crust; the surface is elevated by the thermal expansion of the hot, rising mantle rock and of the crustal rock as it is warmed from below.

Figure 2.19 shows how a continent might rift to form an ocean. The figure shows rifting before uplift, because recent work indicates that this was the sequence for the opening of the Red Sea. The crust is initially stretched and thinned. Numerous normal faults break the crust, and the surface subsides into a central graben (figure 2.19*A*). Shallow earthquakes and basalt eruptions occur in this rift valley, which also has high heat flow. An example of a boundary at this stage is the African Rift Valleys in eastern Africa (figure 2.20). The valleys are grabens that may mark the site of the future breakup of Africa.

As divergence continues, the continental crust on the upper part of the plate clearly separates, and sea water floods into the linear basin between the two divergent continents (figure 2.19*B*). A series of fault blocks have rotated along curved fault planes at the edges of the continents, thinning the continental crust. The rise of hot mantle rock beneath the thinned crust causes continued basalt eruptions that create true oceanic crust between the two continents. The centre of the narrow ocean is marked by a rift valley with its typical high heat flow and shallow earthquakes. The Red Sea is an example of a divergent margin at this stage (figure 2.20).

After modest widening of the new ocean, uplift of the continental edges may occur. As continental crust thins by stretching and faulting, the surface initially subsides. At the same time, hot mantle rock wells up beneath the stretched crust (figure 2.19*B*). The rising diapir of hot mantle rock would cause uplift by thermal expansion and weakening of the crust.

The new ocean is narrow, and the tilt of the adjacent land is away from the new sea, so rivers flow away from the sea (figure 2.19*B*). At this stage the sea water that has flooded into the rift may evaporate, leaving behind a thick layer of rock salt overlying the continental sediments. The likelihood of salt precipitation increases if the continent is in one of the desert belts or if one or both ends of the new ocean should become temporarily blocked, perhaps by volcanism. Not all divergent boundaries contain rock salt, however.

The plates continue to diverge, widening the sea. Thermal uplift creates a mid-oceanic ridge in the centre of the sea (figure 2.19*C*). The flanks of the ridge subside as the sea-floor rock cools as it moves.

The trailing edges of the continents also subside as they are lowered by erosion and as the mantle rock beneath them cools. Subsidence continues until the edges of the continents are under water. A thick sequence of marine sediment blankets the thinned continental rock, forming a *passive continental margin* (figure 2.19*C*). The sediment forms a shallow continental shelf, which may contain a deeply buried salt layer. The deep conti-

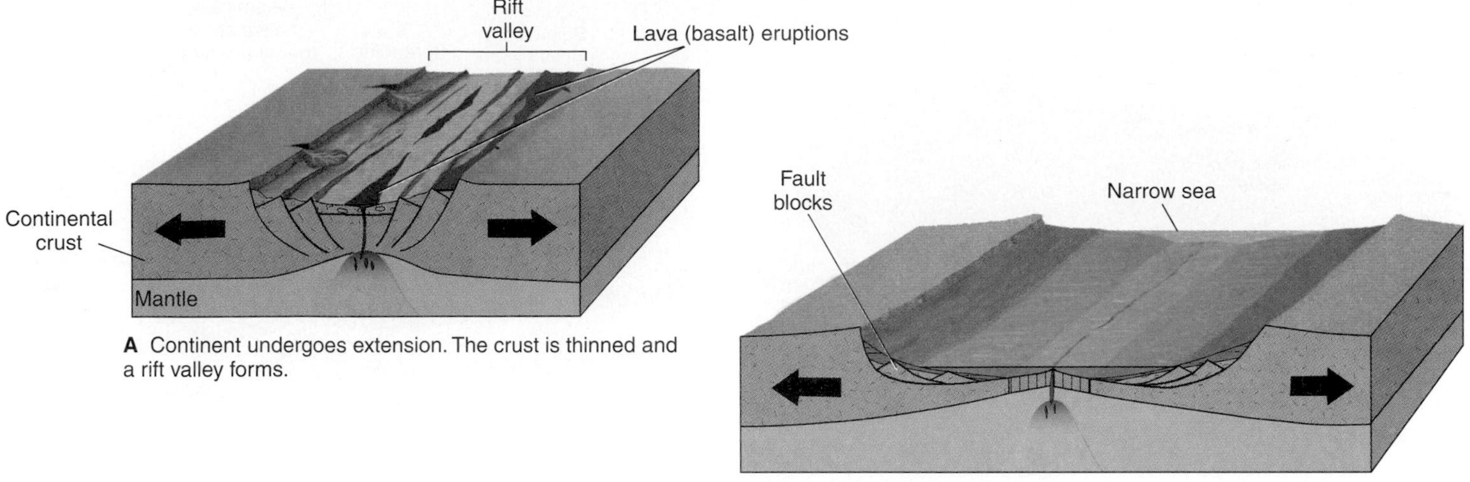

A Continent undergoes extension. The crust is thinned and a rift valley forms.

B Continent tears in two. Continent edges are faulted and uplifted. Basalt eruptions form oceanic crust.

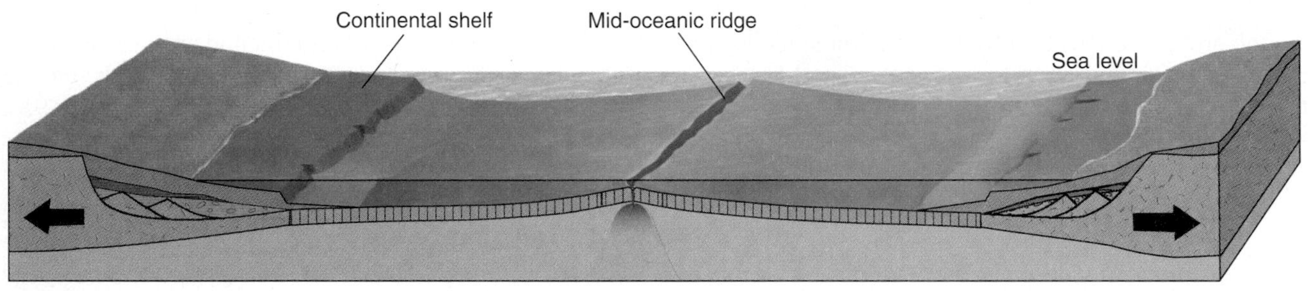

C Continental sediments blanket the subsiding "passive" margins to form continental shelves. The ocean widens and a mid-oceanic ridge develops, as in the Atlantic Ocean.

FIGURE 2.19 

A divergent plate boundary forming in the middle of a continent will eventually create a new ocean.

nental rise is formed as sediment is carried down the continental slope by turbidity currents and other mechanisms. The Atlantic Ocean is currently at this stage of divergence (figure 2.18).

A divergent boundary on the sea floor is located on the crest of a mid-oceanic ridge. If the spreading rate is slow, as it is in the Atlantic Ocean (1 cm/year), the crest has a rift valley. Fast spreading, as along the East Pacific Rise (18 cm/year) and other ridges in the Pacific Ocean, prevents a rift from forming. A divergent boundary at sea is marked by the same features as a divergent boundary on land—tensional cracks, normal faults, shallow earthquakes, high heat flow, and basaltic eruptions. The basalt forms dikes within the cracks and pillow lavas on the sea floor, creating new oceanic crust on the trailing edge of plates.

Mid-Oceanic Ridges

Mid-oceanic ridges are giant undersea mountain ranges that extend around the world like the seams on a baseball (figures 2.8 and 2.25). The ridges, which are made up mostly of basalt, are more than 80,000 km long and 1,500 to 2,500 km wide. They rise 2 to 3 km above the adjacent ocean floor.

A **rift valley** of tensional origin commonly runs down the crest of each ridge (figures 2.8 and 2.21). The rift valley is 1 to 2 km deep and several kilometres wide—about the dimensions of the Grand Canyon in Arizona. Rift valleys are present in the Atlantic and Indian oceans, but are generally absent in the Pacific Ocean.

Geologic Activity on Ridges

Associated with the rift valley at the crest of mid-oceanic ridges, and also with the riftless crest of the ridge in the Pacific Ocean, are *shallow-focus earthquakes* from 0 to 20 km below the sea floor.

Careful measurements of the heat loss from the Earth's interior through the crust have shown a very *high heat flow* on the crest of mid-oceanic ridges. The heat loss at the ridge crest is many times the normal value found elsewhere in the ocean; it decreases away from the ridge crest.

Basalt eruptions occur in and near the rift valley on ridge crests. Sometimes these eruptions build up volcanoes that protrude above sea level as oceanic islands. The large island of Iceland

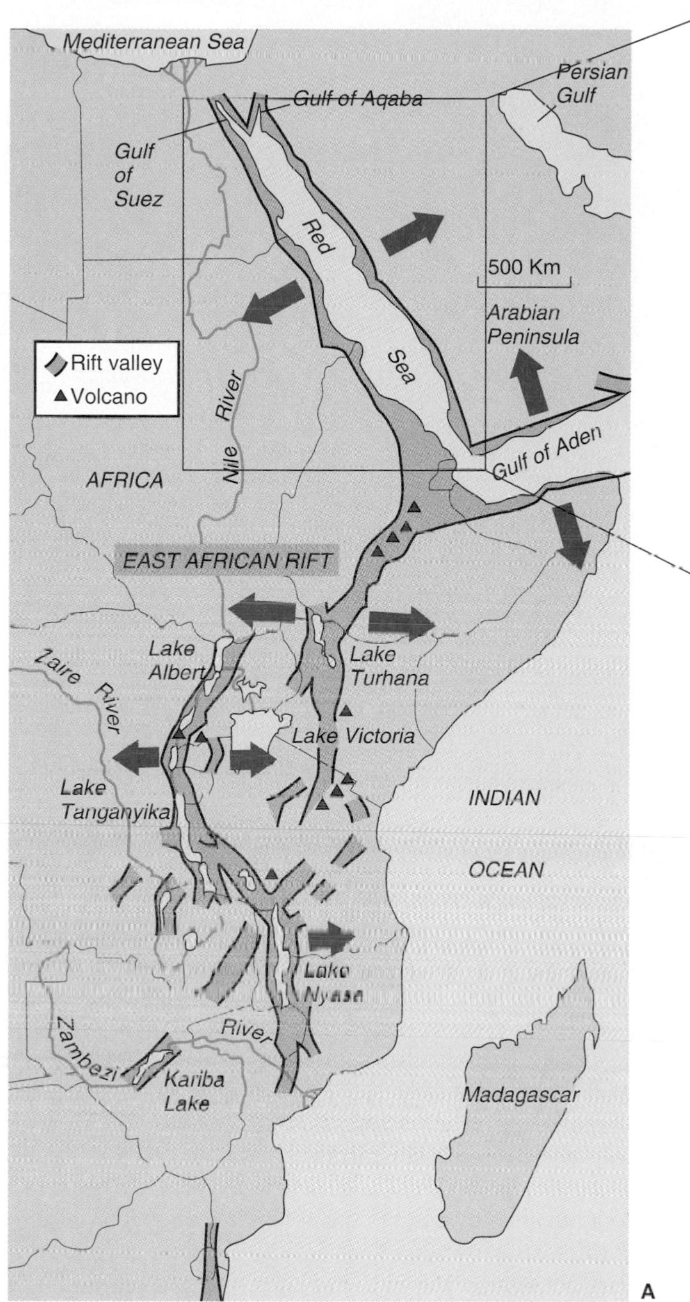

FIGURE 2.20

(*A*) The East African Rift Valleys and the Red Sea. (*B*) Satellite photo of Red Sea. Gulf of Suez is on the upper left and Gulf of Aqaba on upper right. Note the similarities in the shorelines of the Arabian Penninsula (right) and Africa (left) suggesting that the Red Sea was formed by splitting of the continent.

Photo by Jeff Schmaltz, MODIS Rapid Response Team, NASA/GSFC (http://visibleearth.nasa.gov/)

(figure 2.21), which is mostly basaltic, appears to be a section of the mid-Atlantic ridge elevated above sea level. Many geologists have studied the active volcanoes, high heat flow, and central rift valley of Iceland to learn about mid-oceanic ridges. As Iceland is above sea level, however, it may not be a typical portion of a mid-oceanic ridge.

In the summer of 1974 geologists were able to get a first-hand view of parts of submerged ridges and rift valleys. A series of more than 40 dives by submersibles, including ALVIN, carried French and American marine geologists directly into the rift valley in the North Atlantic Ocean. The project (called FAMOUS, for French-American Mid-Ocean Undersea Study) allowed the ridge rock to be seen, photographed, and sampled directly, rather than indirectly from surface ships (figure 2.22).

The geologists on the FAMOUS project saw clear evidence of extensional faults within the rift valley. These run parallel to the axis of the rift valley and range in width from hairline cracks to gaping fissures that ALVIN dived into. Fresh pillow basalts occur in a narrow band along the bottom of the rift valley, suggesting very recent volcanic activity there, although no active eruptions were observed. It appeared to the geologists that the rift was continuous, and that sporadic volcanic activity occurred as a result of the rifting.

Mid-oceanic ridges are often marked by lines of hot springs that carry and precipitate metal sulphides. Geologists in submersibles have observed hot springs in several localities along the rift valleys of the mid-oceanic ridges. The hot springs, caused by the high heat flow and shallow basaltic magma beneath the rift valley, range in temperature from about 20°C up to an estimated 350°C.

As the hot water rises in the rift valley, cold water is drawn in from the sides to take its place. This creates a circulation pattern in which cold sea water is actually drawn *downward* through the cracks in the basaltic crust of the ridge flanks and then moves horizontally toward the rift valley, where it re-emerges on the sea floor after being heated (figure 2.23*A*). As the sea water circulates, it dissolves metals and sulphur from the crustal rocks. When the hot, metal-rich solutions contact the cold sea water, metal sulphide particles are precipitated in the cold sea water at *black smokers,* which build chimney-like mounds around the hot spring (figure 2.23*B*).

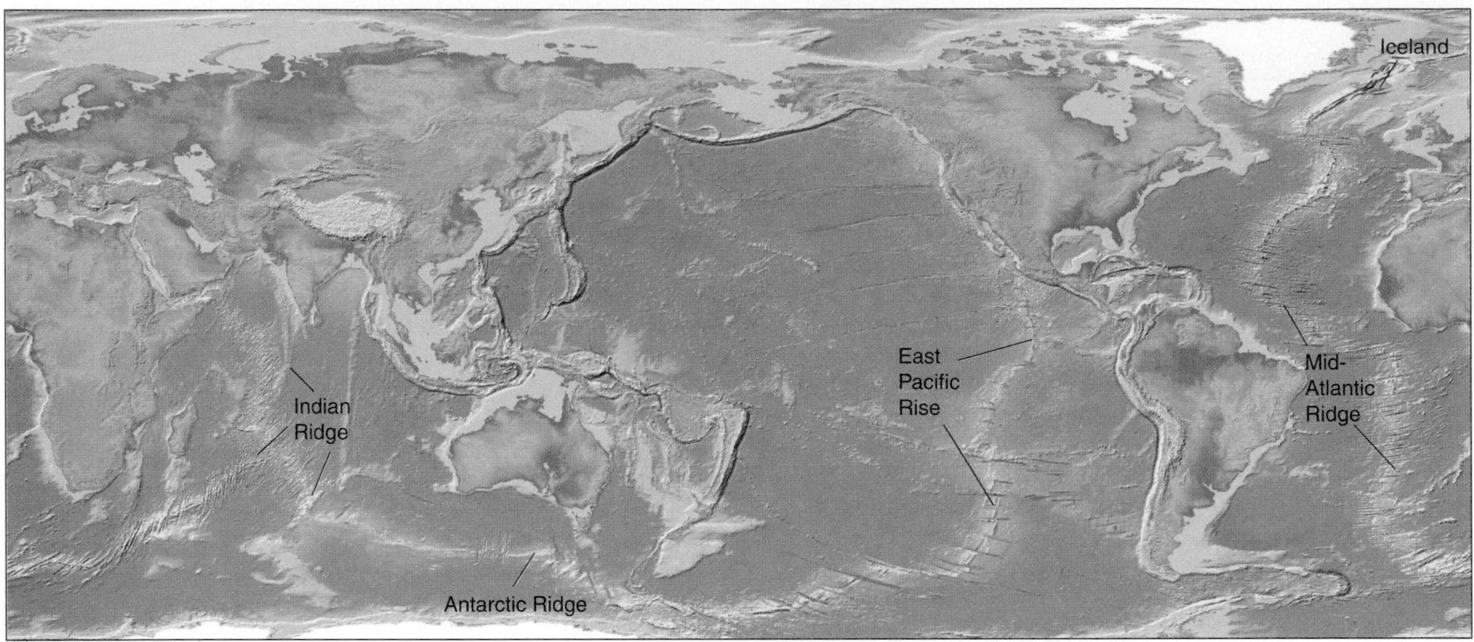

FIGURE 2.21
World map showing oceanic ridges cut by transform faults and fracture zones.

FIGURE 2.22
Underwater photograph of fresh pillow basalt on the floor of the rift valley of the mid-oceanic ridge in the North Atlantic Ocean. A white, tubular sponge grows in the foreground.
Photo courtesy of Woods Hole Oceanographic Institution

Biologic Activity on Ridges

The occurrence of black smokers was a surprise to geologists exploring the mid-oceanic ridges, but an even bigger surprise was the presence of exotic, bottom-dwelling organisms surrounding the hot springs. The exotic organisms, including mussels, crabs, starfish, giant white clams, and giant tube worms, are able to survive toxic chemicals, high temperatures, high pressures, and total darkness at depths of more than 2 km (figure 2.23*C*). The organisms live on bacteria that thrive by oxidizing hydrogen sulphide from the hot springs. It is believed that the heat-loving, or *thermophyllic,* bacteria normally reside beneath the sea floor but are blown out of the hot spring when it erupts. Such sulphur-digesting bacteria have also been found in acidic water in mines containing sulphide minerals. Current research in the new field of *geomicrobiology* is examining the role such bacteria may have had in the precipitation of minerals and in the evolution of early life-forms on Earth and possibly other places in the solar system.

Ridges and Ore Deposits

The metals released by rift-valley hot springs are predominantly iron, copper, and zinc, with smaller amounts of manganese, gold, and silver. Although black smoker mounds are nearly solid metal sulphide, they are small and widely scattered on the sea floor, so commercial mining of them may not be practical. Occasionally, the ore minerals may be concentrated in richer deposits. On the floor of the Red Sea metallic sediments have precipitated in basins filled with hot-spring solutions. Although the solutions

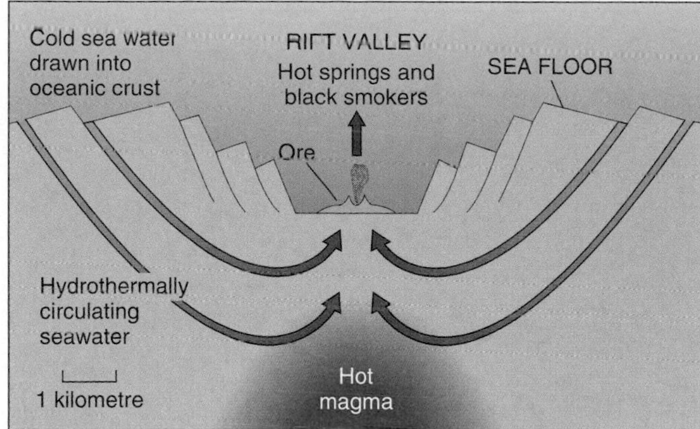

A

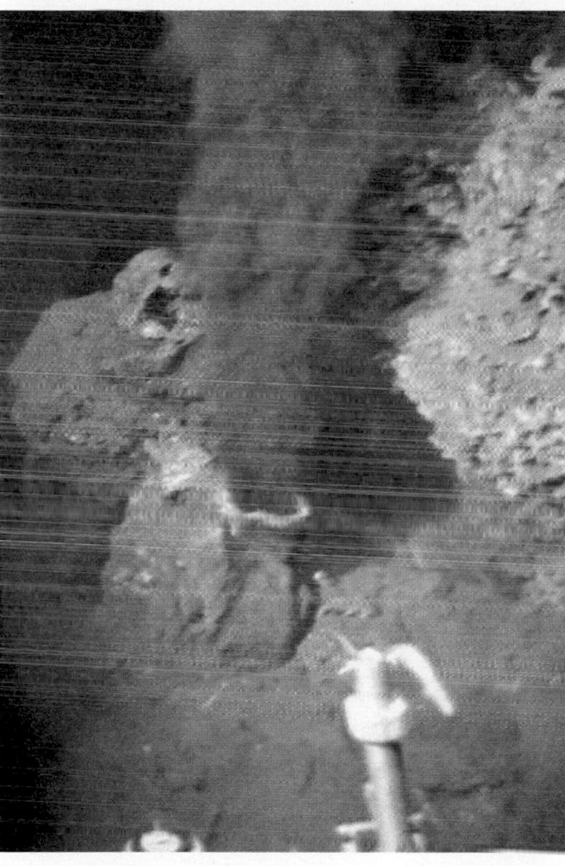

B

C

FIGURE 2.23

(A) Hydrothermal circulation of seawater at ridge crest creates hot springs and metallic deposits in rift valley. Cold seawater is drawn into fractured crust on ridge flanks. (Size of ore deposit is exaggerated.) (B) "Black smoker" or submarine hot spring on the crest of the mid-oceanic ridge in the Pacific Ocean near 21° North latitude. The "smoke" is a hot plume of metallic sulphide minerals being discharged into cold seawater from a chimney 0.5 metres high. The large mounds around the chimney are metallic deposits. The instruments in the foreground are attached to the small submarine from which the picture was taken. (C) Giant worms, crabs, and clams from Galápagos vent.

Photo B by U.S. Geological Survey; Photo C © WHOI/D. Foster/Visuals Unlimited

are hot (up to 60°C), they are very dense because of their high salt content (they are seven times saltier than sea water), so they collect in sea-floor depressions instead of mixing with the overlying sea water. Although not currently mined, the metallic sediments were estimated in 1983 to be worth $25 billion.

Hot metallic solutions are also found along some divergent continental boundaries. Near the Salton Sea in southern California, which lies along the extension of the mid-oceanic ridge inland, hot water very similar to the Red Sea brines has been discovered underground. The hot water is currently being used to run a geothermal power plant. The high salt and metal content is corrosive to equipment, but metals such as copper and silver may one day be recovered as valuable by-products.

Transform Boundaries

At *transform boundaries*, where one plate slides horizontally past another plate, the plate motion can occur on a single fault or on a group of parallel faults. Transform boundaries are marked by shallow-focus earthquakes in a narrow zone for a single fault, or in a broad zone for a group of parallel faults (see chapter 3). First-motion studies of the quakes indicate strike-slip movement parallel to the faults.

The name *transform fault* comes from the fact that the displacement along the fault abruptly ends or transforms into another kind of displacement. The most common type of transform fault

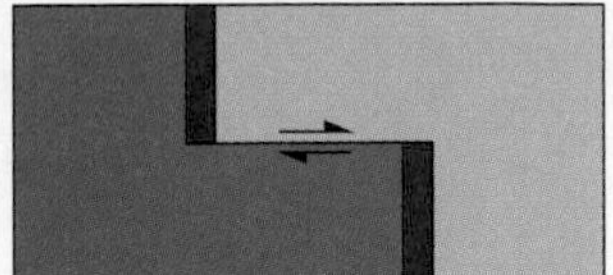

A Ridge–Ridge Transform

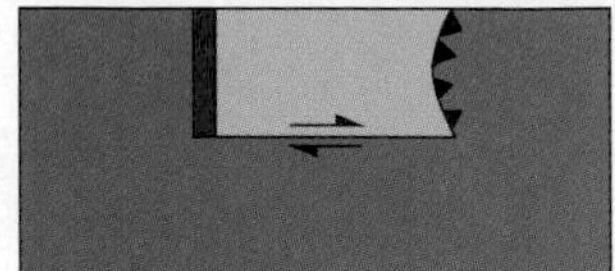

B Ridge–Trench Transform

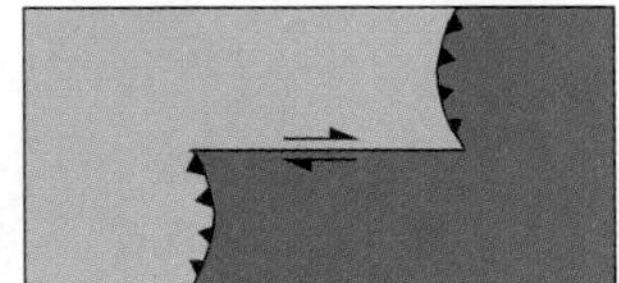

C Trench–Trench Transform

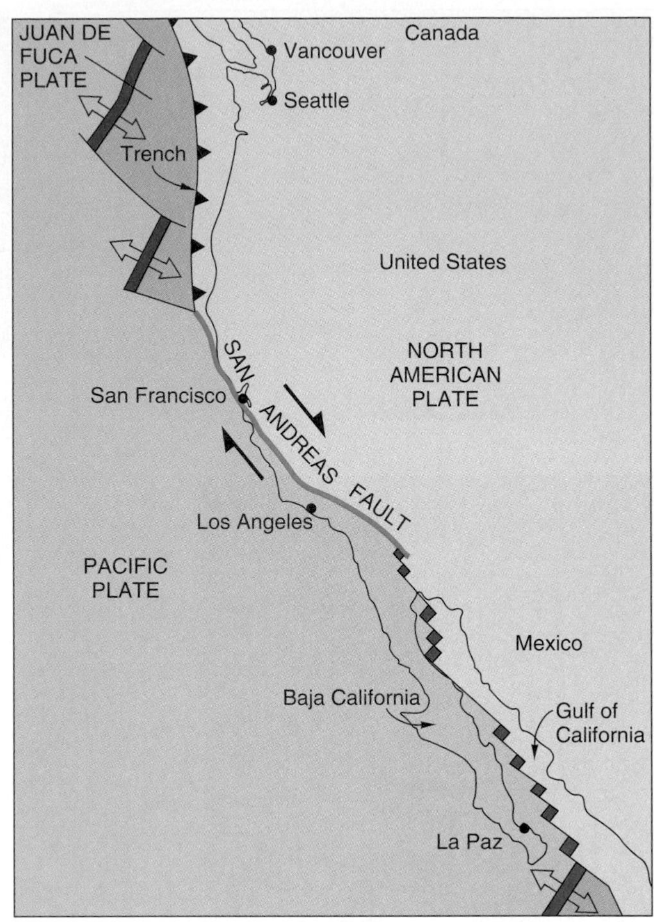

D

FIGURE 2.24

Transform boundaries (*A*) between two ridges; (*B*) between a ridge and a trench; and (*C*) between two trenches. Triangles on trenches point down subduction zones. Trench–trench transform boundaries are common in the southeast Pacific. Colour tones show two plates in each case. (*D*) The San Andreas fault is a ridge–ridge transform plate boundary between the North American plate and the Pacific plate. The south end of the San Andreas fault is a ridge segment (shown in red) near the U.S.–Mexico border. The north end of the fault is a "triple junction" where three plates meet at a point. The relative motion along the San Andreas fault is shown by the large black arrows, as the Pacific plate slides horizontally past the North American plate.

(D) Modified from U.S. Geological Survey

occurs along fracture zones and connects two divergent plate boundaries at the crest of the mid-oceanic ridge (figures 2.24 and 2.18*B*). The spreading motion at one ridge segment is transformed into the spreading motion at the other ridge segment by strike-slip movement along the transform fault.

Not all transform faults connect two ridge segments. As you can see in figure 2.24, a transform fault can connect a ridge to a trench (a divergent boundary to a convergent boundary), or can connect two trenches (two convergent boundaries). The San Andreas fault in California is a transform fault with a complex history (figure 2.24*D*).

What is the origin of the offset in a ridge–ridge transform fault? The offsets appear to be the result of irregularly shaped divergent boundaries (figure 2.25). When two oceanic plates begin to diverge, the boundary may be curved on a sphere. Mechanical constraints prevent divergence along a curved boundary, so the original curves readjust into a series of right-angle bends. The ridge crests align perpendicular to the spreading direction, and the transform faults align parallel to the spreading direction. An old line of weakness in a continent may cause the initial divergent boundary to be oblique to the spreading direction when the continent splits. The boundary will then readjust into a series of transform faults parallel to the spreading direction.

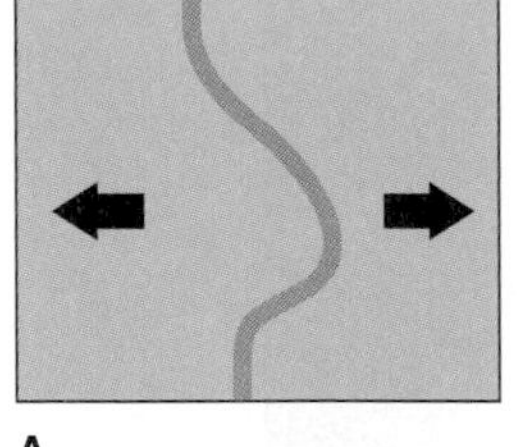

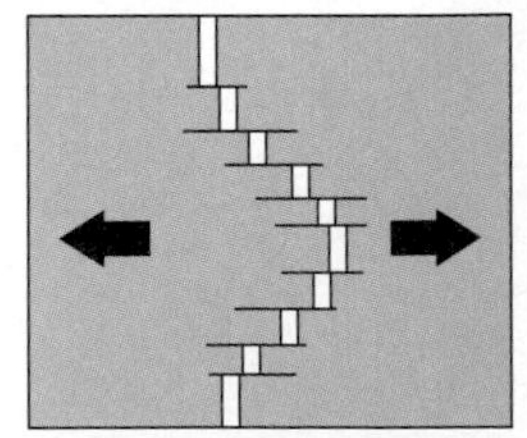

A

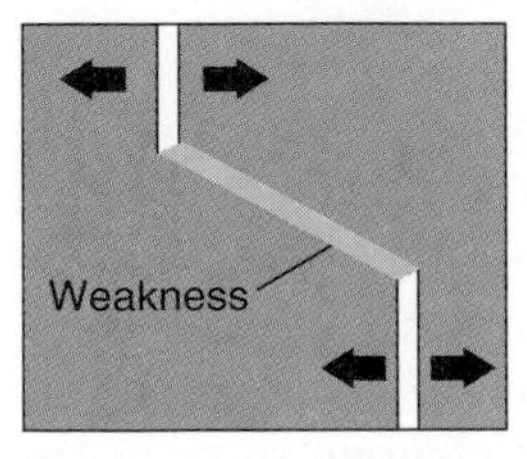

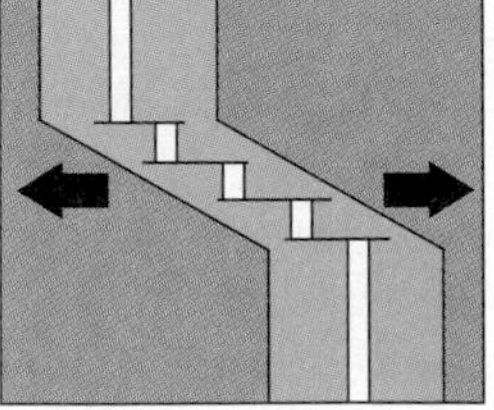

B

FIGURE 2.25

Divergent boundaries form ridge crests perpendicular to the spreading direction and transform faults parallel to the spreading direction. (*A*) Oceanic plates. (*B*) Continental plates.

Convergent Plate Boundaries

At convergent plate boundaries two plates move toward each other (often obliquely). The character of the boundary depends partly on the type of plates that converge. A plate capped by oceanic crust can move toward another plate capped by oceanic crust, in which case one plate dives (subducts) under the other. If an oceanic plate converges with a plate capped by a continent, the dense oceanic plate subducts under the continental plate. If the two approaching plates are both carrying continents, the continents collide and crumple but neither is subducted.

Ocean–Ocean Convergence

Where two plates capped by sea floor converge, one plate subducts under the other (the Pacific plate sliding under the western Aleutian Islands is an example). The subducting plate bends downward, forming the outer wall of an oceanic trench, which usually forms a broad curve convex to the subducting plate (figure 2.26).

An **oceanic trench** is a narrow, deep trough parallel to the edge of a continent or an island arc (a curved line of islands like the Aleutians or Japan), as shown in figure 2.27. The continental slope on an active margin forms the landward wall of the trench, its steepness often increasing with depth. The slope is typically 4 degrees to 5 degrees on the upper part, steepening to 10 degrees to 15 degrees or even more near the bottom of the trench. The elongate oceanic trenches, often 8 to 10 km deep, far exceed the average depth of abyssal plains on passive margins. The deepest spots on Earth, more than 11 km below sea level, are in oceanic trenches in the southwest Pacific Ocean.

As one plate subducts under another, a Benioff zone of shallow-, intermediate-, and deep-focus earthquakes is created within the upper portion of the downgoing lithosphere (figure 2.26). (The reasons for these quakes are discussed in chapter 3.) The existence of deep-focus earthquakes to a depth of 670 km tells us that brittle plates continue to (at least) that depth. The pattern of quakes shows that the angle of subduction changes, usually becoming steeper with depth (figure 2.26). Some plates crumple or break into segments as they descend.

As the descending plate reaches depths of at least 100 km, magma is generated in the overlying asthenosphere (figure 2.26). The magma probably forms by partial melting of the asthenosphere, perhaps triggered by dewatering of the downgoing oceanic crust as it is subducted, as described in chapter 7. Differentiation and assimilation also play important roles in the generation of the magma, which is typically andesitic to basaltic in composition.

The magma works its way upward to erupt as an **island arc,** a curved line of volcanoes that form a string of islands parallel to the oceanic trench (figure 2.26). Beneath the volcanoes are large plutons in the thickened arc crust.

The distance between the island arc and the trench can vary, depending upon where the subducting plate reaches the 100-km depth. If the subduction angle is steep, the plate reaches this magma-generating depth at a location close to the trench, so the horizontal distance between the arc and trench is short. If the subduction angle is gentle, the arc–trench distance is greater.

When a plate subducts far from a mid-oceanic ridge, the plate is cold, with a low heat flow. Oceanic plates form at ridge crests, then cool and sink as they spread toward trenches. Eventually they become cold and dense enough to sink back into the mantle.

The inner wall of a trench (toward the arc) consists of an *accretionary wedge* (or *subduction complex*) of thrust-faulted and folded marine sediment and pieces of oceanic crust (figure 2.26). The sediment is snowploughed off the subducting plate by the overriding plate. New slices of sediment are continually added to the bottom of the accretionary wedge, pushing it upward to form a ridge on the sea floor. A relatively undeformed *forearc basin* lies between the accretionary wedge and the volcanic arc. (The trench side of an arc is the forearc; the other side of the arc is the backarc.)

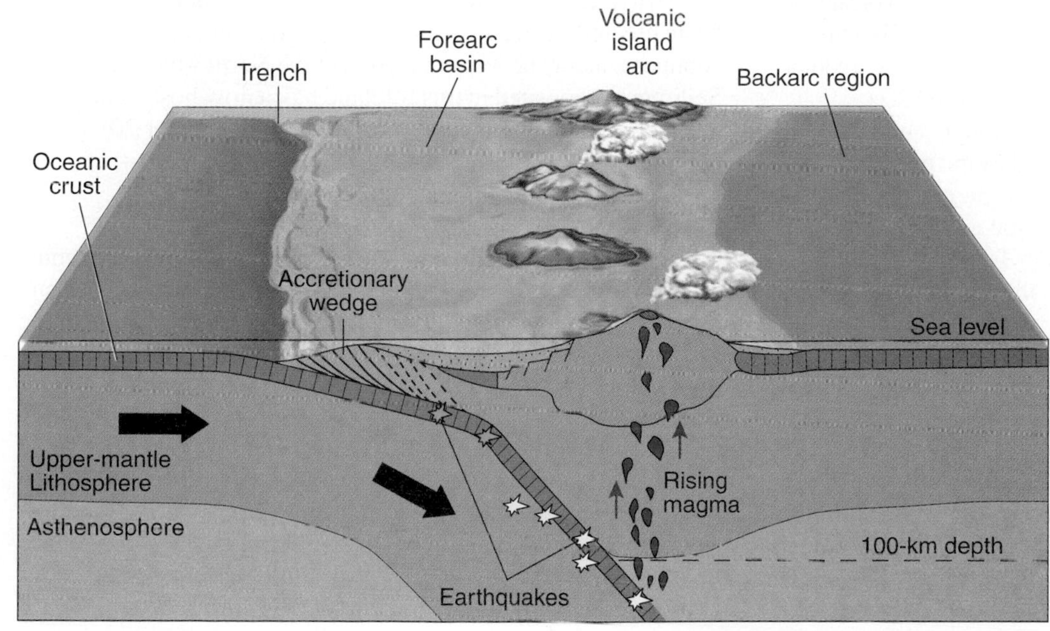

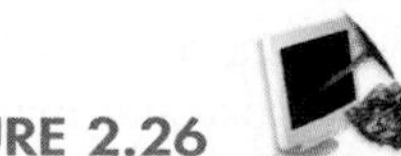

FIGURE 2.26
Ocean–ocean convergence forms a trench, a volcanic island arc, and a Benioff zone of earthquakes.

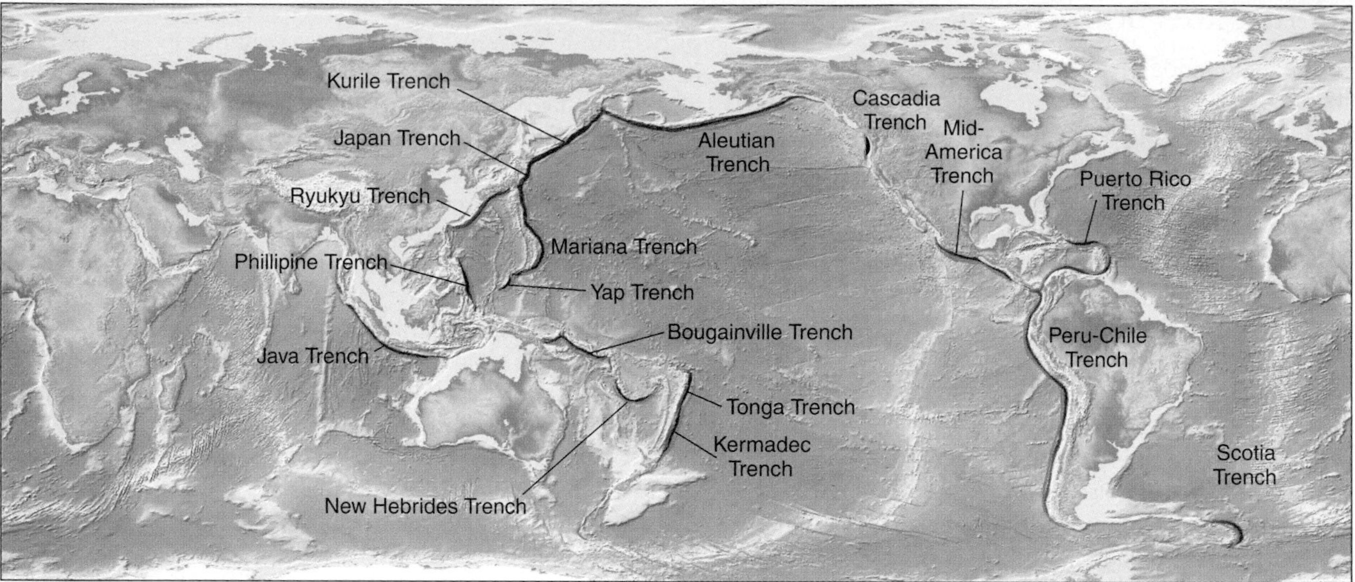

FIGURE 2.27
The distribution of oceanic trenches. Trenches next to continents mark active continental margins and convergent plate boundaries.

Trench positions change with time. As one plate subducts, the overlying plate may be moving toward it. The motion of the leading edge of the overlying plate will force the trench to migrate horizontally over the subducting plate. The Peru–Chile trench is moving over the Nazca plate in this manner as South America moves westward (figure 2.1).

Ocean–Continent Convergence

When a plate capped by oceanic crust is subducted under the *continental* lithosphere, an accretionary wedge and forearc basin form an *active continental margin* between the trench and the continent (figure 2.28). A Benioff zone of earthquakes dips under the edge of the continent, which is marked by andesitic volcanism and a young mountain belt. An example of this type of boundary is the subduction of the Nazca plate under western South America and the Juan de Fuca plate under North America (box 2.2).

The magma that is created by ocean–continent convergence forms a **magmatic arc,** a broad term used both for island arcs at sea and for belts of igneous activity on the edges of continents. The surface expression of a magmatic arc is either a line of andesitic islands (such as the Aleutian Islands) or a line of andesitic continental volcanoes (such as the Cascade volcanoes of the Pacific Northwest). Beneath the volcanoes are large plutons in thickened crust. We see these plutons as *batholiths* on land when they are exposed by deep erosion. The igneous processes that form the granitic and intermediate magmas of batholiths are described in chapter 7.

The hot magma rising from the subduction zone thickens the continental crust and makes it weaker and more mobile than cold crust. Regional metamorphism takes place within this hot, mobile zone. Crustal thickening causes uplift, so a young mountain belt forms here as the thickened crust rises.

Another reason for the growth of the mountain belt is the stacking up of thrust sheets on the continental (backarc) side of the magmatic arc (figure 2.28). The thrust faults, associated with folds, move slivers of mountain-belt rocks landward over the continental interior (the *craton*). Underthrusting of the cold, rigid craton beneath the hot, mobile core of the mountain belt may help form the fold-thrust belt.

Inland of the backarc fold-thrust belt, the craton subsides to form a sedimentary basin (sometimes called a *foreland basin*). The weight of the stacked thrust sheets further depresses the craton. The basin receives sediment, some of which may be marine if the craton is forced below sea level. This basin extends the effect of subduction far inland. In Canada, the Cretaceous Western Interior Seaway developed in an extensive foreland basin created by the thrusting of accreted island arcs and microcontinents along the western margin of the North American craton. Sediments deposited in this foreland basin now host some of the productive oil and gas fields of Alberta (see chapter 20).

Continent–Continent Convergence

Two continents may approach each other and collide. They must be separated by an ocean floor that is being subducted under one continent and that lacks a spreading axis to create new oceanic crust (figure 2.29). The edge of one continent will initially have a magmatic arc and all the other features of ocean– continent convergence.

As the sea floor is subducted, the ocean becomes narrower and narrower until the continents eventually collide and destroy or close the ocean basin. Oceanic lithosphere is heavy and can sink into the mantle, but continental lithosphere is less dense and cannot sink. One continent may slide a short distance under another, but it will not go down a subduction zone.

IN GREATER DEPTH 2.2

British Columbia and the Juan de Fuca Plate/ Cascadia Subduction Zone

The Juan de Fuca plate is a small oceanic plate that lies off the western coast of North America between Queen Charlotte Sound in British Columbia, and Cape Mendocino in northern California (box figure 1). The plate is broken up into three segments, each separated by fracture zones: the Explorer plate in the north, the Central plate offshore from Washington and Oregon, and the Gorda plate in the south.

Ocean-floor basalts that make up the Juan de Fuca plate form at three offset spreading centres located between the mid-Pacific Ocean and the North American land mass: the Explorer Ridge, the Juan de Fuca Ridge, and the Gorda Ridge. Basalt that moves westward from the ridges forms the Pacific plate, while basalt that moves eastward makes up the Juan de Fuca plate.

Northeastward motion of the Juan de Fuca plate causes it to collide obliquely with the westward-moving North American continent along a zone of convergence known as the Cascadia Subduction Zone (box figure 2). The young basaltic materials of the Juan de Fuca plate are only slightly more dense than the North American continental crust and therefore undergo forced subduction with very low rates of movement. Sediments deposited on the relatively buoyant subducting plate are scraped off and thrust against the North American land mass as an accretionary wedge. Magma generated above downgoing crust at great depths rises to the surface to form the volcanoes of the Cascade Mountains of British Columbia, Washington, Oregon, and northern California. The Cascadia Subduction Zone passes northwestward into the Queen Charlotte Fault and southeastward into the San Andreas Fault.

Subduction of the Juan de Fuca plate beneath the North American continent creates the potential for large and frequent earthquakes in the region (see chapter 3). Large earthquakes, on the order of 7+ on the Richter scale, have been estimated to occur along the British Columbia and Washington coasts approximately once every 500 years.

GPS data are now being used to monitor the relative motion of the Earth's surface along the western margin of North America. The Western Canada Deformation Array (WDCA; see box 2.1) of GPS stations has been used to measure annual rates and directions of motion of specific sites relative to a reference site located at the Dominion Radio Astrophysical Observatory (DRAO) south of Penticton (see box figure 3). Sites that overlie the locked portion of the Cascadia Subduction Zone (e.g., UCLU–Ucluelet and NEAH–Neah Bay) move at rates of more than 10 mm/yr in a northeasterly direction, whereas inland sites (e.g., WSLR–Whistler) move at half that rate or less. This suggests that the outer margin is slowly being compressed like a giant spring. If the accumulated strain is totally released during the next great earthquake, the outer coast of southern Vancouver Island will move southwestward by up to 5 metres.

BOX 2.2 ■ FIGURE 1

Sea-floor topography in the Northeast Pacific Ocean.

From www.ocean.washington.edu/people/grads/scottv/exploraquarium/scaletrip/2nepac.html

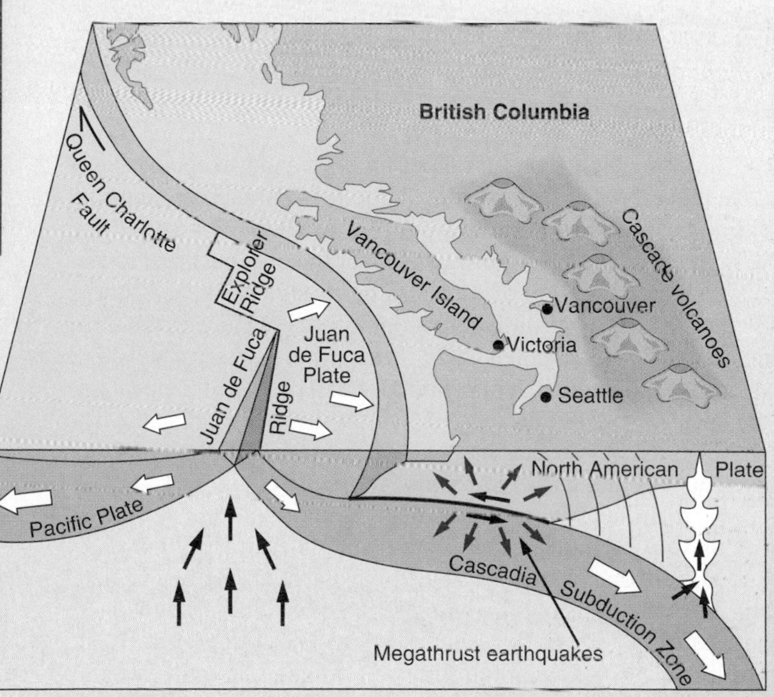

BOX 2.2 ■ FIGURE 2

Schematic cross-section through the Cascadia subduction zone.

Reproduced with the permission of the Minister of Public Works and Government Services Canada, 2006 and Courtesy of Natural Resources Canada, Geological Survey of Canada.

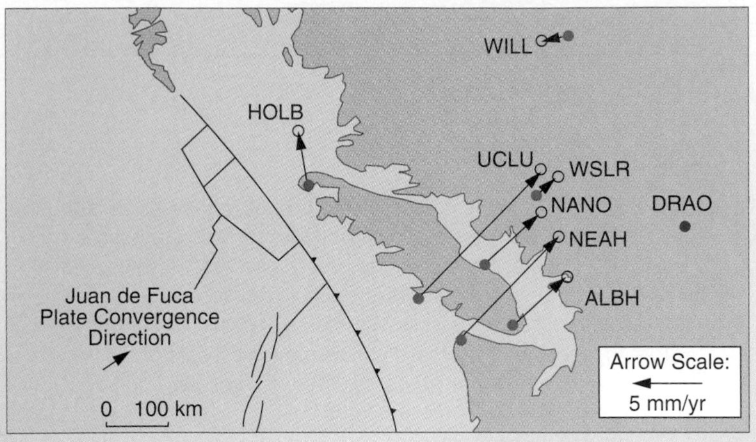

BOX 2.2 ■ FIGURE 3

Annual motion of sites within the WCDA as determined from GPS data. The arrows show the direction of movement; length of arrow is proportional to the amount of annual motion.

From http://gsc.nrcan.gc.ca/geodyn/images/agu99rel.gif
Reproduced with permission of the Minister of Public Works and Government Services Canada, 2006 and Courtesy of Natural Resources Canada, Geological Survey of Canada.

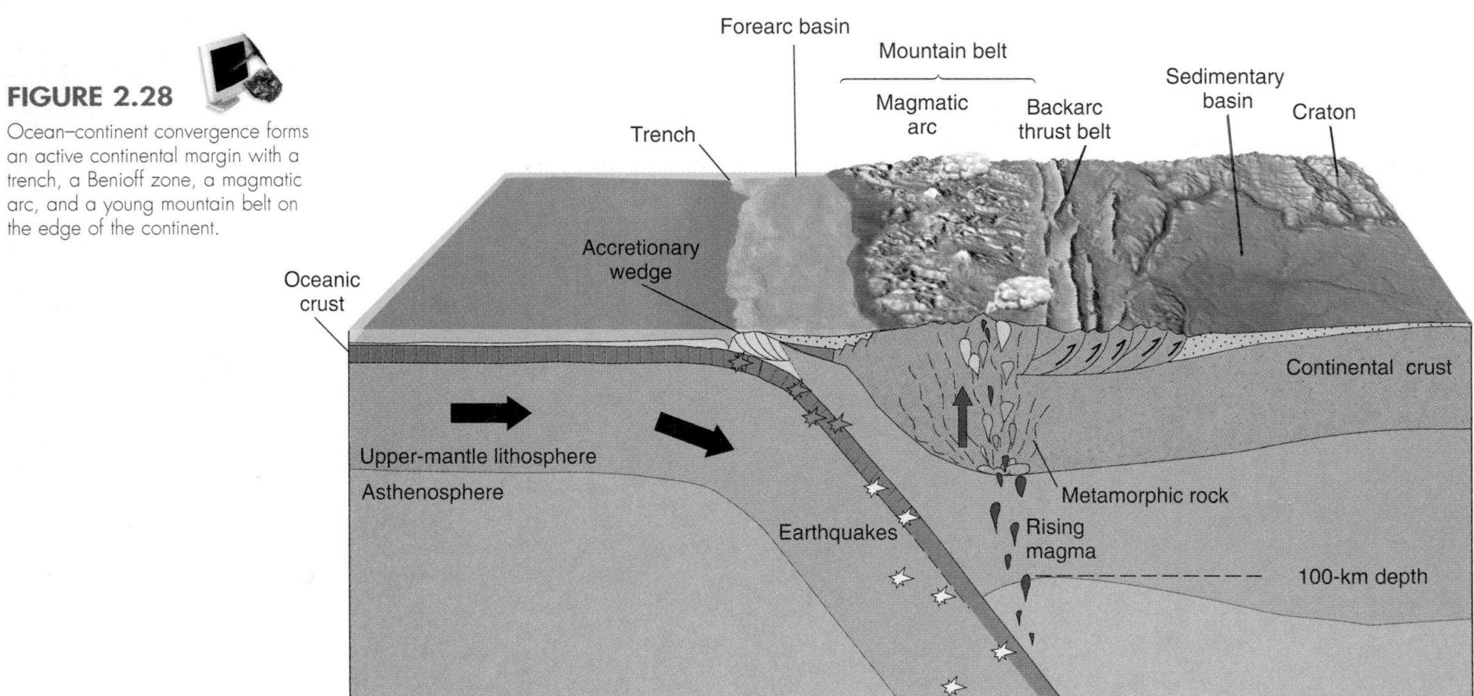

FIGURE 2.28

Ocean–continent convergence forms an active continental margin with a trench, a Benioff zone, a magmatic arc, and a young mountain belt on the edge of the continent.

The two continents are welded together along a dipping *suture zone* that marks the old site of subduction. Thrust belts and subsiding basins occur on both sides of the original magmatic arc, which is now inactive. The presence of the original arc thickens the crust in the region of impact. The crust is thickened further by the shallow underthrusting of one continent beneath the other and also by the stacking of thrust sheets in the two thrust belts. The result is a mountain belt in the interior of a continent (a new large continent formed by the collision of the two smaller continents). The entire region of impact is marked by a broad belt of shallow-focus earthquakes along the numerous faults (see chapter 3).

The Himalayas in central Asia are thought to have formed in this way, as India collided with and underthrust Asia to produce exceptionally thick crust and high elevations. Paleomagnetic studies show that India was once in the southern hemisphere and drifted north to its present position. The collision with Asia began between 40 and 60 million years ago after an intervening ocean was destroyed by subduction (figure 2.29*C*). The stresses generated by the continent–continent collision extend far to the north of the Himalaya, possibly affecting areas up to 5,000 km into Central Asia (figure 2.29*D*).

Backarc Spreading

Regional extension occurs within or behind many arcs. This extension can tear an arc in two, moving the two halves in opposite directions (figure 2.30). If it occurs behind an arc, it can move the arc away from a continent. It can split the edge of a continent, moving a narrow strip of the continent seaward (this is apparently how Japan formed). In each case the spreading creates new oceanic crust that is similar, but not identical, to the oceanic crust formed at the crest of mid-oceanic ridges. This backarc oceanic crust is apparently the type of crust found in most ophiolites (see box 2.3).

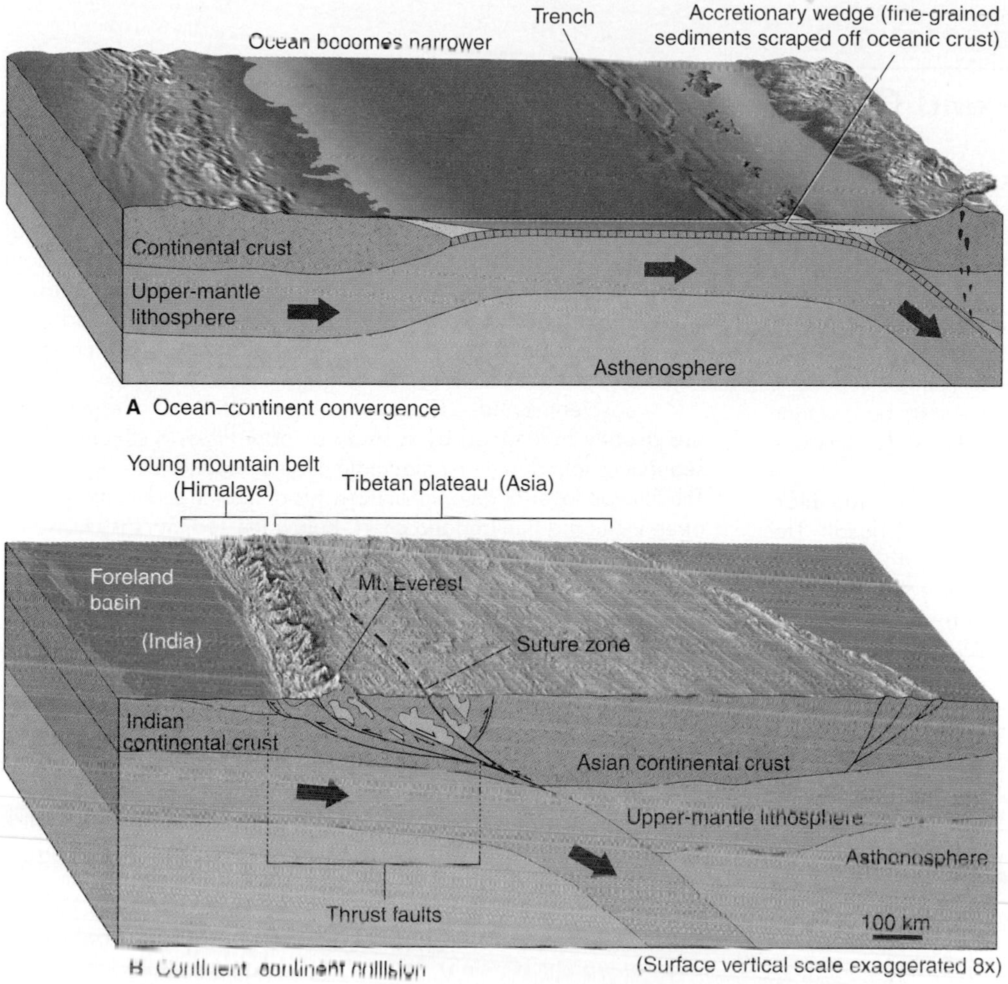

Trench
Accretionary wedge (fine-grained sediments scraped off oceanic crust)
Ocean becomes narrower
Continental crust
Upper-mantle lithosphere
Asthenosphere

A Ocean–continent convergence

Young mountain belt (Himalaya)
Tibetan plateau (Asia)
Foreland basin (India)
Mt. Everest
Suture zone
Indian continental crust
Asian continental crust
Upper-mantle lithosphere
Asthenosphere
Thrust faults
100 km

B Continent–continent collision (Surface vertical scale exaggerated 8x)

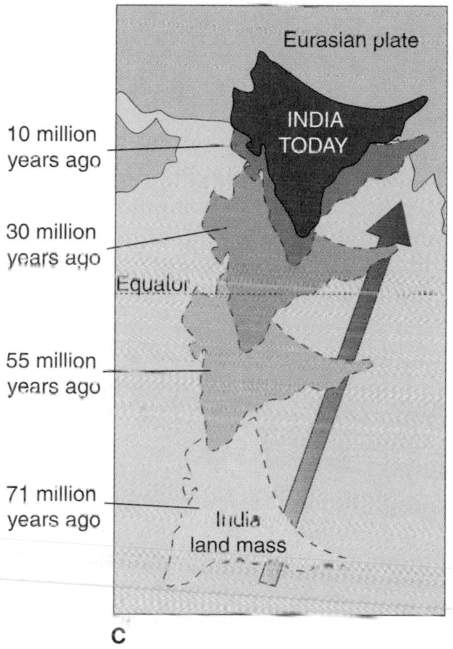

Eurasian plate
INDIA TODAY
10 million years ago
30 million years ago
Equator
55 million years ago
71 million years ago
India land mass

C

D

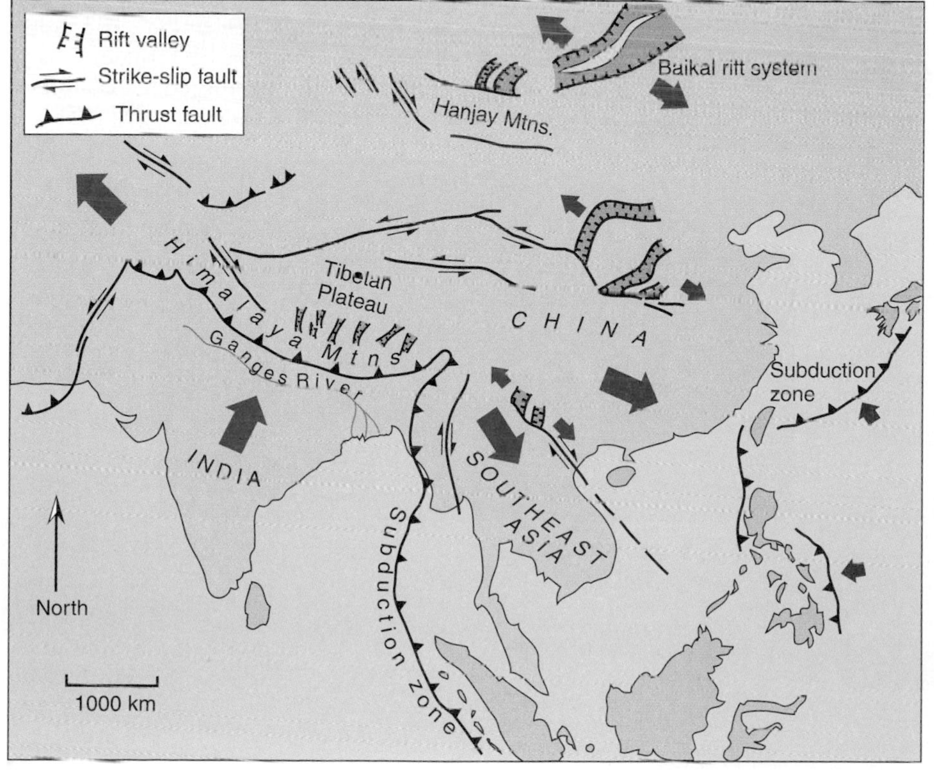

FIGURE 2.29

The collision of two continents forms a young mountain belt in the interior of a new, larger continent. The most famous example of continent–continent collision is the collision of India with Asia. (*A*) India is moving toward Asia due to ocean–continent convergence. (*B*) India collides with Asia to form the Himalayas, the highest mountain range on Earth. (*C*) Map view of the northward movement of India through time. (*D*) Central Asia has adjusted to the broadside collision of India through uplift of the Tibetan plateau, and by stretching and slipping along major "intraplate" faults.

IN GREATER DEPTH 2.3

Oceanic Crust and Ophiolites

Oceanic crust differs significantly from continental crust; it is both thinner and of a different composition than continental crust. Seismic reflection and seismic refraction surveys at sea have shown the oceanic crust to be about 7 km thick and divided into three layers (box figure 1*A*).

The top layer (Layer 1), of variable thickness and character, is marine sediment. In an abyssal fan or on the continental rise, Layer 1 may consist of several kilometres of terrigenous sediment. On the upper flanks of mid-oceanic ridges, there may be less than 100 m of marine sediment. An average thickness for Layer 1 might be 0.5 km.

Beneath the sediment is Layer 2, which is about 1.5 km thick. This layer consists of pillow basalt overlying dikes of basalt. The basalt pillows, rounded masses that form when hot lava erupts into cold water (box figure 1*C*), are highly fractured at the top of Layer 2. The dikes in the lower part of Layer 2 are closely spaced, parallel, vertical dikes ("sheeted dikes"). Widely sampled by drilling and dredging, Layer 2 has also been observed directly by geologists in submersibles diving into the rift valley of the mid-oceanic ridge during the FAMOUS expedition.

The lowest layer in oceanic crust is Layer 3. It is about 5 km thick and thought to consist of sill-like gabbro bodies (see chapter 5). Geologists have drilled through 2 km of pillow basalt and basaltic dikes in oceanic crust, but have not yet been able to reach gabbro by drilling through basalt. Gabbro (and other rocks) are exposed on some fault blocks in rift valleys and on some steep submarine cliffs in fracture zones; 0.5 km of gabbro has been drilled on one such cliff in the southern Indian Ocean. Some geologists presume that the gabbro represents Layer 3 exposed by faulting, but other geologists are not so sure. Even if the drilled gabbro *does* represent the upper 10 percent of Layer 3, no one has yet sampled the lower 90 percent of the layer.

A drillhole in the eastern Pacific Ocean has been reoccupied four times in a 12-year span, and has now reached a total depth of 2,000 m below the sea floor. Seismic evidence suggested that the Layer 2–Layer 3 boundary would be found at a depth of about 1,700 m, but the drill went well past that depth without finding the contact between the dikes of Layer 2 and the expected gabbro of Layer 3. Either the seismic interpretation or the model of Layer 3's composition must be wrong.

Geoscientists' ideas about the composition of oceanic crust are greatly influenced by a study of *ophiolites*, distinctive rock sequences found in many mountain chains on land (box figure 1*B*). The thin top layer of an ophiolite consists of marine sedimentary rock, often including thin-bedded chert. Below the sedimentary rock lies a zone of pillow basalt, which in turn is underlain by a sheeted-dike complex that probably served as feeder dikes for the pillowed lava flows above. The similarities between the upper part of an ophiolite and Layers 1 and 2 of oceanic crust are obvious.

Below the sheeted dikes of an ophiolite is a thick layer of pod-like gabbro intrusions underlain by ultramafic rock such as peridotite, the top part of which has been metamorposed. Geologists have long thought that ophiolites represent slivers of oceanic crust somehow emplaced on land. If this is true, then the gabbro of ophiolites may represent oceanic Layer 3, and the peridotite may represent mantle rock below oceanic crust. The contact between the peridotite and the overlying gabbro would be the Mohorovičić discontinuity.

Recent work has shown that many, if not most, ophiolites are *not* typical sea floor, but a special type of sea floor formed in marginal ocean basins next to continents by the process of backarc spreading. If ophiolites do not represent typical sea floor, then more extensive, deeper sea-floor drilling is needed before a clear picture of oceanic crust can be formed. This is an important goal, for oceanic crust is the most common surface rock, covering 60 percent of the Earth's surface.

7 kilometres
1 Marine sediment
2 Pillow basalt and sheeted dikes
Gabbro intrusions
3
?
Moho
Mantle ?
Ultramafic rock
A

Thin-bedded chert
Pillow basalt and sheeted dikes
Gabbro intrusions
Ultramafic rock
B

C

BOX 2.3 ■ FIGURE 1

A comparison of oceanic crust and an ophiolite sequence. (*A*) Structure of oceanic crust, determined from seismic studies and drilling. Gabbro and ultramafic rock drilled and dredged from sea-floor fault blocks may be from Layer 3 and the mantle. (*B*) Typical ophiolite sequence found in mountain ranges on land. Thickness approximate—sequence is usually highly faulted. (*C*) Pillow basalt from the upper part of an ophiolite, Bett's Cove, Newfoundland. These rocks formed as part of the sea floor, when hot lava cooled quickly in cold sea water.

www.ocean.washington.edu/exploraquarium/scaletrip.2nepac.html

Photo by Nick Eyles.

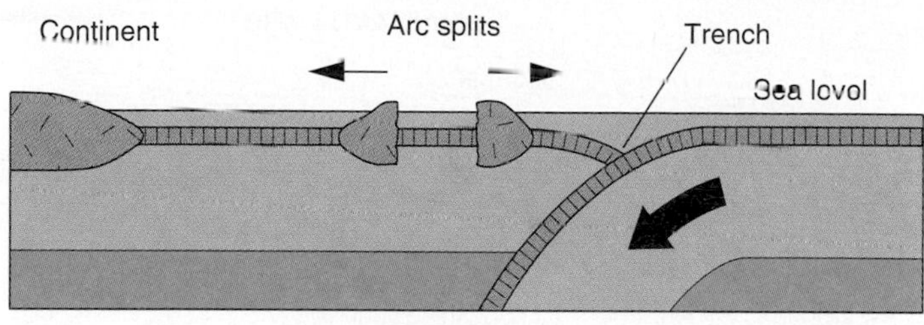

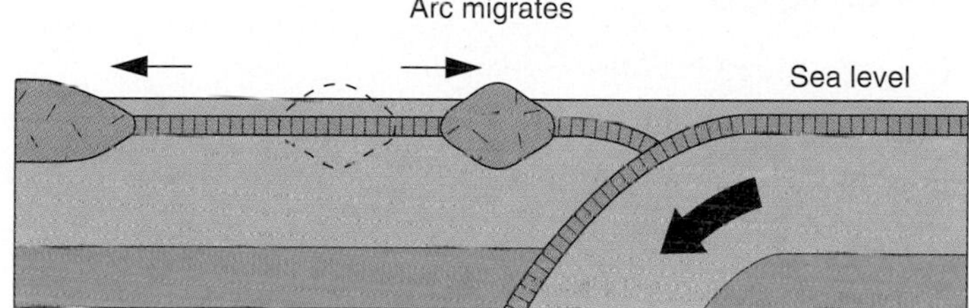

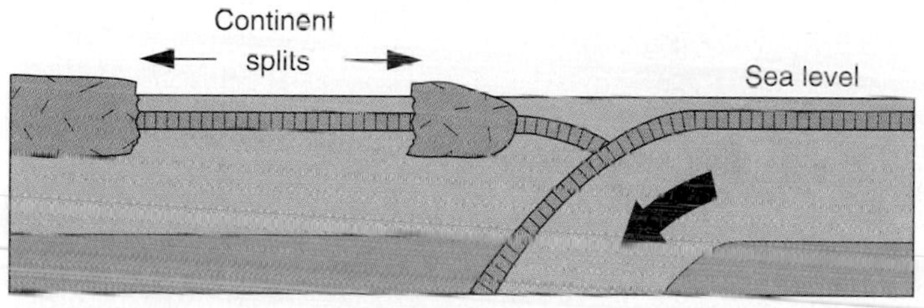

FIGURE 2.30

Backarc spreading. Regional extension in the overlying plate of a subduction zone can split an arc, move an arc offshore, or split a continent.

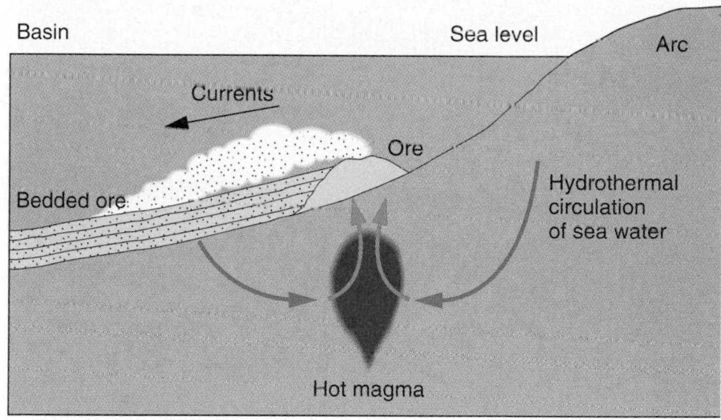

FIGURE 2.31

On island arcs metallic ores can form over hot springs and be redistributed into layers by currents in shallow basins.

Convergent Boundaries and Ore Deposits

Sea-floor spreading carries metallic ores away from ridge crests, perhaps to be subducted beneath island arcs or continents at *convergent plate boundaries.* Slivers of ancient oceanic crust (ophiolite; see box 2.3) exposed on land may contain these rich ore minerals in relatively intact form. A notable example of such ores occurs on the island of Cyprus in the Mediterranean Sea. Banded chromite ores may also be contained in the ancient oceanic crust.

Volcanism at *island arcs* can also produce hot-spring deposits on the flanks of the andesitic volcanoes. Pods of very rich ore collect above local bodies of magma, and the ore is sometimes distributed as sedimentary layers in shallow basins (figure 2.31). The circulation pattern and the ore-forming processes are quite similar to those of spreading centres, but the island arc ores usually contain more lead and gold. Rich *massive sulphide deposits* overlying fractured volcanic rock in the Precambrian shield area of Canada may have formed in this way within ancient island arcs (e.g., Noranda and Kidd Creek in the Abitibi Greenstone Belt of northern Ontario).

Subduction of the sea floor beneath a *continent* produces broad belts of metallic ore deposits near the edge of the continent. The origin of the continental ores above a subduction zone is not clear. Hot-spring deposits from the ridge crest are subducted with oceanic crust and could become remobilized to rise into the continent above. Ores may also "distill" off other parts of the descending oceanic crust or upper mantle. The metals may also derive from the continental crust itself or the mantle below it and may be concentrated somehow by the heat of a rising blob of magma or by hydrothermal circulation.

HOW DO MOUNTAIN RANGES FORM?

Orogenies and Plate Convergence

An **orogeny** (from the Greek words *oros*, meaning "mountain," and *genesis*, meaning "formation") is an episode of mountain building, often characterized by intense deformation of the rocks in a region. Mountain belts form along plate margins, particularly where plate convergence compresses the crust, causing uplift and deformation. By understanding how mountains form in different types of convergent margin settings, we can better reconstruct past mountain building processes.

Ocean–Continent Convergence

The Andes, in which the South American plate is overriding the Nazca plate, is an example of a mountain belt formed by ocean–continent convergence. In this setting, an *accretionary wedge* develops where newly formed layers of marine sediment are folded and faulted as they are snowploughed off the subducting oceanic plate (figures 2.28, 2.32). Rock caught in and pulled down the subduction zone is subjected to intense shearing. If rock is carried further down the subduction zone, it becomes metamorphosed (see chapter 10).

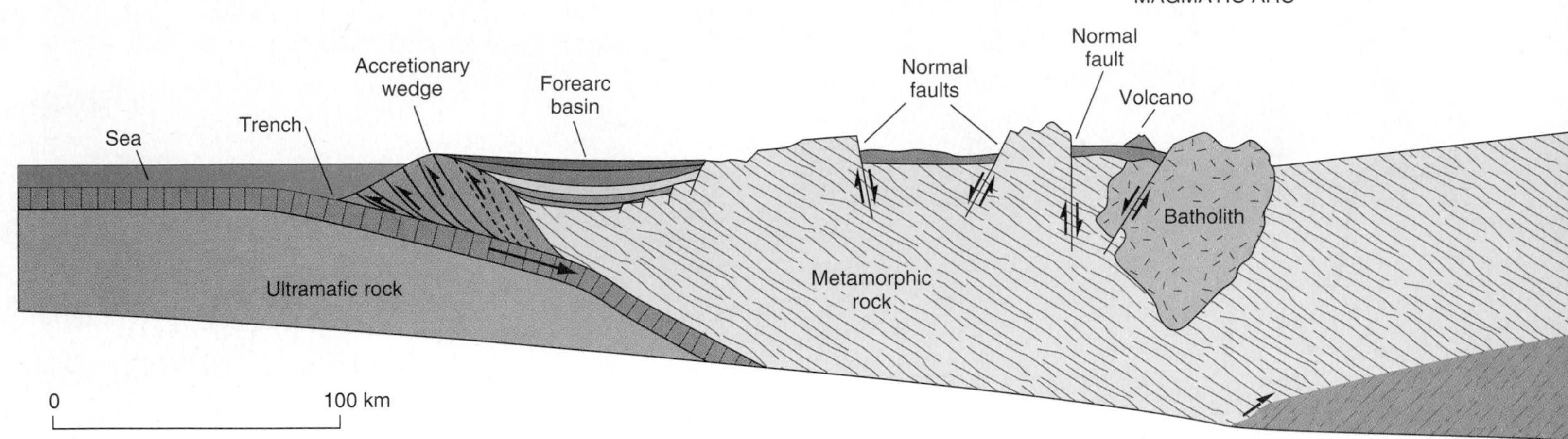

FIGURE 2.32

Cross-section of an "Andean type" mountain belt; that is, one whose orogeny is due to oceanic–continental convergence. For simplicity, only a few of the many layers of sedimentary rock are shown. The size of some features is exaggerated for illustrative purposes.

Fold-and-thrust belts may develop on the craton (backarc) side of the mountain belt (figure 2.32). Thrusting is away from the magmatic arc toward the craton. The magmatic arc is at a high elevation, because the crust is thicker and composed largely of hot igneous and metamorphic rocks. The large-thrust sheets move toward and sometimes over the craton. (In the Rocky Mountains, thrust faulting of the craton itself has taken place; see box 2.4.) The thrusting probably is largely due to the crustal shortening caused by convergence. Normal faulting may take place due to extension in the high, central part of the magmatic arc.

Arc–Continent Convergence

Sometimes an island arc collides with a continent (figure 2.33). As the arc and continent converge, the intervening ocean is destroyed by subduction. When collision occurs, the arc, like a continent, is too buoyant to be subducted. Continued convergence of the two plates may cause the remaining sea floor to break away from the arc and create a new site of subduction and a new trench seaward of the arc (figure 2.33*C*). Note that the direction of the new subduction is opposite to the direction of the original subduction (this is sometimes called a *flipping subduction zone*), but it still may supply the arc with magma. The arc has now become welded to the continent, increasing the size of the continent.

This type of collision apparently occurred in northern New Guinea (north of Australia). A similar collision may have added an island arc to the Sierra Nevada complex in California during Mesozoic time, when a subduction zone may have existed in what now is central California. Many geoscientists think that much of westernmost North America has formed from a series of arcs colliding with North America (discussed in chapter 20).

Continent–Continent Convergence

Some mountain belts form when an ocean basin closes and continents collide. Mountain belts that we find within continents (with cratons on either side) are hypothesized to be products of

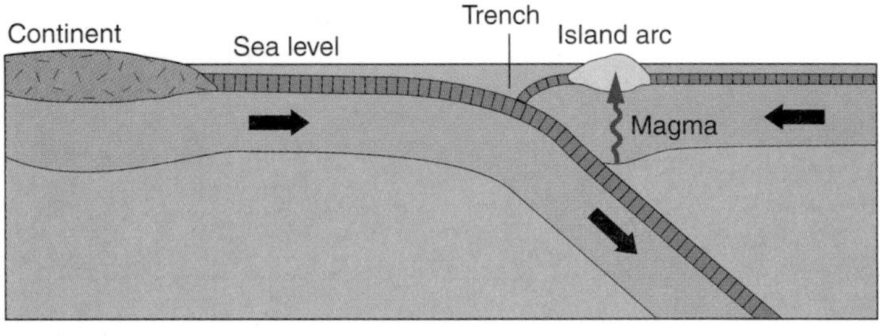

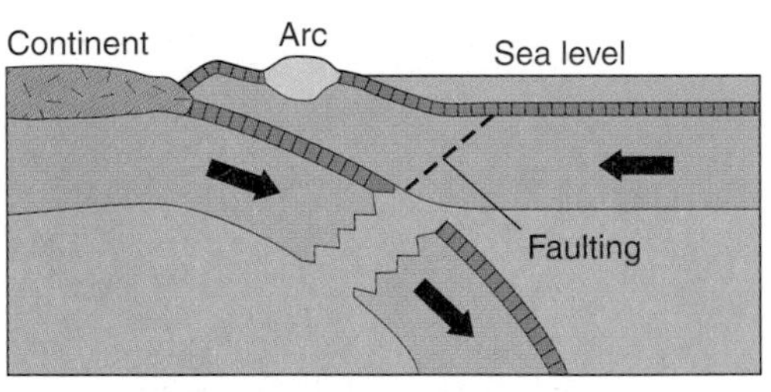

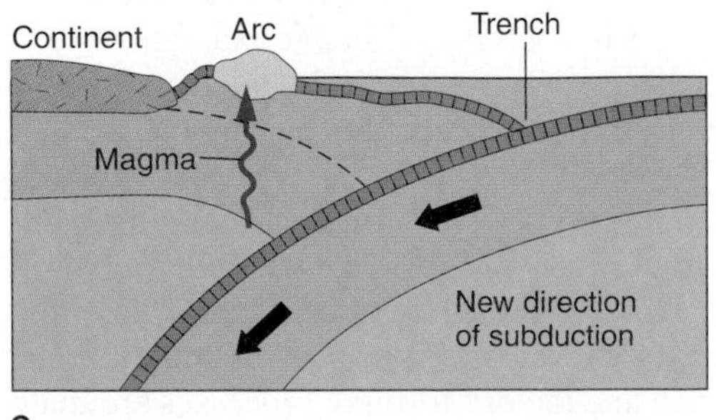

FIGURE 2.33

Arc–continent convergence can weld an island arc onto a continent. The direction of subduction can change after impact.

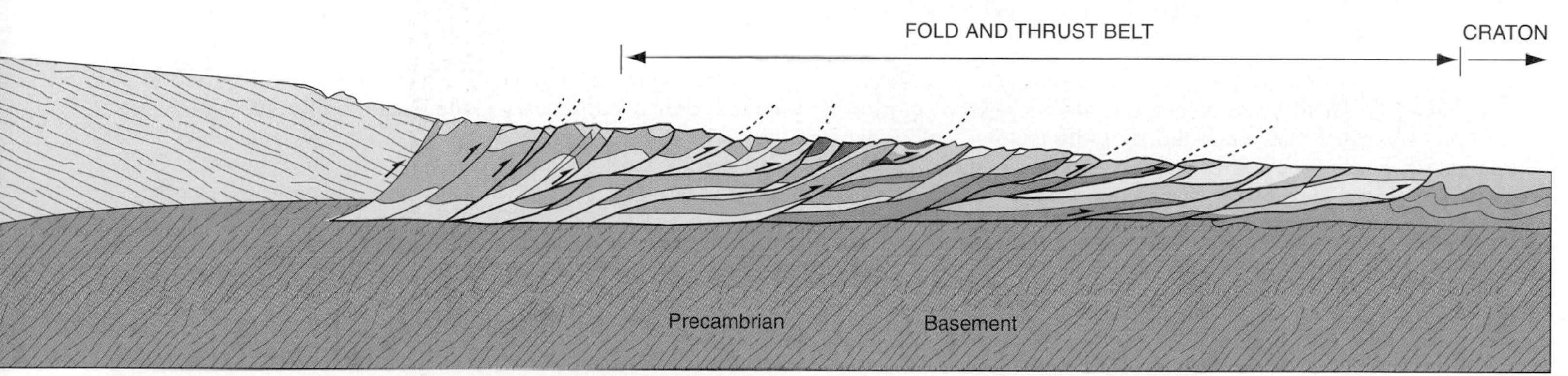

continent–continent convergence. Convergence of the African and European plates created the Alps. Our highest mountains are in the Himalayan belt. The Himalayan orogeny started around 45 million years ago as India began colliding with Asia (figure 2.29). The thick sequences of sedimentary rocks that had built up on both continental margins were intensely faulted and folded. Fold-and-thrust belts developed and were carved by erosion into the mountain ranges that make up the Himalayas. The mountains are still rising, and frequent earthquakes attest to continuing tectonic activity.

The Appalachian Mountains are an example of continent–continent convergence, but with a more complicated history. Arc–continent convergence was also involved and the mountain belt was later split apart by plate divergence.

A condensed version of orogeny in the northern Appalachian mountains of eastern Canada is as follows (figure 2.34). During the Proterozoic, rifting of the supercontinent Rodinia created the ancestral Atlantic Ocean (the Iapetus Ocean; figure 2.34*A*) and sediment began to accumulate on the newly formed continental margin of North America (Laurentia; figure 2.34*B*). During the Cambrian, plate motion shifted, subduction began, and the Iapetus Ocean began to close (figure 2.34*C*). Island arcs developed between the continents, and these were plastered onto the North American craton together with large preserved remnants of ocean crust (forming ophiolites) as the ocean basin closed during the Taconic Orogeny (figure 2.34*D*). A new subduction zone developed to the east and the microcontinent Avalonia (formerly part of Gondwana) collided with North America during the Acadian Orogeny (figure 2.34*E*). By the end of the Carboniferous, the Gondwanan continent had also collided with eastern North America (during the Alleghanian Orogeny) to form the supercontinent Pangea (figure 2.34*F*). At this time, the Appalachian mountain belt formed an extensive suture zone, continuous with the Caledonide mountain belt of Great Britain and Norway, and comparable to the present-day Himalaya.

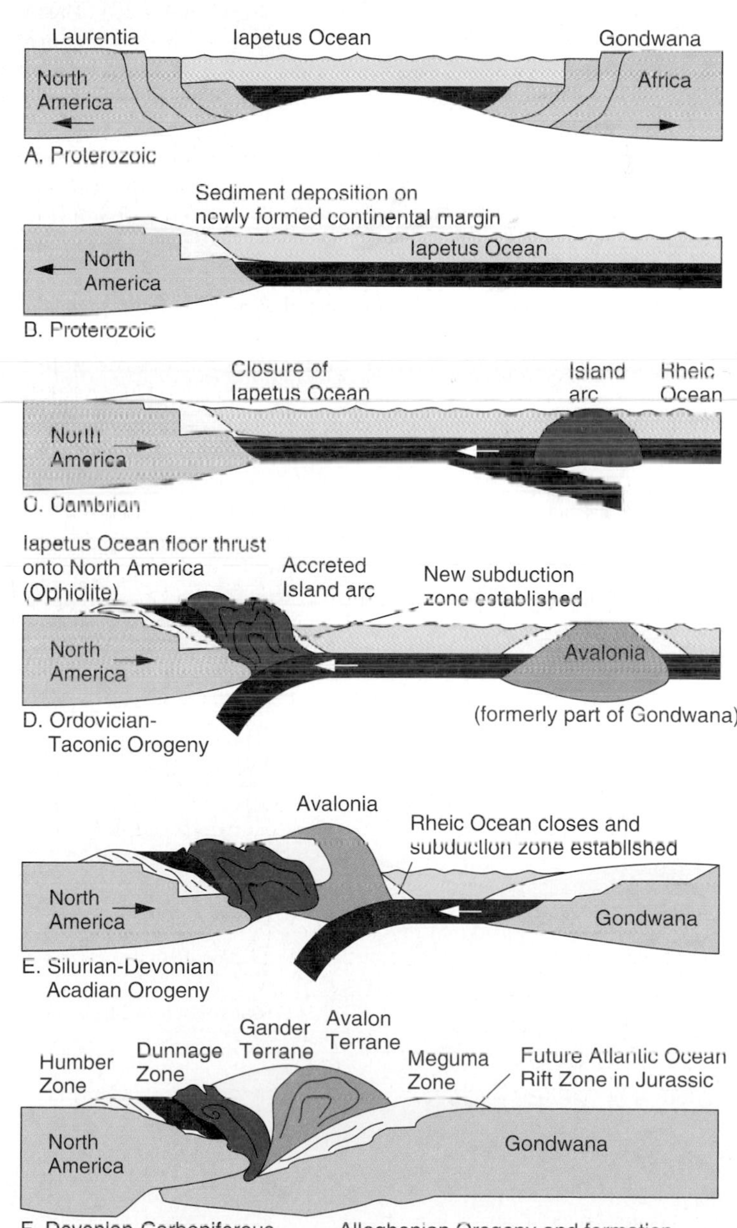

FIGURE 2.34

From Rodinia to Pangea: the geologic history of the Northern Appalachians as recorded in eastern Canada. (see also figure 20.16A)

Modified from Earth: An Introduction to Physical Geography, Tarbuck, 2005, p. 568, Pearson/Prentice Hall. Reprinted with permission by Pearson Education Canada Inc.

IN GREATER DEPTH 2.4

Canadian Rocky Mountains

The Canadian Rocky Mountains are a relatively recent feature of the North American continent, and were formed by erosion and uplift of rock strata deformed by plate tectonic collisions. The bent and folded rocks spectacularly exposed in the mountain ranges record the accretion of several microcontinents (terranes) onto the western seaboard of North America.

The evolution of the Rockies began when the supercontinent Pangea rifted apart around 180 million years ago and North America began to move slowly westward. As it moved, it collided with a number of smaller continents (or microcontinents) similar to modern-day Japan, and the upper parts of these continents were "scraped" off and welded (accreted) to the western margin of North America as distinct "terranes" (see chapter 20). These microcontinents were once probably dispersed in the eastern Pacific Ocean. Fossil and paleomagnetic evidence suggests that large areas of British Columbia, the Yukon, and Alaska are made of terranes that originated much nearer the equator.

The repeated accretion of microcontinents to the western margin of North America created a zone of intense compression and thrusting that created the basic structure that now forms the Rocky Mountains (box figure 1). At least two episodes of intense compression have been recognized by geologists. These episodes are called *orogenies* and occurred during the late Jurassic (Columbian Orogeny) and the early Tertiary (Laramide Orogeny). During compression, rock layers were subject to intense deformation, which created the spectacular thrust faults and contorted bands we see in them today (see chapter 11). Erosional processes then took over. Steep-sided narrow valleys within the mountains were scoured by glaciers that repeatedly covered northern North America during the past 2 million years (chapter 16). Mass wasting processes are currently active on the steep slopes (chapter 13), and rivers carry eroded sediment to the Pacific, Atlantic, or Arctic oceans (chapter 14).

Working in the mid–nineteenth century, Canadian geologists made a dramatic discovery in the Canadian Rockies. They found that older rocks, dating from the Cambrian era, had been thrust eastward several hundred kilometres over much younger Cretaceous strata. The first such thrust to be discovered is named after the geologist McConnell and defines the base of Mount Yamnuska, just 50 km west of Calgary (box figure 2). The steep cliffs above the thrust are made of Cambrian limestone and are very popular with climbers.

BOX 2.4 ■ FIGURE 2

Mount Yamnuska, showing steep cliffs of Cambrian limestone above Cretaceous shales (covered by scree), separated by the McConnell thrust (dashed line). The upper limestones were thrust from left (west) to right (east).

Photo by M. Radomski

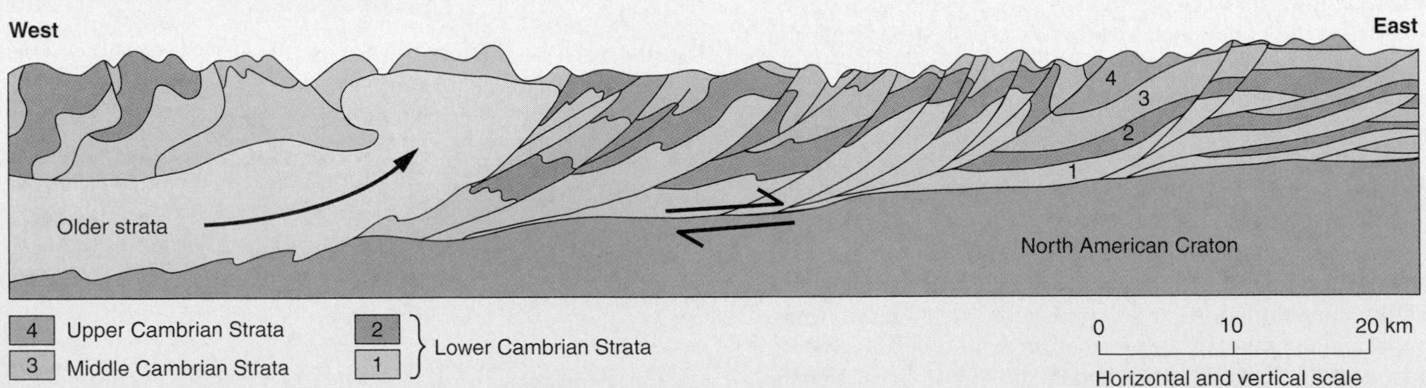

BOX 2.4 ■ FIGURE 1

Geological cross-section through the Canadian Rocky Mountains west of Calgary showing major folds and thrusts. These result from east–west compression during the Columbian and Laramide orogenies.

This process of continental splitting and collision was responsible for adding the Maritime provinces onto the Canadian Shield (see chapter 20). Early in the Mesozoic Pangea spilt roughly parallel to the old suture zone, and the continents of North America, Africa, and Europe separated as the modern Atlantic Ocean began to form.

What happened to the Appalachians would seem too implausible for even a science fiction plot. Yet, if one accepts the principles of plate tectonics theory and examines the rocks and structures in the Appalachians (and their counterparts in Europe and Africa), the argument for this sequence of events becomes not only plausible but also convincing.

The cycle of splitting of a supercontinent, opening of an ocean basin, closing of the basin, and collision of continents is known as the *Wilson Cycle* (box 2.5). Canadian geophysicist J. Tuzo Wilson proposed the cycle in the 1960s for the tectonic history of the Appalachians.

The Wilson Cycle apparently has occurred before. Why would a continent split apart more or less along a suture zone, where one would expect the crust to be thickest? One recently proposed hypothesis is that this is the zone that is weakened and thinned somewhat by outward flow of rock during gravitational collapse and spreading.

HOW DO PLATES CHANGE OVER TIME?

Plate Boundaries

Almost nothing is fixed in plate tectonics. Not only do plates move, but plate boundaries move as well. Plates may move away from each other at a divergent boundary on a ridge crest for tens of millions of years, but the ridge crest can be migrating across Earth's surface as this occurs. Ridge crests can also jump to new positions.

Convergent boundaries migrate, too. As they do, trenches and magmatic arcs migrate along with the boundaries. Convergent boundaries can also jump; subduction can stop in one place and begin suddenly in a new place.

Transform boundaries also change position. California's San Andreas fault has been in its present position about 5 million years. Prior to that, the plate motion was taken up on sea-floor faults parallel to the San Andreas. In the future, the San Andreas may shift eastward again. The 1992 Landers earthquake, on a new fault in the Mojave Desert, and its pattern of aftershocks extending an astonishing 800 km northward suggest that the San Andreas may be trying to jump inland again. Geodetic studies have shown that nearly 25 percent of plate motion between the North American and Pacific plates is accommodated along faults in eastern California and western Nevada (figure 2.35). If more motion is taken up along this zone, most of California will be newly attached to the Pacific plate instead of the North American plate, and California will slide northwestward relative to the rest of North America.

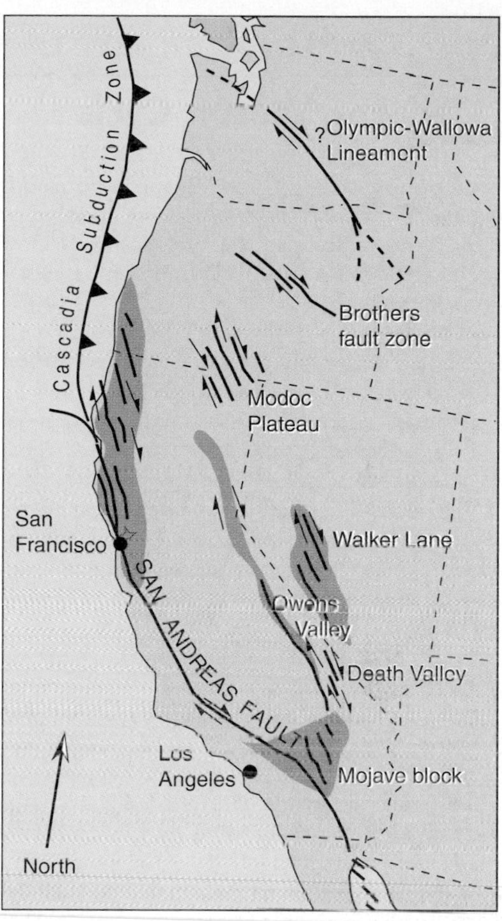

FIGURE 2.35

Many faults participate in easing North America past the Pacific plate along the San Andreas fault system.

Plate Size

Plates can change in size. For example, new sea floor is being added on the trailing edge of the North American plate at the spreading axis in the central Atlantic Ocean. Most of the North American plate is not being subducted along its leading edge because this edge is made up of lightweight continental rock. Thus the North American plate is growing in size as it moves slowly westward.

The Nazca plate is getting smaller. The spreading axis is adding new rock along the western edge of the Nazca plate, but the eastern edge is being subducted down the Peru–Chile Trench. If South America were stationary, the Nazca plate might remain the same size, because the rate of subduction and the rate of spreading are equal. But South America is slowly moving westward because of spreading on the Atlantic Ridge, pushing the Peru–Chile Trench in front of it. This means that the site of subduction of the Nazca plate is gradually coming closer to its spreading axis to the west, and so the Nazca plate is getting smaller. The same thing is probably happening to the Juan de Fuca plate (refer to box 2.2) as a result of westward movement of the North American plate, and to the Pacific plate as the Eurasian plate moves eastward into the Pacific Ocean.

IN GREATER DEPTH 2.5

The Wilson Cycle

Canadian geophysicist J. Tuzo Wilson proposed that ocean basins underwent a cyclic evolution of opening and closing controlled by plate tectonics. The opening and closing of an ocean basin is called a Wilson Cycle and can be subdivided into six stages (box figure 1).

The cycle begins (stage A) with a stable craton under which a hot spot develops, causing the crust to dome upward and fracture creating a rift valley or graben (see figure 2.40). The East African Rift Valley system is an ocean basin in this early stage of development.

The next stage (stage B) is reached when a sea-floor spreading centre with an erupted basaltic ocean floor is flooded to form a new ocean. The separated continental margins on each side of the new ocean cool, become denser, and subside beneath sea level. Sediments accumulating on these subsiding continental margins begin to form continental shelves. The Red Sea is currently at this stage of the Wilson Cycle.

Thereafter, the ocean basin widens (stage C) through addition of igneous mantle material via a mid-ocean ridge system characterized by young volcanoes and steep slopes. Continental shelves are well developed by this stage. The Atlantic Ocean is at this stage of development.

When the ocean basin is enlarged sufficiently to allow the basaltic crust to cool and sink down into the asthenosphere, a subduction zone is formed (stage D). Partial melting of subducted oceanic crust at depth produces rising magma plumes that erupt at the surface to form volcanic mountain chains or island arcs. An example of an ocean basin at this stage is the Pacific Ocean.

When most of the original oceanic crust has been subducted (stage E), the two continents that were originally separated begin to collide. Sea-floor sediments scraped off the descending plates form accretionary wedges characterized by intense folding and faulting.

Collision of the two continents follows complete subduction of the remnant ocean (stage F). The continent oceanward of the subduction zone is obducted onto the other continent, and a mountain range is created. Both continents are sutured together by rising plutonic rocks that originate at depth beneath the dying subduction zone. The Himalayas are the result of collision and suturing of the Indian subcontinent and Asia. Over time, these mountain ranges are peneplained by erosion, creating a new stable craton (stage G).

The Wilson Cycle has operated several times during the Phanerozoic, allowing the opening and closing of several major ocean basins. Breakup of the supercontinent Rodinia allowed development of the Iapetus Ocean, which later closed during formation of Pangea. The Appalachians appear to have formed as a result of such a cycle of ocean basin opening and closing (see figure 2.34). The many suture zones separating distinct geologic terranes in the Canadian Shield also demonstrate operation of the Wilson Cycle in the Precambrian (see chapter 20).

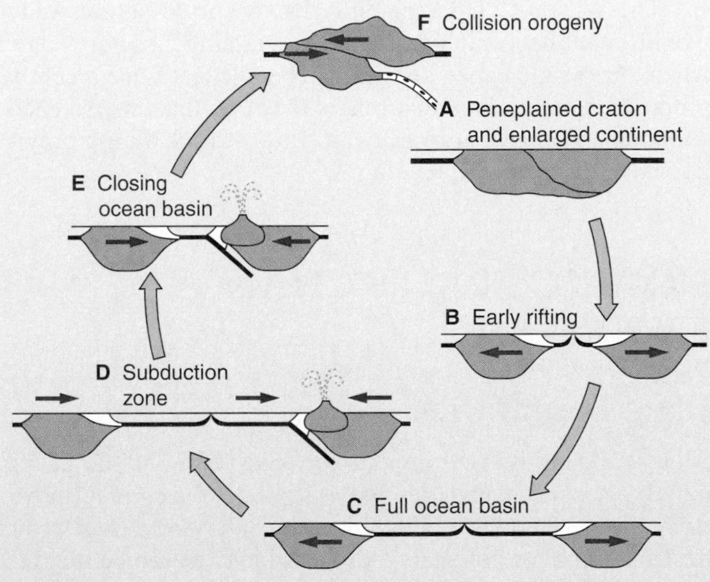

BOX 2.5 ■ FIGURE 1
The Wilson Cycle
From Department of Geology and Environmental Studies, James Madison University, http://brynmawr.edu/geology/102/lectures/cycles_files/frame.htm#slide0071.htm

WHAT CAUSES PLATE MOTIONS?

A great deal of speculation currently exists about why plates move. There may be several reasons for plate motion. Any mechanism for plate motion has to explain why:

1. mid-oceanic ridge crests are hot and elevated, while trenches are cold and deep;
2. ridge crests have tensional cracks;
3. the edges of some plates are subducting sea floor, while the edges of other plates are continents (which cannot subduct).

Convection in the mantle, proposed as a mechanism for sea-floor spreading (figure 2.12), can account for these facts, as we have shown earlier in this chapter. Mantle convection is quite likely because heat loss from the Earth's core should heat the overlying mantle, causing it to overturn. The old sea-floor spreading model assumed mantle-deep convection. Recent studies using seismic tomography and computer modelling of seismic waves suggest that the dynamics of convection in the mantle are not simple (figure 2.36). Cold lithospheric plates may subduct down to the core–mantle boundary, whereas other less dense (younger) plates may only reach the 670-km boundary. One of

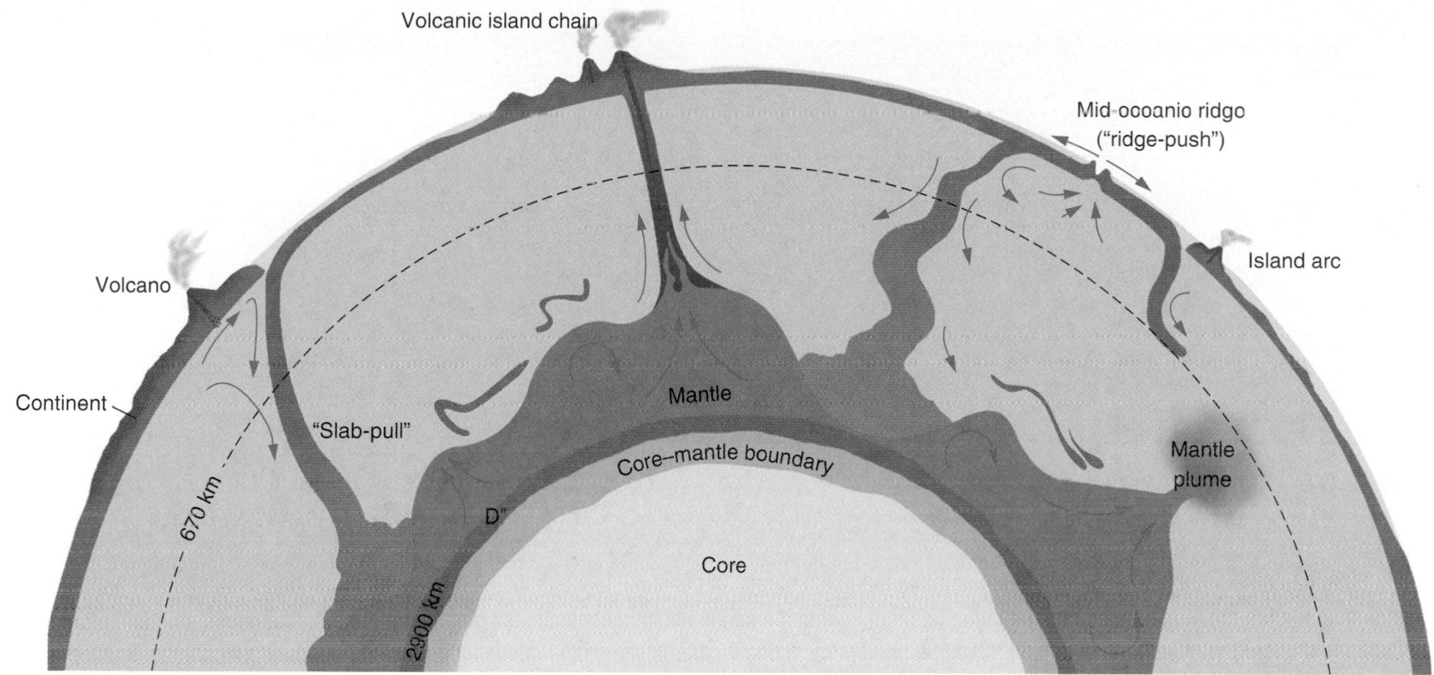

FIGURE 2.36

A possible model of mantle convection.

Modified from L. H. Kellogg, B. H. Hager, and R. D. van der Hilst, 1999, Science, 283:1881–84

the most recent models suggests that the lowermost part of the mantle does not mix with the upper and middle mantle, but acts as a "lava lamp" turned on low, fuelled by internal heating and heat flow across the core–mantle boundary. Variation in the thickness of this dense layer may control where mantle plumes rise and subducted plates ultimately rest.

The basic question in plate motion is why do plates diverge and sink? Two or three different mechanisms may be at work here. One proposal is called *ridge-push.* As a plate moves away from a divergent boundary, it cools and thickens. Cooling sea floor subsides as it moves, and this subsidence forms the broad side slopes of the mid-oceanic ridge. A slope also forms between the lithosphere and the asthenosphere, which may have a relief of 80 to 100 km.

Another mechanism is called *slab-pull* (figure 2.37). Cold lithosphere sinking at a steep angle through hot mantle should pull the surface part of the plate away from the ridge crest and then down into mantle as it cools. Slab-pull is thought to be at least twice as important as ridge-push in moving an oceanic plate away from a ridge crest. Slab-pull is hypothesized to cause rapid plate motion.

If subducting plates fall into the mantle at angles steeper than their dip (figure 2.37), then trenches and the overlying plates are pulled horizontally seaward toward the subducting plates. This mechanism has been termed *trench-suction.* It is probably a minor force, but may be important in moving continents apart.

All three of these mechanisms (ridge-push, slab-pull, and trench-suction), particularly in combination, are compatible with high, hot ridges; cold, deep trenches; and tensional cracks at the ridge crest. They can account for the motion of both oceanic and continental plates. In this scheme, plate motions are controlled by variations in lithosphere density and thickness, which, in turn, are controlled largely by cooling. In other words, the reasons for plate motions are the properties of the plates themselves and the pull of gravity. This idea is in sharp contrast to most convection models, which assume that plates are dragged along by the movement of mantle rock beneath the plates.

HOW ARE MANTLE PLUMES AND HOT SPOTS RELATED?

Mantle plumes are narrow columns of hot mantle rock that rise through the mantle from thermal boundary layers at the base of the mantle (or upper mantle) (figure 2.38). Mantle plumes are now thought to have large spherical or mushroom-shaped heads above a narrow rising tail. They are essentially stationary with respect to moving plates and to each other.

Plumes may form "hot spots" of active volcanism at the Earth's surface. Note in figure 2.39 that many hot spots are located in volcanic regions such as Iceland, Yellowstone, and Hawaii. Recent seismic images of the mantle suggest that not all hot spots are fed by mantle plumes. Of the 45 hot spots identified on Earth, only 12 show evidence of a deep, continuous plume in the underlying mantle. According to one hypothesis, when the large head of the plume nears the surface, it causes uplift and the eruption of vast fields of flood basalts. As the head widens beneath the crust the flood-basalt area widens and the crust is stretched. The tail that follows the head produces a narrow spot of volcanic activity, much smaller than the head.

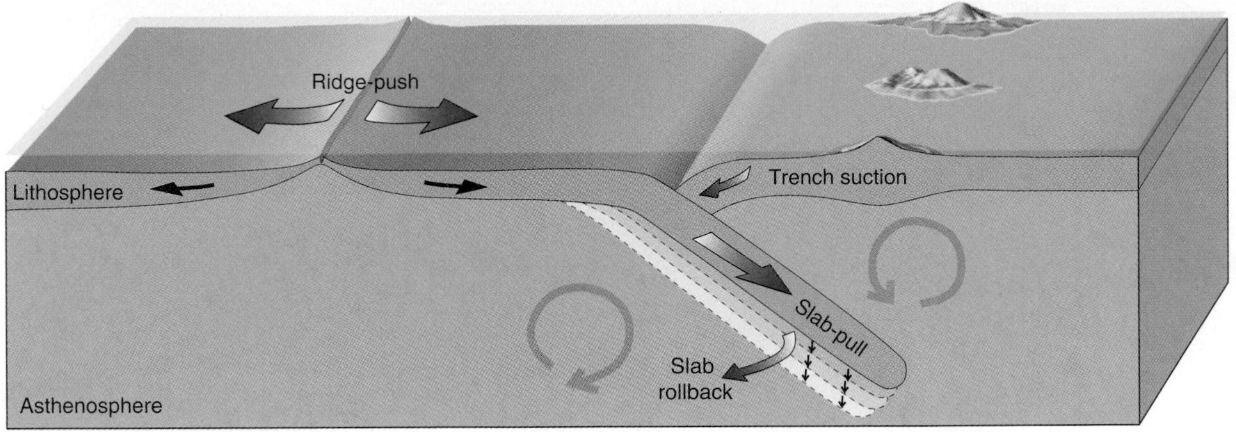

FIGURE 2.37

Other possible mechanisms for plate motion. Plates are pushed apart at the ridge (*ridge-push*) by sliding downhill on the sloping boundary between the lithosphere and asthenosphere. Plates may also be pulled (*slab-pull*) as the dense leading edge of a subducting plate sinks down into the asthenosphere. If the subducting plate falls into the asthenosphere at angles steeper than its dip (*slab rollback*) then the trench and overlying plate are pulled horizontally seaward toward the subducting plate by *trench suction*.

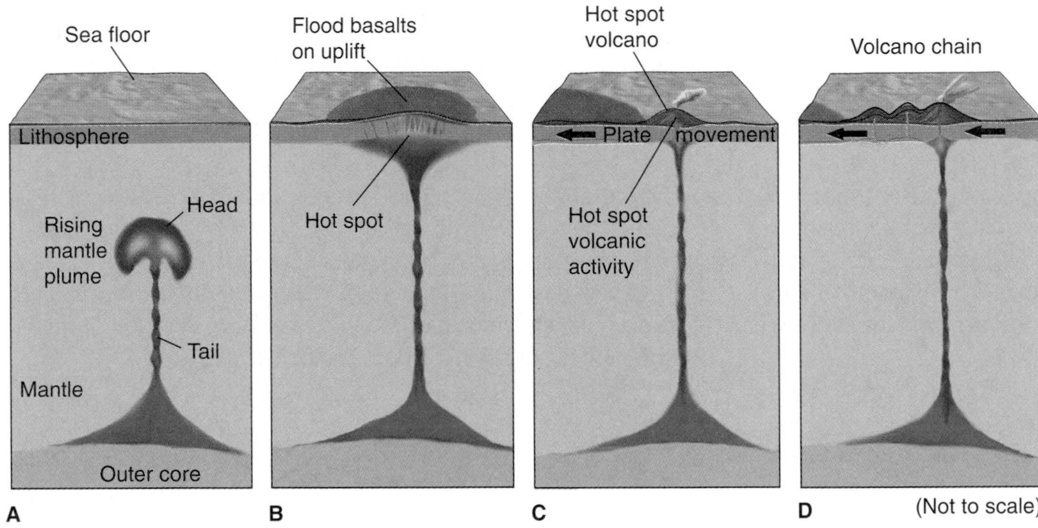

FIGURE 2.38

Model of mantle plume rising upward through the mantle to form a hot spot and associated flood basalts and volcanic chain. (*A*) Rising mantle plume contains a hot, mushroom-shaped plume head and a narrow tail. (*B*) Plume head forms a broad hot spot when it reaches the top of the mantle and causes uplift and stretching of the crust and eruption of flood basalts. (*C*) When the tail rises to the surface, a narrower hot spot forms a volcano. (*D*) Continued plate motion over the hot spot creates a trail or chain of volcanoes.

The outward, radial flow of the expanding head may be strong enough to break the lithosphere and start plates moving. Some geoscientists suggest that a few plumes, such as those underlying hot spots on the mid-oceanic ridge in the Atlantic Ocean in figure 2.39, are enough to drive plates apart (in this case, to push the American plates westward).

A mantle plume rising beneath a continent should heat the land and bulge it upward to form a dome marked by volcanic eruptions. As the dome forms, the stretched crust typically fractures in a three-pronged pattern (figure 2.40). Continued radial flow outward from the rising plume eventually separates the crust along two of the three fractures but leaves the third fracture inactive. In this model of continental breakup, the two active fractures become continental edges as new sea floor forms between the divergent continents. The third fracture is a *failed rift* (or *aulacogen*), an inactive rift that becomes filled with sediment (see box 2.6).

An example of this type of fracturing may exist in the vicinity of the Red Sea (figure 2.40). The Red Sea and the Gulf of Aden are active diverging boundaries along which the Arabian Peninsula is being separated from northeastern Africa. The third, inactive, rift is the northernmost African Rift Valley, lying at an angle of about 120 degrees to each of the narrow seaways.

Figure 2.41 shows how two plumes might split a continent and begin plate divergence. Local uplift *causes* rifting over each plume. The rifts lengthen with time until the land is torn in two. The two halves begin to diverge from being dragged along from below by the outward radial flow of the plume. Along the long rift segments between plumes, rifting occurs *before* uplift.

Some plumes rise beneath the centres of oceanic plates. A plume under Hawaii rises in the centre of the Pacific plate. As the plate moves over the plume, a line of volcanoes forms, creating an aseismic ridge (figure 2.42). The volcanoes are gradually

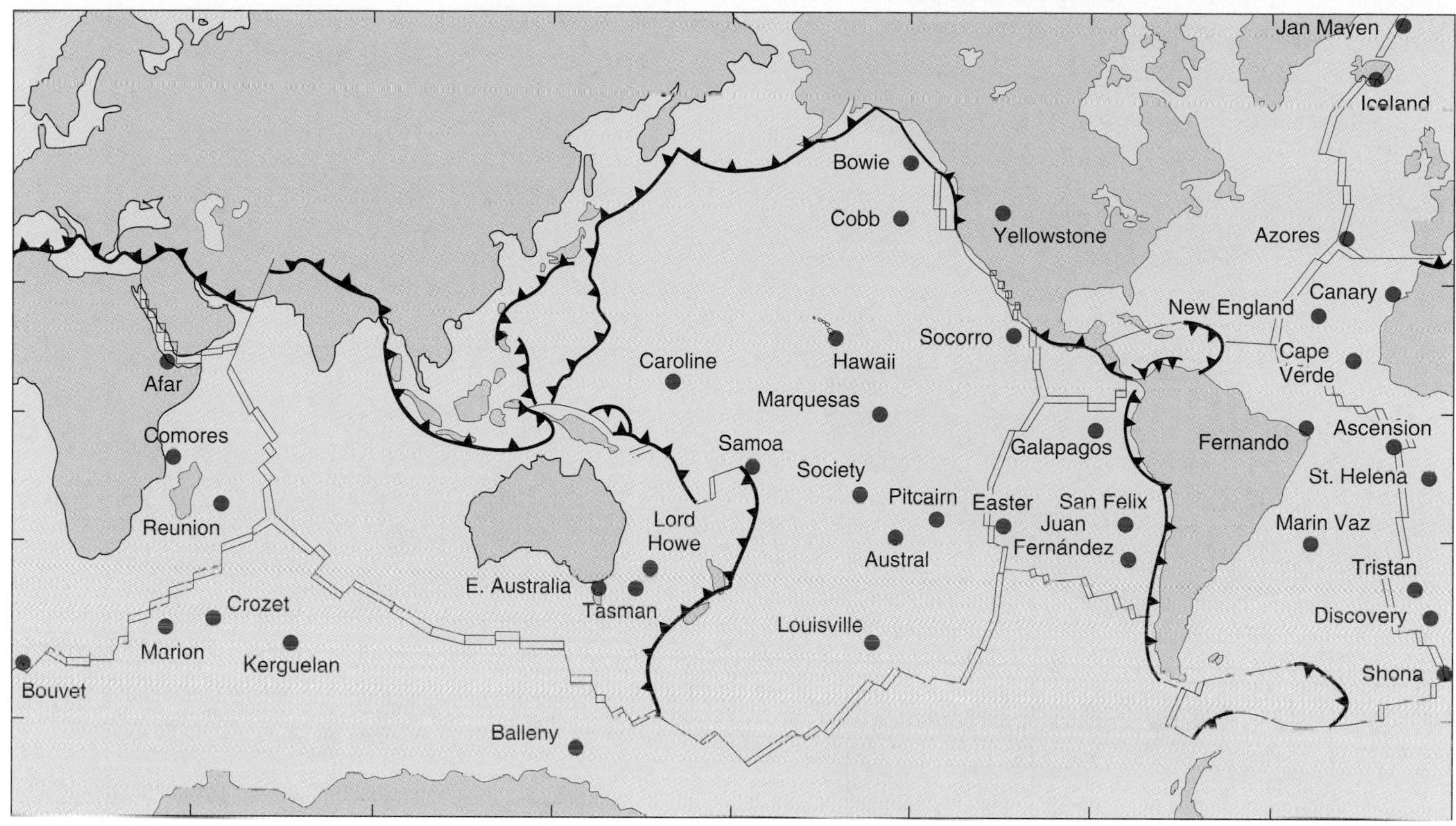

FIGURE 2.39

Distribution of hot spots, identified by volcanic activity and structural uplift within the past few million years. The hot spots near the poles are not shown

carried away from the eruptive centre, sinking as they go because of cooling. The result is a line of extinct volcanoes (seamounts and guyots) increasing in age away from an active volcano directly above the plume.

In the Hawaiian island group, the only two active volcanoes are in the extreme southeastern corner (figure 2.43). The isotopic ages of the Hawaiian basalts increase regularly to the northwest, and a long line of submerged volcanoes forms an aseismic ridge to the northwest of Kauai (figure 2.43). Most aseismic ridges on the sea floor appear to have active volcanoes at one end, with ages increasing away from the eruptive centres. Deep-sea drilling has shown, however, that not all aseismic ridges increase in age along their lengths. This evidence has led to alternate hypotheses for the origin of aseismic ridges. It may pose difficulties for the plume hypothesis itself.

Seamounts, Guyots, and Aseismic Ridges

Conical undersea mountains that rise 1,000 m or more above the sea floor are called **seamounts** (figures 2.8 and 2.44). Some rise above sea level to form islands. They are scattered on the flanks of the mid-oceanic ridge and on other parts of the sea floor, including abyssal plains. One area of the sea floor with a particularly high concentration of seamounts—one estimate is 10,000—is the western Pacific. Rocks dredged from seamounts are nearly always basalt, so it is thought that most seamounts are extinct volcanoes. Of the thousands of seamounts on the sea floor, only a few are active volcanoes. Most of these are on the crests of the mid-oceanic ridges. A few others, such as the two active volcanoes on the island of Hawaii, are found at locations not associated with a ridge.

Guyots are flat-topped seamounts (figure 2.44) found mostly in the western Pacific Ocean. Most geologists think that the flat summits of guyots were cut by wave action. These flat tops are now many hundreds of metres below sea level, well below the level of wave erosion. If the guyot tops were cut by waves, the guyots must have subsided after erosion took place. Evidence of such subsidence comes from the dredging of dead reef corals from guyot tops. Since such corals grow only in shallow water, they must have been carried to their present depths as the guyots sank.

Many of the guyots and seamounts on the sea floor are aligned in chains (figure 2.8). Such volcanic chains, together with some other ridges on the sea floor, are given the name **aseismic ridges**; that is, they are submarine ridges that are not associated with earthquakes. The name *aseismic* is used to distinguish these features from the much larger mid-oceanic ridge, where earthquakes occur along the rift valley.

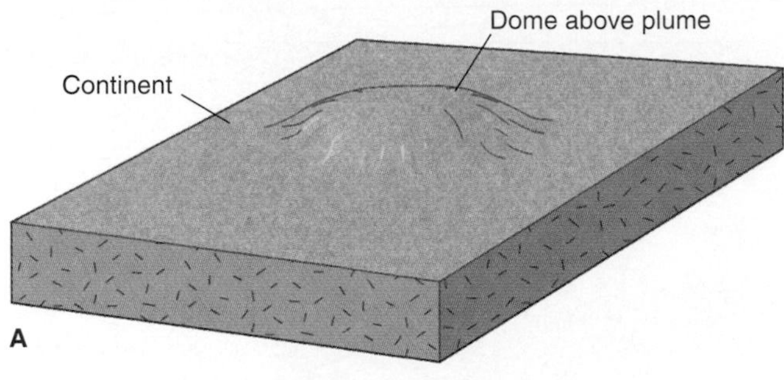

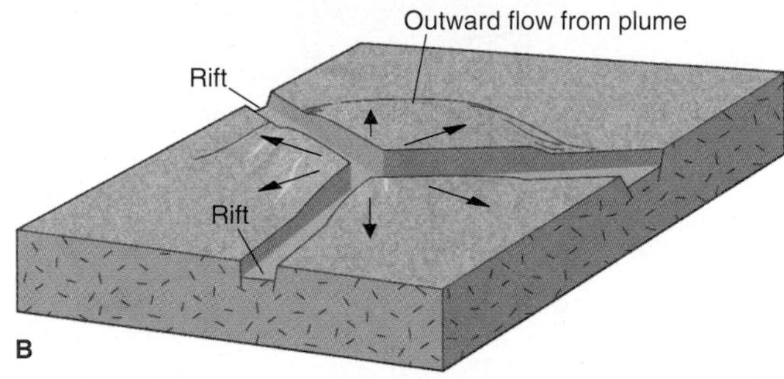

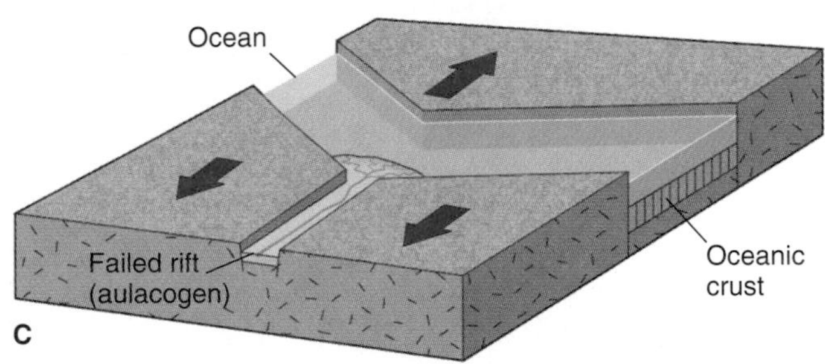

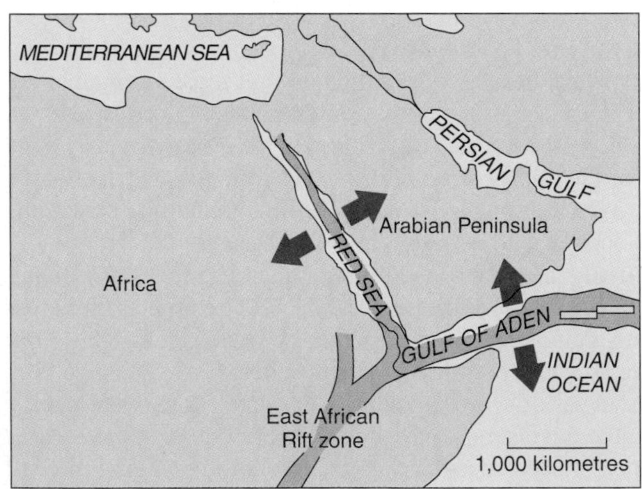

FIGURE 2.40

Continental breakup caused by a mantle plume. (*A*) A dome forms over a mantle plume rising beneath a continent. (*B*) Three radial rifts develop due to outward radial flow from the top of the mantle plume. (*C*) Continent separates into two pieces along two of the three rifts, with new ocean floor forming between the diverging continents. The third rift becomes an inactive "failed rift" (or aulacogen) filled with continental sediment. (*D*) An example of radial rifts. The Red Sea and the Gulf of Aden are the active rifts, as the Arabian peninsula drifts away from Africa. The Gulf of Aden contains a mid-oceanic ridge and central rift valley. The less active, failed rift (aulacogen) is the rift valley shown in Africa.

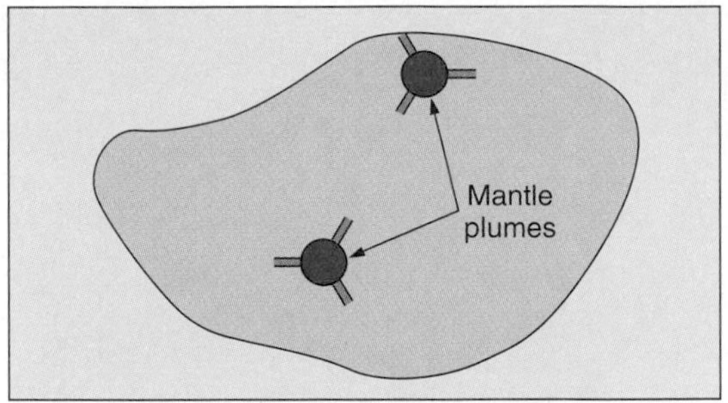

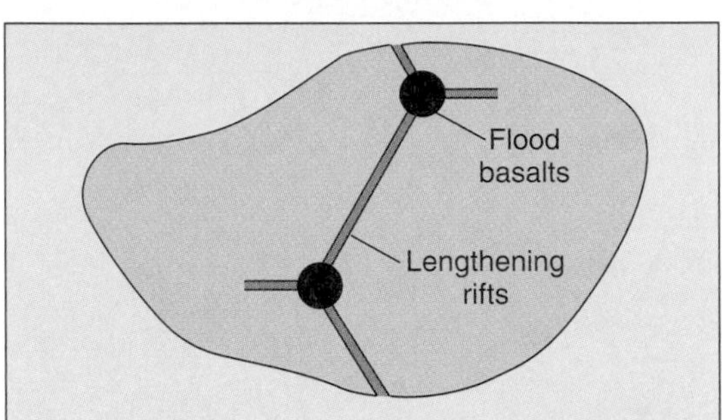

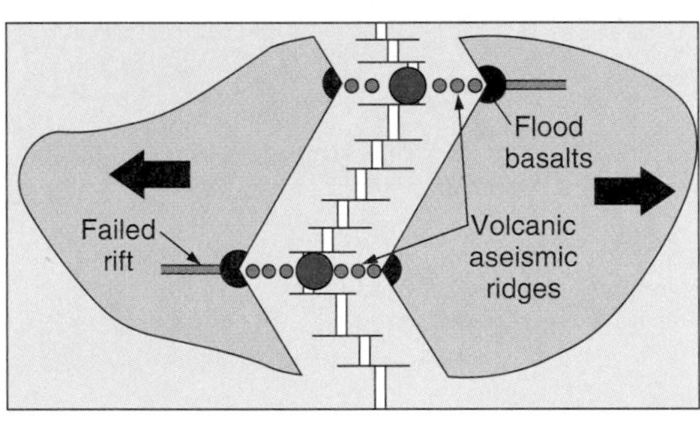

FIGURE 2.41

(*A*) Two mantle plumes beneath a continent. (*B*) The rifts lengthen and flood basalts erupt over the plumes. (*C*) The continent splits and failed rifts form. The new ocean is marked by ridge crests, fracture zones, and aseismic ridges (chains of volcanoes).

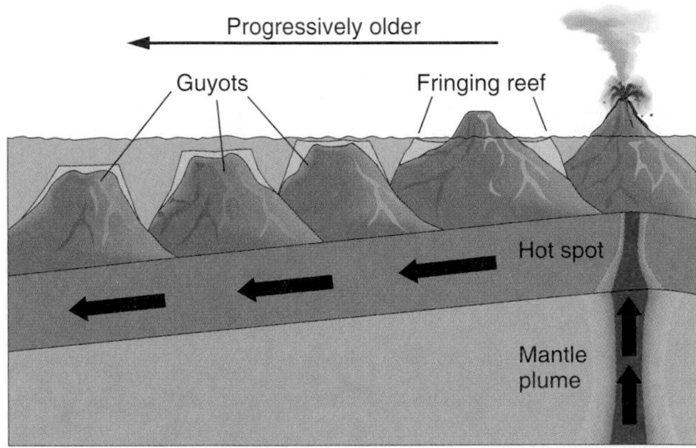

FIGURE 2.42 

Fringing reef forms around volcanic island as it moves off hot spot. Waves erode and flatten top of volcanic islands to form guyots that progressively sink and become submerged away from hot spot.

IN GREATER DEPTH 2.6

The Geological Significance of Aulacogens

Initial doming of a continent prior to rifting produces a three-armed fracture system, or triple junction (see figure 2.40). Each arm of the triple junction forms a rift, or graben. However, only two of the three rifts will continue to widen to form an ocean basin, and the third rift fails to develop further; with time, it infills with many kilometres of sediment. These "failed rifts," or aulacogens, are found on all continents and have considerable environmental significance both as sites of petroleum production and of intraplate earthquake occurrence (see chapter 3).

The North Sea Basin (box figure 1), separating Britain from mainland Europe, is an aulacogen that formed during the Mesozoic opening of the northern part of the Atlantic Ocean. The thick successions of interbedded coarse- and fine-grained sediments deposited within the rift basin contain a wealth of organic material and host some of the world's most productive petroleum reservoirs. Large rivers often develop in failed rift valleys and transport significant volumes of sediment, which are deposited as deltas at the mouth of the valley. The Niger River flows along an aulacogen formed as a result of rifting of the southern part of the Atlantic Ocean and terminates at the Niger Delta; this is also a globally significant petroleum-producing area.

The Ottawa–Bonnechere Graben of southern Ontario (box figure 2) is thought to have formed as an aulacogen when Rodinia broke up more than 600 million years ago to form the Iapetus (or proto-Atlantic) Ocean. The graben was probably reactivated and enlarged as a result of Mesozoic crustal extension during the breakup of Pangea around 150 million years ago. Today, stretching and reactivation of faults in the graben as North America drifts westward may be responsible for the many small to moderate *intraplate* earthquakes felt in this area of Ontario and Quebec (see chapter 3). Lake Nipissing, the Ottawa River, and the city of Ottawa currently lie within the graben, which is 60 km wide and more than 700 km long. The St. Lawrence Rift is also an aulacogen, initially formed during the breakup of Rodinia and later reactivated during the breakup of Pangea. The St. Lawrence Rift is thought to underlie Lake Ontario and may have exerted some control on the development of the Erie and Ontario lake basins (box figure 2).

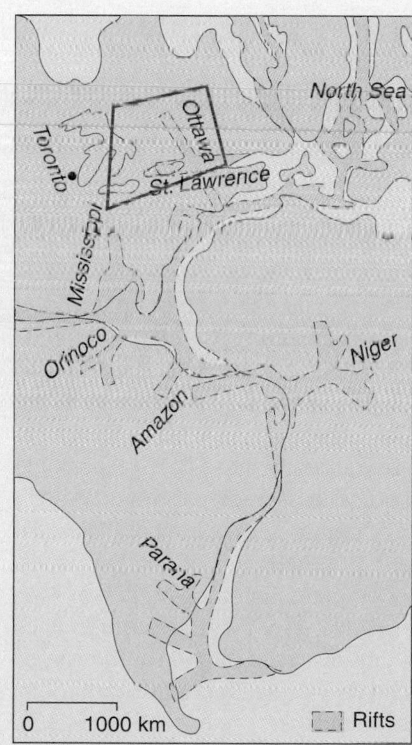

BOX 2.6 ■ FIGURE 1

Formation and reactivation of rifts and grabens during the break up of Pangea around 150 million years ago. The North Sea, Niger, and Ottawa grabens are examples of rifts that failed to open fully (aulacogens).

Modified from Nick Eyles, 2002

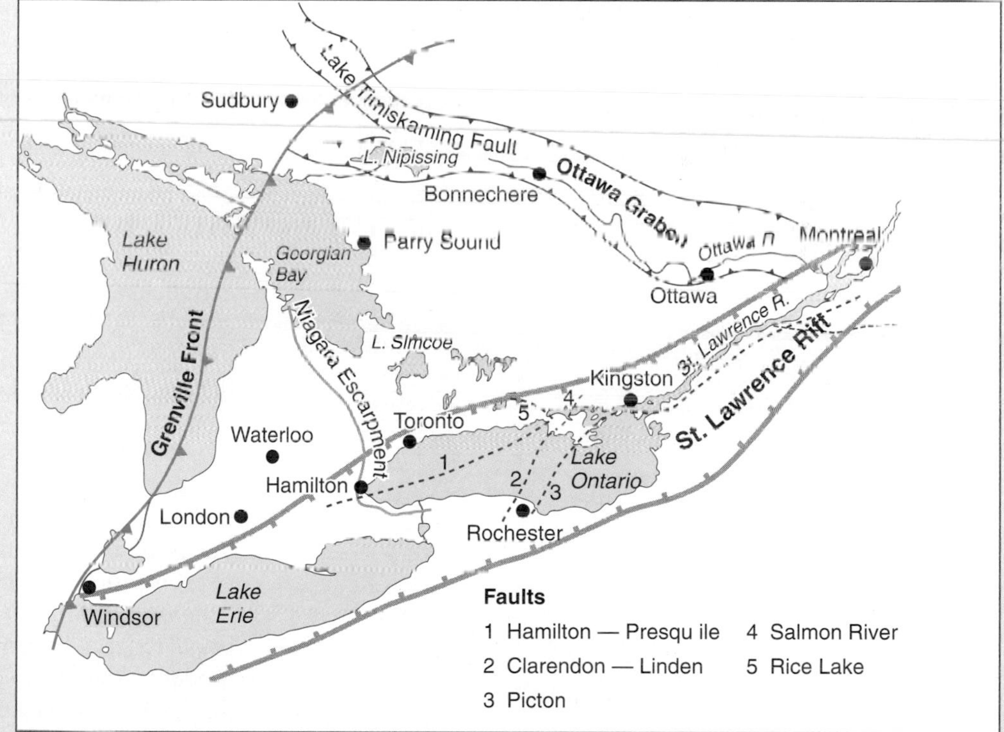

BOX 2.6 ■ FIGURE 2

The Ottawa–Bonnechere Graben and St. Lawrence Rift are both aulacogens (failed rifts) that were reactivated during the breakup of Pangea and are now associated with common small to moderate-level earthquakes. There is also a close relationship between modern surface drainage features (rivers and lakes) and the ancient failed rift systems.

Modified from Nick Eyles, 2002

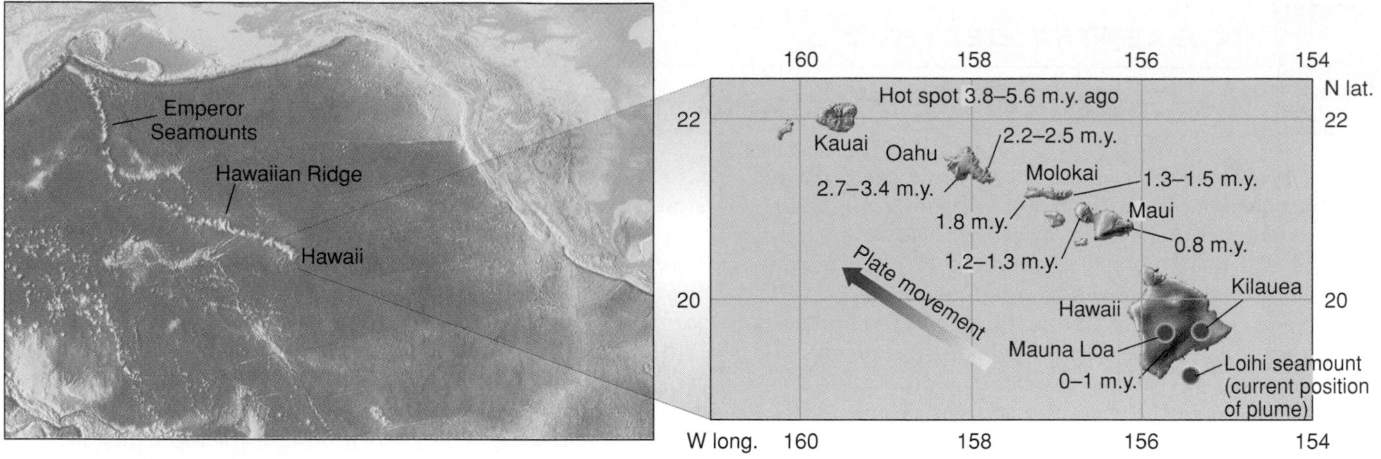

FIGURE 2.43

Ages of volcanic rock of the Hawaiian island group. Ages increase to northwest. Two active volcanoes on Hawaii shown by red dots. The plume is currently offshore under the Loihi seamount (red dot), where recent underwater eruptions have been documented.

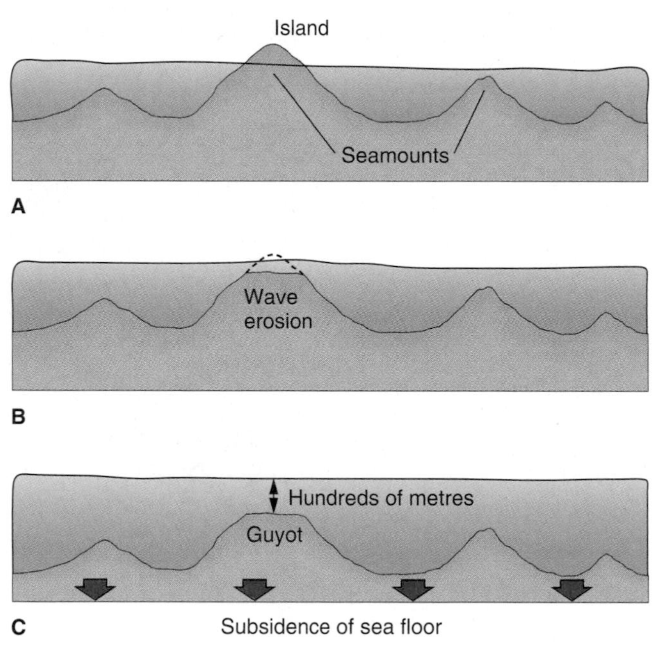

FIGURE 2.44

(A) Seamounts are conical mountains on the sea floor, occasionally rising above sea level to form islands. (B) The flat summit of a guyot was probably eroded by waves when the top of a seamount was above sea level. (C) The present depth of a guyot is due to subsidence.

WHY IS IT IMPORTANT TO UNDERSTAND PLATE TECTONICS?

Plate tectonics is an exciting geological theory that can be used to explain many features of the Earth. The global distribution of volcanoes and earthquakes, sea-floor age and topography, the development of seamounts and hot spots, and the origin of mountain belts can all be related to plate tectonics. The most geologically active areas on Earth today are those associated with plate boundaries where plates are moving relative to each other and new ocean crust may be either created or consumed. These regions are especially prone to violent geologic events such as earthquakes, volcanic eruptions, and landslides (see box 2.2 and chapters 3, 7, and 13). An understanding of plate tectonics processes and their geographic distribution is fundamental to the prediction and preparation for emergency situations arising from such events.

Plate tectonics processes also appear to have been responsible for the repeated creation and breakup of "supercontinents" such as Rodinia and Pangea. Many of our modern continental cratons, including the Canadian Shield, are the result of amalgamation of many smaller land masses by plate tectonics processes (see chapter 20). Ancient plate margins, now forming suture zones or aulacogens in continental cratons, are commonly the site of modern earthquake activity and may threaten urban populations far removed from modern plate boundaries (see boxes 2.6 and 2.7). Hence, reconstruction of ancient plate movements and tectonic processes is necessary to fully evaluate the potential for earthquakes in continental interiors.

Finally, valuable mineral resources are commonly formed at plate margins where geological processes are most active (see chapter 12). Reconstruction of past plate tectonics movements is often required for resource exploration purposes.

IN GREATER DEPTH 2.7

Plate Tectonics and Earthquake Risk in Southern Ontario

The North American craton, the largest exposed portion of which is called the Canadian Shield, formed as a result of the accretion of numerous separate continental masses early in the Earth's history. Each accreted microcontinent has distinctive geological characteristics, and these can be identified as geological *terranes* in areas of the Canadian Shield. The boundaries between individual terranes represent collisional margins or *suture zones* where compression of plate margins created high mountain ranges (box figure 1). The mountains have since been eroded away, but the crust along these suture zones is weakened by extensive fault systems and is commonly the site of earthquake activity. Although most earthquakes occur along plate boundaries (interplate earthquakes), large earthquakes also occur quite distant from such boundaries in continental interiors (intraplate earthquakes; see chapter 3). These earthquakes have the potential to cause significant environmental damage in urban regions of Canada.

Southern Ontario is underlain by the craton, which is covered by younger Paleozoic rocks and Quaternary sediments. The areas of the craton underlying southern Ontario were accreted to ancestral North America around 1.2 billion years ago during the formation of a much larger supercontinent called Rodinia. At this time, rocks of the Central Metasedimentary Belt (CMB) were welded onto the Central Gneiss Belt (CGB) along a suture zone called the Central Metasedimentary Belt Boundary Zone (CMBBZ). A high range of mountains formed along this boundary (the Grenville Mountains, similar to the modern Himalayas), but they have been peneplained by erosion (box figure 1). All that remains of the suture zone is a 10- to 20-km-wide linear zone of intense structural deformation that runs from northeastern Ontario into Pennsylvania and is now mostly buried beneath more recent deposits (box figure 2).

The geological significance of ancient suture zones is that they are zones of crustal weakness that are the preferred site of earthquake activity. As the North American plate moves westward at a rate of about 2.7 cm each year, considerable stresses are built up in crustal rocks. These stresses can be released as earthquakes or may cause joints and buckles ("pop-ups") to develop in sediments and rocks. Only recently have geoscientists recognized that small-magnitude intraplate earthquakes occur frequently in eastern North America, including southern Ontario. Much of this *neotectonic* earthquake activity appears to be concentrated along terrane boundaries or suture zones such as the CMBBZ (box figure 3).

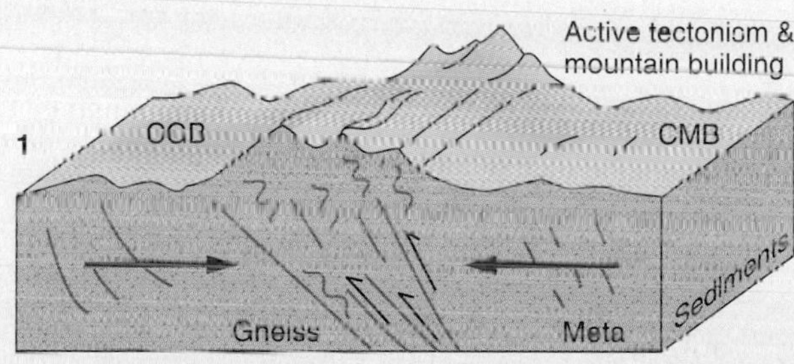

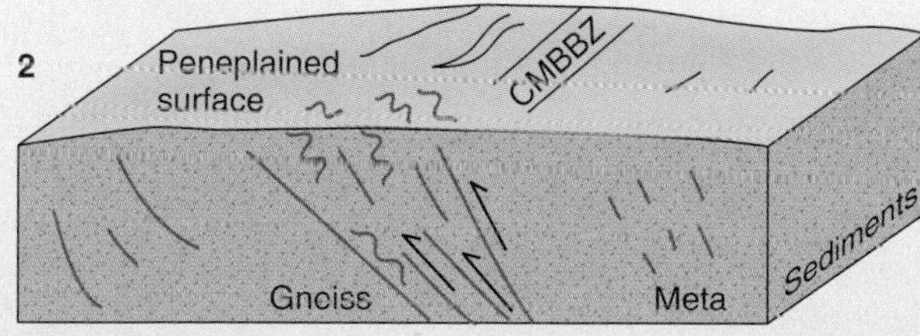

BOX 2.7 ■ FIGURE 1

Collision of two terranes forms a mountain range, which is eroded over time.
From Nick Eyles, 2002

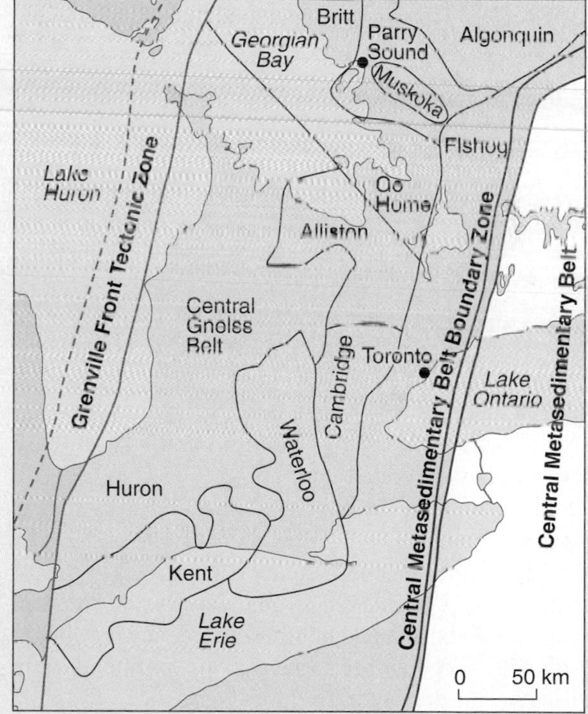

BOX 2.7 ■ FIGURE 2

Buried terranes of the Central gneiss belt (orange) and Central metasedimentary belt (yellow) join along the CMBBZ. The CMBBZ suture zone passes directly below the Toronto area.
From Nick Eyles, 2002

The existence of the CMBBZ and its role as a *seismogenic* structure was not realized when the Pickering Nuclear Generating Station was built on it in the 1960s. Suture zones and faults in buried crustal rocks are difficult to recognize on the ground surface and geoscientists rely on remotely sensed geophysical data to identify significant changes in rock properties. Buried faults and suture zones are commonly identified from changes in the magnetic properties of the rocks (magnetic lineaments; box figure 3).

Today, geologists recognize the potential for earthquake activity along ancient suture zones in continental interiors but are faced with the difficult task of determining the recurrence interval or frequency of these events. It has been suggested that a magnitude 7 earthquake could occur, on average, once every 3,000 years in southern Ontario.

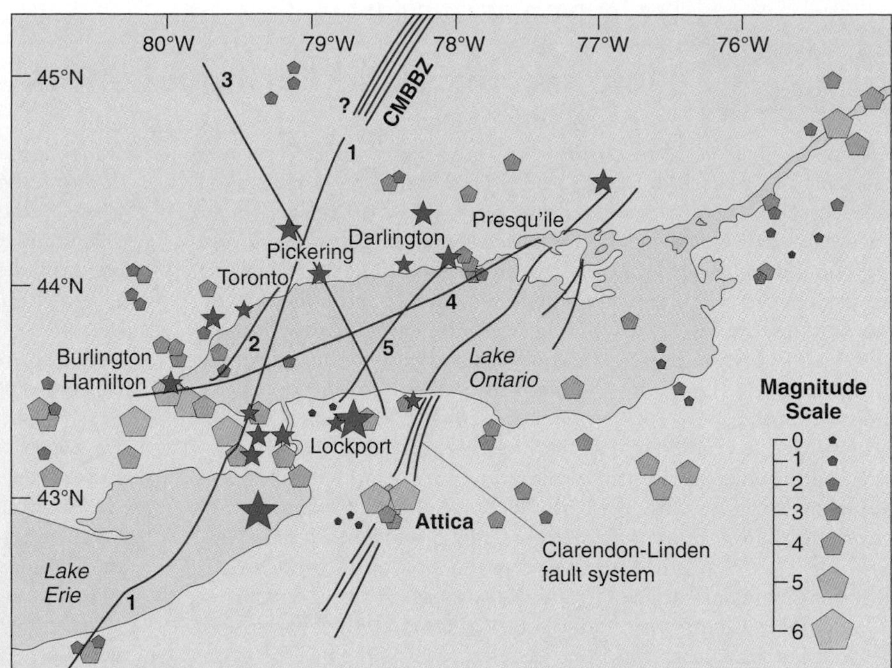

BOX 2.7 ■ FIGURE 3

Earthquake epicentres and bedrock structures in southern Ontario.
After Alex Mohajer, 1977.

SUMMARY

Plate tectonics is a theory that suggests Earth's surface is divided into several large plates that change position and size. Intense geologic activity occurs at plate boundaries.

Plate tectonics combines the concepts of *sea-floor spreading* and *continental drift*.

Alfred Wegener proposed continental drift in the early 1900s. His evidence included coastline fit, similar fossils and rocks in now-separated continents, and paleoclimatic evidence for *apparent polar wandering*. Wegener proposed that all continents were once joined together in the supercontinent *Pangea*.

Wegener's ideas were not widely accepted until the 1950s, when work in paleomagnetism revived interest in polar wandering.

Evidence for continental drift includes careful fits of continental edges and detailed rock matches between now-separated continents. The positions of continents during the past 200 million years have been mapped.

Hess's hypothesis of *sea-floor spreading* suggests that the sea floor moves away from the ridge crest and toward trenches as a result of mantle convection.

According to the concept of sea-floor spreading, the high heat flow and volcanism of the ridge crest are caused by hot mantle rock rising beneath the ridge. Divergent *convection* currents in the mantle cause the rift valley and earthquakes on the ridge crest, which is a *spreading axis* (or *centre*). New sea floor near the rift valley has not yet accumulated pelagic sediment.

Sea-floor spreading explains trenches as sites of sea-floor *subduction*, which causes low heat flow and negative-gravity anomalies. Benioff zones and andesitic volcanism are caused by interaction between the subducting sea floor and the rocks above.

Sea-floor spreading also explains the young age of the rock of the sea floor as caused by the loss of old sea floor through subduction into the mantle.

Plates are composed of blocks of *lithosphere* riding on the plastic upper mantle or *asthenosphere*. Plates move away from spreading axes, which add new sea floor to the trailing edges of the plates.

An apparent confirmation of plate motion came in the 1960s with the correlation of marine *magnetic anomalies* to *magnetic reversals* by Morley, Vine, and Matthews. The origin of magnetic anomalies at sea apparently is due to the recording of normal and reverse magnetization by dikes that intrude the crest of the mid-oceanic ridge, then split and move sideways to give anomaly patterns a mirror symmetry.

The Morley-Vine-Matthews hypothesis gives the rate of plate motion (generally 1 to 6 cm/year) and can predict the age of the sea floor before it is sampled.

Deep-sea drilling has apparently verified plate motions and the age predictions made from magnetic anomalies.

Earthquake distribution and first-motion studies on *transform faults* on fracture zones also verify plate motions.

Divergent plate boundaries are marked by rift valleys, shallow-focus earthquakes, high heat flow, and basaltic volcanism.

Transform boundaries between plates sliding past one another are marked by strike-slip (transform) faults and shallow-focus earthquakes.

Convergent plate boundaries can cause *subduction* or *continental collision*. Subducting plate boundaries are marked by trenches, low heat flow, Benioff zones, andesitic volcanism, and young mountain belts or island arcs. Continental-collision boundaries have shallow-focus earthquakes and form young mountain belts in continental interiors.

The distribution and origin of most volcanoes, earthquakes, young mountain belts, and major sea-floor features can be explained by plate tectonics.

Plate motion was once thought to be caused by *mantle convection*, but is now attributed to the cold, dense, leading edge of a subducting plate pulling the rest of the plate along with it (*slab-pull*). Plates near mid-oceanic ridges also slide down the sloping lithosphere–asthenosphere boundary at the ridge (*ridge-push*). *Trench-suction* may help continents diverge.

Mantle plumes are narrow columns of hot, rising mantle rock. They cause flood basalts and may split continents, causing plate divergence.

An *aseismic ridge* may form as an oceanic plate moves over a mantle plume acting as an eruptive centre (hot spot).

Terms to Remember

anomaly 33
aseismic ridges 57
asthenosphere 33
continental drift 22
convection 29
convergent plate boundary 33
divergent plate boundary 33
guyots 57
island arc 43
lithosphere 32
magmatic arc 44
magnetic polarity time scale 33
magnetic reversals 33
magnetometer 33
mantle plume 55
mid-oceanic ridge 38
normal polarity 33
oceanic trench 43
orogeny 49
paleomagnetism 33
plate 32
plate tectonics 22
polar wandering 28
reversed polarity 33
rift valley 38
sea-floor spreading 22
seamount 57
sidescan sonar 25
subduction 29
transform fault 36
transform plate boundary 33

Testing Your Knowledge

Use the questions below to prepare for exams based on this chapter.

1. What was Wegener's evidence for continental drift?
2. What is polar wandering? What is the paleoclimatic evidence for polar wandering? What is the magnetic evidence for polar wandering? Does polar wandering require the poles to move?
3. What is the evidence that South America and Africa were once joined?
4. In a series of sketches, show how the South Atlantic Ocean might have formed by the movement of South America and Africa.
5. What was Pangea?
6. In a single cross-sectional sketch, show the concept of sea-floor spreading and how it relates to the mid-oceanic ridge and oceanic trenches.
7. How does sea-floor spreading account for the age of the sea floor?
8. What is a plate in the concept of plate tectonics?
9. Define *lithosphere* and *asthenosphere*.
10. What is the origin of marine magnetic anomalies according to Morley, Vine, and Matthews?
11. Why does the pattern of magnetic anomalies at sea match the pattern of magnetic reversals (recorded in lava flows on land)?
12. How has deep-sea drilling tested the concept of plate motion?
13. How has the study of fracture zones tested the concept of plate motion?
14. Explain how plate tectonics can account for the existence of the mid-oceanic ridge and its associated rift valley, earthquakes, high heat flow, and basaltic volcanism.
15. Explain how plate tectonics can account for the existence of oceanic trenches as well as their low heat flow, their negative gravity anomalies, the associated Benioff zones of earthquakes, and andesitic volcanism.

16. What is a transform fault?

17. Discuss possible driving mechanisms for plate tectonics.

18. Describe the various types of plate boundaries and the geologic features associated with them.

19. What is a mantle plume? What is the geologic significance of mantle plumes?

20. The mesozoic southern supercontinent is called
    a. Gondwanaland b. Pangea
    c. Laurasia d. Glossopteris

21. The sliding of the sea floor beneath a continent or island arc is called
    a. rotation b. tension
    c. subduction d. polar wandering

22. In cross-section, the plates are part of a rigid outer shell of the Earth called the
    a. lithosphere b. asthenosphere
    c. crust d. mantle

23. The Morley-Vine-Matthews hypothesis explains the origin of
    a. polar wandering
    b. sea-floor magnetic anomalies
    c. continental drift
    d. mid-ocean ridges

24. The San Andreas fault in California is a
    a. normal fault b. reverse fault
    c. transform fault d. thrust fault

25. What would you most expect to find at ocean–ocean convergence?
    a. suture zone b. island arc
    c. mid-ocean ridge

26. What would you most expect to find at ocean–continent convergence?
    a. magmatic arc b. suture zone
    c. island arc d. mid-ocean ridge

27. What would you most expect to find at continent–continent convergence?
    a. magmatic arc b. mountain belt
    c. island arc d. mid-ocean ridge

28. Passive continental margins are created at
    a. divergent plate boundaries
    b. transform faults
    c. convergent plate boundaries

29. The Hawaiian islands are thought to be the result of
    a. subduction
    b. mid-ocean ridge volcanics
    c. mantle plumes
    d. ocean–ocean convergence

30. Metallic ores are created at diverging plate boundaries
    a. through hydrothermal processes
    b. in lava flows
    c. in sedimentary deposits
    d. through metamorphism

Exploring Web Resources

www.mcgrawhill.ca/olc/plummer

Visit the Online Learning Centre for additional readings, media resources, and some great animations. Check out the videos on plate tectonics, continental drift, and plate dynamics. This site also features interactive quizzes, flashcards, and the answers to the Testing Your Knowledge section. And, from the Online Learning Centre, you can link directly to the sites listed below to further your understanding of plate tectonics.

http://pubs.usgs.gov/publications/text/dynamic.html

This Dynamic Earth: The Story of Plate Tectonics U.S. Geological Survey online book by W. J. Kious and R. Tilling provides general information about plate tectonics.

http://bowie.gsfc.nasa.gov/926/slrtecto.html

Tectonic Plate Motion (explains how plate motion is calculated).

http://vishnu.glg.nau.edu/rcb/globaltext.html

View images of plate tectonic reconstructions by R. Blakely at Northern Arizona University.

Animations

This chapter includes the following animations available on our Online Learning Centre at www.mcgrawhill.ca/olc/plummer.

2.12 Seafloor Spreading
2.15 Magnetic Reversals at MO Ridge
2.16 How Seafloor Spreading Creates Magnetic Polarity Stripes
2.17 Age of Ocean Floor
2.18 Transform Faults
2.19 Continental Rifting and Early Drift
2.26 Convergence of Plates — Ocean–Ocean
2.28 Convergence of Plates — Ocean–Continent
2.29 Convergence of Plates — Continent–Continent
2.42 Formation of Hawaiian Island Chain by Hotspot Volcanism

CHAPTER

Earthquakes

What causes earthquakes?
Why do earthquakes cause so much damage?
How do we know where earthquakes occur?
What kinds of damage can earthquakes cause?
Where do earthquakes occur on a global scale?
What is the relationship between earthquakes and plate tectonics?
Can we predict when earthquakes will occur?

This chapter will help you understand the nature and origin of earthquakes. We discuss the seismic waves created by earthquakes and how the quakes are measured and located by studying these waves. We also describe some effects of earthquakes, such as ground motion and displacement, damage to buildings, and quake-caused fires, landslides, and tsunamis.

Earthquakes are largely confined to a few narrow belts on Earth. This distribution was once puzzling to geologists, but here we show how the concept of plate tectonics neatly explains it.

As geologists learn more about earthquake behaviour, there is the possibility that we will be able to forecast earthquakes. We conclude the chapter with a look at this developing branch of Earth science.

Oil depot at Kreungraya, Indonesia destroyed by the tsunami generated by the December 26, 2005 Indonesian earthquake" *CP/Bullit Marquez*

On February 5, 1663, a violent earthquake struck the Charlevoix-Kamouraska area of Quebec, causing vast landslides along rivers in the St. Lawrence valley. The earthquake was felt over the entire eastern part of North America and, although no loss of life was reported, Jesuit priests reported the event as "so prodigious, so violent, and so terrifying that there are no words strong enough to describe it." This area of Quebec has experienced many seismic events over the past two centuries and is aptly named the Charlevoix-Kamouraska Seismic Zone. The October 20, 1870, Baie-St-Paul earthquake (also in the Charlevoix-Kamouraska Seismic Zone) ripped open the ground along several fissures near the epicentre in Baie-St-Paul, while tremors were felt as far south as Virginia. These are among the earliest earthquakes to be documented in North America.

Fifty years later, the eastern coast of Canada experienced another severe jolt when parts of the sea floor beneath the Atlantic Ocean moved in the Grand Banks earthquake of November 18, 1929. The earthquake triggered a large submarine landslide that ruptured 12 transatlantic cables, allowing scientists to record the speed of movement and erosive capabilities of a turbidity current for the first time (see chapter 18). A large tsunami generated by the quake struck the Burin Peninsula of Newfoundland and claimed the lives of 28 people. The largest earthquake to rock eastern Canada in 53 years jostled the Saguenay region of Quebec on November 25, 1988. This earthquake severely compromised the engineering integrity of structures built on glaciomarine clays ("quickclays"; see chapter 13) that underlie the region.

On April 18, 1906, at 5:12 in the morning, part of California slid abruptly past the rest of the state during a great earthquake. A visible scar 450 km long was left where the Earth was torn along coastal northern California. Rock was displaced horizontally as much as 4.5 m; soil above the rock was displaced up to 6.5 m. The quake, located on a segment of the San Andreas fault near San Francisco, shook the ground for one full minute. Buildings toppled in San Francisco, and broken gas mains fed fires that raged for three days (figure 3.1*A*). Broken water mains hampered fire fighting. The fires were finally extinguished when buildings were dynamited to create a firebreak. At least 3,000 people died and $400 million (in 1906 U.S. dollars) of damage was caused by the earthquake. Perhaps 90 percent of the destruction was caused by the fires.

At 5:04 P.M. on October 17, 1989, San Francisco was again severely shaken for 15 seconds by the Loma Prieta earthquake located to the south, on the San Andreas fault near Santa Cruz. Although the quake did not tear the ground surface, it collapsed some buildings and freeway overpasses built upon the soft "bay fill" sediment in San Francisco and Oakland (figure 3.1*B*). A section of the Bay Bridge collapsed. Just as in 1906, raging fires were fed by broken gas mains in the Marina district of San Francisco, and were difficult to fight because of broken water mains; fire-boats helped extinguish them. Very severe damage occurred in small towns near the centre of the quake. The death toll was 63, and damage was U.S. $6 billion.

At 5:30 P.M. on March 27, 1964, southern Alaska was rocked by an earthquake that lasted for 3 minutes. Although the force of this earthquake was twice as strong as the 1906 San Francisco earthquake, loss of life and property was relatively low because of Alaska's small population—15 people died as a direct result of the shaking, and damage amounted to slightly over $300 million (in 1964 U.S. dollars). The tremor was felt over an area of more than 1 million km^2. A section of the Earth's surface 50 by 200 km was raised as much as 13 m, and a similar block of land sank 1 to 2 m. Horizontal movement was slight. In Anchorage, 150 km from the centre of the earthquake, landslides wrecked parts of the city. The greatest loss of life was caused by large sea waves (or tsunami) generated by land movement associated with the earthquake—almost 100 people drowned in Alaska (figure 3.1*C*), and a few people drowned as far away as Oregon and northern California, as the waves spread over the Pacific Ocean. The 1964 Good Friday earthquake also caused extensive tsunami damage in Port Alberni, British Columbia.

On January 17, 1994, at 4:31 A.M., the Northridge earthquake rocked the San Fernando Valley, just north of Los Angeles, for 40 seconds. The quake, about 3 km from California State University, Northridge, damaged or destroyed all 53 CSUN buildings, and seriously damaged 300 other schools. Numerous freeway overpasses collapsed (including some that had previously collapsed in a nearby 1971 quake), closing four Interstates and seven other highways for months. Damage exceeded $25 billion. The two upper storeys of the Northridge Meadows Apartments collapsed onto the lower storey, killing 16 people (figure 3.1*D*).

On November 3, 2002, the largest earthquake ever recorded in the interior of Alaska ruptured the ground surface for more than 300 km. The rupture (figure 3.1*E*), mainly along the Denali fault, propagated eastward at more than 11,000 km/h, offset streams and glaciers, and triggered thousands of landslides. The earthquake caused no deaths and minimal damage as it occurred in a remote area of south-central Alaska. The Trans-Alaska Pipeline suffered only minor damage and did not break, in large part due to the pipeline's effective engineering design.

At 7:58 A.M. (local time) on December 26, 2004, a magnitude-9.3 earthquake, the second largest recorded since 1900, severely deformed the Indian Ocean floor off the western coast of Northern Sumatra (see box 3.1). NASA scientists calculated that the quake was so severe that it slightly changed the shape of the planet, reduced the length of the day by almost 3 microseconds, and moved the north pole by several centimetres. The tsunami generated by this earthquake caused more casualties than any other in recorded history and was detected on tide gauges in the Indian, Pacific, and Atlantic oceans. The quake epicentre was located to the east of the Sunda Trench, where the India plate is subducted below the Burma plate, and was caused by thrust faulting along the boundary between the two plates. This type of destructive "megathrust" earthquake is also possible along the western coast of Canada and the U.S. Pacific Northwest where the Juan de Fuca plate is being subducted beneath the North American plate. The recent magnitude-6.8 Nisqually earthquake (February 28, 2001), centred 65 km southwest of Seattle, was a deep quake associated with such movement on the downgoing Juan de Fuca plate. Shaking during the Nisqually earthquake was felt over a broad area and caused damage to unreinforced brick and concrete structures (figure 3.1*F*).

A

B

C

D

E

F

FIGURE 3.1

Damage from earthquakes in the United States. (*A*) Damaged buildings and fires in San Francisco after 1906 earthquake. (*B*) Collapsed double-deck Cypress freeway in Oakland after the 1989 Loma Prieta earthquake. (*C*) Tsunami damage from the 1964 Alaska earthquake carried a fishing boat inland in Resurrection Bay at Seward. (*D*) The collapse of the lower story of the Northridge Meadows Apartments killed 16 people in the 1994 earthquake in San Fernando Valley, southern California. (*E*) Scarp formed near the epicentre of the magnitude-7.9 Denali fault earthquake, the largest earthquake to strike the interior of Alaska. This quake resulted in a rupture that broke the surface for over 320 kilometres. (*F*) Damage from falling bricks in downtown Seattle after the February 28, 2001, Nisqually earthquake.

Photo A © Arnold Genthe/Corbis; Photo B © Lloyd Cluff/Corbis; Photo C by National Geophysical Data Center; Photo D © Roger Ressmeyer/Corbis Images; Photo E by Peter Hausler, U.S. Geological Survey; Photo F © AP/Wide World Photos

Although this quake was not "the big one," it served as a reminder that a great earthquake, of similar scale to the December 2004 Sumatra quake, is very probable in the Pacific Northwest.

WHAT CAUSES EARTHQUAKES?

An **earthquake** is a trembling or shaking of the ground caused by the sudden release of energy stored in the rocks beneath Earth's surface. As described in chapter 11, great forces acting deep in the Earth may put a *stress* on the rock, which may bend or change in shape (*strain*). If you bend a stick of wood, your hands put a stress (the force per unit area) on the stick; its bending (a change in shape) is the strain.

Like a bending stick, rock can deform only so far before it breaks. When a rock breaks, waves of energy are released and sent out through the Earth. These are **seismic waves,** the waves of energy produced by an earthquake. It is the seismic waves that cause the ground to tremble and shake during an earthquake.

The sudden release of energy when rock breaks may cause one huge mass of rock to slide past another mass of rock into a different relative position. As you will learn in chapter 11, the break between the two rock masses is a *fault.* The classic explanation of why earthquakes take place is called the **elastic rebound theory** (figure 3.2). It involves the sudden release of progressively stored elastic strain energy in rocks, causing movement along a fault. Deep-seated internal forces (*tectonic forces*) act on a mass of rock over many decades. Initially the rock bends, lifts, or stretches, but does not break. More and more energy is stored in the rock as the bending becomes more severe. Eventually the energy stored in the rock exceeds the breaking strength of the rock, and the rock breaks suddenly, causing an earthquake. Two masses of rock move past one another along a fault. The movement may be vertical, horizontal, or both (figure 3.3). The strain on the rock is released; the energy is expended by moving the rock into new positions and by creating seismic waves.

Recently some modifications have been suggested for the sequence of events shown in figure 3.2. The classic model implies that existing faults are strong; a very large stress must act to break rocks along a fault. The new idea is that faults are weak, and need only a small stress to cause rupture and an earthquake. The evidence for the new idea is suggestive, but not yet conclusive, so we currently have two models for fault behaviour. The weak-fault model poses serious problems for earthquake prediction, as you will see later in the chapter.

The brittle behaviour of breaking rock is characteristic only of rocks near Earth's surface. Rocks at depth are subject to increased temperature and pressure, which tend to reduce brittleness. Deep rocks deform plastically (*ductile* behaviour) instead of breaking (*brittle* behaviour); hence, there is a limit to the depth where faults can occur.

Most earthquakes are associated with movement on faults, but in some quakes the connection with faulting may be difficult to establish. Four recent California quakes, including the 1994 Northridge quake, occurred on buried thrust faults, some of which were unknown and none of which involved surface displacement. Most earthquakes in eastern North America are not associated with surface displacement. Earthquakes also occur during explosive volcanic eruptions and as magma forcibly fills underground magma chambers prior to many eruptions; these quakes may not be associated with fault movement at all.

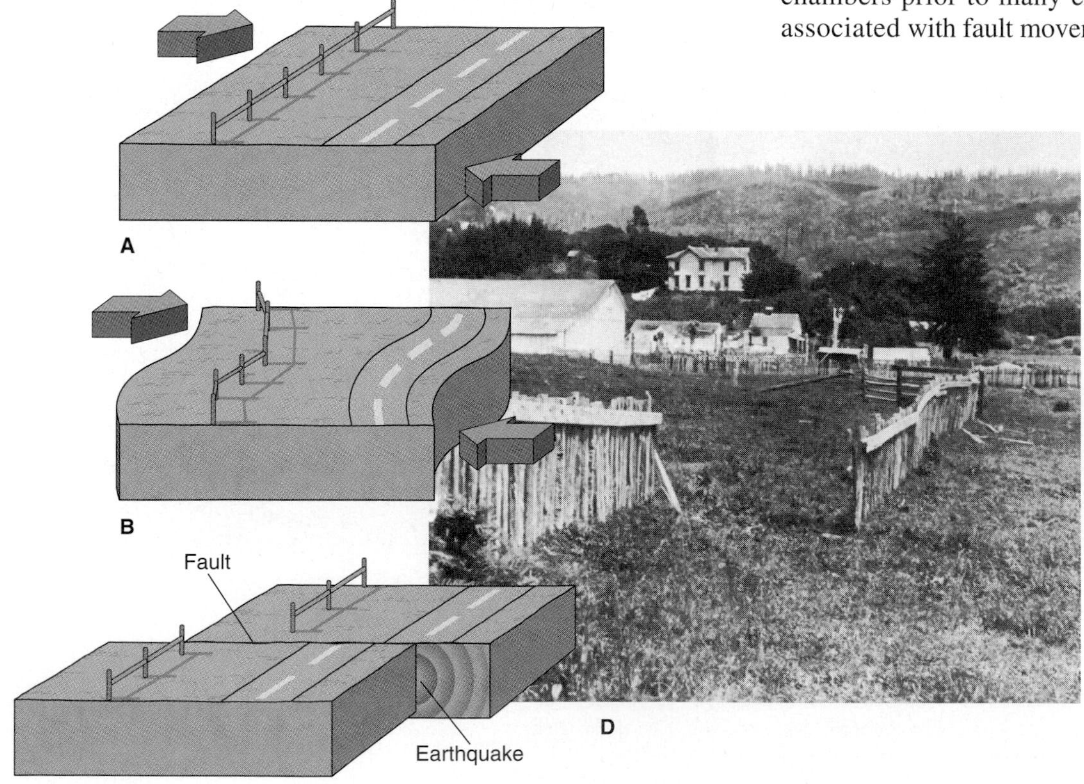

FIGURE 3.2

The elastic rebound theory of the cause of earthquakes. (*A*) Rock with stress acting on it. (*B*) Stress has caused strain in the rock. Strain builds up over a long period of time. (*C*) Rock breaks suddenly, releasing energy, with rock movement along a fault. Horizontal motion is shown; rocks can also move vertically or diagonally. (*D*) Fence offset nearly 3 m after 1906 San Francisco earthquake. *Photo by G. K. Gilbert, U.S. Geological Survey*

IN GREATER DEPTH 3.1

Indonesia/Sumatra Earthquake and Tsunami, December 26, 2004

One of the largest earthquakes ever recorded on Earth occurred on December 26, 2004 and triggered a destructive tsunami that caused the deaths of almost 250,000 people in South Asia and East Africa. The magnitude-9.3 earthquake was caused by the sudden release of large amounts of energy along a portion of the boundary between the India plate and the subducting Burma plate (box figure 1). This boundary essentially forms a large thrust fault that allows movement between the two colliding plates. It is thought that the initial zone of rupture or movement along the boundary was approximately 100 km wide and involved vertical displacement of around 15 m. However, given that the zone of aftershocks that followed the main quake extends a considerable distance to the northwest of the epicentre, it is possible that up to 1,500 km of the plate boundary slipped as a result of the quake. This estimated slip zone is approximately the same size as the entire Cascadia Subduction Zone that extends from northern California to southern British Columbia.

The ocean floor above the thrust fault was uplifted by several metres as a result of the "megathrust" quake and created a massive tsunami wave that spread out in all directions. The wave was detected by a series of Earth-orbiting satellites (including the Jason 1 satellite) and scientists have been able to determine the height of the wave as it spread over the ocean (box figure 2). Wave height reduces over time as it spreads across the ocean and loses energy. At around two hours after the quake the wave height was approximately 60 cm and it was travelling at a speed of 750 km/hr. By nine hours after the quake, the wave height had reduced to between 5 and 10 cm and had spread across most of the Indian Ocean.

Although the tsunami wave had a relatively small amplitude in the open ocean, it was amplified in shallow coastal waters and had devastating effects on low-lying coastal communities of northern Sumatra, Indonesia, Sri Lanka, India, Thailand, Somalia, Myanmar, Maldives, Malaysia, and Tanzania (box figure 3). The tsunami height varied as it inundated different parts of the coastline and was affected by factors such as sea-floor topography and the shape and orientation of the coastline. In Sumatra, wave heights of 20 to 30 m were documented at the island's northwest end and wave heights may have ranged from 15 to 30 m along at least a 100-km stretch of the northwest coast.

The power of the tsunami that affected the region was also documented by measuring two other important factors—inundation distances and run-up elevations (box figure 4). The inundation distance is the distance from the shoreline to the limit of tsunami penetration and ranged from less than 50 metres to more than 1 kilometre in Sri Lanka. Inundation distances on Sumatra were so large they were measured from satellite images (box figure 5). In Sri Lanka, areas that experienced the most extensive inundation were found to be in embayments between rocky headlands where beaches were narrow and backed by flat-lying coastal plains. The run-up elevation of a tsunami is the elevation above sea level of the water surface at the inland limit of inundation (box figure 4) and ranged from less than 3 metres to more than 12 metres in Sri Lanka. Those areas that experienced the greatest run-up elevations were rocky headlands, and lower run-up elevations were measured in low-lying areas. These measurements provide important information for determining how the tsunami lost energy as it moved inland and will be used to model the effects of future tsunamis in the region.

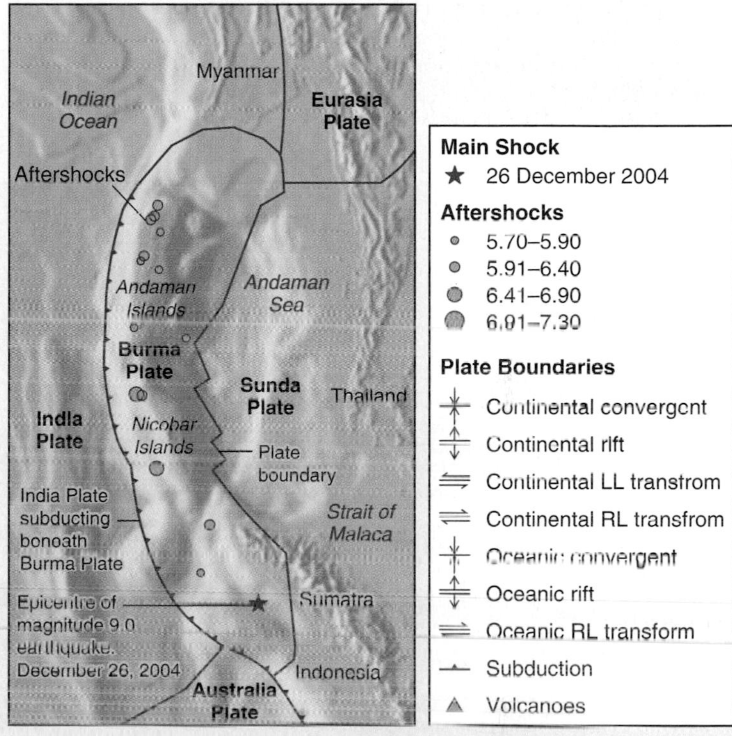

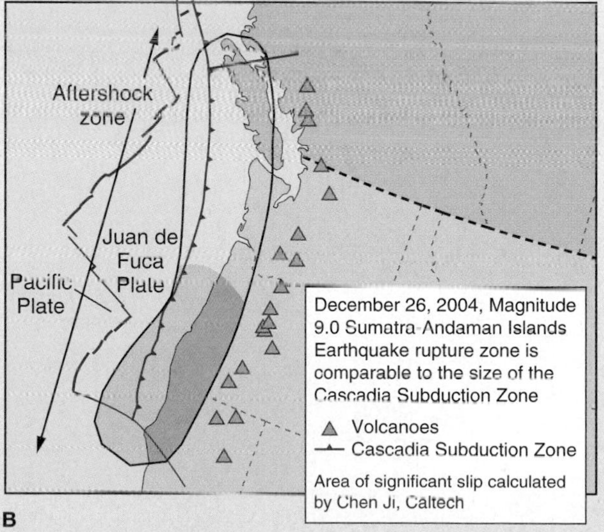

BOX 3.1 ■ FIGURE 1

(A) Location map showing epicentre of the December 26, 2004 earthquake and aftershocks. Plate boundaries are also shown. (B) Rupture area of Indonesian quake compared to Cascadia subduction zone.

M9.0 Sumatra-Andaman Islands Earthquake of Dec. 26, 2004. U.S. Geological Survey., National Earthquake Information Center, Feb. 28, 2005.

Will Such a Disaster Happen Again?

Plates will continue to move and grind past each other on the surface of our planet and will certainly generate great earthquakes and tsunamis in the future. Current efforts are being made to improve our ability to predict such devastating events and to respond to the threats they pose as quickly and effectively as possible. Satellite data are not received by scientists until several hours after a tsunami has developed and cannot be used as a primary warning system. However, satellite images can be used to help improve tsunami hazard forecasts by mapping the shape and form of the ocean floors from space. This will allow scientists to predict where tsunami energy will be focused or dispersed.

Additional Resources

For tsunami animations and information, see:

http://staff.aist.go.jp/kenji.satake/animation.gif
www.earthquake.usgs.gov/eqinthenews/2004/usslav
http://walrus.wr.usgs.gov/tsunami/srilanka05/heights.html
http://gsc.nrcan.gc.ca/index_e.php
http://gisdata.usgs.gov/website/tsunami
http://edc.usgs.gov/Tectonic
www.esa.int/esaEO/SEM3MZW797E_index_2.html
www.noanews.noaa.gov/stories2005/s2365.htm
http://walrus.wr.usgs.gov/tsunami

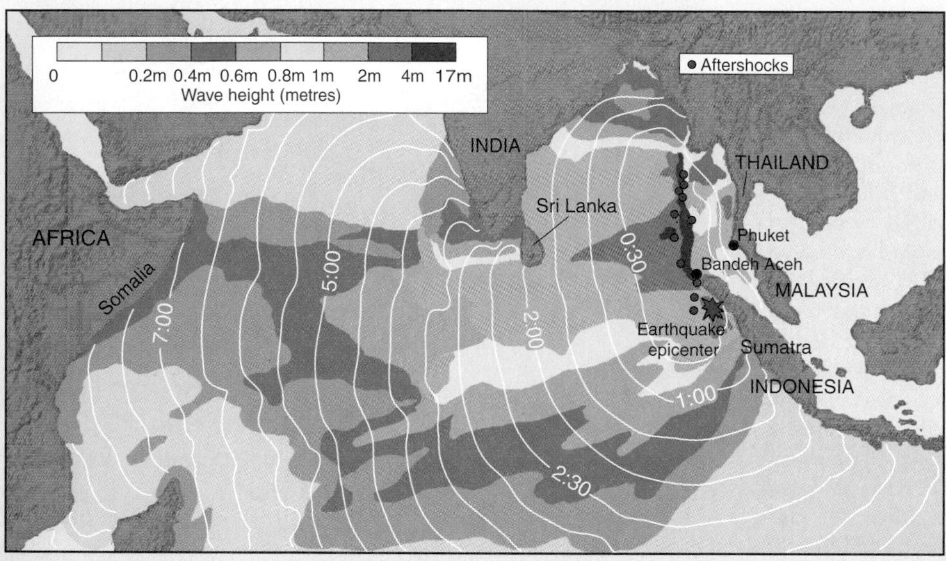

BOX 3.1 ■ FIGURE 2

(A) Tsunami travel time and wave height map from December 26, 2004 Sumatra earthquake. The arrival time (hours) of the first wave of the tsunami are shown by the white contour lines. Bandeh Aceh was hit by the tsunami 15 minutes after the earthquake and it took 7 hours to travel across the Indian Ocean to reach the coast of Africa. (B) Many people are surprised while others run for safety as the tsunami caused by the December 26, 2004 9.3-magnitude Sumantra earthquake comes crashing onshore in Koh Raya, Thailand.

Photo © John Russell/AFP/Getty Images

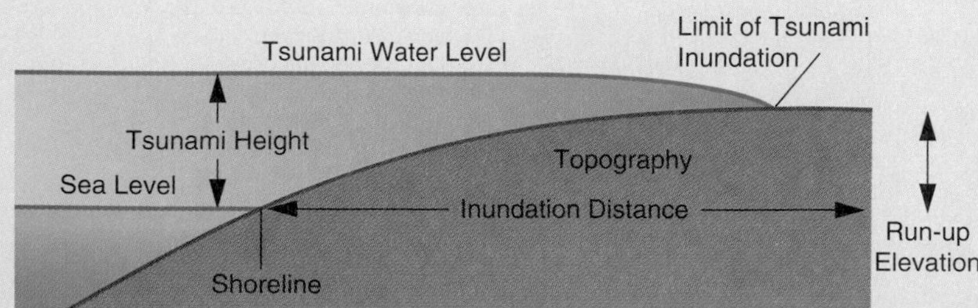

BOX 3.1 ■ FIGURE 4

Tsunami characteristics measured by survey teams.

Bruce Jaffe and Laura Torresan, United States Geological Society

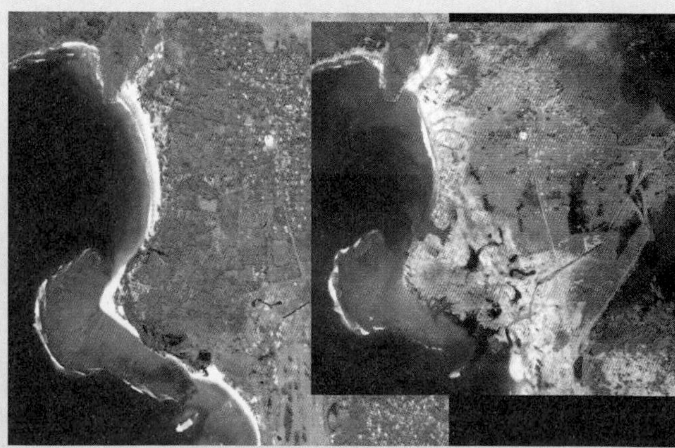

BOX 3.1 ■ FIGURE 5

Erosion of beach sand is evident in this before-and-after pair of satellite images of the coast at Lampuuk, Sumatra. The white structure near the centre of the images is a mosque that survived the tsunami. Satellite image acquired using Space Imaging IKONOS satellite and processed by the Centre for Remote Imaging, Sensing and Procesing (CRISP), National University of Singapore.

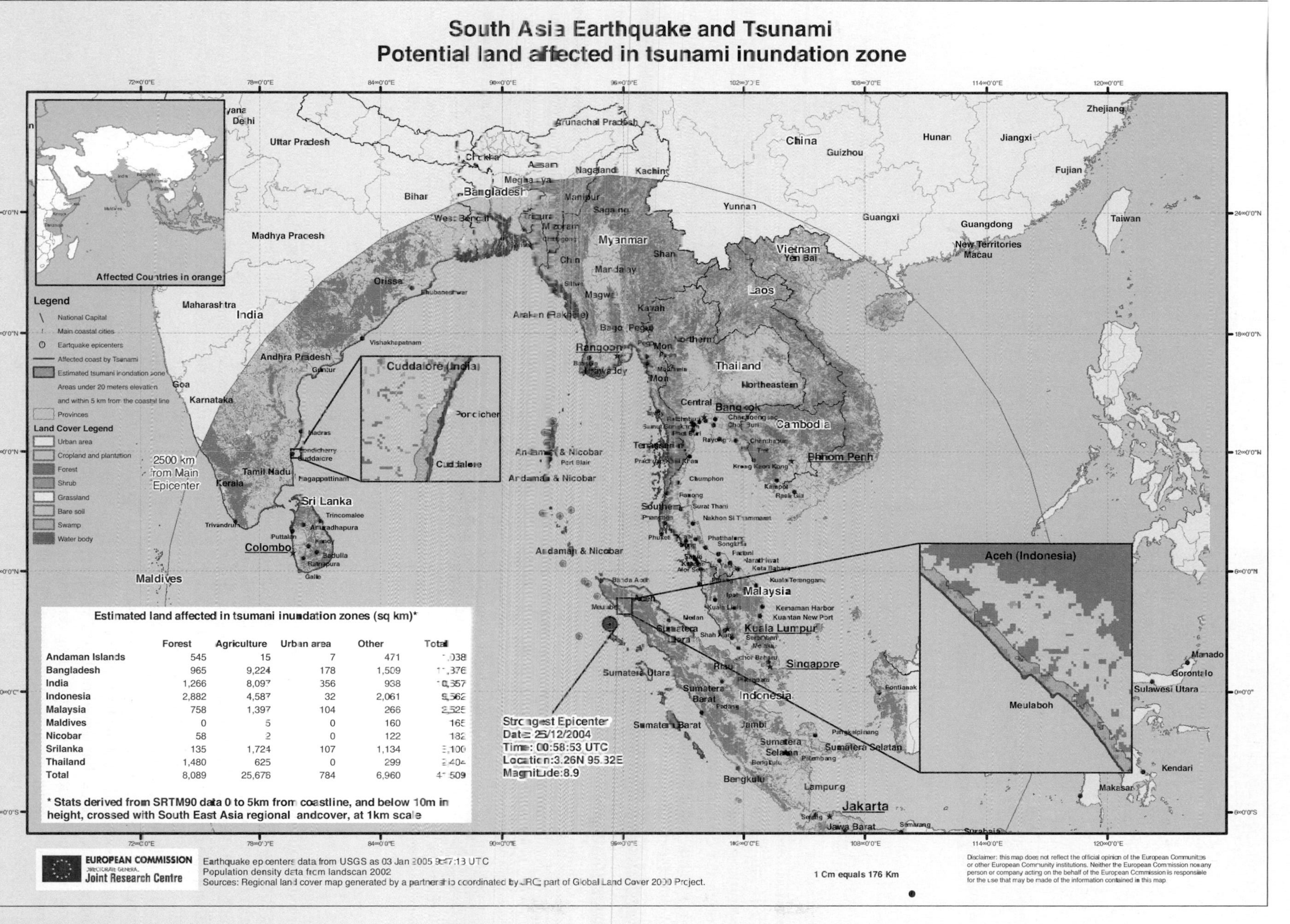

Estimated land affected in tsunami inundation zones (sq km)*

	Forest	Agriculture	Urban area	Other	Total
Andaman Islands	545	15	7	471	1,038
Bangladesh	965	9,224	178	1,509	11,376
India	1,266	8,097	356	938	10,357
Indonesia	2,882	4,587	32	2,061	9,562
Malaysia	758	1,397	104	266	2,525
Maldives	0	5	0	160	165
Nicobar	58	2	0	122	182
Srilanka	135	1,724	107	1,134	3,100
Thailand	1,480	625	0	299	2,404
Total	8,089	25,676	784	6,960	41,509

* Stats derived from SRTM90 data 0 to 5km from coastline, and below 10m in height, crossed with South East Asia regional landcover, at 1km scale

BOX 3.1 ■ FIGURE 3

The earthquake and tsunami of December 26, 2004. Map of the Indian Ocean region showing the epicentre of the quake and the countries and shorelines where people were killed.

Images acquired and processed by CRISP, National University of Singapore IKONOS image © CRISP 2004

FIGURE 3.3

Horizontal offset of trees in an orchard, 1979, El Centro, California.
Photo © John S. Shelton

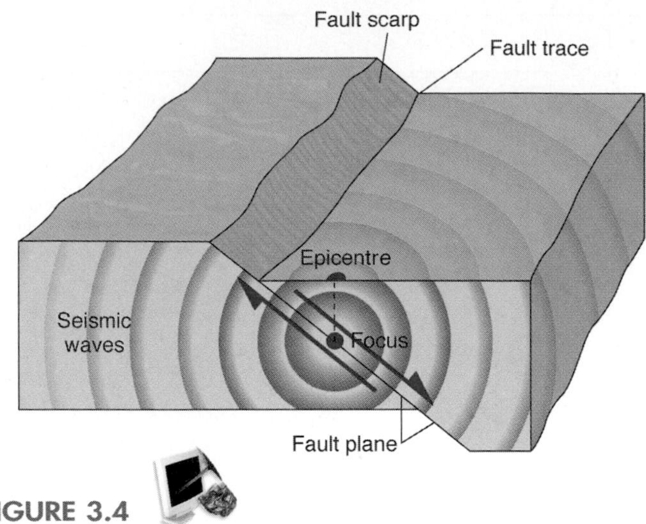

FIGURE 3.4

The focus of an earthquake is the point where rocks first break along a fault; seismic waves radiate from the focus. The epicentre is the point on the Earth's surface directly above the focus.

Another cause has recently been postulated for deep earthquakes (100 to 670 km below the surface), essentially all of which are found on cold, subducting plates sliding down into the mantle. Although the downgoing plates are colder than the surrounding rock, the high temperature and pressure at depth suggests to some geologists that the rock in the plates should behave plastically rather than breaking in the brittle manner of near-surface rocks. The suggested cause of deep quakes is mineral transformations within the downgoing rock, as pressure collapses one mineral into a denser form. Lab experiments have shown bodies of the new, denser minerals along fractures. Whether the process occurs on a large scale to produce large quakes is unknown. Similar suggestions for the cause of deep quakes include the dehydration of water-containing serpentine and the conversion of serpentine into glass. Both these processes occur suddenly on small fractures in lab experiments.

WHY DO EARTHQUAKES CAUSE SO MUCH DAMAGE?

The point within the Earth where seismic waves first originate is called the **focus** (or *hypocentre*) of the earthquakes (figure 3.4). This is the centre of the earthquake, the point of initial breakage and movement on a fault. Rupture begins at the focus and then spreads rapidly along the fault plane. The point on the Earth's surface directly above the focus is the **epicentre.**

Two types of seismic waves are generated during earthquakes. **Body waves** are seismic waves that travel through the Earth's interior, spreading outward from the focus in all directions. **Surface waves** are seismic waves that travel on Earth's surface away from the epicentre, like water waves spreading out from a pebble thrown into a pond. Rock movement associated with seismic surface waves dies out with depth into the Earth, just as water movement in ocean waves dies out with depth.

Body Waves

There are two types of body waves, both shown in figure 3.5. A **P wave** is a compressional (or longitudinal) wave in which rock vibrates back and forth *parallel* to the direction of wave propagation. Because it is a very fast wave, travelling through near-surface rocks at speeds of 4 to 7 km per second (14,400 to more than 25,000 kilometres per hour), a P wave is the first (or *primary*) wave to arrive at a recording station following an earthquake.

The second type of body wave is called an **S wave** (*secondary*) and is a slower, transverse wave that travels through near-surface rocks at 2 to 5 km per second. An S wave is propagated by a shearing motion much like that in a stretched, shaken rope. The rock vibrates *perpendicular* to the direction of wave propagation; that is, crosswise to the direction the waves are moving.

Both P waves and S waves pass easily through solid rock. A P wave can also pass through a fluid (gas or liquid), but an S wave cannot. We discuss the importance of this fact in chapter 4.

Surface Waves

Surface waves are the slowest waves set off by earthquakes. In general, surface waves cause more property damage than body waves because surface waves produce more ground movement and travel more slowly, so they take longer to pass. The two most important types of surface waves are Love waves and Rayleigh waves, named after the geophysicists who discovered them.

Love waves are most like S waves that have no vertical displacement. The ground moves side to side in a horizontal plane that is perpendicular to the direction the wave is travelling or propagating (figure 3.5*C*). Like S waves, Love waves do not travel through liquids and would not be felt on a body of water. Because of the horizontal movement, Love waves tend to knock buildings off their foundations and destroy highway bridge supports.

Body Waves

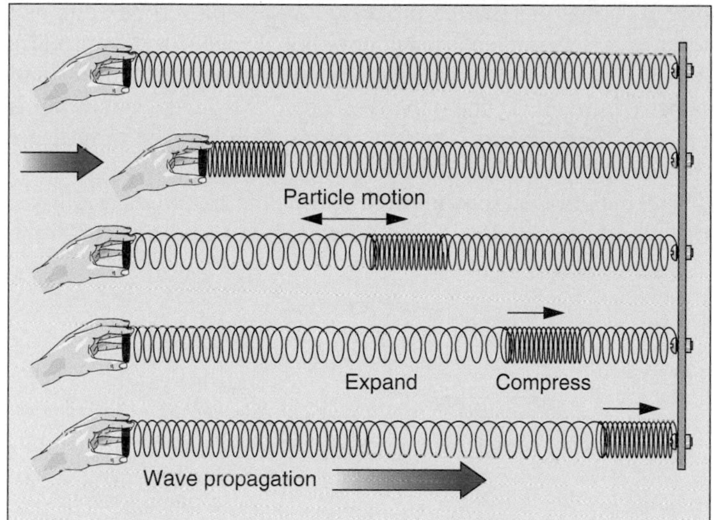

A Primary wave

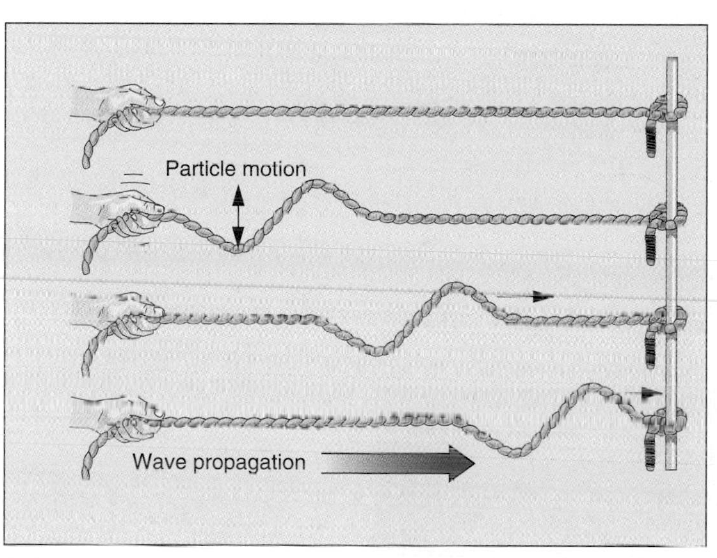

B Secondary wave

Surface Waves

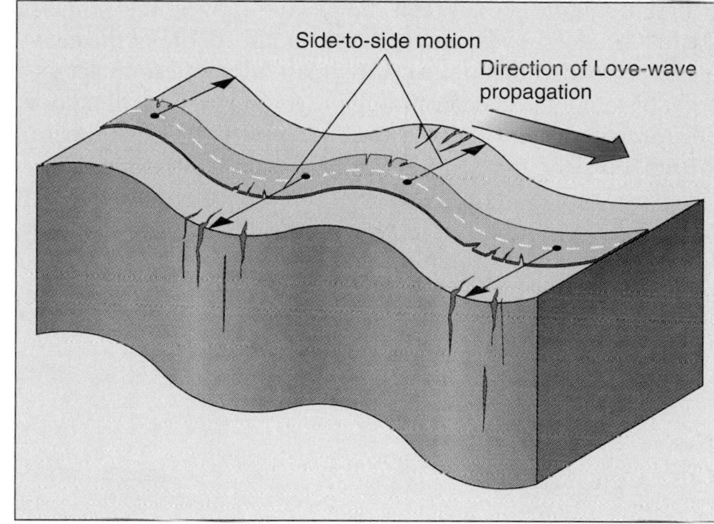

C Love wave

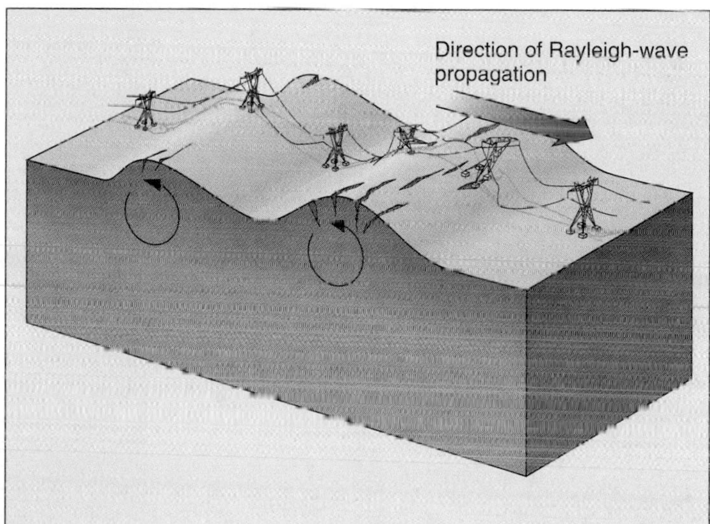

D Rayleigh wave

FIGURE 3.5 

Particle motion in seismic waves. (*A*) A P wave is illustrated by a sudden push on the end of a stretched spring or Slinky. The particles vibrate *parallel* to the direction of wave propagation. (*B*) An S wave is illustrated by shaking a loop along a stretched rope. The particles vibrate *perpendicular* to the direction of wave propagation. (*C*) Love waves behave like S waves in that the particle motion is perpendicular to the direction of wave travel along Earth's surface. (*D*) Rayleigh waves are like ocean waves and cause a rolling motion on Earth's surface. The particle motion is elliptical and opposite (counterclockwise) to the direction of wave propagation.

Rayleigh waves behave like rolling ocean waves. Unlike ocean waves, Rayleigh waves cause the ground to move in an elliptical path opposite to the direction the wave passes (figure 3.5*D*). Rayleigh waves tend to be incredibly destructive to buildings because they produce more ground movement and take longer to pass.

HOW DO WE KNOW WHERE EARTHQUAKES OCCUR?

The invention of instruments that could accurately record seismic waves was an important scientific advance. They measure the amount of ground motion and can be used to find the location, depth, and size of an earthquake.

The instrument used to measure seismic waves is a *seismometer*. The principle of the seismometer is to keep a heavy suspended mass as motionless as possible—suspending it by springs or hanging it as a pendulum from the frame of the instrument (figure 3.6). When the ground moves, the frame of the instrument moves with it; however, the inertia of the heavy mass suspended inside keeps the mass motionless to act as a point of reference in determining the amount of ground motion. Seismometers are usually placed in clusters of three to record the motion along the *x*, *y*, and *z* axes of three-dimensional space.

A seismometer by itself cannot record the motion that it measures. A **seismograph** is a recording device that produces a permanent record of Earth motion detected by a seismometer, in digital format that can be processed and displayed on computer terminals. Seismographs used for public displays usually produce a record in the form of a wiggly line drawn on a moving strip of paper (figure 3.7). The paper record of the Earth's vibration is called a **seismogram.** The seismogram can be used to measure the strength of the earthquake.

A network of seismograph stations is maintained all over the world to record and study earthquakes (and nuclear bomb

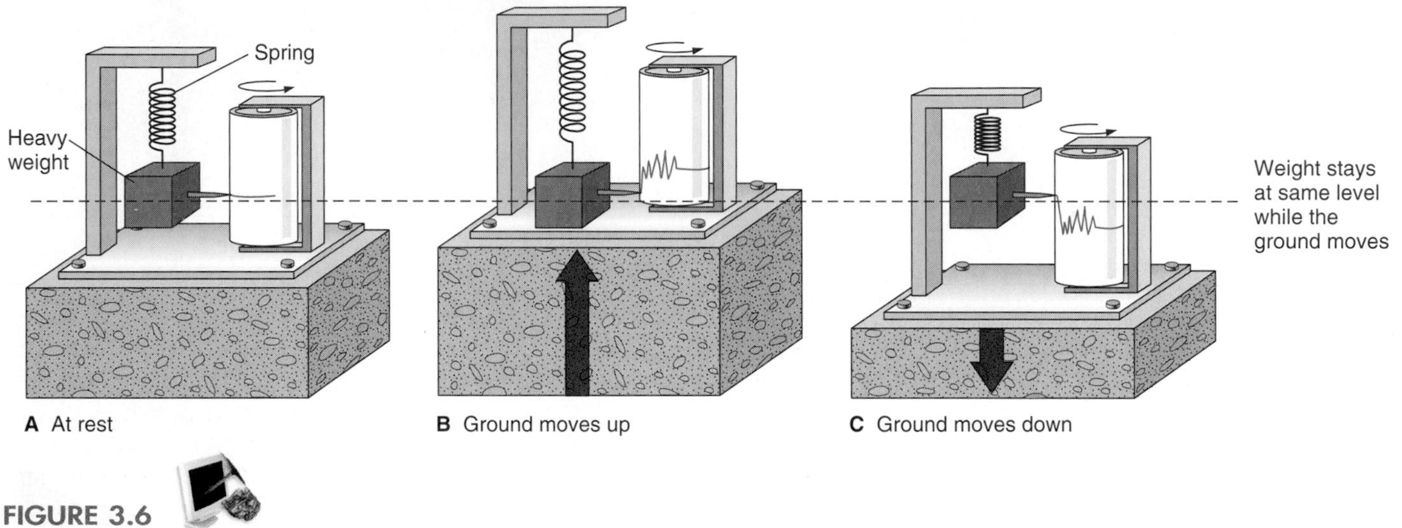

FIGURE 3.6

A simple seismograph for detecting vertical rock motion. The pen records the ground motion on the seismogram as the spring stretches and compresses with up and down movement of the spring. Frame and recording drum move with the ground. Inertia of the weight keeps it and the needle relatively motionless.

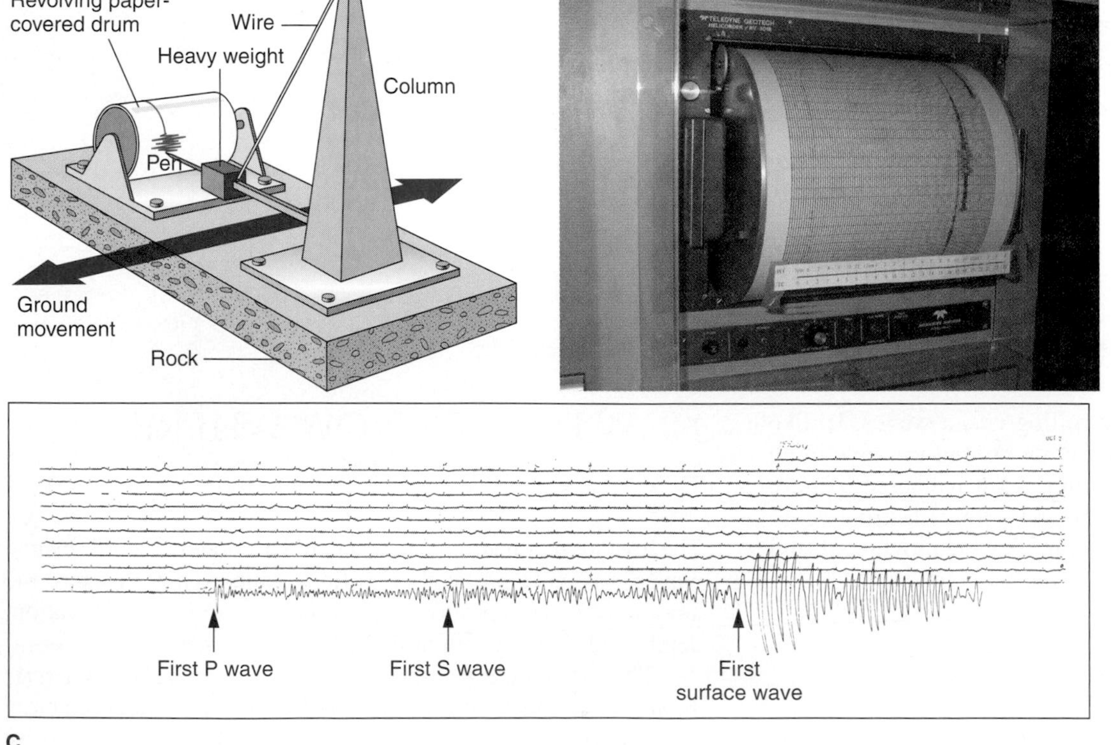

FIGURE 3.7

(A) A seismograph for horizontal motion. Modern seismographs record Earth motion on moving strips of paper. The mass is suspended by a wire from the column and swings like a pendulum when the ground moves horizontally. A pen attached to the mass records the motion on a moving strip of paper. (B) Seismograph at Sunset Crater Volcano National Monument, Arizona. (C) A seismogram of a 1967 earthquake in Taiwan, magnitude 6.2, recorded in Berkeley, California, 10,000 kilometres away. First arrivals of P, S, and surface waves are shown.

A and C courtesy University of California, Berkeley; Photo B by C. H. Eyles

explosions). Within minutes after an earthquake occurs, distant seismographs begin to pick up seismic waves. A large earthquake can be detected by seismographs all over the world.

Because the different types of seismic waves travel at different speeds, they arrive at seismograph stations in a definite order: first the P waves, then the S waves, and finally the surface waves. These three different waves can be distinguished on the seismograms. By analyzing these seismograms, geologists can learn a great deal about an earthquake, including its location and size.

Determining the Location of an Earthquake

P and S waves start out from the focus of an earthquake at essentially the same time. As they travel away from the quake, the two kinds of body waves gradually separate because they are travelling at different speeds. On a seismogram from a station close to the earthquake, the first arrival of the P wave is separated from the first arrival of the S wave by a short distance on the paper record (figure 3.8). At a recording station far from the earthquake, however, the first arrivals of these waves will be recorded much farther apart on the seismogram. The farther the seismic waves travel, the longer the time intervals between the arrivals of P and S waves and the more they are separated on the seismograms.

Because the time interval between the first arrivals of P and S waves increases with distance from the focus of an earthquake, this interval can be used to determine the distance from the seismograph station to a quake. The increase in the P–S interval is regular with increasing distance for several thousand kilometres and so can be graphed in a **travel-time curve,** which plots seismic-wave arrival time against distance (figure 3.9).

In practice, a station records the P and S waves from a quake, then a seismologist matches the interval between the waves to a standard travel-time curve. By reading directly from the graph, one can determine, for example, that an earthquake has occurred 5,300 kilometres away. This determination can often be made very rapidly, even while the ground is still trembling from the quake.

In the past, a single analog station could determine only the distance to a quake, not the direction. A circle was drawn on a globe, with the centre of the circle being the station and its radius the distance to the quake (figure 3.10). The scientists at the station knew that the quake occurred somewhere on that circle, but from the information recorded they were not able to tell where. With information from two or more other stations, however, they could pinpoint the location of the quake. If three or more stations determined the distance to a single quake, a circle was drawn for each station and the intersection of the circles located the epicentre. Modern three-component seismographs now record the intensity of earthquake vibrations in three orthogonal directions. This allows scientists to determine both the direction of the incoming wave and the distance from the epicentre.

Analyses of seismograms can also indicate at what depth beneath the surface the quake occurred. Most earthquakes occur relatively close to Earth's surface, although a few occur much deeper.

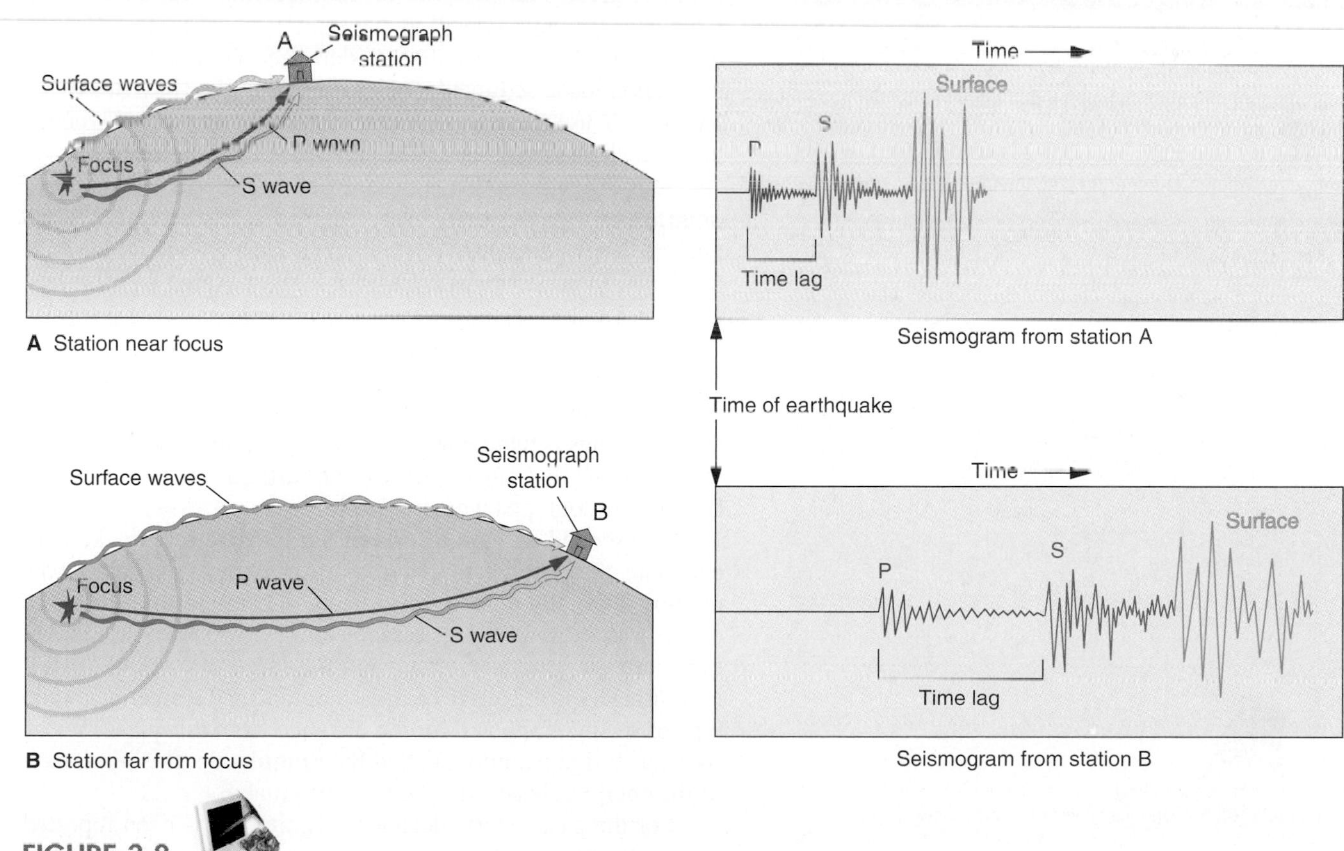

FIGURE 3.8

Because of the difference in travel times, intervals between P waves, S waves, and surface waves increase with distance from the focus.

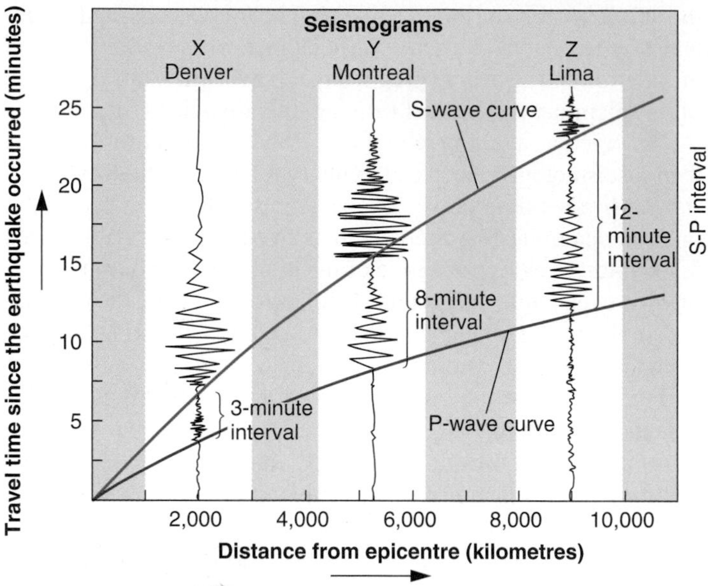

FIGURE 3.9

A travel-time curve is used to determine the distance to an earthquake. Note that the time interval between the first arrival of P and S waves increases with distance from the epicentre. Seismogram X has a 3-minute interval between P and S waves corresponding to a distance of 2,000 km from the epicentre; Y has an interval of 8 minutes, so the earthquake occurred 5,300 km away, and Z an interval of 12 minutes, and is a distance of 9,000 km from the epicentre.

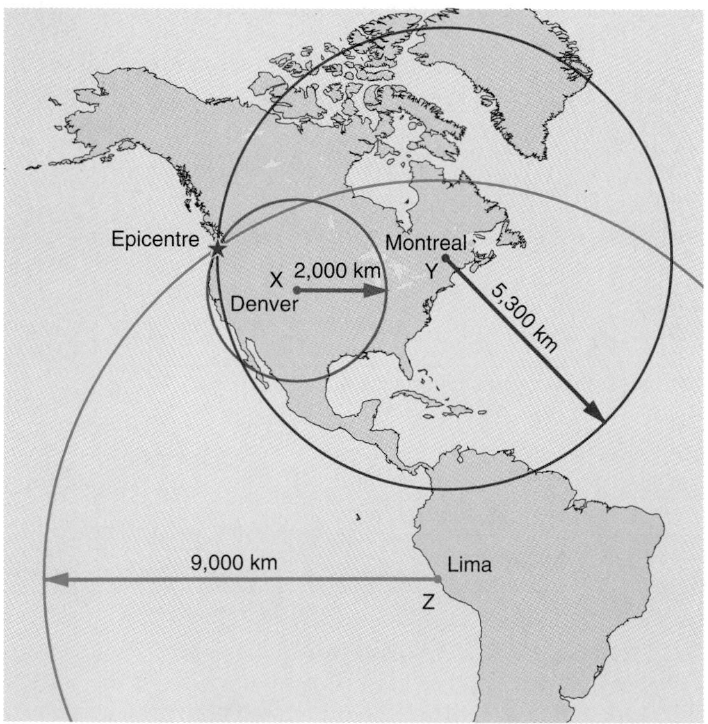

FIGURE 3.10

Locating an earthquake. The distance from each of three stations (Denver, Montreal, and Lima) is determined from seismograms and the travel-time curves shown in figure 3.9. Each distance is used for the radius of a circle about the station. The location of the earthquake is just offshore of Vancouver, British Columbia, where the three circles intersect.

Wood and Neumann, 1931, Bulletin of the Seismology Society of America.

The maximum **depth of focus**—the distance between focus and epicentre—for earthquakes is about 670 km. Quakes are classified into three groups according to their depth of focus:

Shallow focus	0–70 km deep
Intermediate focus	70–350 km deep
Deep focus	350–670 km deep

Shallow-focus earthquakes are most common; they account for 85 percent of total quake energy released. Intermediate (12%) and deep (3%) focus quakes are rarer because most deep rocks flow plastically when stressed or deformed; they are unable to store and suddenly release energy as brittle surface rocks do.

Measuring the Size of an Earthquake

The size of earthquakes is measured in two ways. One method is to find out how much and what kind of damage the quake has caused. This determines the **intensity,** which is a measure of an earthquake's effect on people and buildings. Intensities are expressed as Roman numerals ranging from I to XII on the **modified Mercalli scale** (table 3.1); higher numbers indicate greater damage.

Although intensities are widely reported at earthquake locations throughout the world, using intensity as a measure of earthquake strength has a number of drawbacks. Because damage generally lessens with distance from a quake's epicentre, different locations report different intensities for the same earthquake (figure 3.11). Moreover, damage to buildings and other structures depends greatly on the type of geologic material on which a structure was built as well as the type of construction. Houses built on solid rock normally are damaged far less than houses built upon loose sediment, such as delta mud or bay fill. Brick and stone houses usually suffer much greater damage than wooden houses, which are somewhat flexible. Damage estimates are also subjective: people may exaggerate damage reports consciously or unconsciously. Intensity maps can be drawn for a single earthquake to show the approximate damage over a wide region (figure 3.11). Intensity maps are useful for assessing how different areas respond to seismic waves and provide valuable information for earthquake planning. But such maps cannot be drawn for uninhabited areas (the open ocean, for instance), so not all quakes can be assigned intensities. The one big advantage of intensity ratings is that no instruments are required, which allows seismologists to estimate the size of earthquakes that occurred before seismographs were available.

The second method of measuring the size of a quake is to calculate the amount of energy released by the quake. This method is usually done by measuring the height (amplitude) of one of the wiggles on a seismogram. The larger the quake, the more the ground vibrates and the larger the wiggle. After measuring a specific wave on a seismogram, and correcting for the type of seismograph and for the distance from the quake, scientists can assign a number called the **magnitude.** It is a measure of the energy released during the earthquake.

For the past several decades magnitude has been reported on the **Richter scale,** a numerical scale of magnitudes. The Richter scale is open-ended, meaning there are no earthquakes

TABLE 3.1 Modified Mercalli Intensity Scale of 1931 (Abridged)

I. Not felt except by a very few under especially favourable circumstances.

II. Felt only by a few persons at rest, especially on upper floors of buildings. Delicately suspended objects may swing.

III. Felt quite noticeably indoors, especially on upper floors of buildings, but many people do not recognize it as an earthquake. Standing motor cars may rock slightly. Vibration like passing of truck. Duration estimated.

IV. During the day felt indoors by many, outdoors by few. At night some awakened. Dishes, windows, doors disturbed; walls made cracking sound. Sensation like heavy truck striking building. Standing motor cars rocked noticeably.

V. Felt by nearly everyone; many awakened. Some dishes, windows, etc., broken; a few instances of cracked plaster; unstable objects overturned. Disturbance of trees, poles, and other tall objects sometimes noticed. Pendulum clocks may stop.

VI. Felt by all; many frightened and run outdoors. Some heavy furniture moved; a few instances of fallen plaster or damaged chimneys. Damage slight.

VII. Everybody runs outdoors. Damage *negligible* in buildings of good design and construction; *slight* to moderate in well-built ordinary structures; *considerable* in poorly built or badly designed structures; some chimneys broken. Noticed by persons driving motor cars.

VIII. Damage *slight* in specially designed structures; *considerable* in ordinary substantial buildings with partial collapse; *great* in poorly built structures. Panel walls thrown out of frame structures. Fall of chimneys, factory stacks, columns, monuments, walls. Heavy furniture overturned. Sand and mud ejected in small amounts. Changes in well water. Persons driving motor cars disturbed.

IX. Damage *considerable* in specially designed structures; well-designed frame structures thrown out of plumb; *great* in substantial buildings, with partial collapse. Buildings shifted off foundations. Ground cracked conspicuously. Underground pipes broken.

X. Some well-built wooden structures destroyed; most masonry and frame structures destroyed with foundations; ground badly cracked. Rails bent. Considerable landslides from river banks and steep slopes. Shifted sand and mud. Water splashed (slopped) over banks.

XI. Few, if any (masonry), structures remain standing. Bridges destroyed. Broad fissures in ground. Underground pipelines completely out of service. Earth slumps and land slips in soft ground. Rails bent greatly.

XII. Damage total. Waves seen on ground surface. Lines of sight and level distorted. Objects thrown upward into the air.

From Wood and Neumann, 1931, *Bulletin of the Seismological Society of America*

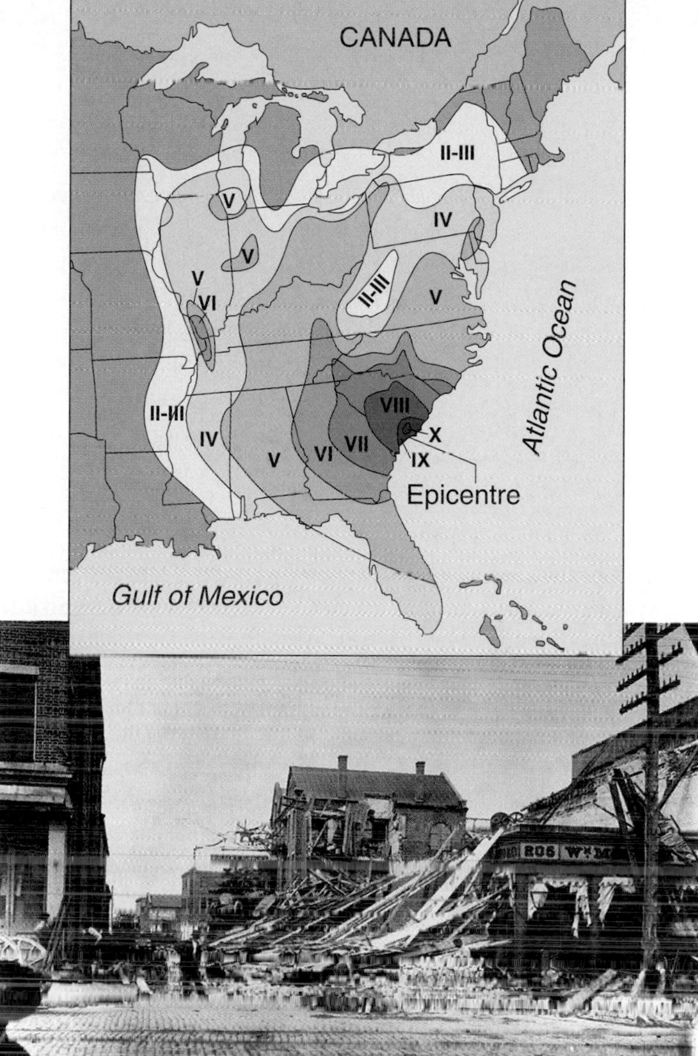

FIGURE 3.11

Zones of different intensity from the 1886 Charleston, South Carolina, earthquake. The map illustrates the general decrease in intensity with increasing distance from the epicentre, as well as the effect of different types of Earth materials. The photo shows damage in Charleston, South Carolina from 1886 earthquake.

Photo by J. K. Hillers, U.S. Geological Survey

too large or too small to fit on the scale. The higher numbers indicate larger earthquakes. Very small earthquakes can have negative magnitudes, but these are seldom reported. The largest Richter magnitude measured so far is 8.6. Smaller earthquakes are much more common than large ones (table 3.2).

There are several methods of measuring magnitude, however. The original Richter scale applied only to shallow earthquakes in southern California. Different seismic waves (body or surface) can be measured to make the scale more useful over larger areas, so several different magnitudes are sometimes reported for a single quake. A further complication is that magnitudes calculated from seismograms tend to be inaccurate (usually too low) above magnitude 7.

A new method of calculating magnitude involves the use of the *seismic moment* of a quake, which is determined from the

TABLE 3.2 Comparison of Earthquake Magnitude, Description, Intensity, and Expected Annual World Occurrence

Richter Magnitude	Description	Maximum Expected Mercalli Intensity at Epicentre	Annual Expected Number
2.0	Very minor	I Usually detected only by instruments	600,000
2.0–2.9	Very minor	I–II Felt by some indoors, especially on upper floors	300,000
3.0–3.9	Minor	III Felt indoors	49,000
4.0–4.9	Light	IV–V Felt by most; slight damage	6,200
5.0–5.9	Moderate	VI–VII Felt by all; damage minor to moderate	800
6.0–6.9	Strong	VII–VIII Everyone runs outdoors; moderate to major damage	266
7.0–7.9	Major	IX–X Major damage	18
8.0 or higher	Great	X–XII Major and total damage	1 or 2

Source: U.S. Geological Survey

strength of the rock, surface area of the rupture, and the amount of rock displacement along the fault. The **moment magnitude** is the most objective way of measuring the energy released by a large earthquake. The 1964 Alaska quake is estimated to have a moment magnitude of 9.2, the 2004 Sumatra, Indonesia quake a moment magnitude of 9.3, and the 1960 Chile quake 9.5. Unfortunately the media rarely indicate which type of magnitude they are reporting, and scientists typically revise magnitudes for several weeks after a quake as they receive more information, so trying to find out the "real" magnitude of a recent quake can be confusing. Table 3.3 lists Richter magnitudes and moment magnitudes for many earthquakes of interest.

Because the Richter scale is logarithmic, the difference between two consecutive whole numbers on the scale means an increase of 10 times in the amplitude of Earth's vibrations, particularly below magnitude 5. This means that if the measured amplitude of vibration for certain rocks is 1 cm during a magnitude-4 quake, these rocks will move 10 cm during a magnitude-5 quake occurring at the same location.

It has been estimated that a tenfold increase in the size of Earth vibrations is caused by an increase of roughly 32 times in terms of energy released. A quake of magnitude 5, for example, releases approximately 32 times more energy than one of magnitude 4. A magnitude-6 quake is about 1,000 times (32×32) more powerful in terms of energy released than a magnitude-4 quake. The actual energy released in earthquakes of varying magnitudes is shown in figure 3.12.

Although a seismograph is usually required to measure magnitude, this measure has many advantages over intensity as an indicator of earthquake strength. A worldwide network of standard seismograph stations now makes determining magnitude a routine matter, and the press reports magnitudes for all earthquakes of interest to Canada and the United States. Eventually, a single magnitude number can be assigned to a single earthquake, whereas intensity varies for a single earthquake depending on the amount and kind of local damage. Magnitudes can be reported for all quakes, even those in distant uninhabited areas where there is no property to affect.

Location and Size of Earthquakes in North America

Figure 3.13 shows the locations of all damaging earthquakes that have occurred in North America from 1977–1997. Note that only a few localities are relatively free of earthquakes.

Most of the large earthquakes occur in western North America. Quakes in the province of British Columbia, and in western states including California, Nevada, Utah, Idaho, Montana, and Washington, are related to known faults and usually (but not always) involve surface rupture of the ground. Earthquakes in Alaska occur mainly below the Aleutian Islands where the Pacific plate is converging with and being subducted beneath the North American plate.

British Columbia is the most seismically active region of Canada and lies close to active plate margins along the Cascadia Subduction Zone (see box 2.2). The southwestern corner of B.C. experiences more than 200 earthquakes a year and nine moderate to large ($M = 6–7$) earthquakes have occurred in the region during historic times. Canada's largest historic earthquake, an $M = 8.1$ event, occurred on August 22, 1949 and was centred offshore of the Queen Charlotte Islands (see box 3.2).

Earthquakes east of the Rocky Mountains are more rare, and are generally smaller and deeper than earthquakes in western North America. They usually are not associated with surface rupture. The quakes may be occurring on the deeply buried, relatively inactive faults of old *divergent plate boundaries* and *failed rifts* (*aulacogens*), both of which are described in chapter 2 (see box 2.6).

Although large quakes are extremely rare in central and eastern North America, when they do occur they can be very destructive and widely felt, because Earth's crust is older, cooler, and more brittle in the east than in the west and seismic waves travel more efficiently. The St. Lawrence River Valley along the Canada–U.S. border has had several intensity IX and X earthquakes, most recently in 1944 centred in Cornwall, Ontario. A series of quakes (intensity XI) that occurred near New Madrid, Missouri, in the winter of 1811–1812 were the most widely felt

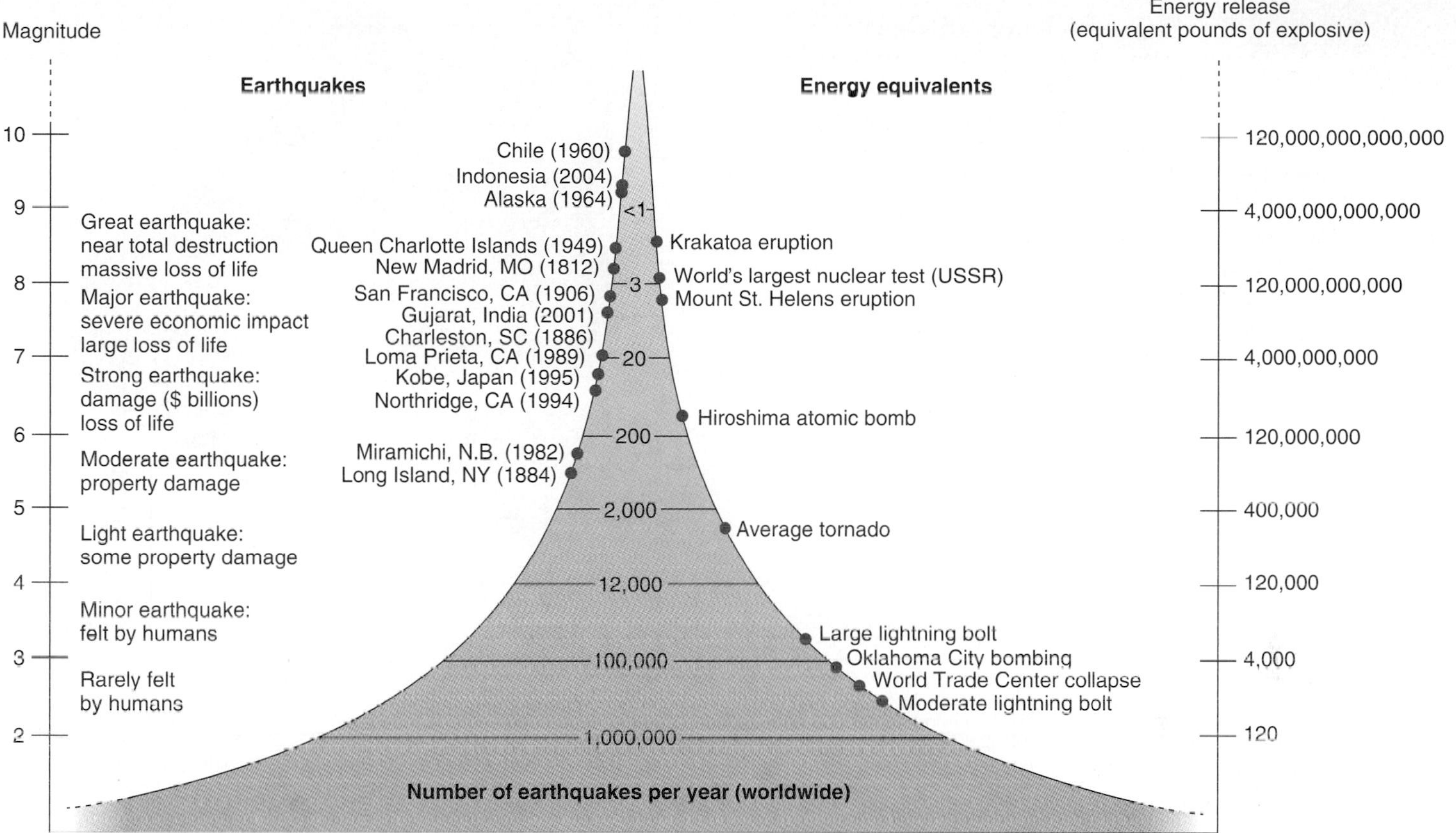

FIGURE 3.12

Diagram showing the relationship between the moment magnitude of an earthquake, the number of earthquakes per year throughout the world, and the energy released during an earthquake.

After IRIS Consortium (www.iris.edu)

earthquakes to occur in North America in recorded history. The quakes knocked over chimneys as far away as Richmond, Virginia, and rang church bells in Boston, 700 km away.

The 1886 quake in Charleston, South Carolina (intensity X) was felt throughout almost half the United States (figure 3.11) and killed 60 people; it was sharply felt in New York City. Moderate quakes hit Arkansas and New Hampshire in 1982, and in 1983 a quake of 5.1 magnitude rocked New York's Adirondack Mountains. A 5.0-magnitude quake in 1987 near Lawrenceville, Illinois, was felt from Kansas to South Carolina to Ontario. In 1988 a 6.0-magnitude quake in the Saguenay region north of Quebec City was felt in Ontario and distant areas of Quebec as well as Indiana and Washington, D.C.

Geologists have mapped regions of seismic risk in North America (figure 3.14) and elsewhere throughout the world, primarily on the assumption that large earthquakes will occur in the future in places where they have occurred in the past.

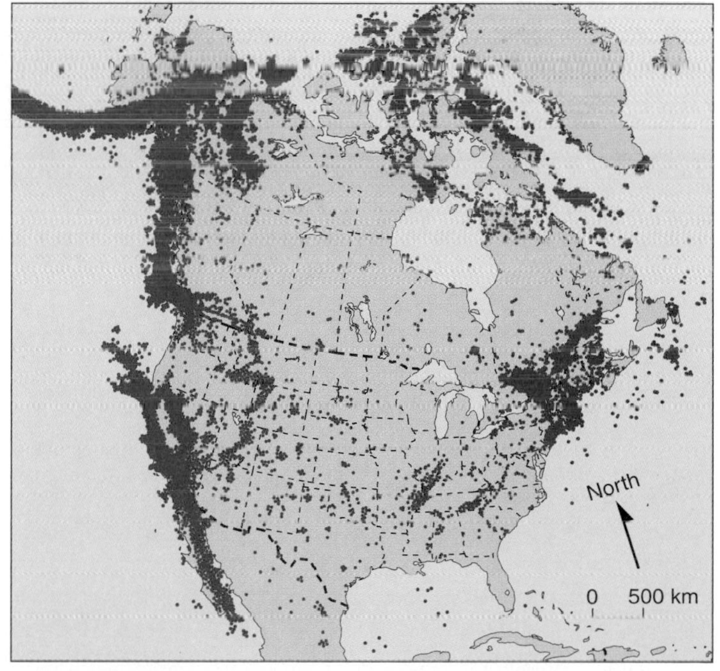

FIGURE 3.13

Locations of earthquakes that have occurred in North America from 1977–1997.

U.S. Geological Survey www.neic.cr.usgs.gov/neis/general/seismicity.us.html and http://earthquakescanada.nrcan.gc.ca/historic_eq/caneqmap_e.php Reproduced with the permission of the Minister of Public Works and Government Services Canada, 2006 and Courtesy of Natural Resources Canada, Geological Survey of Canada.

WHAT KINDS OF DAMAGE CAN EARTHQUAKES CAUSE?

Ground motion is the trembling and shaking of the land that can cause buildings to vibrate. During small quakes, windows and

TABLE 3.3 Earthquake Magnitudes

		Richter Magnitude	Moment Magnitude
2005	Pakistan/Kashmir		7.6
2004	Offshore N. Sumatra, Indonesia		9.3
2002	Denali fault, S-Central Alaska		7.9
2001	Puget Sound (Nisqually), Washington		6.8
2001	Gujarat, India		7.7
2001	El Salvador		7.7
1999	Taipei, Taiwan		7.6
1999	Izmit, Turkey		7.4
1998	Papua, New Guinea		7.1
1995	Kobe, Japan (7.2 on Japanese scale)	6.8	6.9
1994	Northridge, S. Calif.	6.4	6.7
1992	Humboldt County, N. Calif.	7.1, 6.6, 6.7	
1989	Loma Prieta, N. Calif.	7.0	7.2
1989	Ungava, Quebec	6.3	
1988	Saguenay, Quebec	6.0	
1987	Lawrenceville, Illinois	5.0	5.0
1985	Nahanni, N.W.T.	6.9	
1985	Ixtapa, Mexico		8.1, 7.5
1983	Hawaii	6.6	
1983	Challis, Idaho	7.2	7.0
1983	Coalinga, Calif.	6.7	6.2
1982	Miramichi, New Brunswick	5.7	
1970	Peru	7.75	7.9
1965	Aleutian Islands, Alaska	8.2	8.7
1964	near Anchorage, Alaska	8.6	9.2
1964	Nigaka, Japan		
1960	Chile	8.5	9.5
1949	Queen Charlotte Islands, B.C.	8.1	
1946	Vancouver Island, B.C.	7.3	
1929	Grand Banks, Nfld.	7.2	
1906	San Francisco, N. Calif.	8.25	7.7
1886	Charleston, South Carolina	6.7	7.0
1811–12	New Madrid, Missouri area	7.5, 7.3, 7.8	7.7, 7.6, 7.9

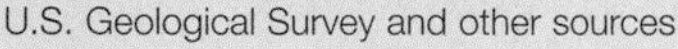
U.S. Geological Survey and other sources

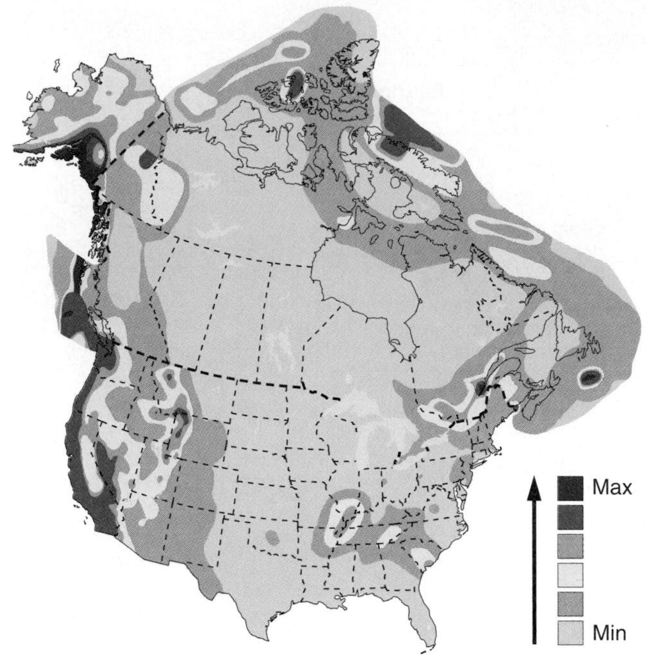

FIGURE 3.14

Map of seismic hazard in North America based on the expected amount of ground shaking and damage.

From USGS-National Seismic Hazard Mapping Project and Natural Resources Canada — National Earthquake Hazards Program (www.earthquakescanada.nrcan.gc.ca/hazard/simphaz_e.php)

walls may crack from such vibration. In a very large quake the ground motion may be visible. It can be strong enough to topple large structures such as bridges and office and apartment buildings (figure 3.15). Most people injured or killed in an earthquake are hit by falling debris from buildings. Because proper building construction can greatly reduce the dangers, building codes need to be both strict and strictly enforced in earthquake-prone areas (see box 3.3). Much of the damage and loss of life in the recent Turkey, El Salvador, and India earthquakes was due to poorly constructed buildings that did not meet building codes. As we have seen, the location of buildings also needs to be controlled; buildings built on soft sediment are damaged more than buildings on hard rock.

Fire is a particularly serious problem just after an earthquake because of broken gas and water mains and fallen electrical wires (figure 3.16). Although fire was the cause of most of the damage to San Francisco in 1906, changes in building construction and improved fire-fighting methods have reduced (but not eliminated) the fire danger to modern cities. The stubborn Marina District fires in San Francisco in 1989 attest to the modern dangers of broken gas and water mains.

Landslides can be triggered by the shaking of the ground (figure 3.17*A*). The 1959 Madison Canyon landslide in Montana was triggered by a nearby quake of magnitude 7.7. More than 300 landslides were triggered by the magnitude-7.3 Vancouver Island earthquake in 1946 over an area of about 20,000 km^2. Landslides and subsidence caused extensive damage in downtown and suburban Anchorage during the 1964 Alaskan quake (magnitude 8.6). The 1970 Peruvian earthquake (magnitude 7.75) set off thousands

IN GREATER DEPTH 3.2

Earthquakes in Canada, Eh?

Several regions of Canada experience frequent earthquake activity (see figure 3.13). Most activity is focused along the Cascadia subduction zone off the western coast of British Columbia (see box 3.6). Earthquakes also occur regularly along the St. Lawrence and Ottawa river valleys, and in 1929 a quake offshore from Newfoundland triggered a tsunami that killed 27 people—Canada's deadliest earthquake to date. One of the earliest recorded earthquakes occurred in 1663 at Charlevoix in Quebec, along the St. Lawrence River valley (estimated M=7). Canada's largest measured earthquake (M=8.1) occurred in 1949 just off the Queen Charlotte Islands in British Columbia. Infrequent earthquake activity occurs along the Mackenzie River valley and in the Canadian Arctic islands. Earthquakes in the Canadian Arctic result from active crustal rebound following the relatively recent retreat of the last ice sheet.

The "Top 10" earthquakes to have affected Canada during historic times are identified in box figure 1. The Cascadia subduction zone averages one earthquake per day, with a damaging earthquake occurring every 20 years or so. The city of Vancouver is at great risk from the next "big one" given that much of the urban area is built on wet silts of the Fraser River delta that may liquefy when shocked.

Eastern Canada lies far from any active plate boundaries, and the majority of earthquakes are caused by movement along ancient basement faults in Precambrian rocks buried below younger sedimentary strata. Southern Ontario is at risk from earthquakes generated along buried faults lying within ancient suture zones (see box 2.7). The St. Lawrence Rift and the Ottawa Graben are both ancient failed rift systems (aulacogens; see box 2.6) and the underlying Precambrian rocks are extensively faulted. Stresses generated as the North American plate moves westward stretch the crust and reactivate some of these ancient faults. Historically, the epicentres of earthquakes felt strongly in Quebec have been located in the Charlevoix/Kamouraska region and, to a lesser extent, in the Saguenay region; both areas lie within the ancient rift systems. Most recently (in January 2000), the Ottawa–Gatineau region was shaken by a magnitude-5.2 earthquake centred at Kipewa, just north of North Bay along the Ontario/Quebec border (box figure 2).

Additional Resources

For further information on Canadian earthquakes, go to
http://earthquakescanada.nrcan.gc.ca/recent_eq/index_e.php

Information regarding earthquake activity for specific regions of Canada can be found at
http://geoscape.nrcan.gc.ca/vancouver/earth_e.php
http://geoscape.nrcan.gc.ca/victoria/eq_e.php
http://geoscape.nrcan.gc.ca/ottawa/earthquakes_e.php

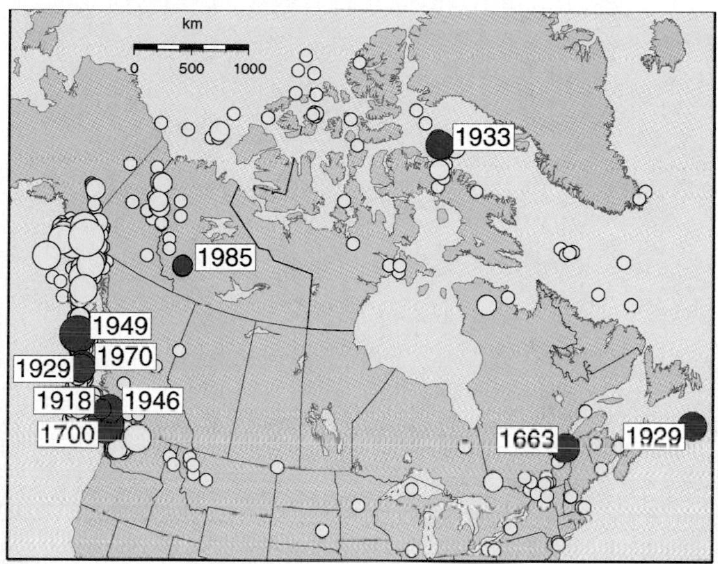

Magnitude
○ 5.0 - 5.9 ○ 6.0 - 6.9 ○ 7.0 - 7.9 ○ ≥8.0 1660-2004

Date	Magnitude	Location
1700/01/26	9.0	Cascadia subduction zone, British Columbia.
1949/08/22	8.1	Offshore Queen Charlotte Islands, British Columbia.
1970/06/24	7.4	South of Queen Charlotte Islands, British Columbia.
1933/11/20	7.3	Baffin Bay, Northwest Territories
1946/06/23	7.3	Vancouver Island, British Columbia.
1929/11/18	7.2	Grand Banks south of Newfoundland.
1929/05/26	7.0	South of Queen Charlotte Islands, British Columbia.
1663/02/05	7.0	Charlevoix, Quebec.
1985/12/23	6.9	Nahanni region, Northwest Territories.
1918/12/06	6.9	Vancouver Island, British Columbia.

BOX 3.2 ■ FIGURE 1

The date, magnitude, and location of the 10 largest-magnitude earthquakes (red circles) recorded in Canada during historical times. Magnitudes for earthquakes prior to the 20th century are estimated from non-instrumental data.

Top 10 earthquakes in Canada[2] Website. Reproduced with the permission of the Minister of Public Works and Government Services Canada, 2006 and Courtesy of Natural Resources Canada, Geological Survey of Canada.

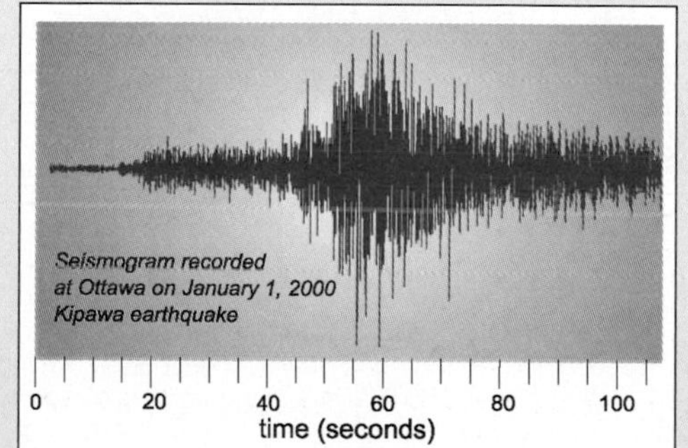

BOX 3.2 ■ FIGURE 2

Seismogram recorded at Ottawa on January 1, 2000, Kipawa earthquake.

Reproduced with permission of the Minister of Public Works and Government Services Canada, 2006 and Courtesy of Natural Resources Canada, Geological Survey of Canada

A

B

C

D

FIGURE 3.15

Earthquake damage to structures from recent major earthquakes throughout the world. (*A*) Elevated highway knocked over by a strong horizontal jolt during the 1995 Kobe, Japan earthquake. Damage exceeded $400 billion and destroyed or severely damaged more than 88,000 buildings. (*B*) Poorly constructed buildings crumbled during the 1999 Izmit, Turkey earthquake while structures built to seismic code and old mosques are left standing. (*C*) Many high-rise buildings collapsed during the 1999 Taiwan earthquake. The M=7.6 quake was the largest to hit central Taiwan in the past 400 years and damage exceeded $14 billion. (*D*) One of the many buildings damaged during the January 2001 Gujarat, India earthquake, which caused more than $1.3 billion in damage.

Photo A by Reuter/Sankei/Shimbun. Photo B by AP/Wide World Photo. Photo C by Smith Glenn/Corbis SYGMA. Photo D by Jaswant Arelekar/IITK, Kanpur, India

IN GREATER DEPTH 3.3

Earthquake Engineering

Damage and loss of life can be substantially reduced by siting structures on solid bedrock or dense compacted soils and by building structures that adhere to strict seismic building codes. In the 7.2-magnitude earthquake that struck Armenia in 1988, 50,000 people lost their lives when poorly constructed buildings crumbled. More recently, on January 26, 2001, the magnitude-7.7 Gujarat, India quake killed 18,000, left 600,000 homeless, and destroyed 332,000 houses. In contrast, earthquake-resistant structures and enforcement of seismic building codes in the San Francisco Bay area resulted in only 63 people dying in the 7.2-magnitude Loma Prieta earthquake.

Buildings that are constructed of strong, flexible, and light materials such as steel, wood, and reinforced concrete (strengthened by steel rebar) are the most resistant to damage by seismic shaking. Houses built with unreinforced concrete block or brick, which are only as strong as the mortar holding the blocks and bricks together, tend to lack flexibility and crumble in large earthquakes. During moderate-sized earthquakes, many houses lose their chimneys or brick facades. Buildings with heavy roofs made of tile or slate also tend to collapse. During the Gujarat, India earthquake, many reinforced concrete buildings failed because the walls, floors, ceilings, and elevator shafts were not well connected to allow the entire building to flex as one (box figure 1*A*).

The 1985 Mexico City earthquake, which killed 5,000 people and caused more than $5 billion in damage, is a classic example of the effect soft soils have on the amplification of earthquake waves. The ground shaking was relatively mild in most parts of Mexico, but it was amplified in some buildings and by the lake-bed sediments beneath Mexico City. Most buildings are not rigid but slightly flexible; they sway gently like large pendulums when struck by wind or seismic waves. The time necessary for a single back-and-forth oscillation is called its *period*, and it varies with a building's height and mass. Natural bodies of rock and sediment vibrate the same way, with periods that vary with the body's size and density.

When earthquake waves struck Mexico City, many were vibrating with a two-second period. The body of lake sediment beneath the city has a natural period of two seconds also, so the wave motion was amplified by the sediments. This type of amplification, or resonance, occurs when you push a child on a swing—if you push gently in time with the swing's natural period, the swing progressively goes higher and higher. The water-saturated sediment began to move like a sloshing waterbed. The ground moved back and forth by 40 centimetres every two seconds, and it did this 15 to 20 times.

This shaking was devastating to buildings with a natural two-second period (generally those 5 to 20 storeys high), and several hundred buildings collapsed as they further resonated with the shaking (box figure 1*B*). Many structures weakened by the main shock collapsed during the large aftershock. Most shorter and taller buildings had other periods of vibration and rode out the shaking with little damage. Although 800 buildings collapsed or were seriously damaged, most of the city's 600,000 structures survived with little or no damage. The Mexico City quake vividly illustrates the fact that proper building design can greatly lessen earthquake damage.

A

B

BOX 3.3 ■ FIGURE 1

(*A*) Insufficient connection between the reinforced-concrete elevator shaft and the rest of the building led to separation and partial collapse during the 2001 Gujarat, India earthquake. (*B*) This 15-storey building collapsed completely during the 1985 Mexico City earthquake, crushing all its occupants as its reinforced-concrete floors "pancaked" together.

Photo A by C.V.R. Marty/IITK, Kanpur, India. Photo B by M. Celebi, U.S. Geological Survey

FIGURE 3.16

Almost 100 homes burned at a Sylmar mobile-home park following the Northridge earthquake, southern California, 1994.

Photo © Ken Lubas/Los Angeles Times

of landslides in the steep Andes Mountains, burying more than 17,000 people (see box 13.1). In 1920 in China more than 100,000 people living in hollowed-out caves in cliffs of loess (described in chapter 17) were killed when a quake collapsed the cliffs. The 2001 El Salvador quake resulted in nearly 500 landslides, the largest of which occurred in Santa Tecla where 1,200 people were missing after tonnes of soil and rock fell on a neighbourhood.

A special type of ground failure caused by earthquakes is *liquefaction.* This occurs when a water-saturated soil or sediment turns from a solid to a liquid as a result of earthquake shaking. Liquefaction may occur several minutes after an earthquake, causing buildings to sink and underground tanks to float as once-solid sediment flows like water (figure 3.17*B*). Liquefaction was responsible for much of the damage in the 1989 Loma Prieta quake, and contributed to the damage in the 1906 San Francisco, the 1964 Alaska, the 1995 Kobe, Japan, and the 2001 Puget Sound, Washington, and Gujarat, India, quakes.

Permanent displacement of the land surface may be the result of movement along a fault. Rocks can move vertically, those on one side of a fault rising while those on the other side drop. Rocks can also move horizontally, those on one side of a fault sliding past those on the other side. Oblique movement with both vertical and horizontal components can also occur during a single quake. Such movement can affect huge areas, although the displacement in a single earthquake seldom exceeds 8 metres. The trace of a fault on Earth's surface may appear as a low cliff, called a *scarp,* or as a closed tear in the ground (figure 3.18). In rare instances small cracks open during a quake (but not to the extent that Hollywood films often portray). Ground displacement during quakes can tear apart buildings, roads, and pipelines that cross faults. Sudden subsidence of land near the sea can cause flooding and drownings.

Aftershocks are small earthquakes that follow the main shock. Although aftershocks are smaller than the main quake, they can cause considerable damage, particularly to structures previously weakened by the powerful main shock. A long period of aftershocks can be extremely unsettling to people who have lived through the main shock. *Foreshocks* are small quakes that precede a main shock. They are usually less common and less damaging than aftershocks, but can sometimes be used to help predict large quakes (although not all large quakes have foreshocks).

A

B

FIGURE 3.17

(*A*) Landslide in Pacific Palisades triggered by the Northridge earthquake, 1994. (*B*) Liquefaction of soil by a 1964 quake in Niigata, Japan caused earthquake-resistant apartment buildings to topple over intact.

Photo A © Al Seib/Los Angeles Times; photo B by National Geophysical Data Center

A

B

C

D

FIGURE 3.18

Varieties of ground displacement caused by earthquakes. (*A*) Five-metre scarp (cliff) formed by vertical ground motion, Alaska, 1964. (*B*) Tearing of the ground near Olema, California, 1906. (*C*) Fence compressed by ground movement, Gallatin County, Montana, 1959. (*D*) Compression of concrete freeway, San Fernando Valley, California, 1971.

Photo A by U.S. Geological Survey; photo B by G. K. Gilbert, U.S. Geological Survey; photo C by I. J. Witkind, U.S. Geological Survey

Tsunami

The sudden movement of the sea floor upward or downward during a submarine earthquake displaces the entire water column and can generate very large sea waves, popularly called "tidal waves." Because the ocean tides have nothing to do with generating these huge waves, the Japanese term *tsunami* is preferred by geologists. **Tsunamis** are also called **seismic sea waves.** They usually are caused by great earthquakes (magnitude 8+) that disturb the sea floor, but they also result from submarine landslides or volcanic explosions. When a large section of sea floor suddenly rises or falls during a quake, all the water over the moving area is lifted or dropped for an instant. As the water returns to sea level, it sets up long, low waves that spread very rapidly over the ocean (figure 3.19). Because vertical motion of the sea floor is most conducive to the formation of a tsunami, most are associated with subduction-zone earthquakes, which tend to be some of the strongest quakes (see box 3.1).

A tsunami is unlike an ordinary water wave on the sea surface. A large wind-generated wave may have a wavelength of 400 m and be moving in deep water at a speed of 90 km per hour. The wave height when it breaks on shore may be only 0.6 to 3 m, although in the middle of hurricanes the waves can be more than 15 m high. A tsunami, however, may have a wavelength of 160 km, and may be moving at 725 km per hour. In deep water the wave height may be only 0.6 to 2 m, but near shore the tsunami may peak up to heights of 15 to 30 m. This great increase in wave height near shore is caused by bottom topography; only a few localities have the combination of gently sloping offshore shelf and funnel-shaped bay that forces tsunamis to awesome heights

FIGURE 3.19

Tsunami waves are generated by a submarine earthquake that displaces the sea floor and entire water column above. Long, low waves are formed above the displaced sea floor to compensate for the momentary rise in sea level and spread very rapidly (at the speed of a jetliner) in the deep ocean. In shallower water, the tsunami slows to highway speeds and builds in height until it breaks and crashes onto the shore with incredible force, causing destructive flooding along low-lying coastal areas.

1. Before the earthquake: In this case, an oceanic plate subducts under a continental plate. The continental plate bends as stresses between the two plates build over time.

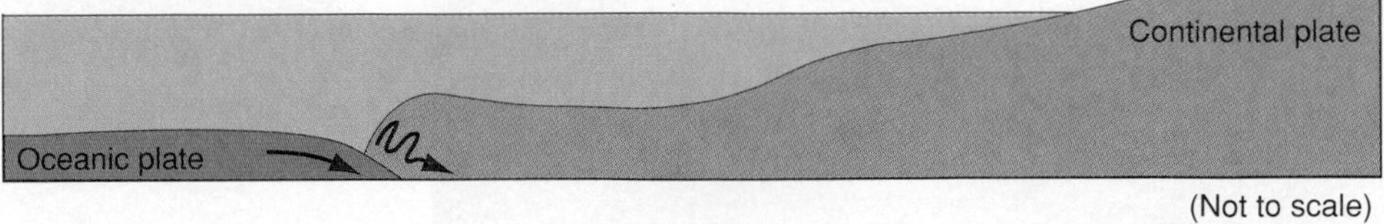

2. Earthquake: Releasing its built-up stress, the continental plate lurches forward over the oceanic crust, lifting the ocean. The displaced water appears as a huge bulge on the sea surface.

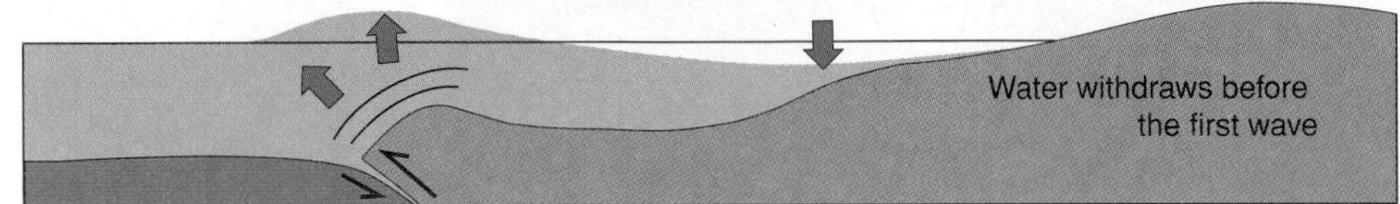

3. After the earthquake, gravity collapses the bulge to start a *succession* of waves, or tsunami. The tsunami waves move away in both directions as the mass of water "bobs" up and down over the source of the earthquake.

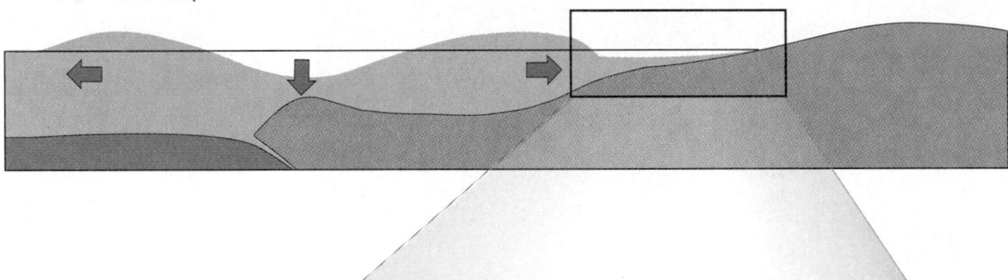

4. Each wave quickly advances over the land as a sediment-filled wall of water. It stops briefly before retreating, carrying sediments and debris back to the sea. Over time, the intensity of the tsunami subsides.

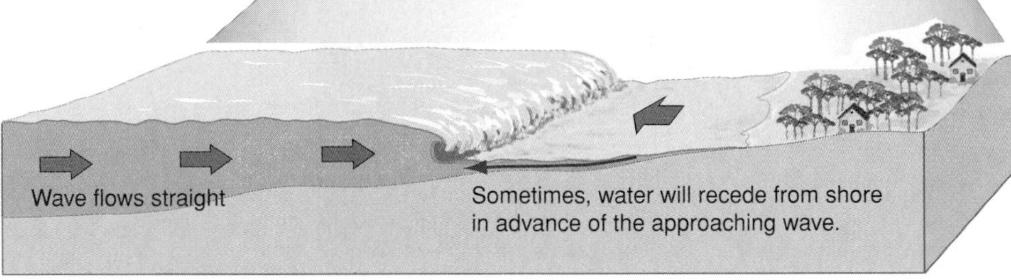

(the record height was 85 m in 1971 in the Ryukyu islands south of Japan). Along most coastlines tsunami height is very small.

Although the speed of the wave slows drastically as it moves through shallow water, a tsunami can still hit some shores as a very large, very fast wave. Because of its extremely long wavelength, a tsunami does not withdraw quickly as normal waves do. The water keeps on rising for five to ten minutes, causing great flooding before the wave withdraws. The long duration and great height of a tsunami can bring widespread destruction to the entire shore zone.

A tsunami formed by the 1960 Chilean quake crossed the Pacific Ocean and did extensive damage in Japan. In 1929, a submarine landslide triggered by the Grand Banks earthquake severed underwater cables and caused a tsunami that killed 27 people on the Burin Peninsula of Newfoundland. A destructive tsunami was generated April 1, 1946, by a 7.3-magnitude earthquake offshore from Alaska that devastated the city of Hilo, Hawaii, causing 159 deaths. The tsunami following the 1964 Alaskan quake drowned 12 people in Crescent City, California, a small coastal town near the Oregon border. Wave damage near an epicentre can be awesome. The 1964 Alaska tsunami destroyed the Scotch Cap lighthouse on nearby Unimak Island, sweeping it off its concrete base, which was 10 m above sea level, and killing its five occupants. The wave also swept away a radio tower, whose base was 31 m above sea level.

By far the most destructive tsunami in recorded history is the Indonesian tsunami of December 26, 2004, generated by a magnitude-9.3 quake off the west coast of Sumatra (figure 3.20). More than 250,000 people were killed and almost 2 million people were displaced in the 10 countries of South Asia and East Africa directly affected by the earthquake and tsunami (see box 3.1). Minor damage was also caused on the west coast of Australia, and water level fluctuations were even recorded in wells in parts of North America.

After the 1946 Hilo, Hawaii tsunami, the U.S. Coast and Geodetic Survey established a Tsunami Early Warning System in an attempt to minimize loss of life in Pacific coastal communities. A network of seismic stations reports large earthquakes that are capable of generating tsunami to the Tsunami Warning Center in Honolulu. An array of tidal gauges and deep-ocean tsunami detectors (figure 3.21) are then read to determine if a tsunami has been generated. Even though tsunamis travel at high speeds, there is usually sufficient time to warn low-lying coastal communities of the impending wave.

A

B

FIGURE 3.20

(A) Marina beach in Madras, India inundated by the tsunami. (B) Tsunami survivors carry items they saved from the rubble of a commercial area of Banda Aceh in northwest Indonesia.
Photo A © AFP/Getty Images; photo B © AP/Wide World Photos

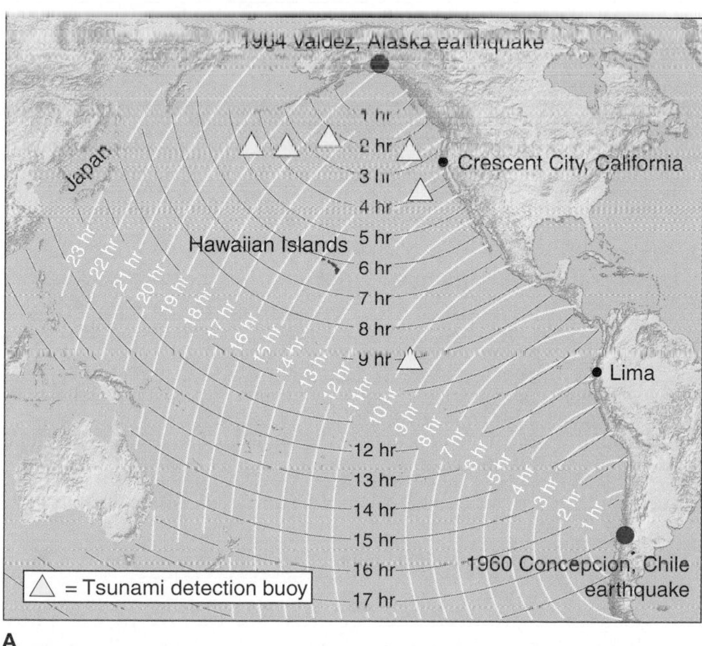

A

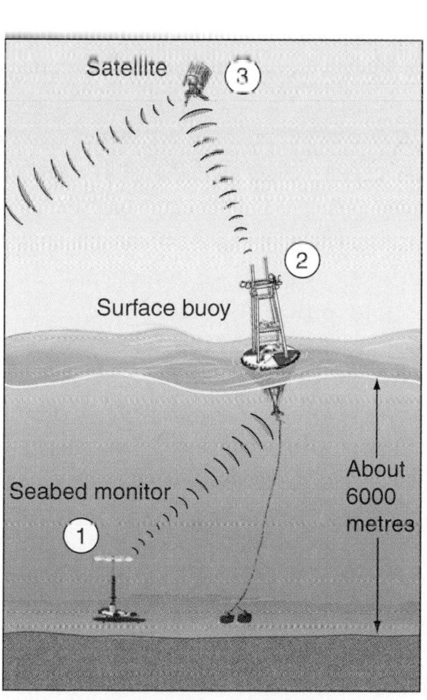

B NOAA's DART mooring system (components not to scale)

FIGURE 3.21

(A) Tsunami travel times from the magnitude-9.2 Alaska earthquake and the magnitude-9.5 Chile earthquake to locations within the Pacific. Yellow triangles show the location of tsunami-detection buoys. (B) Details of the Deep-ocean Assessment and Reporting of Tsunamis (DART) system used by NOAA.
From NOAA

Countries surrounding the Pacific Ocean, including Canada and the United States, share real-time information from seismometers and tidal gauges to provide a tsunami warning network. No tsunami-warning system, or other means of warning communities of the possibility of a tsunami, was in place in the Indian Ocean when the December 2004 tsunami struck. Efforts are now underway to create a global multi-hazard warning system, including a tsunami early warning system for the Indian Ocean. In the months following the Indonesia/Sumatra tsunami disaster, a series of webcast/teleconference presentations were made by the B.C. Provincial Emergency Program to more than 300 coastal communities in British Columbia. These presentations included topics such as earthquake hazards, tsunamis, tsunami warning, responsibilities, and future initiatives, and are viewed as an important step in mitigating losses from future earthquakes and tsunamis along Canada's West Coast.

WHERE DO EARTHQUAKES OCCUR ON A GLOBAL SCALE?

Most earthquakes are concentrated in narrow geographic belts (figure 3.22*A*), although *some* earthquakes have occurred in most regions on Earth. The boundaries of plates in the plate tectonics theory are defined by these earthquake belts (figure 3.22*B*). The most important concentration of earthquakes by far is in the **circum-Pacific belt,** which encircles the rim of the Pacific Ocean. Within this belt occur approximately 80 percent of the world's shallow-focus quakes, 90 percent of the intermediate-focus quakes, and nearly 100 percent of the deep-focus quakes.

Another major concentration of earthquakes is in the **Mediterranean-Himalayan belt,** which runs through the Mediterranean Sea, crosses the Mideast and the Himalayas, and passes through the East Indies to meet the circum-Pacific belt north of Australia.

A number of shallow-focus earthquakes occur in two other significant locations on Earth. One is along the summit or crest of the *mid-oceanic ridges,* huge underwater mountain ranges that run through all the world's oceans (see figure 2.18 and chapter 2). A few earthquakes have also been recorded in isolated spots usually associated with basaltic volcanoes, such as those of Hawaii.

In most parts of the circum-Pacific belt, earthquakes, andesitic volcanoes, and *oceanic trenches* (see chapter 2) appear to be closely associated. Careful determination of the locations and depths of focus of earthquakes has revealed the existence of distinct *earthquake zones* that begin at oceanic trenches and slope landward and downward into Earth at an angle of about 30 degrees to 60 degrees (figure 3.23). Such zones of inclined seismic activity are called **Benioff zones,** after the man who first recognized them.

Benioff zones slope under a continent or a curved line of islands called an **island arc.** Andesitic volcanoes may form the islands of the island arc, or they may be found near the edge of a continent that overlies a Benioff zone.

Most of the circum-Pacific belt is made up of Benioff zones associated in this manner with oceanic trenches and andesitic volcanoes. Parts of the Mediterranean-Himalayan belt represent Benioff zones, too, notably in the eastern Mediterranean Sea and in the East Indies. Essentially all of the world's intermediate- and deep-focus earthquakes occur in Benioff zones.

WHAT IS THE RELATIONSHIP BETWEEN EARTHQUAKES AND PLATE TECTONICS?

One of the great attractions of the concept of plate tectonics is its ability to explain the distribution of earthquakes and the rock motion associated with them.

As described in chapter 2, the concept of plate tectonics is that the Earth's surface is divided into a few giant *plates.* Plates are rigid slabs of rock, thousands of kilometres wide and 70 to 125 (or more) kilometres thick, that move across Earth's surface. Because the plates include continents and sea floors on their upper surfaces, the plate tectonics concept means that the continents and sea floors are moving. The plates change not only position but also size and shape.

Earthquakes occur commonly at the edges of plates (interplate earthquakes) but only occasionally in the middle of a plate (intraplate earthquakes). The close correspondence between plate edges and earthquake belts can be seen by comparing the map of earthquake distribution in figure 3.22*A* with the plate map in figure 3.22*B*.

This correspondence is hardly surprising—plate boundaries are identified and *defined* by earthquakes. According to plate tectonics, earthquakes are caused by the interactions of plates along plate boundaries. Therefore, narrow bands of earthquakes are used to outline plates on plate maps. This can be clearly seen in the east Pacific Ocean off South America, where the Nazca plate (figure 3.22*B*) is almost completely outlined by earthquake epicentres (figure 3.22*A*).

The earthquakes on the western border of the Nazca plate are shallow-focus quakes, and they occur in a narrow belt along the crest of the mid-oceanic ridge here, locally called the East Pacific Rise. The quakes along the eastern boundary occupy a broader belt that lies mostly within South America. This belt includes shallow-, intermediate-, and deep-focus earthquakes in a Benioff zone that begins at the Peru–Chile Trench just offshore and slopes steeply down under South America to the east. The Nazca plate moves eastward, away from the crest of the mid-oceanic ridge and toward the subduction zone at the trench, where the plate plunges down into the mantle. The plate's western boundary is located at the crest of the East Pacific Rise, and its eastern boundary is at the bottom of the Peru–Chile Trench.

Earthquakes at Plate Boundaries

As you learned from chapter 2, there are three types of plate boundaries: *divergent boundaries,* where plates move away from each other; *transform boundaries,* where plates move horizontally past each other; and *convergent boundaries,* where plates move toward each other. Each type of boundary has a characteristic pattern of earthquake distribution and rock motion.

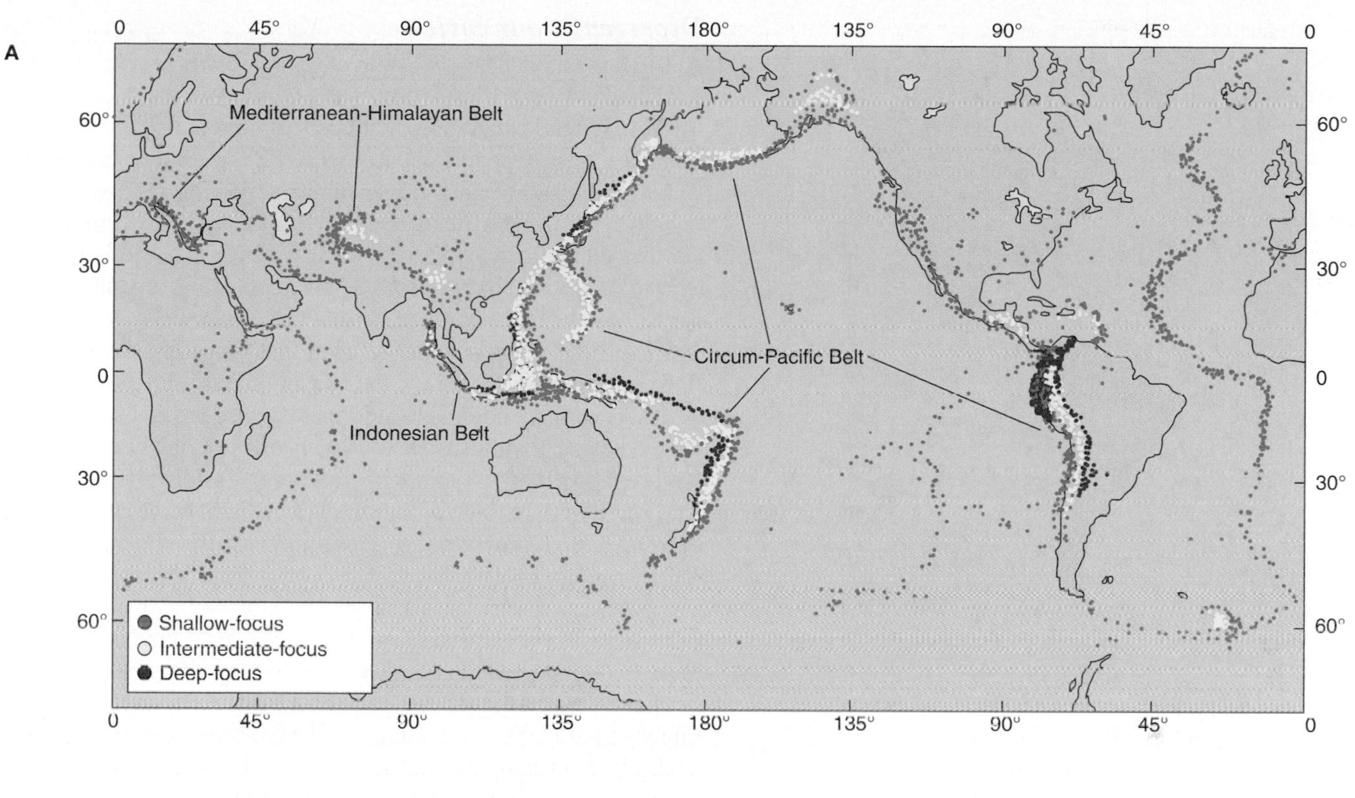

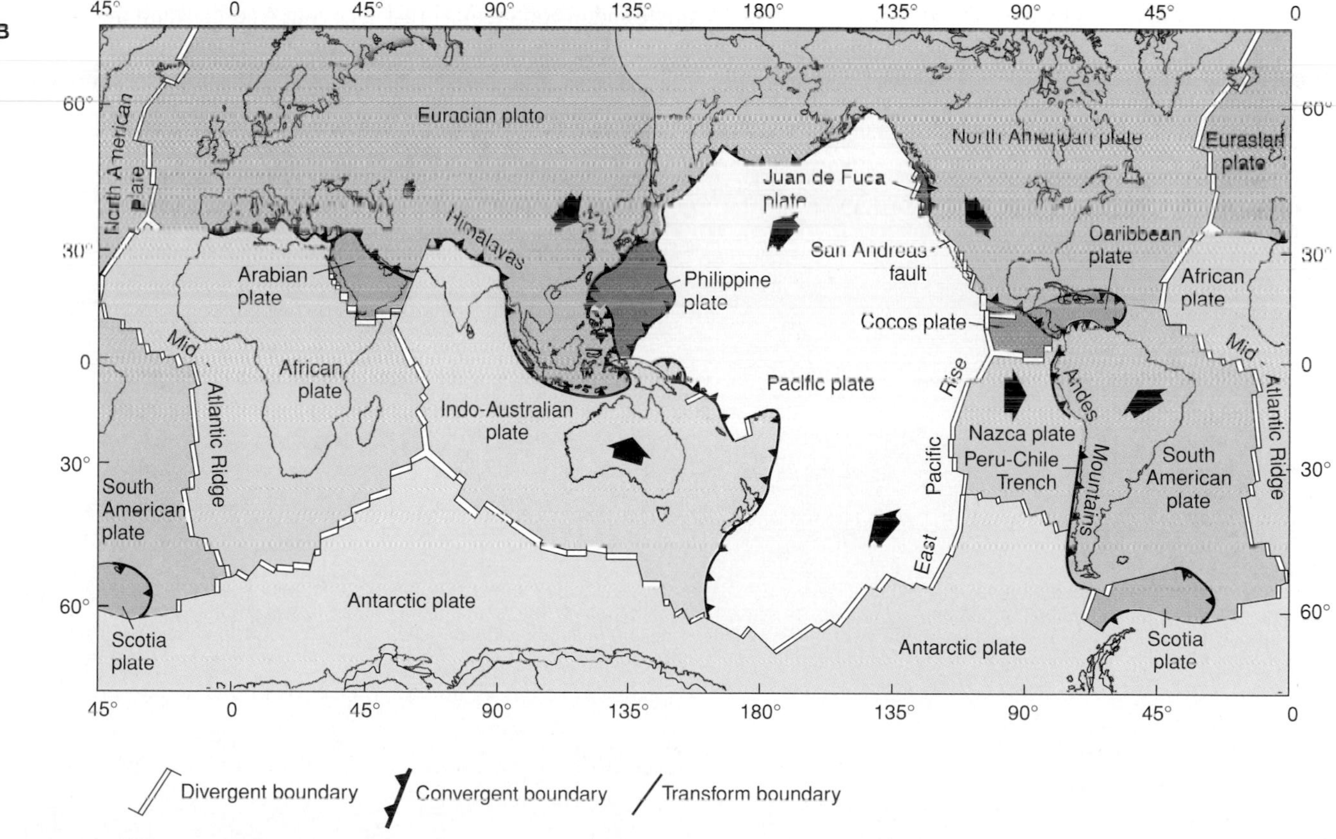

FIGURE 3.22

(*A*) Distribution of shallow-, intermediate-, and deep-focus earthquakes. (*B*) The major plates of the world in the theory of plate tectonics. Compare the locations of plate boundaries with earthquake locations shown in figure 3.22*A*. Double lines show diverging plate boundaries; single lines show transform boundaries. Heavy lines with triangles show converging boundaries; triangles point down subduction zone.

After W. Hamilton, U.S. Geological Survey

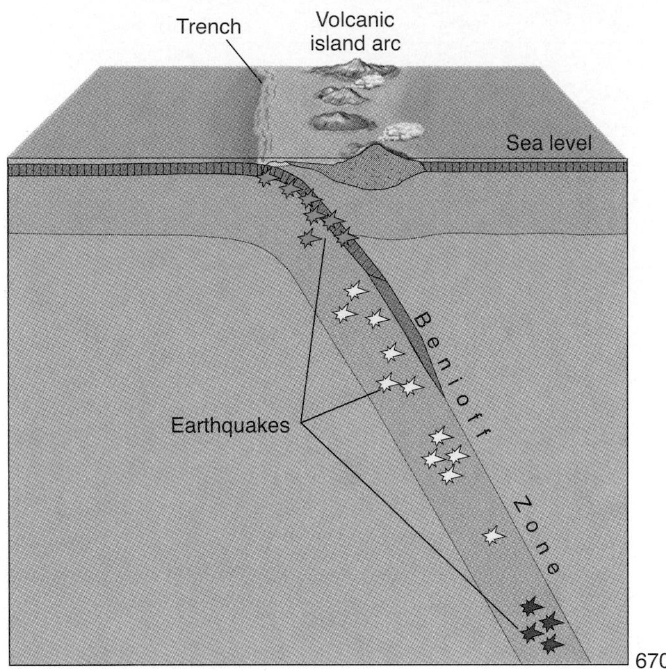

FIGURE 3.23

A Benioff zone of earthquakes begins at an oceanic trench and dips under a continent (such as South America) or a volcanic island arc. Upper part of Benioff zone may extend to a depth of 670 km.

Divergent Boundaries

At a divergent boundary, where plates move away from each other, earthquakes are shallow, restricted to a narrow band, and much lower magnitude than those that occur at convergent or transform boundaries. A divergent boundary on the sea floor is marked by the crest of a mid-oceanic ridge and the *rift valley* that is often (but not always) found on the ridge crest (figure 3.24). The earthquakes are located along the sides of the rift valley and beneath its floor. The rock motion that is deduced from study of seismic signals shows that the faults here are normal faults, parallel to the rift valley. The ridge crest is under tension that is tearing the sea floor open, creating the rift valley and causing the earthquakes.

A divergent boundary within a *continent* is usually also marked by a rift valley, shallow-focus quakes, and normal faults (figure 3.24). The African Rift Valleys in eastern Africa (figure 3.22*B*) seem to be such a boundary. Tensional forces are tearing eastern Africa slowly apart, creating the rift valleys, some of which contain lakes.

Transform Boundaries

Where two plates move past each other along a transform boundary, the earthquakes are shallow. Strike-slip motion occurs on faults parallel to the boundary. The earthquakes are aligned in a narrow band along the transform fault. Although most transform faults occur on the ocean floor and offset ridge segments, some are found in continental crust. The San Andreas fault in California (box 3.6) is the most famous example of a right-lateral transform fault. The Alpine fault in New Zealand is another example of a right-lateral transform fault.

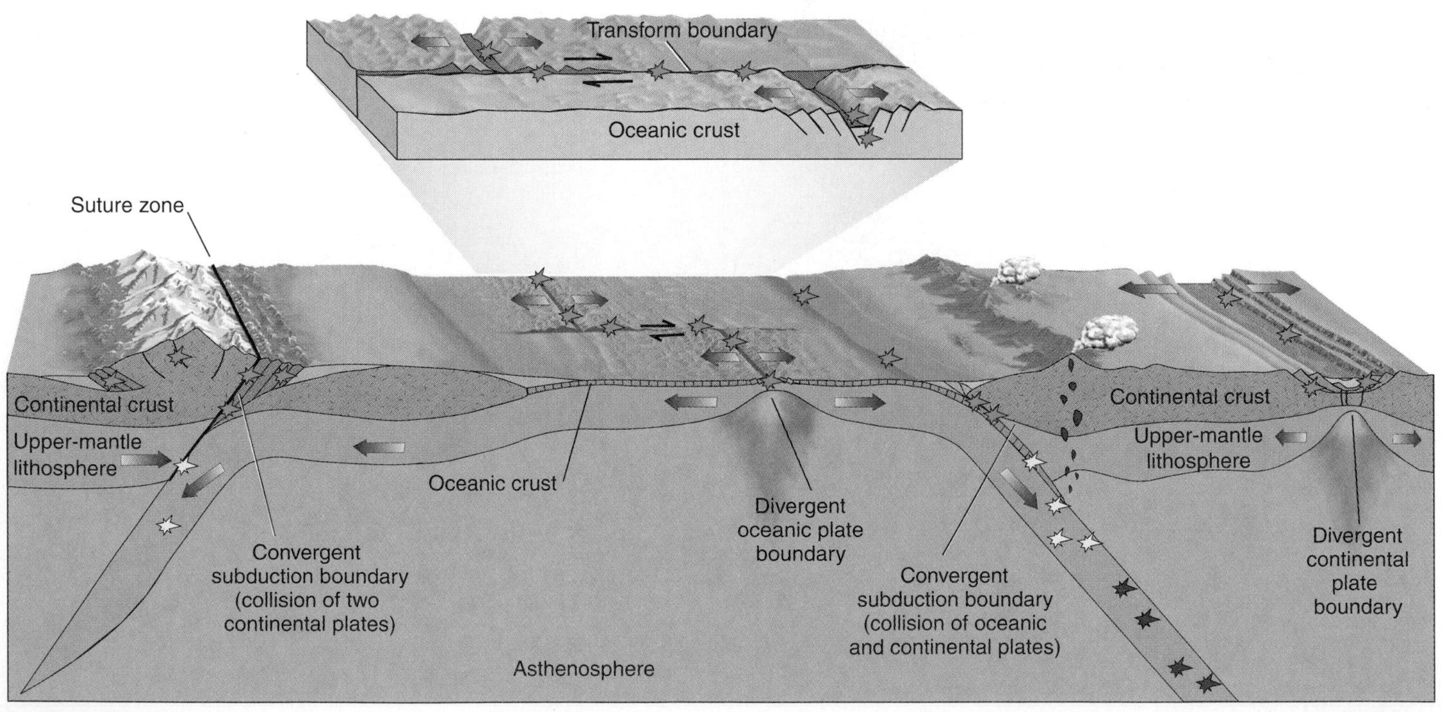

FIGURE 3.24

Distribution of earthquakes at plate boundaries. Shallow-focus earthquakes occur at divergent boundaries where the lithosphere is being pulled apart and also along transform boundaries where slip in the lithosphere accommodates the spreading between oceanic ridges. Shallow- to deep-focus earthquakes occur where a lithosphere subducts during collision of two plates.

ENVIRONMENTAL GEOLOGY 3.4

Waiting for the Big One in British Columbia

The San Andreas fault passes northward into the Cascadia Subduction Zone, where the Juan de Fuca plate is being subducted below the leading edge of the North American plate (see box 2.2). The southwestern corner of British Columbia—which includes Vancouver, Canada's third largest city—lies adjacent to the Cascadia Subduction Zone and is the most seismically active region of Canada. This area experiences more than 200 earthquakes each year (box figure 1). Most of these earthquakes are too small to be felt, although nine moderate to large (M=6–7) earthquakes have occurred within 300 km of Vancouver over the past 150 years. Scientists are more concerned, however, with the geologic record, which suggests that large earthquakes (M>8) have repeatedly shaken tens of thousands of kilometres of the Pacific Northwest during prehistoric times.

Three types of damaging earthquakes could affect southwestern British Columbia. *Crustal* earthquakes may originate within the North American plate; *intraplate* earthquakes within the Juan de Fuca plate; and *subduction* or *plate boundary* earthquakes at the boundary between the two plates. All of the historical earthquakes in the region are thought to have been local crustal or intraplate earthquakes. There is evidence to suggest that the Juan de Fuca–North American plate boundary is currently locked (*aseismic*) and accumulating strain. This strain will release as a great subduction earthquake sometime in the future.

Geologic evidence in the form of buried soils that indicate sudden land subsidence—sheets of sand deposited by tsunamis and liquefaction features in sediments (box figure 2)—suggests that many great earthquakes have occurred in the area during the Holocene. These earthquakes have occurred on average once every 500 to 700 years, although the intervals between them range from a few hundred years to more than 1,000 years, making it difficult to predict the timing of the next great subduction earthquake. The most recent was about 300 years ago.

The locked zone of the Juan de Fuca–North America plate boundary is thought to lie below the continental slope and shelf some 200 km from Vancouver. Rupture of a substantial length of the 1,000-km locked zone could create a subduction earthquake in excess of M=9 and would cause severe damage to Vancouver and Seattle. However, of greater hazard to the residents of southwestern B.C. are the more frequent but lesser magnitude M=6–7 intraplate earthquakes that occur within the North American and Juan de Fuca plates. A "local" earthquake of this magnitude could cause considerable damage depending on the location of the epicentre. An M=7 event could produce significant damage within 50 to 100 km of the epicentre, while an M=6 earthquake would have a damage radius of only 20 to 50 km. Modelling of the economic impact of a M=6.5 crustal earthquake with a focus 10 km below Vancouver indicates the total economic loss would be between $14 billion and $32 billion (Canadian dollars, 1992). These losses are comparable to those from the 1989 Loma Prieta earthquake (US$10 billion) and the 1994 Los Angeles earthquake (US$15 billion).

Earthquake damage in the Vancouver area will be particularly severe due to the thick cover of unconsolidated surficial sediments prone to seismic wave amplification, the presence of waterlogged coastal sediments prone to liquefaction, and the steep slopes in surrounding areas prone to landsliding. The Fraser River delta and its associated floodplain have been identified as a region particularly susceptible to seismic hazards. The delta hosts the city's major airport and is underlain by up to 235 metres of loose, water-saturated clay, silt, and sand, easily liquefied by seismic waves. Landslides occurring in the steep mountain valleys surrounding the city would also cut essential road and rail links.

Further Reading

Clague, John J. 1997. Earthquake Hazard in the Greater Vancouver Area, in N. Eyles (ed.) *Environmental Geology of Urban Areas*. Geological Association of Canada, Newfoundland, pp. 423–437.

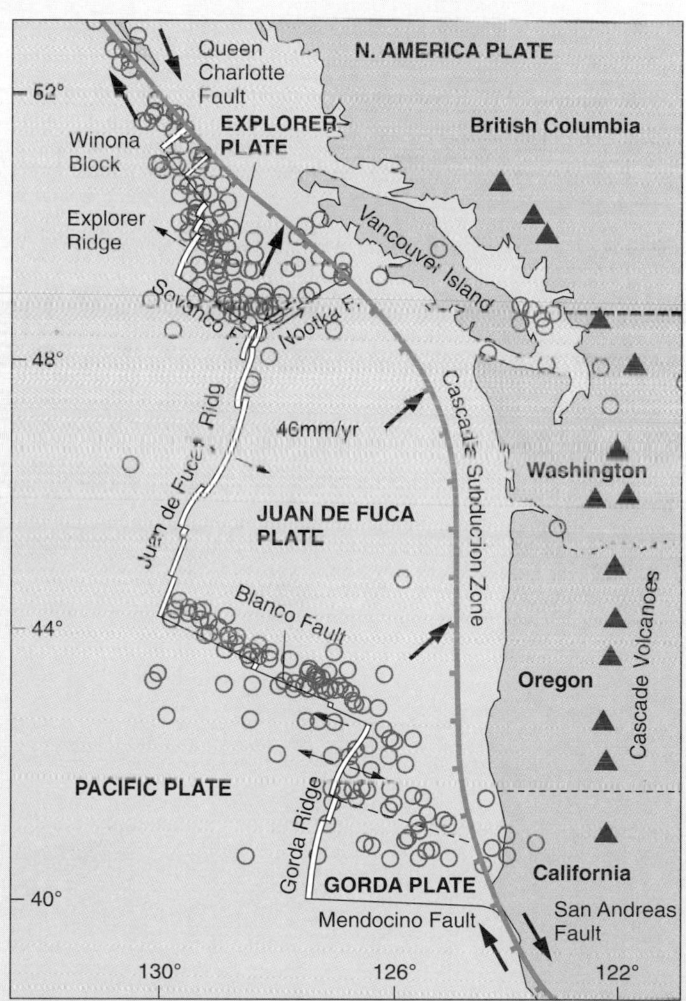

BOX 3.4 ■ FIGURE 1

Plate tectonics regime of the Pacific Northwest. Circles are epicentres of earthquakes of magnitude 5 and larger (data from Canadian Earthquake Epicentre File of the Geological Survey of Canada). All earthquakes prior to 1990 (United States) and 1992 (Canada) are shown. Arrows indicate directions of plate motion. Modified from Dragert et al., 1994.

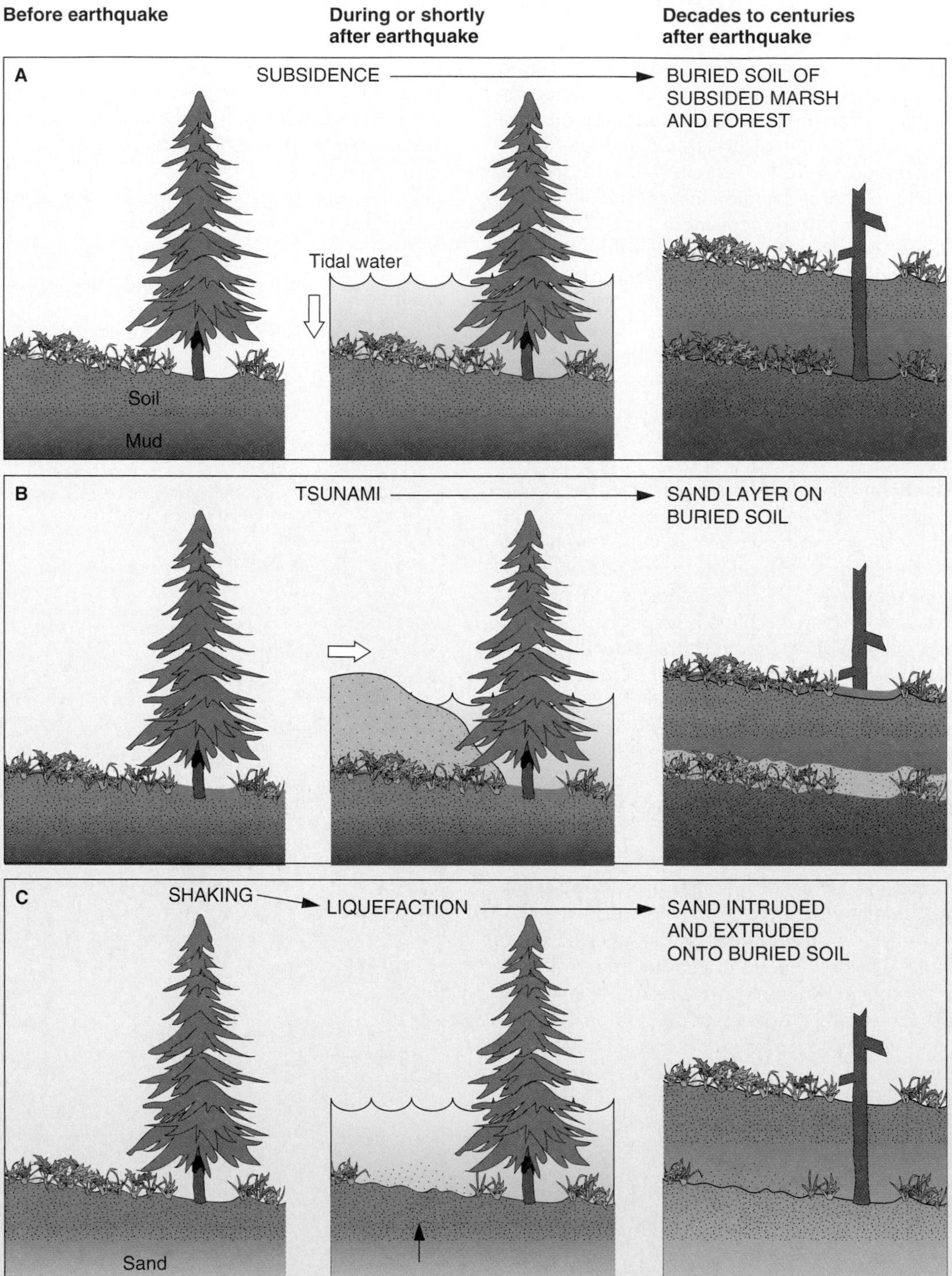

BOX 3.4 ■ FIGURE 2

Inferred origin of the main coastal features cited as evidence for prehistoric earthquakes in southwestern British Columbia and Washington. (*A*) Soil is buried by tidal mud after earthquake-induced subsidence lowers the land into the intertidal zone. (*B*) A sheet of sand is deposited on a subsided surface by a tsunami shortly after an earthquake. (*C*) Liquefied sand is erupted through and onto a subsided surface as a result of shaking during an earthquake Modified from Atwater et al., 1995.

Convergent Boundaries

Convergent boundaries are of two general types, one marked by the *collision* of two continents, the other marked by *subduction* of the ocean floor under a continent (figure 3.24) or another piece of sea floor. Each type has a characteristic pattern of earthquakes.

Collision boundaries are characterized by broad zones of shallow earthquakes on a complex system of faults (figure 3.24). Some of the faults are parallel to the dip of the suture zone that marks the line of collision; some are not. One continent usually overrides the other slightly (continents are not dense enough to be subducted), creating thick crust and a mountain range. The Himalayas are thought to represent such a boundary (figure 3.22*B*). The seismic zone is so broad and complex at such boundaries that other criteria, such as detailed geologic maps, must be used to identify the position of the suture zone at the plate boundary.

During *subduction,* earthquakes occur for several different reasons (figure 3.24). As a dense oceanic plate bends to go down at a trench, it stretches slightly at the top of the bend, and normal faults occur as the rocks are subjected to *tension.* This gives a block-faulted character to the outer (seaward) wall of a trench. For some distance below the trench, the subducting plate is in contact with the overlying plate. Studies of earthquakes at these shallow depths show that the quakes are caused by shallow-angle thrust-faulting. This is the motion expected as one plate slides beneath another, a process commonly called *underthrusting.*

At greater depths, where the descending plate is not in direct contact with the overlying plate, earthquakes are common but the reasons for them are not obvious. The quakes are confined to a thin zone, only 20 to 30 km thick, within the lithosphere of the descending plate, which is about 100 km thick. This zone is thought to be near the top of the lithosphere, where the rock is colder and more brittle.

Subduction Angle

The horizontal and vertical distributions of earthquakes can be used to determine the angle of subduction of a downgoing plate. Subduction angles vary considerably from trench to trench. Many plates start subducting at a gentle angle, which becomes much steeper with depth. At a few trenches in the open Pacific, subduction begins (and continues) at almost a vertical angle. Subduction angle is probably controlled by plate density and the rate of plate convergence. Older oceanic lithosphere, such as that in the southeast Pacific, tends to be colder and more dense and therefore subducts at a steeper angle; younger oceanic plates in close proximity to the oceanic ridge are warmer and more buoyant and subduct at a shallower angle. A faster rate of convergence may also result in a shallower angle of subduction.

Earthquakes Away from Plate Boundaries

Not all earthquakes occur along the boundaries of plates (figure 3.13). Many earthquakes also occur within plates, although with less frequency than those at plate boundaries, and account for only about 5 percent of earthquake energy released in a year. These are termed *intraplate* earthquakes.

Most intraplate earthquakes occur in areas of thinned or weakened crust such as continental margins, aulacogens, or ancient suture zones (see boxes 2.6 and 2.7). They may be triggered by a buildup of stress between the lithosphere and the underlying asthenosphere as plates move, or by isostatic adjustments of the crust due to loading and unloading by ice or sediment. The North American continent is moving westward over the asthenosphere at a rate of several centimetres per year, and the buildup of strain has the potential to reactivate ancient or buried faults in the crust.

Analysis of the distribution of intraplate earthquakes in North America shows that they occur in distinct zones. The areas around New Madrid, Missouri, and the St. Lawrence Valley/ Charlevoix-Kamouraska area are particularly prone to earthquake activity (refer to figure 3.13). Many of these areas are underlain by ancient rift systems dating back to the Precambrian (figure 3.25). Other seismically active areas in eastern North America are located on ancient suture zones that record the collision of former continents. The densely populated region of southern Ontario was not thought at risk from severe earthquake activity until recently. However, many small-magnitude earthquakes occur in the region, mostly less than M=4.3 (see box 2.7), and are located along buried

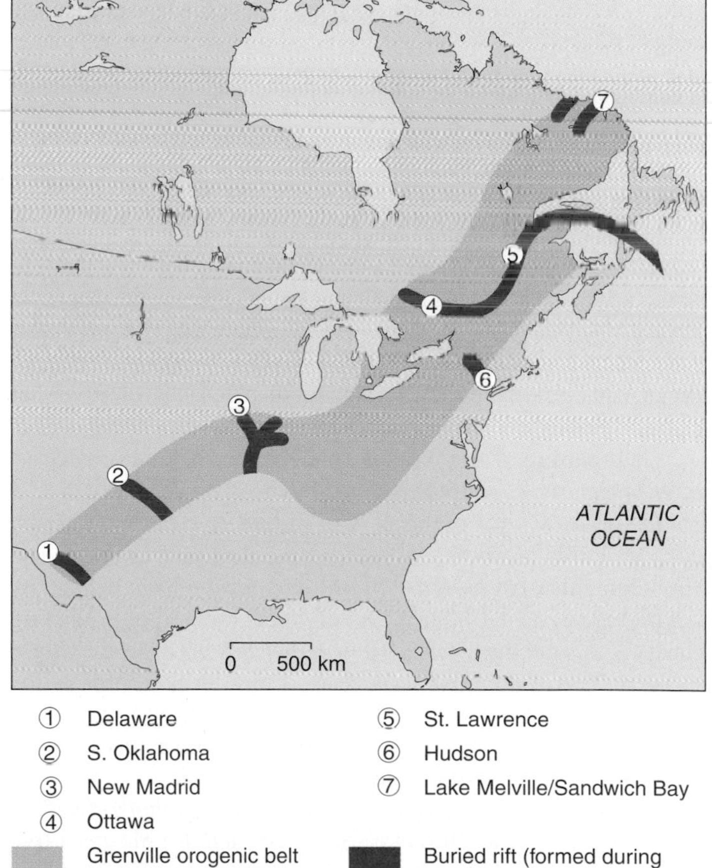

FIGURE 3.25

Seismically active zones of eastern North America.

bedrock structures marking the position of former suture zones. Future reactivation of these buried structures creates the potential for moderate to large intraplate earthquakes in southern Ontario. The probabilities of earthquakes of M=5, M=6, and M=7 in the next 50 years have been calculated as 57 percent, 6 percent, and less than 1 percent, respectively.

Residents of New Brunswick were made aware of the potential for damaging earthquakes in 1982 when two earthquakes (M=5.7 and M=5.1), centred in the Miramichi region, occurred within hours of each other and were followed by many smaller aftershocks. No surface faulting was observed following the quakes and they are also attributed to reactivation of ancient fault systems in buried Precambrian basement rocks.

Intraplate earthquakes can be particularly damaging as they occur on areas of relatively rigid plate that allow the efficient transmission of seismic waves. It is therefore important to understand the distribution and causes of such earthquakes in order to better predict their occurrence.

CAN WE PREDICT WHEN EARTHQUAKES WILL OCCUR?

People who live in earthquake-prone regions are plagued by unscientific predictions of impending earthquakes by popular writers and self-proclaimed prophets. Several techniques are being explored for *scientifically* forecasting a coming earthquake. One group of methods involves monitoring slight changes, or precursors, that occur in rock next to a fault before the rock breaks and moves; these methods assume that large amounts of strain are stored in rock before it breaks (figure 3.2). Box 3.5 describes some ways that people can prepare for—and react to—an earthquake.

Just as a bent stick may crackle and pop before it breaks with a loud snap, a rock may give warning signals that it is about to break. Before a large quake, small cracks may open within the rock, causing small tremors, or *microseisms,* to increase. The *properties of the rock* next to the fault may be changed by the opening of such cracks. Changes in the rock's magnetism, electrical resistivity, or seismic velocity may give some warning of an impending quake.

The opening of tiny cracks changes the rock's porosity, so *water levels in wells* often rise or fall before quakes. The cracks provide pathways for the release of radioactive radon gas from rocks (radon is a product of radioactive decay of uranium and other elements). An *increase in radon emission from wells* may be a prelude to an earthquake. A very local method of predicting quakes is to time the *interval between eruptions of Old Faithful geyser* in Yellowstone National Park. Long-term records of the time between eruptions have shown that this interval changes in a regular way before a large local earthquake, probably because of porosity changes within the surrounding rock.

In some areas the *surface of Earth tilts and changes elevation* slightly before an earthquake. Scientists use highly sensitive instruments to measure this increasing strain, in hopes of predicting quakes.

Chinese scientists claim successful short-range predictions by watching *animal behaviour*—horses become skittish and snakes leave their holes shortly before a quake. North American scientists are conducting a few pilot programs along these lines, but many are skeptical of the Chinese claims. It is interesting that very few animals were killed by the 2004 Indian Ocean tsunami. Apparently before the tsunami hit, elephants were seen running to higher ground, flamingoes left low-lying breeding areas, and dogs refused to go outdoors.

Japanese and Russian geologists were the first to predict earthquakes successfully, and Chinese geologists have made some very accurate predictions. In 1975 a 7.3-magnitude earthquake near Haicheng in northeastern China was predicted five hours before it happened. Alerted by a series of *foreshocks,* authorities evacuated about a million people from their homes; many watched outdoor movies in the open town square. Half the buildings in Haicheng were destroyed, along with many entire villages, but only a few hundred lives were lost. In grim contrast, however, the Chinese program failed to predict the 1976 Tangshan earthquake (magnitude 7.6), which struck with no warning and killed an estimated 250,000 people.

Most of these methods were once considered very promising, but have since proved to be of little real help in predicting quakes. A typical quake predictor, such as tilt of the land surface, may precede one quake, and then be absent for the next 10 quakes. In addition, each precursor can be caused by forces unrelated to earthquakes (land tilt is also caused by mass wasting and wetting and drying of the land).

A fundamentally different method of determining the probability of an earthquake occurring relies on the history of earthquakes along a fault and the amount of tectonic stress building in the rock. Geoscientists look at the geologic record for evidence of past earthquakes using the techniques of *paleoseismology.* One technique involves digging a trench across the fault zone to examine sedimentary layers that have been offset and disrupted during past earthquakes (figures 3.26*A* and *B*). If the offset layers contain material such as volcanic ash, pollen, or organic material such as tree roots that can give a numerical age, then the average length of time between earthquakes (*recurrence interval*) can be determined. If the length of time since the last recorded earthquake far exceeds the recurrence interval, the fault is given a high probability of generating an earthquake.

Along some long-active faults are short, inactive segments called *seismic gaps* where earthquakes have not occurred for a long time. These gaps form as part of the seismic cycle and result in a zone of lowered stress, or stress shadow zone, where earthquake activity sharply decreases after a major seismic event. Such was the case after the 1906 San Francisco earthquake and after the 1857 break along the southern section of the San Andreas fault (see box 3.6).

The recurrence interval and likelihood of future earthquakes are also determined by measuring the slip rate along plate boundaries. Exciting new satellite-based techniques such as InSAR (interferometric synthetic aperture radar; see box 3.7), in addition to GPS, have allowed seismologists to measure the vertical and horizontal movement along active faults and to determine how long it would take for sufficient stress to build up along the plate boundary to generate rupturing and slip along

A

FIGURE 3.26

To determine the likelihood of a large earthquake occurring again along an active fault, geoscientists need to know how often quakes have occurred in the past and how large the last one was. By using the techniques of paleoseismology, geoscientists dig a trench (*A*) across or alongside a fault and very carefully map disturbed layers of sediment and soil exposed in the upper few metres of the trench (*B*). (*A*) Trench being dug across the southern Hayward fault near Fremont, California, by the U.S. Geological Survey to reevaluate the seismic risk for the San Francisco Bay area. (*B*) Photomosaic of the wall of a trench dug across the Coachella Valley section of the San Andreas fault near Thousand Palms Oasis, California reveals evidence for three of the five past earthquakes that struck this area since 825 A.D. (Events TP 1, 2, and 5). Evidence for the three separate earthquakes is shown in the trench either by the fault displacing different channel deposits against one another (*TP 1 offsets channel IV against VI and TP5 offsets channel I against II*) or the fault being buried or terminated by younger channel deposits (*TP2 cuts channel IV but not the overlying channel V sediments*). Based on these relations and on radiocarbon dates obtained from the disrupted layers (shown by small yellow boxes), it has been determined that the average time between earthquakes for this section of the San Andreas fault is 215 ± 25 years. Because the last earthquake (TP 1) occurred sometime after 1520–1680 A.D., more than 233 years have elapsed since the most recent earthquake, and geologists are concerned that the southernmost San Andreas fault zone is overdue for a large earthquake.

Photo A by Jennifer Adleman, U.S. Geological Survey; Photo B from Bulletin Seismological Society of America, 2002, v. 92, no. 7, p. 2851, courtesy of T. E. Fumal, U.S. Geological Survey

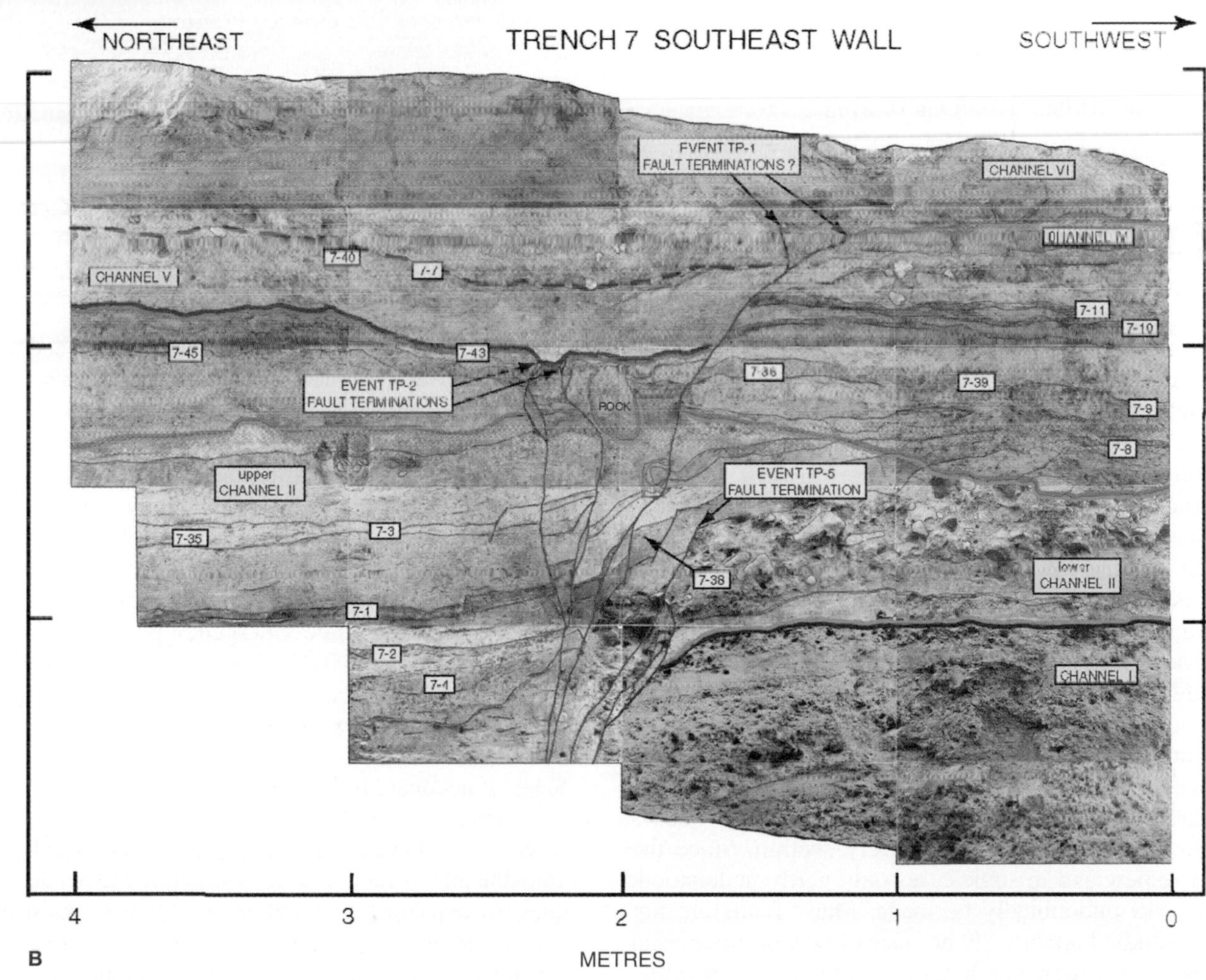

B

IN GREATER DEPTH 3.5

What to Do Before, During, and After an Earthquake

Before

- Develop a family emergency plan and practice it regularly.
- Identify an out-of-area phone contact person to call and check in with.
- Choose a couple of family meeting places; pick easy to identify, open and accessible places that you can likely walk to.
- Prepare to be self-sufficient for a minimum of three days.
- Assemble an emergency supply kit; include food, water, prescription medications and first aid supplies, a battery operated radio, flashlight, extra batteries, shelter, clothing, sturdy shoes, and personal toiletries.
- Assemble similar emergency kits for your workplace and vehicle.
- Take an approved first aid course.
- Quake-proof your house by securing heavy furniture and objects.

During

- Remain calm! The shaking usually lasts no more than a minute.
- If inside, stay inside.... "DROP, COVER, and HOLD!" Duck under sturdy furniture. Cover as much of your head and torso as you can. Hold onto the furniture. If you cannot get under sturdy furniture, move to an inside wall or archway and sit with your back to the wall, bring your knees to your chest and cover your head.
- Stay away from mirrors and windows.
- Do not exit the building during the shaking.
- If outdoors, move to an open area away from all structures, especially buildings, bridges, and overhead power lines.
- If driving, stop in an open area away from all structures especially bridges, overpasses, tunnels, and overhead power lines. Stay as low as possible inside the vehicle.

After

- STAY CALM! Count to 60 to allow time for objects to fall before moving.
- Move cautiously, and check for unstable objects and other hazards above and around you.
- Check yourself for injuries.
- Help those around you and provide first aid, if you are qualified.
- Hang up all phones. Only use phones (including cell phones) if a life is at stake.
- Inspect gas, water, and electric lines. If there are leaks or if there is any doubt about leaks, shut off mains; evacuate immediately if you hear or smell gas and can't shut it off. Report leaks to the authorities.
- Anticipate aftershocks, especially if the shaking lasted longer than two minutes.
- Stay out of damaged buildings.
- Listen to the radio or watch local TV for emergency information and additional safety instructions.

Related Web Resources

www.pep.bc.ca
B.C. Provincial Emergency Program

www.seismo.nrcan.gc.ca
National Resource Canada's National Earthquake Hazards Program is a comprehensive resource for Canadian eathquake information

For advice on all types of emergency preparation go to: www.psepc.gc.ca (Public Safety and Emergency Preparedness Canada).

Source: www.pep.bc.ca/hazard_preparedness/Quake_survival_guide.pdf

a fault (see box 2.1). For example, if the slip rate along the boundary is determined to be 5 centimetres per year and the last earthquake resulted in 5 metres of slip, then you would expect the next large earthquake to occur in 100 years. Just as a rubber band will break if stretched too far, rock will also break or rupture if a critical level of stress is exceeded. In other cases, the accumulating stress is released aseismically by so-called silent earthquakes where a fault slips very slowly or creeps to gradually relieve the stress. Slip rates and recurrence intervals are used to determine the statistical probability of an earthquake occurring over a given amount of time.

By studying the seismic history of faults, geoscientists are sometimes able to forecast earthquakes along some segments of some faults. In 1988, the U.S. Geological Survey estimated a 50 percent chance of a magnitude-7 quake along the segment of the San Andreas fault near Santa Cruz. In 1989, the magnitude-7 Loma Prieta quake occurred on this very section. Since the techniques are new and in some cases only partly understood, some errors will undoubtedly be made. Many faults are not monitored or studied historically because of lack of money and personnel, so we will never have a warning of impending quakes in some regions. For large urban areas near active faults such as the San Andreas, however, earthquake risk analysis may reduce damage and loss of life.

Another more recent approach to minimize loss of life and reduce damage in a major earthquake is to closely monitor the amount and location of strong shaking by using a dense network of broadband seismometers that digitally relay information via satellites to a central location. At this location, maps showing where the greatest amount of shaking occurred can be generated within minutes to guide emergency personnel to the areas of most damage (figure 3.27). Such a system has been developed in southern California, and there are plans for integrating other regional seismic networks into an Advanced National Seismic System (ANSS) to monitor earthquakes throughout the United States if adequate funding can be obtained.

A major goal of the ANSS program is to locate strong-motion seismometers in buildings, bridges, canals, and pipelines to provide valuable information on how a structure moves during an earthquake to help engineers build more earthquake-resistant structures. One key to reducing damage and loss of life is to create stronger structures that resist catastrophic damage during a major earthquake.

ENVIRONMENTAL GEOLOGY 3.6

Waiting for the Big One in California

The San Andreas fault, running north–south for 1,300 kilometres through California, is a right-lateral fault capable of generating great earthquakes of magnitude 8 or more. The 1906 earthquake near San Francisco caused a 450-kilometre scar in northern California (box figure 1). The portion of the fault nearest Los Angeles last broke in 1857 in a quake that was probably of comparable size. The ground has not broken in either of these regions since these quakes. Each old break is now a seismic gap, where rock strain is being stored prior to the next giant quake.

Recent California quakes were considerably smaller than the "big one" long predicted by geologists to be in the magnitude-8 range. The 1906 quake in the north had an estimated Richter magnitude of 8.25, and the southern break in 1857 near Fort Tejon was estimated to have a moment magnitude (M) of 7.8. In contrast, the 1989 Loma Prieta quake on the San Andreas fault near San Francisco was a M7.2 and the 1994 Northridge quake (not on the San Andreas fault) was M6.7. So, recent California quakes have been about magnitude 7 or less, and the "big one" should be 8.

A great earthquake of magnitude 8 could strike either the northern section or the southern section of the San Andreas fault tomorrow. Which section will break first? Because the southern section has been inactive longer, it may be the more likely candidate. A magnitude-8 quake here could cause hundreds of billions of dollars in damage and kill thousands of people if it struck during weekday business hours.

Detailed paleoseismology studies suggest that great earthquakes have a recurrence interval of about 105 years on the southern portion of the San Andreas fault near San Bernardino. A 1,500-year record of earthquake activity is well-preserved in the sedimentary layers along the southern part of the 1857 break and reveals evidence for 14 separate earthquakes. Because the time elapsed since the most recent 1857 earthquake is much longer than the 105-year average between quakes, geologists are concerned that the southern part of the fault may rupture again in a M7.6–7.8 earthquake within the next few decades putting the urban San Bernardino–Riverside area at great risk.

But the northern portion of the San Andreas fault is dangerous, too. Prior to 1906, this section of the fault broke in another giant quake in 1838. These quakes were only 68 years apart, and 1906 plus 68 equals 1974, so the northern section may actually be overdue for a big quake.

Another way of estimating the recurrence interval is by rock displacement. In 1906, rocks were displaced about 4.5 metres at the epicentre, and we know that plate motion across the San Andreas fault is about 5 centimetres per year. It should therefore take about 90 years to store enough strain to move rocks 5 metres, so the next quake should have occurred in 1996.

The probability of a repeat of the 1906 quake (8+) on the locked northern section of the San Andreas fault may be very low, less than 21 percent for the next 30 years. The latest probability studies estimate the chance of a 6.7 or greater quake in the San Francisco Bay area to be 62 percent from 2003–2032. A likely candidate for the quake is not the San Andreas but the Hayward fault across the bay from San Francisco.

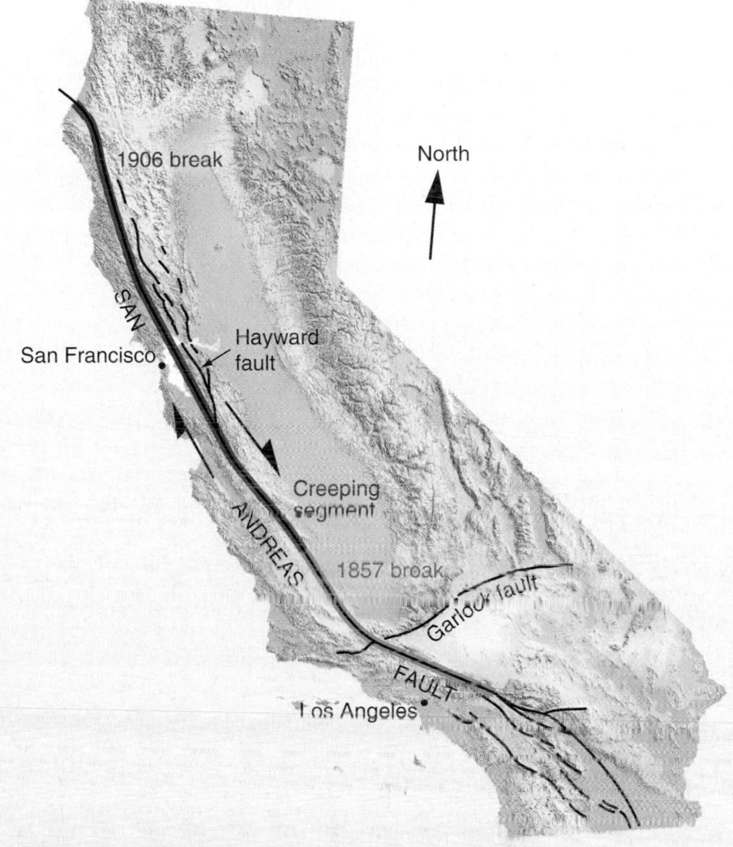

BOX 3.6 ■ FIGURE 1

The two major breaks on the San Andreas fault in California. Each break occurred during a giant earthquake (break from the 1857 earthquake is shown in green and the 1906 earthquake is shown in red). Each old break is now a seismic gap where the fault is locked and may be the future site for another major earthquake. A creeping segment (blue) separates the two locked portions.

From U.S. Geological Survey

The U.S. Geological Survey has estimated that the southern portion of the fault has about a 60 percent chance of an earthquake of magnitude 7.5 to 8.3 within the next 30 years. The 1992 Landers (M7.3), 1994 Northridge (M6.7), and 1999 Hector Mine (M7.1) quakes have occurred since that prediction, and geologists disagree as to whether those quakes increase or decrease the likelihood of the "big one" in southern California. Even more likely (85 percent chance) is a 7+-magnitude quake on any one of several faults that parallel the San Andreas fault and lie closer to (or even under) Los Angeles.

Geoscientists are concerned that a blind thrust fault (fault that cannot be seen at the surface) might rupture close to downtown Los Angeles. To determine the underground configuration of the blind thrust faults and to investigate how deep sedimentary basins are that will amplify shaking in the region, the Los Angeles Regional Seismic Experiment (LARSE) was undertaken to predict where the strongest shaking will occur during future earthquakes. The LARSE project involved setting off underground explosive charges across the Los Angeles Basin northward toward the San

Andreas fault to generate sound waves that could be analyzed by powerful computers to produce images of the subsurface. The experiment revealed a main blind thrust fault 20 kilometres beneath the surface that extends from near the San Andreas fault and transfers stress and strain upward and southward under the San Gabriel Valley and the Los Angeles Basin (box figure 2). The images also show that the sedimentary basin under the San Gabriel Valley is nearly 5 km deep—much deeper than originally thought—which will increase the potential for strong shaking during the next earthquake in this highly populated area.

Although the chance of a magnitude-8+ "big one" along the San Andreas fault may be higher in southern California, the chance of a 7+-quake killing thousands of people and causing extensive damage is about equal in the north and the south.

Additional Resources

U.S. Geological Survey. 1990. *The San Andreas fault system.* Professional Paper 1515.

Special issue on the paleoseismology of the San Andreas fault. 2002. *Bulletin of the Seismological Society of America,* vol. 92, no. 7.

For more information about the San Andreas fault and the likelihood of it creating a large earthquake, visit U.S. Geological Survey websites:
http://pubs.usgs.gov/gip/earthq3/safaultgip.html and
http://geopubs.wr.usgs.gov/fact-sheet/fs152-99/

For more details on the Los Angeles Regional Seismic Experiment, visit:
http://geopubs.wr.usgs.gov/fact-sheet/fs110-99/

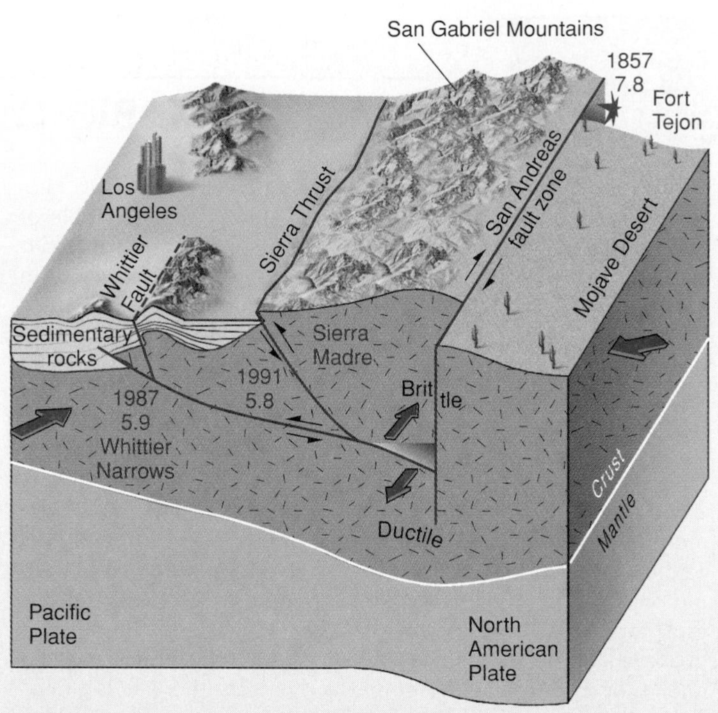

BOX 3.6 ■ FIGURE 2

Diagram drawn from a subsurface image generated by the Los Angeles Regional Seismic Experiment (LARSE) shows an interpretation of the subsurface structures under the San Andreas fault zone westward under the San Gabriel and Los Angeles Basins.

From U.S. Geological Survey

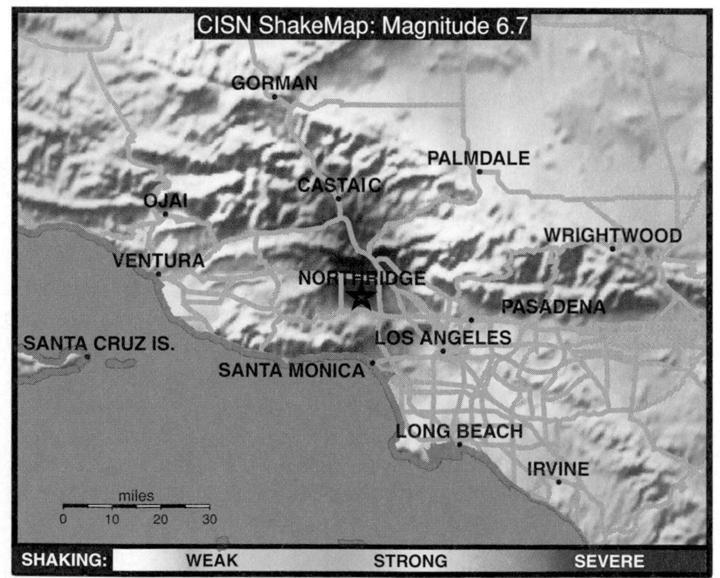

FIGURE 3.27

Map shows the amount of shaking that occurred after the 1994 Northridge earthquake. The ability to create maps within minutes after an earthquake that show the location and severity of maximum ground shaking (ShakeMap) was developed in 1995 by the U.S. Geological Survey. Had this ShakeMap been available minutes after the 1994 Northridge earthquake, emergency personnel could have been immediately directed to the most damaged areas.

Image courtesy David Wald, U.S. Geological Survey

A future goal of the program is to minimize risk by developing an early warning system. With a wide enough distribution of real-time seismometers, it is technically possible for an urban area to get an early warning of an impending earthquake if the earthquake's epicentre is far enough away from the city. For example, if an earthquake occurred 100 kilometres from downtown Los Angeles and its waves are moving at 4 kilometres per second, the system would have 25 seconds to process and analyze the data and broadcast it as an early warning. Even seconds of warning could be enough to shut off main gas pipelines, shut down subway trains, and give schoolchildren time to get under their desks. Japan has successfully used such a system for detecting offshore earthquakes that will shut down the Bullet Train; it is also trying to pursue other ways to use the system to give early warnings to save lives in a major earthquake.

GEOMATICS 3.7

Measuring Ground Displacement Caused by Earthquakes

The spatial analysis of satellite data is beginning to play a significant role in earthquake science as our ability to observe vertical and horizontal deformation on the ground surface improves. InSAR (Interferometric Synthetic Aperture Radar) is a space-borne system that emits electromagnetic radiation (EMR) and records the strength and time delay of the returning signal to produce images of the ground surface. These high-resolution images (with centimetre-scale accuracy) can be compared over time and subtle changes in the ground surface can be identified. InSAR is particularly useful for measuring movement along faults and has been used to compare actual movement caused by earthquakes with that predicted by theoretical models. InSAR combined with detailed GPS surveys is also one of the most accurate methods for determining the location of earthquakes in remote areas. InSAR has been used to create striking images showing fault displacement following a number of recent earthquakes including the magnitude-7.3 Landers earthquake in 1992 and the nearby magnitude-7.1 Hector Mine earthquake in 1999 (box figure 1), as well as the Nenana Mountain earthquake of 2002.

Additional Web Resource

www.gisdevelopment.net/application/natural_hazards/earthquakes/nhmeq0012a.htm.

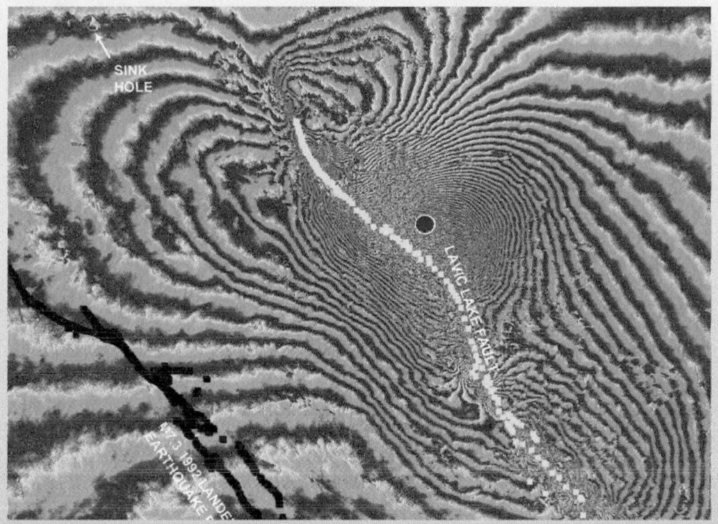

BOX 3.7 ■ FIGURE 1

Interferogram of 1999 Hector Mine, magnitude-7.1 earthquake. Each fringe corresponds to 28 mm of ground displacement away from the radar. The earthquake ruptured 45 km of faults in the Mojave Desert.

Near Real-Time Synthetic Aperture Radar Interferometry of Hector Mine 7.1 M Earthquake http://topex.ucsd.edu/hector/David Sandwell, L. Sichoix, A. Jacobs, R. Scharroo, B. Minster, Y. Bock, P. Jamason, E. Price, and H. Zebker

SUMMARY

Earthquakes usually occur when rocks break and move along a fault to release strain that has gradually built up in the rock. Volcanic activity can also cause earthquakes. Deep quakes may be caused by mineral transformations.

Seismic waves move out from the earthquake's *focus. Body waves* (P waves and S waves) move through Earth's interior, and *surface waves (Love* and *Rayleigh waves)* move on Earth's surface.

Seismographs record seismic waves on *seismograms,* which can be used to determine an earthquake's strength, location, and depth of focus. Most earthquakes are shallow-focus quakes, but some occur as deep as 670 kilometres below Earth's surface.

The time interval between first arrivals of P and S waves is used to determine the distance between the seismograph and the *epicentre.* Three or more stations are needed to determine the location of earthquakes.

Earthquake *intensity* is determined by assessing damage and is measured on the *modified Mercalli scale.*

Earthquake *magnitude,* determined by the amplitude of seismic waves on a seismogram, is measured on the *Richter scale. Moment magnitudes,* determined by field work, are widely used today and often are larger than Richter magnitudes.

The most noticeable effects of earthquakes are ground motion and displacement (which destroy buildings and thereby injure or kill people), fire, landslides, and *tsunamis. Aftershocks* can continue to cause damage months after the main shock.

Earthquakes are generally distributed in belts. The *circum-Pacific belt* contains most of the world's earthquakes. Earthquakes also occur on the Mediterranean-Himalayan belt, the crest of the mid-oceanic ridge, and in association with basaltic volcanoes.

Benioff zones of shallow-, intermediate-, and deep-focus earthquakes are associated with andesitic volcanoes, oceanic trenches, and the edges of continents or island arcs.

The concept of plate tectonics explains most earthquakes as being caused by interactions between two plates at their boundaries. Plate boundaries are generally defined by bands of earthquakes.

Divergent plate boundaries are marked by a narrow zone of shallow earthquakes along normal faults, usually in a rift valley. Transform boundaries are marked by shallow quakes caused by strike-slip motion along one or more faults.

Convergent boundaries where continents collide are marked by a very broad zone of shallow quakes. Convergent boundaries involving deep subduction are marked by Benioff zones of quakes caused by tension, underthrusting, and compression.

The distribution of quakes indicates subduction angles of a down-going plate. The subduction angle is probably controlled by plate density and rate of plate convergence.

Determining the probability of an earthquake occurring uses the measurement of rock properties near faults, slip rate studies, and paleoseismology investigations to determine the recurrence interval of quakes along individual faults.

Terms to Remember

aftershock 84
Benioff zone 88
body wave 72
circum-Pacific belt 88
depth of focus 76
earthquake 68
elastic rebound theory 68
epicentre 72
focus 72
intensity 76
island arc 88
Love wave 72
magnitude 76
Mediterranean-Himalayan belt 88
modified Mercalli scale 76
moment magnitude 78
P wave 72
Rayleigh wave 73
Richter scale 76
seismic sea wave 85
seismic wave 68
seismogram 74
seismograph 74
surface wave 72
S wave 72
travel-time curve 75
tsunami (seismic sea wave) 85

Testing Your Knowledge

Use the questions below to prepare for exams based on this chapter.

1. Describe in detail how earthquake epicentres are located by seismograph stations.
2. What causes earthquakes?
3. Compare and contrast the concepts of intensity and magnitude of earthquakes.
4. Name and describe the various types of seismic waves.
5. Discuss the distribution of earthquakes with regard to location and depth of focus.
6. Show with a sketch how the concept of plate tectonics can explain the distribution of earthquakes in a Benioff zone and on the crest of the mid-oceanic ridge.
7. Describe several techniques that may help scientists predict earthquakes.
8. How may the timing of earthquakes someday be controlled?
9. Describe several ways that earthquakes cause damage.
10. How do earthquakes cause tsunami?
11. What are aftershocks?
12. The elastic rebound theory
    a. explains folding of rocks
    b. explains the behaviour of seismic waves
    c. involves the sudden release of progressively stored strain in rocks, causing movement along a fault
    d. none of the preceding
13. The point within Earth where seismic waves originate is called the
    a. focus b. epicentre
    c. fault scarp d. fold
14. P waves are
    a. compressional b. transverse
    c. tensional
15. What is the minimum number of seismic stations needed to determine the location of the epicentre of an earthquake?
    a. 1 b. 2 e. 10
    c. 3 d. 5
16. The Richter scale measures
    a. intensity
    b. magnitude
    c. damage and destruction caused by the earthquake
    d. the number of people killed by the earthquake
17. Benioff zones are found near
    a. mid-ocean ridges b. ancient mountain chains
    c. interiors of continents d. oceanic trenches
18. Most earthquakes at divergent plate boundaries are
    a. shallow focus b. intermediate focus
    c. deep focus d. all of the preceding
19. Most earthquakes at convergent plate boundaries are
    a. shallow focus b. intermediate focus
    c. deep focus d. all of the preceding
20. A zone of shallow earthquakes along normal faults is typical of
    a. diverging boundaries b. transform boundaries
    c. subduction zones d. collision boundaries

21. A seismic gap is
    a. the time between large earthquakes
    b. a segment of an active fault where earthquakes have not occurred for a long time
    c. the centre of a plate where earthquakes rarely happen

22. Which of the following is not true of tsunamis?
    a. very long wave length
    b. high wave height in deep water
    c. very fast moving
    d. continued flooding after wave crest hits shore

Exploring Web Resources

www.mcgrawhill.ca/college/plummer
Visit our Online Learning Centre for additional readings and media resources. Check your answers for Testing Your Knowledge and click on the links to the other great Websites listed below.

http://quake.wr.usgs.gov/prepare/future
U.S. Geological Survey, 1990. *The Next Big Quake.*

http://pubs.usgs.gov/gip/earthq3/safaultgip.html
U.S. Geological Survey, 1990. *The San Andreas Fault.* Professional Paper 1515. Online version.

www.geophys.washington.edu/seismosurfing.html
Exhaustive list of worldwide Internet sites for information about earthquakes.

http://quake.wr.usgs.gov/
U.S. Geological Survey Earthquake Information. Gives information on reducing earthquake hazards, earthquake preparedness, latest quake information, historical earthquakes, and how earthquakes are studied. Also a good starting place for links to other earthquake sites.

http://quake.wr.usgs.gov/recenteqs/faq.html
Frequently Asked Questions about recent earthquakes maintained by the U.S. Geological Survey.

http://earthquakescanada.nrcan.gc.ca
Gives information on recent earthquakes, earthquake hazards, and earthquake research in Canada.

www.seismo.unr.edu/
University of Nevada, Reno Seismological Laboratory site contains information about recent earthquakes, earthquake preparedness, and links to other earthquake sites.

www.seismo.berkeley.edu/seismo/Homepage.html
Seismographic information page maintained by U. C.–Berkeley that has many links to other earthquake sites (particularly in California), 3-D earthquake movie, Northridge earthquake rupture movies, and information on earthquake preparedness.

http://vquake.calstatela.edu/
California State University, Los Angeles *Virtual Earthquake.* Create and analyze an earthquake.

http://pubs.usgs.gov/gip/earthq4/severitygip.html
General information about the size of an earthquake. Discussion of Richter and Mercalli scales.

http://pubs.usgs.gov/publications/text/dynamic.html
General information about plate tectonics.

http://geopubs.wr.usgs.gov/circular/c1187/
U.S. Geological online version of Tsunami Circular.

http://walrus.wr.usgs.gov/tsunami/PNGhome.html
U.S. Geological Survey Web page gives information about the devastating July 17, 1998 tsunami at Papua, New Guinea and links to other sites.

www.iris.edu
Seismic monitor views global earthquakes in real time.

http://pasadena.wr.usgs.gov/step
Real time forecast of earthquake hazards.

Animations

This chapter includes the following animations available on our Online Learning Centre at www.mcgrawhill.ca/olc/plummer.

3.4 Focus of an Earthquake
3.5 Particle Motion in Seismic Waves
3.6 Seismograph for Detecting Vertical Rock Motion
3.7 Seismograph for Detecting Horizontal Rock Motion
3.8, 3.9, 3.10 Locating Earthquake Epicentre

CHAPTER

4

The Earth's Interior

What can we learn from the study of seismic waves?
What is inside the Earth?
How does the elevation of continents change?
What can gravity tell us about the Earth's crust?
How does the Earth's magnetic field change through time?
How hot is the Earth's core? What is the origin of the Earth's heat?

The only rocks that geoscientists can study directly in place are those of the Earth's crust; the Earth's crust is but a thin skin of rock, making up less than 1 percent of the Earth's total volume. Mantle rocks brought to the Earth's surface in basalt flows, in diamond-bearing kimberlite pipes, and also by the tectonic attachment of lower parts of the oceanic lithosphere to the continental crust give geoscientists a glimpse of what the underlying mantle might look like. Meteorites also give clues about the possible composition of the core of the Earth. But, to learn more about the deep interior of the Earth, geoscientists must study it indirectly, largely by using the tools of geophysics—that is, seismic waves and the measurement of gravity, heat flow, and Earth magnetism.

The evidence from geophysics suggests that the Earth is divided into three major layers—the crust on the surface, the rocky mantle beneath the crust, and the metallic core at the centre. The study of plate tectonics has shown that the crust and uppermost mantle can be conveniently divided into the brittle lithosphere and the plastic asthenosphere.

Diamonds form in the mantle and are brought to the surface in kimberlite pipes, giving geologists a glimpse of the Earth's interior. *Photo © Reuters NewMedia Inc./Corbis*

You will learn in this chapter how gravity measurements can indicate where certain regions of the crust and upper mantle are being held up or held down out of their natural position of equilibrium. We will also discuss Earth's magnetic field and its history of reversals. We will show how magnetic anomalies can indicate hidden ore and geologic structures. We close with a discussion of the distribution and loss of Earth's heat.

What *do* geoscientists know about the Earth's interior? How do they obtain information about the parts of the Earth beneath the surface? Geoscientists, in fact, are not able to sample rocks very far below the surface. Some deep mines penetrate 3 km into the Earth, and a deep oil well may go as far as 8 km beneath the surface; the deepest scientific well has reached 12 km in Russia (see box 4.1). Rock samples can be brought up from a mine or a well for geoscientists to study.

IN GREATER DEPTH 4.1

Deep Drilling on Continents

The structure and composition of most of the continental crust is unknown. Surface mapping and seismic reflection and refraction suggest that the continents are largely igneous and metamorphic rock, such as granite and gneiss, overlain by a veneer of sedimentary rocks. This sedimentary cover is generally thin, like icing on a cake, but it may thicken to 10 km or more in giant sedimentary basins where the underlying "basement rock" has subsided. Although oil companies have drilled as deep as 8 km on land, they drill in the sedimentary basins. The igneous and metamorphic basement, which averages 40 km thick and makes up most of the continental crust, has rarely been sampled deeper than 2 or 3 km (although uplift and erosion have exposed some rocks widely thought to have been formed much deeper in the crust).

Russia has drilled the world's deepest hole, on the Kola Peninsula near Murmansk north of the Arctic Circle. The 12-km-deep hole took 15 years to drill and penetrated ancient Precambrian basement rocks. The second deepest well drilled is the KTB hole in southeastern Germany, which reached a depth of 10 km and cost more than a billion dollars (box figure 1). Deep drilling is as technically complex as space exploration. High pressures and 300°C temperatures require special equipment and techniques. The Russians used a turbodrill that rotated under the pressure of circulating drilling mud. Unlike normal drilling operations, the lightweight aluminum drill pipe does not turn. Because the Kola drilling operation resulted in a crooked hole, the Germans advanced deep-drilling technology by developing a system to keep the hole straight while being drilled.

The drilling at Kola shows that seismic models for this area are wrong. The Russians expected 4.7 km of metamorphosed sedimentary and volcanic rock, then a granitic layer to a depth of 7 km, and a "basaltic" layer below that. The granite, however, appeared at 6.8 km and extends to more than 12 km; the "basalt" has not yet been found. These results, and data from the other deep holes, show that seismic surveys of continental crust are being systematically misinterpreted.

The Russians and Germans unexpectedly found open fractures and circulating fluids throughout the borehole. The fluids include hydrogen, helium, and methane (natural gas), as well as mineralized waters forming ore bodies. Copper–nickel ore was found deeper than theory predicted, and gold mineralization was present from 9.5 to 11 km down. These results will change geoscientists' models of ore formation and fluid circulation underground.

BOX 4.1 ■ FIGURE 1

The KTB drilling operation in southeastern Germany reached a depth of 10 km and has advanced the technology of deep drilling.

Photo courtesy of ICDP, GeoForschungsZentrum Potsdam

Additional Resources

Kerr, R. A. 1993, Looking—deeply—into the Earth's crust in Europe. *Science* 261:295.

Kozlovsky Y. A. 1987, The Superdeep well of the Kola Peninsula, Springer-Verlag, 558 p.

Scientific Information System for the world's deepest borehole, Kola SDB-3.

IGCP408: Rocks and minerals at great depth and on the surface.

http://icdp.gfz-potsdam.de/html/kola/IGCP408.html

A direct look at rocks from deeper levels can be had where mantle rocks have been brought up to the surface by basalt flows, by the intrusion and erosion of diamond-bearing kimberlite pipes (see box 4.4 later in this chapter), or where the lower part of the oceanic lithosphere (see chapter 2) has been tectonically attached to the continental crust at a convergent plate boundary. However, Earth has a radius of about 6,370 km, so it is obvious that geoscientists can only scratch the surface when they try to study *directly* the rocks beneath their feet.

Deep parts of the Earth are studied *indirectly,* however, largely through the branch of geology called **geophysics,** which is the application of physical laws and principles to a study of the Earth. Geophysics includes the study of seismic waves and Earth's magnetic field, gravity, and heat. All these things tell us something about the nature of the deeper parts of the Earth. Together they create a convincing picture of what makes up the Earth's interior.

WHAT CAN WE LEARN FROM THE STUDY OF SEISMIC WAVES?

Seismic waves from a large earthquake may pass through the entire Earth. A nuclear bomb explosion also generates seismic waves. Geoscientists obtain new information about the Earth's interior after every large earthquake and bomb test.

One important way of learning about the Earth's interior is the study of **seismic reflection,** the return of some of the energy of seismic waves to the Earth's surface after the waves bounce off a rock boundary. If two rock layers of differing densities are separated by a fairly sharp boundary, seismic waves reflect off that boundary just as light reflects off a mirror (figure 4.1). These reflected waves are recorded on a seismogram, which shows the amount of time the waves took to travel down to the boundary, reflect off it, and return to the surface. From the amount of time necessary for the round trip, geoscientists calculate the depth of the boundary. The Canadian Lithoprobe Project has been applying seismic reflection techniques to map crustal structures at the base of the crust, up to 50 kilometres below the ground surface (see box 4.2).

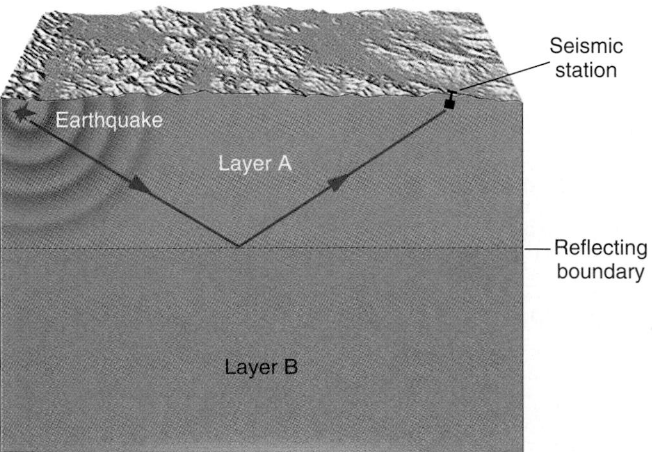

FIGURE 4.1
Seismic reflection. Seismic waves reflect from a rock boundary deep within the Earth and return to a seismograph station on the surface.

Another method used to locate rock boundaries is the study of **seismic refraction,** the bending of seismic waves as they pass from one material to another, which is similar to the way that light waves bend when they pass through the lenses of eyeglasses. As a seismic wave strikes a rock boundary, much of the energy of the wave passes across the boundary. As the wave crosses from one rock layer to another, it changes direction (figure 4.2). This change of direction, or refraction, occurs only if the velocity of seismic waves is different in each layer (which is generally true if the rock layers differ in density or strength).

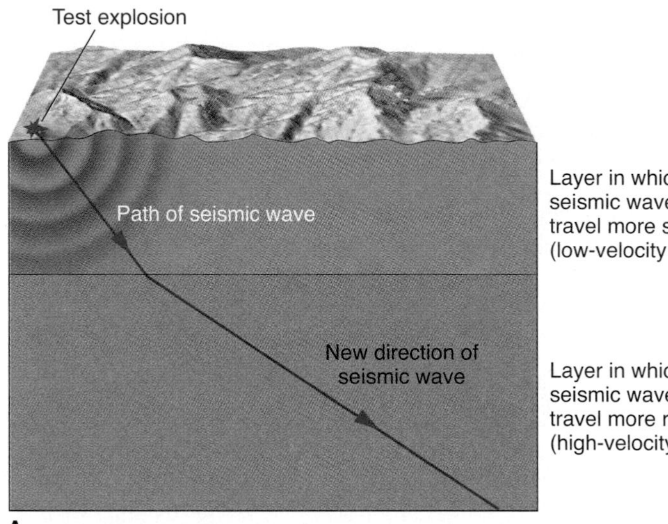

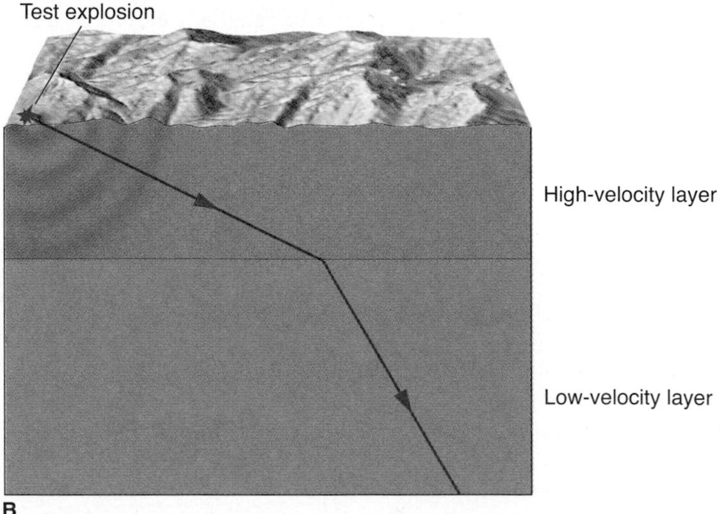

FIGURE 4.2
Seismic refraction occurs when seismic waves bend as they cross rock boundaries. At an interface, seismic (or sound or light) waves will bend toward the lower-velocity material. (*A*) Low-velocity layer above high-velocity layer. (*B*) High-velocity layer above low-velocity layer. Some of the seismic waves will also return to the surface by reflecting off the rock boundary.

IN GREATER DEPTH 4.2

Canadian Lithoprobe Project

The Lithoprobe Project is a groundbreaking Canadian scientific effort that has been investigating the composition and structure of the Canadian Shield and surrounding orogenic belts since 1984. This enormous undertaking, involving no less than 900 scientists, has provided an exceptional opportunity to pioneer new perspectives on understanding continental evolution. No other nation has conducted such a focused study of the evolution of a continent.

The aim of the Lithoprobe Project is to develop a comprehensive understanding of the geological evolution of the North American continent, specifically the Canadian Shield. The shield is formed of distinct geological terranes that were once separate land masses but were brought together by the forces of plate tectonics (see chapter 2). By gaining information about the characteristics of major terranes and their boundaries within the Canadian Shield, scientists will be able to answer fundamental questions on how current continental configuration came to be and what tectonic processes were involved. This information will also help to evaluate earthquake risk across the shield and will identify geological structures that may be associated with oil and gas reservoirs or mineral plays.

One of the techniques used by the Lithoprobe Project to collect data about the Earth's crust is *seismic reflection*. This involves sending sound waves into the ground and recording them when they bounce back. In order for sound waves to pass deep into the crust, the Lithoprobe Project has used a series of large *vibroseis* trucks to generate sufficiently large vibrations (box figure 1). Four or five high-tech vibrating trucks, which have been termed "dancing elephants," work together to stamp in unison along a road bed or road shoulder (box figure 2). The energy waves they create pass through the Earth and *reflect* or *refract* when they encounter a boundary between rocks with different physical properties. Such boundaries may be created by changes in rock types, buried fault systems, or intrusive bodies such as plutons. Geophones (cup-sized motion sensors) laid out on the ground surface detect the reflected sound waves, and the data are recorded on computers. As many as 10,000 geophones may be connected over a distance of 20 km on the ground surface and allow geologic features to be identified at depths of up to 50 km (essentially the base of the crust). The dancing elephants move relatively slowly and can cover a distance of only approximately 10 km a day when vibrating every 100 m.

Seismic reflection data received by the geophones is computer analyzed and two-dimensional images of the subsurface geology are created (box figure 3). These images can be analyzed further by adding data regarding characteristics of the gravitational and magnetic fields and changes in electrical conductivity with depth and isotopic ages of geologic units. Since 1984, the Lithoprobe Project has collected more than 10,000 km worth of seismic reflection data from the Canadian Shield, which are now being used to create multidimensional maps of the Earth's crust (box figure 3). These maps will be used to unravel the puzzle of how the variety of ancient continents, oceans, and islands assembled together to form the core of the North American continent.

For more information: www.lithoprobe.ca

BOX 4.2 ■ FIGURE 1
Vibroseis trucks ("dancing elephants") on the Stewart-Cassiar Highway, B.C.
Photo by Philip Hammer, Lithoprobe

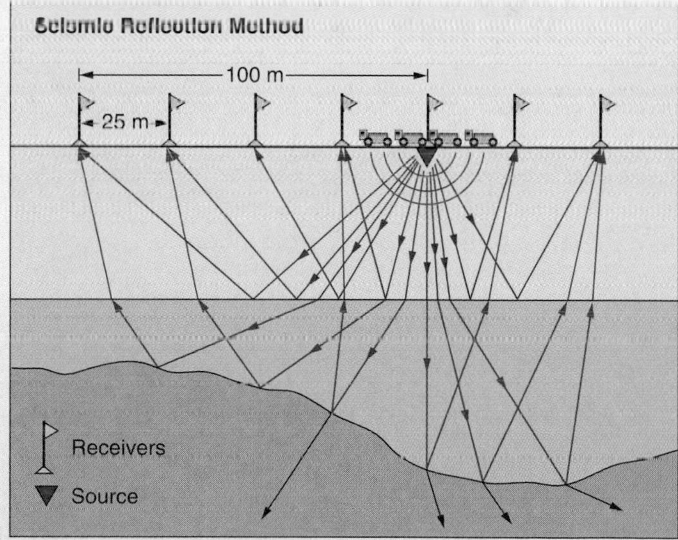

BOX 4.2 ■ FIGURE 2
Seismic energy waves generated by four vibroseis trucks propagate through the Earth and reflect or refract when reaching a boundary between rocks of different physical properties. The reflected energy waves are detected by a series of geophones on the surface (marked by yellow flags) and recorded on a computer.
Modified from Calvert et al., Nature, 1995. Courtesy of Lithoprobe www.lithoprobe.ca

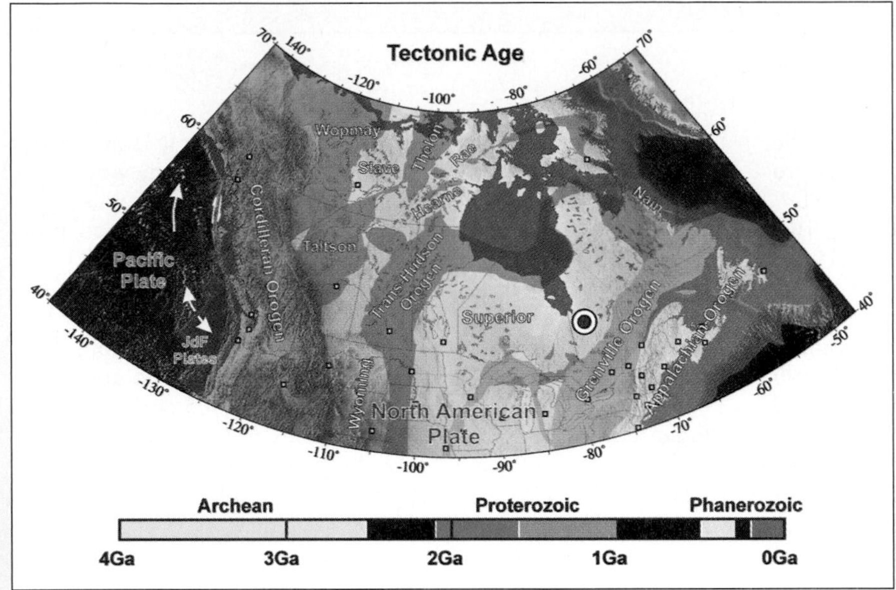

BOX 4.2 ■ FIGURE 3A

Map showing geologic provinces of Canada. The continent consists of ancient cratons (green-blue) joined together along suture zones (black/white patterns) by plate collisions during the Proterozoic. Pink areas have been added by more recent collisions. The position of seismic reflection line AG48 crossing part of the Archean Superior Province is shown. The Superior Province itself was formed by collision and amalgamation of a number of "subprovinces" (see chapter 20), and line AG48 was designed to study the transition between the Opatica plutonic belt and Abitibi greenstone belt.

Modified from Calvertet al., Nature, *1995; www.lithoprobe.ca*

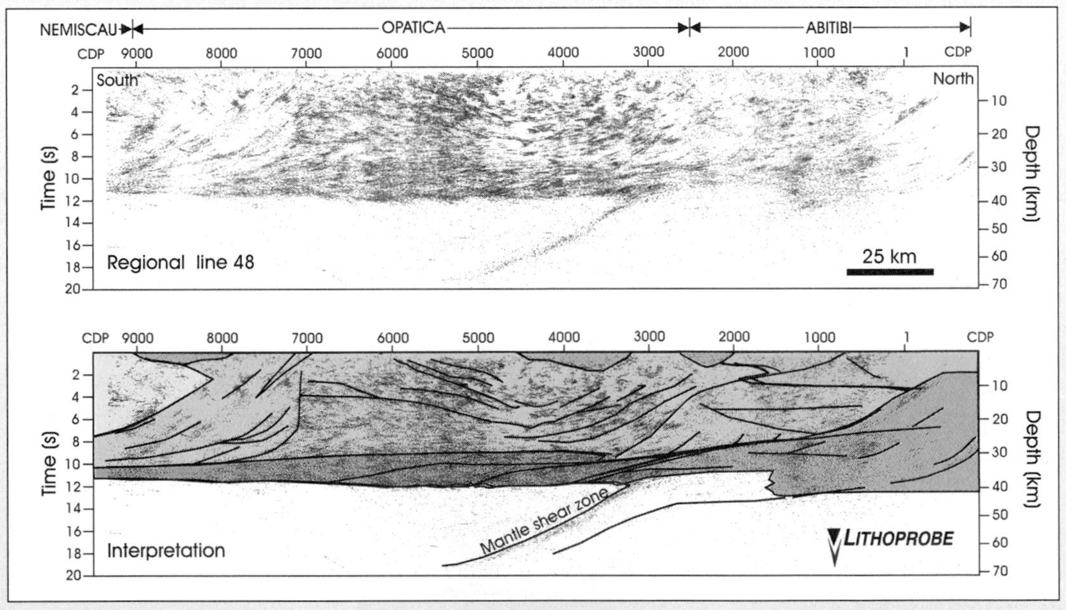

BOX 4.2 ■ FIGURE 3B

Top: Seismic reflection data from line AG48.

Bottom: Interpretation of crustal structure identified from seismic data (colours represent different geologic units). This section clearly shows a steeply dipping structure deep below the ground surface, which is interpreted as the remains of a subducted oceanic plate. The seismic line is interpreted to represent an ancient subduction zone along which two continental masses collided around 2.7 billion years ago and is the oldest record of plate tectonics yet recognized on Earth.

Modified from Calvert et al., Nature, *1995; www.lithoprobe.ca*

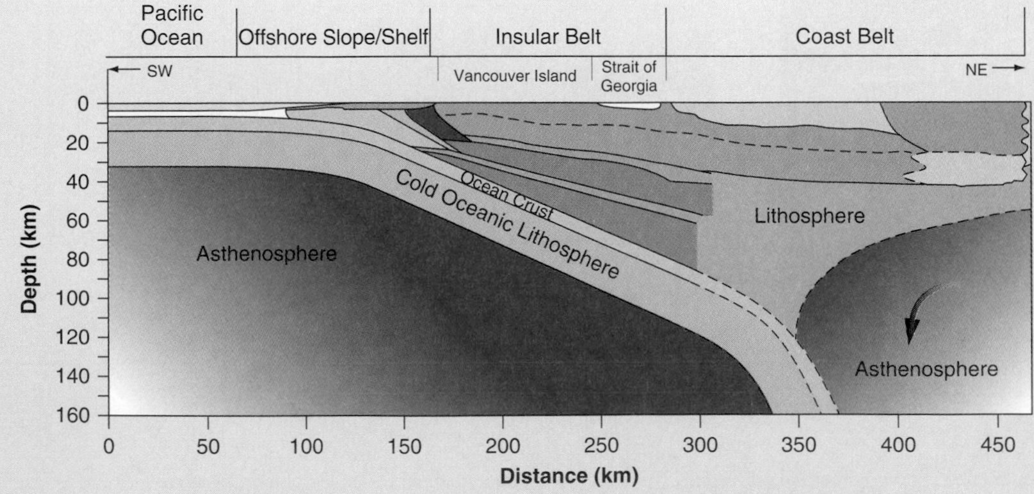

BOX 4.2 ■ FIGURE 3C

Schematic cross-section through the modern subduction zone beneath southwest British Columbia. The crustal structure identified in line AG48 looks remarkably similar to this.

Modified from Calvert et al., Nature, 1995. Courtesy of Lithoprobe www.lithoprobe.ca

The boundaries between such rock layers are usually distinct enough to be located by seismic refraction techniques, as shown in figure 4.3. Seismograph station 1 is receiving seismic waves that pass directly through the upper layer *A*. Stations farther from the epicentre, such as station 2, receive seismic waves from two pathways: (1) a direct path straight through layer *A* and (2) a refracted path through layer *A* to a higher-velocity layer *B* and back to layer *A*. Station 2 therefore receives the same wave twice.

Seismograph stations close to station 1 receive only the direct wave or possibly two waves, the direct (upper) wave arriving before the refracted (lower) wave. Stations near station 2 receive both the direct and the refracted waves. At some point between station 1 and station 2 there is a transformation from receiving the direct wave first to receiving the *refracted* wave first. Even though the refracted wave travels farther, it can arrive at a station first because most of its path is in the high-velocity layer *B*.

The distance between this point of transformation and the epicentre of the earthquake is a function of the depth to the rock boundary between layers *A* and *B*. A series of portable seismographs can be set up in a line away from an explosion (a *seismic shot*) to find this distance, and the depth to the boundary can then be calculated. The velocities of seismic waves within the layers can also be found.

Figure 4.2 shows how waves bend as they travel downward into higher-velocity layers. But why do waves return to the surface, as shown in figure 4.3? The answer is that advancing waves give off energy in all directions. Much of this energy continues to travel horizontally within layer *B* (figure 4.3). This energy passes beneath station 2 and out of the figure toward the right. A small part of the energy "leaks" upward into layer *A*, and it is this pathway that is shown in the figure. There are many other pathways for this wave's energy that are not shown here.

A sharp rock boundary is not necessary for the refraction of seismic waves. Even in a thick layer of uniform rock, the increasing pressure with depth tends to increase the velocity of the waves. The waves follow curved paths through such a layer, as shown in figure 4.4. To understand the reason for the curving path, visualize the thick rock layer as a stack of very thin layers, each with a slightly higher velocity than the one above. The curved path results from many small changes in direction as the wave passes through the many layers.

FIGURE 4.3

Seismic refraction can be used to detect boundaries between rock layers. See text for explanation.

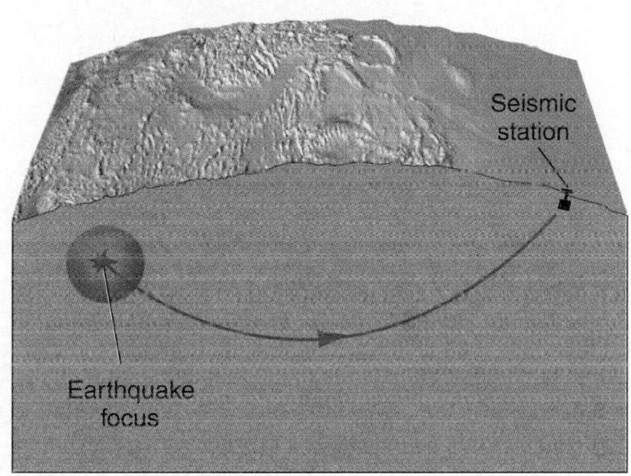

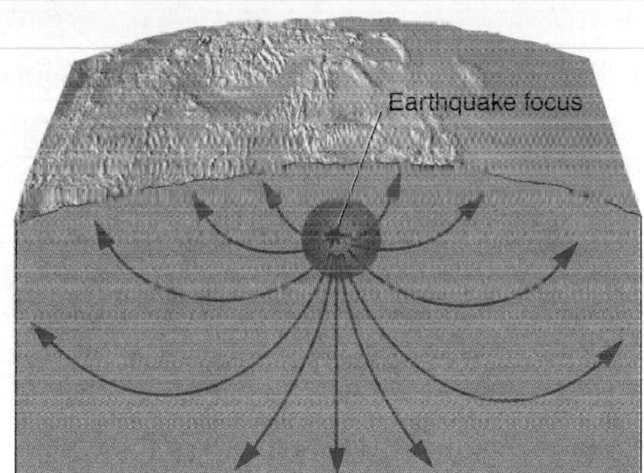

FIGURE 4.4

Curved paths of seismic waves caused by uniform rock with increasing seismic velocity with depth. (*A*) Path between earthquake and recording station. (*B*) Waves spreading out in all directions from earthquake focus.

WHAT IS INSIDE THE EARTH?

It was the study of seismic refraction and seismic reflection that enabled scientists to plot the three main zones of the Earth's interior (figure 4.5). The **crust** is the outer layer of rock, which forms a thin skin on Earth's surface. Below the crust lies the **mantle,** a thick shell of rock that separates the crust above from the core below. The **core** is the central zone of Earth. It is probably metallic and the source of Earth's magnetic field.

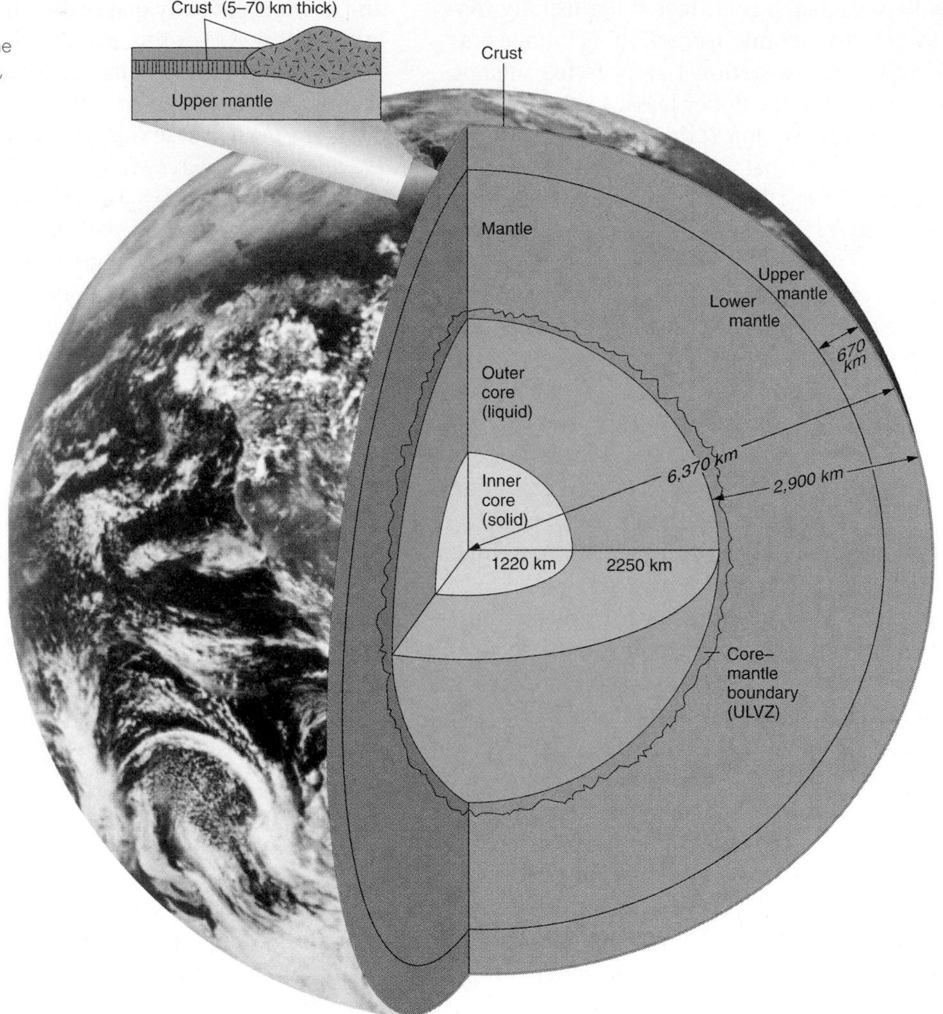

FIGURE 4.5
Earth's interior. Seismic waves show the three main divisions of Earth: the crust, the mantle, and the core.
Photo by NASA

The Crust

Studies of seismic waves have shown (1) that the crust is thinner beneath the oceans than beneath the continents (figure 4.6) and (2) that seismic waves travel faster in oceanic crust than in continental crust. Because of this velocity difference, it is assumed that the two types of crust are made up of different kinds of rock.

Seismic P waves travel through oceanic crust at about 7 km per second, which is also the speed at which they travel through basalt and gabbro (the coarse-grained equivalent of basalt). Samples of rocks taken from the sea floor by oceanographic ships verify that the upper part of the oceanic crust is basalt and suggest that the lower part is gabbro. The oceanic crust averages 7 km in thickness, varying from 5 to 8 km (table 4.1).

Seismic P waves travel more slowly through continental crust—about 6 km per second, the same speed at which they travel through granite and gneiss. Continental crust is often called "granitic," but the term should be put in quotation marks because most of the rocks exposed on land are not granite. The continental crust is highly variable and complex, consisting of a crystalline basement composed of granite, other plutonic rocks, gneisses, and schists, all capped by a layer of sedimentary rocks, like icing on a cake. Since a single rock term cannot accurately describe crust that varies so greatly in composition, some geoscientists use the term *felsic*—rocks high in *feldspar* and *silicon*—for continental crust and *mafic*—rocks high in magnesium and iron (ferric)—for oceanic crust.

Continental crust is much thicker than oceanic crust, averaging 30 to 50 km in thickness, though it varies from 10 to 70 km. Seismic waves show that the crust is thickest under geologically young mountain ranges, such as the Andes and the Himalayas, bulging downward as a *mountain root* into the mantle (figure 4.6). The continental crust is also less dense than oceanic crust, a fact that is important in plate tectonics (table 4.1).

The boundary that separates the crust from the mantle beneath it is called the **Mohorovičić discontinuity** (**Moho** for short). Note from figure 4.6 that the mantle lies closer to the Earth's surface beneath the ocean than it does beneath continents. The idea behind an ambitious program called Project Mohole (begun during the early 1960s) was to use specially equipped ships to drill through the oceanic crust and obtain samples from the mantle. Although the project was abandoned because of high costs, ocean-floor drilling has become routine since then, but not to the great depth necessary to sample the mantle. Perhaps in the future the original concept of drilling to the mantle through oceanic crust will be revived.

TABLE 4.1 Characteristics of Oceanic Crust and Continental Crust

	Oceanic Crust	Continental Crust
Average thickness	7 km	20 to 70 km (thickest under mountains)
Seismic P-wave velocity	7 km/second	6 km/second (higher in lower crust)
Density	3.0 gm/cm^3	2.7 gm/cm^3
Probable composition	Basalt underlain by gabbro	Granite, other plutonic rocks, schist, gneiss (with sedimentary rock cover)

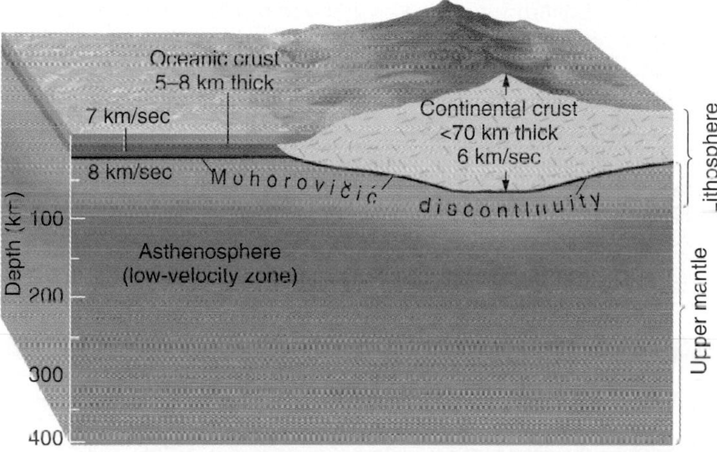

FIGURE 4.6

Thin oceanic crust has a P-wave velocity of 7 km per second, whereas thick continental crust has a lower velocity. Mantle velocities are about 8 km per second. The oceanic and continental crust along with the upper rigid part of the upper mantle form the lithosphere. The asthenosphere underlies the lithosphere and is defined by a decrease in P-wave velocities.

The Mantle

Because of the way seismic waves pass through the mantle, geoscientists interpret it to be made of solid rock. Localized magma chambers of melted rock may occur as isolated pockets of liquid in both the crust and the upper mantle, but most of the mantle seems to be solid. Because P waves travel at about 8 km per second in the upper mantle, it appears that the mantle is a different type of rock from either oceanic crust or continental crust. The best hypothesis that geoscientists can make about the composition of the upper mantle is that it consists of ultramafic rock such as peridotite. *Ultramafic rock* is dense igneous rock made up chiefly of ferromagnesian minerals such as olivine and pyroxene (see chapter 5). Some ultramafic rocks contain garnet, and all of them lack feldspar.

The crust and uppermost mantle together form the **lithosphere,** the outer shell of Earth that is relatively strong and brittle. The lithosphere makes up the plates of plate tectonics theory. The lithosphere averages about 70 km thick beneath oceans and may be 125 to 250 km thick beneath continents. Its lower boundary is marked by a curious mantle layer in which seismic waves slow down (figure 4.6).

Generally, seismic waves increase in velocity with depth as increasing pressure alters the properties of the rock. Beginning at a depth of 70 to 125 km, however, seismic waves travel more slowly than they do in shallower layers, and so this zone has been called the *low-velocity zone* (figure 4.6). This zone, extending to a depth of perhaps 200 km, is called the **asthenosphere.** The rocks in this zone may be closer to their melting point than the rocks above or below the zone. (The rocks are probably not *hotter* than the rocks below—melting points are controlled by pressure as well as temperature.) Some geoscientists think that these rocks may actually be partially melted, forming a crystal-and-liquid slush; a very small percentage of liquid in the asthenosphere could help explain some of its physical properties.

If the rocks of the asthenosphere are close to their melting point, this zone may be important for two reasons: (1) it may represent a zone where magma is likely to be generated; and (2) the rocks here may have relatively little strength and therefore are likely to flow. If mantle rocks in the asthenosphere are weaker than they are in the overlying lithosphere, then the asthenosphere can deform easily by plastic flow. Plates of brittle lithosphere probably move easily over the asthenosphere, which may act as a lubricating layer below.

There is widespread agreement on the existence and depth of the asthenosphere under oceanic crust, but considerable disagreement about asthenosphere under continental crust. Figure 4.6 shows asthenosphere at a depth of 125 km below the continents. Some geoscientists think that the lithosphere is much thicker beneath continents than shown in the figure, and that the asthenosphere begins at a depth of 250 km (or even more). A few geoscientists say that there is *no* asthenosphere beneath continents at all. The reasons for this disagreement are the results of the rapidly developing field of seismic tomography, which is described in box 4.3.

Data from seismic reflection and refraction indicate several concentric layers in the mantle (figure 4.7), with prominent boundaries at 400 and 670 km (670 km is also the depth of the deepest earthquakes). It is doubtful that the layering is due to the presence of several different kinds of rock. Most geoscientists think that the chemical composition of the mantle rock is about the same throughout the mantle. Because pressure increases with depth into the Earth, the boundaries between mantle layers possibly represent depths at which pressure collapses the internal structure of certain minerals into denser minerals. For example, at a pressure equivalent to a depth of about 670 km, the mineral *olivine* should collapse into the denser structure of the mineral *perovskite.* If the boundaries between mantle layers represent pressure-caused transformations of minerals, the entire mantle may have the same *chemical* composition throughout, although not the same *mineral* composition (see box 4.4). However, some geoscientists think that the 670-km boundary represents a chemical change as well as a physical change and separates the *upper mantle* from the chemically different *lower mantle* below.

IN GREATER DEPTH 4.3

A CAT Scan of the Mantle

A new technique for looking at the mantle is similar to the medical technique of CAT scanning (CAT stands for computed axial tomography), which builds up a three-dimensional picture of soft body tissues such as the brain by taking a series of X-ray pictures along successive planes in the body.

Seismic tomography uses earthquake waves and powerful computers to study planar cross-sections of the mantle following large earthquakes. Slight variations from expected arrival times at distant seismograph stations can be used to find temperature variations in the mantle. Hot rock slows down seismic waves, so a late arrival of a seismic wave shows that the wave went through hot rock. Cold rock is dense and strong, so it speeds up seismic waves, resulting in early arrivals. Sophisticated computer analysis of hundreds of sections through the mantle allows maps of seismic-wave velocity (and therefore mantle rock temperature) to be drawn for various depths.

Box figure 1, top shows mantle velocities at a depth of 100 km. Red areas show low velocities (probably caused by hot rock) in generally expected positions—along the crest of the mid-oceanic ridge and beneath hot spots. Blue areas show high-velocity (probably cold) rock under continents and old sea floor such as the western Pacific. Box figure 1, bottom shows that these patterns are dramatically different at a depth of 300 km. High-velocity rock extends to this depth below most continents, implying that continents have very deep roots. Some areas that appear hot at 100 km are cold at 300 km, such as the ridge crest just south of Australia. Areas such as the central Pacific and the Red Sea region appear cold at 100 km and hot at 300 km.

In box figure 2, vertical cross-sections of seismic velocity are shown to a depth of 670 km for two regions. Note that high-velocity (cold) roots beneath North America, Asia, and Antarctica extend 400 to 600 km downward. This finding casts doubt on our simple lithosphere–asthenosphere model of plate behaviour—continental plates here seem to be hundreds of kilometres thick. Notice, too, how some low-velocity hot spots near Greenland (box figure 2, top) and in the south Atlantic and south Pacific (box figure 2, bottom) are underlain by apparently cold rock. This pattern suggests to some geologists that mantle plumes may be quite shallow and may not extend vertically throughout the mantle. On the other hand, plume tails may be too narrow to be detected by this technique.

More recent, deeper CAT scans of the mantle (box figure 3) indicate that some mantle plumes emanate from the core–mantle boundary and are fed by heat loss from the core. The plume under the Hawaiian hot spot was recently found to contain material from the crust, mantle, and core. It is likely that the hot plumes originate from a hot thermal boundary larger either at the core–mantle boundary or at the base of the upper mantle.

The new tomographic images also reveal that high-velocity areas, which are interpreted as cold sinking slabs of subducted plates, also extend all the way to the core–mantle boundary (box figures 3 and 4).

BOX 4.3 ■ FIGURE 1

Map views of seismic-wave velocities in the mantle at depths of 100 and 300 km, as determined by seismic tomography. Blue indicates high velocity (cold rock), red indicates low velocity (hot rock). White lines outline plates; white circles are major hot spots.

From Dziewonski and Anderson, American Scientist, 1984, 72:483–94

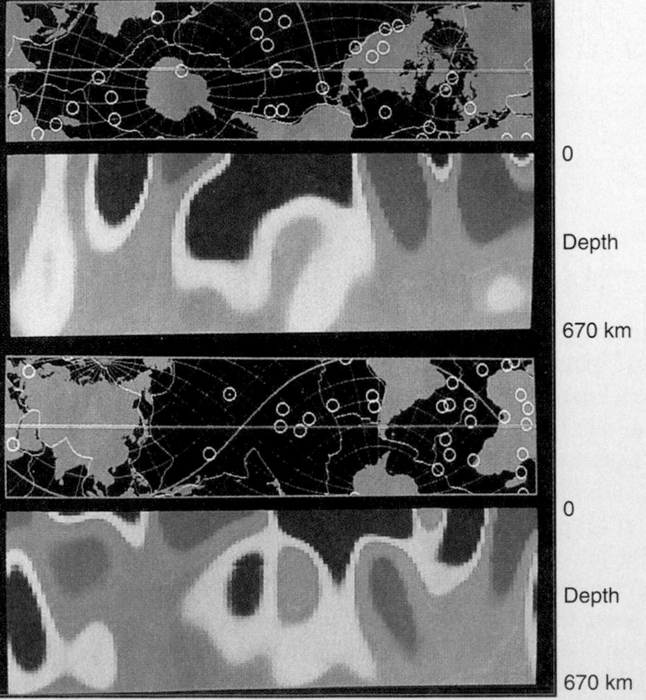

BOX 4.3 ■ FIGURE 2

Vertical cross-sections of seismic-wave velocities to a depth of 670 km in the mantle. The orange lines show the locations of the cross-sections.

From Dziewonski and Anderson, American Scientist, 1984, 72:483–94

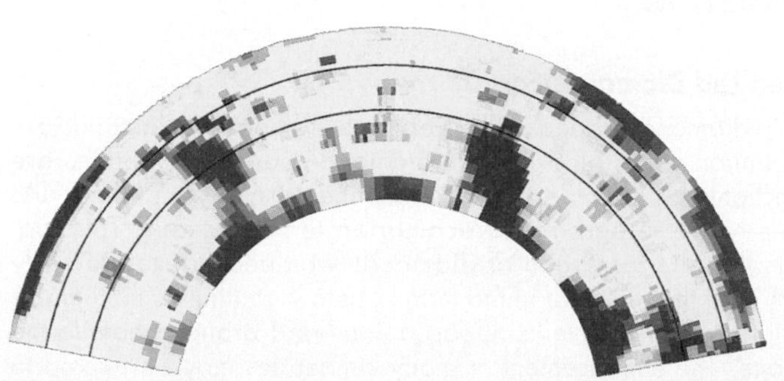

BOX 4.3 ■ FIGURE 3

Cross-section of seismic-wave velocities from the Earth's surface (upper curve) to core. Blue indicates fast seismic velocities (cold rock), and red indicates low velocities (hot rock). There is a presumed cold slab of rock, shown on the left side of the cross-section, that is sinking into the lower mantle into other slabs that rest on the core–mantle boundary. Hot rocks, believed to represent mantle plumes, also emanate from the core–mantle boundary, on the right side of the cross-section.

Photo courtesy of Stephen Grand, University of Texas at Austin

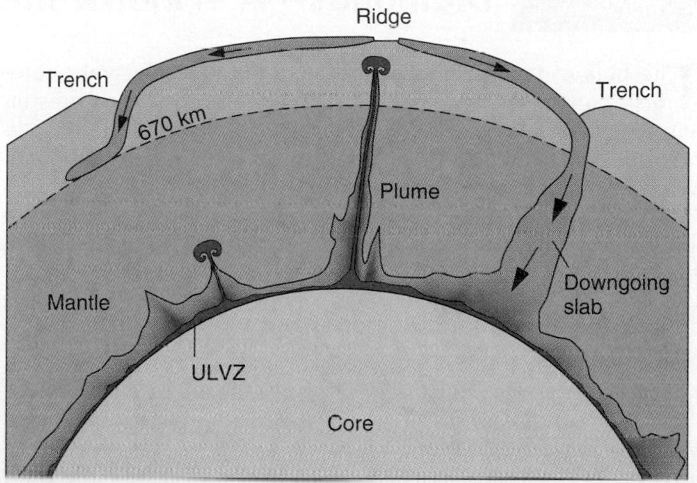

BOX 4.3 ■ FIGURE 4

Seismic data suggest some plates sink to the base of the mantle, whereas other plates are impeded by the increase in density of the mantle at 670 km. Deep mantle plumes emanating from the core mantle boundary are underlain by an ultra-low-velocity zone (ULVZ).

Other plates stop descending at the 670-kilometre boundary within the mantle. Perhaps the depth of sinking is controlled by plate density. The older the subducting rock is, the colder and denser it is. Old, dense plates may sink to the base of the mantle, while younger plates, being less dense, stop at a depth of 670 kilometres (box figure 4).

It is becoming increasingly apparent that the core–mantle boundary may play an important role in the overall mechanism of plate movement.

Additional Readings

Kerr, R. A. 1991. Do plumes stir earth's entire mantle? *Science* vol. 252: 1068–1069.

———. 1997. Deep-sinking slabs stir the mantle. *Science* vol. 275: 613–615.

Grand, S. P., Van der Hilst, R. D., and Widiyantoro, S. 1997. Global seismic tomography: A snapshot of convection in the Earth. *GSA Today* 7:1–7.

The Core

Seismic wave data provide the primary evidence for the existence of the core of the Earth. (See chapter 3 for a discussion of seismic P and S waves.) Seismic waves do not reach certain areas on the opposite side of the Earth from a large earthquake. Figure 4.8 shows how seismic P waves spread out from a quake until, at 103 degrees of arc (11,500 km) from the epicentre, they suddenly disappear from seismograms. At more than 142 degrees (15,500 km) from the epicentre, P waves reappear on seismograms. The region between 103 degrees and 142 degrees, which lacks P waves, is called the **P-wave shadow zone.**

The P-wave shadow zone can be explained by the refraction of P waves when they encounter the core boundary deep within Earth's interior. Because the paths of P waves can be accurately calculated, the size and shape of the core can be determined also. In figure 4.8, notice that Earth's core deflects the P waves and, in effect, "casts a shadow" where their energy does not reach the surface. In other words, P waves are missing within the shadow zone because they have been bent (refracted) by the core.

The chapter on earthquakes explains that while P waves can travel through solids and fluids, S waves can travel only through solids. As figure 4.9 shows, an **S-wave shadow zone** also exists and is larger than the P-wave shadow zone. Direct S waves are not recorded in the entire region more than 103 degrees away from the epicentre. The S-wave shadow zone seems to indicate that S waves do not travel through the core at all. If this is true, it implies that the core of Earth is a liquid, or at least acts like a liquid.

The way in which P waves are refracted within the Earth's core (as shown by careful analysis of seismograms) suggests that the core has two parts, a *liquid outer core* and a *solid inner core* (figure 4.7).

IN GREATER DEPTH 4.4

Diamonds—A Window Into the Mantle

The bulk of the Earth's volume lies in the mantle, yet geologists cannot see or sample this material directly and know little about its exact composition. Occasionally, fragments of mantle materials are brought to the Earth's surface via volcanoes. Mantle materials are also found in diamond-bearing igneous rocks called kimberlites (named for Kimberley, South Africa), which form carrot-shaped bodies (up to 200 m across and more than 1 km deep) called *kimberlite pipes* (box figure 1).

How Do We Know That Diamonds Form in the Mantle?

Diamonds are made of a high-pressure form of crystalline carbon that can be produced experimentally only at extreme temperatures (1500°C) and pressures (55 kilobars). In nature, these conditions would be found at depths of more than 150 km below the Earth's crust, in the mantle. Formed under such immense temperatures and pressures, diamond is the hardest known natural material on Earth.

How Do Diamonds Form?

The carbon that forms diamonds is thought to have originated from carbon-bearing rocks on oceanic plates that were subducted at collisional plate margins sometime in the past. This carbon transformed into diamond under extreme heat and pressure and was trapped in the mantle below continents. Subsequent eruption of the kimberlite magmas through the volcano-like vents of kimberlite pipes brought diamonds to the surface of continents. Kimberlite magmas are viscous and contain large amounts of dissolved gas, making them frothy and able to rise to the surface quickly—this rapid ascent from the mantle did not allow the diamonds to break down on their journey through the crust. Contrary to popular belief, diamonds are not stable at the Earth's surface and will eventually, over geologic time, break down to form graphite.

When Did Diamonds Form?

Many diamonds appear to be very old—Kimberley diamonds are 3.3 billion years old—and all diamond-bearing kimberlites are found only on the oldest parts of continents. However, kimberlite pipes range in age from Precambrian to Cretaceous. This suggests that the formation of diamonds was peculiar to the early Earth and that they sat in the mantle below continents until major continental rifting events released them and brought them to the surface. The emplacement of many kimberlites may be related to the rifting and breakup of the supercontinents of Rodinia and Pangea (see chapter 2) in the late Precambrian and the Mesozoic.

Where Can We Find Diamonds in North America?

Diamonds are found on most of the ancient continental cratons—southern and central Africa, Siberia, India, Australia, and the Canadian Shield. The first commercial diamond mine in North America is the Ekati mine, which lies about 300 km northeast of Yellowknife in the Northwest Territories (box figure 2). A total of 121 kimberlite pipes have been identified in the Ekati claim area—five are currently being developed, and others will be mined in the future. The Ekati mine has a projected lifespan of around 25 years and will produce more than 2 million carats in diamonds each year. Other diamond-bearing kimberlite pipes are known in Ontario and Quebec and in the U.S. Rocky Mountains.

BOX 4.4 ■ FIGURE 2

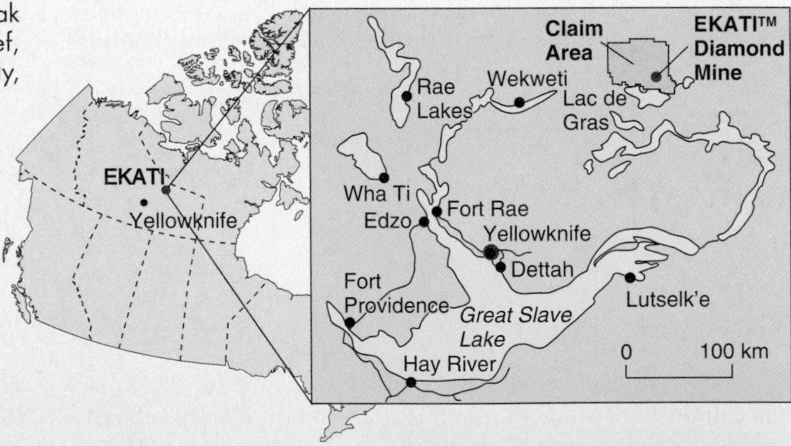

BOX 4.4 ■ FIGURE 1

The "big hole" at Kimberley, South Africa shows what a kimberlite pipe looks like after it has been mined out for diamonds.

Photo by Nick Eyles

BOX 4.4 ■ FIGURE 3

Diamonds from the Ekati mine.

Photo by CP (Dave Buston)

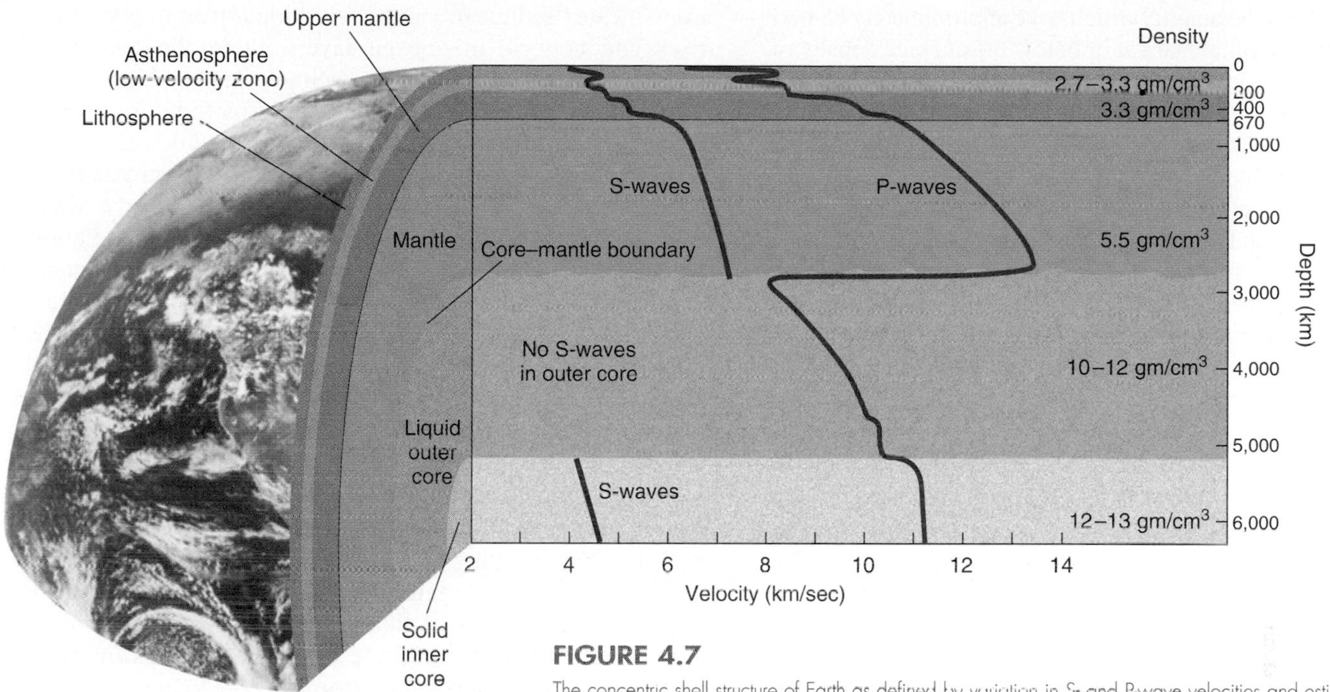

FIGURE 4.7

The concentric shell structure of Earth as defined by variation in S- and P-wave velocities and estimates of density. The velocity of seismic P- and S-waves generally increases with depth except in the low-velocity zone. The plastic asthenosphere slows down seismic waves. Velocity increases at 400 and 670 km may be caused by mineral collapse. S-waves do not pass through the outer core, but are thought to travel through the solid inner core.

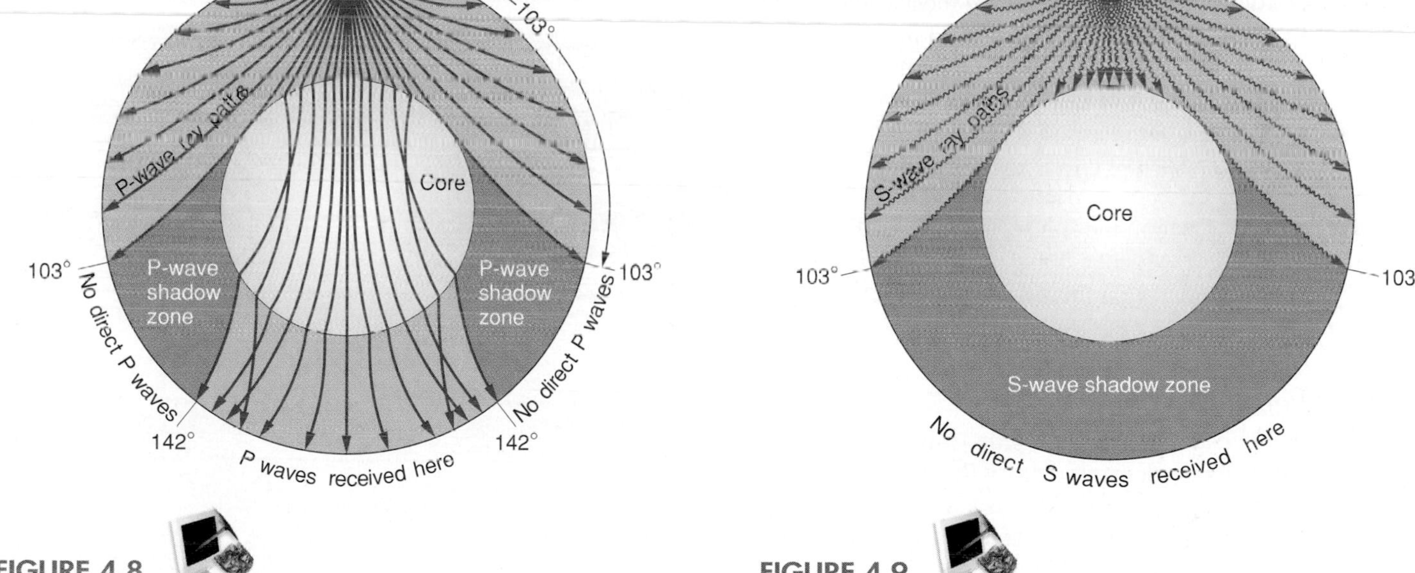

FIGURE 4.8

The P-wave shadow zone, caused by refraction of P waves within the Earth's core.

FIGURE 4.9

The S-wave shadow zone. Because no S waves pass through the core, the outer core is apparently a liquid (or acts like a liquid).

Composition of the Core

When evidence from astronomy and seismic-wave studies is combined with what we know about the properties of materials, it appears that the Earth's core is made of metal—not silicate rock—and that this metal is probably iron (along with a minor amount of oxygen, silicon, sulphur, or nickel). How did geoscientists arrive at this conclusion?

The overall density of the Earth is 5.5 gm/cm^3, based on calculations from Newton's law of gravitational attraction. The crustal rocks are relatively low density, from 2.7 gm/cm^3 for granite to 3.0 gm/cm^3 for basalt. The ultramafic rock thought to make up the mantle probably has a density of 3.3 gm/cm^3 in the upper mantle, although rock pressure should raise this value to about 5.5 gm/cm^3 at the base of the mantle (figure 4.7).

If the crust and the mantle, which have approximately 85 percent of the Earth's volume, are at or below the average density of the Earth, then the core must be very heavy to bring the average up to 5.5 gm/cm^3.

Calculations show that the core has to have a density of about 10 gm/cm^3 at the core–mantle boundary, increasing to 12 or 13 gm/cm^3 at the centre of Earth (figure 4.7). This great density would be enough to give the Earth an average density of 5.5 gm/cm^3.

Under the great pressures existing in the core, iron would have a density slightly greater than that required in the core. Iron mixed with a small amount of a lighter element, such as oxygen, sulphur, or silicon, would have the required density. Therefore, many geoscientists think that such a mixture makes up the core.

But a study of density by itself is hardly convincing evidence that the core is mostly iron, for many other heavy substances could be there instead. The choice of iron as the major component of the core comes from looking at meteorites (see box 1.10). Meteorites are thought by some scientists to be remnants of the basic material that created our own solar system. An estimated 10 percent of meteorites are composed of iron mixed with small amounts of nickel. Material similar to these meteorites may have helped create the Earth, perhaps settling to the centre of Earth because of metal's high density. The composition of these meteorites, then, may tell us what is in the core. Nickel is denser than iron, however, so a mixture of just iron and nickel would have a density greater than that required in the core. (The other 90 percent of meteorites are mostly ultramafic rock and perhaps represent material that formed the mantle.)

Seismic and density data, together with assumptions based on meteorite composition, point to a core that is largely iron, with at least the outer part being liquid. The existence of Earth's magnetic field, which is discussed later in this chapter, also suggests a metallic core. Of course, no geoscientist has seen the core, nor is anyone likely to in the foreseeable future. But since so many lines of indirect evidence point to a liquid metal outer core, most scientists accept this theory as the best conclusion that can be made about the core's composition.

The Core–Mantle Boundary

The boundary between the core and mantle is marked by great changes in seismic velocity, density (figure 4.7), and temperature, as we see later in the chapter. Here there is a transition zone up to 200 km thick, known as the *D″ layer,* at the base of the mantle where P-wave velocities decrease dramatically. The *ultra-low-velocity zone* (ULVZ) (figure 4.10) that forms the undulating border at the core–mantle boundary may be due to hot core partially melting overlying mantle rock or could be due to part of the liquid outer core reacting chemically with the adjacent mantle. The latest seismic and geodetic studies have hinted that lighter iron alloys from the liquid outer core may react with silicates in the lower mantle to form iron silicates. The less dense iron silicate "sediment," along with liquid iron in pore spaces, rises and collects in uneven layers along the core–mantle boundary. The pressure of the accumulating "sediment" along the boundary causes some of the liquid iron to be squeezed out of the pore spaces to form an electrically conductive layer that connects the core and mantle and explains the decrease in seismic velocities at the ULVZ. It may be difficult to prove whether the lowermost mantle is being partially melted by the core or whether the core is instead chemically reacting with the mantle.

Both the mantle and the core are undergoing **convection,** a circulation pattern in which low-density material rises and high-density material sinks. Based on seismic tomography studies, heavy portions of the mantle (including subducted plates) sink to its base, but are unable to penetrate the denser core. Light portions of the core may rise to its top, and may be incorporated into the mantle above. This is suggested by recent isotopic studies of the mantle plume that feeds the Hawaiian hot spot. The resulting Hawaiian volcanic rocks (basalts) contain a light isotope signature that is characteristic of the core. Continent-sized blobs of liquid and liquid-crystal slush may accumulate at the core–mantle boundary, perhaps interfering with or helping cause heat loss from the core to help drive mantle convection and transfer of heat to the surface, and also causing changes in Earth's magnetic field. This boundary is an exciting frontier for geologic study, but data, of course, are sparse and hard to obtain.

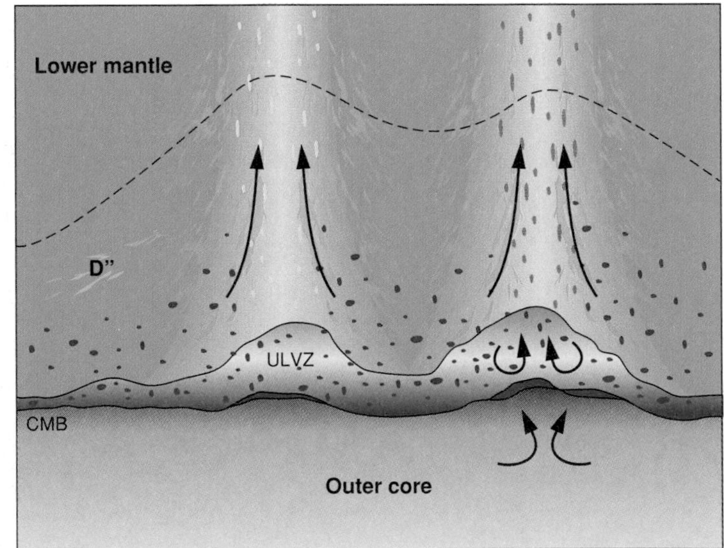

FIGURE 4.10

Recent seismic and geodetic studies are redefining the boundary between the lower 200 km of the mantle (D" layer) and the outer core. Iron silicate "sediments" (shown in brown) may rise from the underlying liquid core and fill pockets or inverted basins at the core–mantle boundary (CMB). Alternatively, the outer core material (shown in red) may be melting the lowermost mantle (shown in yellow) to form the ultra-low-velocity zone (ULVZ).

Modified from Garnero and Jeanloz, Science, *2000*

HOW DOES THE ELEVATION OF CONTINENTS CHANGE?

Isostasy is a balance or *equilibrium* of adjacent blocks of brittle crust "floating" on the upper mantle. Since crustal rocks weigh less than mantle rocks, the crust can be thought of as floating on the denser mantle much as wood floats on water.

Blocks of wood floating on water rise or sink until they displace an amount of water equal to their own mass weight. In a greatly simplified way, crustal rocks can be thought of as tending to rise or sink gradually until they are balanced by the weight of displaced mantle rocks. This concept of vertical movement to reach equilibrium is called **isostatic adjustment.** Once crustal blocks have come into isostatic balance, a tall block (a mountain range) extends deep into the mantle (a *mountain root,* as shown in figure 4.11). A column of thick continental crust (a mountain and its root) has the same weight as a column containing thin continental crust and some of the upper mantle. A column containing sea water, thin oceanic crust, and a thick section of heavy mantle weighs the same as the other two columns.

Let us look at some examples of isostatic balance (equilibrium) in crustal rocks. Suppose that two sections of crust of unequal thickness are next to each other, as in figure 4.11. Sediment from the higher part, which is subject to more rapid erosion, is deposited on the lower part. The decrease in weight from the high part causes it to rise, while the increase in weight on the low part causes it to sink. These vertical movements (isostatic adjustment) do in fact take place whenever large volumes of material are eroded from or deposited on parts of the crust.

Rising or sinking of the crust, of course, requires plastic flow of the mantle to accommodate the motion. By measuring the rate of rising or sinking, the viscosity of the mantle can be calculated. The plastic flow of the mantle probably takes place within the asthenosphere in the upper mantle.

Another example of isostatic adjustment, caused by plastic mantle flow, is the upward movement of large areas of the crust since the glacial ages. The weight of the thick continental ice sheets during the Quaternary depressed the crust underneath the ice (figure 4.12). Depression of the crust also caused local upwarping around the ice sheet margin (called a forebulge; see figure 4.12*B*). After the melting of the ice, the crust rose back upward, a process that is still going on in some areas (figure 4.13). This rise of the crust after the removal of the ice is known as **crustal rebound.** Upwarped areas also subside to their original position during the rebound process (figure 4.12).

Recent geophysical studies have shown that some mountains, such as the Rockies and southern Sierra Nevada, do not have thick roots and are instead buoyed by warm, less dense mantle. It appears that the upper mantle beneath some continents is not homogeneous, but has zones that are quite buoyant due to higher temperatures and less dense mineral phases.

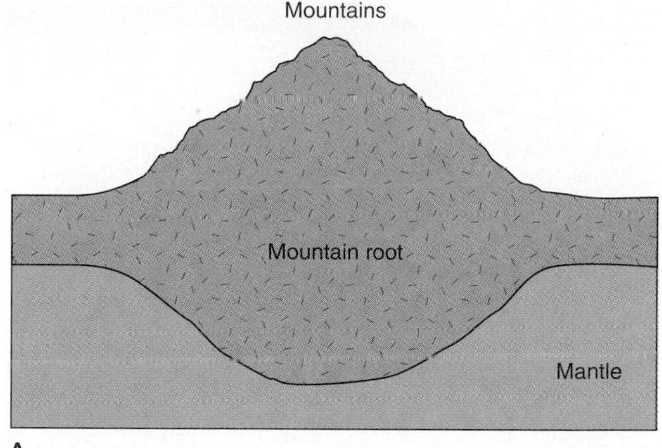

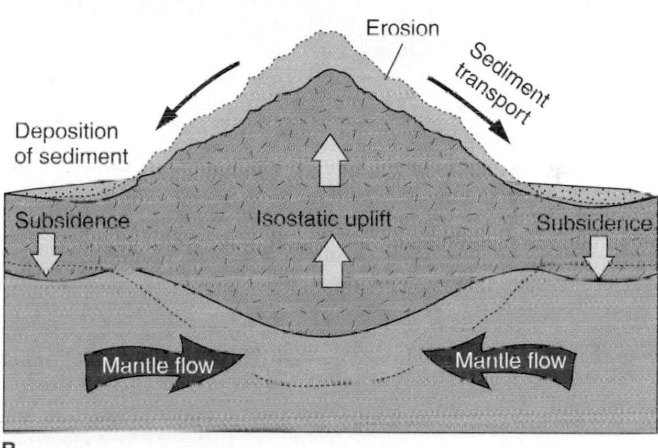

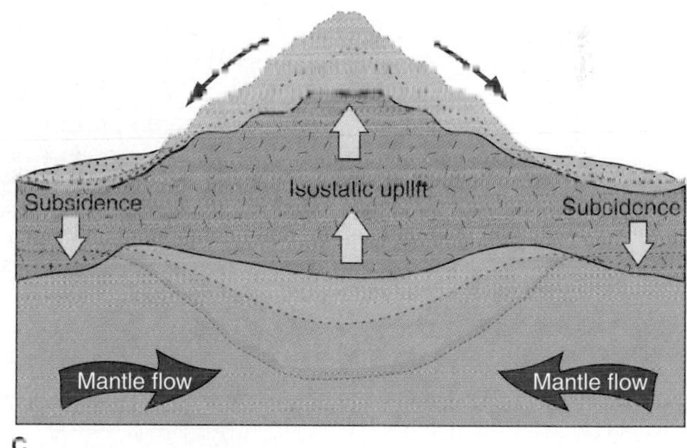

FIGURE 4.11

Isostatic adjustment due to erosion and deposition of sediment. Rock within the mantle must flow to accommodate vertical motion of crustal blocks. Mantle flow occurs in the asthenosphere, deeper than shown in *C*.

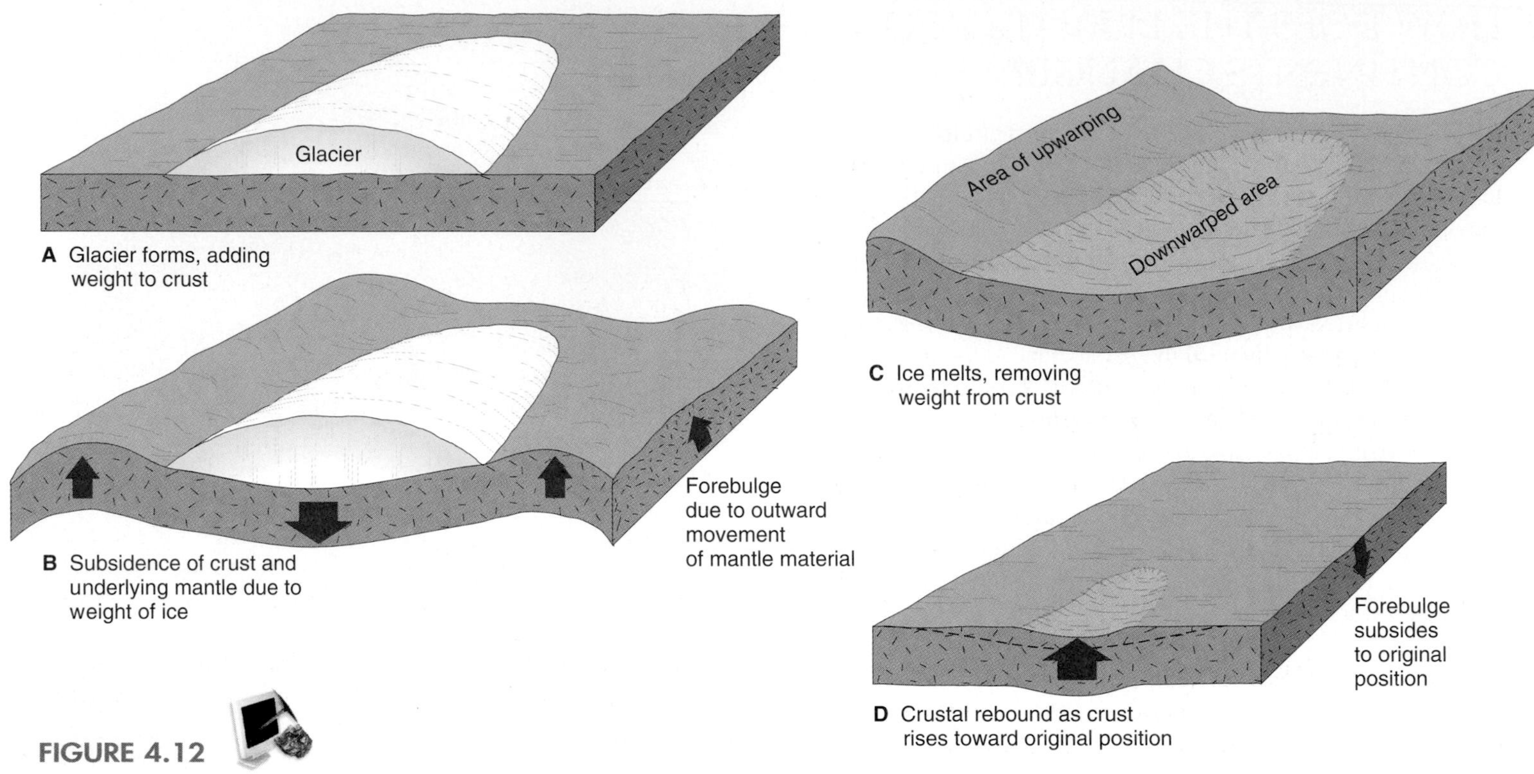

FIGURE 4.12

The weight of glaciers depresses the crust, and the crust rebounds when the ice melts. Downwarped areas may be flooded by short-lived lakes or seas.

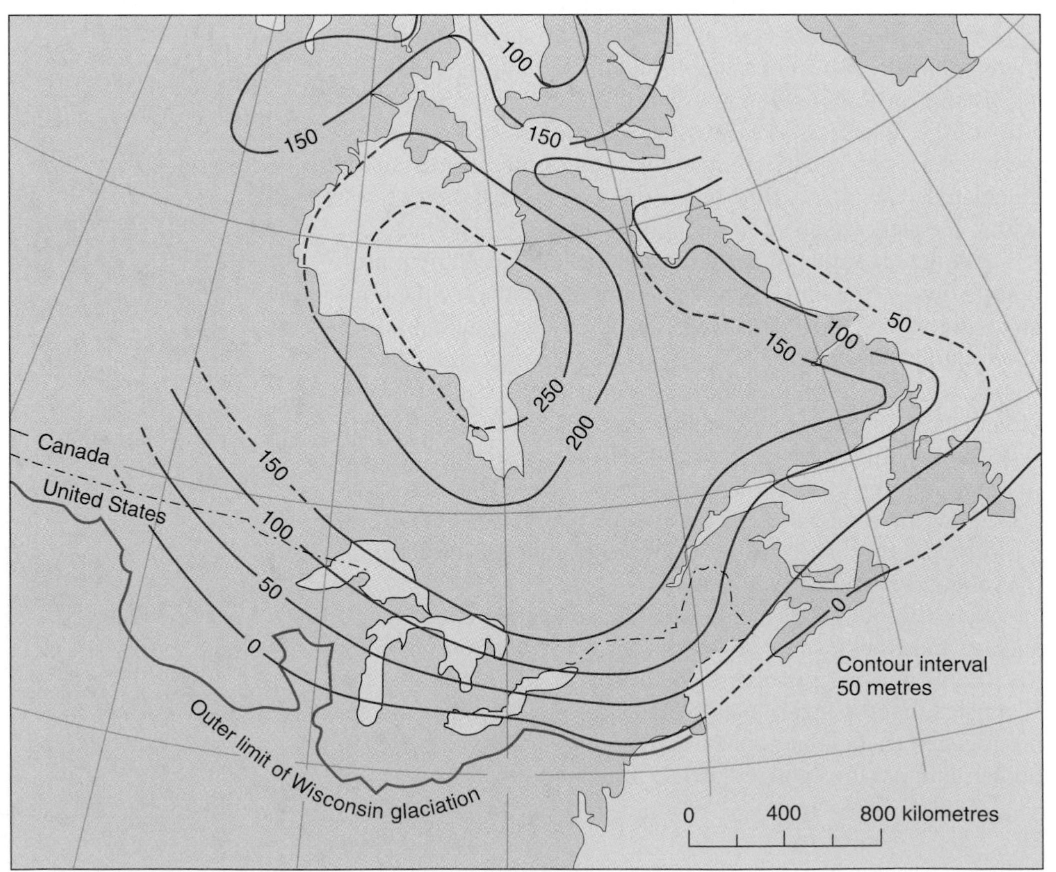

FIGURE 4.13

Uplift of land surface in Canada and the northern United States caused by crustal rebound after glaciers melted. Coloured lines show the amount of uplift in metres since the ice disappeared.

From Phillip B. King, "Tectonics of Quaternary Time in Middle North America," in The Quaternary of the United States, *H. D. Wright, Jr. and David G. Frey, eds., fig. 4A, p. 836. Reprinted by permission of Princeton University Press*

WHAT CAN GRAVITY TELL US ABOUT THE EARTH'S CRUST?

The force of gravity between two objects varies with the masses of the objects and the distance between them (figure 4.14):

$$\text{Force of gravity between } A \text{ and } B = \text{Constant}\left(\frac{\text{mass}_A \times \text{mass}_B}{\text{distance}^2}\right)$$

The force increases with an increase in either mass. The gravitational attraction between Earth and the moon, for example, is vastly greater than the extremely small attraction that exists between two bowling balls. The equation also shows that force decreases with the square of the distance between the two objects. The farther two objects are apart, the less gravitational attraction there is between them.

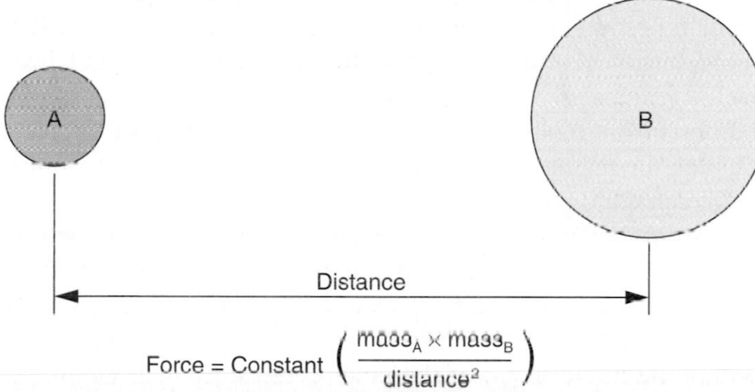

FIGURE 4.14

The force of gravitational attraction between two objects is a function of the masses of the objects and the distance between the centres of the objects.

A useful tool for studying the crust and upper mantle is the **gravity meter,** which measures the gravitational attraction between the Earth and a mass within the instrument. Geoscientists use gravity meters to identify relative changes in gravity that may indicate local variations in rock density (mass = density × volume). Dense rock such as metal ores and ultramafic rock pulls strongly on the mass inside the meter (figure 4.15). The strong pull stretches a spring, and the amount of stretching can be very precisely determined. So, a gravity meter can be used to explore for metallic ore deposits. A cavity or a body of low-density material such as sediment causes a much weaker pull on the meter's mass (figure 4.15). (The use of a gravity meter to explore for salt domes and their associated traps for oil and gas is shown in box 12.1.)

Another important use of a gravity meter is to discover whether regions are in isostatic equilibrium. If a region is in isostatic balance, as in figure 4.16*A*, each column of rock has the same mass. If a gravity meter were carried across the rock columns, it would register the same amount of gravitational attraction for each column (after correcting for differences in elevation—gravitational attraction is less on a mountaintop than at sea level because the mountaintop is farther from the centre of the Earth).

Some regions, however, are held up out of isostatic equilibrium by deep tectonic forces. Figure 4.16*B* shows a region with uniformly thick crust. Tectonic forces are holding the centre of the region up. This uplift creates a mountain range without a mountain root. There is a thicker section of heavy mantle rock under the mountain range than there is on either side of the mountain range. Therefore, the central "column" has more mass than the neighbouring columns, and a gravity meter shows that the gravitational attraction is correspondingly greater over the central than over the side columns.

A gravity reading higher than the normal regional gravity is called a **positive gravity anomaly** (figure 4.16*B*). It can indicate that tectonic forces are holding a region up out of isostatic equilibrium, as shown in figure 4.16*B*. When the forces stop acting, the land surface sinks until it re-establishes isostatic balance. The gravity anomaly then disappears. For the region shown in figure 4.16*B*, equilibrium will be established when the land surface becomes level.

Positive gravity anomalies, particularly small ones, are also caused by local concentrations of dense rock such as metal ore. The gravity meter in figure 4.15 is registering a positive gravity anomaly over ore (the spring inside the meter is stretched). Since there can be more than one cause of a positive gravity anomaly, geoscientists may disagree about the interpretation of anomalies. Drilling into a region with a gravity anomaly usually discloses the reason for the anomaly.

A region can also be held down out of isostatic equilibrium, as shown in figure 4.16*C*. The mass deficiency in such a region produces a **negative gravity anomaly**—a gravity reading lower than the normal regional gravity. Negative gravity anomalies indicate either that a region is being held down (figure 4.16*C*) or that local mass deficiencies exist for other reasons (figure 4.15).

The greatest negative gravity anomalies in the world are found over oceanic trenches (see chapter 2). These negative anomalies are interpreted to mean that trenches are actively being held down and are out of isostatic balance.

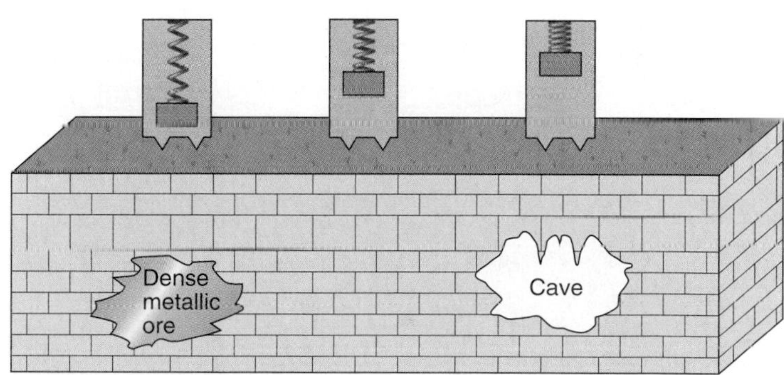

FIGURE 4.15

A gravity-meter reading is affected by the density of the rocks beneath it. Dense rock pulls strongly on the mass within the meter, stretching a spring; a cavity exerts a weak pull on the mass. A gravity meter can be used to explore for hidden ore bodies, caves, and other features that have density contrasts with the surrounding rock.

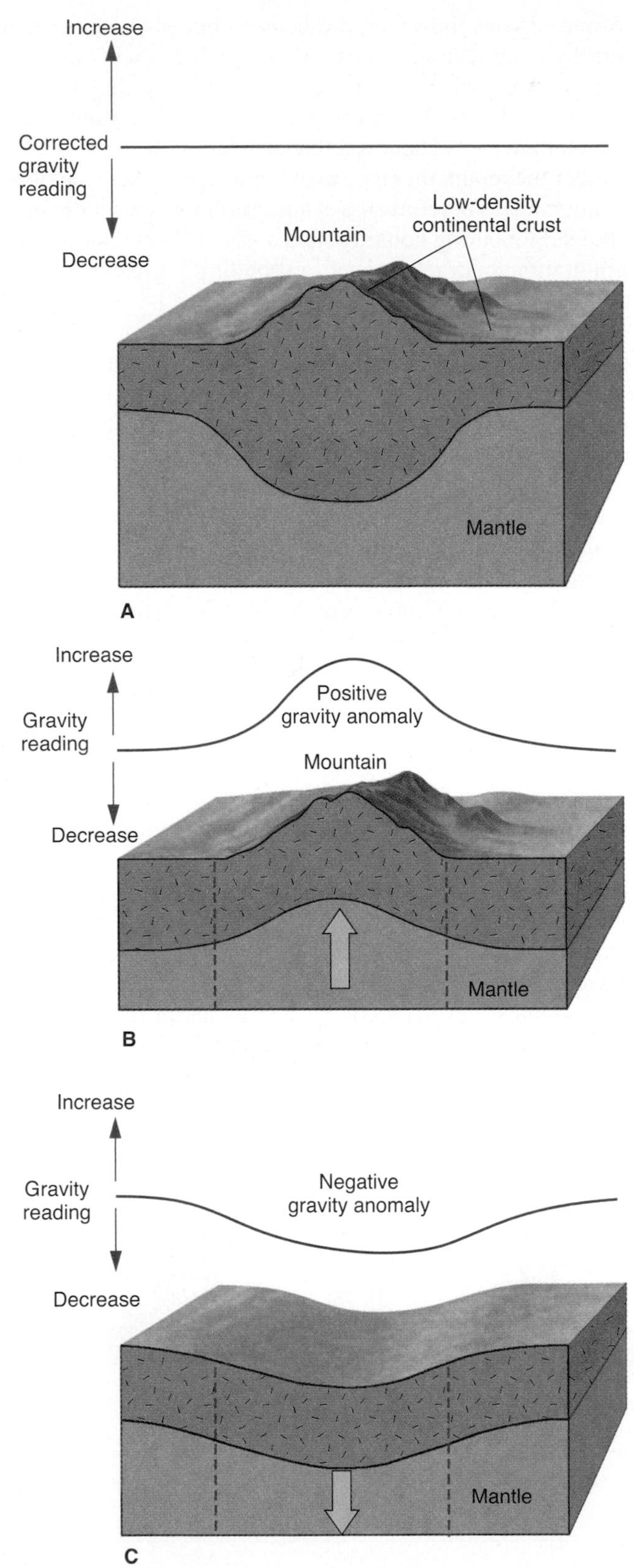

FIGURE 4.16

(*A*) A region in isostatic balance gives a uniform gravity reading (no gravity anomalies), after correcting for differences in elevation. (*B*) A region being held up out of isostatic equilibrium gives a positive gravity anomaly. (*C*) A region being held down out of isostatic equilibrium gives a negative gravity anomaly.

HOW DOES THE EARTH'S MAGNETIC FIELD CHANGE THROUGH TIME?

A region of magnetic force—a **magnetic field**—surrounds Earth. The invisible lines of magnetic force surrounding Earth deflect magnetized objects, such as compass needles, that are free to move. The field has north and south **magnetic poles,** one near the geographic North Pole, the other near the geographic South Pole. (Because it has two poles, Earth's field is called *dipolar.*) The strength of the magnetic field is greatest at the magnetic poles where magnetic lines of force appear to leave and enter Earth vertically (figure 4.17).

Because the compass is important in navigation, Earth's magnetism has been observed for centuries. It has long been known that the magnetic poles are displaced about 11.5 degrees from the geographic poles (about which Earth rotates). Furthermore, changes in the position of the magnetic poles have been well documented, especially since the time of the great explorations of the globe. Because Earth's field is not 100 percent dipolar, the magnetic poles appear to be moving slowly around the geographic poles.

More recently, geophysical studies have been directed toward the *source* of Earth's magnetism. The rate of the poles' changes in position, together with the strength of the magnetic field, strongly suggest that the magnetic field is generated within the liquid metal of the outer core rather than within the solid rock of the crust or the mantle (see box 4.5).

How is Earth's magnetic field generated? A number of hypotheses have been put forth. One widely accepted hypothesis suggests that the magnetic field is created by electric currents within the liquid outer core. The outer core is extremely hot and flows at a rate of several kilometres per year in large convection currents, about one million times faster than mantle convection above it. Convecting metal, in the presence of an existing magnetic field, creates electric currents, which in turn could sustain Earth's magnetic field. This hypothesis requires the core to be an electrical conductor. Metals are good conductors of electricity, whereas silicate rock is generally a poor electrical conductor. Indirectly, this is evidence that the core is metallic.

Magnetic Reversals

In the 1950s evidence began to accumulate that Earth's magnetic field has periodically reversed its polarity in the past. Such a change in the polarity of the magnetic field is a **magnetic reversal.** During a time of *normal polarity,* magnetic lines of force leave Earth near the geographic South Pole and re-enter near the geographic North Pole (figure 4.17). This orientation is called "normal" polarity because it is the same as the present polarity. During a time of *reversed polarity,* the magnetic lines of force run the other way, leaving Earth near the North Pole and entering near the South Pole (figure 4.17). In other words, during a magnetic reversal, the north magnetic pole and the south magnetic pole exchange positions.

IN GREATER DEPTH 4.5

Earth's Spinning Inner Core

Recent studies have led to a new understanding of the dynamics of Earth's inner core and generation of Earth's magnetic field and periodic magnetic reversals. Gary A. Glatzmaier of Los Alamos National Laboratory in New Mexico, and Paul H. Roberts of the University of California, Los Angeles developed a very sophisticated computer model of convection in the outer core that has been successful in simulating a magnetic field very similar to that measured on Earth. The model utilizes circulating metallic fluids in the outer core, caused by cooling and heat loss, as the driving force of Earth's magnetic field. The circulation of metallic fluids in the outer core has been theorized for many years, and the computer model was successful in simulating and maintaining a magnetic field similar to that measured on Earth. The model also predicted that Earth's solid inner core spins faster than the rest of the planet, gaining a full lap on the rest of the planet every 150 years. Because the magnetic lines of force penetrate and connect both the inner and outer core, a faster rate of rotation of the inner core would play an important role in the generation of Earth's magnetic field and may also influence periodic magnetic reversals. Interestingly, Glatzmaier and Roberts' model produced a magnetic reversal on its own without any additional input from the experimenters after about 35,000 years of simulated time (box figure 1).

The results of this computer model inspired seismologists Xiao Dong Song and Paul Richards from Columbia's Lamont-Doherty Earth Observatory to look for evidence that the inner core actually spins at a more rapid rate than the rest of the planet. The seismologists knew that previous studies suggested seismic waves pass through the inner core faster along a nearly north–south route. This faster route, or high-velocity pathway, is similar to the grain in a piece of wood. This pathway is not aligned directly with the inner core's spin axis but is tilted about 10 degrees from it (box figure 2). Seismic waves tend to travel slower along other paths, such as in an east–west direction parallel to the inner core's equator.

The seismologists studied seismic wave records from 38 separate, closely spaced earthquakes from 1967 to 1995 near the Sandwich Islands off Argentina to determine how long it took them to reach a monitoring station in College, Alaska. The waves all took about the same amount of time to reach Alaska; however, the seismic waves in the 1990s arrived in Alaska about 0.3 seconds faster than the seismic waves in the 1960s. Since the seismic waves would have travelled through the inner core, the seismologists have explained the difference in travel time as indicating that the inner core had changed its position relative to the monitoring station in Alaska. That is, the inner core and the high-velocity pathway had rotated slightly with respect to the rest of the planet.

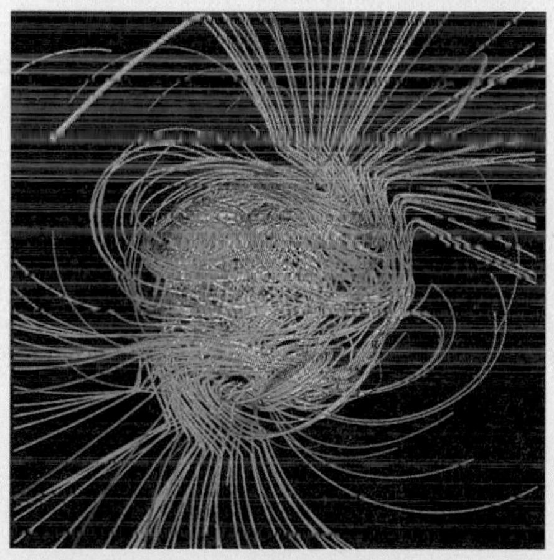

A

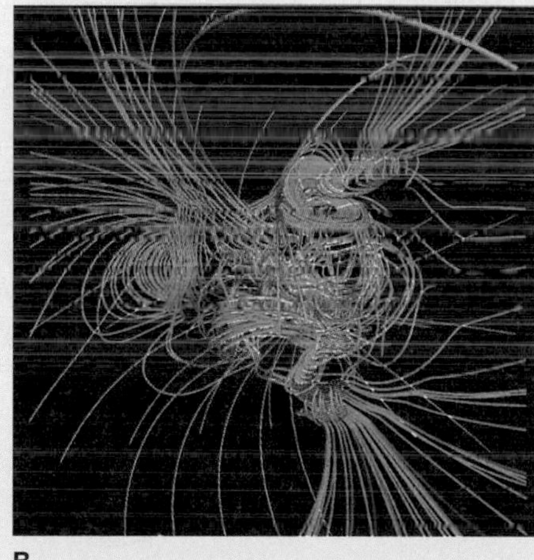

B

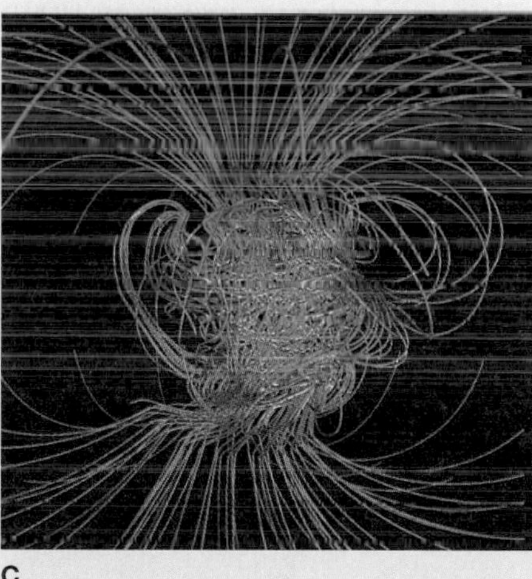

C

BOX 4.5 ■ FIGURE 1

Computer simulation of Earth's magnetic field and magnetic reversal. (*A*) Reversed magnetic polarity with magnetic field lines leaving the north magnetic pole (orange) and re-entering at the south pole. (*B*) Transitional magnetic field. (*C*) Normal magnetic field.

Photos from the Geodynamo Computer Simulation, courtesy of G. A. Glatzmaier, Los Alamos National Laboratory, and P. H. Roberts, University of California, Los Angeles

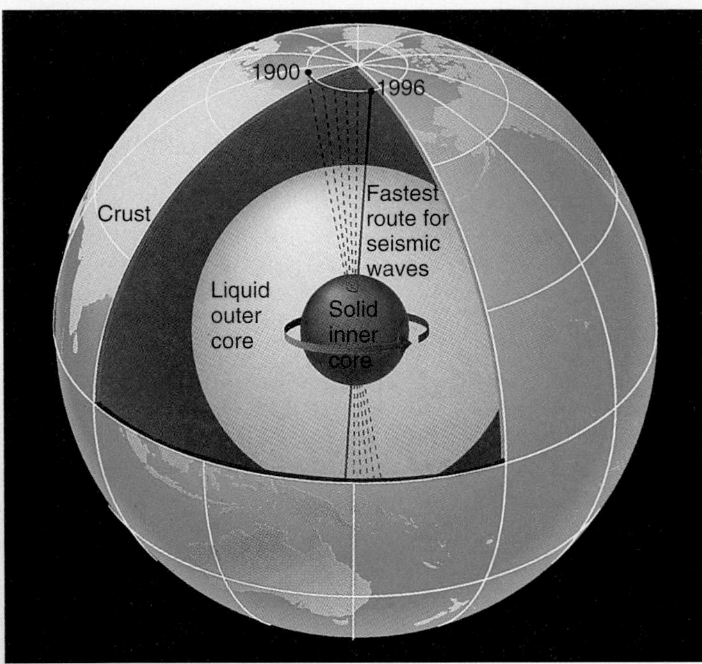

BOX 4.5 ■ FIGURE 2

Seismic waves indicate that Earth's core rotates faster than the rest of the planet by about a degree per year. The solid line indicates the 1996 position of a point in the core relative to the surface of Earth, and the dashed line indicates where the point was in 1900.

Courtesy of Lamont-Doherty Earth Observatory, Columbia University. Data from Michael Carlowicz, Earth Magazine, *p. 21, 1996*

Seismologists at Harvard looked at additional earthquake records and calculated that the inner core is rotating at approximately the same rate as Glatzmaier and Roberts' model predicted. Future studies to examine earthquake records over a longer period of time are needed to confirm whether the inner core has been spinning faster than the rest of the Earth. This is an exciting time for Earth scientists since we may now have a better idea about the inner motion of the core and the generation of Earth's magnetic field.

Additional Resources

Carlowicz, M. 1996. Spin control. *Earth* 12(21): 62–63.

Core convection and the Geodynamo Website discusses the recent model for reversals of Earth's magnetic field:

http://ees5-www.lanl.gov/IGPP/Geodynamo.html

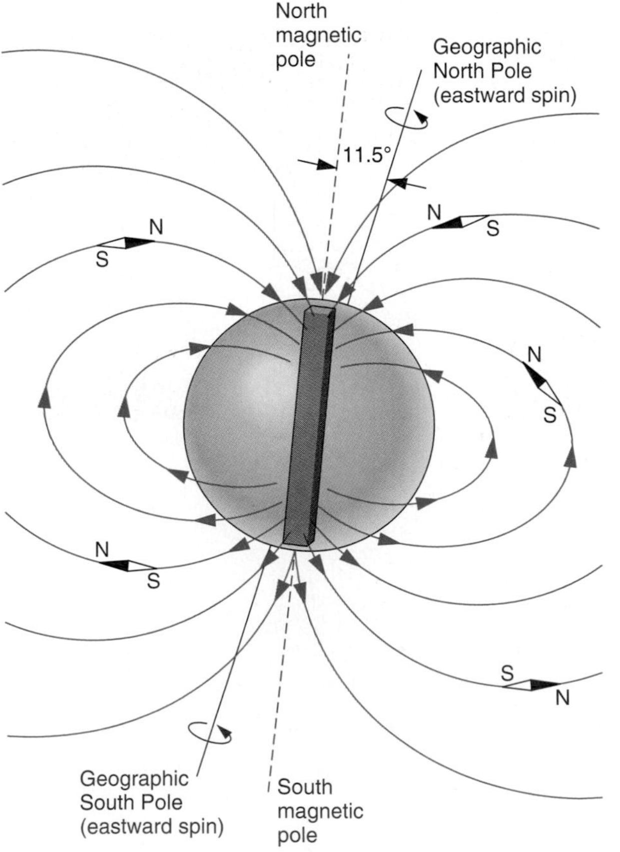

FIGURE 4.17

The Earth's magnetic field. The depiction of the internal field as a large bar magnet is a simplification of the real field, which is more complex. N and S in the two small figures indicate the *geographic poles.*

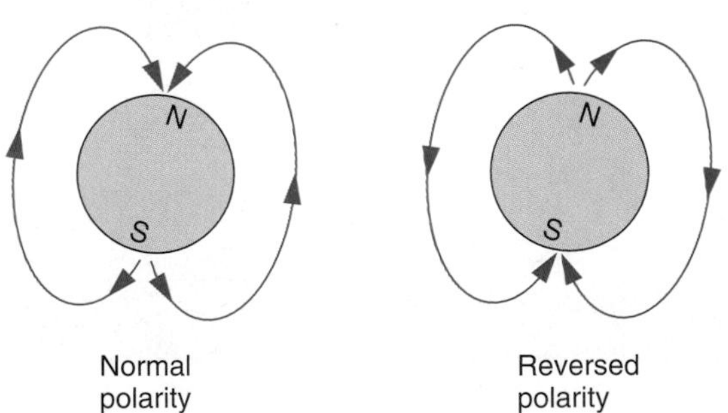

Many rocks contain a record of the strength and direction of the magnetic field *at the time the rocks formed.* When the mineral magnetite, for example, is crystallizing in a cooling lava flow, the atoms within the crystals respond to Earth's magnetic field and form magnetic alignments that "point" toward the north magnetic pole. As the lava cools slowly below the **Curie point** (580°C for magnetite), this magnetic record is permanently trapped in the rock (figure 4.18*A*). Unless the rock is heated again above the Curie point or temperature this magnetic record is retained, and when studied reveals the direction of Earth's magnetic field at the time the lava cooled. Other rock types, including sedimentary rocks stained red by iron compounds, also record former magnetic-field directions. The study of ancient magnetic fields is called **paleomagnetism.**

Most of the evidence for magnetic reversals comes from lava flows on the continents. Paleomagnetic studies of a series of stacked lava flows often show that some of the lava flows have a magnetic orientation directly opposite to Earth's present orientation (figure 4.18*B*). That is, at the time these lava flows cooled, the magnetic poles had exchanged positions. During this time of magnetic reversal, a compass needle would have pointed south rather than north. Many periods of normal and reverse magnetization are recorded in continental lava flows. They are worldwide events. Since lava flows can be dated isotopically, the time of these reversals in the Earth's past can be determined. Although reversals appear to occur randomly (figure 4.19), records for tens of millions of years suggest that Earth's field reverses on average about once every 500,000 years. The present normal orientation has lasted for the past 700,000 years. It takes time for one magnetic orientation to die out and the reverse orientation to build up. Most geoscientists think that it takes 10,000 years for a reversal to develop, although new evidence suggests that a reversal can occur much faster than that.

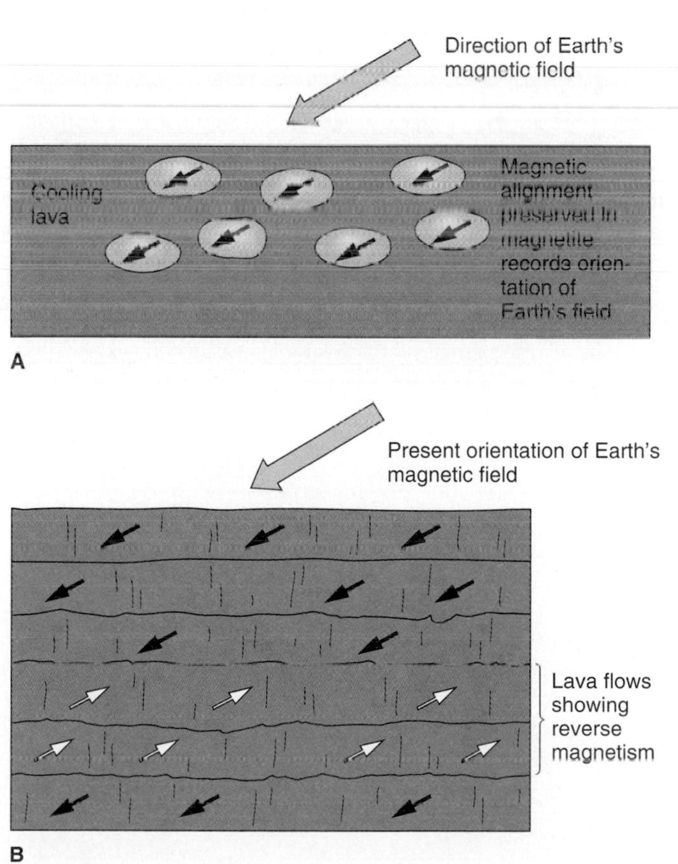

FIGURE 4.18

(*A*) Some rocks preserve a record of Earth's magnetic field. (*B*) Cross-section of stacked lava flows showing evidence of magnetic reversals.

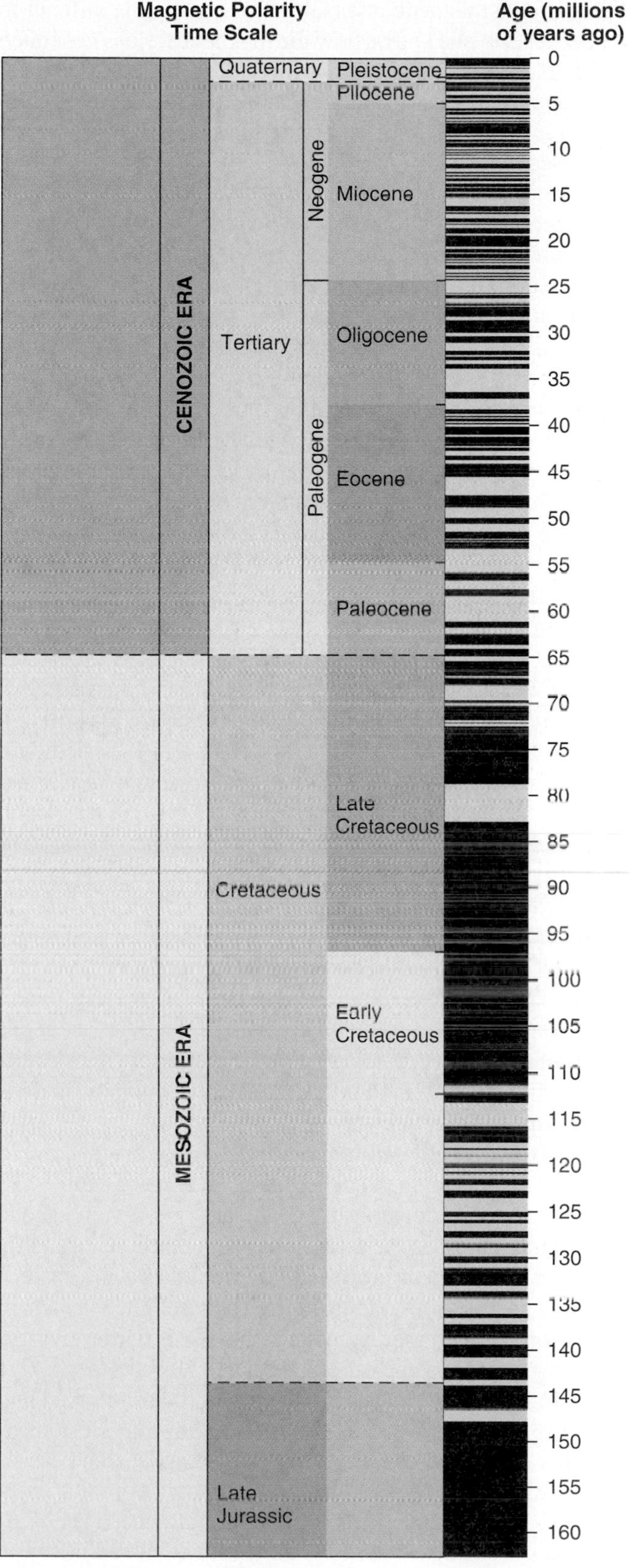

FIGURE 4.19

Worldwide magnetic polarity time scale for the Cenozoic and Mesozoic Eras. Black indicates positive anomalies (and therefore normal polarity). Tan indicates negative anomalies (reverse polarity).

Modified from R. L. Larson and W. C. Pitman, III, 1972, Geological Society of America Bulletin

What causes magnetic reversals? The question is difficult to answer because no one knows how the magnetic field is generated in the first place. Recent computer modelling and seismological research support the theory that the magnetic field is generated by convection currents in the liquid outer core (see box 4.5). *If* the field is caused by convection currents within the liquid outer core, perhaps a reversal is caused when the currents change direction, or by a temporary current building up and then dying out.

A magnetic reversal can have some profound effects on the Earth. The strength of the Earth's magnetic field probably declines to near zero before the orientation reverses; then the field strength increases to its usual values, but in the opposite orientation. This collapse of the magnetic field means that deadly cosmic radiation from the sun would be much more intense at the surface. When the magnetic field is at its usual strength, it shields the Earth from these rays, but when the field collapses, this shielding is lost. Cosmic radiation affects organisms; the extinction of some species and the appearance of new species by mutation have been correlated with some magnetic reversals; however, there have been far more reversals than extinctions.

Magnetic Anomalies

A **magnetometer** is an instrument used to measure the strength of Earth's magnetic field. A magnetometer can be carried over the land surface or flown over land or sea. At sea, magnetometers can also be towed behind ships. They are also used as metal detectors in airports.

The strength of Earth's magnetic field varies from place to place. As with gravity, a deviation from average readings is called an *anomaly.* Very broad regional magnetic anomalies may be due to *circulation patterns in the liquid outer core* or to other deep-seated causes. Smaller anomalies generally reflect *variations in rock type,* for the magnetism of near-surface rocks adds to the main magnetic field generated in the core. Rocks differ in their magnetism, depending upon their content of iron-containing mineral, particularly magnetite.

A **positive magnetic anomaly** is a reading of magnetic-field strength that is higher than the regional average. Figure 4.20 shows three geologic situations that can cause positive magnetic anomalies. In figure 4.20*A,* a body of magnetite ore (a highly magnetic ore of the metal iron) has been emplaced in a bed of limestone by hot solutions rising along a fracture. The magnetism of the iron ore adds to the magnetic field of the Earth, giving a stronger magnetic-field measurement at the surface (a positive anomaly). In figure 4.20*B* a large dike of gabbro has intruded into granitic basement rock. Because gabbro contains more ferromagnesian minerals than granite, gabbro is more magnetic and causes a positive magnetic anomaly. Figure 4.20*C* shows a granitic basement high (perhaps originally a hill) that has influenced later sediment deposits, causing a draping of the layers as the sediments on the hilltop compacted less than the thicker sediments to the sides. Such a structure can form an *oil trap* (see chapter 12). The granite in the hill contains more iron in its ferromagnesian minerals than the surrounding sedimentary rocks, so a small positive magnetic anomaly occurs where the granite is closer to the surface. Note how each example shows horizontal sedimentary rocks at the surface, with no surface hint of the subsurface geology. The magnetometer helps find hidden ores and geologic structures.

A **negative magnetic anomaly** is a reading of magnetic-field strength that is lower than the regional average. Figure 4.21 shows how a negative anomaly can be produced by a down-

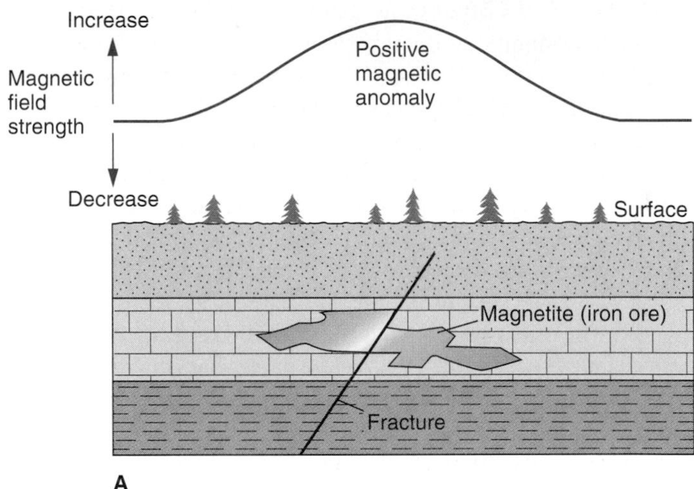

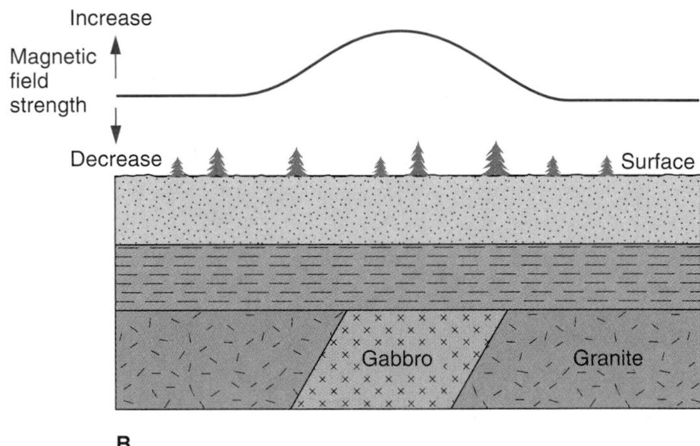

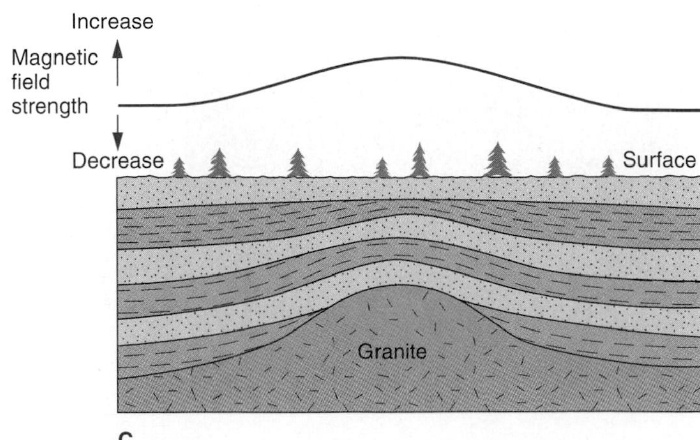

FIGURE 4.20
Positive magnetic anomalies can indicate hidden ore and geologic structures.

dropped fault block (a *graben*) in igneous rock. The thick sedimentary fill above the graben is less magnetic than is the igneous rock, so a weaker field (a negative magnetic anomaly) develops over the thick sediment.

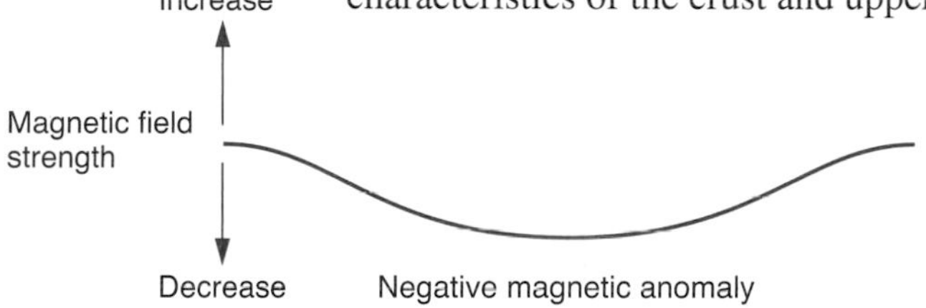

FIGURE 4.21

A graben filled with sediment can give a negative magnetic anomaly if the sediment contains fewer magnetic minerals than the rock beneath it.

Not all local magnetic anomalies are caused by variations in rock type. The linear magnetic anomalies found in oceanic crust are caused by a *variation in the direction of magnetism,* as discussed in chapter 2. Extremely small variations in the Earth's electromagnetic field are now being used to identify structural characteristics of the crust and upper mantle (see box 4.6).

GEOMATICS 4.6

Magnetotellurics: A New Tool for Investigating the Earth's Interior

Spatial analysis of digital data is extensively used in geophysical studies of the Earth's Interior. **Magnetotellurics** is a new geophysical approach that is being used in remote regions of the Canadian Arctic to investigate and map structures within the underlying crust and mantle. The approach is based on the measurement of very small variations in the Earth's electrical and magnetic fields (box figure 1). Electromagnetic energy released from the sun penetrates the Earth and as it does so interacts with the materials thorough which it passes. These interactions can be measured on the surface of the Earth using very sensitive equipment. The Earth's magnetic field is roughly 50,000 nano Tesla (a Tesla is a unit of magnetic field strength), and the equipment used in magnetotellurics can measure changes smaller than 0.03 nano Tesla. Sophisticated processing and modelling of the data collected allow scientists to produce maps showing the electrical resistivity characteristics of the subsurface (box figure 2).

The maps produced by magnetotelluric surveys are used for a variety of purposes including the search for economic ore bodies, to find geothermal energy, to look for fluids and for groundwater contamination, and to aid in the assessment of earthquake risk. Scientists from the Geological Survey of Canada are conducting magnetotellurics surveys of Baffin Island to learn more about the structure of ancient Canadian Shield and the underlying crust and upper mantle.

BOX 4.6 ■ FIGURE 1

Geoscientist setting up magnetotelluric equipment on Baffin Island.

Photo by Alan Jones

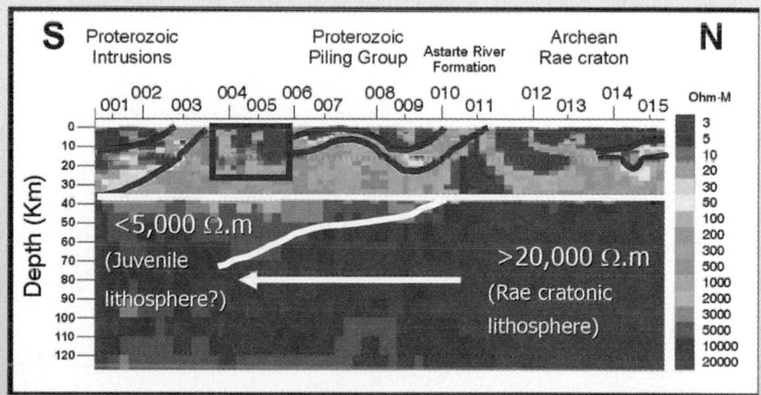

BOX 4.6 ■ FIGURE 2

Preliminary magnetotelluric model of Baffin Island.

From http://gsc-cgd.nrcan.gc.ca/baffin4d/proj_e.asp?id=5
Courtesy of Alan Jones

HOW HOT IS THE EARTH'S CORE? WHAT IS THE ORIGIN OF THE EARTH'S HEAT?

Geothermal Gradient

The rate of temperature increase with depth into Earth is called the **geothermal gradient.** The geothermal gradient can be measured on land in abandoned wells or on the sea floor by dropping specially designed probes into the mud. The average temperature increase is 25°C per kilometre of depth. Some regions have a much higher gradient, indicating concentrations of heat at shallow depths. Such regions have a potential for generating *geothermal energy* (discussed in chapter 15).

The temperature increase with depth creates a problem in deep mines, such as in a 3-kilometre-deep gold mine in South Africa, where the temperature is close to the boiling point of water. Deep mines must be cooled by air-conditioning for the miners to survive. High temperatures at depth also complicate the drilling of deep oil wells. A well drilled to a depth of 7 or 8 kilometres must pass through rock with a temperature of 200°C. At such high temperatures, a tough steel drilling pipe will become soft and flexible unless it is cooled with a special mud solution pumped down the hole.

Geoscientists hypothesize that the geothermal gradient must taper off sharply a short distance into the Earth. The high values of 25°C/kilometre recorded near the Earth's surface could not continue very far into the Earth. If they did, the temperature would be 2,500°C at the shallow depth of 100 km. This temperature is above the melting point of all rocks at that depth—even though the increased pressure with depth into the Earth increases the melting point of rocks. Seismic evidence seems to indicate a solid, not molten, mantle, so the geothermal gradient must drop to values as low as 0.3°C/kilometre within the mantle (figure 4.22*A*).

At the boundary between the inner core and the outer core, there would be some constraints on possible temperatures if the core is molten metal above the boundary and solid metal below. The weight of the thick rock layer of the mantle and the liquid metal of the outer core raises the pressure at this boundary (figure 4.22*B*) to about 3 million atmospheres. (An *atmosphere* of pressure is the force per unit area caused by the weight of the air in the atmosphere. It is about 1 kilogram per square centimetre.)

Using geophysical and geochemical data, in addition to computer modelling and high-pressure experiments, the internal temperature of the Earth can be estimated. Recent laboratory experiments with pressure anvils and giant guns have created (for a millionth of a second) the enormous pressures found at the centre of the Earth. The measured temperature at this pressure was far higher than expected. New estimates of the Earth's internal temperatures have resulted: 3,800°C at the core–mantle boundary, 6,300°C ± 800°C at the inner-core/outer-core boundary, and 6,400°C ± 600°C at Earth's centre (hotter than the surface of the sun!).

Heat Flow

A small but measurable amount of heat from the Earth's interior is being lost gradually through the surface. This gradual loss of heat through Earth's surface is called **heat flow.** What is the origin of the heat? It could be "original" heat from the time that Earth formed; that is, *if* the Earth formed as a mass of planetesimals that coalesced and compressed the inner material. Or the heat

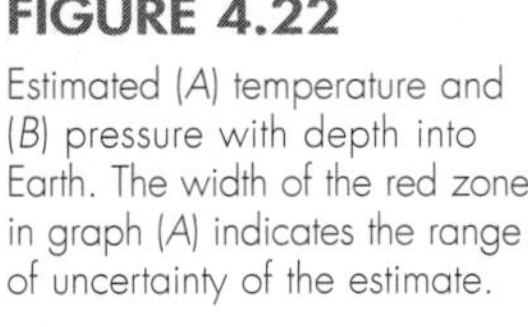

FIGURE 4.22

Estimated (*A*) temperature and (*B*) pressure with depth into Earth. The width of the red zone in graph (*A*) indicates the range of uncertainty of the estimate.

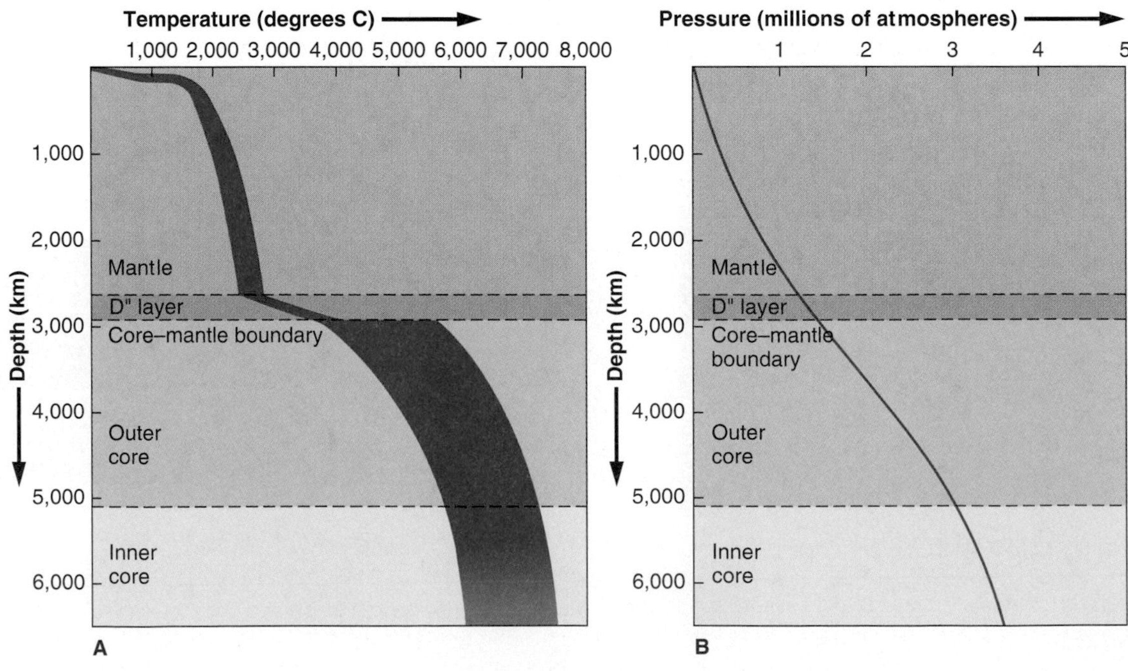

could be a by-product of the decay of radioactive isotopes inside Earth. Radioactive decay *may* actually be warming up the planet. Geoscientists are not sure whether Earth formed as a hot or cold mass, or whether the planet is now cooling off or warming up. Changes in Earth's internal temperature are extremely slow (on the order of 100 million years), and trying to work out its thermal history is a slow, often frustrating job.

Some regions on Earth have a high heat flow. More heat is being lost through the surface in these regions than is normal. High heat flow is usually caused by the presence of a magma body or still-cooling pluton near the surface (figure 4.23). An old body of igneous rock that is rich in uranium and other radioactive isotopes can cause a high heat flow, too, because radioactive decay produces heat as it occurs. High heat flow over an extensive area may be due to the rise of warm mantle rock beneath abnormally thin crust.

The average heat flow from continents is the same as the average heat flow from the sea floor, a surprising fact if you consider the greater concentration of radioactive material in continental rock (figure 4.24). The unexpectedly high average heat flow under the ocean may be due to hot mantle rock rising slowly by convection under parts of the ocean. Regional patterns of high heat flow and low heat flow on the sea floor (heat flow decreases away from the crest of the mid-oceanic ridge) may also be explained by convection of mantle rock, as we discussed in chapter 2.

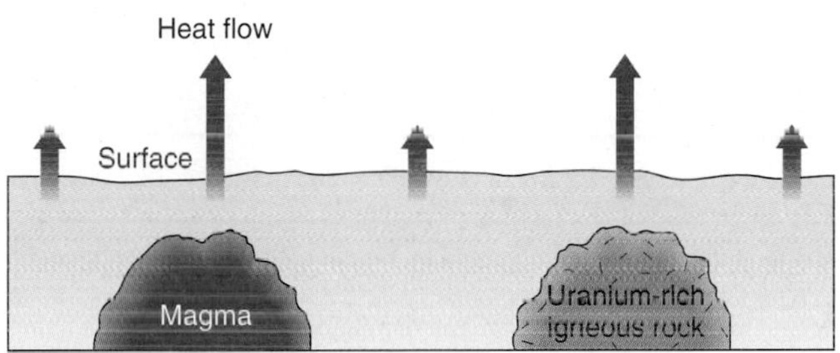

FIGURE 4.23

Some regions have higher heat flow than others; the amount of heat flow is indicated by the length of the arrow. Regions of high heat flow may be underlain by cooling magma or uranium-rich igneous rock.

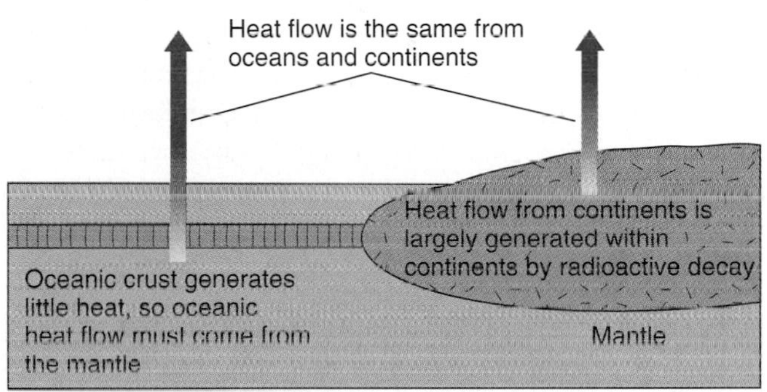

FIGURE 4.24

The average heat flow from oceans and continents is the same, but the origin of the heat differs from the ocean to continents.

SUMMARY

The interior of Earth is studied indirectly by *geophysics*—a study of seismic waves, gravity, Earth magnetism, and Earth heat.

Seismic reflection and *seismic refraction* can indicate the presence of boundaries between rock layers.

Earth is divided into three major units—the *crust*, the *mantle*, and the *core*.

The crust beneath oceans is 7 km thick and made of basalt on top of gabbro. Continental crust is 30 to 50 km thick and consists of a crystalline basement of granite and gneiss (and other rocks) capped by sedimentary rocks.

The *Mohorovičić discontinuity* separates the crust from the mantle.

The mantle is a layer of solid rock 2,900 km thick and is probably composed of an ultramafic rock such as peridotite. Seismic waves show the mantle has a structure of concentric shells, perhaps caused by pressure transformations of minerals.

The *lithosphere*, which forms plates, is made up of brittle crust and upper mantle. It is 70 to 125 (or more) km thick and moves over the plastic asthenosphere.

The *asthenosphere* lies below the lithosphere and may represent rock close to its melting point (seismic waves slow down here). It is probably the region of most magma generation and isostatic adjustment.

Seismic-wave shadow zones show the core has a radius of 3,450 km and is divided into a liquid outer core and a solid inner core. A core composition of mostly iron is suggested by Earth's density, the composition of meteorites, and the existence of Earth's magnetic field.

Isostasy is the equilibrium of crustal columns floating on plastic mantle. *Isostatic adjustment* occurs when weight is added to or subtracted from a column of rock. *Crustal rebound* is isostatic adjustment that occurs after the melting of glacial ice.

A *gravity meter* can be used to study variations in rock density or to find regions that are out of isostatic equilibrium.

A *positive gravity anomaly* forms over dense rock or over regions being held up out of isostatic balance. A *negative gravity anomaly* indicates low-density rock or a region being held down.

Earth's *magnetic field* has two *magnetic poles*, probably generated by convection circulation and electric currents in the outer core.

Some rocks record Earth's magnetism at the time they form. *Paleomagnetism* is the study of ancient magnetic fields.

Magnetic reversals of polarity occurred in the past, with the north magnetic pole and south magnetic pole exchanging positions. Isotopic dating of rocks shows the ages of the reversals.

A *magnetometer* measures the strength of the magnetic field.

A *positive magnetic anomaly* develops over rock that is more magnetic than neighbouring rock. A *negative magnetic anomaly* indicates rock with low magnetism.

Magnetic anomalies can also be caused by circulation patterns in Earth's core and by variations in the direction of rock magnetism.

The *geothermal gradient* is about 25°C/km near the surface but decreases rapidly at depth. The temperature at the centre of the Earth may be 6,900°C. *Heat flow* measurements show that heat loss per unit area from continents and oceans is about the same, perhaps because of convection of hot mantle rock beneath the oceans.

Terms to Remember

asthenosphere 109
convection 114
core 107
crust 107
crustal rebound 115
Curie point 120
geophysics 104
geothermal gradient 124
gravity meter 117
heat flow 124
isostasy 115
isostatic adjustment 115
lithosphere 109
magnetic field 118
magnetic pole 118
magnetic reversal 118
magnetometer 122
magnetotellurics 123
mantle 107
Mohorovičić discontinuity (Moho) 108
negative gravity anomaly 117
negative magnetic anomaly 122
paleomagnetism 121
positive gravity anomaly 117
positive magnetic anomaly 122
P-wave shadow zone 111
seismic reflection 104
seismic refraction 104
S-wave shadow zone 111

Testing Your Knowledge

Use the questions below to prepare for exams based on this chapter.

1. Describe how seismic reflection and seismic refraction show the presence of layers within Earth.
2. Sketch a cross-section of the entire Earth showing the main subdivisions of Earth's interior and giving the name, thickness, and probable composition of each.
3. What facts make it probable that the Earth's core is composed of mostly iron?
4. Describe the differences between continental crust and oceanic crust.
5. What is a gravity anomaly, and what does it generally indicate about the rocks in the region where it is found?
6. Discuss seismic-wave shadow zones and what they indicate about the Earth's interior.
7. Describe the Earth's magnetic field. Where is it generated?
8. What is the temperature distribution with depth into the Earth?
9. Heat flow has been found to be about equal through continents and the sea floor. Why was this unexpected? What might cause this equality?
10. What is the Mohorovičić discontinuity?
11. What is the asthenosphere? Why is it important?
12. How does the lithosphere differ from the asthenosphere?
13. What is a magnetic reversal? What is the evidence for magnetic reversals?
14. What is a magnetic anomaly? How are magnetic anomalies measured at sea?
15. *Felsic* and *mafic* are terms used by some geoscientists to describe
    a. composition of continental and oceanic crust
    b. behaviour of earthquake waves
    c. regions in the mantle
16. The boundary that separates the crust from the mantle is called the
    a. lithosphere
    b. asthenosphere
    c. Mohorovičić discontinuity
    d. none of the above
17. The core is probably composed mainly of
    a. silicon b. sulphur
    c. oxygen d. iron
18. The principle of continents being in a buoyant equilibrium is called
    a. subsidence b. isostasy
    c. convection d. rebound
19. A positive gravity anomaly indicates that
    a. tectonic forces are holding a region up out of isostatic equilibrium
    b. the land is sinking
    c. local mass deficiencies exist in the crust
    d. all of the above
20. A positive magnetic anomaly could indicate
    a. a body of magnetic ore
    b. the magnetic field strength is higher than the regional average
    c. an intrusion of gabbro
    d. the presence of a granitic basement high
    e. all of the above

21. Which of the following is not an example of the effects of isostasy?
    a. deep mountain roots
    b. magnetic reversals
    c. the post-glacial rise of northeastern North America
    d. mountain ranges at subduction zones

22. The S-wave shadow zone is evidence that
    a. the core is made of iron and nickel
    b. the inner core is solid
    c. the outer core is fluid
    d. the mantle is plastic

Exploring Web Resources

www.mcgrawhill.ca/olc/plummer
Visit the Online Learning Centre for Mantle Xenoliths—A Peek at the Deep. Here you can read what geoscientists believe the upper mantle looks like. You can also check your answers for the Testing Your Knowledge section and click on the direct links to the Websites listed below.

http://rses.anu.edu.au/gfd/Gfd_other_pages/Convection_demo/Demo_page_1.html
The Geophysical Fluid Dynamics Group Website contains images of mantle convection models.

http://ees5-www.lanl.gov/IGPP/Geodynamo.html
Core Convection and Geodynamo Website discusses the recent model for reversals of Earth's magnetic field.

4.2 Seismic Refraction
4.12 Isostatic Adjustment of Mountains and Sedimentary Basins
4.13 Isostatic Adjustment by Glacial Rebound

Animations

This chapter includes the following animations available on our Online Learning Centre at www.mcgrawhill.ca/olc/plummer.

4.8, 4.9 P and S wave shadow zones
4.11 How Isostacy, Orogeny, and Metamorphism are Interrelated
4.12 Isostatic Adjustment of Mountains and Sedimentary Basins

CHAPTER

5

Atoms, Elements, and Minerals

What are minerals?
What are atoms and elements?
How do minerals vary in structure and composition?
How do we identify minerals?
How do minerals form?

This chapter is the first of eight on the materials of which the Earth is made. The following chapters are mostly about rocks. Nearly all rocks are made of minerals. Therefore, to be ready to learn about rocks, you must first understand what minerals are as well as the characteristics of some of the most common minerals.

In this chapter, you are introduced to some basic principles of chemistry. This will help you understand material covered in the chapters on rocks, weathering, and the composition of Earth's crust and its interior. You will discover that each mineral is composed of specific chemical elements, the atoms of which are in a remarkably orderly arrangement. A mineral's chemistry and the architecture of its internal structure determine the physical properties used to distinguish it from other minerals. You should learn how to readily determine physical properties and use them to identify common minerals. (Appendix A is a further guide to identifying minerals.)

Crystals of tourmaline (variety: elbaite). Differences in colour within each crystal are due to small changes in chemical composition incorporated into the minerals as they grew. *Photo © Parvinder Sethi*

WHAT ARE MINERALS?

Introduction

A quick glance at a rock may show some colour or pattern, perhaps a speckled appearance that may be attractive to a collector but otherwise holds no meaning for most people. But these colours and patterns, made up of packages of matter called *minerals,* tell a very important story about the origin of our world, and indeed about all Earth-like planets. The considerable information conveyed by minerals enriches our appreciation for nature, and perhaps gives us more reason to take good care of it.

Minerals are compositionally and physically distinctive substances (figure 5.1). Each mineral type develops in a particular way under natural conditions, guided by the principles of chemistry. Factors such as heat, pressure, oxygen, available atoms, and acid content all play a role in determining how minerals form.

There are about 4,500 kinds of minerals in the world, with only a couple hundred that are really common and only a couple dozen that form the majority of all rocks. Each type of mineral is distinguished by a combination of properties, some of which we can see with the unaided eye, others that are discernable only at the microscopic and atomic levels. Examples of these properties include colour, lustre, hardness, chemical composition, and the transmission of light under a microscope. Minerals are so important and so easily distinguishable that geologists use them as the basis for classifying almost all rocks.

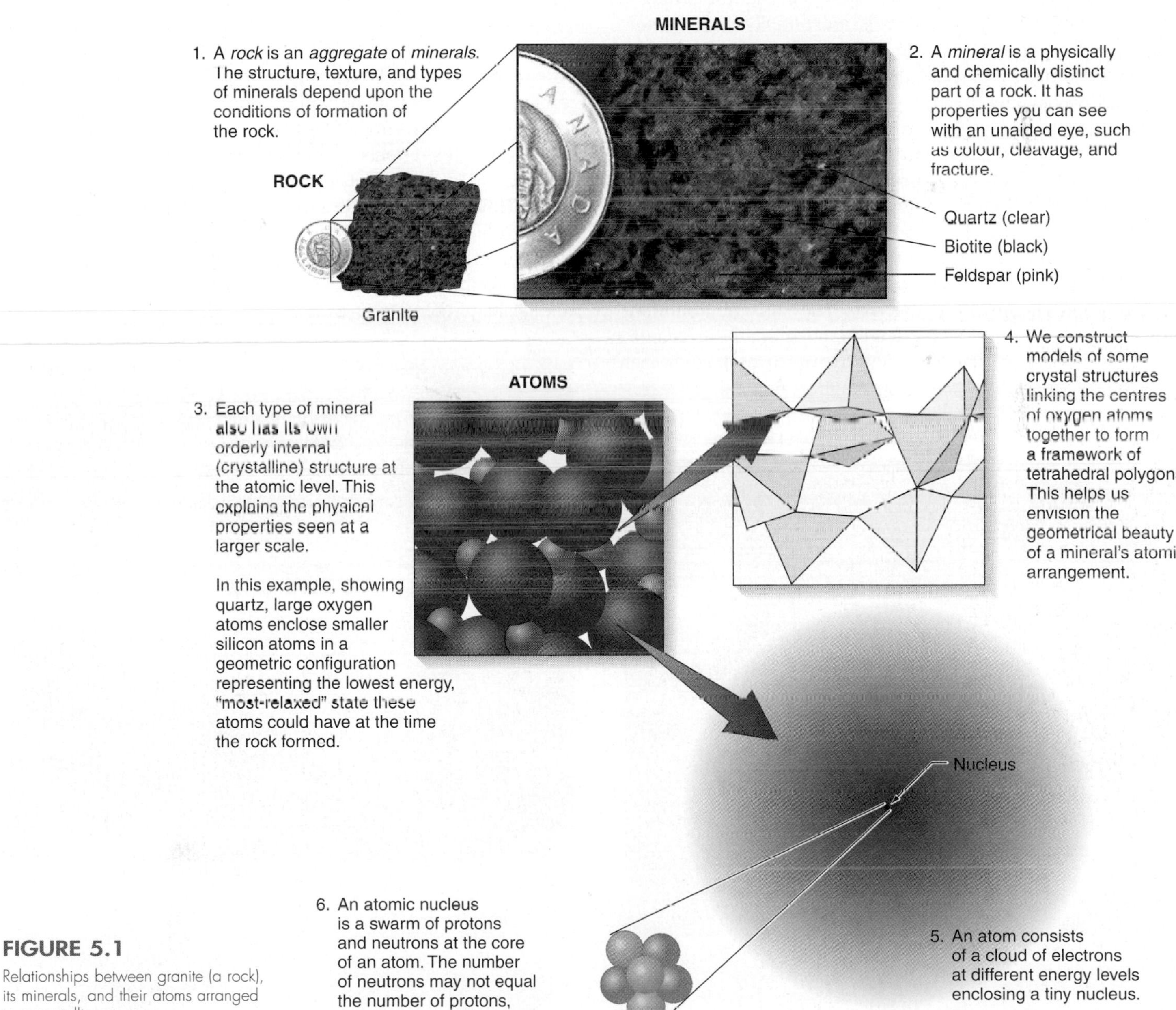

FIGURE 5.1

Relationships between granite (a rock), its minerals, and their atoms arranged in a crystalline structure.

Photo by S. Fletcher

What most people call *crystals* essentially are just "perfectly formed" minerals. They are mineral specimens whose surfaces consist of faces, edges, and corners that show beautiful geometrical symmetry. For example, the quartz you can see lining the interior of a geode (a natural pocket in certain kinds of volcanic rocks; figure 5.2) forms crystals whose smooth faces sparkle brightly in reflected light. Most mineral grains in rocks do not display the faces that we associate with crystals. Yet, geoscientists often call them *crystals* anyway because they have an orderly arrangement of their constituent atoms. It would be less ambiguous to say they are *crystalline* substances. The irregular outlines of most mineral grains seen in rocks results from the fact that these minerals form simultaneously under conditions of close confinement, or grow to fill the gaps left between earlier-forming minerals.

The existence of crystals reflects the fact that the arrangements of atoms within *all* minerals are *orderly and regular.* Detailed X-ray study of minerals shows that there are 230 different kinds of symmetrical atomic arrangements possible in nature and each mineral type exhibits one or several of these 230 arrangements. Minerals are said to be *crystalline* because of this universal internal property. A **crystalline** substance is one in which the atoms are arranged in a three-dimensional, regularly repeating, orderly pattern. Figure 5.3 is a model of the **crystal structure** of one mineral.

Putting it all together, then, **minerals** can be defined as a family of naturally occurring, crystalline substances that are physically and chemically distinctive. They are the "building blocks" of rocks. Minerals also are compositionally inorganic; that is, they don't consist of carbon-hydrogen molecules that also form crystalline substances through biological processes (sugars, for example).

FIGURE 5.2

Quartz crystals line the inside of a geode. This purple variety of quartz is known as amethyst.

Photo © Parvinder Sethi

When vitamin advertisers and nutritional specialists talk about "minerals," they are not, of course, referring to the strictly geologic definition, but rather to single elements, such as calcium or magnesium, that have certain dietary benefits. Commercial processing of true minerals yields these popular ingredients. In reality, most true minerals are complex assemblages of multiple elements.

The chemical formula of a mineral represents not only the types of atoms in the mineral, but their relative proportions as well, expressed in lowest whole number ratios. For example, quartz is made up exclusively of oxygen and silicon atoms. More precisely, quartz contains twice as many oxygen (O) atoms as silicon (Si) atoms. Therefore, the chemical formula for quartz is SiO_2, its specific composition.

Another common mineral is halite (rock salt), whose chemical composition is NaCl. This means that halite is composed of *equal numbers* of sodium (Na) and chlorine (Cl) atoms. These atoms are arranged in an orderly, three-dimensional lattice that resembles a stack of boxes (figure 5.4). This imparts an overall cubic shape to crystals of halite.

Consider the formula of a more common mineral, feldspar—$KAlSi_3O_8$. This reflects not only a more complex composition, but also a less symmetric atomic arrangement than that of halite. How do the atoms in a mineral like feldspar stick together? Why are minerals crystalline at all? Science reveals an underlying order to physical reality that is breathtaking and largely hidden from view when we look at the apparent randomness and chaos of the natural world.

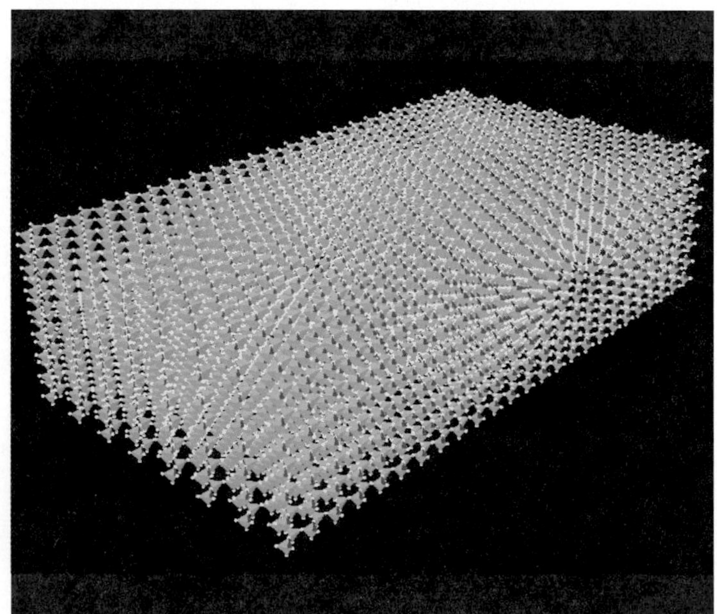

FIGURE 5.3

Three-dimensional image of the structure of a framework silicate mineral such as quartz. It is particularly important to understand the structure of silicate minerals as they are the most common rock forming minerals found on Earth and account for nearly 75 percent (by weight) of the continental crust. The image was created using the simple schematic representation of silicon–oxygen tetrahedrons and has been modelled in ArcGIS with the 3D Analyst extension.

Image created by Patrick Kennelly, Long Island University

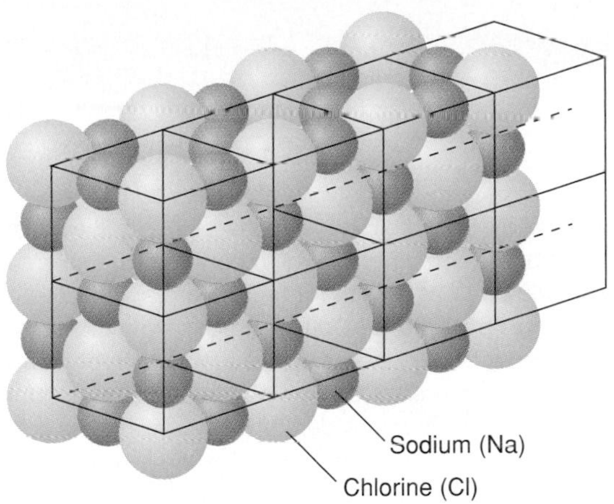

FIGURE 5.4

Model of the atomic structure of halite. The alternating three-dimensional stacking of atoms creates a box-like grid that is expressed in the cubic form of halite crystals seen in hand samples.

WHAT ARE ATOMS AND ELEMENTS?

To answer the questions just posed, we need to take a look at what is happening at an infinitesimally small scale.

Atoms are the smallest, electrically neutral assemblies of energy and matter that we know exist in the universe. It is important to understand what "electrically neutral" means. Many of us have the misfortune of knowing electrical force as a sharp jolt that occurs when we accidentally touch a live wire. This force results when tiny, charged particles called **electrons** flow from one place to another; for example, along a wire. Physicists say that the electrons carry a negative charge—the electrical force that we exploit to power the world.

IN GREATER DEPTH 5.1

Atomic Number, Atomic Mass Number, Isotopes, and Atomic Weight

The number of protons in an atom controls the "behaviour" of an element more than does the number of other subatomic particles. The **atomic number** of an element is the number of protons in each atom. As per our earlier definition of an element, each atom of an element has the *same number of protons*.

The **atomic mass number** is the total number of neutrons and protons in an atom. The atomic mass number of the oxygen atom shown in box figure 1 is 16 (8 protons plus 8 neutrons) and is indicated by the symbol ^{16}O. Heavier elements have more neutrons and protons than do lighter ones. For example, the heavy element gold has an atomic mass number of 197, whereas helium has an atomic mass number of only 4.

Isotopes of an element are atoms containing different numbers of neutrons but the same number of protons. Isotopes are either stable or unstable. An unstable, or *radioactive,* isotope is one in which protons or neutrons are, over time, spontaneously lost from the nucleus. The subatomic particles that unstable isotopes emit are what Geiger counters detect. This is *radioactivity,* which we know can be hazardous in high doses. Unstable isotopes of uranium and a few other elements are very important to geology because they are used to determine the ages of rocks. These isotopes decay at a known rate and, as described in chapter 19, are used as a kind of geologic stopwatch that starts running at the time some rocks form.

A *stable isotope* is an isotope that will retain all of its protons and neutrons through time. During recent years, stable isotopes have become increasingly important to geology and related sciences. The isotopes most commonly studied in geology are those of carbon, nitrogen, oxygen, sulphur, and hydrogen. Their usefulness in scientific investigations is due to the tendency of isotopes of a given element to partition (distribute preferentially between substances) in different proportions due to their minute weight difference. For instance, oxygen and hydrogen isotopes can be used as a proxy for the surface temperature of the Earth because when water vapour evaporates from liquid water, the vapour will have a slightly higher ratio of lighter to heavier isotopes compared to the isotopes that remain in the liquid. Box 5.2 describes one of the applications of oxygen isotopes.

An element's atomic weight is closely related to the mass number. **Atomic weight** is the weight of an *average* atom of an element, given in atomic mass units. Because sodium has only one naturally occurring isotope, its atomic mass number and its atomic weight are the same—23. On the other hand, chlorine has two common isotopes, with mass numbers of 35 and 37. The atomic weight of chlorine, which takes into account the abundance of each isotope, is about 35.5 because the lighter isotope is more common than the heavier one.

BOX 5.1 ■ FIGURE 1
Model of an oxygen atom and its nucleus.

IN GREATER DEPTH 5.2

Oxygen Isotopes and Climate Change

Oxygen has three stable isotopes. ^{16}O (the 16 tells us there are 16 protons and neutrons in the nucleus) is most abundant, making up 99.762% of Earth's oxygen. ^{17}O constitutes 0.038%, and ^{18}O, 0.200%. The ratio of ^{18}O to ^{16}O in a substance is determined using very accurate instruments called *mass spectrometers.* The ratio of ^{18}O to ^{16}O is 0.0020:1. If partitioning did not take place, we would expect to find the same ratio of isotopes in any substance containing oxygen. However, there is considerable deviation because of the tendency of lighter and heavier atoms to partition.

Water that evaporates or is given off by respiration of plants or animals will have a slightly higher abundance of the lighter isotope (^{16}O) relative to the heavier isotope (^{18}O). ^{16}O-rich water evaporated from the oceans will fall as precipitation on land, either as rain or snow, and will become concentrated in large ice sheets during times of extensive glaciation (box figure 1). When large amounts of ^{16}O are "locked up" in ice masses on land during glacial periods, ocean waters become relatively enriched in the heavier ^{18}O isotope. As glacial ice melts during warm or interglacial periods, the ^{16}O-rich water returns to the oceans, changing the oxygen isotope ratios in the oceans, atmosphere, plants, and animals.

Scientists have been able to reconstruct changes in the volume of ice on Earth during the recent geologic past by studying the record of oxygen isotopes preserved in fossils found in ocean-floor deposits. *Foraminifera* are microscopic organisms that live in ocean waters. While they are alive, they grow shells of calcite ($CaCO_3$), incorporating oxygen from the seawater. The oxygen in the shells has the same $^{18}O/^{16}O$ ratio as that of the seawater in which they live and reflects the amount of glacial ice on land. When foraminifera die, their shells settle to the deep ocean floor and accumulate along with layers of mud. Deep-sea drilling retrieves cores of these mud layers and foraminifera are extracted from each of the layers. Foraminifera are analyzed and their $^{18}O/^{16}O$ ratios are determined. These data can then be used to infer global ice cover for the times the foraminifera were alive. Mud layers in the cores are also dated using paleomagnetic reversal stratigraphy (see chapter 4), and the oxygen isotope record can be used to reconstruct the timing of glacial and interglacial periods that occurred during the recent past (box figure 2). The oxygen isotope record is well established for the past 800,000 years and indicates that at least 11 glacial/interglacial cycles occurred during that time. The interesting question is what causes global climate to change so dramatically.

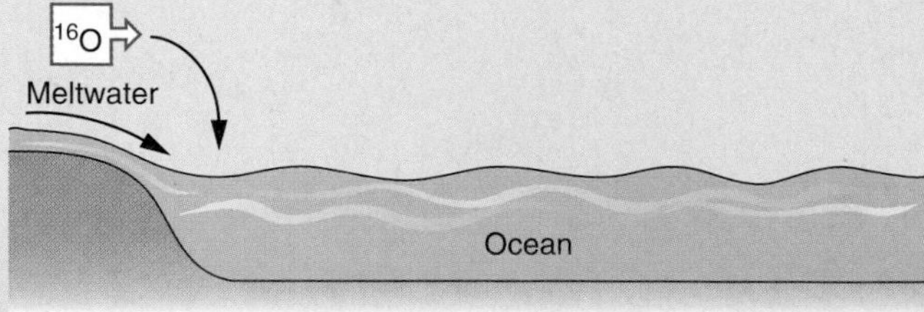

BOX 5.2 ■ FIGURE 1
Changing isotopic composition of ocean waters related to changing ice volumes.

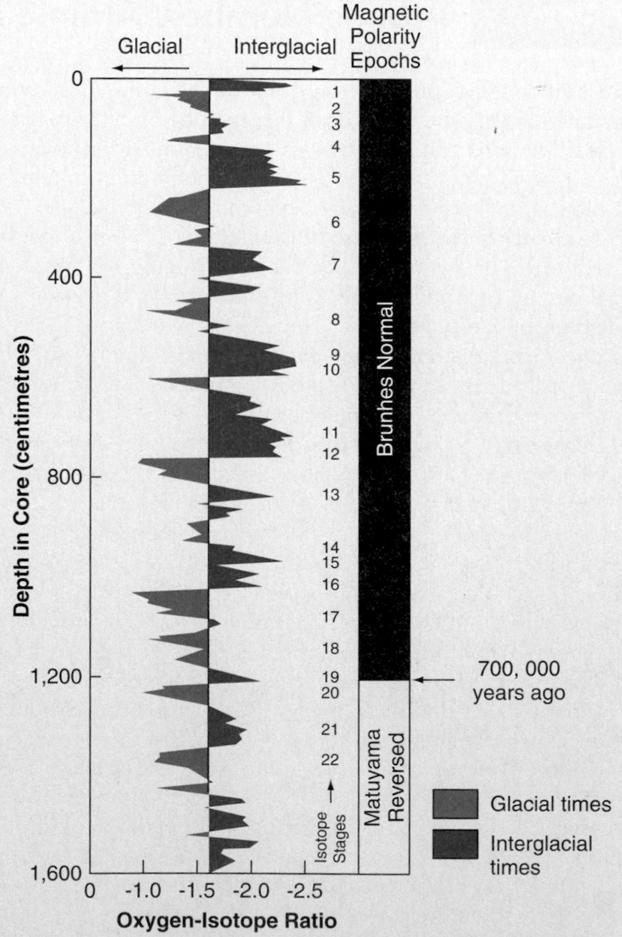

BOX 5.2 ■ FIGURE 2
Oxygen isotope record and glacial/interglacial periods identified from core taken from the Pacific Ocean. Oxygen isotope stages are numbered: odd numbers are interglacials, even numbers are glacials.

Electrons move in directions that allow them to balance out, or "neutralize," their charges. In atoms, electron charges are neutralized as the electrons crowd in around a central core of "positively charged" **protons.** The core, or **nucleus** of the atom, also contains **neutrons,** which are neutrally charged particles of some importance to geology (figure 5.1).

There are 92 different kinds of naturally occurring atoms arranged in order of increasing size and complexity on the periodic table (see appendix D) used by chemists. We call each "species" of atom an *element.* An **element** is defined by the number of protons in its nucleus. For example, oxygen has 8 protons. The number of protons in an atom is that element's *atomic number* (see box 5.1). For example, sodium, potassium, chlorine, and oxygen are all different kinds of elements. In addition to having 8 protons, each atom of oxygen contains 8 electrons and, in its most stable form, 8 neutrons. Chlorine, in contrast, brings together 17 electrons, 17 protons, and 18 neutrons in each atom. Notice that the number of neutrons need not match the numbers of protons and electrons in each atom (see box 5.1).

The electrons in an atom are continuously on the move, like bees buzzing around a hive. Some are more energetic than others and move farther away from the nucleus as they move in the space around it. Although each electron moves throughout the space surrounding the nucleus, it will spend most of its time as part of an *energy level.* (Energy levels used to be shown as concentric spherical shells, but chemists regard that as misleading.) The most stable configuration is to have complete energy levels. The first energy level is complete with 2 electrons. Helium has a complete energy level because it has 2 electrons that balance the 2 protons in its nucleus. The second and third energy levels are each complete with 8 electrons (see figure 5.5). For a more thorough explanation of atomic theory from a chemist's perspective, go to *Understanding Chemistry,* www.chemguide.co.uk/atommenu.html#top.

The linking together (**bonding**) of atoms to form minerals largely takes place because most individual atoms in a free state have a deficit or surplus of electrons in their outermost energy levels. They are not fully charge-neutral, in other words, and those with negative net charges are drawn toward those with positive net charges to address this imbalance. An electrically charged atom is called an **ion.** It is largely the combination of negative and positive ions in a stable, charge-neutral arrangement that makes up the crystalline structure of minerals (but also see box 5.3).

Ions and Crystalline Structures

The most important reason why ions exist is that a typical atom is most stable when each of its energy levels is *completely filled* with electrons—that is, when the electrical energy around the nucleus is compact and concentrated. The innermost energy level in the standard model of an atom is full when it possesses 2 electrons. The second and third orbital energy levels each require 8 electrons for an atom to be nonreactive. (Elements having additional energy levels are more complicated.)

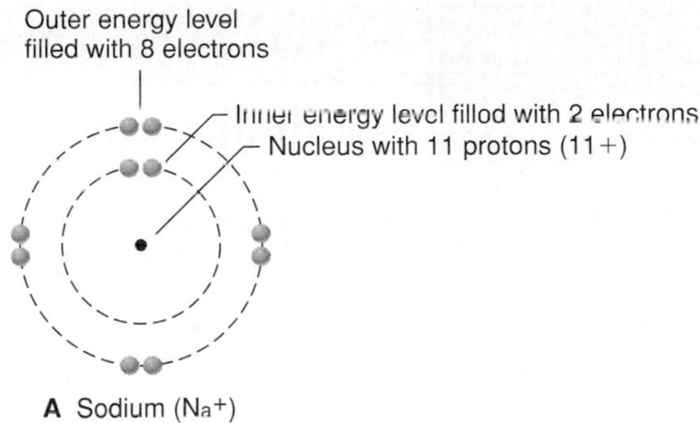

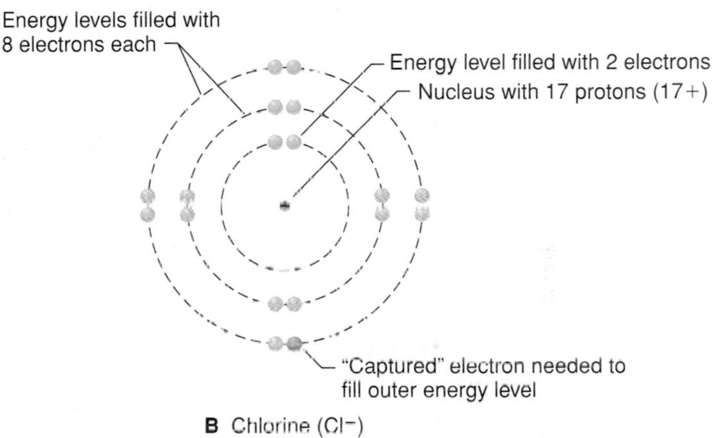

FIGURE 5.5

Diagrammatic representation of (*A*) sodium and (*B*) chlorine ions. The dots represent electrons in energy levels within an ion. Sodium has lost the electron that would have made it electronically neutral because a single electron in a higher energy level would be unstable. Chlorine has gained an electron to complete its outer energy level and make it stable.

Consider the two ions that bond to form halite—chlorine and sodium. Note that sodium, shown in figure 5.5, has a complete inner energy level with 2 electrons and a second energy level, also filled with 8 electrons. One more electron would neutralize the positive charge of the 11 protons in the nucleus, but an eleventh electron alone in an energy level is unstable, so the sodium atom gives it up if it can be taken up by *other* electron-deficient atoms. In each sodium ion, then, the 11 protons (11^+) and 10 electrons (10^-) add up to a single excess positive charge ($^+1$). Positively charged ions like this are called *cations.* Chemists customarily abbreviate the sodium cation as Na^+.

Chlorine, with an atomic number of 17, has a complete inner energy level with 2 electrons and a complete second energy level of 8 electrons around the inner level. A neutral chlorine atom would have only 7 electrons in the third level, but this level requires 8 electrons, so an extra electron is captured and incorporated. The chlorine ion then contains 18 electrons and 17 protons, and so have a single excess negative charge (Cl^-). Chlorine is an example of a negatively charged ion, or *anion.*

Several factors explain why ions such as chlorine and sodium bond together to form a crystalline solid. One is the need for different ions of like-charge to be as widely separated as possible.

IN GREATER DEPTH 5.3

Bonding

Ions may be regarded as tiny spheres that behave much like magnets. Positively charged ions attract negatively charged ions so that their electrical charges can be neutralized. In saltwater, equal numbers of sodium ions (Na^+) and chlorine ions (Cl^-) move about freely. The electrical neutrality of the water is maintained because positive sodium ions exactly balance negative chlorine ions. If the water evaporates, the sodium and chlorine are electrically attracted to each other and crystallize into halite. The crystalline structure is the most orderly way for chlorine and sodium ions to pack themselves together and neutralize their collective charges.

A chlorine ion and a sodium ion are fixed in place by their electrical attraction to each other. This is called **ionic bonding** because it is brought about by an attraction between positively and negatively charged ions (box figure 1).

Ionic bonding is the most common type of bonding in minerals. However, in most minerals the bonds between atoms are not purely ionic. Atoms are also commonly bonded together by **covalent bonding,** or bonding in which adjacent atoms *share* electrons. Diamond is composed exclusively of covalently bonded carbon atoms (box figure 2). Carbon has an atomic number of 6, which means that the innermost energy level is full with 2 electrons. Four more electrons are required to maintain electrical neutrality. In a diamond, each carbon atom has 4 electrons in the outer energy level to maintain neutrality, while the need for 8 electrons in that shell is satisfied by electrons that are shared with adjacent carbon atoms. Neighbouring carbon atoms are so close together that each of the outer-shell electrons spends half its time in one atom and half in an adjacent atom. Electrical neutrality is maintained, and each atom, in a sense, has 8 electrons in the outer energy level (even though they are not all there at the same time). Covalent bonds in the diamond are extremely strong, and diamond is the hardest natural substance on Earth. However, covalent bonds are not necessarily stronger than ionic bonds.

Graphite, like diamond, is pure carbon. Graphite is used in pencils and as a lubricant. Amazingly, the hardest mineral and one of the softest have the same composition. The distinction is in the bonding.

A third type of bonding, *metallic bonding,* is not as important to geology. In metals, such as iron or gold, the atoms are closely packed and the electrons move freely throughout the crystal so as to hold the atoms together. The ease with which electrons move accounts for the high electrical conductivity of metals.

Finally, after all atoms have bonded together, there may be weak, attractive forces remaining. This is the very weak force that holds adjacent sheets of mica or graphite together.

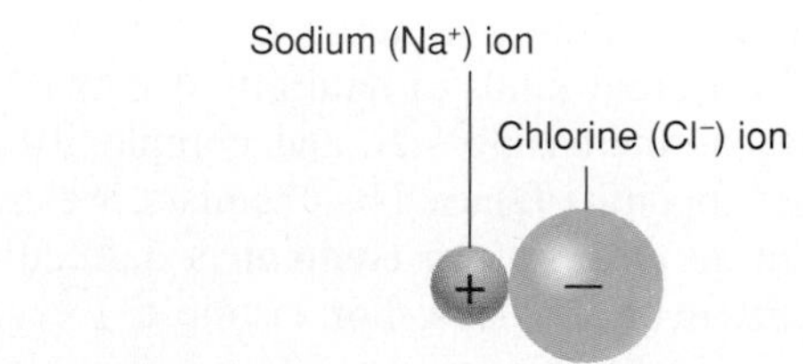

BOX 5.3 ■ FIGURE 1
Ionic bonding between sodium (Na^+) and chlorine (Cl^-).

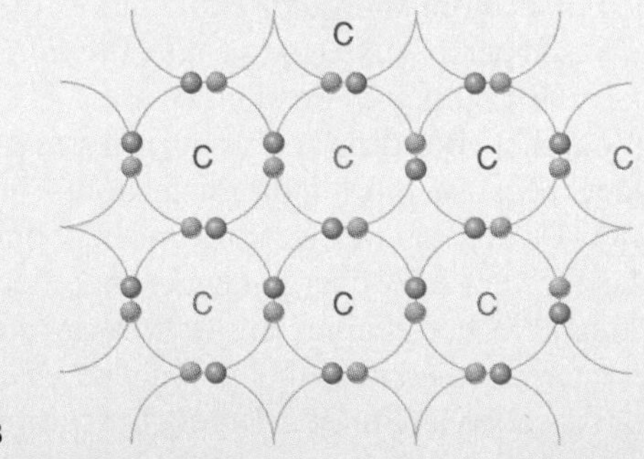

BOX 5.3 ■ FIGURE 2
Covalent bonding in diamond. (*A*) Three-dimensional arrangement of carbon atoms. The rods represent bonds between adjacent carbon atoms. (*B*) Schematic representation of carbon's covalent bonding. Each C represents a carbon nucleus with 6 protons and the 2 inner energy level electrons. Dots represent electrons in the outer energy level. For clarity, the colour of the "C" corresponds to the colour of the dots "belonging" to that atom.
Photo © Japack Company/Corbis

Recommended Web Investigation

To see how diamond and graphite differ, go to www.minweb.co.uk/Mineral_Web.html to see rotating, crystal structures in 3-D. (First you must download and install Chime software, which is easy to do from the Website.) From the nonsilicates pull-down menu, select graphite. Note the rotating crystal structure. The short rods connecting carbon atoms represent strong bonds (ignore the long, thin rods). You can use your mouse to stop the rotation and view the structure from any perspective. Then go to diamond. How do the two crystal structures differ? Can you see why plates of tightly bonded graphite slide easily past one another?

Under ordinary circumstances, like-charged ions repel one another and quickly move apart. They only come close to form a stable mineral structure because they are "glued" into place by bonding with ions of the opposite charge. In other words, the need to neutralize electrical charges while at the same time keeping like-charges apart works to create a regular arrangement of atoms.

The field of swarming electrons extends farther out from the atomic nuclei of some elements than it does for others. The "size" of an atom (or an ion) is essentially the radius of its electron field; its *ionic radius,* in other words. Ionic radii play an important role in the arrangement of atoms in a crystalline structure as well. When ions come together they tend to pack as efficiently as possible. No irregular holes may exist in the arrangement. A large number of small *anions* (negatively charged ions) may crowd around a single, large *cation* (positively charged ion), while only a few, large anions may cluster about a small cation (as in figure 5.6).

Of particular importance in this respect are the crystal structures derived from the two most common elements in the Earth's crust—oxygen and silicon (box 5.4).

Silicon is the *element used to make computer chips.* **Silica** is a term for *oxygen combined with silicon.* Because silicon is the second most abundant element in the crust, most minerals contain silica. The common mineral quartz (SiO_2) is pure silica that has crystallized. Quartz is one of many minerals that are **silicates,** substances that contain silica (as indicated by their chemical formulas). Most silicate minerals also contain one or more other elements.

The Silicon–Oxygen Tetrahedron

Silicon and oxygen combine to form the atomic framework for most common minerals on Earth. The basic structural unit consists of 4 oxygen atoms (anions) packed together around a single, much smaller silicon atom, as shown in figure 5.6*A*. The four-sided, pyramidal, geometric shape called a *tetrahedron* is used to represent the 4 oxygen atoms surrounding a silicon atom. Each *corner* of the tetrahedron represents the *centre* of an oxygen atom (figure 5.6*B*). This basic building block of a crystal is called a **silicon–oxygen tetrahedron** (also known as a *silica tetrahedron*); see figure 5.3.

The atoms of the tetrahedron are strongly bonded together. Within a silicon–oxygen tetrahedron, the negative charges exceed the positive charges (see figure 5.7*A*). A single silicon–oxygen tetrahedron is a complex ion with a formula of SiO_4^{-4} because silicon has a charge of +4 and the 4 oxygen ions have 8 negative charges (−2 for each oxygen atom).

A silicon–oxygen tetrahedron can either bond with positively charged ions, such as iron or aluminum, or with other silicon–oxygen tetrahedrons. In other words, for the silicon–oxygen tetrahedron to be stable within a crystal structure, it must either (1) be balanced by enough positively charged ions or (2) share oxygen atoms with adjacent tetrahedrons (as shown in figures 5.7*C* and *D*) and therefore reduce the need for extra, positively charged ions. The structures of silicate minerals range from an *isolated silicate structure,* which depends entirely on positively charged ions to hold the tetrahedrons together, to *framework silicates* (quartz, for example), in which all oxygen atoms are shared by adjacent tetrahedrons. The most common types of silicate structures are shown diagrammatically in figure 5.8 and are discussed next.

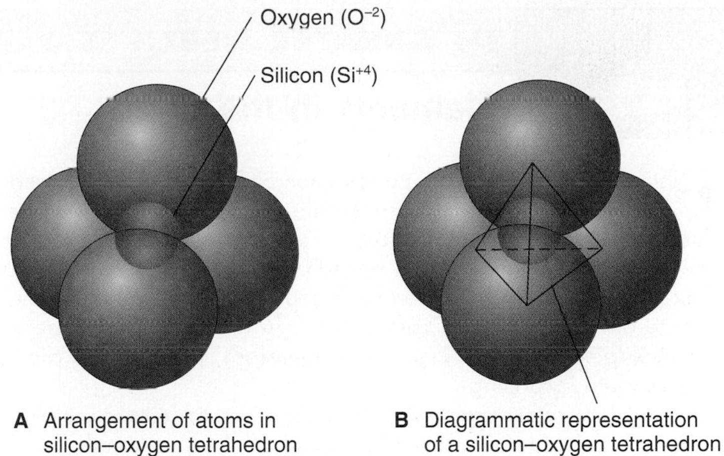

FIGURE 5.6

(*A*) The silicon–oxygen tetrahedron. (*B*) The silicon–oxygen tetrahedron showing the corners of the tetrahedron coinciding with the centres of oxygen ions.

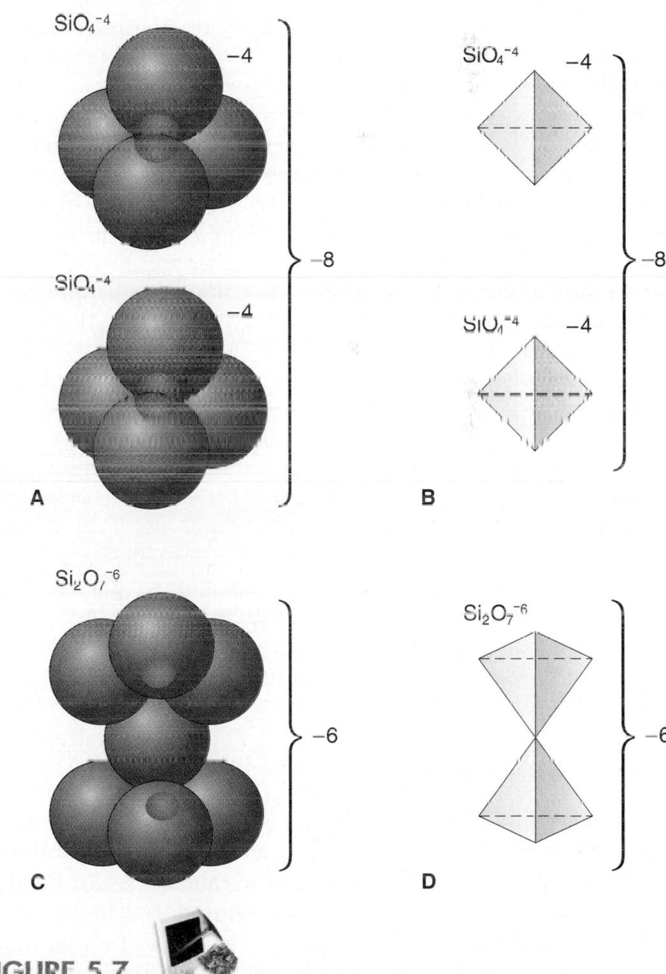

FIGURE 5.7

Two single tetrahedrons (*A* and *B*) require more positively charged ions to maintain electrical neutrality than two tetrahedrons sharing an oxygen atom (*C* and *D*). *B* and *D* are the schematic representations of *A* and *C*, respectively.

IN GREATER DEPTH 5.4

Elements in the Earth

Estimates of the chemical composition of Earth's crust are based on many chemical analyses of the rocks exposed on Earth's surface. (Models for the composition of the interior of the Earth—the core and the mantle—are based on more indirect evidence.) Box figure 1 shows the generally accepted estimates of the abundance of elements in the Earth's crust. At first glance, the chemical composition of the crust (and, therefore, the average rock) seems quite surprising.

We think of oxygen as the O_2 molecules in the air we breathe. Yet most rocks are composed largely of oxygen, as it is the most abundant element in the Earth's crust. Unlike the oxygen gas in air, oxygen in minerals is strongly bonded to other elements. By weight, oxygen accounts for almost half the crust, but it takes up 93 percent of the volume of an average rock. This is because the oxygen atom takes up a large amount of space relative to its weight. (Note how much bigger oxygen atoms are relative to other atoms in figure 5.6 and others.) It is not an exaggeration to regard the crust as a mass of oxygen with other elements occupying positions in crystalline structures between oxygen atoms.

Note that the third most abundant element is aluminum, which is more common in rocks than iron. Knowing this, one might assume that aluminum would be less expensive than iron, but of course this is not the case. Common rocks are not mined for aluminum because it is so strongly bonded to oxygen and other elements. The amount of energy required to break these bonds and separate the aluminum makes the process too costly for commercial production. Aluminum is mined from the uncommon deposits where aluminum-bearing rocks have been weathered, producing compounds in which the crystalline bonds are not so strong.

Collectively, the eight elements listed in box figure 1 account for more than 98 percent of the weight of the crust. All the other elements total only about 1.5 percent. Absent from the top eight elements are such vital elements as hydrogen (tenth by weight) and carbon (seventeenth by weight).

The element copper is only 27th in abundance, but our industrialized society is highly dependent on this metal. Most of the wiring in electronic equipment is copper, as are many of the telephone and power cables that crisscross the continent. However, the Earth's crust is not homogeneous, and geological processes have created concentrations of elements such as copper in a few places. Exploration geologists are employed by mining companies to discover where (as well as why) ore deposits of copper and other metals occur (see chapter 12).

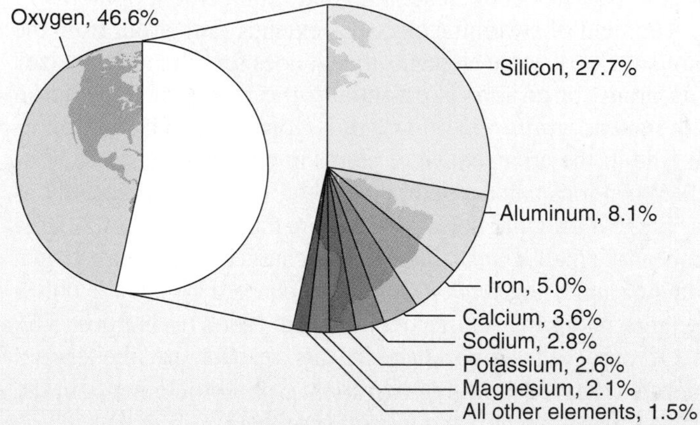

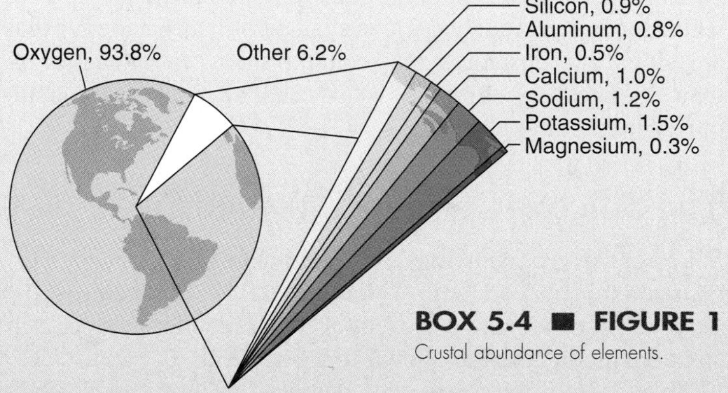

BOX 5.4 ■ FIGURE 1
Crustal abundance of elements.

Isolated Silicate Structure

Silicate minerals that are structured so that none of the oxygen atoms is shared by tetrahedrons have an **isolated silicate structure.** The individual silicon–oxygen tetrahedrons are bonded together by positively charged ions (figure 5.9). The common mineral *olivine,* for example, contains two ions of either magnesium (Mg^{+2}) or iron (Fe^{+2}) for each silicon–oxygen tetrahedron. The formula for olivine is $(Mg,Fe)_2SiO_4$.

Chain Silicates

A **chain silicate structure** forms when two of a tetrahedron's oxygen atoms are shared with adjacent tetrahedrons to form a chain (figures 5.8 and 5.10). Each chain, which extends indefinitely, has a net excess of negative charges. Minerals may have a *single-* or *double-chain structure.* For single-chain silicate structures, the ratio of silicon to oxygen (as figure 5.10 shows) is 1:3; therefore, each mineral in this group (the *pyroxene* group) incorporates SiO_3^{-2} in its formula, and it must be electrically balanced by the positive ions (e.g., Mg^{+2}) that hold the parallel chains together. If a pyroxene has magnesium, as the +2 ions bonding the chains shown in figure 5.10*A,* it has a formula of $MgSiO_3$.

A double-chain silicate is essentially two adjacent single chains that are sharing oxygen atoms. The *amphibole* group is characterized by two parallel chains in which every other tetrahedron shares an oxygen atom with the adjacent chain's tetrahedron (figure 5.8). In even a small amphibole crystal, millions of parallel double chains are bonded together by positively charged ions.

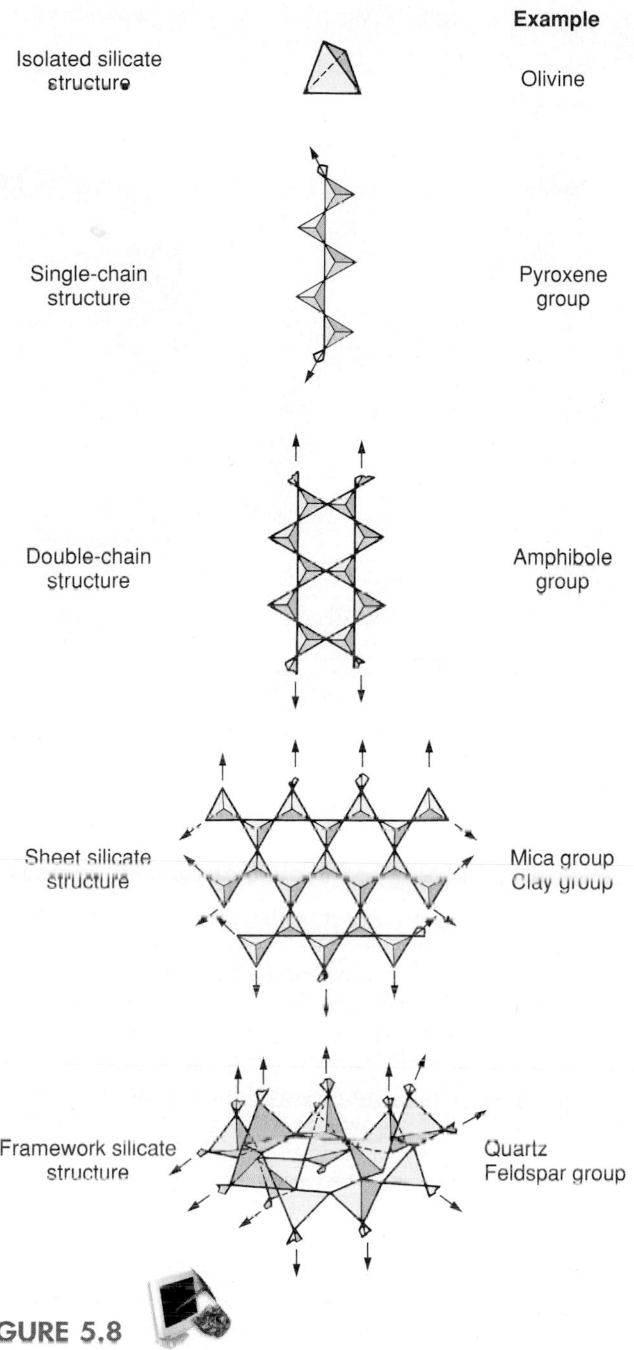

FIGURE 5.8

Common silicate structures. Arrows indicate directions in which structure repeats indefinitely.

Chain silicates tend to be shaped like columns, needles, or even fibres. The long structure of the external form corresponds to the linear dimension of the chain structure. Fibrous aggregates of certain minerals are called *asbestos* (see box 5.5)

Sheet Silicates

In a **sheet silicate structure** each tetrahedron shares three oxygen atoms to form a sheet (figure 5.8). The *mica* group and the *clay* group of minerals are sheet silicates. The positive ions that hold the sheets together are "sandwiched" between the silicate sheets (box 5.6).

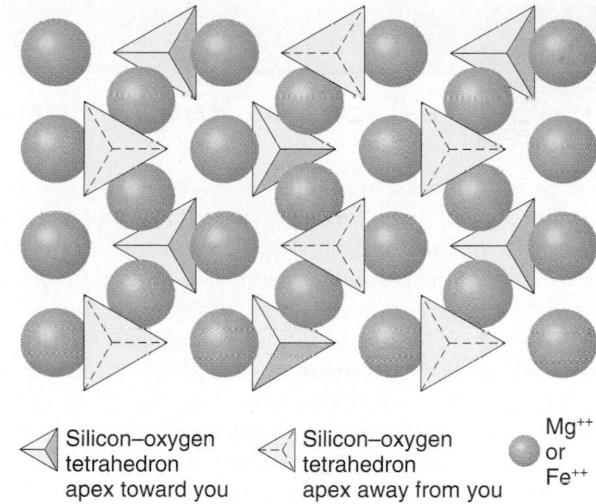

FIGURE 5.9

Diagram of the crystal structure of olivine, as seen from one side of the crystal.

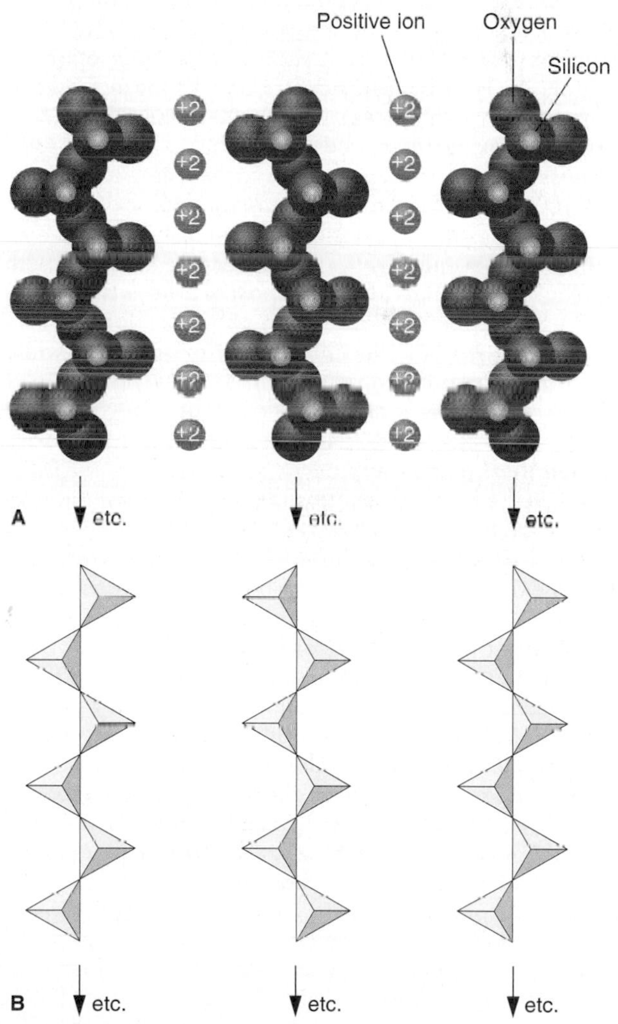

FIGURE 5.10

Single-chain silicate structure. (*A*) Model of a single-chain silicate mineral. (*B*) The same chain silicate shown diagrammatically as linked tetrahedrons; positive ions between the chains are not shown.

ENVIRONMENTAL GEOLOGY 5.5

Is Asbestos Really Harmful to Humans?

Asbestos is a generic name for fibrous aggregates of minerals (box figure 1). Because it does not ignite or melt in fire, asbestos has a number of valuable industrial applications. Woven into cloth, it may be used to make suits for firefighters. It can also be used as a fireproof insulation for homes and other buildings and has commonly been used in plaster for ceilings. Five of the six commercial varieties of asbestos are amphiboles, known commercially as "brown" and "blue" asbestos. The sixth variety is *chrysotile,* which is not a chain silicate and belongs to the *serpentine* family of minerals, and is more commonly known as "white asbestos." White asbestos is, by far, the most commonly used in North America (about 95 percent of that used in the United States).

Public fear of asbestos in North America has resulted in its being virtually outlawed by the federal government. Tens of billions of dollars have been spent (probably unnecessarily) to remove or seal off asbestos from schools and other public buildings.

Asbestos' bad reputation comes from the high death rate among asbestos workers exposed, without protective attire, to extremely high levels of asbestos dust. Some of these workers, who were covered with fibres, were called "snowmen." In Manville, New Jersey, children would catch the "snow" (asbestos particles released from a nearby asbestos factory) in their mouths. The high death rates among asbestos workers are attributed to *asbestosis* and lung cancer. Asbestosis is similar to silicosis contracted by miners; essentially, the lungs become clogged with asbestos dust after prolonged heavy exposure. The incidence of cancer has been especially high among asbestos workers who were also smokers. It's not clear that heavy exposure to white asbestos caused cancer among nonsmoking asbestos workers. However, brown and blue (amphibole) varieties, which are not mined in North America, have been linked to cancer for heavy exposure (even if for a short term).

What are the hazards of asbestos to an individual in a building where walls or ceilings contain asbestos? Recent studies from a wide range of scientific disciplines indicate that the risks are minimal to nonexistent, at least for exposure to white asbestos. The largest asbestos mines in the world are at Thetford Mines, Quebec. A study of longtime Thetford Mines residents, whose houses border the waste piles from the asbestos mines, indicated that their incidence of cancer was no higher than that of Canadians overall. Nor have studies in the United States been able to link nonoccupational exposure to asbestos and cancer. One estimate of the risk of death from cancer due to exposure to asbestos dust is one per 100,000 lifetimes. (Compare this to the risk of death from lightning of 4 per 100,000 lifetimes or automobile travel—1,600 deaths per 100,000 lifetimes.)

Following the collapse of New York's World Trade Center towers on September 11, 2001, the dust in the air from destroyed buildings contained high levels of asbestos—much higher than the safety levels set by the Environmental Protection Agency (EPA). Faced with widespread panic and a mass exodus from the city, the EPA reversed itself and declared the air safe. In doing so, the agency admitted that its standards were too stringent and were based on long-term exposure.

BOX 5.5 ■ FIGURE 1
Chrysotile asbestos.
Photo © Parvinder Sethi

In California, a closed-down white asbestos mining site designated for EPA Superfund cleanup is a short distance from where asbestos is being mined cleanly and efficiently. It is packaged and shipped to Japan. It cannot be used in the United States, because the United States is the only industrialized nation whose laws do not distinguish among asbestos types and permit the use of chrysotile. In 1999, Canada was the third largest producer of chrysotile asbestos in the world. Approximately 70 percent of the chrysotile exported by Canada is destined for Asian countries.

A reason chrysotile is less hazardous than amphibole asbestos is that chrysotile fibres will dissolve in lungs and amphibole will not. Experiments by scientists at Virginia Polytechnic Institute indicate that it takes about a year for chrysotile fibres to dissolve in lung fluids, whereas glass fibres of the same size will dissolve only after several hundred years. Yet fibreglass is being used increasingly as a substitute for asbestos.

Additional Resources

Chrysotile Institute
www.chrysotile.com

National Cancer Institute
www.ncic.cancer.ca

ENVIRONMENTAL GEOLOGY 5.6

Clay Minerals That Swell

Clay minerals are very common at the Earth's surface; they are a major component of soil. There are a great number of different clay minerals. What they all have in common is that they are sheet silicates. They differ by which ions hold sheets together and by the number of sheets "sandwiched" together.

Ceramic products and bricks are made from clay. Surprisingly, some clay minerals are edible; some are used in the manufacturing of pills. *Kaolinite,* a clay mineral, is the main ingredient in Kaopectate, a remedy for intestinal distress. Popular fast-food chains use clay minerals as a thickener for shakes (you can tell which ones, because the chains do not call them "milk shakes"—they do not use milk).

Montmorillonite is one of the more interesting clay minerals. It is better known as *expansive clay* or *swelling clay.* If water is added to the montmorillonite, the water molecules are adsorbed into the spaces between silicate layers (box figure 1). This results in a large increase in volume, sometimes up to several hundred percent. The pressure generated can be up to 50,000 kilograms per square metre. This is sufficient to lift a good-sized building.

If a building is erected on expansive clay that subsequently gets wet, a portion of the building will be shoved upward. In all likelihood, the foundation will break. Some people think that expansive soils have caused more damage than earthquakes and landslides combined. Damage in the United States is estimated to cost $2 billion a year.

BOX 5.6 ■ FIGURE 1

Expansive clays. (The orange ion represents aluminum in the clay layers and is not drawn to scale.)

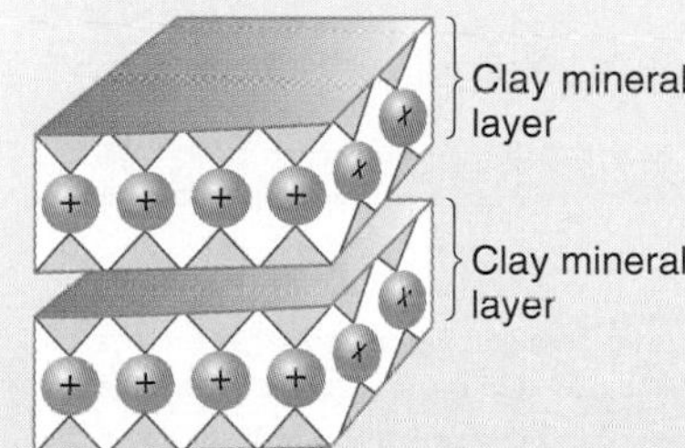

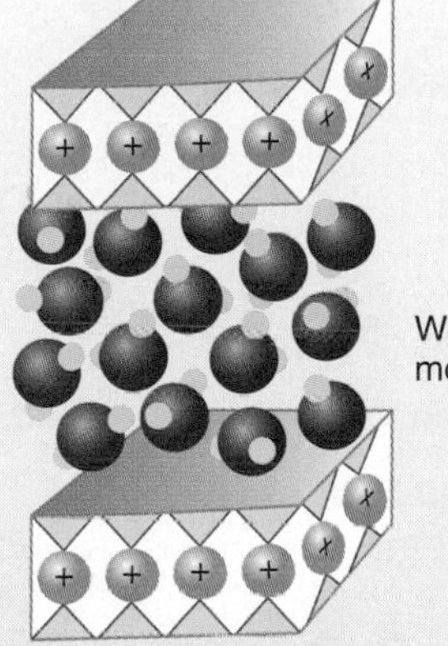

On the other hand, swelling clays can be put to use. Montmorillonite, mixed with water, can be pumped into fractured rock or concrete. When the water is adsorbed, swelling clay expands to fill and seal the crack. The technique is particularly useful where dams have been built against fractured bedrock. Sealing the cracks with expansive clays ensures that water will stay in the reservoir behind the dam. Swelling clays are also used to thicken drilling fluids when deep wells are drilled. This helps keep the borehole open during drilling and prevents the escape of pressurized gases encountered in resevoir beds.

Framework Silicates

When all four oxygen ions are shared by adjacent tetrahedrons, a **framework silicate structure** is formed (figure 5.3). *Quartz* is a framework silicate mineral. A *feldspar* is a framework silicate as well. However, its structure is slightly more complex because aluminum substitutes for some of the silicon atoms in some of the tetrahedrons. The same kind of substitution also takes place in amphiboles and micas, which helps account for the wide variety of silicate minerals.

Nonsilicate Minerals

Although not as abundant in Earth, *nonsilicates,* minerals that do not contain silica, are nevertheless important. The *carbonates* have CO_3 in their formulas. Calcite, $CaCO_3$, is a member of this group and is one of the most abundant minerals at the Earth's surface, where it occurs mainly in limestone. In dolomite, also a carbonate, magnesium replaces some calcium in the calcite formula. Gypsum is a *sulphate* (containing SO_4). *Sulphides* have S but not O in their formulas (pyrite, FeS_2, is an example). Hematite (Fe_2O_3) is an *oxide*—that is, it contains oxygen not bonded to Si, C, or S. Halite, NaCl, is a member of the *chloride* group. *Native elements* have only one element in their formulas. Some examples are gold (Au), copper (Cu), and the two minerals that are composed of pure carbon (C), diamond and graphite.

HOW DO MINERALS VARY IN STRUCTURE AND COMPOSITION?

It stands to reason that only a limited number of mineral compositions exist in nature because atoms cannot be combined randomly and they can come together to form only a restricted number of crystalline structures. This does not mean, however, that each kind of mineral is compositionally different, or that individual mineral types can't show some internal compositional variation.

Ions of like size and charge may freely substitute for one another in the atomic structures of minerals. Iron (Fe^{2+}) and magnesium (Mg^{2+}), for example, interchangeably substitute to create a range of compositions in the common silicate mineral olivine. This is represented by the parentheses in the formula of olivine—$(Mg,Fe)_2SiO_4$. Olivine is an example of a *solid solution,* with pure magnesium olivine, Mg_2SiO_4, forming the bright green variety forsterite (or peridot, as a gem), and pure iron olivine forming the jet black variety fayalite, Fe_2SiO_4. The crystal structures of forsterite and fayalite are virtually identical.

Some minerals that show solid solution, like plagioclase feldspar and augite (a pyroxene), also show *compositional zoning,* with the centres of crystals dominated by one type of cation and the rims dominated by another. The grains of plagioclase in certain igneous rocks typically have calcium-rich centres and sodium-rich rims (figure 5.11). The change is due to the cooling of

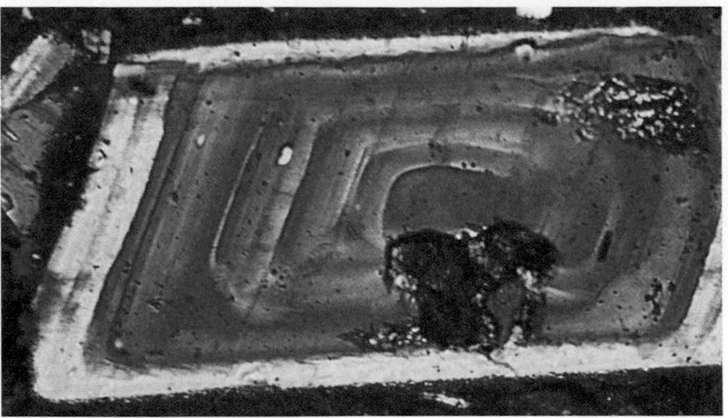

FIGURE 5.11

Zoning in plagioclase feldspar, as seen under a polarizing microscope. The concentric colour bands each indicate different amounts of Ca and Na in the crystal structure.

Photographed using cross-polarizers and a red-1 (550 nm) retardation plate. Photo by C. C. Plummer

the molten rock from which the plagioclase crystallizes. Calcium-rich plagioclase is more stable at the high temperatures in which the crystals start growing. The crystals then develop sodium-rich rims as the remaining melt crystallizes.

Some minerals can have the same chemical composition but have different crystalline structures—a phenomenon termed *polymorphism.* For example, calcite and aragonite both have the same formula $CaCO_3$. Their atomic crystal structures differ greatly, however. As you might expect, these two similar but distinctive mineral types result from separate conditions and processes of formation, with aragonite usually being an indicator of high-pressure crystallization.

Graphite and diamond are another, particularly spectacular example of polymorphism. Both minerals are made up of elemental carbon. They are unusual in that there is no other element involved in their structures. Dull-looking graphite, however, has a sheetlike structure that makes the mineral quite soft—useful as pencil lead; while the structure of shiny diamond is much more compact, making this the hardest substance on Earth. Graphite forms within the crust, while diamond originates much deeper, in the higher-pressure conditions of the mantle.

It is important to note that the physical characteristics of minerals we can observe without fancy laboratory equipment, such as colour, hardness, and lustre, are linked closely with the atomic structures and compositions of the minerals.

HOW DO WE IDENTIFY MINERALS?

To identify an unknown mineral, you should first determine its physical properties, then match the properties with the appropriate mineral, using a mineral identification key or chart such as the ones included in appendix A of this book. With a bit of experience, you may get to know the few diagnostic tests for each common mineral and no longer need to refer to an identification table. The most common minerals found in the Earth's crust are listed in table 5.1.

TABLE 5.1 Minerals of the Earth's Crust

Mineral Name	Chemical Composition	Abundance	Mineral Hardness
Silicates			
Framework Silicates			
Plagioclase	Ca and Na Al silicate	Very abundant	6
Potassium Feldspar	K Al silicate	Very abundant	6
Quartz	silica	Very abundant	7
Single-chain Silicates			
Pyroxene (augite most common)	Fe, Mg silicate (with some Al, Na, Ca)	Very abundant	6
Double-chain Silicates			
Amphibole	Complex Fe, Mg, Al silicate hydroxide	Very abundant	5–6
Sheet Silicates			
Muscovite	K Al silicate hydroxide	Very abundant	2–3
Biotite	K Fe, Mg Al silicate hydroxide	Very abundant	2–3
Clay minerals	Complex Al silicate hydroxides	Abundant	
Isolated Silicates			
Olivine	Mg, Fe silicate	Abundant	6–7
Garnet	Complex silicates	Abundant	7
Non-Silicates			
Carbonates			
Calcite	$CaCO_3$	Abundant	3
Dolomite	$CaMg(CO_3)_2$	Abundant	3
Sulphides			
Chalcopyrite	Cu, Fe sulphide	Not abundant	3–4
Sphalerite	Zn sulphide	Not abundant	3–4
Galena	Pb sulphide	Not abundant	2.5
Oxides			
Hematite	Iron Oxide (Fe_2O_3)	Not abundant	5–6
Magnetite	Iron Oxide (Fe_3O_2)	Not abundant	5–6
Native Elements			
Gold	Au (gold)	Not abundant	3
Diamond	C	Not abundant	10
Sulphates			
Gypsum	$CaSO_4 \cdot 2H_2O$	Not abundant	2
Chlorides			
Halite	NaCl	Not abundant	2.5

Colour

The first thing most people notice about a mineral is its colour. For some minerals, colour is a useful property. *Muscovite* mica is silvery white or colourless, whereas *biotite* is black or dark brown. Most of the **ferromagnesian minerals** (iron/magnesium-bearing), such as *augite, hornblende,* olivine, and biotite, are either green or black.

Because colour is so obvious, beginning students tend to rely too heavily on it as a key to mineral identification. Unfortunately, colour is also apt to be the most ambiguous of physical properties (figure 5.12). If you look at a number of quartz crystals, for instance, you may find specimens that are white, pink, black, yellow, or purple. Colour is extremely variable in quartz and many other minerals because even minute chemical impurities can strongly influence it. Obviously, it is poor procedure to attempt to identify quartz strictly on the basis of colour.

Streak

A pulverized mineral gives a colour, called a **streak,** that usually is more reliable than the colour of the specimen itself. Scraping the edge of a mineral sample across an unglazed porcelain plate leaves a streak that may be diagnostic of the mineral. For instance, hematite always leaves a reddish brown streak though the sample may be brown or red or silver.

Unfortunately, few of the silicate minerals—the most common minerals—leave an identifying streak because most are harder than the porcelain streak plate.

Lustre

The quality and intensity of *light* that is reflected from the surface of a mineral is termed **lustre.** (A photograph cannot always show this quality.) The lustre of a mineral is described by comparing it to familiar substances.

FIGURE 5.12

Why colour may be a poor way of identifying minerals. These are all corundum gems, including ruby and sapphire (see box 5.7 for more on gems).
Photo © Parvinder Sethi

Lustre is either *metallic* or *nonmetallic.* A **metallic lustre** gives a substance the appearance of being made of metal. Metallic lustre may be very shiny, like a chrome car part, or less shiny, like the surface of a broken piece of iron.

Nonmetallic lustre is more common. The most important type is **glassy** (also called **vitreous**) **lustre,** which gives a substance a glazed appearance, like glass or porcelain. Most silicate minerals have this characteristic. The feldspars, quartz, the micas, and the pyroxenes and amphiboles all have a glassy lustre.

Less common is an **earthy lustre.** This resembles the surface of unglazed pottery and is characteristic of the various clay minerals. Some uncommon lustres include *resinous* lustre (appearance of resin), *silky* lustre, and *pearly* lustre.

Hardness

The property of "scratchability," or **hardness,** can be tested fairly reliably. For a true test of hardness, the harder mineral or substance must be able to make a groove or scratch on a smooth, fresh surface of the softer mineral. For example, quartz can always scratch calcite or feldspar and is thus said to be harder than both of these minerals. Substances can be compared to **Mohs' hardness scale,** in which ten minerals are designated as standards of hardness (figure 5.13). The softest mineral, talc (used for talcum powder because of its softness), is designated as 1. Diamond, the hardest natural substance on Earth, is 10 on the scale (for more on diamonds, see box 5.7). Mohs' scale is a relative hardness scale. Figure 5.13 shows the absolute hardness for the ten minerals. Hardness values for other common minerals in the Earth's crust are given in table 5.1. The absolute hardness is obtained using an instrument that measures how much pressure is required to indent a mineral. Note that the difference in absolute hardness between corundum (9) and diamond (10) is around six times the difference between corundum and topaz (8).

Rather than carry samples of the ten standard minerals, a geologist doing field work usually relies on common objects to test for hardness (figure 5.13). A fingernail usually has a hardness of about 2-1/2. If you can scratch the smooth surface of a mineral with your fingernail, the hardness of the mineral must be less than 2-1/2 (figure 5.14). A copper coin or a penny has a hardness between 3 and 4; however, the brown, oxidized surface of most pennies is much softer, so check for a groove into the coin. A knife blade or a steel nail generally has a hardness slightly greater than 5, but it depends on the particular steel alloy used. A geologist uses a knife blade to distinguish between softer minerals, such as calcite, and similarly appearing harder minerals, such as quartz. Ordinary window glass, usually slightly harder than a knife blade (although some glass, such as that containing lead, is much softer), can be used in the same way as a knife blade for hardness tests. A file (one made of tempered steel for filing metal, not a fingernail file) can be used for a hardness of between 6 and 7. A porcelain streak plate also has a hardness of around 6-1/2.

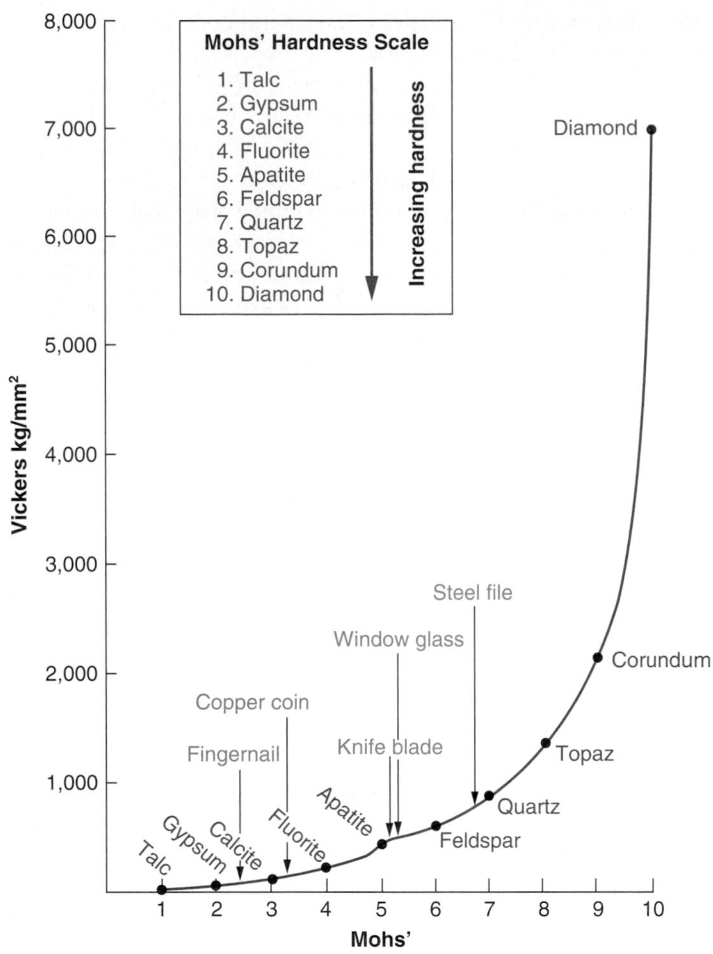

FIGURE 5.13

Mohs' hardness scale plotted against Vickers indentation values (kg/mm^2). Indentation values are obtained by an instrument that measures the force necessary to make a small indentation into a substance.

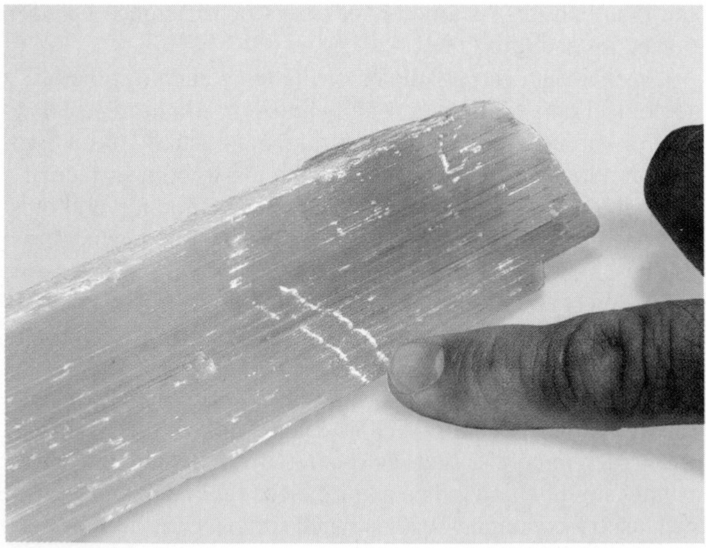

FIGURE 5.14

Fingernail (hardness of 2-1/2) easily scratches gypsum (hardness of 2).
Photo © Parvinder Sethi

External Crystal Form

The **crystal form** of a mineral is a set of faces that have a definite geometric relationship to one another. A well-formed crystal of halite, for example, consists of six faces all square and joined at right angles. The crystal form of halite is a *cube,* in other words. Other kinds of minerals whose crystals commonly consist of single forms include magnetite (octahedron ◆) and garnet (dodecahedron ⬟).

Crystals consisting of single forms, like those mentioned previously, are the most highly symmetrical shapes in all of nature other than spheres, which single minerals cannot develop. Crystals more commonly consist of several types of forms combined together to generate the full body of each specimen. As a rule of thumb, if two or more faces on a crystal are identical in shape and size, they belong to the same crystal form (figure 5.15).

Minerals displaying well-developed crystal faces have played an important role in the development of chemistry and physics. Steno, a Danish naturalist of the seventeenth century, first noted that the angle between two adjacent faces of quartz is always exactly the same, no matter what part of the world the quartz sample comes from or the colour or size of the quartz. As shown in figure 5.16, the angle between any two adjacent sides of the six-sided "pillar" (which is called a prism by mineralogists) is always exactly 120 degrees, while between a face of the "pillar" and one of the "pyramid" faces (actually part of a rhombohedron) the angle is always exactly 141°45′.

The discovery of such regularity in nature usually has profound implications. When minerals other than quartz were studied, they too were found to have sets of angles for adjacent faces that never varied from sample to sample. This observation became formalized as the *law of constancy of interfacial angles.* Later the discovery of X-ray beams and their behaviour in crystals confirmed Steno's theory about the structure of crystals.

Steno suspected that each type of mineral was composed of many tiny, identical building blocks, with the geometric shape of the crystal being a function of how these building blocks are put together. If you are stacking cubes, you can build a structure having only a limited variety of planar forms. Likewise, stacking rhombohedrons in three dimensions limits you to other geometric forms (figure 5.17).

Steno's law was really a precursor of atomic theory, developed centuries later. Our present concept of crystallinity is that atoms are clustered into geometric forms—cubes, bricks, hexagons, and so on—and that a crystal is essentially an orderly, three-dimensional stacking of these tiny geometric forms. Halite, for example, may be regarded as a series of cubes stacked in three dimensions (see figure 5.4). Because of the cubic "building block," its usual crystal form is a cube with crystal faces at 90-degree angles to each other.

Cleavage

The internal order of a crystal may be expressed externally by crystal faces, or it may be indicated by the mineral's tendency to split apart along certain preferred directions. **Cleavage** is the

FIGURE 5.15

Characteristic crystal forms of three common minerals: (*A*) Cluster of quartz crystals. (*B*) Crystals of potassium feldspar. (*C*) Intergrown cubic crystals of fluorite.
Photos © Parvinder Sethi

ability of a mineral to break, when struck or split, along preferred planar directions.

A mineral tends to break along certain planes because the bonding between atoms is weaker there. In quartz, the bonds are equally strong in all directions; therefore, quartz has no cleavage. The micas, however, are easily split apart into sheets (figure 5.18). If we could look at the arrangement of atoms in the crystalline structure of micas, we would see that the individual silicon–oxygen tetrahedrons are strongly bonded to one another within each of the silicate sheets. The bonding *between* adjacent sheets, however, is very weak. Therefore, it is easy to split the mineral apart parallel to the plane of the sheets.

Cleavage is one of the most useful diagnostic tools because it is identical for a given mineral from one sample to another. Cleavage is especially useful for identifying minerals when they are small grains in rocks.

The wide variety of combinations of cleavage and *quality* of cleavage also increases the diagnostic value of this property. Mica has a single direction of cleavage, and its quality is perfect (figures 5.18*A* and 5.19*A*). Other minerals are characterized by one, two, or more cleavage directions; the quality can range from perfect to poor (poor cleavage is very hard for anyone but a well-trained mineralogist to detect).

Three of the most common mineral groups—the feldspars, the amphiboles, and the pyroxenes—have two directions of cleavage (figure 5.19*B* and *C*). In feldspars, the two directions are at angles of about 90 degrees to each other, and both directions are of very good quality. In pyroxenes, the two directions are also at about right angles, but the quality is only fair. In amphiboles (figure 5.20), the quality of the cleavage is very good and the two directions are at an angle of 56 degrees (or 124 degrees for the obtuse angle).

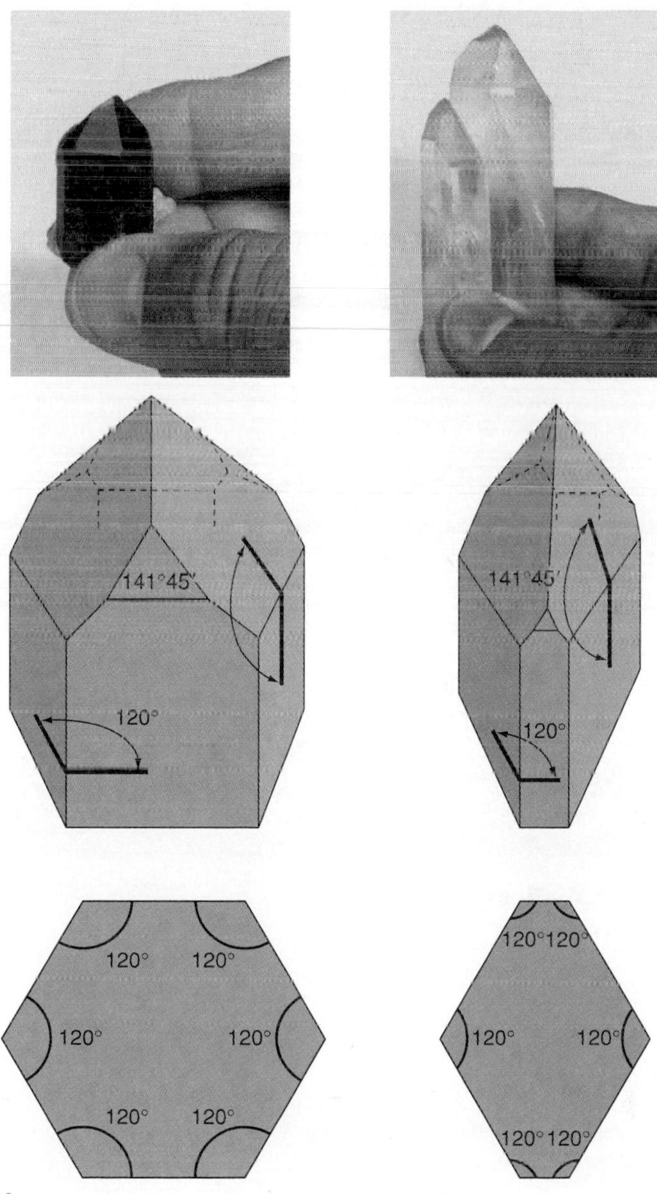

FIGURE 5.16

Quartz crystals showing how interfacial angles remain the same in perfectly proportioned (*A*) and misshapen (*B*) crystals. Cuts perpendicular to the prisms show that all angles are exactly 120°.
Photos © Parvinder Sethi

IN GREATER DEPTH 5.7

Gemstones

Gemstones are varieties of certain minerals that are valuable because of their beauty. Precious stones such as diamond, sapphire, ruby, emerald, and aquamarine are particularly valuable. These precious stones have a number of characteristics in common—they are transparent, with even colouration, and have a hardness greater than quartz (7 on Mohs' hardness scale). Their hardness ensures that they are durable.

Diamonds are usually clear, although some tinted varieties are particularly valuable (the famous Hope Diamond is blue). The appeal of diamond is due to its unique, brilliant lustre (called *adamantine* lustre), which causes light to be reflected from within the crystal and dispersed into rainbow-like colours. The facets on a diamond have been cut (or more correctly, ground, using diamond dust) to enhance the gem's brilliance. Diamonds are found extensively on the Canadian Shield (refer to box 4.4) and Canada is soon expected to be the third largest producer of diamonds in the world.

Emerald and aquamarine are varieties of beryl (hardness of 7.5). Emerald is the most expensive of these and owes its green colour to chromium impurities. Aquamarine's blue colour is due to iron impurities in the crystal structure. Gem-quality emeralds have been discovered in the Yukon (box figure 1) and there is potential for additional emerald finds in the Northwest Territories and British Columbia.

Sapphire and ruby are both varieties of corundum (9 on Mohs' scale). Sapphire can be various colours (except red), but blue sapphires are most valuable. Minute amounts of titanium and iron in the crystal structure give sapphire its blue colouration. Rubies are red due to trace amounts of chromium in corundum. Recently, gem-quality sapphires were discovered from the "Beluga" property near Kimmirut on Baffin Island, Nunavut (box figure 2).

Recommended Web Investigation

The Image: www.theimage.com/
Click on the "Gemstone Gallery," then "Beryl." You can read about properties of beryl and details about emerald and aquamarine. Below the description, you can access images of these gems. This site also contains information on how gems are faceted.

A

B

BOX 5.7 ■ FIGURE 1

(*A*) A collection of rough emeralds. (*B*) A collection of cut emeralds from Regal Ridge, Yukon. The largest stone is about 0.5 carats.

Photos by True North Gems.

BOX 5.7 ■ FIGURE 2

Sapphires from the Beluga property, Nunavut. The blue stone on the left weighs 1.17 ct, and the colourless stone on the right weighs 0.63 ct. Both were collected in July 2004.

Photo by Bradley S. Wilson

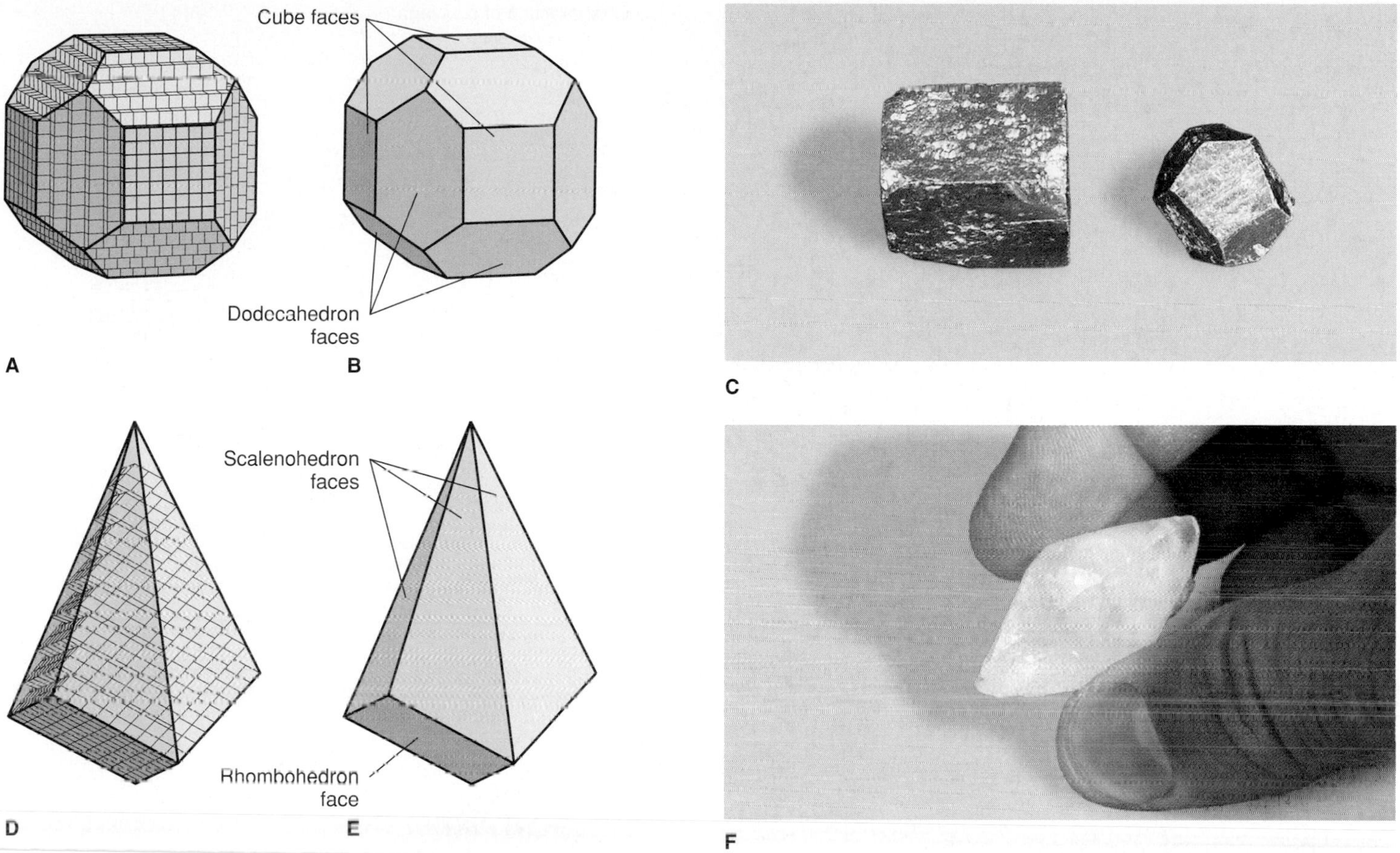

Cube faces

Dodecahedron faces

A B

C

Scalenohedron faces

Rhombohedron face

D E

F

FIGURE 5.17

Geometric forms built by stacking cubes (*A*, *B*, and *C*) and rhombohedrons (*D*, *E*, *F*). *A* and *D* are from a diagram published in 1801 by Haüy, a French mathematician. *A* and *B* show how cubes can be stacked for cubic and dodecahedral (12-sided) crystal forms. *C* is a photo showing a cube and a dodecahedron of pyrite. *D* and *E* show the relationship of stacked rhombohedrons to a "dog tooth" (scalenohedron) form and a rhombohedral face. *F* is a calcite "dog tooth" without a rhombohedral face.

Photos © Parvinder Sethi

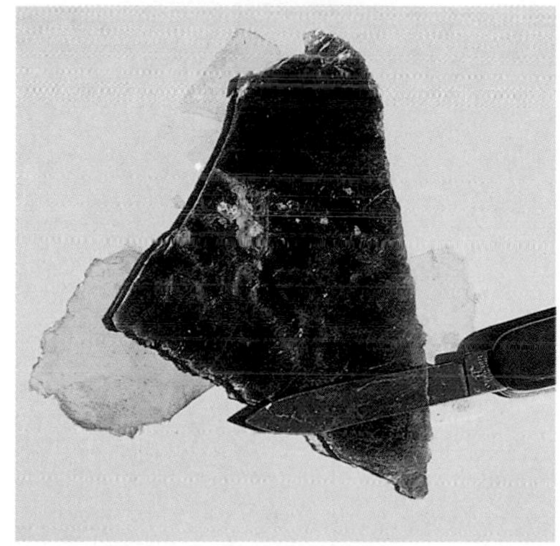

A

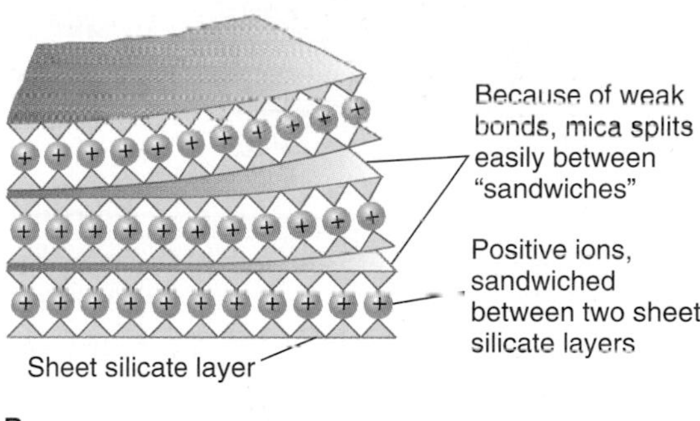

B

FIGURE 5.18

(*A*) Mica cleaves easily parallel to the knife blade. (*B*) Relationship of mica to cleavage. Mica crystal structure is simplified in this diagram.

Photo by C. C. Plummer

Halite is an example of a mineral with three excellent cleavage directions, all at 90 degrees to each other. This is called *cubic cleavage* (figure 5.19*D*). Halite's cleavage tells us that the bonds are weak in the planes parallel to the cube faces shown in figure 5.4.

Calcite also has three cleavage directions, each excellent. But the angles between them are clearly not right angles. Calcite's cleavage is known as *rhombohedral* cleavage (figure 5.19*E* and figure 5.21).

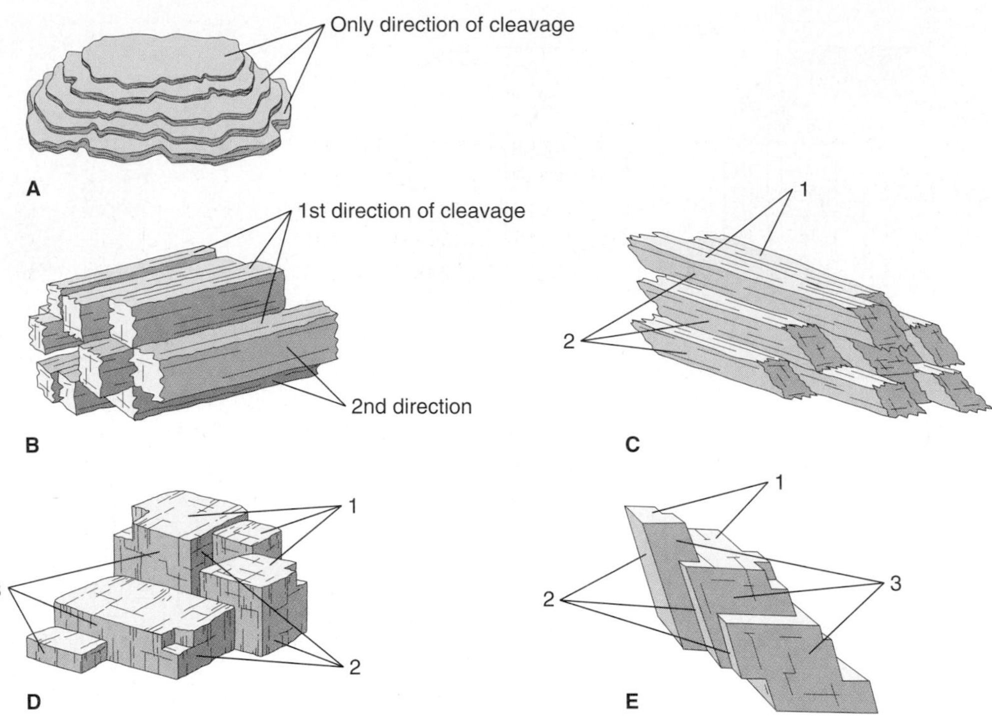

FIGURE 5.19

Most common types of mineral cleavage. Straight lines and flat planes represent cleavage. (*A*) One direction of cleavage. Mica is an example. (*B*) Two directions of cleavage that intersect at 90-degree angles. Feldspar is an example. (*C*) Two directions of cleavage that do not intersect at 90-degree angles. Amphibole is an example. (*D*) Three directions of cleavage that intersect at 90-degree angles. Halite is an example. (*E*) Three directions of cleavage that do not intersect at 90-degree angles. Calcite is an example. Not shown are the two other possible types of cleavage—four directions (such as in diamond) and six directions (as in sphalerite).

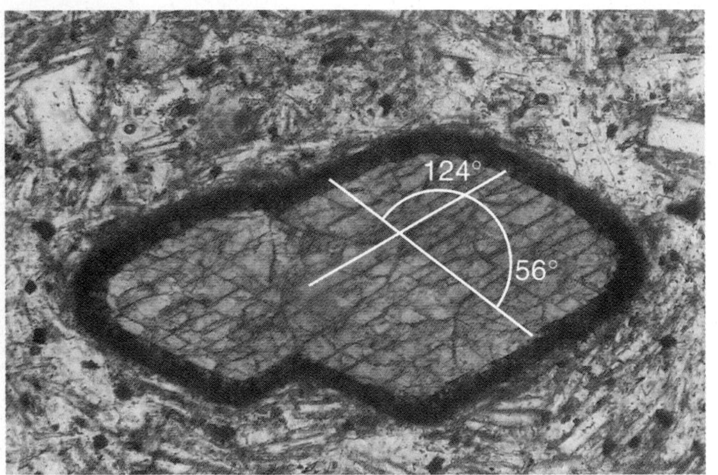

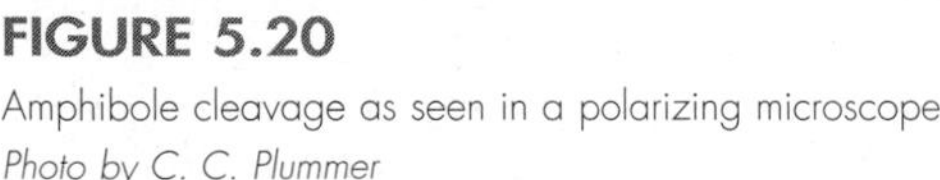

FIGURE 5.20

Amphibole cleavage as seen in a polarizing microscope.
Photo by C. C. Plummer

FIGURE 5.21

Cleavage fragments of calcite.
Photo © Parvinder Sethi

Some minerals have more than three directions of cleavage. Diamond has very good cleavage in four directions (ironically, the hardest natural substance on Earth can be easily shattered into small cleavage fragments). Sphalerite, the principal ore of zinc, has six directions.

Recognizing cleavage and determining angular relationships between cleavage directions take some practice. Students new to mineral identification tend to ignore cleavage because it is not as immediately apparent to the eye as colour. But determining cleavage is frequently the key to identifying a mineral, so the small amount of practice needed to develop this skill is worthwhile.

Fracture

Fracture is the way a substance breaks where not controlled by cleavage. Minerals that have no cleavage commonly have an *irregular fracture.*

Some minerals break along curved fracture surfaces known as *conchoidal fractures* (figure 5.22). These look like the inside of a clam shell. This type of fracture is commonly observed in quartz and garnet (but these minerals also show irregular fractures). Conchoidal fracture is particularly common in glass, including obsidian (volcanic glass).

FIGURE 5.22
Conchoidal fracture in volcanic glass (obsidian).
Photo by C. H. Eyles

FIGURE 5.23
Plagioclase striations.
Photo by C. C. Plummer

Minerals that have cleavage can fracture along directions other than that of the cleavage. The mica in figure 5.18*A* has irregular edges, which are fractures due to being torn perpendicular to the cleavage direction.

Specific Gravity

It is easy to tell that a brick is heavier than a loaf of bread just by hefting each of them. The brick has a higher **density**, weight per given volume, than the bread. Density is commonly expressed as **specific gravity,** the ratio of a mass of a substance to the mass of an equal volume of water.

Liquid water has a specific gravity of 1. (Ice, being lighter, has a specific gravity of about 0.9.) Most of the common silicate minerals are about two and a half to three times as dense as equal volumes of water: quartz has a specific gravity of 2.65; the feldspars range from 2.56 to 2.76. Special scales are needed to determine specific gravity precisely. However, a person can easily distinguish by hand very dense minerals such as galena (a lead sulphide with a specific gravity of 7.5) from the much less dense silicate minerals.

Gold, with a specific gravity of 19.3, is much heavier than galena. Because of its high density, gold can be collected by "panning." While the lighter clay and silt particles in the pan are sloshed out with the water, the gold lags behind in the bottom of the pan.

Special Properties

Some properties apply to only one or a few minerals. Smell is one. Some clay minerals have a characteristic "earthy" smell when they are moistened. A few minerals have a distinctive taste. If you lick halite, it tastes salty—because it is, of course, table salt.

Plagioclase feldspar commonly exhibits **striations**—straight, parallel lines on the flat surfaces of one of the two cleavage directions (figure 5.23). The lines appear to be etched by a delicate scriber. In plagioclase, they are caused by a systematic change within the pattern of crystalline structure.

The mineral *magnetite* (an iron oxide) owes its name to its characteristic physical property of being attracted to a magnet. Where large bodies of magnetite are found in the Earth's crust, compass needles point toward the magnetite body rather than to magnetic north. Airplanes navigating by compass have become lost because of the influence of large magnetite bodies. Some other minerals are weakly magnetic; their magnetism can only be detected by specialized magnetometers, similar to metal detectors in airports. Magnetism is important to modern civilization. We use magnetic tape (coated with magnetite or other magnetic material) for sound and video recordings as well as for magnetic memory disks in our computers. In later chapters, you will see how magnetite in igneous rocks has preserved a record of Earth's magnetic field through geologic time; this has been an important part of the verification of plate tectonic theory. Some bacteria create magnetite, and this has been used to support the hypothesis that life has existed on Mars (as described in NASA's Mars micromagnet site, http://science.nasa.gov/headlines/y2000/ast20dec_1.htm).

Quartz has the property of generating electricity when squeezed in a certain crystallographic direction. This property relates to its use in quartz watches.

A mineral has numerous other properties, including its melting point, electrical and heat conductivity, and so on. Most are not relevant to introductory geology. Two categories of properties that are important are optical properties and the effects of X-rays on minerals.

A clear crystal of calcite exhibits an unusual property. If you place transparent calcite over an image on paper, you will see two images (figure 5.24). This phenomenon is known as *double refraction* and is caused by light splitting into two components when it enters some crystalline materials. Each of the components is travelling through the mineral at different velocities. Most minerals possess double refraction, but it is usually slight and can be observed using polarizing filters, notably

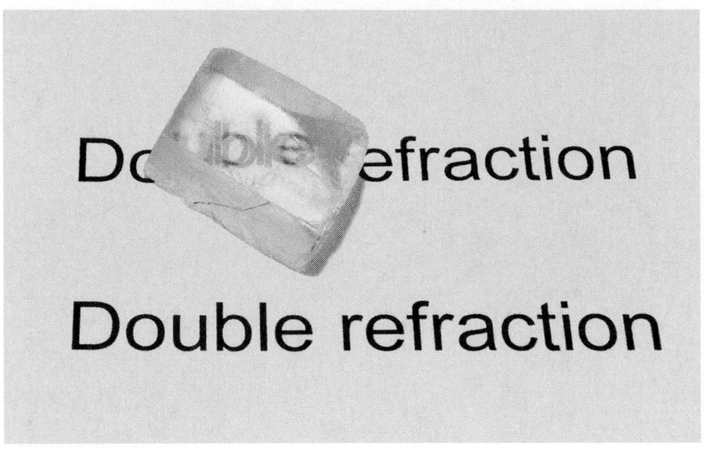

FIGURE 5.24

Double refraction in calcite. Two images of the letters are seen through the transparent calcite crystal.

Photo © Parvinder Sethi

in polarizing microscopes. Polarizing microscopes are very useful to professional geologists and advanced students for identifying minerals and interpreting how rocks formed. Photomicrographs elsewhere in this book were taken through polarizing microscopes (for example, figures 5.11 and 6.3). Explaining optical phenomena such as this is beyond the scope of this book, but if interested you can go to the Molecular Expressions Microscopy Primer site at http://micro.magnet.fsu.edu/primer/virtual/virtualpolarized.html.

Specialized equipment is needed to determine some properties. Perhaps most important are the characteristic effects of minerals on X-rays, which we can explain only briefly here. X-rays entering a crystalline substance are deflected by planes of atoms within the crystal. The X-rays leave the crystal at precise and measurable angles controlled by the orientation of the planes of atoms that make up the internal crystalline structure (figure 5.25). The pattern of X-rays exiting can be recorded on photographic film or by various recording instruments. Each mineral has its own pattern of reflected X-rays, which serves as an identifying "fingerprint."

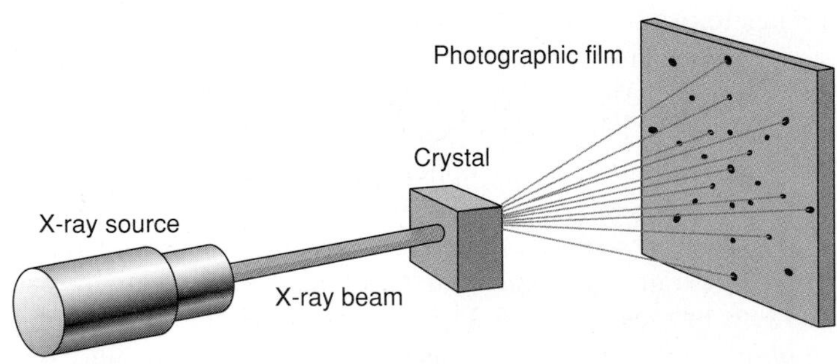

FIGURE 5.25

An X-ray beam passes through a crystal and is deflected by the rows of atoms into a pattern of beams. The dots exposed on the film are an orderly pattern used to identify the particular mineral.

Chemical Tests

One chemical reaction is routinely used for identifying minerals. The mineral calcite, as well as some other carbonate minerals (those containing CO_3^{-2}), reacts with a weak acid to produce carbon dioxide gas. In this test, a drop of dilute hydrochloric acid applied to the sample of calcite bubbles vigorously, indicating that CO_2 gas is being formed. Normally, this is the only chemical test that geologists do during field research.

Chemical analyses of minerals and rocks are done in labs using a wide range of techniques. A chemical analysis can accurately tell us the amount of each element present in a mineral. However, chemical analysis alone cannot be used to conclusively identify a mineral. We also need to know about the mineral's crystalline structure. As we have seen, diamond and graphite have an identical composition but very different crystalline structures.

HOW DO MINERALS FORM?

Minerals form under an enormously wide variety of conditions—most purely geological; others biological in nature. Some form tens of kilometres beneath the surface; others right at the surface and virtually out of the atmosphere itself.

The most common minerals are silicates, which incorporate the most abundant elements on Earth. Silicate minerals such as quartz, olivine, and the feldspars (plagioclase and potassium feldspar) crystallize primarily from molten rock (magma). They are *precipitates*—products of crystallizing liquid. Other precipitates include the carbonates calcite and aragonite, which grow in spring and cave waters.

Some minerals precipitate due to evaporation (e.g., halite). The very thick salt deposits underlying central Europe and the southern Great Plains exist because of the evaporation of seas millions of years ago.

Ice may be regarded as a very transient mineral at all but the coldest parts of the Earth's surface. (Ice is a major crust-forming mineral on planets of the outer solar system, where it cannot melt; box 5.8).

Some minerals result from biological activity; for example, the building of coral reefs creates huge masses of calcite-rich limestone. Many organisms, including human beings, create magnetite within their skull cases. Some researchers believe that birds and whales exploit the properties of this mineral to assist them in migratory navigation (why do we make magnetite in our own heads?). Bacteria also form huge amounts of sulphur by processing preexisting sulphate minerals. Most of our commercial supply of sulphur, in fact, comes from the mining of these *biogenic* deposits.

Some minerals crystallize directly from volcanic gases around volcanic vents—a process termed *sublimation.* Examples include ordinary sulphur, ralstonite, and thenardite (used as a natural rat poison). Sublimates are much less common than precipitates, though on planets and moons with intense volcanic activity, like Venus and Io, they cover wide swaths of planetary surface in thick beds. See box 5.9 for a look at mineral deposits in Ontario.

IN GREATER DEPTH 5.8

Water and Ice—Molecules and Crystals

Earth is often called the *blue planet* because oceans cover 70 percent of its surface. Ice, which is a mineral, dominates our planet's polar regions. Perhaps "Aqua" rather than "Earth" would be a more appropriate name for our planet. It is fortunate that water is so abundant because life would be impossible without it. In fact, we humans are made up mostly of water. The nature and behaviour of water molecules helps explain why water is vital to life on Earth.

In a water molecule, the two hydrogen atoms are tightly bonded to the oxygen atom. However, the shape of the molecule is asymmetrical, with the two hydrogen atoms on the same side of the atom (box figure 1). This means the molecule is polarized, with a slight excessive positive charge at the hydrogen side of the molecule and a slight excess negative charge at the opposite side. Because of the slight electrical attraction of water molecules, other substances are readily attracted to the molecules and dissolved or carried away by water. Water has been called the universal solvent. Dirt washes out of clothing; water, in blood, carries nutrients to our muscles and transports waste to our kidneys and out of our bodies.

When water is in its liquid state, the molecules are moving about. Because of the polarity, molecules are slightly attracted to one another. For this reason, water molecules are closer together than they are in most other liquids. However, in ice the water molecules are not as tightly packed together as in liquid water.

When water freezes, the hydrogen atoms are attracted to oxygen atoms in adjacent water molecules (box figure 2), resulting in an orderly, three-dimensional pattern that is hexagonal, as in a honeycomb (this explains the hexagonal shape of snowflakes). The openness of the honeycomblike, crystalline structure of ice contrasts with the more closely packed molecules in liquid water. This is the reason why ice is less dense than liquid water. This is an unusual solid–liquid relationship. For most substances, the solid is denser than its liquid phase.

The fact that ice is less dense than liquid water has profound implications. Ice floats rather than sinks in liquid water. Icebergs float in the ocean. Lakes freeze from the top down. Ice on a lake surface acts as an insulating layer that retards the freezing of underlying water. If ice sank, lakes would freeze much more readily and thaw much more slowly. Our climate would be very different if ice sank. The Arctic Ocean surface freezes during the winter but only at its surface. If the ice were to sink, more ocean water would be exposed to the cold atmosphere and would freeze and sink. Eventually, the entire Arctic Ocean would freeze and would not thaw during the summer. If this were the case, life as we know it probably would not exist.

When water freezes, it expands. A bottled beverage placed in a freezer breaks its container upon freezing. When water trapped in cracks in rock freezes, it will expand and will help break up the rock (as explained in the chapter on weathering).

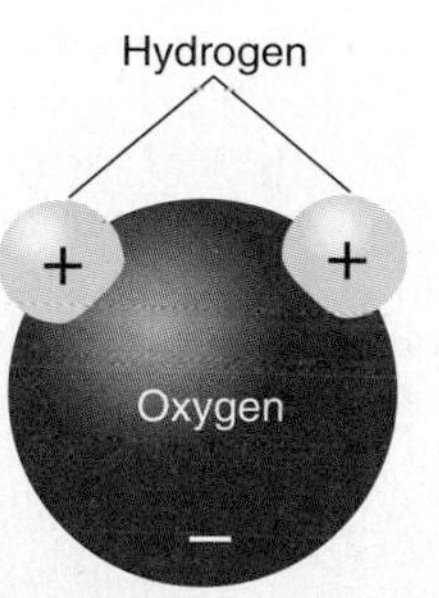

BOX 5.8 ■ FIGURE 1
Water molecule.

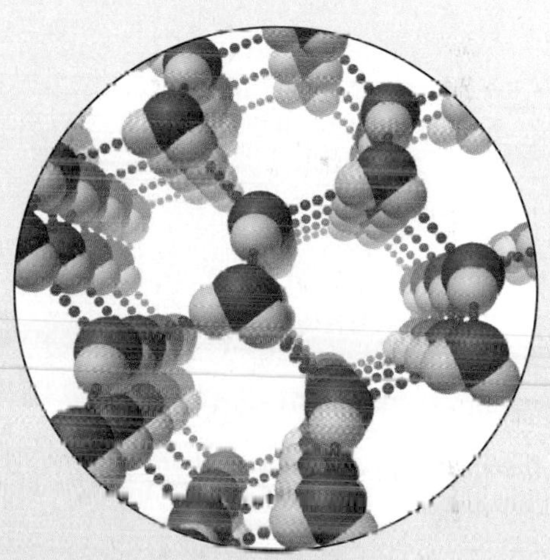

BOX 5.8 ■ FIGURE 2
Hexagonal structure of ice. Small, black dots represent the attraction between hydrogen atoms and oxygen atoms for adjacent water molecules.

Additional Resources

Snow Crystal Research
Nice images taken with an electron microscope.
http://emu.arsusda.gov/snowsite/default.html

Snow Crystals
Caltech's site. More about ice and nice pictures of snow crystals. Click on "Ice Properties" under "Snowflake Physics" to see a model of the arrangement of oxygen and hydrogen atoms in the crystal structure of ice.
www.its.caltech.edu/~atomic/snowcrystals/

We are able to understand the conditions of formation of most minerals with varying degrees of accuracy and precision using the tools of chemistry, especially with an understanding of thermodynamics and solutions. In fact, as implied at the beginning of this chapter, a good grasp of chemistry is a necessity for any advanced study of minerals.

IN GREATER DEPTH 5.9

Bancroft: Mineral Capital of Canada

Bancroft, in southern Ontario, is known as the Mineral Capital of Canada. The Bancroft area contains many localities where corundum, feldspar, uranium, graphite, iron, nephaline syenite, mica, granite, gold, marble, lead, molybdenite, apatite, beryl, fluorite, talc, and sodalite have been mined since the mid-1800s. Bancroft's mineral wealth is due to the Grenvillian orogeny, which occurred some 1.1 billion years ago. During this orogeny (see chapter 2), far-travelled land masses (terranes) collided with ancestal North America to form a larger continent. Marine sediments such as limestone and volcanic rocks formed in the seaways that existed between the colliding land masses were crushed and deformed during the orogenic event. Metamorphism of the sedimentary and volcanic rocks deep below the Grenville mountains created mineral-rich veins and narrow ore bodies. These are now exposed on and close to the ground surface as a result of millions of years of erosion that subdued the Grenville mountains into the low-relief surface we know as the Canadian Shield.

BOX 5.9 ■ FIGURE 1

Mineral collectors visiting Bancroft today can find a wealth of beautiful specimens at a number of abandoned mine sites. Each summer Canada's largest mineral show, "The Rockhound Jamboree," in Bancroft attracts visitors from around the world.

Photo by Nick Eyles

BOX 5.9 ■ FIGURE 2

Sodalite: a blue-coloured silicate mineral containing sodium, chlorine, and aluminum. It is used as a gemstone.

Photo by Nick Eyles

SUMMARY

Atoms are composed of *protons* (+), *neutrons,* and *electrons* (−). A given element always has the same number of protons. An atom in which the positive and negative electric charges do not balance is an *ion.*

Ions or atoms bond together in very orderly, three-dimensional structures that are *crystalline.*

A *mineral* is a crystalline substance that is naturally occurring and is chemically and physically distinctive. Minerals are the building blocks of rocks.

The two most abundant elements in the Earth's crust are oxygen and silicon. Most minerals are silicates, having the silicon–oxygen tetrahedron as their basic building block.

Minerals are usually identified by their physical properties. Cleavage is perhaps the most useful physical property for identification purposes. Other important physical properties are colour, streak, lustre, hardness, external crystal form, fracture, and specific gravity.

Terms to Remember

atom 131
atomic mass number 131
atomic number 131
atomic weight 131
bonding 133
chain silicate structure 136
cleavage 142
covalent bonding 134
crystal form 142
crystal structure 130
crystalline 130
density 147
earthy lustre 141
electron 131
element 133
ferromagnesian mineral 141
fracture 146
framework silicate structure 139
glassy (vitreous) lustre 141
hardness 141
ion 133
ionic bonding 134
isolated silicate structure 136
isotope 131
lustre 141
metallic lustre 141
mineral 130
Mohs' hardness scale 141
neutron 133
nonmetallic lustre 141
nucleus 133
proton 133
sheet silicate structure 137
silica 135
silicates 135
silicon–oxygen tetrahedron 135
specific gravity 147
streak 141
striations 147

Testing Your Knowledge

Use the following questions to prepare for exams based on this chapter.

1. Compare feldspar and quartz.
   a. How do they differ chemically?
   b. What type of silicate structure does each have?
   c. How would you distinguish between them on the basis of cleavage?
2. How do the crystal structures of pyroxenes and amphiboles differ from one another?
3. How do the various feldspars differ from one another chemically?
4. Distinguish between the following pairs of terms:
   silica/silicate
   silicon/silicon–oxygen tetrahedron
5. What is the distinction between cleavage and external crystal form?
6. How would you distinguish the following on the basis of physical properties? (You might refer to appendix A.)
   feldspar/quartz calcite/feldspar
   muscovite/feldspar pyroxene/feldspar
7. Using triangles to represent tetrahedrons, start with a single triangle (to represent isolated silicate structure) and, by drawing more triangles, build on the triangle to show a single-chain silicate structure. By adding more triangles, convert that to a double-chain structure. Turn your double-chain structure into a sheet silicate structure.
8. What major factor controls chemical activity between atoms?
9. What are the three most common elements (by number and approximate percentage) in the Earth's crust?
10. What are the next five most common elements?
11. A substance that cannot be broken down into other substances by ordinary chemical methods is a(n)
    a. crystal b. element
    c. molecule d. compound
12. The subatomic particle that contributes mass and a single positive electrical charge is the
    a. proton b. neutron
    c. electron

13. Atoms containing different numbers of neutrons but the same number of protons are called
    a. compounds b. ions
    c. elements d. isotopes
14. Atoms with either a positive or negative charge are called
    a. compounds b. ions
    c. elements d. isotopes
15. The bonding between Cl and Na in halite is
    a. ionic b. covalent
    c. metallic d. male
16. Which is not true of a single silicon–oxygen tetrahedron?
    a. The atoms of the tetrahedron are strongly bonded together.
    b. It has a net negative charge.
    c. The formula is SiO_4.
    d. It has four silicon atoms.
17. Which is not a type of silicate structure?
    a. isolated b. single chain
    c. double chain d. sheet
    e. framework f. pentagonal
18. Which of the common minerals is not a silicate?
    a. quartz b. calcite
    c. pyroxene d. feldspar
    e. biotite
19. On Mohs' hardness scale, ordinary window glass has a hardness of about
    a. 2–3 b. 3–4
    c. 5–6 d. 7–8
20. The ability of a mineral to break along preferred directions is called
    a. fracture b. crystal form
    c. hardness d. cleavage
21. Striations are associated with
    a. quartz b. mica
    c. potassium feldspar d. plagioclase
22. Glass is
    a. atoms randomly arranged b. crystalline
    c. ionically bonded d. covalently bonded
23. Crystalline substances are always
    a. ionically bonded b. minerals
    c. made of repeating patterns of atoms d. made of glass

Exploring Web Resources

www.mcgrawhill.ca/olc/plummer

This is the dedicated Website for this book. You can go to it for new and updated information. The universal resource locators (URL) listed in this book are also given as links on the Website, making it easy to go to those Websites without typing in the URL. When you visit our Online Learning Centre, go to the Student Centre and then to the chapter of interest. Here you will find additional readings and media resources, as well as answers to the Testing Your Knowledge section, more quizzes, animation, and video clips. Links to additional Websites can also be found. We have added questions for some of the sites to allow you to get the most of your exploration of the web. Using the web is an enjoyable way of enhancing your knowledge of geology.

www.rockhounds.com/

Bob's Rock Shop. Contains a great amount of information for mineral collectors. Find and go to "crystallography and mineral crystal systems" for a more in-depth study of crystallography than presented in this book.

www.minweb.co.uk/

Mineral Web. Crystal structures are displayed in 3-D. The structures rotate, and you can manipulate the rotation using a mouse. You must install Chime, a program for viewing the structures. Chime can be downloaded easily from this site. Once installed, take a look at the crystal structures of diamond, olivine, muscovite, and other minerals. You can also observe the various silicon–oxygen tetrahedron structures.

http://webmineral.com/

Mineralogy Database. There are descriptions of close to 4,000 mineral species. The descriptions include mineral properties beyond the scope of an introductory geology course; however, there are links to other sites that include pictures of minerals.

www.mindat.org/

The *online mineralogy resource.* Another comprehensive source for mineral information. You can search for a mineral by name or by locality.

www.theimage.com/

The Image. Photos of minerals and gems. Click on Mineral Gallery and choose a mineral to view photos and properties of that mineral. The Gemstone Gallery has photos of gem minerals.

www.webelements.com/

WebElements periodic table. The periodic table of elements. You can click on an element to determine its properties.

www.uky.edu/Projects/Chemcomics/

The comic book periodic table of elements. An entertaining site in which you click on an element and see examples of comic book stories about that element.

www.canadianrockhound.ca

Free online magazine with interesting articles about Canadian minerals and rocks.

Animations

This chapter includes the following animation available on our Online Learning Centre at **www.mcgrawhill.ca/olc/plummer**.

5.7, 5.8 Silicate mineral structures

C H A P T E R

6

Igneous Rocks, Intrusive Activity, and the Origin of Igneous Rocks

What are igneous rocks?
How are igneous rocks classified?
What happens when magma cools underground?
How do different types of magma form?
How do magmas of different compositions evolve?
How does igneous activity relate to plate tectonics?

Chapters 6 and 7 are about igneous rocks and igneous processes. Chapter 6 describes igneous processes, particularly those that take place underground. Chapter 7 focuses on volcanoes and igneous activity that takes place at the Earth's surface.

We begin the chapter by focusing on igneous rocks. After the section on igneous rock classification, we describe structural relationships between bodies of intrusive rock and other rocks in the Earth's crust. This is followed by a discussion of how magmas form and are altered. We conclude by discussing various hypotheses that relate igneous activity to plate tectonics theory.

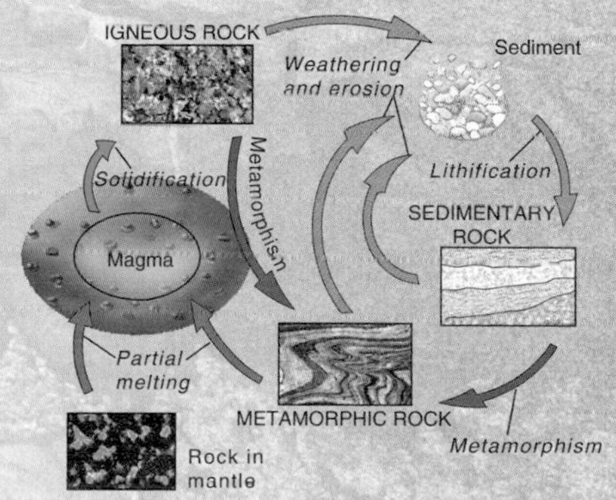

Pale coloured dikes of igneous rock intruded into darker coloured country rock, British Columbia. *Photo by Nick Eyles*

WHAT ARE IGNEOUS ROCKS?

If you go to the island of Hawaii, you might observe red hot lava flowing over the land, and, as it cools, solidifying into the fine-grained (the grains are less than 1 millimetre across), black rock we call basalt. Basalt is an **igneous rock,** rock that has solidified from magma. **Magma** is molten rock, usually rich in silica and containing dissolved gases. (**Lava** is magma on the Earth's surface.) Igneous rocks may be either **extrusive** if they form at the Earth's surface (e.g., basalt) or **intrusive** if magma solidifies underground. **Granite,** a coarse-grained (the grains are larger than 1 millimetre) rock composed predominantly of feldspar and quartz, is an intrusive rock. In fact, granite is the most abundant intrusive rock found in the continents.

Unlike the volcanic rock in Hawaii, nobody has ever seen magma solidify into intrusive rock. So what evidence suggests that bodies of granite (and other intrusive rocks) solidified underground form magma?

- Mineralogically and chemically, intrusive rocks are essentially identical to volcanic rocks.
- Volcanic rocks are fine-grained (or glass) due to their rapid solidification; intrusive rocks are generally coarse-grained, which is inferred to mean that the magma crystallized slowly underground.
- Experiments have confirmed that most of the minerals in these rocks can form only at high temperatures. Other experiments indicate that some of the minerals could have formed only under high pressures, implying they were deeply buried. More evidence comes from examining *intrusive contacts,* such as shown in figures 6.1 and 6.2. (A **contact** is a surface separating different rock types. Other types of contacts are described elsewhere in this book.)
- Pre-existing solid rock, *country rock,* appears to have been forcibly broken by an intruding liquid, with the magma flowing into the fractures that developed. **Country rock,** incidentally, is an accepted term for any older rock into which an igneous body intruded.
- Close examination of the country rock immediately adjacent to the intrusive rock usually indicates that it appears "baked," or *metamorphosed,* close to the contact with the intrusive rock.
- Rock types of the country rock often match **xenoliths,** fragments of rock that are distinct from the body of igneous rocks in which they are enclosed.
- In the intrusive rock adjacent to contacts with country rock are **chill zones,** finer-grained rocks that indicate magma solidified more quickly here because of the rapid loss of heat to cooler rock.

FIGURE 6.1
Granite (light-coloured rock) solidified from magma that intruded dark-coloured country rock in Torres del Paine, Chile. The dark-coloured country rock is shale deposited in a marine environment. The spires are erosional remnants of rock that were once deep underground.
Photo by Kay Kepler

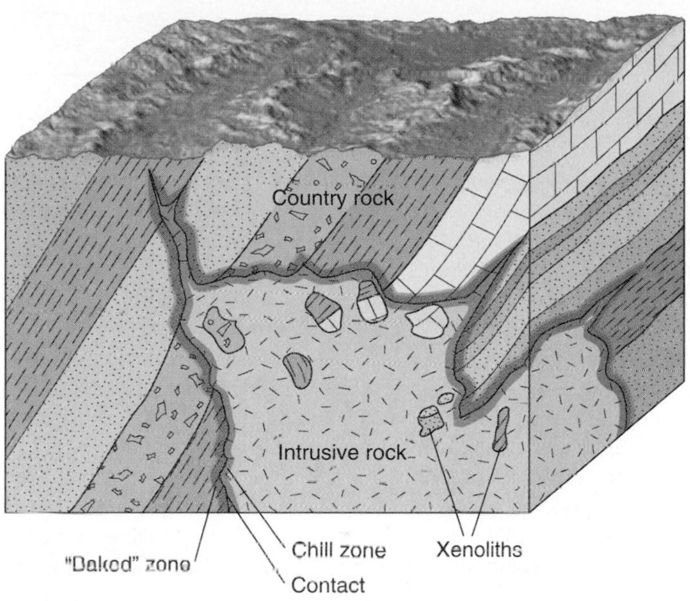

FIGURE 6.2

Igneous rock intruded pre-existing rock (country rock) as a liquid. (Xenoliths are usually much smaller than indicated.)

Laboratory experiments have greatly increased our understanding of how igneous rocks form. However, geologists have not been able to artificially make coarsely crystalline granite. Only very fine-grained rocks containing the minerals of granite have been made from artificial magmas, or "melts." The temperature and pressure at which granite apparently forms can be duplicated in the laboratory—but not the time element. According to calculations, a large body of magma requires over a million years to solidify completely. This very gradual cooling causes the coarse-grained texture of most intrusive rocks. Chemical processes involving silicates are known to be exceedingly slow. Yet another problem in trying to apply experimental procedures to real rocks is determining the role of water and other gases in the crystallization of rocks such as granite. Only a small amount of gases is retained in rock crystallized underground from a magma, but large amounts of gas (especially water vapour) are released during volcanic eruptions. No one has seen an intrusive rock forming; hence, we can only speculate about the role these gases might have played before they escaped. One example shows why gases are important. Laboratory studies have shown that granite can melt at temperatures as low as 650°C if water is present and under high pressure. Without the water, the melting temperature is several hundred degrees higher. Not knowing how much water was present during crystallization makes accurate determination of temperatures difficult and speculative.

Igneous Rock Textures

Texture refers to a rock's appearance with respect to the size, shape, and arrangement of its grains or other constituents. Most (but not all) igneous rocks are *crystalline;* that is, they are made of interlocking crystals (of, for instance, quartz and feldspar). The most significant aspect of texture in igneous rocks is grain (or crystal) size. Extrusive rocks typically are **fine-grained rocks,** in which most of the grains are smaller than 1 millimetre. The grains, if they are crystals, are small because magma cools rapidly at the Earth's surface, and so they have less time to form. Some intrusive rocks are also fine-grained; these occur as smaller bodies that apparently solidified near the surface upon intrusion into relatively cold country rock (probably within a couple kilometres of the Earth's surface). *Basalt, andesite,* and *rhyolite* are the common fine-grained igneous rocks. Igneous rocks that formed at considerable depth—usually more than several kilometres—are called **plutonic rocks** (after Pluto, the Roman god of the underworld). Characteristically, these rocks are coarse-grained, reflecting the slow cooling and solidification of magma. For our purposes, **coarse-grained** (or **coarsely crystalline**) **rocks** are defined as those in which most of the grains are larger than 1 millimetre. The crystalline grains of plutonic rocks are commonly interlocked in a granular pattern (figure 6.3). An extremely coarse-grained (grains over 5 centimetres) igneous rock is called a *pegmatite* (see box 6.1).

A

B

FIGURE 6.3

(A) Coarse-grained texture characteristic of plutonic rock. Feldspars are white and pink. Quartz is transparent. Biotite mica is black. *(B)* A similar rock seen through a polarizing microscope. Note the interlocking crystal grains of individual minerals.
Photo A by S. Fletcher; Photo B by C. C. Plummer

IN GREATER DEPTH 6.1

Pegmatite—A Rock Made of Giant Crystals

Pegmatites are extremely coarse-grained igneous rocks. In some pegmatites, crystals are as large as 10 m across. Strictly speaking, a pegmatite can be of diorite, gabbro, or granite. However, the vast majority of pegmatites are silicic, with very large crystals of potassium feldspar, sodium-rich plagioclase feldspars, and quartz. Hence, the term *pegmatite* generally refers to a rock of granitic composition (if otherwise, a term such as *gabbroic pegmatite* is used). Pegmatites are interesting as geological phenomena and important as minable resources (see box figure 1).

The extremely coarse texture of pegmatites is attributed to both slow cooling and the low viscosity (resistance to flow) of the fluid from which they form. Lava solidifying to rhyolite is very viscous. Magma solidifying to granite, being chemically similar, should be equally viscous.

Pegmatites, however, probably crystallize from a fluid composed largely of water under high pressure. Water molecules and ions from the parent, granitic magma make up a residual magma. Geoscientists hypothesize the following sequence of events accounts for most pegmatites.

As a granite pluton cools, increasingly more of the magma solidifies into the minerals of a granite. By the time the pluton is well over 90 percent solid, the residual magma contains a very high amount of silica and ions of elements that will crystallize into potassium and sodium feldspars. Also present are elements that could not be accommodated into the crystal structures of the common minerals that formed during the normal solidification phase of the pluton. Fluids, notably water, that were in the original magma are left over as well. If no fracture above the pluton permits the fluids to escape, they are sealed in, as in a pressure cooker. The watery residual magma has a low viscosity, which allows appropriate atoms to migrate easily toward growing crystals. The crystals add more and more atoms and grow very large.

Pegmatite bodies are generally quite small. Many are podlike structures, located either within the upper portion of a granite pluton or within the overlying country rock near the contact with granite, the fluid body evidently having squeezed into the country rock before solidifying. Pegmatite dikes are fairly common, especially within granite plutons, where they apparently filled cracks that developed in the already solid granite. Some pegmatites form small dikes along contacts between granite and country rock, filling cracks that developed as the cooling granite pluton contracted.

Most pegmatites contain only quartz, feldspar, and perhaps mica. Minerals of considerable commercial value are found in a few pegmatites. Large crystals of muscovite mica are mined from pegmatites. These crystals are called "books" because the cleavage flakes (tens of centimetres across) look like pages. Because muscovite is an excellent insulator, the cleavage sheets are used in electrical devices, such as toasters, to separate uninsulated electrical wires. Even the large feldspar crystals in pegmatites are mined for various industrial uses, notably the manufacture of ceramics.

Many rare elements are mined from pegmatites. These elements were not absorbed by the minerals of the main pluton and so were concentrated in the residual pegmatitic magma, where they crystallized as constituents of unusual minerals. Minerals containing the element lithium are mined from pegmatites. Lithium becomes part of a sheet silicate structure to form a pink or purple variety of mica (called lepidolite). Uranium ores, similarly concentrated in the residual melt of magmas, are also extracted from pegmatites. Tantalum, a rare metal used in the manufacture of miniaturized electronic circuits, is mined from pegmatites at the Tanco Mine in Manitoba.

Some pegmatites are mined for gemstones. Emerald and aquamarine, varieties of the mineral beryl, occur in pegmatites that crystallized from a solution containing the element beryllium. A large number of the world's very rare minerals are found only in pegmatites, many of these in only one known pegmatite body. These rare minerals are mainly of interest to collectors and museums.

Hydrothermal veins (described in chapter 7) are closely related to pegmatites. Veins of quartz are common in country rock near granite. Many of these are interpreted to have formed from water that escapes from the magma. Silica dissolved in the very hot water cakes on the walls of cracks as the water cools while travelling surfaceward. Sometimes valuable metals such as gold, silver, lead, zinc, and copper are deposited with the quartz in veins.

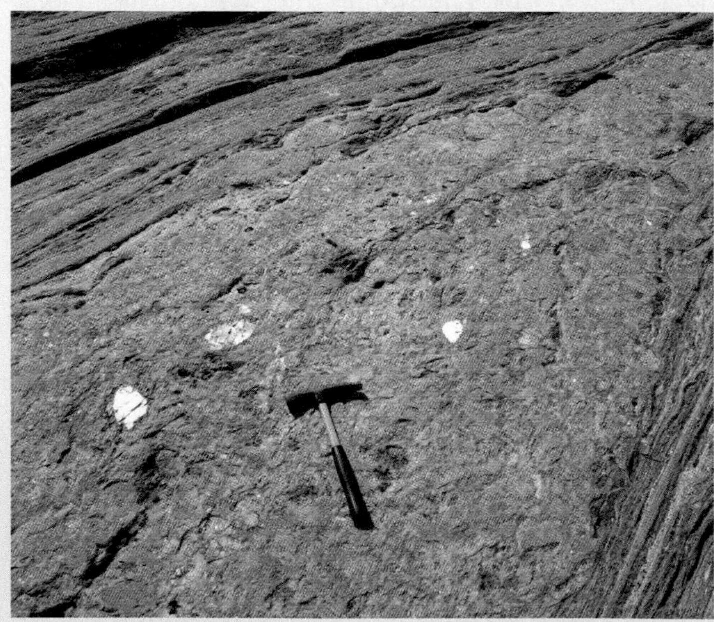

BOX 6.1 ■ FIGURE 1

Pegmatite in gneiss exposed on the Canadian Shield. Large crystals are feldspar (pink) and quartz (white).

Photo by Nick Eyles

The crystals or grains of most fine-grained rocks are considerably smaller than 1 millimetre and cannot be distinguished by the unaided eye. So, for practical purposes, if you can discern the individual grains, regard the rock as coarse-grained; if not, consider it fine-grained.

Some rocks are **porphyritic;** that is, large crystals are enclosed in a *groundmass* of finer-grained crystals or glass. An analogy for porphyritic texture is a milk chocolate bar containing whole almonds. If the groundmass is fine-grained, extrusive rock names are used. (For instance, figure 6.5 shows a *porphyritic andesite.*) Porphyritic extrusive rocks are usually interpreted as having begun crystallizing slowly underground followed by eruption and rapid solidification of the remaining magma at the Earth's surface. Some porphyritic rocks have a coarse-grained groundmass in which the individual grains are more than 1 millimetre. The larger crystals enclosed in the groundmass are much bigger, usually two or more centimetres across. *Porphyritic granite* is an example.

HOW ARE IGNEOUS ROCKS CLASSIFIED?

Identification of Igneous Rocks

Igneous rock names are based on texture (notably grain size) and mineralogical composition (which reflects chemical composition). Mineralogically (and chemically) equivalent rocks are *granite-rhyolite, diorite-andesite,* and *gabbro-basalt.* The relationships between igneous rocks are shown in figure 6.4.

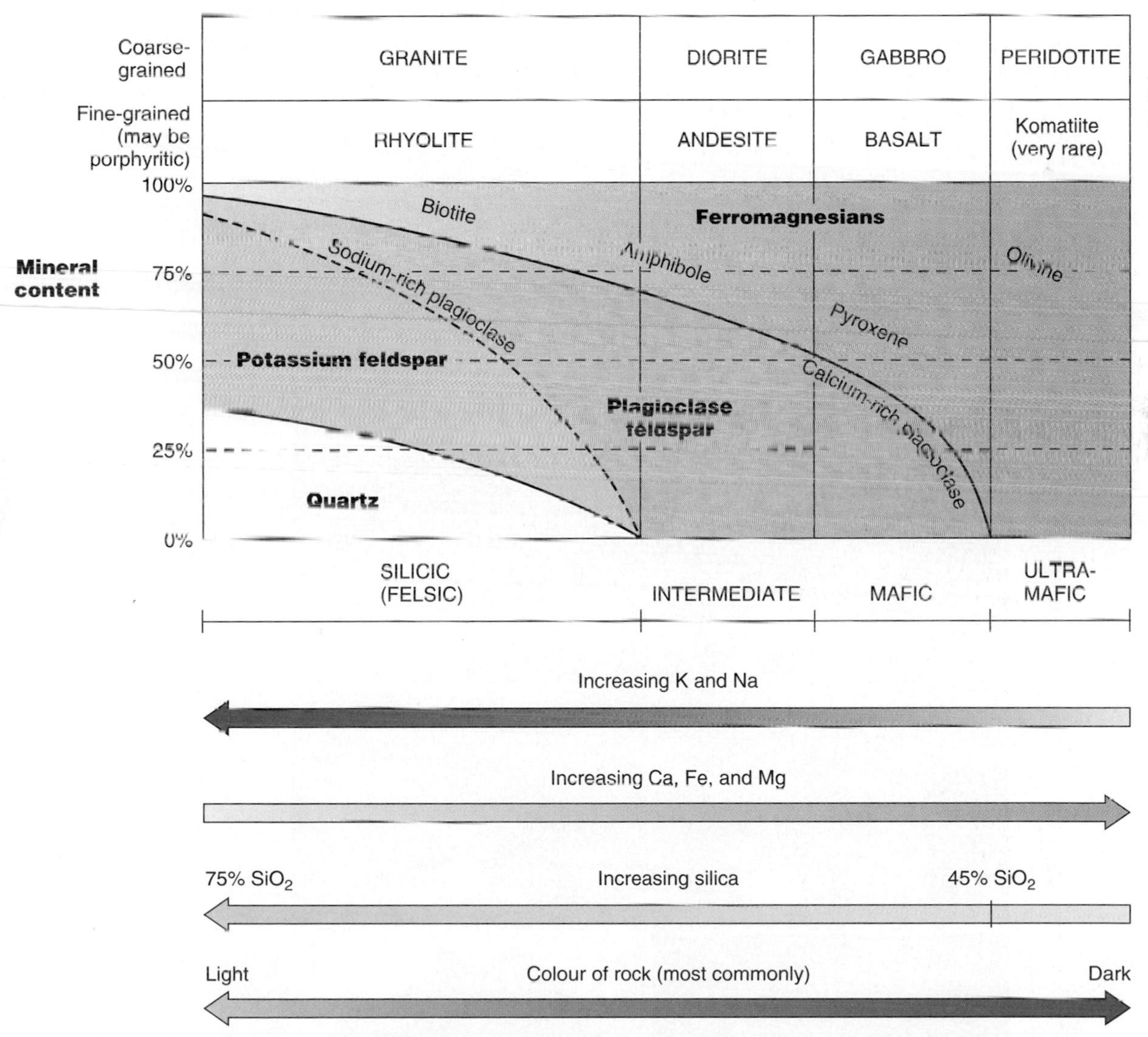

FIGURE 6.4

Classification chart for the most common igneous rocks. Rock names based on special textures are not shown. Sodium-rich plagioclase is associated with silicic rocks, whereas calcium-rich plagioclase is associated with mafic rocks. The names of the particular ferromagnesian minerals (biotite, etc.) are placed in the diagram at the approximate composition of the rocks in which they are most likely to be found.

Because of their larger mineral grains, plutonic rocks are easier to identify than extrusive rocks. The physical properties of each mineral in a plutonic rock can be determined more readily. And, of course, knowing what minerals are present makes rock identification a simpler task. For instance, **gabbro** is formed of coarse-grained ferromagnesian minerals and grey, plagioclase feldspar. (Recall from the mineral chapter that ferromagnesian minerals are silicates that contain iron and magnesium—amphibole, pyroxene, olivine, and biotite.) One can positively identify the feldspar on the basis of cleavage and, with practice, verify that no quartz is present. Gabbro's fine-grained counterpart is **basalt,** which is also composed of ferromagnesian minerals and plagioclase. The individual minerals cannot be identified by the naked eye, however, and one must use the less reliable attribute of colour—basalt is usually dark grey to black.

As you can see from figure 6.4, granite and **rhyolite** are composed predominantly of feldspars (usually white or pink) and quartz. Granite, being coarse-grained, can be positively identified by verifying that quartz is present. Rhyolite is usually cream-coloured, tan, or pink. Its light colour indicates that ferromagnesian minerals are not abundant. **Diorite** and **andesite** are composed of feldspars and significant amounts of ferromagnesian minerals (30–50%). The minerals can be identified and their percentages estimated to indicate diorite. Andesite, being fine-grained, can usually be identified by its medium-grey or medium-green colour. Its appearance is intermediate between light-coloured rhyolite and dark basalt.

Use the chart in figure 6.4 to identify common igneous rocks. You may also find it helpful to turn to appendix B, which includes a key for identifying common igneous rocks. (Photos of typical igneous rocks are shown in figure 6.5.)

Varieties of Granite

Granite and rhyolite occupy a larger area in the classification chart than do the other rocks. This reflects their greater variation in composition. Figure 6.6 reproduces the granite-rhyolite field shown in figure 6.4. Rocks X and Y represent two quite different granites (or rhyolites). Figure 6.6 shows the mineral composition of each of these rocks as determined by using the

FIGURE 6.5
Common igneous rocks
Photos by S. Fletcher

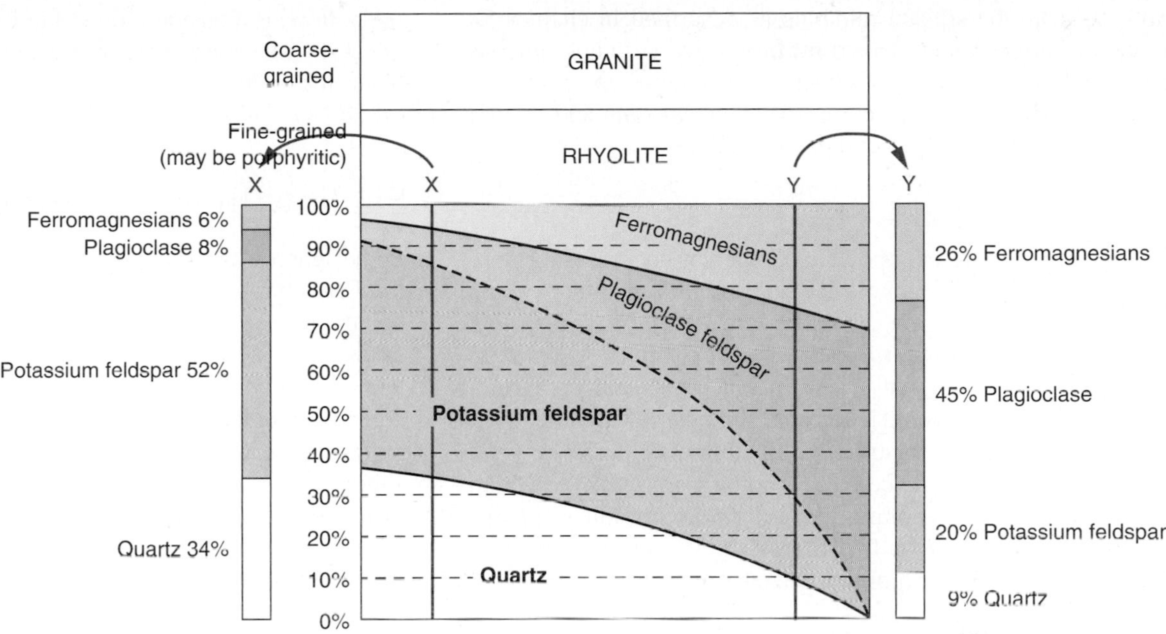

FIGURE 6.6

The compositional variation of granite shown in figure 6.4. Granite x and granite y are quite different in the proportions of minerals present. The bar graphs for each show the percentage of minerals in each rock.

percentage scale on the left of the diagram. Rock X has 34 percent quartz and 52 percent potassium feldspar (together, 86 percent of the rock). Only 8 percent is sodium-rich plagioclase feldspar and 6 percent ferromagnesian minerals (most likely biotite mica). This is a very silica-rich (*silicic*) rock. By contrast, granite (or rhyolite) Y is close to the diorite-andesite field and has a composition that has only 11 percent quartz and 21 percent potassium feldspar. Plagioclase feldspar is, at 44 percent, the most abundant mineral and 24 percent of the rock is ferromagnesian minerals (likely a mixture of biotite and amphibole). Geologists have subdivided the field of granite and named each of the varieties; a rock in the right portion of the field, rock Y, for example, is called granodiorite.

Any classification system is, of course, a human device, and for this reason, classification systems differ somewhat among groups of geologists. We define the boundary between granite and diorite by the presence or absence of quartz; but we could just as easily have placed the boundary slightly to the left, so that a rock with 10 percent or less quartz would be diorite.

Chemistry of Igneous Rocks

The chemical composition of the magma determines which minerals and how much of each will crystallize when an igneous rock forms. For instance, the presence of quartz in a rock indicates that the magma was enriched in silica (SiO_2). The lower part of figure 6.4 shows the relationship of chemical composition to rock type. Chemical analyses of rocks are reported as weight percentages of oxides (e.g., SiO_2, MgO, Na_2O, etc.) rather than as separate elements (e.g., Si, O, Mg, Na). Figure 6.7 shows the chemical composition of average rocks. For virtually all igneous rocks, SiO_2 (silica) is the most abundant component. The amount of SiO_2 varies from about 45 percent to 75 percent of the total weight of common volcanic rocks. The variations between these extremes account for striking differences in the appearance (figure 6.5) and mineral content (figure 6.4) of the rocks.

Mafic Rocks

Rocks with a silica content close to 50 percent (by weight) are considered *silica-deficient*, even though SiO_2 is, by far, the most abundant constituent (figure 6.7). Chemical analyses show that the remainder is composed mostly of the oxides of aluminum (Al_2O_3), calcium (CaO), magnesium (MgO), and iron (FeO and Fe_2O_3). (These oxides generally combine with

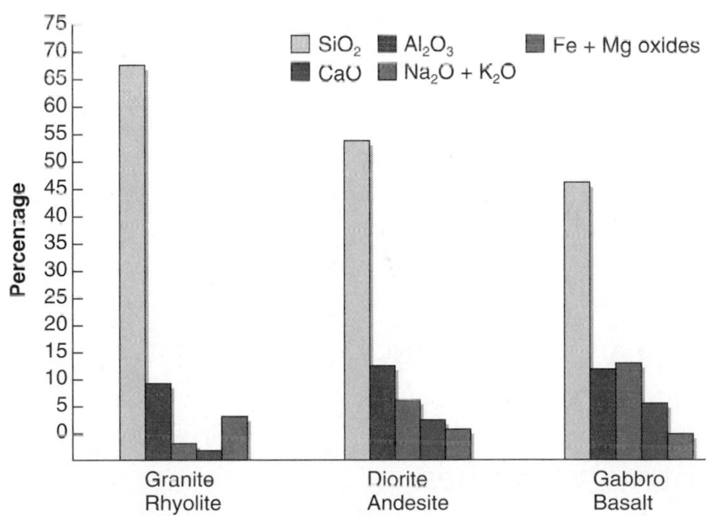

FIGURE 6.7

The average chemical composition of silicic, intermediate, and mafic rocks. Composition is given in weight percent of oxides. Note that as the amount of silica decreases, the oxides of Na and K decrease, and the oxides of Ca, Fe, and Mg increase. Al oxide does not vary significantly.

SiO_2 to form the silicate minerals as described in chapter 5.) Rocks in this group are called **mafic**—silica-deficient igneous rocks with a relatively high content of magnesium, iron, and calcium. (The term *mafic* comes from *magnesium* and *ferric.*) Basalt and gabbro are, of course, mafic rocks.

Silicic (Felsic) Rocks

At the other extreme, the *silica-rich* (65 percent or more SiO_2) rocks tend to have only very small amounts of the oxides of calcium, magnesium, and iron. The remaining 25 percent to 35 percent of these rocks is mostly aluminum oxide (Al_2O_3) and oxides of sodium (Na_2O) and potassium (K_2O). These are called **silicic** or **felsic** rocks—silica-rich igneous rocks with a relatively high content of potassium and sodium (the *fel* part of the name comes from *feldspar,* which crystallizes from the potassium, sodium, aluminum, and silicon oxides; *si* in *felsic* is for silica). The silicic rocks rhyolite and granite are light-coloured because of the low amount of ferromagnesian minerals.

Intermediate Rocks

Rocks with a chemical content between that of felsic and mafic are classified as **intermediate rocks.** *Andesite,* which is usually green or medium grey, is the most common intermediate volcanic rock.

Ultramafic Rocks

An **ultramafic rock** is composed entirely or almost entirely of ferromagnesian minerals. No feldspars are present and, of course, no quartz. **Peridotite,** a coarse-grained rock composed of pyroxene and olivine, is the most abundant ultramafic rock. Chemically, these rocks contain less than 45 percent silica.

Note from the chart (figure 6.4) that komatiite, the volcanic ultramafic rock, is very rare. Ultramafic extrusive rocks are mostly restricted to the very early history of the Earth. For our purposes, they need not be discussed further.

Some ultramafic rocks form from differentiation (explained later in this chapter) of a basaltic magma at very high temperatures. Most ultramafic rocks come from the mantle, rather than from the Earth's crust. Where we find large bodies of ultramafic rocks, the usual interpretation is that a part of the mantle has travelled upward as solid rock.

WHAT HAPPENS WHEN MAGMA COOLS UNDERGROUND?

Intrusive Bodies

Intrusions, or **intrusive structures,** are bodies of intrusive rock whose names are based on their size and shape, as well as their relationship to surrounding rocks. They are important aspects of the architecture, or *structure,* of the Earth's crust. The various intrusions are named and classified on the basis of the following considerations: (1) Is the body large or small? (2) Does it have a particular geometric shape? (3) Did the rock form at a considerable depth, or was it a shallow intrusion? (4) Does it follow layering in the country rock or not?

Shallow Intrusive Structures

Some igneous bodies apparently solidified near the surface of the Earth (probably at depths of less than 2 kilometres). These bodies appear to have solidified in the subsurface "plumbing systems" of volcanoes or lava flows. Shallow intrusive structures tend to be relatively small compared with those that formed at considerable depth. Because the country rock near the Earth's surface generally is cool, intruded magma tends to chill and solidify relatively rapidly. Also, smaller magma bodies will cool faster than larger bodies, regardless of depth. For both of these reasons, shallow intrusive bodies are likely to be fine-grained.

A **volcanic neck** is an intrusive structure apparently formed from magma that solidified within the throat of a volcano. One of the best examples is Ship Rock in New Mexico (figure 6.8). Here is how geologists interpret the history of this feature. A volcano formed above what is now Ship Rock. The magma for the volcano moved upward through a more or less cylindrical "plumbing system." Eruptions ceased and the magma underground solidified into what is now Ship Rock. In time, the volcano and its underlying rock—the country rock around Ship Rock—eroded away. The more resistant igneous body eroded more slowly into its present shape. Weathering and erosion are continuing (falling rock has been a serious hazard to rock climbers). Black Tusk in the Garibaldi Park area of British Columbia is also an example of an exposed volcanic neck.

Dikes and Sills

Another, and far more common, intrusive structure can also be seen at Ship Rock. The low, wall-like ridge extending outward from Ship Rock is an eroded dike. A **dike** is a tabular (shaped like a tabletop), discordant, intrusive structure (figure 6.9). *Discordant* means that the body is not parallel to any layering in the country rock. (Think of a dike as cutting across layers of country rock.) Dikes may form at shallow depths and be fine-grained, such as those at Ship Rock, or form at greater depths and be coarser-grained. Dikes need not appear as walls protruding from the ground (figure 6.10). The ones at Ship Rock do so only because they are more resistant to weathering and erosion than the country rock. One of the larger dike swarms in the world is found in northern Canada. The Mackenzie dike swarm formed around 1.3 billion years ago and is related to injection of magma from an underlying mantle plume.

A **sill** is also a tabular intrusive structure, but it is *concordant.* That is, sills, unlike dikes, are parallel to any planes or layering in the country rock (figures 6.9 and 6.11). Typically, the country rock bounding a sill is layered sedimentary rock. As magma squeezes into a crack between two layers, it solidifies into a sill.

If the country rock is not layered, a tabular intrusion is regarded as a dike.

A

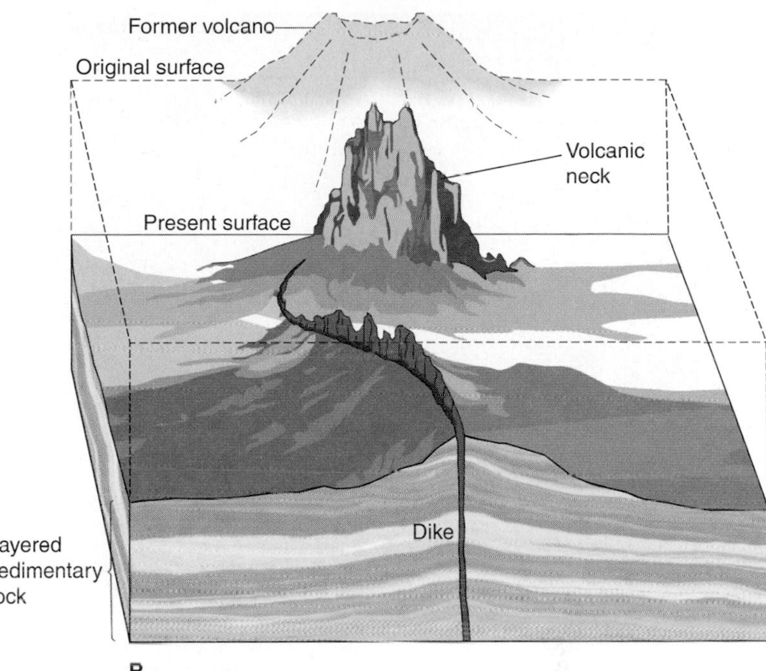

B

FIGURE 6.8

(*A*) Ship Rock in New Mexico, which rises 420 metres above the desert floor. (*B*) Relationship to the former volcano. *Photo by Frank M. Hanna*

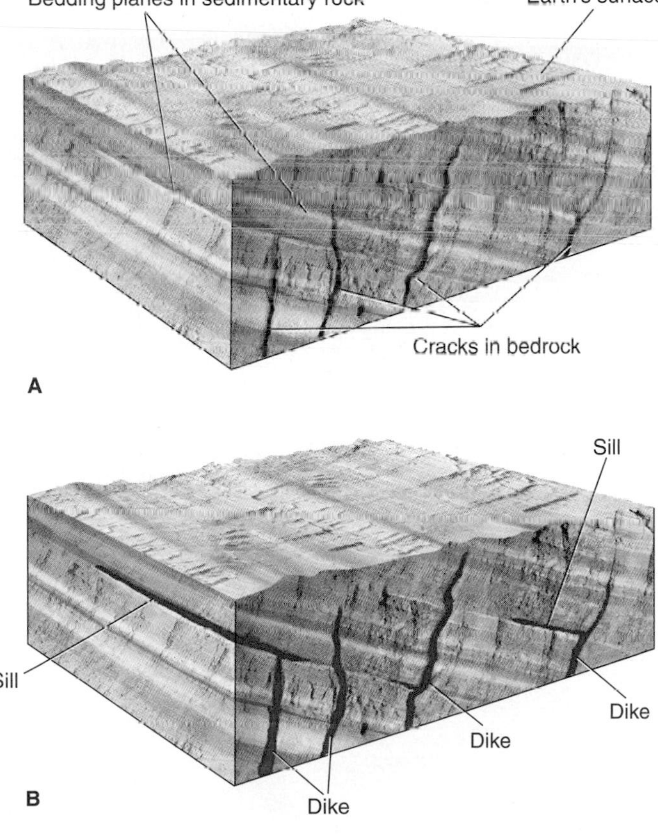

FIGURE 6.9

(*A*) Cracks and bedding planes are planes of weakness. (*B*) Concordant intrusions where magma has intruded between sedimentary layers are sills; discordant intrusions are dikes.

FIGURE 6.10

Dikes (pink-coloured rocks) intruded into banded gneiss (dark-coloured), Ontario, Canada.

Photo by Nick Eyles

FIGURE 6.11

A sill (dark layer) intruded between limestone layers, Glacier National Park, Montana. The limestone adjacent to the sill has been contact metamorphosed into light-coloured marble (explained in chapter 10).

Photo © William E. Ferguson

Intrusives That Crystallize at Depth

A **pluton** is a body of magma or igneous rock that crystallized at considerable depth within the crust. Where plutons are exposed at the Earth's surface, they are arbitrarily distinguished by size. A **stock** is a small discordant pluton with an outcrop area (i.e., the area over which it is exposed to the atmosphere) of less than 100 square kilometres. If the outcrop area is greater than 100 square kilometres, the body is called a **batholith** (figure 6.12). Most batholiths crop out over areas vastly greater than the minimum 100 square kilometres.

Although batholiths may contain mafic and intermediate rocks, they almost always are predominantly composed of granite. Detailed studies of batholiths indicate that they are formed of numerous, coalesced plutons. Apparently, large blobs of magma worked their way upward through the lower crust and collected 5 to 30 kilometres below the surface, where they solidified (figure 6.12). These blobs of magma, known as **diapirs,** are less dense than the surrounding rock that is pliable and shouldered aside as the magma rises. Batholiths occupy large portions of North America, particularly in the west. Over half of California's Sierra Nevada mountains is a batholith whose individual plutons were emplaced during a period of over 100 million years. An even larger batholith extends almost the entire length of the mountain ranges of Canada's west coast and southeastern Alaska—a distance of 1,800 kilometres (figure 6.13). Smaller batholiths are also found in eastern North

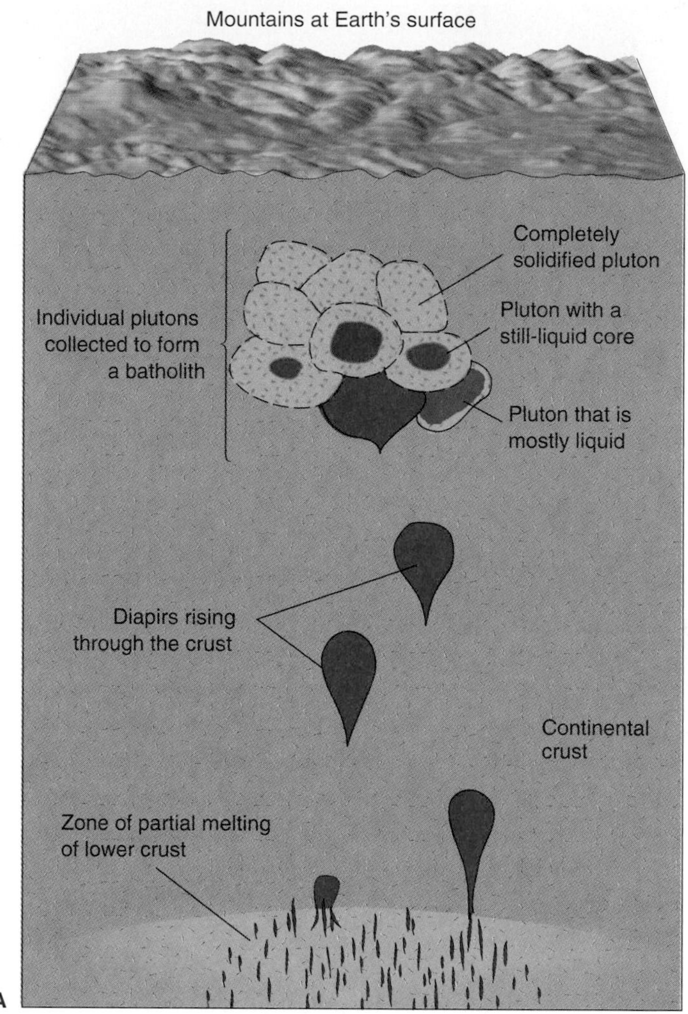

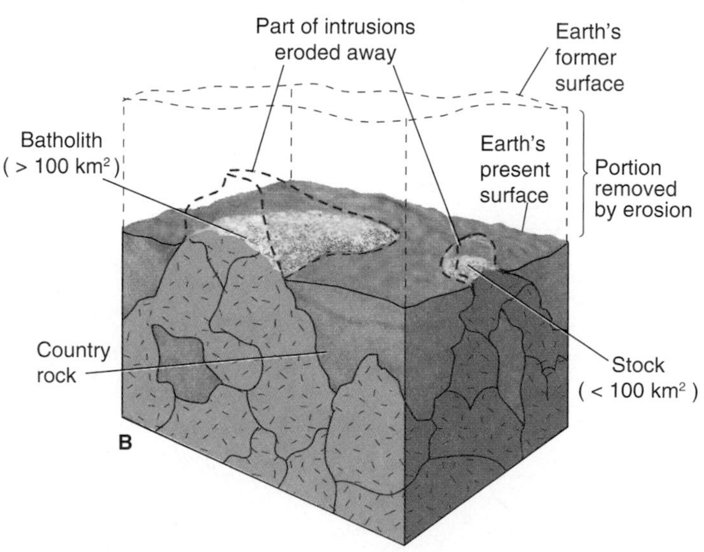

FIGURE 6.12

(A) Diapirs of magma travel upward from the lower crust and solidify in the upper crust. Coalescence of many magma diapirs forms a solid mass of plutonic rock. (B) After uplift and erosion, surface exposures of plutonic rock form a batholith and a stock.

FIGURE 6.13

Exhumed granite batholith at Squamish, B.C. The batholith forms part of the Coast Range Batholith that extends along Canada's west coast.

Photo by Enlightened Photography

America in the Piedmont east of the Appalachian Mountains and in New England and the coastal provinces of Canada. (The extent and location of North American batholiths are shown on the geologic map on the inside front cover.)

Granite is considerably more common than rhyolite, its volcanic counterpart. Why is this? Silicic magma is much more *viscous* (that is, more resistant to flow) than mafic magma. Therefore, a silicic magma body will travel upward through the crust more slowly and with more difficulty than mafic or intermediate magma. Unless it is exceptionally hot, a silicic magma will not be able to work its way through the relatively cool and rigid rocks of the upper few kilometres of crust. Instead, it is much more likely to solidify slowly into a pluton.

Abundance and Distribution of Plutonic Rocks

Granite is the most abundant igneous rock in mountain ranges. It is also the most commonly found igneous rock in the interior lowlands of continents. Throughout the lowlands of much of Canada, very old plutons have intruded even older metamorphic rock. Metamorphic and plutonic rocks similar in age and complexity to those in Canada are found in the Great Plains of the United States. Here, however, they are mostly covered by a veneer (a kilometre or so) of younger, sedimentary rock. These "basement" rocks are exposed to us in only a few places. In Grand Canyon, Arizona, the Colorado River has eroded through the layers of sedimentary rock to expose the ancient plutonic and metamorphic basement. In the Black Hills of South Dakota, local uplift and subsequent erosion have exposed similar rocks.

Granite, then, is the predominant igneous rock of the continents. As described in chapter 7, basalt and gabbro are the predominant rocks underlying the oceans. Andesite (usually along continental margins) is the building material of most young volcanic mountains. Underneath the crust, ultramafic rocks make up the upper mantle.

HOW DO DIFFERENT TYPES OF MAGMA FORM?

If a rock is heated sufficiently, it begins melting to form magma. Under ideal conditions, rock can melt and yield a granitic magma at temperatures as low as 650°C. Temperatures over 1,000°C are required to create basaltic magma. However, several factors control the melting temperature of rock. Pressure, amount of gas (particularly water) present, and the particular mix of minerals all influence when melting takes place. These factors are discussed in the following sections.

Heat for Melting Rock

Most of the heat that contributes to the generation of magma comes from the very hot Earth's core (where temperatures are estimated to be greater than 5,000°C). Heat is conducted toward the Earth's surface through the mantle and crust. This is comparable to the way heat is conducted through the wall from a hot room into a cooler room or through the metal of a frying pan. Heat is also brought from the lower mantle when part of the mantle flows upward, either through convection (described in chapters 1 and 19) or by hot mantle plumes. The geothermal gradient, described next, is a manifestation of heat transfer in the mantle.

Geothermal Gradient

A miner descending a mine shaft notices a rise in temperature. This is due to the **geothermal gradient,** the rate at which temperature increases with increasing depth beneath the surface. Data show the geothermal gradient, on the average, to be about 3°C for each 100 metres (30°C/km) of depth in the upper part of the crust. The geothermal gradient is not the same everywhere. Figure 6.14 shows geothermal gradients for two regions. The curve for the volcanic region indicates a higher geothermal gradient than that for the continental interior. Temperatures high enough to melt rock would be expected at a relatively shallow depth beneath the volcanic region. You would have to go deeper in the continental interior to reach the same temperature; however, the rock there does not melt because of the increased pressure at that depth.

One reason for a higher geothermal gradient is that deeper, and therefore hotter, mantle rock has worked its way upward closer to the Earth's surface due either to mantle convection or mantle plumes. "Hot spots" in the crust (where the geothermal gradient is locally very high) have been hypothesized to be due to hot **mantle plumes,** which are narrow upwellings of hot material within the mantle. Mantle plumes have been used to explain some igneous activity, notably that which takes place in the interior of tectonic plates, far from a plate boundary. Examples include the long-lasting volcanic activity that built the Hawaiian Islands and the eruptions at Yellowstone National Park in Wyoming. The silicic eruptions at Yellowstone that took place some 600,000 years ago were much larger and more violent than any eruptions that have occurred in historical time. Geoscientists attribute these eruptions to a hot mantle plume that caused melting of the crust beneath this area.

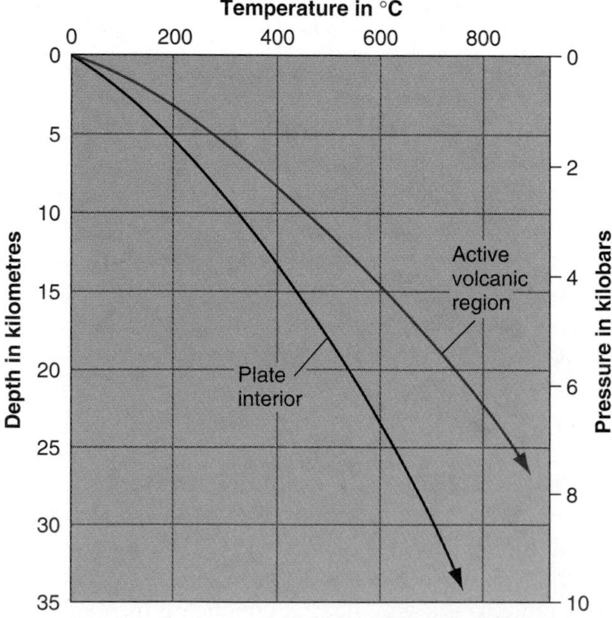

FIGURE 6.14

Geothermal gradients at two parts of the Earth's crust.

Factors That Control Melting Temperatures

Pressure

The melting point of a mineral generally *increases* with increasing pressure. Pressure increases with depth in the Earth's crust, just as temperature does. So a rock that melts at a given temperature at the surface of the Earth requires a higher temperature to melt deep underground. Rock will not melt where the geothermal gradient is close to that of the plate interior because at all depths, the melting temperature will always be higher than the temperature of the rock. Thus, we need mechanisms that raise the rock to a higher temperature or lower the melting temperature for the rock.

Decompression melting takes place when a body of hot mantle rock moves upward and the pressure is reduced to the extent that the melting point drops to the temperature of the body. Hawaiian volcanic activity, perhaps attributable to an underlying mantle plume, illustrates how *reduced* pressure contributes to the creation of magma. Solid rock that was once deeper in the mantle (and, therefore, very hot) has worked its way upward. Most of its heat has been retained during the upward journey. However, the pressure decreases as the rock body travels upward. As it approaches 50 kilometres or so from the Earth's surface, pressure is sufficiently reduced so that melting takes place.

Water under Pressure

If enough gas, especially water vapour, is present and under high pressure, a dramatic change occurs in the melting process. Water sealed in under high pressure helps break the silicon–oxygen bonds, causing the crystal to liquify. A mineral's melting temperature is significantly lowered by water under high pressure (figure 6.15).

Experiments have shown that, under moderately high pressure, water mixed with granite lowers the melting point of granite from over 900°C (when dry) to as low as 650°C when saturated with water under pressure equivalent to that of 10,000 atmospheres, or *bars.* (Pressure at depth is usually expressed in *kilobars;* one kilobar is equal to 1,000 bars.) Ten kilobars corresponds to a depth of approximately 35 kilometres.

Effect of Mixed Minerals

Two metals—as in solder, which, typically, is a mixture of tin and lead—can be mixed in a ratio that lowers their melting temperature far below that of the melting points of the pure metals. Minerals behave similarly. Experiments have shown that in some cases, mixed fragments of two minerals melt at a lower temperature than either mineral alone. Figure 6.16 shows

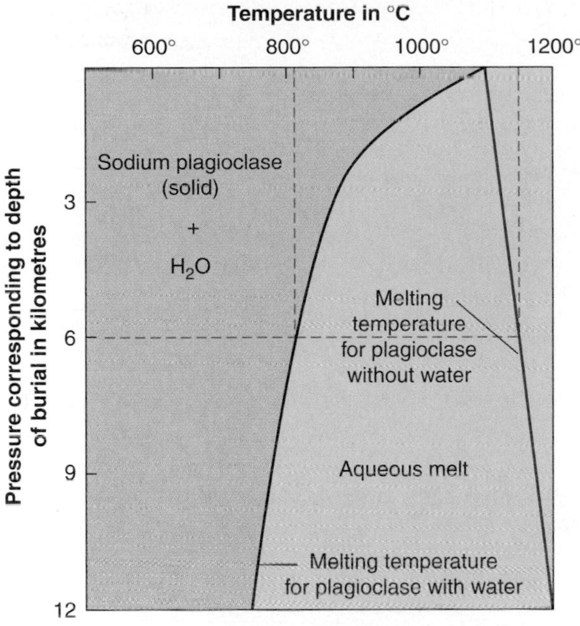

FIGURE 6.15

Melting temperature of a mineral with and without water present. The curve on the left is for melting of plagioclase saturated with water under the pressure corresponding to depth of burial. The line on the right corresponds to melting of dry plagioclase. The dashed, red line indicates that at pressures corresponding to 6 kilometres, plagioclase with water melts at just over 800°C, whereas dry plagioclase melts at around 1150°C.

After Tuttle and Brown, Geologic Society of America, 1958

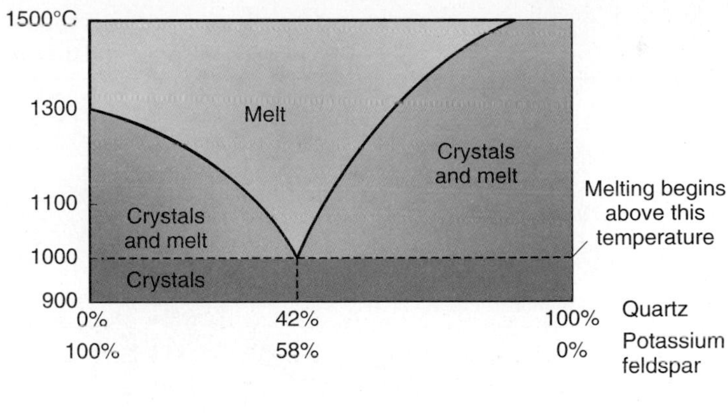

FIGURE 6.16

Melting temperatures for mixtures of quartz and potassium feldspar at atmospheric pressure.

Modified from Schairer and Bowen, 1956. V 254, p. 16. American Journal of Science

the melting temperatures for quartz and potassium feldspar mixed in various proportions. If the mixture is 42 percent quartz (58 percent potassium feldspar), complete melting takes place at just above 1,000°C. For the composition 80 percent quartz (20 percent potassium feldspar), melting will start at 1,000°C with a drop of liquid, but for complete melting the temperature must be raised to 1,500°C. Pure quartz requires even higher temperatures to melt.

HOW DO MAGMAS OF DIFFERENT COMPOSITIONS EVOLVE?

A major topic of investigation for geoscientists is why igneous rocks are so varied in composition. On a global scale, magma composition is clearly controlled by geologic setting. But why? Why are basaltic magmas associated with oceanic crust, whereas granitic magmas are common in the continental crust? On a local scale, igneous bodies often show considerable variation in rock type. For instance, individual plutons typically display a considerable range of compositions, mostly varieties of granite, but many also will contain minor amounts of gabbro or diorite. In this section, we describe processes that result in differences in composition of magmas. The final section of this chapter relates these processes to plate tectonics for the larger view of igneous activity.

Sequence of Crystallization and Melting

Early in the twentieth century, N. L. Bowen conducted a series of experiments that determined the sequence in which minerals crystallize in a cooling magma. The sequence became known as **Bowen's reaction series** and is shown in figure 6.17. A simplified explanation of the series and its importance to igneous rocks is presented next. For a more in-depth presentation, go to our Website at **www.mcgrawhill.ca/olc/plummer**.

Bowen's experiments showed that in a cooling magma, certain minerals are stable at higher melting temperatures and crystallize before those stable at lower temperatures. Looking at the *discontinuous branch,* which contains only ferromagnesian minerals, we can see that olivine crystallizes before pyroxene and pyroxene crystallizes before amphibole. A complication is that early formed crystals *react* with the remaining melt and recrystallize as cooling proceeds. For instance, early formed olivine crystals react with the melt and recrystallize to pyroxene when pyroxene's temperature of crystallization is reached. Upon further cooling, pyroxene continues to crystallize until all of the melt is used up or the melting temperature of amphibole is reached. At this point, pyroxene reacts with the remaining melt and amphibole forms at its expense. If all of the iron and magnesium in the melt is used up before all of the pyroxene recrystallizes to amphibole, then the ferromagnesian minerals in the solid rock would be amphibole and pyroxene. (The rock would not contain olivine or biotite.)

Crystallization in the discontinuous and the *continuous branch* takes place at the same time. The continuous branch contains only plagioclase feldspar. Plagioclase is a *solid-solution* mineral (discussed in chapter 5 on minerals) in which either sodium or calcium atoms can be accommodated in its crystal structure, along with aluminum, silicon, and oxygen. The composition of plagioclase changes as magma is cooled

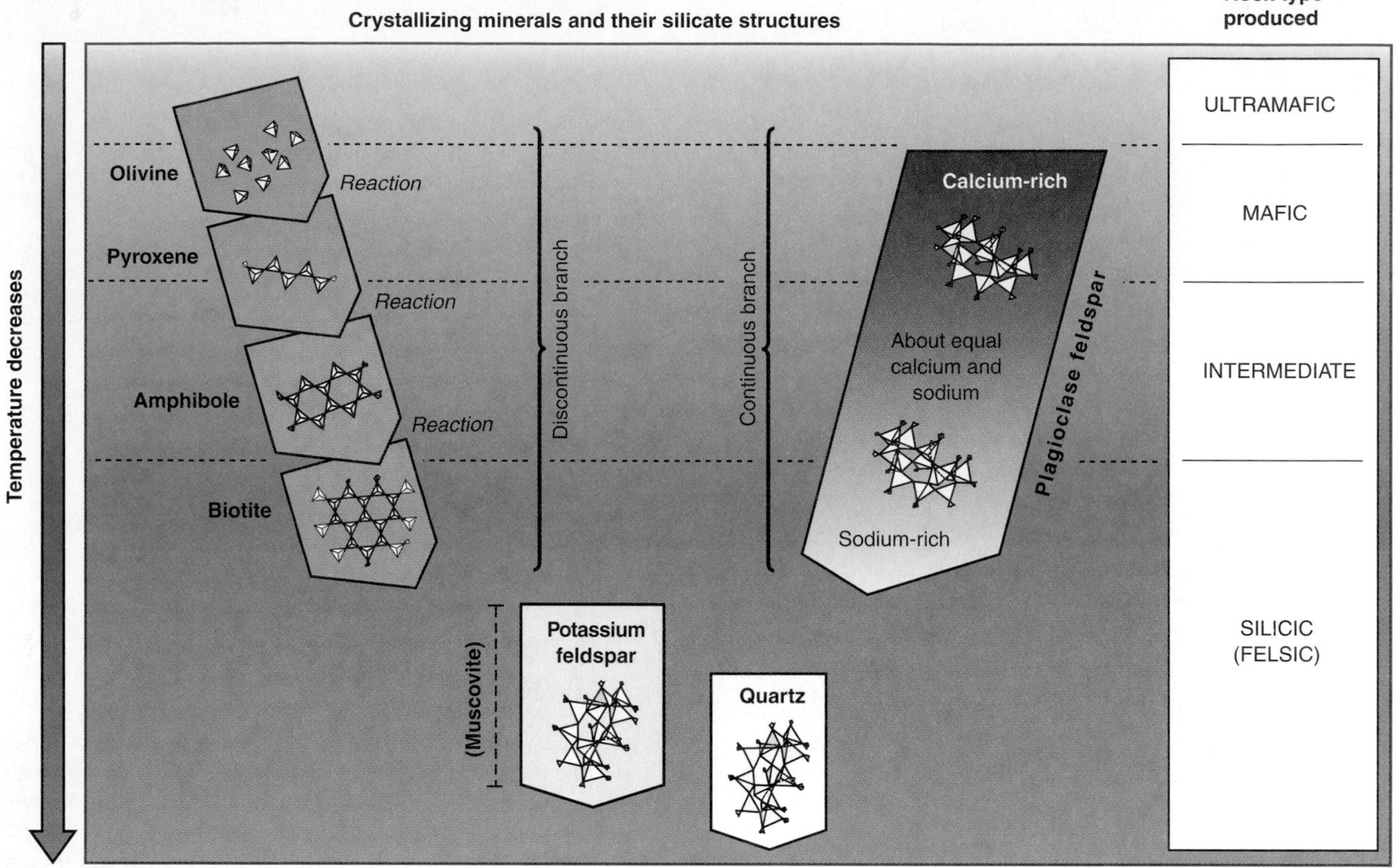

FIGURE 6.17

Bowen's reaction series. The reaction series as shown is very generalized. Moreover, it represents Bowen's experiments that involved melting a relatively silica-rich variety of basalt.

and earlier formed crystals react with the melt. The first plagioclase crystals to form as a hot melt cools contain calcium but little or no sodium. As cooling continues, the early formed crystals grow and incorporate progressively more sodium into their crystal structures.

Any magma left after the crystallization is completed along the two branches is richer in silicon than the original magma and also contains abundant potassium and aluminum. The potassium and aluminum combine with silicon to form *potassium feldspar.* (If the water pressure is high, *muscovite* may also form at this stage.) Excess SiO_2 crystallizes as *quartz.*

From Bowen's reaction series, we can derive several important concepts that are necessary to understand igneous rocks and processes:

- A mafic magma will crystallize into pyroxene (with or without olivine) and calcium-rich plagioclase—that is, basalt or gabbro—if the early formed crystals are not removed from the remaining magma. Similarly, an intermediate magma will crystallize into diorite or andesite, if early formed minerals are not removed.
- If minerals are separated from a magma, the remaining magma is more silicic than the original magma. For example, if olivine and calcium-rich plagioclase are removed, the residual melt would be richer in silicon and sodium and poorer in iron and magnesium.
- If you heat a rock, the minerals will melt in reverse order. In other words, you would be going up the series as diagrammed in figure 6.17. Quartz and potassium feldspar would melt first. If the temperature is raised further, biotite and sodium-rich plagioclase would contribute to the melt. Any minerals higher in the series would remain solid unless the temperature is raised further.
- Bowen's reaction series can be used to show how two important processes that create and modify magma composition work. These are *differentiation* and *partial melting.*

Differentiation

The process by which different ingredients separate from an originally homogenous mixture is **differentiation.** An example is the separation of whole milk into cream and nonfat milk.

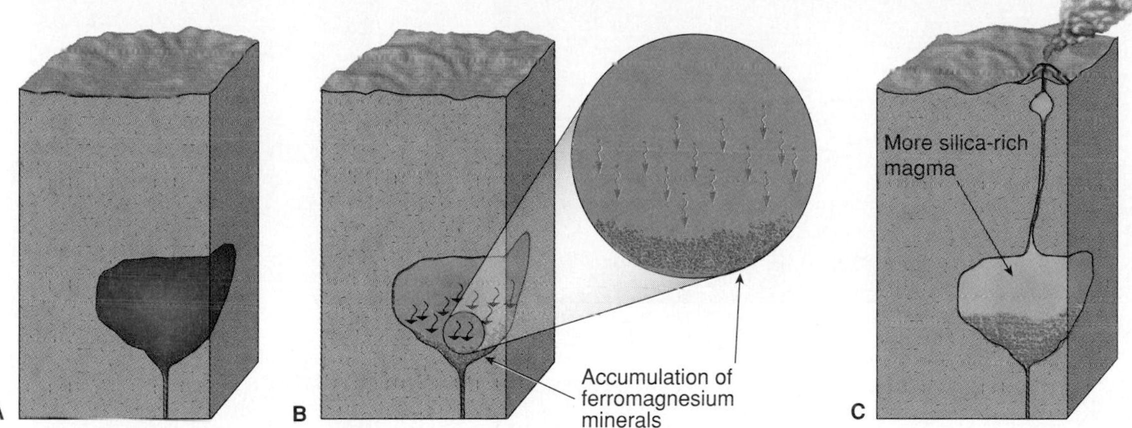

FIGURE 6.18
Differentiation of a magma body. (*A*) Recently intruded mafic magma is completely liquid. (*B*) Upon slow cooling, ferromagnesian minerals, such as olivine, crystallize and sink to the bottom of the magma chamber. The remaining liquid is now an intermediate magma. (*C*) Some of the intermediate magma moves upward to form a smaller magma chamber at a higher level that feeds a volcano.

Differentiation in magmas takes place mainly through **crystal settling,** the downward movement of minerals that are denser (heavier) than the magma from which they crystallized.

If crystal settling takes place in a mafic magma chamber, olivine and, perhaps, pyroxene crystallize and settle to the bottom of the magma chamber (figure 6.18). This makes the remaining magma more silicic. Calcium-rich plagioclase also separates as it forms. The remaining magma is, therefore, depleted of calcium, iron, and magnesium. Because these minerals were economical in using the relatively abundant silica, the remaining magma becomes richer in silica as well as in sodium and potassium.

It is possible that by removing enough mafic components, the residual magma would be silicic enough to solidify into granite (or rhyolite). But it is more likely only enough mafic components would be removed to allow an intermediate residual magma, which would solidify into diorite or andesite. The lowermost portions of some large sills are composed predominantly of olivine and pyroxene, whereas upper levels are considerably less mafic. Even in large sills, however, differentiation has rarely progressed far enough to produce granite within the sill.

Partial Melting

As mentioned earlier, progressing upward through Bowen's reaction series (going from cool to hot) gives us the sequence in which minerals in a rock melt. As might be expected, the first portion of a rock to melt as temperatures rise forms a liquid with the chemical composition of quartz and potassium feldspar. The oxides of silicon plus potassium and aluminum "sweated out" of the solid rock could accumulate into a pocket of silicic magma. If higher temperatures prevailed, more mafic magmas would be created. Small pockets of magma could merge and form a large enough mass to rise as a diapir. In nature, temperatures rarely rise high enough to entirely melt a rock.

Partial melting of the lower continental crust likely produces silicic magma. The magma rises and eventually solidifies at a higher level in the crust into granite, or rhyolite if it reaches Earth's surface.

Geoscientists generally regard basaltic magma (Hawaiian lava, for example) as the product of partial melting of ultramafic rock in the mantle, at temperatures hotter than those in the crust. The solid residue left behind in the mantle when the basaltic magma is removed is an even more silica-deficient ultramafic rock.

Assimilation

A very hot magma may melt some of the country rock and *assimilate* the newly molten material into the magma (figure 6.19). This is like putting a few ice cubes into a cup of hot coffee. The ice melts and the coffee cools as it becomes diluted. Similarly, if a hot basaltic magma, perhaps generated from the mantle, melts portions of the continental crust, the magma simultaneously becomes richer in silica and cooler. Possibly intermediate magmas such as those associated with circum-Pacific andesite volcanoes may derive from assimilation of some crustal rocks by a basaltic magma.

Mixing of Magmas

Some of our igneous rocks may be "cocktails" of different magmas. The concept is quite simple. If two magmas meet and merge within the crust, the combined magma will be compositionally intermediate (figure 6.20). If you had approximately equal amounts of a granitic magma mixing with a basaltic magma, the resulting magma should crystallize underground as diorite or erupt on the surface to solidify as andesite.

HOW DOES IGNEOUS ACTIVITY RELATE TO PLATE TECTONICS?

One of the appealing aspects of the theory of plate tectonics is that it accounts reasonably well for the variety of igneous rocks and their distribution patterns. Divergent boundaries are associated with creation of basalt and gabbro of the oceanic crust. Andesite and granite are associated with convergent boundaries. Table 6.1 summarizes the relationships.

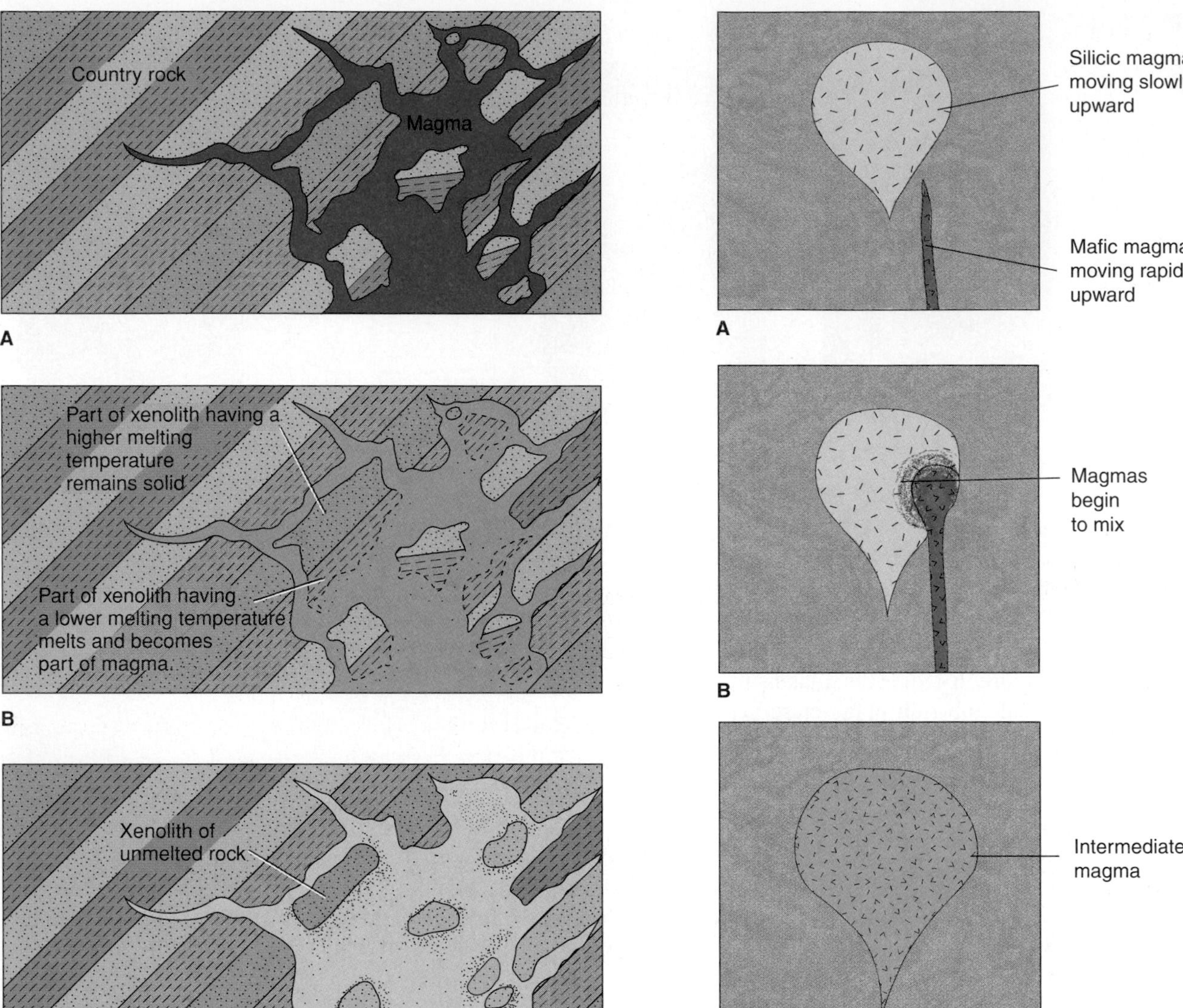

FIGURE 6.19

Assimilation. Magma formed is intermediate in composition between the original magma and the absorbed country rock. (*A*) Ascending magma breaks off blocks of country rock (the process is called stoping). (*B*) Xenoliths of country rock with melting temperatures lower than the magma melt. (*C*) The molten country rock blends with the original magma, leaving unmelted portions as inclusions.

FIGURE 6.20

Mixing of magmas. (*A*) Two bodies of magma moving surfaceward. (*B*) The mafic magma catches up with the silicic magma. (*C*) The two magmas combine and become an intermediate magma.

Igneous Processes at Divergent Boundaries

The crust beneath the world's oceans (more than 70 percent of Earth's surface) is mafic volcanic and intrusive rock, covered to a varying extent by sediment and sedimentary rock. Most of this basalt and gabbro was created at mid-oceanic ridges, which also are divergent plate boundaries. Geoscientists agree that the mafic magma produced at divergent boundaries is due to partial melting of the asthenosphere. The asthenosphere beneath divergent boundaries probably is mantle material that has welled upward from deeper levels of the mantle. As the hot asthenosphere gets close to the surface, decrease in pressure results in partial melting. In other words, *decompression melting* takes place. The magma that forms is mafic and will solidify as basalt or gabbro. The portion that did not melt remains behind as a silica-depleted, iron-and-magnesium-enriched ultramafic rock.

Some of the basaltic magma erupts along a submarine ridge to form pillow basalts (described in chapter 7), while some fills near-surface fissures to create dikes. Deeper down, magma solidifies more slowly into gabbro. The newly solidified rock is pulled apart by spreading plates; more magma fills the new fracture and some erupts on the sea floor. The process is repeated, resulting in a continuous production of mafic crust (see chapter 2).

TABLE 6.1 Relationships between Igneous Rock Types and Their Usual Plate Tectonic Setting

Rock	Original Magma	Final Magma	Processes	Plate Tectonic Setting
Basalt and gabbro	Mafic	Mafic	Partial melting of mantle (asthenosphere)	1. Divergent boundary—oceanic crust created 2. Intraplate • plateau basalt • volcanic island chains (e.g., Hawaii)
Andesite and diorite	Mafic (usually)	Intermediate	Partial melting of mantle (asthenosphere) followed by: • differentiation or • assimilation or • magma mixing	Convergent boundary
Granite and rhyolite	Silicic	Silicic	Partial melting of lower crust	1. Convergent boundary 2. Intraplate • over mantle plume

Intraplate Igneous Activity

Igneous activity within a plate, a long distance from a plate boundary, is unusual. As described earlier, Hawaii and Yellowstone eruptions represent intraplate igneous activity. The ongoing volcanic eruptions in Hawaii take place on oceanic crust, whereas eruptions at Yellowstone represent continental intraplate activity.

The huge volume of mafic magma that erupted to form the Columbia plateau basalts of Washington and Oregon (described in chapter 7) is attributed to a past hot mantle plume, according to a recent hypothesis (figure 6.21). In this case, the large volume of basalt is due to the arrival beneath the lithosphere and decompression melting of a mantle plume with a large head on it.

Igneous Processes at Convergent Boundaries

Intermediate and silicic magmas are clearly related to the convergence of two plates and subduction. However, exactly what takes place is debated by geoscientists. Compared to divergent boundaries, there is less agreement about how magmas are generated at convergent boundaries. The scenarios that follow are currently argued by geoscientists to be the best explanations of the data.

The Origin of Andesite

Magma for most of our andesitic composite volcanoes (such as those found along the west coast of the Americas) seems to originate from a depth of about 100 kilometres. This coincides with the depth at which the subducted oceanic plate is sliding under the asthenosphere (figure 6.22). Partial melting of the asthenosphere takes place, resulting in a mafic magma. In most cases, melting occurs because the subducted oceanic crust releases water into the asthenosphere. The water collected in the oceanic crust when it was beneath the ocean is driven out as the descending plate is heated. The water lowers the melting temperature of the ultramafic rocks in this part of the mantle. Partial melting produces a mafic magma.

But how can we keep producing magma from ultramafic rock after those rocks have been depleted of the constituents of the mafic magma? The answer is that hot asthenospheric rock continues to flow into the zone of partial melting. As shown in figure 6.22, asthenospheric ultramafic rock is dragged downward by the descending lithospheric slab. More ultra-

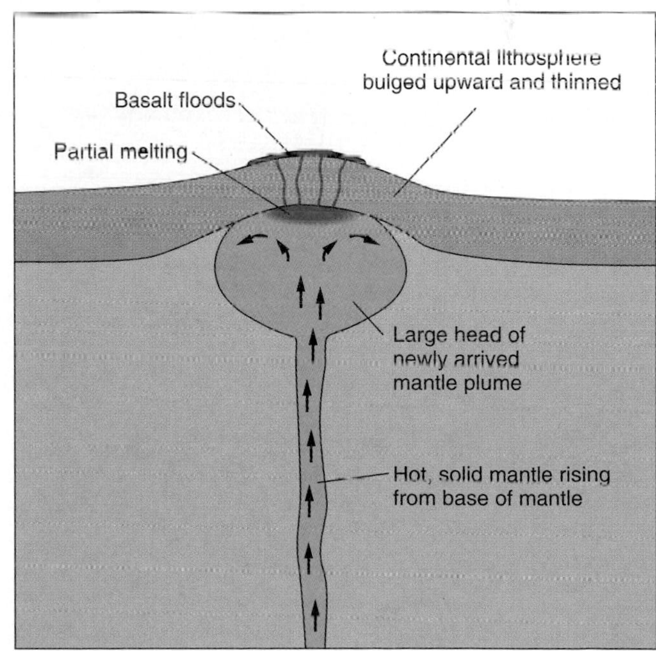

FIGURE 6.21

A hot mantle plume with a large head rises from the lower mantle. When it reaches the base of the lithosphere, it uplifts and stretches the overlying lithosphere. The reduced pressure results in decompression melting, producing basaltic magma. Large volumes of magma travel through fissures and flood the Earth's surface.

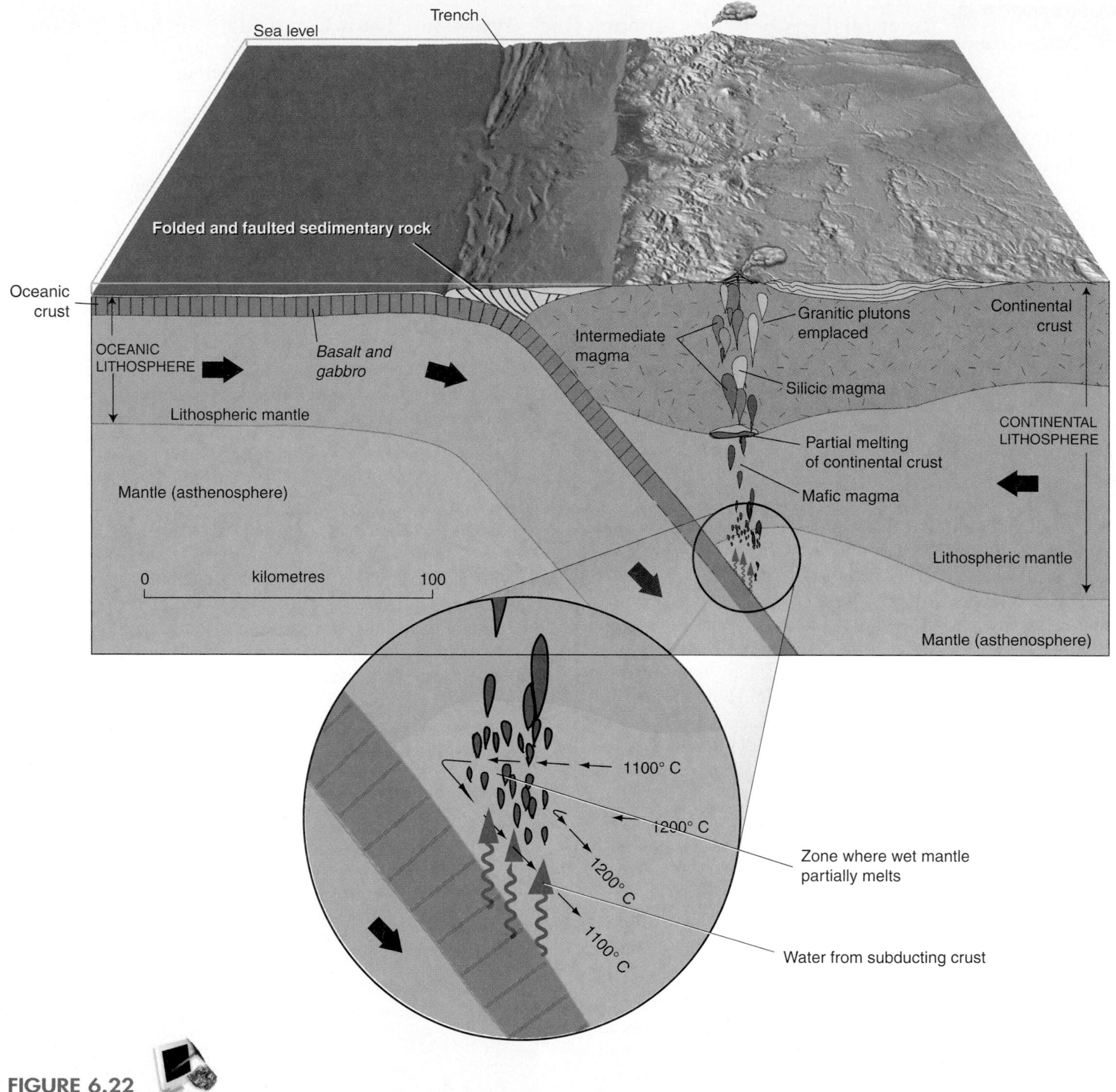

FIGURE 6.22

Generation of magma at a convergent boundary. Mafic magma is generated in the asthenosphere above the subducting oceanic lithosphere, and silicic magma is created in the lower crust. The insert shows the circulation of asthenosphere and lines of equal temperature (isotherms). Partial melting of "wet" ultramafic rock takes place in the zone where it is between 1100 and 1200°C.

mafic rock flows laterally to replace the descending material. A continuous flow of hot, "fertile" (containing the constituents of basalt) ultramafic rock is brought into the zone where water, moving upward from the descending slab, lowers the melting temperature. After being depleted of basaltic magma, the solid, residual, ultramafic rock continues to sink deeper into the mantle.

On its slow journey through the crust, the mafic magma evolves into an intermediate magma by differentiation, assimilation of silicic crustal rocks, and by magma mixing.

The Origin of Granite

To explain the great volumes of granitic plutonic rocks, most geologists think that partial melting of the lower continental crust must take place. The continental crust contains the high amount of silica needed for a silicic magma. As the silicic rocks of the continental crust have relatively low melting temperatures (especially if water is present), partial melting of the lower continental crust is likely. However, calculations indicate that the temperatures we would expect from a normal geothermal gradient are too low for melting to take place. Therefore, we need an additional heat source.

Currently, geologists think that the additional heat is provided by mafic magma that was generated in the asthenosphere and moved upward. The process of *magmatic underplating* involves mafic magma pooling at the base of the continental crust, supplying the extra heat necessary to partially melt the overlying, silica-rich crustal rocks (figure 6.23). Mafic magma generated in the asthenosphere rises to the base of the crust. The mafic magma is denser than the overlying silica rich crust; therefore, it collects as a liquid mass that is much hotter than the crust. The continental crust becomes heated (as if by a giant hotplate). When the temperature of the lower crust rises sufficiently, partial melting takes place, creating silicic magma. The silicic magma collects and forms diapirs, which rise to a higher level in the crust and solidify as granitic plutons (or, on occasion, reach the surface and erupt violently).

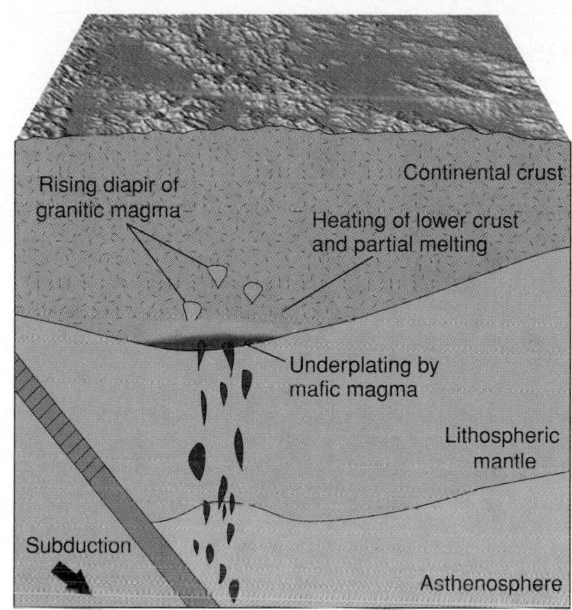

FIGURE 6.23

How mafic magma could add heat to the lower crust and result in partial melting to form a granitic magma. Mafic magma from the asthenosphere rises to underplate the continental crust

SUMMARY

Igneous rocks form from solidification of magma. If the rock forms at the Earth's surface it is *extrusive. Intrusive rocks* are igneous rocks that formed underground. Some intrusive rocks have solidified near the surface as a direct result of volcanic activity. Volcanic *necks* solidified within volcanoes. Fine-grained *dikes* and *sills* may also have formed in cracks during local extrusive activity. A sill is *concordant*—parallel to the planes within the country rock. A dike is *discordant*—not parallel to planes in the country rock. Both are tabular bodies. Coarser grains in either a dike or a sill indicate that it probably formed at considerable depth.

Most intrusive rock is *plutonic*—that is, coarse-grained rock that solidified slowly at considerable depth. Most plutonic rock exposed at the Earth's surface is in *batholiths*—large plutonic bodies. A smaller body is called a *stock.*

Silicic (or felsic) rocks are rich in silica, whereas mafic rocks are silica deficient. Most igneous rocks are named on the basis of their mineral content, which in turn reflects the chemical composition of the magmas from which they formed, and on grain sizes. *Granite, diorite,* and *gabbro* are the coarse-grained equivalents of *rhyolite, andesite,* and *basalt,* respectively. *Peridotite* is an *ultramafic* rock made entirely of ferromagnesian minerals and is mostly associated with the mantle.

Basalt and gabbro are predominant in the oceanic crust. Granite predominates in the continental crust. Younger granite batholiths occur mostly within younger mountain belts. Andesite is largely restricted to narrow zones along convergent plate boundaries.

The *geothermal gradient* is the increase in temperature with increase in depth. Hot *mantle plumes* and magma at shallow depths in volcanic regions locally raise the geothermal gradient.

No single process can satisfactorily account for all igneous rocks. In the process of *differentiation,* based on *Bowen's reaction series,* a residual magma more silicic than the original mafic magma is created when the early-forming minerals separate out of the magma. In *assimilation,* a hot, original magma is contaminated by picking up and absorbing rock of a different composition. *Magma mixing* produces a magma whose composition is intermediate, between that of the two types of magma that were mixed.

Partial melting of the mantle usually produces basaltic magma, whereas granitic magma is most likely produced by partial melting of the lower continental crust.

The theory of *plate tectonics* incorporates the preceding concepts. Basalt is generated where hot mantle rock partially melts, most notably along divergent boundaries. The fluid magma rises easily through fissures, if present. Granite and andesite are associated with subduction. Differentiation, assimilation, partial melting, and mixing of magmas may each play a part in creating the observed variety of rocks.

Terms to Remember

andesite 158
basalt 158
batholith 162
Bowen's reaction series 165
chill zone 154
coarse-grained (coarsely crystalline) rock 155
contact 154
country rock 154
crystal settling 167
decompression melting 164
diapir 162
differentiation 166
dike 160
diorite 158
extrusive rock 154
fine-grained rock 155
gabbro 158
geothermal gradient 164
granite 154
igneous rock 154
intermediate rock or magma 160
intrusion (intrusive structure) 160
intrusive rock 154
lava 154
mafic rock or magma 160
magma 154
mantle plume 164
peridotite 160
pluton 162
plutonic rock 155
porphyritic 157
rhyolite 158
silicic (felsic) rock or magma 160
sill 160
stock 162
texture 155
ultramafic rock 160
volcanic neck 160
xenolith 154

Testing Your Knowledge

Use the following questions to prepare for exams based on this chapter.

1. Why do mafic magmas tend to reach the surface much more often than silicic magmas?
2. What role does the asthenosphere play in generating magma at (a) a convergent boundary; (b) a divergent boundary?
3. How do batholiths form?
4. How would you distinguish, on the basis of minerals present, among granite, gabbro, and diorite?
5. How would you distinguish andesite from a diorite?
6. What rock would probably form if magma that was feeding volcanoes above subduction zones solidified at considerable depth?
7. Why is a higher temperature required to form magma at the oceanic ridges than in the continental crust?
8. What is the difference between feldspar found in gabbro and feldspar found in granite?
9. What is the difference between a dike and a sill?
10. Describe the differences between the continuous and the discontinuous branches of Bowen's reaction series.
11. A surface separating different rock types is called a
    a. xenolith b. contact
    c. chill zone d. none of the preceding
12. The major difference between intrusive igneous rocks and extrusive igneous rocks is
    a. where they solidify b. chemical composition
    c. type of minerals d. all of the preceding
13. Which is not an intrusive igneous rock?
    a. gabbro b. diorite
    c. granite d. andesite
14. By definition, stocks differ from batholiths in
    a. size b. shape
    c. chemical composition d. all of the preceding
15. Which is not a source of heat for melting rock?
    a. geothermal gradient b. the hotter mantle
    c. mantle plumes d. water under pressure
16. The geothermal gradient is, on the average, about
    a. 1°C/km b. 10°C/km
    c. 30°C/km d. 50°C/km
17. The continuous branch of Bowen's reaction series contains the mineral
    a. pyroxene b. plagioclase
    c. amphibole d. biotite
18. The discontinuous branch of Bowen's reaction series contains the mineral
    a. pyroxene b. amphibole
    c. biotite d. all of the preceding
19. The most common igneous rock of the continents is
    a. basalt b. granite
    c. rhyolite d. ultramafic
20. Granitic magmas are associated with
    a. convergent boundaries and magmatic underplating
    b. divergent boundaries and differentiation
    c. convergent boundaries and decompression melting
    d. divergent boundaries and water release
21. The difference in texture between intrusive and extrusive rocks is primarily due to
    a. different mineralogy
    b. different rates of cooling and crystallization
    c. different amounts of water in the magma
22. Mafic magma is generated at divergent boundaries because of
    a. water under pressure b. decompression melting
    c. magmatic underplating d. melting of the lithosphere

23. A change in magma composition due to melting of surrounding country rock is called
    a. magma mixing
    b. assimilation
    c. crystal setting
    d. partial melting

Exploring Web Resources

www.mcgrawhill.ca/olc/plummer

This is the dedicated website for this book. You can go to it for new and updated information. The universal resource locators (URLs) listed in this book are also given as links on the website, making it easy to go to those websites without typing in the URL. When you visit our Online Learning Centre, go to the Student Centre and then to the chapter of interest. Here you will find additional readings and media resources, as well as answers to the Testing Your Knowledge section, more quizzes, animation and video clips, and direct links to the sites listed in this book. Links to additional websites can also be found. We have added questions for some of the sites to allow you to get the most of your exploration of the web. Using the web is an enjoyable way of enhancing your knowledge of geology.

http://uts.cc.utexas.edu/~rmr/

Rob's Granite Page. This site has a lot of information on granite and related igneous activity. The site is useful for people new to geology as well as for professionals. There are numerous images of granite. Click on "Granite is like ice cream?" for an interesting comparison. The page also has photos of various granites and links to other sites that have more images.

www.geolab.unc.edu/Petunia/IgMetAtlas/mainmenu.html

Atlas of Rocks, Minerals, and Textures (from University of North Carolina). This site contains some photomicrographs of plutonic and volcanic rocks. The images are thin sections (slices of rock so thin that most minerals are transparent) seen in a polarizing microscope. Most images are taken from cross-polarized light, which causes many minerals to appear in distinctive, bright colours. For some of the rocks (gabbro, for instance), you can also see what they look like under plain polarized light by clicking the circle with the horizontal, grey lines.

http://seis.natsci.csulb.edu/basicgeo/IGNEOUS_TOUR.html

Igneous Rocks Tour. This site has some hand specimen images of common igneous rocks and should provide a useful review for rock identification.

www.gpc.edu/~pgore/stonemtn/text.html

Stone Mountain, Georgia, Virtual Field Trip. Stone Mountain is an exposure of granite in Georgia that is a famous landmark. Begin by reading the geologic summary, then take the virtual tour.

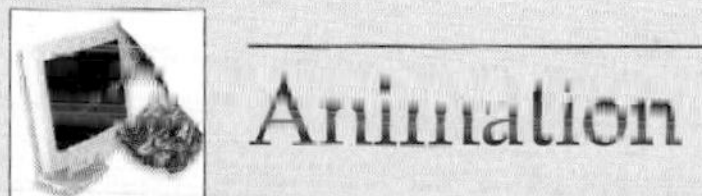

Animation

This chapter includes the following animations available on our Online Learning Center at **www.mcgrawhill.ca/olc/plummer**.

6.22 How subduction causes volcanism

CHAPTER

7

Volcanism and Extrusive Rocks

How does volcanic activity affect humans?
What determines the degree of violence associated with volcanic activity?
What kinds of rocks do volcanoes produce?
What are the characteristics of different types of volcanoes?
What are lava floods?

Volcanic eruptions, while awesome natural spectacles, also provide important information on the workings of the Earth's interior. Volcanic eruptions vary in nature and in degree of explosive violence. A strong correlation exists between the chemical composition of magma (or lava) and the violence of an eruption. The size and shape of volcanoes and lava flows and their pattern of distribution on the Earth's surface also correspond to the composition of their lavas.

Understanding volcanism provides a background for theories relating to mountain building, the development and evolution of continental and oceanic crust, and plate tectonic theory.

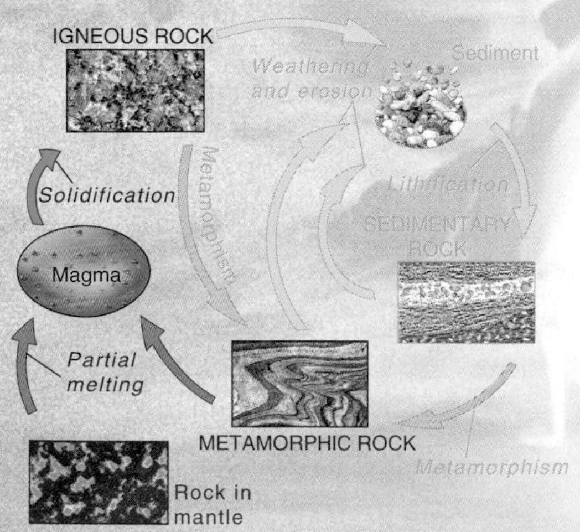

Lava flows into the sea at Hawaii during the eruption of Kilavea that has been ongoing since 1983. *Photo © Parvinder Sethi*

A

B

FIGURE 7.1

Contrasting styles of volcanic eruptions. (A) May 18, 1980. Exploding magma blasts out of the side of Mount St. Helens. (B) Lava flow in Hawaii, 1989. A lava fountain is at the source of lava cascading over a cliff.

Photo A © 1980 Keith Ronnholm; Photo B by D. A. Swanson, U.S. Geological Survey

The May 18, 1980 eruption of Mount St. Helens (figure 7.1*A* and box 7.1) was a spectacular release of energy from the Earth's interior. The plate tectonic explanation is that North America is overriding a portion of the Pacific Ocean floor. Melting of previously solid rock takes place at depth, just above the subducting plate. Some of the **magma** (molten rock or liquid that is mostly silica) worked its way upward to the Earth's surface to erupt. Magma does not always reach the Earth's surface before solidifying, but when it does it is called **lava.**

ENVIRONMENTAL GEOLOGY 7.1

Mount St. Helens Reawakens

Before 1980, Mount St. Helens, in southern Washington, had not erupted since 1857. On March 27, 1980, minor ash and steam eruptions began and continued for the next six weeks. These eruptions were due to exploding gas blasting out the volcano's previously formed rock, but the pattern of earthquakes recorded at the volcano indicated magma was working its way upward.

After several weeks, the peak began swelling—like a balloon being inflated—indicating magma was now inside the volcano. The northern flank of the volcano bulged outward at a rate of 1.5 metres per day. Bulging continued until the surface of the northern slope was displaced outward over a hundred metres from its original position. The bulge was too steep to be stable, and the U.S. Geological Survey warned of another hazard—a mammoth landslide.

On May 18, a monumental blast destroyed the summit and north flank of Mount St. Helens (see figure 7.1). Seconds after the eruption began, an area extending northward 10 kilometres was stripped of all vegetation and soil.

Although the sequence of events was exceedingly rapid, it is now clear what happened (box figure 1). A fairly strong earthquake loosened the bulging north slope, triggering a landslide. The landslide, known as a *debris avalanche,* moved at speeds of over 160 kilometres per hour. It was one of the largest landslides ever to occur, but it was eclipsed by the huge eruption that followed. The landslide stripped away the lid on the magma chamber, and because of the reduced pressure, the previously dissolved gases in the magma exploded (figure 7.1*A*). The violent froth of gas and magma blasted away the mountain's north flank and roared outward at up to 1,000 kilometres per hour. The huge lateral blast of hot gas and volcanic rock debris killed everything near the volcano and, beyond the 10-kilometre scorched zone, knocked down every tree in the forest.

For the next 30 hours, exploding gases propelled frothing magma and volcanic ash vertically into the high atmosphere. The mushroom-shaped cloud of ash was blown northeastward by winds. A rain of ash went on for days, causing damage as far away as Montana. Volcanic mudflows caused enormous damage during and after the eruption. The mudflows resulted from water from melted snow and glacier ice mixing with volcanic debris to form a slurry having the consistency of wet cement. Mudflows flowed down river valleys, carrying away steel bridges and other structures (see chapter 9, notably figure 9.13).

Damage was in the hundreds of millions of dollars, and 63 people were killed. The death toll might have been much worse had not scientists warned public officials about the potential hazards, causing them to evacuate the danger zone before the eruption. For comparison, 29,000 people were killed during an eruption of Mount Pelée (described later in this chapter), and 23,000 lives were lost in a 1985 volcanic mudflow in Colombia.

Between 1980 and 1986 a lava dome grew in the crater of Mount St. Helens. Volcanic activity levels in and around the volcano were generally low until September 2004, when seismic activity beneath the lava dome increased dramatically. Swarms of small, shallow earthquakes indicated upward movement of lava beneath the volcano; frequent small steam and ash eruptions occurred, and the lava dome began to grow once more (box figure 2). Scientists are now carefully monitoring the volcano and are recording changes in the form of the lava dome through analysis of remote camera images and digital elevation models created from aerial photographs. Fumarole activity and gas emissions from vents are also monitored and the composition of erupted lavas compared to those from the 1980 eruption. Volcano alerts are issued if scientists feel there is any possibility of a violent and hazardous eruption.

Mount St. Helens is not the only potentially active volcano in the Pacific Northwest, and several cities lie dangerously close to volcanoes that have remained quiescent for decades and centuries (see figure 7.4). Vancouver is threatened by potential eruptions from Mount Garibaldi to the north or Mount Baker to the south. Seattle and Tacoma are close to Mount Rainier, and Mount Hood lies almost within the suburbs of Portland, Washington.

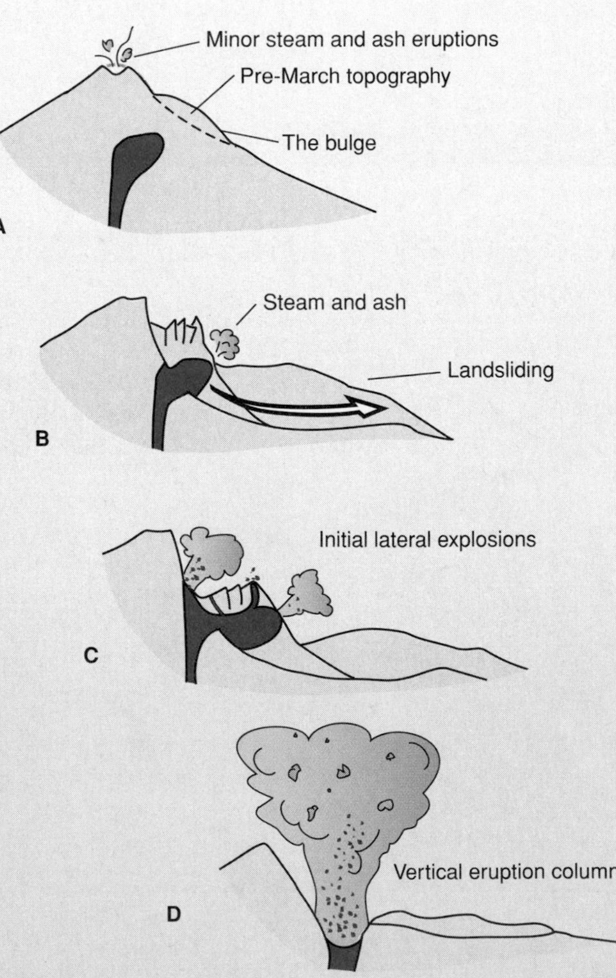

BOX 7.1 ■ FIGURE 1

Sequence of events at Mount St. Helens, May 18, 1980. (*A*) Just before the eruption. (*B*) The landslide relieves the pressure on the underlying magma. (*C*) Magma blasts outward. (*D*) Full vertical eruption.

Additional Resource

USGS Cascade Volcano Observatory—Mount St. Helens

http://vulcan.wr.usgs.gov/Volcanoes/MSH/framework.html

For more information on volcanic activity at Mount St. Helens and the Cascade volcanoes, go to:

http://vulcan.wr.usgs.gov/

www.volcano.und.edu/vwdocs

Excellent images from the recent activity of Mount St. Helens can be found at:

http://vulcan.wr.usgs.gov/Volcanoes/MSH/Images/MSH04/framework.html

BOX 7.1 ■ FIGURE 2

Aerial view, Mount St. Helens crater and dome as seen from the northeast.

USGS photograph taken on February 15, 2006, by Jim Vallance and Matt Logan.

At Mount St. Helens the lava solidified quickly as it was blasted explosively by gases into the air, producing rock fragments known as **pyroclasts** (from the Greek *pyro,* "fire," and *clast,* "broken"). Pyroclastic debris is also known as *tephra.* In Hawaii, lava extrudes out of fissures in the ground as **lava flows** (figure 7.1*B*). Pyroclastic debris and rock formed by solidification of lava are collectively regarded as **extrusive rock,** surface rock resulting from volcanic activity.

The most obvious landform created by **volcanism** is a **volcano,** a hill or mountain formed by the extrusion of lava or ejection of rock fragments from a vent; however, volcanoes are not the only volcanic landforms. Very fluid lava may flow out of the Earth and flood an area, solidifying into a nearly horizontal layer of extrusive rock. Successive layers of lava flows may accumulate, building a lava plateau.

HOW DOES VOLCANIC ACTIVITY AFFECT HUMANS?

Supernatural Beliefs

Not surprisingly, myths and religions relating gods to volcanoes flourish in cultures that live with volcanoes. In Iceland, Loki, of Norse mythology, is regarded as imprisoned underground, blowing steam and lava up through fissures. Pacific Northwest Indians regarded the Cascade volcanoes as warrior gods who would sometimes throw red-hot boulders at each other. They also had a romantic side: Mount Hood and Mount Adams fought over Mount St. Helens, the youngest and prettiest of the volcano gods. In British Columbia, a lava flow killed about 2,000 members of the Nisga'a tribe around A.D. 1700 (see box 7.2). According to the Nisga'a, children were harassing salmon, including putting flaming sticks in a fish's back and watching the smoking fish swim upstream. Disrespect of fish is a major taboo and was believed to have brought on the lava eruption. In Hawaii, Madame Pele is regarded as a goddess who controls eruptions. According to legend, Pele and her sister tore up the ocean floor to produce the Hawaiian island chain. Today, many fervently believe that Pele dictates when and where an eruption will take place. In the 1970s, when Kilauea began erupting near a village, residents chartered an airplane and dropped flowers and a bottle of gin into the lava vent to appease Pele.

Volcanism is also relevant to human affairs in very tangible ways. Its effects can be catastrophic or, surprisingly, beneficial.

The Growth of an Island

Although occasionally a highway or village is overrun by outpourings of lava, the overall effects of volcanism have been favourable to humans in Hawaii. Lava flowing into the sea and solidifying adds real estate to the island of Hawaii. Kilauea Volcano has been erupting since 1983, spewing out an average of 325,000 cubic metres of lava a day. This is the equivalent of 40,000 dumptruck loads of material. In 20 years, 2.5 billion cubic metres of lava were produced—enough to build a highway that circles the world over five times. The downside is that during the 1980s and 1990s 181 houses were destroyed by lava flows.

Were it not for volcanic activity, Hawaii would not exist. The islands are the crests of a series of volcanoes that have been built up from the bottom of the Pacific Ocean over millions of years (the vertical distance from the summit of Mauna Loa Volcano to the ocean floor greatly exceeds the height above sea level of Mount Everest). When lava flows into the sea and solidifies, more land is added to the islands. Hawaii is, quite literally, growing.

IN GREATER DEPTH 7.2

Volcanoes and Volcanic Hazards in Canada

Western Canada is home to many volcanoes, none of which are erupting at present, but many have the potential to erupt in the near future. Volcanoes that have been active during the past 5 million years form seven major volcanic belts in the Canadian Cordillera (box figure 1). The Garibaldi volcanic belt extends northward from the Cascade volcanic belt (figure 7.4) and lies adjacent to the densely populated areas of southwestern British Columbia. The volcanoes of the Garibaldi belt (including Mount Garibaldi, Mount Cayley, and Mount Meagre) are large stratovolcanoes formed as a result of subduction of the Juan de Fuca plate below the North American plate (refer to box 2.2) and are the most explosive young volcanoes in Canada. Eruption of any of these volcanoes could cause pyroclastic flows, lahars (volcanic debris flows), widespread ash falls, landslides, and damming of river valleys. An explosive eruption of Mount Meagre more than 2,300 years ago deposited ash as far as Alberta and generated pyroclastic flows that extended 7 km down the Lillooet River. Volcanic debris from the eruption dammed up a large lake that failed catastrophically carrying large blocks of volcanic debris several kilometres downstream. An eruption occurring anywhere along the corridor between Vancouver and Pemberton could have similar effects.

The Stikine volcanic belt in northern B.C. is the most active volcanic region in Canada with more than 100 volcanoes, and extends from the area of Prince Rupert into the Yukon and the Alaska border. This volcanic belt includes (among many others) Volcano Mountain, Mount Edziza, and three young volcanoes (Tseax Cone, Lava Fork, and Ruby Mountain) that have erupted in the last few hundred years. Eruptions from these volcanoes tend to be small and generate fluid basaltic lavas (box figure 2) that do not pose serious risk to the relatively sparse local population. However, several of these volcanoes are covered by glaciers and eruptions could cause rapid melting of snow and ice and generate debris flows (lahars) and floods that may threaten local villages in the surrounding valleys.

Many volcanoes in Canada were ice-covered at the time of former eruptions and have a flat-topped form (box figure 3). These flat-topped, subglacially formed volcanoes are called "tuyas" after Tuya Butte in northern B.C. Similar flat-topped volcanoes have formed in Antarctica and Iceland; in Iceland these volcanoes are called "table mountains."

Additional Resource

An excellent source of information on volcanoes and volcanic hazards in Canada is:

http://gsc.nrcan.gc.ca/volcanoes/index_e.php

BOX 7.2 ■ FIGURE 1

Volcanoes younger than 5 million years can be grouped into seven major volcanic belts in western Canada.

http://gsc.nrcan.gc.ca/volcanoes/map/index_e.php

Reproduced with the permission of the Minister of Public Works and Government Services, 2006 and Courtesy of Natural Resources Canada, Geological Survey of Canada.

BOX 7.2 ■ FIGURE 2

Lava flows from eruptions of the Tseax Cone (220 and 650 years ago) within the Nisga'a Memorial Lava Beds Provincial Park (approximately 60 km north of Terrace, B.C.). The fluid lava travelled more than 22 km from the cone.

Photograph by C. J. Hickson, Geological Survey of Canada

BOX 7.2 ■ FIGURE 3

Tuya Butte in nothern B.C. is a flat-topped tuya or subglacial volcano.

Photo by Ben Edwards

In addition to gaining more land, Hawaii benefits in other ways from its volcanoes. Weathered volcanic ash and lava produce excellent fertile soils (think pineapples and papayas). Moreover, Hawaii's periodically erupting volcanoes (which are relatively safe to watch) are great spectacles that attract both tourists and scientists, benefiting the island's economy (figure 7.1*B*).

Geothermal Energy

In other areas of recent volcanic activity, underground heat generated by igneous activity is harnessed for human needs. In Iceland, Italy, Mexico, New Zealand, Argentina, Japan, and California, geothermal installations produce electric power. Steam or superheated water trapped in layers of hot volcanic rock is tapped by drilling and then piped out of the ground to power turbines that generate electricity. Naturally heated geothermal fluids can also be tapped for space or domestic water heating or industrial use, as in paper manufacturing. In Iceland, more than 85 percent of homes are heated with geothermal energy, and geothermal greenhouses are used to grow most of the nation's produce, including bananas. Geothermal energy resources are currently being developed in the Garibaldi volcanic belt of British Columbia. (For more information, go to http://geothermal.marin.org/ and Chapter 15 on groundwater.)

Effect on Weather

Occasionally, a volcano will spew large amounts of fine, volcanic dust and gas into the high atmosphere. Winds can keep fine particles suspended over the Earth for years. The 1991 eruption of Mount Pinatubo in the Philippines produced noticeably more colourful sunsets worldwide. More significantly, it reduced solar radiation that penetrates the atmosphere. Measurements indicated that the worldwide average temperature dropped approximately half a degree Celsius for a couple years. While this may not seem like much, it was enough to temporarily offset the global warming trend of the past 100 years.

The 1815 eruption of Tambora in Indonesia was the largest single eruption in a millennium—40 cubic kilometres of material was blasted out of a volcanic island, leaving a six-kilometre-wide depression. The following year, 1816, became known as "the year without summer." In New England, snow in June was widespread and frosts throughout the summer ruined crops. Parts of Europe suffered famine because of the cold-weather effects on agriculture. However, these effects were relatively short lived and appear to have been most severe in areas lying to the north of the jet stream during the summer of 1816.

Volcanic Catastrophes

While the eruption of Mount St. Helens in 1980 was indeed awesome, its effects were not nearly as disastrous as a number of historical eruptions elsewhere in the world. For instance, the Roman city of Pompeii and at least four other towns near Naples in Italy were destroyed in A.D. 79 when Mount Vesuvius erupted (figure 7.2). Before the eruption, vineyards on the flanks of the apparently "dead" volcano extended to the summit. After it erupted without warning, Pompeii was buried under 5 to 8 metres of hot ash. Seventeen centuries later, the town was rediscovered.

FIGURE 7.2

(A) Pompeii with Mount Vesuvius in the background. *(B)* Casts of bodies of people who died in Pompeii, buried by ash from the eruption of Vesuvius, A.D. 79. The casts were made by pouring plaster into voids in the ash left by the dead.

Photo A by R. W. Decker; Photo B © Bettmann/Corbis

Excavation revealed moulds of people suffocated by the ashfall, many with facial expressions of terror. This eruption was not the end of Vesuvius's activity. The volcano was active almost continually from 1631 to 1944, with major twentieth-century eruptions in 1906, 1929, and 1944.

The island of Krakatoa in the western Pacific, composed of three apparently inactive volcanoes, erupted in 1883 with the force of several hydrogen bombs. This Indonesian island, which formerly rose 800 metres above sea level, was blown apart. Only one-third of the island remained after the eruption. An estimated 13 cubic kilometres of rock collapsed into the subsurface magma chamber that had been emptied by the eruption, leaving an underwater depression 300 metres deep where the major part of the island had been. The huge explosion was heard 5,000 km away. More than 34,000 people died as a result of the giant sea waves (tsunamis) generated by the explosion.

A similar series of explosions in prehistoric time (about 6,600 years ago) was at least partially responsible for creating the depression occupied by Crater Lake in Oregon (figure 7.3). Volcanic debris covering more than a million square kilometres in Oregon and neighbouring states has been traced to those eruptions. The original volcano, named Mount Mazama, is estimated to have been about 2,000 metres higher than the present rim of Crater Lake. For more information on Crater Lake and Mount Mazama, go to http://pubs.usgs.gov/fs/2002/fs092-02/.

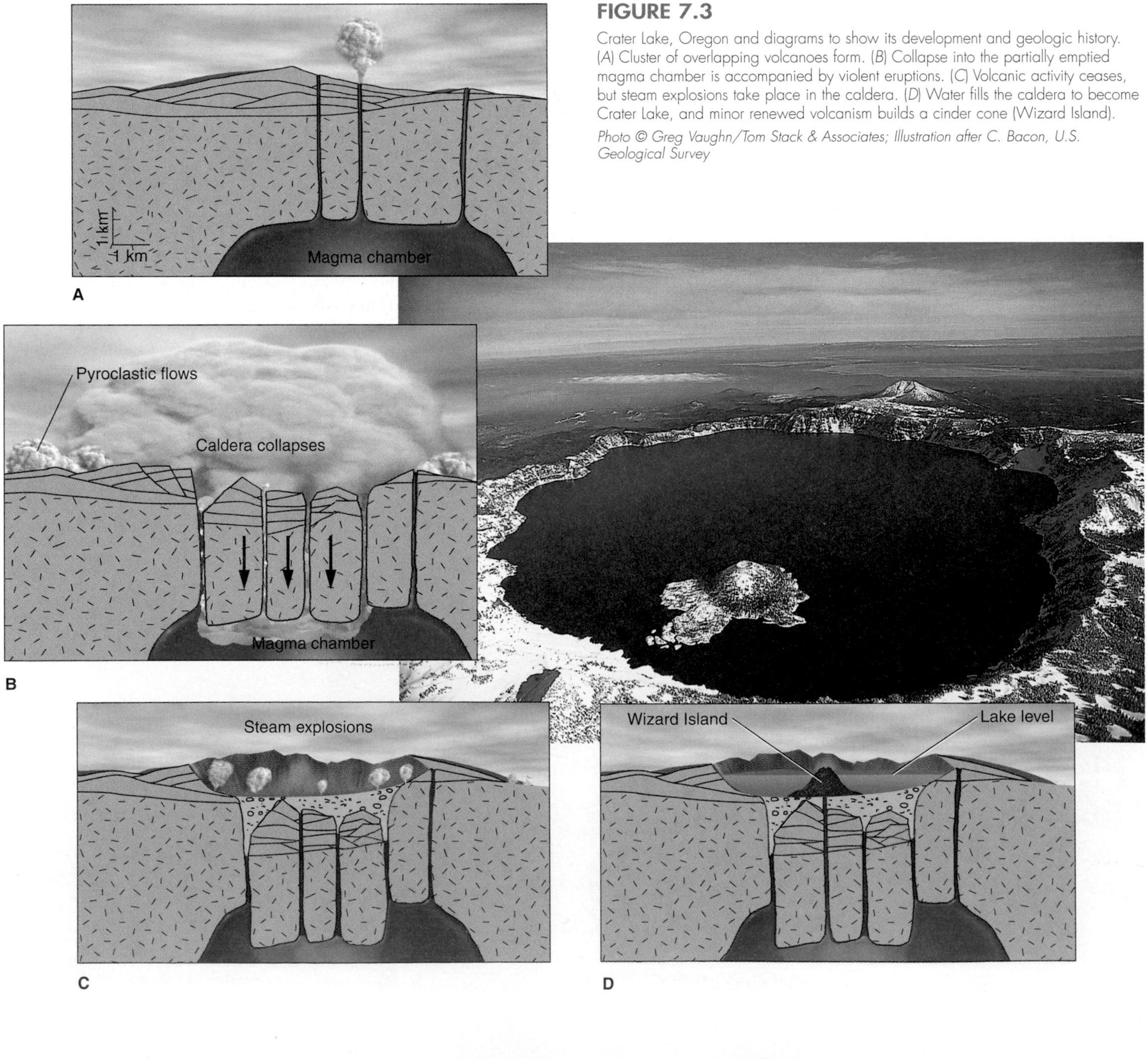

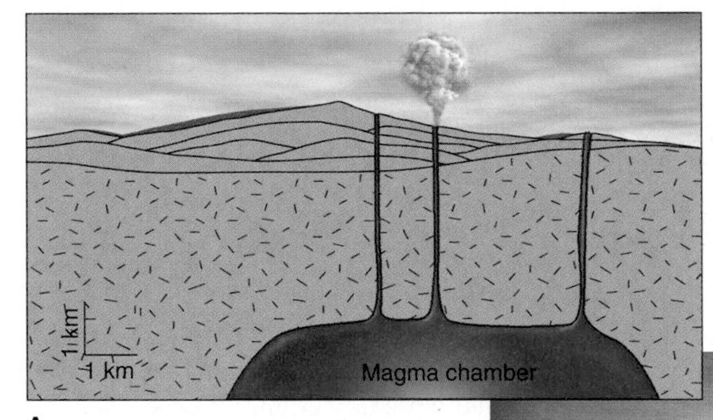

A

FIGURE 7.3

Crater Lake, Oregon and diagrams to show its development and geologic history. (*A*) Cluster of overlapping volcanoes form. (*B*) Collapse into the partially emptied magma chamber is accompanied by violent eruptions. (*C*) Volcanic activity ceases, but steam explosions take place in the caldera. (*D*) Water fills the caldera to become Crater Lake, and minor renewed volcanism builds a cinder cone (Wizard Island).

Photo © Greg Vaughn/Tom Stack & Associates; Illustration after C. Bacon, U.S. Geological Survey

The southern Cascade Mountains, where Crater Lake is located, have been built up by eruptions over the past 30 to 40 million years (figure 7.4; see also the geologic map, inside cover). Only the youngest peaks (those built within the past 2 million years), such as Mount St. Helens, Mount Rainier, Mount Shasta, and Mount Hood, still stand out as cones. As Mount St. Helens has demonstrated, any of these could again erupt (see box 7.1).

Although none of Canada's volcanoes are currently erupting, many have the potential to erupt in the near future and pose significant risks to western Canada (see box 7.2).

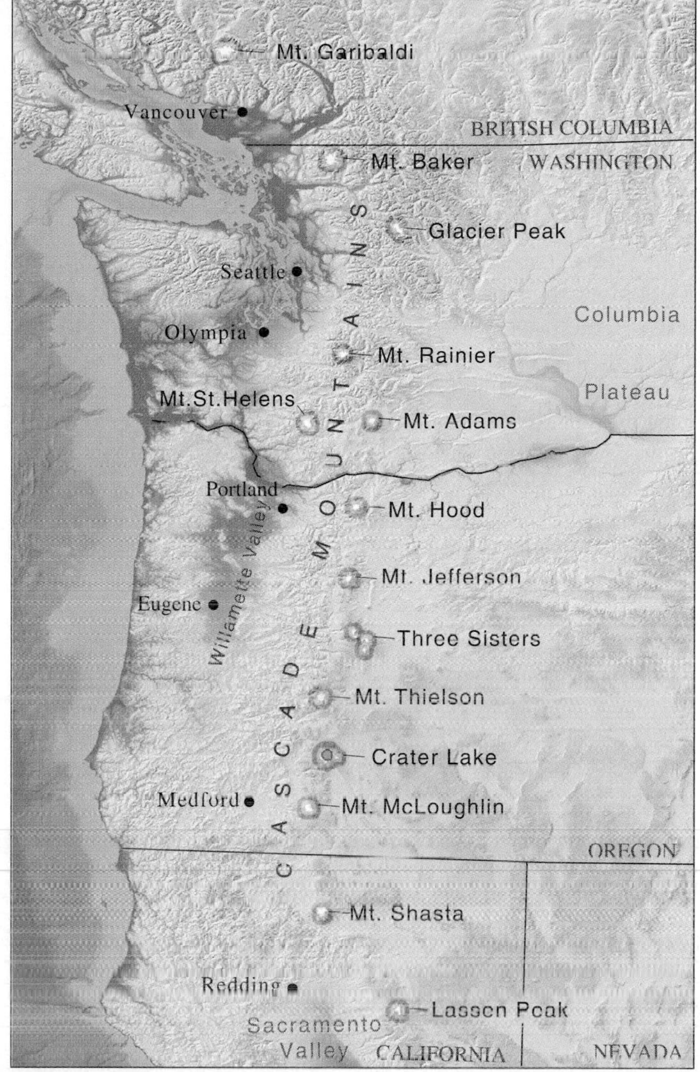

FIGURE 7.4

The Cascade volcanoes. The named volcanoes are ones that have erupted in geologically recent time.

Adapted from U.S. Geological Survey

The Record of Fatalities

Figure 7.5 shows the results of research at the Smithsonian Institute and Macquarie University, Australia. Note the dramatic increase in fatalities during the recent centuries (figure 7.5*A*). This is not due to increasing volcanic activity but to increasing population, more people living near volcanoes, and more intensive reporting of such events. Figure 7.5*B*, which shows the cumulative number of deaths during the last seven centuries, also shows that most of the fatalities have been caused by seven major eruptions.

Volcanoes can kill in a number of ways. Figure 7.5*C* indicates that pyroclastic flows account for the most fatalities. A *pyroclastic flow,* described later in this chapter, is a mixture of hot gas and pyroclastic debris that rapidly flows down a volcano's flanks. Famine and other indirect causes account for the next greatest number of fatalities. Widespread destruction of crops and farm animals can cause regional famine (as occurred with the eruption of Tambora in 1815). Note that relatively few events (specific eruptions) have caused the large number of deaths attributable to famine.

Pyroclastic material accounts for the largest number of deadly events; however, few people die in each event, so the total number of deaths is not great. Most of the deaths due to pyroclastic material are caused by collapse of ash-covered roofs or by being hit by falling, pyroclastic fragments.

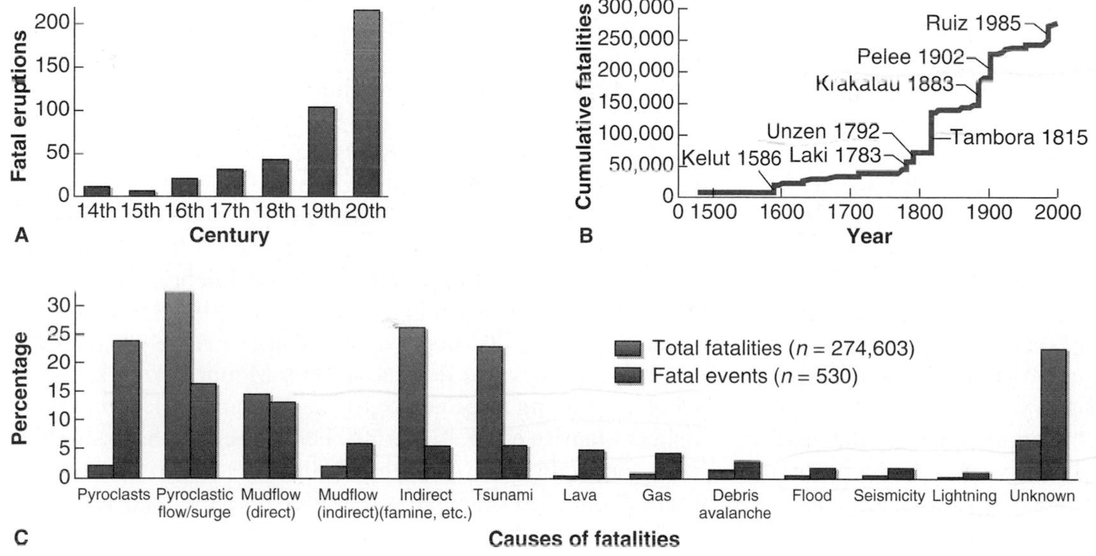

FIGURE 7.5

Volcano fatalities. (*A*) Fatal volcano eruptions per century. (*B*) Cumulative volcano fatalities. Note the big jumps with the seven most deadly eruptions. These were eruptions that killed more than 10,000 people and account for two-thirds of the total. (*C*) The causes of volcano fatalities.

Reprinted with permission from "Volcano Fatalities" by T. Simkin, L. Siebert, and R. Blong, Science, v. 291: p. 255. Copyright © 2001 American Association for the Advancement of Science

WHAT DETERMINES THE DEGREE OF VIOLENCE ASSOCIATED WITH VOLCANIC ACTIVITY?

Eruptive Violence and Physical Characteristics of Lava

Why can we state confidently that active volcanism in Hawaii poses only slight danger to humans, but expect violent explosions to occur in the Cascade Mountains? Whether eruptions are very explosive or relatively "quiet" is largely determined by two factors: (1) the amount of gas in the lava or magma and (2) the ease or difficulty with which the gas can escape to the atmosphere. The **viscosity,** or resistance to flow, of a lava determines how easily the gas escapes. The more viscous the lava and the greater the volume of gas trying to escape, the more violent the eruption. Later we will show how these factors not only determine the degree of violence of an eruption but also influence the shape and height of a volcano.

The three most important factors that influence viscosity are (1) the silica (SiO_2) content of the lava, (2) the temperature of the lava relative to the cooler temperature at which it solidifies, and (3) the gas dissolved in magma—the greater the dissolved gas content, the more fluid the lava. If the lava being extruded is considerably hotter than its solidification temperature, the lava is less viscous (more fluid) than when its temperature is near its solidification point. Temperatures at which lavas solidify range from about 700°C for silicic rocks to 1,200°C for mafic rocks.

Volcanic rocks, and the magma from which they formed, have a silica content that ranges from 45 percent to 75 percent by weight. **Silicic** (or felsic) rocks are silica-rich (65% or more SiO_2) rocks. *Rhyolite* is the most abundant silicic volcanic rock. **Mafic rocks** are *silica-deficient* rocks. Their silica content is close to 50 percent. *Basalt* is the most common mafic rock. **Intermediate rocks** have a chemical content between that of felsic and mafic rocks. The most common intermediate rock is *andesite.* Chapter 6 contains a more complete description of the chemistry of igneous rocks and their relationship to the mineral content of rocks.

Mafic lavas, which are relatively low in SiO_2, tend to flow easily. Conversely, felsic lavas are much more viscous and flow sluggishly. Mafic lava is around 10,000 times as viscous as water, whereas silicic magma is around 100 million times the viscosity of water. Lavas rich in silica are more viscous because even before they have cooled enough to allow crystallization of minerals, silicon–oxygen tetrahedrons have begun to form small frameworks in the lava. Although too few atoms are involved for the structures to be considered crystals, the total effect of these silicate structures is to make the liquid lava more viscous, much like the way that flour or cornstarch thickens gravy.

Because silicic magmas are the most viscous, they are associated with the most violent eruptions. Mafic magmas are the least viscous and commonly erupt as lava flows (such as in Hawaii). Eruption associated with intermediate magma can be violent or can produce lava flows. The Cascade volcanoes in North America and the Andes in South America are predominantly composed of intermediate rock.

Scientific Investigation of Volcanism

Volcanoes and lava flows, unlike many other geologic phenomena, can be observed directly, and samples can be collected without great difficulty (at least for the quiet, Hawaiian type of eruption). The growth and deformation of volcanoes can now be measured directly using satellite-based radar images (http://volcanoes.usgs.gov/insar/more_insar.html). We can measure the temperature of lava flows, collect samples of gases being given off, observe the lava solidifying into rock, and take newly formed rock samples into the laboratory for analysis and study. By comparing rocks observed solidifying from lava with similar ones from other areas of the world (and even with samples from the moon) where volcanism is no longer active, we can infer the nature of volcanic activity that took place in the geologic past.

Gases

From active volcanoes we have learned that most of the gas released during eruptions is water vapour, which condenses as steam. Other gases, such as carbon dioxide, sulphur dioxide, hydrogen sulphide (which smells like rotten eggs), and hydrochloric acid, are given off in lesser amounts with the steam.

Surface water introduced into a volcanic system can greatly increase the explosivity of an eruption, as exemplified by the devastation of the island of Krakatoa (described earlier).

Gases and Pyroclastics

During an eruption, expanding, hot gases may propel pyroclastics high into the atmosphere as a column rising from a volcano. At high altitudes, the pyroclastics often spread out into a dark mushroom cloud. The fine particles are transported by high-atmosphere winds (see box 7.3). Eventually, debris settles back to Earth under gravity's influence as *ashfall* (or sometimes *pumice fall*) deposits.

A **pyroclastic flow** is a mixture of gas and pyroclastic debris that is so dense that it hugs the ground as it flows rapidly into low areas (figure 7.6). As figure 7.6 shows, there are two ways in which pyroclastic flows develop. An exploding froth of gas and magma can blast out from under a solid or very viscous plug capping a volcano. Or it may be caused by gravitational collapse of a column of gas and pyroclastic debris that was initially blasted vertically into the air. These turbulent masses can travel more than 100 km per hour and are extremely dangerous. In 1991, a pyroclastic flow at Japan's Mount Unzen killed 31 people, including three geologists and famous volcano photographers Maurice and Katia Krafft. Far worse was the destruction of St. Pierre (figure 7.7) on the Caribbean island of Martinique, where about 29,000 people were killed by a pyroclastic flow in 1902 (see box 7.4).

IN GREATER DEPTH 7.3

Volcanoes and Flying

There have been several occasions in which jumbo jets have flown into volcanic ash clouds with nearly disastrous results. In 1989, a KLM Boeing 747 unknowingly entered an ash plume over Mount Redoubt, a volcano in Alaska, at an altitude of 8,000 m. The pilot applied full power hoping to climb out of the plume. After climbing a thousand metres, all four engines stopped. The plane dropped to an altitude of 4,000 m in eight tension-filled minutes before the flight crew was able to restart the engines. Although the plane landed safely in Anchorage, the cost to repair it was $80 million. Its engines, which had to be replaced, contained glassy coatings that turned out to be melted and resolidified ash. When full power was applied, the engines became very hot—hotter than the melting temperature of the ash. Following this discovery, the standard procedure is to reduce power to keep the engine temperature well below the melting point of volcanic ash and lessen the chances of engine failure. It is, of course, preferable to fly *around* pyroclastics.

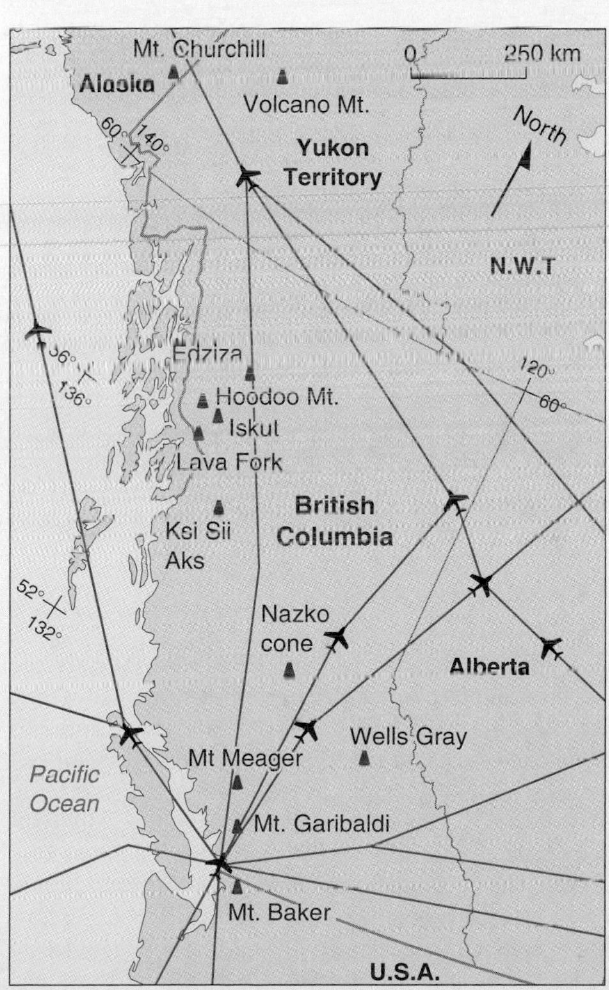

BOX 7.3 ■ FIGURE 1

Air routes in western Canada and their proximity to volcanoes.

Reproduced with permission of the Minister of Public Works and Government Services Canada, 2006 and Courtesy of Natural Resources Canada, Geological Survey of Canada.

Another less serious problem is what appears to be extensive scratching of airplane windows. The enormous amount of sulphuric acid aerosol that was belched into the atmosphere by Mount Pinatubo in 1991 caused scratching. Acid attacks the windows, made of acrylic, and etches fine lines in them.

The threat from airborne ash is one of the most important short-term impacts of volcanic activity on the Canadian public, as future eruption of Canadian or Alaskan volcanoes could severely impact air travel in western Canada (box figure 1). Ash released during the 1992 eruption of Mount Spurr in Alaska was blown by upper atmospheric winds southeastward across the Yukon into B.C., Alberta, and eventually Ontario and Quebec before dispersing over the North Atlantic (box figure 2).

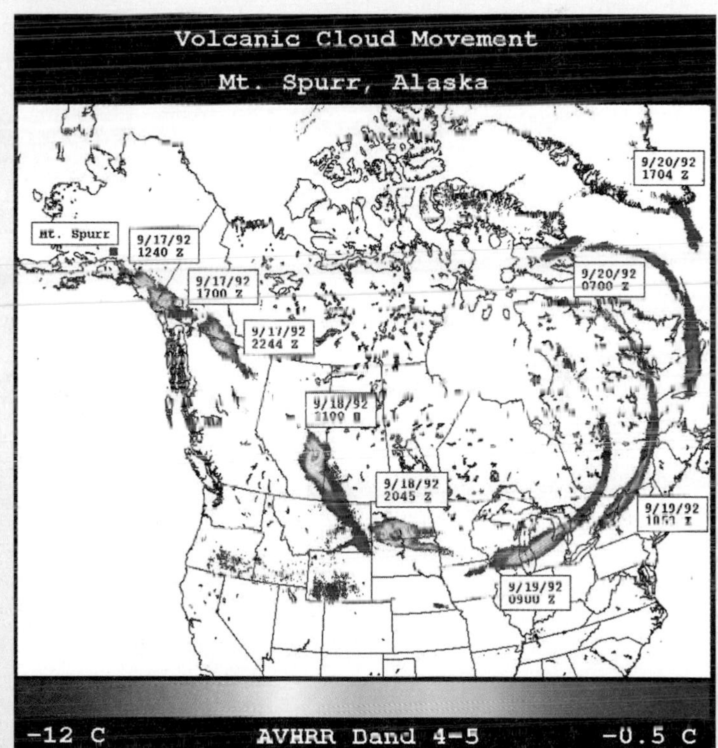

BOX 7.3 ■ FIGURE 2

Composite satellite radar image of ash cloud from the 1992 eruption of Mount Spurr.

Image courtesy of D. Schneider, United States Geological Survey

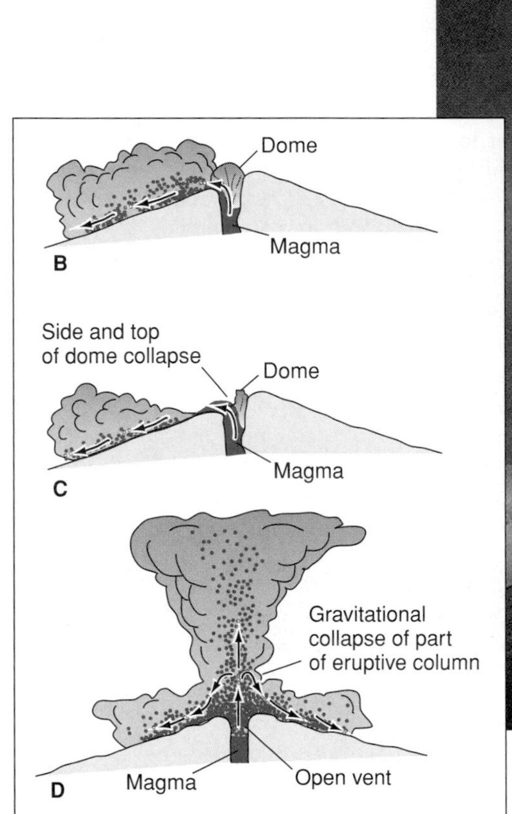

FIGURE 7.6

(*A*) Pyroclastic flow descending Mayon Volcano, Philippines, in 1984. Ways in which pyroclastic flows can form: (*B*) Blasting out from under a plug capping a volcano. (*C*) Collapse of part of a steep-sided dome. (*D*) Gravitational collapse of an eruptive column.

Photo by Chris Newhall, U.S. Geological Survey

FIGURE 7.7

The ruins of St. Pierre in 1902. Mount Pelée is in the clouds.

Photo by Underwood & Underwood, courtesy U.S. Library of Congress

WHAT KINDS OF ROCKS DO VOLCANOES PRODUCE?

Extrusive Rocks

Most extrusive rocks are named and identified on the basis of their composition and texture. But some names are based solely on texture (e.g., pumice).

Composition

The amount of silica in a lava largely controls not only the viscosity of lava and the violence of eruptions but also which particular rock is formed. Chapter 6 describes how igneous rocks are identified based on the minerals present and their relative abundance in the rock (for photos and diagrams, refer to figure 6.4). Because extrusive igneous rocks are generally fine-grained, a specialized microscope is usually needed for precise identification of the component minerals. In most cases, however, we can guess the probable mineral content by noting how dark or light in colour an extrusive rock is. Most silicic rocks are light-coloured because they contain abundant feldspar and quartz (both of which are silica-rich) and few dark minerals (which contain iron and magnesium and are silica-deficient). Mafic rocks, on the other hand, tend to be dark because of the abundance of ferromagnesian minerals.

ENVIRONMENTAL GEOLOGY 7.4

A Tale of Two Volcanoes—Lives Lost and Lives Saved in the Caribbean

Montserrat and Martinique are two of the tropical islands that are part of a volcanic island arc (box figure 1). During the twentieth century, both islands had major eruptions that destroyed towns. Violent and deadly pyroclastic flows associated with growth of volcanic domes caused most of the destruction. For one island, the death toll was huge, and for the other, it was minimal.

In 1902, the port city of St. Pierre on the island of Martinique was destroyed after a period of dome growth and pyroclastic flows on Mount Pelée (no relationship to Pele, Hawaii's goddess of volcanoes). A series of pyroclastic flows broke out of a volcanic dome and flowed down the sides of the volcano. Searingly hot pyroclastic flows can travel at up to 200 kilometres per hour and will destroy any living things in their paths. After the pyroclastic flows began, the residents of St. Pierre became fearful and many wanted to leave the island. The authorities claimed there was no danger and prevented evacuation. There was an election coming up, and the governor felt that most of his supporters lived in the city. He did not want to lose their votes, but neither the governor nor any of the city's residents would ever vote. The climax came on the morning of May 8, when great fiery, exploding clouds descended like an avalanche down the mountainside, raced down a stream valley, through the port city and onto the harbour. St. Pierre and the ships anchored in the harbor were incinerated (see figure 7.7). Temperatures within the pyroclastic flow were estimated at 700°C. Some of the dead had faces that appeared unaffected by the incinerating storm. However, the backs of their skulls were blasted open by their boiling brains. About 29,000 people were burned to death or suffocated (of the two survivors in St. Pierre, one was a condemned prisoner in a poorly ventilated dungeon).

Ninety-three years later, in July 1995, small steam-ash eruptions began at Soufrière Hills volcano on the neighbouring island of Montserrat. As a major eruption looked increasingly likely, teams of volcanologists from France, the United Kingdom, the United States, Canada, and elsewhere flew in to study the volcano and help assess the hazards. An unprecedented array of modern instruments (including seismographs, tiltmeters, and gas analyzers) were deployed around the volcano. In November 1995, viscous, andesitic lava built a dome over the vent. Pyroclastic flows began when the dome collapsed in March 1996. Pyroclastic flows continued with more dome building and collapsing. By 1997, nearly all of the people in the southern part of the island were evacuated, following advice from the scientific teams. In June 1997, large eruptions took place and pyroclastic flows destroyed the evacuated capital city of Plymouth. In contrast to the tragedy of St. Pierre, only 19 people were killed in the region.

In August 1997, major eruptions resumed. This time, the northern part of the island, previously considered safe, was faced with pyroclastic flows (box figure 1), and more people were evacuated from the island. Activity continued, at least into the mid-2000s, but with decreasing intensity. In May 2004, a volcanic mudflow went through the already uninhabitable town of Plymouth. Up to 6 metres of debris were deposited, partially burying buildings still left in the town.

Additional Resources

Mount Pelée, West Indies (Volcano World site)

This site contains some excellent photos from the 1902 eruptions. The second page has photos of the famous spine that grew in Mount Pelée after the tragic eruption.

www.volcano.und.edu/vwdocs/volc_images/img_mt_pelee.html

Montserrat Volcano Observatory

Includes up-to-date reports on volcanic activity.

www.mvo.ms

BOX 7.4 ■ FIGURE 1

Eruption of Soufriére Hills volcano on Montserrat, August 4, 1997. An ash cloud billows upward above a ground-hugging pyroclastic flow. Map of the West Indies showing location of Montserrat, Martinique, and Soufrière Hills volcano.

Photo by AP/Kevin West

Rhyolite, a silicic rock, is usually cream-coloured, tan, or pink; it is made up mostly of feldspar but always includes some quartz. Note that the rhyolite (and granite) portion of figure 6.4 is larger than the areas shown for andesite and basalt. Geologists commonly subdivide this portion of the classification system. For example, *dacite,* the rock associated with the 1980 Mount St. Helens eruptions, contains more ferromagnesian minerals and plagioclase but less potassium feldspar and quartz than the average rhyolite. In our classification system, dacite corresponds to the right portion of the area in figure 6.4 assigned to rhyolite.

A **basalt** has a relatively low amount (about 50 percent by weight) of SiO_2. Much of that silica is bonded to iron and magnesium to form ferromagnesian minerals, such as *olivine* or *pyroxene,* which are green or black. The remaining silica plus aluminum is bonded predominantly with calcium to form calcium-rich *plagioclase feldspar* (which tends to be darker grey than the white or pink potassium or sodium feldspars associated with felsic rocks). Basalt does not contain quartz because no silica is left over after the other minerals have formed. Because of the preponderance of dark minerals in basalt, this rock is usually dark grey to black.

Andesite, which crystallizes from an intermediate lava, can be recognized by its moderately grey or green colour. It is this colour because a little over half the rock is light- to medium-grey plagioclase feldspar, while the rest is ferromagnesian minerals (usually *pyroxene* or *amphibole*).

Extrusive Textures

Texture refers to a rock's appearance with respect to the size, shape, and arrangement of its grains or other constituents. Table 7.1 is a summary of the textures described on the following pages. Some extrusive rocks (such as obsidian and pumice) are classified solely on the basis of their textures, but most are classified by composition *and* texture. *Grain size* is a rock's most important textural characteristic. For the most part, extrusive rocks are fine-grained or else made of glass.

A **fine-grained rock** is one in which most of the mineral grains are smaller than 1 millimetre. In some, the individual minerals are distinguishable only with a microscope. **Obsidian** (figure 7.8), which is volcanic glass, is one of the few rocks that is not composed of minerals. A fine-grained or glassy texture distinguishes extrusive rocks from most intrusive rocks.

Two critical factors determine grain size during the solidification of igneous rocks: rate of cooling and viscosity. If lava cools rapidly, the atoms have time to move only a short distance; they bond with nearby atoms, forming only small crystals. With extremely rapid or almost instantaneous cooling, individual atoms in the lava are "frozen" in place, forming glass rather than crystals.

Grain size is controlled to a lesser extent by the viscosity of the lava. Atoms in a highly viscous lava cannot move as freely as those in a more fluid lava. Hence, a rock formed from viscous lava is more likely to be obsidian or of finer grains than one formed from more fluid lava. Most obsidian, when chemically analyzed, has a very high silica content and is silicic, the chemical equivalent of rhyolite.

TABLE 7.1 Summary of Textures in Volcanic Rocks

Name	Description
Fine-grained (adjective)	Mosaic of interlocking minerals that are smaller than 1 mm.
Porphyritic (adjective)	Some crystals, phenocrysts, are larger than 1 mm (usually considerably larger). Most grains are smaller than 1 mm. Or phenocrysts are enclosed in glass.
Obsidian	Glass. Arrangement of atoms is disordered.
Vesicular (adjective)	Holes (vesicles) in rock due to trapped gas.
Pumice	Frothy glass.
Tuff	Consolidated fine pyroclastic material.
Volcanic breccia	Consolidated pyroclastic debris that includes blocks or bombs

Porphyritic Textures

Extrusive rock that does not have a uniformly fine-grained texture throughout is described as porphyritic. A **porphyritic rock** is one in which larger crystals are enclosed in a *matrix* (or *groundmass*) of much finer-grained minerals or obsidian. The larger crystals are termed **phenocrysts.** A porphyritic rock looks rather like raisin bread; the matrix or groundmass is the bread, the phenocrysts are the raisins. In the porphyritic andesite shown in figure 7.9, phenocrysts of feldspar and ferromagnesian minerals are enclosed in a matrix of crystals too fine-grained to distinguish with the naked eye but visible under a microscope.

Porphyritic texture usually indicates two stages of solidification. Slow cooling takes place while the magma is underground. Minerals that form at higher temperatures crystallize and grow to form phenocrysts in the still partly fluid magma. If the entire mass is then erupted, the remaining liquid portion cools rapidly and forms the fine-grained matrix.

FIGURE 7.8
Obsidian.
Photo by Nick Eyles

A

B

FIGURE 7.9

Porphyritic andesite. A few large crystals (phenocrysts) are surrounded by a great number of fine grains. (*A*) Hand specimen. (*B*) Photomicrograph (using polarized light) of the same rock. The black and white striped phenocrysts are plagioclase and the green ones are ferromagnesian minerals.

Photo A by S. Fletcher; Photo B by C. C. Plummer

Textures Due to Trapped Gas

A magma deep underground is under high pressure, generally high enough to keep all its gases in a dissolved state. On eruption, the pressure is suddenly released and the gases come out of solution. This is analogous to what happens when a bottle of beer or soda is opened. Because the drink was bottled under pressure, the gas (carbon dioxide) is in solution. Uncapping the drink relieves the pressure, and the carbon dioxide separates from the liquid as gas bubbles. If you freeze the newly opened drink very quickly, you have a piece of ice with small, bubble-shaped holes. Similarly, when a lava solidifies while gas is bubbling through it, holes are trapped in the rock, creating a distinctive vesicular texture. **Vesicles** are cavities in extrusive rock resulting from gas bubbles that were in lava and the texture is called *vesicular*. A vesicular rock has the appearance of Swiss cheese (whose texture is caused by trapped carbon dioxide gas). *Vesicular basalt* is quite common (figure 7.10). *Scoria,* a highly vesicular basalt, actually contains more gas space than rock.

In more viscous lavas, where the gas cannot escape as easily, the lava is churned into a froth (like the head in a glass of beer). When cooled quickly, it forms **pumice** (figure 7.11), a frothy glass with so much void space that it floats in water. Powdered pumice is used as an abrasive because it can scratch metal or glass.

FIGURE 7.10

Vesicular basalt.

Photo by S. Fletcher

FIGURE 7.11

Pumice contains so many air spaces that it can float on water.

Photos by Nick Eyles

A

B

FIGURE 7.12

(*A*) Volcanic bomb. (*B*) Night-time eruption at Cerro Negro, a cinder cone in Nicaragua. Magma blobs that solidify in the air will land as bombs. If they are still molten upon landing, they will spatter.

Photo A by C. H. Eyles; Photo B by R. W. Decker

Fragmental Textures

Pyroclasts, the fragments formed by volcanic explosion, can be almost any size. Pyroclasts are named on the basis of their size. *Dust* and *ash* are less than 2 millimetres diameter, *cinders* are between 2 and 64 millimetres, and *blocks* and *bombs* are greater than 64 millimetres diameter. When solid rock has been blasted apart by a volcanic explosion, the pyroclastic fragments are *angular,* with no rounded edges or corners, and are called **blocks.** If lava is ejected into the air, a molten blob becomes streamlined during flight, solidifies, and falls to the ground as a **bomb,** a spindle- or lens-shaped pyroclast (figure 7.12).

When pyroclastic material (ash, bombs, etc.) accumulates and is cemented or otherwise consolidated, the new rock is called *tuff* or *volcanic breccia,* depending on the size of the fragments. A **tuff** (figure 7.13) is a rock composed of fine-grained pyroclastic particles. A **volcanic breccia** is a rock that includes larger pieces of volcanic rock (blocks, bombs).

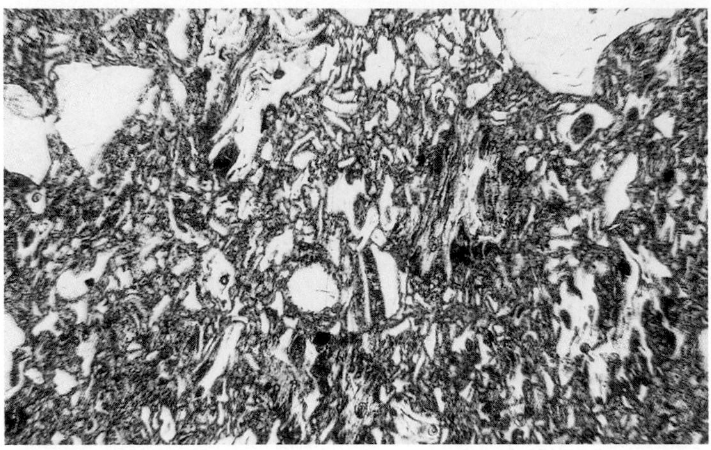

FIGURE 7.13

Photomicrograph of a tuff. Fragments of different rocks, mainly obsidian and pumice, are angular and variously coloured.

Photo by C. C. Plummer

WHAT ARE THE CHARACTERISTICS OF DIFFERENT TYPES OF VOLCANOES?

Volcanic material that is ejected from and deposited around a central vent produces the conical shape typical of volcanoes. The **vent** is the opening through which an eruption takes place. The **crater** of a volcano is a basinlike depression over a vent at the summit of the cone (figure 7.14). Material is not always ejected from the central vent. In a **flank eruption,** lava pours from a vent on the side of a volcano.

A **caldera** is a volcanic depression much larger than the original crater, having a diameter of at least one kilometre. (The most famous caldera in the United States is misnamed "Crater Lake.") A caldera can be created when a volcano's summit is blown off by exploding gases, as occurred at Mount St. Helens in May 1980, or, as in the case of Crater Lake, when a volcano (or several volcanoes) collapses into a vacated magma chamber (see figure 7.3). A few calderas may exist buried beneath ice caps in Canada including at Mount Edziza and Mount Silverthorne (see box 7.2).

The three major types of volcanoes (shield, pyroclastic cone, and composite) are compared in table 7.2; they are markedly distinct from one another in size, shape, and, usually, composition. Although volcanic domes are not cones, they are associated with volcanoes and are also examined in this section.

Shield Volcanoes

Shield volcanoes are broad, gently sloping volcanoes constructed of solidified lava flows (see box 7.5). During eruptions, the lava spreads widely and thinly due to its low viscosity. Because the lava flows from a central vent, without building up much near the vent, the slopes are usually between 2 degrees and 10 degrees from the horizontal, producing a volcano in the shape of a flattened dome or "shield" (figure 7.15).

FIGURE 7.14

Crater and caldera in Kamchatka, Russia. In the foreground is the crater on Karymsky volcano. In the background is a lake-filled caldera.

Photo by C. Dan Miller, U.S. Geological Survey

TABLE 7.2 Comparison of the Three Types of Volcanoes

Profile of Volcano	Description	Composition
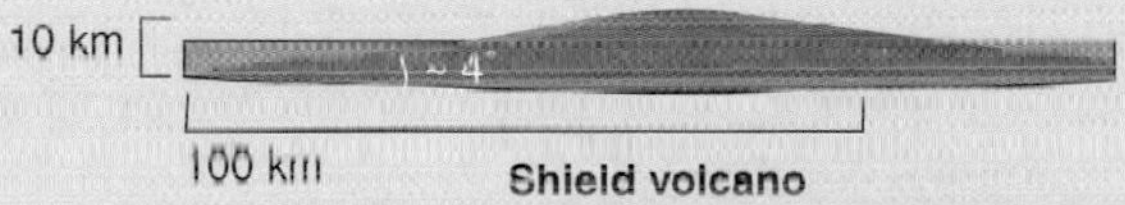 10 km; ~ 4°; 100 km; Shield volcano	**Shield Volcano** Gentle slopes – between 2° and 10°. The Hawaiian example rises 10 kilometres from the sea floor.	Basalt. Layers of solidified lava flows.
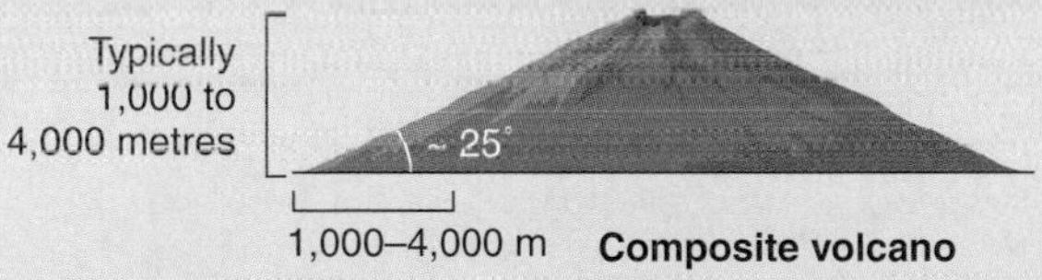 Typically 1,000 to 4,000 metres; ~ 25°; 1,000–4,000 m; Composite volcano	**Composite Volcano** Slopes less than 33°. Considerably larger than cinder cones.	Layers of pyroclastic fragments and lava flows. Mostly andesite.
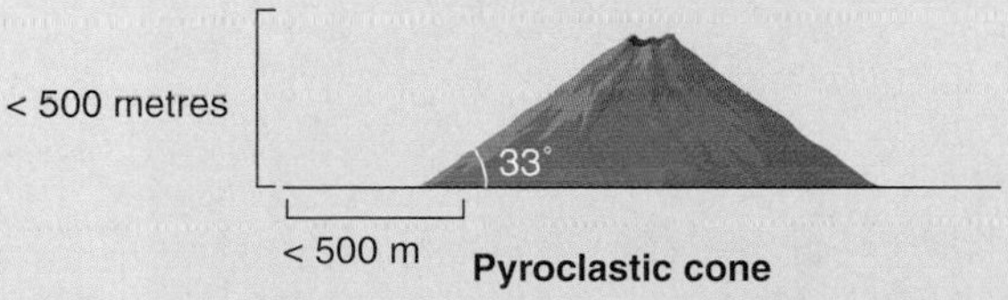 < 500 metres; 33°; < 500 m; Pyroclastic cone	**Pyroclastic Cone** Steep slopes – 33°. Smallest of the three types.	Pyroclastic fragments of any composition. Basalt is most common.

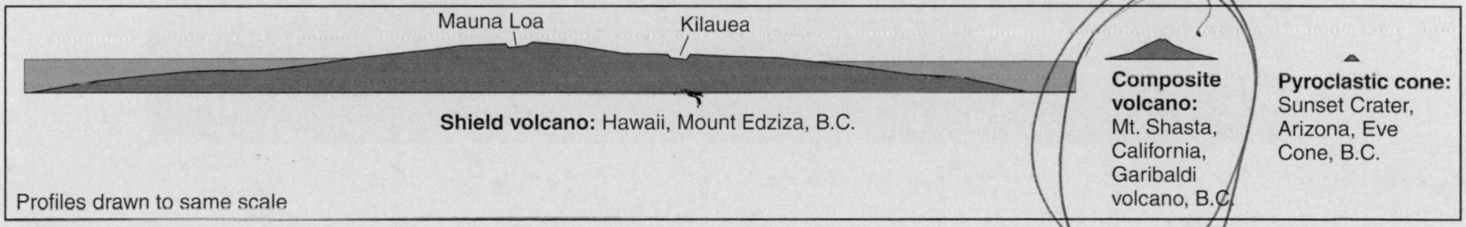

The islands of Hawaii are essentially a series of shield volcanoes built upward from the ocean floor by intermittent eruptions over millions of years (figure 7.15*B*). Although spectacular to observe, the eruptions are relatively nonviolent because the lavas are fairly fluid (less viscous). By implication, then, the shield volcanoes of the Hawaiian Islands are composed of a series of layers of basalt. Examples of shield volcanoes in Canada include Mount Edziza and the Ilgachuz Range of central B.C.

Hawaiian names have been given to two distinctive surfaces of basalt flows. *Pahoehoe* (pronounced *pah-hoy-hoy*) is characterized by a ropy or billowy surface (figure 7.16). The surface is formed by the quick cooling and solidification from the surface downward of a lava flow or pool of lava that was fully liquid. By contrast, basalt that is cool enough to have partially solidified moves as a slow, pasty mass. Its largely solidified front is shoved forward as a pile of rubble. A flow such as this is called *aa* (pronounced *ah-ah*) and has a jagged, rubbly surface (figure 7.17).

A minor feature called a *spatter cone,* a small, steep-sided cone built from lava sputtering out of a vent (figure 7.18), will occasionally develop on a solidifying lava flow. When a small concentration of gas is trapped in a cooling lava flow, lava is belched out of a vent through the solidified surface of the flow. Falling lava plasters itself onto the developing cone and solidifies. The sides of a spatter cone can be very steep, but they are rarely more than 10 metres high. An exception to this is Pu'u 'O'o, the 250-metre-high, combined spatter and cinder cone on the eastern flank of Kilauea shield volcano. It is located at the vent for the ongoing (1983–onward) lava eruptions.

Much of the lava in the ongoing Hawaiian eruptions flows underground in a lava tube, travelling about 7 kilometres from Pu'u 'O'o to the sea. A *lava tube* is a tunnel-like conduit for lava that develops after most of a flow has solidified (figure 7.19). The tube's roof and walls solidified along with the earlier, broader flow. The tube provides insulation so that the rapidly flowing lava loses little heat and remains fluid.

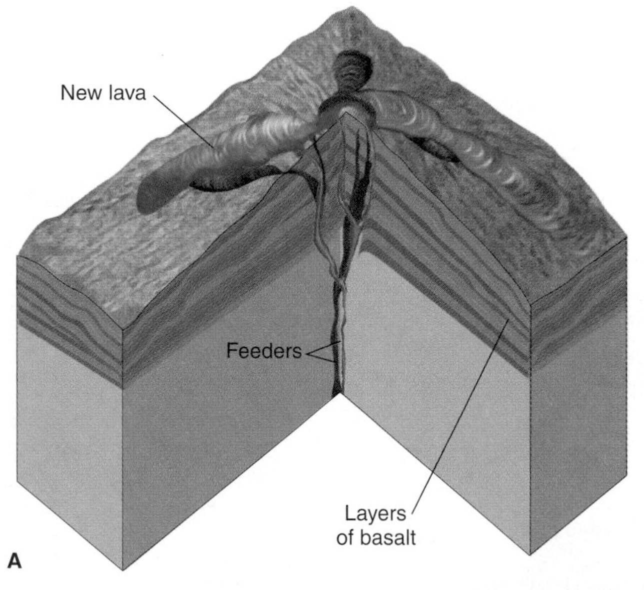

FIGURE 7.15

(*A*) Cutaway view of a shield volcano. (*B*) The top of Mauna Loa, a shield volcano in Hawaii, and its summit caldera. The smaller depressions are pit craters. (*C*) The Illgachuz Range in western B.C. is a shield volcano several million years old.

Photo B © James L. Amos/Corbis Images; Photo C by C. J. Hickson (Geological Survey of Canada).

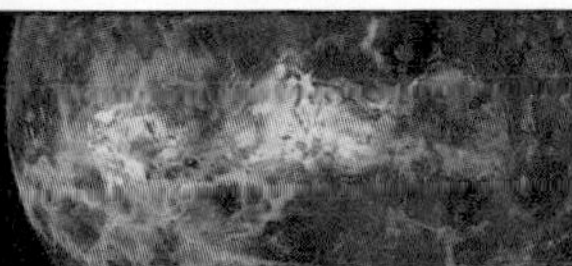

PLANETARY GEOLOGY 7.5

Extraterrestrial Volcanic Activity

Volcanic activity has been a common geologic process operating on the moon and several other bodies in the solar system. Approximately one-sixth of the moon's surface consists of nearly circular, dark-coloured, smooth, relatively flat lava plains. The lava plains, found mostly on the near side of the moon, are called *maria* (singular, *mare;* literally, "seas"). They are interpreted to be huge meteorite impact craters that were flooded with basaltic lava during the moon's early history. There are also a few extinct shield volcanoes on the moon.

Elongate trenches or cracklike valleys called *rilles* are found mainly in the smoother portions of the lunar maria. They range in length from a few kilometres to hundreds of kilometres. Some are arc-shaped or crooked and are regarded as drained basaltic lava channels.

Mercury, the innermost planet, also has areas of smooth plains, suspected to be volcanic in origin.

Radar images of Venus show a surface that is young and probably still volcanically active. More than three-fourths of that surface is covered by continuous plains formed by enormous floods of lava. Close examination of these plains reveals extensive networks of lava channels and individual lava flows thousands of kilometres long.

Large-shield volcanoes, some in chains along a great fault, have been identified on Venus, and molten lava lakes may exist. In other places, thick lavas have oozed out to form kilometre-high, pancake-shaped domes. Radar studies have shown that some of these domes are composed of a glassy substance mixed with bubbles of trapped gas. Fan-shaped deposits adjacent to some volcanoes may be pyroclastic debris.

Several of Venus's volcanoes emit large amounts of sulphur gases, causing the almost continuous lightning that has been observed by spacecraft. It is strongly suspected that the planet is still volcanically active.

Nearly half of the planet Mars may be covered with volcanic material. There are areas of extensive lava flows similar to the lunar maria and a number of volcanoes, some with associated lava flows.

Mars has at least 19 large-shield volcanoes, probably composed of basalt. The largest one, Olympus Mons (box figure 1), is three times the height of Mount Everest and wider than Arizona. Its caldera is more than 90 kilometres across.

Hundreds of volcanoes have been discovered on Jupiter's moon Io (box figure 2), and some of those have erupted for periods of at least four months. Material rich in sulphur compounds is thrown at least 500 kilometres into space at speeds of up to 3,200 kilometres per hour. This material often forms umbrella-shaped clouds as it spreads out and falls back to the surface. Lakes of very hot silicate lava, perhaps mafic or ultramafic, are common. More than 100 calderas larger than 25 kilometres across have been observed, including one that vents sulphur gases. The energy source for Io's volcanoes may be the gravitational pulls of Jupiter and two of its other larger satellites, causing Io to heat up much as a piece of wire will do if it is flexed continuously.

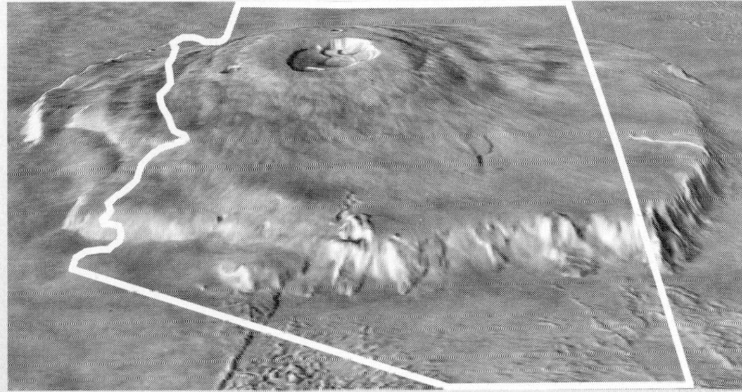

BOX 7.5 ■ FIGURE 1

Perspective view of Olympus Mons, the largest volcano and tallest mountain in the solar system. This Martian volcano is over 650 km wide and 24 km high. Note the outline of the state of Arizona for size comparison.

Photo by NASA/MOLA Science Team

BOX 7.5 ■ FIGURE 2

Two volcanic plumes on Jupiter's moon Io. The plume on left horizon (and upper insert) is 140 km high; the one in the centre (and lower insert) is 75 km high. For details go to photojournal.jpl.nasa.gov/catalog/PIA00703.

Photo by JPL/NASA

Neptune's moon Triton is the third object in the solar system that has active volcanoes. There, "ice volcanoes" erupt what is probably nitrogen frost.

Related Web Resource

The Nine Planets

www.nineplanets.org/

FIGURE 7.16
Flow of lava solidifying to pahoehoe in Hawaii.
Photo © Parvinder Sethi

FIGURE 7.17
An *aa* flow in Hawaii, 1983.
Photo by J. D. Griggs, U.S. Geological Survey

FIGURE 7.18
A spatter cone (approximately 1 metre high) erupting in Hawaii.
Photo by B. Judd, U.S. Geological Survey

FIGURE 7.19
Lava stream seen through a collapsed roof of a lava tube during a 1970 eruption of Kilauea Volcano, Hawaii. Note the ledges within the tube, indicating different levels of flows.
Photo by B. Judd, U.S. Geological Survey

Pyroclastic Cones

A **pyroclastic cone** (also called a *cinder cone* or *lapilli cone* depending on the particle sizes involved) is a volcano constructed of pyroclastic fragments ejected from a central vent (figure 7.20). In contrast to the gentle slopes of shield volcanoes, pyroclastic cones commonly have slopes of about 30 degrees. Most of the ejected material lands near the vent during an eruption, building up the cone to a peak. The steepness of slopes of accumulating loose material is limited by gravity to about 33 degrees. Pyroclastic cones tend to be very much smaller than shield volcanoes. In fact, pyroclastic cones are commonly found on the flanks and in the calderas of Hawaii's shield volcanoes. Few pyroclastic cones exceed a height of 500 m. Pyroclastic cones are common in western Canada and can be found on the flanks of shield volcanoes such as Mount Edziza (Eve Cone, figure 7.20).

Pyroclastic cones form by ejected material (such as ash, blocks, and bombs) accumulating around a vent. They form because of a buildup of gases and are independent of composition. Most cinder cones are associated with mafic or intermediate lava. Silicic pyroclastic cones, which are made of fragments of pumice, are also known as pumice cones.

The lifespan of an active pyroclastic cone tends to be short. The local concentration of gas is depleted rather quickly during the eruptive periods. Moreover, as landforms, pyroclastic cones are temporary features in terms of geologic time. The unconsolidated pyroclasts are eroded relatively easily.

FIGURE 7.20

Pyroclastic cone, Eve Cone in the Stikine volcanic belt, British Columbia. *Photo by C. J. Hickson (Geological Survey of Canada).*

Composite Volcanoes

A **composite volcano** (also called **stratovolcano**) is one constructed of alternating layers of pyroclastic fragments and solidified lava flows (figure 7.21*A*). The slopes are intermediate in steepness compared with cinder cones and shield volcanoes. Pyroclastic layers build steep slopes as debris collects near the vent, just as in pyroclastic cones; however, subsequent lava flows partially flatten the profile of the cone as the downward flow builds up the height of the flanks more than the summit area. The solidified lava acts as a protective cover over the loose pyroclastic layers, making composite volcanoes less vulnerable to erosion than pyroclastic cones. Composite volcanoes are also strengthened from intrusion of dikes and sills of igneous rock that are feeders to the lava flows (figure 7.21).

Composite volcanoes are built over long spans of time. Eruption is intermittent, with hundreds or thousands of years of inactivity separating a few years of intense activity. During the quiet intervals between eruptions, composite volcanoes may be eroded by running water, landslides, or glaciers (figure 7.21*B*). These surficial processes tend to alter the surface, shape, and form of the cone. But because of their long lives and relative resistance to erosion, composite cones can become very large. Aconcagua, a composite volcano in the Andes, is 6,960 m above sea level and the highest peak in the western hemisphere.

The extrusive material that builds composite cones is predominantly of intermediate composition, although there may be local minor silicic and mafic eruptions. Therefore, *andesite* is the rock most associated with composite volcanoes. If the lava is especially hot, the relatively low viscosity fluid flows easily from the crater down the slopes. On the other hand, if enough gas pressure exists, an explosion may litter the slopes with pyroclastic andesite, particularly if the lava has fully or partially solidified and clogged the volcano's vent.

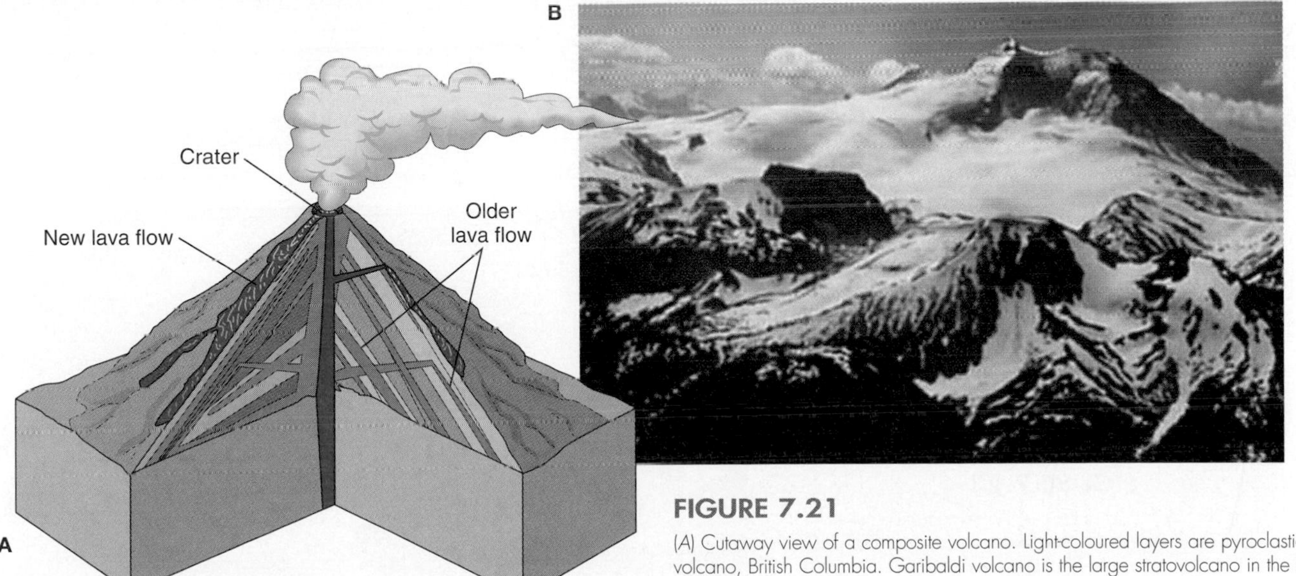

FIGURE 7.21

(*A*) Cutaway view of a composite volcano. Light-coloured layers are pyroclastics. (*B*) Garibaldi volcano, British Columbia. Garibaldi volcano is the large stratovolcano in the background. Table Mountain is the flat-topped, steep sided volcano that erupted under glacial ice. Prominent levees on Mount Price (foreground) identify the edges of a lava flow that erupted about 10,000 years ago. *Photo by C. J. Hickson (Geological Survey of Canada)*

The composition as well as eruptive history of individual volcanoes can vary considerably. For instance, Mount Rainier is composed of 90 percent lava flows and only 10 percent pyroclastic layers. Conversely, Mount St. Helens was built mostly from pyroclastic eruptions—reflecting a more violent history. As would be expected, the composition of the rocks formed during the 1980 eruptions of Mount St. Helens is somewhat higher in silica than the average for Cascade volcanoes.

Distribution of Composite Volcanoes

Nearly all the larger and better known volcanoes of the world are composite volcanoes. They tend to align along two major belts on the Earth (figure 7.22). The **circum-Pacific belt,** or "Ring of Fire," is the larger. The Cascade Range volcanoes and the Garibaldi Volcanic Belt (figure 7.4 and box 7.2) make up a small segment of the circum-Pacific belt.

Several composite volcanoes in Mexico rise higher than 5,000 m, including Orizaba (third highest peak in North America) and Popocatépetl (see box 7.6).

The circum-Pacific belt includes many volcanoes in Central America, western South America (including Nevado del Ruiz in Colombia), and Antarctica. Mount Erebus, in Antarctica, is the southernmost active volcano in the world (figure 7.23).

The western portion of the Pacific belt includes volcanoes in New Zealand, Indonesia, the Philippines (with Pinatubo, Mayon, whose 1993 eruptions killed more than 30 people, and many other volcanoes), and Japan. The beautifully symmetrical Fujiyama, in Japan, is probably the most frequently painted volcano in the world. The northernmost part of the circum-Pacific belt includes active volcanoes in Russia (see figure 7.14) and on Alaska's Aleutian Islands (figure 7.24).

The second major volcanic belt is the **Mediterranean belt,** which includes Mount Vesuvius. An exceptionally violent eruption of Mount Thera, an island in the Mediterranean, may have destroyed an important site of early Greek civilization. (Some archaeologists consider Thera the original "lost continent" of Atlantis.) Mount Etna, on the island of Sicily, has been called the "lighthouse of the Mediterranean" because of its frequent

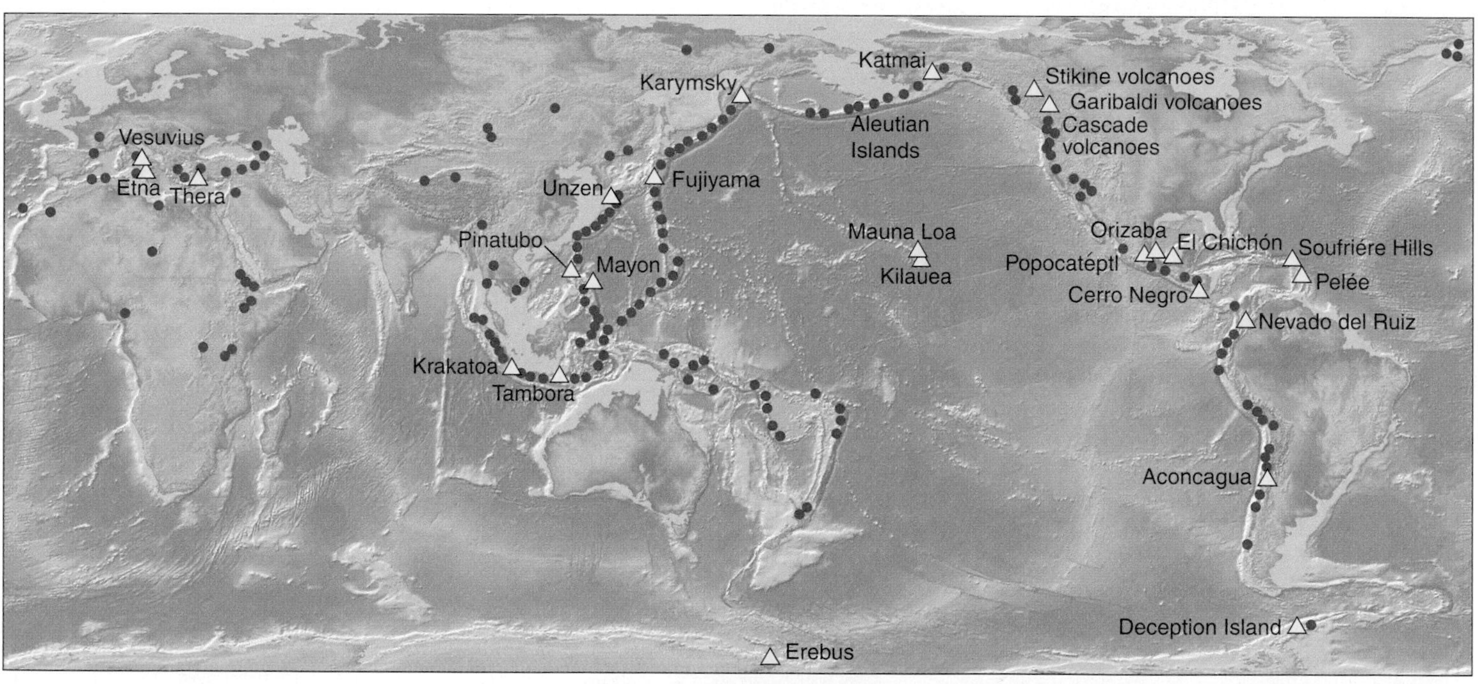

FIGURE 7.22

Map of the world showing recently active major volcanoes. Red dots represent individual volcanoes. Yellow triangles represent volcanoes mentioned in this chapter.

FIGURE 7.23

Mount Erebus, Antarctica, the southernmost active volcano in the world. The photo is taken on sea ice. The summit is 3,794 metres above sea level. A summit crater contains a convecting lava lake.

Photo by Philip R. Kyle

ENVIRONMENTAL GEOLOGY 7.6

Popocatépetl—Will it Erupt Big Time?

Popocatépetl, located 55 km east of Mexico City, one of the world's largest cities, and 70 km west of the city of Puebla, began erupting in 1994. Some 30 million people live within view of Popocatépetl (Aztec for "smoking mountain"). A major eruption could endanger hundreds of thousands of those people.

Popocatépetl, affectionately called "Popo," at 5,484 m above sea level, is one of North America's highest mountains. Not only does Popocatépetl provide a majestic scenic presence (box figure 1*A*), but it figures prominently in Mexico's history, art, and culture. According to Aztec legend, Popocatépetl is a warrior eternally guarding his sleeping lover, the neighbouring mountain Ixtaccihuatl (Aztec for "white lady"), whose outline resembles that of a supine woman. Cortez sent his men to climb Popocatépetl during the Spanish conquest of Mexico in 1521. They were lowered into the smoking crater and returned with sulphur used to make gunpowder. (This was the first recorded ascent in the world of a major mountain.)

The volcano began awakening from a long period of dormancy in December 1994 with a minor dusting of ash on Puebla. Some 75,000 people living on the eastern flank of the volcano were temporarily evacuated. An extensive monitoring network of instruments was deployed and teams of Mexican scientists assisted by members of the U.S. Geological Survey Volcanic Disaster Assistance Program began assessing the potential hazards.

The threat of a disaster is taken very seriously, because in 1982 an apparently insignificant, jungle-covered, 1,000-metre-high volcano in southern Mexico called *El Chichon* erupted with a series of violent explosions. Towns near the previously inactive volcano were buried by heavy ashfall or blasted by searing, gas-charged pyroclastic flows. The death toll could be only roughly estimated to be in the thousands.

By determining the size and extent of ancient pyroclastic deposits and dating them, geologists have determined that Popocatépetl produced major explosive eruptions every 1,000 to 3,000 years for the last 10,000 years. Each has produced widespread pumice falls, pyroclastic flows, and mudflows. Pre-conquest population centres were repeatedly destroyed by these catastrophic eruptions. Since the year 1345, records indicate that there have been some 30 small eruptions before the present activity. Volcanologists consider it one of the world's most dangerous volcanoes. Will the current activity culminate in a colossal event that takes place every thousand years or so?

If such an event were to occur, pyroclastic flows and mudflows would destroy villages and could kill thousands of people if they are not evacuated in time. Heavy ashfall would cause further damage. Mexico City is not likely to be affected by pyroclastic flows or mudflows, but if ash is blown over and deposited in the city there could be serious consequences. Air traffic to and from Mexico City International Airport would be threatened. Water supplies, electrical power grids, and sewer systems could be damaged or destroyed.

Minor steam and ash eruptions from Popocatépetl continued through 1995, 1996, and into 1997. In late 2000 activity increased, and on December 18 Popo's largest eruption in more than 1,000 years took place with spectacular night-time displays of incandescent lava expelled from the mountain. On January 31, 2001 a pyroclastic flow descended the volcano to within 8 km of a town.

A

B

BOX 7.6 ■ FIGURE 1

Popocatépetl in 1960 (*A*) and during the December 19, 2000 eruption (*B*). Snow-covered glaciers that stand out in the 1960 photo are now covered with ash.

Photo A by C. C. Plummer; Photo B © AP/Wide World Photos

Related Web Resource

CENAPRED Volcano Site

www.cenapred.unam.mx/mvolcan.html

Although a summary of reports can be accessed in English, you can get a daily report in Spanish. You can also see the volcano live by clicking on *Tamano B* under *Imagen del volcan.*

eruptions throughout the centuries. The largest eruption in 300 years began in 1991 and lasted for 473 days. Some 250 million cubic metres of lava covered 7 km^2 of land. A town was saved from the lava by heroic efforts that included building a dam to retain the lava (the lava quickly overtopped it), plugging some natural channels, and diverting the lava into other, newly constructed channels. More recent eruptions of Mount Etna occurred in 2002.

Volcanic Domes

Volcanic domes are steep-sided, dome- or spine-shaped masses of volcanic rock formed from viscous lava that solidifies in or immediately above a volcanic vent. A volcanic dome grew within the caldera of Mount St. Helens after the climactic eruption of May 1980 (box 7.1). This was expected because of the high viscosity of the lava from the St. Helens eruptions. In 1983 alone, the dome increased its elevation by 200 metres. Growth of the dome resumed in 2004 due to addition of lava at a rate of about 1.5 m^3/sec (figure 7.25 and http://vulcan.wr.usgs.gov/Volcanoes/MSH/framework.html). Most of the viscous lavas that form volcanic domes are high in silica. They solidify as rhyolite or, less commonly, andesite if minerals crystallize, or as obsidian if no minerals crystallize.

Because the thick, pasty lava that squeezes from a vent is too viscous to flow, it builds up a steep-sided dome or spine (figure 7.25). Some volcanic domes act like champagne corks, keeping gases from escaping. If the plug is removed or broken, the gas escapes suddenly and violently. Some of the most destructive volcanic explosions known have been associated with volcanic domes (see box 7.4).

FIGURE 7.24
Augustine Volcano, Alaska.
Photo by G. McGimsey, Alaska Volcano Observatory/U.S. Geological Survey

FIGURE 7.25
Mount St. Helens crater, new dome and lava spine.
USGS Photograph taken on March 17, 1988, by Lyn Topinka

WHAT ARE LAVA FLOODS?

Not all extrusive rocks are associated with volcanoes. Lava that is very nonviscous and flows almost as easily as water does not build a cone around its vents. Such lava is, of course, mafic (low in silica).

Plateau basalts were produced during the geologic past by vast outpourings of lava (figure 7.26). The Columbia plateau area of Washington, Idaho, and Oregon (see inside cover), for example, is an area of over 400,000 square kilometres constructed of layer upon layer of basalt, in places as thick as 3,000 metres. Each individual flood of lava added a layer usually between 15 and 100 metres thick and sometimes thousands of square kilometres in extent. The outpourings of lava that built the Columbia Plateau took place from 17.5 to 6 million years ago but 95 percent erupted between 17 and 15.5 million years ago. Similar huge, lava plateau-building events have not occurred since then. (The hypothesis that these are due to the arrival of huge mantle plumes beneath the lithosphere is described in chapter 6.) Even larger basalt plateaus are found in India and Siberia. Their times of eruption coincide with the two largest mass extinctions of life on Earth. The one in Siberia occurred about 250 million years ago, around the time of the largest mass extinction, when over 90 percent of living species were wiped out. The eruptions are a prime suspect because of the enormous amount of gases that must have been emitted. These would have changed the atmosphere and worldwide climate. The Indian eruptions occurred about 65 million years ago and coincided with the mass extinction in which the last of the dinosaurs died. Although this mass extinction is generally blamed on a large asteroid hitting Earth (see chapter 19), the intense volcanic activity may have been a contributing factor.

Basalt layers give the landscape a striking appearance in most places where they are exposed. Instead of stacked-up slabs or tablets of solid, unbroken rock, the individual layers may appear to be formed of parallel, vertical columns, mostly six-sided. This characteristic of basalt is called **columnar structure** or **columnar jointing** (figure 7.27). The columns can be explained by the way in which basalt contracts as it cools *after* solidifying. Basalt solidifies completely at temperatures below about 1,200°C. The hot layer of rock then continues to cool to temperatures normal for the Earth's surface. Like most solids, basalt contracts as it cools. The layer of basalt is easily able to accommodate the shrinkage in the narrow vertical dimension; but the cooling rock cannot "pull in" its edges, which may be many kilometres away. The tension fractures the rock into an orderly hexagonal pattern. Instead, the rock contracts toward evenly spaced centres of contraction. Tension cracks develop halfway between neighbouring centres. A hexagonal fracture pattern is the most efficient way in which a set of contraction centres can share fractures. Although most columns are six-sided, some are five- or seven-sided.

FIGURE 7.26

Ancient lava flows (bottom left), Yukon Territory. Volcano Mountain can be seen in distance.

GSC Photo 2002-324. Photo by Lionel Jackson. Reproduced with permission of the Minister of Public Works and Government Services Canada, 2006 and Courtesy of Natural Resources Canada, Geological Survey of Canada.

A

B

FIGURE 7.27

Columnar jointing at Devil's Postpile, California. (*A*) The columns as seen from above. (Scratches were caused by glacial erosion as described in chapter 16.) (*B*) Side view.

Photos by C. C. Plummer

Submarine Eruptions

Submarine eruptions, notably those occurring along mid-oceanic ridges, almost always consist of mafic lavas that create basalt. As described in chapter 6, basaltic rock, thought to have been formed from lava erupting along mid-oceanic ridges, or solidifying underground beneath the ridges, makes up virtually the entire crust underlying the oceans. In a few places—Iceland, for example—volcanic islands rise above the otherwise submerged system (see box 7.7).

Pillow Basalts

Figure 7.28 shows **pillow structure**—rocks, generally basalt, occurring as pillow-shaped rounded masses closely fitted together. From observations of submarine eruptions by divers, we know how the pillow structure is produced. Elongate blobs of lava break out of a thin skin of solid basalt over the top of a flow that is submerged in water. Each blob is squeezed out like toothpaste, and its surface is chilled to rock within seconds. A new blob forms as more lava inside breaks out. Each new pillow settles down on the pile, with little space left in between. Some pillow basalt forms in lakes and rivers; however, most forms at mid-oceanic ridge crests (figure 7.29). According to plate tectonic theory, basalt magma flows up the fracture that develops at a divergent boundary (explained in chapter 6). The magma that reaches the sea floor solidifies as pillow basalt. The rest solidifies in the fracture as a dike. Pillow basalt that is overlying a series of dikes is sometimes found in mountain ranges. These probably formed during sea-floor spreading in the distant past, followed much later by uplift. Metamorphosed pillow basalts more than 1 billion years old (figure 7.28*B*) can be found in the eroded remnants of mountain chains that now form parts of the Canadian Shield (see chapter 20).

A

B

FIGURE 7.28

(*A*) Pillow basalt in Iceland. (*B*) One-billion-year-old metamorphosed pillow basalts of the Canadian Shield, Ontario.

Photo A by R. W. Decker; Photo B by N. Eyles

FIGURE 7.29

Pillow basalt on a mid-oceanic ridge. Photo taken from a submersible vessel.

Courtesy of Woods Hole Oceanographic Institution

ENVIRONMENTAL GEOLOGY 7.7

Fighting a Volcano in Iceland—and Winning

In 1973 a volcano began erupting on a small island in Iceland. Go to the book's Online Learning Centre at www.mcgrawhill.ca/olc/plummer to learn about:

- how a town was almost buried by ash;
- what volunteers did to keep roofs from collapsing from heavy ash deposits;
- a lava flow that threatened to seal off the harbour and end the town's thriving fishing industry;
- an unprecedented effort to halt the lava flow;
- the cleanup and rebuilding of the town;
- how the residents get heat and hot water from the lava flow.

BOX 7.7 ■ FIGURE 1
Lava fountaining behind the town on Heimaey. The glow behind the town in the left part of the photo is the lava flow advancing to the harbour.
Photo © Solarfilma

SUMMARY

Lava is molten rock that reaches the Earth's surface, having been formed as magma from rock within the Earth's crust or from the uppermost part of the mantle.

More people have been killed by pyroclastic flows and, indirectly, by famine than by other volcanic hazards.

Lava contains 45 percent to 75 percent *silica* (SiO_2). The more silica, the more viscous the lava. Viscosity is also determined by the temperature of the lava. Viscous lavas are associated with more violent eruptions than are fluid lavas. *Volcanic domes* form from the extrusion of very viscous lavas.

A *mafic* lava, relatively low in silica, crystallizes into *basalt*, the most abundant extrusive igneous rock. Basalt, which is dark in colour, is composed of minerals that are relatively high in iron, magnesium, and calcium.

Rhyolite, a light-coloured rock, forms from *silicic* lavas that are high in silica but contain little iron, magnesium, or calcium. Because potassium and sodium are important elements in rhyolite, its constituent minerals are mostly potassium- and sodium-rich feldspars and quartz.

A lava with a composition between mafic and silicic crystallizes to *andesite*, a moderately dark rock. Andesite contains about equal amounts of ferromagnesian minerals and sodium- and calcium-rich feldspars.

Extrusive rocks are characteristically fine-grained. *Porphyritic* rock contains some larger crystals in an otherwise fine-grained rock. Rocks that solidified too rapidly for crystals to develop form a natural glass called *obsidian*. Gas trapped in rock forms *vesicles*.

Pyroclasts are the result of volcanic explosions. *Tuff* is volcanic ash that has consolidated into a rock. If large pyroclastic fragments have reconsolidated, the rock is a *volcanic breccia*.

A *pyroclastic cone* is composed of loose pyroclastic material that forms steep slopes as it falls back around the crater. *Cinder cones* are not as large as the other two major types of cones.

A *shield volcano* is built up by successive eruptions of mafic lava. Its slopes are gentle but its volume generally large.

Composite cones are made of alternating layers of pyroclastic material and solidified lava flows. They are not as steep as pyroclastic cones but steeper than shield volcanoes. Young composite volcanoes, predominantly composed of andesite, are aligned along the circum-Pacific belt and, less extensively, in the Mediterranean belt.

Plateau basalts are thick sequences of lava floods. *Columnar jointing* develops in solidified basalt flows. Basalt that erupts underwater forms a *pillow structure*. Pillow basalts are common along the crests of mid-oceanic ridges.

Terms to Remember

andesite 186
basalt 186
block 188
bomb 188
caldera 188
circum-Pacific belt 194
columnar structure (columnar jointing) 197
composite volcano (stratovolcano) 193
crater 188
extrusive rock 177
fine-grained rock 186
flank eruption 188
intermediate rock 182
lava 175
lava flows 177
mafic rock 182
magma 175
Mediterranean belt 194
obsidian 186
phenocryst 186
pillow structure (pillow basalts) 198
plateau basalts 197
porphyritic rock 186
pumice 187
pyroclastic cone 192
pyroclastic flow 182
pyroclasts 177
rhyolite 186
shield volcano 188
silicic (felsic) 182
texture 186
tuff 188
vent 188
vesicle 187
viscosity 182
volcanic breccia 188
volcanic dome 196
volcanism 177
volcano 177

Testing Your Knowledge

Use the questions below to prepare for exams based on this chapter.

1. Compare the hazards of lava flows to pyroclastic flows.
2. What roles do gases play in volcanism?
3. What do pillow structures indicate about the environment of volcanism?
4. Name the minerals, and the approximate percentage of each, that you would expect to be present in each of the following rocks: andesite, rhyolite, basalt.
5. What property (or characteristic) of obsidian makes it an exception to the usual geologic definition of *rock?*
6. What determines the viscosity of a lava?
7. What determines whether a series of volcanic eruptions builds a shield volcano, a composite volcano, or a pyroclastic cone? Describe each type of volcanic cone.
8. Explain how a vesicular porphyritic andesite might have formed.
9. Why are extrusive igneous rocks fine-grained?
10. Why don't flood basalts build volcanic cones?
11. Mount St. Helens
    a. last erupted violently in 1980
    b. is part of the Cascade Range
    c. is located in southern Washington
    d. all of the preceding
12. Volcanic eruptions can affect the climate because
    a. they heat the atmosphere
    b. volcanic dust and gas can reduce the amount of solar radiation that penetrates the atmosphere
    c. they change the elevation of the land
    d. all of the preceding
13. Whether volcanic eruptions are very explosive or relatively quiet is largely determined by
    a. the amount of gas in the lava or magma
    b. the ease or difficulty with which the gas escapes to the atmosphere
    c. the viscosity of the lava
    d. all of the preceding
14. Temperatures at which lavas solidify range from about ____°C for silicic rocks to ____°C for mafic rocks.
    a. 100, 200 b. 300, 1,000
    c. 700, 1,200 d. 1,000, 2,000
15. One gas typically not released during a volcanic eruption is
    a. water vapour b. carbon dioxide
    c. sulphur dioxide d. hydrogen sulphide
    e. oxygen
16. Mafic rocks contain about ____% silica.
    a. 10 b. 25
    c. 50 d. 65
    e. 80
17. Silicic rocks contain about ____% silica.
    a. 10 b. 25
    c. 50 d. 70
    e. 80
18. Which is not an extrusive igneous rock?
    a. granite b. rhyolite
    c. basalt d. andesite
19. Which is not a major type of volcano?
    a. shield b. pyroclastic cone
    c. composite d. stratovolcano
    e. spatter cone
20. A typical example of a shield volcano is
    a. Mount St. Helens b. Kilauea in Hawaii
    c. El Chichón d. Mount Vesuvius

21. An example of a composite volcano is
    a. Mount St. Helens
    b. El Chichón
    c. Mount Vesuvius
    d. all of the preceding
22. Which volcano is not usually made of basalt?
    a. shield
    b. composite cone
    c. spatter cone
    d. pyroclastic cone
23. An igneous rock made of pyroclasts has a texture called
    a. fragmental
    b. vesicular
    c. porphyritic
    d. fine-grained

Exploring Web Resources

www.mcgrawhill.ca/college/plummer
Visit the Online Learning Centre for additional readings and article updates, direct links to the sites listed below, and other media resources.

http://volcano.und.nodak.edu/
Volcano World. This is an excellent site to learn about volcanoes. At the home page you may click on "Volcano of the Week" or "Volcano World Starting Points." "Starting Points" presents a menu that includes currently active volcanoes, volcano video clips, interviews with volcanologists, and more. You can also subscribe to being alerted to new eruptions by e-mail.

www.geo.mtu.edu/volcanoes/
Michigan Tech Volcanoes Page. The focus for this site is on scientific and educational information relative to volcanic hazard mitigation. Clicking on "volcanic humour" will show the lighter side of volcanology.

http://gsc.nrcan.gc.ca/volcanoes/index_e.php
For information on Canadian volcanoes.

http://volcanoes.usgs.gov/Products/Pglossary/
A photo glossary of volcano terms.

CHAPTER

8

Weathering and Soil

How do rocks break down into smaller particles?
How do rocks decompose?
How does climate affect weathering?
What is soil and how does it form?
What are soil horizons?
How does climate affect soil?

In this chapter you see how minerals and rocks change when they are subjected to the physical and chemical conditions existing at Earth's surface. Rocks undergo mechanical weathering (physical disintegration) and chemical weathering (decomposition) as they are attacked by air and water. Your knowledge of the chemical composition and atomic structure of minerals will help you understand the reactions that occur during chemical weathering.

Weathering processes create sediments (primarily mud and sand) and soil. Weathering prepares rocks for erosion and is a fundamental part of the rock cycle, transforming rocks into the raw material that eventually becomes sedimentary rocks.

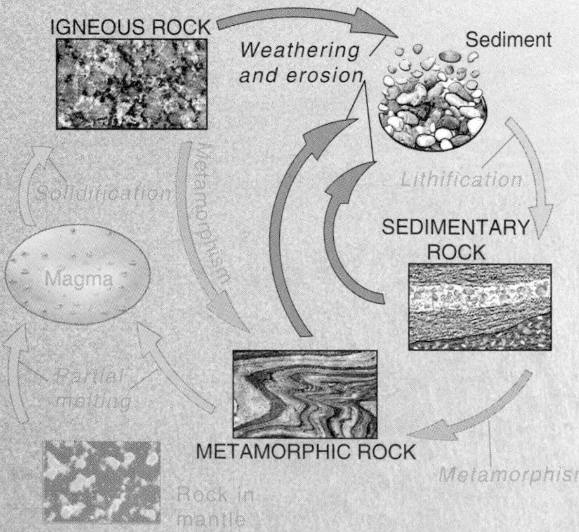

Weathering reactions between sulphide minerals and oxygen-rich water at an abandoned mine site near Sudbury, Ontario produce acidic surface waters (acid mine drainage). *Photo by Nick Eyles*

HOW DO ROCKS BREAK DOWN INTO SMALLER PARTICLES?

Rocks exposed at Earth's surface are constantly being changed by water, air, varying temperature, and other environmental factors. Granite may seem indestructible, but given time and exposure to air and water, it can decompose and disintegrate into soil. The processes that alter rock are *weathering, erosion,* and *transportation.*

The term **weathering** refers to the group of destructive processes that change the physical and chemical character of rock at or near the surface. For example, if you abandon a car, particularly in a wet climate, eventually the paint will flake off and the metal will rust. The car weathers. Similarly, the tightly bound crystals of any rock can be loosened and altered to new minerals when exposed to air and water during weathering. Weathering breaks down rocks that are either stationary or moving.

Erosion is the picking up or *physical removal* of rock particles by an agent such as running water or glaciers. Weathering helps break down a solid rock into loose particles that are easily eroded. Rainwater flowing down a cliff or hillside removes the loose particles produced by weathering. Similarly, if you sand blast rust off of a car, erosion takes place.

After a rock fragment is picked up (eroded), it is transported. **Transportation** is the movement of eroded particles by agents such as rivers, waves, glaciers, or wind. Weathering processes continue during transportation. A boulder being transported by a stream can be physically worn down and chemically altered as it is carried along by the water. In the car analogy, transportation would take place when a stream of rust-bearing water flows away from a car in which rust is being hosed off.

How Weathering Alters Rocks

Rocks undergo both mechanical weathering and chemical weathering. **Mechanical weathering** (physical disintegration) includes several processes that break rock into smaller pieces. The change in the rock is physical; there is little or no chemical change. For example, water freezing and expanding in cracks can cause rocks to disintegrate physically. **Chemical weathering** is the decomposition of rock from exposure to water and atmospheric gases (principally carbon dioxide, oxygen, and water vapour). As rock is decomposed by these agents, new chemical compounds form.

Mechanical weathering breaks up rock but does not change the composition. A large mass of granite may be broken into smaller pieces by frost action, but its original crystals of quartz, feldspar, and ferromagnesian minerals are unchanged. On the other hand, if the granite is being chemically weathered, some of the original minerals are chemically changed into different minerals. Feldspar, for example, will change into a clay mineral (with a crystal structure similar to mica). In nature, mechanical and chemical weathering usually occur together, and the effects are interrelated.

Weathering is a relatively long, slow process that takes hundreds to thousands of years. Typically, cracks in rock are enlarged gradually by frost action or plant growth (as roots pry into rock crevices), and, as a result, more surfaces are exposed to attack by chemical agents. Chemical weathering initially works along contacts between mineral grains. Tightly bound crystals are loosened as weathering products form at their contacts. Mechanical and chemical weathering then proceed together, until a once-tough rock slowly crumbles into individual grains.

Solid minerals are not the only products of chemical weathering. Some minerals—calcite, for example—dissolve when chemically weathered. We can expect limestone and marble, rocks consisting mainly of calcite, to weather chemically in quite a different way than granite.

Effects of Weathering

The results of chemical weathering are easy to find. Look along the edges or corners of old stone structures for evidence. The inscriptions on statues and gravestones that have stood for several decades may no longer be sharp (figure 8.1). Building blocks of limestone or marble exposed to rain and atmospheric gases may show solution effects of chemical weathering in a surprisingly short time. Granite and slate gravestones and building materials are much more resistant to weathering due to the strong silicon–oxygen bonds in the silicate minerals. However, after centuries the mineral grains in granite may be loosened, cracks enlarged, and the surface discoloured and dulled by the products of weathering. Surface discolouration is also common on rock *outcrops,* where rock is exposed to view, with no plant or soil cover. That is why field geologists carry rock hammers—to break rocks to examine unweathered surfaces.

We tend to think of weathering as destructive because it mars statues and building fronts. As rock is destroyed, however, valuable products can be created. Soil is produced by rock weathering, so most plants depend on weathering for the soil they need in order to grow. In a sense, then, all agriculture depends upon weathering. Weathering products transported to the sea by rivers as dissolved solids make seawater salty and serve as nutrients for many marine organisms. Some metallic ores, such as those of copper and aluminum, are concentrated into economic deposits by chemical weathering

Many weathered rocks display interesting shapes. **Spheroidal weathering** occurs where rock has been rounded by weathering from an initial blocky shape. It is rounded because chemical weathering acts more rapidly or intensely on the corners and edges of a rock than on the smooth rock faces (figure 8.2).

Differential weathering is the term for varying rates of weathering in an area where some rocks are more resistant to weathering than others. Resistant rocks weather slowly, and may protrude above softer rocks that weather rapidly. Figures 8.3 and 8.4 show some striking landforms produced by erosion of rocks that weather at different rates. Layers of resistant rock (such as sandstone) weather to form steep cliffs while less resistant rocks (such as shale) form shallow slopes covered by eroded rock debris called **talus.**

A

B

FIGURE 8.1

(*A*) The effects of chemical weathering are obvious in the marble gravestone on the right but not in the slate gravestone on the left, which still retains its detail. Both gravestones date back to the 1780s. (*B*) This marble statue has lost most of the fine detail on the face by chemical weathering.

Photos by C. C. Plummer

A

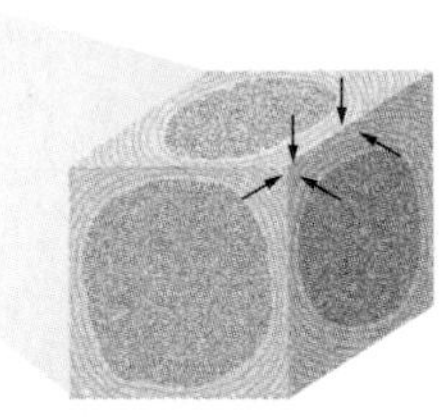

B

C

D

FIGURE 8.2

(*A*) Water penetrating along cracks at right angles to one another in an igneous rock produces spheroidal weathering of once-angular blocks. (*B*) Because of the increased surface area, chemical weathering attacks the corners and edges more rapidly than the flat faces, creating the spheroidal shapes shown in (*C*). (*D*) Spheroidally weathered granite exposed along the Salt River Canyon, Arizona.

FIGURE 8.3

Pedestal rock near Lees Ferry, Arizona. Resistant sandstone cap protects weak shale pedestal from weathering and erosion. Hammer for scale is barely visible at base of pedestal.

Photo by David McGeary

FIGURE 8.4

Tilted sedimentary rocks exposed at Joggins, Nova Scotia. Pale-coloured sandstones resist weathering and form prominent beds and steep sections of the cliff. Large blocks of sandstone released by weathering can be seen at the foot of the cliff. Less resistant layers of shale weather to form more gently sloping sections and finer-grained talus. Cliff is approximately 8 m high.

Photo by C. H. Eyles

Mechanical Weathering

Of the many processes that cause rocks to disintegrate, the most effective are pressure release and frost action.

Pressure Release

The reduction of pressure on a body of rock can cause it to crack as it expands; **pressure release** is a significant type of mechanical weathering. A large mass of rock, such as a batholith, may originally form under great pressure from the weight of several kilometres of rock above it. This batholith is gradually exposed by tectonic uplift of the region followed by erosion of the overlying rock (figure 8.5). The removal of the great weight of rock above the batholith, usually termed *unloading,* allows the granite to expand upward.

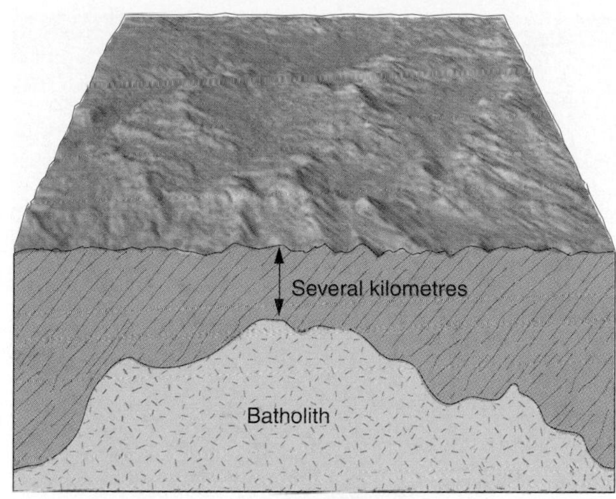

A

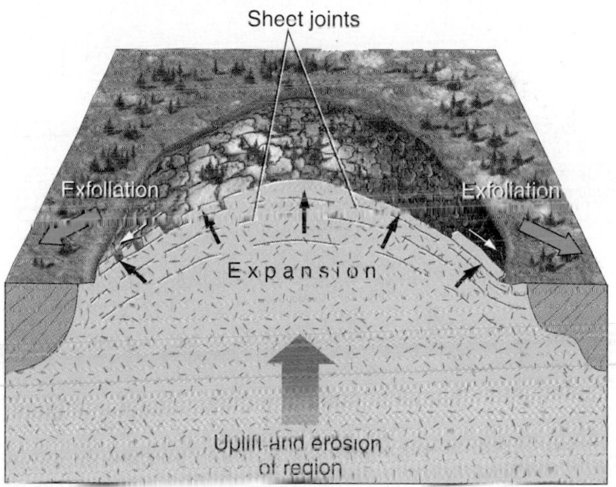

B

C

FIGURE 8.5

Sheet joints caused by pressure release. A granite batholith (*A*) is exposed by regional uplift followed by the erosion of the overlying rock (*B*). Unloading reduces pressure on the granite and causes outward expansion. Sheet joints are closely spaced at the surface where expansion is greatest. Exfoliation of rock layers produces rounded exfoliation domes. (*C*) Sheet joints in a granite outcrop near the top of the Sierra Nevada, California. The granite formed several kilometres below the surface and expanded outward when it was exposed by uplift and erosion.

Photo by David McGeary

FIGURE 8.6
Sheet joints in gneiss of the Canadian Shield, Ontario.
Photo by Nick Eyles

FIGURE 8.7
Exfoliation dome, Yosemite National Park, California. Onionlike layers of rock are peeling off the dome.
Photo by David McGeary

Cracks called **sheet joints** develop parallel to the outer surface of the rock as the outer part of the rock expands more than the inner part. Sheet joints are also common in gneisses and granites of the Canadian Shield (figure 8.6). On slopes, gravity may cause the rock between such joints to break loose in concentric slabs from the underlying granite mass. This process of spalling off of rock layers is called **exfoliation;** it is somewhat similar to peeling layers from an onion. **Exfoliation domes** (figure 8.7) are large, rounded landforms developed in massive rock, such as granite, by exfoliation. Some famous examples of exfoliation domes include Stone Mountain in Georgia and Half Dome in Yosemite.

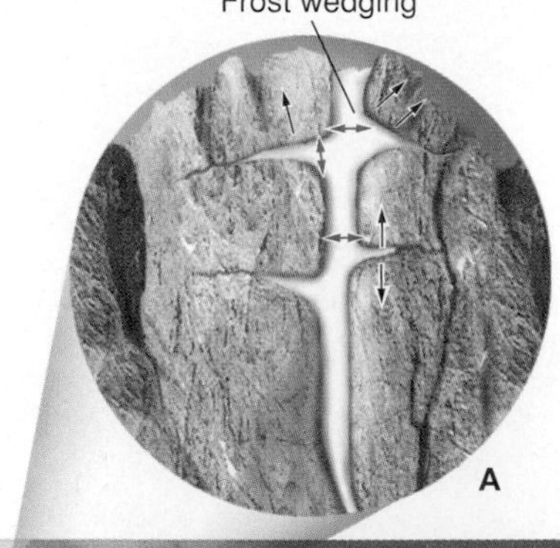

FIGURE 8.8
(*A*) Frost wedging occurs when water fills joints (cracks) in a rock and then freezes. The expanding ice wedges the rock apart. (*B*) Frost wedging has broken the rock and sculpted Crawford Mountain in Banff National Park, Alberta. The broken rock forms cone-shaped piles of debris (talus) at the base of the mountains.
Photo © Martin G. Miller/Visuals Unlimited

Frost Action

Did you ever leave a bottle of water in the freezer, coming back later to find the water frozen and the bottle burst open? When water freezes at 0°C, the individual water molecules jumbled together in the liquid align into an ordered crystal structure, forming ice. Because the crystal structure of ice takes up more space than the liquid, water expands 9 percent in volume when it freezes. This unique property makes water a potent agent of mechanical weathering in any climate where the temperature falls below freezing.

Frost action—the mechanical effect of freezing water on rocks—commonly occurs as frost wedging or frost heaving. In **frost wedging** the expansion of freezing water pries rock apart.

FIGURE 8.9

Growth of ice pedestals (pipraker) below stones in soil lifts them vertically. This process is called frost heaving.

Photo by Nick Eyles

Most rock contains a system of cracks called *joints,* caused by the slow flexing of brittle rock by deep-seated Earth forces (see chapter 11). Water that has trickled into a joint in a rock can freeze and expand when the temperature drops below 0°C. The expanding ice wedges the rock apart, extending the joint or even breaking the rock into pieces (figure 8.8). Frost wedging is most effective in regions with many days of freezing and thawing (mountaintops and midlatitude regions with pronounced seasons). Partial thawing during the day adds new water to the ice in the crack; refreezing at night adds new ice to the old ice.

Frost heaving lifts rock and soil vertically. Solid rock conducts heat better than soil, so on a cold winter day the bottom of a partially buried rock will be much colder than soil at the same depth. As the ground freezes in winter, ice forms first under large rock fragments in the soil. The expanding ice layers elevate pebbles and boulders out of the ground, a process well known to Canadian farmers and other residents of rocky soils (figure 8.9). Frost heaving bulges the ground surface upward in winter, breaking up roads and leaving lawns spongy and misshapen after the spring thaw.

Other Processes

Several other processes mechanically weather rock but in most environments are less effective than frost action and pressure release. *Plant growth,* particularly roots growing in cracks (figure 8.10*A*), can break up rocks, as can *burrowing animals.* Such activities help to speed up chemical weathering by enlarging passageways for water and air. The *pressure of salt crystals* formed as water evaporates inside small spaces in rock also helps to disintegrate desert rocks (figure 8.10*B*). *Extreme changes in temperature,* as in a forest fire, can cause a rock to expand or contract until it cracks. Whatever processes of mechanical weathering are at work, as rocks disintegrate into smaller fragments the total surface area increases (figure 8.11), allowing more extensive chemical weathering by water and air.

A

B

FIGURE 8.10

(*A*) Tree roots pry rocks apart as they grow within the rock joints, Niagara Escarpment, Ontario. (*B*) This rock is being broken by the growth of salt crystals, which precipitate as water evaporates within the cracks in the rock, Mojave Desert, California. Note the tremendous increase in surface area that results from the rock being split into layers.

Photo A by Nick Eyles; Photo B by Crystal Hootman and Diane Carlson

HOW DO ROCKS DECOMPOSE?

Chemical Weathering

The processes of *chemical weathering*, or *rock decomposition,* transform rocks and minerals exposed to water and air into new chemical products. A mineral that crystallized deep underground from a water-deficient magma may eventually be exposed at the surface, where it can react with the abundant water there to form

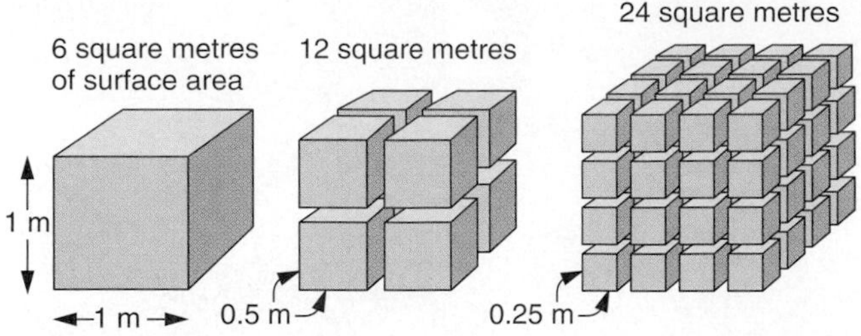

FIGURE 8.11

Mechanical weathering can increase the surface area of a rock, increasing the rate of chemical weathering. As a cube breaks up into smaller pieces, its volume remains the same but its surface area increases.

a new, different mineral. A mineral containing very little oxygen may react with oxygen in the air, extracting oxygen atoms from the atmosphere and incorporating them into its own crystal structure, thus forming a different mineral. These new minerals are weathering products. They have adjusted to physical and chemical conditions at (or near) Earth's surface. Minerals change gradually at the surface until they come into *equilibrium,* or balance, with the surrounding conditions.

Role of Oxygen

Oxygen is abundant in the atmosphere and quite active chemically, so it often combines with minerals or with elements within minerals that are exposed at the Earth's surface.

The rusting of an iron nail exposed to air is a simple example of chemical weathering. Oxygen from the atmosphere combines with the iron to form iron oxide, the reaction being expressed as follows:

$4Fe$	+	$3O_2$	→	$2Fe_2O_3$
iron	+	oxygen	→	iron oxide

Iron oxide formed in this way is a weathering product of numerous minerals containing iron, such as the ferromagnesian group (pyroxenes, amphiboles, biotite, and olivine). The iron in the ferromagnesian silicate minerals must first be separated from the silica in the crystal structure before it can oxidize. The iron oxide (Fe_2O_3) formed is the mineral **hematite,** which has a brick-red colour when powdered. If water is present, as it usually is at the Earth's surface, the iron oxide combines with water to form **limonite,** which is the name for a group of mostly amorphous, hydrated iron oxides (often including the mineral *goethite*), which are yellowish-brown when powdered. The general formula for this group is $Fe_2O_3 \cdot nH_2O$ (the n represents a small, whole number such as 1, 2, or 3 to show a variable amount of water). The brown, yellow, or red colour of soil and many kinds of sedimentary rock is commonly the result of small amounts of hematite and limonite released by the weathering of iron-containing minerals (figure 8.12).

FIGURE 8.12

Shale that has been coloured red by weathering of iron-rich minerals. Red Rock Canyon, Waterton Lakes National Park, Alberta.

Photo by Nick Eyles

How Acids Affect Rocks

The most effective agent of chemical weathering is acid. Acids are chemical compounds that give off hydrogen ions (H^+) when they dissociate, or break down, in water. Strong acids produce a great number of hydrogen ions when they dissociate, and weak acids produce relatively few such ions.

The hydrogen ions given off by natural acids disrupt the orderly arrangement of atoms within most minerals. Because a hydrogen ion has a positive electrical charge and a very small size, it can substitute for other positive ions (such as Ca^{++}, Na^+, or K^+) within minerals. This substitution changes the chemical composition of the mineral and disrupts its atomic structure. The mineral decomposes, often into a different mineral, when it is exposed to acid.

Some strong acids occur naturally on Earth's surface, but they are relatively rare. Sulphuric acid and hydrofluoric acid are strong acids emitted during many volcanic eruptions. They can kill trees and cause intense chemical weathering of rocks near volcanic vents. The bubbling mud of Yellowstone National Park's mudpots (figure 8.13) is produced by rapid weathering caused by acidic sulphur gases that are given off by some hot springs. Strong acids also drain from some mines as sulphur-containing minerals such as pyrite oxidize and form acids at the surface (figure 8.14). Uncontrolled mine drainage can kill fish and plants downstream and accelerate rock weathering.

FIGURE 8.13

A mudpot of boiling mud is created by intense chemical weathering of the surrounding rock by the acid gases dissolved in a hot spring. Yellowstone National Park, Wyoming.
Photo by David McGeary

The most important natural source of acid for rock weathering at Earth's surface is dissolved carbon dioxide (CO_2) in water. Water and carbon dioxide form *carbonic acid* (H_2CO_3), a weak acid that dissociates into the hydrogen ion and the bicarbonate ion (see equation *A* in table 8.1). Even though carbonic acid is a weak acid, it is so abundant at Earth's surface that it is the single most effective agent of chemical weathering.

Earth's atmosphere (mostly oxygen and nitrogen) contains 0.03 percent carbon dioxide. Some of this carbon dioxide dissolves in rain as it falls, so most rain is slightly acidic when it hits the ground (see box 8.1). Large amounts of carbon dioxide also dissolve in water that percolates through soil. The openings in soil are filled with a gas mixture that differs from air. Soil gas has a much higher content of carbon dioxide (up to 10%) than does air, because carbon dioxide is produced by the decay of organic matter and the respiration of soil organisms, such as worms. Rainwater that has trickled through soil is therefore usually acidic and readily attacks minerals in the unweathered rock below the soil (figure 8.15).

Solution Weathering

Some minerals are completely dissolved by chemical weathering. *Calcite,* for instance, goes into solution when exposed to carbon dioxide and water, as shown in equation *B* in table 8.1. The carbon dioxide and water combine to form carbonic acid, which dissociates into the hydrogen ion and the bicarbonate ion, as you have seen, so the equation for the solution of calcite can also be written as equation *C* in table 8.1.

There are no solid products in the last part of the equation, indicating that complete solution of the calcite has occurred. Caves can form underground when flowing groundwater dissolves the sedimentary rock limestone, which is mostly calcite.

A

B

FIGURE 8.14

(*A*) Acid rock drainage assocated with nickel-mining operations. Sudbury, Ontario.
(*B*) Acidified waters flowing from an abandoned coal mining shaft at Joggins, Nova Scotia.
Photo A by L. Warren; Photo B by C.H. Eyles

ENVIRONMENTAL GEOLOGY 8.1

Acid Rain

The burning of coal, oil, and natural gas (the *fossil fuels*) adds a great deal of carbon dioxide to the atmosphere (box figure 1). As you have seen in table 8.1, this carbon dioxide combines with water to form carbonic acid in rain. Coal and oil can also contain nitrogen and sulphur, which are given off as gases (NO_2 and SO_2) when these fuels are burned, forming nitric acid and sulphuric acid in rain. These two acids are much stronger than carbonic acid.

The strength of an acidic solution is measured on the pH scale from 0 to 14 (box figure 2). A solution of pH 7 is chemically neutral, neither acidic nor alkaline. Values below 7 are acidic; the lower the number the more acidic the solution. Values above 7 are alkaline or basic. The pH scale is logarithmic, so a change of 1 on the scale means a change of 10 in the concentration of H^+ ions that make a solution acidic.

Ordinary rain has a pH of about 5.5 to 6 from the small amount of carbon dioxide given off during respiration (every time we exhale, we add a little CO_2 to the atmosphere), and from natural sources of acidic sulphur gases such as volcanoes and coastal marshes. Ordinary rain is about as acidic as milk—hardly a strong acid.

In cities downwind of industrial smokestacks (often for hundreds of kilometres), the increased amount of acid gases can reduce the pH of rain to 4, 3, or even 2 (the pH of lemon juice or vinegar), creating "acid rain." This rain in turn can lower the pH of streams, lakes, and soils. Such low pH values are hard on organisms; fish may die in streams and lakes polluted by acid rain, and forests suffer under acid rain. This is particularly a problem where rocks and soil do not buffer the acid.

For example, in Canada the areas most prone to the damaging effects of acid rain are those underlain by silicic igneous and metamorphic rocks of the Canadian Shield, such as Ontario, Quebec, Nova Scotia, and New Brunswick. These provinces have a large number of industries but are also located in areas where acid gases are carried by prevailing winds from the U.S. (box figure 3). It is estimated that around 75 percent of the acid deposition affecting southern Ontario and Quebec comes from sources in the U.S.

More than 300,000 lakes in eastern Canada underlain by silicic rocks are susceptible to the damaging effects of acid rain; more than 14,000 of these lakes have been acidified to the point where fish populations are significantly reduced. In eastern Canada, spring snowmelt is a particular problem as it can cause acid shock, when large amounts of acid are quickly released into streams and lakes. Acid rain also accelerates chemical weathering processes and causes statues and stone buildings in cities to weather many times faster than those in rural areas free of acid rain (refer to figure 8.1).

Related Web Resources

http://minerals.er.usgs.gov/acid1.html
http://bqs.usgs.gov/acidrain/index.htm
www.ec.gc.ca/acidrain/acidfact.html

BOX 8.1 ■ FIGURE 1
The burning of fossil fuels releases carbon dioxide and other acid gases, such as sulphur dioxide and nitrous oxide, that combine with water to produce acid rain.

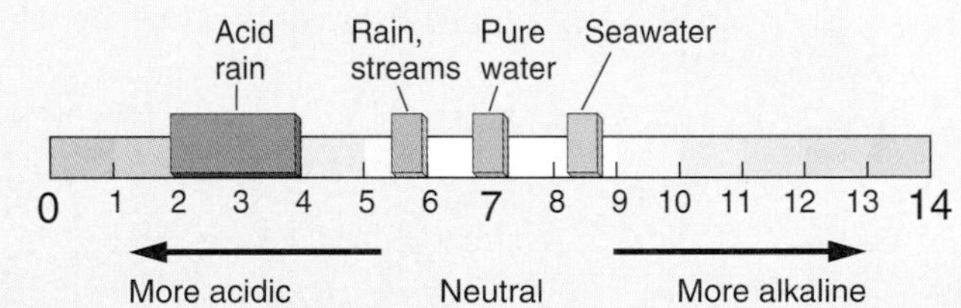

BOX 8.1 ■ FIGURE 2
The pH scale.

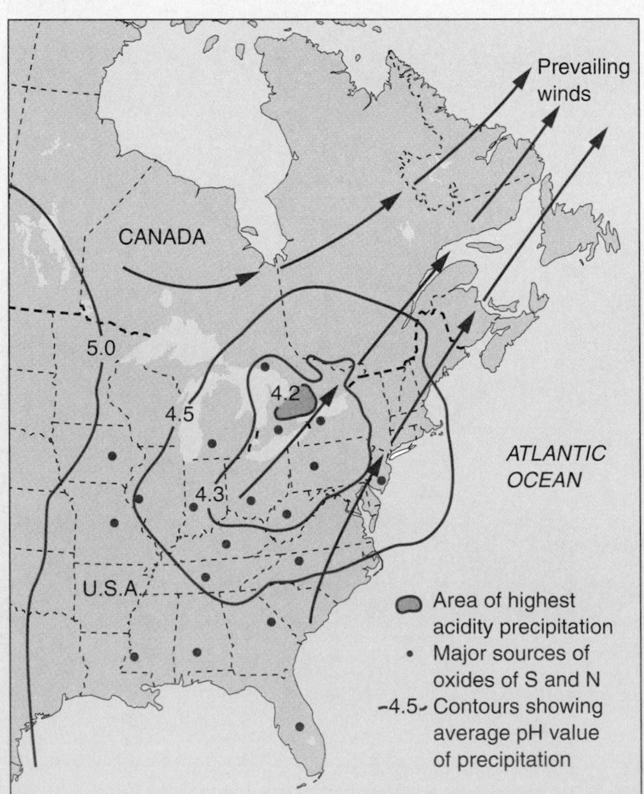

BOX 8.1 ■ FIGURE 3
Acid precipitation in eastern North America.
Based on B. W. Murck, B. J. Skinner, S. Porter. Environmental Geology. New York: Wiley and Sons Inc., 1996, p. 483.

TABLE 8.1 Chemical Equations Important to Weathering

A. Solution of Carbon Dioxide in Water to Form Acid

CO_2 carbon dioxide	+	H_2O water	⇄	H_2CO_3 carbonic acid	⇄	H^+ hydrogen ion	+	HCO_3^- bicarbonate ion

B. Solution of Calcite

$CaCO_3$ calcite	+	CO_2 carbon dioxide	+	H_2O water	⇄	Ca^{++} calcium ion	+	$2HCO_3^-$ bicarbonate ion

C. Solution of Calcite

$CaCO_3$	+	H^+	+	HCO_3^-	⇄	Ca^{++}	+	$2HCO_3^-$

D. Chemical Weathering of Feldspar to Form a Clay Mineral

$2KAlSi_3O_8$ potassium feldspar	+	$2H^+ + 2HCO_3^-$ (from CO_2 and H_2O)	+	H_2O	→	$Al_2Si_2O_5(OH)_4$ clay mineral	+	$2K^+ + 2HCO_3^-$ (soluble ions)	+	$4\,SiO_2$ silica in solution or as fine solid particles

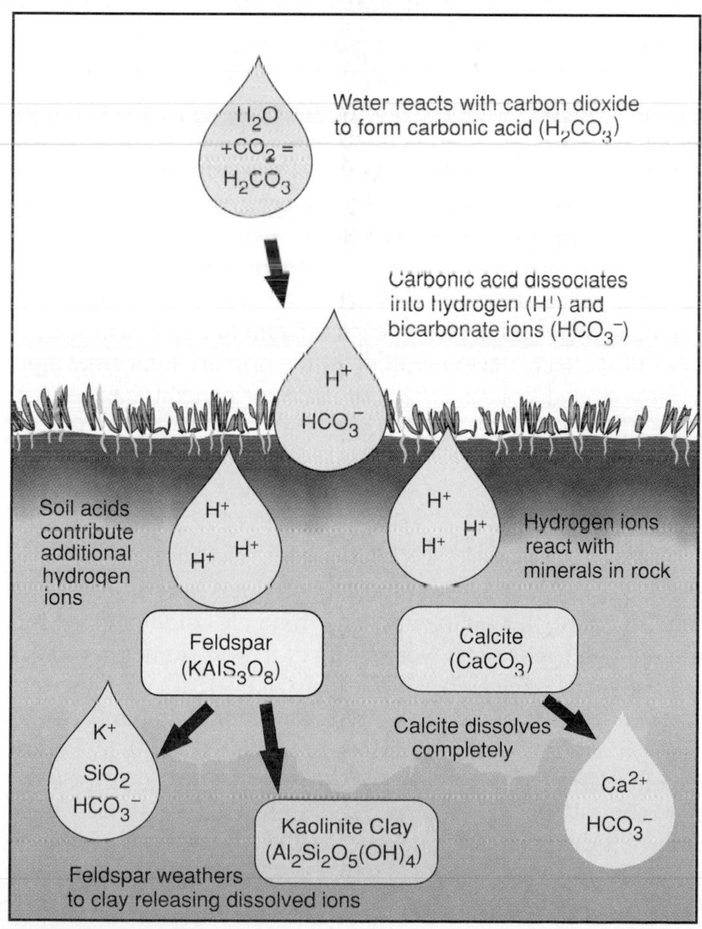

FIGURE 8.15

Chemical weathering of feldspar and calcite by carbonic and soil acids. Water percolating through soil weathers feldspar to clay and completely dissolves calcite. Soluble ions and soluble silica weathering products are washed away.

Rain can discolour and dissolve statues and tombstones carved from the metamorphic rock marble, which is also mostly calcite (see figure 8.1). Another product of solution weathering in areas underlain by limestone is hard water. Hard water contains relatively large amounts of dissolved calcium and magnesium, which prevent soap from lathering and may precipitate as a scaly deposit inside pipes and kettles.

Chemical Weathering of Feldspar

The weathering of feldspar is an example of the alteration of an original mineral to an entirely different type of mineral as the weathered product. When feldspar is attacked by the hydrogen ion of carbonic acid (from carbon dioxide and water), it forms clay minerals. In general, a **clay mineral** is a hydrous aluminum silicate with a sheet-silicate structure like that of mica. Therefore, the entire silicate structure of the feldspar crystal is altered by weathering: feldspar is a framework silicate, but the clay mineral product is a sheet silicate, differing both chemically and physically from feldspar. Partly because of the complexity of the reaction, the chemical weathering of feldspar proceeds at a much slower rate than the solution weathering illustrated by calcite.

Let us look in more detail at the weathering of feldspar (equation *D* in table 8.1). Rainwater percolates down through soil, picking up carbon dioxide from the atmosphere and the upper part of the soil. The water, now slightly acidic, comes in contact with feldspar in the lower part of the soil (figure 8.15), as shown in the first part of the equation. The acidic water reacts with the feldspar and alters it to a clay mineral.

The hydrogen ion (H^+) attacks the feldspar structure, becoming incorporated into the clay mineral product. When the hydrogen moves into the crystal structure, it releases potassium (K) from the feldspar. The potassium is carried away in solution as a dissolved ion (K^+). The bicarbonate ion from the original carbonic acid does not enter into the reaction; it reappears on the right side of the equation. The soluble potassium and bicarbonate ions are carried away by water (groundwater or streams).

All the silicon from the feldspar cannot fit into the clay mineral, so some is left over and is carried away as silica (SiO_2) by the moving water. This excess silica may be carried in solution or as extremely small solid particles.

The weathering process is the same regardless of the type of feldspar: K-feldspar forms potassium ions; Na-feldspar and Ca-feldspar (plagioclase) form sodium ions and calcium ions, respectively. The ions that result from the weathering of Ca-feldspar are calcium ions (Ca^{++}) and bicarbonate ions (HCO_3^-), both of which are very common in rivers and in underground water, particularly in humid regions.

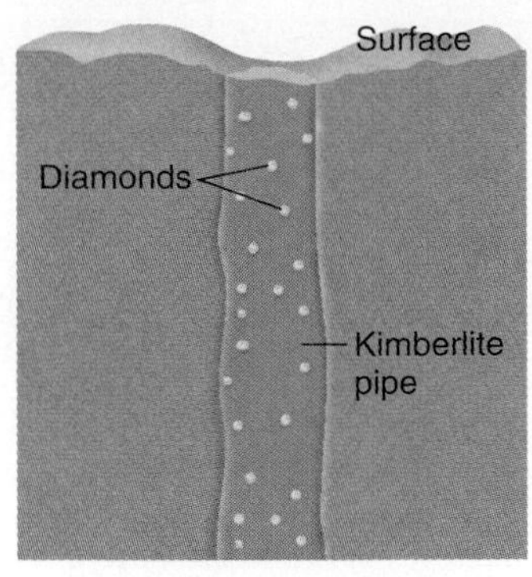

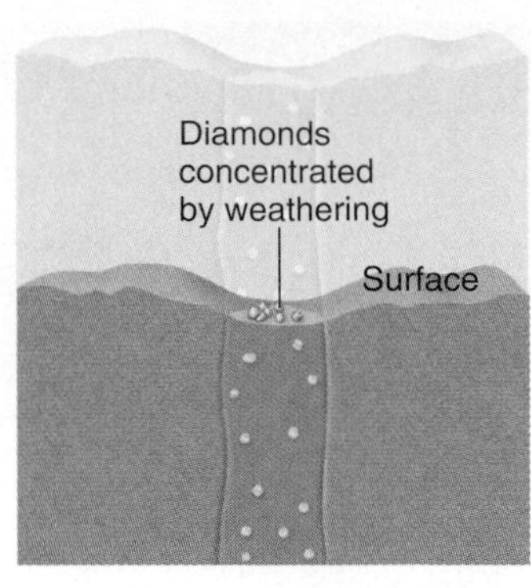

FIGURE 8.16

Residual concentration by weathering. (*A*) Cross-sectional view of diamonds widely scattered within kimberlite pipe. (*B*) Diamonds concentrated on surface by removal of rock by weathering and erosion.

Chemical Weathering of Other Minerals

The weathering of ferromagnesian or dark minerals is much the same as that of feldspars. Two additional products are found on the right side of the equation—magnesium ions and iron oxides (hematite, limonite, and goethite).

The susceptibility of the rock-forming minerals to chemical weathering is dependent upon the strength of the mineral's chemical bonding within the crystal framework. Because of the strength of the silicon–oxygen bond, quartz is quite resistant to chemical weathering. Thus, quartz (SiO_2) is the rock-forming mineral least susceptible to chemical attack at Earth's surface. Ferromagnesian minerals such as olivine, pyroxene, and amphibole include other positively charged ions such as Al, Fe, Mg, and Ca. The presence of these positively charged ions in the crystal framework makes these minerals vulnerable to chemical attack due to the weaker chemical bonding between these ions and oxygen, as compared to the much stronger silicon–oxygen bonds. For example, olivine $(Fe, Mg)_2SiO_4$ weathers rapidly because its isolated silicon–oxygen tetrahedra are held together by relatively weak ionic bonds between oxygen and iron and magnesium. These ions are replaced by H^+ ions during chemical weathering similar to that described for the feldspars.

Weathering and Diamond Concentration

Diamond is the hardest mineral known and is also extremely resistant to weathering. This is due to the very strong covalent bonding of carbon, as described in chapter 5. But diamonds are often concentrated by weathering, as illustrated in figure 8.16. Diamonds are brought to the surface of the Earth in *kimberlite pipes,* columns of brecciated or broken ultramafic rock that have risen from the upper mantle (refer to box 4.4). Diamonds are widely scattered in kimberlite pipes when they form. At the Earth's surface the ultramafic rock in the pipe is preferentially weathered and eroded away. The diamonds, being more resistant to weathering, are left behind, concentrated in rich deposits on top of the pipes. Rivers may redistribute and reconcentrate the diamonds, as in South Africa and India. In Canada, diamond pipes have been eroded by glaciers and diamonds may be found dispersed in glacial deposits in the direction of ice flow. Studying past ice flow directions allows prospectors to trace the trail of diamonds back to the source area (see chapter 12).

As diamond is extremely rare in kimberlite (<1 ppm), exploration for primary diamond deposits is carried out by looking for other, more common minerals that weather out with diamonds and are useful indicators of kimberlite. These minerals are called *indicator minerals* and include Cr-pyrope, eclogitic garnet, Cr-diopside, Mg-ilmenite, Cr-spinel, and olivine. Surface soils and sediments such as glacial tills, sands, and gravels are sampled in and around a potential source area and the medium-sand-sized fraction is analyzed by electron microprobe to determine concentrations of indicator minerals. Kimberlite indicator minerals can survive long distances of glacial transport, and the relative abundance of each mineral in a sediment sample generally reflects the mineralogy of the primary kimberlite pipe. Areas with high concentrations of indicator minerals can be used to identify *indicator mineral trains* that can be traced up-glacier to the kimberlite source.

Weathering Products

Table 8.2 summarizes weathering products for the common minerals. Note that quartz and clay minerals commonly are left after complete chemical weathering of a rock. Sometimes other solid products, such as iron oxides, also are left after weathering.

The solution of calcite supplies substantial amounts of calcium ions (Ca^{++}) and bicarbonate ions (HCO_3^-) to underground water. The weathering of Ca-feldspars (plagioclase) into clay minerals can also supply Ca^{++} and HCO_3^- ions, as well as silica (SiO_2), to water. Under ordinary chemical circumstances, the dissolved Ca^{++} and HCO_3^- can combine to form solid $CaCO_3$ (calcium carbonate), the mineral calcite. Dissolved silica can also precipitate as a solid from underground water. This is significant because calcite and silica are the most common materials precipitated as *cement,* which binds loose particles of sand, silt, and clay into solid sedimentary rock (see chapter 9).

The weathering of calcite, feldspars, and other minerals is a likely source for such cement.

If the soluble ions and silica are not precipitated as solids, they remain in solution and may eventually find their way into a stream and then into the ocean. Enormous quantities of dissolved material are carried by rivers into the sea (one estimate is 4 billion tons per year). This is the main reason why seawater is salty.

HOW DOES CLIMATE AFFECT WEATHERING?

The intensity of both mechanical and chemical weathering is affected by a variety of factors, including climate. Chemical weathering is largely a function of the availability of liquid water. Rock chemically weathers much faster in humid climates than in arid climates. Limestone, which is extremely susceptible to dissolution, weathers quickly and tends to form valleys in wet regions such as the Appalachian Mountains. However, in the arid west, limestone is a resistant rock that forms ridges and cliffs. Temperature is also a factor in chemical weathering. The most intense chemical weathering occurs in the tropics, which are both wet and hot. Polar regions of Canada experience very little chemical weathering because of the frigid temperatures and the absence of liquid water. Mechanical weathering intensity is also related to climate (temperature and humidity), as well as to slope. Cool temperate climates, where abundant water repeatedly freezes and thaws, promote extensive frost weathering. Steep slopes cause rock to fall and break up under the influence of gravity. The most intense mechanical weathering probably occurs in high mountain peaks where the combination of steep slopes, precipitation, freezing, and thawing promote rapid disintegration of solid rock. Climate change may have a significant effect on weathering processes in Canada. Climate warming in the Great Lakes–St. Lawrence region and western Canada, where limestones are exposed, may significantly increase rates of weathering. Climate change also affects weathering processes by changing the frequency of freeze–thaw cycles. The changing intensity of weathering processes over time is considered to have affected global climate through the variable release of carbon dioxide, a "greenhouse gas" (see box 8.2).

WHAT IS SOIL AND HOW DOES IT FORM?

In civil engineering and construction, soil is the usual name for any kind of loose, unconsolidated earth material; but most geologists commonly use the term **soil** for a layer of *weathered,* unconsolidated material on top of bedrock (a general term for rock beneath soil). Soil scientists further restrict the term *soil* to horizons of weathered, unconsolidated material that contains organic matter and is capable of supporting plant growth. (If this definition is used, then the term *regolith* can be applied to any loose surface sediment; soil would be the upper part of the regolith.) A mature, fertile soil is the product of centuries of mechanical and chemical weathering of rock, combined with the addition and decay of plant and other organic matter.

The term **loam** refers to a soil of approximately equal amounts of sand, silt, and clay. (*Clay-sized* particles usually consist of *clay minerals.*) Loamy soils are often well-drained, may contain organic matter, and are often very fertile. *Topsoil* is the upper part of the soil and is more fertile than the underlying *subsoil,* which is often stony and lacks organic matter.

Clay minerals and quartz, the two minerals usually remaining after complete weathering of rock (table 8.2), have important roles in soil development and plant growth. Quartz crystals form sand grains that help keep soil loose and aerated, allowing good water drainage. (Partially weathered crystals of feldspar and other minerals can also form sand-sized grains.)

Clay minerals help to hold water and plant nutrients in a soil. Clay minerals occur as microscopic plates. Because of ion substitution within their sheet silicate structure, most clay minerals have a negative electrical charge on the flat faces of the plates. This negative charge attracts water and nutrient ions to the clay mineral.

The water molecule, made up of two hydrogen atoms and one oxygen atom, is neutral in charge but has a positive end and a negative end. The negative charge on the flat faces of the clay mineral attracts the positive ends of the water molecules to the clay flake (figure 8.17). The clay holds the water loosely enough that most of it is available for uptake by plant roots.

Plant nutrients, such as Ca^{++} and K^{+}, commonly supplied by the weathering of minerals such as feldspar, are also held loosely on the surface of clay minerals. A plant root is able to release H^{+} from organic acids and exchange it for the Ca^{++} and K^{+} that the plant needs for healthy growth (figure 8.18).

WHAT ARE SOIL HORIZONS?

Most soils take a long time to form. The rate of soil formation is controlled by rainfall, temperature, slope, and to some extent the type of bedrock that weathers to form soil. High temperatures and abundant rainfall speed up soil formation, but in most places a fully developed soil that can support plant growth takes hundreds or thousands of years to form.

As soils mature, distinct layers appear in them (figure 8.19). Soil layers are called **soil horizons,** and can be distinguished from one another by appearance and chemical composition. Boundaries between soil horizons are usually transitional rather than sharp. By observing a vertical cross-section, or *soil profile,* various horizons can be identified.

In Canada the uppermost layer that consists entirely of non-decomposed and highly decomposed organic material is given the designation **LFH:** L(leaf), F(needle), or H(humus). (This is the **O horizon** of the U.S. soil classification system.)

The **A horizon** is the dark-coloured soil layer that is rich in organic material and forms just below surface vegetation. This horizon contains decomposed plant material, or *humus,* and contributes to the formation of organic acids that accelerate leaching in the lower part of the A horizon (called *Ae horizon*). The lower part of the A horizon, or **zone of leaching,** is characterized by the downward movement of water. Part of the rain falling on

IN GREATER DEPTH 8.2

Weathering, the Carbon Cycle, and Global Climate

Weathering has affected the long-term climate of Earth by changing the carbon dioxide content of the atmosphere through the inorganic carbon cycle (see box figure 1). Carbon dioxide is a "greenhouse gas" that traps solar heat near the surface, warming the Earth. The planet Venus has a dense atmosphere composed mostly of CO_2, which traps so much solar heat that the surface temperature averages a scorching 480°C. Earth has comparatively very little CO_2 in its atmosphere (see box table 1)—enough to keep most of the surface above freezing but not too hot to support life. However, when Earth first formed, its atmosphere was probably very much like that of Venus, with much more CO_2. What happened to most of the original carbon dioxide in Earth's atmosphere? Geoscientists hypothesize believe that a quantity of CO_2 equal to approximately 65,000 times the mass of CO_2 in the present atmosphere lies buried in the crust and upper mantle of Earth. Some of this CO_2 was used to make organic molecules during photosynthesis and is now trapped as buried organic matter and fossil fuels in sedimentary rocks. However, the majority of the missing CO_2 was converted to bicarbonate ion (HCO_3^-) during chemical weathering and is locked away in carbonate minerals (primarily $CaCO_3$) that formed layers of limestone rock.

The inorganic carbon cycle helps to regulate the climate of Earth because CO_2 is a greenhouse gas, chemical weathering accelerates with warming, and the formation of limestone occurs mostly in warm, tropical oceans. When Earth's climate is warm, chemical weathering and the formation of limestone increase, drawing CO_2 from the atmosphere, which cools the climate. When the global climate cools, chemical weathering and limestone formation slow down, allowing CO_2 to accumulate in the atmosphere from volcanism, which warms the Earth. An increase in chemical weathering can also lead to global cooling by removing more CO_2 from the atmosphere. For example, the Cenozoic uplift and weathering of large regions of high mountains such as the Alps and the Himalaya may have triggered the global cooling that culminated in the glaciations of the Pleistocene epoch.

BOX 8.2 ■ TABLE 1

Carbon Dioxide in the Atmospheres of Earth, Mars, and Venus

	Earth	Mars	Venus
CO_2%	0.33	95.3	96.5
Total surface pressure, bars	1.0[a]	.006	92

[a]Approximately 50 bars of CO_2 is buried in the crust of the Earth as limestone and organic carbon.

Additional Resources

http://earthobservatory.nasa.gov/Library/CarbonCycle/

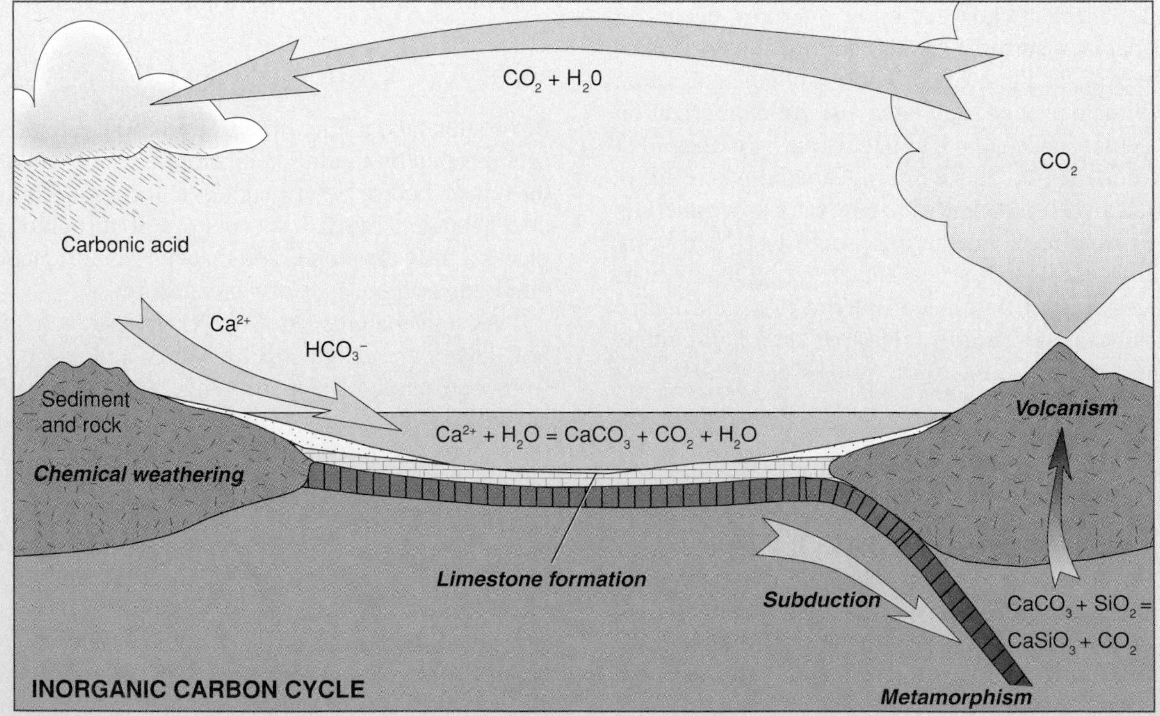

BOX 8.2 ■ FIGURE 1

Carbon dioxide dissolves in water to form carbonic acid in the atmosphere. Carbonic acid reacts with sediment and rocks during chemical weathering, releasing calcium ions and bicarbonate ions (HCO_3^-), which are carried by rivers into the sea. The precipitation of $CaCO_3$ mineral in the oceans (see chapter 9) forms layers of limestone rock. Deep burial of limestone leads to metamorphism, which reacts silica and calcite to form calcium silicate minerals and carbon dioxide. The CO_2 remains trapped in Earth's interior until it is released during volcanic eruptions.

TABLE 8.2 Weathering Products of Common Rock-Forming Minerals

Original Mineral	Under Influence of CO_2 and H_2O	Main Solid Product		Other Products (Mostly Soluble)
Feldspar	→	Clay mineral	+	Ions (Na^+, Ca^{++}, K^+), SiO_2
Ferromagnesian minerals (including biotite mica)	→	Clay mineral	+	Ions (Na^+, Ca^{++}, K^+, Mg^{++}), SiO_2, Fe oxides
Muscovite mica	→	Clay mineral	+	Ions (K^+), SiO_2
Quartz	→	Quartz grains (sand)		
Calcite	→	—		Ions (Ca^{++}, HCO_3^-)

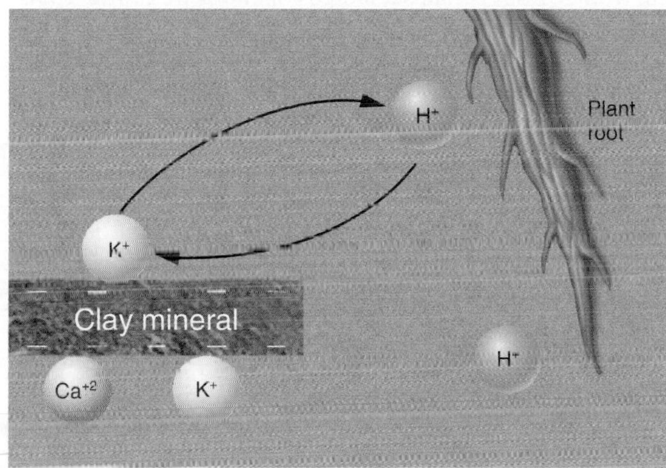

FIGURE 8.17

Ion exchange between plant root and clay mineral.

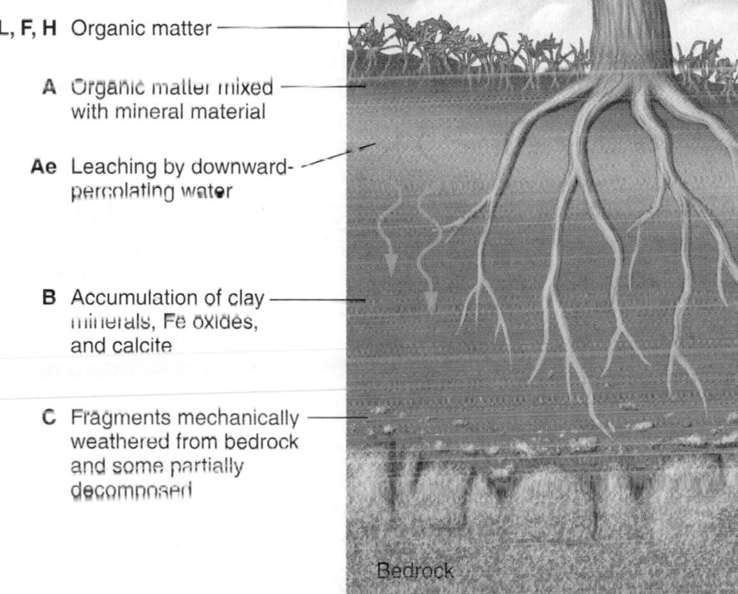

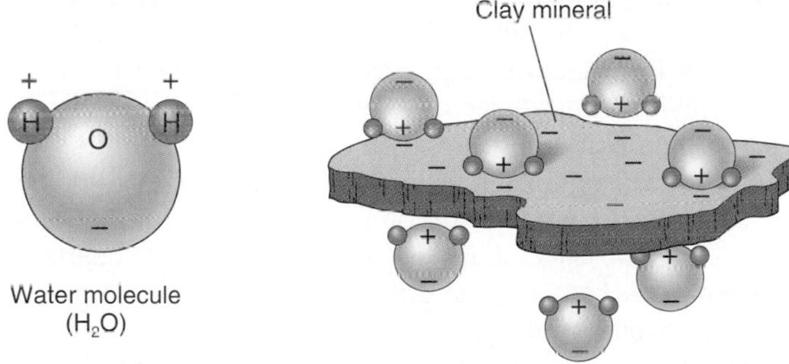

FIGURE 8.18

Negative charges on a clay mineral attract positive ends of water molecules.

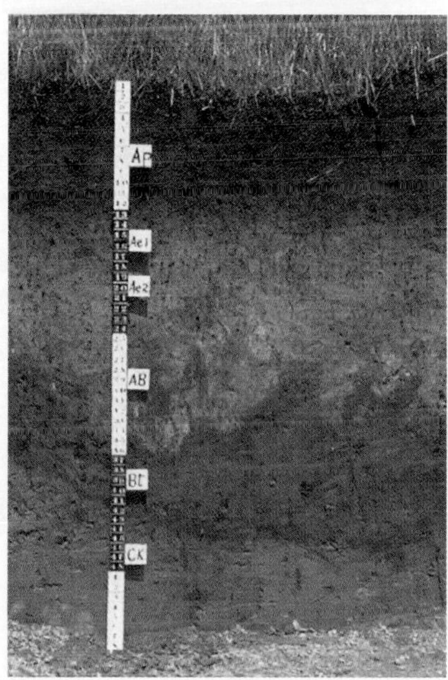

FIGURE 8.19

(*A*) Horizons (LFH, A, Ae, B, and C) in a soil profile that forms in a humid climate. (*B*) Soil profile through a grey luvisol in southern Ontario that shows an organic-rich A horizon, a paler-coloured and leached Ae horizon, clayey B horizon, and calcareous parent material (C horizon).

Photo A Reproduced with the permission of the Minister of Public Works and Government Services Canada, 2006. Agriculture and Agri-Food Canada, Manitoba

Photo B Soil Landscape Illustration of Southern Ontario and Quebec. Photo by Robert G. Eilers. Reproduced with the permission of the Minister of Public Works and Government Services Canada, 2006. Agriculture and Agri-Foods Canada, Manitoba

the ground percolates downward through the soil. This tends to *leach*, or carry dissolved chemicals downward to lower levels in the soil profile. In a humid (wet) climate, iron oxides and dissolved calcite are most typically leached downward; clays are also transported downward. Leaching may make the lower part of the A horizon pale and sandy, but the uppermost part is often darkened by humus.

The **B horizon,** or **zone of accumulation,** is a soil layer characterized by the accumulation of material leached downward from the A horizon above. This layer is often quite clayey and stained red or brown by hematite and limonite. Calcite may also build up in B horizons.

The **C horizon** is incompletely weathered parent material, lying below the B horizon. The parent material is commonly the underlying bedrock, which is subjected to mechanical and chemical weathering from frost action, roots, plant acids, and other agents. In such a case, the C horizon is transitional between unweathered bedrock below and developing soil above.

Soil Classification

The Soil Conservation Service of the U.S. Department of Agriculture and the Canadian Soil Information System have developed soil classification systems to group soils with similar properties so that soils can be mapped in a systematic way. There are several large groups, called *orders,* that are distinguished by the characteristics of the horizons present in soil profiles. Brief descriptions of the orders are given in table 8.3, and their distribution in the continental United States and Canada is shown on the map in figure 8.20. Note that soil orders are named differently in the U.S. and Canada; equivalent names for soil orders are given in table 8.3. Each order can be further subdivided into many subdivisions, or suborders, which are defined by even more specific diagnostic physical and chemical properties observed in the soil.

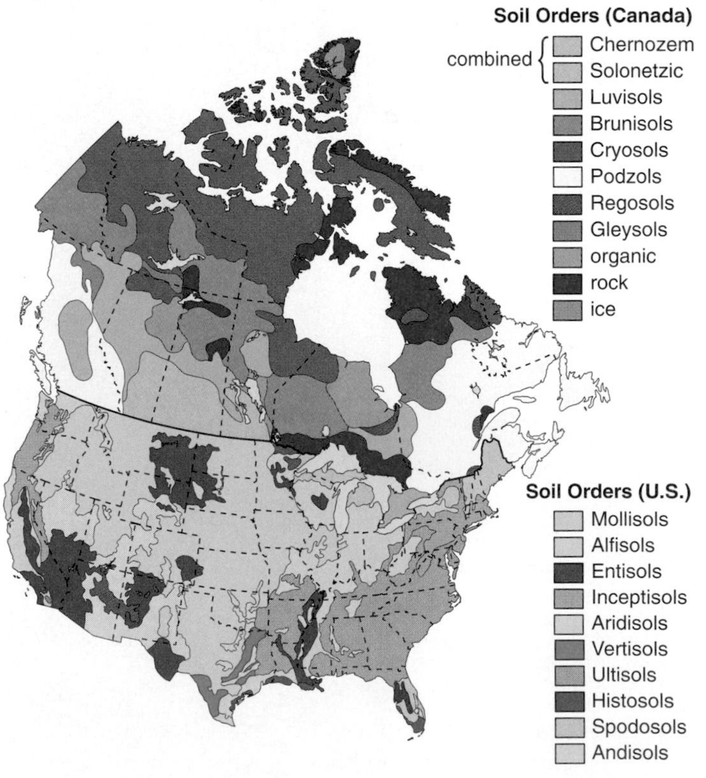

FIGURE 8.20

Distribution of soil orders in the United States and Canada (U.S.D.A. Soil Conservation Map; Canadian Soil Information Map)

Residual and Transported Soils

A **residual soil** is one that develops from weathering of the rock directly beneath it. Figure 8.19*A* is a diagram of a residual soil developing in a humid climate from a bedrock source. Although this is a typical situation, a number of important agricultural regions in the United States and Canada have developed on **transported soils,** which did not form from the local rock but from regolith brought in from some other region. Transported soils usually form on sediment deposited by running water, wind, or glacial ice. For example, mud deposited by a river during times of flooding can form an excellent agricultural soil next to the river after floodwaters recede. The soil-forming mud was not weathered from the rock beneath its present location but was carried downstream from regions perhaps hundreds of kilometres away. Transported wind deposits, called *loess* (see chapter 17), are the parent material for some of the most valuable food-producing soils in the U.S. Midwest and Pacific Northwest.

Soils, Parent Material, Time, and Slope

The character of a soil depends partly on the parent material from which it develops. A soil developing on weathering granite will be sandy, as sand-sized particles of quartz and partially weathered feldspar are released from the granite. As time passes, the partially weathered feldspar grains weather completely, forming fine-grained clay minerals. The quartz does not weather, so the resulting soil has both sand and clay (and perhaps silt) in it.

A soil forming on basalt may never be sandy, even in its early stages of development (this depends on the relative rates of chemical weathering versus mechanical weathering). The fine-grained feldspars in the basalt weather to fine-grained clay minerals. Since the parent rock had no coarse-grained minerals and no quartz to start with, the resulting soil may lack sand. Such a soil may not drain well, although it can be quite fertile.

Note that the character of a soil changes with time. A soil developing from granite begins as a sandy soil and becomes more clayey with time. Over very long periods, the type of parent rock becomes less and less important. Given enough time, soils forming from many different kinds of igneous, metamorphic, and sedimentary rocks can become quite similar (in the same climate). The presence or absence of coarse grains of quartz in the parent rock becomes the only characteristic of the parent rock to have long-term significance.

With time, soils tend to become thicker (figure 8.21*A*); most modern soils have taken centuries to form. A new deposit of volcanic ash, which is very fine-grained and rich in plant nutrients, may be covered with grass and other low plants in just a few years, but a new lava flow, which weathers much more slowly than ash, may not have enough soil to support grass for many decades. It may take centuries for the lava-flow soil to thicken enough to support shrubs, and thousands of years to support trees. The fertile agricultural soils of the Canadian plains and the northern United States took more than 10,000 years to develop on glacial deposits after the thick continental ice sheets melted (see figure 16.33).

TABLE 8.3 Soil Orders

Canadian Soil Order	Description	U.S. Soil Order
Brunisol	Immature soils, lack distinct B horizon. Often found in forested ecosystems. Occupy abut 10% area of Canada.	Some inceptisols and entisols
Chernozem	Humus-rich A horizon at least 10 cm thick, calcium rich. Develop only under prairie climates. Occupy about 5% area of Canada.	Some mollisols
Cryosol	High-latitude or high-altitude soils with permafrost layer within 1 m of the surface. Occupy 10–15% area of Canada.	Gelisol
Gleysol	Found in areas that are frequently flooded or permanently waterlogged. Horizons show chemical signs of oxidation and reduction. Occupy about 5% area of Canada.	Wet inceptisol
Luvisol	Soil often has calcareous parent material. A horizon has high pH and strong eluviation of clay. Occur throughout boreal areas and range into taiga areas. Occupy about 9% area of Canada.	Some mollisols, vertisols, inceptisols, and alfisols
Organic	Soil mostly composed of organic matter, which may be in various stages of decomposition. Common in bogs and fens. Occupy about 11% area of Canada (41% of Manitoba).	Histosols
Podzol	Acidic soils with an eluviated A horizon; B horizon enriched in organic matter, aluminum, and iron. Common in coniferous forests. Occupy about 22% area of Canada.	Spodosol
Regosol	Any young and undeveloped soil. May have thin A horizon and C horizon but no B horizon. Occupy about 10% area of Canada (mostly in north).	Entisol
Solonetz	Dry soils experiencing high levels of evapotranspiration, deposition of salts at or near soil surface. High sodium and relatively low calcium content. Common in dry, prairie regions. Occupy about 1% area of Canada.	Some aridisols, vertisols, and mollisols
n/a	Heavily weathered soils low in plant nutrient ions and rich in aluminum and iron oxides. Common in moist, tropical regions.	Oxisols
n/a	Strongly weathered soils, low in plant nutrient ions with clay accumulation in the subsurface. Moist.	Ultisols
n/a	Soils formed in volcanic ash.	Andisols

Soil orders of Canada and approximate equivalents in the United States. (Note: matching Canadian and U.S. soil classifications is difficult as there are not exact equivalencies between the units.)

Another factor controlling soil thickness is the slope of the land surface (figure 8.21*B*). Soils tend to be thick on flat land where erosion is slow and water can collect, and thin on steep slopes where gravity pulls water and soil particles downhill.

Organic Activity

Organisms contribute to soil development. Plant roots break up rocks, and burrowing organisms such as ants, worms, and rodents bring soil particles to the surface and create passageways for water and air to get underground, thus speeding up chemical weathering. Respiration of soil organisms and decay of plant and animal material adds carbon dioxide gas to soil, creating carbonic acid. Plants and humus also release organic acids that increase chemical weathering. Once soil begins to develop on a newly exposed rock, it attracts plants and soil organisms that increase chemical weathering, accelerating the rate of soil development. Partially decayed organic matter provides plant nutrients, increasing soil fertility.

HOW DOES CLIMATE AFFECT SOIL?

Climate affects soil thickness and character. Soils in many wet climates, as in Europe and the eastern United States, tend to be thick and are generally characterized by downward movement of water through Earth materials (figure 8.19 shows such a soil). In general these soils tend to have a high content of aluminum and iron oxides, and are marked by effective downward leaching due to high rainfall and to the acids produced by the decay of abundant humus. In much of Canada, soil thickness is more closely related to temperature, time since deglaciation, and parent material than precipitation.

In arid (dry) climates, as in many parts of the western United States, the Prairie regions of Canada, and the Okanagan Valley of south central B.C., soils tend to be thin and are characterized by little leaching, scant humus, and the *upward* movement of soil water beneath the land surface. The water is drawn up by subsurface evaporation and capillary action.

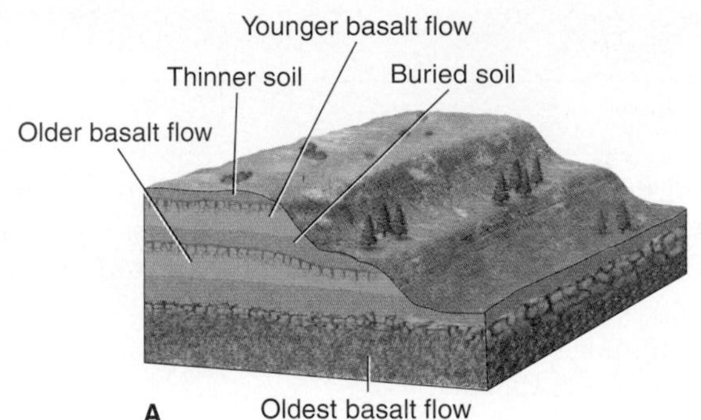

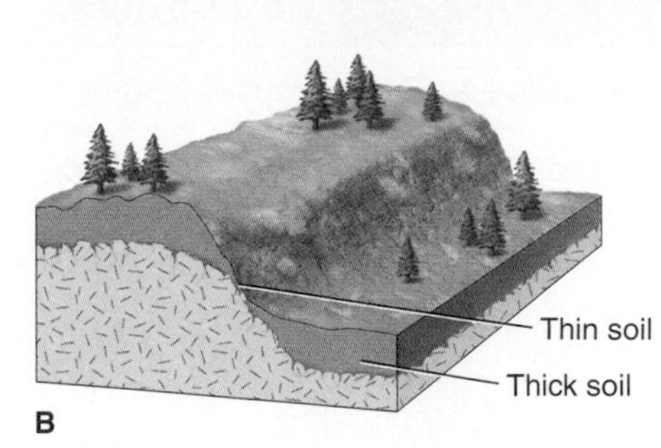

FIGURE 8.21

Soil thickness. (*A*) Soil thickens with time. Oldest basalt flow was exposed to soil-forming processes for a longer time and has a thicker soil developed on it than younger flows. The soils developed below the younger basalt flow are examples of a buried soil. (*B*) Steep slopes have thin soil.

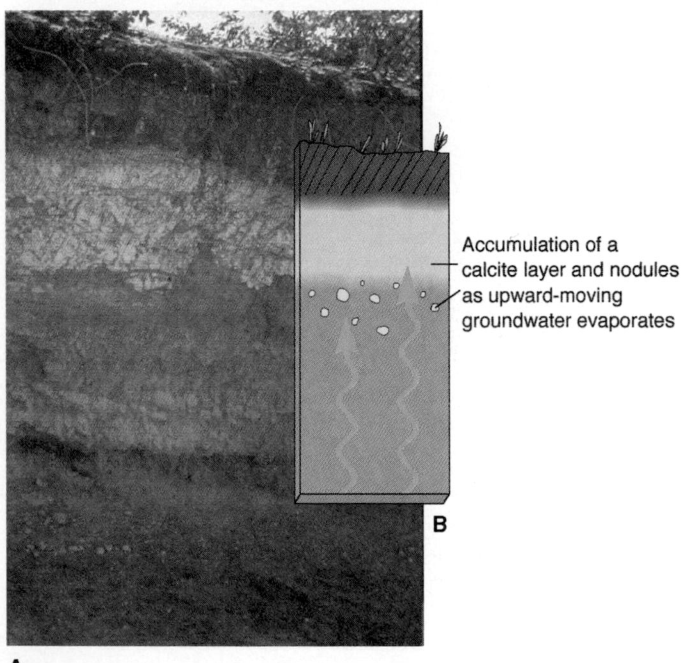

FIGURE 8.22

Soil profile marked by upward-moving groundwater that evaporates underground in a drier climate, precipitating salts within the soil, often forming a light-coloured layer. *Photo by D. Yost, USDA-Soil Conservation Service*

The evaporation of water beneath the land surface can cause the precipitation of salts within the soil. These salts are usually calcium salts such as calcite (figure 8.22). An extreme example of salt buildup can be found in desert *alkali soils,* in which heavy concentrations of toxic sodium salts may prevent plant growth.

Hardpan

"Hardpan" is a general term for a hard layer of Earth material that is difficult to dig or drill. Geoscientists usually restrict the term to a hard, often clayey, layer of cemented soil particles. Such a layer may be too hard for even backhoes to dig through; planting a tree in a lawn with a hardpan layer may require a jackhammer. Hardpan layers in wet climates are usually formed of clay minerals, silica, and iron compounds that have accumulated in the B horizon. In arid climates a different type of hardpan forms from the cementing of soil by calcium carbonate and other salts that precipitate in the soil as water evaporates. Both types of hardpan are really layers of rock within loose soil. A hardpan layer can break ploughs, prevent water drainage through the soil, and act as a barrier to plant roots. Tree roots may grow laterally along rather than down through hardpan; such shallow-rooted trees are easily uprooted by wind.

Laterites

In tropical regions where temperatures are high and rainfall is abundant, highly leached soils called **laterites** (*oxisols*) form. Under such conditions weathering is deep and intense. Laterites are usually red and are composed almost entirely of iron and aluminum oxides, generally the least soluble products of rock weathering in tropical climates (figure 8.23). If the soil is rich in hematite it can be mined as iron ore, but tropical rainfall usually hydrates the hematite to limonite, which is seldom rich enough to mine. However, aluminum is sometimes found in nearly pure layers of *bauxite* ($Al_2O_3 \cdot nH_2O$, the principal ore of aluminum), particularly in laterites formed by the weathering of aluminum-rich volcanic tuffs.

Under tropical conditions of high rainfall and high temperature, most weathering products are soluble—even silica. The least soluble product is aluminum oxide, which remains on top of the weathering rocks, forming bauxite in a soil very rich in aluminum. Like the diamonds, the aluminum has been concentrated residually by the removal of everything else. The aluminum ores may be redistributed slightly by running water (figure 8.24). Because bauxite forms under conditions of tropical weathering, the United States has very little aluminum ore (Canada has none) and depends almost entirely on recycling and tropical countries for its aluminum supply.

Laterites are relatively nonproductive soils. This may seem strange when you think of the lush jungle growth that often exists on tropical lateritic soils. Jungle vegetation, though, is nourished largely by a layer of humus on top of the soil. If the jungle and the humus layer are cleared away or burned—an increasingly common practice in tropical regions—the laterite quickly becomes incapable of sustaining plant growth, making tropical agriculture very difficult. Laterite exposed to the sun is apt to bake into a permanent, bricklike layer that makes digging nearly impossible. This hard layer can be quarried, however, and makes a durable building material.

FIGURE 8.23

Laterite soil (oxisol) develops in very wet climates, where intense, downward leaching carries away all but iron and aluminum oxides. Many laterites are a rusty orange to deep red colour from the oxidation of the iron oxides.

Photo by USDA Natural Resources Conservation Service

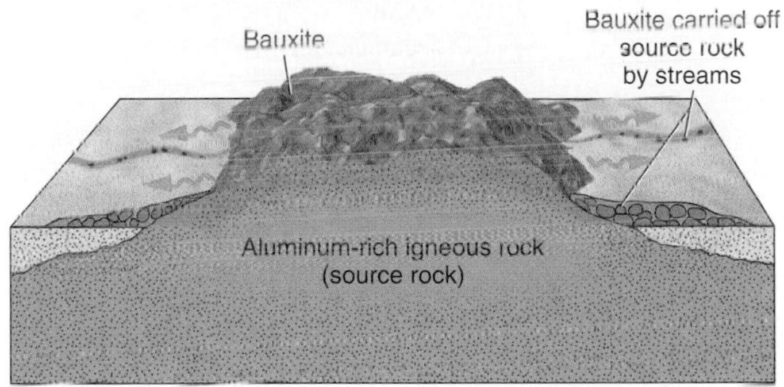

FIGURE 8.24

Bauxite forms by intense tropical weathering of an aluminum-rich source rock such as a volcanic tuff.

Buried Soils

A soil may become buried by volcanic ash, windblown dust, glacial deposits, other sediment, or lava (see figure 8.21). A buried soil is called a *paleosol* (paleo means ancient). Such soils may be distinctive and traceable over wide regions, making them useful for dating rocks and sediments, and for interpreting past climates and topography (figure 8.25).

FIGURE 8.25

Series of paleosols (dark layers) in alluvial sediments of the Peace River, Alberta. Lowermost layer is 7,500 years old; top layers were deposited within the past few centuries.

Photo by P.T. Bobrowsky

SUMMARY

When rocks that formed deep in the Earth become exposed at the Earth's surface, they are altered by *mechanical* and *chemical weathering*.

Weathering processes produce *spheroidal weathering, differentially weathered* landforms, *sheet joints*, and *exfoliation domes*.

Mechanical weathering, largely caused by *frost action* and *pressure release* after unloading, disintegrates (breaks) rocks into smaller pieces.

By increasing the exposed surface area of rocks, mechanical weathering helps speed chemical weathering.

Chemical weathering results when a mineral is unstable in the presence of water and atmospheric gases. As chemical weathering proceeds, the mineral's components recombine into new minerals that are more in equilibrium.

Weak acid, primarily from the solution of carbon dioxide in water, is the most effective agent of chemical weathering.

Calcite dissolves when it is chemically weathered. Most of the silicate minerals form *clay minerals* when they chemically weather. Quartz is very resistant to chemical weathering.

Soil develops by chemical and mechanical weathering of a parent material. Some definitions of soil require that it contain organic matter and be able to support plant growth.

Soils, which can be *residual* or *transported*, usually have distinguishable layers, or *horizons*, caused in part by water movement within the soil.

Climate is the most important factor determining soil type. Other factors in soil development are parent material, time, slope, and organic activity.

Laterites form under conditions of intense tropical weathering; they are usually red from concentrated iron oxides. Bauxite, the ore of aluminum, may be found in laterites.

Terms to Remember

A horizon (zone of leaching) 213
B horizon (zone of accumulation) 216
C horizon 216
chemical weathering 203
clay mineral 211
differential weathering 203
erosion 203
exfoliation 206
exfoliation dome 206
frost action 206
frost heaving 207
frost wedging 206
hematite 208
laterite 218
LFH 213
limonite 208
loam 213
mechanical weathering 203
O horizon 213
pressure release 205
residual soil 216
sheet joints 206
soil 213
soil horizon 213
spheroidal weathering 203
talus 203
transportation 203
transported soil 216
weathering 203
zone of accumulation 216
zone of leaching 213

Testing Your Knowledge

Use the questions below to prepare for exams based on this chapter.

1. Why are some minerals stable several kilometres underground but unstable at Earth's surface?
2. Describe what happens to each mineral within granite during the complete chemical weathering of granite in a humid climate. List the final products for each mineral.
3. Explain what happens chemically when calcite dissolves. Show the reaction in a chemical equation.
4. Why do stone buildings tend to weather more rapidly in cities than in rural areas?
5. Describe at least three processes that mechanically weather rock.
6. How can mechanical weathering speed up chemical weathering?
7. Name at least three natural sources of acid in solution. Which one is most important for chemical weathering?
8. What is the difference between a residual soil and a transported soil?
9. What is a laterite and how does it form?
10. What are the soil horizons? How do they form?
11. Physical disintegration of rock into smaller pieces is called
    a. chemical weathering
    b. transportation
    c. deposition
    d. mechanical weathering

12. The decomposition of rock from exposure to water and atmospheric gases is called
    a. chemical weathering
    b. transportation
    c. deposition
    d. mechanical weathering
13. Which is not a type of mechanical weathering?
    a. frost wedging
    b. frost heaving
    c. pressure release
    d. oxidation
14. The single most effective agent of chemical weathering at Earth's surface is
    a. carbonic acid H_2CO_3
    b. water H_2O
    c. carbon dioxide CO_2
    d. hydrochloric acid HCl
15. The most common end product of the chemical weathering of feldspar is
    a. clay minerals
    b. pyroxene
    c. amphibole
    d. calcite
16. The most common end product of the chemical weathering of quartz is
    a. clay minerals
    b. pyroxene
    c. amphibole
    d. calcite
    e. quartz does not usually weather chemically
17. Soil with approximately equal amounts of sand, silt, and clay along with a generous amount of organic matter is called
    a. loam
    b. inorganic
    c. humus
    d. caliche
18. Which is characteristic of soil horizons?
    a. they can be distinguished from one another by appearance and chemical composition
    b. boundaries between soil horizons are usually transitional rather than sharp
    c. they are classified by letters
    d. all of the preceding
19. The soil horizon containing only organic material is the
    a. A horizon
    b. B horizon
    c. C horizon
    d. LFH horizon
20. Hardpan forms in the
    a. A horizon
    b. B horizon
    c. C horizon
    d. LFH horizon
21. Tropical soils are typically
    a. rich in organic material
    b. very fertile
    c. deeply leached
    d. easily replenished

Exploring Web Resources

www.mcgrawhill.ca/olc/plummer
Visit the Online Learning Centre to review your Testing Your Knowledge answers. This Website also has additional quizzing, direct links to the sites listed below, as well as reading articles and other media resources.

http://soils.ag.uidaho.edu/soilorders/
University of Idaho Soil Science Division. Web page contains photos, descriptions, and surveys of the 12 major soil orders.

http://res.agr.ca/cansis/_overview.html
Canadian Soil Information System provides links to detailed soil surveys and land inventories.

CHAPTER

9

Sediments and Sedimentary Rocks

What is sediment?
What happens when sediment is moved?
How does sediment change into a sedimentary rock?
What are the main types of sedimentary rocks?
How do oil and gas form?
What do sedimentary structures tell us about depositional conditions?

You saw in chapter 8 how weathering produces sediment. In this chapter, we explain more about sediment origin, as well as the erosion, transportation, sorting, deposition, and eventual transformation of sediments to sedimentary rock. Because they have such diverse origins, sedimentary rocks are difficult to classify. We divide them into detrital, chemical, and organic sedimentary rocks, but this classification does not do justice to the great variety of sedimentary rock types.

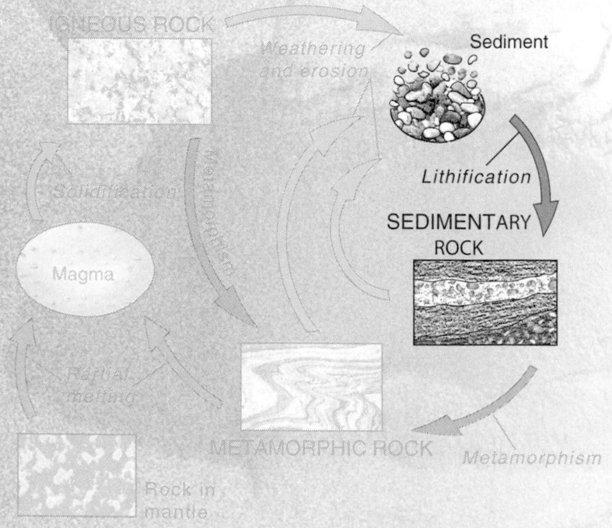

Layered sedimentary rock. *Photo by Nick Eyles.*

Sedimentary rocks contain sedimentary structures such as ripple marks, crossbeds, and mud cracks, as well as the fossilized remains of extinct organisms. These features, combined with knowledge of the sediment types and the sequence of rock layers, allow geoscientists to interpret the environments in which the rocks were deposited. About 75 percent of the surface of the continents is blanketed by sedimentary rock, providing geologists with the information they need to reconstruct a detailed history of the surface of Earth.

Sedimentary rocks are also economically important. Most building materials such as stone, concrete, silica (glass), gypsum (plaster), and iron are quarried and mined from sedimentary rock. Salt is also a sedimentary product and, in many places in the world, supplies of fresh water are pumped from sedimentary layers. Coal, crude oil, and natural gas, the fossil fuels that drove the industrial revolution and that power our technological society, are all formed within and extracted from sedimentary rock.

WHAT IS SEDIMENT?

Most sedimentary rocks form from loose grains or chemical precipitation of sediment. Sediment includes such particles as sand on beaches, mud on a lake bottom, boulders frozen into glaciers, pebbles in streams, and dust particles settling out of the air. An accumulation of clam shells on the sea bottom offshore is sediment, as are coral fragments broken from a reef by large storm waves.

Sediment is the collective name for loose, solid particles that originate from:

1. Weathering and erosion of pre-existing rocks.
2. Chemical precipitation from solution, including secretion by organisms in water.

These particles usually collect in layers on the Earth's surface. An important part of the definition is that the particles are loose. Sediments are said to be *unconsolidated*, which means that the grains are separate, or unattached to one another.

Sediment particles are classified and defined according to the size of individual fragments. Table 9.1 shows the precise definitions of particles by size.

Gravel includes all rounded particles coarser than 2 mm in diameter, the thickness of a nickel. (Angular fragments of this size are called *rubble*.) *Pebbles* range from 2 to 64 mm (about the size of a tennis ball). *Cobbles* range from 64 to 256 mm (about the size of a basketball), and *boulders* are coarser than 256 mm.

Sand grains are from 1/16 mm (about the thickness of a human hair) to 2 mm in diameter. Grains of this size are visible and feel gritty between the fingers. **Silt** grains are from 1/256 to 1/16 mm. They are too small to see without a magnifying device, such as a geoscientist's hand lens. Silt does not feel gritty between the fingers, but it does feel gritty between the teeth (geoscientists often bite sediments to test their grain size). **Clay** is the finest sediment, at less than 1/256 mm, too fine to feel gritty to fingers or teeth. *Mud* is a term loosely used for a mixture of silt and clay.

TABLE 9.1 Sediment Particles and Clastic Sedimentary Rocks

Diameter (mm)	Sediment		Sedimentary Rock
	Boulder	Gravel	**Breccia** (angular particles) or **Conglomerate** (rounded particles)
256	Cobble		
64	Pebble		
2	Sand		**Sandstone**
0.0625	Silt	"Mud"	Siltstone (mostly silt)
0.004	Clay		**Shale** or mudstone (mostly clay)

Sandstone and shale are quite common; the others are relatively rare.

Note that we have two different uses of the word *clay*—a *clay-sized particle* (table 9.1) and a *clay mineral*. A clay-sized particle can be composed of any mineral at all provided its diameter is less than 1/256 mm. A clay mineral, on the other hand, is one of a small group of silicate minerals with a sheet-silicate structure. Clay minerals usually fall into the clay-size range.

Quite often the composition of sediment in the clay-size range turns out to be mostly clay minerals, but this is not always the case. Because of its resistance to chemical weathering, quartz may show up in this fine-size grade. (Most silt is quartz.) Intense mechanical weathering can break down a wide variety of minerals to clay size, and these extremely fine particles may retain their mineral identity for a long time if chemical weathering is slow. The great weight of glaciers is particularly effective at grinding minerals down to the clay-size range, producing "rock flour," which gives a milky appearance to glacial meltwater streams (see chapter 16).

Weathering, erosion, and transportation are some of the processes that affect the character of sediment. Both mechanically weathered and chemically weathered rock and sediment can be eroded, and weathering continues as erosion takes place. Sand being transported by a river also can be actively weathering, as can mud on a lake bottom. The character of sediment can also be altered by *rounding* and *sorting* during transportation, and by eventual *deposition*.

WHAT HAPPENS WHEN SEDIMENT IS MOVED?

Transportation

Most sediment is **transported** some distance by gravity, wind, water, or ice before coming to rest and settling into layers. During transportation, sediment continues to weather and change in character in proportion to the distance the sediment is moved. **Rounding** is the grinding away of sharp edges and corners of rock fragments during transportation. Rounding occurs in sand

and gravel as rivers, glaciers, or waves cause particles to hit and scrape against one another (figure 9.1) or against a rock surface, such as a rocky streambed. Boulders in a stream may show substantial rounding in less than two kilometres of travel. Because rounding during transportation is so rapid, it is a much more important process than spheroidal weathering (see chapter 8), which also tends to round off sharp edges.

Sorting is the process by which sediment grains are selected and separated according to grain size (or grain shape or specific gravity) by the agents of transportation, especially by running water. Because of their high viscosity and manner of flow, glaciers are poor sorting agents. Glaciers deposit all sediment sizes in the same place, so glacial sediment usually consists of a mixture of clay, silt, sand, and gravel. Such glacial sediment is considered *poorly sorted.* Sediment is considered *well-sorted* when the grains are nearly all the same size. A river, for example, is a good sorting agent, separating sand from gravel, and silt and clay from sand. Sorting takes place because of the greater weight of larger particles. Boulders weigh more than pebbles and are more difficult for the river to transport, so a river must flow more rapidly to move boulders than to move pebbles. Similarly, pebbles are harder to move than sand, and sand is harder to move than silt and clay.

Figure 9.2 shows the sorting of sediment by a river as it flows out of steep mountains onto a gentle plain, where the water loses energy and slows down. As the river loses energy, the heaviest particles of sediment are deposited. The boulders come to rest first (figure 9.3). As the river continues to slow and becomes less turbulent, cobbles and then pebbles are deposited. Sand comes to rest as the river loses still more energy (figure 9.4). Finally, the river is carrying only the finest sediment—silt and clay (figure 9.5). The river has sorted the original sediment mix by grain size.

FIGURE 9.1
These boulders have been rounded by abrasion as wave action rolled them against one another on this beach.

Deposition

When transported material settles or comes to rest, **deposition** occurs. Sediment is deposited when running water, glacial ice, waves, or wind loses energy and can no longer transport its load.

Deposition also refers to the accumulation of chemical or organic sediment, such as clam shells on the sea floor, or plant material on the floor of a swamp. Such sediments may form as organisms die and their remains accumulate, perhaps with no transportation at all. Deposition of salt crystals can take place as sea water evaporates. A change in the temperature, pressure, or chemistry of a solution may also cause precipitation—hot springs may deposit calcite or silica as the warm water cools.

The **environment of deposition** is the location in which deposition occurs. A few examples of environments of deposition are the deep sea floor, beneath a glacier, a river channel, a coral reef, a lake bottom, a beach, and a delta. Each environment is marked by characteristic physical, chemical, and biological

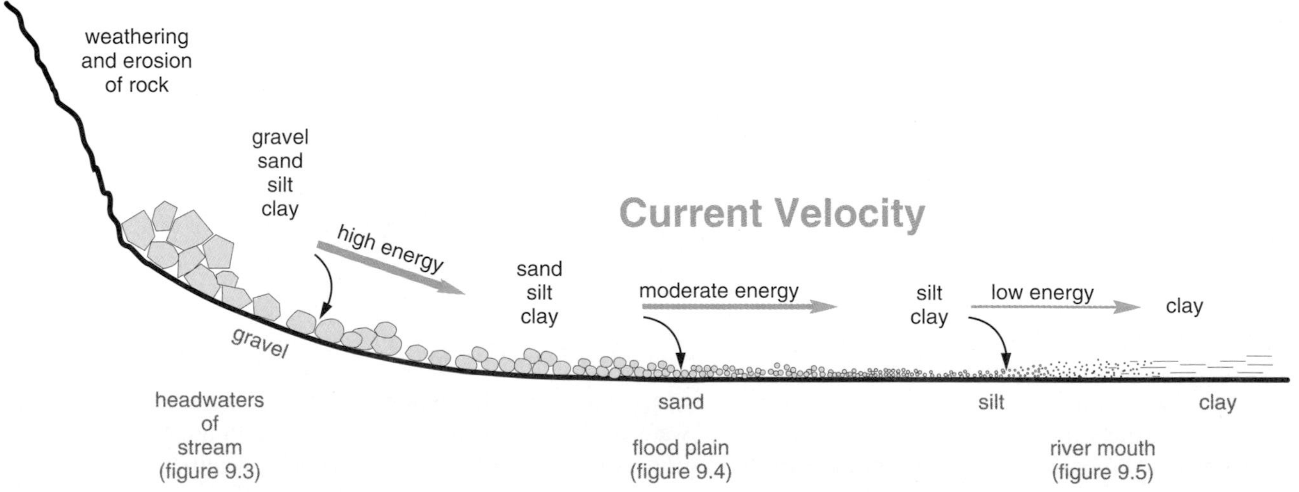

FIGURE 9.2
Sorting of sediment by a river. The coarse sediment is deposited first, and the finest sediment is carried in farthest.

FIGURE 9.3
Coarse gravel (boulder size) is deposited first along a river's course as the river sorts out the various sediment sizes. River gravel is usually deposited in or near steep mountains. Red Rock Canyon, Waterton Lakes National Park, Alberta.
Photo by M. Radomski

conditions. You might expect mud on the sea floor to differ from mud on a lake bottom. Sand on a beach may differ from sand in a river channel. Some differences are due to varying sediment sources and transporting agents, but some are the result of conditions in the environments of deposition themselves.

One of the most important jobs of geoscientists studying sedimentary rocks is to try to determine the ancient environment of deposition of the sediment in which the rock formed. Factors that can help in determining this are a detailed knowledge of modern environments, the shape and vertical sequences of rock layers in the field, the features (including fossils) found within the rock, the mineral composition of the rock, and the size, shape, and surface texture of the individual sediment grains. Later in the chapter we give a few examples of interpreting sedimentary rocks.

Preservation

Not all sediments are preserved as sedimentary rock. Gravel in a river may be deposited when a river is low, but then may be eroded and transported by the next flood on the river. Many sediments on land, particularly those well above sea level, are easily eroded and carried away, so they are not commonly preserved. Sediments on the sea floor are easier to preserve. In general, continental and marine sediments are most likely to be preserved if they are deposited in a *subsiding* (sinking) *basin* and if they are covered or *buried by later sediments*.

FIGURE 9.4
Deposition of sand occurs as a river loses energy as it flows across a gentle plain.

FIGURE 9.5

The river on the right is carrying only silt and clay as it enters the clear river on the left. This fine sediment may come to rest at the mouth of a river where it enters a lake or the sea.

Photo by C. W. Montgomery

HOW DOES SEDIMENT CHANGE INTO A SEDIMENTARY ROCK?

Lithification is the general term for the processes that convert loose sediment into sedimentary rock. Most sedimentary rocks are lithified by a combination of *compaction,* which packs loose sediment grains tightly together, and *cementation,* in which the precipitation of cement around sediment grains binds them into a firm, coherent rock. *Crystallization* of minerals from solution, without passing through the loose-sediment stage, is another way that rocks may be lithified. Some layers of sediment persist for tens of millions of years without becoming fully lithified. Usually, layers of *partially lithified sediment* have been buried deep enough to become compacted, but have not experienced the conditions required for cementation.

As sediment grains settle slowly in a quiet environment such as a lake bottom, they form an arrangement with a great deal of open space between the grains (figure 9.6*A*). The open spaces between grains are called *pores,* and in a quiet environment, a deposit of sand may have 40 percent to 50 percent of its volume as open **pore space.** (If the grains were travelling rapidly and impacting one another just before deposition, the percentage of pore space will be less.) As more and more sediment grains are deposited on top of the original grains, the increasing weight of this *overburden* packs the original grains together, reducing the amount of pore space. This shift to a tighter packing, with a resulting decrease in pore space, is called **compaction** (figure 9.6*B*). As pore space decreases, some of the interstitial water that usually fills sediment pores is driven out of the sediment.

As underground water moves through the remaining pore space, solid material called **cement** can precipitate in the pore space and bind the loose sediment grains together to form a solid rock. The cement attaches very tightly to the grains, holding them in a rigid framework. As cement partially or completely fills the pores, the total amount of pore space is further reduced (figure 9.6*C*), and the loose sand forms a hard, coherent sandstone by **cementation.**

Sedimentary rock cement is often composed of the mineral calcite or of other carbonate minerals. Dissolved calcium and bicarbonate ions are common in surface and underground waters. If the chemical conditions are right, these ions may recombine to form solid calcite, as shown in the following reaction.

$$\underbrace{Ca^{++} + 2HCO_3^{-}}_{\text{dissolved ions}} \rightarrow \underbrace{CaCO_3}_{\text{calcite}} + H_2O + CO_2$$

Silica is another common cement. Iron oxides and clay minerals can also act as cement but are less common than calcite and silica. The dissolved ions that precipitate as cement originate from the chemical weathering of minerals such as feldspar and calcite. This weathering may occur within the sediments being cemented, or at a very distant site, with the ions being transported tens or even hundreds of kilometres by water before precipitating as solid cement.

A sedimentary rock that consists of sediment grains bound by cement into a rigid framework is said to have a **clastic texture.** Usually such a rock still has some pore space because cement rarely fills the pores completely (figure 9.6*C*).

Some sedimentary rocks form by **crystallization,** the development and growth of crystals by precipitation from solution at or near Earth's surface (the term is also used for igneous rocks that crystallize as magma cools). These rocks have a **crystalline texture,** an arrangement of interlocking crystals that develops as crystals grow and interfere with each other. Crystalline rocks lack cement. They are held together by the interlocking of crystals. Such rocks have no pore space because the crystals have grown until they fill all available space. Some sedimentary rocks with a crystalline texture are the result of *recrystallization,* the growth of new crystals that form from and then destroy the original clastic grains of a rock that has been buried (figure 9.7).

WHAT ARE THE MAIN TYPES OF SEDIMENTARY ROCKS?

Sedimentary rocks are formed from (1) lithification of sediment, (2) precipitation from solution, or (3) consolidation of the remains of plants or animals. These different types of sedimentary rocks are called, respectively, *clastic, chemical,* and *organic* rocks. (Some kinds of sedimentary rocks have commercial value; see box 9.1.)

Most sedimentary rocks are **clastic sedimentary rocks,** formed from cemented sediment grains that are fragments of pre-existing rocks. The rock fragments can be either identifiable

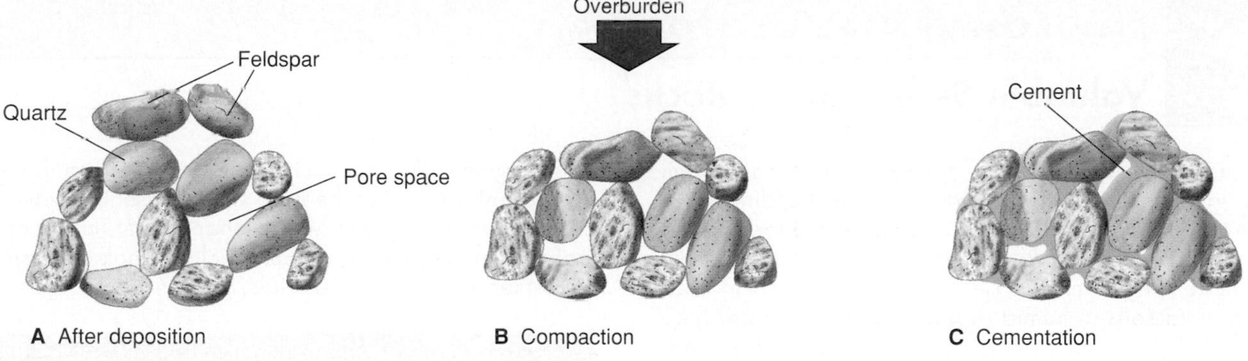

FIGURE 9.6
Lithification of sand grains to become sandstone. (*A*) Loose sand grains are deposited with open pore space between the grains. (*B*) The weight of overburden compacts the sand into a tighter arrangement, reducing pore space. (*C*) Precipitation of cement in the pores by groundwater binds the sand into the rock sandstone, which has a clastic texture.

FIGURE 9.7
Crystalline dolostone as seen through a polarizing microscope. Note the interlocking crystals of dolomite mineral that grew as they precipitated during recrystallization. Such crystalline sedimentary textures have no cement or pore spaces.
Photo by Bret Bennington

pieces of rock, such as pebbles of granite or shale, or individual mineral grains, such as sand-sized quartz and feldspar crystals loosened from rocks by weathering and erosion. Clay minerals formed by chemical weathering are also considered fragments of pre-existing rocks. During transportation the grains may have been rounded and sorted. Table 9.1 shows the clastic rocks, such as conglomerate, sandstone, and shale, and shows how these rocks vary in grain size.

Chemical sedimentary rocks are deposited by precipitation of minerals from solution. An example of inorganic precipitation is the formation of *rock salt* as seawater evaporates. Chemical precipitation can also be caused by organisms. The sedimentary rock *limestone,* for instance, can form by the precipitation of calcite within a coral reef by corals and algae. Such a rock is classified as a *biochemical* limestone.

Not all chemical sedimentary rocks accumulate as sediment. Some limestones are crystallized as solid rock by corals and coralline algae in reefs. Chert crystallizes in solid masses within some layers of limestone. Rock salt may crystallize directly as a solid mass or it may form from the crystallization of individual salt crystals that behave as sedimentary particles until they grow large enough to interlock into solid rock.

Organic sedimentary rocks are rocks that are composed of organic carbon compounds. *Coal* is an organic rock that forms from the compression of plant remains, such as moss, leaves, twigs, roots, and tree trunks.

Appendix B describes and helps you identify the common sedimentary rocks. The standard geologic symbols for these rocks (such as dots for sandstone, and a "brick-wall" symbol for limestone) are shown in appendix F and will be used in the remainder of the book.

Clastic Sedimentary Rocks

Breccia and Conglomerate

Sedimentary breccia is a coarse-grained sedimentary rock formed by the cementation of coarse, angular fragments of rubble (figure 9.8). Because grains are rounded so rapidly during transport, it is unlikely that the angular fragments within breccia have moved very far from their source. Sedimentary breccia might form from fragments that have accumulated at the base of a steep slope of rock that is being mechanically weathered. Landslide deposits also might lithify into sedimentary breccia. This type of rock is not particularly common.

Conglomerate is a coarse-grained sedimentary rock formed by the cementation of rounded gravel. It can be distinguished from breccia by the definite roundness of its particles (figure 9.9). Because conglomerates are coarse-grained, the particles may not have travelled far; but some transport was necessary to round the particles. Angular fragments that fall from a cliff and then are carried a few kilometres by a river or pounded by waves crashing in the surf along a beach are quickly rounded. Gravel that is transported down steep submarine canyons, or carried by glacial ice, however, can be transported tens or even hundreds of kilometres before deposition.

ENVIRONMENTAL GEOLOGY 9.1

Valuable Sedimentary Rocks

Many sedimentary rocks have uses that make them valuable. *Limestone* is widely used as building stone and is also the main rock type quarried for crushed rock for road construction. Pulverized limestone is the main ingredient of cement for mortar and concrete and is also used to neutralize acid soils in humid regions. *Coal* is a major fuel, used widely for generating electrical power and for heating. Plaster and plasterboard for home construction are manufactured from *gypsum,* which is also used to stabilize the shrink–swell characteristics of clay-rich soils in some areas. Huge quantities of *rock salt* (box figure 1) are consumed by industry, primarily for the manufacture of hydrochloric acid. More familiar uses of rock salt are for table salt and for melting ice on roads.

Some *chalk* is used in the manufacture of blackboard chalk, although most classroom chalk is now made from pulverized limestone. The filtering agent for beer brewing and for swimming pools is likely to be made of *diatomite,* an accumulation of the siliceous remains of microscope diatoms.

Clay from *shale* and other deposits supplies the basic material for ceramics of all sorts, from hand-thrown pottery and fine porcelain to sewer pipe. *Sulphur* is used for matches, fungicides, and sulphuric acid; and *phosphates* and *nitrates* for fertilizers are extracted from natural occurrences of special sedimentary rocks (although other sources also are used). Potassium for soap manufacture comes largely from *evaporites,* as does boron for heat-resistant cookware and fibreglass, and sodium for baking soda, washing soda, and soap. *Quartz sandstone* is used in glass manufacturing and for building stone.

Many *metallic ores,* such as the most common iron ores, have a sedimentary origin. The pore space of sedimentary rocks acts as a reservoir for groundwater (chapter 15), crude oil, and natural gas. In chapter 12 we take a closer look at these resources and other useful Earth materials.

BOX 9.1 ■ FIGURE 1

Underground salt-mining operations near Goderich, Ontario. Pillars of salt are left in place to support overlying strata, a mining method called "pillar and stall."

Photo by Nick Eyles

FIGURE 9.8

Breccia is characterized by coarse, angular fragments. The cement in this rock is coloured by hematite. The wide black and white bars on the scale are 1 cm long, the small divisions are 1 mm. Note that most grains exceed 2 mm (table 9.1).

Photo by David McGeary

FIGURE 9.9

An outcrop of conglomerate. Note the rounding of cobbles, which vary in composition. Long scale bar 10 cm, short bars 1 cm.

Photo by David McGeary

Sandstone

Sandstone is formed by the cementation of sand grains (figure 9.10). Any deposit of sand can lithify to sandstone. Rivers deposit sand in their channels, and wind piles up sand into dunes. Waves deposit sand on beaches and in shallow water. Deep-sea currents spread sand over the sea floor. As you might imagine, sandstones show a great deal of variation in mineral composition, degree of sorting, and degree of rounding.

Quartz sandstone is a sandstone in which more than 90 percent of the grains are quartz (figure 9.10*A*). Because quartz is resistant to chemical weathering, it tends to concentrate in sand deposits as the less resistant minerals such as feldspar are weathered away. The quartz grains in a quartz sandstone are usually well-sorted and well-rounded because they have been transported for great distances (figure 9.11*A*). Most quartz sandstone was deposited as fluvial, beach, or dune sand.

A sandstone with more than 25 percent of the grains consisting of feldspar is called *arkose* (figure 9.10*B*). Because feldspar grains are preserved in the rock, the original sediment obviously did not undergo severe chemical weathering, or the feldspar would have been destroyed. Mountains of granite in a desert could be a source for such a sediment, for the rapid erosion associated with rugged terrain would allow feldspar to be mechanically weathered and eroded before it is chemically weathered (a dry climate slows chemical weathering). Most arkoses contain coarse, angular grains (figure 9.11*B*), so transportation distances were probably short. An arkose may have been deposited on an alluvial fan, a fan shaped accumulation of sediment that usually forms where a stream emerges from a narrow canyon onto a flat plain at the foot of a mountain range (figure 9.12).

Sandstones may contain a substantial amount of **matrix** in the form of fine-grained silt and clay in the space between larger sand grains (figure 9.13). A matrix-rich sandstone is poorly sorted and often dark in colour. It is sometimes called a "dirty sandstone."

Greywacke (pronounced "grey-wacky") is a type of sandstone in which more than 15 percent of the rock's volume consists of fine-grained matrix (figure 9.10*C* and 9.11*C*). Greywackes are often tough and dense and are generally dark grey or green. The sand grains may be so coated with matrix that they are hard to see, but they typically consist of quartz, feldspar, and sand-sized fragments of other fine-grained sedimentary, volcanic, and metamorphic rocks.

Most greywackes probably formed from sediments transported by **turbidity currents,** dense masses of sediment-laden water that flow downslope along the sea floor. The sediment–water mixture is more dense than clear water, so it is pulled down-slope by gravity until it comes to rest on the sea floor at the base of the slope (figure 9.14). Turbidity currents may be generated by underwater landslides, perhaps triggered by earthquakes, or by violent surface storms such as hurricanes, which stir up bottom sediment.

FIGURE 9.10

Types of sandstone. (*A*) Quartz sandstone; more than 90 percent of the grains are quartz. (*B*) Arkose; the grains are mostly feldspar and quartz. (*C*) Greywacke; the grains are surrounded by dark, fine-grained matrix. (Small scale divisions are 1 millimetre; most of the sand grains are about 1 millimetre in diameter.)

Photos by David McGeary

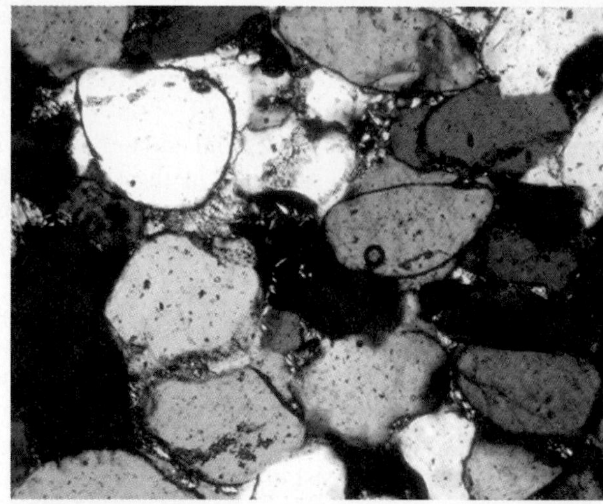
A

B

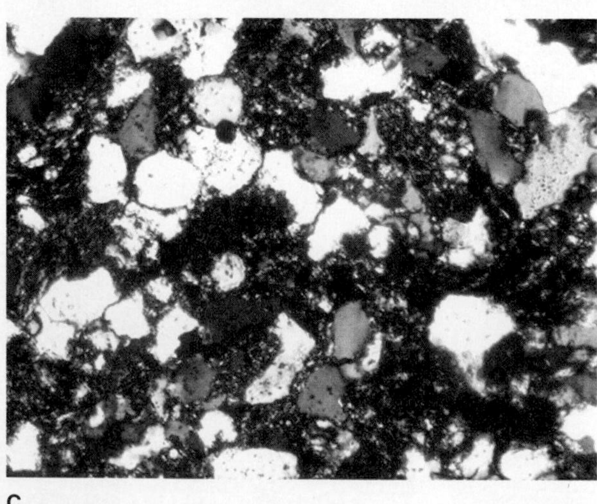
C

FIGURE 9.11

Detrital sedimentary rocks viewed through a polarizing microscope. (*A*) Quartz sandstone; note the well-rounded and well-sorted grains. (*B*) Arkose; large feldspar grain in centre surrounded by angular quartz grains. (*C*) Greywacke; quartz grains surrounded by brownish matrix of mud.

Photos by Bret Bennington

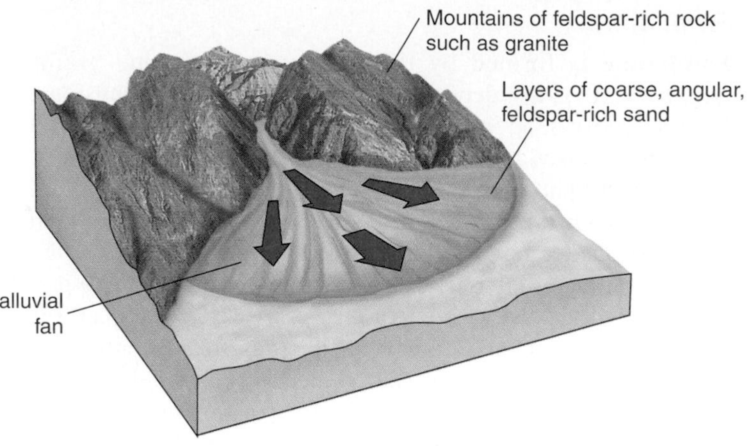

FIGURE 9.12

Feldspar-rich sand (arkose) may accumulate from the rapid erosion of feldspar-containing rock such as granite. Steep terrain accelerates erosion rates so that feldspar may be eroded before it is completely chemically weathered into clay minerals.

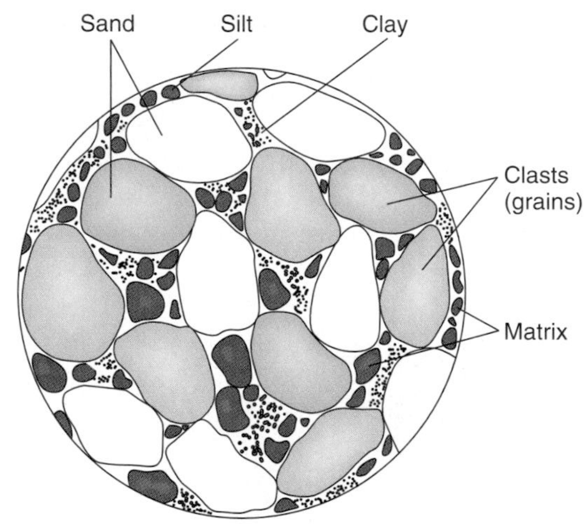

FIGURE 9.13

A poorly sorted sediment of sand grains (clasts) surrounded by a matrix of silt and clay grains. Lithification of such a sediment would produce a "dirty sandstone."

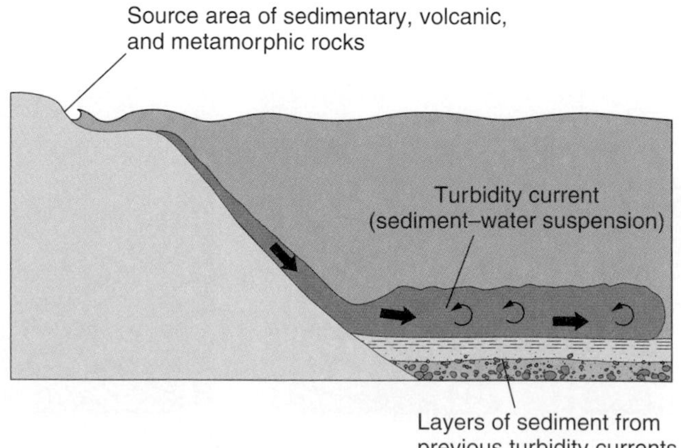

FIGURE 9.14

A turbidity current flow downslope along the sea floor. Sediment-laden water is more dense than the clear water that it flows beneath.

The Fine-Grained Rocks

Rocks consisting of fine-grained silt and clay are called *shale, siltstone, claystone,* and *mudstone.*

Shale is a fine-grained sedimentary rock notable for its ability to split into layers (called *fissility*). Splitting takes place along the surfaces of very thin layers (called *laminations*) within the shale (figure 9.15). Most shales contain both silt and clay (averaging about two-thirds clay-sized clay minerals, one-third silt-sized quartz) and are so fine-grained that the surface of the rock feels very smooth. The silt and clay deposits that lithify as shale accumulate on lake bottoms, at the ends of rivers in deltas, on river flood plains, and on quiet parts of the deep ocean floor.

Fine-grained rocks such as shale typically undergo pronounced compaction as they lithify. Figure 9.16 shows the role of compaction in the lithification of shale from wet mud. Before compaction, as much as 80 percent of the volume of the wet mud may have been pore space filled with water. The flakelike clay minerals were randomly arranged within the mud. Pressure from overlying material packs the sediment grains together and reduces the overall volume by squeezing water out of the pores. The clay minerals are reoriented perpendicular to the pressure, becoming parallel to one another like a deck of cards. The fissility of shale is due to weaknesses between these parallel clay flakes.

Compaction by itself does not generally lithify sediment into sedimentary rock. It does help consolidate clayey sediments by pressing the microscopic clay minerals so closely together that attractional forces at the atomic level tend to bind them together. Even in shale, however, the primary method of lithification is cementation.

A

B

FIGURE 9.15

(A) An outcrop of Queenston Formation shale near Toronto, Ontario. Note how this fine-grained rock tends to split into very thin layers. (B) Shale pieces; note the very fine grain size, very thin layers (laminations) on the edge of the large piece, and tendency to break into small, flat pieces (fissility).

Photo A by Nick Eyles, Photo B by David McGeary

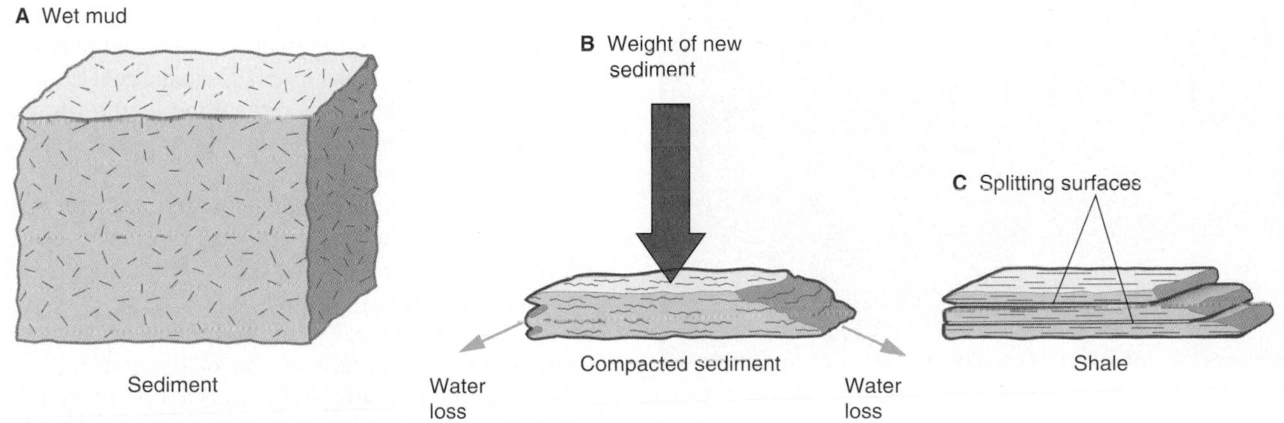

FIGURE 9.16

Lithification of shale from the compaction and cementation of wet mud. (A) Randomly oriented silt and clay particles in wet mud. (B) Particles reorient, water is lost, and pore space decreases during compaction caused by the weight of new sediment deposited on top of the wet mud. (C) Splitting surfaces in cemented shale form parallel to the oriented mineral grains.

A rock consisting mostly of silt grains is called *siltstone.* Somewhat coarser-grained than most shales, siltstones lack the fissility and laminations of shale. *Claystone* is a rock composed predominantly of clay-sized particles, but lacking the fissility of shale. *Mudstone* contains both silt and clay, having the same grain size and smooth feel of shale but lacking shale's laminations and fissility. Mudstone is massive and blocky, while shale is visibly layered and fissile.

Chemical Sedimentary Rocks

Chemical sedimentary rocks are precipitated from an aqueous environment. Chemical sedimentary rocks are precipitated directly either by inorganic processes or by the actions of organisms. Chemical rocks include *carbonates, chert,* and *evaporites.*

Carbonate Rocks

Carbonate rocks contain the CO_3^{2-} ion as part of their chemical composition. The two main types of carbonates are limestone and dolomite.

LIMESTONE

Limestone is a sedimentary rock composed mostly of calcite ($CaCO_3$). Limestones are either precipitated by the actions of organisms or are precipitated directly as the result of inorganic processes. Thus, the two major types of limestone can be classified as either *biochemical* or *inorganic limestone.*

FIGURE 9.17
Corals precipitate calcium carbonate to form limestone in a reef. Water depth about 8 metres, San Salvador Island, Bahamas.
Photo by David McGeary

Biochemical limestones are precipitated through the actions of organisms. Most biochemical limestones are formed on continental shelves in warm, shallow water. Biochemical limestone may be precipitated directly as a solid rock in the core of a reef by corals, encrusting algae, or by other shell-forming organisms (figure 9.17). Such a rock would have a crystalline texture and would contain the fossil remains of organisms preserved in growth position.

The great majority of limestones are biochemical limestones formed of wave-broken fragments of algae, corals, and shells. The fragments may be of any size (gravel, sand, silt, and clay) and are often sorted and rounded as they are transported by waves and currents across the sea floor (figure 9.18). The action of these waves and currents and subsequent cementation of these fragments into rock give these limestones a clastic texture. These *bioclastic* (or *skeletal*) *limestones* take a great variety of appearances. They may be relatively coarse-grained with recognizable fossils (figure 9.19) or uniformly fine-grained and dense from the accumulation of microscopic fragments of coralline algae (figures 9.19 and 9.20). A variety of limestone called *coquina* forms from the cementation of shells and shell fragments that accumulated on the shallow sea floor near shore (figure 9.21). It has a clastic texture and is usually coarse-grained, with easily recognizable shells and shell fragments in it. *Chalk* is a light-coloured, porous, very fine-grained variety of bioclastic limestone that forms from the sea-floor accumulation of microscopic marine organisms that drift near the sea surface (figure 9.22).

Inorganic limestones are precipitated directly as the result of inorganic processes. *Oolitic limestone* is a distinctive variety of inorganic limestone formed by the cementation of sand-sized *oolites* (or *ooids*), small spheres of calcite inorganically precipitated in warm, shallow sea water (figure 9.23). Strong tidal currents roll the oolites back and forth, allowing them to maintain a nearly spherical shape as they grow. Wave action may also contribute to their shape.

Oolitic limestone has a clastic texture. *Tufa* and *travertine* are inorganic limestones that form from fresh water. Tufa is precipitated from solution in the water of a continental spring, lake, or from percolating groundwater. Travertine may form in caves when carbonate-rich water loses CO_2 to the cave atmosphere. Tufa and travertine both have a crystalline texture; however, tufa is generally more porous, cellular, or open than travertine, which tends to be more dense.

Limestones are particularly susceptible to **recrystallization,** the process by which new crystals, often of the same composition as the original grains, develop in a rock. Calcite grains recrystallize easily, particularly in the presence of water and under the weight of overlying sediment. The new crystals that form are often large and can be easily seen in a rock as light reflects off their broad, flat faces. Because recrystallization often destroys the original clastic texture and fossils of a rock, replacing them with a new crystalline texture, the geologic history of such a rock can be very difficult to determine.

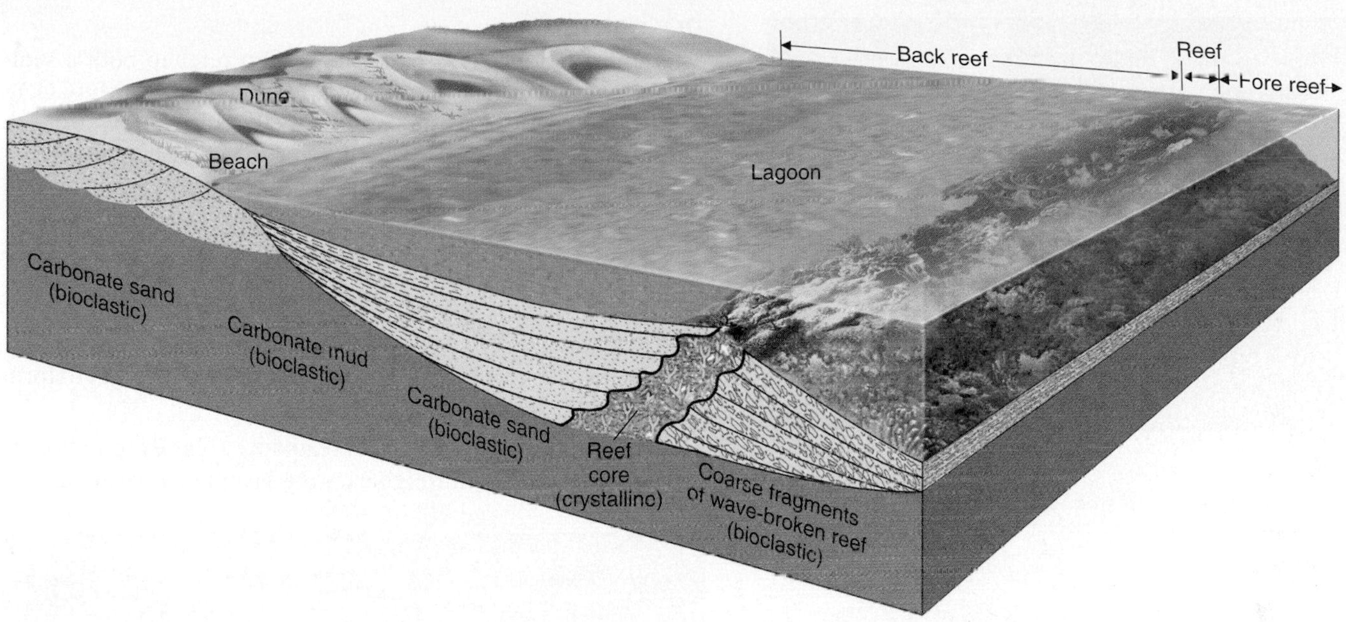

FIGURE 9.18

A living coral-algal reef sheds bioclastic sediment into the fore-reef and back-reef environments. The fore reef consists of coarse, angular fragments of reef. Coralline algae are the major contributors of carbonate sand and mud in the back reef. Beaches and dunes are often bioclastic sand. The sediments in each environment can lithify to form highly varied limestones.

FIGURE 9.19

Bioclastic limestones. The two on the left are coarse-grained and contain visible fossils of corals and shells. The limestone on the right consists of fine-grained carbonate mud formed by coralline algae.

Photo by David McGeary

FIGURE 9.20

Coralline algae on the sea floor in 3 m of water on the Bahama Banks. The "shaving brush" alga is Penicillus, which produces great quantities of fine-grained carbonate mud.

Photo by David McGeary

FIGURE 9.21

Coquina, a variety of bioclastic limestone, is formed by the cementation of shells and coarse shell fragments.

Photo by David McGeary

FIGURE 9.22

Chalk is a fine-grained variety of bioclastic limestone formed of the remains of microscopic marine organisms that live near the sea surface.

Photo by David McGeary

DOLOMITE

The term **dolomite** (table 9.2) is used to refer to both a sedimentary rock and the mineral that composes it, $CaMg(CO_3)_2$. (Some geoscientists use *dolostone* for the rock.) Dolomite often forms from limestone as the calcium in calcite is partially replaced by magnesium, usually as water solutions move through the limestone.

$$\underset{\text{magnesium in solution}}{Mg^{++}} + \underset{\text{calcite}}{2CaCO_3} \rightarrow \underset{\text{dolomite}}{CaMg(CO_3)_2} + \underset{\text{calcium in solution}}{Ca^{++}}$$

Regionally extensive layers of dolomite are thought to form in one of two ways:

1. As magnesium-rich brines created by solar evaporation of sea water trickle through existing layers of limestone.

A

B

FIGURE 9.23

(*A*) Aerial photo of underwater dunes of oolites (ooids) chemically precipitated from seawater on the shallow Bahama Banks, south of Bimini. Tidal currents move the dunes. (*B*) An oolitic limestone formed by the cementation of oolites (small spheres). Small divisions on scale are 1 mm wide.

Photos by David McGeary

TABLE 9.2 Chemical Sedimentary Rocks

Inorganic Sedimentary Rocks			
Rock	**Composition**	**Texture**	**Origin**
Limestone	$CaCO_3$	Crystalline	May be precipitated directly from seawater. Cementation of oolites (ooids) precipitated chemically from warm shallow seawater (*oolitic limestone*). Also forms in caves as *travertine* and in springs, lakes, or percolating groundwater as *tufa*.
Dolomite	$CaMg(CO_3)_2$	Crystalline	Alteration of limestone by Mg-rich solutions (usually)
Evaporites			Evaporation of seawater or a saline lake.
Rock salt	NaCl	Crystalline	
Rock gypsum	$CaSO_4 \cdot 2H_2O$	Crystalline	

Biochemical Sedimentary Rocks			
Rock	**Composition**	**Texture**	**Origin**
Limestone	$CaCO_3$ (calcite)	Clastic or crystalline	Cementation of fragments of shells, corals, and coralline algae (*bioclastic limestone* such as coquina and chalk). Also precipitated directly by organisms in reefs.
Chert	SiO_2 (silica)	Crystalline (usually)	Cementation of microscopic marine organisms; rock usually recrystallized.

2. As chemical reactions take place at the boundary between fresh underground water and sea water; the Mg ions could migrate through layers of limestone as sea level rises or falls.

The dolomitization process causes recrystallization of the pre-existing limestone, resulting in dolomite rock that is hard and very finely crystalline. Original features such as grain size, fossils, and sedimentary structures are often destroyed during recrystallization, making it difficult to interpret the environment of deposition of the original limestone. Dolomitization also involves a reduction in volume of the rock and creates small cavities (vugs) which are often enlarged by weathering (figure 9.24). These cavities may be later filled with oil and gas and vuggy dolostones can form important reservoir rocks.

FIGURE 9.24

Vugs (cavities) in dolostone block from the Niagara Escarpment, Ontario. Block is approximately 1.5 m across.

Photo by C.H. Eyles

Chert

A hard, compact, fine-grained sedimentary rock formed almost entirely of silica, **chert** occurs in two principal forms—as irregular, lumpy nodules within other rocks and as layered deposits like other sedimentary rocks (figure 9.25). The nodules, often found in limestone or dolomite, probably formed from inorganic precipitation as underground water replaced part of the original rock with silica. The layered deposits typically form from the accumulation of hard, shell-like parts of microscopic marine organisms on the sea floor.

Microscopic fossils composed of silica are abundant in some cherts. But because chert is susceptible to recrystallization, the original fossils are easily destroyed, and the origin of many cherts is uncertain.

Evaporites

Rocks formed from crystals that precipitate during evaporation of water are called **evaporites.** They form from the evaporation of seawater or a saline lake (figure 9.26), such as Great Salt Lake in Utah. *Rock gypsum,* formed from the mineral gypsum ($CaSO_4 \cdot 2H_2O$), is a common evaporite. *Rock salt,* composed of the mineral halite (NaCl), may also form if evaporation continues. Other less common evaporites include the borates, potassium salts, and magnesium salts. All evaporites have a crystalline texture.

Organic Sedimentary Rocks

Coal

Coal is a sedimentary rock that forms from the compaction of plant material that has not completely decayed (figure 9.27; see chapter 12). Rapid plant growth and deposition in water with a

A

B

FIGURE 9.25
(*A*) Chert nodules in limestone near Bluefield, West Virginia. (*B*) Bedded chert from the Coast Ranges, California. Camera lens cap (5.5 cm) for scale.
Photo A by Parvinder Sethi; Photo B by David McGeary

FIGURE 9.26
Salt deposited on the floor of a dried-up desert lake, Bonneville salt flats, Utah.
Photo by Diane Carlson

FIGURE 9.27
Coal seam interbedded with shales and sandstones, Joggins, Nova Scotia. Keys (circled) for scale.
Photo by C.H. Eyles

low oxygen content are needed, so shallow swamps or bogs in a temperate or tropical climate are likely environments of deposition. The plant fossils in coal beds include leaves, stems, tree trunks, and stumps with roots often extending into the underlying shales, so apparently most coal formed right at the place where the plants grew. Coal usually develops from *peat,* a brown, lightweight, unconsolidated or semiconsolidated deposit of moss and other plant remains that accumulates in wet bogs. Peat is transformed into coal largely by compaction after it has been buried by sediments.

Partial decay of the abundant plant material uses up any oxygen in the swamp water, so the decay stops and the remaining organic matter is preserved. Burial by sediment compresses the plant material, gradually driving out any water or other volatile compounds. The coal changes from brown to black as the amount of carbon in it increases. Several varieties of coal are recognized on the basis of the type of original plant material and the degree of compaction (chapter 12).

HOW DO OIL AND GAS FORM?

Oil and natural gas seem to originate from organic matter in marine sediment. Microscopic organisms, such as diatoms and other single-celled algae, settle to the sea floor and accumulate in marine mud. The most likely environments for this are restricted basins with poor water circulation, particularly on continental shelves. The organic matter may partially decompose, using up the dissolved oxygen in the sediment. As soon as the oxygen is gone, decay stops and the remaining organic matter is preserved.

Continued sedimentation buries the organic matter and subjects it to higher temperatures and pressures, which convert the organic matter to oil and gas. As muddy sediments compact, the gas and small droplets of oil may be squeezed out of the mud and may move into more porous and permeable sandy layers nearby. Over long periods of time large accumulations of gas and oil can collect in the sandy layers. Both oil and gas are less dense than water, so they generally tend to rise upward through water-saturated rock and sediment. Natural gas represents the end point in petroleum maturation.

Details of the origin of coal, oil, and gas are discussed in chapter 12.

WHAT DO SEDIMENTARY STRUCTURES TELL US ABOUT DEPOSITIONAL CONDITIONS?

Sedimentary structures are features found within sedimentary rock. They usually form during or shortly after deposition of the sediment, but before lithification. Structures found in sedimentary rocks are important because they provide clues that help geologists determine the means by which sediment was transported and also its eventual resting place, or environment of deposition (see box 9.2). Sedimentary structures may also reveal the orientation or upward direction of the deposit, which helps geologists unravel the geometry of rocks that have been folded and faulted in tectonically active regions.

One of the most prominent structures, seen in most large bodies of sedimentary rock, is **bedding,** a series of visible layers within rock (figure 9.28). Most bedding is horizontal because the sediments from which the sedimentary rocks formed were originally deposited as horizontal layers (figure 9.28*A*). The principle of **original horizontality** states that most water-laid sediment is deposited in horizontal or near-horizontal layers that are essentially parallel to Earth's surface. In many cases this is also true for sediments deposited by ice or wind. If each new layer of sediment buries previous layers, a stack of horizontal layers will develop with the oldest layer on the bottom and the layers becoming younger upward. This is the principle of **superposition.** Sedimentary rocks formed from such sediments preserve the horizontal layering in the form of beds (figure 9.28*B*). A **bedding plane** is a nearly flat surface of deposition separating two layers of rock. A change in the grain size or composition of the particles being deposited, or a pause during deposition, can create bedding planes (figure 9.28*B*).

A

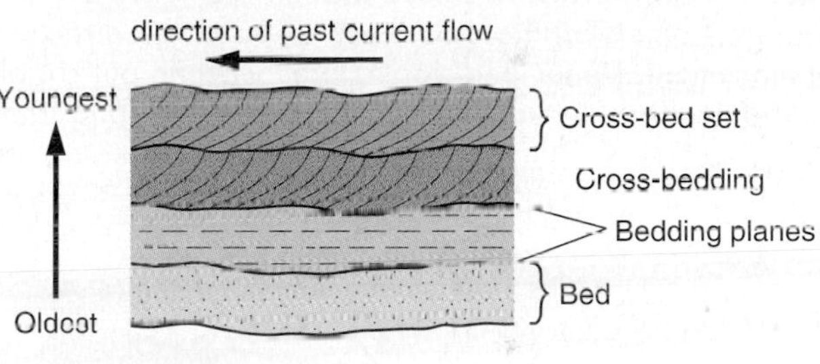

B

FIGURE 9.28

(*A*) Bedding in sandstone and shale, Horseshoe Canyon, Alberta. The horizontal layers formed as one type of sediment buried another in the geologic past. The layers get younger upward. (*B*) Bedding and cross-bedding.
Photo by M. Radomski

In sandstone, a thicker bed of rock will often consist of a series of thinner, inclined beds called **cross-beds** (figure 9.29). Cross-beds form because in flowing air and water, sand grains move as migrating ripples and dunes. Sand is pushed up the shallow side of the ripple to the crest, where it then avalanches down the steep side, forming a cross-bed. Cross-beds form one after the other as the ripple migrates downstream (figure 9.30). Ripples can also be preserved on the surface of a bed of sandstone, forming **ripple marks,** if they are buried by another layer of sediment (figure 9.31). Ripple marks produced by currents flowing in a single direction are asymmetrical (as discussed previously and in figures 9.31*C* and *D*). In waves, water moves back and forth, producing symmetrical wave ripples (figures 9.31*A* and *B*). Ripple marks and cross-beds can form in conglomerates, sandstones, siltstones, and limestones, and in environments such as deserts, river channels, river deltas, and shorelines.

IN GREATER DEPTH 9.2

How Fast Did The Current Flow?

Geoscientists can determine the approximate velocity of a flowing current by examining the grain size of the sediment (generally, the coarser the grain size, the faster the current) and also by examining the sedimentary structures preserved in the deposit. A bedform stability diagram (box figure 1) shows the types of bedforms that form in sediments of different grain sizes at different current velocities and in different water depths. Box figure 1 shows the relationship between current velocity, sediment grain size, and bedforms for a flow approximately 1 m deep. For example, sediment will not move at very low flow velocities on the bed of a sandy stream. When the flow velocities increase to between 20 and 60 cm/sec, small bedforms (ripples: see box figure 2) form in very fine to coarse-grained sands and migrate downstream, forming ripple cross-bedded sands. If flow velocities exceed 50 cm/sec, ripples grow larger and form sand waves or dunes in fine to coarse-grained sands (box figure 3). These larger bedforms produce cross-bedded sandstones and the steeply dipping *forsets* on the cross-beds slope in the direction of current movement. Dunes and ripples are "washed out" in all grain sizes when current velocities exceed 60 cm/sec and form a flat bed (box figure 4). Sand deposition on a flat bed produces horizontally laminated sandstone with bed surfaces marked by elongate ridges (only a few grains high); these ridges form *parting lineation* and may also be used to indicate current direction. Current velocities in excess of 100 cm/sec allow the development of ridges on the bed called *antidunes*; antidunes migrate upstream but are rarely preserved as they are reworked into other bedforms as current velocities decrease.

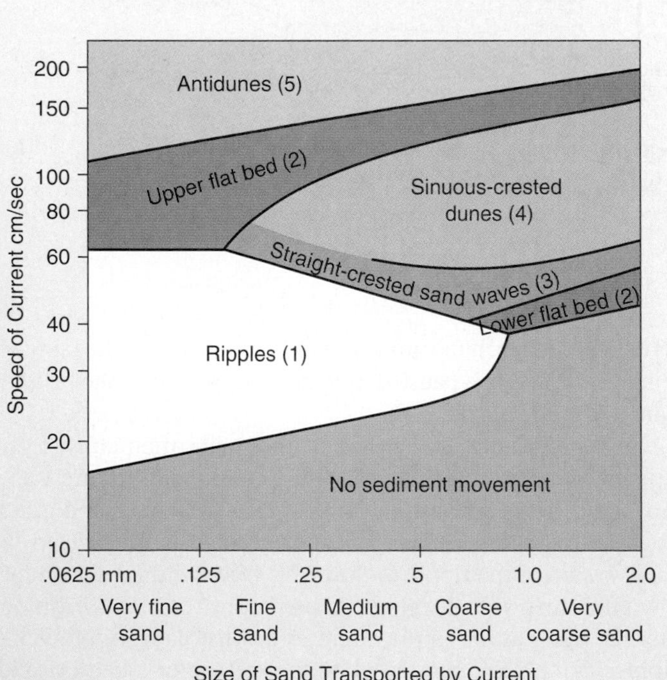

BOX 9.2 ■ FIGURE 1

Bedform stability diagram for flows approximately 1 m deep.

A

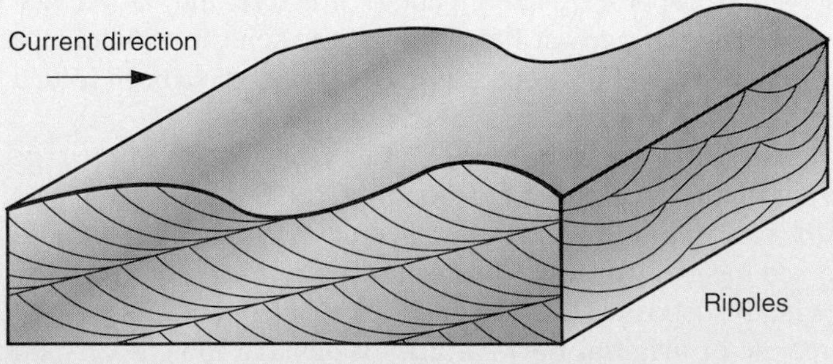

B

BOX 9.2 ■ FIGURE 2

Ripples formed in fine sand.

Photo by Nick Eyles

A

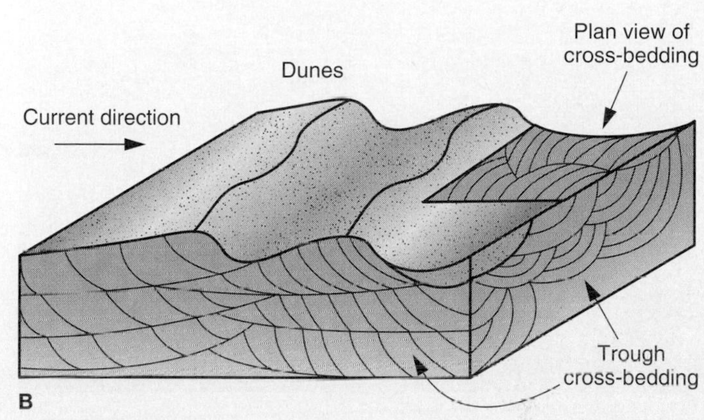

B

BOX 9.2 ■ FIGURE 3

Dunes formed in coarse-grained sands. *Photo by Nick Eyles*

A

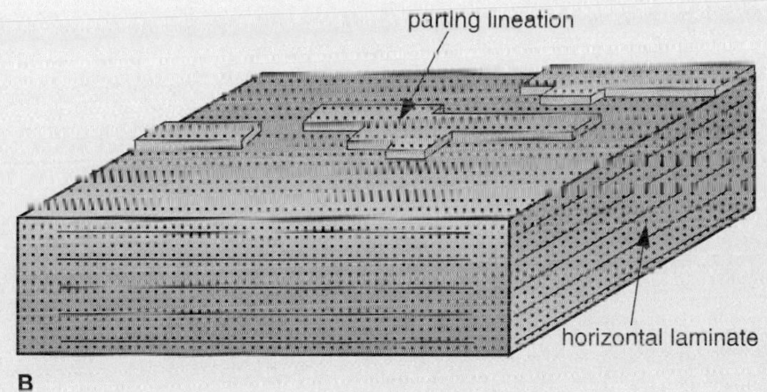

B

BOX 9.2 ■ FIGURE 4

Horizontally laminated coarse-grained sands. *Photo by Nick Eyles*

A **graded bed** is a layer with a vertical change in particle size, usually from coarse grains at the bottom of the bed to progressively finer grains toward the top (figure 9.32). A single bed may have gravel at its base and grade upward through sand and silt to fine clay at the top. A graded bed may build up as sediment is deposited by a gradually slowing current. This seems particularly likely to happen during deposition from a *turbidity current* on the deep sea floor (figure 9.14). Figure 9.33 shows the development of a graded bed by turbidity-current deposition.

Mud cracks are a polygonal pattern of cracks formed in very fine-grained sediment as it dries (figure 9.34). Because drying requires air, mud cracks form only in sediment exposed above water. Mud cracks may form in lake-bottom sediment as the lake dries up; in flood-deposited sediment as a river level drops; or in marine sediment exposed to the air, perhaps temporarily by a falling tide. Cracked mud can lithify to form shale, preserving the cracks. The filling of mud cracks by sand can form casts of the cracks in an overlying sandstone.

Fossils

Fossils are the remains of organisms preserved in sedimentary rock. Most sedimentary layers contain some type of fossil and some limestones are composed entirely of fossils. Most fossils are preserved by the rapid burial in sediment of bones, shells, or teeth, which are the mineralized hard parts of

FIGURE 9.29

Cross-bedded sandstone in Zion National Park, Utah. Note how the thin layers have formed at an angle to the more extensive bedding planes (also tilted) in the rock. This cross-bedding was formed in sand dunes deposited by the wind.

Photo by David McGeary

animals most resistant to decay (figure 9.35). The original bone or shell is rarely preserved unaltered; the original mineral is often recrystallized or replaced by a different mineral such as pyrite or silica. Bone and wood may be *petrified* as organic material is replaced and pore spaces filled with mineral. Shells entombed within rock are commonly dissolved away by pore waters, leaving only impressions or *moulds* of the original fossil. Leaves and undecayed organic tissue can also be preserved as a thin film of carbon (figure 9.35*C*). *Trace fossils* are a type of sedimentary structure produced by the impact of an organism's activities on the sediment. Footprints, trackways, and burrows are the most common trace fossils (figure 9.35*B*).

Many *paleontologists* study fossils to learn about the evolution of life on Earth, but fossils are also very useful for interpreting depositional environments and for reconstructing the climates of the past. Fossils can be used to distinguish fresh water from marine environments and to infer the water depth at which a particular sedimentary layer was deposited. Tropical, temperate, and arid climates can be associated with distinctive types of fossil plants. Marine *microfossils,* the tiny shells produced by ocean-dwelling plankton, can be analyzed to determine the water temperature that surrounded the shell when it formed. Much of our detailed knowledge of Earth's climate

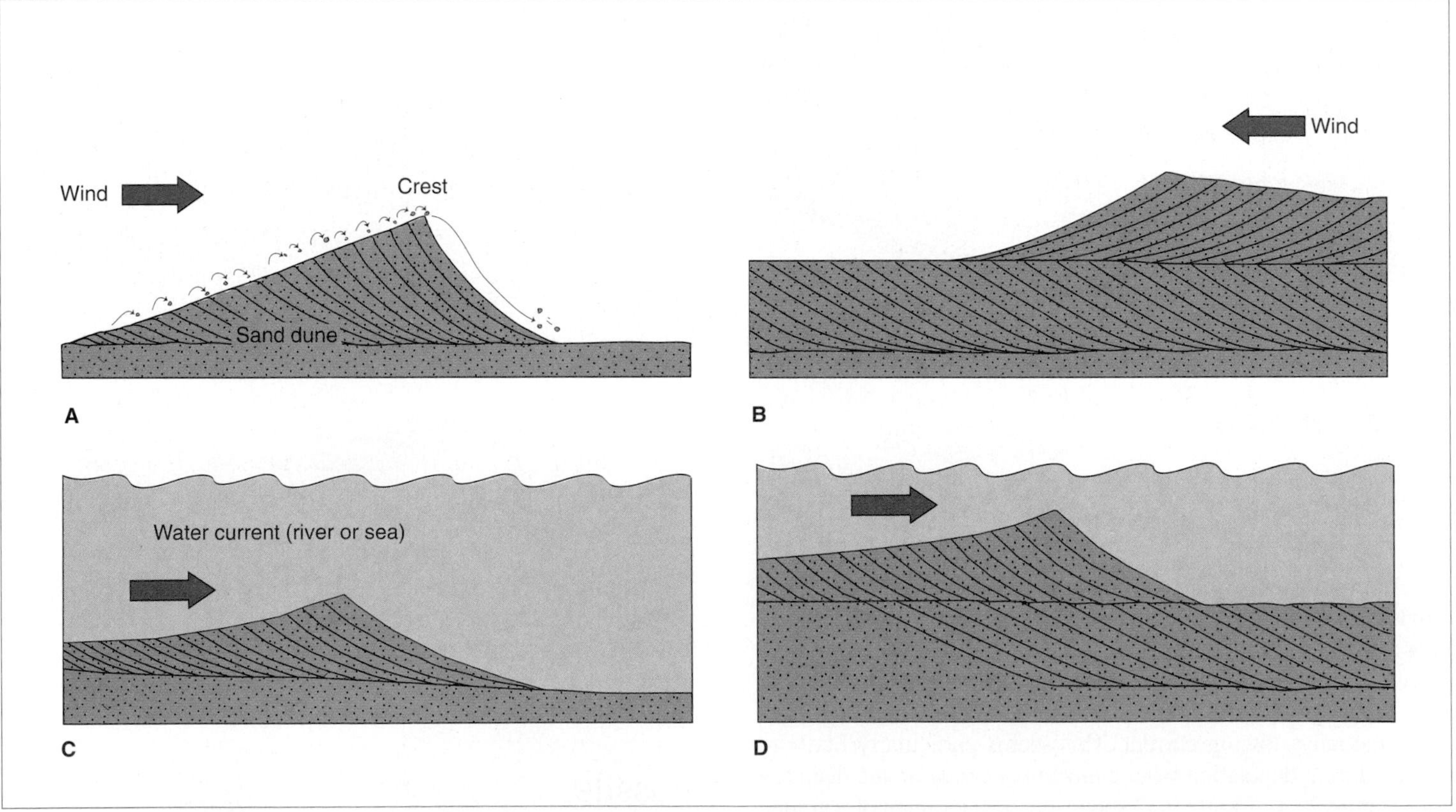

FIGURE 9.30

The development of cross-beds in wind-blown sand (*A* and *B*) and water-deposited sand (*C* and *D*). (*A*) Sand grains migrate up the shallow side of the dune and avalanche down the steep side, forming cross-beds. (*B*) Second layer of cross-beds forms as wind shifts and a dune migrates from the opposite direction. (*C*) Underwater current deposits cross-beds as ripple migrates downstream. (*D*) Continued deposition and migration of ripples produces multiple layers of cross-beds.

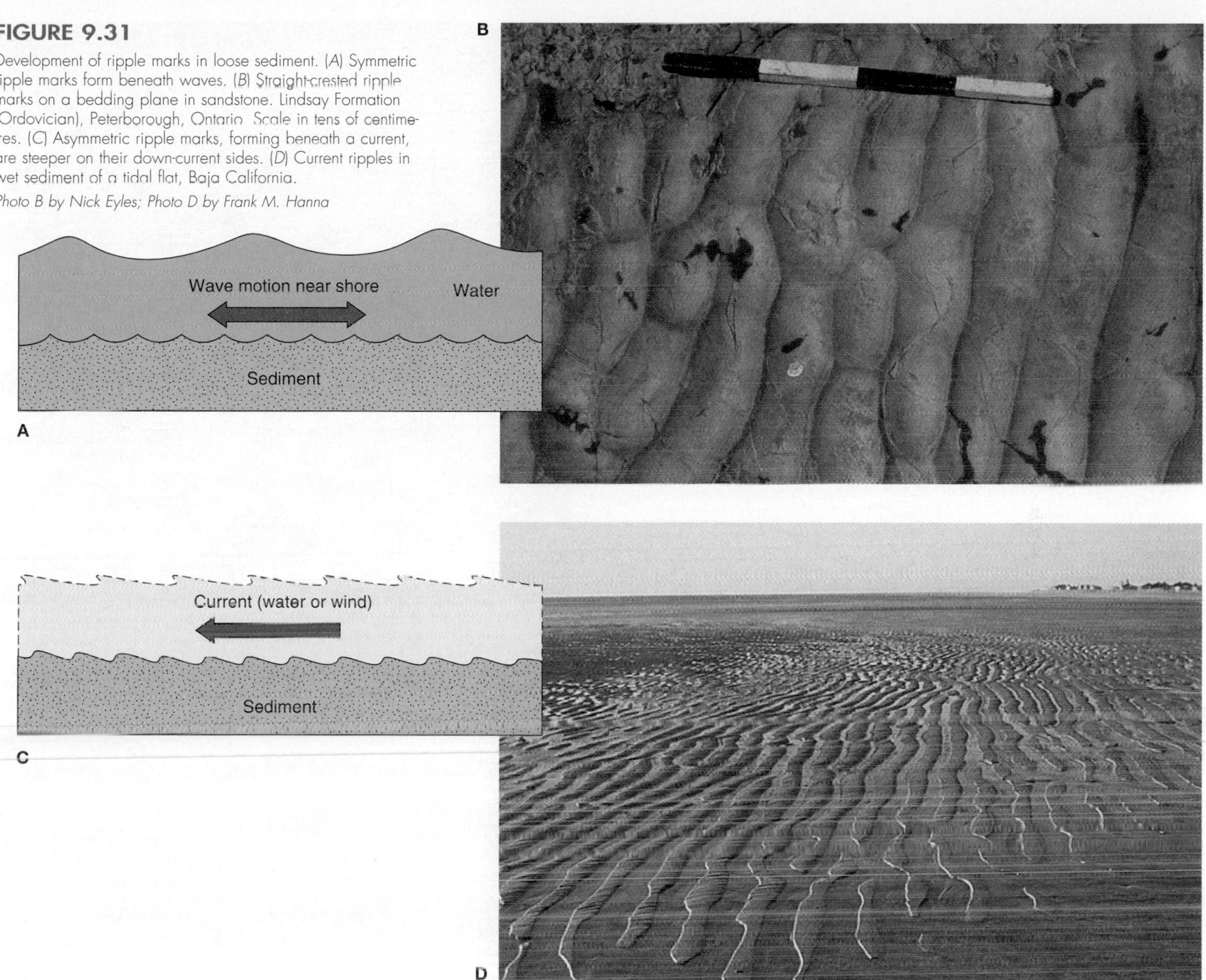

FIGURE 9.31

Development of ripple marks in loose sediment. (*A*) Symmetric ripple marks form beneath waves. (*B*) Straight-crested ripple marks on a bedding plane in sandstone. Lindsay Formation (Ordovician), Peterborough, Ontario. Scale in tens of centimetres. (*C*) Asymmetric ripple marks, forming beneath a current, are steeper on their down-current sides. (*D*) Current ripples in wet sediment of a tidal flat, Baja California.

Photo B by Nick Eyles; Photo D by Frank M. Hanna

changes over the last 150 million years has come from the study of microfossils extracted from layers of mud deposited on the deep-ocean floor.

Formations

A **formation** is a body of rock of considerable thickness with characteristics that distinguish it from adjacent rock units and that is large enough to be mappable. Although a formation is usually composed of one or more beds of sedimentary rock, units of metamorphic and igneous rock are also called formations. It is a convenient unit for mapping, describing, or interpreting the geology of a region.

Formations are often based on rock type. A formation may be a single thick bed of rock such as sandstone. A sequence of several thin sandstone beds could also be called a formation, as could a sequence of alternating limestone and shale beds.

The main criterion for distinguishing and naming a formation is some visible characteristic that makes it a recognizable unit. This characteristic may be rock type or sedimentary structures or both. For example, a thick sequence of shale may be overlain by basalt flows and underlain by sandstone. The shale, the basalt, and the sandstone are each a different formation. Or a sequence of thin limestone beds, with a total thickness of hundreds of metres, may have recognizable fossils in the lower half and distinctly different fossils in the upper half. The limestone sequence is divided into two formations on the basis of its fossil content.

Formations are given proper names: the first name is often a geographic location where the rock is well exposed, and the second the name of a rock type, such as Mt. Wilson Quartzite, Dunvegan Sandstone, Owen Creek Dolostone, or Queenston Shale. If the formation has a mixture of rock types, so that one rock name does not accurately describe it, it is called simply "formation," as in the Waterton Formation or the Banff Formation.

FIGURE 9.32

A graded bed has coarse grains at the bottom of the bed and progressively finer grains toward the top. Coin for scale.

Photo by David McGeary

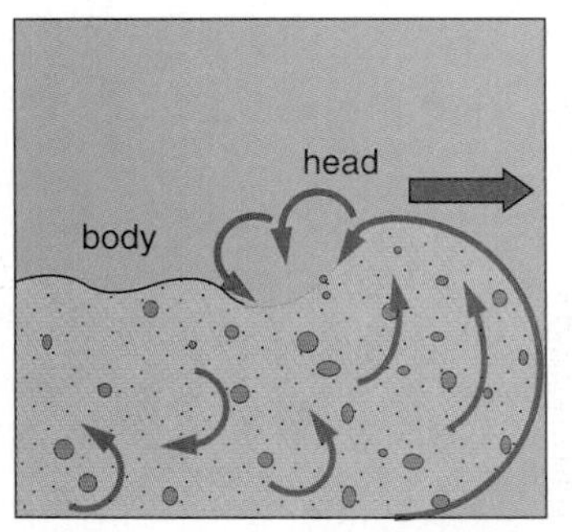

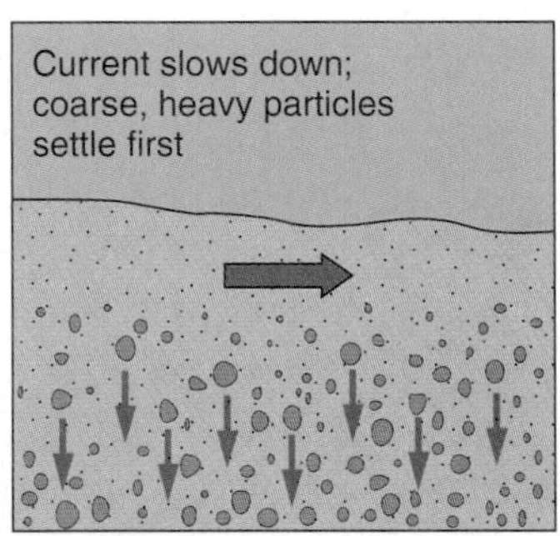

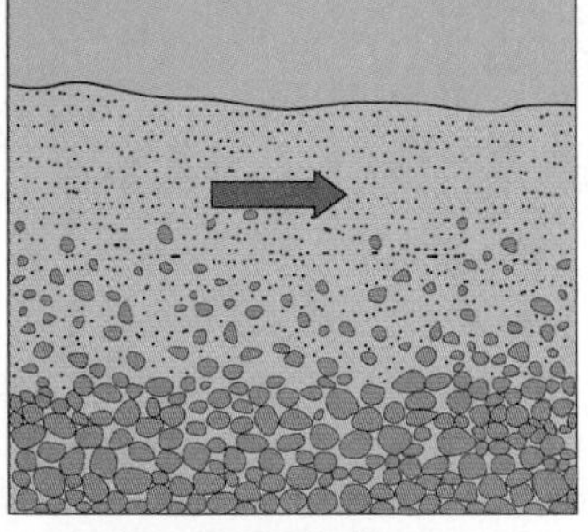

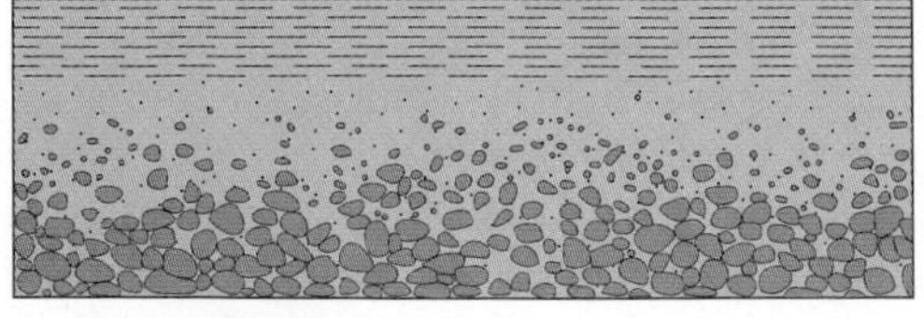

FIGURE 9.33

Development of a graded bed of sediment deposited by a turbidity current.

A

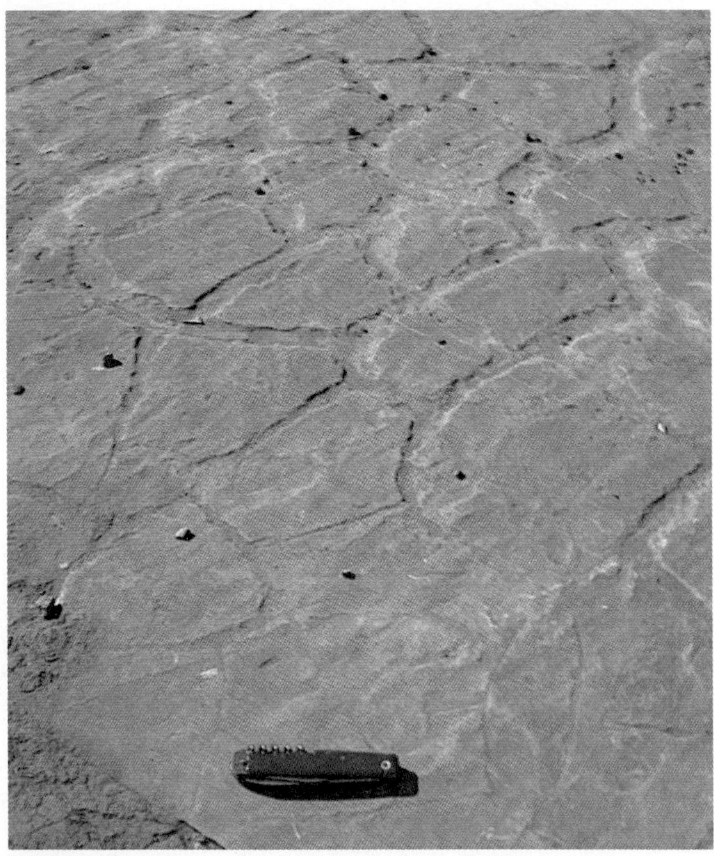

B

FIGURE 9.34

(A) Mud cracks in recently dried mud. *(B)* Mud cracks preserved in shale; they have been partially filled with sediment.

Photos by David McGeary

A

B

C

FIGURE 9.35

(A) Excavation of fossilized leg bone of a duckbilled dinosaur (Edmontosaurus), Drumheller Valley, Alberta. (B) Footprint of plant-eating dinosaur, Dinosaur Provincial Park, southeastern Alberta. (C) Fossil plant stems preserved as thin films of carbon within sandstone, Joggins, Nova Scotia.

Photo A © Royal Tyrrell Museum/Alberta Community Development; Photo C by C. H. Eyles

A **contact** is the boundary surface between two different rock types or ages of rocks. In sedimentary rock formations, the contacts are usually bedding planes (figure 9.28*B*).

Figure 9.36 shows the geological formations exposed along the Niagara Gorge that separates Canada and the United States. The contacts between formations are also shown.

Interpretation of Sedimentary Rocks

Sedimentary rocks are important in interpreting the geologic history of an area (see box 9.3). Geoscientists examine sedimentary formations to look for clues such as fossils; sedimentary structures; grain shape, size, and composition; and the overall shape and extent of the formation. These clues are useful in determining the source area of the sediment, environment of deposition, and the possible plate tectonics setting at the time of deposition.

Source Area

The **source area** of a sediment is the locality from which the sediment originated. The most important things to determine about a source area are the type of rocks that were present and its location and distance from the site of eventual deposition.

The *rock type* exposed in the source area determines the character of the resulting sediment. The composition of a sediment can indicate the source area rock type, even if the source area has been completely eroded away. A conglomerate may contain cobbles of basalt, granite, and chert; these rock types

FIGURE 9.36

Geological formations of Upper Ordovician to Silurian age exposed in the Niagara Gorge, Ontario and New York State. The Lockport Dolostone is resistant to erosion and forms the "cap rock" of the Niagara Escarpment over which the Niagara Falls flow. The black lines are drawn to show the boundaries (contacts) between formations.

Photo by C. H. Eyles

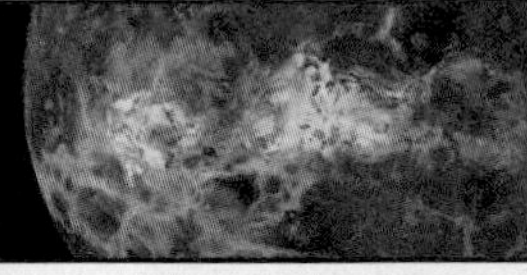

ASTROGEOLOGY 9.3

Sedimentary Rocks: The Key To Mars' Past

Sedimentary rocks on Mars will one day allow planetary geologists to decipher its early history and determine if Mars was once a warmer, wetter planet. Currently, the atmosphere on Mars is too thin and its surface too cold to allow liquid water to exist (see chapter 22). But was Mars wet enough to host lakes and seas long ago? New observations from robotic spacecraft exploring Mars show evidence for extensive deposits of water-lain sedimentary rock. In orbit around Mars, the *Mars Global Surveyor* and *Mars Express* spacecraft have taken thousands of high-resolution photographs, many of which reveal widespread, laterally continuous layers that appear to be sedimentary rock. For example, hundreds of layers of rock are exposed in parts of the walls of the Valles Marineris, a large chasm on Mars that resembles the Grand Canyon but is almost 4,000 kilometres long! Recently, robotic rovers have explored two regions of Mars that may have once been covered by liquid water. The Mars Exploration Rover named *Opportunity* landed inside of a small crater with exposures of layered rock and later traversed the Martian surface to enter a larger crater with more layered rock (box figure 1). Detailed photographic and spectrographic analysis of these layered rocks has revealed sedimentary features such as cross-beds, hematite mineral concretions, and the presence of minerals such as jarosite that typically form in water. Halfway around the planet in Gusev Crater, the Mars Exploration Rover named *Spirit* has discovered exposures of bedrock in hills near its landing site. Although most rocks in this area appear to be of volcanic origin, some have characteristics that indicate they have been chemically and texturally altered by exposure to liquid water. Although the accomplishments of these robotic geologists are remarkable, confirmation of the presence of water-deposited sedimentary rock on Mars will have to wait until samples can be returned to Earth for detailed study or until human geologists arrive on Mars.

Scientists are particularly excited by the prospect of Mars being a wet planet early in its history because of the possibility that life could have evolved there. If, as now seems likely, the early history of Mars is recorded in layers of sedimentary rock, perhaps fossils of Martian microbes remain to be discovered. Mars continues to be the most promising place to look for evidence of extraterrestrial life in our solar system.

BOX 9.3 ■ FIGURE 1
Layers of sedimentary rock exposed inside the rim of Endurance Crater photographed by the Mars Exploration Rover *Opportunity*.
Photo by NASA/JPL/Cornell

Additional Resources

M. C. Malin and K. S. Edgett. 2000. Sedimentary rocks of early Mars. *Science* 290 (5498): 1927–37.

Information about the Mars exploration program at NASA, including images and updates from ongoing missions such as the Mars Global Surveyor and the Mars Exploration Rovers, is available from the Jet Propulsion Laboratory/NASA Mars Program website.

http://marsprogram.jpl.nasa.gov/

Visit the European Space Agency's Mars Express website for information about this ongoing mission.

www.esa.int/SPECIALS/Mars_Express/

Visit the Malin Space Science Systems website for an extensive collection of archived Mars Orbiter Camera images.

www.msss.com

were obviously in its source area. An arkose containing coarse feldspar, quartz, and biotite may have come from a granitic source area. Furthermore, the presence of feldspar indicates the source area was not subjected to extensive chemical weathering and that erosion probably took place in an arid environment with high relief. A quartz sandstone containing well-rounded quartz grains, on the other hand, probably represents the erosion and deposition of quartz grains from pre-existing sandstone. Quartz is a hard, tough mineral very resistant to rounding by abrasion, so if quartz grains are well-rounded they have undergone many cycles of erosion, transportation, and deposition, probably over tens of millions of years.

Sedimentary rocks are also studied to determine the *direction* and *distance* to the source area. Figure 9.2 shows how several characteristics of sediment may vary with distance from a source area. Many sediment deposits get thinner away from the source, and the sediment grains themselves usually become better sorted, finer, and more rounded.

Sedimentary structures often give clues about the directions of past currents (*paleocurrents*) that deposited sediments. Refer

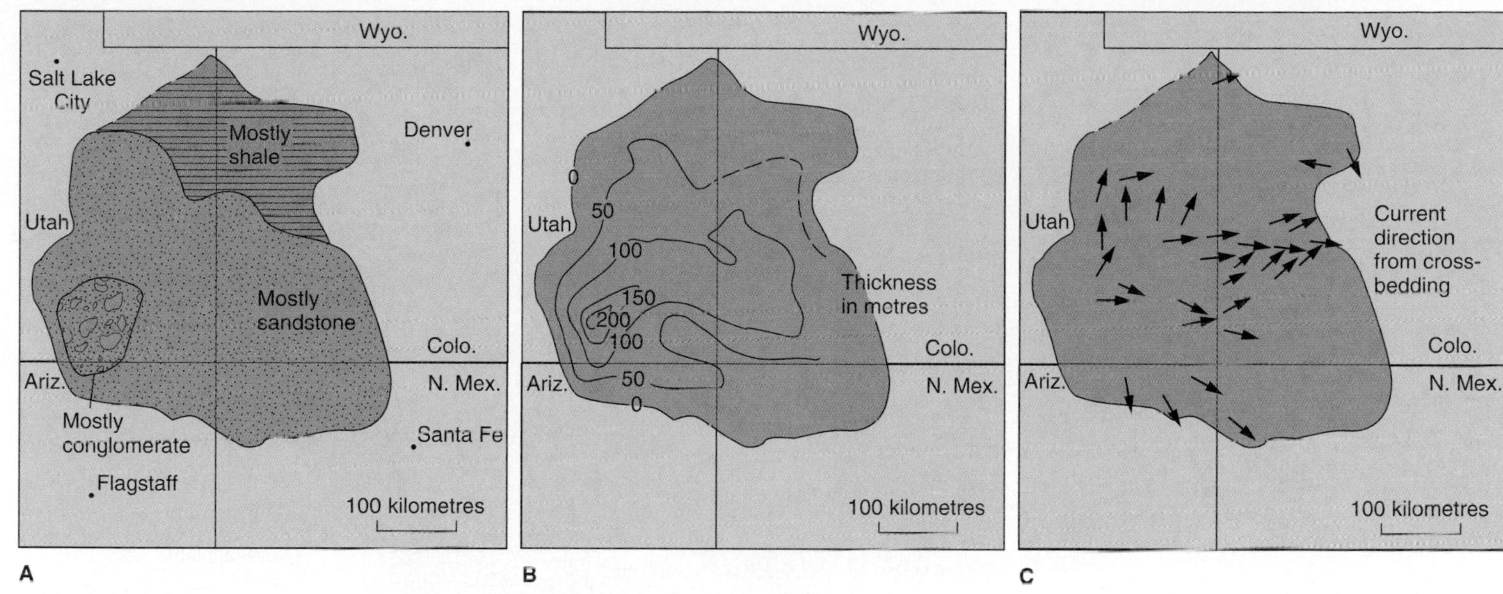

FIGURE 9.37

Characteristics of the Salt Wash Member of the Morrison Formation that help locate its source area. (*A*) The sediment grains become coarser to the southwest. (*B*) The deposit becomes thicker to the southwest. The contour lines show the thickness of the Salt Wash Member in metres. (*C*) Cross-bedding shows that the depositing currents came mostly from the southwest (arrows point down current).

Redrawn and simplified from Craig and others, 1955, U.S. Geological Survey Bulletin

back to figures 9.28*B* and 9.30 and notice how cross-beds dip (slope) downward in the direction of current flow. Paleocurrent direction can also be determined from asymmetric ripple marks (figure 9.31).

Figure 9.37 shows how three of these characteristics were used to determine the location of the source area for a particular rock unit in the southwestern United States. The unit is the Salt Wash Member of the Morrison Formation (a *member* is a subdivision of a formation). It is an important rock unit, for it contains a great deal of uranium, deposited within the rock by groundwater long after the rock formed. The unit thickens and coarsens to the southwest, and cross-beds show that the old currents that deposited the sediment came from rivers that flowed from the southwest. These three facts strongly suggest that the source area was to the southwest. This information helps exploration geologists search for uranium within the Salt Wash Member. The Morrison Formation is also world famous for its dinosaur fossils.

Environment of Deposition

Figure 9.38 shows the common environments in which sediments are deposited. Geoscientists study modern environments in great detail so that they can interpret ancient rocks. Clues to the ancient environment of deposition come from a rock's composition, the size and sorting and rounding of the grains, the sedimentary structures and fossils present, and the shape and vertical sequence of the sedimentary layers.

Continental environments include alluvial fans, river channels, flood plains, lakes, and dunes. Sediments deposited on land are subject to erosion, so they often are destroyed. The great bulk of sedimentary rocks come from the more easily preserved shallow marine environments, such as deltas, beaches, lagoons, shelves, and reefs. The characteristics of major environments are covered in detail in later chapters (14, 16–18). In this section we describe the main sediment types and sedimentary structures found in each environment.

GLACIAL ENVIRONMENTS

Glacial ice can transport sediment ranging in size from large boulders to silt and clay and typically deposits very poorly sorted sediments. **Till** is the name given to unsorted glacial sediment deposited beneath glaciers. Many areas of Canada consist of extensive till plains formed beneath the huge ice sheets that covered northern North America during the last glaciation (see chapter 16). Till commonly contains boulders and cobbles that have been scratched and shaped by grinding over one another under the great weight of the ice. Glaciers also supply coarse- and fine-grained sediment to adjacent fluvial, lacustrine, and marine environments (figure 9.39).

ALLUVIAL FAN

Rivers flowing from steep mountain slopes onto flatter plains often deposit broad, fan-shaped piles of sediment called *alluvial fans*. The sediment consists of coarse, arkosic sandstones and conglomerates, marked by coarse cross-bedding and lens-like channel deposits (figure 9.40).

RIVER CHANNEL AND FLOOD PLAIN

Rivers deposit elongate lenses of conglomerate or sandstone in their channels (figure 9.40). The sandstones may be arkoses or may consist of sand-sized fragments of fine-grained rocks. River channel deposits typically contain cross-beds and current ripple marks. Broad, flat flood plains are covered by periodic

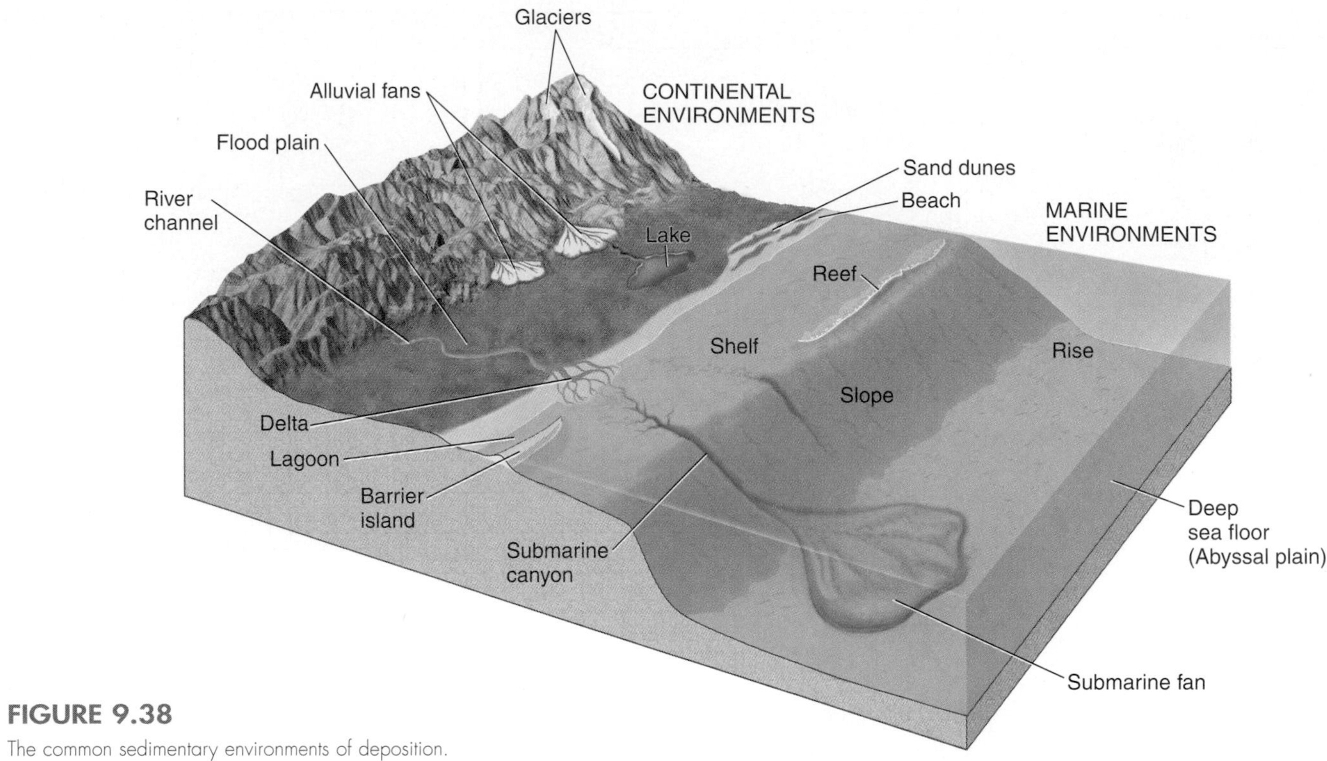

FIGURE 9.38
The common sedimentary environments of deposition.

FIGURE 9.39
Saskatchewan Glacier, Alberta, with lake in foreground.
Photo by L.K. Dekeyser

FIGURE 9.40
Alluvial fan developed at mouth of stream entering the Mackenzie River, near Tulita, NWT.
Photo by Nick Eyles

floodwaters, which deposit thin-bedded shales characterized by mud cracks and fossil footprints of animals. Hematite may colour flood-plain deposits red.

LAKE

Thin-bedded shale, perhaps containing fish fossils, is deposited on lake bottoms. If the lake periodically dries up, the shales will be mud-cracked and perhaps interbedded with evaporites such as gypsum or rock salt. Lake deposits are often associated with glacial sediments in northern parts of North America (figure 9.39). Sediments accumulating in lakes adjacent to large urbanized regions are often contaminated with a variety of industrial and organic pollutants (see box 9.4).

DELTA

A delta is a body of sediment deposited when a river flows into standing water, such as the sea or a lake. Most deltas contain a great variety of subenvironments but are generally made up of thick successions of siltstone and shale, marked by low-angle cross-bedding and cut by coarser channel deposits. Deltaic successions may contain beds of peat or coal, as well as marine fossils such as clam shells.

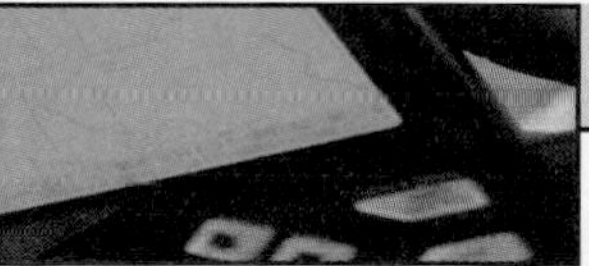

GEOMATICS 9.4

Mapping Contaminated Sediments in an Urban Harbour

The clean-up and remediation of contaminated sediments in urban water bodies is a major issue in Canada and is important in both freshwater (e.g., Lake Ontario) and saltwater (e.g., Halifax Harbour) environments. Effective remediation of contaminated sediments, which involves either their removal or burial, requires the compilation of accurate maps showing their thickness and distribution. However, the creation of sediment distribution maps is not an easy task and all too often inaccurate maps are drawn on the basis of widely spaced core data.

Recent work in Hamilton Harbour, a highly polluted harbour surrounded by steel manufacturing and other industrial activities at the western end of Lake Ontario, has shown the effective use of geophysical methods to map lake-floor sediments contaminated with magnetic minerals. Data from a number of cores taken in the harbour showed that contaminated sediment contained high amounts of magnetic oxides and had high magnetic susceptibility. A study was conducted in which a magnetometer was towed behind a boat along closely spaced survey grid lines criss-crossing the whole of the harbour area. The magnetometer measured changes in magnetic susceptibility of the lake floor sediments and in effect provided a rapid means of mapping areas of highly magnetized (and contaminated) sediment (box figure 1*A*). This magnetic susceptibility map was then used to produce a qualitative map showing the estimated contaminant impact levels in the harbour (box figure 1*B*). Digital maps such as this can be used to design and guide remediation plans and are likely to be more accurate than those based on widely spaced core data.

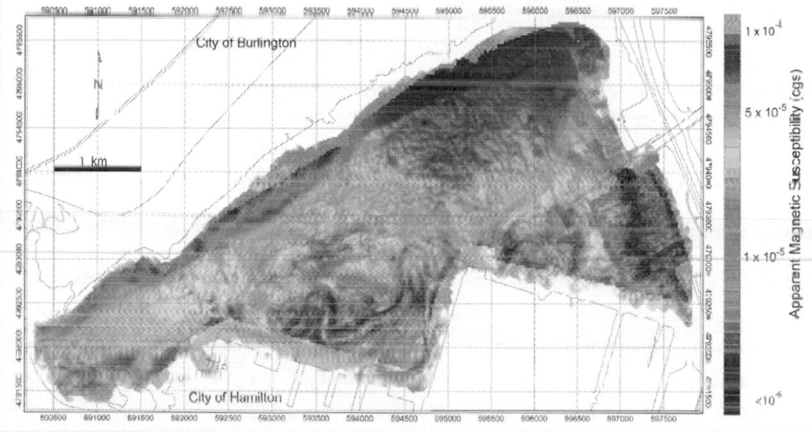

A

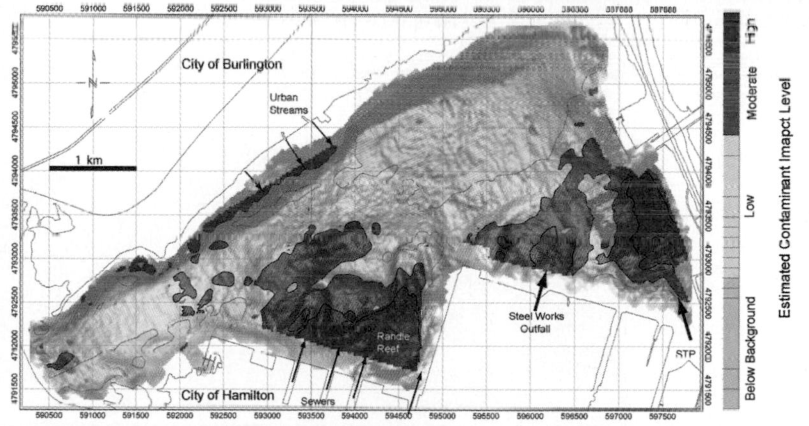

BOX 9.4 ■ FIGURE 1

(*A*) Apparent magnetic susceptibility map of Hamilton Harbour. Hot colours represent areas of high magnetic susceptibility and cool colours represent areas of low magnetic susceptibility. (*B*) Qualitative sediment classification map showing estimated contaminant impact levels in the harbour.

Source: Reprinted from Journal of Applied Geophysics, Vol. 57, 2004, Pages 23–41. Pozza et al, Lake Based Magnetic Mapping of... with permission of Elsevier.

BEACH, BARRIER ISLAND, DUNE

A barrier island is an elongate bar of sand built by wave action. Well-sorted quartz sandstone with well-rounded grains is deposited on beaches, barrier islands, and dunes. Beaches and barrier islands are characterized by cross-bedding (often low-angle) and marine fossils. Dunes have both high-angle and low-angle cross-bedding and occasionally contain fossil footprints of land animals such as lizards. All three environments can also contain carbonate sand in tropical regions, thus yielding cross-bedded clastic limestones.

LAGOON

A semi-enclosed, quiet body of water between a barrier island and the mainland is a lagoon. Fine-grained dark shale, cut by tidal channels filled with coarse sand and containing fossil oysters and other marine organisms, is formed in lagoons. Limestones may also form in lagoons adjacent to reefs (refer to figure 9.18).

SHALLOW MARINE SHELVES

On the broad, shallow shelves adjacent to most shorelines, sediment grain size decreases offshore. Widespread deposits of sandstone, siltstone, and shale can be deposited on such shelves. The sandstone and siltstone contain symmetrical ripple marks, low-angle cross-beds, and marine fossils such as clams and snails. If fine-grained *tidal flats* near shore are alternately covered and exposed by the rise and fall of tides, mud-cracked marine shale will result.

REEFS

Massive limestone is deposited in reef cores, with steep beds of limestone breccia forming seaward of the reef, and horizontal beds of sand-sized and finer-grained limestones forming landward (refer to figure 9.18). All these limestones are full of fossil fragments of corals, coralline algae, and numerous other marine organisms.

DEEP MARINE ENVIRONMENTS

On the deep sea floor shale and greywacke sandstones are deposited on submarine fans (figure 9.38). The greywackes are deposited by turbidity currents (refer to figures 9.14 and 9.32) and typically contain graded bedding and current ripple marks.

Plate Tectonics and Sedimentary Rocks

The dynamic forces that move plates on Earth are also responsible for the distribution of many sedimentary rocks. As such, the distribution of sedimentary rocks often provides information that helps geologists reconstruct past plate tectonics settings.

In tectonically active areas, such as *convergent plate boundaries*, sedimentary basins are formed both oceanward and inland of mountain belts (figure 9.41). Rapid erosion of the rising mountains produces enormous quantities of sediment that are transported by streams and turbidities to accumulate offshore in a **forearc basin**. Continued subsidence of the forearc basin results in the formation of great thicknesses of sedimentary rock that record the history of uplift and erosion in the mountain belt. On the landward side of the mountain belt, loading of the crust by the growing mountains causes depression and the development of a **foreland basin** (figure 9.41). Foreland basins are often flooded to form inland seas, in which shallow marine sediments accumulate.

Thick successions of interbedded sandstones and mudstones accumulated in the Western Interior Sedimentary Basin, which formed as a foreland basin across parts of western Canada around 100 million years ago (see chapter 20). Organic materials trapped within these shallow marine sediments have been transformed into hydrocarbons and now form the rich oil and gas fields of Alberta. The Appalachian Foreland Basin in eastern Canada formed in a similar fashion during the Paleozoic and contains sedimentary rocks that host important coal deposits.

Successions of shallow marine sedimentary rocks that accumulate in foreland basins record changes in shoreline positions caused by uplift of the mountains and/or climate changes.

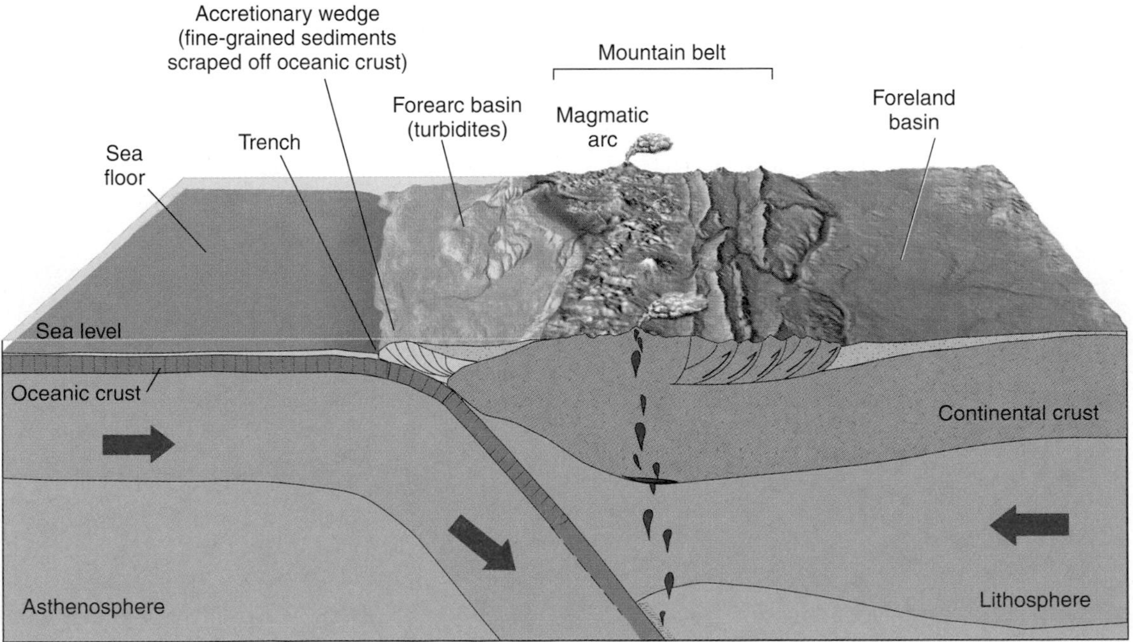

FIGURE 9.41

Sedimentary basins associated with a convergent plate boundary include a forearc basin on the oceanward side that contains mainly clastic sediments deposited by streams and turbidity currents from an eroding magmatic arc. Toward the craton (continent) a backarc basin also collects clastic sediment derived from the uplifted mountain belt and craton.

When water depths increase, the shoreline moves landward (**transgression**), and as water depths reduce, the shoreline moves basinward (**regression**).

It is not uncommon for rugged mountain ranges that stand several thousand metres above sea level, such as the Canadian Rockies, European Alps, and Himalayas, to contain sedimentary rocks of marine origin that were originally deposited below sea level. The presence of marine sedimentary rocks such as limestone, chert, and shale containing marine fossils at high elevations attests to the tremendous uplift associated with mountain building at convergent plate boundaries (see chapter 2).

Transform plate boundaries are also characterized by rapid rates of erosion and deposition of sediments as fault-bounded basins open and subside rapidly with continued plate motion. Because of the rapid rate of deposition and the rapid burial of organic material, fault-bounded basins are good places to explore for petroleum. Many of the petroleum occurrences in California are related to basins that formed as the San Andreas transform fault developed.

A *divergent plate boundary* may result in the splitting apart of a continent and formation of a new ocean basin. In the initial stages of continental divergence, a rift valley forms and fills with thick wedges of gravel and coarse sand along its fault-bounded margins; lake bed deposits and associated evaporite rocks may form in the bottom of the rift valley (figure 9.42). In the early stages, continental rifts will have extensive volcanics that contribute to the sediments in the rift. The Red Sea is located along the East African Rift Zone and is a good example of the features and sedimentary rocks formed during the initial stages of continental rifting.

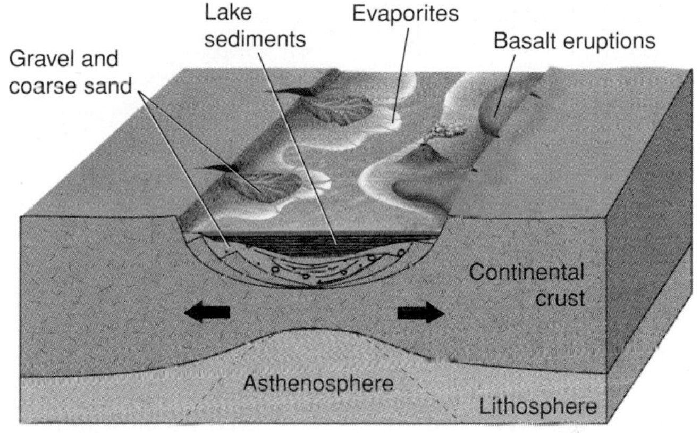

FIGURE 9.42

Divergent plate boundary showing thick wedges of gravel and coarse sand along fault-bounded margins of developing rift valley. Lake bed deposits and evaporite rocks are located on the floor of the rift valley.

SUMMARY

Sediment forms by the weathering and erosion of pre-existing rocks and by chemical precipitation, sometimes by organisms.

Gravel, sand, silt, and *clay* are sediment particles defined by grain size.

The composition of sediment is governed by the rates of chemical weathering, mechanical weathering, and erosion. During transportation, grains can become rounded and sorted.

Sedimentary rocks form by *lithification* of sediment, by *crystallization* from solution, or by consolidation of remains of organisms. Sedimentary rocks may be *clastic, chemical,* or *organic.*

Clastic sedimentary rocks form mostly by *compaction* and *cementation* of grains. *Matrix* can partially fill the *pore space* of clastic rocks.

Conglomerate forms from coarse, rounded sediment grains that often have been transported only a short distance by a river or waves. *Sandstone* forms from sand deposited by rivers, wind, waves, or turbidity currents. *Shale* forms from river, lake, or ocean mud.

Limestone consists of calcite, formed either as a chemical precipitate in a reef or, more commonly, by the cementation of shell and coral fragments or of oolites. *Dolomite* usually forms from the alteration of limestone by magnesium-rich solutions.

Chert consists of silica and usually forms from the accumulation of microscopic marine organisms. *Recrystallization* often destroys the original texture of chert (and some limestones).

Evaporites, such as rock salt and gypsum, form as water evaporates. *Coal,* a major fuel, is consolidated plant material.

Sedimentary rocks are usually found in *beds* separated by *bedding planes* because the original sediments are deposited in horizontal layers.

Cross-beds and *ripple marks* develop as moving sediment forms ripples and dunes during transport by wind, underwater currents, and waves.

A *graded bed* forms as coarse particles fall from suspension before fine particles, perhaps in a turbidity current.

Mud cracks form in drying mud.

Fossils are the traces of an organism's hard parts or tracks preserved in rock.

A *formation* is a convenient rock unit for mapping and describing rock. Formations are lithologically distinguishable from adjacent rocks; their boundaries are *contacts.*

Geoscientists try to determine the *source area* of a sedimentary rock by studying its grain size, composition, and sedimentary structures. The source area's rock type and location are important to determine.

The *environment of deposition* of a sedimentary rock is determined by studying bed sequence, grain composition and rounding, and sedimentary structures. Typical environments include glaciers, alluvial fans, river channels, flood plains, lakes, dunes, deltas, beaches, shallow marine shelves, reefs, and the deep sea floor.

Plate tectonics plays an important role in the distribution of sedimentary rocks; the occurrence of certain types of sedimentary rocks is used by geoscientists to reconstruct past plate tectonic settings.

Terms to Remember

bedding 237
bedding plane 237
cement 226
cementation 226
chemical sedimentary rock 227
chert 235
clastic sedimentary rock 226
clastic texture 226
clay 223
coal 235
compaction 226
conglomerate 227
contact 243
cross-beds 237
crystalline texture 226
crystallization 226
deposition 224
dolomite 234
environment of deposition 224
evaporite 235
forearc basin 248
foreland basin 248
formation 241
fossil 239
graded bed 239
gravel 223
limestone 232
lithification 226
matrix 229
mud crack 239
organic sedimentary rock 227
original horizontality 237
pore space 226
recrystallization 232
regression 249
ripple mark 237
rounding 223
sand 223
sandstone 229
sediment 223
sedimentary breccia 227
sedimentary rock 226
sedimentary structure 237
shale 231
silt 223
sorting 224
source area 243
superposition 237
till 245
transgression 249
transported 223
turbidity current 229

Testing Your Knowledge

Use the questions below to prepare for exams based on this chapter.

1. Quartz is a common mineral in sandstone. Under certain circumstances, feldspar is common in sandstone, even though it normally weathers rapidly to clay. What conditions of climate, weathering rate, and erosion rate could lead to a feldspar-rich sandstone? Explain your answer.
2. Describe with sketches how wet mud compacts before it becomes shale.
3. What do mud cracks tell about the environment of deposition of a sedimentary rock?
4. How does a graded bed form?
5. List the clastic sediment particles in order of decreasing grain size.
6. How does a sedimentary breccia differ in appearance and origin from a conglomerate?
7. Describe three different origins for limestone.
8. How does dolomite usually form?
9. What is the origin of coal?
10. Sketch the cementation of sand to form sandstone.
11. How do evaporites form? Name two evaporites.
12. Name the three most common sedimentary rocks.
13. What is a formation?
14. Explain two ways that cross-bedding can form.
15. Particles of sediment from 1/16 mm to 2 mm in diameter are of what type?
    a. gravel b. sand
    c. silt d. clay
16. Rounding is
    a. the rounding of a grain to a spherical shape
    b. the grinding away of sharp edges and corners of rock fragments during transportation
    c. a type of mineral
    d. none of the preceding
17. Compaction and cementation are two common processes of
    a. erosion b. transportation
    c. deposition d. lithification
18. Which is not a chemical or organic sedimentary rock?
    a. rock salt b. shale
    c. limestone d. gypsum
19. The major difference between breccia and conglomerate is
    a. size of grains b. rounding of the grains
    c. composition of grains d. all of the preceding

20. Which is not a type of sandstone?
    a. quartz sandstone b. arkose
    c. greywacke d. coal

21. Shale differs from mudstone in that
    a. shale has larger grains
    b. shale is visibly layered and fissile; mudstone is massive and blocky
    c. shale has smaller grains
    d. there is no difference between shale and mudstone

22. The chemical element found in dolomite not found in limestone is
    a. Ca b. Mg
    c. C d. O
    e. Al

23. In a graded bed the particle size decreases
    a. upward
    b. downward
    c. in the direction of the current
    d. particle size stays the same

24. A body or rock of considerable thickness with characteristics that distinguish it from adjacent rock units is called a/an
    a. formation b. contact
    c. bedding plane d. outcrop

25. If sea level drops or the land rises, what is likely to occur?
    a. a flood b. a regression
    c. a transgression d. no geologic change will take place

26. Thick accumulations of greywacke and volcanic sediments can indicate an ancient
    a. divergent plate boundary
    b. convergent boundary
    c. transform boundary

27. A sedimentary rock made of fragments of pre-existing rocks is
    a. organic b. chemical
    c. clastic

28. Clues to the nature of the source area of sediment can be found in
    a. the composition of the sediment
    b. sedimentary structures
    c. rounding of sediment
    d. all of the preceding

Exploring Web Resources

www.mcgrawhill.ca/olc/plummer

Visit the Online Learning Centre Website for an animation and discussion on how sediments are deposited along a broad, shallow marine shelf with the rising and falling sea level. The Online Learning Centre also has Testing Your Knowledge answers, additional quizzes, and reading and media resources relating to sedimentary rocks.

http://darkwing.uoregon.edu/%7Erdorsey/SedResources.html

The *Web Resources for Sedimentary Geologists* site contains a comprehensive listing of resources available on the Web.

www.lib.utexas.edu/geo/folkready/entirefolkpdf.pdf

Online version of *Petrology of Sedimentary Rocks* by Professor Robert Folk at the University of Texas at Austin.

http://walrus.wr.usgs.gov/seds/

Visit the *U.S. Geological Survey Bedform and Sedimentology* site for computer and photographic images and movies of sedimentary structures.

http://zircon.geology.union.edu/Gildner/stack.html

For a virtual field trip of the Ordovician age rocks in the Mohawk Valley of New York state, visit Union College's geology Website.

www.palaeo.de/edu/JRP/index.html

For a virtual field trip of the Jurassic Reef Park in Germany. Learn how fossil reefs preserved in sedimentary rocks are used to interpret past ecosystems.

Animations

This chapter includes the following animations available on our Online Learning Centre at www.mcgrawhill.ca/olc/plummer.

9.30 Migration of Sand Grains to Form Ripples, Dunes, and Crossbeds

9.33 Formation of a Graded Bed

CHAPTER

10

Metamorphism, Metamorphic Rocks, and Hydrothermal Rocks

How do metamorphic rocks form?
What factors control the characteristics of metamorphic rocks?
How do we classify metamorphic rocks?
Where does metamorphism occur?
What is the relationship between plate tectonics and metamorphism?
What is the role of hot water in metamorphic processes?

This chapter on metamorphic rocks, the third major category of rocks in the rock cycle, completes our description of Earth materials (rocks and minerals). The information on igneous and sedimentary processes in previous chapters should help you understand metamorphic rocks, which form from pre-existing rocks.

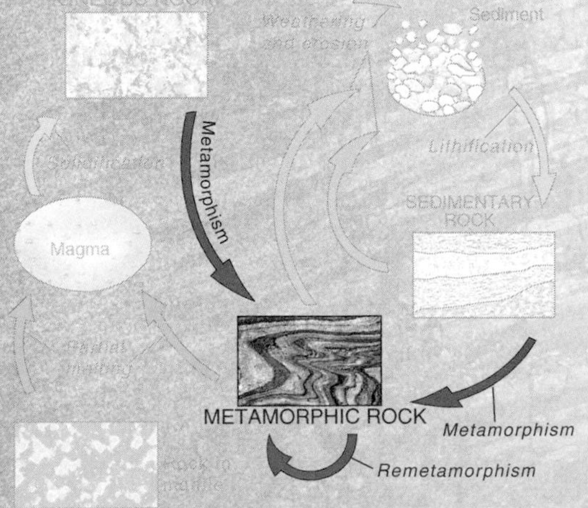

Metamorphic rocks of the Canadian Shield near Parry Sound, Ontario. The banded grey rock (mylonite) has been intensively sheared and intruded by pink granite dykes. *Photo by Nick Eyles*

After reading chapter 8 on weathering, you know how rocks are altered when exposed at Earth's surface. *Metamorphism* (a word from Latin and Greek that means literally "changing of form") also involves alterations, but the changes are due to deep burial, tectonic forces, and/or high temperature rather than surface conditions.

Because nearly all metamorphic rocks form deep within the Earth's crust, they provide geologists with many clues about conditions at depth. Therefore, understanding metamorphism will help you when we consider geologic processes involving Earth's internal forces. Metamorphic rocks are a feature of the oldest exposed rocks of the continents and of major mountain belts. They are especially important in providing evidence of what happens during subduction and plate convergence.

We also discuss hydrothermally deposited rocks and minerals, which are usually found in association with both igneous and metamorphic rocks. Hydrothermal ore deposits, while not volumetrically significant, are of great importance to the world's supply of metals.

HOW DO METAMORPHIC ROCKS FORM?

From your study so far of Earth materials and the rock cycle, you know that rocks change, given enough time, when their physical environment changes radically. In chapter 6, you saw how deeply buried rocks melt (or partially melt) to form magma when temperatures are high enough. What happens to rocks that are deeply buried but are not hot enough to melt? They become metamorphosed. **Metamorphism** refers to changes to rocks that take place in the Earth's interior. The changes may be new textures, new mineral assemblages, or both. Transformations occur in the solid state (meaning the rock does not melt). The new rock is a **metamorphic rock.**

As most metamorphism takes place in moderate to great depths in the Earth's crust, metamorphic rocks provide us with a window to processes that take place deep underground, beyond our direct observation. Metamorphic rocks are exposed over large regions because of erosion of mountain belts along with uplift due to isostatic adjustment. In fact, the cores of the continents are largely metamorphic rocks and granitic plutons. These form the stable interior of North America, the central lowlands between the Appalachians and the Rocky Mountains, and other ranges of western North America. Very ancient (Precambrian) complexes of metamorphic and intrusive igneous rocks are exposed over much of Canada (known as the *Canadian Shield*; see box 10.1). The inside front cover shows the Canadian Shield as the region underlain by Precambrian rocks. In the Great Plains of the United States similar rocks form the *basement* underlying a veneer of younger sedimentary rocks (see the brown area on the inside front cover map that the legend indicates is "Platform deposits on Precambrian basement"). Similar ancient metamorphic and plutonic rocks also form the stable cores or *cratons* of the other continental landmasses (e.g., Africa, Antarctica, Australia).

In nearly all cases, a metamorphic rock has a texture clearly different from that of the *original* rock, or **parent rock.** When limestone is metamorphosed to marble, for example, the fine grains of calcite coalesce and recrystallize into larger calcite crystals. The calcite crystals form a mosaic pattern that gives marble a texture distinctly different from that of the parent limestone. If the limestone is composed entirely of calcite, then metamorphism into marble involves no new minerals, only a change in texture.

More commonly, the various elements of a parent rock react chemically and crystallize into new minerals, thus making the metamorphic rock distinct both mineralogically and texturally from the parent rock. This is because the parent rock is unstable in its new environment. The old minerals recrystallize into new ones that are at *equilibrium* in the new environment. For example, clay minerals form at Earth's surface (see chapter 8). Therefore, they are stable at the low temperature and pressure conditions both at and just below the Earth's surface. When subjected to the temperatures and pressures deep within the Earth's crust, the clay minerals of a shale can recrystallize into coarse-grained mica. Everyone who has fashioned clay into pots and fired them knows that clay minerals will recrystallize at temperatures between 200 and 300°C. This is the temperature range of stability of the clay-bearing mudstones; as the temperature is increased by burial they are metamorphosed into shale and schist. There is no melting in these reactions, so it is a solid-state reaction—there are new minerals formed, a new texture to the rock, a loss of water, and an increase in density. Another example is that under appropriate temperature and pressure conditions, a quartz sandstone with a calcite cement metamorphoses as follows:

$$\underset{\text{calcite}}{CaCO_3} + \underset{\text{quartz}}{SiO_2} \rightarrow \underset{\substack{\text{wollastonite}\\ \text{(a mineral)}}}{CaSiO_3} + \underset{\substack{\text{carbon}\\ \text{dioxide}}}{CO_2}$$

This reaction was first interpreted from observations of textures of metamorphosed quartz sandstones. It has been reproduced experimentally, so it is known that the reaction takes place in the temperature range 400 to 700°C (depending on pressure), and there is no sign of melting. As the sedimentary rock is subjected to increasing temperatures, by burial or by igneous intrusion, this reaction to form *wollastonite* will take place. Many metamorphic rocks found on the Earth's surface exhibit contorted banding (figure 10.1). Commonly, banding in metamorphic rocks can be demonstrated to have originally been flat-lying sedimentary layering (even though the rock has since recrystallized). These rocks, now hard and brittle, would shatter if smashed with a hammer. But they must have been **ductile** (or *plastic*), capable of being bent and moulded under stress, to have been folded into such contorted patterns. Because high temperature and pressure are necessary to make rocks ductile, a reasonable conclusion is that these rocks formed at considerable depth, where such conditions exist. Moreover, crystallization of a magma would not produce contorted layering.

IN GREATER DEPTH 10.1

The Oldest Rocks on Earth

The oldest known rocks on Earth are metamorphic rocks and are found in ancient cratons, such as the Canadian Shield, that now form the centre of our continents. One of the oldest rocks dated is the Acasta Gneiss (box figure 1), a metamorphosed sedimentary rock found in the Northwest Territories. Zircon crystals contained within the Acasta Gneiss have been dated using radiometric techniques that measure uranium and thorium isotopes and their radioactive decay products, such as lead (see chapter 19). The Acasta Gneiss is thought to have formed approximately 3.96 billion years ago. Other gneisses from the same area of Canada are even older and have been dated at around 4.03 billion years ago.

Earth rocks of this age are rare because the dynamic geological processes that operate on our planet destroyed the rocks created during the first 500 million years of the Earth's history. However, metamorphic rocks such as the Acasta Gneiss can tell us much about early conditions on the Earth. Contorted white bands of quartz and feldspar in the gneiss are also common in immature sediments derived from granite, an igneous rock found in continental crust. Granites usually form when older basaltic crust melts in the presence of water rather than directly from mantle materials. This suggests that when the Acasta Gneiss formed around 4 billion years ago water was present on the Earth, and basaltic crust (which forms the floor of ocean basins) existed even before this time.

BOX 10.1 ■ FIGURE 1
Polished slab of Acasta Gneiss, Northwest Territories.
Photo by Nick Eyles

Related Web Resources

www.nmnh.si.edu/minsci/outreach/monthly/archive/rocks/acasta/

FIGURE 10.1
Contorted banding in metamorphic rocks of the Canadian Shield, Georgian Bay, Ontario.
Photo by Nick Eyles

WHAT FACTORS CONTROL THE CHARACTERISTICS OF METAMORPHIC ROCKS?

A metamorphic rock owes its characteristic texture and particular mineral content to several factors, the most important being (1) the composition of the parent rock before metamorphism, (2) temperature and pressure during metamorphism, (3) the effects of tectonic forces, and (4) the effects of fluids, such as water.

Composition of the Parent Rock

Usually no new elements or chemical compounds are added to the rock during metamorphism, except perhaps water. (Metasomatism, discussed later in this chapter, does involve the addition of other elements.) Therefore, the mineral content of the metamorphic rock is controlled by the chemical composition of the parent rock. For example, a basalt always metamorphoses into a rock in which the new minerals can collectively accommodate the approximately 50-percent silica and relatively high amounts of the oxides of iron, magnesium, calcium, and aluminum in the original rock. On the other hand, a limestone, composed essentially of calcite ($CaCO_3$), cannot metamorphose into a silica-rich rock.

Temperature

Heat, necessary for metamorphic reactions, comes primarily from the outward flow of geothermal energy from the Earth's deep interior. The deeper a rock is beneath the surface, the hotter it will be. The particular temperature for rock at a given depth depends on the local *geothermal gradient* (described in chapter 4). Additional heat could be derived from magma, if magma bodies are locally present.

A mineral is said to be *stable* if, given enough time, it does not react with another substance or convert to a new mineral or substance. Any mineral is stable only within a given temperature range. The stability temperature range of a mineral varies with factors such as pressure and the presence or absence of other substances. Some minerals are stable over a wide temperature range. Quartz, if not mixed with other minerals, is stable at atmospheric pressure (i.e., at the Earth's surface) up to about 800°C. At higher pressures, quartz remains stable to even higher temperatures. Other minerals are stable over a temperature range of only 100 or 200°C.

By knowing (from results of laboratory experiments) the particular temperature range in which a mineral is stable, a geologist may be able to deduce the temperature of metamorphism for a rock that includes that *index mineral* (see box 10.4 later in this chapter).

The upper limit on temperature in metamorphism overlaps the temperature of partial melting of a rock. If partial melting takes place, the component that melts becomes a magma; the solid residue remains a metamorphic rock. Temperatures at which the igneous and metamorphic realms can coexist vary considerably. For an ultramafic rock (containing only ferromagnesian silicate minerals) the temperature will be more than 1,200°C; for a metamorphosed shale under high water pressure, a granitic melt component can form in the metamorphic rock at temperatures as low as 650°C.

Pressure

Usually, when we talk about pressure, we mean **confining pressure;** that is, pressure applied equally on all surfaces of a substance as a result of burial or submergence. A diver senses confining pressure (known as *hydrostatic pressure*) proportional to the weight of the overlying water (figure 10.2). The pressure uniformly squeezes the diver's entire body surface. Likewise, an object buried deeply within Earth's crust is compressed by strong confining pressure, called *lithostatic pressure,* which forces grains closer together and eliminates pore space. For metamorphism, pressure is usually given in *kilobars.* A kilobar is 1,000 bars. A *bar* is very close (0.99) to standard atmospheric pressure, so that, for all practical purposes, a kilobar is the pressure equivalent of a thousand times the pressure of the atmosphere at sea level. The *pressure gradient,* the increase in lithostatic pressure with depth, is approximately 1 kilobar per each 3.3 kilometres of burial in crustal rock.

Any new mineral that has crystallized under high-pressure conditions tends to occupy less space than did the mineral or minerals from which it formed. The new mineral is denser than its low-pressure counterparts because the pressure forces atoms closer together into a more closely packed crystal structure.

But what if pressure and temperature both increase, as is commonly the case with increasing depth into the Earth? If the effect of higher temperature is greater than the effect of higher pressure, the new mineral will likely be less dense. A denser new mineral is likely if increasing pressure effects are greater than increasing temperature effects.

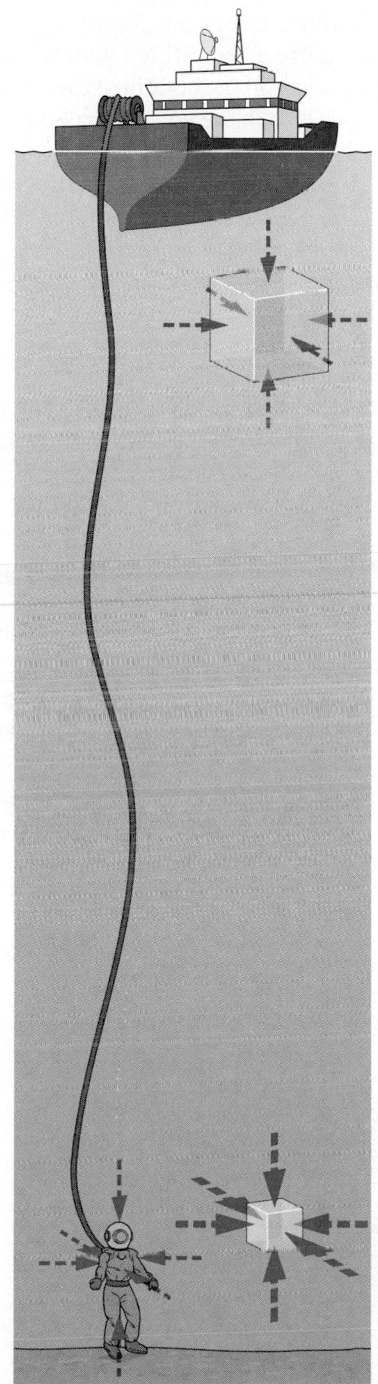

FIGURE 10.2

Confining pressure. The diver's suit is pressurized to counteract hydrostatic pressure. Object (cube) has a greater volume at low pressure than at high pressure.

Differential Stress

Most metamorphic rocks show the effects of tectonic forces. When forces are applied to an object, the object is under **stress.** If the forces on a body differ in different directions, a body is subjected to **differential stress.** Differential stress tends to deform objects into oblong or flattened forms. If you squeeze a rubber ball between your thumb and forefinger, the ball is under differential stress. If you squeeze a ball of dough, it will remain flattened after you stop squeezing, because dough is plastic (figure 10.3*A*). To illustrate the difference between confining pressure and differential stress, visualize a drum filled with water. If you place a ball of putty underwater in the bottom of the drum, the ball will not change its shape (its volume will decrease slightly due to the weight of the overlying water). Now take the putty ball out of the water and place it under the drum. The putty will be flattened into the shape of a pancake due to the differential stress. In this case, the putty is subjected to *compressive differential stress* or, more simply, **compressive stress** (as is the dough ball shown in figure 10.3*A*).

Differential stress also causes **shearing,** which causes parts of a body to move or slide relative to one another across a plane; an example is when you spread out a deck of cards on a table with your hand moving parallel to the table. Shearing often takes place perpendicular to, or nearly perpendicular to, the direction of compressive stress. If you put a ball of putty between your hands and slide your hands while compressing the putty, as shown in figure 10.3*B,* the putty flattens parallel to the shearing (the moving hands) as well as perpendicular to the compressive stress.

Some rocks can be attributed exclusively to shearing during faulting (movement of bedrock along a fracture, described in chapter 15) in a process sometimes called *dynamic metamorphism.* Rocks in contact along the fault are broken and crushed when movement takes place. A *mylonite* is an unusual rock that is formed from pulverized rock in a fault zone (see box 10.2). The rock is streaked out parallel to the fault in darker and lighter components due to shearing. Mylonites are hypothesized to form at a depth of around a kilometre or so, where the rock is still cool and brittle (rather than ductile), but the pressure is sufficient to compress the pulverized rock into a compact, hard rock. Where found, they occupy zones that are only about a metre or so wide. In the Parry Sound region of Ontario, mylonite is quarried for use as a decorative building stone.

Foliation

Differential stress has a very important influence on the texture of a metamorphic rock because it forces the constituents of the rock to become parallel to one another. For instance, the pebbles in the metamorphosed conglomerate shown in figure 10.4 were originally more spherical but have been flattened by differential stress. When a rock has a planar texture, it is said to be **foliated.** Foliation is manifested in various ways. If a platy mineral (such as mica) is crystallizing within a rock that is undergoing differential stress, the mineral grows in such a way that it remains parallel to the direction of shearing or perpendicular to the direction of compressive stress (figure 10.5). Any platy mineral attempting to grow against shearing is either ground up or forced into alignment. Minerals that crystallize in needlelike shapes (for example, hornblende) behave similarly, growing with their long axes parallel to the plane of shearing or perpendicular to compressive stress. The three very different textures described below (from lowest to highest degree of metamorphism) are all variations of foliation and are important in classifying metamorphic rocks:

1. If the rock splits easily along nearly flat and parallel planes, indicating that pre-existing, microscopic, platy minerals were pushed into alignment during metamorphism, we say the rock is **slaty,** or that it possesses **slaty cleavage.**
2. If visible platy or needle-shaped minerals have grown essentially parallel to a plane due to differential stress, the rock is **schistose** (figure 10.6).
3. If the rock became very ductile and the new minerals separated into distinct (light and dark) layers or lenses, the rock has a layered or **gneissic** texture, such as in figures 10.1 and 10.14.

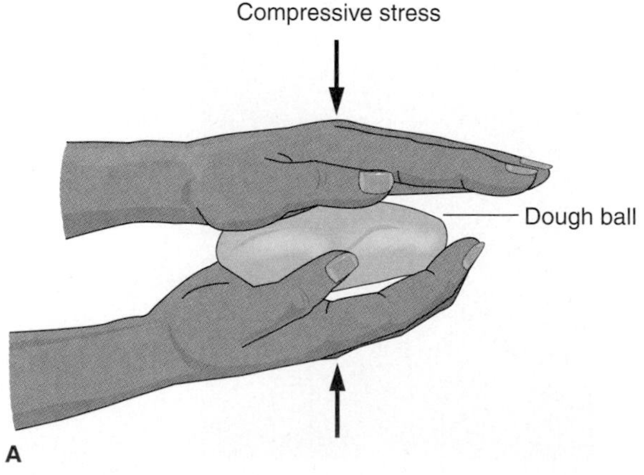

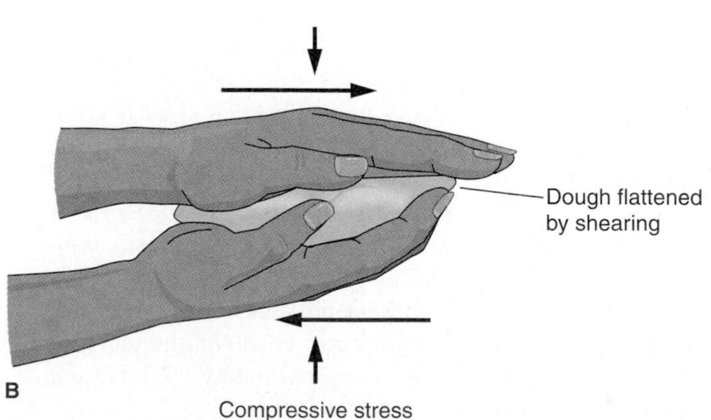

FIGURE 10.3

(*A*) Compressive stress exerted on a ball of putty by two hands. More force is exerted in the direction of arrows than elsewhere on the putty.
(*B*) Shearing takes place as two hands move parallel to each other at the same time that some compressive force is exerted perpendicular to the flattening putty.

GEOMATICS 10.2

3-D Visualization and Analysis of Metamorphic Deformation in Mylonite

Mylonite is a metamorphic rock formed under conditions of intense shearing in fault zones (box figure 1). The rock has a streaky or banded structure formed as rock components were stretched and sheared during metamorphism (box figure 2). Analysis of the internal structure of mylonite using newly developed geomatics techniques can help geoscientists reconstruct the type of deformation that affected the rock during metamorphism and can provide valuable information regarding the geologic history of the area in which it formed. Hand specimens of mylonite are cut into a series of thin slices, which are scanned and digitized. The digitized data are then used to create a virtual 3-D reconstruction of the internal composition of the sample showing the orientation and form of individual recrystalized mineral grains called porphyroclasts. Analysis of the distribution, shape, and orientation of grains within a mylonite can provide information regarding the type of strain that affected the rock and its history of deformation (box figure 3).

BOX 10.2 ■ FIGURE 1
Outcrop of mylonite (approximately 1.5 m high) showing sheared dark and light layers in the Parry Sound Shear Zone, Ontario. Pink coloured rock in upper right forms part of a pegmatite dike.
Photo by Nick Eyles

BOX 10.2 ■ FIGURE 2
Close-up photograph of mylonite in the Parry Sound area, Ontario.
Photo by Nick Eyles

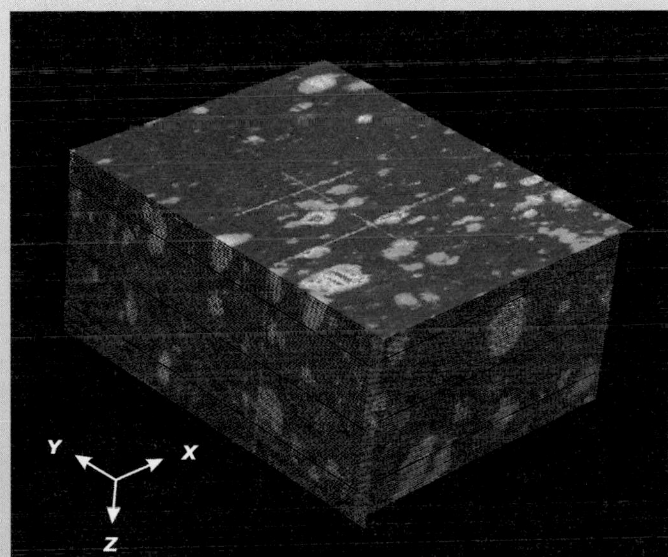

BOX 10.2 ■ FIGURE 3
3-D image of a mylonite sample showing position and orientation of mineral grains within the sample. Analysis of the fabric patterns shows that the grains exhibit orthorhombic symmetry and are characterized by three mutually perpendicular axes of symmetry.
Image from Eric Grunsky and Eric de Kemp, Geological Survey of Canada

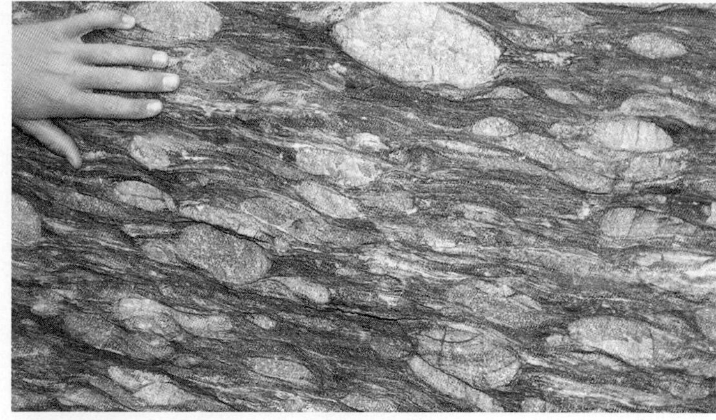

A

B

FIGURE 10.4

(*A*) Metamorphosed pebble conglomerate in which the pebbles have been flattened (sometimes called a stretched pebble conglomerate). Compare with Figure 9.9. (*B*) Pebbles have been stretched and squeezed further to form a banded gneiss. The elongate form of the original pebbles can still be seen. Both photos at Hwy 41, north of Kaladar (Grenville Province), Ontario.

Photos by Nick Eyles

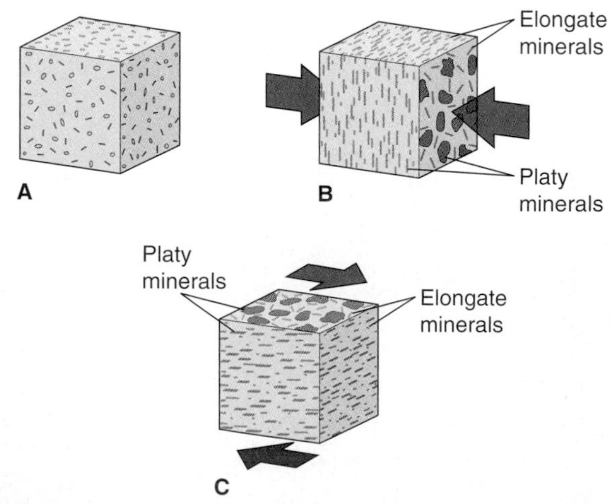

FIGURE 10.5

Orientation of platy and elongate minerals in metamorphic rock. (*A*) Platy minerals randomly oriented (e.g., clay minerals before metamorphism). No differential stress involved. (*B*) Platy minerals (e.g., mica) and elongate minerals (e.g., amphibole) have crystallized under the influence of compressive stress. (*C*) Platy and elongate minerals developed under the influence of shearing.

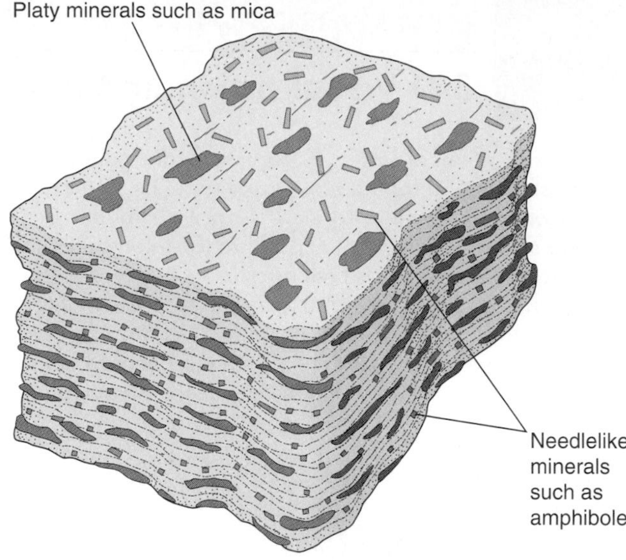

FIGURE 10.6

Schistose texture.

Fluids

Hot water (as vapour) is the most important fluid involved in metamorphic processes, although other gases, such as carbon dioxide, sometimes play a role. The water may have been trapped in a parent sedimentary rock or given off by a cooling pluton. Water may also be given off from minerals that have water in their crystal structure (e.g., clay, mica). As temperature rises during metamorphism and a mineral becomes unstable, its water is released.

Water is thought to help trigger metamorphic chemical reactions. Water, moving through fractures and along grain margins, is a sort of intra-rock rapid transit for ions. Under high pressure, it moves between grains, dissolves ions from one mineral, and then carries these ions elsewhere in the rock where they can react with the ions of a second mineral. The new mineral that forms is stable under the existing conditions. The role of fluids in metamorphism is discussed in more detail in the section on hydrothermal processes at the end of this chapter.

Time

The effect of time on metamorphism is hard to comprehend. Most metamorphic rocks are composed predominantly of silicate minerals, and silicate compounds are notorious for their sluggish chemical reaction rates. Recently, garnet crystals taken from a metamorphic rock collected in Vermont were analyzed and scientists calculated a growth rate of 1.4 millimetres per million years. The garnets' growth was sustained over a 10.5-million-year period. Many laboratory attempts to duplicate metamorphic reactions that occur in nature have been frustrated by the time element. The several million years during which a particular combination of temperature and pressure may have prevailed in nature are impossible to duplicate.

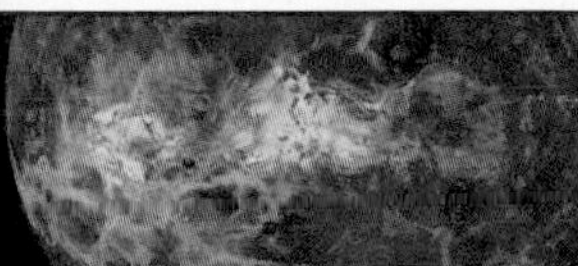

PLANETARY GEOLOGY 10.3

Impact Craters and Shock Metamorphism

The spectacular collision of the comet Shoemaker-Levy with Jupiter in 1994 served to remind us that asteroids and comets occasionally collide with a planet. Earth is not exempt from collisions. Large meteorites have produced impact craters when they have collided with the Earth's surface. One well-known meteorite crater is Manicouagan Crater in Quebec, which is almost 100 kilometres in diameter (box figure 1). Many other large craters are known in Canada, Germany, Australia, and other places. The Sudbury Structure in Ontario is a large-impact crater that formed around 1.85 billion years ago and is thought to have had an original diameter of between 150 and 225 km (box figure 2). Later deformation of the circular crater during an orogenic event created the oval shape of the structure that we see today.

Impact craters display an unusual type of metamorphism called *shock metamorphism.* The sudden impact of a large extraterrestrial body results in brief, but extremely high pressures. Quartz may recrystallize into the rare SiO_2 mineral coesite. Quartz that is not as intensely impacted suffers damage (detectable under a microscope) to its crystal lattice (box figure 3).

The impact of a meteorite also may generate enough heat to locally melt rock. Molten blobs of rock are thrown into the air and become streamlined in the Earth's atmosphere before solidifying into what are called *tektites.* Tektites may be found hundreds of kilometres from the point of meteorite impact.

The intense shock caused by a meteorite impact also creates large faults that can be filled with crushed and partially melted rocks, called *pseudotachylytes.* The broken and fragmented rocks around the Sudbury Structure in Ontario form the largest known pseudotachylyte in the world and are host to rich ore deposits.

A large meteorite would also throw large quantities of dust high into the atmosphere. According to one theory, the change in global climate due to a meteorite impact around 65 million years ago caused extinctions of many varieties of creatures (refer to box 8.2 on the extinction of dinosaurs). Evidence for this impact includes finding tiny fragments of shock metamorphosed quartz and tektites in sedimentary rock that is 65 million years old.

Shock metamorphosed rock fragments are much more common on the moon than on Earth. There may be as many as 400,000 craters larger than a kilometre in diameter on the moon. Mercury's surface is remarkably similar to that of the moon. Our two neighbouring planets, Venus and Mars, are not as extensively cratered as is the moon.

Related Web Resources

Meteor Crater

www.meteorcrater.com/

See an animation of the meteorite impact. Go to "reference information" for details about the meteorite impact.

BOX 10.3 ■ FIGURE 1

Landsat image of Manicouagan Crater in Quebec.

U.S. Geological Survey

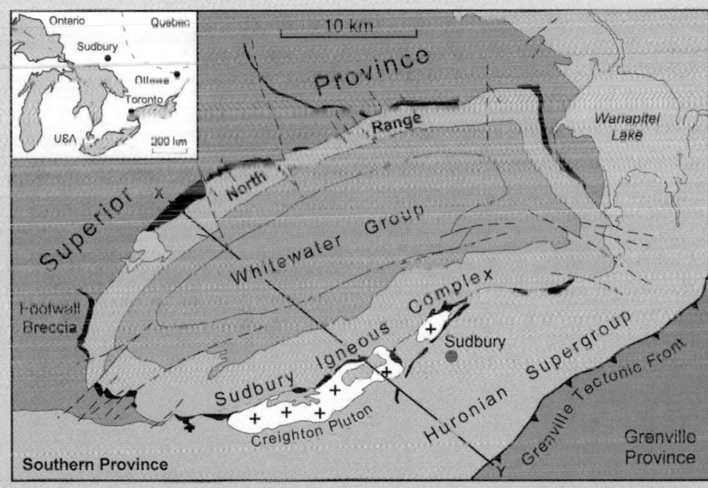

BOX 10.3 ■ FIGURE 2

The Sudbury Structure in Ontario is an ancient meteorite-impact crater. The Sudbury Igneous Complex formed by the melting of rocks during impact and shocked and partially melted rocks form the Footwall Breccia (black).

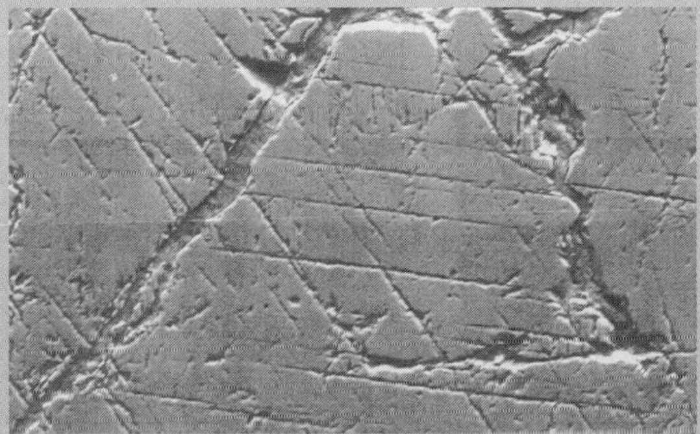

BOX 10.3 ■ FIGURE 3

Shocked quartz crystal with veins of pseudotachylyte and criss-crossing fractures.

Glikson et al. 2005 figure 2c (published with the permission of the Geological Society of Australia).—A. Y. Glikson, S. Eggins, D. S. Golding, P. W. Haines, R. P. Iasky, T. P. Mernagh, A. J. Mory, F. Pirajno & I. T. Uysal 2005. Microchemistry and microstructures of hydrothermally altered shock-metamorphosed basement gneiss, Woodleigh impact structure, Southern Carnarvon Basin, Western Australia. Australian Journal of Earth Sciences 52.555–573.

HOW DO WE CLASSIFY METAMORPHIC ROCKS?

As we noted before, the kind of metamorphic rock that forms is determined by the metamorphic environment (primarily the particular combination of pressure, stress, and temperature) and by the chemical constituents of the parent rock. Many kinds of metamorphic rocks exist because of the many possible combinations of these factors. These rocks are classified based on broad similarities. (Appendix B contains a systematic procedure for identifying common metamorphic rocks.) The relationship of texture to rock name is summarized in table 10.1.

First, consider the texture of a metamorphic rock. Is it *nonfoliated* or *foliated* (figure 10.7)?

Nonfoliated Rocks

If the rock is nonfoliated, it is named on the basis of its composition. The two most common nonfoliated rocks are marble and quartzite, composed, respectively, of calcite and quartz.

Marble, a coarse-grained rock composed of interlocking calcite crystals (figure 10.8), forms when limestone recrystallizes during metamorphism. If the parent rock is dolomite, the recrystallized rock is a *dolomite marble.* Marble has long been valued as a building material and as a material for sculpture, partly because it is easily cut and polished and partly because it reflects light in a shimmering pattern, a result of the excellent cleavage of the individual calcite crystals. Marble is, however, highly susceptible to chemical weathering (see chapter 8).

TABLE 10.1 Classification and Naming of Metamorphic Rocks (Based Primarily on Texture)

NONFOLIATED			
Name Based on Mineral Content of Rock			
Usual Parent Rock	**Rock Name**	**Predominant Minerals**	**Identifying Characteristics**
Limestone	Marble	Calcite	Coarse interlocking grains of calcite (or, less commonly, dolomite)
Dolomite	Dolomite marble	Dolomite	Calcite (or dolomite) has rhombohedral cleavage; hardness intermediate between glass and fingernail. Calcite effervesces in weak acid
Quartz sandstone	Quartzite	Quartz	Rock composed of interlocking small granules of quartz. Has a sugary appearance and vitreous lustre; scratches glass
Shale	Hornfels	Fine-grained micas	A fine-grained, dark rock that generally will scratch glass. May have a few coarser minerals present
Basalt	Hornfels	Fine-grained ferromagnesian minerals, plagioclase	
FOLIATED			
Name Based Principally on Kind of Foliation Regardless of Parent Rock. Adjectives Describe the Composition (e.g., biotite-garnet schist)			
Texture	**Rock Name**	**Typical Characteristic Minerals**	**Identifying Characteristics**
Slaty	Slate	Clay and other sheet silicates	A very fine-grained rock with an earthy lustre. Splits easily into thin, flat sheets
Intermediate between slaty and schistose	Phyllite	Mica	Fine-grained rock with a silky lustre. Generally splits along wavy surfaces
Schistose	Schist	Biotite and muscovite amphibole	Composed of visible platy or elongated minerals that show planar alignment. A wide variety of minerals can be found in various types of schist (e.g., garnet-mica schist, hornblende schist, etc.).
Gneissic	Gneiss	Feldspar	Light and dark minerals are found in separate, parallel layers or lenses. Commonly, the dark layers include biotite and hornblende; the light-coloured layers are composed of feldspars and quartz. The layers may be folded or appear contorted

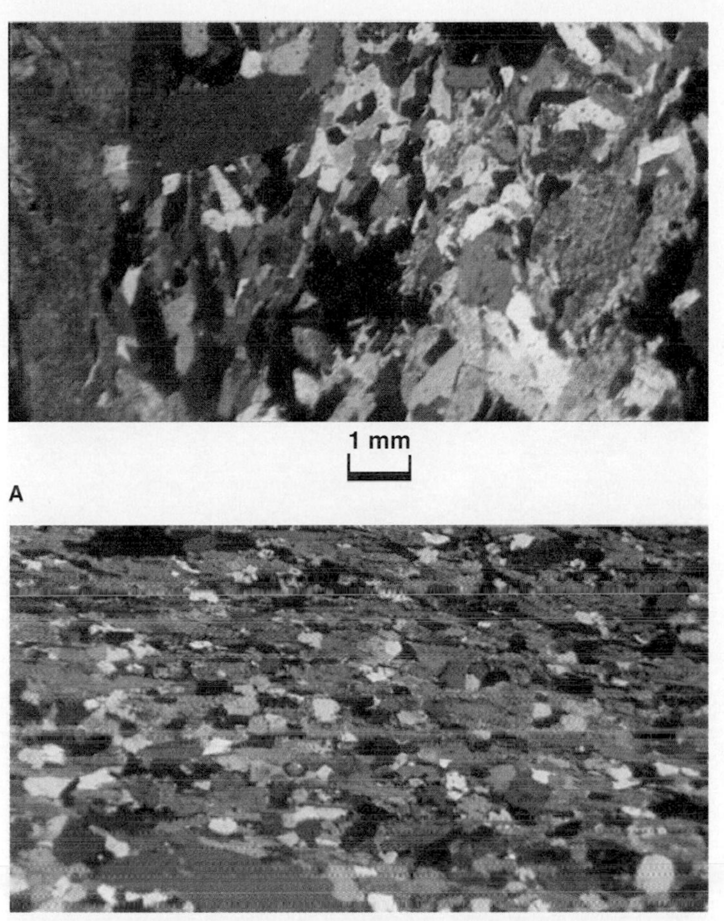

FIGURE 10.7

Photomicrographs of metamorphic rocks taken through a polarizing microscope. (*A*) nonfoliated rock and (*B*) foliated rock.

Photos by C. C. Plummer

Quartzite (figure 10.9) is produced when grains of quartz in sandstone are welded together while the rock is subjected to high temperature. This makes it as difficult to break along grain boundaries as through the grains. Therefore, quartzite, being as hard as a single quartz crystal, is difficult to crush or break. It is the most durable of common rocks used for construction, both because of its hardness and because quartz is not susceptible to chemical weathering.

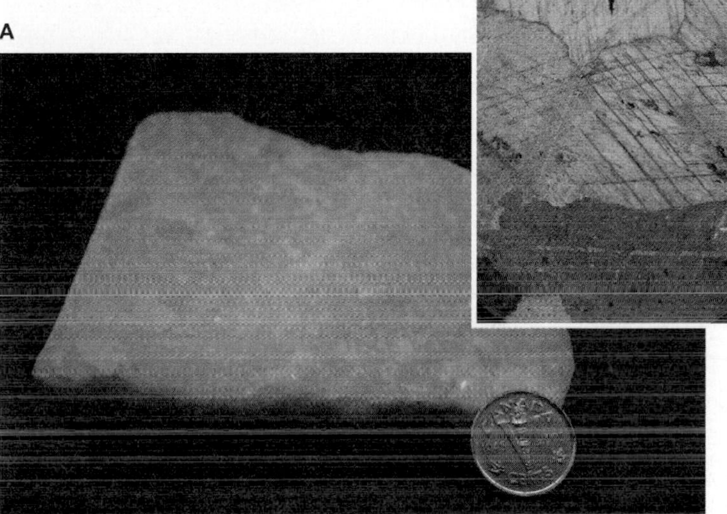

FIGURE 10.8

(*A*) Hard specimen of marble. (*B*) Photomicrograph showing interlocking crystals of calcite. Each crystal is approximately 2 millimetres across.

Photo A by S. Fletcher; Photo B by C. C. Plummer

FIGURE 10.9

Quartzite. (*A*) Cross-bedded quartzite at Whitefish Falls, Ontario. The quartzite originally formed as a quartz sand fluvial deposit and was later metamorphosed. (*B*) Hand specimen of quartzite. (*C*) Photomicrograph taken using a polarizing microscope. Interlocking quartz crystals are about 0.5 mm across.

Photo A by C. H. Eyles; Photo B by S. Fletcher; Photo C by C. C. Plummer

Hornfels is a very fine-grained, nonfoliated, metamorphic rock whose parent rock is either shale or basalt. If it forms from shale, characteristically only microscopically visible micas form from the shale's clay minerals. Sometimes a few minerals grow large enough to be seen with the naked eye; these are minerals that are especially capable of crystallizing under the particular temperature attained during metamorphism. If hornfels forms from basalt, amphibole, rather than mica, is the predominant fine-grained mineral produced.

Foliated Rocks

If the rock is foliated, you need to determine the type of foliation to name the rock. For example, a schistose rock is called a *schist.* But this name tells us nothing about what minerals are in this rock, so we add adjectives to describe the composition—for example, *garnet-mica schist.* The following are the most common foliated rocks progressing from lower grade (they usually form at lower temperatures) to higher grade:

Slate is a very fine-grained rock that splits easily along flat, parallel planes (figure 10.10). Although some slate forms from volcanic ash, the usual parent rock is shale. Slate develops under temperatures and pressures only slightly greater than those found in the sedimentary realm. The temperatures are not high enough for the rock to thoroughly recrystallize. The important controlling factor is differential stress. The original clay minerals partially recrystallize into equally fine-grained, platy minerals. Under differential stress, the old and new platy minerals are aligned, creating slaty cleavage in the rock. A slate indicates that a relatively cool and brittle rock has been subjected to intense tectonic activity.

Because of the ease with which it can be split into thin, flat sheets, slate is used for making chalkboards, pool tables, and roofs.

Phyllite is a rock in which the newly formed micas are larger than the platy minerals in slate but still cannot be seen with the naked eye. This requires a further increase in temperature over that needed for slate to form. The very fine-grained mica imparts a satin sheen to the rock, which may otherwise closely resemble slate (figure 10.11). The slaty cleavage may be crinkled in the process of conversion of slate to phyllite.

A **schist** is characterized by megascopically visible, approximately parallel-oriented minerals. Platy or elongate minerals that crystallize from the parent rock are clearly visible to the naked eye. Which minerals form depends on the particular combination of temperature and pressure prevailing during recrystallization as well as the composition of the parent rock. Two, of several, schists that form from shale are *mica schist* and *garnet-mica schist* (figure 10.12). Although they both have the same parent rock, they form under different combinations of temperature and pressure. If the parent rock is basalt, the schists that form are quite different. If the predominant ferromagnesian mineral that forms during metamorphism of basalt is amphibole, it is an *amphibole schist* (also called an *amphibolite*). At a lower grade, the predominant mineral is chlorite, a green micaceous mineral, in a *chlorite schist* (or *greenschist*).

FIGURE 10.10
(*A*) Slate outcrop in Antarctica. (*B*) Hard specimen of slate.
Photo A by P. D. Rowley, U.S. Geological Survey; Photo B by C. H. Eyles

FIGURE 10.11
Phyllite, exhibiting a crinkled, silky-looking surface.
Photo by C. H. Eyles

FIGURE 10.12
Garnet-mica schist. Red garnet crystals give the rock a "raisin bread" appearance.
Photo by Nick Eyles

FIGURE 10.13
Gneiss.
Photo by C. H. Eyles

Gneiss is a rock consisting of light and dark mineral layers or lenses. The highest temperatures and pressures have changed the rock so that minerals have separated into layers. Platy or elongate minerals (such as mica or amphibole) in dark layers alternate with layers of light-coloured minerals of no particular shape. Within the light-coloured layers, coarse feldspars have crystallized. In composition, a gneiss may resemble granite or diorite, but it is distinguishable from those plutonic rocks by its foliation (figures 10.1 and 10.13).

Temperature conditions under which a gneiss develops approach those at which granite solidifies. It is not surprising, then, that the same minerals are found in gneiss and in granite. In fact, a previously solidified granite can be converted to a gneiss under appropriate pressure and temperature conditions and if the rock is under differential stress.

WHERE DOES METAMORPHISM OCCUR?

The two most common types of metamorphism are *contact metamorphism* and *regional metamorphism*. Hydrothermal processes, in which hot water plays a major role during metamorphism, are discussed later in this chapter. See box 10.2 for a discussion of shock metamorphism.

Contact Metamorphism

Contact metamorphism (also known as *thermal* metamorphism) is metamorphism in which high temperature is the dominant factor. Confining pressure may influence which new minerals crystallize; however, the confining pressure is usually relatively low. This is because contact metamorphism mostly takes place not too far beneath Earth's surface (less than 10 kilometres). Contact metamorphism occurs adjacent to a pluton when a body of magma intrudes relatively cool country rock. The process can be thought of as the "baking" of country rock adjacent to an intrusive contact; hence the term *contact metamorphism*. The zone of contact metamorphism (also called an *aureole*) is usually quite narrow—generally from 1 to 100 metres wide. Differential stress is rarely significant. Therefore, the most common rocks found in an aureole are the nonfoliated rocks: marble when igneous rock intrudes limestone; quartzite when quartz sandstone is metamorphosed; hornfels when shale is scorched.

Marble and quartzite also form under conditions of regional metamorphism. When grains of calcite or quartz recrystallize, they tend to be equidimensional, rather than elongate or platy. For this reason, marble and quartzite do not usually exhibit foliation, even though subjected to differential stress during metamorphism.

Regional Metamorphism

The great majority of the metamorphic rocks found on Earth's surface are products of **regional metamorphism** (also known as *dynamothermal* metamorphism), which is metamorphism that takes place at considerable depth underground (generally greater than 5 kilometres). Regional metamorphic rocks are almost always foliated, indicating differential stress during recrystallization. Metamorphic rocks are prevalent in the most intensely deformed portions of mountain ranges. They are visible where once deeply buried cores of mountain ranges are exposed by erosion. Furthermore, large regions of the continents are underlain by metamorphic rocks, thought to be the roots of ancient mountains long since eroded down to plains or rolling hills. The Grenville Province of the Canadian Shield is the eroded remnant of the once-majestic Grenville Mountains, which formed around 1.3 billion years ago (see box 10.4).

Temperatures during regional metamorphism vary widely. Usually, the temperatures are in the range of 300 to 800°C. Temperature at a particular place depends to a large extent on depth of

IN GREATER DEPTH 10.4

Index Minerals

Certain minerals can form only under a restricted range of pressure and temperature. Stability ranges of these minerals have been determined in laboratories. When found in metamorphic rocks, these minerals can help us infer, within limits, what the pressure and temperature conditions were during metamorphism. For this reason, they are known as *index minerals.* Among the best known are *andalusite, kyanite,* and *sillimanite.* All three have an identical chemical composition (Al_2SiO_5), but different crystal structures. They are found in metamorphosed shales that have an abundance of aluminum. Box figure 1 is a phase diagram showing the pressure/temperature fields in which each is stable. Box figure 2 is a map showing metamorphic patterns across the Grenville Province of the Canadian Shield. These patterns were established using the minerals andalusite-sillimanite-kyanite.

If andalusite is found in a rock, this indicates that pressures were relatively low. Andalusite is often found in contact-metamorphosed shales (hornfels). Kyanite, when found in schists, is regarded as an indicator of high pressure; but note that the higher the temperature of the rock, the greater the pressure needed for kyanite to form. Sillimanite is an indicator of high temperature and can be found in some contact metamorphic rocks adjacent to very hot intrusions as well as in regionally metamorphosed schists and gneisses that formed at considerable depths.

Note that if you find all three minerals in the same rock, and could determine that they were mutually stable, you could infer that the temperature was close to 500°C and the confining pressure was almost 4 kilobars during metamorphism. (A kilobar is equivalent to the pressure of approximately 1,000 atmospheres.)

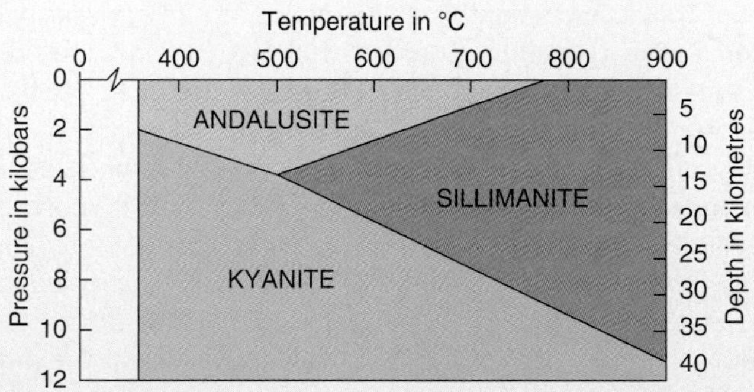

BOX 10.4 ■ FIGURE 1

Phase diagram showing the stability relationships for the Al_2SiO_5 minerals.

M. J. Holdaway, 1971, American Journal of Science, v. 271. Reprinted by permission of American Journal of Science and Michael J. Holdaway.

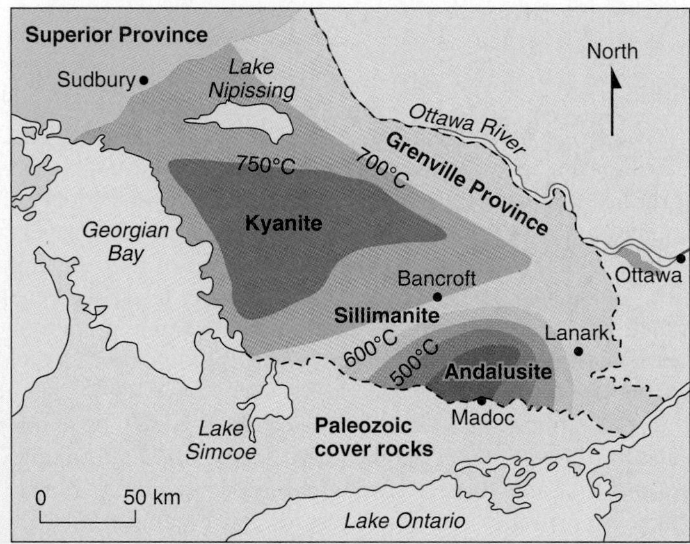

BOX 10.4 ■ FIGURE 2

Regional metamorphic patterns across the Grenville Province of the Canadian Shield. Coloured bands represent reconstructed burial temperatures based on minerals present in the metamorphic rocks. Higher grades of metamorphism occur in the west of the Grenville Province and indicate deeper burial and higher temperatures in that area.

burial and the geothermal gradient of the region (see box 10.5). Locally, temperature may also increase because of heat given off by nearby magma bodies. The high confining pressure is due to burial under 5 or more kilometres of rock. The differential stress is due to tectonism; that is, the constant movement and squeezing of the crust during mountain-building episodes.

Temperatures and pressures during metamorphism can be estimated through the results of laboratory experimental studies of minerals. In many cases, we can estimate temperature and pressure by determining the conditions under which an assemblage of several minerals can coexist. In some instances, a single mineral, or *index mineral,* suffices for determining the pressure and temperature combination under which a rock recrystallized (box 10.4).

Depending on the pressure and temperature conditions during metamorphism, a particular parent rock may recrystallize into one of several metamorphic rocks. For example, if basalt is metamorphosed at relatively low temperatures and pressures, it will recrystallize into a *greenschist,* a schistose rock containing chlorite (a green sheet-silicate), actinolite (a green amphibole), and sodium-rich plagioclase or it will recrystallize into a *greenstone,* a rock that has similar minerals but is not foliated. (A greenstone would indicate that the tectonic forces were not strong enough to induce foliation while the basalt was recrystallizing.) At higher temperatures and pressures, the same basalt would recrystallize into an *amphibole schist* (also called *amphibolite*), a rock composed of hornblende, plagioclase feldspar, and, perhaps, garnet. Metamorphism of other parent rocks under conditions similar to those that produce amphibole schist from basalt should produce the metamorphic rocks shown in table 10.2.

IN GREATER DEPTH 10.5

Metamorphic Facies and the Relationship to Plate Tectonics

During the early part of the twentieth century, geologists in Scandinavia introduced the concept of metamorphic facies. They noted that metamorphic rocks that were once basalts had one set of minerals in some parts of Europe, but in other areas, metabasalts had quite different mineral assemblages. As these rocks were *chemically* similar, the different *mineral assemblages* were regarded as indicating significantly different pressure and temperature conditions during metamorphism. Rocks having the same mineral assemblage are regarded as belonging to the same *metamorphic facies,* implying that they formed under broadly similar pressure and temperature conditions. The name for each facies is based on the assemblage of minerals or the name of a rock common to that facies. For instance, a schistose metabasalt composed mostly of the minerals chlorite, actinolite, and epidote (all of which are green minerals) belongs to the *greenschist facies.* On the other hand, rocks of the same chemical composition (metabasalts) belonging to the *amphibolite facies* are largely made up of hornblende and garnet. (Do not try to remember the names of the facies or their compositions; your aim should be to understand the concept.)

Based on the geologic setting, early workers inferred that the temperature conditions during metamorphism were lower for the greenschist facies than for the amphibolite facies. Laboratory work has since confirmed this as well as determined the pressure and temperature *stability fields* for each of the facies (box figure 1).

The concept of metamorphic facies is analogous to defining climatic zones by the combinations of plants found in each zone. A place where ferns, palm trees, and vines flourish corresponds to a climate with warm temperatures and abundant rainfall. On the other hand, a combination of palm trees, cactus, and sagebrush implies a hot, dry climate.

By identifying the metamorphic facies of rocks currently cropping out on the surface, geologists can infer, within broad limits, the depth at which metamorphism took place. They may also (again, within broad limits) be able to determine the corresponding temperature.

The concept of metamorphic facies preceded plate tectonics theory by several decades. Although earlier geologists were able to relate the individual facies to pressure and temperature combinations, they had no satisfactory explanation for the environments that produced the various combinations. Figure 10.16, which relates the temperatures of regional metamorphism to plate tectonics, may be used to infer the environment for each of the metamorphic facies shown in box figure 1. Box figure 2 shows the likely distribution of metamorphic facies across the same converging boundary as in figure 10.16. To understand the relationship, study box figures 1 and 2 as well as figure 10.16.

If one were to determine the geothermal gradient represented by the three vertical lines marked *A, B,* and *C* on figure 10.16 and box figure 2, the temperatures for particular depths should plot on the corresponding arrows shown in box figure 1. Follow arrow A in box figure 1. This means that if you were able to drill vertically downward along line *A,* you would find, beneath unmetamorphosed rocks, rocks of the zeolite facies. At a greater depth would be the boundary between the zeolite and prehnite-pumpellyite facies. You would reach this boundary at a depth where the pressure is about 4 kilobars and the temperature is approximately 150°C. Your drill would penetrate rocks of the prehnite-pumpellyite facies until you reached the blueschist facies. In hole *B,* in the interior of the plate, the progression would be from zeolite facies to prehnite-pumpellyite facies to greenschist facies to amphibolite facies to granulite facies. Hole *C,* in the volcanic-plutonic complex, would not pass through the prehnite-pumpellyite facies but would go from zeolite facies to greenschist facies to amphibolite facies to granulite facies.

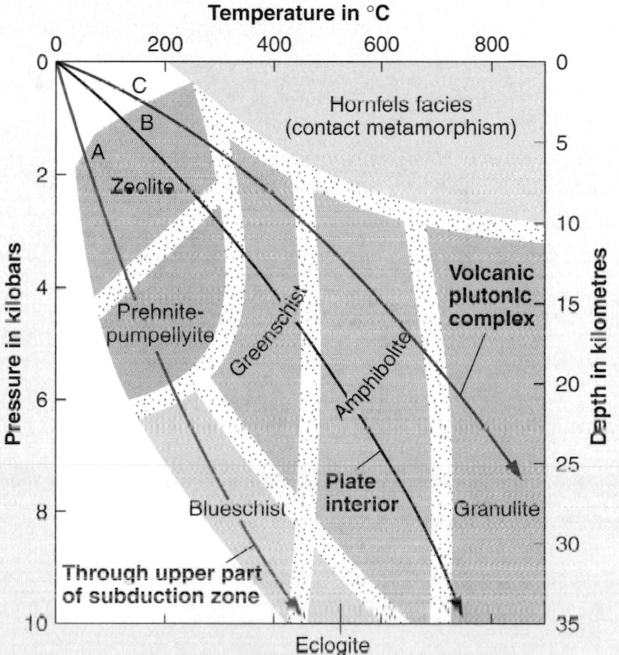

BOX 10.5 ■ FIGURE 1

The metamorphic facies. Facies are named after minerals (prehnite, zeolite, pumpellyite) or rock types (e.g., blueschist, granulite). Boundaries between facies are approximate. The arrows represent increases in temperature with depth for the three lines labelled *A, B, C* in figure 2 and in figure 10.16.

From W. G. Ernst, Metamorphism and Plate Tectonic Regimes. *Stroudsberg, Pa.: Dowden, Hutchinson & Ross, 1975; p. 425. Reprinted by permission of the author*

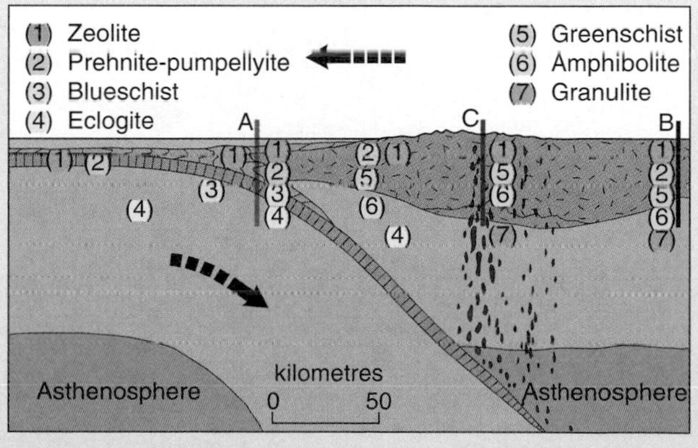

BOX 10.5 ■ FIGURE 2

Schematic representation of the distribution of facies across a convergent plate boundary.

From W. G. Ernst, Metamorphism and Plate Tectonic Regimes. *Stroudsburg, Pa.: Dowden, Hutchinson & Ross, 1975; p. 426. Reprinted by permission of the publisher*

TABLE 10.2 Regional Metamorphic Rocks That Form under Approximately Similar Pressure and Temperature Conditions

Parent Rock	Rock Name	Predominant Minerals
Basalt	Amphibole schist (amphibolite)	Hornblende, plagioclase, garnet
Shale	Mica schist	Biotite, muscovite, quartz, garnet
Quartz sandstone	Quartzite	Quartz
Limestone or dolomite	Marble	Calcite or dolomite

The minerals present in a rock indicate its *metamorphic grade*. Low-grade rocks formed under relatively cool temperatures and high-grade rocks at high temperatures, whereas medium-grade rocks recrystallized at around the middle of the range of metamorphic temperatures. Greenschist and greenstone are regarded as low-grade rocks, while amphibole schist is regarded as a medium-grade rock.

Prograde Metamorphism

When a rock becomes buried to increasingly greater depths, it is subjected to increasingly greater temperatures and pressures and will undergo *prograde metamorphism*—that is, it recrystallizes into a higher-grade rock. To show how rocks are changed by regional metamorphism, we look at what happens to shale during prograde metamorphism as progressively greater pressure and temperature act on a rock type with increasing depth in Earth's crust (figure 10.14).

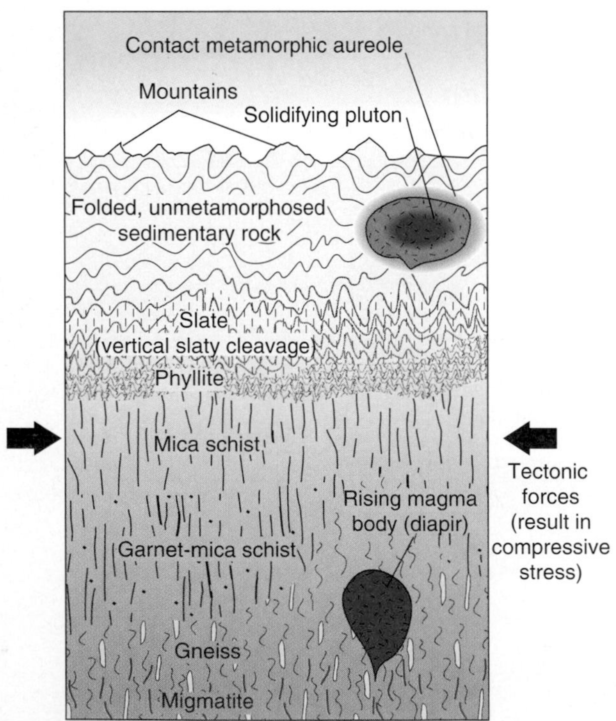

FIGURE 10.14

Schematic cross-section representing an approximately 30-kilometre vertical section through the Earth's crust during metamorphism. Rock names given are those produced from shale.

Slate, which looks quite similar to the shale from which it forms, is the lowest-grade rock in progressive metamorphism. Its slaty cleavage develops as a result of differential stress during incipient recrystallization of clay minerals to other platy minerals. As described earlier, phyllite is a rock that is transitional between slate and schist and, as such, we expect it to have formed at a depth between where slate and schist form.

Schist forms at higher temperatures and usually higher pressures than does phyllite. However, schist with shale as a parent rock forms over a wide range of temperatures and pressures. Figure 10.14 indicates the metamorphic setting for two varieties of schist (there are a number of others) that form from shale. *Mica schist* indicates a grade of metamorphism slightly higher than that of phyllite. Garnet requires higher temperatures to crystallize in a schist, so the *garnet-mica schist* probably formed at a deeper level than that of mica schist.

If schist is subjected to high enough temperatures, its constituents become more mobile and the rock recrystallizes into gneiss. The constituents of feldspar migrate (probably as ions) into planes of weakness caused by differential stress where feldspars, along with quartz, crystallize to form light-coloured layers. The ferromagnesian minerals remain behind as the dark layers.

If the temperature is high enough, partial melting of rock may take place, and a magma collects in layers within the foliation planes of the solid rock. After the magma solidifies, the rock becomes a **migmatite,** a mixed igneous and metamorphic rock (figure 10.15). A migmatite can be thought of as a "twilight zone" rock that is neither fully igneous nor entirely metamorphic.

The metamorphic rocks that we see usually have minerals that formed at or near the highest temperature reached during metamorphism. But why doesn't a rock recrystallize to one stable at lower temperature and pressure conditions during its long journey to the surface, where we now find it? The answer is that water is usually available during prograde metamorphism and the rock is relatively dry after reaching its peak temperatures. The absence of water means that chemical reaction will be prohibitively slow at the cooler temperatures. Substantial *retrograde metamorphism* occurs only if additional water is introduced to the rock after peak metamorphism. Tectonic forces at work during the peak of metamorphism fracture the rock extensively and permit water to get to the mineral grains. After tectonic forces are relaxed, the rocks move upward as a large block as isostatic adjustment takes place. It is unusual to find rocks that indicate retrograde metamorphism. These are rocks that recrystallized under lower temperature and pressure

FIGURE 10.15

Complexly folded migmatite exposed near Gravenhurst, Ontario. These rocks record the partial melting of gneiss below the Grenville mountains.
Photo by Nick Eyles

conditions than during the peak of metamorphism. They were fractured during their ascent, permitting water to trigger reactions to new, lower-grade minerals.

Pressure and Temperature Paths in Time

Index minerals and mineral assemblages in a rock can be used to determine the approximate temperature and pressure conditions that prevailed during metamorphism. Precise determination of the chemical composition of some minerals can determine the temperature or pressure present during the growth of a particular mineral. The usual basis for determining temperature (*geothermometry*) or pressure (*geobarometry*) during mineral growth is the ratio of pairs of elements (e.g., Fe and Al) within the crystal structure of the mineral.

Modern techniques allow us to determine chemical compositional changes across a grain of a mineral in a rock. An *electron microprobe* is a microscope that allows the user to focus on a tiny portion of a mineral in a rock, then shoot a very narrow beam of electrons into that point in the mineral. The extent and manner in which the beam is absorbed by the mineral are translated (by computer) into the precise chemical composition of the mineral at that point. If the mineral is *zoned* (that is, the chemical composition changes within the mineral, as described in chapter 5), the electron microprobe will indicate the differing composition within the mineral grain.

A mineral will grow from the centre outward, adding layers of atoms as it becomes larger. If pressure and temperature conditions change as the mineral grows, the concentric zoning will reflect those changes. Figure 10.16 shows the results of one such study. The diagram shows the changes of temperature and pressure in time, with the line showing the temperature-pressure-time path. If pressure and temperature are both increasing, this indicates the rock is being buried deeper while becoming hotter. If temperature and pressure are both decreasing, the rock is cooling down at the same time that pressure is being reduced because of erosion at Earth's surface.

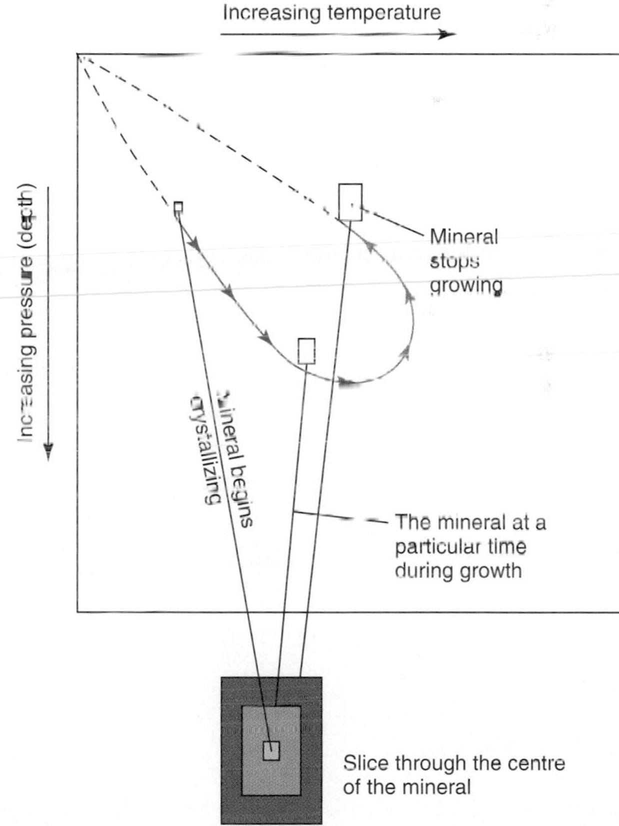

FIGURE 10.16

Pressure-temperature-time path for growth of a mineral during metamorphism. An electron microprobe is used to determine the precise chemical composition of the concentric zones of the mineral. The data are used to determine the pressure and temperature during the growth of the mineral. Three stages during the growth of the mineral are correlated to the graph—beginning of growth (centre of crystal), an arbitrary point during its growth, and the end of crystallization (the outermost part of the crystal).

The green segment of the path indicates increasing pressure and temperature during metamorphism. The orange segment indicates that pressure was decreasing while temperature continued to rise. The blue segment indicates temperature and pressure were both decreasing. The decrease in pressure is likely to be the result of uplift and erosion at the surface. The dashed lines are inferred pressure and temperature paths before and after metamorphism.

WHAT IS THE RELATIONSHIP BETWEEN PLATE TECTONICS AND METAMORPHISM?

Studies of metamorphic rocks have provided important information on conditions and processes within the lithosphere and have aided our understanding of plate tectonics. Conversely, plate tectonic theory has provided models that allow us to explain many of the observed characteristics of metamorphic rocks.

Foliation and Plate Tectonics

Figure 10.17 shows an oceanic-continental boundary (oceanic lithosphere is subducted beneath continental lithosphere). One of the things the diagram shows is where differential stress that is responsible for foliation is taking place. Shearing takes place in the subduction zone where the oceanic crust slides beneath continental lithosphere. For here, we infer that the sedimentary rocks and some of the basalt becomes foliated, during metamorphism, roughly parallel to the subduction zone (parallel to the lines in the diagram).

Within the thickest part of the continental crust shown in figure 10.17, flowage of rock is indicated by the purple arrows. The crust is thickest here beneath a growing mountain belt. The thickening is due to the compression caused by the two colliding plates. Within this part of the crust, rocks flow downward and then outward (as indicated by the arrows) in a process of *gravitational collapse and spreading.* Under this concept, the central part of a mountain belt becomes too high after plate convergence and is gravitationally unstable. This forces the rock downward and outward. Regional metamorphism takes place throughout and we expect foliation in the recrystallizing rocks to be approximately parallel to the arrows.

Pressure–Temperature Regimes

Before the advent of plate tectonics, geologists were hard-pressed to explain how some rocks apparently were metamorphosed at relatively cool temperatures yet high pressures. We expect rocks to be hotter as they become more deeply buried. How could rocks stay cool, yet be deeply buried?

Figure 10.18 shows experimentally determined stability fields for a few metamorphic minerals. Line x indicates a common geothermal gradient during metamorphism. At the appropriate pressure and temperature, kyanite begins to crystallize in the rock. If it is buried deeper, its pressure and temperature would change along line x. Eventually, it would cross the stability boundary and sillimanite would crystallize rather than kyanite. By contrast, if a rock contains glaucophane (an amphibole), rather than hornblende, the rock must have formed under high pressure but abnormally low temperature for its depth of burial. Line y represents a possible geothermal gradient that must have been very low and the increase in temperature was small with respect to the increase in pressure.

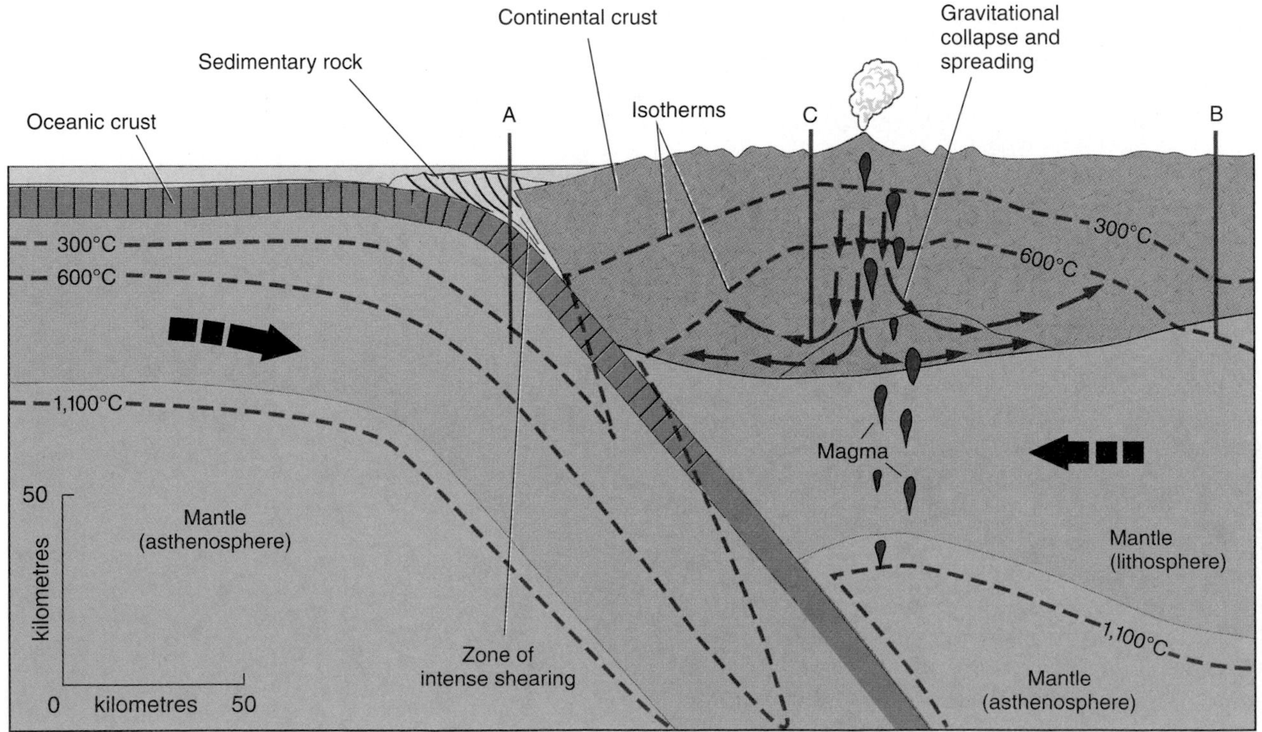

FIGURE 10.17

Metamorphism across a convergent plate boundary. All rock that is hotter than 300 degrees or deeper than 5 kilometres is likely to be undergoing metamorphism.

Modified from W. G. Ernst. Metamorphism and Plate Tectonic Regimes. Stroudsberg, Pa.: Dowden, Hutchinson & Ross, 1975; p. 425. Reprinted by permission of the publisher

If we return to figure 10.17, we can use it to see how plate tectonics explains these very different pressure-temperature regimes at a convergent boundary. Confining pressure is directly related to depth. For this reason, we expect the same pressure at any given depth. For example, the pressure corresponding to 20 kilometres is the same under a hot volcanic area as it is within the relatively cool rocks of a plate's interior. Temperature, however, is quite variable as indicated by the dashed red lines. Each of these lines is an **isotherm,** a line connecting points of equal temperature.

Each of the three places (*A, B,* and *C*) in figure 10.17 would have a different geothermal gradient. If you were somehow able to push a thermometer through the lithosphere, you would find the rock is hotter at shallower depths in areas with higher geothermal gradients than at places where the geothermal gradient is low. As indicated in figure 10.17, the geothermal gradient is higher progressing downward through an active volcanic-plutonic complex (for instance, the Cascade Mountains of British Columbia, Washington, and Oregon) than it is in the interior of a plate (beneath the Great Plains of North America, for example). The isotherms are bowed upward in the region of the volcanic-plutonic complex because magma created at lower levels works its way upward and brings heat from the asthenosphere into the mantle and crust of the continental lithosphere. At point *C* we would expect the metamorphism that takes place to result in minerals that reflect the high temperature relative to pressure conditions such as those along line *x* in figure 10.18.

If we focus our attention at the line at *A* in figure 10.17, we can understand how minerals can form under high pressure but relatively low temperature conditions. You may observe that the bottom of line *A* is at a depth of about 50 kilometres, and if a hypothetical thermometer were here, it would read just over 300 degrees because it would be just below the 300-degree isotherm. Compare this to vertical line *C* in the volcanic-plutonic complex. The confining pressure at the base of this line would be the same as at the base of line *A,* yet the temperature at the base of line *C* would be well over 600 degrees. The minerals that could form at the base of line *A* would not be the same as those that could form at line *C*. Therefore, we would expect quite different metamorphic rocks in the two places, even if the parent rock had been the same (box 10.5).

So when we find high-pressure/low-temperature minerals (such as glaucophane) in a rock, we can infer that metamorphism took place while subduction carried basalt and overlying sedimentary rocks downward. Thus, plate tectonics accounts for the abnormally high-pressure/low-temperature geothermal gradients (such as line *y* in figure 10.18).

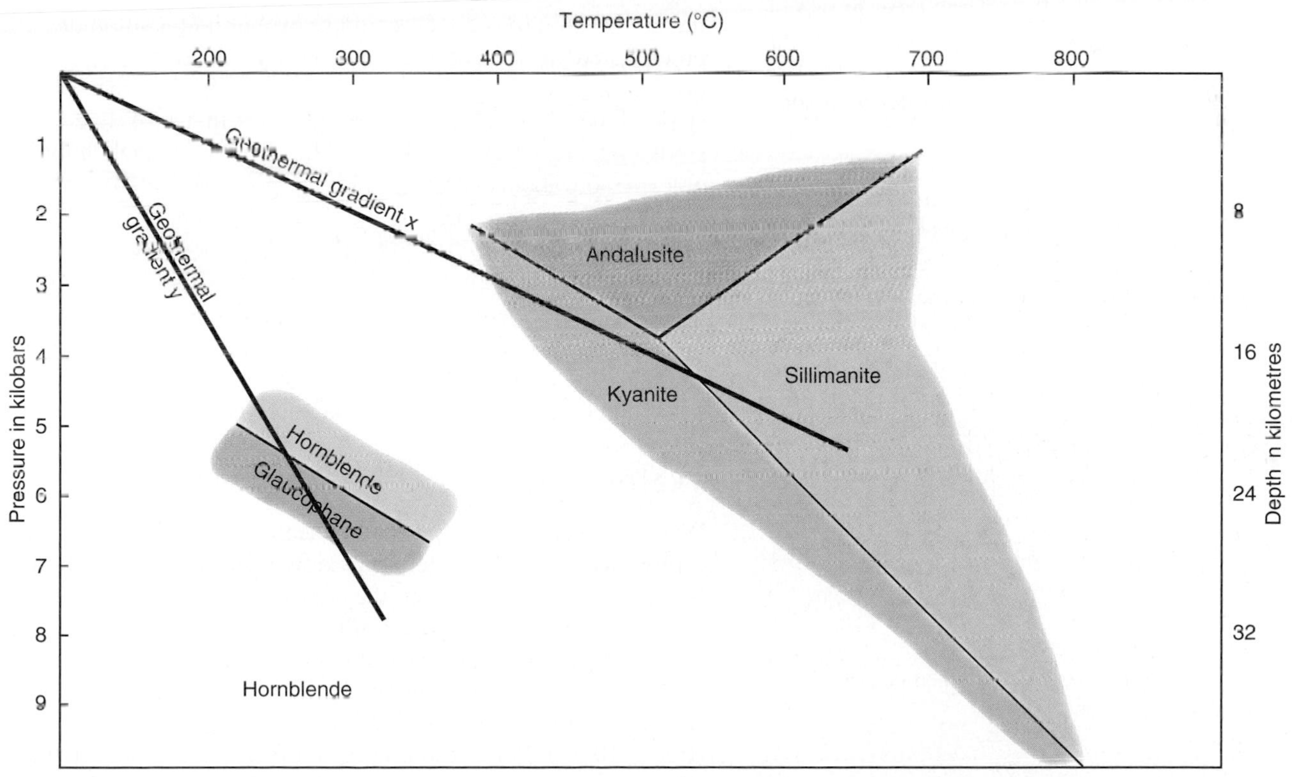

FIGURE 10.18

Stability fields for a few minerals. (Many more mineral stability fields can be used for increased accuracy.) The fields are based on laboratory research. Prograde metamorphism taking place with a geothermal gradient *x* involves a high temperature increase with increasing pressure. Prograde metamorphism under conditions of geothermal gradient *y* involves low temperature increase with increasing pressure.

WHAT IS THE ROLE OF HOT WATER IN METAMORPHIC PROCESSES?

Rocks that have precipitated from hot water or have been altered by hot water passing through are hard to classify. As described earlier, hot water is involved to some extent in most metamorphic processes. Beyond metamorphism, hot water also plays an important role creating new rocks and minerals. These form entirely by precipitation of ions derived from hydrothermal solutions. *Hydrothermal minerals* can form in void spaces or between the grains of a host rock. An aggregate of hydrothermal minerals, a **hydrothermal rock,** may crystallize within a pre-existing fracture in a rock to form a hydrothermal **vein.**

Hydrothermal processes are summarized in table 10.3. As we have seen, water is important for metamorphic processes not only because water transports ions from one mineral to another, but also because many of the minerals (the micas, for instance) that crystallize during metamorphism incorporate water into their crystal structures.

Hydrothermal Activity at Divergent Plate Boundaries

Hydrothermal processes are particularly important at midoceanic ridges (which are also divergent plate boundaries). As shown in figure 10.19, cold seawater moves downward through cracks in the basaltic crust and is cycled upward by heat from magma beneath the ridge crest. Very hot water returns to the ocean at submarine hot springs (hydrothermal vents; figure 10.21).

Hot water travelling through the basalt and gabbro of the oceanic lithosphere helps metamorphose these rocks while they are close to the divergent boundary. This is sometimes called *seafloor metamorphism*. During metamorphism the ferromagnesian igneous minerals, olivine and pyroxene, become converted to *hydrous* (water-bearing) minerals such as amphibole. An important consequence of this is that the hydrous minerals may eventually

TABLE 10.3 Hydrothermal Processes

Role of Water	Name of Process or Product
Water transports ions between grains in a rock. Some water may be incorporated into crystal structures.	Metamorphism
Water brings ions from outside the rock, and they are added to the rock during metamorphism. Other ions may be dissolved and removed.	Metasomatism
Water passes through cracks or pore rocks spaces in rock and precipitates minerals on the walls of cracks and within pore spaces.	Hydrothermal

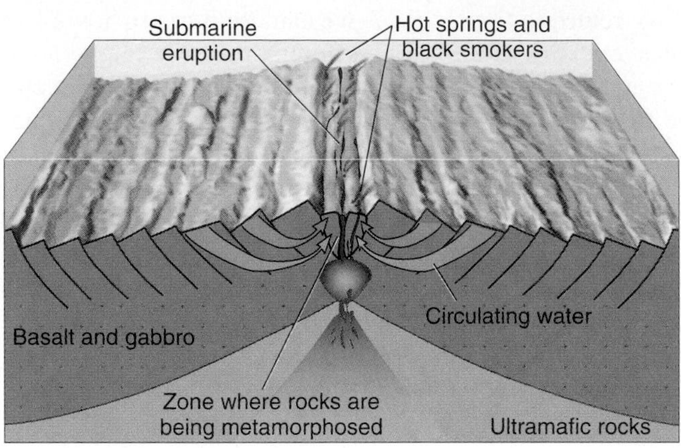

FIGURE 10.19

Cross-section of a mid-oceanic ridge (divergent plate boundary). Water descends through fractures in the oceanic crust, is heated by magma and hot igneous rocks, and rises.

contribute to magma generation at convergent boundaries. After oceanic crust is subducted the minerals are dehydrated deep in a subduction zone (see figure 10.20). The water produced moves upward into the overlying asthenosphere and contributes to melting and magma generation, as described in chapter 6.

Ore Deposits at Divergent Plate Boundaries

As the seawater moves through the crust, it dissolves metals and sulphur from the crustal rocks and magma. When the hot, metal-rich solutions contact cold seawater, metal sulphides are precipitated in a mound around the hydrothermal vent. This process has been filmed in the Pacific, where some springs spew clouds of fine-grained ore minerals that look like black smoke (figure 10.21). (To see a video clip of a "black smoker" or learn more about hydrothermal vents, go to the seafloor geology page of *Voyage to the Deep,* www.ocean.udel.edu/deepsea/level-1/geology/geology.html.)

The metals in rift-valley hydrothermal vents are predominantly iron, copper, and zinc, with smaller amounts of manganese, gold, and silver. Although the mounds are nearly solid metal sulphide, they are small and widely scattered on the sea floor, so commercial mining of them may not be practical.

Water at Convergent Boundaries

Water that percolates from the surface into the ground becomes *groundwater.* Groundwater seeps downward through pores and fractures in rocks. However, the depth to which surface-derived water can penetrate is quite limited.

Plate tectonics can account for water at deeper levels in the lithosphere as seawater trapped in the oceanic crust can be carried to depths of up to 100 kilometres through subduction (figure 10.20). Water trapped in sediment and in sedimentary rocks lying on basalt may be carried down with the descending crust. It is driven out by pressure at depths up to around 30 kilometres. However, studies indicate that most of the water is carried by hydrous minerals (amphibole, for example) in the basaltic crust. When the rocks get hot enough, the hydrous

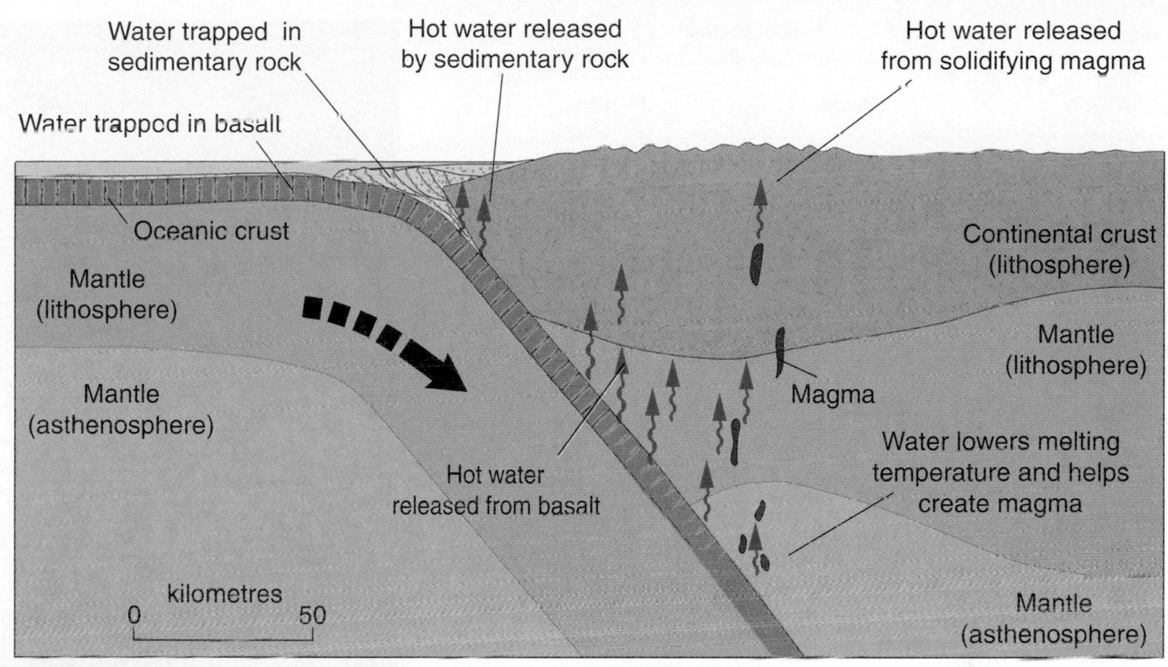

FIGURE 10.20

Water at a convergent boundary. Seawater trapped in the oceanic crust is carried downward and released upon heating at various depths within the subduction zone.

FIGURE 10.21

"Black smoker" or hydrothermal vent on the crest of the mid-oceanic ridge in the Pacific Ocean near 21 degrees North latitude. The "smoke" is a hot plume of metallic sulphide minerals being discharged into cold sea water from a chimney 0.5 metres high. The large mounds around the chimney are metal deposits. The instruments in the foreground are attached to the small submarine from which the picture was taken.

Photo © WHOI-D. Foster/Visuals Unlimited

minerals recrystallize, releasing water. The water vapour works its way upward through the overlying continental lithosphere through fissures. In the process of ascending, water assists in the metamorphism of rocks, dissolves minerals, and carries the ions to interact during metasomatism, or it deposits quartz and other minerals in fissures as veins. The water can also lower the melting points of rocks at depth, allowing magma to form (as described in chapter 6 on igneous rocks).

Metasomatism

Metasomatism is metamorphism coupled with the introduction of *ions* from an external source. The ions are brought in by water from outside the immediate environment and are incorporated into the newly crystallizing minerals. Often, metasomatism involves ion exchange. Newly crystallizing minerals replace pre-existing ones as water simultaneously dissolves and replaces ions.

When metasomatism takes place during regional metamorphism, very hot water travels through a rock while gneiss or schist is crystallizing. Ions (typically K^+, Na^+, and SiO_4^{-4}) are carried by the water and participate in metamorphic reactions. Large feldspar crystals may grow in schist due to the addition of potassium or sodium ions.

If metasomatism is associated with contact metamorphism, the ions are introduced from a cooling magma. Some important commercially mined deposits of metals such as iron, tungsten, copper, lead, zinc, and silver are attributed to metasomatism. Figure 10.22 shows how magnetite (iron oxide) ore bodies have formed through metasomatism. Ions of the metal are transported by water and react with minerals in the host rock. Elements within the host rock are simultaneously dissolved out of the host rock and replaced by the metal ions brought in by the fluid. Because of the solubility of calcite, marble commonly serves as a host for metasomatic ore deposits.

Hydrothermal Rocks and Minerals

Quartz veins (figure 10.23) are especially common where igneous activity has occurred. These can form from hot water given off by a cooling magma. They also are produced by groundwater heated by a pluton and circulated by convection, as shown in figure 10.24. Where the water is hottest, the most material (notably silica) is dissolved. As the hydrous fluid continues upward through increasingly cooler rocks during its journey toward the Earth's surface, pressure decreases and heat is lost. Fewer ions can be carried in solution, and so the silicon and oxygen leave the water and cake onto the walls of the crack as silica (SiO_2), forming a quartz vein.

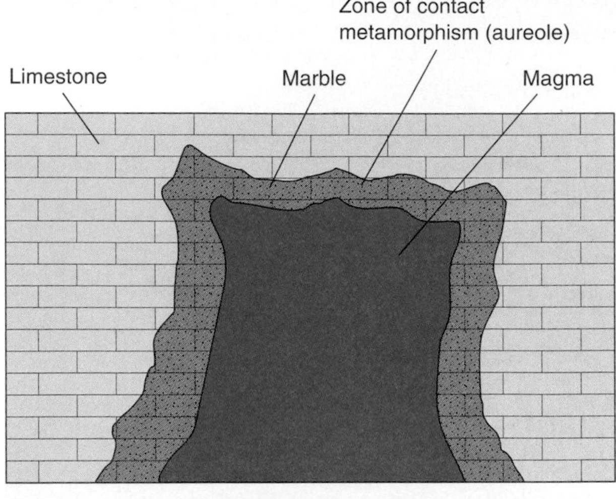

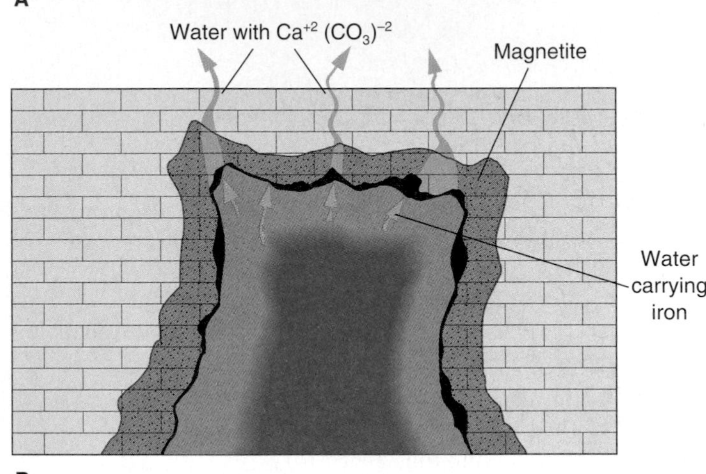

FIGURE 10.22

Development of a contact metasomatic deposit of iron (magnetite). (*A*) Magma intrudes country rock (limestone), and marble forms along contact. (*B*) As magma solidifies, gases bearing ions of iron leave the magma, dissolve some of the marble, and deposit iron as magnetite.

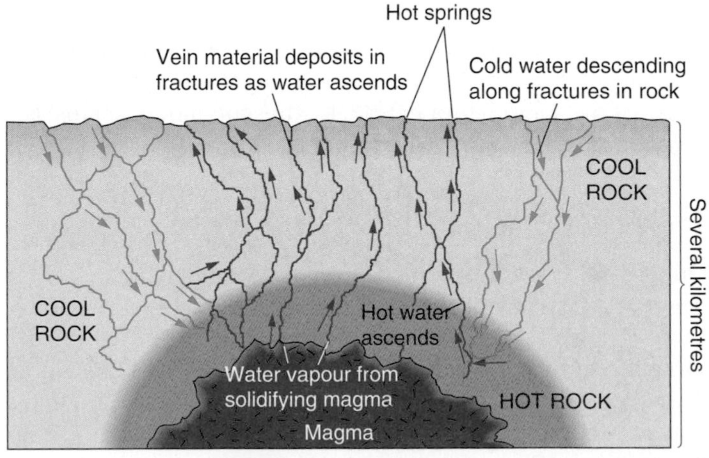

FIGURE 10.24

How veins form. Cold water descends, is heated, dissolves material, ascends, and deposits material as water cools and pressure drops upon ascending.

FIGURE 10.23

Gold-bearing syenite containing 2 per cent disseminated syenite and cut by quartz-carbonate veinlets, Young-Davidson property, Matachewan, Ontario.

Photo by R. Kelly. Reproduced with the permission of the Minister of Public Works and Government Services Canada, 2007 and Courtesy of Natural Resources Canada, Geological Survey of Canada

Veins consisting only of quartz are the most widespread, although some quartz veins contain other minerals. Veins with no quartz are not as common and are composed of calcite or some other minerals.

Hydrothermal veins are very important economically. In them we find most of the world's great deposits of zinc, lead, silver, gold, tungsten, tin, mercury, and, to some extent, copper. Ore minerals containing these metals are usually found in quartz veins. Veins containing commercially extractable amounts of metals are by no means common.

Some ore-bearing solutions percolate upward between the grains of the rock and deposit very fine grains of ore mineral throughout. These are called *disseminated ore deposits.* Usually, metallic sulphide ore minerals are distributed in very low concentration through large volumes of rock, both above and within a pluton. The ore in the pluton is in the upper part, which solidified earliest. As crystallization continued in the underlying magma, hydrothermal solutions were given off, and ore minerals crystallized in the tiny fractures and between grains in the overlying rock. Most of the world's copper comes from disseminated deposits, also called *porphyry copper deposits,* because the associated pluton is usually porphyritic (see box 10.6). Along with the copper are deposited many other metals, such as lead, zinc, molybdenum, silver, and gold (and iron, though not in commercial quantities). Some very large gold mines are also in disseminated ore deposits.

ENVIRONMENTAL GEOLOGY 10.6

The World's Largest Human-made Hole— The Bingham Canyon Copper Mine

The biggest human-made hole in the world is located at Bingham Canyon near Salt Lake City, Utah. The 800-metre-deep open pit mine is 4 km wide at the top and continues to be enlarged. The reason for this hole is copper. Similar large open-pit copper mines are common in the Cordillera of the United States and Canada (see chapter 12).

About 40,000 kilograms of explosives are used per day to blast apart more than 60,000 tons of ore (copper-bearing rock) and an equal amount of waste rock. A conveyor belt system moves up to 10,000 tons of crushed rock per hour through a tunnel out of the pit for processing.

Mining began here as a typical underground operation in 1863. The shafts and tunnels of the mine followed a series of veins. Originally ores of silver and lead were mined. Later it was discovered that fine-grained, copper-bearing minerals (chalcopyrite and other copper sulphide minerals) were disseminated in tiny veinlets throughout a granite stock. Although the percentage of copper in the rock was small, the total volume of copper was recognized as huge. With efficient earth-moving techniques, large volumes of ore-bearing rock can be moved and processed. Today mining is still going on, and the company is able to make a profit even though 0.6 percent of the rock being mined is copper. Since 1904, more than 12 million tons of copper have been mined, processed, and sold. The mine has also produced impressive amounts of gold, silver, and other metals.

Such an operation is not without environmental problems. Some people regard the huge hole in the mountains as an eyesore (but it is a popular tourist attraction). Disposing of the waste—more than 99 percent of the rock material mined—creates problems. Wind stirs up dust storms from the piles of finely crushed waste rock unless it is kept wet. The nearby smelter that extracts the pure copper from the sulphide minerals has created a toxic smoke containing sulphuric acid fumes. During most of the twentieth century the toxic smoke was released into the atmosphere; occasionally wind blew polluted air to Salt Lake City. Now, more than 99 percent of the sulphur fumes are removed at the smelter, recovered as sulphuric acid and sold as a by-product.

BOX 10.6 ■ FIGURE 1
Bingham Canyon copper mine in Utah.
Courtesy Kennecott Copper Company

Related Web Resource

Mining Technology—Bingham Canyon

www.mining-technology.com/projects/bingham/

SUMMARY

Metamorphic rocks form from other rocks that are subjected to high temperature generally accompanied by high confining pressure. Although recrystallization takes place in the solid state, water, which is usually present, aids metamorphic reactions. Foliation in metamorphic rocks is due to *differential stress* (either *compressive stress* or *shearing*). Slate, phyllite, schist, and gneiss are foliated rocks that indicate increasing grade of regional metamorphism. They are distinguished from one another by the type of foliation.

Contact metamorphic rocks are produced during metamorphism usually without significant differential stress but with high temperature. Contact metamorphism occurs in rocks immediately adjacent to intruded magmas.

Regional metamorphism, which involves heat, confining pressure, and differential stress, has created most of the metamorphic rock of Earth's crust. Different parent rocks as well as widely varying combinations of pressure and temperature result in a large variety of metamorphic rocks. Combinations of minerals in a rock can indicate what the pressure and temperature conditions were during metamorphism. Extreme metamorphism, where the rock partially melts, can result in *migmatites*.

Hydrothermal veins form when hot water precipitates material that crystallizes into minerals. During *metasomatism*, hot water introduces ions into a rock being metamorphosed, changing the chemical composition of the metasomatized rock from that of the parent rock.

Plate tectonic theory accounts for the features observed in metamorphic rocks and relates their development to other activities in Earth. In particular, plate tectonics explains (1) the deep burial of rocks originally formed at or near the Earth's surface; (2) the intense squeezing necessary for the differential stress, implied by foliated rocks; (3) the presence of water deep within the lithosphere; and (4) the wide variety of pressures and temperatures hypothesized to be present during metamorphism.

Terms to Remember

Testing Your Knowledge

Use the questions below to prepare for exams based on this chapter.

1. What are the effects on metamorphic minerals and textures of temperature, confining pressure, and differential stress?
2. What are the various sources of heat for metamorphism?
3. How do regional metamorphic rocks differ in texture from contact metamorphic rocks?
4. Why is such a variety of combinations of pressure and temperature environments possible during metamorphism?
5. How would you distinguish
   a. schist and gneiss? b. slate and phyllite?
   c. quartzite and marble? d. granite and gneiss?
6. Why is an edifice built with blocks of quartzite more durable than one built of marble blocks?
7. Metamorphism of limestone may contribute to global warming by the release of
   a. oxygen b. sulphuric acid
   c. nitrogen d. CO_2
8. Shearing is a type of
   a. compressive stress b. confining pressure
   c. lithostatic pressure d. differential stress
9. Metamorphic rocks with a planar texture (the constituents of the rock are parallel to one another) are said to be
   a. concordant b. foliated
   c. discordant d. nonfoliated
10. Metamorphic rocks are classified primarily on
    a. texture—the presence or absence of foliation
    b. mineralogy—the presence or absence of quartz
    c. environment of deposition
    d. chemical composition
11. Which is not a foliated metamorphic rock?
    a. gneiss b. schist
    c. quartzite d. slate
12. Limestone recrystallizes during metamorphism into
    a. hornfels b. marble
    c. quartzite d. schist
13. Quartz sandstone is changed during metamorphism into
    a. hornfels b. marble
    c. quartzite d. schist
14. The correct sequence of rocks that are formed when shale undergoes progressive metamorphism is
    a. slate, gneiss, schist, phyllite b. phyllite, slate, schist, gneiss
    c. slate, phyllite, schist, gneiss d. schist, phyllite, slate, gneiss

15. The major difference between metamorphism and metasomatism is
    a. the temperature at which each takes place
    b. the minerals involved
    c. the area or region involved
    d. metasomatism is metamorphism coupled with the introduction of ions from an external source

16. Ore bodies at divergent plate boundaries can be created through
    a. contact metamorphism
    b. regional metamorphism
    c. hydrothermal processes

17. A schist that developed in a high-pressure, low-temperature environment likely formed
    a. in the lower part of the continental crust
    b. in a subduction zone
    c. in a mid-oceanic ridge
    d. near a contact with a magma body

18. A metamorphic rock that has undergone partial melting to produce a mixed igneous-metamorphic rock is a
    a. gneiss
    b. hornfels
    c. schist
    d. migmatite

Exploring Web Resources

www.mcgrawhill.ca/olc/plummer

Visit this Website for answers to Testing Your Knowledge, as well as additional quizzes, readings, media resources, and direct links to the sites listed below.

www.eos.ubc.ca/courses/eosc221/meta/metamorphic.html

University of British Columbia's *Metamorphic Rocks Home Page.* This site is meant for a geology course on the study of rocks. Although it is at a more advanced level, it can be used to reinforce some of the concepts covered in this chapter.

www.geolab.unc.edu/Petunia/IgMetAtlas/mainmenu.html

University of North Carolina's *Atlas of Rocks, Minerals, and Textures.* Click on "Metamorphic microtextures." Click on terms covered in this chapter (e.g., foliation, gneiss, phyllite, marble, quartzite, slate) to see excellent photomicrographs taken through a polarizing microscope.

Animation

This chapter includes the following animations available on our Online Learning Centre at www.mcgrawhill.ca/olc/plummer.

10.24 Hydrothermal ore vein formation

CHAPTER

11

Geologic Structures

What causes rocks to bend and break?
How do rocks behave when stressed?
How do we measure and record geologic structures?
What do folded rocks tell us about geologic processes?
What are fractures and faults?

In previous chapters we have discussed how rock at the Earth's surface is affected by erosional agents such as wind and water. We now shift our focus to changes in bedrock caused by powerful forces originating deep within the Earth. In this chapter we explain how rocks respond to these tectonic forces and how geoscientists study the resulting geologic structures or architecture of the Earth's crust.

The main purpose of this chapter is to help you recognize certain geologic structures, understand the forces that caused them, and thus determine the geologic history of an area. Recognition of unconformities as well as the principles of original horizontality, superposition, and cross-cutting relationships are as important to structural geology as they are to determining relative time (see chapter 19).

Many areas of geoscience require an understanding and knowledge of structural geology. To understand earthquakes, for instance, one must know about faults. Appreciating how major mountain belts and the continents have evolved (chapter 20) calls for a comprehension of faulting and folding. In areas of active tectonics, the location of geologic structures is important in the selection of safe sites for schools, hospitals, dams, bridges, and nuclear power facilities.

Folded sedimentary rocks in the Canadian Rocky Mountains.
Photo by M. Radomski

Also, understanding structural geology can help us more fully appreciate the problem of finding more of the Earth's dwindling natural resources. Chapter 12 discusses the association of certain geologic structures with petroleum deposits and other valuable resources. In a broad sense structural geology can be thought of as the study of the architecture of the Earth's crust, its deformational features, and their mutual relations and origins. For our purposes, **structural geology** can be defined as the branch of geology concerned with the shapes, arrangement, and interrelationships of bedrock units and the forces that cause them.

WHAT CAUSES ROCKS TO BEND AND BREAK?

Stress and Strain in the Earth's Lithosphere

Tectonic forces move and deform parts of the lithosphere, particularly along plate margins. Deformation may cause a change in orientation, location, and shape of a rock body. In figure 11.1, originally horizontal rock layers have been deformed into wave-like folds that are broken by faults. The layers have been deformed, probably by tectonic forces that pushed or compressed the layers together until they were shortened by buckling and breaking.

When studying deformed rocks, structural geologists typically refer to **stress,** a force per unit area. Where stress can be measured, it is expressed as the force per unit area at a particular point; however, it is difficult to measure stress in rocks that are currently buried. We can observe the effects of past stress (caused by tectonic forces and confining pressure from burial) when rock bodies are exposed after uplift and erosion. From our observations we may be able to infer the principal directions of stress that prevailed. We also can observe in exposed rocks the effect of forces on a rock that was stressed. **Strain** is the change in size (volume) or shape, or both, in response to stress.

The relationship between stress and strain can be illustrated by deforming a piece of pizza dough (figure 11.2). If the pizza dough is pushed together or squeezed from opposite directions, we say the stress is **compressive.** Compressive stress is common along convergent plate boundaries and typically results in rocks being deformed by a *shortening strain.* In figure 11.2*A,* an elongate piece of pizza dough may shorten by bending, or folding, whereas a ball of pizza dough will flatten by shortening in the direction parallel to the compressive stress and elongating or stretching in the direction perpendicular to it. Rocks that have been shortened or flattened are typically found along convergent plate boundaries where rocks have been pushed or shoved together.

A **tensional stress** is caused by forces pulling away from one another in opposite directions (figure 11.2*B*). Tensional stress results in a *stretching or extension* of material. If we apply a tensional stress on a ball of pizza dough, it will elongate or stretch parallel to the applied stress. If the tensional stress is applied rapidly, the pizza dough will first stretch and then break apart (figure 11.2*B*). At divergent plate boundaries, the lithosphere is undergoing extension as the plates move away from one another. Because rocks are very weak when pulled apart, fractures and faults are common structures.

FIGURE 11.1
Geoscientists examine folded sedimentary rocks of Moose Mountain, Alberta.
Photo courtesy of Moose Oils Ltd. Calgary, Alberta

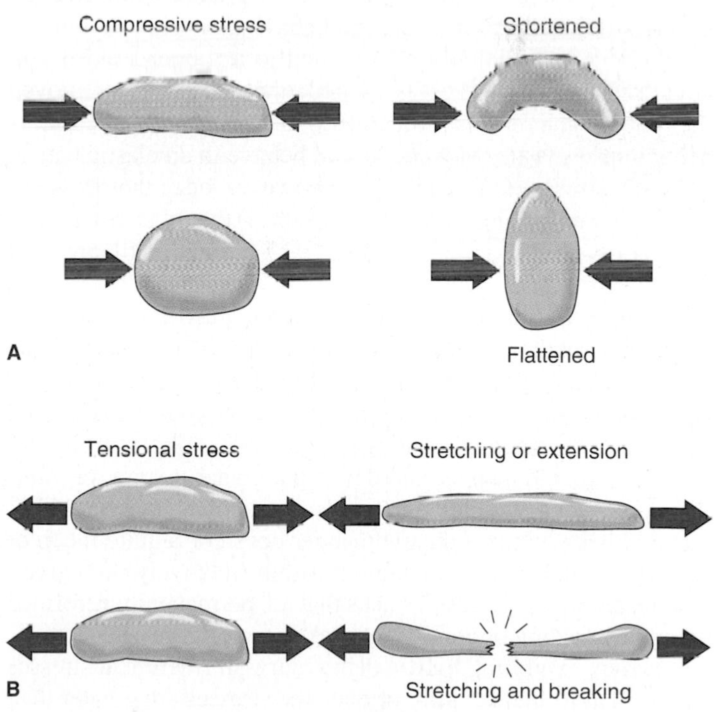

FIGURE 11.2
The effects of compressional and tensional stresses on pizza dough. (*A*) Compressing pizza dough results in shortening either by folding or flattening. (*B*) Pulling (tensional stress) pizza dough causes stretching or extension; if pulled (strained) too fast, or chilled, the pizza dough will break after first stretching.

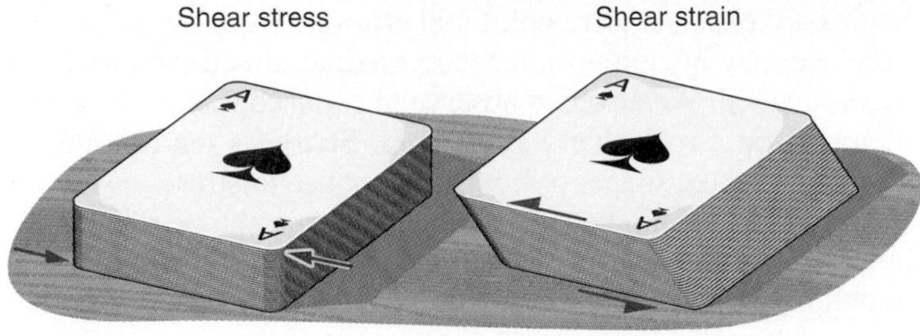

FIGURE 11.3
Shear stress can be modelled by shearing a deck of cards.

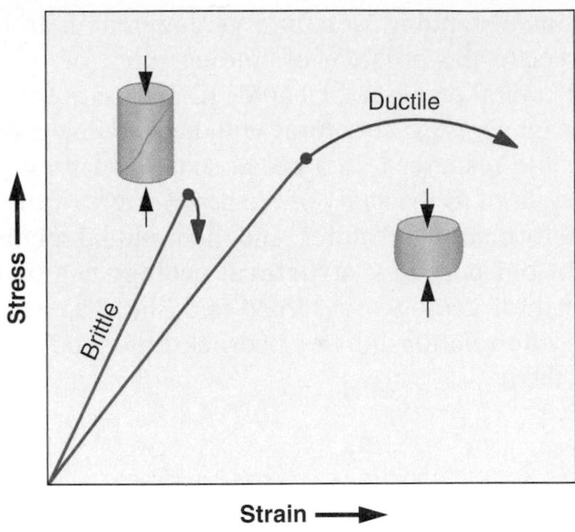

FIGURE 11.4
Graph shows the behaviour of rocks with increasing stress and strain. Elastic behaviour occurs along the straight line portions (shown in blue) of the graph. At stresses greater than the elastic limit (red points), the rock will either deform as a ductile material or break, as shown in the deformed rock cylinders.

When stresses act parallel to a plane, **shear stress** is produced. It is much like putting a deck of cards in your hands and shearing the deck by moving your hands in opposite directions (figure 11.3). A shear stress results in a *shear strain* parallel to the direction of the stresses. Shear stresses occur along actively moving faults.

HOW DO ROCKS BEHAVE WHEN STRESSED?

Rocks behave as elastic, ductile, or brittle materials depending on the amount and rate of stress applied, the type of rock, and the temperature and pressure under which the rock is strained.

If a deformed body recovers its original shape after the stress is reduced or removed, the behaviour is **elastic.** For example, if a tensional stress is applied to a rubber band it will stretch as long as the stress is applied; once the stress is removed the rubber band returns (or recovers) to its original shape and its behaviour is elastic. Most rocks can behave in an elastic way at very low stresses (a few kilobars); however, once the stress applied exceeds the **elastic limit** (figure 11.4) the rock will deform in a permanent way, just as the rubber band will break if stretched too far.

A rock that behaves in a **ductile** or plastic manner will bend while under stress and does not return to its original shape after relaxation of the stress. Pizza dough behaves as a ductile material unless the rate of strain is rapid. Rocks exposed to elevated pressure and temperature during regional metamorphism also behave in a ductile manner and develop a planar texture, or *foliation*, due to alignment of minerals. As shown in figure 11.4, material behaving in a ductile manner does not require much of an increase in stress to continue to strain (relatively flat curve). Ductile behaviour results in rocks that are permanently deformed mainly by folding or bending of rock layers (figure 11.1).

A rock exhibiting **brittle** behaviour will fracture at stresses higher than its elastic limit, or once the stresses are greater than the strength of the rock. Rocks typically exhibit brittle behaviour at or near the Earth's surface where temperatures and pressure are low. Under these conditions, rocks favour breaking rather than bending. Faults and joints are examples of structures that form by brittle behaviour of the crust.

A sedimentary rock exposed at the Earth's surface is brittle; it will fracture if you hit it with a hammer. How then do sedimentary rocks, such as those shown in figure 11.1, become bent (or deformed in a ductile way)? The answer is that either stress increased very slowly or the rock was deformed under considerable confining pressure (buried under more rock) and higher temperatures.

Note, however, that there are some fractures (faults) disrupting the bent layers in figure 11.1. This tells us that although the rock was plastic initially, the stress became too intense or the rate of strain increased and the rock fractured.

HOW DO WE MEASURE AND RECORD GEOLOGIC STRUCTURES?

Some geologic structures that give us clues to the past have been described in earlier chapters. Batholiths, stocks, dikes, and sills, for example, are keys to past igneous activity (see chapter 6 on intrusive activity). In this chapter we are mainly concerned with types of structures that can provide a record of crustal deformation that is no longer active. Very old structures that are now visible at the Earth's surface were once buried and are exposed through erosion.

The study of geologic structures is of more than academic interest. The petroleum and mining industries, for example, employ geologists to look for and map geologic structures associated with oil and metallic ore deposits. Understanding and mapping geologic structures is also important for evaluating problems related to engineering decisions and environmental planning, such as determining the most appropriate sites for dams, large bridges, or nuclear reactors, and even the building of houses, schools, and hospitals.

Geologic Maps and Field Methods

In an ideal situation, a geoscientist studying structures would use remotely sensed data, collected using satellites or airplanes, to identify local and regional patterns of bedrock from above (figure 11.5*A* and box 11.1). These data would be "ground truthed" with observations from a number of individual *outcrops* (exposures of bedrock at the surface) to determine the patterns of geologic structures (figure 11.5*B*). The characteristics of rock at each outcrop in an area are plotted on a map by means of appropriate symbols. With the data that can be collected, a geologist can make inferences about those parts of the area he or she cannot observe. A **geologic map,** which uses standardized symbols and patterns to represent rock types and geologic structures, is typically produced from the field map for a given area (for example, see the geologic map of North America inside the front cover). On such a map are plotted the type and distribution of rock units, the occurrence of structural features (folds, faults, joints, etc.), ore deposits, and so forth. Sometimes surficial features, such as deposits by former glaciers, are included, but these may be shown separately on a different type of geologic map.

Anyone trained in the use of geologic maps can gain considerable information about local geologic structures because standard symbols and terms are used on the maps and the accompanying reports. For example, the symbol ⊕ on a geologic map denotes horizontal bedding in an outcrop. Different colours and patterns on a geologic map represent distinct rock units.

Strike and Dip

Many sedimentary rocks and some lava flows and ash falls are deposited as horizontal beds or strata. Where these originally horizontal rocks are found tilted, it indicates that tilting must have occurred after deposition and lithification (figure 11.6). Someone studying a geologic map of the area would want to know the extent and direction of tilting. By convention, this is determined by plotting the relationship between a surface of an inclined bed and an imaginary horizontal plane. You can understand the relationship by looking carefully at figure 11.7, which represents sedimentary beds cropping out alongside a lake (the lake surface provides a convenient horizontal plane for this discussion).

A

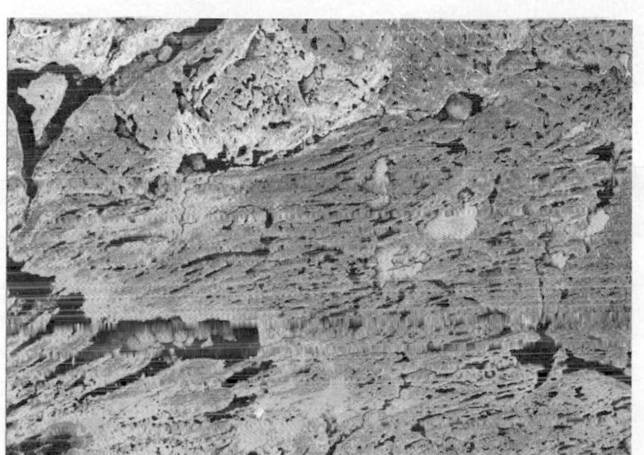

B

FIGURE 11.5

(*A*) Satellite image of tightly folded sedimentary rocks of the Labrador Fold Belt on the Canadian Shield. (*B*) Students mapping regional structure from a ridge-top vantage point, Whitefish Falls, Ontario.

Photo A courtesy of Jeff Harris, Geological Survey of Canada; Photo B by M. Radomski

FIGURE 11.6

Tilted sedimentary beds along the coast of northern California near Pt. Arena. Here, the strike is the line formed by the intersection of the tilted sedimentary beds and the horizontal layer of sand in the foreground. The direction of dip is toward the left.

Photo by Diane Carlson

GEOMATICS 11.1

Remote Predictive Mapping and 3-D Modelling of Geologic Structures on Baffin Island

The understanding of geologic structures is a 3-D science and requires analysis of both the surface distribution of geologic units and also interpretation of their underground form. Knowledge of the structural geology of an area is important not only for reconstruction of its geologic and plate tectonic history but also for resource exploration, particularly in complex areas like the Canadian Shield. The Geological Survey of Canada (GSC) is investigating the geology of part of the Canadian Shield in remote areas of Baffin Island and is developing methods for both remotely sensing geologic data and visualizing and analyzing geologic structures in 3-D. Analysis of hyperspectral satellite images has allowed different rock units to be identified in intensely folded and contorted areas of southern Baffin Island (box figure 1). These images allow discrimination of potentially economically valuable rock types that cannot be easily identified by ground-based mapping alone.

Structural geologists are also conducting detailed field mapping of geologic units and structures in central Baffin Island in order to reconstruct their 3-D distribution (box figure 2). Field data are collected in digital form using hand-held Palm Pilot devices and integrated directly into GIS for analysis. Digital elevation models (DEMs) showing the form of the land surface are constructed from regional topographic maps and combined with elevation data collected using GPS during ground surveys (box figure 2). Field-based geological map and DEM data are then analyzed using 3-D computer modelling software in order to create images showing the distribution of rock types (box figure 3). These detailed 3-D structural models can be used to better interpret the geologic history and resource potential of Baffin Island and the methodologies developed may be applied to other structurally complex regions of Canada.

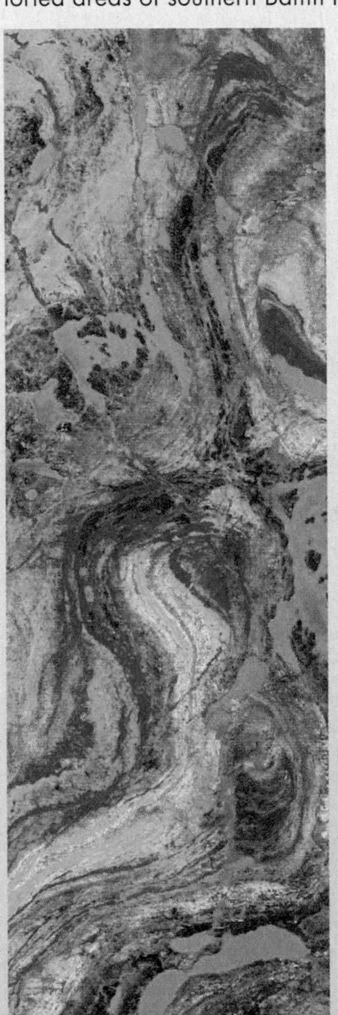

BOX 11.1 ■ FIGURE 1

Hyperspectral image showing various lithologic units and fold structures in southern Baffin Island. Blue = quartzite, green = iron-rich metamorphic rocks, dark red = vegetation and metagabbro, and pink = iron-poor metamorphic rocks. Traditional field mapping of this area did not discriminate the iron-rich and iron-poor units shown here.

Image courtesy of Jeff Harris, Geological Survey of Canada, Earth Sciences Sector NRCan

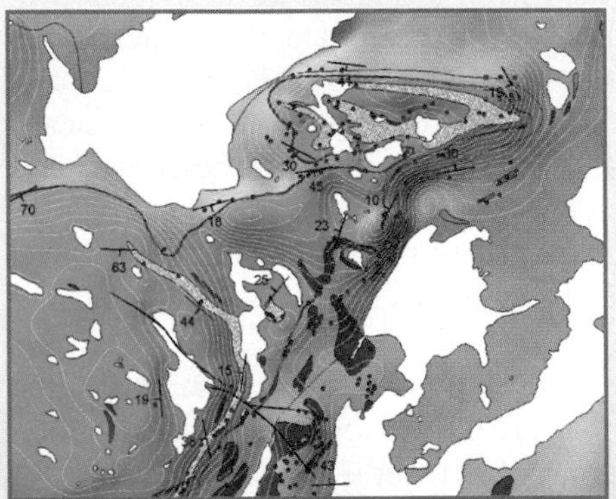

BOX 11.1 ■ FIGURE 2

(*A*) Bedrock geology superimposed upon DEM map of part of central Baffin Island showing structural field observations of geologic bedding (strike and dips). Area shown is approximately 10 × 15 km. (*B*) Strike and dip measurements displayed in 3-D can be used to create 3-D structural maps.

Image courtesy of Eric de Kemp, Natural Resources of Canada

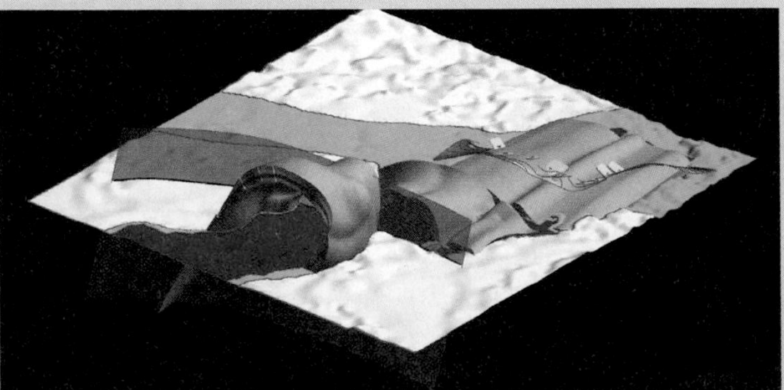

BOX 11.1 ■ FIGURE 3

3-D interpretive map of regional geologic data. The nature of folding changes from gentle upright structures on the right-hand side to tight, slightly overturned folding on the left. This may indicate proximity to a deformation front to the left of the image.

Image courtesy of Eric de Kemp, Natural Resources of Canada

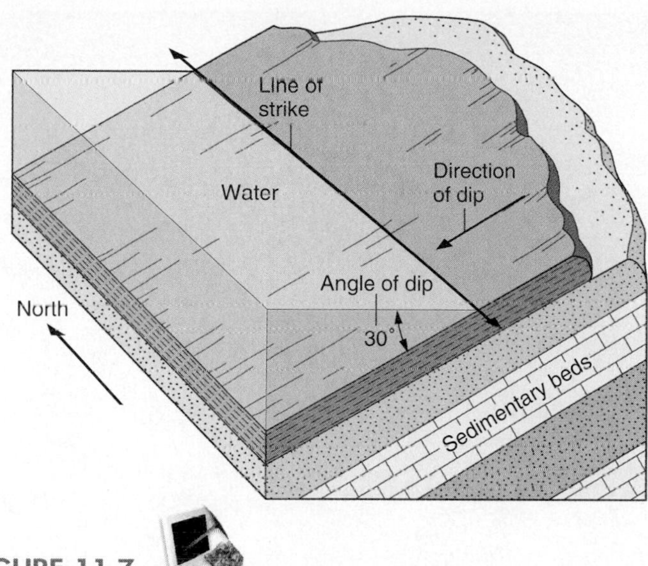

FIGURE 11.7

Strike, direction of dip, and angle of dip. The line of strike is found where an inclined bed intersects a horizontal plane (as shown here by the water). The dip direction is always perpendicular to the strike and in the direction the bed slopes (or a ball would roll down). The dip angle is the vertical angle of the inclined bed as measured from the horizontal.

Strike is the compass direction of a line formed by the intersection of an inclined plane with a horizontal plane. In this example, the inclined plane is a bedding plane. You can see from figure 11.7 that the beds are striking from north to south. Customarily, only the northerly direction of the strike line is given, so we simply say that beds strike north a certain number of degrees east or west (such as N50°E).

Observe that the **angle of dip** is measured downward from the horizontal plane to the bedding plane (an inclined plane). Note that the angle of dip (30° in the figure) is measured within a vertical plane that is perpendicular to both the bedding and the horizontal planes.

The **direction of dip** is the compass direction in which the angle of dip is measured. If you could roll a ball down a bedding surface, the compass direction in which the ball rolled would be the direction of dip.

The dip angle is always measured at a right angle to the strike; that is, perpendicular to the strike line as shown in figure 11.7. Because the beds could dip away from the strike line in either of two possible directions, the general direction of dip is also specified—in this example, west. In North America, it is now conventional to use the "right-hand rule" to record *azimuthal* strikes and dips. The right-hand rule states that strike is recorded in the direction such that the dip is down to the right. Thus, a bed oriented N45°E, 25°NW would be recorded as 225/25, and a bed oriented N0°, 30°W as 180/30. Azimuthal strikes and dips are necessary for input into computer mapping programs.

A specially designed instrument called a Brunton pocket transit (after the inventor) is used by geologists for measuring the strike and dip (figure 11.8). The Brunton pocket transit contains a compass, a level, and a device for measuring angles of inclination. Besides recording strike and dip measurements in a field notebook, a geologist who is mapping an area draws strike and dip symbols on the field map, such as ⅄ or ⅄ for each outcrop with dipping or tilted beds. On the map, the intersection of the two lines at the centre of each strike and dip symbol represents the location of the outcrop where the strike and dip of the bedrock were measured. The long line of the symbol is aligned with the compass direction of the strike. The small tick, which is always drawn perpendicular to the strike line, is put on one side or the other, depending on which of the two directions the beds actually dip. The angle of dip is given as a number next to the appropriate symbol on the map. Thus, ^{25}⅄ indicates that the bed is dipping 25 degrees from the horizontal toward the northwest and the strike is northeast (assuming that the top of the page is north). The orientation of the bed would be written 225/25. Figure 11.9 is a geologic map that shows all the sedimentary layers striking north and dipping 30 degrees to the west (180, 30).

Beds with vertical dip require a unique symbol because they dip neither to the left nor to the right of the direction of strike. The symbol used is ⨯, which indicates that the beds are striking northeast and that they are vertical (045/90). Strike and dip measurements can now be recorded digitally in the field using hand-held devices and the data later displayed as 3-D images (see box 11.1).

FIGURE 11.8

Geoscientist measuring the strike and dip of sedimentary rocks exposed along the Niagara Escarpment, Ontario.

Photo by J. Maclachlan

Geologic Cross-Sections

A **geologic cross-section** represents a vertical slice through a portion of the Earth. It is much like a roadcut or the wall of a quarry in that it shows the orientation of rock units and structures in the vertical dimension. Geologic cross-sections are constructed from geologic maps by projecting the dip of rock units into the subsurface (figure 11.9), and are quite useful in helping visualize geology in three dimensions. They are used extensively throughout this book as well as in professional publications. A number of software packages are also available to create 3-D geological models and images of subsurface data.

WHAT DO FOLDED ROCKS TELL US ABOUT GEOLOGIC PROCESSES?

Folds are bends or wavelike features in layered rock. Folded rock can be compared to several layers of rugs or blankets that have been pushed into a series of arches and troughs. Folds in rock can often be seen in roadcuts or other exposures (figure 11.10). When the arches and troughs of folds are concealed (or when they exist on a grand scale), geologists can still determine the presence of folds by noticing repeated reversals in the direction of dip taken on outcrops in the field or shown on a geologic map.

The fact that the rock is folded or bent shows that it behaved as a ductile material. Yet the rock exposed in outcrops is generally brittle and shatters when struck with a hammer. The rock is not metamorphosed (most metamorphic rock is intensely folded because it is ductile under the high pressure and temperature environment of deep burial and tectonic stresses). Perhaps folding took place when the rock was buried at a moderate depth where high temperature and confining pressure favour ductile behaviour. Alternatively, folding could have taken place close to the surface under a very low rate of strain.

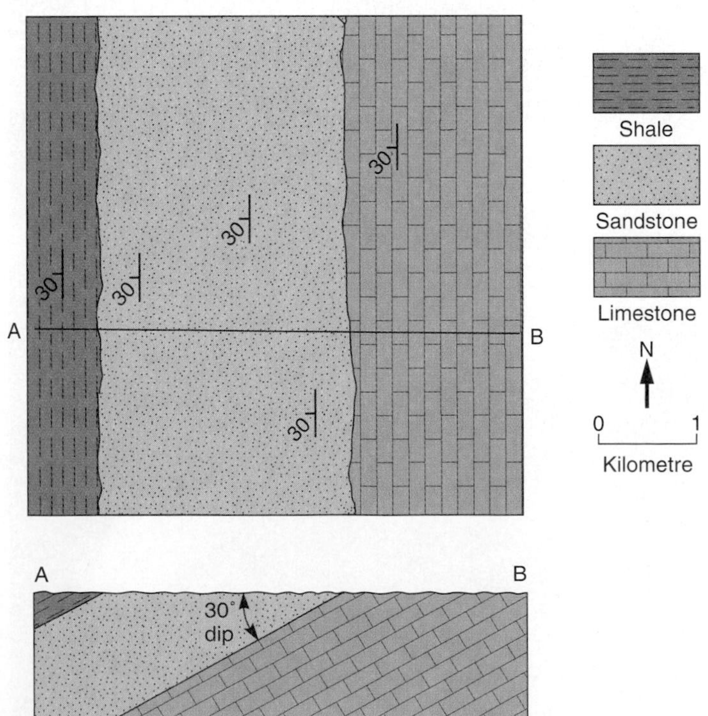

FIGURE 11.9

A geologic map and cross-section of an area with three sedimentary formations. (Each formation may contain many individual sedimentary layers, as explained in chapter 9.) Beds strike north–south and dip 30° to the west. The geologic cross-section (vertical cut) is constructed between points A and B on the map.

FIGURE 11.10

Folded sedimentary rock layers exposed at Lulworth Cove, Dorset, England. *Photo © Tom Bean*

Geometry of Folds

Determining the geometry or shape of folds may have important economic implications because many oil and gas deposits (see boxes 11.2 and 11.3) and also some metallic mineral deposits are localized in folded rocks (see chapter 12). The geometry of folds is also important in unravelling how a rock was strained and how it might be related to the movement of tectonic plates. Folds are usually associated with compressive stresses along convergent plate boundaries, but are also commonly formed where rock has been sheared along a fault.

Because folds are wave-like forms, two basic fold geometries are common—anticlines and synclines (figure 11.11).

IN GREATER DEPTH 11.2

Structures Associated with Salt Domes

Most folding is the product of horizontal compression; what we would expect if rock layers were placed in a giant vise. However, we get vertical compressive stresses when bodies of salt rise through the crust. Domes and normal faults are the characteristic structures that form above the rising bodies of salt.

Blobs (more correctly, *diapirs*) of rock salt rise through layers of sedimentary rock. Sometimes these bodies, which are a kilometre (or more) wide, travel upward as much as 10 kilometres. They originate from buried layers of evaporites. Because rock salt is easily deformed plastically, it squeezes out of the original beds. It works its way upward because salt is less dense than the overlying rocks (box figure 1). Where it breaks through rock layers, those layers are upturned adjacent to the rock salt (box figure 2).

The rising salt pushes the overlying rock into a dome or, less commonly, an anticline. The compressive stresses caused by the rising salt diapirs may also create normal faults that trap oil (box figure 2).

Many important oil and gas fields are associated with salt domes and diapirs. Oil fields along the Gulf Coast of Texas and Louisiana are located at salt domes. Triassic-Jurassic salt diapirs along the Canadian East Coast are the focus of renewed exploration interest, and salt diapirs in the Sverdrup Basin of the Canadian Arctic form the cores of many petroleum plays where drag folds around the diapirs have trapped hydrocarbons against the salt. On Axel Heiberg Island, geologic evidence suggests that salt diapirs made predominantly of gypsum and anhydrite are still rising and have moved upward by several hundreds of metres since glacial erosion of the valleys in which they are found (box figure 3).

Related Web Resources

Visit the Bureau of Economic Geology Website at the University of Texas–Austin for interesting 3-D images and animations of salt domes.
www.beg.utexas.edu/indassoc/agl/agl_if.html

A description of the research conducted on salt domes and diapirs in the Canadian Arctic can be found at
http://earthsciences.dal.ca/research/facility/ftl/ftl-strand.html

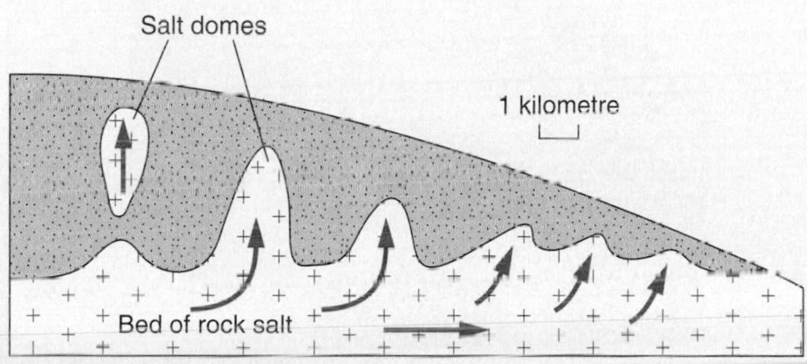

BOX 11.2 ■ FIGURE 1
Salt domes may form as a bed of rock salt is loaded unevenly by a thick wedge of sediment. The salt flows toward the thin part of the sediment wedge and also flows upward.

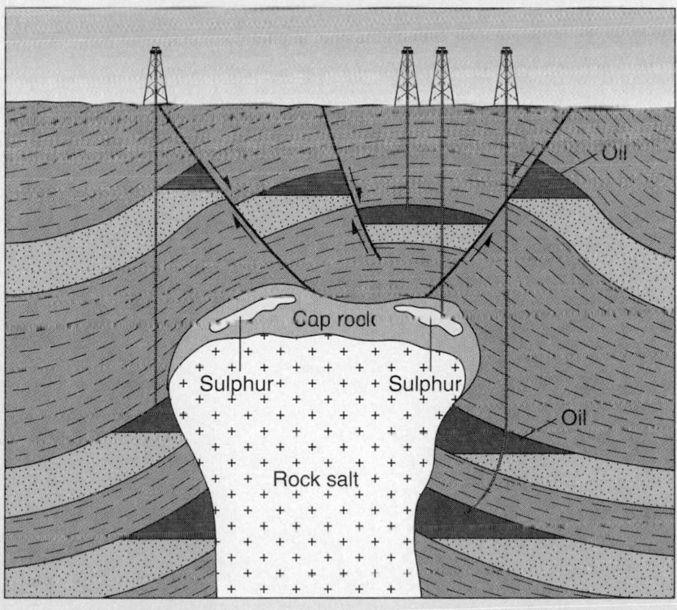

BOX 11.2 ■ FIGURE 2
A salt dome. Oil and gas are trapped in folds and along faults above the dome and within upturned sandstones along the flanks of the dome. Insoluble cap rock may contain recoverable sulphur.

BOX 11.2 ■ FIGURE 3
Deformed shale (grey unit to right of figure) within an anhydrite-gypsum diapir, Axel Heiberg Island.
Photo by M. Zentilli, Dalhousie University http://earthsciences.dal.ca/research/facility/ftl

IN GREATER DEPTH 11.3

Is There Oil Beneath My Property—First Check the Geologic Structure

An "oil pool" can exist only under certain conditions. Crude oil does not fill caves underground, as the term *pool* may suggest; rather, it simply occupies the pore spaces of certain sedimentary rocks, such as poorly cemented sandstone, in which void space exists between grains. Natural gas (less dense) often occupies the pore spaces above the crude oil, while water (more dense) is generally found saturating the rock below the oil pool (box figure 1).

A *source rock,* which is always a sedimentary rock, must be present for oil to form (see chapter 12). The sediment of the source rock has to include remains of organisms buried during sedimentation. This organic matter partially decomposes into petroleum and natural gas. Once formed, the droplets of petroleum tend to migrate, following fractures and interconnecting pore spaces. Being less dense than the rock, the petroleum usually migrates upward, although horizontal migration does occur.

If it is not blocked by impermeable rock, the oil may migrate all the way to the surface, where it is dissipated and permanently lost. Natural oil seeps, where leakage of petroleum is taking place, exist both on land and offshore. Where impermeable rock blocks the oil droplets' path of migration, an oil pool may accumulate below the rock, much like helium-filled balloons might collect under a domed ceiling. For any significant amount of oil to collect, the rock below the impermeable rock must be porous as well as permeable. Such a rock, when it contains oil, is called a *reservoir rock.*

Another necessary condition is that the geologic structure must be one that favours the accumulation and retention of petroleum. An "anticlinal trap" is one of the best structures for holding oil. As oil became a major energy source and the demand for it increased, most of the newly discovered wells penetrated anticlinal traps. Geologists discovered these by looking for indication of anticlines exposed at the surface. As time went by, other types of structures were also found to be oil traps. Many of these were difficult to find because of the lack of telltale surface patterns indicating favourable underground structures. Box figure 1 also illustrates structures other than anticlinal traps that might have a potential for oil production.

A major area of focus for gas exploration in Canada is the Foothills region of the Western Canadian Sedimentary Basin. Almost 12 billion cubic feet of gas per year are produced from wells penetrating complex fold and fault structures at Moose Mountain, Alberta (box figure 2). The sedimentary rocks in this area have been horizontally compressed or "shortened" into folds that are commonly cut by low-angle thrust faults ("overthrust tectonics"). This style of deformation dominates in parts of Alberta, British Columbia, Yukon, and the western Northwest Territories and results from compression or shortening of the crust caused by accretion of terranes to the western margin of the North American craton between 180 and 80 million years ago (see chapter 20). In these areas, effective oil and gas traps are formed by the juxtaposition of permeable and impermeable beds in anticlines and along thrust faults. One of the major challenges for future oil and gas exploration in the Foothills region lies in the amount of data that needs to be processed and interpreted in order to develop the natural gas reserves within such complex geological structures.

For further information on oil and gas exploration in the Foothills region, go to www.mooseoils.com.

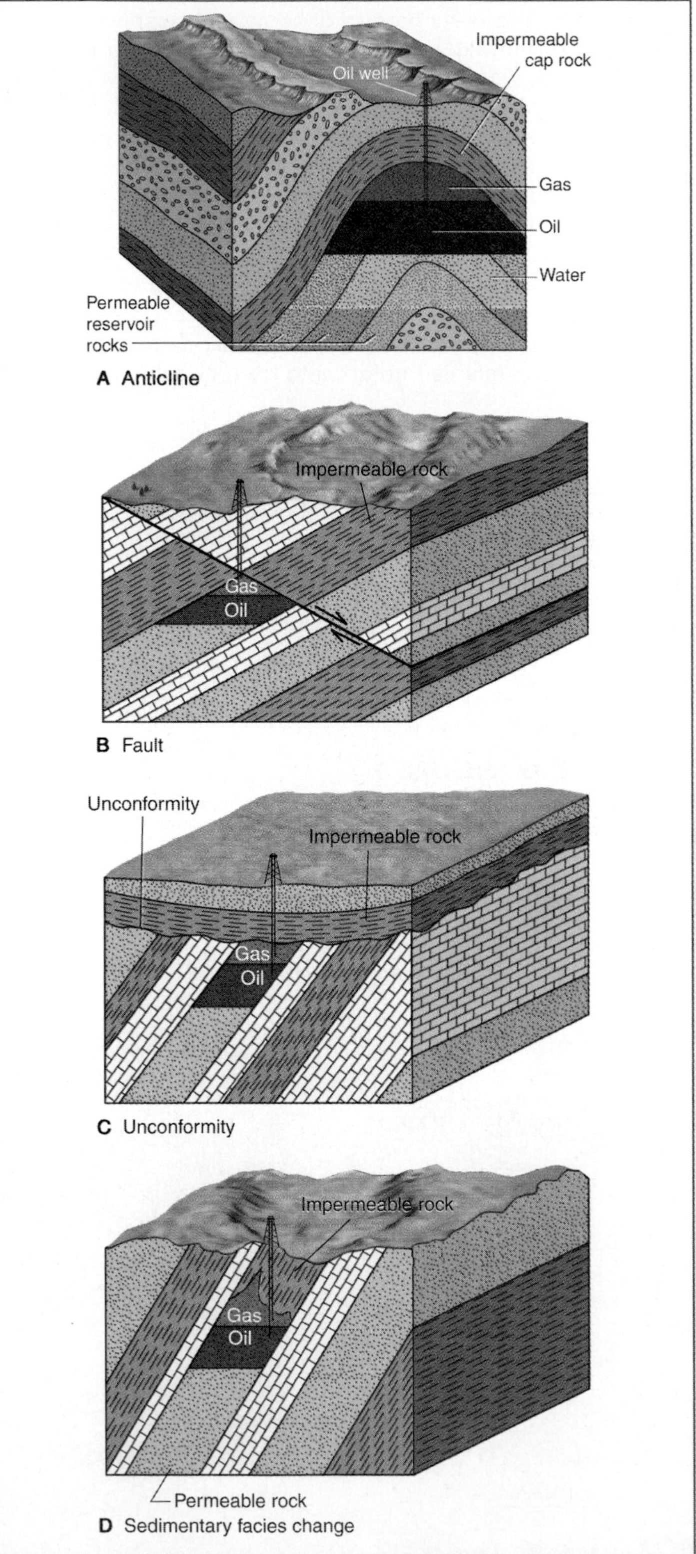

BOX 11.3 ■ FIGURE 1
Geological structures that may trap oil and gas.

A

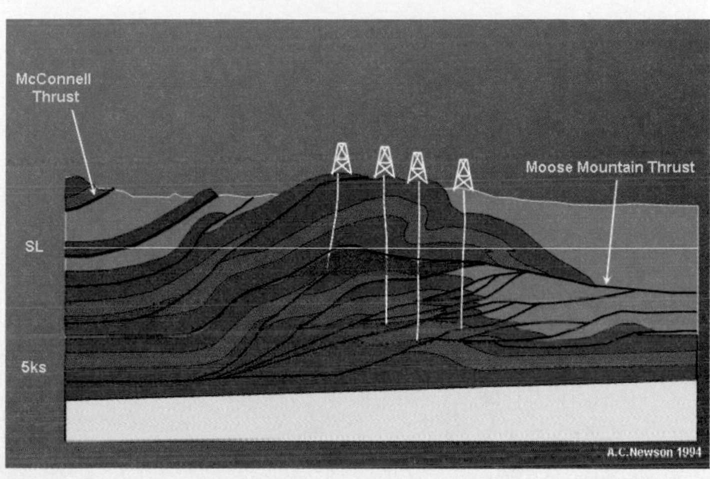

B

BOX 11.3 ■ FIGURE 2

(A) Production well at Moose Mountain, Alberta
(B) Structural cross-section through Paleozoic rocks within the Moose Mountain thrust structure showing location of gas-producing wells.
Photo A by Nick Eyles; Photo B by Andrew C. Neson, Moose Oils Ltd., Calgary

An **anticline** is an upward-arching fold. Usually the rock layers dip away from the **hinge line** (or *axis*) of the fold. The downward-arching counterpart of an anticline is a **syncline,** a troughlike fold. The layered rock usually dips toward the syncline's hinge line. In the series of folds shown in figure 11.11, two anticlines are separated by a syncline. Each anticline and adjacent syncline share a **limb.** Note the hinge lines on the crests of the two anticlines and bottom of the syncline. Similar hinge lines could be located in the hinge areas at the contacts between any two adjacent folded layers. For each anticline and the syncline, the hinge lines are contained within the shaded vertical planes. Each of these planes is an **axial plane,** an imaginary plane containing all of the hinge lines of a fold. The axial plane divides the fold into its two limbs.

It is important to remember that anticlines are not necessarily related to ridges nor synclines to valleys, because valleys and ridges are nearly always erosional features. In an area that has been eroded to a plain, the presence of underlying anticlines and synclines is determined by the direction of dipping beds in exposed bedrock, as shown in figure 11.12. (In the field, of course, the cross-sections are not exposed to view as they are in the diagram.)

Figure 11.12 also illustrates how determining the relative ages of the rock layers, or beds, can tell us whether a structure is an anticline or a syncline. Observe that the oldest exposed rocks are along the hinge line of the anticline. This is because lower layers in the originally flat-lying sedimentary or volcanic rock were moved upward and are now in the core of the anticline. The youngest rocks, on the other hand, which were originally in the upper layers, were folded downward and are now exposed along the synclinal hinge line.

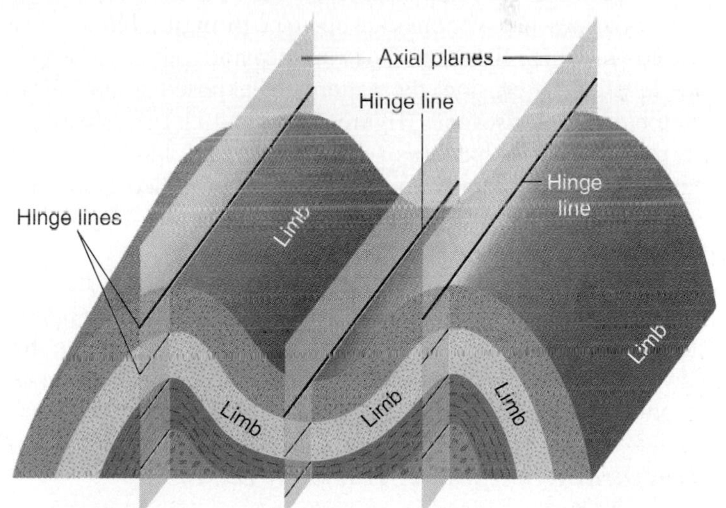

FIGURE 11.11

Diagrammatic sketch of two anticlines and a syncline illustrating the axial planes, hinge lines, and fold limbs.

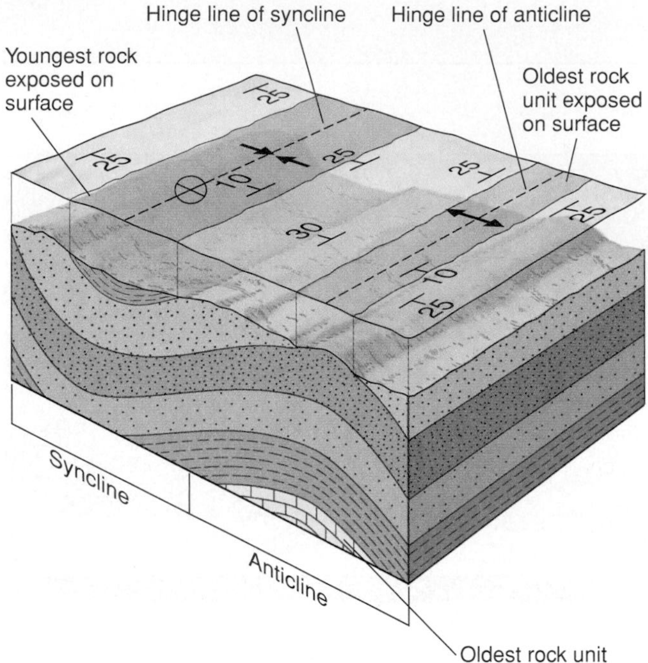

FIGURE 11.12

By measuring the strike and dip of exposed sedimentary beds in the field and plotting them on a geologic map (top surface) geologists can interpret the geometry of the geologic structure below the ground surface.

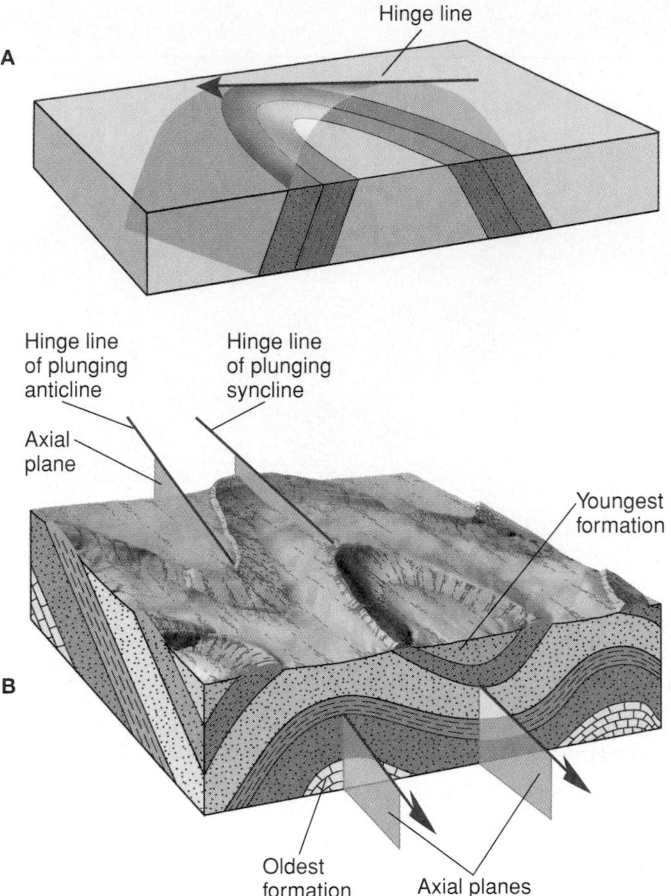

FIGURE 11.13

(*A*) Plunging fold that is cut by a horizontal plane has a V-shaped pattern. (*B*) Plunging anticline on left and right and plunging syncline in centre. The hinge lines plunge toward front of block diagram and lie within the axial planes of the folds.

Plunging Fold

The examples shown so far have been of folds with horizontal hinge lines. These are the easiest to visualize. In nature, however, anticlines and synclines are apt to be **plunging folds**—that is, folds in which the hinge lines are not horizontal. On a surface levelled by erosion, the patterns of exposed strata (beds) resemble Vs or horseshoes (figures 11.13 and 11.14) rather than the parallel striped patterns of nonplunging folds. However, plunging anticlines and synclines are distinguished from one another in the same way as are nonplunging folds—by directions of dip or by relative ages of beds.

A plunging syncline contains the youngest rocks in its centre or core, and the V or horseshoe points in the direction opposite the plunge. Conversely, a plunging anticline contains the oldest rocks in its core and the V points in the same direction as the plunge of the fold.

Structural Domes and Structural Basins

A **structural dome** is a structure in which the beds dip away from a central point. In cross-section, a dome resembles an anticline and is sometimes called a *doubly plunging anticline*. In a **structural basin,** the beds dip toward a central point (figure 11.15); in cross-section, it is comparable to a syncline (doubly plunging syncline). A structural basin is like a set of nested bowls. If the set of bowls is turned upside down, it is analogous to a structural dome.

FIGURE 11.14

Rock layers dip away from centre of a plunging anticline exposed in Utah. Anticline plunges in the direction of the upper part of the photo.

Photo by Frank M. Hanna

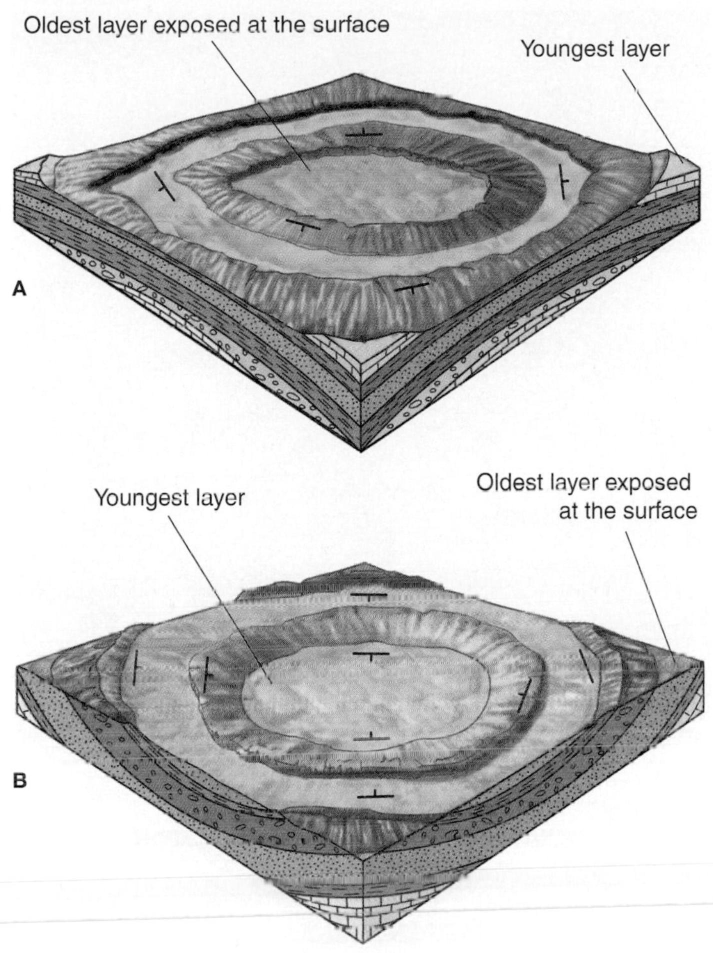

FIGURE 11.15
(A) Structural dome. (B) Structural basin.

FIGURE 11.16
Dome near Casper, Wyoming. The ridges are sedimentary layers that are resistant to erosion. Beds dip away from the centre of the dome, where the oldest layers are exposed.
Photo by D. A. Rahm, courtesy of Rahm Memorial Collection, Western Washington University

Domes and basins tend to be features on a grand scale (some are more than a hundred kilometres across), formed by uplift somewhat greater (for domes) or less (for basins) than that of the rest of a region. Michigan's lower peninsula and parts of adjoining states and Ontario are on a large structural basin (see map on the inside front cover). Domes of similar size are found in other parts of the Middle West. Smaller domes are found in the Rocky Mountains (figure 11.16).

Domes and anticlines (as well as some other structures) are important to the world's petroleum resources, as described in chapter 12.

Interpreting Folds

Folds occur in many varieties and sizes. Some are studied under the microscope, while others can have adjacent hinge lines tens of kilometres apart. Some folds are a kilometre or more in height. Figure 11.17 shows several of the more common types of folds. **Open folds** (figure 11.17*A*) have limbs that dip gently. All other factors being equal, the more open the fold, the less it has been strained by shortening. By contrast, an **isoclinal fold,** one in which limbs are parallel to one another, implies larger shortening strain or shear strain (figure 11.17*B*).

Folds that have vertical axial planes are referred to as upright folds. However, where the axial plane of a fold is not vertical but is inclined or tipped over, the fold may be classified as *asymmetric*. If the axial plane is inclined to such a degree that the fold limbs dip in the same direction, the fold is classified as an **overturned fold** (figure 11.17*C*). Looking at an outcrop where only the overturned limb of a fold is exposed, you would probably conclude that the youngest bed is at the top. The principles of *superposition* (see chapter 19), however, cannot be applied to determine top and bottom for overturned beds. You must either see the rest of the fold or find primary sedimentary structures within the beds such as mudcracks that indicate the original top or upward direction.

Recumbent folds (figure 11.17*D*) are overturned to such an extent that the limbs are essentially horizontal. Recumbent folds are found in the cores of mountain ranges such as the Canadian Rockies, Alps, and Himalayas and record extreme shortening and shearing of the crust typically associated with plate convergence.

WHAT ARE FRACTURES AND FAULTS?

If a rock is brittle, it will fracture. Commonly there is some movement or displacement. If essentially no displacement occurs, a fracture or crack in bedrock is called a **joint.** If the rock on either side of a fracture moves parallel to the fracture surface, the fracture is a *fault.* Most rock at or near the surface is brittle, so nearly all exposed bedrock is jointed to some extent.

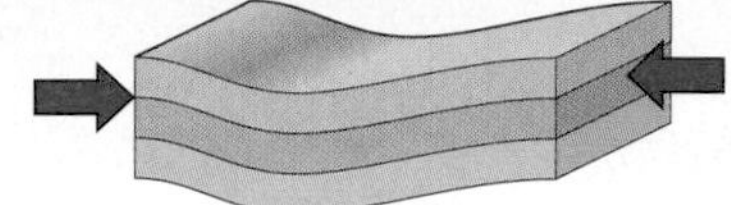

A Open folds

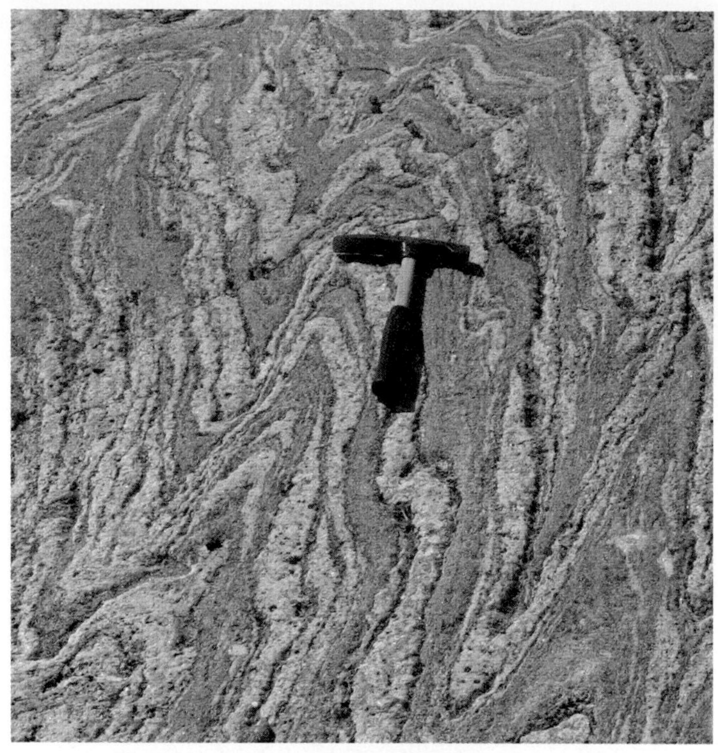

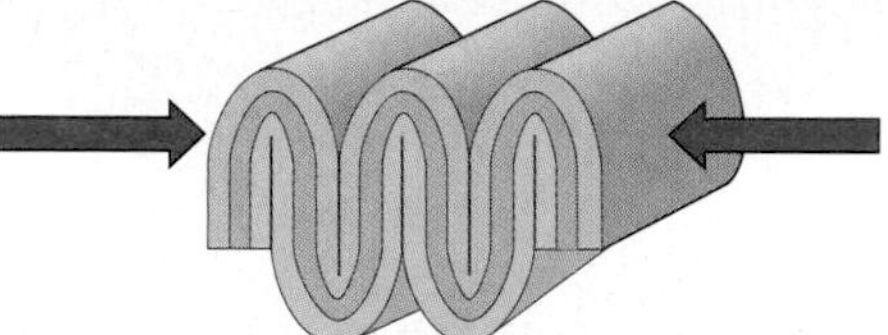

B Isoclinal ("hairpin") folds

C Overturned folds

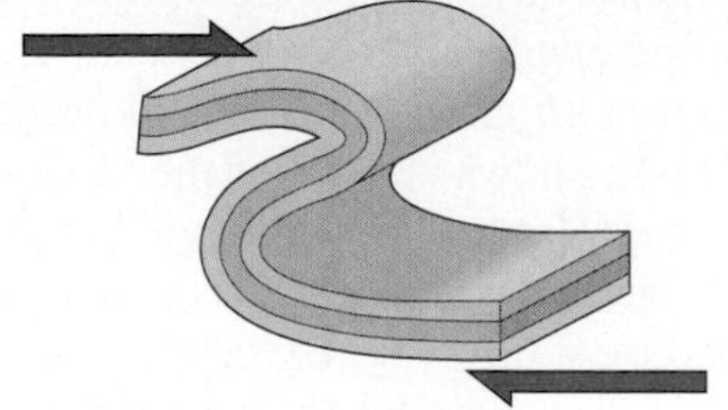

D Recumbent folds

FIGURE 11.17

Various types of folds. The length of the arrows in *A* through *D* is proportional to the amount and direction of shortening and shearing that caused folding. (*A*) Open folds in Spain (they are plunging away from the people). (*B*) Tight isoclinal folds in gneiss of the Canadian Shield. (*C*) Overturned anticline from northern California. (*D*) Recumbent folds, Port au Peninsula, Newfoundland.

Photo A by C. C. Plummer; Photo B by Nick Eyles; Photo C by Diane Carlson; Photo D Saint Mary's University: www.stmarys.ca/academic/science/geology/structural/folds.html

Joints

In discussing volcanoes, we described *columnar jointing,* in which hexagonal columns form as the result of tension and contraction of a cooling, solidified lava flow. *Sheet jointing,* a type of jointing due to expansion (discussed along with weathering in chapter 8) is caused by pressure release due to the removal of overlying rock and has the effect of creating tensional stress perpendicular to the land surface.

Columnar and sheet joints are examples of fractures that form from nontectonic stresses and are therefore referred to as primary joints. In this chapter we are concerned with joints that form not from cooling or unloading but from tectonic stresses.

Joints are one of the most commonly observed structures in rocks (figure 11.18). A joint is a fracture or crack in a rock body along which essentially no displacement has occurred. Joints form at shallow depths in the crust where rock breaks in a brittle way and is pulled apart slightly by tensional stresses caused by bending or regional uplift. Where joints are oriented approximately parallel to one another, a **joint set** can be defined.

FIGURE 11.18

Vertical joints in sedimentary rock of the Colorado Plateau formed in response to tectonic uplift of the region.
Photo by Frank M. Hanna

FIGURE 11.19

Geoscientists examining joints (spray-painted red) in clay-rich glacial deposits in a test pit at a potential landfill site in Ontario. The joints would facilitate the movement of groundwater and the site was removed from further consideration as a landfill.
Photo by Nick Eyles

Geologists sometimes find valuable ore deposits by studying the orientation of joints. For example, hydrothermal solutions may migrate upward through a set of joints and deposit quartz and economically important minerals such as gold, silver, copper, and zinc in the cracks. Accurate information about joints also is important in the planning and construction of large engineering projects, particularly dams and reservoirs. If the bedrock at a proposed location is intensely jointed, the possibility of dam failure or reservoir leakage may make that site too hazardous. The movement of contaminated groundwater from unlined landfills and abandoned mines may also be controlled by joints (figure 11.19), which results in difficult and costly cleanups.

Faults

Faults are fractures in bedrock along which sliding has taken place. The displacement may be only several centimetres or may involve hundreds of kilometres. For many geoscientists, an active fault is regarded as one along which movement has taken place during the last 11,000 years. Most faults, however, are no longer active.

The nature of past movement ordinarily can be discerned where a fault is exposed in an outcrop (figure 11.20). The geologist looks for dislocated beds or other features of the rock that might show how much displacement has occurred and the relative direction of movement. In some faults, the contact between the two displaced sides is very narrow. In others the rock has been broken or ground to a fractured or pulverized mass sandwiched between the displaced sides.

Geologists describe fault movement in terms of direction of slippage: dip-slip, strike-slip, or oblique-slip (figure 11.21). In a **dip-slip fault,** movement is parallel to the dip of the fault surface. A **strike-slip fault** indicates *horizontal* motion parallel to the strike of the fault surface. An **oblique-slip fault** has both strike-slip and dip-slip components.

FIGURE 11.20

Fault in Quaternary sediments of the Rouge Valley, Ontario. Note downward displacement of sediment layers to right of fault.

Photo by Nick Eyles

Dip-Slip Faults

In a dip-slip fault, the movement is up or down parallel to the dip of the inclined fault surface. The side of the fault above the inclined fault surface is called the **hanging wall,** whereas the side below the fault is called the **footwall** (figure 11.22). These terms came from miners who tunnelled along the fault looking for veins of mineralized rock (ore). As they tunnelled, their feet were on the lower *footwall block* and they could hang their lanterns on the upper surface, or *hanging-wall block.*

Normal and reverse faults, the most common types of dip-slip faults, are distinguished from each other on the basis of the relative movement of the footwall block and the hanging-wall block.

In a **normal fault** (figures 11.23 and 11.24), the hanging-wall block has moved downward relative to the footwall block. The relative movement is represented on a geological cross-section by a pair of arrows, because geodetic measurement of normal faults indicate that both blocks move during slip. As shown in figure 11.23, a normal fault results in extension or lengthening of the crust. When there is extension of the crust, the hanging-wall block moves downward along the fault to compensate for the pulling apart of the rocks. Sometimes a block bounded by normal faults will drop down, creating a *graben,* as shown in figure 11.23*C*. (*Graben* is the German word for "ditch.")

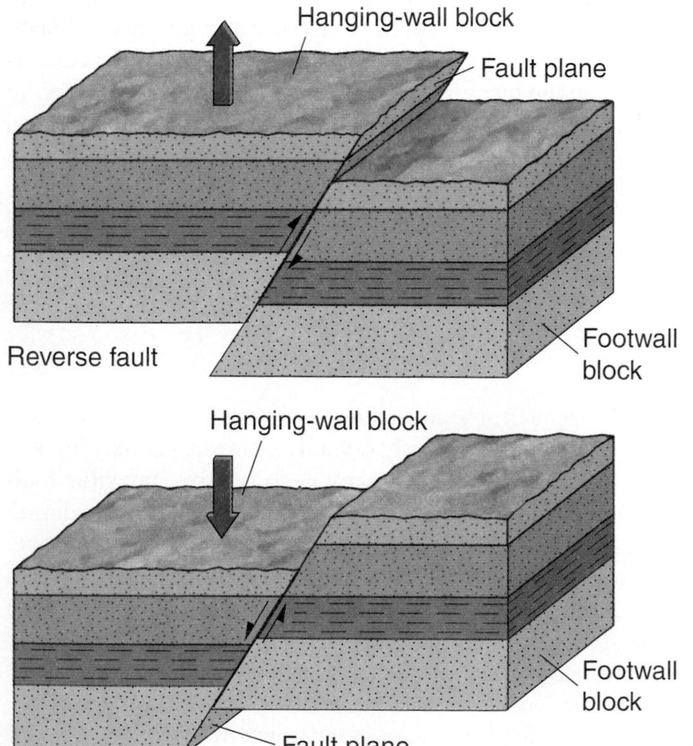

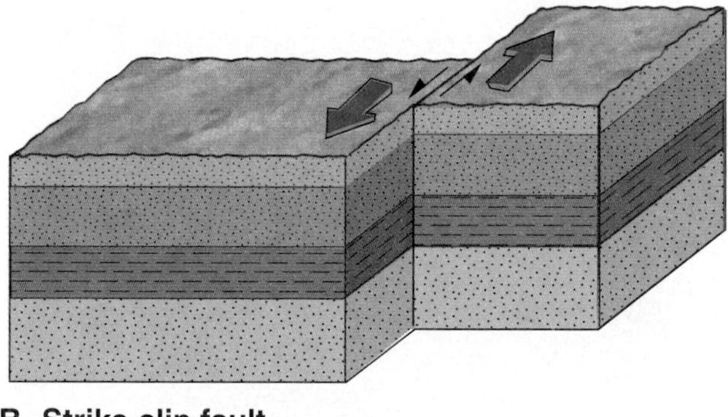

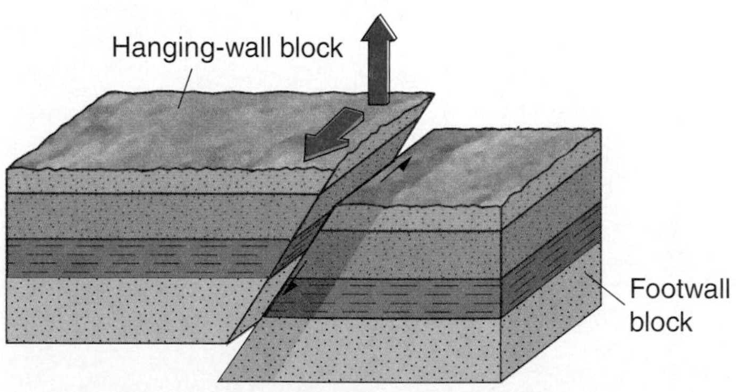

FIGURE 11.21

Three types of faults illustrated by displaced blocks. Although both blocks probably move when the fault slips, the heavier arrows show only the direction of movement on the left. (*A*) Dip-slip movement. (*B*) Strike-slip movement. (*C*) Oblique-slip movement. Black arrows show dip-slip and strike-slip components of movement.

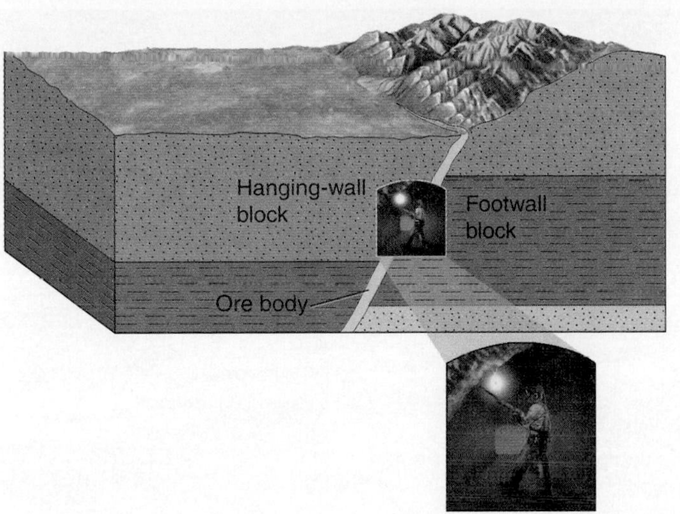

FIGURE 11.22

Relationship between the hanging wall block and footwall block of a fault. The upper surface where a miner can hang a lantern is the hanging wall. The lower surface below the fault is the footwall.

Rifts are grabens associated with diverging plate boundaries, either along mid-oceanic ridges or on continents (see chapter 2). The Rhine Valley in Germany and the Red Sea Rift are examples of grabens.

If a block bounded by normal faults is uplifted sufficiently, it becomes a fault-block mountain range. (This is also called a *horst,* the opposite of a graben.) The Basin and Range province of Nevada and portions of adjoining states are characterized by numerous mountain ranges (horsts) separated from adjoining valleys by normal faults. Normal fault planes typically dip at steep angles (60°) at shallow depths but may become curved or even horizontal at depth.

In a **reverse fault,** the hanging-wall block has moved up relative to the footwall block (figures 11.25 and 11.26). As shown in figure 11.25, horizontal compressive stresses cause reverse faults. Reverse faults tend to shorten the crust.

A **thrust fault** is a reverse fault in which the dip of the fault plane is at a low angle (<30°) or even horizontal (figures 11.25*C* and 11.27). In some mountain regions, it is not uncommon for the upper plate (or hanging-wall block) of a thrust fault to have overridden the lower plate (footwall block) for several tens of kilometres.

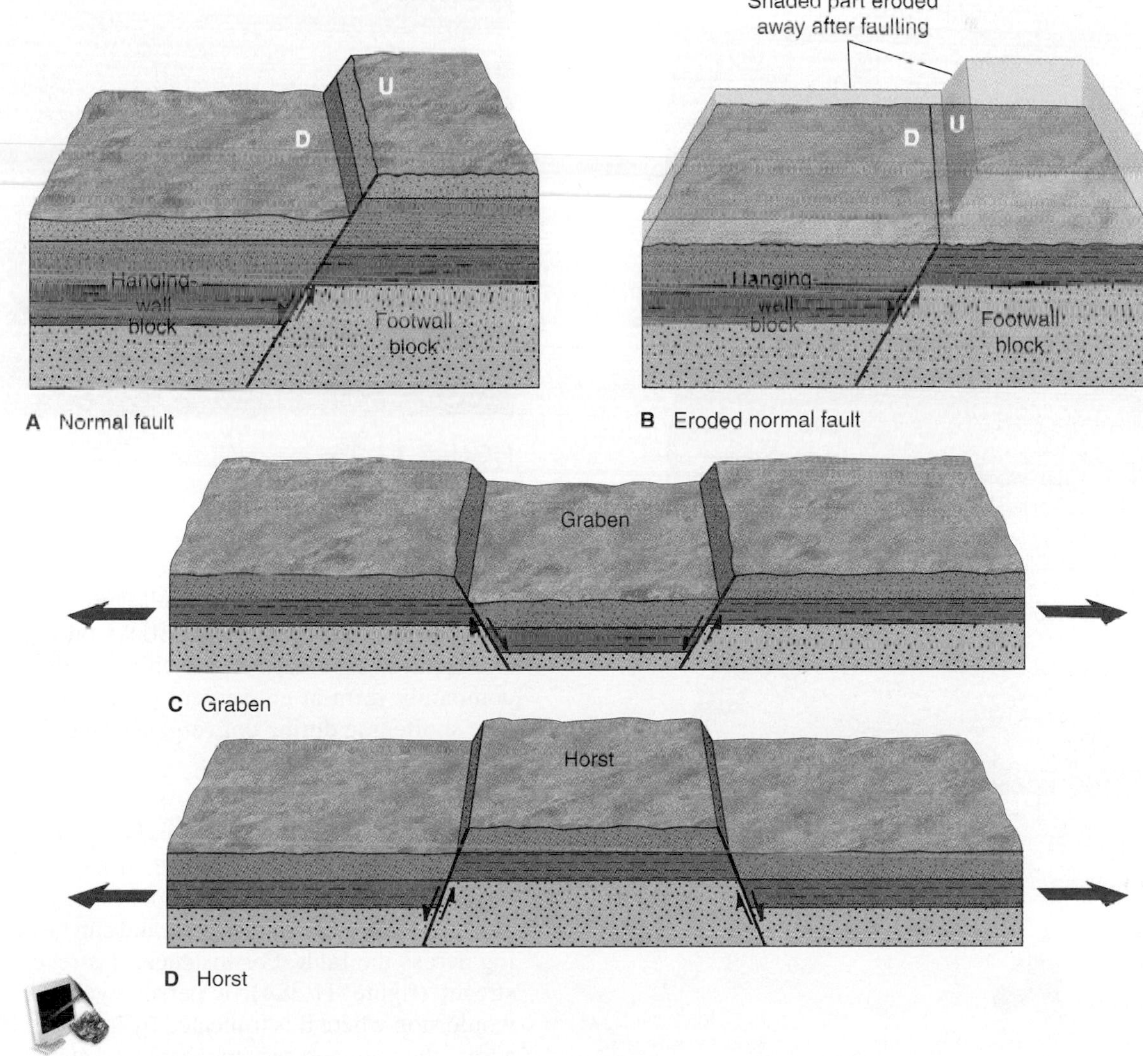

FIGURE 11.23

Normal faults. (*A*) Diagram shows the fault before erosion and the geometric relationships of the fault. (*B*) The same area after erosion. (*C*) A graben. (*D*) A horst. Arrows in *C* and *D* indicate horizontal extension of the crust.

FIGURE 11.24
Normal faults with prominent horst block offset volcanic ash layers in southern Oregon.
Photo by Diane Carlson

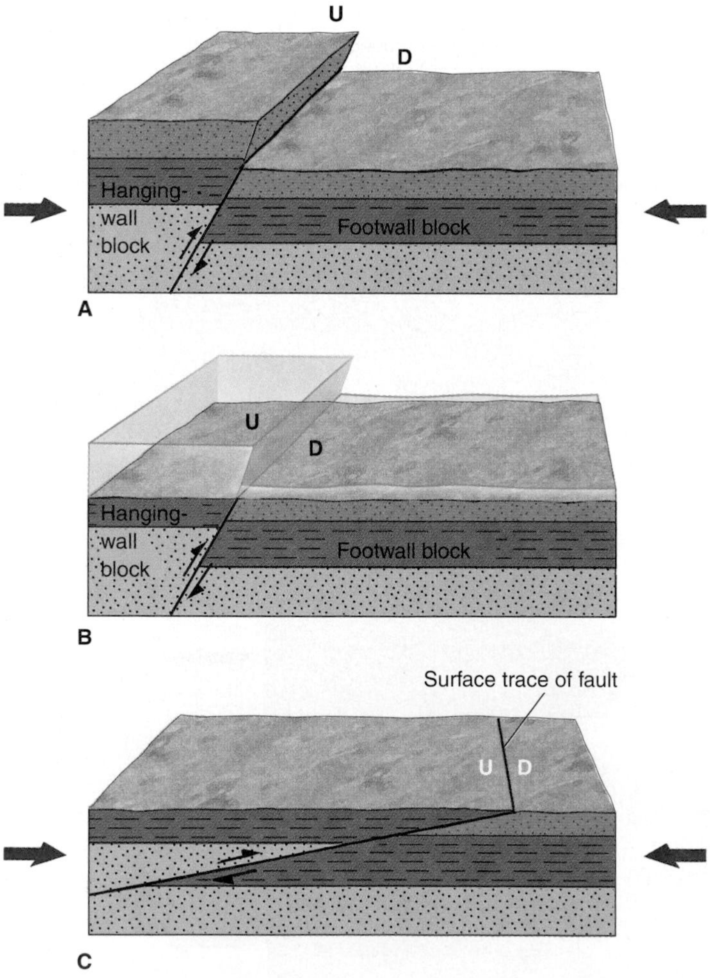

FIGURE 11.25
(*A*) A reverse fault. The fault is unaffected by erosion. Arrows indicate shortening direction. (*B*) Diagram shows area after erosion. (*C*) Thrust fault has a lower angle of dip and accommodates more shortening by stacking rock layers on top of one another.

FIGURE 11.26
Reverse fault in shales, Waterton National Park, Alberta.
Photo by Nick Eyles

Thrust faults such as the McConnell Thrust in Alberta (figure 11.27) typically move or thrust older rocks on top of younger rocks, and result in an extreme shortening of the crust. Thrust faults commonly form at convergent plate boundaries to accommodate shortening during convergence (see chapter 20).

Strike-Slip Faults

A fault where the movement (or *slip*) is predominantly horizontal and therefore parallel to the strike of the fault is called a **strike-slip fault.** The displacement along a strike-slip fault is either left-lateral or right-lateral and can be determined by looking across the fault. For instance, if a recent fault displaced a stream (figure 11.28*A*), a person walking along the stream would stop where it is truncated by the fault. If the person looks across the fault and sees the displaced stream to the right, it is a **right-lateral fault.** In a **left-lateral fault,** a stream or other dis-

A

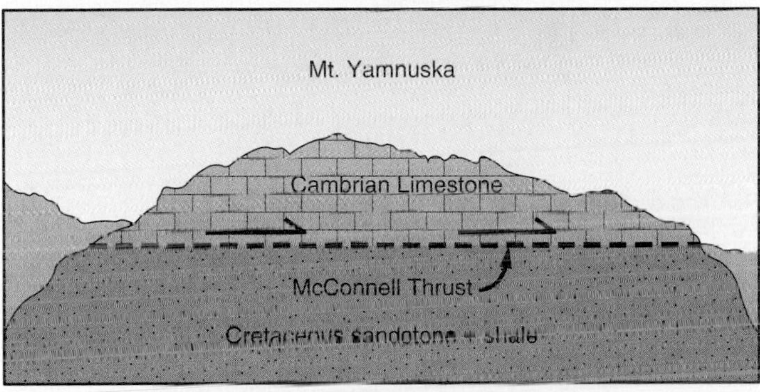

B

FIGURE 11.27

(A) Mt. Yamnuska in the Canadian Rockies. The steep cliff is 350 m high. (B) Sketch of Yamnuska showing older, resistant Paleozoic limestones thrust from the southwest to the northeast, along the McConnell Thrust, up and over younger and softer Cretaceous sandstones and shales.

Photo by Nick Eyles

placed feature would appear to the left across the fault. Again, we cannot tell which side actually moved, so pairs of arrows are used to indicate relative movement.

Large strike-slip faults, like the San Andreas fault in California, typically define a zone of faulting that may be several kilometres wide and hundreds of kilometres long (see box 11.4). The surface trace of an active strike-slip fault is usually defined by a prominent linear valley that has been more easily eroded where the rock has been ground up along the fault during movement. The linear valley may contain lakes or sag ponds where the highly permeable fault rock allows groundwater to pond at the surface. The trace of the fault may also be marked by offset surface features such as streams, fences, and roads or by distinctive rock units (see box 11.4, figure 3).

Strike-slip faults that have experienced a large amount of offset typically do not remain straight for long distances. They may either bend or step over to another fault that is parallel. Depending on the direction of the bend or stepover, the lithosphere is either pulled apart (*releasing bend*) or pushed together (*restraining bend*) (figure 11.28*B*). Normal faults and grabens form in response to the pulling apart at the releasing bends and folds and thrust faults form at the restraining bends to accommodate the pushing or pinching together of the lithosphere.

Strike-slip faults accommodate shearing strain in the brittle, uppermost lithosphere, and may also represent transform plate boundaries where plates slide past one another. One of the most famous examples of a transform fault is the San Andreas fault. The San Andreas fault is a right-lateral strike-slip fault that forms part of the boundary between the North American and Pacific plates.

A

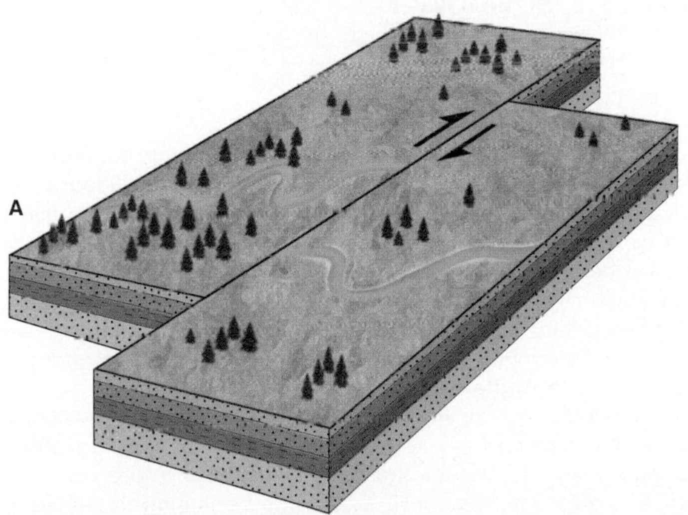

B

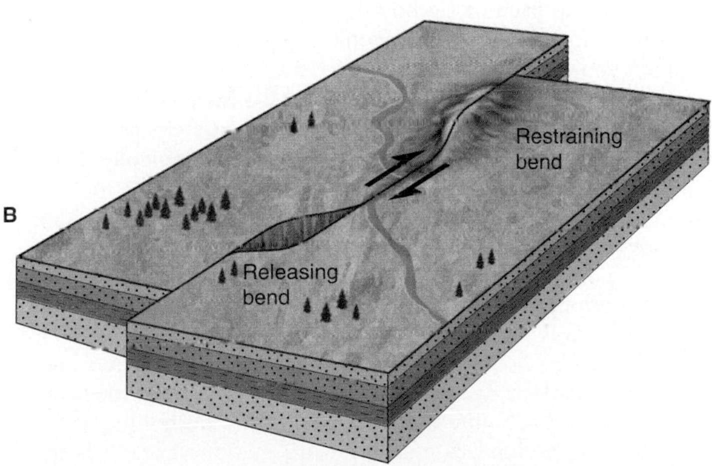

FIGURE 11.28

(A) Right-lateral strike-slip fault offsets a stream channel. Looking across the fault, you would need to walk to the right to find the continuation of the stream. (B) Strike-slip movement along curved faults produces gaps or basins at releasing bends where the lithosphere is pulled apart or shortening and hills where it is pushed together at restraining bends.

IN GREATER DEPTH 11.4

The World's Most Famous Fault—The San Andreas

The San Andreas fault in California is one of the best known geologic structures in North America, but the geologists and seismologists who study it admit that our knowledge of its activity and its history is far from complete. Actually, the San Andreas is the longest of several, subparallel faults that transect western California (box figure 1). Collectively, these right-lateral faults are known as the San Andreas fault system. The system is in a belt approximately 100 kilometres wide and 1,300 kilometres long that extends into Mexico, ending at the Gulf of California.

Los Angeles is slowly moving toward San Francisco because of San Andreas fault motion. At an average rate of movement of about 2 centimetres per year, Los Angeles could be a western suburb of San Francisco (or San Francisco an eastern suburb of Los Angeles) in some 25 million years. Earthquakes are produced by sudden movement within the fault system, as explained in chapter 3. Bedrock along the San Andreas fault shifted as much as 4.5 metres in association with the 1906 quake that destroyed much of San Francisco.

The San Andreas fault is not a simple crack, but a belt of broken and ground-up rock, usually a hundred metres or more wide. Its presence is easy to determine throughout most of its length. Along the fault trace are long, straight valleys (formed by erosion and subsidence) that show quite different terrain on either side. Stream channels follow much of the fault zone because the weak, ground-up material along the fault is easily eroded. Locally, elongate lakes (called sag ponds) are found where the ground-up material has settled more than the surface of adjacent parts of the fault zone. The fault was named after one of these ponds, San Andreas Lake, just south of San Francisco (box figure 2).

One can visually follow the fault northward from San Andreas Lake into the southwestern suburbs of San Francisco. There the fault zone is hidden by recently built housing tracts. Apparently the builders and residents have chosen to ignore the hazards of living on the nation's most famous fault.

Geoscientists have been unable to agree on the total displacement of the fault or on how long it has been active. Some hypothesize that movement began in the Mesozoic Era (more than 65 million years ago); most think that it began later. The difficulty in establishing an age for the inception of the faulting lies in finding clear evidence of displaced bedrock. What geoscientists would like to find, if it exists, is a rock unit that can be isotopically dated and that was formed across the fault zone just before the faulting began. Currently displaced rock on both sides of the fault zone would have to be clearly identifiable as having been the same unit.

Geologically young features that cross the fault, such as displaced stream channels (box figure 3), are common. Similarly, ancient rocks that undoubtedly were there before faulting began are recognized as having been displaced. Many California geoscientists think that the belt of granitic rock just west of the fault was once the southern continuation of the granitic batholiths of the Sierra Nevada, which are more than 80 million years old. But these extremes tell us only that the age of the San Andreas is somewhere between approximately 80 million years and a few thousand years, when the stream channel in box figure 3 carved its course across the fault.

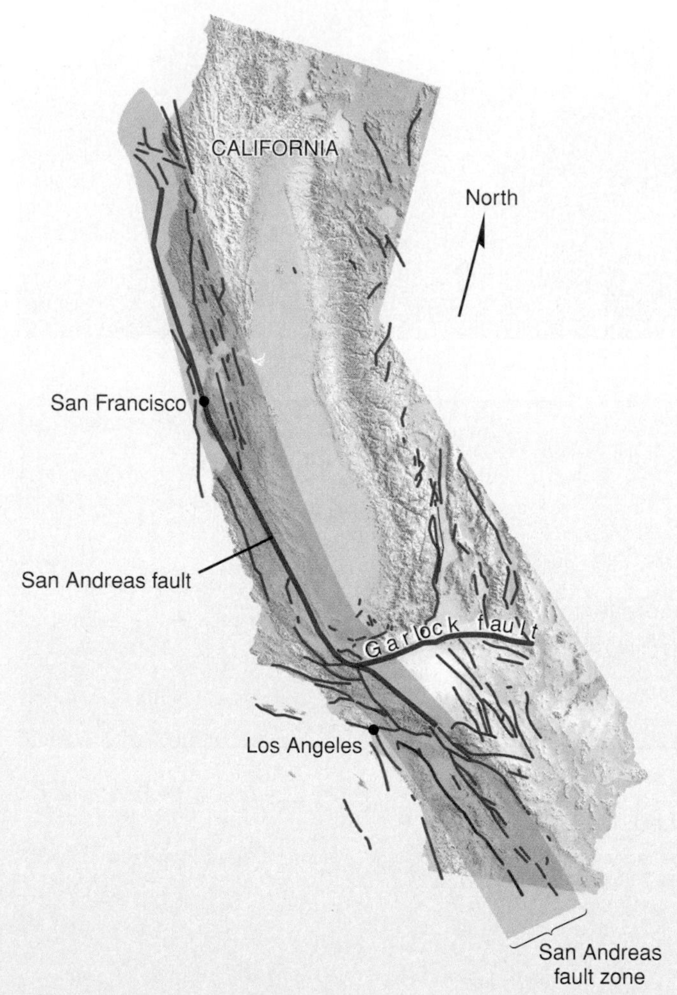

BOX 11.4 ■ FIGURE 1

California has its faults. Red lines indicate faults that have been active within the last 200 years, and blue lines indicate faults that have been active over the last 2 million years.

From California Division of Mines and Geology

The strongest evidence for long-term faulting comes from almost identical volcanic sequences now 315 kilometres apart. The volcanic activity took place 23.5 million years ago. Using these figures, we can calculate the average rate of motion as 1.3 centimetres per year for the San Andreas fault. (But, movement along other faults means the rate of motion is higher for the fault system.) Older rocks that appear to have been offset 560 kilometres have been correlated with less certainty, suggesting that the total offset for the San Andreas fault is at least 560 kilometres.

How long ago faulting began remains controversial. According to plate tectonics theory, the San Andreas fault is a transform boundary that separates the North American plate from the Pacific plate. One hypothesis places the beginning of strike-slip movement at about 30 million years ago. According to this hypothesis, the Baja California peninsula split from mainland Mexico as sea-floor spreading began. As the Gulf of California widens, the block of crust west of the San Andreas is pushed northward.

BOX 11.4 ■ FIGURE 2

Part of the San Andreas fault. View northward toward San Francisco. Lakes occupy the fault zone. Hills to the left of the fault are moving northward.

Photo by B. Amundson

BOX 11.4 ■ FIGURE 3

Stream channel displaced by the San Andreas fault. The arrows on either side of the fault trace indicate relative motion.

Photo by C. C. Plummer

Resources

Wallace, R. E., ed. 1990. *The San Andreas fault system, California*: U.S. Geological Survey Professional Paper 1515.

Related Web Resources

http://pubs.usgs.gov/gip/earthq3/

Internet version of the San Andreas fault system.

http://quake.usgs.gov/info/1906/index.html

"What have we learned about the San Andreas fault since 1906" Web page shows research sites along the San Andreas fault.

SUMMARY

Tectonic forces result in deformation of the Earth's crust. *Stress* (force per unit area) is a measure of the tectonic force and confining pressure acting on bedrock. Stress can be *compressive*, *tensional*, or *shearing*. *Strained* (changed in size or shape) rock records past stresses, usually as joints, faults, or folds.

A geologic map shows the structural characteristics of a region. *Strike* and *dip* symbols on geologic maps indicate the attitudes of inclined surfaces such as bedding planes. The strike and dip of a bedding surface indicate the relationship between the inclined plane and a horizontal plane.

If rock layers bend (ductile behaviour) rather than break, they become folded. Rock layers are folded into *anticlines* and *synclines* and recumbent folds. If the hinge line of a fold is not horizontal, the fold is *plunging*. Older beds exposed in the core of a fold indicate an anticline, whereas younger beds in the centre of the structure indicate a syncline. In places where folded rock has been eroded to a plain, an anticline can usually be distinguished from a syncline by whether the beds dip toward the centre (syncline) or away from the centre (anticline). Also, the oldest rocks are found in the centre of an eroded anticline whereas the youngest rocks are found in the centre or core of a syncline.

Fractures in rock are either *joints* or *faults*. A joint indicates that movement has not occurred on either side of the fracture; displaced rock along a fracture indicates a fault. *Dip-slip* faults are either *normal* or *reverse*, depending on the motion of the hanging-wall block relative to the footwall block. The relative motion of the hanging wall is upward in a reverse fault and downward in a normal fault. A reverse fault with a low angle of dip for the fault plane is a *thrust fault*. Reverse faults accommodate horizontal shortening of the crust, whereas normal faults accommodate horizontal stretching or extension.

In a *strike-slip* fault, which can be either left-lateral or right-lateral, horizontal movement parallel to the strike has occurred.

Terms to Remember

angle of dip 281
anticline 285
axial plane 285
brittle 278
compressive stress 277
dip-slip fault 289
direction of dip 281
ductile 278
elastic 278
elastic limit 278
fold 282
footwall 290
geologic cross-section 282
geologic map 279
hanging wall 290
hinge line 285
isoclinal fold 287
joint 287
joint set 289
left-lateral fault 292
limb 285
normal fault 290
oblique-slip fault 289
open fold 287
overturned fold 287
plunging fold 286
recumbent fold 287
reverse fault 291
right-lateral fault 292
shear stress 278
strain 277
stress 277
strike 281
strike-slip fault, 289, 292
structural basin 286
structural dome 286
structural geology 277
syncline 285
tensional stress 277
thrust fault 291

Testing Your Knowledge

Use the questions below to prepare for exams based on this chapter.

1. Most anticlines have both limbs dipping away from their hinge lines. For which kind of fold is this not the case?
2. What are the four main types of contacts and how would you distinguish among them if you were a geologist doing field work?
3. On a geologic map, if no cross-sections were available, how could you distinguish an anticline from a syncline?
4. If you locate a dip-slip fault while doing field work, what kind of evidence would you look for to determine whether the fault is normal or reverse?
5. Name several geologic structures described in earlier chapters.
6. What is the difference between strike, direction of dip, and angle of dip?
7. Draw a simple geologic map, using strike and dip symbols for a syncline plunging to the west.
8. How does a structural dome differ from a plunging anticline?

9. Which of the following statements is true?
   a. when forces are applied to an object, the object is under stress
   b. strain is the change in size (volume) or shape, or both, while an object is undergoing stress
   c. stresses can be compressive, tensional, or shear
   d. all of the above

10. The compass direction of a line formed by the intersection of an inclined plane with a horizontal plane is called
    a. strike b. direction of dip
    c. angle of dip

11. Folds in a rock show that the rock behaved in a _____ way.
    a. ductile b. elastic
    c. brittle d. all of the above

12. An anticline is
    a. any fold
    b. overturned fold
    c. an upward-arched fold
    d. a downward-arched fold

13. A syncline is
    a. an upward-arched fold
    b. overturned fold
    c. a downward-arched fold
    d. horizontal beds

14. A structure in which the beds dip away from a central point is called a
    a. basin b. anticline
    c. structural dome d. syncline

15. Which is not a type of fold?
    a. open b. isoclinal
    c. overturned d. recumbent
    e. thrust

16. Fractures in bedrock along which movement has taken place are called
    a. joints b. faults
    c. cracks d. crevasses

17. In a normal fault, the hanging-wall block has moved _____ relative to the footwall block.
    a. upward b. downward
    c. sideways

18. Normal faults accommodate what kind of strain?
    a. shortening b. extensional
    c. ductile

19. Faults that typically move older rock on top of younger rock are
    a. normal faults b. thrust faults
    c. strike-slip faults

Exploring Web Resources

www.mcgrawhill.ca/olc/plummer
Visit our Online Learning Centre for additional readings and media resources. Check your Testing Your Knowledge answers, and click on the links to go directly to the sites listed below.

www.science.smith.edu/departments/Geology/Structure_Resources/
Structural geology site maintained by Kevin J. Smart at the University of Oklahoma contains many links to online courses, computer software, bibliographies, and research projects dealing with structural geology.

www.geo.cornell.edu/geology/classes/geol326/326.html
Website for structural geology course taught by the Department of Geological Sciences at Cornell University contains images showing structural features and models of thrust–fault movement.

www.brocku.ca/ctg
Canadian Tectonics Group Website contains structural geology images, computer software, and a newsletter outlining research projects.

Animations

This chapter includes the following animations available on our Online Learning Centre at www.mcgrawhill.ca/olc/plummer.

11.7 Strike and Dip
11.17 Styles of Folding
11.21 Styles of Faulting
11.23 Normal Faulting
11.25 Reverse and Thrust Faults

CHAPTER

12

Geologic Resources

What are geologic resources?
How do resources differ from reserves?
What kinds of energy resources are used in North America?
How do oil and gas form?
Where do we find oil and gas?
How much oil and gas do we have left?
What are renewable sources of energy?
Where do metals come from?
Where do we find important metals?
What are some important nonmetallic resources?
What are some possible future trends?

Our examination of Earth materials would not be complete without a discussion of human use of Earth materials, most of which are nonrenewable. Water is a renewable resource that is an important exception and will be discussed in chapters 14 and 15. Our purpose in this chapter is to survey briefly some important geologic resources, other than water, that are of economic value.

We first look at energy resources to see which ones might help reduce our dependence on hydrocarbons such as oil. Then we discuss metals and their relation to igneous rocks and plate tectonics and conclude with nonmetallic resources such as sand and gravel.

Raw and sorted diamonds from the Jericho Diamond Mine near Carat Lake, Nunavut. The Jericho Mine, is Canada's third and Nunavut's first diamond mine. Canada is now the third largest producer of diamonds in the world. *Photo courtesy of CP/Jeff McIntosh*

Nearly every manufactured object requires geologic resources. An automobile contains substantial amounts of iron, chromium, manganese, nickel, platinum, tin, copper, lead, and aluminum in its body and engine and quartz sand in its window glass. It consumes petroleum in several forms—as fuel and lubricants, as synthetic rubber for tires, and as plastic for electrical parts, upholstery, and steering wheels. Dozens of other geologic resources go into automobiles, from tungsten in lightbulb filaments to sulphur in battery acid.

People are beginning to realize how heavily dependent on geologic resources they are. Some have tried to limit their consumption of resources, or at least the rate at which their consumption *increases*. But it is impossible to stop consuming geologic resources. Think about some of the objects near you as you are reading this chapter. If you're wearing jeans, they may be made of fabric that is 100-percent cotton, but many jeans are made of shrinkproof fabrics that blend cotton with petroleum-based synthetic fibres. The brass zipper is made of copper and zinc. Some brands of jeans have pocket rivets made of copper. To make the leather tags on the back of some jeans, either aluminum or chromium was used during tanning. The fabric dye almost certainly came from petroleum. A cell phone, which weighs on average 218 grams, uses many geologic resources (figure 12.1) and a computer many more. Computer chips require silicon from quartz sand. More than 40 percent of gold that is mined is used for computers. Dental fillings are made of mercury, silver, and other metals. Every one of us uses geologic resources, and therefore is indirectly responsible for the existence of mines and oil wells.

FIGURE 12.1

Geologic resources used to make a cell phone. A cell phone contains approximately 40 percent metals, 40 percent plastics, and 20 percent ceramics.

Photo by Nick Eyles

WHAT ARE GEOLOGIC RESOURCES?

Geologic resources are valuable materials of geologic origin that can be extracted from the Earth. There are three main categories of geologic resources:

1. Energy resources—petroleum (oil and natural gas), coal, uranium, and a few others, such as geothermal resources.
2. Metallic resources—iron, copper, aluminum, lead, zinc, gold, silver, and many more.
3. Nonmetallic resources—sand and gravel, building stone, limestone (for cement), sulphur, gems, gypsum, fertilizers, and many more. Groundwater (chapter 15) should also be regarded as an important geologic resource.

Geologic resources are sometimes called *mineral resources*, but the term, though widely used, is not accurate. Oil, natural gas, and coal, for example, are not formed of solid, crystalline minerals.

Some geologic resources are *renewable*; that is, they are replenished by natural processes fast enough that people can use them continuously. Water is the best example. Under sustainable conditions, the supply of water is never-ending, provided that we extract water no faster than it is replenished naturally by precipitation, runoff, and infiltration. Most geologic resources, however, are **nonrenewable resources.** They form very slowly, often over millions of years under unusual conditions in restricted geographic settings. They are "happy accidents" of nature. Humans extract nonrenewable resources much faster than nature replaces them. The annual rate of extraction of crude oil, for example, is on the order of a million times faster than natural rates of replenishment.

HOW DO RESOURCES DIFFER FROM RESERVES?

Two different terms are used to describe the amount of geologic material not yet extracted from Earth. **Resources** is a broad term used for the total amount of a geologic material in all deposits, discovered and undiscovered. It includes both the deposits that can be economically extracted under present conditions and those that may be extracted economically in the future (figure 12.2). It is very difficult to estimate resources because educated guesses must be made about the existence and sizes of deposits yet undiscovered as well as about what type of deposit might someday be economical to extract. **Reserves** are a small part of resources. They are the *discovered* deposits that can be extracted economically and legally under present conditions. That is, they are the short-term supply of a geologic material.

Once *resources* have been carefully estimated the amount should not change from year to year, for an estimate of resources is basically an estimate of the total future supply. Estimates of *reserves,* however, change all the time. The extraction and use of a substance lowers reserves. New discoveries add to reserves, for part of the definition of reserves is that we have to know the

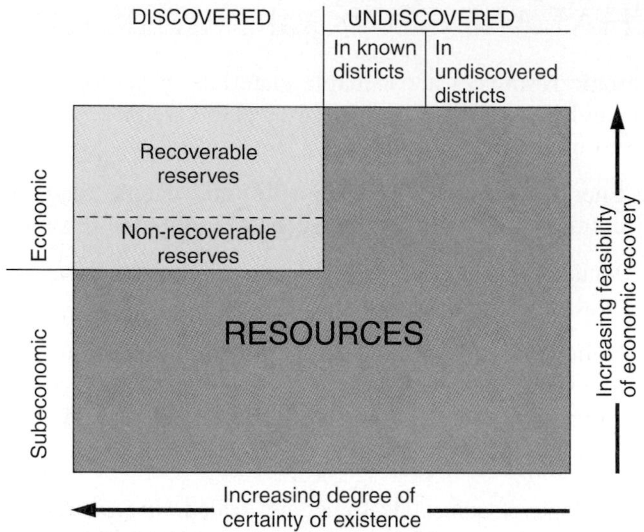

FIGURE 12.2

The difference between reserves and resources. *Reserves* (discovered deposits that are economically and legally extractable) form a small part of total *resources* (discovered and undiscovered deposits that are economical to extract now or may be economical in the future).

deposit is there. (*Reserves* are like your bank balance, rising and falling with time; *resources* are like your future, lifetime income.)

A deposit also has to be profitable to extract, and many factors determine whether a profit can be made. The costs of extraction, including workers' salaries and the fuel used to run equipment, are important. New inventions that make extraction cheaper can increase reserves. The final price a company can receive for its product is also important. When prices are low, reserves are naturally low because few deposits can be extracted profitably at that price. When prices rise, more deposits become economical and reserves increase, even without new discoveries. Reducing the amount of taxes that extraction companies pay can also increase reserves. However, economic factors are not the only controls on the size of reserves. Some deposits that could be economically exploited are not exploited due to political or social preferences.

Changes in laws can affect reserves. Large areas of government-owned land are off-limits to mining and drilling, so any geologic materials under these areas are not legally extractable and cannot be included in reserves. Opening more land to extraction can therefore increase reserves.

WHAT KINDS OF ENERGY RESOURCES ARE USED IN NORTH AMERICA?

The United States and Canada consume huge amounts of energy. The sources of energy used in the U.S. have been changing in recent years in response to changing prices and availability of imported fuels.

Oil and natural gas account for two-thirds of the U.S.'s energy supply; the fossil fuels (coal, oil, and gas) provide almost 90 percent of energy used in the U.S.

On a per-capita basis, Canadians are some of the highest consumers of energy on Earth. This is largely due to the country's cold climate conditions, vast continental area requiring large transport distances, energy-intensive industrial base, and high standard of living. However, Canada is also the world's fifth largest energy producer (after the U.S., Russia, China, and Saudi Arabia), and exports approximately 30 percent of the energy it produces, mostly to the United States. Canada is the major source of U.S. oil imports, and is the world's second largest gas exporter (after Russia). More than two-thirds of Canada's energy is produced in Alberta.

Most of Canada's energy consumption is from fossil fuels (more than 65%; see figure 12.3*A*), with the remainder from hydroelectricity (28%) and nuclear energy (7%). Despite concern over the use of nonrenewable energy resources, alternative energy sources—such as solar and wind power—make up less than one ten-thousandth of Canada's total energy consumption. Canada is the largest producer of hydro power in the world, and almost 60 percent of the electricity generated in Canada comes from hydroelectric sources (figure 12.3*B*)

Most of the energy consumed in Canada is accounted for by industry (39%) and transportation (27%); the remaining 34 percent is consumed by agriculture, homes, and businesses. In the industrial sector, 70 percent of the energy is used by only five major industries: pulp and paper, metal smelting, steel making, mining, and petrochemicals.

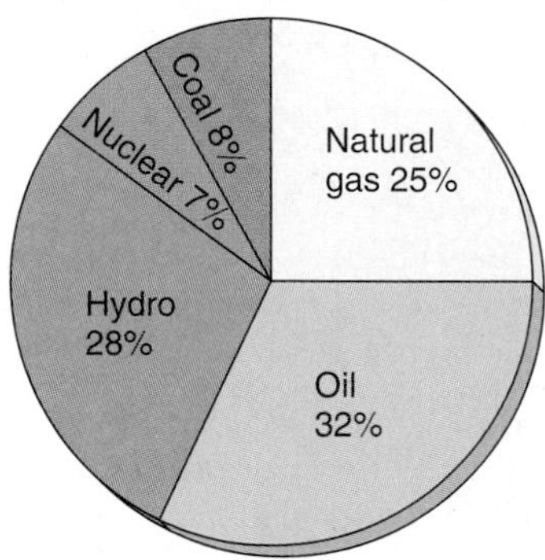

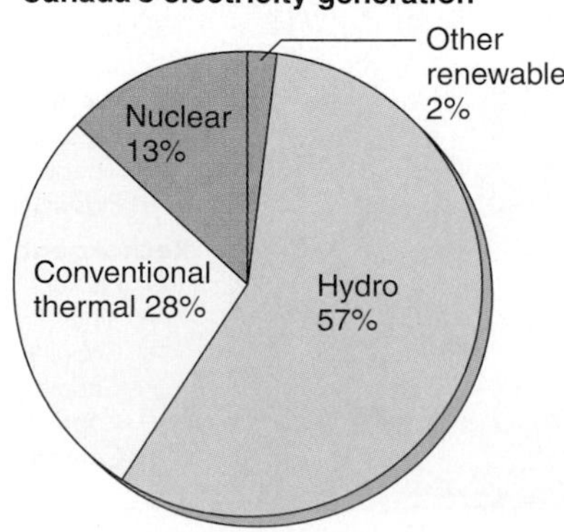

FIGURE 12.3

Canada's energy consumption and electricity generation.

Sources: www.eia.doe.gov/emeu/cabs/canada.html; www.ec.gc.ca

HOW DO OIL AND GAS FORM?

Within the petroleum industry, **petroleum** is a broad term that includes both crude oil and natural gas. That will be our usage. (In common usage *petroleum* is synonymous with crude oil; many geoscientists also use it this way.)

Crude oil is a liquid mixture of naturally occurring hydrocarbons (compounds containing hydrogen and carbon), which can be distilled to yield a great variety of products.

Natural gas is a *gaseous* mixture of naturally occurring hydrocarbons. Its origin and occurrence closely parallel that of oil. Many wells that recover oil also recover natural gas, although either may exist alone.

Hydrocarbons form from organic matter that accumulates together with fine-grained sediment in marine or lake basins. Microscopic life forms such as plankton, foraminifera, diatoms, and other organisms thrive in well-lit, nutrient-rich coastal waters (often fed by rivers or upwelling currents) and fall to the basin floor when they die. These organisms may be preserved from decay if basin floor conditions are anoxic and bacterial decomposition is prevented. The accumulation of organic rich muds in sufficient quantities in anoxic basins may eventually allow the formation of *petroleum source rocks*.

Organic-rich muds on the basin floor are progressively buried and compressed by younger sediments that accumulate on top. During burial, the carbohydrates and proteins of the organic material are destroyed and the remaining organic compounds form a material called *kerogen*. As the sediments and the kerogen they contain are buried more deeply they heat up, at a rate of approximately 30°C/1,000 m of burial (the geothermal gradient). When kerogen heats up to temperatures between 80 and 100°C, the complex organic molecules it contains break down into long chains of hydrogen and carbon atoms and form oil. The process of breaking down complex organic molecules into simpler forms is called *cracking* (also known as *catagenesis*). With increased heating (or cooking) oil breaks down further to form *natural gas*. Thus, oil will form only under certain temperature conditions beneath the Earth's surface. The *petroleum window* is the zone from about 2,300 m to 4,600 m below the ground surface in which conditions are ideal for the formation of oil (figure 12.4). At shallower depths the kerogen has not "cooked" sufficiently to form oil, and in deeper zones the high temperatures cause oil to change into gas. When temperatures exceed 200°C any useful hydrocarbon products are combusted.

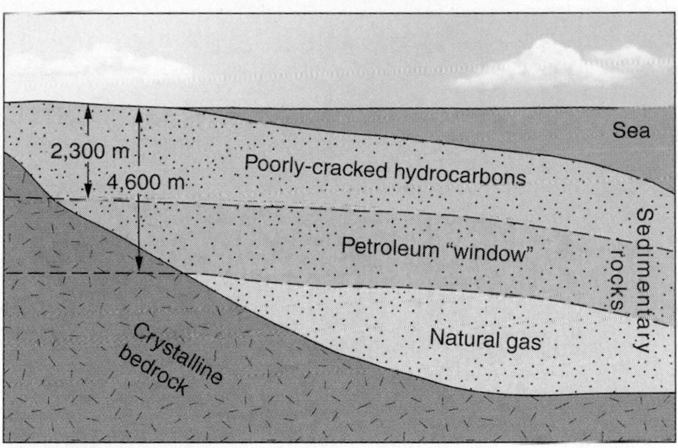

FIGURE 12.4

Typical depths of hydrocarbon cracking. The "petroleum window" lies at depths of between 2,300 and 4,600 metres. Depth will vary somewhat, depending on the geothermal gradient.

WHERE DO WE FIND OIL AND GAS?

Once oil and gas have formed they tend to migrate away from the source rock, through microscopic cracks and pores, into overlying permeable layers that are under lower pressure. The hydrocarbons may migrate all the way to the Earth's surface to form oil or tar seeps (see box 12.1), but often become trapped by impermeable rocks (*seals* or *trap rocks*) to form oil or gas pools (figure 12.5).

Oil pools are valuable underground accumulations of oil. They are found only where three specific conditions occur together figure 12.6): (1) a **source rock** (such as shale) containing organic matter that is converted to petroleum; (2) a **reservoir rock** (usually sandstone or limestone) that is sufficiently porous and permeable to store and transmit the petroleum; (3) an **oil trap**, a set of conditions to hold petroleum in a reservoir rock and prevent its escape by migration. Natural gas requires the same conditions as oil for accumulation and formation of gas pools.

Figure 12.5 depicts several types of *structural traps* for oil and gas. Some types of traps are more abundant in particular regions than in others. For example, anticlines (described in chapter 11) create the most common oil traps in the Persian Gulf; thrust faults are important trap formers in the foothills of Alberta, and salt domes account for most of the petroleum reserve in the Gulf of Mexico.

Where oil and water occur together in folded sandstone beds, the oil droplets, being less dense than water, rise within the permeable sandstones toward the top of the fold. There the oil may be trapped by impermeable shale overlying the sandstone reservoir rocks. Because natural gas is less dense than oil, the gas collects in a pocket, under fairly high pressure, on the top of the oil.

Faults may create oil traps when permeable reservoir rocks break and slide next to impermeable rocks. Thrust faults are often associated with folds because both are caused by compression. The backarc thrust belts on the landward side of both the Cordilleran and Appalachian mountain belts have been intensively explored for oil and gas.

A *stratigraphic trap* is a result of natural sedimentation rather than folding or faulting. It may be a lens of sandstone within a larger bed of shale. Another such trap is the narrow edge of a gently dipping sandstone layer where it pinches out within a shale sequence. Oil can collect under *unconformities* if shale seals off a reservoir rock. Limestone *reefs* can form a variety of traps. The core of a reef is usually full of large openings formed by the irregular growth of coral and algae. Oil can collect both in a reef core and in the dipping beds of wave-broken debris on the reef flanks. *Salt domes* create a variety of traps (refer to box 11.2).

IN GREATER DEPTH 12.1

The Development of Canada's Oil and Gas Industry

The first commercial oil well in North America was dug in 1858 at Oil Springs in Ontario. This well consisted of a deep hole into which oil seeped and was then pumped to the surface. Oil was originally discovered in the area by local First Nations bands, who used the sticky oil that oozed out onto the ground surface (as "gum") to waterproof canoes and for medicinal purposes. In 1862, the world's first "gusher" was encountered near Petrolia, Ontario (box figure 1) when drilling penetrated a body of oil at a depth of 49 m below the ground surface. Oil is still pumped from shallow reservoirs in Devonian limestones at a depth of 140 m.

Oil Springs and Petrolia supplied 90 percent of Canada's needs before 1900, and total production to date is around 100 million barrels. Southern Ontario currently produces approximately 2 million barrels of oil per year, mostly from Ordovician and Silurian limestones and dolostones.

In 1890, Eugene Coste realized the potential of "nuisance gas" flared from the producing oil wells in southern Ontario as an energy resource and began to supply to local communities. Most of the natural gas now produced in Ontario comes from wells drilled into Silurian sandstones beneath Lake Erie. Permeable limestones and sandstones that have had their original gas removed are now used to store natural gas piped from Alberta for use during peak consumption periods.

In Alberta, the village of Medicine Hat was the first to supply local energy needs from a gas reservoir found by CPR workers drilling for water. A major gas reservoir was subsequently discovered in 1908 (also by Coste) at Bow Island, and a 270-kilometre pipeline was built to transport the gas to Calgary. The first oil discovery was in 1902 near Pincher Creek in the southwestern part of Alberta, and in 1914 a small oil pool at Turner Valley southwest of Calgary was discovered. However, it was not until Imperial Oil drilled 133 dry holes in a row across Western Canada that a significant oil find was made at the Leduc #1 well just south of Edmonton in 1947. This find opened up exploration of the Western Canada Sedimentary Basin and paved the way for large discoveries in the 1950s and 60s. By 1953, oil supply pipelines had been built to Sarnia and Vancouver, and the city of Calgary began to grow as an administrative centre for the petroleum industry. Today, Alberta is Canada's largest producer of petroleum and natural gas, accounting for 55 percent of Canada's oil production.

Exploitation of Canada's oil sands began in 1964 when the Great Canadian Oil Sands consortium, later Suncor Energy, began construction of the first successful open-pit oil sands mining operation. An even larger facility was built by the Syncrude consortium in 1978. Oil sands are an abundant and valuable resource for Canadians, but extraction of the oil is difficult and costly (box figure 2). Exciting new developments in technologies used to extract the heavy oil involve in situ extraction by stream injection through horizontal wells (the Firebag Project at Suncor: see box figure 3). These new extraction techniques are more efficient and will reduce environmental impacts of oil sand mining. Currently, oil sands account for 30 percent of Canada's oil production.

Canada's oil and gas industry has also recently expanded into offshore areas of the Atlantic coast. The Hibernia (see box 12.2), Terra Nova, and White Rose oil fields are located on the Grand Banks offshore from Newfoundland, and the Sable Island gas field lies offshore from Nova Scotia. Exploration of Canada's east coast waters began in 1959; to date, more than $15 billion has been spent on offshore Newfoundland targets (such as Hibernia) and $5 billion on offshore Nova Scotia targets.

BOX 12.1 ■ FIGURE 1
The first oil "gusher," drilled at Petrolia, Ontario in 1862.
Courtesy of Oil Springs Museum, Ontario

BOX 12.1 ■ FIGURE 2
Truck and shovel used to mine oil sands, Suncor Mine, Fort McMurray, Alberta.
Courtesy Suncor Energy Inc., www.suncor.com

BOX 12.1 ■ FIGURE 3
Construction of well pad for Firebag Project, a system for in situ extraction of oil by high-pressure steam injection. Suncor Mine, Fort McMurray, Alberta.
Courtesy Suncor Energy Inc., www.suncor.com

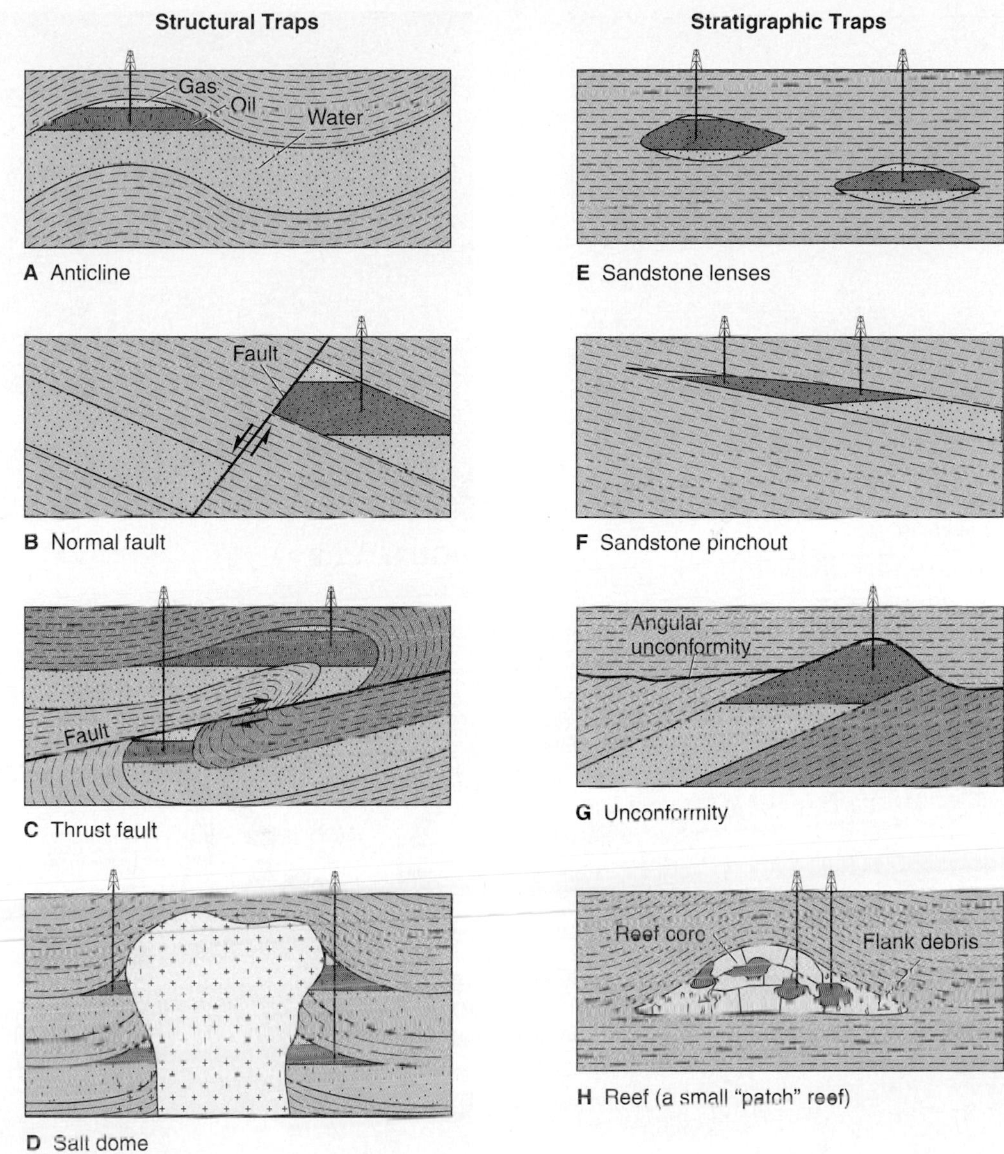

FIGURE 12.5

Major types of hydrocarbon traps. *A–D* are structural traps. *D–H* are stratigraphic traps. In all cases, an impermeable barrier encloses or caps the hydrocarbons.

It is important to bear in mind that the layered pools of fluids that form oil and gas pools do not fill hollow, underground chambers, like flooded caves, but merely *fill all of the pore spaces* in a highly permeable sedimentary rock (figure 12.5). Because of the surface tension of fluids clinging to the walls of the pores within the rock, only a certain fraction of the petroleum may ever be extracted and, in most cases, this is less than half of the oil present.

Oil fields are regions underlain by one or more oil pools. Figure 12.7 shows the location of most of the major North American oil fields. In Canada, the largest producing oil fields are in Alberta, although oil fields such as Hibernia on Canada's east coast are becoming important oil producers. The two largest oil fields within the United States are in eastern Texas and on Alaska's North Slope. Most of the world's remaining oil lies in giant fields in the Middle East (especially in Saudi Arabia and Kuwait), Russia and Azerbaijan, Venezuela, and Mexico.

Recovering the Oil

When an oil pool or field has been discovered, wells are drilled into the ground. Permanent derricks used to be built to handle the long sections of drilling pipe. Now portable drilling rigs are set up to drill (figure 12.8), and are then dismantled and moved. When the well reaches a pool, oil usually rises up the well because of its density difference with water or because of the pressure of the expanding gas cap above the oil. Although this rise of oil is almost always carefully controlled today, spouts of oil, or *gushers,* were common in the past (box 12.1). In time,

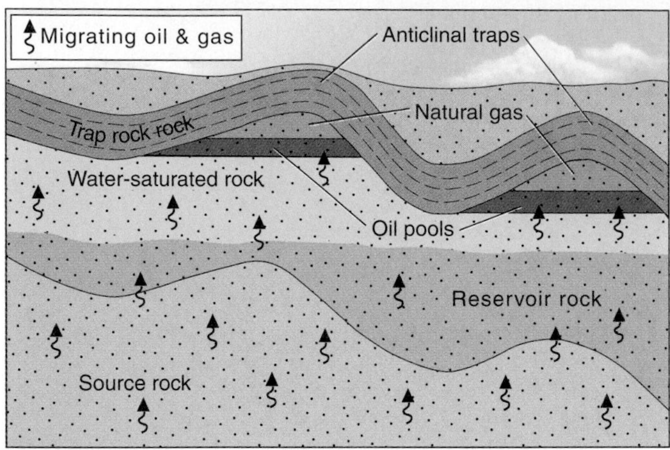

FIGURE 12.6
Migration of hydrocarbons into reservoirs.

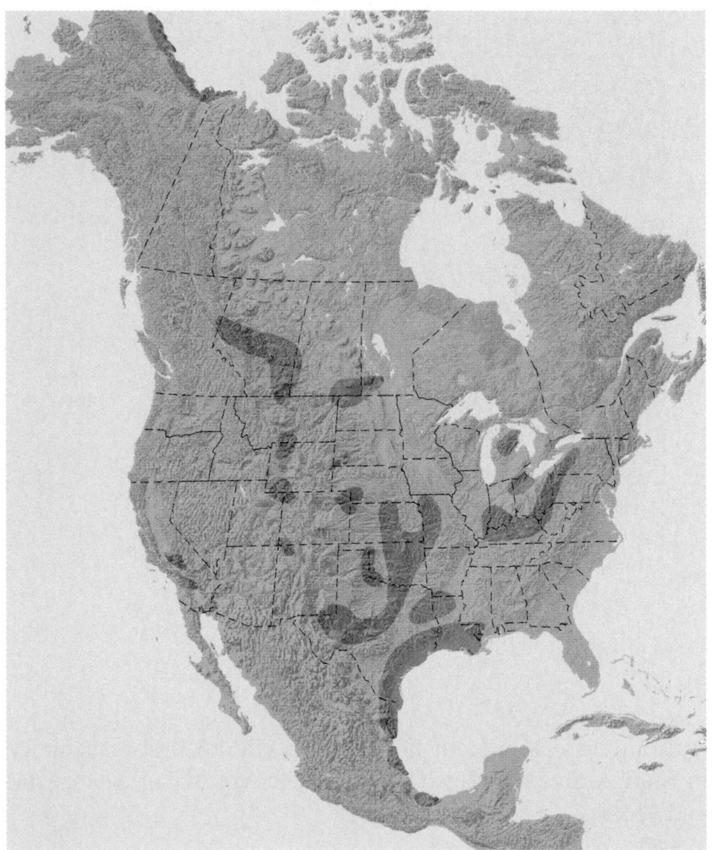

FIGURE 12.7
Major oil fields in North America. The amount of oil in a field is not necessarily related to its areal extent on a map. It is also governed by the vertical "thickness" of the oil pools in a field. The fields with the most oil are in Alaska and east Texas.
From U.S. Geological Survey and other sources

FIGURE 12.8
Drilling rig in Tuktoyaktuk, Northwest Territories.
Courtesy www.canadian-wellsite.com, "Home page for the Canadian Oilpatch"

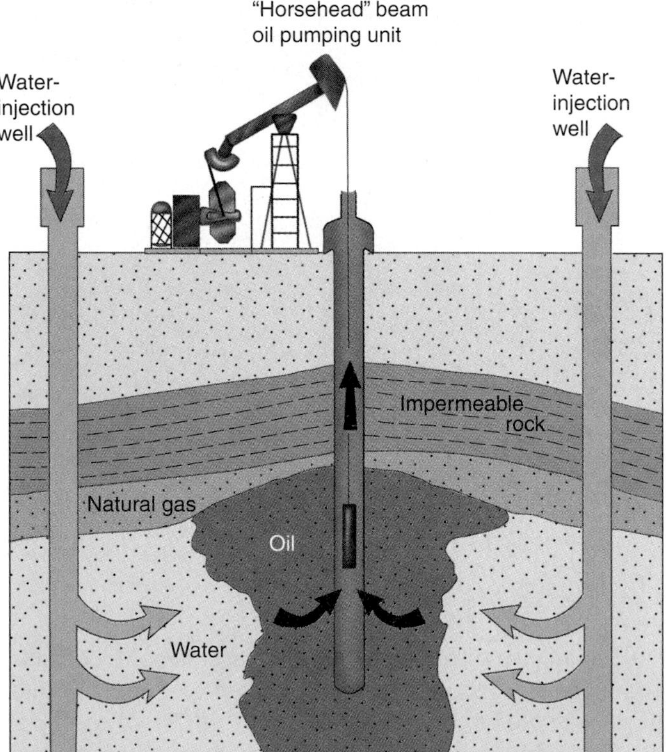

FIGURE 12.9
Water is often injected to drive additional, hard-to-get oil out of the ground (secondary recovery).

fluid pressure diminishes and an oil pool becomes less economical to operate. Remaining oil may be flushed out of the ground by "flooding" the reservoir with injected groundwater. The groundwater drives the petroleum ahead of it from the area of injection wells toward oil wells for removal (figure 12.9). Developers have also used steam to drive out the oil. As much as one-third of the original reserve in an oil pool may be extracted using these *secondary recovery* methods. At a refinery the crude oil from underground is separated into natural gas, gasoline, kerosene, lubricating oil, fuel oil, grease, asphalt, and paraffin.

Petrochemicals, manufactured from petroleum, include dyes, fertilizers, medicines, synthetic rubber, explosives, perfumes, paints, saccharin, solvents, synthetic fibres, and plastics used for such varied products as compact discs, cassettes, floor tile, and garbage bags.

As oil becomes increasingly difficult to find, the search for it is extended into more hostile environments. The development of the oil field on the North Slope of Alaska and the construction of the Alaska pipeline are examples of the great expense and difficulty involved in new oil discoveries. Offshore platforms extend the search for oil to the ocean's continental shelves (box 12.2) and even beyond. More than one-quarter of the world's oil and almost one-fifth of the world's natural gas comes from offshore, even though offshore drilling is six to seven times as expensive as drilling on land.

Environmental Effects

Getting petroleum out of the ground and to the consumer can create environmental problems anywhere along the line. Salty *brine,* a by-product of most oil wells, can pollute surface water. Pipelines carrying oil can be broken by faulting, landsliding, or acts of war, causing an *oil spill.* Tanker spillage from collisions or groundings (such as the *Exxon Valdez* off Alaska in 1989) can create oil slicks at sea. Offshore platforms may also lose oil (such as the Santa Barbara blowout off southern California in 1969). Oil slicks can drift ashore, fouling the beaches. It has been estimated that spills and oil well blowouts release more than 10,000 cubic metres of oil into the Canadian environment each year. Natural oil seeps can also cause pollution. Tar-like lumps collect on some California beaches as a result of natural seepage of petroleum from the ground. At one place in Alaska, oil seeps build a crude oil lake behind ice during the winter; in the summer, when the ice melts, the oil flows out to a nearby bay.

Subsidence of the ground can occur as oil is removed. The Wilmington field near Long Beach, California, has subsided 9 metres in 50 years; dikes have had to be built to prevent sea water from flooding the area. The refining and burning of petroleum and its products can cause *air pollution.* Advancing technology and strict laws are helping control some of these adverse environmental effects.

HOW MUCH OIL AND GAS DO WE HAVE LEFT?

The World Situation

The U.S. Geological Survey (USGS) assessed the world's recoverable oil for the year 2000 and estimated the *reserves* to be 870 billion barrels of oil (a barrel contains 159 litres of oil). More recent estimates (2004) increase this value to 1,265 billion barrels (figure 12.10*A*). Canada has 15 percent of the world's proven crude oil reserves (mostly in Alberta's oil sands), second only to Saudi Arabia (figure 12.10*A*). Total world *resources,* of course, are much larger than reserves. The USGS estimate is around 2,300 billion barrels (including reserves). Since the world's present consumption of oil is 1 billion barrels of oil every 10 days, known *reserves* will last about 30 to 40 years. Forecasts like this, however, fail to take into account changing reserve sizes and consumption rates.

There is far more oil underground than is listed in the resources estimate; some oil cannot be recovered. The oil may be in a pool too small, or lie too far from market, to justify the expense of drilling. Some oil lies under regions where drilling is forbidden, such as national parks or other public lands. Even using the best extraction techniques, only about 30 percent to 40 percent of the oil in a given pool can be brought to the surface. The rest is far too difficult to extract, and has to remain underground.

The global recoverable reserve of natural gas is currently thought to be on the order of 6,079 trillion cubic feet (tcf), most of which lies in the Middle East, Europe, and Russia and its neighbours (figure 12.10*B*). Unfortunately, it is much harder to transport natural gas than petroleum, requiring pipelines and LNG (liquid-natural gas) tankers for widespread distribution. Nevertheless, natural gas has become vital in the modern world. It is used to heat homes, cook food, make synthetic NPK (nitrogen-phosphorus-potassium) fertilizers for agriculture, produce electricity, and power fuel cells. Recent estimates of worldwide gas in hydrate form range from 100,000s to 1,000,000s of trillion cubic feet.

The Outlook in North America

Canada is the ninth largest petroleum producer and the third largest natural gas producer in the world. Oil reserves in Canada are estimated at almost 180 billion barrels (figure 12.11). Natural gas resources are estimated at almost 60 trillion cubic feet (figure 12.10*B*). Recent discoveries of large volumes of gas hydrates off the west coast (box 12.3) may also provide valuable alternate sources of energy for the future.

With 6 percent of the world's population, the U.S. uses 29 percent of the world's oil production each year. Sixty-five percent of U.S. consumption is used for transportation. The United States imports 55 percent of the oil it uses, and at present rates of extraction will run out of domestic oil within 20 years.

The U.S. Energy Information Agency estimates that nearly 1,300 trillion cubic feet of recoverable natural gas exist in the United States. At current levels of consumption, this provides a 75-year reserve of natural gas. Unconventional gas sources such as coal beds, tightly compacted sandstones, organic-rich shales, and gas hydrates have been the focus of renewed exploration interest in the U.S. since the decontrol of natural gas prices.

IN GREATER DEPTH 12.2

Offshore Oil: The Hibernia Project

A

In 1979 a major offshore oil discovery was made at the Hibernia field on the Grand Banks off Canada's east coast. The Hibernia field is located in the Jeanne d'Arc Basin, 315 km east of St. John's, Newfoundland, in water 80 m deep. The field consists principally of two sandstone reservoirs—Hibernia and Avalon—located at average depths of 3,700 m and 2,400 m, respectively. Hibernia oil is a light, sweet crude and has a recoverable reserve estimate of 884 million barrels. Oil production from the Hibernia field began in 1997, and an average of 150,000 barrels of oil are being extracted per day from nine wells. The Hibernia field has a projected lifespan of more than 20 years.

Development of the Hibernia field was a mega project that required collaboration among industry and government partners. Severe environmental conditions on the Grand Banks, including high storm waves and frequent icebergs (many weighing up to several million tonnes), required the development of a unique offshore drilling and production platform. The Hibernia facility uses a \$5.8-million floating platform that sits on top of a fixed base structure (box figure 1). Tugboats are used to tow icebergs away from the platform, but in the event of a potential collision the production platform can be disconnected from the base and floated away. The base itself is designed to withstand the impact of a 6-million-tonne iceberg travelling at one knot. Two tankers were also specially designed to transport the oil from Hibernia to ports on the eastern coast of Newfoundland. These tankers, built at a cost of \$150 million, each hold up to 850,000 barrels of oil and are designed to withstand the stormy waters of the Grand Banks.

Offshore oil production is a difficult and expensive task, but with improved drilling and oil recovery techniques it will become a more economically viable proposition in the future.

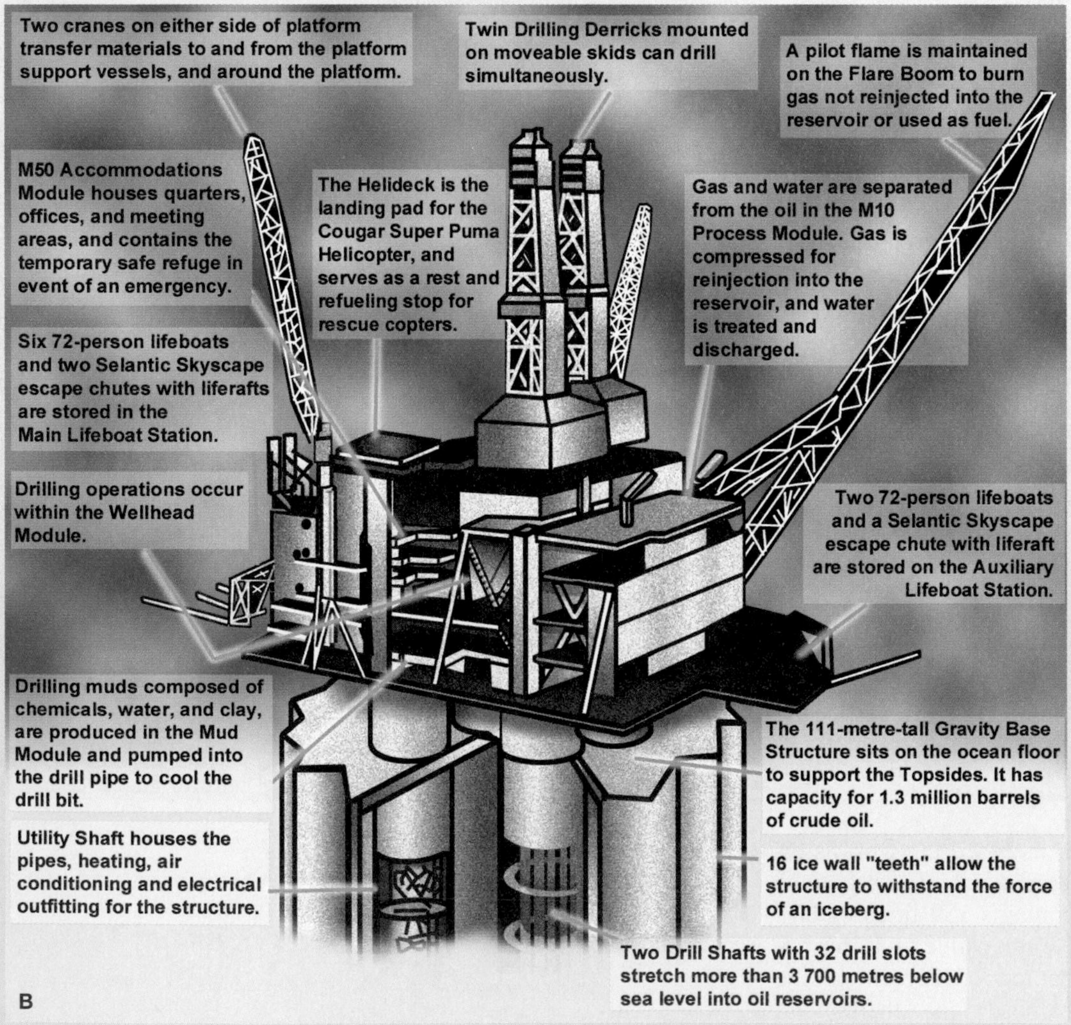

Two cranes on either side of platform transfer materials to and from the platform support vessels, and around the platform.

Twin Drilling Derricks mounted on moveable skids can drill simultaneously.

A pilot flame is maintained on the Flare Boom to burn gas not reinjected into the reservoir or used as fuel.

M50 Accommodations Module houses quarters, offices, and meeting areas, and contains the temporary safe refuge in event of an emergency.

The Helideck is the landing pad for the Cougar Super Puma Helicopter, and serves as a rest and refueling stop for rescue copters.

Gas and water are separated from the oil in the M10 Process Module. Gas is compressed for reinjection into the reservoir, and water is treated and discharged.

Six 72-person lifeboats and two Selantic Skyscape escape chutes with liferafts are stored in the Main Lifeboat Station.

Drilling operations occur within the Wellhead Module.

Two 72-person lifeboats and a Selantic Skyscape escape chute with liferaft are stored on the Auxiliary Lifeboat Station.

Drilling muds composed of chemicals, water, and clay, are produced in the Mud Module and pumped into the drill pipe to cool the drill bit.

The 111-metre-tall Gravity Base Structure sits on the ocean floor to support the Topsides. It has capacity for 1.3 million barrels of crude oil.

Utility Shaft houses the pipes, heating, air conditioning and electrical outfitting for the structure.

16 ice wall "teeth" allow the structure to withstand the force of an iceberg.

Two Drill Shafts with 32 drill slots stretch more than 3 700 metres below sea level into oil reservoirs.

B

Additional Reading

Draper, D., 1998. *Our Environment – A Canadian Perspective*

Related Web Resources

www.offshore-technology.com/projects/hibernia

www.gov.nf.ca/mines&en/Exploration/offshore.stm

BOX 12.2 — FIGURE 1

Photo (*A*) and sketch (*B*) of the Hibernia oil platform, Grand Banks, Newfoundland.

Photo A from www.gov.nf.ca/mines&en/EXPLORATION/GrandBanks/Hibernia.htmbarrels; Illustration B after www.hibernia.ca

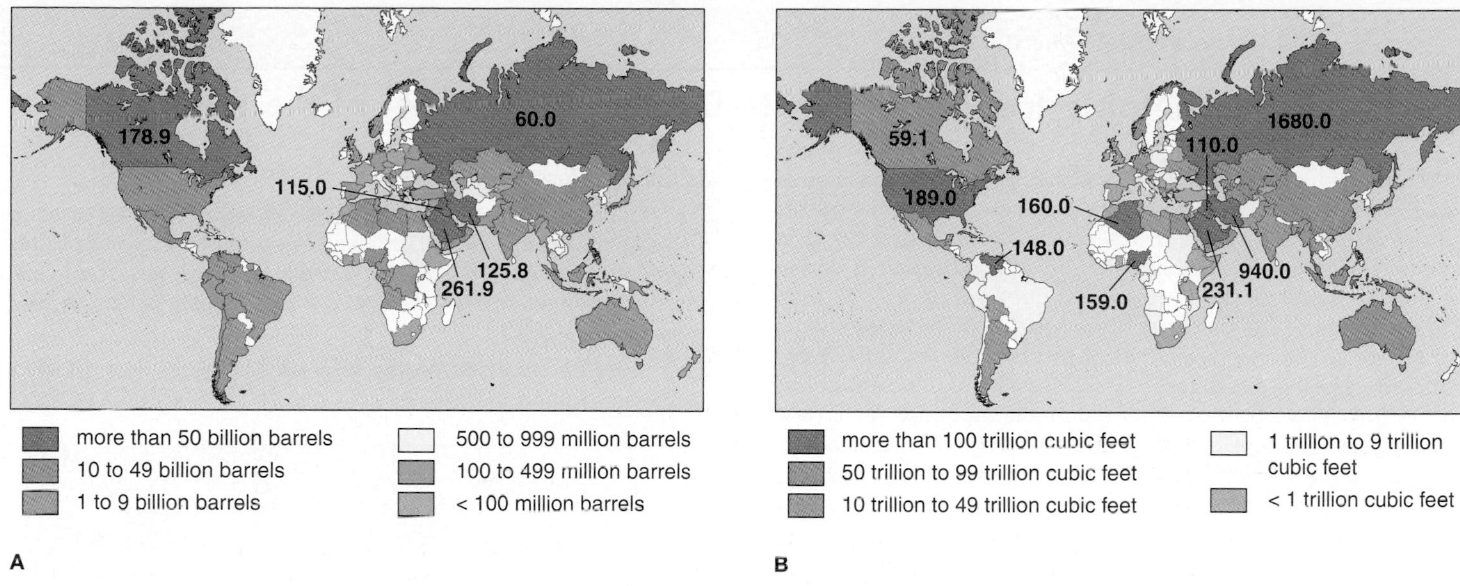

FIGURE 12.10

(A) World oil reserves, 2004. (B) World gas reserves, 2004.

From Oil & Gas Journal, "Worldwide Report," December 22, 2003. Used with permission.

Note: Reserve figures are proved reserves recoverable with present technology (except former USSR and Canada gas figures, which include some probable reserves).

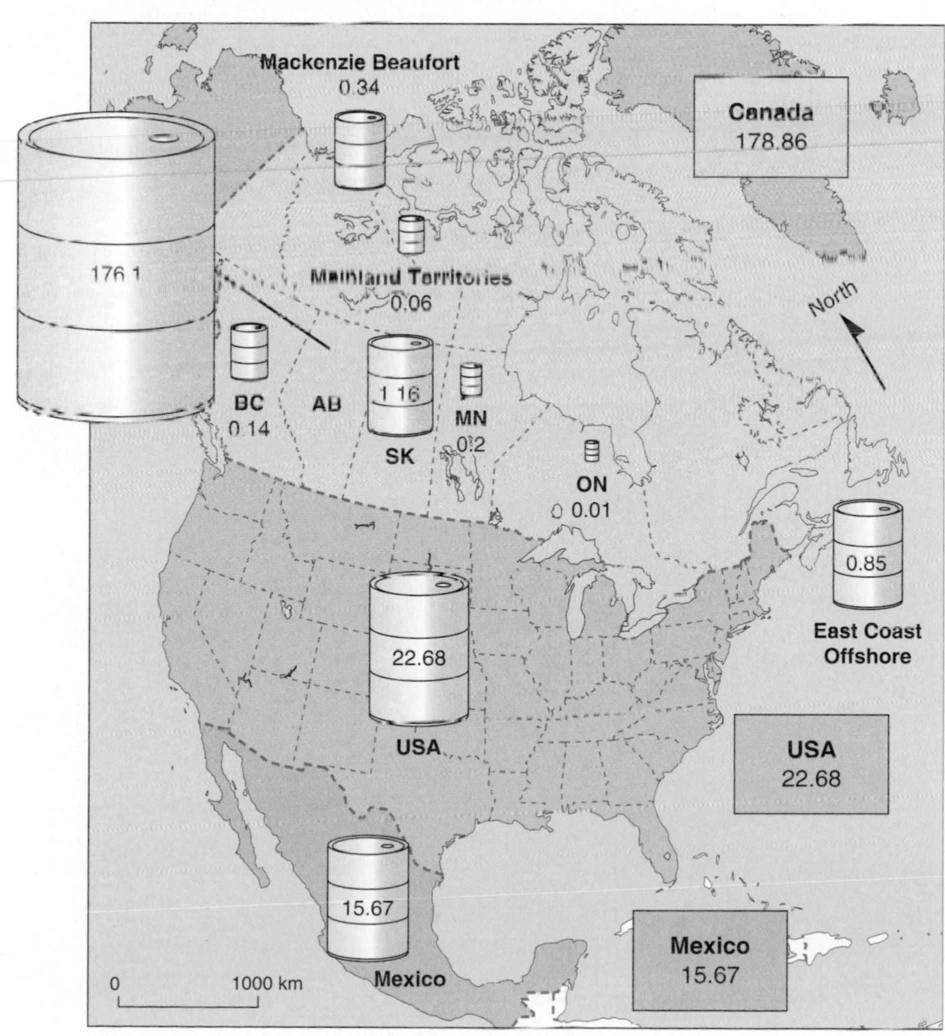

FIGURE 12.11

Oil reserves of North America (2003 values).

Courtesy of The Alberta Department of Energy

ENVIRONMENTAL GEOLOGY 12.3

Flammable Ice. Gas Hydrate Deposits—The Fuel of the Future?

Gas hydrates (also called methane hydrates) are an unusual mixture of ice and gas in which methane (one of the gasses in natural gas) is trapped in ice crystals. These are found in extreme environments, notably permafrost in polar regions and in the deep ocean floor. If lit, a piece of gas hydrate ice will burn with a red flame. The amount of gas hydrate in the ocean floors is staggering. Although estimates of gas hydrate resources vary, it appears that there is at least twice the amount of potential fuel tied up in gas hydrates as in all petroleum and coal combined.

The recent discovery of huge amounts of gas hydrates on the sea floor off the western coast of Vancouver Island (box figure 1) may provide a potential alternative energy source for Canadians. The methane hydrates were identified in water depths of 850 m using a remotely operated submersible and are visible as large mounds 3 to 4 metres high and up to 10 metres wide (box figure 2). More than 1200 m of core from gas hydrate deposits were collected by geoscientists working on the Joides Resolution drilling ship during the fall of 2005.

A world research site for the study of natural gas hydrates has recently been established at the Mallik gas hydrate field in the Mackenzie Delta of northwestern Canada. A 1,150-m-deep research well was drilled at the site (box figure 3) and the first terrestrial gas hydrate cores in the world were collected (box figure 4). Initial analyses indicate that the Mallik field is one of the most concentrated gas hydrate reservoirs in the world.

Commercially exploiting gas hydrate deposits presents formidable challenges. Most of the deposits are in lenses frozen in sediment at deep ocean floors, beneath a kilometre or more of water. There, methane hydrate is stable because of the cold and high pressure. If pressure is reduced or the substance heated, it becomes unstable and the methane escapes. Mining anything at this depth is difficult, but trying to remove the icy substance from the sediment and get it to the surface without losing the methane is an extreme technological problem.

Gas hydrate could significantly exasperate global warming. Methane, like carbon dioxide, is a greenhouse gas and is at least 15 times as effective at trapping UV radiation as CO_2. Unlike CO_2, methane will stay in the atmosphere for only around 10 years. However, methane reacts with oxygen in the atmosphere and produces CO_2, which remains in the atmosphere indefinitely. Significant volumes of methane could be released if methane hydrate sediments are disrupted by submarine landslides, earthquakes, or other means.

Related Web Resources

http://geophys.seos.uvic.ca/hydrates.html

http://gsc.nrcan.gc.ca/gashydrates/index_e.php

Additional Reading

E. Suess, G. Bohrmann, J. Greinert and E. Lausch, 1999, Flammable Ice. *Scientific American*, Nov. 1999, p. 76–83.

Additional Resources

The November 2004 issue of *Geotimes* features three articles about gas hydrates: G. R. Dickens. Methane hydrate and abrupt climate change: pp. 18–22. T. S. Collett. Gas hydrates as a future energy resource: pp. 24–27. N. Lubick. Detecting marine gas hydrates: pp. 28–30.

These may be accessed on the web at www.geoltimes.org Go to "Archives" to find the articles.

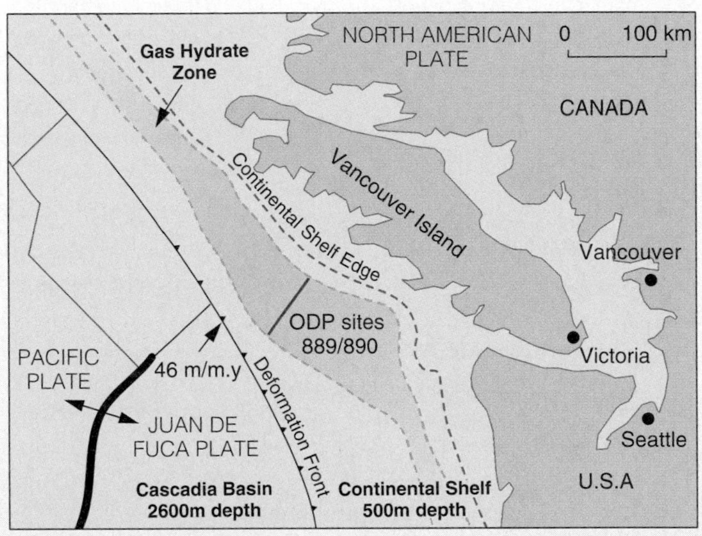

BOX 12.3 ■ FIGURE 1

Location of area of gas hydrate occurrence beneath the continental slope off Vancouver Island and the position of the International Ocean Drilling Program (ODP) research wells.

Reproduced with the permission of the Minister of Public Works and Government Services Canada, 2006 and Courtesy of Natural Resources Canada, Geological Survey of Canada.

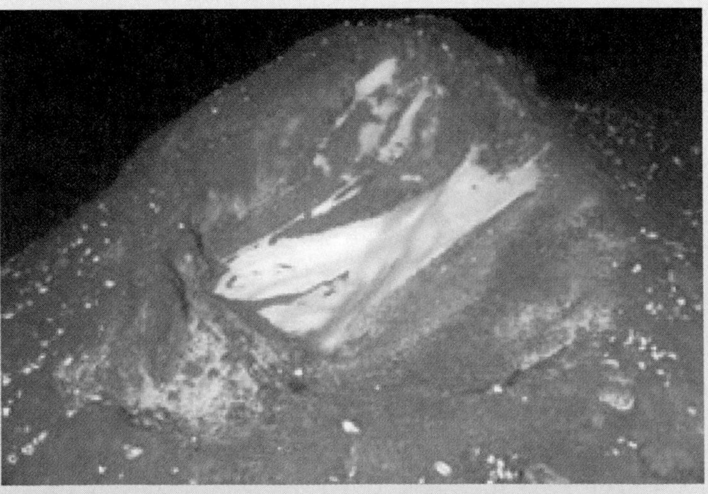

BOX 12.3 ■ FIGURE 2

A solid gas hydrate cap (about the size of a Volkswagen Beetle) on the ocean floor, offshore Vancouver Island.

Courtesy Ross Chapman, Chief Scientist. Hydrate research supported by NSERC, with participation by the Canadian Scientific Submersible Facility

BOX 12.3 ■ FIGURE 3
Drilling at the Mallik gas hydrate field, Mackenzie Delta, December 2002.
Photo by Hideaki Takahashi. http://gsc.nrcan.gc.ca/gashydrates/mallik2002/photo_e.php Core Studies at the Rig. Reproduced with the permission of the Minister of Public Works and Government Services Canada, 2005 and Courtesy of Natural Resources Canada, Geological Survey of Canada.

BOX 12.3 ■ FIGURE 4
White gas hydrate crystals visible in core from the Mallik field, Mackenzie Delta
Photo from Mallik 2002.http://www.gsc.nrcan.gc.ca/gashydrates/mallik2002/images/site_studies_1.jpg Reproduced with the permission of the Minister of Public Works and Government Services Canada, 2005 and Courtesy of Natural Resources Canada, Geological Survey of Canada.

Coal Bed Methane

Coal beds may also prove to be a major source of natural gas in the future. When coal forms, water and natural gas in the form of methane are trapped in the fine pores, pockets, and fractures that speckle and lace the interior of the coal. Pumping the water out lowers pressure and releases gas in huge quantities. Coal can store six to seven times more gas than an equivalent amount of rock in an ordinary natural gas field. A problem arises with respect to the water removed during pumping. Coal-water gets saltier the deeper the deposit, and disposal of salty water into surface watersheds seriously degrades water quality. Groundwater supplies may also be contaminated during gas extraction. In any event, there is a considerable amount of **coal bed methane** in Canada and the United States. Canadian coal bed methane resources are estimated to be between 187 tcf and 460 tcf, with the majority in Alberta and Nova Scotia. The U.S. Geological Survey estimates that the total coalbed methane resource worldwide might be as high as 7,500 tcf.

Heavy Crude and Oil Sands

Heavy crude is dense, viscous petroleum. It may flow into a well, but its rate of flow is too slow to be economical. As a result, heavy crude is left out of reserve and resource estimates of less viscous "light oil," or regular oil. Heavy crude can be made to flow faster by injecting steam or solvents down wells, and if it can be recovered, it can be refined into gasoline and many other products just as light oil is. Most California oil is heavy crude.

Oil sands (or *tar sands*) are asphalt-cemented sand or sandstone deposits. The asphalt is solid, so oil sands are often mined rather than drilled into, although the techniques for reducing the viscosity of heavy crude often work on oil sands as well.

The origin of heavy crude and oil sands is uncertain. They may form from regular oil if the lighter components are lost by evaporation or other processes. Oil sands and asphalt seeps at the Earth's surface (such as the Rancho La Brea Tar Pits in Los Angeles and Oil Springs, Ontario) probably formed from evaporating oil. But some heavy crudes and oil sands are found as far as 4,000 metres underground. Most of them have much higher concentrations of sulphur and metals, such as nickel and vanadium, than does regular oil. These facts suggest that heavy crude and oil sands may have a somewhat different origin than light oil.

The best-known oil sand deposit in the world is the Athabasca Oil Sand in northern Alberta (figure 12.12). About 30 percent of Canada's present oil production comes from these oil sands and in 2004 oil sands production averaged more than 1 million barrels per day. Recently, Canada's economically recoverable reserves took a big jump because of technological breakthroughs (figure 12.11). Traditionally, the oil sands near the surface have been strip-mined and trucked to a processing plant. New technology involves mixing the oil sands in place with hot water and then pumping the slurry through a pipeline. This permits economically feasible extraction of deeper deposits. The cost savings mean that the price of a barrel of this oil is now competitive with oil produced elsewhere. The *Oil and Gas Journal* added the new reserves of oil sands to Canada's "conventional" oil reserves, boosting the country's total oil reserves in 2003 from 4.9 billion barrels to 180 billion barrels.

With the new reserves, Canada now has the second-highest oil reserves in the world (behind Saudi Arabia). Venezuela has even more oil sand than Canada. The United States has more than 100 billion barrels of heavy crude and oil in oil sands, including about 30 billion in the form of oil sands in Utah. Half of it may be ultimately recoverable. This means that heavy crude and oil sands may supply as much oil in the future as our light oil reserves.

Oil Shale

Oil shale is a black or brown shale with a high content of solid organic matter from which oil may be extracted by distillation. Oil shales were burned in the Collingwood area of Ontario to produce oil in the 1850s (figure 12.12), but the costly and inefficient processing methods were uneconomical. The best-known oil shale in the United States is the Green River Formation, which covers more than 40,000 square kilometres in Colorado, Wyoming, and Utah, with deposits up to 650 metres thick (figure 12.12). The Green River Formation includes more than 400 billion barrels of oil in rich beds that yield more than 25 gallons of oil per ton of rock.

The mining of oil shale can create environmental problems. During distillation the shale expands, creating a space problem. Spent shale could be piled in valleys and compacted, but land reclamation would be troublesome. A great amount of water is required, both for distillation and for reclamation, and water supply is always a problem in the arid West. New processing techniques that extract the oil in place without bringing the shale to the surface may eventually help solve some of the problems and lower the water requirements.

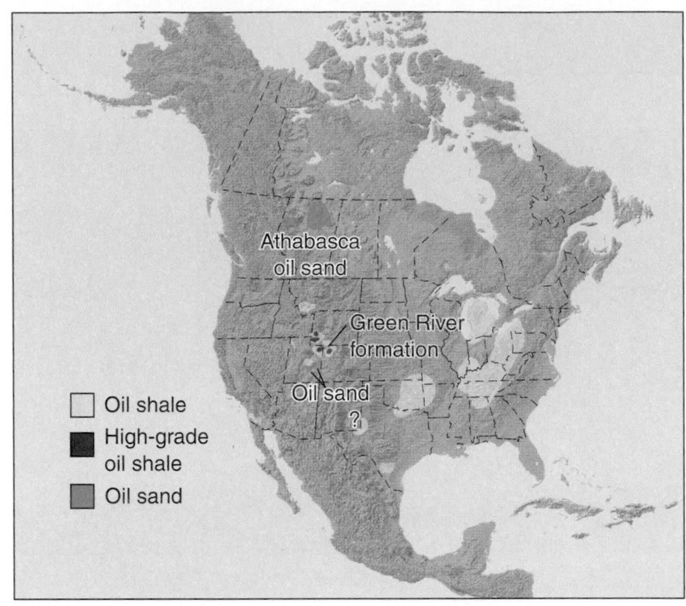

A

B

FIGURE 12.12

(*A*) Distribution of major deposits of oil sand and oil shale in the United States and Canada. (*B*) Organic-rich shales exposed along the shores of Georgian Bay (Lake Huron), Craigleith Provincial Park, Collingwood, Ontario.

Illustration from U.S. Geological Survey and other sources; Photo by Nick Eyles

Coal

Coal, as described in chapter 9, is a sedimentary rock that forms from the compaction of plant material that has not completely decayed. Coal is the world's most abundant fossil fuel with proven reserves of more than 1 trillion tonnes. After oil and natural gas and hydroelectricity, coal is the third major energy resource in the United States; it is the fourth most important energy resource in Canada (figure 12.3).

About 88 percent of the present use of coal in the United States, but less than 30 percent of its use in Canada, is for generating electricity. Coal is also used to make *coke,* which is used in steel making. In the future, coal may be used instead of petroleum in the manufacture of some chemicals. *Coal bed methane* and *coal oil* made from coal are reasonable approximations of natural gas and oil and can be used for some of the same purposes, although they are still very expensive to produce. Coal can also be powdered and mixed with water to form a *slurry,* which can be transported through pipelines and burned as a liquid fuel.

Varieties of Coal

Table 12.1 shows the common varieties (ranks) of coal. *Peat,* a mat of unconsolidated plant material, is not coal but probably represents the initial stage of coal development (figure 12.13*A*). When dry, it can be burned as a fuel (peat fires used to dry malted barley give Scotch whisky its smoky flavour). With compaction, peat can become *lignite* (*brown coal*), which may still contain visible pieces of wood. Lignite is soft and often crumbles as it dries in air. It is subject to spontaneous combustion as it oxidizes in air, and this somewhat limits its use as a fuel. *Subbituminous coal* and *bituminous coal* (*soft coal*) are black and often banded with layers of different plant material (figure 12.13*B*). They are dusty to handle, ignite readily, and burn with a smoky flame. *Anthracite* (*hard coal*) is actually a metamorphic rock, generally formed only under the regional compression associated with folding. It is hard to ignite but is dust-free and smokeless.

TABLE 12.1 Varieties (Ranks) of Coal

	Colour	Water Content (%)	Other Volatiles (%)	Fixed Carbon[2] (%)	Approximate Heat Value (BTUs of heat per pound of dry coal)
Peat[1]	Brown	75	10	15	Varies
Lignite	Brown to brownish-black	45	25	30	7,000
Subbituminous coal	Black	25	35	40	10,000
Bituminous coal (soft coal)	Black	5 to 15	20 to 30	50 to 75	12,000 to 15,000
Anthracite (hard coal)	Black	5	5	90	14,000

1. Peat is not a coal.
2. "Fixed carbon" means solid combustible material left after water, volatiles, and ash (noncombustible solids) are removed.

A

B

FIGURE 12.13

(A) A layer of peat being cut and dried for fuel, island of Mull, Scotland. Coal often forms from peat. (B) Excavation of bituminous coal at the Fording River Mine, B.C.

Photo A by David McGeary; Photo B by Coal Association of Canada

Occurrence of Coal

Coal beds are typically found interlayered with sedimentary rocks such as sandstone or shale. Coal bed thickness ranges from a few centimetres to 30 metres or more. If the beds are deeply buried, underground mines are dug to extract the coal (figure 12.14). If the coal beds are close to the land surface, the coal is mined in a **strip mine,** in which the overburden is removed to expose coal at the surface (figure 12.15). When a strip of coal has been uncovered and removed, the resulting trench is filled in with the overburden from the adjoining strip.

Figure 12.16 shows the coal fields of Canada and the lower 48 U.S. states. We will discuss three major regions, the Appalachian and eastern Canadian fields, the interior fields, and the far western fields.

The major coal-producing fields in the United States are the *Appalachian fields,* which stretch from Pennsylvania to Alabama and contain extensive beds of bituminous coal. Bituminous coals are also found in the Maritime provinces of Canada (Nova Scotia and New Brunswick; figure 12.16), although little is mined at present. The coals are mostly of Pennsylvanian age; a few are Mississippian and Permian. The coal beds, which thin westward, were included in the late Paleozoic folding and faulting of the Appalachian orogeny, so they are strongly deformed in the eastern part of the belt. Steeply dipping coal beds here are mostly mined underground. The folds are gentler to the west, where the coals can be extracted by both underground and strip mining.

There are 25 to 50 coal beds over most of this region; each bed generally is 2 m or less in thickness, although some are locally thicker. The coal lies within repeated successions of sandstone, shale, and limestone called *cyclothems,* which indicate deposition under alternating continental and marine conditions.

FIGURE 12.14

An underground coal mine.

Photo by Larry Lee/West Light

FIGURE 12.15

Coal strip mining at the Bienfait Mine, Saskatchewan. Ninety-five percent of Canada's coal is mined from the surface using either open-pit or strip mines.

http://spans.gscc.nrcan.gc.ca/~ren/coal/wall2.html Bienfait Mine, Saskatchewan. Reproduced with the permission of the Minister of Public Works and Government Services Canada, 2007 and Courtesy of Natural Resources Canada, Geological Survey of Canada.

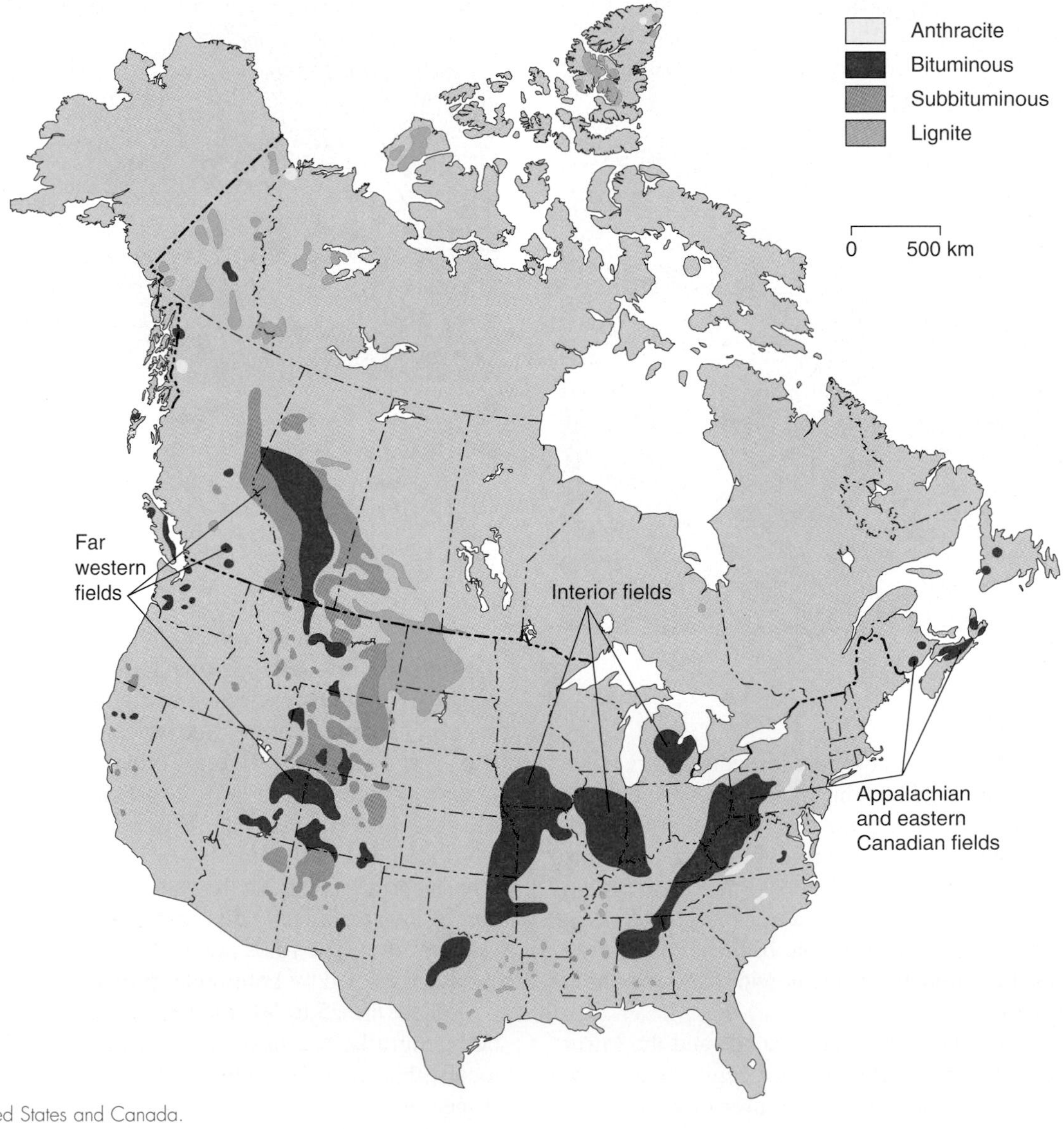

FIGURE 12.16

Coal fields of the United States and Canada.

From U.S. Geological Survey and www.coal.ca

This implies a low-lying environment near the sea, such as lagoons, large deltas, and swampy coastal plains similar to those that exist in present-day Florida, Georgia, and South Carolina.

The *interior fields* (figure 12.16) extend from Michigan through Illinois to Texas and are extensions of the Appalachian rocks westward onto the continental interior. In Michigan and Illinois the rocks are preserved in large basins; the fields from Iowa to Texas are mostly horizontal. The coals are usually strip-mined, particularly near major industrial centres around the basin's edge in Illinois, Indiana, and Kentucky.

The *far western fields* extend from New Mexico northward through the Rocky Mountains to Montana and into Canada. These coal beds are thicker and younger than eastern coals. They are generally of Cretaceous or Eocene age and are up to 30 m thick. The coals occur in large basins and are generally of low rank, being either lignite or subbituminous coal, although some good-quality bituminous coals occur in Colorado and Utah. Anthracite occurs in remote fields in British Columbia and the Yukon but is not being mined at present. Many thick beds are very close to the surface and are strip-mined (figure 12.15), although underground mining is common in some states. Western coals are attractive fuels, and are currently in very high demand, because they typically have less sulphur than eastern coals (sulphur compounds can pollute the air when coal burns).

Environmental Effects of Coal

Extracting and using coal creates environmental problems. The presence of a mine usually lowers the local water table as groundwater is pumped out of the mine. Mine drainage tends to be highly acidic, polluting surface streams and water supplies. In the past, strip mines have been refilled as barren, unsightly piles, but new techniques of recontouring the overburden and restoring the topsoil and vegetation help reclaim mined-out areas for other uses. Uncontrolled dumping of toxic by products created during the process of making coke from coal has caused serious environmental damage in many areas (box 12.4). When coal is burned, uranium, ash, and sulphur gases are released into the air, causing widespread air pollution. Carbon is also released from burning coal and forms CO_2, a "greenhouse gas" that contributes to global warming. Solving environmental problems associated with coal, of course, raises the immediate cost of extraction and the price to the consumer. Long-term costs may be *decreased,* however; preventing environmental damage is usually less expensive than cleaning it up.

Reserves and Resources

About 25 percent of the world's total coal reserves lie in the United States. Russia, China, and India also have substantial coal reserves. Canada is the world's 12th largest coal producer, with huge reserves in Alberta, B.C., and the Maritime provinces. Coal production in the United States is about 1 billion tons per year (the United States uses 0.9 billion tons and exports the rest). Sixty percent of the production is from surface strip mines. In recent years, the production of western, low-sulphur coal has surpassed that of Appalachian coal. In 2002, the western states provided 50 percent of U.S. coal compared to 36 percent from the Appalachian states.

Recoverable *reserves* in the United States are about 270 billion tons (40% of this is in Montana and Wyoming). Canadian coal reserves are almost 9 billion tons. As you can see, there are centuries of coal left at the present rate of production.

Coal *resources* within 1,000 m of the surface are 1,700 billion tons. A greater amount is estimated to lie deeper, so the total U.S. resources are an impressive 3,900 billion tons (much of which, of course, is not currently usable).

Uranium

The metal *uranium,* which powers nuclear reactors, occurs as *pitchblende,* a black uranium oxide found in hydrothermal veins, or as yellow *carnotite,* a complex hydrated oxide found as incrustations in sedimentary rocks. Groundwater easily transports oxidized uranium, which is highly soluble. Organic matter reduces uranium, making it relatively insoluble, so uranium precipitates in association with organic matter.

Canada is the largest producer of uranium in the world, supplying about one-third of the world's mined uranium; most of this comes from large, high-grade uranium resources located in Saskatchewan. Although Canadian uranium mining began around Elliot Lake in Ontario during the 1950s, the richest uranium deposits now mined occur near the base of the Athabasca Basin sandstone in Saskatchewan, close to an erosional unconformity with underlying crystalline rocks of the North American Craton. The uranium ores occur predominantly as pitchblende and are mined by both underground and open-pit mining methods. Extracted ore is crushed and milled to separate the uranium, which is then formed into a product known as yellowcake (U^3O^8) for shipment out of the province. Yellowcake is further refined and enriched to produce fuel for electric generation reactors. At current rates of production, Saskatchewan's uranium resources are sufficient to supply electric power utilities in Canada, the United States, Europe, and the Far East for more than 40 years.

Most of the easily recoverable uranium in the United States is found as carnotite in sandstone in New Mexico, Utah, Colorado, and Wyoming, some of it in and near petrified wood. In the 1950s uranium boom, western prospectors looked for petrified logs and checked them with Geiger counters. Some individual logs contained tens of thousands of dollars worth of uranium. Most of the uranium, however, is in sandstone channels that contain plant fragments.

Organic phosphorite deposits of marine origin in Idaho and Florida also contain uranium. The uranium is not very concentrated, but the deposits are so large that overall they contain a substantial amount of uranium. The black Devonian shales of the eastern United States also contain uranium. These shales are really low-grade oil shales (see figure 12.12), and they also contain large amounts of natural gas. Uranium may be recovered from phosphorites or shales as a by-product of another mining operation.

ENVIRONMENTAL GEOLOGY 12.4

The Sydney Tar Ponds—Canada's Largest Toxic Dump

The Sydney Tar Ponds in Nova Scotia is Canada's most contaminated site. The ponds contain more than 700,000 tonnes of toxic coal-tar deposits derived from coking processes associated with Sydney's steel-making industry (box figure 1). Contaminants contained within the coal tar deposits include PAHs (polyaromatic hydrocarbons), PCBs (polychlorinated biphenyls), and heavy metals. Other contaminants released into the local environment by the coking plant include benzene, toluene, kerosene, naphthalene, and tar. Toxic effluent was dumped untreated into Muggah Creek from the coking plant from the early 1900s and 1988, infilling a former tidal estuary to form the shallow tar "ponds" (box figure 2). These ponds are still open to tidal circulation and contaminants are flushed into Sydney harbour with each tide. It is estimated that about 800 kilograms of PAHs are released annually to the harbour, and local fisheries have been closed due to contamination.

In 1986, federal and provincial government officials signed an agreement to clean up the tar ponds by excavating and incinerating the contaminated fill. Unfortunately, the dredging operation failed because the fill contained PCBs at 10 times the level that had been estimated and the low-temperature incinerators available could not destroy these contaminants. The dredging plan was abandoned and in 1996 the provincial government announced a second plan, to encase the contaminated waste in slag and pave it over. This plan was also abandoned due to widespread opposition, and the future of the tar ponds was placed in the hands of a joint action group that included local community members. In 2004, $400 million in funding from federal and provincial sources was committed to a cleanup process that will destroy the worst contaminants and treat the remainder before enclosing the site in an engineered containment system.

BOX 12.4 ■ FIGURE 1
Sydney Tar Ponds, with steel-making industry in background, Nova Scotia.
CP (Cape Breton Post)

BOX 12.4 ■ FIGURE 2
Aerial view of the Sydney Tar Ponds, Nova Scotia. Area in background was once a deep tidal inlet and has been infilled with contaminated sediment.
CP (Cape Breton Post)

Uranium is used in nuclear reactors to produce electricity, and in nuclear weapons and some naval craft. At present, nuclear reactors produce 13 percent of the energy needs of Canada and 10 percent of the energy needs of the United States. In France, 75 percent of electricity is produced by nuclear power. Use of nuclear power may not rise appreciably in the future, depending on public acceptance. Nuclear plants produce long-lived waste products that remain dangerous for centuries; no suitable storage site yet exists, although Yucca Mountain in Nevada is being intensively studied as a possible nuclear waste repository (see chapter 15). Accidents at the Three Mile Island reactor in Pennsylvania in 1979 and the Chernobyl reactor near Kiev in Ukraine in 1986 caused a major rethinking of the desirability of nuclear power.

There are six nuclear power generation sites in Canada (four in Ontario, one in Quebec, one in New Brunswick). Canadian reactors use ^{238}U in the CANDU process. All of the United States' nuclear reactors use ^{235}U, which is much less abundant than ^{238}U (only 0.7% of uranium is ^{235}U). Half of France's nuclear reactors are breeder reactors, in which some of the neutrons given off during a chain reaction of ^{235}U bombard nuclei of ^{238}U, converting them into plutonium. Plutonium, like ^{235}U, is also capable of nuclear chain reactions. Thus, breeder reactors can greatly extend the amount of uranium that can be used as fuel.

Recoverable reserves of nearly 500,000 tons of uranium oxide in western sandstones of the United States and "known" low-cost recoverable reserves of more than 450,000 tons of uranium in Canada (about 14% of the world total) seem adequate to power the existing reactors in North America well into the century.

Approximately 65,000 tons of uranium ore must be produced every year to satisfy the needs of the world's 430 nuclear reactors, which supply about 7 percent of the world's energy needs.

Geothermal Power

The world's largest geothermal power plant is at the Geysers in the Coast Range of northern California. This 1,000-megawatt facility provides the energy needs for a million people, though its level of production has been declining steadily since 1980, underscoring the point that geothermal energy is a somewhat fragile resource. The most important reason for this is that groundwater supplies are withdrawn faster than nature can replenish—and reheat—them.

It is difficult to know whether to call geothermal energy a "renewable" form of energy or not. Given that this resource is exhausted readily as most plants worldwide, geothermal energy must be regarded as nonrenewable. On the other hand, groundwater and heat are both readily replenished on a time scale of centuries or millennia, and in a few places there seems to be a never-ending supply of extractable Earth-heat. Unlike fossil fuels, however, geothermal power is not transportable worldwide. Geothermal energy is likely to have only local or regional applications.

Geothermal power (chapter 15) may contribute to power needs, particularly if successful techniques are developed for tapping the heat of areas not marked by surface hot springs. Currently, geothermal power is being developed as a renewable energy source in British Columbia, but contributes only 0.2 percent to U.S. energy use.

WHAT ARE SOME RENEWABLE SOURCES OF ENERGY?

Some energy resources are unquestionably renewable and easily tapped. These include solar, wind, wave (tidal), and hydroelectric power, which together may provide the world with as much as 9 percent of its electricity by 2010. At present, the growth in renewable energy supply is about 2.3 percent per year, compared to an increase of 1.6 percent in nonrenewable resource extraction.

Unlike fossil fuels, which require huge industries to mine, transport, and distribute to users, solar and wind power can be generated locally. Solar collectors and windmills require large amounts of steel, copper, and other mineral resources to make. They are also rather inefficient sources of energy (5–15% for solar, usually around 20% for wind farms). These problems will diminish with improving technology and rising demand.

Wind power is one of the fastest growing sources of energy in the world. Canada is developing a series of "wind farms" across the country (figure 12.17), and the goal is to provide 5 percent of Canada's energy from wind-powered turbines by 2010. Wind turbines are relatively expensive to install, and the most cost-effective wind turbines are located on tall towers in the windiest areas.

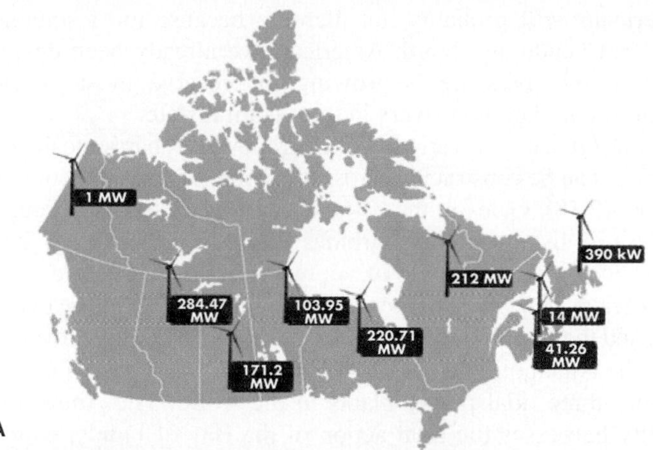

FIGURE 12.17

(*A*) Canada's installed capacity for wind power generation (2006).
(*B*) Wind Farm, Pincher Creek, Alberta.

Photo A from http://www.canwea.ca/en/CanadianWindFarms.html; Photo B courtesy of TransAlta Energy Corporation

Hydroelectric power facilities transform gravitational energy in the form of falling water into electrical energy. Most hydroelectric power stations lie at the foot of dams where water spilling from reservoirs into rivers downstream spins turbines. These stations produce electricity somewhat more cheaply than fossil fuels, especially in regions where oil, coal, or gas has to be imported. Downstream bank erosion, disruption of fish migrations, the flooding of land, and displacement of populations by filling reservoirs are major environmental concerns.

Hydroelectric power is by far the largest of the developed renewable energy resources at present, accounting for 96 percent of all renewable energy production worldwide—about half of it in Europe and North America. Hydropower provides 28 percent of the energy needs of Canada (see figure 12.3) but only 3 percent of the energy consumed in the United States. It can be generated locally; a small station placed on a creek or stream can provide the power needs of a home, farm, or ranch. The contribution of hydropower to the energy needs of North

Americans will probably not increase because most suitable sites in Canada and North America have already been developed. Public pressure is growing to preserve most of the remaining undammed rivers in their natural state.

Tidal power is a variation of hydropower. A barrier, called a *barrage,* can be constructed across the mouth of an estuary or bay (figure 12.18). Gates in the barrage allow water to pass through as the tide rises, spinning turbines to produce electricity. The gates close when the tide is in, capturing the water inshore from the barrage. The gates reopen after the tide falls on the seaward side, and the water pouring out spins the turbines again in reverse.

The Annapolis Power Tidal Facility in Nova Scotia is one of only three tidal power plants in the world. The Annapolis facility harnesses the tidal action of the Bay of Fundy, which experiences the largest tides in the world (see chapter 18) and generates up to 20 megawatts of power daily. The world's largest tidal power facility is at Rance, in France, and generates 320 megawatts daily. In order to produce effective amounts of tidal power, the difference between high and low tides needs to be at least 5 metres. Relatively few coastal areas in the world experience such high tidal ranges and this, together with the high costs of tidal power station construction and concerns about impacts on fish and bird life, has prevented the construction of more tidal power facilities. New technologies are currently being developed to harness tidal energy, including the use of floating or anchored offshore tidal turbines (figure 12.18*C*) that will offer significant advantages over barrage and fence tidal systems and reduce environmental effects.

WHERE DO METALS COME FROM?

The successful search for metals depends on finding **ores,** which are naturally occurring materials that can be profitably mined (see box 12.5 and table 12.2). It is important to recognize that the local concentration of a metal must be greater (usually much greater) than its average crustal abundance to be a potential ore body. Metals must be concentrated in a particular place in a large enough amount to be viable ore bodies. Take gold in sea water. You could become fabulously wealthy if you could extract a fraction of the gold in sea water, because there are more than 10^{11} troy ounces of gold—around $52 trillion worth—in the world's oceans. But the concentration is 4 grams per 1 million tons of water. It would cost you far more to remove that gold than you could sell it for.

Whether or not a mineral (or rock) is considered a metal ore depends on its chemical composition, the percentage of extractable metal, and the market value of the metal. The mineral *hematite* (Fe_2O_3), for example, is usually a good *iron ore* because it contains 70 percent iron by weight; this high percentage is profitable to extract at current prices for iron. Limonite ($Fe_2O_3{\bullet}nH_2O$) contains less iron than hematite and hence is not as extensively mined. Even a mineral containing a high percentage of metal is not described as an *ore* if the metal is too difficult to extract or the site is too far from a market; profit is part of what defines an ore. Many different kinds of geologic processes can accumulate ores, from weathering and sedimentation to the settling of crystals deep within magma chambers (table 12.3).

A

Power house

Bay

Incoming tide

Turbines in tidal gate

Barrage

B

FIGURE 12.18

(*A*) and (*B*) Generation of tidal power. (*C*) Artist's impression of an offshore tidal turbine designed for use in water depths of approximately 60 metres.

Illustration C from TidalStream Partnership

IN GREATER DEPTH 12.5

Canada's Mineral Wealth

Canada is one of the world's leading exporters of minerals and metals. In 2005, the value of mineral exports reached almost $60 billion and the mining industry contributed approximately 4 percent of the national gross domestic product. Gold, nickel, zinc, copper, and iron ore were the major metals produced, and potash was the most important nonmetal mined (box figure 1). Canada is the world's largest producer of potash and uranium and ranks in the top five for the production of nickel, asbestos, zinc, cadmium, titanium concentrate, aluminum, platinum-group metals, salt, gold, molybdenum, copper, gypsum, cobalt, and diamonds. In 2004 there were 190 metal, nonmetal, and coal mines; 3,000 stone quarries and sand and gravel pits; and about 50 nonferrous smelters, refineries, and steel mills operating in Canada.

Despite all of this activity, less than 0.03 percent of Canada's land area (less than half the size of Prince Edward Island) has been used for mining since metal mining began more than 150 years ago. Many of Canada's communities were founded on the basis of mineral extraction industries—beginning with the Klondike gold rush in the Yukon and continuing today with the development of diamond mines in the Northwest Territories (see box 12.7).

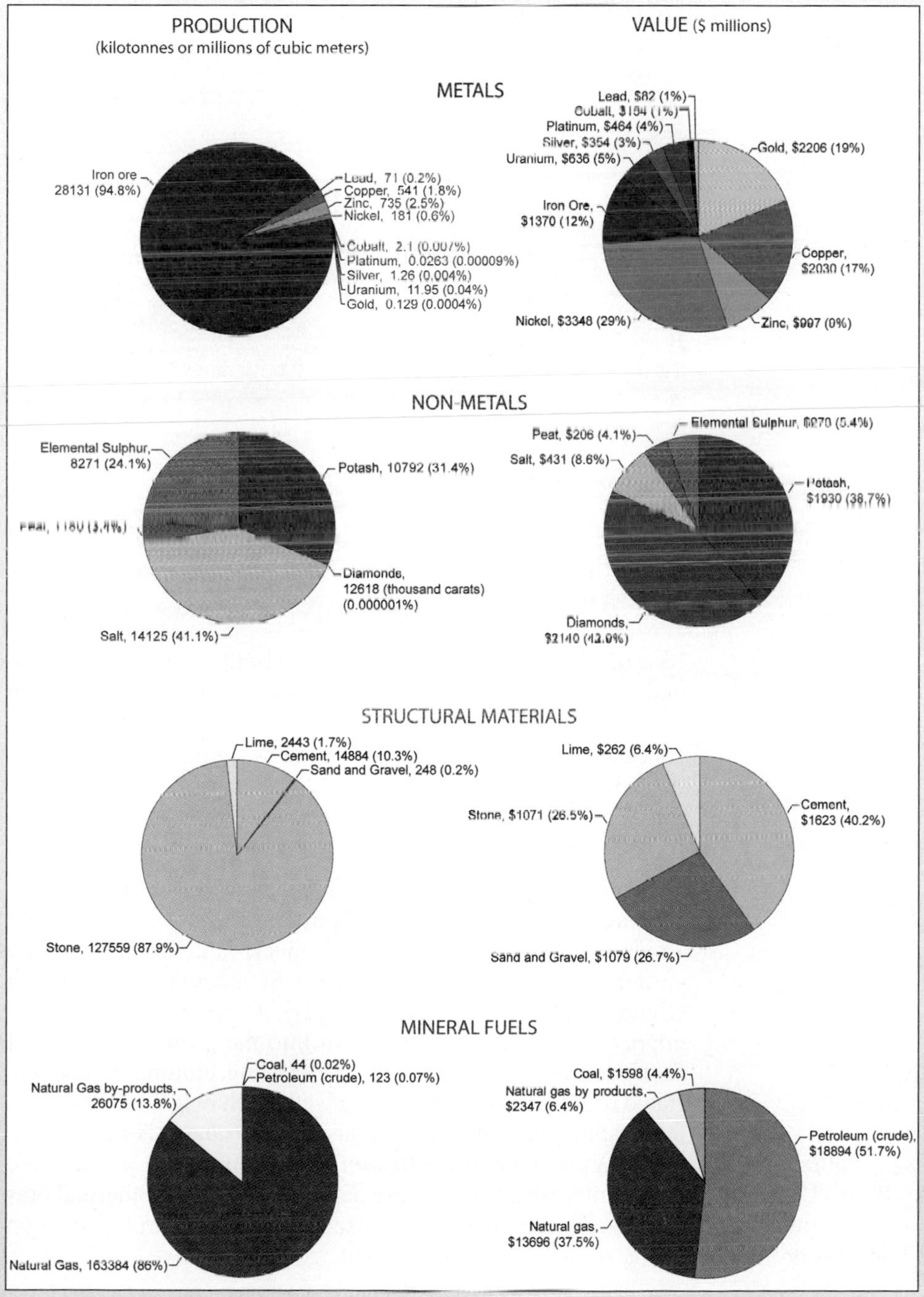

BOX 12.5 ■ FIGURE 1

Production and value of Canada's mineral resources (2004).

Note: For resource production of metals, non-metals, and structural minerals, amounts are given in kilotonnes; for mineral fuels, amounts are given in millions of cubic metres. Values are given in millions of dollars.

TABLE 12.2 Common Ore Minerals

Metal	Ore Mineral	Composition
Aluminum	Bauxite (a mineral mixture)	$AlO(OH)$ and $Al(OH)_3$
Chromium	Chromite	$FeCr_2O_4$
Copper	Native copper	Cu
	Chalcocite	Cu_2S
	Chalcopyrite	$CuFeS_2$
Gold	Native gold	Au
Iron	Hematite	Fe_2O_3
	Magnetite	Fe_3O_4
Lead	Galena	PbS
Manganese	Pyrolusite	MnO_2
Mercury	Cinnabar	HgS
Nickel	Pentlandite	(Fe, Ni)S
Silver	Native silver	Ag
	Argentite	Ag_2S
Tin	Cassiterite	SnO_2
Uranium	Pitchblende	U_3O_8
	Carnotite	$K(UO_2)_2(VO_4)_2 \cdot 3H_2O$
Zinc	Sphalerite	ZnS

TABLE 12.3 Some Ways Ore Deposits Form

Type of Ore Deposit	Some Metals Found in This Type of Ore Deposit
Crystal settling within cooling magma	Chromium, platinum, iron
Hydrothermal deposits (contact metamorphism, hydrothermal veins, disseminated deposits, hot-spring deposits)	Copper, lead, zinc, gold, silver, iron, molybdenum, tungsten, tin, mercury, cobalt
Pegmatites	Lithium, rare metals
Chemical precipitation as sediment	Iron, manganese, copper
Placer deposits	Gold, tin, platinum, titanium
Concentration by weathering and groundwater	Aluminum, nickel, copper, silver, uranium, iron, manganese, lead, tin, mercury

Ores Formed by Igneous Processes

Crystal Settling

Crystal settling occurs as early-forming minerals crystallize and settle to the bottom of a cooling body of magma (figure 12.19). This process was described under differentiation in chapter 6. The metal *chromium* comes from chromite ore bodies near the base of sills and other intrusions. Most of the world's chromium and platinum come from a single intrusion, the huge Bushveldt Complex in South Africa. In Montana, another huge Precambrian sill called the Stillwater Complex contains similar but lower-grade deposits of these two metals.

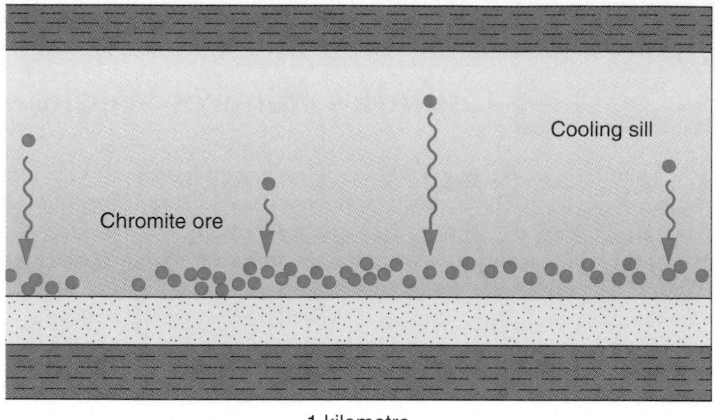

FIGURE 12.19

Early-forming minerals such as chromite may settle through magma to collect in layers near the bottom of a cooling sill.

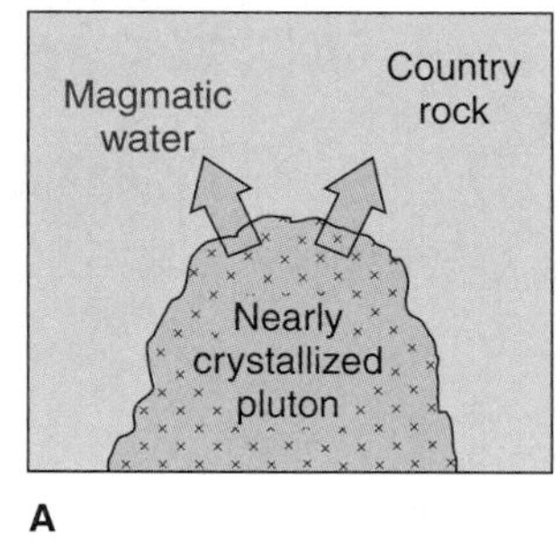

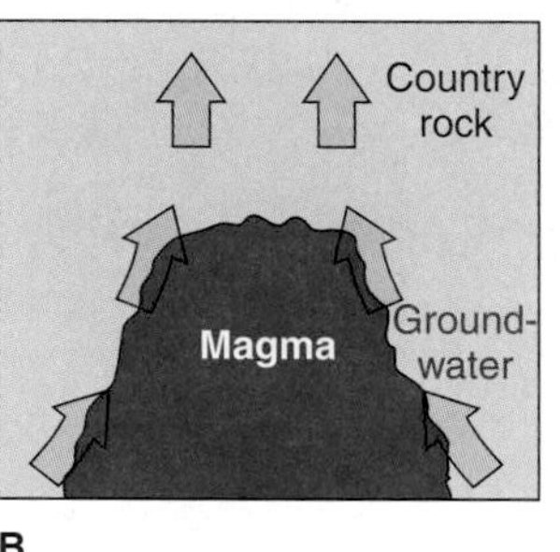

FIGURE 12.20

Two possible origins of hydrothermal fluids. (*A*) Residually concentrated magmatic water moves into country rock when magma is nearly all crystallized. (*B*) Groundwater becomes heated by magma (or by a cooling solid pluton), and a convective circulation is set up.

Hydrothermal Fluids

Hydrothermal fluids, discussed in chapter 10, are the most important source of metallic ore deposits. The hot water and other fluids are part of the magma itself, injected into the surrounding country rock during the last stages of magma crystallization (figure 12.20*A*; see also figure 10.21). Atoms of metals such as copper and gold, which do not fit into the growing crystals of feldspar and other minerals in the cooling pluton, are concentrated residually in the remaining water-rich magma. Eventually, a hot solution, rich in metals and silica (quartz is the lowest-temperature mineral on Bowen's reaction series), moves into the country rock to create ore deposits. Most hydrothermal ores are metallic sulphides, often mixed with milky quartz. The origin of the sulphur is widely debated.

A magma body or hot rock may heat groundwater and cause convection circulation. This water may mix with water given off from solidifying magma or it may leech metals from solid rock and deposit metallic minerals elsewhere as the water cools. However the hydrothermal solutions form, they tend to create four general types of hydrothermal ore deposits: (1) contact metamorphic deposits, (2) hydrothermal veins, (3) disseminated deposits, and (4) hot-spring deposits.

Contact metamorphism can create ores of iron, tungsten, copper, lead, zinc, silver, and other metals in country rock. The country rock may be completely or partially removed and replaced by ore (figure 12.21*A*). This is particularly true of limestone beds, which react readily with hydrothermal solutions. (The metasomatic addition of ions to country rock is described in chapter 10.) The ore bodies can be quite large and very rich.

Hydrothermal veins are narrow ore bodies formed along joints and faults (figure 12.21*B*). They can extend great distances from their apparent plutonic sources. Some extend so far that it is questionable whether they are even associated with plutons. The fluids can precipitate ore (and quartz) within cavities along the fractures (figure 12.21*D*) and may also replace the wall rock of the fractures with ore. Hydrothermal veins form most of the world's great deposits of lead, zinc, silver, gold, tungsten, tin, mercury, and, to some extent, copper.

Hot solutions can also form *disseminated deposits* in which metallic sulphide ore minerals are distributed in very low concentration through large volumes of rock, both above and within a pluton (figure 12.21*C* and box 10.4). Most of the world's copper comes from disseminated deposits (also called *porphyry copper deposits* because the associated pluton is usually porphyritic). Along with the copper are deposited many other metals, such as lead, zinc, molybdenum, silver, and gold (and iron, though not in commercial quantities).

Where hot solutions rise to the Earth's surface, *hot springs* form. Hot springs on land may contain large amounts of dissolved metals. More impressive are hot springs on the sea floor (figure 12.21*D*), which can precipitate large mounds of metallic sulphides, sometimes in commercial quantities.

Pegmatites (refer to box 6.1) are another type of ore deposit associated with igneous rocks. They may contain important concentrations of minerals containing lithium, beryllium, and other rare metals, as well as gemstones such as emeralds and sapphires.

Ores Formed by Surface Processes

Chemical precipitation in layers is the most common origin for ores of iron and manganese. A few copper ores form in this way too. Banded iron ores, usually composed of alternating layers of iron minerals and chert, formed as sedimentary rocks in many parts of the world during the Precambrian, in shallow, water-filled basins (figure 12.22). Later folding, faulting, metamorphism, and solution have destroyed many of its original features so the origin of the ore is difficult to interpret. The water may have been fresh or marine, and the iron may have come from volcanic activity or deep weathering of the surrounding continents. The alternating bands may have been created by some rhythmic variation in volcanic activity, river runoff, basin water circulation, growth of organisms, or some other factor. Since banded iron ores are all Precambrian, their origin might be connected to an ancient atmosphere or ocean chemically different from today's.

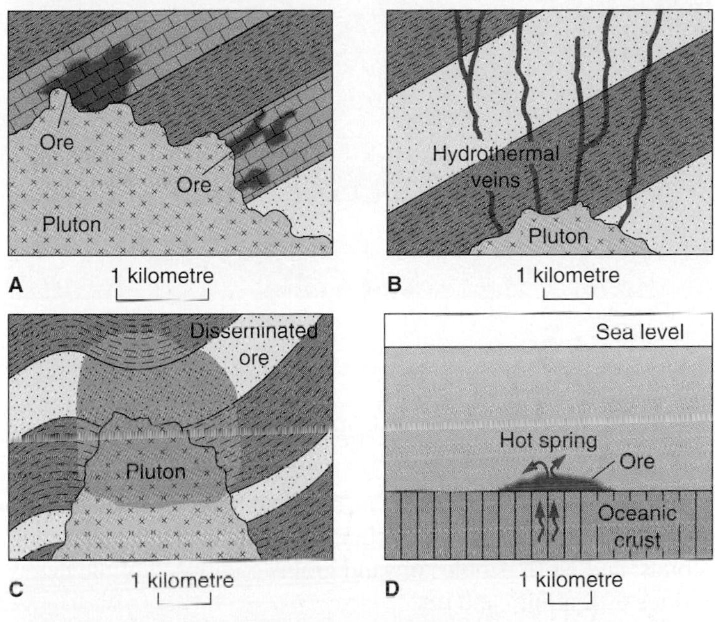

FIGURE 12.21

Hydrothermal ore deposits. (*A*) Contact metamorphism in which ore replaces limestone. (*B*) Ore emplaced in hydrothermal veins. (*C*) Disseminated ore within and above a pluton (porphyry copper deposits, for example). (*D*) Ore precipitated around a submarine hot spring (size of ore deposit exaggerated). (*E*) Hydrothermal vein of milky quartz in gneiss, Canadian Shield, Ontario.

Photo by C.H. Eyles

FIGURE 12.22

Banded iron ores of Precambrian age probably accumulated in shallow basins.

Placer deposits in which streams have concentrated heavy sediment grains in a river bar are described in chapter 14. Wave action can also form placers at beaches. Placers include gold nuggets and dust, native platinum, diamonds and other gemstones, and worn pebbles or sand grains composed of the heavy oxides of titanium and tin.

Ore deposits due to *concentration by weathering* were described in chapter 8. Aluminum (in bauxite) forms through weathering in tropical climates.

Another type of concentration by weathering is the *supergene enrichment* of disseminated ore deposits. Through supergene enrichment, low-grade ores of 0.3 percent copper in rock can be enriched to a mineable 1 percent copper. The major ore mineral in a disseminated copper deposit is chalcopyrite, a copper-iron sulphide containing about 35 percent copper. Near the Earth's surface, downward-moving groundwater can leach copper and sulphur from the ore, leaving the iron behind (figure 12.23). At or below the water table the dissolved copper can react with chalcopyrite in the lower part of the disseminated deposit, forming a richer ore mineral such as chalcocite, which is about 80 percent copper:

$3\ Cu^{++}$	+	$CuFeS_2$	→	$2\ Cu_2S$	+	Fe^{++}
Copper dissolved in groundwater		Chalcopyrite		Chalcocite		Iron in solution

In this way copper is removed from the top of the deposit and added to the lower part (figure 12.23). The ore below the water table may be several times richer than in the rest of the deposit (silver can move with the copper). The iron left behind at the surface forms a rusty cap called a *gossan,* or "iron hat," which is a visible clue to the ore below.

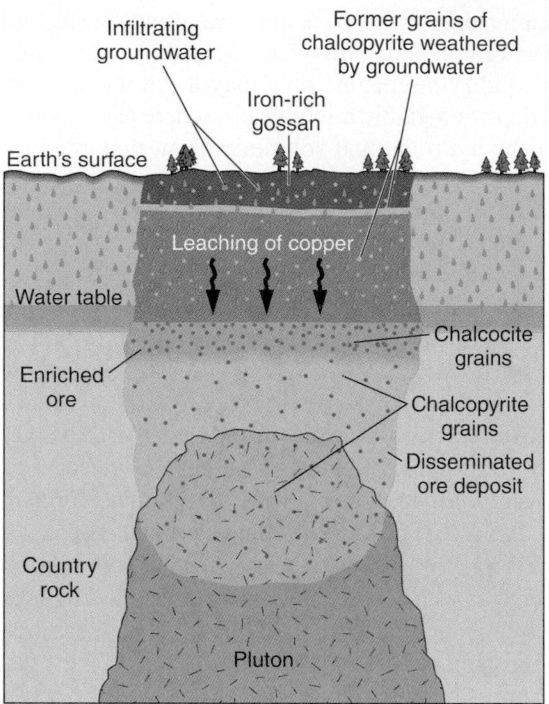

FIGURE 12.23

Supergene enrichment. Groundwater leaches copper from upper part of disseminated deposit and precipitates it at or below the water table, forming rich ore.

Mining

As in the case of coal, miners use both surface and underground techniques to extract ore minerals (figure 12.24). Strip mining—the wholesale removal of large areas of soil and shallow rock cover—has already been mentioned in connection with coal beds. Aluminum ore (bauxite), which forms in weathered soil beds under tropical conditions, is often most easily extracted this way. *Open pit mining* is related to strip mining, but concentrates on the removal of valuable deposits from a specific, relatively small area. Open-pit mines often dig much deeper than strip mines. *Placer mines* are localized to ancient or modern river bar or beach deposits.

Underground, or *bedrock mining* must be done to excavate many valuable mineral deposits. Bedrock mining of ores typically extends to much greater depths than ordinary coal mines, and this presents its own set of technical challenges. The world's deepest mines, in South Africa, extend to depths of 1,500–2,500 metres. The walls grow hot to the touch so deep underground, and pumping of fresh, cool air and water must be done to make working conditions tolerable. Mines have notoriously poor air circulation, and the use of dynamite to blast openings releases toxic gases that must be removed. Ammonium nitrate (NH_4NO_3) mixed with fuel oil (CH_2) is a typical blasting agent. The explosive reaction generates poisonous carbon monoxide whenever there is a slight excess of oil in the mixture. Carbon monoxide (CO) is a heavier-than-air gas that sinks deep into the mine:

$$3\ NH_4NO_3 + CH_2 \rightarrow CO_2\ \text{(or CO plus oxygen)} + 7\ H_2O + 3\ N_2 + \text{heat}$$

This is one of the reasons why abandoned mines should never be explored.

GEOMATICS 12.6

Planning Effective Mining Strategies Using 3-D Models

There are many applications of geomatics in the mineral mining industry. Remotely sensed digital data from satellite images and airborne geophysical surveys are combined with ground-based geologic and geophysical surveys to identify potential areas for exploration purposes—most ore bodies and diamond-bearing kimberlite pipes identified on the Canadian Shield have been located in this way. The spatial analysis of digital data can also be used to effectively design and plan mining operations once economic deposits have been found. Metallic ore bodies generally have complex subsurface geometries, and planning the most profitable and least environmentally damaging mining operation to extract the ores is difficult. 3-D models of the subsurface form of ore bodies can now be created to guide mine planning operations, including the location of shafts and drifts to extract the ore (box figure 1). These models allow mine operators not only to effectively plan the construction of their mine but also to communicate information to the public regarding potential hazards or environmental effects.

BOX 12.6 ■ FIGURE 1

(*A*) and (*B*): Two different views of a mine plan. The large red structure is a model of the ore body to be mined, while the white and green lines are the planned shafts, the purple lines are the planned drifts, and the orange line is the planned ramp.

Images: INCO Ltd.

Where the mines extend beneath the water table, the water that seeps in must be pumped out to avoid flooding. An active mine consumes large amounts of energy as well as material resources.

The design of a mine takes into consideration three vital factors: (1) the geometry of the underground ore body (see box 12.6); (2) the need for safety; and (3) the need to maximize profit. It is typically easiest for miners to construct a set of vertical and horizontal passages to access and remove the ore (figure 12.24*A*). The vertical openings, called *shafts* (or *winzes,* if they do not open all the way to the surface) allow elevators to take miners underground and bring ore up to the surface. Shafts are also conduits for electrical cables, water hoses, and air lines. The horizontal tunnels, termed *adits* (or *drifts,* if they do not open to the surface), are the pathways through which ore is directly excavated. In larger mines, multiple drifts radiate off of shafts (see box 12.6). *Ramps* are slanted tunnels, many of which have tracks for winching ore carts. In some places, the ore may be so rich that miners excavate a giant underground chamber called a *stope.* To avoid collapse, the walls of the stope may have to be shored up with timbers or other construction material.

Today, more sophisticated—and expensive—prospecting techniques are applied to locating and determining the shape of an underground ore body. Exploration geoscientists must study the structural geology (stratigraphy and deformation of the surrounding rocks), examine the evidence brought up in preliminary boreholes, and conduct geophysical surveys, including the use of gravimeters, magnetometers, and electrical-resistivity equipment. Some ore bodies are excellent electrical conductors and may be highly magnetic. Exploration geoscientists have the ultimate say on whether or not a company should proceed with mining.

Environmental Effects of Mining

Some of the environmental problems associated with mining can be partially solved if care is taken. For example, in the past *waste rock* (or *tailings*) was routinely left in unsightly heaps and dumps. The excavations for strip mines and placer mines can be filled in with waste rock, levelled or graded, and then covered over with topsoil to restore the land to usable condition. In some cases crops can be grown on reclaimed land within one year after mining operations are completed. Underground mines are sometimes back-filled with waste rock to prevent land *subsidence* after ore is removed.

One of the more difficult problems to deal with is *acid drainage* from mines, caused by groundwater running or being pumped out of a mine. Sulphide ore minerals (table 12.2) and pyrite (FeS_2) are most often the source of the trouble. Groundwater transports oxygen to the sulphides where they are oxidized to iron oxide and sulphuric acid. In some mines, expensive programs of holding and neutralizing drainage water in ponds or artificial wetlands prevent pollution of surface streams and harm to forests and wildlife. The worst problem is with long-abandoned mines that are still draining acid waters. Many of these may never be neutralized.

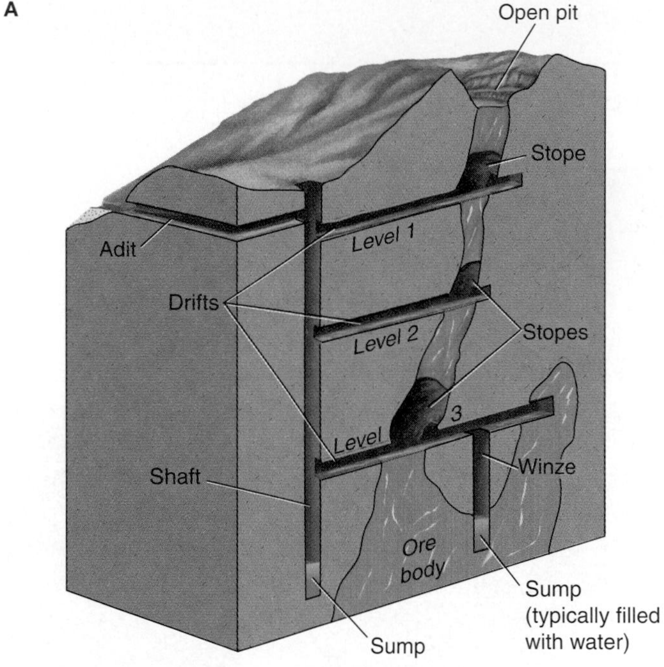

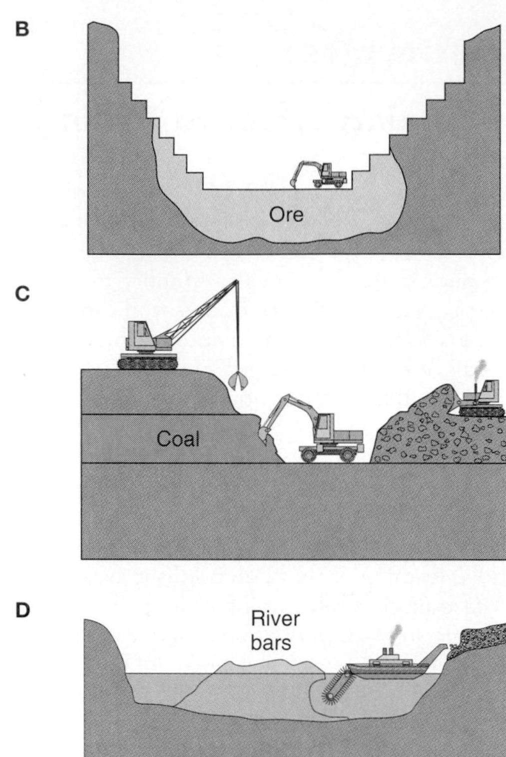

FIGURE 12.24

Types of mines: (*A*) Underground. (*B*) Open pit. (*C*) Strip. (*D*) Placer (being mined by a floating dredge).

WHERE DO WE FIND IMPORTANT METALS?

Iron

Iron is one of the three essential ingredients required to make steel. Coal (coke) and limestone, the other two, are needed to melt the iron in furnaces before moulding. Iron can be mixed with other metals ("alloyed"), including silicon, chromium, and nickel, to make special kinds of steel. In fact, there are more than 2,000 kinds of steel used in the world today, representing various mixtures of many metals. But in all, iron is dominant. Steel is used to make everything from skillets to locomotives.

The major iron ore minerals are hematite and magnetite. In Canada, iron ore is produced in Newfoundland and Labrador and in Quebec (figure 12.25). Most of the iron in the United States comes from the region around Lake Superior, most notably Minnesota. The ores are banded iron ores of Precambrian age, typical of most iron ores in the world. Mining is done mostly by open-pit methods.

Copper

Less abundant is *copper,* another important metal for industry. More than half the copper used in the United States and Canada goes into electrical wire and equipment and one-third into the manufacture of brass, a copper–zinc alloy.

Most copper ores are sulphides. Chalcopyrite is the most important copper ore mineral. Some vein deposits of copper exist (as at Butte, Montana), but most major deposits are disseminated through large volumes of rock; so, most copper mines are open pits (figure 12.26). British Columbia, Ontario, Quebec, and Manitoba are the major copper producing provinces (figure 12.25). Copper will also be produced as a by-product to nickel and cobalt when the Voisey's Bay mine of Newfoundland and Labrador begins production. Arizona, Utah, and a few other western states are the major producers in the United States (figure 12.25). The concentration of copper averages about 0.5 percent in most currently worked deposits; that is, 1 kg of copper is recovered for every 200 kg of rock processed.

Aluminum

Aluminum is consumed in the manufacture of beer and soft drink cans, airplanes, electrical cable, and many other products. The use of aluminum is increasing rapidly.

The ore of aluminum is bauxite, which forms under tropical weathering conditions. Canada has no bauxite resources but is a major producer of aluminum because of the availability of low-cost hydroelectric power required for smelting. The United States has very little bauxite, so it imports 90 percent of its aluminum ore from tropical countries. The largest aluminum mine in the United States is in Arkansas (figure 12.25). Open-pit mining is the usual technique for extracting bauxite. Aluminum was the first metal to be subject to widespread recycling, in part because so much of North America's aluminum ore is imported.

Lead

The main use of *lead* (79 percent) is in batteries. Substantial amounts are recycled, largely from automobile batteries.

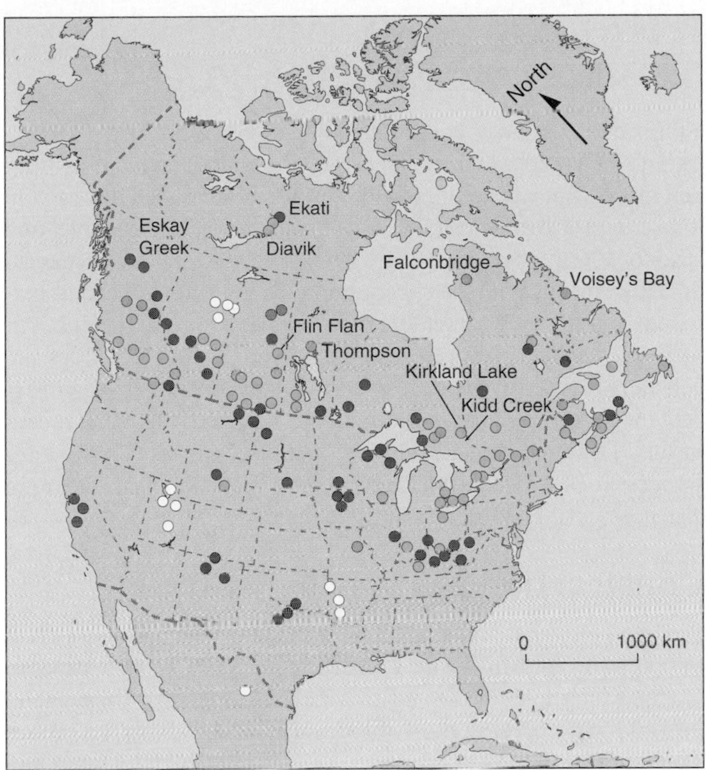

FIGURE 12.25

Geologic resources of North America.

Canadian data from geological survey of Canada, 2004. Canadian Resources of North America, Geological Survey of Canada, Map 900A, 2000. Reproduced with the permission of the Minister of Public Works and Government Services Canada, 2007 and Courtesy of Natural Resources Canada, Geological Survey of Canada.

FIGURE 12.26

Open-pit workings at the Highland Valley Copper Mine, British Columbia.

The most important ore of lead is galena. In Canada, lead is produced as a by-product of zinc in New Brunswick, British Columbia, and Nunavut (figure 12.25). Major lead deposits occur in Missouri, Idaho, Utah, and Colorado (figure 12.25). The Missouri deposits are mostly found in limestone beds; their precise origin is a matter of some controversy. The ore is mined both underground and from open pits. Deposits in Idaho occur mostly as veins and are usually mined underground.

Zinc

Widely used in industry, *zinc* is necessary for galvanizing and the manufacture of brass and other alloys.

The major zinc ore is sphalerite. As sphalerite usually is found closely associated with galena, most lead mines also extract zinc. Zinc occurs without lead in some areas, however. Canada is one of the largest producers and exporters of zinc, which is found in Ontario and Quebec (figure 12.25).

Silver

Coins, tableware, jewellery, photographic film, and many other products are made of *silver.*

Silver, found as a native metal and in sulphide ores, is a common by-product of lead and copper mining. Ontario, Quebec, British Columbia, and Manitoba produce silver (figure 12.25). The lead–zinc mines of Idaho (the Coeur d' Alene district) are the largest silver producers in the United States.

Gold

The rare and valuable metal *gold* is used in coins, jewellery, decoration, dentistry, electronics, and the space program. Gold bars are stored to back national currency, although this use is rapidly disappearing.

Gold is found as a native element in the form of nuggets and grains. In some parts of the world these are concentrated in placer deposits (California's Gold Rush of 1849 was triggered by discoveries of placer gold). Gold nuggets, flakes, and dust can be separated from the other sediments by (1) *panning;* (2) *sluice boxes*, which catch the heavy gold on the bottom of a box as gravel is washed through it; (3) *hydraulic mining* (figure 12.27), which washes gold-bearing gravel from a hillside into a sluice box; or (4) floating *dredges* (figure 12.28), which separate gold from gravel aboard a large barge, piling the spent gravel behind. When gold is found in hydrothermal veins associated with milky quartz, as it is in parts of Colorado and in California's Mother Lode (figure 12.25), it is mined underground. Canada is one of the world's leading producers of gold. More than half of the gold mined in Canada comes from rich lode deposits in the Hemlo and Red Lake areas of Ontario (figure 12.25). The largest underground gold mine in the United States is the Homestake Mine in South Dakota, where finely disseminated gold is extracted from folded metamorphic rock. A large amount of disseminated gold is mined in open pits near Carlin, Nevada, and at several other localities in Nevada and California.

FIGURE 12.27
Hydraulic mining for gold, Dawson City, Yukon.
Photo by Nick Eyles

FIGURE 12.28
A gold dredge separates gold from gravel.
Photo by David McGeary

The world has large reserves of iron and aluminum, moderate reserves of copper, lead, and zinc, and scanty reserves of gold and silver.

Other Metals

Many other metals are vital to our economy. *Chromium, nickel, cobalt, manganese, molybdenum, tungsten,* and *vanadium* are all important in the steel industry, particularly in the manufacture of specialty products such as stainless steel and armour plate. Most of these metals have other uses as well. *Tin* is used in solder and for plating steel in tin cans. *Mercury* is used in thermometers, silent electrical switches, medical compounds, and batteries. *Magnesium* is used in aircraft. *Titanium,* as strong as steel but weighing half as much, is used in aircraft. *Platinum* is used in catalytic converters to clean automobile exhaust.

WHAT ARE SOME IMPORTANT NONMETALLIC RESOURCES?

Nonmetallic resources are Earth resources that are not mined to extract a metal or as a source of energy. Most rocks and minerals contain metals, but when nonmetallic resources are mined it is usually to use the rock (or mineral) as-is (for example, using gravel and sand for construction projects), whereas metallic ores are processed to extract metal. With the exception of the gemstones, such as diamonds and rubies, nonmetallic resources do not have the glamour of many metals or energy resources. Nonmetallic resources are generally inexpensive and are needed in large quantities (again, except for gemstones); however, their value exceeds that of all mined metals. The large demand and low unit price means that these resources are best taken from local sources. Transportation over long distances would add significantly to the cost.

Construction Materials

Sand and *gravel* are both needed for the manufacture of concrete for building and highway construction. Sand is also used in mortar, which holds bricks and cement blocks together. The demand for sand and gravel in the United States and Canada has more than doubled in the last 25 years. Sand dunes, river channel and bar deposits, glacial outwash, and beach deposits are common sources for sand and gravel. Cinder cones are mined for "gravel" in some areas. Sand and gravel are ordinarily mined in open pits (figure 12.29).

Stone refers to rock used in blocks to construct buildings or crushed to form roadbed. Most stone in buildings is limestone or granite, and most crushed stone is limestone. Huge quantities of stone are used each year in North America. Stone is removed from open pits called *quarries* (figure 12.30).

Limestone has many uses other than building stone or crushed roadbed. Cement, used in concrete and mortar, is made from limestone. Pulverized limestone is in demand as a soil conditioner and is the principal ingredient of many chemical products.

FIGURE 12.29
Sand and gravel pit in the Oak Ridges Moraine, Ontario.
Photo by Nick Eyles

FIGURE 12.30

Aerial view of quarry extracting dolostone, Vineland, Ontario. Width of rock face in centre of photo approximately 500 m.

Photo by S. Fletcher

FIGURE 12.31

Arial view of Daivik diamond mine, which lies approximately 300 kilometres northeast of Yellowknife, Northwest Territories.

Photo courtesy Diavik Diamond Mines Inc.

Fertilizers and Evaporites

Fertilizers (phosphate, nitrate, and potassium compounds) are extremely important to agriculture today, so much so that they are one of the few nonmetallic resources transported across the sea. *Phosphate* is produced from phosphorite, a sedimentary rock formed by the accumulation and alteration of the remains of marine organisms. Major phosphate deposits in the United States are in Idaho, Wyoming, and Florida. *Nitrate* can form directly as an evaporite deposit but today is usually made from atmospheric nitrogen. *Potassium* compounds (e.g., potash) are often found as evaporites.

Rock salt is coarsely crystalline halite formed as an evaporite. Salt beds are mined underground in Canada, Ohio, and Michigan; underground salt domes are mined in Texas and Louisiana. (Some salt is also extracted from sea water by evaporation.) Rock salt is used in many ways de-icing roads in winter, preserving food, as table salt, and in manufacturing hydrochloric acid and sodium compounds for baking soda, soap, and other products. Rock salt is heavily used by the chemical industry.

Gypsum forms as an evaporite. Beds of gypsum are mined in many states in the U.S., notably California, Michigan, Iowa, and Texas, and in Canada in Ontario, British Columbia, Manitoba, and Nova Scotia. Gypsum, the essential ingredient of plaster and wallboard (Sheetrock), is used mainly by the construction industry, although there are other uses.

Sulphur occurs in bright-yellow deposits of elemental sulphur. Most of its commercial production comes from the cap rock of salt domes. Sulphur is widely used in agriculture as a fungicide and fertilizer and by industry to manufacture sulphuric acid, matches, and many other products.

Other Nonmetallics

Gemstones (called *gems* when cut and polished) include precious stones such as diamonds, rubies, emeralds, and sapphires and semiprecious stones such as beryl, garnet, jade, spinel, topaz, turquoise, and zircon. Gems are used for jewellery, bearings, and abrasives (most are above 7 on Mohs' scale of hardness). Diamond drills and diamond saws are used to drill and cut rock. Old watches and other instruments often have hard gems at bearing points of friction ("17-jewel watches"). Gemstones are often found in pegmatites or in close association with other igneous intrusives. Some are recovered from placer deposits. The Canadian Shield is host to many kimberlite pipes that are now being mined for diamonds in Canada (figure 12.31 and box 12.7).

Asbestos is a fibrous variety of serpentine or chain silicate minerals. The fibre can be separated and woven into fireproof fabric used for firefighters' clothes and theatre curtains. It is also used in manufacturing ceiling and sound insulation, shingles, and brake linings, although the use of asbestos is being rapidly curtailed because of concern about its connection with lung cancer (box 5.5). The United States produces little asbestos, mostly from belts of serpentinized ultramafic rocks in the Appalachians and the Pacific Coast states. Large amounts are mined in Canada, chiefly by surface mining of narrow veins of asbestos in igneous rocks of the Eastern Townships region of Quebec. *Talc,* used in talcum powder and other products, is often found associated with asbestos.

Other nonmetallic resources are important. *Mica* is used in electrical insulators. *Barite* ($BaSO_4$), because of its high specific gravity, is used to make heavy drilling mud to prevent oil gushers. *Borates* are boron-containing evaporites used in fibreglass, cleaning compounds, and ceramics. *Fluorite* (CaF_2) is used in toothpaste, Teflon finishes, and steel-smelting. *Clays* are used in ceramics, manufacturing paper, and as filters and absorbents. *Diatomite* is used in swimming-pool filters and to filter out yeast in beer and wine. *Glass sand,* which is more than 95 percent quartz, is the main component of glass. *Graphite* is used in foundries, lubricants, steel-making, batteries, and pencil "lead."

IN GREATER DEPTH 12.7

Canada's Diamond Rush

Canada's diamond industry is booming. In 2004 production from Canada's two diamond-producing mines—EKATI and Diavik, both in the Northwest Territories (box figure 1)—increased to 12.8 million carats (valued at $2.1 billion). Canada is now the third largest producer of diamonds in the world (after Botswana and Russia) and accounts for approximately 15 percent of world diamond production. Diamond exploration in Canada began in the 1960s, but the true "diamond rush" began in the 1990s when several major kimberlite (diamond-hosting) discoveries were made in the Lac de Gras area of the Northwest Territories by geologists Chuck Fipke and Stuart Blusson. The EKATI diamond mine came into production in October 1998, and by April 1999 had delivered 1 million carats. In 2004, production at EKATI was 5.11 million carats. A new underground mining project at EKATI, termed the Panda Project, is expected to produce more than 4.5 million carats of high-value Panda diamonds over a six-year time span (2006–12).

Diavik, Canada's second diamond mine, began production in January 2003. The Diavik mine has estimated reserves of 95.6 million carats of rough diamonds (box figure 2), found in three extremely high-grade kimberlite pipes. With a projected 20-year lifespan, the Diavik mine is expected to produce 6 to 8 million carats a year at its peak, approximately 5 percent of the world's total supply. By 2008, another three diamond mines are scheduled to be operational: at Snap Lake (NWT), Jericho (Nunavut), and Viktor (near James Bay, Ontario).

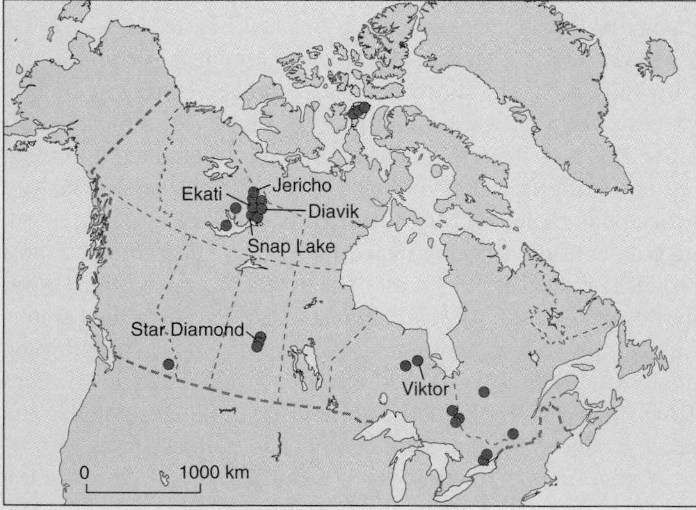

BOX 12.7 ■ FIGURE 1

Location of some of Canada's known kimberlite pipes (red diamonds).

The Atlas of Canada, Earth Sciences Sector, Natural Resources Canada. Reproduced with the permission of the Minister of Public Works and Government Services Canada, 2006, and Courtesy of Natural Resources Canada, Geological Survey of Canada.

The "diamond rush" continues in Canada with more than 120 companies involved in exploration work in Alberta, Saskatchewan, Manitoba, Ontario, Quebec, and Newfoundland and Labrador (see box figure 1). More than 1,200 people are currently employed in diamond mine operations, and approximately 4,000 indirect jobs have been created by the industry. Canada now has a small but growing diamond manufacturing industry with four factories operating in Yellowknife and other manufacturers in Vancouver, Toronto, Montreal, and Matane (Quebec). The largest of these facilities has an output of 2,500 carats a month.

Additional Web Resources

For information on the Canadian diamond industry:

www.nrcan-rncan.gc.ca/mms/diam/index_e.htm;
http://gsc.nrcan.gc.ca/diamonds/index_e.php;
www.nrcan.gc.ca/mms/cmy/2004revu/diam_e.htm.

Animation showing the formation of a kimberlite pipe and an "interactive" kimberlite pipe:

www.nrcan-rncan.gc.ca/mms/diam/Kimberlite-EN/Kimberlite.swf

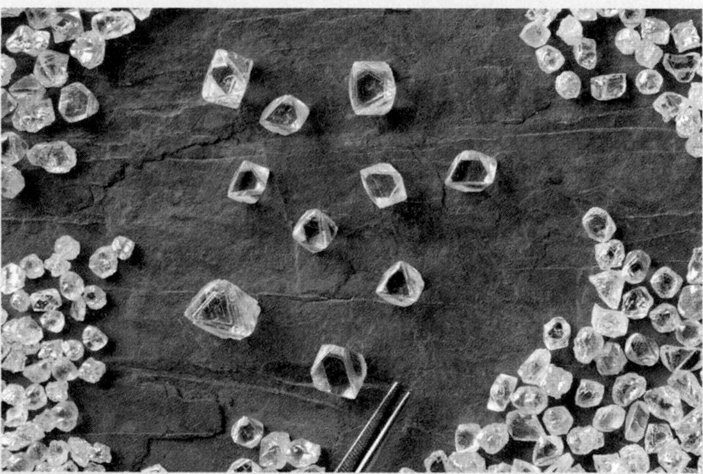

A

B

BOX 12.7 ■ FIGURE 2

(A) Rough diamonds from the Diavik mine. The large diamond on the lower left weighs approximately 8 carats and has a value of approximately $30,000.

(B) Cut and polished Diavik diamonds

Photos courtesy of Diavik Diamond Mines Inc.

WHAT ARE SOME POSSIBLE FUTURE TRENDS?

Ocean mining, now rather uncommon, will increase in the near future. Mining of *manganese nodules* from the deep sea floor (see chapter 18) could provide substantial amounts of copper and nickel, together with far more manganese than industry can consume. The copper content of many nodules is four to five times higher than most deposits on land. *Metallic brines and deposits* of the Red Sea type may be a source of several metals in the future.

Several tools are of great help in mineral exploration on land. Highly sophisticated *geochemical tests* of soils and soil gases point to ore bodies underground. *Geophysical techniques* continue to be refined for resource exploration. Satellites and space-shuttle crews photograph the Earth's surface in many different wavelengths of energy, and careful analysis of satellite imagery applying geomatics techniques is proving to be of great help in prospecting. The economic returns from the satellite photography program will far outweigh its costs.

As the relation of plate tectonics to the distribution of resources such as metals and petroleum becomes clearer, selecting areas for exploring for these materials should become more sophisticated.

The Human Perspective

There is a tendency for humans to take a one-sided view toward geologic resources and the problems created by the extraction and transportation of those resources. The conflicts are between (1) maintaining or increasing our standard of living and raising the quality of life not only for ourselves, but also for poverty-stricken people throughout the world; (2) maintaining the environment; and (3) making sure that we do not deprive future generations of the resources that sustain us.

The extreme position for each of the three concerns could be stated as follows: (1) Extreme exploiter: "Let's mine what we can and make ourselves as rich as possible now. Technological breakthroughs will provide for future generations. Damage to the environment due to extraction is insignificant (or it's where it doesn't bother me)." (2) Extreme environmentalist: "Any mine or oil well does environmental damage and therefore should not be permitted. We can maintain our lifestyles by recycling or by leading less technologically dependent lifestyles." (3) Extreme conservationist: "Let's not mine anything now because there are many future generations that need to rely on these resources."

You probably agree that none of the three extreme positions is reasonable. The middle ground among the three is where almost everyone thinks we should be. The challenge is deciding where is the middle ground. Should we lean toward more exploitation and away from environmental concerns in order that underdeveloped countries can raise their standards of living? Should we minimize mining and energy consumption so as to reduce any impact on air pollution or wildlife? Our hope is that we can at least understand the perspective of those who may disagree to strike a balance and try to deal with each case with enlightenment. Your understanding of geoscience is an important step in your being able to help resolve moral dilemmas that we face in which there is no ideal solution.

SUMMARY

Geologic resources include energy resources, metals, and non metallic resources. All are nonrenewable, except groundwater.

Reserves are known deposits that can be legally and economically recovered now—the short-term supply. *Resources* include reserves as well as other known and undiscovered deposits that may be economically extracted in the future.

Petroleum (*oil* and *natural gas*) supplies almost two-thirds of the energy used by the United States.

The occurrence of oil and natural gas is limited to regions having these conditions together: *source rocks*, *reservoir rocks*, *traps*, such as anticlines, faults, stratigraphic traps, and salt domes, and *thermal maturity* from burial.

Natural gas, *heavy crude*, *oil sand*, and *oil shale* may all help replace liquid petroleum in the future. Most of these resources are in western North America.

The United States and Canada have huge *coal* reserves, enough for centuries of use at the present rate. Coal, now used mostly for generating electricity, will probably be used more widely in the future as oil runs out.

The United States and Canada have ample *uranium* for their reactor programs, mostly in sandstones in western states and prairie provinces.

Metallic ores, which can be profitably mined, are often associated with igneous rocks, particularly their hydrothermal fluids, which can form *contact metamorphic deposits*, *hydrothermal veins*, *disseminated deposits*, and submarine *hot-spring deposits*. Iron occurs in sedimentary layers, and aluminum ores form from weathering.

Metallic ores form from hot springs at divergent plate boundaries, on the flanks of island arcs, and in belts on the edges of continents above subduction zones.

Ores are mined *underground* and also at the Earth's surface in *strip mines*, *open-pit mines*, and *placer mines*.

Metals are vital to an industrial economy, particularly *iron* for steel production and *copper* for electrical equipment.

Nonmetallic resources such as *sand and gravel* and *limestone* for crushed rock and cement are used in huge quantities. *Fertilizers*, *rock salt*, *gypsum*, *sulphur*, and *clays* are also widely used.

Substitutes, recycling, and conservation can help cut consumption of some resources but will not eliminate the need for finding new deposits.

Deep-ocean mining and increasingly sophisticated exploration techniques will help supply some of our future resource needs.

Terms to Remember

coal bed methane 309
crude oil 301
gas hydrates 308
geologic resources 299
heavy crude 309
natural gas 301
nonrenewable resources 299
oil field 303
oil pool 301
oil sand 309
oil shale 310
oil trap 301
ore 316
petroleum 301
reserves 299
reservoir rock 301
resources 299
source rock 301
strip mine 311

Testing Your Knowledge

Use the questions below to prepare for exams based on this chapter.

1. Name the three major classes of geologic resources. Give four examples of each class.
2. Discuss Canada's supplies and potential use of natural gas, heavy crude, oil sand, and uranium.
3. List in decreasing order of use the energy resources used in Canada. Discuss possible future trends in this ranking.
4. What geologic conditions are necessary for the accumulation of oil and natural gas?
5. Differentiate between *reserves* and *resources.* Can reserves be increased? Can resources be increased?
6. Compare oil reserves with coal reserves. What might this indicate for the future use of each?
7. Describe several ways in which ore deposits form. Which are the most important?
8. Describe four ways in which resources are mined.
9. Discuss the environmental effects of oil extraction and coal mining.
10. Discuss common uses for iron, copper, lead, zinc, and aluminum.
11. Describe the potential of substitutes, conservation, recycling, and deep-ocean mining for meeting an increasing need for geologic resources.
12. All of the following are nonrenewable resources except
    a. groundwater b. oil
    c. coal d. iron
13. The major source of energy for Canada is
    a. natural gas b. coal
    c. oil d. nuclear power
14. The major source of energy for the United States is
    a. natural gas b. coal
    c. oil d. nuclear power
15. Which is not a type of coal?
    a. peat b. lignite
    c. bituminous d. anthracite
16. Which metal would most likely be found in an ore deposit formed by crystal settling?
    a. copper b. gold
    c. silver d. chromium
17. Which metal would not be found in hydrothermal veins?
    a. aluminum b. lead
    c. zinc d. silver
    e. gold
18. Metal ore deposits have been found at all these tectonic settings except
    a. mid-oceanic ridges b. island arcs
    c. subduction zones d. mantle plumes
19. The main use of lead is in
    a. coins b. gasoline
    c. batteries d. pencils
20. What factors can increase reserves of the Earth's resources (choose all that apply)
    a. extraction of the resource b. new discoveries
    c. price controls d. new mining technology
21. The largest use of sand and gravel is
    a. glassmaking b. extraction of quartz
    c. construction d. ceramics
22. Oil accumulates when the following conditions are met (choose all that apply)
    a. source rock where oil forms b. permeable reservoir rock
    c. impermeable oil trap d. shallow burial
23. Coal forms
    a. by crystal settling b. through hydrothermal processes
    c. by compaction of plant material d. on the ocean floor

Exploring Web Resources

www.mcgrawhill.ca/olc/plummer

This is the dedicated Website for this book. You can go to it for new and updated information. The universal resource locators (URL) listed in this book are also given as links on the Website, making it easy to go to those Websites without typing in the URL. When you visit our Online Learning Centre, go to the Student Centre and then to the chapter of interest. Here you will find additional readings and media resources, as well as answers to the Testing Your Knowledge section, more quizzes, animation and video clips, and direct links to the sites listed in this book. Links to additional Websites can also be found. We have added questions for some of the sites to allow you to get the most of your exploration of the Web. Using the Web is an enjoyable way of enhancing your knowledge of geology.

www.klws.com/gold/gold.html

Gold prospecting site. This site is mainly aimed at the amateur gold prospector, but contains facts and information on gold.

www.NRCan.gc.ca/

Natural Resources Canada. Use this site to get information on Canada's mineral and energy resources.

http://minerals.usgs.gov/

U.S. Geological Survey Mineral Resources Program. Provides current information on occurrence, quality, quantity, and availability of mineral resources.

www.eia.doe.gov/

U.S. Energy Information Administration. Provides data, analysis, and forecasts of energy and issues related to energy.

www.api.org

American Petroleum Institute. Information on all aspects of petroleum from the industry's perspective.

www.energy.gov.ab.ca

Information on Alberta's energy resources including oil, natural gas, oil sands, coal, and minerals.

CHAPTER

13

Mass Wasting

What is mass wasting?
What makes a slope fail?
How does material move downslope?
What happens when landslides occur underwater?
How can we prevent landslides?

This chapter and chapters 14 through 18 are concerned with surficial processes, the interaction of rock, air, and water in response to gravity at or near the Earth's surface. Nearly all of the features we see as landforms—rounded or rugged mountains, river valleys, cliffs and beaches along seashores, caves, sand dunes, and so on—are products of surficial processes. Surficial processes involve weathering, erosion, transportation, and deposition. Subsequent chapters address the work of running water, groundwater (water that is beneath the surface), glaciers, wind, and ocean waves. This chapter is about the downward movement of masses of rock or loose material. When material on a hillside has weathered (the process described in chapter 8), it is likely to move downslope because of the pull of gravity. Soil or rock moving in bulk at the Earth's surface is called mass wasting. Landsliding is the best known type of mass wasting. In this chapter we describe how different types of mass wasting shape the land and alter the environment and what factors control the rate at which the processes operate. Understanding mass wasting and its possible hazards is particularly important in hilly or mountainous regions, but can also be important in areas of low topographic relief where quickclays are present.

Fallen rock fragments form talus fans at the base of steep mountain slopes surrounding Moraine Lake, Banff National Park, Canadian Rocky Mountains. *Photo by C.H. Eyles*

WHAT IS MASS WASTING?

You may recall from previous chapters that mountains are products of tectonic forces. Most mountains are associated with present or past converging plate boundaries. If tectonism were not at work, the surfaces of the continents would long ago have been reduced to featureless plains due to weathering and erosion. We consider the material on mountain slopes or hillsides to be out of equilibrium with respect to gravity. Because of the force of gravity, the various agents of erosion (moving water, ice, and wind) work to make slopes gentler and therefore increasingly more stable. The process of rock and sediment movement discussed in this chapter is *mass wasting*.

Mass wasting (also called *mass movement*) is movement in which bedrock, rock debris, or soil moves downslope in bulk, or as a mass, because of the pull of gravity. Mass wasting includes movement so slow that it is almost imperceptible (called *creep*) as well as **landslides,** a general term for the slow to very rapid descent of rock or soil.

Mass wasting affects humans in many ways. Its effects range from the devastation of a killer landslide (see box 13.1) to the nuisance of having a fence slowly pulled apart by soil creep. The cost in lives and property from landslides is surprisingly high; landslides are Canada's most destructive geological hazard. According to the U.S. Geological Survey, more people in the United States died from landslides during the final three months of 1985 than were killed during the last 20 years by all other geologic hazards, such as earthquakes and volcanic eruptions. Over time, landslides have cost North Americans triple the combined costs of earthquakes, hurricanes, floods, and tornadoes. In Canada, landslides cause between $100 million and $200 million in damage annually and have caused more than 600 deaths since 1840. One person was killed, two houses were destroyed, and 70 homes were evacuated as a result of a landslide in North Vancouver on January 19, 2005. In 2003, damage from landslides in the U.S. reached $2.3 billion and an average of 25 to 50 lives are lost annually. In many cases of mass wasting, a little knowledge of geology, along with appropriate preventive action, could have averted destruction.

Classification of Mass Wasting

A number of systems are used by geologists, engineers, and others for classifying mass wasting, but none has been universally accepted. Some are very complex and useful only to the specialist.

The classification system used here and summarized in table 13.1 is based on (1) rate of movement, (2) type of material, and (3) nature of the movement.

Rate of Movement

A landslide (debris avalanche) like the one in Peru described in box 13.1 clearly involves rapid movement. Just as clearly, movement of soil at a rate of less than a centimetre a year is slow movement. There is a wide range of velocities between these two extremes.

Type of Material

Mass wasting processes are usually distinguished on the basis of whether the descending mass started as bedrock (as in a rockslide) or as debris. The term **debris,** as applied to mass wasting processes, means any unconsolidated material at the Earth's surface, such as soil and rock fragments (weathered or unweathered) of any size (see box 13.2 for a description of the movement of a different type of material). Mud has a high content of water, clay, and silt. The amount of water (or ice and snow) in a descending mass strongly influences the rate and type of movement.

TABLE 13.1 Some Types of Mass Wasting[1]

	Slowest ——→	Increasing Velocities ——→		Fastest
Type of Movement	Less than 1 centimetre/year	1 millimetre/day to 1 kilometre/hour	1 to 5 kilometre/hour	*Velocities generally greater than 4 kilometres/hour*
		←—— Debris Flow ——→		
Flow	Creep ——→ (debris)	Earthflow ——→ Gelifluction	Mudflow ——→ Quickclay flow	Debris avalanche (debris) ——→ Rock avalanche (bedrock)
Slide		←—— Debris slide or earthslide ——→		
		←—— Rockslide (bedrock) ——→		
Fall				Rockfall (bedrock)
		←—— "Landslides" ——→		

1. The type of material at the start of movement is shown in parentheses. Rates given are typical velocities for each type of movement.

ENVIRONMENTAL GEOLOGY 13.1

Disaster in the Andes

As a result of a tragic combination of geological conditions and human ignorance of geologic hazards, one of the most devastating landslides (a debris avalanche) in history destroyed the town of Yungay in Peru in 1970. Yungay was one of the most picturesque towns in the Santa River Valley, which runs along the base of the highest peaks of the Peruvian Andes. Heavily glaciated Nevado Huascarán, 6,663 metres above sea level, rises steeply above the populated narrow plains along the Santa River.

In May 1970, an earthquake, centred offshore from Peru about 100 kilometres from Yungay, occurred. Although the tremors in this part of the Andes were no stronger than those that have caused only light damage to cities in North America, many poorly constructed homes collapsed. Because of the steepness of the slopes, thousands of small rockfalls and rockslides were triggered.

The greatest tragedy began when a slab of glacier ice about 800 metres wide, perched near the top of Huascarán, was dislodged by the shaking. (A few years earlier American climbers returning from the peak had warned that the ice looked highly unstable. The Peruvian press briefly noted the danger to the towns below, but the warning was soon forgotten.)

The mass of ice rapidly avalanched down the extremely steep slopes, breaking off large masses of rock debris, scooping out small lakes and loose rock that lay in its path. Eye-witnesses described the mass as a rapidly moving wall the size of a ten-storey building. The sound was deafening. More than 50 million cubic metres of muddy debris travelled 3.7 kilometres vertically and 14.5 kilometres horizontally in less than four minutes, attaining speeds between 200 and 435 kilometres per hour. The main mass of material travelled down a steep valley until it came to rest blocking the Santa River and burying about 1,800 people in the small village of Ranrahirca (box figure 1). A relatively small part of the mass of mud and debris that was moving especially rapidly shot up the valley sidewall at a curve and overtopped a ridge. The mass was momentarily airborne before it fell on the town of Yungay, completely burying it under several metres of mud and loose rock. Only the top of the church and tops of palm trees were visible, marking where the town centre was buried (box figure 2). Ironically, the cemetery was not buried because it occupied the high ground. The few survivors were people who managed to run to the cemetery.

The estimated death toll at Yungay was 17,000. This was considerably more than the town's normal population because it was Sunday, a market day, and many families had come in from the country.

For several days after the slide the debris was too muddy for people to walk on, but within three years grass had grown over the site. Except for the church steeple and the tops of palm trees that still protrude above the ground, and the crosses erected by families of those buried in the landslide, the former site of Yungay appears to be a scenic meadow overlooking the Santa River. The U.S. Geological Survey and Peruvian geologists found evidence that Yungay itself had been built on top of debris left by an even bigger slide in the recent geologic past. More slides will almost surely occur here in the future.

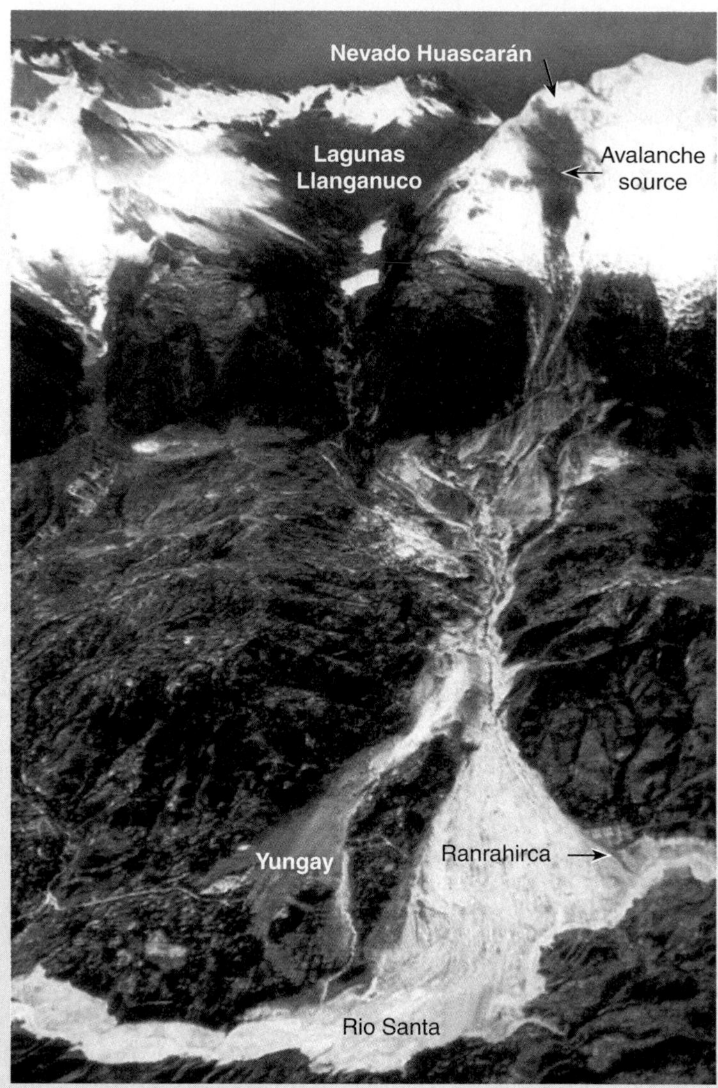

BOX 13.1 ■ FIGURE 1

Air photo showing the 1970 debris avalanche in Peru, which buried Yungay. The main mass of debris destroyed the small village of Ranrahirca.

Photo by Servicio Aerofotografico de Peru, courtesy of U.S. Geological Survey

Further Reading

Ericksen, G. E., G. Plafker, and J. Fernandez Concha, 1970. *Preliminary report on the geologic events associated with the May 31, 1970, Peru earthquake.* U.S. Geological Survey Circular 639.

BOX 13.1 ■ FIGURE 2

(A) Yungay is completely buried, except for the cemetery and a few houses on the small hill in the lower right of the photograph. (B) Behind the palm trees is the top of a church buried under 5 metres of debris at Yungay's central plaza. (C) Three years later.
Photos A and B by George Plafker, U.S. Geological Survey; photo C by C. C. Plummer

Type of Movement

In general, the type of movement in mass wasting can be classified as flow, slide, or fall (figure 13.1). A **flow** implies that the descending mass is moving downslope as a viscous fluid. **Slide** means the descending mass remains relatively intact, moving along one or more well-defined surfaces. A **fall** occurs when material free-falls or bounces down a cliff.

Two kinds of slide are shown in figure 13.1. In a **translational slide,** the descending mass moves along a plane approximately parallel to the slope of the surface. A **rotational slide** (also called a *slump*) involves movement along a curved surface, the upper part moving downward while the lower part moves outward.

WHAT MAKES A SLOPE FAIL?

Controlling Factors in Mass Wasting

Table 13.2 summarizes the factors that influence the likelihood and the rate of movement of mass wasting. The table makes apparent some of the reasons why the landslide (a debris avalanche) in Peru (box 13.1) occurred and why it moved so rapidly:

ENVIRONMENTAL GEOLOGY 13.2

Garbage on the Move!

Natural sediments are not the only materials prone to mass movement. In April 1991, a massive landslide occurred on the margin of a roadway leading to the Scarborough Bluffs on the northern shore of Lake Ontario, blocking the roadway for several weeks (box figure 1). The slide involved the downslope movement of a 25-metre-thick pile of municipal garbage that had been dumped into the Brimley Road Ravine between 1960 and 1968. The material was originally dumped there to both solve a garbage disposal problem and to help slow coastal erosion by covering exposed sediments. However, oversteepening of the ravine sideslopes occurred in the 1970s, when a road was built along the ravine to access the Lake Ontario shoreline. Slow downslope creep of the sideslopes was a persistent problem but mass failure did not occur until heavy rainfall following snowmelt in the spring of 1991 saturated the sideslopes to the point of failure. Previously installed drainage measures were found to be inoperative! The slide materials consisted predominantly of garbage, and significant volumes of contaminated leachate were released during the failure (box figure 2). The cost of cleaning up this garbage slide was more than $2.1 million. Landfill consisting of garbage is particularly prone to mass movement as it is poorly consolidated and contains large pore spaces (voids) that fill with water. Even on low-angle slopes, landfill materials can easily liquefy in the event of earthquakes.

BOX 13.2 ■ FIGURE 1

Oblique air photo of the 1991 landslide on the margins of the Brimley Road Ravine (area of slide is shown by dashed lines).

Photo by Nick Eyles

BOX 13.2 ■ FIGURE 2

Geoscientist examines landslide material, consisting primarily of garbage dumped into the ravine during the 1960s. Pieces of newspaper within the landslide debris were not affected by decomposition and could still be read.

Photo by Nick Eyles

FIGURE 13.1
Flow, slide, and fall.

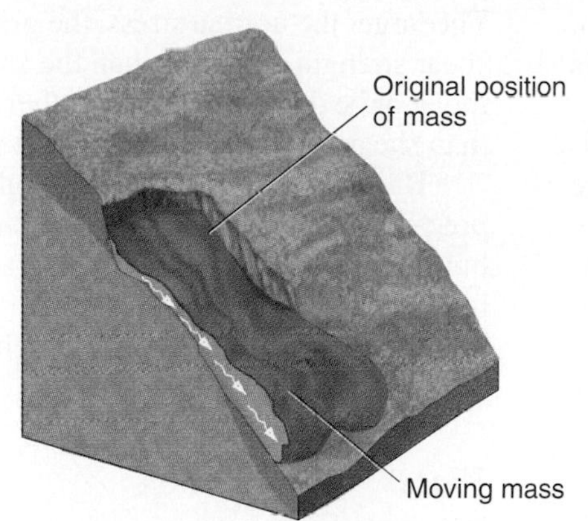

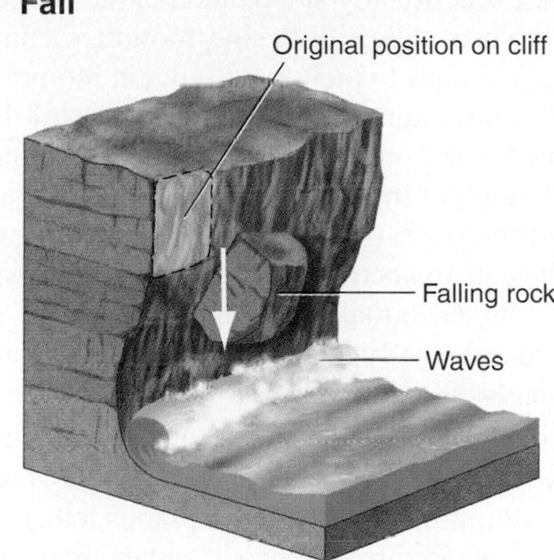

TABLE 13.2 Summary of Controls of Mass Wasting

Driving Force: Gravity

Contributing Factors	Most Stable Situation	Most Unstable Situation
Slope angle	Gentle slopes or horizontal surface	Steep or vertical
Local relief	Low	High
Thickness of debris over bedrock	Thin debris	Thick debris
Bedrock type	Massive, unjointed	Layered, stratified, jointed
Orientation of planes of weakness in bedrock	Planes at right angles to hillside slopes	Planes parallel to hillside slopes
Climatic factors:		
Ice	Temperature stays above freezing	Repeated freezing and thawing
Water in soil or debris	Film of water around fine particles	Episodic saturation of debris with water
Precipitation	Frequent but light rainfall or snow	Long periods of drought with rare episodes of heavy precipitation, snow melt
Vegetation	Heavily vegetated	Sparsely vegetated

Triggering Mechanisms: (1) earthquakes; (2) weight added to upper part of a slope; (3) undercutting of bottom of slope; (4) heavy rainfall.

(1) The slopes were exceptionally steep, and (2) the **relief** (the vertical distance between valley floor and mountain summit) was great, allowing the mass to pick up speed and momentum. (3) Water and ice not only added weight to the mass of debris but they also saturated it and reduced the shear strength, making it more fluid. (4) Abundant loose rock and debris were available in the course of the slide. (5) Where the slide began, there were no plants with roots to anchor loose material on the slope. Finally, (6) the area is earthquake prone. Although the slide would have occurred eventually even without an earthquake, it was triggered by an earthquake.

Other factors influence susceptibility to mass wasting as well as its rate of movement. The orientation of planes of weakness in bedrock (bedding planes, foliation planes, etc.) is important if the movement involves bedrock rather than debris. Fractures or bedding planes oriented so that slabs of rock can slide easily along these surfaces greatly increase the likelihood of mass wasting.

Climatic controls inhibit some types of mass wasting and aid others (table 13.2). Climate influences how much and what kinds of vegetation grow in an area and what type of weathering occurs. Infrequent but heavy rainfall aids mass wasting because it quickly saturates debris that lacks the protective vegetation found in wetter climates. By contrast, a climate in which rain drizzles intermittently much of the year results in vegetation that tends to inhibit mass wasting. In cold climates, freezing and thawing contribute to downslope movement.

Gravity

The driving force for mass wasting is gravity. Figures 13.2*A* to *C* show gravity acting on a block on a slope. The length of the red vertical arrow is proportional to the force—the heavier the material, the longer the arrow. When discussing the effect of a force on a rock body, it is important to consider the area over which the force is distributed. We can therefore resolve the effect of gravity into two component stresses (stress = force per unit area), indicated by the black arrows. One, the *normal stress,* is perpendicular to the slope and is the component of gravity that tends to hold the block in place. The other, called the **shear stress,** is parallel to the slope, and indicates the block's ability to move. The length of the arrows is proportional to the strength of each stress. The steeper the slope (and the heavier the block), the greater is the shear stress and the greater the tendency of the block to slide. Friction counteracts the shear stress. *Shear resistance* (represented by the brown arrow) is the stress that would be needed to move the block. If that arrow is larger than the arrow representing shear stress (as in figure 13.2*A*), the block will not move. The magnitude of the shear resistance (and the length of the brown arrow) is a function of friction and the size of the normal stress. The brown arrow will be shorter (and the shear resistance lower) if water or ice reduces the friction beneath the block. If the shear resistance becomes lower than the shear stress, the block will slide. Similar forces act on debris on a hillside (figure 13.2*C*). The resistance to movement or deformation of that debris is its **shear strength.** Shear strength is controlled by factors such as the cohesiveness of the material, friction between particles, pore pressure of water, and the anchoring effect of plant roots. Shear strength is also related to the normal stress. The larger the normal stress, the greater the shear strength. If the shear strength is greater than the shear stress, the debris will not move or be deformed. On the other hand, if shear strength is less than shear stress, the debris will flow or slide.

Building a heavy structure high on a slope demands special precautions. To prevent movement of both the slope and the building, pilings may have to be sunk through the debris, perhaps even into bedrock. Developers may have to settle for fewer buildings than planned if the weight of too many structures will make the slope unsafe.

Water

Water is a critical factor in mass wasting. When debris is saturated with water (as from heavy rain or melting snow), it becomes heavier and less viscous and is more likely to flow downslope. The added gravitational shear stress from the increased weight, however, is probably less important than the reduction in shear strength. This is due to increased *pore pressure* in which water forces grains of debris apart.

Paradoxically, a small amount of water in soil can actually prevent downslope movement. When water does not completely fill the pore spaces between the grains of soil, it forms a thin film around each grain (as shown in figure 13.3). Loose grains adhere to one another because of the *surface tension* created by the film of water, and shear strength increases. Surface tension of water between sand grains is what allows you to build a sand castle. The sides of the castle can be steep or even vertical because surface tension holds the moist sand grains in place. Dry sand cannot be shaped into a sand castle because the sand grains slide back into a pile that generally slopes at an angle of about 30 to 35 degrees from the horizontal. On the other hand, an experienced sand-castle builder also knows that it is impossible to build anything with sand that is too wet. In this case the water completely occupies the pore space between sand grains, forcing them apart and allowing them to slide easily past one another. When the tide comes in, or someone pours a pail of water on your sand castle, all you have is a puddle of wet sand.

Similarly, as the amount of water in debris increases, rate of movement tends to increase. Damp debris may not move at all, whereas moderately wet debris moves slowly downslope. Slow types of mass wasting, such as creep, are generally characterized by a relatively low ratio of water to debris. Mudflows always have high ratios of water to debris. A mudflow that continues to gain water eventually becomes a muddy stream.

Triggering Mechanisms

A sudden event may trigger mass wasting of a hillside that is unstable. Eventually, movement would occur without the triggering event if conditions slowly became more unstable.

Earthquakes commonly trigger landslides. The 1970 debris avalanche in Peru (see box 13.1) was one of thousands of landslides, mostly small ones, triggered by a quake. The worst damage

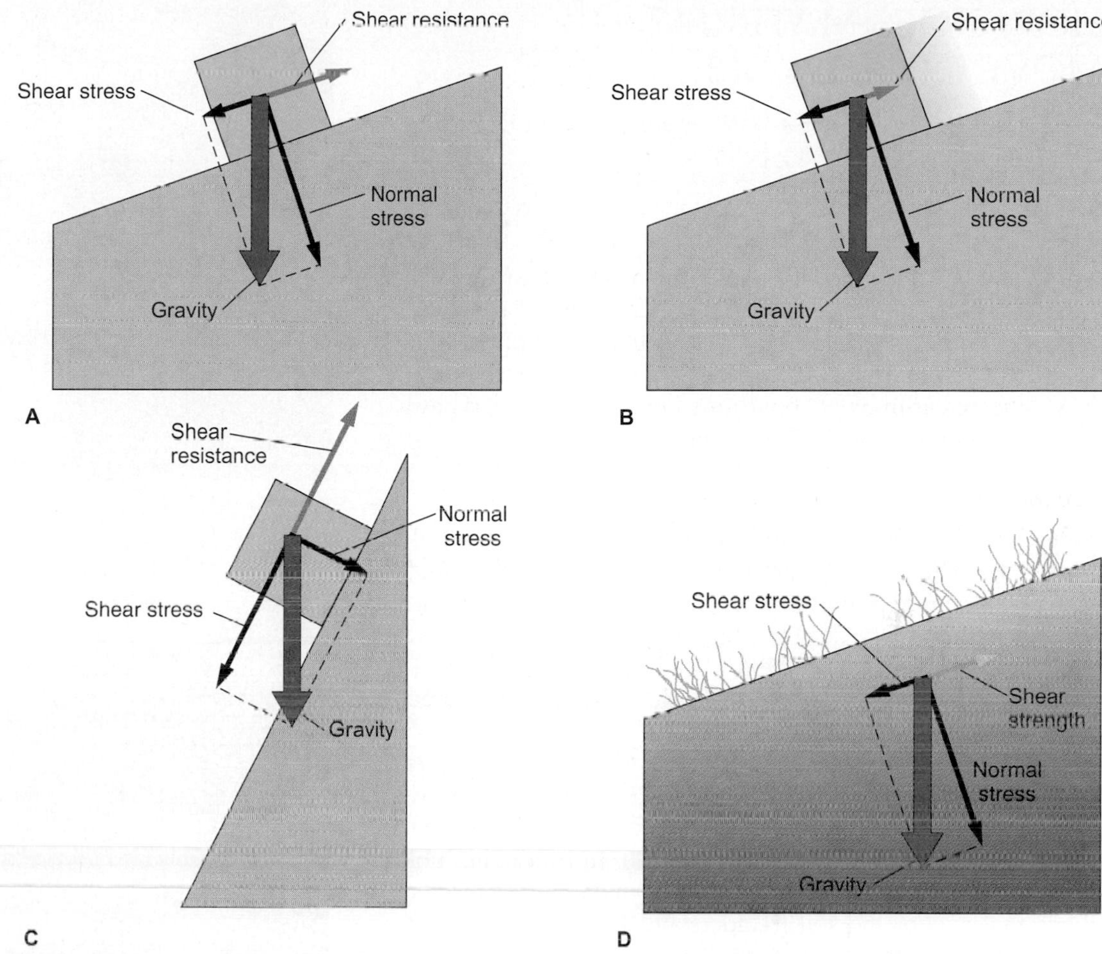

FIGURE 13.2

Relationship of shear stress and normal stress to gravity. (*A*) A block on a gently inclined slope in which the shear resistance (brown arrow) is greater than the shear stress; therefore, the block will not move. (*B*) The same situation as in A, except that the shear resistance is less than the shear stress; therefore, the block will be moving. (*C*) A block on a steep slope. Note how much greater the shear stress is and how much larger the shear resistance has to be to prevent the block from moving. (*D*) Forces acting at a point in debris (soil). Shear strength is represented by a yellow arrow. If that arrow is longer than the one represented by shear stress, debris at that point will not slide or be deformed.

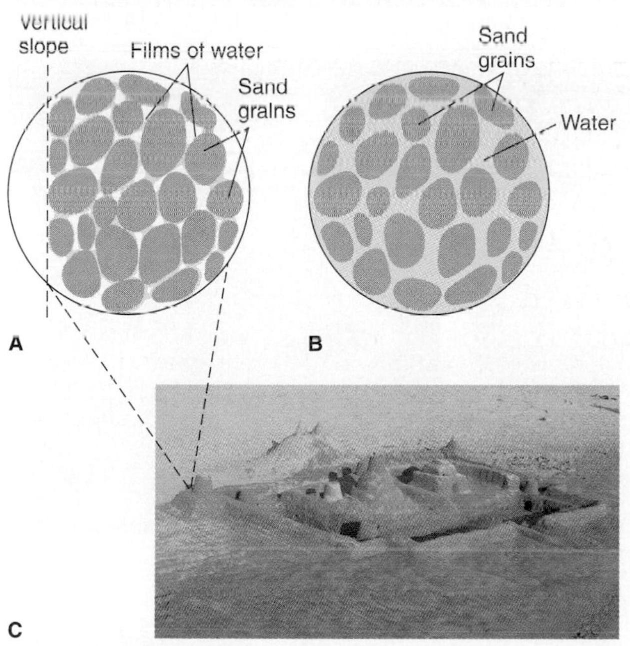

FIGURE 13.3

The effect of water in sand. (*A*) Unsaturated sand held together by surface tension of water. (*B*) Saturated sand grains forced apart by water; mixture flows easily. (*C*) A sand castle in Acapulco, Mexico.

Photo by C. C. Plummer

from California's 1989 Loma Prieta earthquake was in the nearly flat areas of San Francisco's Marina District (where fires from broken gas mains ravaged buildings) and across the bay in Oakland. In Oakland, ground failure occurred beneath a two-tiered freeway, the upper level of which collapsed onto the lower level, crushing many vehicles (see chapter 3). Without the earthquake, movement may not have taken place for decades or centuries, and then mass wasting would have been slow enough that corrective measures could have been taken to stabilize the ground.

Landslides often are triggered by heavy rainfall. The sudden influx of voluminous water quickly adds weight and increases pore pressure in material. Heavy persistent rainfall in British Columbia and in normally dry southern California caused numerous landslides in January 2005. Quickclay slides, discussed later in this chapter, are also commonly triggered by heavy rainfall.

Construction sometimes triggers mass wasting. The extra weight of buildings on a hillside can cause a landslide, as can bulldozing a road cut at the base of a slope.

HOW DOES MATERIAL MOVE DOWNSLOPE?

The common types of mass wasting are shown in table 13.1. Here we will describe each type in detail.

Flow

Flow occurs when motion is taking place within a moving mass of unconsolidated or weakly consolidated material. Grains move relative to adjacent grains, or motion takes place along closely spaced, discrete fractures. The common varieties of flow—creep, earthflow, debris flow, mudflow, and debris avalanche—are described in this section.

Creep

Creep is very slow, downslope movement of soil or unconsolidated debris. Shear forces are only slightly greater than shear strengths. The rate of movement is usually less than a centimetre per year and can be detected only by observations taken over months or years. When conditions are right, creep can take place along nearly horizontal slopes. Some indicators of creep are illustrated in figures 13.4 and 13.5.

Two factors that contribute significantly to creep are water in the soil and daily cycles of freezing and thawing. As we have said, water-saturated ground facilitates movement of soil downhill. What keeps downslope movement from becoming more rapid in most areas is the presence of abundant grass or other plants that anchor the soil. (Understandably, overgrazing can severely damage sloping pastures.)

Several processes contribute to soil creep. Particles are displaced in cycles of wetting and drying. The soil tends to swell

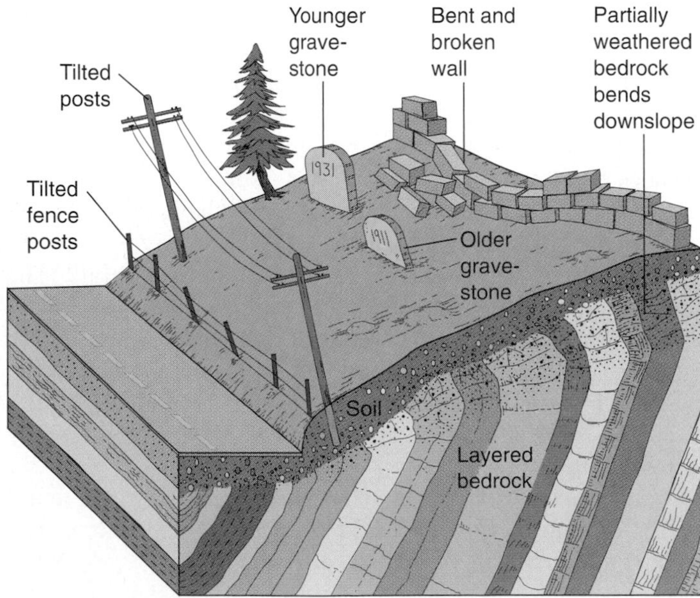

FIGURE 13.4
Indicators of creep.
After C. F. S. Sharpe

A

B

C

FIGURE 13.5
(*A*) Tilted gravestones in a churchyard at Lyme Regis, England (someone probably straightened the one upright gravestone). Grassy slope is inclined gently to the left. (*B*) Soil and partially weathered, nearly vertical sedimentary strata have crept downslope. (*C*) As a young tree grew, it grew vertically but was tilted by creeping soil. As it continued to grow, its new, upper part would grow vertically but in turn would be tilted.
Photo A by C. C. Plummer; Photo B by Frank M. Hanna; Photo C © Parvinder Sethi

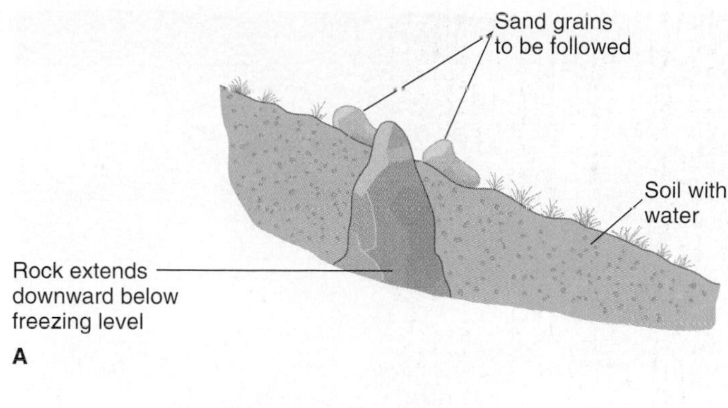

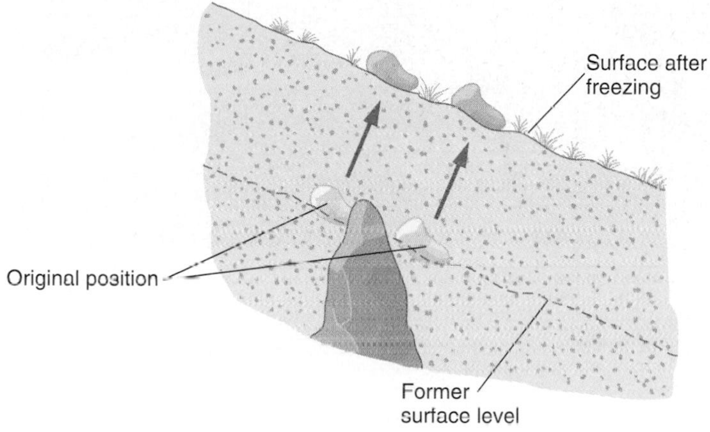

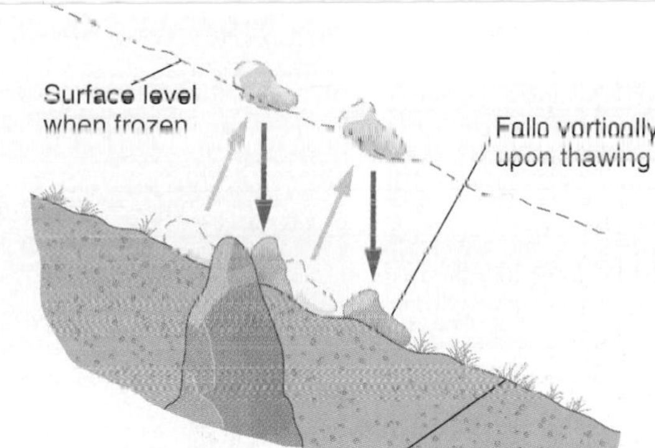

FIGURE 13.6

Downslope movement of soil, illustrated by following two sand grains (each less than a millimetre in size) during a freeze–thaw cycle. Movement downward might not be precisely vertical if adjacent grains interfere with each other.

when wet and contract when dry so that movement takes place in a manner similar to that of a freeze–thaw cycle. Burrowing worms and other creatures "stir" the soil and facilitate movement under gravity's influence. The process is more active where the soil freezes and thaws during part of the year. During the winter over most of Canada and the northeastern United States, the temperature may rise above and fall below freezing once a day. When there is moisture in the soil, each freeze–thaw cycle moves soil particles a minute amount downhill, as shown in figure 13.6 (see also figure 8.22).

Earthflow

In an **earthflow,** debris moves downslope as a viscous fluid; the process can be slow or rapid. Earthflows usually occur on hillsides that have a thick cover of debris, often after heavy rains have saturated the soil. Typically, the flowing mass remains covered by a blanket of vegetation, with a *scarp* (steep cut) developing where the moving debris has pulled away from the stationary upper slope.

A landslide may be entirely an earthflow, as in figure 13.7*A*, with debris particles moving past one another roughly parallel to the slope. Commonly, however, rotational sliding (slumping) takes place above the earthflow, as in figure 13.8. This example is a *debris slide* (upper part) and an *earthflow* (lower part). In such cases, debris remains in a relatively coherent block or blocks that rotate downward and outward, forcing the debris below to flow.

A hummocky lobe usually forms at the toe or front of the earthflow where debris has accumulated (figure 13.7*B*). An earthflow can be active over a period of hours, days, or months; in some earthflows intermittent slow movement continues for years. In March 1995, following an extraordinarily wet year, a slump-earthflow destroyed or severely damaged 14 homes in the southern California coastal community of La Conchita (figure 13.8*B*). In January 2005, following 15 days of record-breaking rainfall, around 15 percent of the 1995 landslide remobilized (figure 13.8*C*). Rapidly moving flow of soil killed 10 people and severely damaged or destroyed 36 houses. Because future landslides are likely, the town of La Conchita was abandoned. For details of the La Conchita landslides go to http://pubs.usgs.gov/of/2005/1067/pdf/OF2005-1067.pdf.

People can trigger earthflows by adding too much water to soil from septic tank systems or by overwatering lawns. In one case, in Los Angeles, a man departing on a long trip forgot to turn off the sprinkler system for his hillside lawn. The soil became saturated, and both house and lawn were carried downward on an earthflow whose lobe spread out over the highway below.

Earthflows, like other kinds of landslides, can be triggered by undercutting at the base of a slope. The undercutting can be caused by waves breaking along shorelines or streams eroding and steepening the base of a slope. Along coastlines, mass wasting commonly destroys buildings (figure 13.9). Entire housing developments and expensive homes built for a view of the ocean are lost. A home buyer who knows nothing of geology may not realize that the sea cliff is there because of the relentless erosion of waves along the shoreline. Nor is the person likely to be aware that a steepened slope creates the potential for landslides.

Bulldozers can undercut the base of a slope more rapidly than wave erosion, and such oversteepening of slopes by human activity has caused many landslides. Unless careful engineering measures are taken at the time a cut is made, road-cuts or platforms carved into hillsides for houses may bring about disaster (see figure 13.18).

GELIFLUCTION

Another type of earthflow commonly associated with colder climates is gelifluction. **Gelifluction** is the flow of water-saturated debris over impermeable material. Because the impermeable

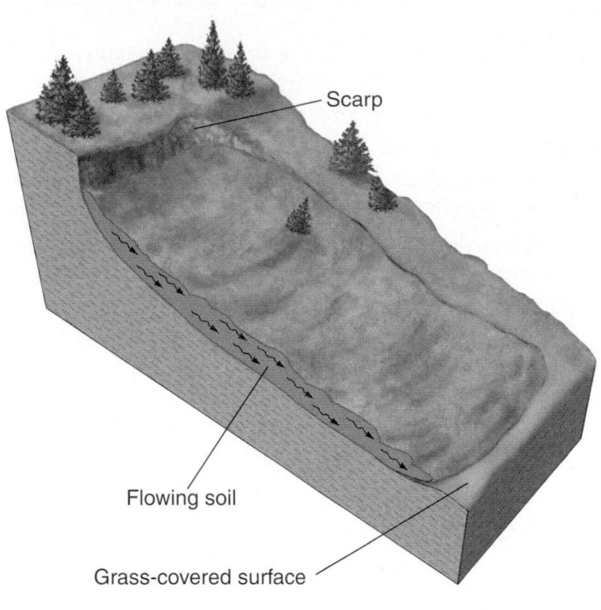

FIGURE 13.7

(*A*) Earthflow. Soil flows beneath a blanket of vegetation. (*B*) Hummocky surface of an earthflow at Pavillon, in British Columbia.

Photo by Nick Eyles

FIGURE 13.8

Earthflow with rotational sliding (slumping). (*A*) Debris in the upper part of the diagram remained mostly intact as it rotated downward in blocks (debris slide). Debris in the lower portion flowed (earthflow). (*B*) A slump-earthflow destroyed several houses in March, 1995, at La Conchita, California. (*C*) After heavy rains in January 2005 part of the landslide remobilized and flowed into the town of La Conchita.

Photo B by Robert L. Schuster, U.S. Geological Survey; Photo C © Kevork Djansezian/AP/Wide World Photos

FIGURE 13.9

Coastal erosion threatens urban development along the northern shore of Lake Ontario.

Photo by Nick Eyles

material beneath the debris prevents water from draining freely, the debris between the vegetation cover and the impermeable material becomes saturated (figure 13.10). Even a gentle slope is susceptible to movement under these conditions.

The impermeable material beneath the saturated soil can be either impenetrable bedrock or, as is more common, **permafrost,** ground that remains frozen for many years (see chapter 16). Most gelifluction takes place in areas of permanently frozen ground, such as in northern Canada and Alaska. Permafrost occurs at depths ranging from a few centimetres to a few metres beneath the surface. The ice in permafrost is a cementing agent for the debris. Permafrost is as solid as concrete.

Above the permafrost is a zone that, if the debris is saturated, is frozen during the winter and indistinguishable from the underlying permafrost. When this zone (called the *active layer*) thaws during the summer, the water, along with water from rain and runoff, cannot percolate downward through the permafrost, and so the slopes become susceptible to gelifluction.

As the water-saturated debris moves downslope, it may form gelifluction lobes or terraces (figure 13.11). This process eventually produces a landscape of smoothed gentle slopes as debris is stripped from the upper parts of the slopes and accumulates in low areas.

Debris Flow

A **debris flow** is a flow in which coarse material (gravel, boulders) is predominant. A debris flow can be like an earthflow and travel relatively short distances to the base of a slope or, if there is a lot of water, a debris flow can behave like a mudflow and flow rapidly, travelling considerable distance in a channel. Rapidly moving debris flows can be extremely devastating.

Debris flows (and mudflows) are a major problem in the Canadian Cordillera, where heavy rainfall mobilizes unstable debris on steep slopes. The 1983 Alberta Creek disaster was cause by a debris flow triggered by an intense rainstorm that swept though the community of Lions Bay on the eastern shore of Howe Sound, B.C., killing two people. Forest harvesting and the improper construction of logging roads, which increases surface water runoff and pore water pressure in soils, has also been blamed for triggering debris flows in mountainous areas of

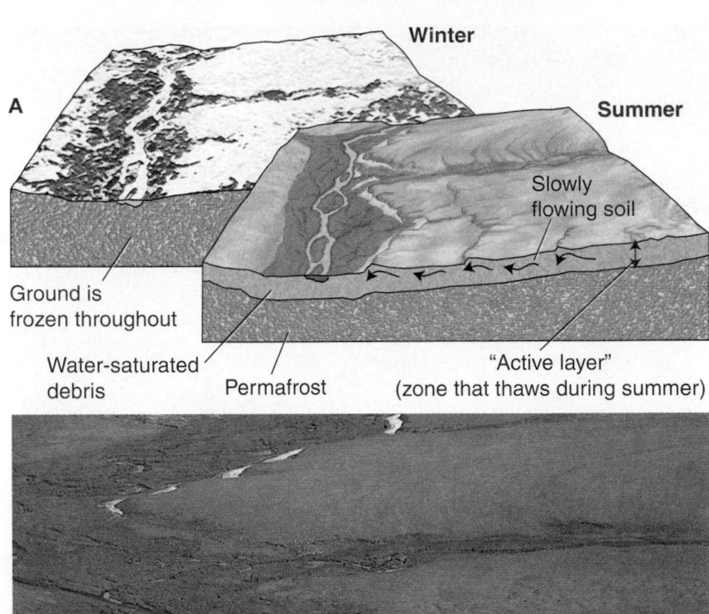

FIGURE 13.10

(A) Gelifluction due to downslope movement of water-saturated debris in the "active layer" that thaws during the summer. (B) Gelifluction lobes in northwestern Alaska.
Photo by C. C. Plummer

FIGURE 13.12

Debris flow scars on a steep mountain slope near Wahleach Lake, Chilliwack, B.C. The debris flows were triggered by heavy rains that caused failure of embankments along the logging road.
Photo by John Clague, Simon Fraser University, 2002

FIGURE 13.11

Gelifluction lobes creeping downslope and covering sandy marine raised beaches, south shore of Melville Sound, NWT.
Photo by C. H. Eyles

British Columbia (figure 13.12). In areas of northern Canada, debris flows are commonly caused by disturbance of permafrost (permanently frozen ground) along stream and river banks (see chapter 16).

The steep mountains that rise above Los Angeles and other southern California urban centres are also sources of sometimes catastrophic debris flows. In December 2003, dozens of debris flows took place in the San Bernadino Mountains. Widespread forest fires scorched southern California during the hot, dry summer, killing trees and ground cover. Heavy rains in late December saturated the soil, triggering a series of debris flows that destroyed a church camp, killing 14 people.

Mudflow

A **mudflow** is a flowing mixture of debris and water, usually moving down a channel (figure 13.13). It can be visualized as a stream with the consistency of a thick milkshake. Most of the solid particles in the slurry are clay and silt (hence the muddy appearance), but coarser sediment commonly is part of the mixture. A slurry of debris and water forms after a heavy rainfall or other influx of water, and begins moving down a slope. Most mudflows quickly become channelled into valleys. They then

FIGURE 13.13

Man examining a 75-metre-long bridge on Washington State highway 504, across the North Fork of the Toutle River. The bridge was washed out by mudflow during the May 18, 1980 eruption of Mount St. Helens. The steel structure was carried about 0.5 km downstream and partially buried by the mudflow

Photo by Robert L. Schuster, U.S. Geological Survey

FIGURE 13.14

Mudflows active on steep slopes of unconsolidated Quaternary-age sediment. Note fan shape of main mudflow lobe. Ice pick is approximately 1 metre.

Photo by Nick Eyles

move downvalley like a stream—except that, because of the heavy load of debris, they are more viscous. Mud moves more slowly than a stream but, because of its high viscosity, can transport boulders, automobiles, and even locomotives. Houses in the path of a mudflow will be filled with mud, if not broken apart and carried away.

Mudflows are most likely to occur in places where debris is not protected by a vegetative cover. For this reason, mudflows are more likely to occur in arid regions than in wet climates. A hillside in a desert environment, where it may not have rained for many years, may be covered with a blanket of loose material. With sparse desert vegetation offering little protection, a sudden thunderstorm with drenching rain can rapidly saturate the loose debris and create a mudflow in minutes.

Mudflows frequently occur on young volcanoes that are littered with ash. Water from heavy rains mixes with pyroclastic debris, as at Mount Pinatubo in 1991. (More than a decade after the big eruption, mudflows near Mount Pinatubo continue to cost lives and destroy property.) Or the water can come from glaciers that are melted by lava or hot pyroclastic debris, as occurred at Mount St. Helens in 1980 (figure 13.13) and at Colombia's Nevado del Ruiz in 1985, which cost 23,000 lives. Like debris flows, mudflows also occur after forest fires destroy slope vegetation that normally anchors soil in place.

Steep, unvegetated cliffs composed of unconsolidated, Quaternary-age sediments are common along the eastern and western seaboards of North America and around the shorelines of the Great Lakes basins. These cliffs are particularly prone to mass movement by mudflow processes (refer to figure 13.14).

The year 1978 was particularly bad for debris flows in southern California. One mudflow roared through a Los Angeles suburb carrying almost as many cars as large boulders. A sturdily built house withstood the onslaught but began filling with muddy debris. Two of its occupants were pinned to the wall of a bedroom and could do nothing as the room filled slowly with mud. The mud stopped rising just as it was reaching their heads. Hours later they were rescued. (John McPhee's *The Control of Nature,* listed in *Exploring Web Resources* on this book's Website [www.mcgrawhill.ca/olc/plummer], has a highly readable account of the 1978 debris flows in southern California.) In 2005, heavy winter rains caused additional extensive and damaging debris flows in southern California (see http://pubs.usgs.gov/fs/2005/3107/.

QUICKCLAYS

Clays deposited in marine environments in northern latitudes are particularly susceptible to downslope movement by a combination of flow and slide processes when exposed on the land surface. These clays are called "quickclays" or sensitive clays and are common in coastal lowlands of both eastern and western Canada (figure 13.15*A*). **Quickclays** are originally deposited in saline waters and the salt forms an important part of the clay structure. When these clays are exposed on the land surface, fresh groundwaters remove the salt from the clays by leaching and cause dramatic weakening of the sediment structure. If the weakened clays are "shocked" (i.e., by earthquakes), overloaded, or contain extremely large amounts of water, they collapse and liquefy, flowing downslope quite rapidly (up to tens of metres per second). A number of quickclay failures have occurred in the St. Lawrence Lowlands of eastern Canada, causing millions of dollars of damage and numerous fatalities. In 1971, 31 people were killed when their houses were engulfed by quickclays mobilized during the St-Jean-Vianney landslide in Quebec. The Lemieux quickclay landslide of 1993 involved the

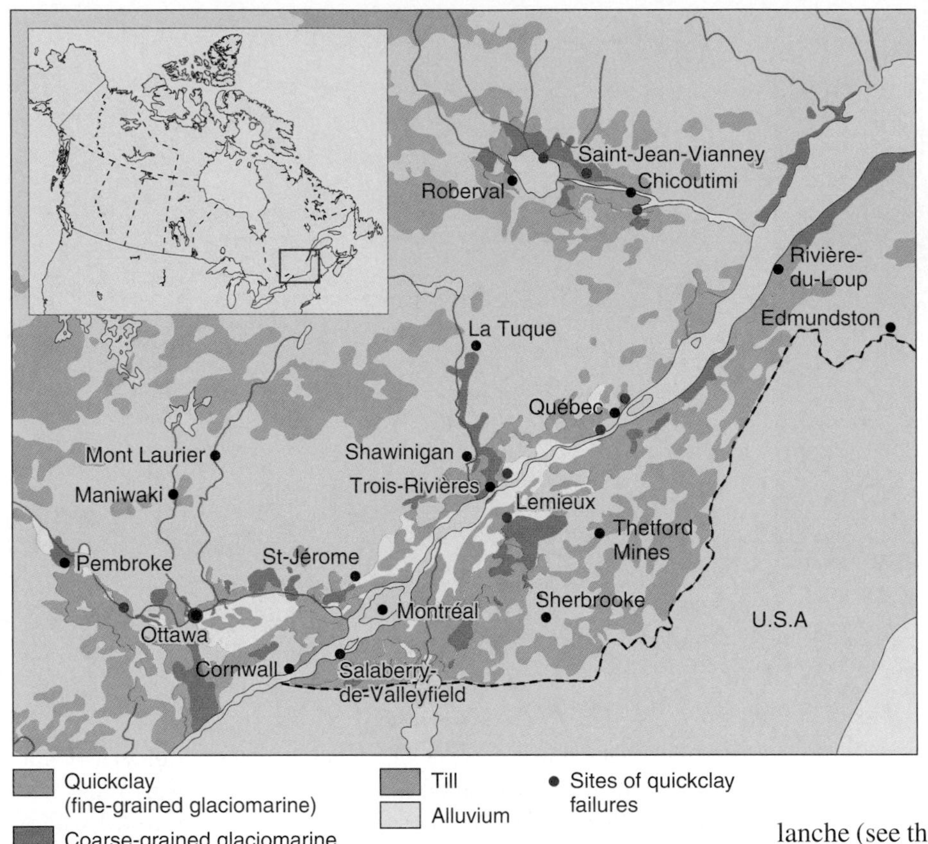

A

B

FIGURE 13.15

(A) Location of quickclay deposits and major quickclay failures in eastern Canada. (B) St. Boniface quickclay slide near Trois-Riviéres, Quebec, April 1996. Approximately 7 million cubic metres of sediment slid into the Machiche River valley, damming the river. Quickclay slide debris forms a series of concentric ridges visible in the centre of the photograph.

Illustration A modified from http://atlas.gc.ca/site/english/maps/environment/land/surflcia/materials. Photo B GSC Photo number 2002-703. St. Boniface landslide, Machiche River valley, near Trois-Riveres, Quebec. Photo by Greg Brooks. http://gsc.nrcan.gc.ca/landscapes/details_e.php?photoID=890 Reproduced with the permission of the Minister of Public Works and Government Services Canada, 2007 and Courtesy of Natural Resources Canada, Geological Survey of Canada

movement of between 2.5 and 3.5 million cubic metres of silt and clay that flowed into the valley of the South Nation River, blocking the river for several months (figure 13.15*B*). Similar large-scale quickclay failures have also occurred in coastal British Columbia, notably at Haney (1880) and Terrace (1993).

Avalanches

The fastest type of mass movement by flow is the **avalanche.** Snow avalanches are rapidly moving mass movements of snow and air that create a serious hazard to skiers and hikers in mountainous regions such as the Canadian Cordillera, where they have been the cause of many deaths (see box 13.3). *Debris avalanches* consist of turbulent mixtures of rock, debris, air, and water that move downslope at speeds of up to several hundred kilometres per hour. The best modern example is the one that buried the Peruvian town of Yungay in 1970 (see box 13.1). Debris avalanches are also common in the Cordillera of western Canada. The 1903 Frank Slide, which killed more than 70 people and was Canada's worst landslide disaster, was probably a rock avalanche (see the section on rockslides below). It is estimated that the 1959 Pandemonium Creek rock avalanche in British Columbia travelled at speeds up to 360 kilometres per hour.

Rockfalls and Rockslides

Rockfalls

When a block of bedrock breaks off and falls freely or bounces down a cliff, it is a **rockfall** (figure 13.16). Cliffs may form naturally by the undercutting action of a river, wave action, or glacial erosion. Highway or other construction projects may also oversteepen slopes. Bedrock commonly has cracks (joints) or other planes of weakness such as foliation (in metamorphic rocks) or sedimentary bedding planes. Blocks of rock will break off along these planes. In colder climates rock is effectively broken apart by frost wedging (as explained in chapter 8).

Commonly, an apron of fallen rock fragments, called **talus,** accumulates at the base of a cliff (figure 13.17).

A spectacular rockfall took place in Yosemite National Park in the summer of 1996, killing one man and injuring several other people. The rockfall originated from near Glacier Point. Two huge slabs (weighing approximately 80,000 tons) of an overhanging arch broke loose just seconds apart. (The arch was a product of exfoliation, and broke loose along a sheet joint—see chapter 8.) The slabs slid a short distance over steep rock from which they were launched outward, as if from a ski jump, away from the vertical cliffs. The slabs fell free for around 500 metres and hit the valley floor 30 metres out from the base of the cliff (you would not have been hit if you were standing at the base of the cliff). They shattered upon impact and created a dust cloud (figure 13.18) that obscured visibility for hours.

ENVIRONMENTAL GEOLOGY 13.3

Snow Avalanches

Snow avalanches, which involve the extremely rapid downslope movement of snow and air, have constantly threatened transportation routes and have been responsible for the deaths of many people in the Rocky Mountains of Canada. Fourteen people were killed in a period of less than two weeks during the winter of 2003 when unstable masses of snow swept down the mountainsides of Glacier National Park in British Columbia. An average of 13 people are killed by avalanches each year in Canada. The worst avalanche disaster ever recorded was the burial of more than 10,000 soldiers on the Austro-Italian front in the Alps in a single day during the First World War.

A snow avalanche starts when a slab of snow begins to move downslope. This slab may remain intact (a slab avalanche) or may fragment into blocks that shatter further as the flow moves downslope. Most slab avalanches involve failure of a weak layer of surface hoar frost that has been buried by subsequent snowfalls. The hoar frost contains large, loosely packed ice crystals that fail easily under stress and can persist as a weak layer in the snow pack several weeks after their burial. The forces that trigger avalanches are of particular interest and include loud noises, thunder, and human disturbance. More than 85 percent of the 1.5 million avalanches (many of them small) that occur in western Canada every year are triggered by some form of human disturbance.

The five-point Canadian Avalanche Size Classification scale rates avalanches according to the mass and volume of snow moved. Avalanches can range from moving only a few tonnes to more than 500,000 tonnes of snow. A size 3 avalanche moves 1,000 tonnes of snow over a kilometre and can destroy cars, trees, and small buildings. A size 4 avalanche moves 10,000 tonnes of snow and can destroy a small forest or overturn railroad cars as it travels over a distance of two kilometres.

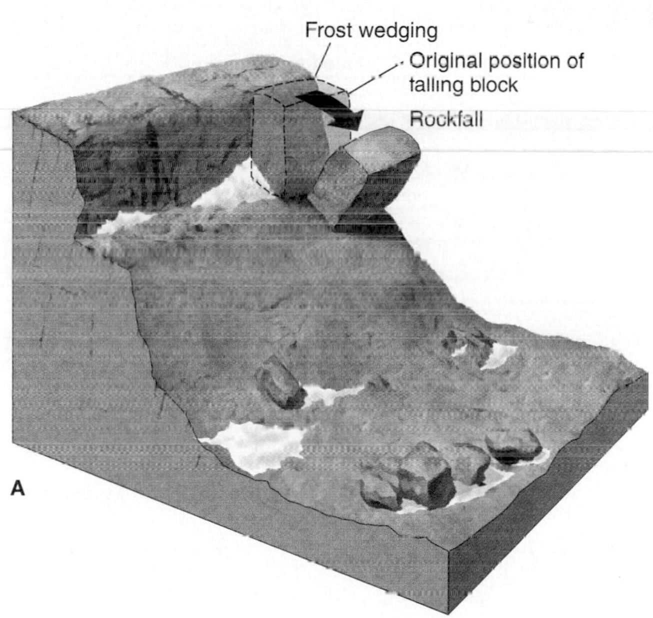

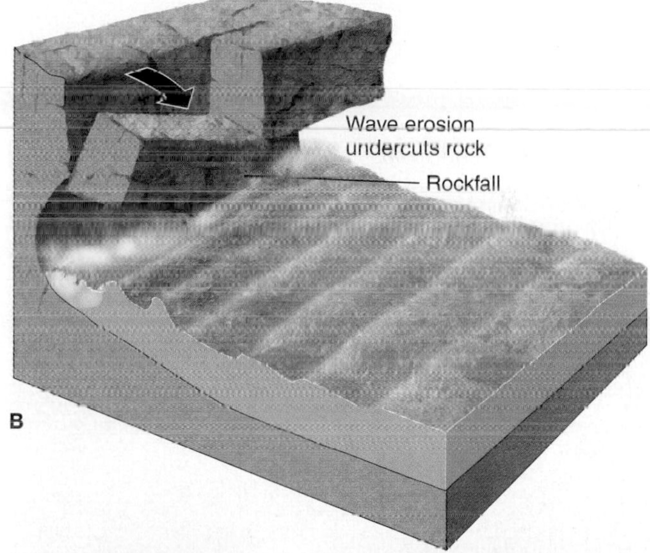

FIGURE 13.16

(A) and (B) Two examples of rockfall. (C) Hazards associated with rockfalls. This car was crushed by an eight-tonne block that fell from the top of a 100-m slope in Upper Island Cove, Newfoundland, February 14, 1999.

Photo by David Liverman. Reproduced by permission of the Government of Newfoundland and Labrador © 1999

FIGURE 13.17

Talus.

Photo by C. C. Plummer

FIGURE 13.18

Small dust clouds linger high above Yosemite Valley, where rock slabs broke loose and fell to the valley floor which, upon impact, created the debris-laden blast of air climbing up the other side of the valley. The photo was taken by a rock climber on a nearby cliff.

Photo by Ed Youmans

A powerful air blast was created as air between the rapidly falling rock and the ground was compressed. The debris-laden wind felled a swath of trees between the newly deposited talus and a nature centre building. In 1999, another rockfall in the same area killed one rock climber and injured three others. For an excellent and thorough report go to the U.S. Geological Survey site "Rockfall in Yosemite" at http://greenwood.cr.usgs.gov/ pub/open-file-reports/ofr-99-0385/.

FIGURE 13.19

(*A*) Frank Slide, Alberta. (*B*) Cross-section of Turtle Mountain showing internal geologic structure.

Photo A courtesy of EUB/AGS or Alberta Energy and Utilities Board/Alberta Geological Survey; Illustration B after D. M. Cruden and C. B. Beaty

Rockslides and Rock Avalanches

A **rockslide** is, as the term suggests, the rapid sliding of a mass of bedrock along an inclined surface of weakness, such as a bedding plane (figure 13.19), a major fracture in the rock, or a foliation plane (box 13.4). Once sliding begins, a rock slab usually breaks up into rubble. Like rockfalls, rockslides can be caused by undercutting at the base of the slope from erosion or construction.

ENVIRONMENTAL GEOLOGY 13.4

Landslides in the Canadian Cordillera

The Frank Slide was Canada's worst landslide disaster but was not the largest landslide to have occurred in the Canadian Cordillera. In southeastern British Columbia, the floor of the Valley of the Rocks is covered by more than 1 billion cubic metres of debris resulting from a prehistoric landslide. More than 30 high-velocity rock avalanches have occurred in the Canadian Cordillera since 1855, the largest being the 1965 Hope Slide (box figure 1) in which 48 million cubic metres of rock crashed down into the valley below. More than 1,000 rockslides are estimated to have occurred during the past 10,000 years in the Canadian Rockies.

Volcanic rocks in the Garibaldi Volcanic belt of southwestern British Columbia are particularly prone to catastrophic downslope movement. In 1975, the Devastation Glacier rock avalanche moved approximately 12 million cubic metres of volcanic rock. Many other volcanic rock avalanches and flank collapse events have occurred on the Quaternary volcanoes of the Garibaldi belt, involving the mass movement of up to 3 billion cubic metres of material, comparable in scale to the rockslide on the flanks of Mount St. Helens in 1980.

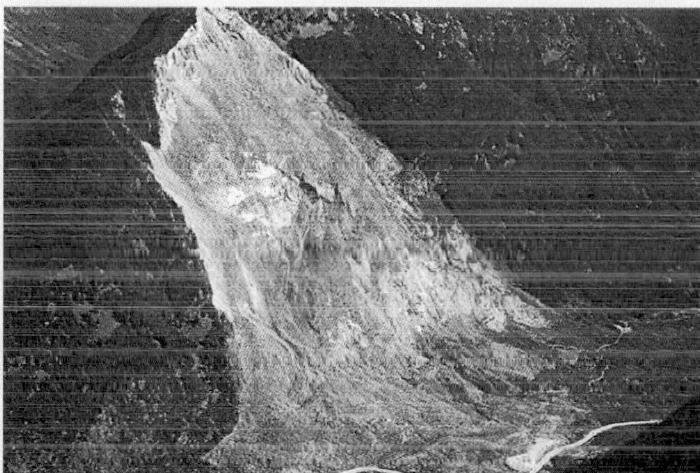

A

THE HOPE SLIDE

EARLY SATURDAY MORNING JAN. 9TH 1965 AN ENORMOUS LAND SLIDE DESCENDED INTO THIS VALLEY OF THE NICOLUM CREEK. THE SLIDE DESTROYED ABOUT TWO MILES OF THE HOPE-PRINCETON HWY. FILLING UP THE BOTTOM OF THE VALLEY WITH ROCK AND MUD TO DEPTHS UP TO 275 FT. APPARENTLY TRIGGERED BY A MINOR EARTHQUAKE. THE SLIDE CONSISTED OF 60 MILLION CUBIC YDS. OF ROCK, OR MORE THAN 100 MILLION TONS, TOGETHER WITH SOME EARTH AND SNOW. IT CRASHED DOWN IN SECONDS FROM THE 6500 FT. HIGH MOUNTAIN RIDGE FORMING THE NORTH SIDE OF THE VALLEY. OUTRAM LAKE AT THE FOOT OF THE SLIDE WAS COMPLETELY FILLED WITH DEBRIS. THE WATER AND SOFT CLAY OF THE LAKE BED AND ADJOINING LAND WERE DISPLACED AND PROJECTED VIOLENTLY UP THE OPPOSITE MOUNTAIN SIDE, AND BACK INTO THE VALLEY SPREADING OUT IN A SOUTH EASTERLY DIRECTION AND BACK UP THE NORTH SLOPE TO A HEIGHT OF 100 TO 200 FT. THE BOUNDARIES OF THE AREA SWEPT BY MUD AND SLIDE DEBRIS ARE VISIBLE ALONG THE SOUTH SIDE OF THE VALLEY WHERE THE TREES HAVE BEEN COMPLETELY REMOVED LEAVING A CLEAN SCAR MARKING THE PATH. UNFORTUNATELY FOUR PEOPLE IN THREE VEHICLES, STOPPED BY AN EARLIER SMALL SNOW SLIDE, WERE CAUGHT IN A WAVE OF MUD WHICH SWEPT BACK INTO THE VALLEY FROM THE SOUTH SIDE. TWO OF THE VICTIMS WERE NEVER FOUND. SEISMOGRAPHS RECORDED TWO EARTHQUAKES THAT MORNING WITH EPICENTRES IN THE NICOLUM VALLEY AREA. THE SECOND OF THESE WAS AT 6:58 A.M. THE APPROXIMATE TIME THE BIG SLIDE OCCURRED. THE NEW HIGHWAY AND THIS VIEWPOINT AND PARKING AREA ARE BUILT ON THE SLIDE DEBRIS 150 FT. ABOVE THE ORIGINAL GROUND LEVEL.

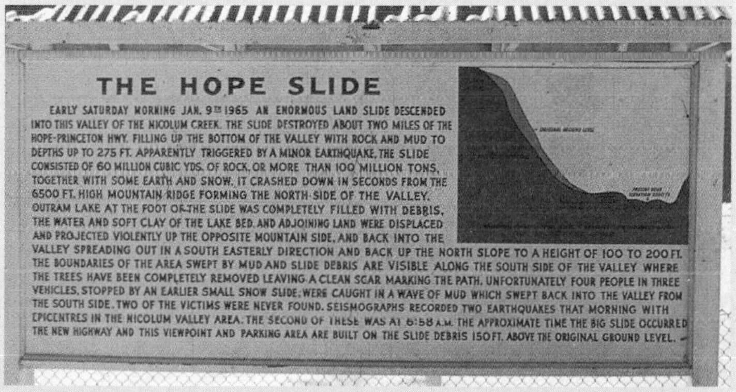

B

BOX 13.4 ■ FIGURE 1

The Hope Slide, British Columbia.

Photo B by Nick Eyles

Landslides and other mass wasting processes have a significant impact on populations living in mountainous areas. In British Columbia, the Sea-to-Sky Highway (Hwy 99; box figure 2*A*), which links Vancouver and the tourist areas of Whistler, Pemberton, and the Cariboo, is repeatedly closed due to rockslides and debris flows. The highway is particularly prone to debris flows because of heavy rains and steep mountain slopes covered in loose debris (box figure 2*B*). A series of debris flows in the area north of Horseshoe Bay claimed the lives of 11 people in the 1980s and stimulated the introduction of zoning regulations to control development on debris fans. Flow-deflection barriers, dams, and stream channel liners can protect existing structures but are costly to build and may be insufficient to protect against large failures. A recent commitment was made by the B.C. government to upgrade the Sea-to-Sky Highway at a cost of more than $600 million.

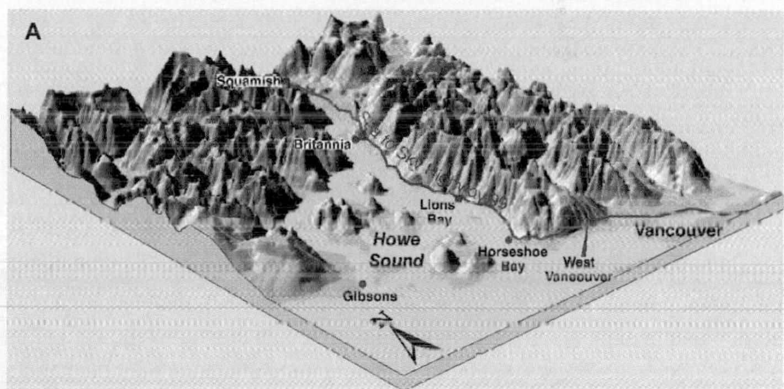

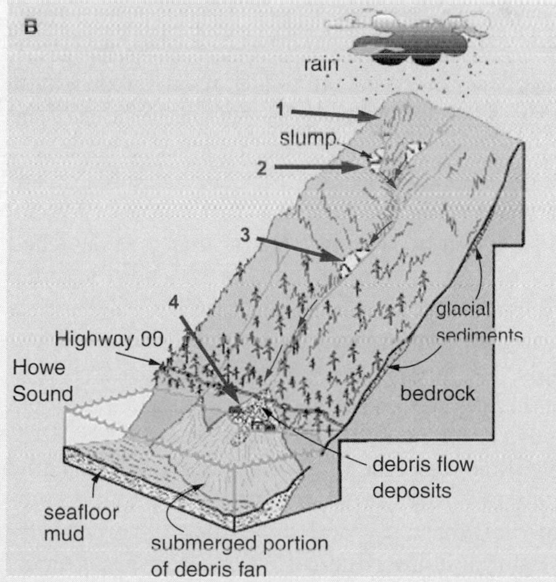

BOX 13.4 ■ FIGURE 2

(*A*) Sea-to-Sky Highway. (*B*) Formation of debris flows on steep mountain slopes in the area of Horseshoe Bay: **1.** Torrential rainfall swells streams along the mountain crest. **2.** Sediment slumps into a raging stream, forming a slurry (debris flow) that surges down the channel. **3.** The debris flow swells in volume as it picks up additional sediment and trees from the channel and canyon walls. **4.** The debris flow emerges onto a fan where it damages houses, roads, bridges, and a rail line.

From http://geoscape.nrcan.gc.ca/vancouver/sea_e.php

FIGURE 13.20

Multiple rock avalanches from steep valley walls in Jasper National Park, Alberta.
Photo by Nick Eyles

Some rockslides travel only a few metres before halting at the base of a slope. In country with high relief, however, a rockslide may travel hundreds or thousands of metres before reaching a valley floor (see box 13.4). If movement becomes very rapid, the rockslide may break up and become a rock avalanche. A **rock avalanche** is a very rapidly moving, turbulent mass of broken-up bedrock. Movement in a rock avalanche is flowage on a grand scale. The only difference between a rock avalanche and a debris avalanche is that a rock avalanche begins its journey as bedrock.

Ultimately, a rockslide or rock avalanche comes to rest as the terrain becomes less steep (figure 13.20). Sometimes the mass of rock fills the bottom of a valley and creates a natural dam. If the rock mass suddenly enters a lake or bay, it can create a huge wave that destroys lives and property far beyond the area of the original landslide.

One of the largest landslides to affect populated regions of North America was the Frank Slide in Alberta (figure 13.19*A*). On April 29, 1903, at 4:10 a.m., a large mass of limestone detached from the summit of Turtle Mountain and buried a portion of the town of Frank, killing 76 people. More than 36 million cubic metres of rock buried an area of more than 3 square kilometres in less than 100 seconds. The average velocity of this landslide is estimated at around 112 kilometres per hour, suggesting that it should be more correctly considered a rock avalanche.

The Frank Slide was essentially caused by the unstable geologic structure of Turtle Mountain. Turtle Mountain is composed of Paleozoic limestones and shales that have been deformed into a tight anticline (figure 13.19*B*). These sediments were thrust to the northeast along the Turtle Mountain Thrust Fault, which now underlies the whole mountain. Fractures developed in the folded mountaintop rocks allow water to penetrate to the core of the mountain and have caused extensive dissolution of the limestone and lubrication of contacts between sedimentary layers. This—together with weakening of the mountain structure caused by oversteepening of the eastern face by glaciers, coal mining operations at the base of the mountain, and severe weather conditions—caused the failure of the summit in the spring of 1903. There is still considerable concern over the stability of Turtle Mountain, and a state-of-the-art predictive monitoring system has been established to provide early warning of any future rockfall hazard (see box 13.5).

Many other areas of the Canadian Cordillera are also prone to landslides and pose considerable problems to engineers maintaining road and rail links through the mountains (box 13.4).

Another example is a disastrous landslide that took place in northern Italy in 1963. A huge layer of limestone broke loose parallel to its bedding planes. The translational slide involved around 250 million cubic metres that slid into the Vaiont Reservoir at up to 100 km/hr, creating a giant wave. The 175-metre-high-wave overtopped the Vaiont Dam (it was the world's highest dam, rising 265 metres above the valley floor). Three thousand people were killed in the villages that it flooded in the valleys below. The dam was not destroyed, a tribute to excellent engineering, but the men in charge of the building project were convicted of criminal negligence for ignoring the landslide hazards.

Debris Slides and Debris Falls

As the names suggest, debris slides and debris falls behave similarly to rockslides and rockfalls, except that they involve debris that moves as a coherent mass (at least initially). A **debris fall** is a free-falling mass of debris.

A **debris slide** is a coherent mass of debris moving along a well-defined surface (or surfaces). If the movement is along a curved surface the landslide is a *rotational debris slide.* Debris slides were mentioned earlier with earthflows, with which they are commonly associated (see figure 13.8). Debris may slide, however, even without an earthflow taking place (figure 13.21).

FIGURE 13.21

Debris slide blocking Cecil Lake Road in the Fort St. John area of British Columbia, July 2001. The slide occurred in clay-rich glacial lake deposits and was triggered in part by heavy rainfall.
Photo by Department of Transportation and Highways, British Columbia. Reproduced with the permission of the Minister of Public Works and Government Services Canada, 2007 and Courtesy of Natural Resources Canada, Geological Survey of Canada
http://gsc.nrcan.gc.ca/landslides/clp/photos_high/cecil_lake_rd_bc_2001.jpg
GSC Photo number: 2002–585

GEOMATICS 13.5

Monitoring an Unstable Mountain

When a large section of Turtle Mountain collapsed on April 29, 1903 it produced the most deadly landslide in Canadian history. The landslide was essentially caused by the unstable structure of Turtle Mountain and there are serious concerns that more of the mountain will collapse in the future. As a result, a state-of-the-art monitoring system, administered by the Alberta Geologic Survey, has been established at the mountain that records and analyses many types of digital data that may provide early warning of a rockslide event. The monitoring system consists of several sub-systems that aim to identify any movement or deformation of the highly cracked and fractured rock on the mountain. *Crack meters* are used to detect movements as small as 10 microns (the width of a human hair) across cracks and fissures. Displacement over larger distances (tens of metres) is measured with *surface extensometers*, which are long steel wires anchored at one end and hung over a pulley with a weight at the other end. Any movement across cracks or fissures spanned by the wire is recorded by a transducer on the pulley. *Tilt meters* have been positioned in several places on the mountain and measure any tilting or rotation of the rock to which they are attached. *Laser ranging systems* and *GPS stations* have also been established to obtain high precision distance measurements from key points on the mountain. Comparison of elevation data obtained from satellite-borne radar images taken at different times (*satellite interferometry*) is also used to detect subtle changes in the mountain's surface topography. Mass movements are often triggered by stress release events that emit seismic waves and a highly sensitive *seismic monitoring system* has been constructed on and around Turtle Mountain to continuously record both local and more regional earth movements. Water pressure in cracks and fissures within a rock can also greatly affect slope stability and a series of *piezometers* (water pressure gauges) have been located at the top of the mountain as well as a *flow monitoring station* at the base of the mountain. Accurate meteorological data are collected at the Turtle Mountain weather station, a small automated weather station located close to the mountain peak and at a station at the Frank Slide Interpretive Centre close to the valley floor.

All of the digital data collected by the various elements of the monitoring system are telemetered in near real time to the Frank Slide Interpretive Centre, where they undergo automatic preliminary analysis. In the event of any data indicating a potential mass movement threat, a warning management system automatically contacts appropriate officials.

The Turtle Mountain Monitoring Project is intended to provide an early warning system for any potential landslide threat but will also provide a wealth of scientific data that will greatly enhance our understanding of rock stability issues.

A

B

C

D

E

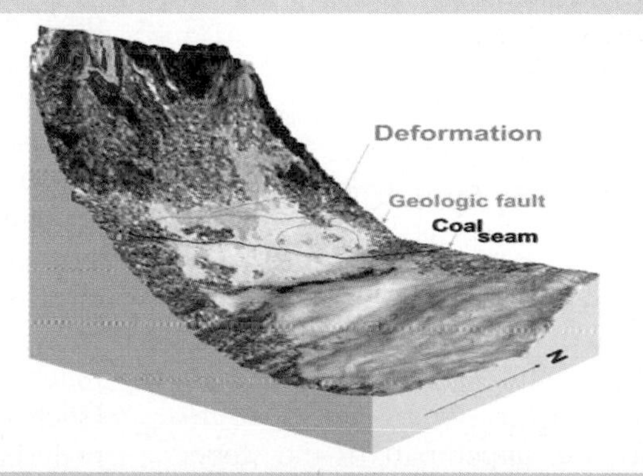

BOX 13.5 ■ FIGURE 1

Turtle Mountain Monitoring Project. (*A*) Tilt meter (*B*) Weather station (*C*) Installing surface extensometers (*D*) Laser prism and GPS station (*E*) Surface deformation map of east face of Turtle Mountain generated from ERS interferometry data (information from the Canada Centre for Remote Sensing)

All data and photos from: Alberta Geological Society; http://www.ags.gov.ab.ca/activities/Turtle_Mountain/mainpage.htm

WHAT HAPPENS WHEN LANDSLIDES OCCUR UNDERWATER?

The steeper parts of the ocean floors sometimes have very large landslides. Prehistoric ones are indicated by large masses of jumbled debris on the deep-ocean floor. One, off the coast of the Hawaiian Islands, is much larger than any landslide mass on land. The debris from what is called the Nuuanu debris avalanche covers an area of 5,000 square kilometres, larger than all of the present Hawaiian Islands combined, and includes volcanic rock blocks several kilometres across.

One very large underwater landslide took place off the coast of northeastern Canada in 1929 following an earthquake. It systematically cut a series of trans-Atlantic telephone and telegraph lines. The existence and extent of the event were inferred decades later by analyzing the timing of the telephone conversation cutoffs and the distance of cables from the earthquake's epicentre. The underwater debris avalanche, described as a *turbidity current* in chapter 18 (see figure 18.28), travelled more than 700 kilometres in 13 hours at speeds from 15 to 60 kilometres per hour. The lengths of the sections of cable carried away indicate that the debris flow was up to 100 kilometres wide.

Scientists have recently found that a very large area of thick sediment off the central part of the East Coast of the United States is unstable and could become a giant submarine landslide. If it does fall, it very likely will generate a giant *tsunami* (discussed in chapter 3) that could be disastrous to coastal communities in Europe as well as North America.

Underwater landslides are particularly common at the fronts of active deltas where large loads of sediment are deposited rapidly at the mouths of rivers entering the sea. The buildup of sediment on such delta fronts creates steep slopes that often fail, triggering debris flows and slides. The Fraser Delta in B.C. has experienced several failures in the the past and future submarine landslides could seriously damage ferry terminals, port facilities, and submarine cables supplying power to Vancouver Island. Earthquake activity along the B.C. coast significantly increases the risk of submarine landslides.

HOW CAN WE PREVENT LANDSLIDES?

Preventing Mass Wasting of Debris

Mass movements of debris can often be prevented. Proper engineering is essential when the natural environment of a hillside is altered by construction. As shown in figure 13.22, construction generally makes a slope more susceptible to mass wasting of debris in several ways: (1) the base of the slope is undercut, removing the natural support for the upper part of the slope; (2) vegetation is removed during construction; (3) buildings constructed on the upper part of a slope add weight to the potential slide; and (4) extra water may be allowed to seep into the debris.

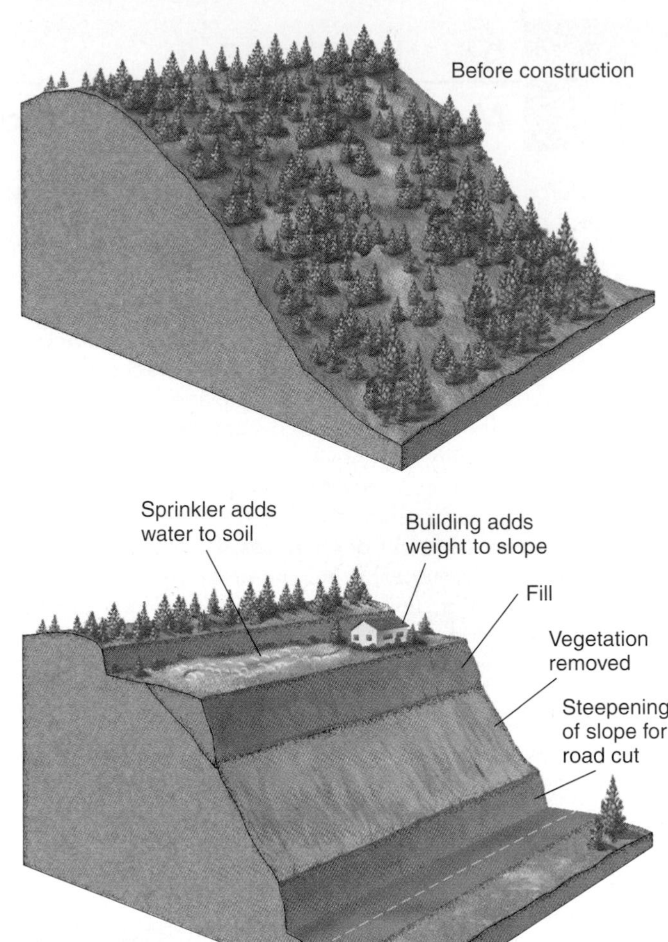

FIGURE 13.22
A hillside becomes vulnerable to mass wasting due to construction activities.

Some preventive measures can be taken during construction. A retaining wall is usually built where a cut has been made in the slope, but this alone is seldom as effective a deterrent to downslope movement as people hope. If, in addition, drain pipes are put through the retaining wall and into the hillside, water can percolate through and drain away rather than collecting in the debris behind the wall (figure 13.23). Without drains, excess water results in decreased shear strength and the whole soggy mass can easily burst through the wall.

Another practical preventive measure is to avoid oversteepening the slope. The hillside can be cut back in a series of terraces rather than in a single steep cut. This not only reduces the slope angle but also reduces the shear force by removing much of the overlying material. It also prevents loose material (such as boulders dislodged from the top of the cut) from rolling to the base. Road cuts constructed in this way are usually reseeded with rapidly growing grass or plants, the roots of which help anchor the slope. A vegetation cover also minimizes erosion from running water.

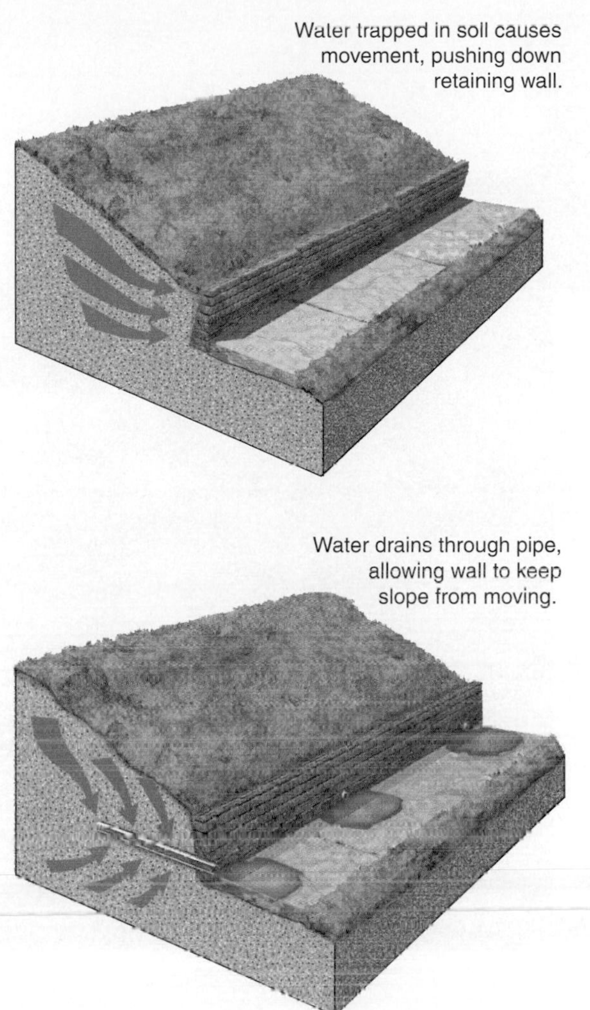

FIGURE 13.23

Use of drains to help prevent mass wasting.

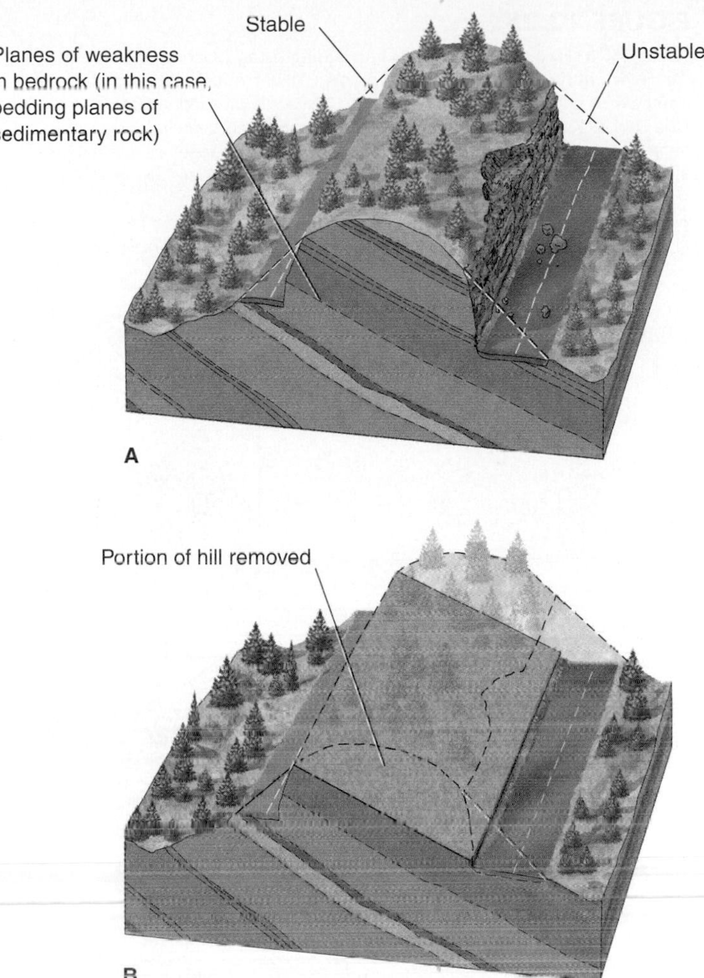

FIGURE 13.24

(*A*) Cross-section of a hill showing a relatively safe road cut on the left and a hazardous road cut on the right. (*B*) The same hazardous road cut after removal of rock that might slide.

Preventing Rockfalls and Rockslides on Highways

Rockslides and rockfalls are a major problem on highways built through mountainous country. Steep slopes and cliffs are created when road cuts are blasted and bulldozed into mountain sides. If the bedrock has planes of weakness (such as joints, bedding planes, or foliation planes), the orientation of these planes relative to the road cut determines whether there is a rockslide hazard (as in figure 13.24*A*). If the planes of weakness are inclined into the hill, there is no chance of a rockslide. On the other hand, where the planes of weakness are approximately parallel to the slope of the hillside, a rockslide may occur.

Various techniques are used to prevent rockslides. By doing a detailed geologic study of an area before a road is built, builders might avoid a hazard by choosing the least dangerous route for the road. If a road cut must be made through bedrock that appears prone to sliding, all of the rock that might slide could be removed (sometimes at great expense), as shown in figure 13.24*B*.

In some instances, slopes prone to rock sliding have been "stitched" in place by the technique shown in figure 13.25. Spraying a roadside exposure with gunnite may retard a landslide in some instances. Fences, railings, or construction of covered tunnels can prevent minor rockfalls from blocking road and railway routes (figure 13.26)

FIGURE 13.25

"Stitching" a slope to keep bedrock from sliding along planes of weakness. (*A*) Holes are drilled through unstable layers into stable rock. (*B*) Expanded view of one hole. A cable is fed into the hole and cement is pumped into the bottom of the hole and allowed to harden. (*C*) A steel plate is placed over the cable and a nut tightened. (*D*) Tightening all the nuts pulls unstable layers together and anchors them in stable bedrock. (*E*) Road cut in Acapulco, Mexico stabilized by "stitching" and sprayed concrete.

Photo by C. C. Plummer

A

B

FIGURE 13.26

(*A*) Strategically placed fencing and (*B*) covered tunnels protecting road and railway routes from rockfalls and avalanches near Lytton, British Columbia.

Photos by M. Radomski

SUMMARY

Mass wasting is the movement of a mass of debris (soil and loose rock fragments) or bedrock toward the base of a slope. Movement can take place as a flow, slide, or fall. Gravity is the driving force. The component of gravitational force that propels mass wasting is the *shear force*, which occurs parallel to the slopes. The resistance to that force is the *shear strength* of rock or debris. If shear force exceeds shear strength, mass wasting takes place. Water is an important factor in mass wasting.

A number of other factors determine whether movement will occur and, if it does, the rate of movement.

The slowest type of movement, *creep*, occurs mostly on relatively gentle slopes, usually aided by water in the soil. In colder climates, repeated freezing and thawing of water within the soil contributes to creep. *Landsliding* is a general term for more rapid mass wasting of rock, debris, or both. *Flows* include creep, earthflows, mudflows, and debris avalanches. *Earthflows* vary greatly in velocity although they are not as rapid as *debris avalanches*, which are turbulent masses of debris, water, and air. *Gelifluction*, a special variety of earthflow, usually takes place in arctic or subarctic climates where ground is permanently frozen (*permafrost*). A *mudflow* is a slurry of debris and water. Most mudflows flow in channels much as streams do.

Rockfall is the fall of broken rock down a vertical or near-vertical slope. A *rockslide* is a slab of rock sliding down a less-than-vertical surface. *Debris falls* and *debris slides* involve unconsolidated material rather than bedrock. Landslides also take place underwater. The larger ones of these are vastly bigger than any that have occurred on land.

Terms to Remember

avalanche 344
creep 338
debris 331
debris fall 348
debris flow 341
debris slide 348
earthflow 339
fall 333
flow 333
gelifluction 339
landslide 331
mass wasting 331
mudflow 342
permafrost 341
quickclays 343
relief 336
rock avalanche 348
rockfall 344
rockslide 346
rotational slide (slump) 333
shear strength 336
shear stress 336
slide 333
talus 344
translational slide 333

Testing Your Knowledge

Use the questions below to prepare for exams based on this chapter.

1. Describe the effect on shear strength of the following:
   a. thickness of debris;
   b. orientation of planes of weakness;
   c. water in debris; and
   d. vegetation.
2. Compare the shear force to the force of gravity (drawing diagrams similar to figure 13.2) for the following situations:
   a. a vertical cliff;
   b. a flat horizontal plane; and
   c. a 45-degree slope.
3. How does a rotational slide differ from a translational slide?
4. What role does water play in each of the types of mass wasting?
5. Why is gelifluction more common in colder climates than in temperate climates?
6. List and explain the key factors that control mass wasting.
7. What is the slowest type of mass wasting process?
   a. debris flow
   b. rockslide
   c. creep
   d. rockfall
   e. avalanche
8. The largest landslide has taken place
   a. on the sea floor
   b. in the Andes
   c. on active volcanoes
   d. in the Himalaya
9. A descending mass moving downslope as a viscous fluid is referred to as a
   a. fall
   b. landslide
   c. flow
   d. slide
10. The driving force behind all mass wasting processes is
    a. gravity
    b. slope angle
    c. type of bedrock material
    d. presence of water
    e. vegetation

11. The resistance to movement or deformation of debris is its
    a. mass
    b. shear strength
    c. shear force
    d. density
12. Flow of water-saturated debris over impermeable material is called
    a. gelifluction
    b. flow
    c. slide
    d. fall
13. A flowing mixture of debris and water, usually moving down a channel, is called a
    a. mudflow
    b. slide
    c. fall
    d. debris flow
14. An apron of fallen rock fragments that accumulates at the base of a cliff is called
    a. debris
    b. sediment
    c. soil
    d. talus
15. How does construction destabilize a slope?
    a. adds weight to the top of the slope
    b. decreases water content of the slope
    c. adds weight to the bottom of the slope
    d. increases the shear strength of the slope
16. How can landslides be prevented during construction? (choose all that apply)
    a. install retaining walls
    b. cut steeper slopes
    c. install water drainage systems
    d. add vegetation

Exploring Web Resources

www.mcgrawhill.ca/olc/plummer

Go to the Online Learning Centre to access the answers for the Testing Your Knowledge section. This site also has additional quizzes, readings, and media resources to further your understanding of mass wasting. The Online Learning Centre also provides you with direct links to all the sites listed below.

http://landslides.usgs.gov/

Geologic hazards, landslides, U.S. Geological Survey. You can get to several useful sites from here. Reports on recent landslides can be accessed by clicking on the ones listed. Click on "National Landslide Information Center" for photos of landslides, including some described in this chapter. Watch animation of a landslide. You can access sources of information on landslides and other geologic features for any U.S. state, usually from the state's geologic survey.

http://sts.gsc.nrcan.gc.ca/landslides/index_e.php

Landslides and snow avalanches in Canada. The Geological Survey of Canada's site has generalized descriptions and some photos of significant Canadian landslides.

www.ags.gov.ab.ca/activities/Turtle_Mountain/mainpage.htm

Detailed descriptions and excellent photographs of instruments used to monitor Turtle Mountain, Alberta.

Animation

This chapter includes the following animation available on our Online Learning Centre at www.mcgrawhill.ca/olc/plummer.

13.1 Types of Earth Movements

CHAPTER 14

Streams and Floods

What is the hydrologic cycle?
What is a stream?
How do drainage patterns reflect geological conditions?
What factors affect stream erosion and deposition?
How does a stream erode?
How do streams transport sediment?
Where do streams deposit sediment?
Can we predict when floods will occur?
How do stream valleys change with time?

Running water is one of the most important geologic agents in eroding, transporting, and depositing sediment. Almost every landscape on Earth shows the results of stream erosion or deposition. Although other agents—groundwater, glaciers, wind, and waves—can be locally important in sculpturing the land, stream action and mass wasting are the dominant processes of landscape development.

The first part of this chapter deals with the various ways that streams erode, transport, and deposit sediment. The second part describes landforms produced by stream action, such as valleys, flood plains, deltas, and alluvial fans, and shows how each of these is related to changes in stream characteristics. The chapter also includes a discussion of the causes and effects of flooding, and various measures used to control flooding.

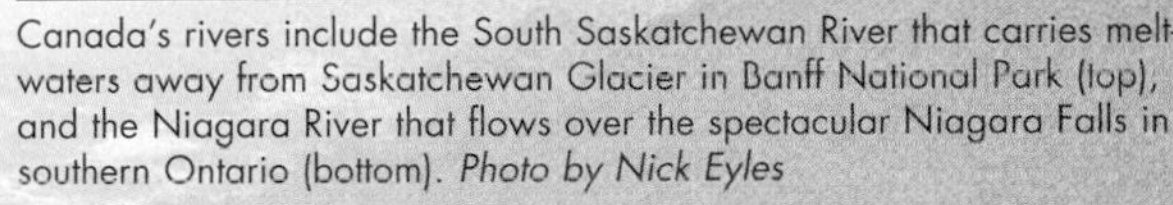

Canada's rivers include the South Saskatchewan River that carries meltwaters away from Saskatchewan Glacier in Banff National Park (top), and the Niagara River that flows over the spectacular Niagara Falls in southern Ontario (bottom). *Photo by Nick Eyles*

WHAT IS THE HYDROLOGIC CYCLE?

The **hydrologic cycle** describes the movement and interchange of water among the sea, air, and land (figure 14.1). Solar radiation provides the necessary energy for *evaporation* of water vapour from the land and sea. When air becomes saturated with water (100 percent relative humidity), rises, and cools in the atmosphere, liquid droplets condense to form clouds. These droplets grow larger as more water leaves the gaseous state to form rain or snow, depending on the temperature. When rain (or snow) falls on the land surface as *precipitation,* more than half the water returns rather rapidly to the atmosphere by evaporation or *transpiration* from plants. Some of the water is held as ice in glaciers and snow pack. The remainder either flows over the land surface as *runoff* in streams, is held temporarily in lakes, or soaks into the ground by *infiltration* to form groundwater. Groundwater (the subject of chapter 15) moves, usually very slowly, underground and may flow back onto the surface a long distance from where it seeped into the ground. Most water eventually reaches the sea, where ongoing evaporation completes the cycle.

Only about 15 percent to 20 percent of rainfall normally ends up as surface runoff in rivers, although the amount of runoff can range from 2 percent to more than 25 percent with variations in climate, steepness of slope, soil and rock type, and vegetation. Steady, continuous rains can saturate the ground and the atmosphere, however, and lead to floods as runoff approaches 100 percent of rainfall.

WHAT IS A STREAM?

A **stream** is a body of running water that is confined in a channel and moves downhill under the influence of gravity. In some parts of the country, *stream* implies size: rivers are large, streams somewhat smaller, and brooks or creeks even smaller. Geoscientists, however, use *stream* for any body of running water, from a small trickle to a huge river.

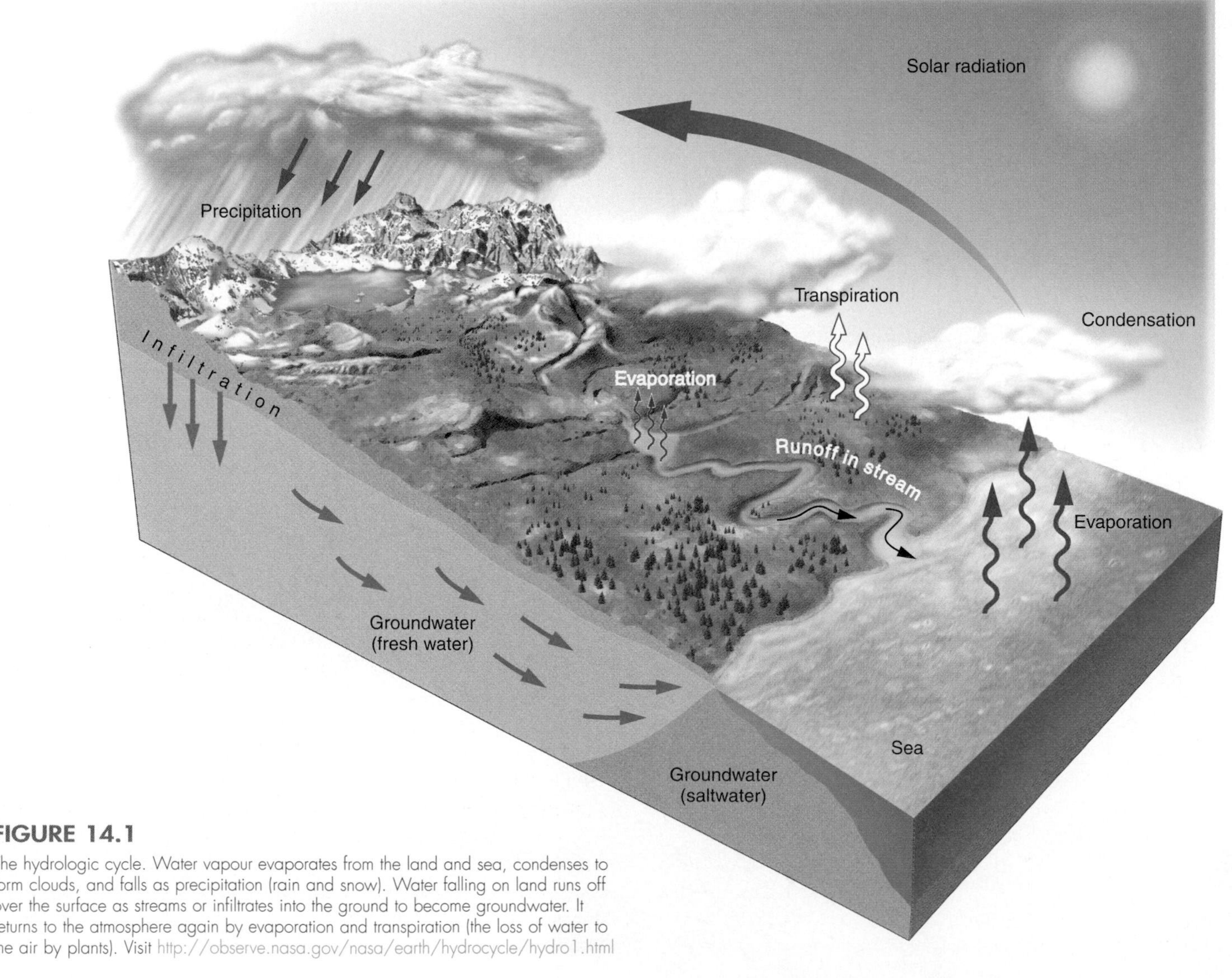

FIGURE 14.1

The hydrologic cycle. Water vapour evaporates from the land and sea, condenses to form clouds, and falls as precipitation (rain and snow). Water falling on land runs off over the surface as streams or infiltrates into the ground to become groundwater. It returns to the atmosphere again by evaporation and transpiration (the loss of water to the air by plants). Visit http://observe.nasa.gov/nasa/earth/hydrocycle/hydro1.html

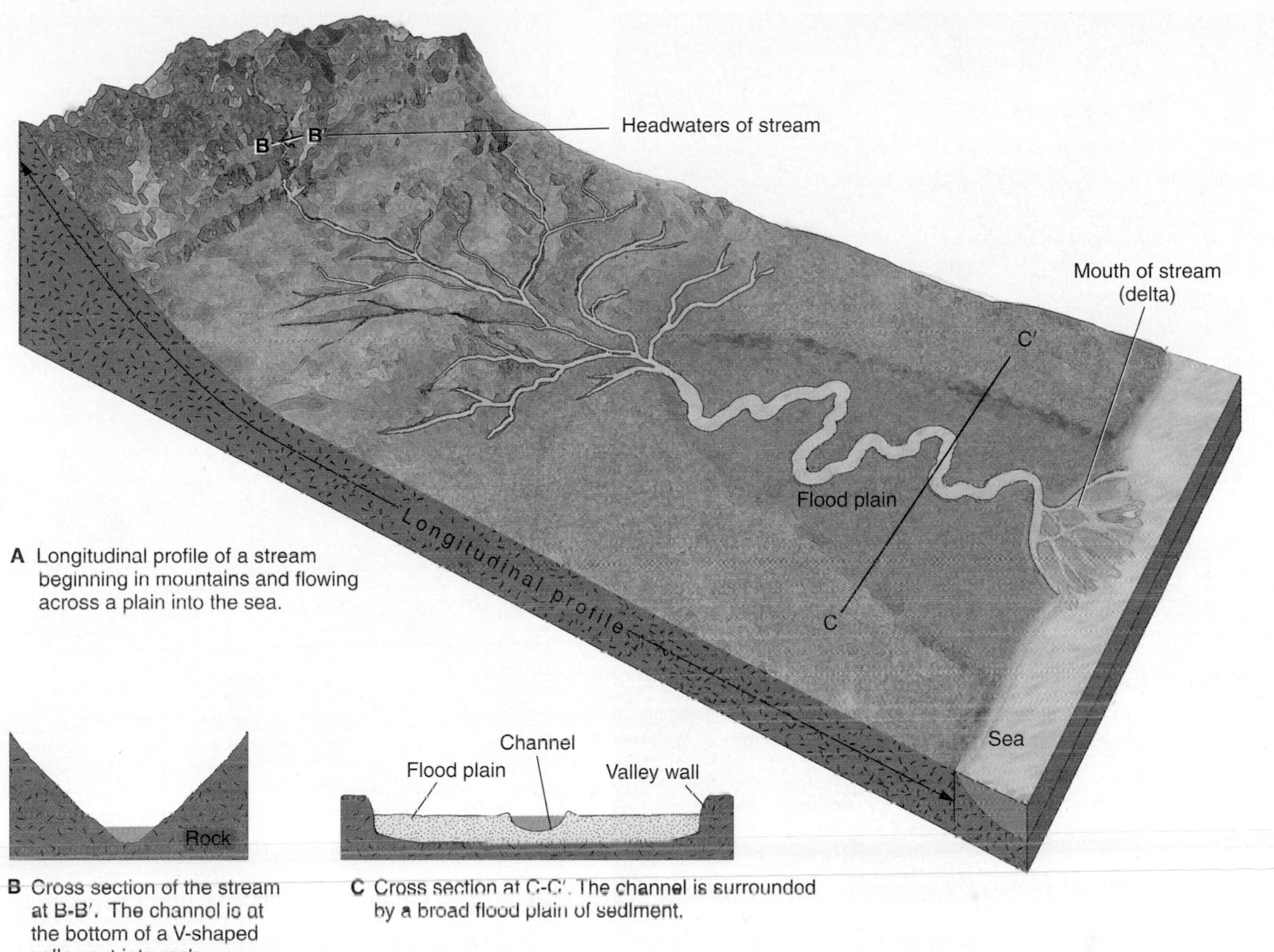

FIGURE 14.2

Longitudinal profile and cross-sections of a typical stream.

Figure 14.2*A* shows a *longitudinal profile* of a typical stream viewed from the side. The stream begins in steep mountains and flows out across a gentle plain into the sea. The *headwaters* of a stream are the upper part of the stream near its source in the mountains. The *mouth* is the place where a stream enters the sea, a lake, or a larger stream. A *cross-section* of a stream in steep mountains is usually a V-shaped valley cut into solid rock, with the stream channel occupying the narrow bottom of the valley; there is little or no flat land next to the stream on the valley bottom (figure 14.2*B*). Near its mouth a stream usually flows within a broad, flat-floored valley. The stream channel is surrounded by a flat *flood plain* of sediment deposited by the stream (figure 14.2*C*).

A stream normally stays in its **stream channel,** a long narrow depression eroded by the stream into rock or sediment. The stream *banks* are the sides of the channel; the stream *bed* is the bottom of the channel. During a flood the waters of a stream may rise and spill over the banks onto the flat flood plain of the valley floor (figure 14.3).

Not all water that moves over the land surface is confined to channels. Sometimes, particularly during heavy rains, water runs off as **sheetwash,** a thin layer of unchannelled water flowing downhill. Sheetwash is particularly common in deserts, where the lack of vegetation allows rainwater to spread quickly over the land surface. It also occurs in humid regions during heavy thunderstorms when water falls faster than it can soak into the ground. A series of closely spaced storms can also promote sheetwash; as the ground becomes saturated, more water runs over the surface.

Sheetwash, along with the violent impact of raindrops on the land surface, can produce considerable *sheet erosion,* in which a thin layer of surface material, usually topsoil, is removed by the flowing sheet of water. This gravity-driven movement of sediment is a process intermediate between mass wasting and stream erosion.

Overland sheetwash becomes concentrated in small channels, forming tiny streams called *rills.* Rills merge to form small streams, and small streams join to form larger streams. Most regions are drained by networks of coalescing streams.

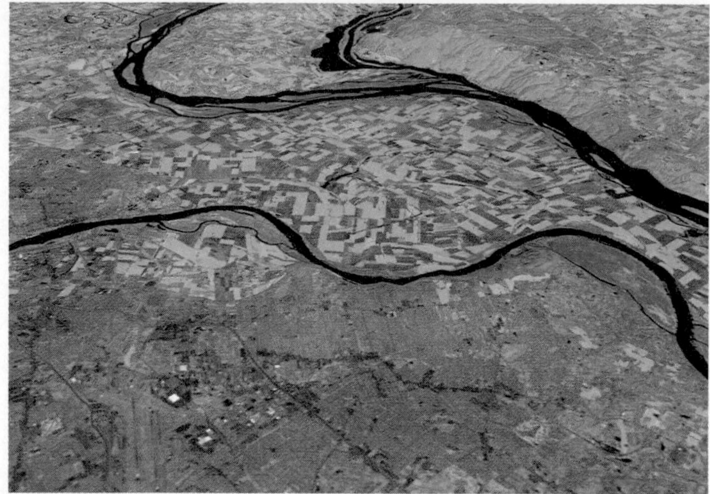

A

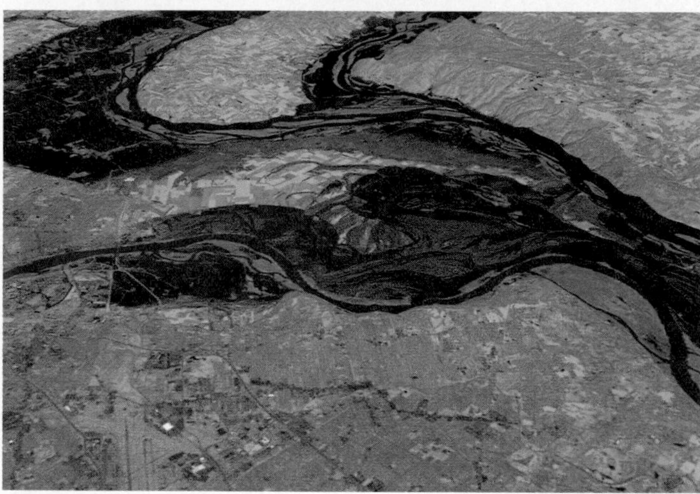

B

FIGURE 14.3

A stream normally stays in its channel, but during a flood it can spill over its banks onto the adjacent flatland (flood plain) as shown in these three-dimensional satellite images. (*A*) Before flooding image (August 14, 1991) of Missouri River (bottom), Mississippi River (upper left), and Illinois River (upper right). Vegetation is shown in green and red indicates recently plowed fields (bare soil). (*B*) Image taken on November 7 after the huge floods of 1993 showing how the rivers spilled over their banks onto the flat flood plains.

Photos © NASA/GSFC/Photo Researchers

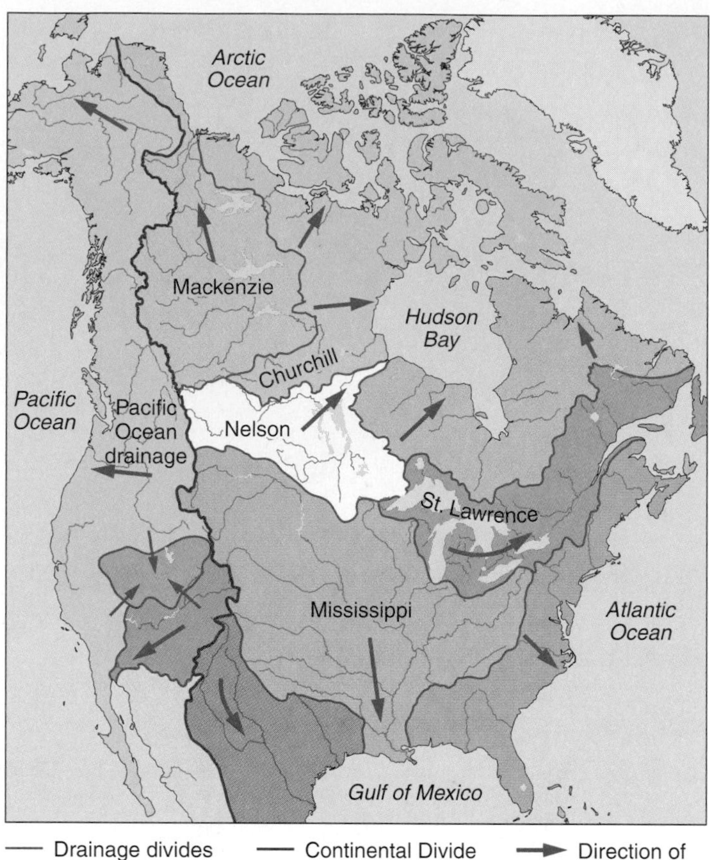

FIGURE 14.4

Drainage basins of North America. The drainage basin of the Mississippi River covers more than 2.5 million square kilometres. The Continental Divide separates rivers that flow into the Pacific Ocean from rivers that flow into the Arctic Ocean, Hudson Bay, the Atlantic Ocean, and the Gulf of Mexico. The Mackenzie River system drains into the Arctic Ocean and includes three major lakes (Great Slave, Great Bear, and Athabasca) and numerous rivers including the Peace, Athabasca, Liard, Peel, Hay, and South Nahanni.

Drainage Basins

Each stream, small or large, has a **drainage basin,** the total area drained by a stream and its tributaries (a **tributary** is a small stream flowing into a larger one). A drainage basin can be outlined on a map by drawing a line around the region drained by all the tributaries to a river (figure 14.4). The Mississippi River's drainage basin, for example, includes all the land area drained by the Mississippi River itself and by all its tributaries, including the Ohio and Missouri Rivers. This great drainage system includes more than one-third the land area of the contiguous 48 states. The second largest river system in North America is the Mackenzie River System, which drains over 1.8 million square kilometres of central and western Canada.

A ridge or strip of high ground dividing one drainage basin from another is termed a **divide** (figure 14.4). The best known in North America is the Continental Divide, a line separating streams that flow to the Pacific Ocean from those that flow to the Atlantic and the Gulf of Mexico. The Continental Divide, which extends from the Yukon Territory down into Mexico, crosses Montana, Idaho, Wyoming, Colorado, and New Mexico in the United States. Road signs indicating the crossing of the Continental Divide have been placed at numerous points where major highways intersect the divide.

HOW DO DRAINAGE PATTERNS REFLECT GEOLOGICAL CONDITIONS?

The arrangement, in map view, of a river and its tributaries is a **drainage pattern.** A drainage pattern can, in many cases, reveal the nature and structure of the rocks underneath it.

Most tributaries join the main stream at an acute angle, forming a V (or Y) pointing downstream. If the pattern resembles branches of a tree or nerve dendrites, it is called **dendritic** (figures 14.4 and 14.5*A*). Dendritic drainage patterns develop on uniformly erodible rock or regolith, and are the most common type of pattern. A **radial pattern,** in which streams diverge outward like spokes of a wheel, forms on high conical mountains, such as composite volcanoes and domes (figure 14.5*B*). A **rectangular pattern,** in which tributaries have frequent 90-degree bends and tend to join other streams at right angles, develops on regularly fractured rock (figure 14.5*C*). A network of fractures meeting at right angles forms pathways for streams because fractures are eroded more easily than unbroken rock. A **trellis pattern** consists of parallel main streams with short tributaries meeting them at right angles (figure 14.5*D*). A trellis pattern forms in a region where tiled layers of resistant rock such as sandstone alternate with nonresistant rock such as shale. Erosion of such a region results in a surface topography of parallel ridges and valleys.

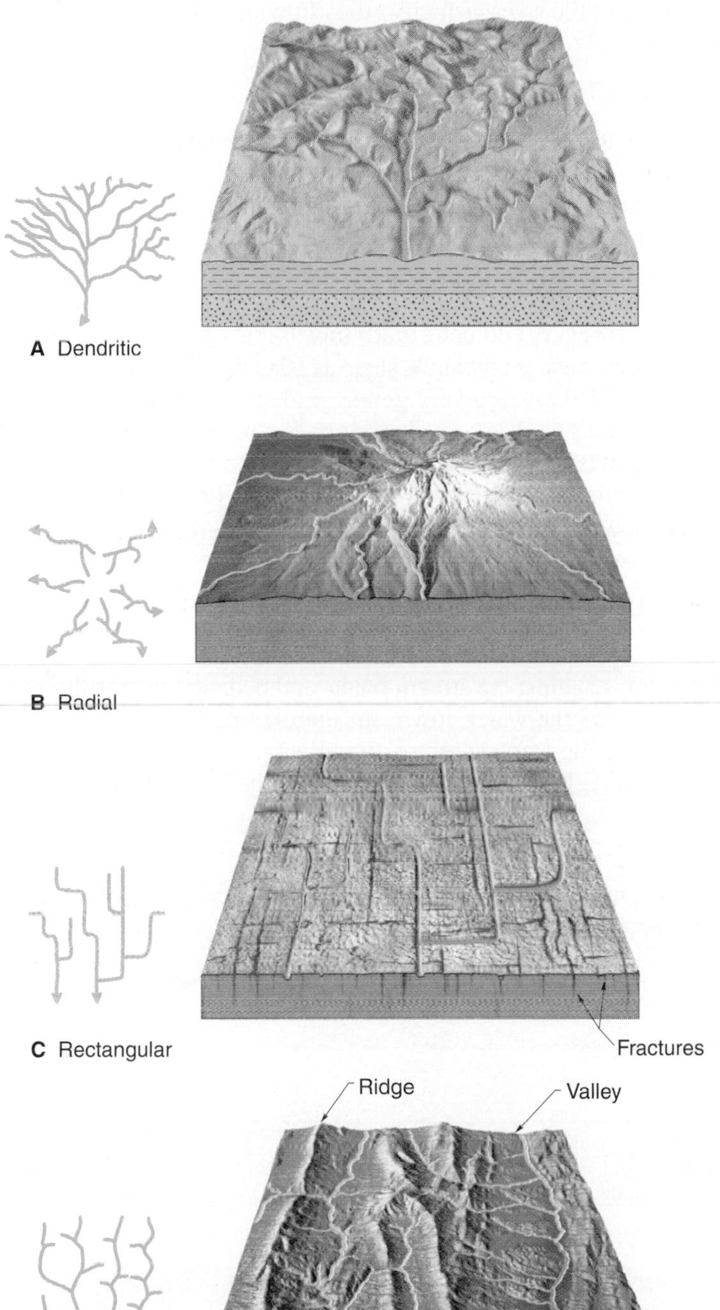

FIGURE 14.5

Drainage patterns can reveal something about the rocks underneath. (*A*) Dendritic pattern develops on uniformly erodible rock. (*B*) A radial pattern develops on a conical mountain or dome. (*C*) A rectangular pattern develops on regularly fractured rock. (*D*) A trellis pattern develops on alternating ridges and valleys caused by the erosion of resistant and nonresistant tilted rock layers.

WHAT FACTORS AFFECT STREAM EROSION AND DEPOSITION?

Stream erosion and deposition are controlled primarily by a river's *velocity* and, to a lesser extent, by its *discharge.* Velocity is largely controlled by the stream *gradient,* channel shape, and channel roughness.

Velocity

The distance water travels in a stream per unit time is called the **stream velocity.** A moderately fast river flows at about 5 kilometres per hour. Rivers flow much faster during flood, sometimes exceeding 25 kilometres per hour.

The cross-sectional views of a stream in figure 14.6 show that a stream reaches its maximum velocity near the middle of the channel. When a stream goes around a curve, the region of maximum velocity is displaced by inertia toward the outside of the curve. Velocity is the key factor in a stream's ability to erode, transport, and deposit. High velocity (meaning greater energy) generally results in erosion and transportation; low velocity causes sediment deposition. Slight changes in velocity can cause great changes in the sediment load carried by the river.

Figure 14.7 shows the stream velocities at which sediments are eroded, transported, and deposited. For each grain size, these velocities are different. The upper curve represents the minimum velocity needed to erode sediment grains. This curve shows the velocity at which previously stationary grains are first picked up by moving water. The lower curve represents the velocity below which deposition occurs, when moving grains come to rest. Between the two curves the water is moving fast enough to transport grains that have already been eroded. Note that it takes a higher stream velocity to erode grains (set them in motion) than to transport grains (keep them in motion).

Point A on figure 14.7 represents fine sand on the bed of a stream that is barely moving. The vertical red arrows represent a flood with gradually increasing stream velocity. No sediment moves until the velocity is high enough to intersect the *upper* curve and move into the area marked "erosion." As the flood recedes, the velocity drops below the upper curve and into the transportation area. Under these conditions the sand that was

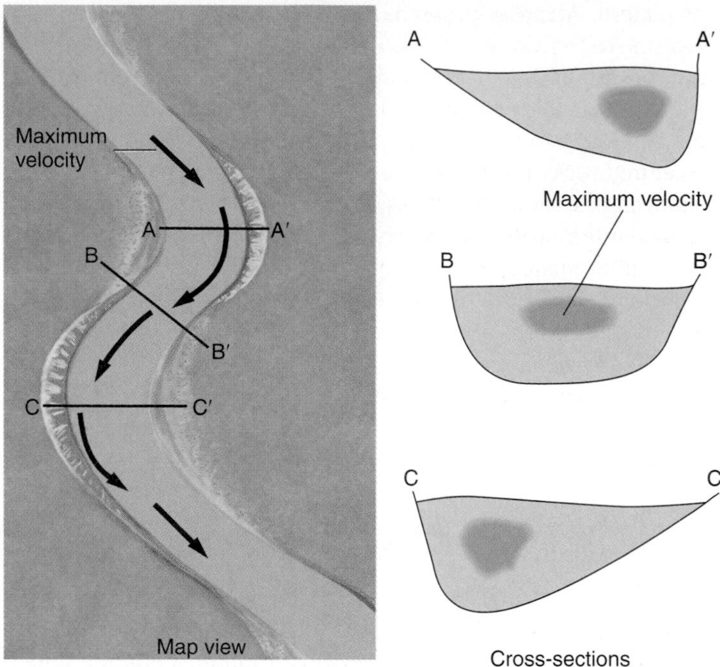

FIGURE 14.6

Regions of maximum velocity in a stream. Arrows on the map show how the maximum velocity shifts to the outside of curves. Sections show maximum velocity on outside of curves and in the centre of the channel on a straight stretch of stream.

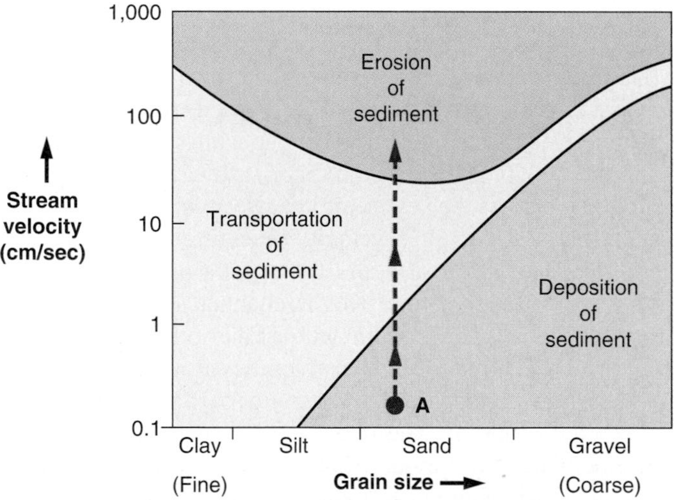

FIGURE 14.7

Logarithmic graph showing the stream velocities at which erosion and deposition of sediment occur. These velocities vary with the grain size of the sediment. See text for a discussion of point A and the dashed red line above it.

already eroded continues to be transported, but no new sand is eroded. As the velocity falls below the lower curve, all the sand is deposited again, coming to rest on the stream bed.

The right half of the diagram shows that coarser particles require progressively higher velocities for erosion and transportation, as you might expect (boulders are harder to move than sand grains). The erosion curve also rises toward the left of the diagram, however. This shows that fine-grained silt and clay are actually harder to erode than sand. The reason is that molecular forces tend to bind silt and clay into a smooth, cohesive mass that resists erosion. Once silt and clay are eroded, however, they are easily transported. As you can see from the lower curve, the silt and clay in a river's suspended load are not deposited until the river virtually stops flowing.

Gradient

One factor that controls a stream's velocity is the **stream gradient,** the downhill slope of the bed (or of the water surface, if the stream is very large). A stream gradient is usually measured in metres per kilometre. (In the United States, gradients are expressed in feet per mile because these units are used on U.S. maps.) A gradient of 5 metres per kilometre means that the river drops 5 metres vertically for every kilometre that it travels horizontally. Mountain streams may have gradients as steep as 10 to 40 m/km. The lower Mississippi River has a very gentle gradient, 0.1 m/km or less.

A stream's gradient usually decreases downstream. Typically, the gradient is greatest in the headwater region and decreases toward the mouth of the stream (see figure 14.2). Local increases in the gradient of a stream are usually marked by rapids.

Channel Shape and Roughness

The *shape of the channel* also controls stream velocity. Flowing water drags against the stream banks and bed, and the resulting friction slows the water down. In figure 14.8, the streams in *A* and *B* have the same cross-sectional area, but stream *B* flows slower than *A* because the wide, shallow channel in *B* has more surface for the moving water to drag against.

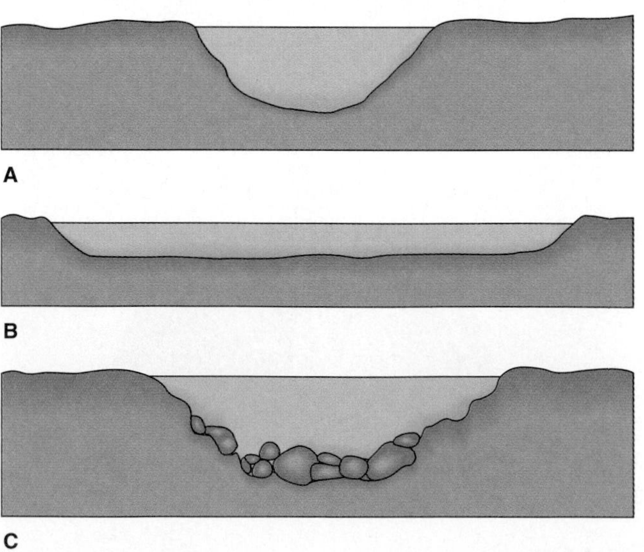

FIGURE 14.8

Channel shape and roughness influence stream velocity. (*A*) Semi-circular channel allows stream to flow rapidly. (*B*) Wide, shallow channel increases friction, slowing river down. (*C*) Rough, boulder-strewn channel slows river.

A stream may change its channel width as it flows across different rock types. Hard, resistant rock is difficult to erode, so a stream may have a relatively narrow channel in such rock. As a result it flows rapidly (figure 14.9*A*). If the stream flows onto a softer rock that is easier to erode, the channel may widen, and the river will slow down because of the increased surface area dragging on the flowing water. Sediment may be deposited as the velocity decreases.

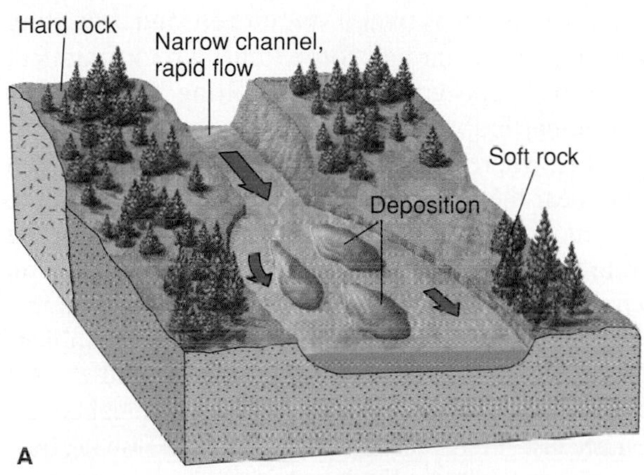

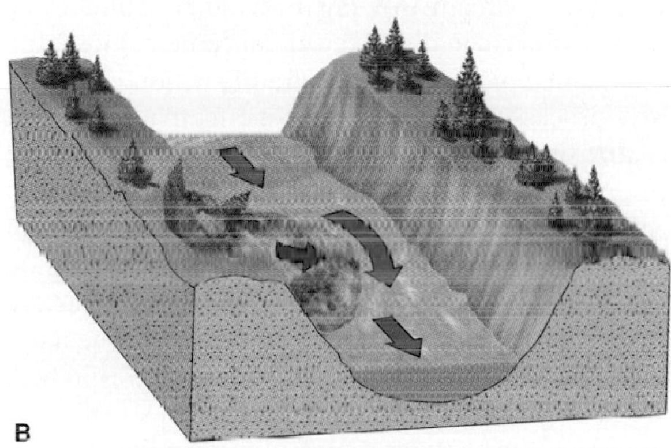

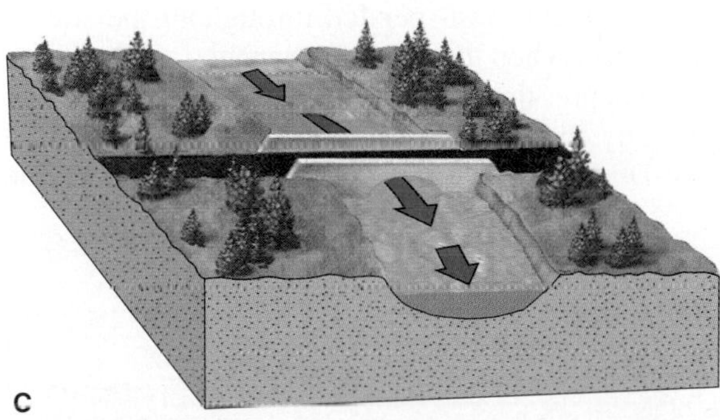

FIGURE 14.9

Channel width variations caused by rock type and obstructions. Length of arrow indicates velocity. (*A*) A channel may widen in soft rock. Deposition may result as stream velocity drops. (*B*) Landslide may narrow a channel, increasing stream velocity. Resulting erosion usually removes landslide debris. (*C*) Bridge piers (or other obstructions) will increase velocity and sometimes erosion next to the piers.

The width of a stream may be controlled by factors external to the stream. A landslide may carry debris onto a valley floor, partially blocking a stream's channel (figure 14.9*B*). The constriction causes the stream to speed up as it flows past the slide, and the increased velocity may quickly erode the landslide debris, carrying it away downstream. Human interference with a river can promote erosion and deposition. Construction of a culvert or bridge can partially block a channel, increasing the stream's velocity (figure 14.9*C*). If the bridge was poorly designed, it may increase velocity to the point where erosion may cause the bridge to collapse.

The *roughness of the channel* also controls velocity. A stream can flow rapidly over a smooth channel, but a rough, boulder-strewn channel floor creates more friction and slows the flow (see figure 14.8*C*). Coarse particles increase the roughness more than fine particles, and a rippled or wavy sand bottom is rougher than a smooth sand bottom.

Discharge

The **discharge** of a stream is the volume of water that flows past a given point in a unit of time. It is found by multiplying the cross-sectional area of a stream by its velocity (or width × depth × velocity). Discharge can be reported in cubic metres per second (m^3/sec), or cubic feet per second (cfs), which is standard in the United States.

$$\text{Discharge } (m^3/\text{sec}) = \text{average stream width (m)} \times \text{average depth (m)} \times \text{average velocity (m/sec)}$$

A stream 30 m wide and 5 m deep flowing at 2 metres per second has a discharge of 300 cubic metres per second (m^3/sec). In streams in humid climates, discharge increases downstream for two reasons: (1) water flows out of the ground into the river through the stream bed, and (2) small tributary streams flow into a larger stream along its length, adding water to the stream as it travels.

To handle the increased discharge, these streams increase in width and depth downstream. Some streams surprisingly increase slightly in velocity downstream, as a result of the increased discharge (the increase in discharge and channel size, and the typical downstream smoothness of the channel, override the effect of a lessening gradient).

During floods a stream's discharge and velocity increase, usually as a result of heavy rains over the stream's drainage basin. Flood discharge may be 50 to 100 times normal flow. Stream erosion and transportation generally increase enormously as a result of a flood's velocity and discharge. Swift mountain streams in flood can sometimes move boulders the size of automobiles (figure 14.10*A*). Flooded areas may be intensely scoured, with river banks and adjacent lawns and fields washed away (figure 14.10*B*). As floodwaters recede, both velocity and discharge decrease, leading to the deposition of a blanket of sediment, usually mud, over the flooded area.

In a dry climate, a river's discharge can decrease in a downstream direction as river water evaporates into the air and soaks into the dry ground (or is used for irrigation). As the discharge decreases, the load of sediment is gradually deposited.

A

B

FIGURE 14.10

(A) These large boulders of granite in a mountain stream are moved only during floods. Note the rounding of the boulders and the scoured high-water mark of floods on the valley walls. Note people for scale. (B) Property damage and sediment deposition resulting from flash floods along DesRosiers Creek, August 2003. Flooding was caused by intense rainfall in the Bois Francs Region, Quebec.

Photo A by David McGeary; Photo B by G. R. Brooks. GSC Photo 1997-42NN Reproduced with the permission of the Minister of Public Works and Government Services Canada, 2005 and Courtesy of Natural Resources Canada, Geological Survey of Canada

HOW DOES A STREAM ERODE?

A stream usually erodes the rock and sediment over which it flows. In fact, streams are one of the most effective sculptors of the land. Streams cut their own valleys, deepening and widening them over long periods of time and carrying away the sediment that mass wasting delivers to valley floors. The particles of rock and sediment that a stream picks up are carried along to be deposited farther downstream. Streams erode rock and sediment in three ways—*hydraulic action, solution,* and *abrasion*

Hydraulic action refers to the ability of flowing water to pick up and move rock and sediment (figure 14.11). The force of running water swirling into a crevice in a rock can crack the rock and break loose a fragment to be carried away by the stream. Hydraulic force can also erode loose material from a stream bank on the outside of a curve. The pressure of flowing water can roll or slide grains over a stream bed, and a swirling eddy of water may exert enough force to lift a rock fragment above a stream bed. The great force of falling water makes hydraulic action particularly effective at the base of a waterfall (figure 14.12), where it may erode a deep plunge pool. You may be able to hear the results of hydraulic action by standing beside a swift mountain stream and listening to boulders and cobbles hitting one another as they tumble along downstream.

From what you have learned about weathering, you know that some rocks can be dissolved by water. **Solution,** although ordinarily slow, can be an effective process of weathering and erosion (weathering because it is a response to surface chemical conditions; erosion because it removes material). A stream flowing over limestone, for example, gradually dissolves the rock, deepening the stream channel. A stream flowing over other sedimentary rocks, such as sandstone, can dissolve calcite cement, loosening grains that can then be picked up by hydraulic action.

The erosive process that is usually most effective on a rocky stream bed is **abrasion,** the grinding away of the stream channel by the friction and impact of the sediment load. Sand and gravel tumbling along near the bottom of a stream wear away the stream bed much as moving sandpaper wears away wood. The abrasion of sediment on the stream bed is generally much more effective in wearing away the rock than hydraulic action alone. The more sediment a stream carries, the faster it is likely to wear away its bed.

The coarsest sediment is the most effective in stream erosion. Sand and gravel strike the stream bed frequently and with great force, while the finer-grained silt and clay particles weigh so little that they are easily suspended throughout the stream and have little impact when they hit the channel.

Potholes are depressions that are eroded into the hard rock of a stream bed by the abrasive action of the sediment load (figure 14.13). During high water when a stream is full, the

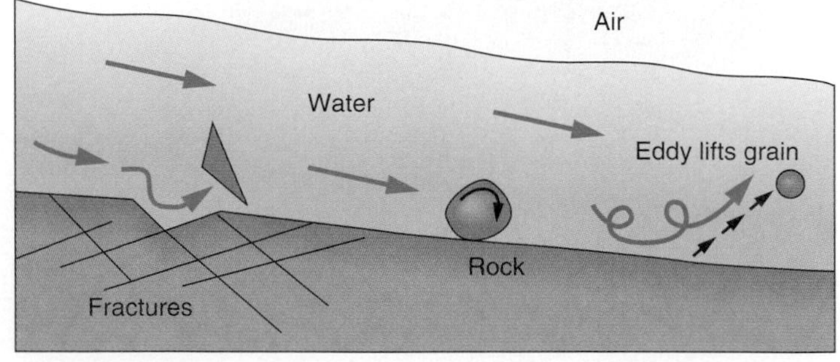

FIGURE 14.11

Hydraulic action can loosen, roll, and lift grains from the stream bed.

FIGURE 14.12

Hydraulic action is particularly effective at the base of Niagara Falls, along the Canada–U.S. border.

Photo by Nick Eyles

swirling water can cause sand and pebbles to scour out smooth, bowl-shaped depressions in hard rock. Potholes tend to form in spots where the rock is a little weaker than the surrounding rock. Although potholes are fairly uncommon, you can see them on the beds of some streams at low water level. Potholes may contain sand or an assortment of beautifully rounded pebbles (figure 14.13).

FIGURE 14.13

Potholes and grooves scoured along the bed of the South Saskatchewan River, Alberta.

Photos by Nick Eyles

HOW DO STREAMS TRANSPORT SEDIMENT?

The sediment load transported by a stream can be subdivided into *bed load, suspended load,* and *dissolved load.* Most of a stream's load is carried in suspension and in solution.

The **bed load** is the large or heavy sediment particles that travel on the stream bed (figure 14.14). Sand and gravel, which form the usual bed load of streams, move by either *traction* or *saltation.*

Large, heavy particles of sediment, such as cobbles and boulders, may never lose contact with the stream bed as they move along in the flowing water. They roll or slide along the stream bottom, eroding the stream bed and each other by abrasion. Movement by rolling, sliding, or dragging is called **traction.**

Sand grains move by traction, but they also move downstream by **saltation,** a series of short leaps or bounces off the bottom (see figure 14.14). Saltation begins when sand grains are momentarily lifted off the bottom by *turbulent* water (eddying, swirling flow). The force of the turbulence temporarily counteracts the downward force of gravity, suspending the grains in water above the stream bed. The water soon slows down because the velocity of water in an eddy is not constant; then gravity overcomes the lift of the water, and the sand grain once again falls to the bed of the stream. While it is suspended, the grain moves downstream with the flowing water. After it lands on the bottom, it may be picked up again if turbulence increases, or it may be thrown up into the water by the impact of another falling sand grain. In this way sand grains saltate downstream in leaps and jumps, partly in contact with the bottom and partly suspended in the water.

The **suspended load** is sediment that is light enough to remain lifted indefinitely above the bottom by water turbulence (see figure 14.14). The muddy appearance of a stream during a flood or after a heavy rain is due to a large suspended load. Silt and clay usually are suspended throughout the water, while the coarser-bed load moves on the stream bottom. Suspended load has less effect on erosion than the less visible bed load, which causes most of the abrasion of the stream bed. Vast quantities of sediment, however, are transported in suspension.

Soluble products of chemical weathering processes can make up a substantial **dissolved load** in a stream. Most streams contain numerous ions in solution, such as bicarbonate, calcium, potassium, sodium, chloride, and sulphate. The ions may precipitate out of water as evaporite minerals if the stream dries up, or they may eventually reach the ocean. Very clear water may in fact be carrying a large load of material in solution, for the dissolved load is invisible. Only if the water evaporates does the material become visible as crystals begin to form.

One estimate is that rivers in the United States carry about 250 million tons of solid load and 300 million tons of dissolved load each year. (It would take a freight train eight times as long as the distance from Boston to Los Angeles to carry 250 million tons.) The Mackenzie River in Canada transports more than 100 million tons of suspended sediment load each year.

WHERE DO STREAMS DEPOSIT SEDIMENT?

The sediments transported by a stream are often deposited temporarily along the stream's course (particularly the bed load sediments). Such sediments move sporadically downstream in

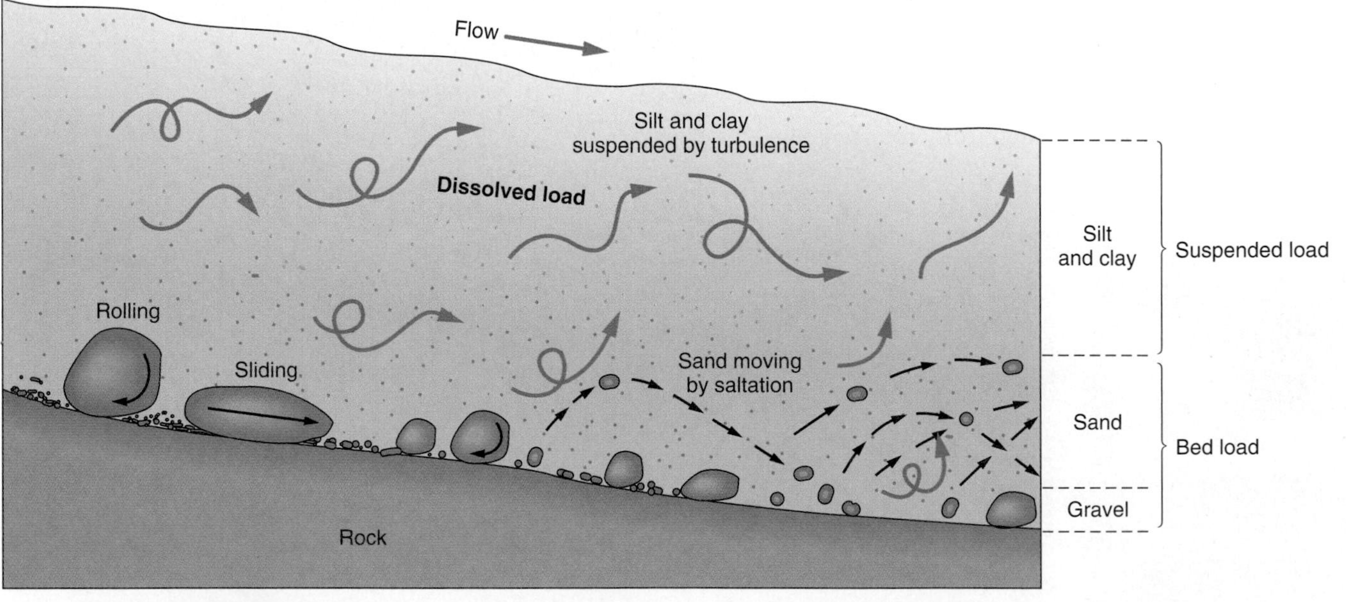

FIGURE 14.14

A stream's bed load consists of sand and gravel moving on or near the stream bed by traction and saltation. Finer silt and clay form the suspended load of the stream. The dissolved load of soluble ions is invisible.

repeated cycles of erosion and deposition, forming *bars* and *flood-plain deposits.* At or near the end of a stream, sediments may be deposited more permanently in a *delta* or an *alluvial fan.*

Bars

Stream deposits may take the form of a **bar,** a ridge of sediment, usually sand and gravel, deposited in the middle or along the banks of a stream (figure 14.15). Bars are formed by deposition when a stream's discharge or velocity decreases. During a flood, a river can move all sizes of sediment, from silt and clay up to huge boulders, because the greatly increased volume of water is moving very rapidly. As the flood begins to recede, the water level in the stream falls and the velocity drops. With the stream no longer able to carry all its sediment load, the larger boulders drop down on the stream bed, slowing the water locally even more. Finer gravel and sand are deposited between the boulders and downstream from them. In this way, deposition builds up a sand and gravel bar that may become exposed as the water level falls.

The next flood on the river may erode most of the sediment in this bar and move it farther downstream. But as the flood slows, it may deposit new gravel in approximately the same place, forming a new bar. After each flood, river fishers and boat operators must relearn the size and position of the bars. Sometimes gold panners discover fresh gold in a mined-out river bar after a flood has shifted sediment downstream. A dramatic example of the shifting of sand bars occurred during the planned flood on the Colorado River downstream from the Glen Canyon Dam (see box 14.1).

Placer Deposits

Placer deposits are found in streams where the running water has mechanically concentrated heavy sediment. The heavy sediment is concentrated in the stream where the velocity of the water is high enough to carry away lighter material but not the heavy sediment. Such places include river bars on the inside of meanders, plunge pools below waterfalls, and depressions on a stream bed (figure 14.16). Grains concentrated in this manner include gold dust and nuggets, native platinum, diamonds and other gemstones, and worn pebble or sand grains composed of the heavy oxides of titanium and tin.

FIGURE 14.15

Gravel bars in the South Saskatchewan River, Alberta. Water and sediment are supplied by the Saskatchewan Glacier, which lies to the left of the snow-covered mountain in the background.

Photo by Nick Eyles

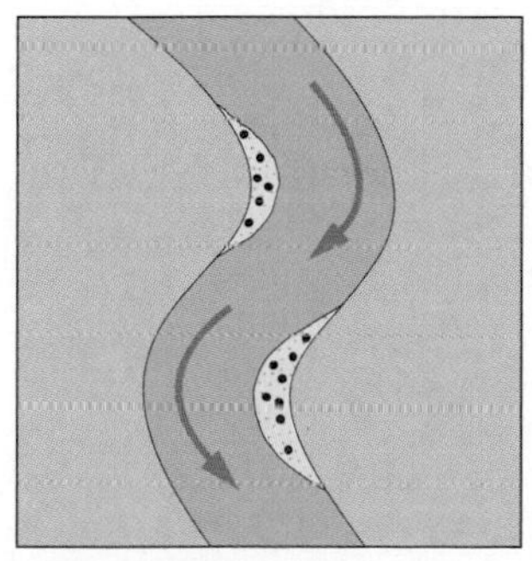

A Map view

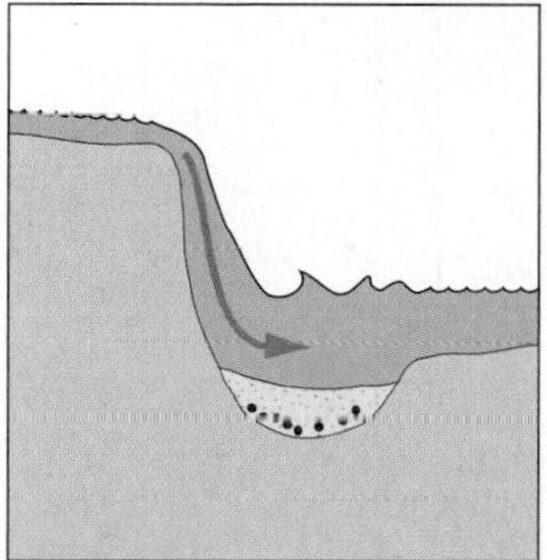

B Side view

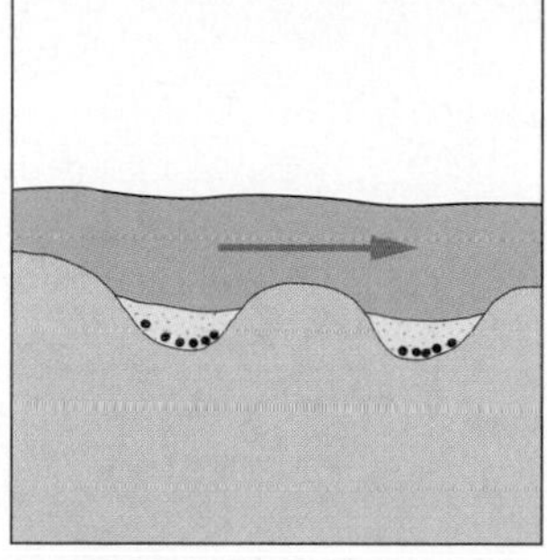

C Side view

D

FIGURE 14.16

Types of placer deposits. (*A*) Stream bar. (*B*) Below waterfall. (*C*) Depressions on stream bed. Valuable grains shown in black. (*D*) Mining for placer gold on gravel bars along the Fraser River, B.C.

Photo by Nick Eyles

ENVIRONMENTAL GEOLOGY 14.1

A Controlled Flood in the Grand Canyon: A Bold Experiment to Restore Sediment Movement in the Colorado River

On March 26, 1996, one of the largest experiments ever conducted on a river took place along the Colorado River below the Glen Canyon Dam (box figure 1). For six days the discharge from the Glen Canyon Dam was increased from 230 m^3/sec to 1,270 m^3/sec (a spike flow) to emulate the effects of a flood on the Colorado River (box figure 2). One of the main goals of this controlled flooding experiment was to determine whether the higher flows would result in bed scour and redeposition of sand bars and beaches along the sides of the channel. Another goal was to measure and observe how rocks move along the bed of the river with increasing discharge and velocity of floodwaters.

The Colorado River had not experienced its usual summertime floods since the Glen Canyon Dam was completed in 1963. The construction of the dam controlled peak discharges or flows on the Colorado River, which resulted in sand being deposited mainly along the bed or bottom of the river and erosion of beaches along the banks of the river. The Glen Canyon Dam cuts off a significant percentage of the sand supply to the lower Colorado River such that most of the downstream sand is supplied by two tributary streams, the Paria and Little Colorado Rivers. In August 1992 the Paria River flooded and deposited 330,000 tons of sand into the Colorado River, and in January 1993 a flood on the Little Colorado River deposited 10 million tons of sediment below its confluence with the Colorado River. The influx of sediment, coupled with the relatively low discharges from the dam (230–570 m^3/sec), resulted in sand being concentrated along the bed of the Colorado River.

One of the main predictions of the experiment proved true. That is, sand caught in deep pools in the bottom of the main channel was scoured and carried in suspension downstream, where it was redeposited along the river banks as beaches (box figure 3). Deposition of sand along the banks occurs due to back eddies that upwell and move upstream along the river banks and decrease the velocity of the downstream flow so that deposition can occur. Most of the scouring and deposition occurred in the first three days of the experiment; however, when flows were reduced back to 230 m^3/sec beaches began to erode and redeposition occurred once again in the deep pools on the bottom of the river.

Downstream at Lava Falls, an experiment was set up to determine how and if large boulders deposited in the main channel from a debris flow would move with the increased discharge and velocity of the floodwater. Holes were drilled into 150 basalt boulders and radio tags were inserted (box figure 4) so their movement could be monitored and correlated with the increase in discharge and velocity of the river. Surface velocity measurements were taken by kayaking the river and charting the speed at which floating balls moved. The surface velocities were used to calculate the velocity of the water close to the river bed where the boulders were positioned. Dye was also injected into the river at peak flows to determine the average velocity of the water. The dye indicated that the velocity of the water increased downstream, particularly at the Lava Falls debris flow. This is because the floodwater accelerated as it flowed downstream, pushing the river water in front of it, which increased the downstream velocity. The first crest of water actually arrived behind Hoover Dam at Lake Mead a day ahead of the floodwater marked with a red dye.

The experiment was deemed a success and for the first time a flood was studied as it happened. The experimental flood, even though smaller than the size of a naturally occurring flood (box figure 2), showed that beaches could be restored below a dam and that boulders could be moved out of rapids much like what occurs on an undammed river during a seasonal flood. It is proposed that other dammed rivers would benefit from periodic floods to help restore their natural conditions and thus minimize the adverse effects of damming a river.

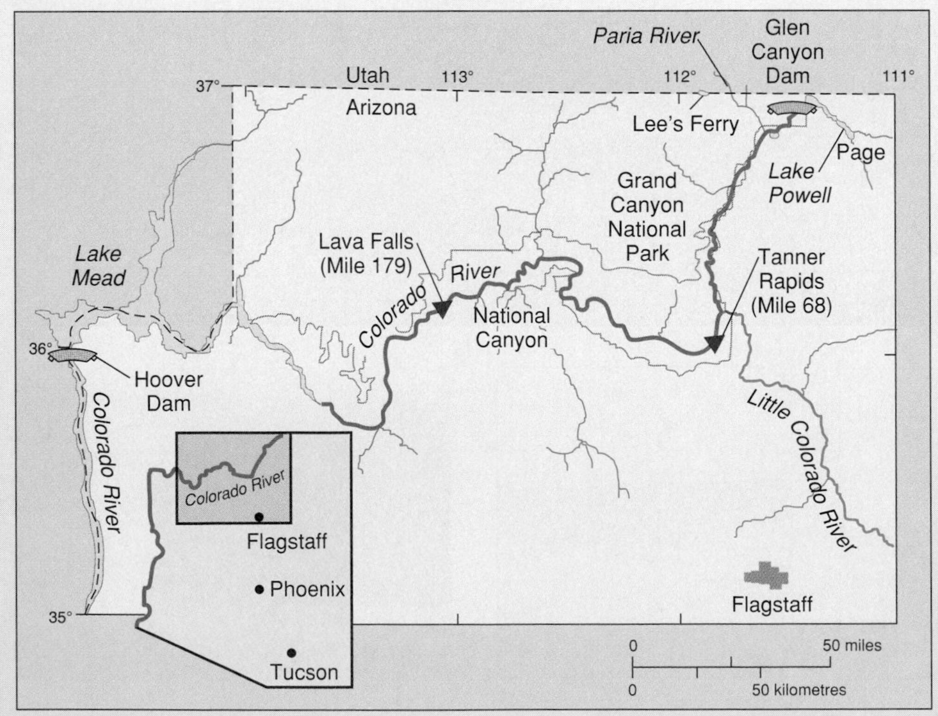

BOX 14.1 ■ FIGURE 1
Location map of the Grand Canyon controlled-flood experiment.
U.S. Geological Survey

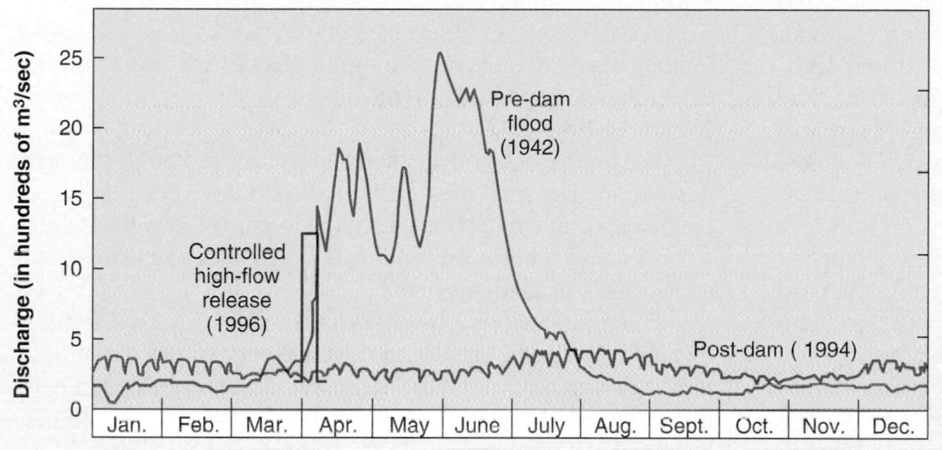

BOX 14.1 ■ FIGURE 2

Graph of annual peak discharges before (black) and after Glen Canyon Dam (blue) and the 1996 controlled high-flow release.

U.S. Geological Survey

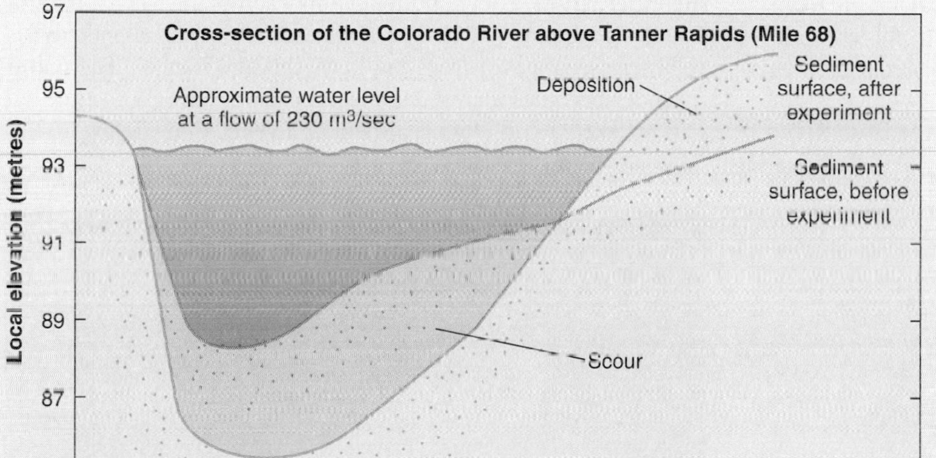

BOX 14.1 ■ FIGURE 3

Cross-section of the channel downstream of the confluence with the Little Colorado River. Increased flows have scoured the bottom sediment and redeposited it as a beach along the river bank.

U.S. Geological Survey

A

B

BOX 14.1 ■ FIGURE 4

(A) Hole being drilled into a basalt boulder and (B) radio tag installed to track the movement of boulders as the discharge and velocity of the Colorado River increases at the Lava Falls debris flow locality.

Photos courtesy of KUAT-TV. University of Arizona, photo by Dan Duncan

Additional Information

Flooding in Grand Canyon. *Scientific American,* January 1997, pp. 82–89.

Grand Canyon Flood! *NOVA Video,* 1997.

Webb, R. H., J. C. Schmidt, G. R. Marzolf, and R. A. Valdez, eds. 1999. *The Controlled Flood in Grand Canyon.* Geophysical Monograph Series 110.

Related Web Resources

For an overview and details of the specific experiments conducted during the planned flood, visit the following Websites:

http://water.usgs.gov/nrp/grandcanyon.html

www.pbs.org/wnet/nature/grandcanyonflood

Braided Streams

Deposition of a bar in the centre of a stream (a *midchannel bar*) diverts the water toward the sides where it washes against the stream banks with greater force, eroding the banks and widening the stream (figure 14.17*A*). A stream heavily loaded with sediment may deposit many bars in its channel, causing the stream to widen continually as more bars are deposited. Such a stream typically goes through many stages of deposition, erosion, deposition, and erosion, especially if its discharge fluctuates. The stream may fill its main channel and become a **braided stream,** flowing in a network of interconnected rivulets around numerous bars (figures 14.17*B* and *C*). A braided stream characteristically has a series of wide, shallow channels.

A stream tends to become braided when it is heavily loaded with sediment (particularly bed load) and has banks that are easily eroded. The braided pattern develops in deserts as a sediment-laden stream loses water through evaporation and through percolation into the ground. In meltwater streams flowing off glaciers, braided patterns tend to develop when the discharge from the melting glaciers is low relative to the great amount and ranges of size of sediment the stream has to carry.

In Canada, braided streams are much more common than meandering streams due to high sediment supply available from glaciers in Arctic and mountainous areas, and strong seasonal variability in discharge. Major Canadian river systems such as the Mackenzie River, the Yukon River, the St. Lawrence River (at Montreal), the Ottawa River (at Ottawa), and the Bow River (at Calgary) are braided.

Meandering Streams and Point Bars

Rivers that carry fine-grained silt and clay in suspension tend to be narrow and deep and to develop pronounced, sinuous curves called **meanders** (figure 14.18). In a long river, sediment tends to become finer downstream, so meandering is common in the lower reaches of a river.

You have seen in figure 14.6 that a river's velocity is higher on the outside of a curve than on the inside. This high velocity can erode the river bank on the outside of a curve, often rapidly (figure 14.19).

The low velocity on the inside of a curve promotes sediment deposition. The sand bars in figure 14.20 have been deposited on the inside of curves because of the lower velocity there. Such a bar is called a **point bar** and usually consists of a series of arcuate ridges of sand or gravel.

The simultaneous erosion on the outside of a curve and deposition on the inside can deepen a gentle curve into a hairpin-like meander (see figure 14.20). Meanders are rarely fixed in position. Continued erosion and deposition cause them to migrate back and forth across a flat valley floor as well as downstream, leaving scars and arcuate point bars to mark their former positions.

At times, particularly during floods, a river may form a **meander cutoff,** a new, shorter channel across the narrow neck of a meander (figure 14.21). The old meander may be abandoned as sediment separates it from the new, shorter channel. The cutoff meander becomes a crescent-shaped **oxbow lake**. With time, an oxbow lake may fill with sediment and vegetation (figure 14.22).

A
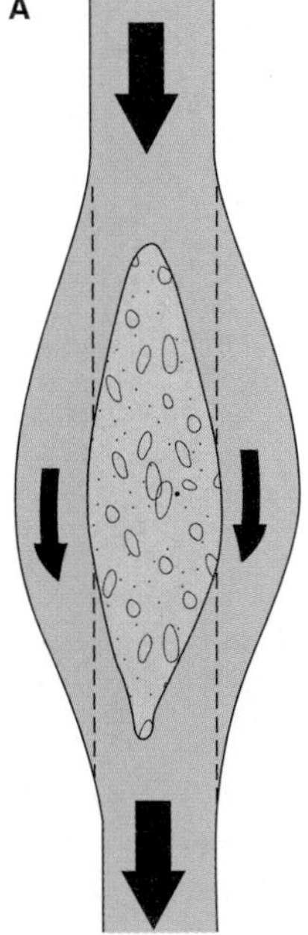

B
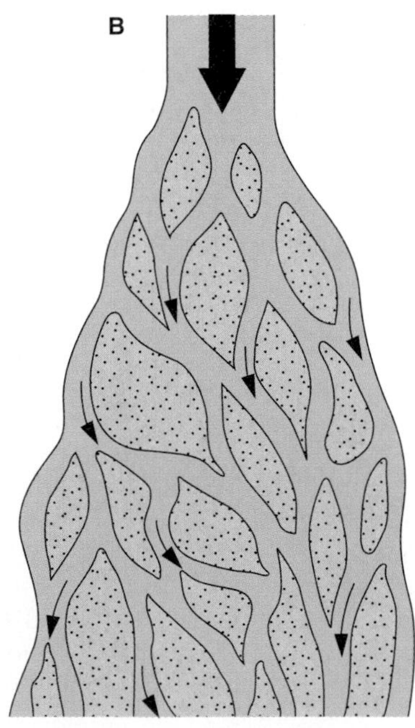

C
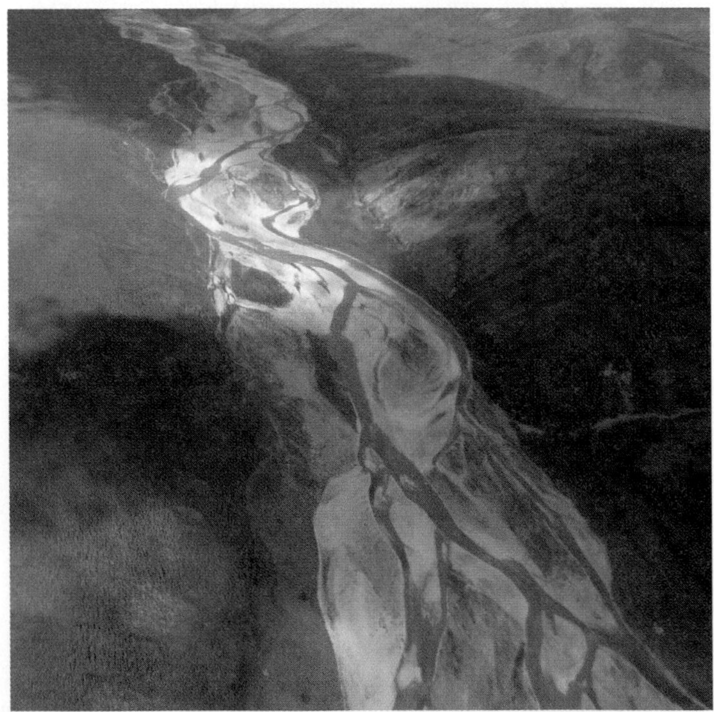

FIGURE 14.17

(*A*) A midchannel bar can divert a stream around it, widening the stream. (*B*) Braided stream occurs where there is an excess of sediment load. Bars split main channel into many smaller channels, greatly widening the stream. (*C*) Braided river in the Mackenzie Mountains, NWT. The bars and islands consist of sand and gravel deposited by the river.
Photo by Nick Eyles

FIGURE 14.18

Meanders in a stream. These sinuous curves develop because a stream's velocity is highest on the outside of curves, promoting erosion there.

Photo © Glenn M. Oliver/Visuals Unlimited

FIGURE 14.19

Erosion along the outside (left side of stream) of a meander curve on DesRosiers Creek, Quebec. Much of this erosion occurred as a result of flooding following heavy rainstorms in 2003. Trees undercut by erosion of cutbanks upstream have been deposited on the point bar on the inside of the meander curve together with boulders and gravel transported by the flood.

Photo from http://gsc.nrcan.gc.ca/floods/boisfrancs/photos_e.php. Reproduced with the permission of the Minister of Public Works and Government Services Canada, 2005 and Courtesy of Natural Resources Canada, Geological Survey of Canada.

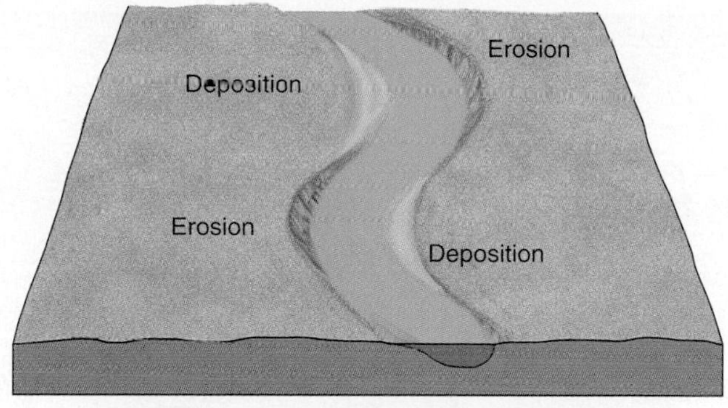

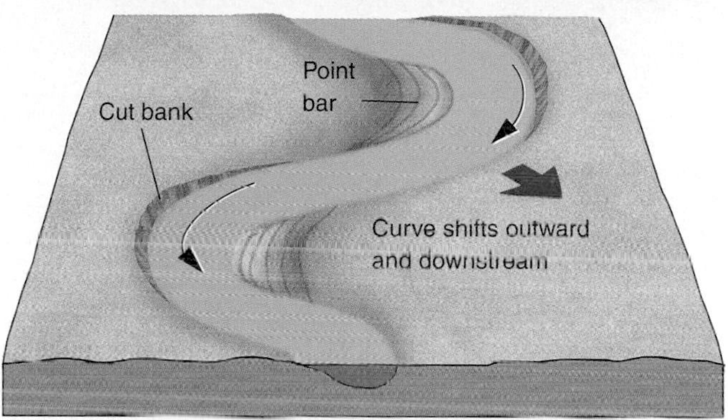

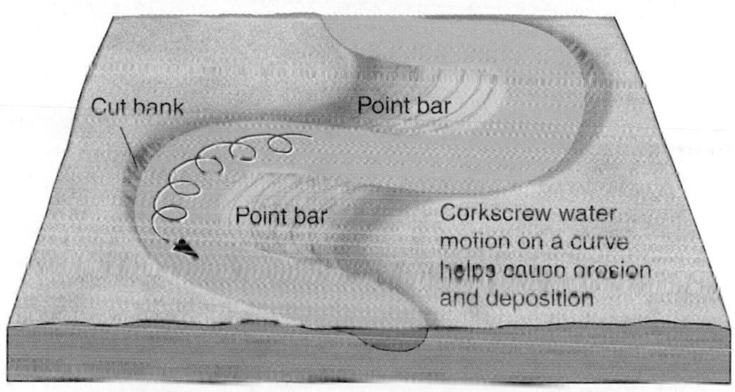

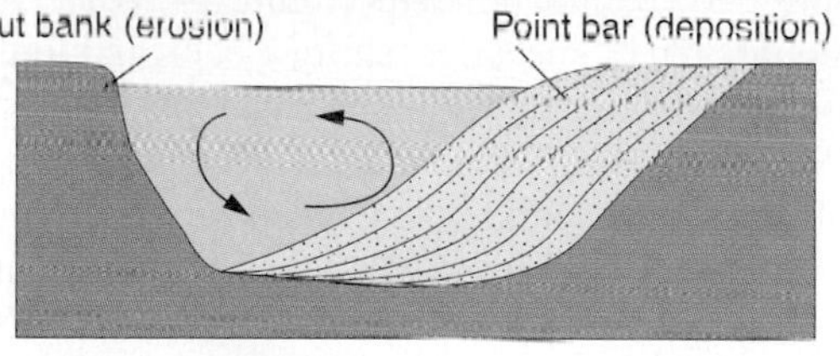

FIGURE 14.20

Development of river meanders and point bars by erosion and deposition on curves. Arrows indicate direction of water motion.

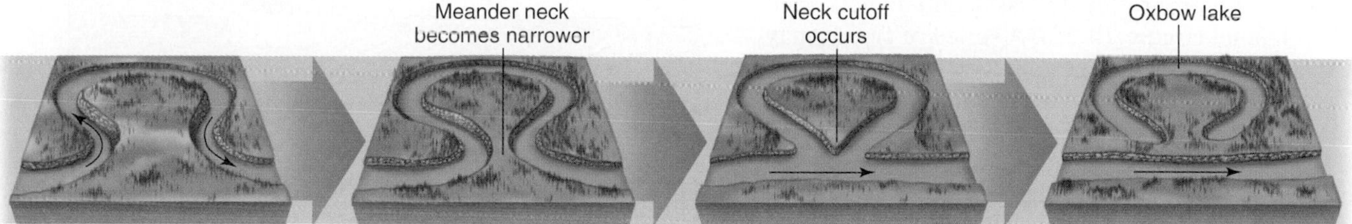

FIGURE 14.21

Creation of an oxbow lake by a meander neck cutoff. Old channel is separated from river by sediment deposition.

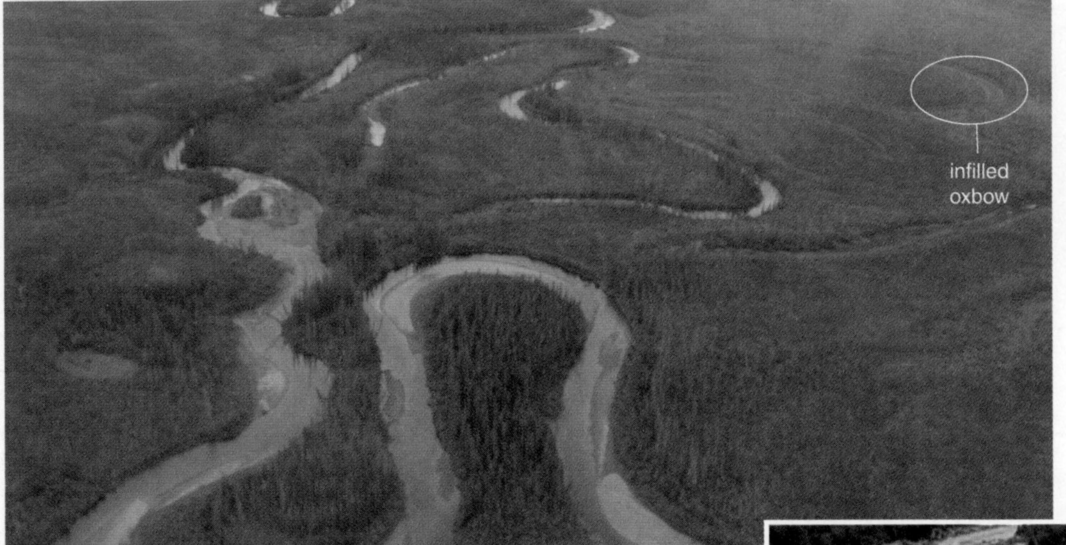

FIGURE 14.22

Well-developed meanders in tributary to the Mackenzie River, NWT. Infilled oxbow lakes can also be seen (horseshoe-shaped areas of brighter green colour).

Photo by Nick Eyles

FIGURE 14.23

Sediment deposited during the 1996 floods on the flood plain of the Rivière des Ha! Ha!, Quebec covered several properties with up to about 2 metres of sand derived from extensive erosion upstream.

Photo by G. R. Brooks. GSC Photo 1997-42KK Reproduced with permission of the Minister of Public Works and Government Services Canada, 2005, and Courtesy of Natural Resources Canada, Geological Survey of Canada.

Flood Plains

A **flood plain** is a broad strip of land built up by sedimentation on either side of a stream channel. During floods, flood plains may be covered with water carrying suspended silt and clay (figure 14.23). When the floodwaters recede, these fine-grained sediments are left behind as a horizontal deposit on the flood plain.

Some flood plains are constructed almost entirely of horizontal layers of fine-grained sediment, interrupted here and there by coarse-grained channel deposits (figure 14.24*A*). Other flood plains are dominated by meanders shifting back and forth over the valley floor and leaving sandy point-bar deposits on the inside of curves. Such a river will deposit a characteristic fining-upward sequence of sediments: coarse channel deposits are gradually covered by medium-grained point-bar deposits, which in turn are overlain by fine-grained flood deposits (figure 14.24*B*).

As a flooding river spreads over a flood plain, it slows down. The velocity of the water is abruptly decreased by friction as the water leaves the deep channel and moves in a thin sheet over the flat valley floor. The sudden decrease in velocity of the water causes the river to deposit most of its sediment near the main channel, with progressively less sediment deposited away from the channel (figure 14.25). A series of floods may build up **natural levees**—low ridges of flood-deposited sediment that form on either side of a stream channel and thin away from the channel. The sediment near the river is coarsest, often sand and silt, while the finer clay is carried farther from the river into the flat, lowland area (the back swamp). However, levee development is limited on many rivers due to available sediment supply and flooding patterns. The Red River in Manitoba, for example, has no natural levees.

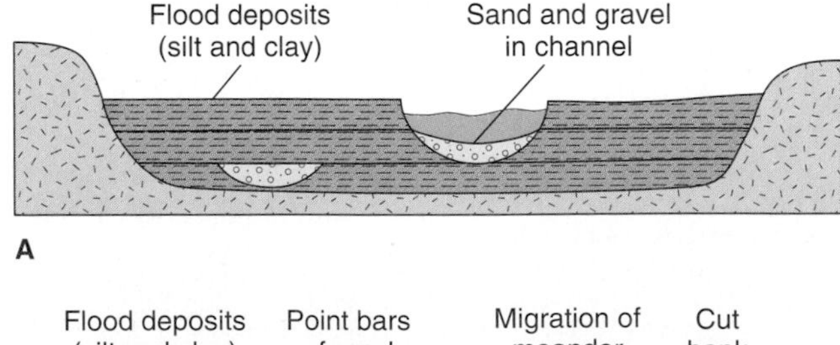

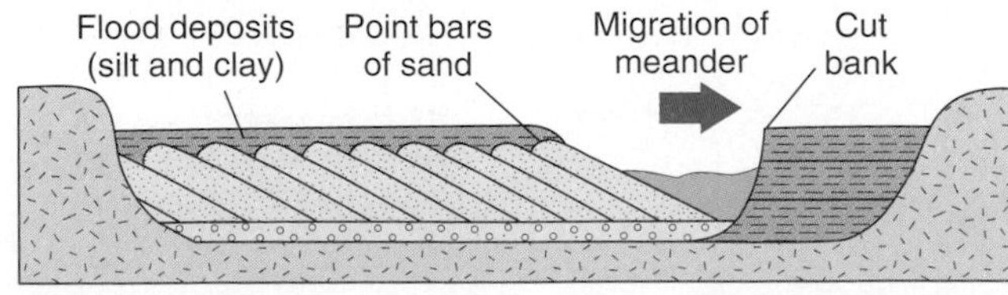

FIGURE 14.24

Flood plains. (*A*) Horizontal layers of fine-grained flood deposits with lenses of coarse-grained channel deposits. (*B*) A fining-upward succession deposited by a migrating meander. Channel gravel is overlain by sandy point bars, which are overlain by fine-grained flood deposits.

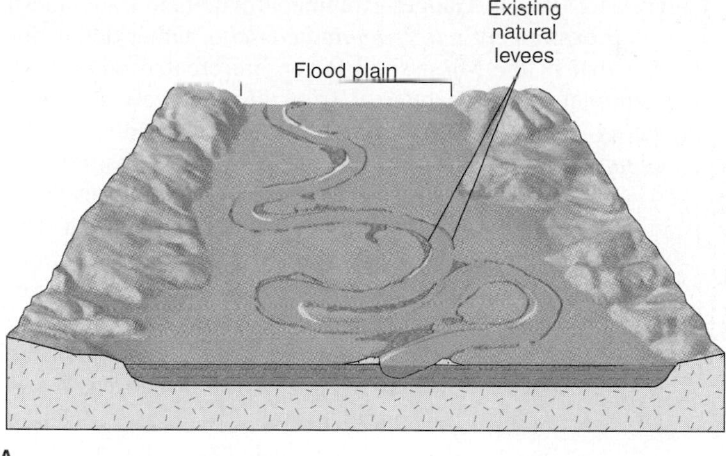

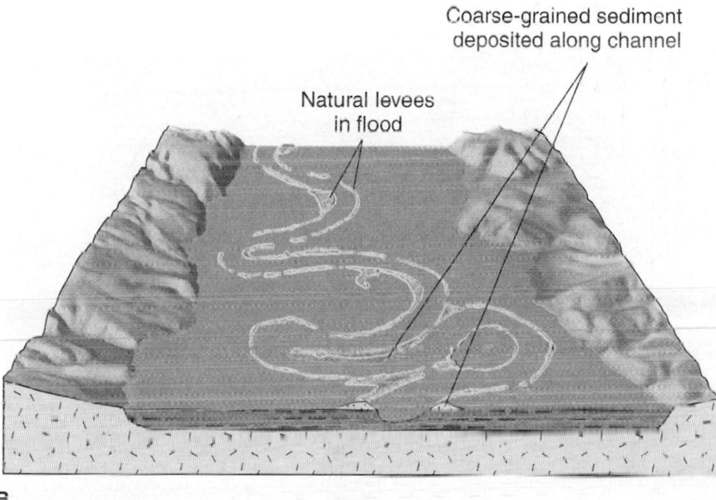

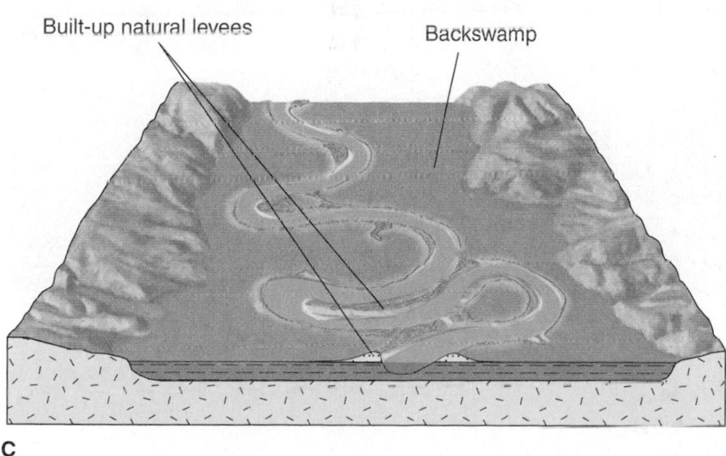

FIGURE 14.25

Natural levee deposition during a flood. Levees are thickest and coarsest next to the river channel and build up from many floods, not just one. (Relief of levees is exaggerated.) (*A*) Normal flow. (*B*) Flood. (*C*) After flood.

Deltas

Most streams ultimately flow into the sea or into large lakes. A stream flowing into quiet water usually builds a **delta,** a body of sediment deposited at the mouth of a river when the river's velocity decreases (figure 14.26)

The surface of most deltas is marked by **distributaries**—small, shifting channels that carry water away from the main river channel and distribute it over the surface of the delta (figure 14.27). Sediment deposited at the end of a distributary tends to block the water flow, causing distributaries and their sites of sediment deposition to shift periodically.

The shape of a marine delta in map view depends on the balance between sediment supply from the stream and the erosive power of waves and tides (figure 14.28). Some deltas, like that of the Nile River, are broadly triangular; this delta's resemblance to the Greek letter *delta* (Δ) is the origin of the name.

The Nile delta is a *wave-dominated delta* that contains barrier islands along its oceanward side (figure 14.28*A*); the barrier islands form by waves actively reworking the deltaic sediments. Wave reworking of delta fronts also occurs in large freshwater lakes. The Slave River delta in Great Slave Lake and the William River delta in Lake Athabasca are good examples of wave-dominated deltas. Some deltas form along a coast that is dominated by strong tides and the sediment is reshaped into tidal bars that are aligned parallel to a tidal current

FIGURE 14.26

Delta building out into Tanquary Fiord, Ellesmere Island, Nunavut. The river divides into several channels (distributaries) as it enters the fiord.

Photo by D. G. F. Long

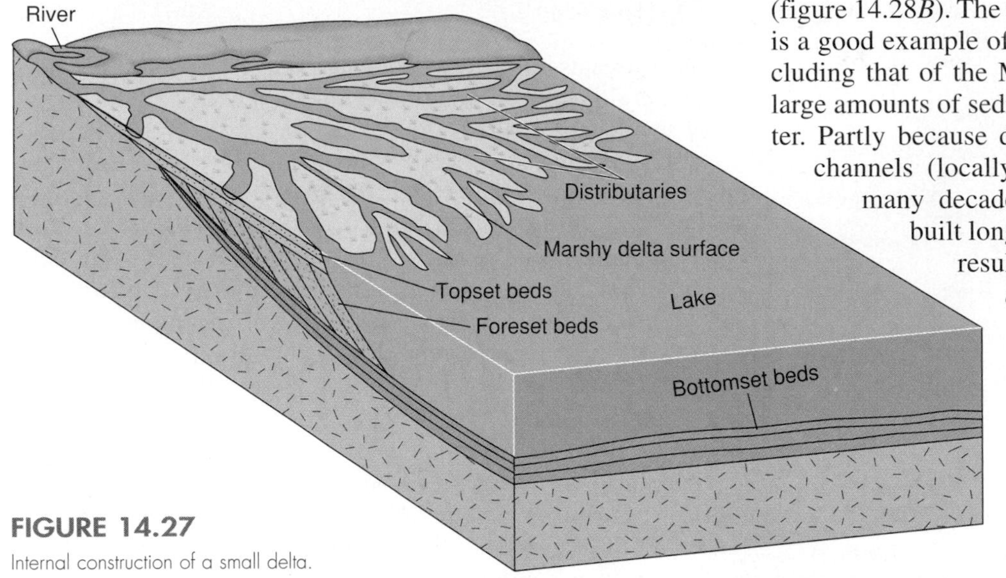

FIGURE 14.27

Internal construction of a small delta.

(figure 14.28*B*). The Ganges-Brahmaputra delta in Bangladesh is a good example of a *tide-dominated delta.* Other deltas, including that of the Mississippi River, are created when very large amounts of sediment are carried into relatively quiet water. Partly because dredging has kept the major distributary channels (locally called "passes") fixed in position for many decades, the Mississippi's distributaries have built long fingers of sediment out into the sea. The resulting shape has been termed a *birdfoot delta.* Because of the dominance of river sedimentation that forms the finger-like distributaries, birdfoot deltas like the Mississippi are also referred to as *river-dominated deltas.* The Fraser delta in B.C. and the Mackenzie delta in N.W.T. are river-dominated deltas but do not have a true birdfoot shape as they are also

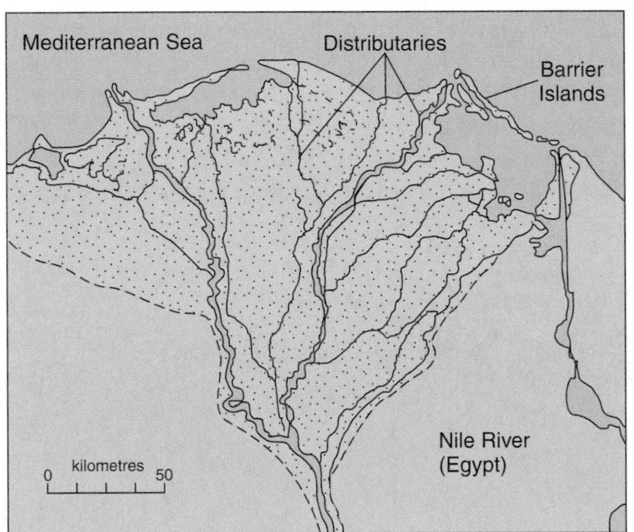

A Wave-dominated delta

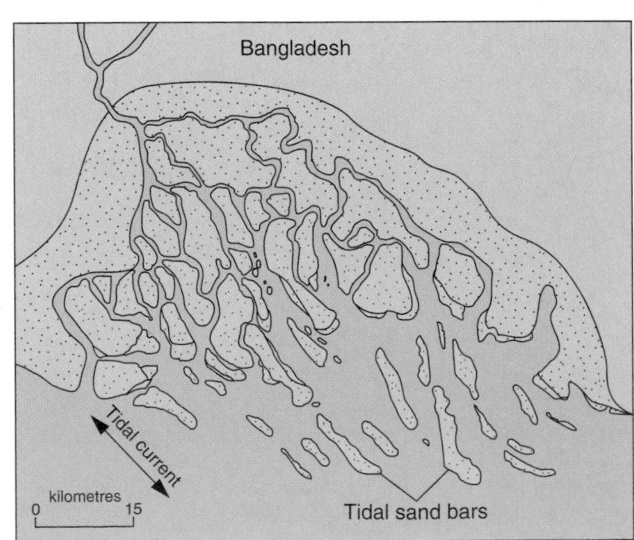

B Tide-dominated delta

C River-dominated delta

FIGURE 14.28

The shape of a delta depends on the amount of sediment being carried by the river and on the vigour of waves and tides in the sea. (*A*) The Nile delta is a wave-dominated delta with prominent barrier islands. (*B*) The Ganges-Brahmaputra delta in Bangladesh contains tidal sand bars formed by strong tidal currents. (*C*) A Landsat image of the Mississippi River delta. Note how sediment (light yellow) is carried by both river and ocean currents.

Figure A and B After R. G. Walker, "Facies Models," Geoscience Canada, *fig. 7, p. 109, 1979; Photo C by NASA*

influenced by wave activity. Marine ice, pushed onshore by winds and tides, also has an effect on delta morphology, producing ice-dominated deltas such as the Yukon River delta near Nome, Alaska. The Mackenzie delta in N.W.T. also shows the effects of seasonal ice.

Many deltas, particularly small ones in freshwater lakes, are built up from three types of deposits, shown in the diagram in figure 14.27. *Foreset beds* form the main body of the delta. They are deposited at an angle to the horizontal. This angle can be as great as 20 to 25 degrees in a small delta where the foreset beds are sandy, or less than 5 degrees in large deltas with fine-grained sediment. On top of the foreset beds are the *topset beds,* nearly horizontal beds of varying grain size formed by distributaries shifting across the delta surface. Out in front of the foreset beds are the *bottomset beds,* deposits of the finest silt and clay carried out into the lake by the river water flow or by sediments sliding downhill on the lake floor. Many of the world's great deltas in the ocean are far more complex than the simplified diagram shown in the figure. Shifting river mouths, wave energy, currents, and other factors produce many different internal structures.

The persistence of large deltas as relatively "dry" land depends upon a balance between the rate of sedimentation and the rates of tectonic subsidence and of compaction of water-saturated sediment. Many deltas are sinking, with seawater encroaching on once-dry land. The Mississippi River delta in Louisiana is sinking, as upstream dams catch sediment, reducing the delta's supply, and as extraction of oil and gas from beneath the delta accelerates subsidence. The Peace River delta of northern Alberta is also subsiding and adjacent ponds and wetlands are drying as a result of upstream dam development and climate changes, which have reduced water flows. The flat surface of a delta is a risky place to live or farm, particularly in regions threatened by the high waves and storm surges associated with hurricanes, such as the U.S. Gulf Coast and the country of Bangladesh (the Ganges-Brahmaputra delta) on the Indian Ocean. Low lying areas of the U.S. Gulf Coast were severely affected by repeated hurricanes during the 2005 hurricane season (see box 14.4 later in this chapter).

Alluvial Fans

Some streams, particularly in dry climates, do not reach the sea or any other body of water. They build *alluvial fans* instead of deltas. An **alluvial fan** is a large, fan- or cone-shaped pile of sediment that usually forms where a stream's velocity decreases as it emerges from a narrow mountain canyon onto a flat plain (figure 14.29). Alluvial fans are particularly well developed and exposed in the southwestern desert of the United States and in other desert regions such as parts of the Canadian Arctic.

An alluvial fan builds up its characteristic fan shape gradually as streams shift back and forth across the fan surface and deposit sediment, usually in a braided pattern. Deposition on an alluvial fan in the desert is discontinuous because streams typically flow for only a short time after the infrequent rainstorms. When rain does come, the amount of sediment to be moved is often greater than the available water and material is moved as a debris flow before it comes to rest and is deposited.

The sudden loss of velocity when a stream flows from steep mountains onto a broad plain causes the sediment to deposit on an alluvial fan. The loss of velocity is due to the widening or branching of the channel as it leaves the narrow mountain canyon. The gradual loss of water as it infiltrates into the fan also promotes sediment deposition. On large fans, deposits are graded in size within the fan, with the coarsest sediment dropped nearest the mountains and the finer material deposited progressively farther away. Small fans do not usually show such grading.

CAN WE PREDICT WHEN FLOODS WILL OCCUR?

Many of the world's cities, such as New Orleans, Prague, Florence, Bangkok, and Winnipeg, are built beside rivers and therefore can be threatened by floods. Rivers are important transportation routes for ships and barges, and flat flood plains have excellent agricultural soil and offer attractive building sites for houses and industry.

Flooding does not occur every year on every river, but flooding is a natural process on all rivers and must be prepared for by river cities and towns. Heavy rains and the rapid melting of snow in the springtime are the usual causes of floods (box 14.2).

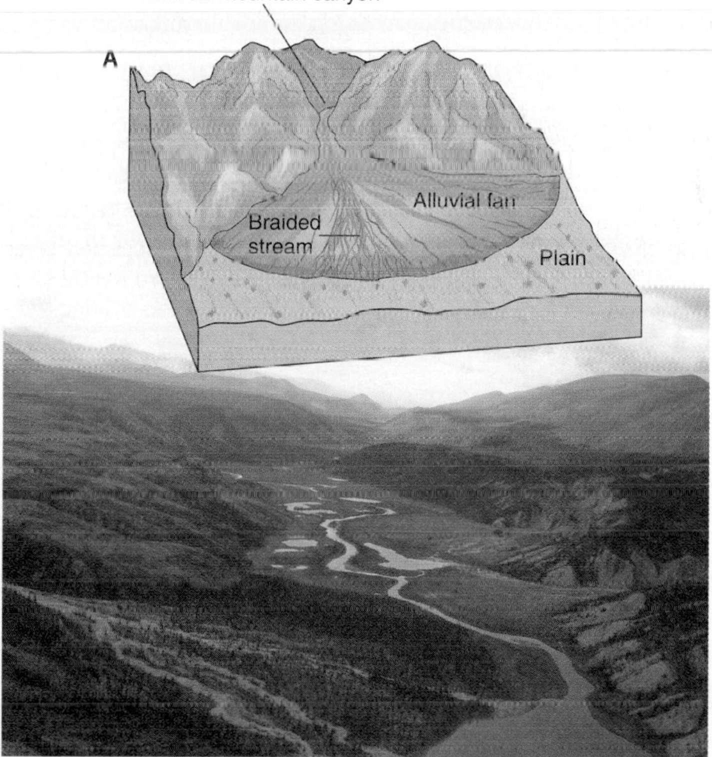

FIGURE 14.29

(*A*) An alluvial fan at the mouth of a desert canyon. (*B*) Alluvial fans formed by streams issuing from narrow tributary valleys onto the flat floor of the main valley. Near Norman Wells, Northwest Territories.

Photo by Nick Fyles

GEOMATICS 14.2

Monitoring Flood Events and Rehabilitation Efforts

In order to effectively delineate flood hazard lands and plan effective flood relief operations, accurate and timely information on flood extent is required. Analysis of remotely sensed data, including that obtained from satellite and airborne sensors, is now being used to monitor flood events and to assist in the planning and implementation of rehabilitation efforts. The most usual causes of flooding in Canada are heavy rainfall over short time periods and rapid melting of snow in the springtime.

A major flood event caused by heavy rainfall occurred in July 1996, when more than 29 cm of rain fell in the Saguenay region of eastern Quebec in a period of less than 36 hours. Severe flooding of many rivers in the area caused extensive bank and channel erosion, damage to bridges and roads, and breaching of dams and dikes. Sixteen thousand people were evacuated from their homes, approximately 1,350 homes were destroyed, and 10 people were killed in the floods, which caused an estimated $1 billion in damage. Damage to industrial facilities including pulp and paper mills and aluminum plants also allowed the release of more than 100 tonnes of toxic waste and chemicals into the Saguenay River. Following the flood, the Canada Centre for Remote Sensing analyzed and interpreted remotely sensed images obtained from Landsat Thematic Mapper (TM), SPOT Haute Résolution Visible (HRV), Radarsat Synthetic Aperture Radar (SAR), and multispectral airborne videography. The images allowed evaluation of the extent and nature of flood damage at a variety of scales, including geomorphological changes (box figure 1*B*), property damage, and disruption of transportation and communication infrastructure.

Extensive flooding can also be caused by rapid spring snowmelt or a winter thaw. The 877-km-long Red River, which flows from North Dakota/Minnesota northward into Manitoba, regularly floods its valley in the spring. Flooding of the Red River is primarily caused by melting of snow in the southern headwaters of the river before the northern reaches in Manitoba are ice-free. Severe floods in 1948 and 1950 caused 80,000 inhabitants of Winnipeg to evacuate, forced 20,000 rural residents to flee their homes, and flooded more than 13,000 homes. As a result of these events a $63-million floodway was constructed to divert the river around the city of Winnipeg (box figure 2*B*). This floodway has saved Winnipeg from flooding 18 times since it was opened in 1968. However, in the early spring of 1997, extensive flooding of the Red River was caused by a combination of large amounts

1. New bar deposit
2. Displaced piece of concrete dam
3. New channel eroded between west dam abutment and valleyside; erosion of valleyside
4. New channel (now dry) formed by overflow of dam
5. Eroded bank
6. Damaged concrete dam overtopped during flood
7. Breached wing-wall of dam
8. Eroded bank
9. Bar deposits within former reservoir basin
10. Minor failures of valleysides above former reservoir
11. Severely damaged powerhouse
12. Area of major lateral valleyside erosion that caused the collapse by undermining of two apartment buildings and damage to others
13. Former spillway below dam
14. Intact dam overtopped during flood, but no longer functional due to adjacent lateral erosion and deep incision of the river
15. Part of former reservoir basin
16. New, deeply-incised channel (down to bedrock) adjacent to dam
17. Minor scouring of the bank of former reservoir
18. Wide, shallow retrogressive failure in bank (possibly caused by rapid drawdown of reservoir)
19. Extensive erosion of the west approach of rail bridge; western bridge support undermined by fluvial scouring of foundation
20. New bar deposition (side, point and mid-channel bars)

BOX 14.2 ■ FIGURE 1

Geomorphic interpretation of 1996 videograph image of the area of Saguenay River near Jonquière after the flood (interpretation by Dr. Greg Brooks and Dr. Ted Lawrence, GSC).

Image from: Kung, Yatabe and Pulz, 1998: Application of Remote Sensing to Environmental Monitoring: A Case Study of the 1996 Saguenay Flood in Quebec. GAC/MAC Annual Meeting

of water supplied by melting of a near–record breaking snowpack, ground that had been fully saturated with rainfall in the fall, and a major storm that rapidly dumped 50 to 75 cm of snow and freezing rain. More than 5 percent of Manitoba's farmland was covered by floodwater, and the losses to dairy farmers alone were estimated to be almost $2 billion. The flood protection measures effectively reduced the impact of the floods in the urban areas, but still 28,000 Manitobans were evacuated and 2,500 homes were damaged. Without the floodway, it is estimated that more than 80 percent of Winnipeg would have been inundated by water and 550,000 people evacuated from the city.

The Province of Manitoba established a flood monitoring system (MFMS) to generate digital and hard-copy maps showing the extent and evolution of any flood along the Red River and provided the types of statistics necessary to determine impact on infrastructure and land uses. RADARSAT data were used to generate the information on the 1997 Red River flood in near-real-time. Thirteen days of imagery were acquired, starting on April 21, 1997 and ending June 7, 1997—the data were received by MFMS within four hours of acquisition by the satellite. By the time the 1997 flood peaked on May 4, RADARSAT imagery was readily available for use by emergency measures personnel responsible for the effective planning of evacuation orders, infrastructure protection, and search and rescue operations (box figure 2*A*).

A

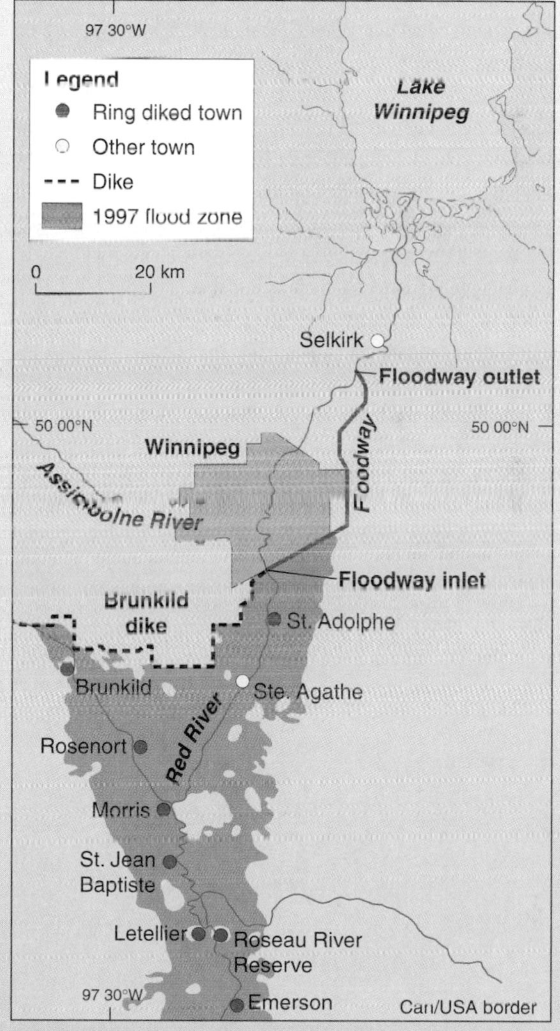

B

BOX 14.2 ■ FIGURE 2

(*A*) Central part of the Red River Valley at the peak of the flood on May 4, 1997.
The image was created by combining flood information generated from the RADARSAT image with LANDSAT imagery.

(*B*) Map of the Red River valley in Manitoba showing location of floodway around the city of Winnipeg.

Used with permission of Manitoba Remote Sensing Centre; http://www.gov.mb.ca/conservation/geomatics/remote_sensing/floodmonitor.html

The rate and volume of rainfall, and the geographic path of rainstorms or hurricanes, often determine whether flooding will occur. The most devastating flooding event to affect Canadians was that caused by Hurricane Hazel in 1954. Hurricane Hazel dumped 18 cm of rain on the Toronto region in less than 24 hours, causing the extensive flooding of creeks and rivers that led to the drowning deaths of 81 people. Land zoning practices were introduced in the region following this storm, and potential flood hazard lands were identified along creeks and rivers.

Floods are described by *recurrence interval,* the average time between floods of a given size. A "100-year flood" is one that can occur, on the *average,* every 100 years (box 14.3). A 100-year flood has a 1-in-100, or one percent, chance of occurring in any given year. It is perfectly possible to have two 100-year floods in successive years—or even in the same year. If a 100-year flood occurs this year on the river you live beside, you should not assume that there will be a 99-year period of safety before the next one.

Flood erosion is caused by the high velocity and large volume of water in a flood. Although relatively harmless on an uninhabited flood plain, flood erosion can be devastating to a city. As a river undercuts its banks, particularly on the outside of curves where water velocity is high, buildings, piers, and bridges may fall into the river (figure 14.30*A*). As sections of flood plain are washed away, highways and railroads are cut (figure 14.30*B*).

High water covers streets and agricultural fields and invades buildings, shorting out electrical lines and backing up sewers. Water-supply systems may fail or be contaminated. Water in your living room will be drawn upward in your walls by capillary action in wall plasterboard and insulation, creating a soggy mess that has to be torn out and replaced. High water on flat flood plains often drains away very slowly; street travel may be by boat for weeks. If floodwaters are deep enough, houses may float away.

Flood deposits are usually silt and clay. A new layer of wet mud on a flood plain in an agricultural region can be beneficial in that it renews the fields with topsoil from upstream—as used to be the case with the Nile River until the Aswan dam was built. The same mud in a city will destroy lawns, furniture, and machinery. Cleanup is slow; imagine shovelling 10 centimetres of worm-filled mud that smells like sewage out of your house.

Urban Flooding

Urbanization contributes to severe flooding. Paved areas and storm sewers increase the amount and rate of surface runoff of water, making river levels higher during storms (figure 14.31). Such a rapid increase in runoff or discharge to a river is called a "flashy" discharge. Storm sewers are usually designed for a 100-year storm; however, large storms that drop a lot of rain in a short period of time (cloudburst) may overwhelm sewer systems and cause localized flooding. Rising river levels may block storm sewer outlets and also add to localized flooding problems.

Bridges, docks, and buildings built on flood plains can also constrict the flow of floodwaters, increasing the water height and velocity and promoting erosion.

Flash Floods

Some floods occur rapidly and die out just as quickly. *Flash floods* are local, sudden floods of large volume and short duration, often triggered by heavy thunderstorms. A startling example occurred in 1976 in north-central Colorado along the Big

A

B

FIGURE 14.30

(*A*) Townhouses undermined by erosion of the river bank during the 1996 flooding of the Saguenay region of Quebec.
(*B*) Undermining of railway tracks along the Rivière à Mars, Quebec by floodwaters in 1996.
Photo A CP (Jacques Boissinot); Photo B CP (Jacques Boissinot)

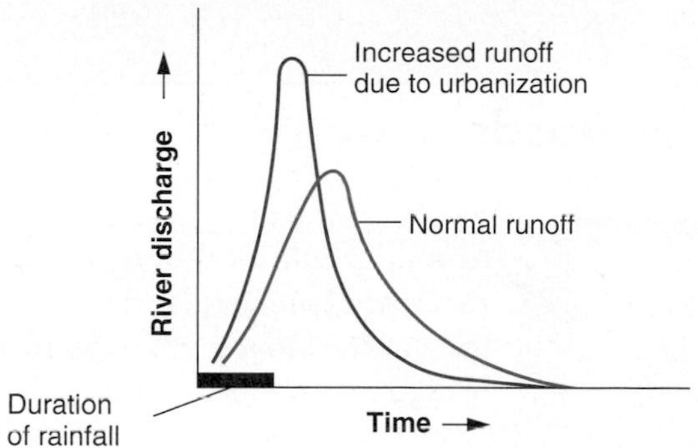

FIGURE 14.31

Schematic diagram to show that urbanization can increase the chance of floods. The blue curve shows the normal increase in a river's discharge following a rainstorm (black bar). The red curve shows the great increase in runoff rate and amount caused by pavement and storm sewers in a city.

Thompson River. Strong winds from the east pushed moist air up the front of the Colorado Rockies, causing thunderstorms in the steep mountains. The storms were unusually stationary, allowing as much as 30 cm of rain to fall in 2 days. The volume of water in the Big Thompson River swelled to four times the previously recorded maximum, and the river's velocity rose to an impressive 25 kilometres per hour for a few hours on the night of July 31. By the next morning the flood was over, and the appalling toll became apparent—139 people dead, 5 missing, and more than $35 million in damages.

Controlling Floods

Flood-control structures can partially reduce the dangers of flood waters and sedimentation to river cities (figure 14.32). Upstream dams can trap water and release it slowly after the storm. (A dam also catches sediment, which eventually fills its reservoir and ends its life as a flood-control structure.) Artificial levees are embankments built along the banks of a river channel to contain floodwaters within the channel. Protective walls of stone (rip-rap) or concrete are often constructed along river banks, particularly on the outside of curves, to slow erosion. Floodwalls, walls of concrete, may be used to protect cities from flooding; however, these flood-control structures may constrict the channel and cause water to flow faster with more erosive power downstream. Bypasses are also used along the Mississippi and other rivers to reduce the discharge in the main channel by diverting water into designated basins in the flood plain. The bypasses serve to give part of the natural flood plain back to the river.

Dams and levees are designed to control certain specified floods. If the flood-control structures on your river were designed for 75-year floods, then a much larger 100-year flood will likely overtop these structures, and may destroy your home. The disastrous floods along the Missouri and Mississippi Rivers and their tributaries north of Cairo, Illinois in 1993 resulted from many such failures in flood control. Extensive flooding of New Orleans in 2005 in the wake of Hurricane Katrina was in part due to failure of improperly designed, constructed, and maintained levee systems (see box 14.4).

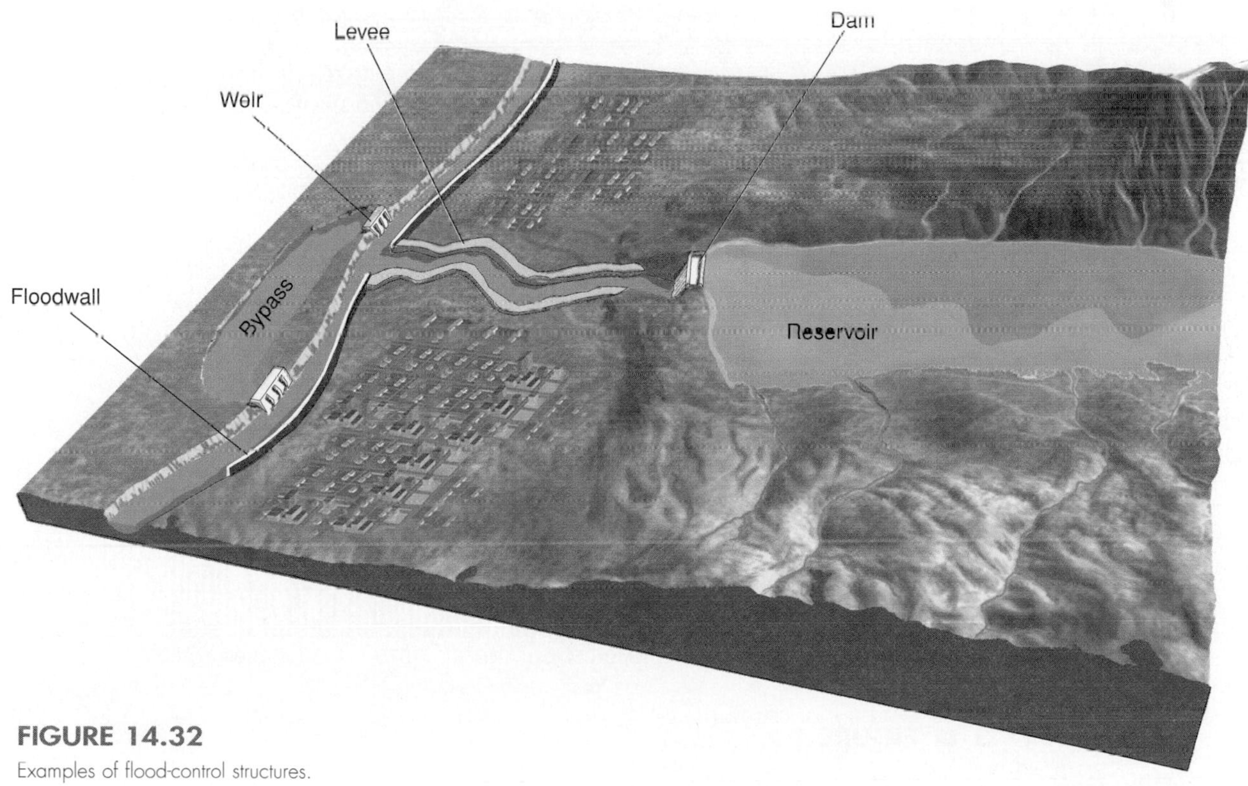

FIGURE 14.32

Examples of flood-control structures.

IN GREATER DEPTH 14.3

Estimating the Size and Frequency of Floods

Because people have encroached on the flood plains of many rivers, flooding is one of the most universally experienced geologic hazards. To minimize flood damage and loss of life, it is useful to know the potential size of large floods and how often they might occur. This is often a difficult task because of the lack of long-term records for most rivers. Environment Canada monitors the stage (water elevation) and discharge of rivers and streams throughout Canada in order to collect data that can be used to attempt to predict the size and frequency of flooding and to make estimates of water supply.

Hydrologists designate floods based on their *recurrence interval*, or *return period*. For example, a 100-year flood is the largest flood expected to occur within a period of 100 years. This does not mean that a 100-year flood occurs once every century, but that there is a 1-in-100 chance, or a one-percent probability, each year that a flood of this size will occur. Usually flood-control systems are built to accommodate a 100-year flood.

To calculate the recurrence interval of flooding for a river, the annual peak discharges (largest discharge of the year) are collected and ranked according to size (box figure 1 and table 1). The largest annual peak discharge is assigned a rank (m) of one, the second a two, and so on until all of the discharges are assigned a rank number. The *recurrence interval* (**R**) of each annual peak discharge is then calculated by adding one to the *number of years of record* (**n**) and dividing by its *rank* (**m**).

$$\mathbf{R} = \frac{\mathbf{n+1}}{\mathbf{m}}$$

BOX 14.3 TABLE 1 **Annual peak discharge and recurrence intervals in rank order for the Red River at Redwood Bridge, Manitoba**

Year	Peak Discharge (m^3/sec)	Magnitude Rank (m)	Recurrence Interval
1997	4,600	1	109.00
1950	3,060	2	54.50
1979	3,030	3	36.33
1996	2,960	4	27.25
1974	2,720	5	21.80
1904	2,210	10	10.90
1995	1,913	20	5.45
1962	1,684	30	3.63
1898	1,475	40	2.73
1964	1,146	50	2.18
1930	1,083	60	1.82

For example, the Red River in Manitoba has 108 years of record (n = 108), and in 1950 the second largest peak discharge (m = 2) of 3,060 m^3/sec occurred. The recurrence interval (R), or expected frequency of occurrence, for a discharge this large is 54.5 years:

$$\mathbf{R} = \frac{\mathbf{108+1}}{\mathbf{2}} = \mathbf{54.5\ years}$$

That is, there is a 1-in-54—or less than 2-percent—chance each year of a peak discharge of 3,060 m^3/sec or greater occurring on the Red River.

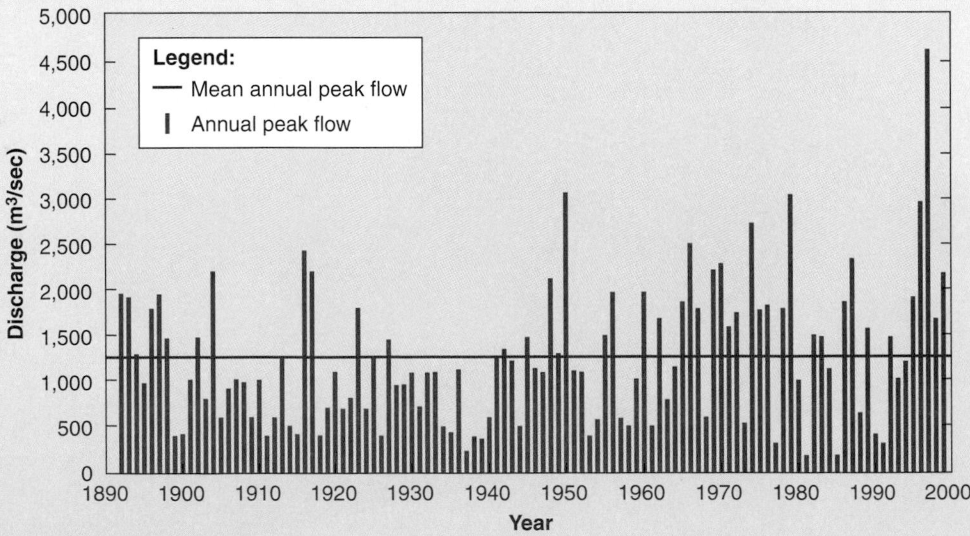

BOX 14.3 ■ FIGURE 1

Annual peak discharge for the Red River, Manitoba, from 1892–1999 (data from Manitoba Water Resources)

BOX 14.3 ■ FIGURE 2
The town of Morris, Manitoba, during the 1997 flooding of the Red River. The ground surface is lower than the water, as the town is protected by a ring dike.
http://gsc.nrcan.gc.ca/floods/redriver/images_e.php Reproduced with permission of the Minister of Public Works and Government Services Canada, 2007 and Courtesy of Natural Resources Canada, Geological Survey of Canada.

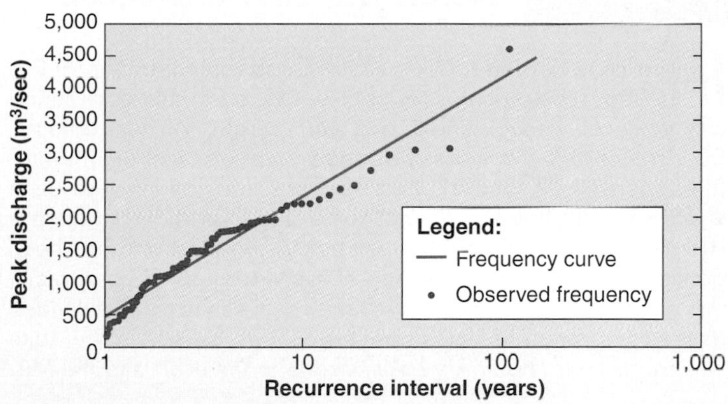

BOX 14.3 ■ FIGURE 3
Flood frequency curves for the Red River.

The flood of record (largest recorded discharge) occurred in the early spring of 1997, when melting of a record amount of snow and additional precipitation from a severe storm caused flooding over large areas of Manitoba. A peak discharge of 4,600 m^3/sec resulted in widespread flooding of homes and agricultural areas (box figure 2). The recurrence interval for the 1997 flood (4,600 m^3/sec) is 109 years.

$$R = \frac{108 + 1}{1} = 109 \text{ years}$$

A *flood-frequency curve* can be useful in providing an estimate of the discharge and the frequency of floods. The flood-frequency curve is generated by plotting the annual peak discharges against the calculated recurrence intervals (box figure 3). Because most of the data points defining the curve plot in the lower range of discharge and recurrence interval, there is some uncertainty in projecting larger flood events. Large floods do not occur as often as small floods, and the rare large flood can have a dramatic effect on the shape of the flood-frequency curve and the estimate of a 100-year event.

Related Web Resources

To find data sets to calculate the recurrence interval for rivers throughout Canada, access the Environment Canada National Water Data Archive Website:

www.wsc.ec.gc.ca/climate/data_archives/water/Index_e.cfm.

For rivers in the United States, access the U.S. Geological Survey Water Data Retrieval Website:

http://water.usgs.gov/usa/nwis/

Wise land-use planning and zoning for flood plains should go hand in hand with flood control. Wherever possible, buildings should be kept out of areas that might someday be flooded by 100-year floods.

HOW DO STREAM VALLEYS CHANGE WITH TIME?

Valleys, the most common landforms on the Earth's surface, are usually cut by streams. By removing rock and sediment from the stream channel, a stream deepens, widens, and lengthens its own valley.

Downcutting and Base Level

The process of deepening a valley by erosion of the stream bed is called **downcutting.** If a stream removes rock from its bed, it can cut a narrow *slot canyon* down through rock (figures 14.33*A* and *B*). Such narrow canyons do not commonly form, because mass wasting and sheet erosion usually remove rock from the valley walls. These processes widen the valley from a narrow, vertical-walled canyon to a broader, open, V-shaped canyon (figures 14.33*C* and *D*). Slot canyons persist, however, in very resistant rock with favourably oriented fractures or in regions where downcutting is rapid.

Downcutting cannot continue indefinitely, because the headwaters of a stream cannot cut below the level of the stream bed at the mouth. If a river flows into the ocean, sea level becomes the lower limit of downcutting. The river cannot cut below sea level, or it would have to flow uphill to get to the sea. For most streams, sea level controls the level to which the land can be eroded.

The limit of downcutting is known as **base level;** it is a theoretical limit for erosion of the Earth's surface (figure 14.34*A*). Downcutting will proceed until the stream bed reaches base level. If the stream is well above base level, downcutting can be quite rapid; but as the stream approaches base level, the rate of

IN GREATER DEPTH 14.4

When the Levee Breaks: Flooding of New Orleans, 2005

Hurricane Katrina struck the Gulf Coast region on August 29, 2005, subjecting the city of New Orleans to a lengthy period of high winds, heavy rainfall, and storm surges. Within 36 hours, 80 percent of the city was flooded by up to 6 metres of water (box figures 1, 2, and 3), and most of its inhabitants had been evacuated. This flooding did not come as a surprise. New Orleans is built on a delta marsh that lies below sea level and has been protected from seasonal flooding of the Mississippi River by a system of levees initially built in the nineteenth century and extended as the city grew. The city is now surrounded by water, with Lake Pontchartrain to the north, Lake Borgne to the east, and the Mississippi River and numerous drainage and shipping canals running through its streets. The problem of flooding has increased with the growth of New Orleans as drainage of swampy grounds for construction has caused land subsidence, and increased traffic along the river and canals enhances bank erosion and loss of protective vegetation. Hurricane Katrina weakened the city's levee system with heavy winds and powerful storm surges, causing the Mississippi River to breach its levees in at least 20 places, and levees along the 17th Street Canal (box figure 4), the London Avenue Canal, and the Industrial Canal to fail. Flooding from these breaches inundated the city with water for many weeks.

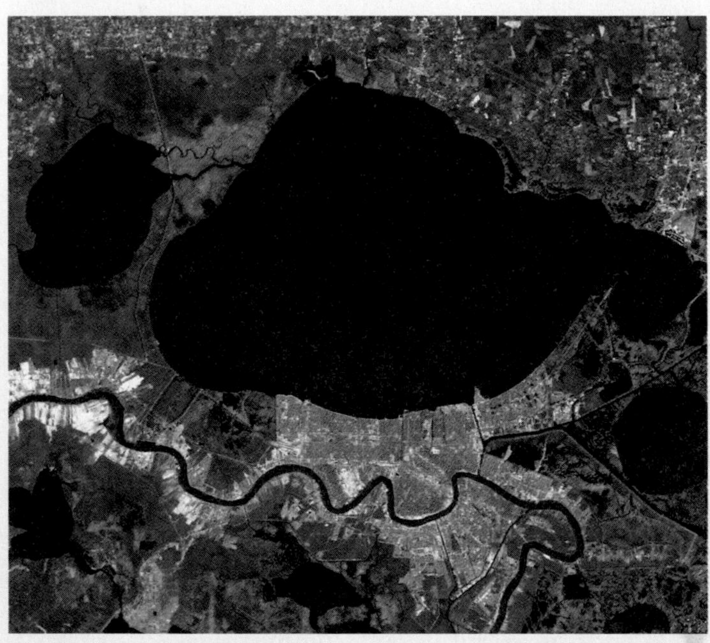

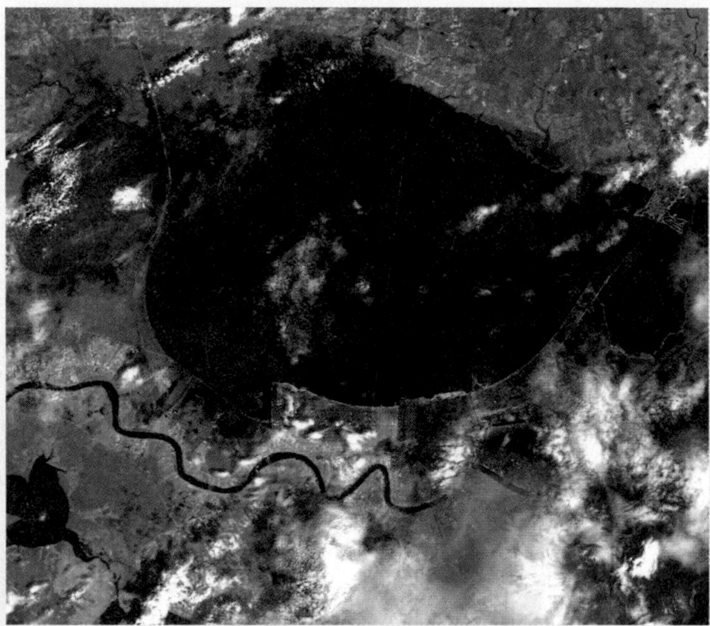

BOX 14.1 ■ FIGURE 1

Top: Landsat 7 image of New Orleans, August 24, 2005.
Bottom: Landsat 7 image of New Orleans, August 30, 2005. Flooded areas of the city are shown in dark green.
It is estimated that approximately 80 percent of New Orleans was underwater when this image was taken.
From U.S. Geological Survey, U.S. Department of the Interior/USGS. http://landsat.usgs.gov/gallery/detail/412/.

BOX 14.4 ■ FIGURE 2

Homes surrounded by floodwaters from Hurricane Katrina as fires burn downtown New Orleans, September 2, 2005
Photo CP/AP (David J. Phillip)

Investigations of the levee breaches immediately following Hurricane Katrina suggested that the levees had been overtopped by floodwaters beyond their design strength. However, subsequent investigations have shown that in places the levees were built improperly—lower than recommended by appropriate guidelines, and using materials that were easily erodible.

BOX 14.4 ■ FIGURE 3
Flooding of homes and roadways near New Orleans (as seen from Air Force One), August 31, 2005
Photo CP/AP (Susan Walsh)

BOX 14.4 ■ FIGURE 4
Workers repair the broken 17th Street Levee in New Orleans, September 3, 2005
Photo CP/AP (David J. Phillip)

downcutting slows down. For streams that reach the ocean base level is close to sea level, but since streams need at least a gentle gradient in order to flow, base level slopes gently upward in an inland direction.

During the glacial fluctuations of the Quaternary (see chapter 16), sea level rose and fell as water was removed from the sea to form the glaciers on the continents and returned to the sea when the glaciers melted. This means that base level rose and fell for streams flowing into the sea. As a result, the lower reaches of such rivers alternated between erosion (caused by low sea level) and deposition (caused by high sea level). Since the glaciers advanced and retreated several times, the cycle of erosion and deposition was repeated many times, resulting in a complex history of cutting and filling near the mouths of most old rivers.

Base levels for streams that do not flow into the ocean are not related to sea level. In Death Valley in California (figure 14.34*B*), base level for in-flowing streams corresponds to the lowest point in the valley, 86 metres *below* sea level (the valley has been dropped below sea level by tectonic movement along faults). On the other hand, base level for a stream above a high reservoir or a mountain lake can be hundreds of metres above sea level. The surface of the lake or reservoir serves as temporary base level for all the water upstream (figure 14.34*C*). The base level of a tributary stream is governed by the level of its junction with the main stream. A ledge of resistant rock may act as a temporary base level if a stream has difficulty eroding through it.

The Concept of a Graded Stream

As a stream begins downcutting into the land, its longitudinal profile is usually irregular, with rapids and waterfalls along its course (see figure 14.33*D*). Such a stream, termed *ungraded,* is using most of its erosional energy in downcutting to smooth out these irregularities in gradient.

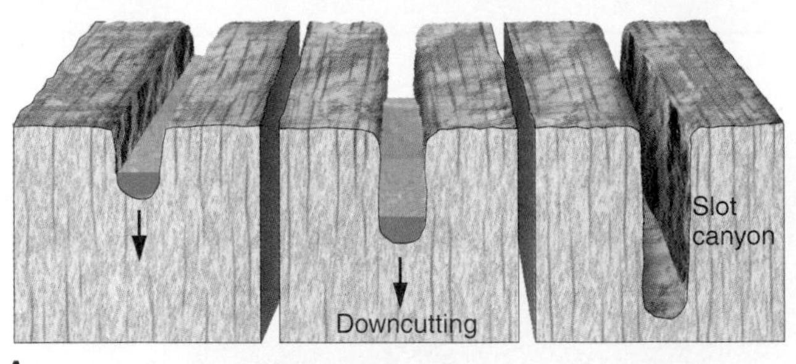

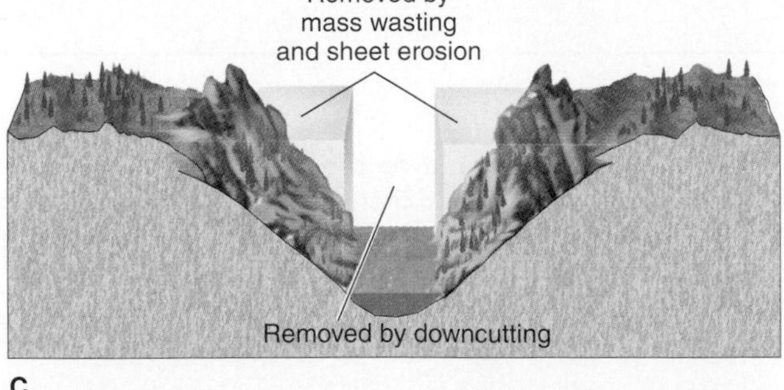

B

D

FIGURE 14.33

Downcutting, mass wasting, and sheet erosion shape canyons and valleys. *(A)* Downcutting can create slot canyons in resistant rock, particularly where downcutting is rapid during flash floods and fractures in the rock are favourably oriented. *(B)* Stream erosion has cut this unusual slot canyon through porous sandstone, Zion National Park, Utah. *(C)* Downslope movement of rock and soil on valley walls widens most canyons into V-shaped valleys. *(D)* The waterfall and rapids on the Yellowstone River in Wyoming indicate that the river is ungraded and actively downcutting. Note the V-shaped cross-profile and lack of flood plain due to the downslope movement of volcanic rock.

Photo B by Allen Hagood, Zion Natural History Association; Photo D by David McGeary

As the stream smooths out its longitudinal profile to a characteristic concave-upward shape (figure 14.35), it becomes graded. A **graded stream** is one that exhibits a delicate balance between its transporting capacity and the sediment load available to it. This balance is maintained by cutting and filling any irregularities in the smooth longitudinal profile of the stream.

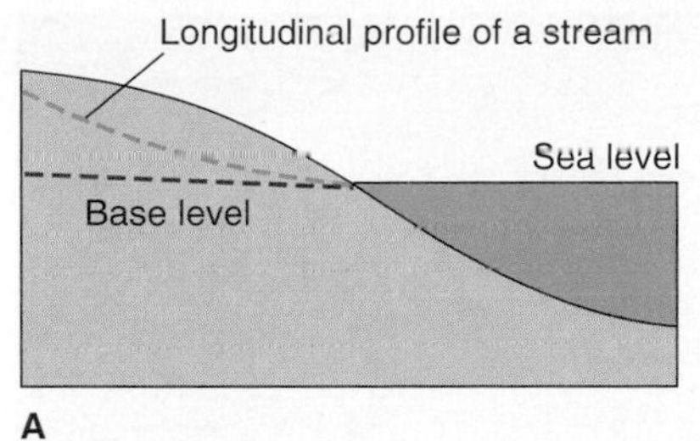

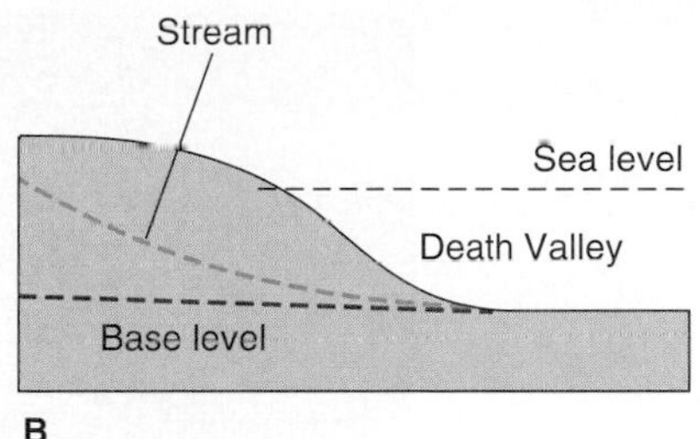

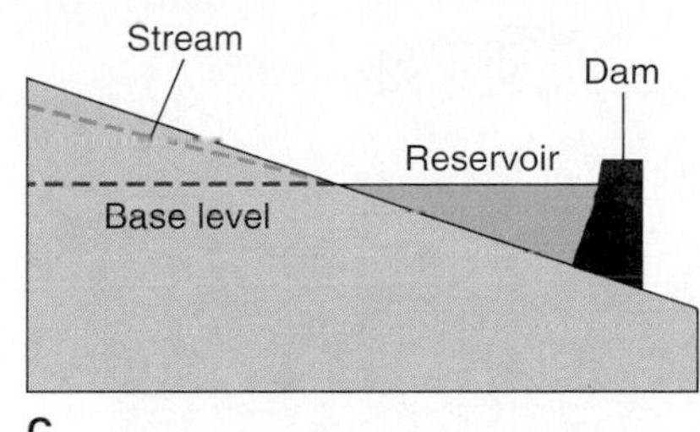

FIGURE 14.34

Base level is the lowest level of downcutting.

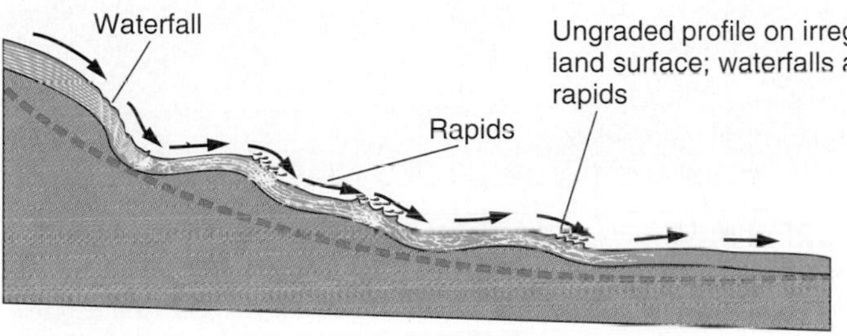

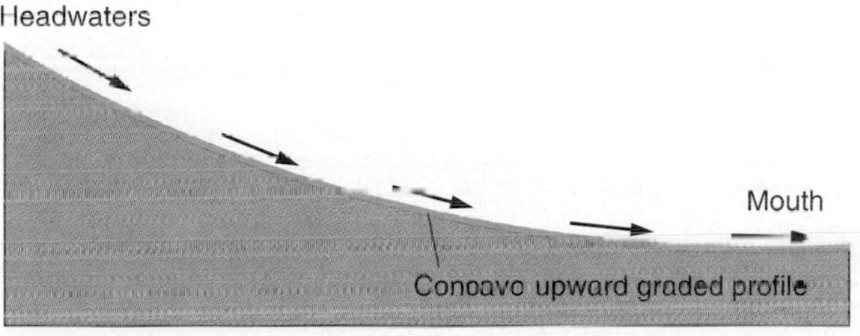

FIGURE 14.35

An ungraded stream has an irregular longitudinal profile with many waterfalls and rapids. A graded stream has smoothed out its longitudinal profile to a smooth, concave-upward curve.

Earlier in this chapter you learned how changes in a stream's gradient can cause changes in its sediment load. An increase in gradient causes an increase in a stream's velocity, allowing the stream to erode and carry more sediment. A balance is maintained—the greater load is a result of the greater transporting capacity caused by the steeper gradient.

The relationship also works in reverse—a change in sediment load can cause a change in gradient. For example, a decrease in sediment load may bring about erosion of the stream's channel, thus lowering the gradient. Because dams trap sediment in the calm reservoirs behind them, most streams are almost completely sediment-free just downstream from dams. In some streams this loss of sediment has caused severe channel erosion below a dam, as the stream adjusts to its new, reduced load.

A river's energy is used for two things—transporting sediment and overcoming resistance to flow. If the sediment load decreases, the river has more energy for other things. It may use this energy to erode more sediment, deepening its valley. Or it may change its channel shape or length, increasing resistance to flow, so that the excess energy is used to overcome friction. Or the river may increase the roughness of its channel, also increasing friction. The response of a river is not always predictable, and construction of a dam can sometimes have unexpected and perhaps harmful results.

Lateral Erosion

A graded stream can be deepening its channel by downcutting while part of its energy is also widening the valley by **lateral erosion,** the erosion and undercutting of a stream's banks and valley walls as the stream swings from side to side across its valley floor. The stream channel remains the same width as it moves across the flood plain, but the valley widens by erosion, particularly on the outside of curves, and meanders where the stream impinges against the valley walls (figure 14.36). The valley widens as its walls are eroded by the stream and as its walls retreat by mass wasting triggered by stream undercutting. As a valley widens, the stream's flood plain increases in width also.

Headward Erosion

Building a delta or alluvial fan at its mouth is one way a river can extend its length. A stream can also lengthen its valley by **headward erosion,** the slow uphill growth of a valley above its original source through gullying, mass wasting, and sheet erosion (figure 14.37). This type of erosion is particularly difficult to stop. When farmland is being lost to gullies that are eroding headward into fields and pastures, farmers must divert sheet flow and fill the gully heads with brush and other debris to stop, or at least retard, the loss of topsoil.

Stream Terraces

Stream terraces are steplike landforms found above a stream and its flood plain (figure 14.38). Terraces may be benches cut in rock (sometimes sediment-covered), or they may be steps formed in sediment by deposition and subsequent erosion.

Figure 14.39 shows how one type of terrace forms as a river cuts downward into a thick sequence of its own flood-plain deposits. Originally the river deposited a thick section of flood-plain sediments. Then the river changed from deposition to erosion and cut into its old flood plain, parts of which remain as terraces above the river.

Why might a river change from deposition to erosion? One reason might be regional uplift, raising a river that was once meandering near base level to an elevation well above base level. Uplift would steepen a river's gradient, causing the river to speed up and begin erosion. But there are several other

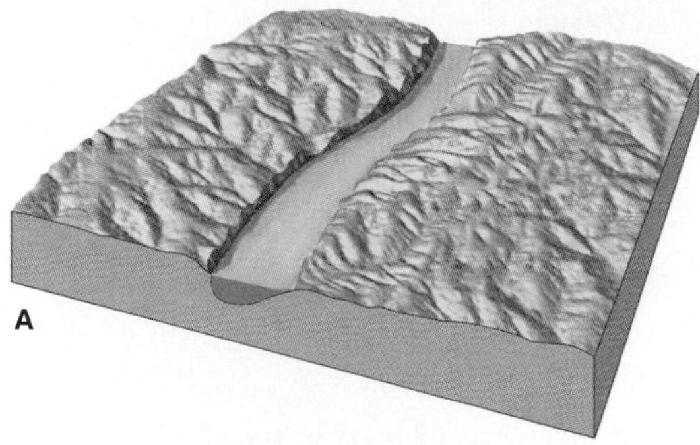

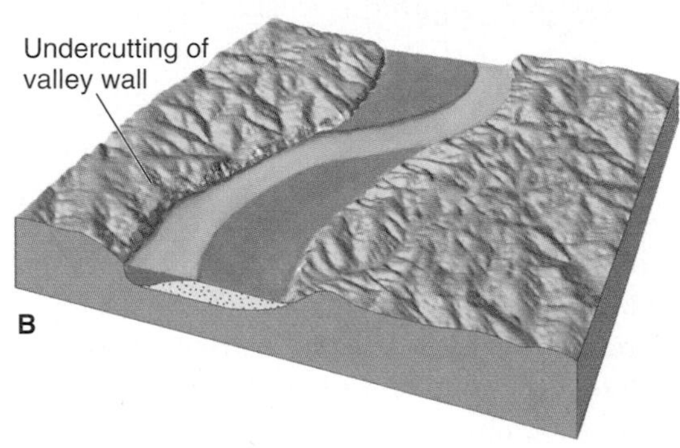

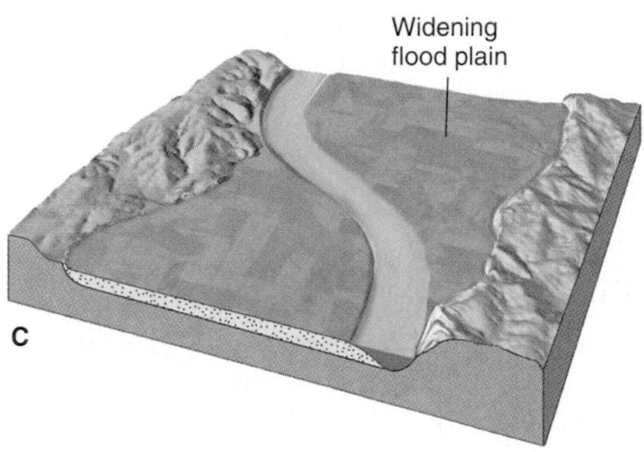

FIGURE 14.36

Lateral erosion can widen a valley by undercutting and eroding valley walls.

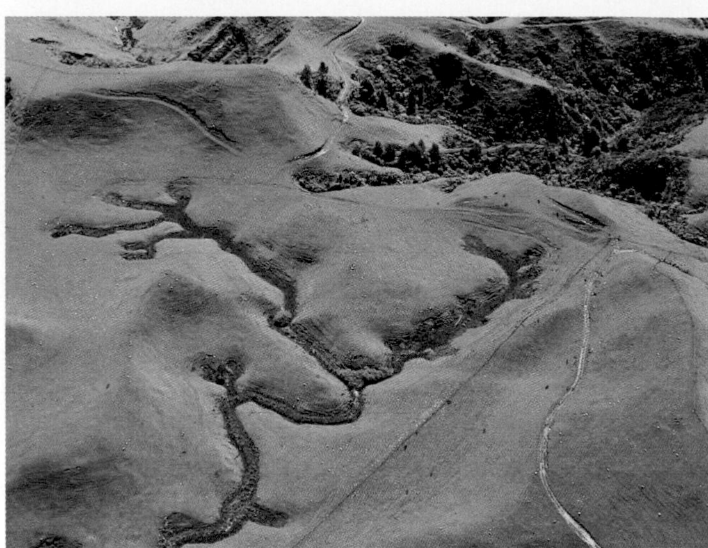

FIGURE 14.37

Headward erosion is lengthening this stream channel. Note the dendritic drainage pattern that is developing in the headwaters of the streams, New Plymouth, New Zealand.
Photo © G. R. "Dick" Roberts

FIGURE 14.38

Stream terraces near Jackson Hole, Wyoming. The stream has cut downward into its old flood plain.
Photo by Diane Carlson

reasons why a river might change from deposition to erosion. A change from a dry to a wet climate may increase discharge and cause a river to begin eroding. A drop in base level (such as lowering of sea level) can have the same effect. A situation like that shown in figure 14.39 can develop in a recently glaciated region. Thick valley fill such as glacial outwash (see chapter 16) may be deposited in a stream valley and later, after the glacier stops producing large amounts of sediment, be dissected into terraces by the river.

Terraces can also develop from erosion of a bedrock valley floor. Bedrock benches are usually capped by a thin layer of flood-plain deposits.

Incised Meanders

Incised meanders are meanders that retain their sinuous pattern as they cut vertically downward below the level at which they originally formed. The result is a meandering *valley* with essentially no flood plain, cut into the land as a steep-sided canyon (figure 14.40*A*).

Some incised meanders may be due to the profound effects of a change in base level. They may originally have been formed as meanders in a laterally eroding river flowing over a flat flood plain, perhaps near base level (figure 14.40*B*). If regional uplift elevated the land high above base level, the river would begin downcutting and might be able to maintain its characteristic meander pattern while deepening its valley (figure 14.40*C*). A drop in base level without land uplift (possibly because of a lowering of sea level) could bring about the same result.

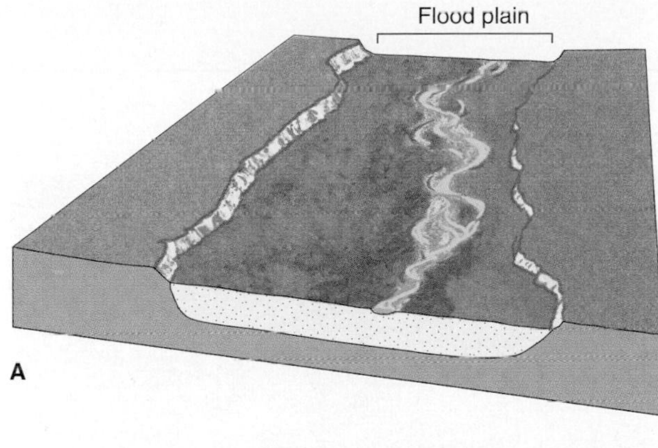

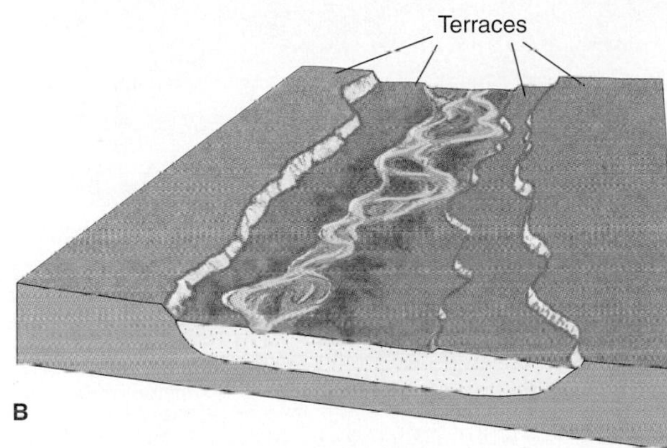

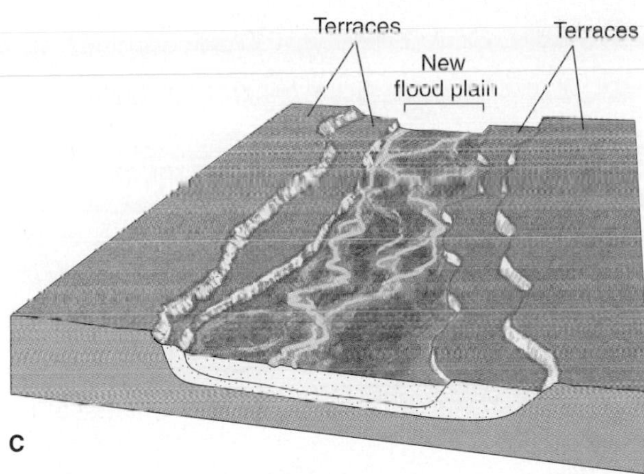

FIGURE 14.39

Terraces formed by a stream cutting downward into its own flood-plain deposits. (A) Stream deposits thick, coarse, flood-plain deposits. (B) Stream erodes its flood plain by downcutting. Old flood-plain surface forms terraces. (C) Lateral erosion forms new flood plain below terraces.

Although uplift may be a key factor in the formation of many incised meanders, it may not be *required* to produce them. Lateral erosion certainly seems to become more prominent as a river approaches base level, but some meandering can occur as soon as a river develops a graded profile. A river flowing on a flat surface high above base level may develop meanders early in its erosional history, and these meanders may become incised by subsequent downcutting. In such a case uplift is not necessary. See box 14.5 for a discussion of stream features on another planet.

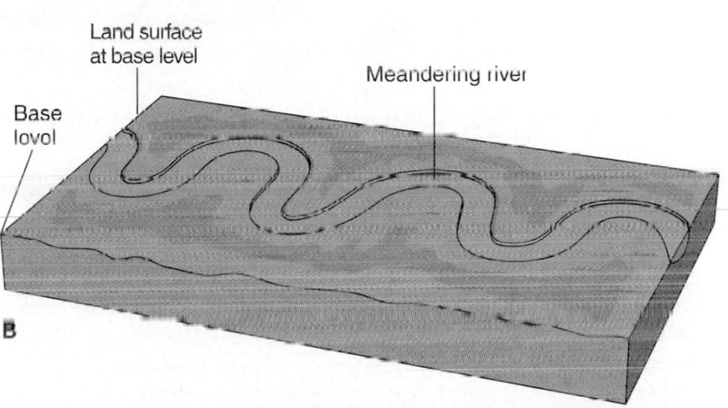

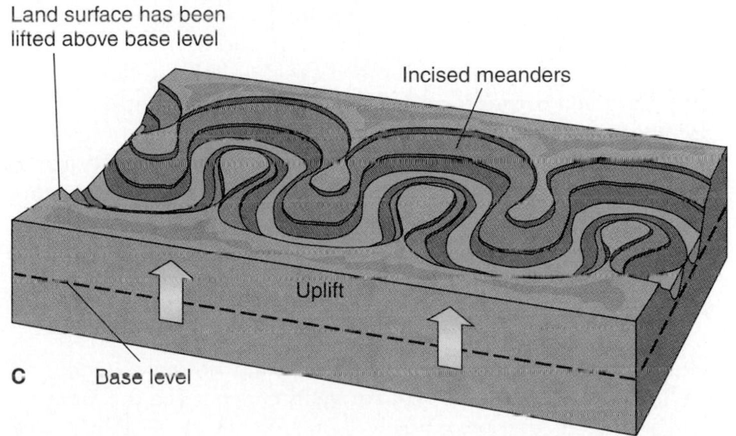

FIGURE 14.40

(A) Incised meanders of the Colorado River ("The Loop"), Canyonlands National Park, southwestern Utah. (B) Meandering river flowing over a flat plain cut to base level. (C) Regional uplift of land surface allows river to downcut and incise its meanders. *Photo by Frank M. Hanna*

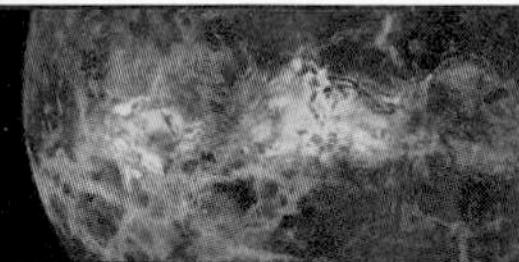

ASTROGEOLOGY 14.5

Stream Features on the Planet Mars

There is probably no liquid water on the surface of Mars today. With the present surface temperatures, atmospheric pressures, and water content in the Martian atmosphere, any liquid water would immediately evaporate. Recent evidence collected from the 2003 Mars Exploration Rovers *Spirit* and *Opportunity* indicate that conditions may have been different in the past and that liquid water existed on Mars, at least temporarily.

Certain features on Mars, called *channels,* closely resemble certain types of stream channels on Earth. They have tributary systems and meanders and are sometimes braided. The channels trend downslope and tend to get wider toward their mouths. These Martian channels are restricted to certain areas and appear to have been formed by intermittent episodes of erosion.

BOX 14.5 ■ FIGURE 1
View from the Mars Pathfinder Lander showing the Sojourner Rover and a variety of rocks from Ares Vallis.
Photo courtesy of Jet Propulsion Laboratory/NASA

One type of channel on Mars appears to have formed by large flooding events and is similar in appearance to those observed in the Channeled Scablands of Washington. The Channeled Scablands were formed by extensive flooding during the Pleistocene glacial ages, when a naturally formed ice dam broke and released water from a large lake. The mouth of Ares Vallis, an ancient Martian flood channel similar to those observed in the Channeled Scablands, was selected for the July 4, 1997, landing of the *Pathfinder* spacecraft and Sojourner Rover. It was postulated that a variety of rock types should be present in the mouth of an ancient flood channel. The first photos from the Mars *Pathfinder* Lander and *Sojourner* Rover, a "robotic field geologist," revealed a variety of rock types (box figure 1) in what does appear to be an ancient outflow channel.

A second kind of Martian channel (box figure 2), a meandering streamlike feature, occurs on the older surfaces of Mars (more than 3.5 billion years old) and may indicate that early in the history of Mars, temperature and atmospheric conditions were such that rainfall could have occurred and long-lived river systems could have existed. On June 8, 1998, the Mars Orbiter Camera aboard the Mars *Global Surveyor* spacecraft captured an image of what appears to be a meandering stream channel and stream terraces inside the Nanedi Vallis canyon. The 2.5-km-wide canyon is located approximately 1,600 kilometres south of where the *Pathfinder* landed. The presence of the narrow, streamlike channel and associated terraces within the Nanedi Vallis canyon strongly suggest that a river of water repeatedly flowed down the canyon. But the lack of smaller tributary channels may argue that the channel, like many others on Mars, may have instead formed by surface collapse caused by frozen water underground.

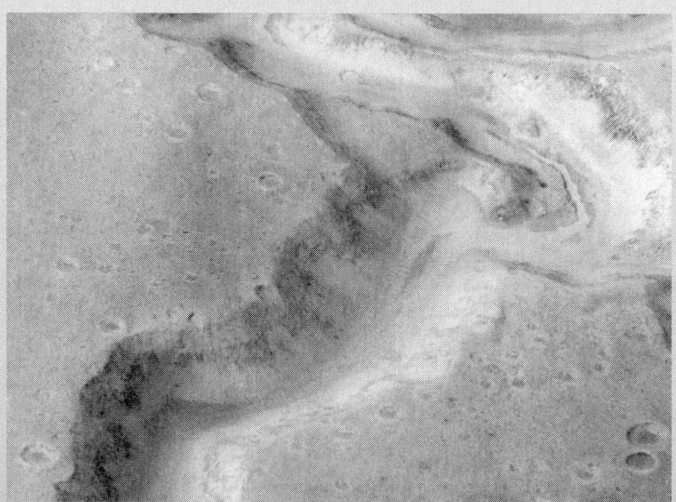

BOX 14.5 ■ FIGURE 2
Meandering channel and flat terraces within the Nanedi Vallis canyon, which resemble stream features cut by running water on Earth.
Photo courtesy of NASA

Additional Resources

M. P. Golombek. 1998. The Mars Pathfinder Mission. *Scientific American 7.*

For more information on the possibility of water on Mars, visit the NASA Goddard Institute for Space Studies research site:

www.giss.nasa.gov/research/intro/gornitz.03/

Information about the Mars Global Surveyor can be found at the Jet Propulsion Laboratory/NASA site:

http://mars.jpl.nasa.gov/mgs

SUMMARY

Normally *stream channels* are eroded and shaped by the streams that flow in them. Unconfined sheet flow can cause significant erosion.

Drainage basins are separated by *divides*.

A river and its tributaries form a *drainage pattern*. A *dendritic* drainage pattern develops on uniform rock, a *rectangular* pattern on regularly jointed rock. A *radial* pattern forms on conical mountains, while a *trellis* pattern usually indicates erosion of folded sedimentary rock.

Stream velocity is the key factor controlling sediment erosion, transportation, and deposition. Velocity is in turn controlled by several factors.

An increase in a *stream's gradient* increases the stream's velocity. *Channel shape* and *roughness* affect velocity by increasing or lessening friction. As tributaries join a stream, the stream's *discharge* increases downstream. Floods increase stream discharge and velocity.

Streams erode by *hydraulic action*, *abrasion*, and *solution*. They carry coarse sediment by *traction* and *saltation* as *bed load*. Finer-grained sediment is carried in *suspension*. A stream can also have a substantial *dissolved load*.

Streams create features by erosion and deposition. *Potholes* form by abrasion of hard rock on a stream bed. *Bars* form in the middle of streams or on stream banks, particularly on the inside of curves where velocity is low (*point bars*). A *braided pattern* can develop in streams with a large amount of bed load.

Meanders are created when a laterally eroding stream shifts across the flood plain, sometimes creating cutoffs and oxbow lakes.

A *flood plain* develops by both lateral and vertical deposition. *Natural levees* are built up beside streams by flood deposition.

A *delta* forms when a stream flows into standing water. The shape and internal structure of deltas are governed by river deposition and wave and current erosion. *Alluvial fans* form, particularly in dry climates, at the base of mountains as a stream's channel widens and its velocity decreases.

Rivers deepen their valleys by *downcutting* until they reach *base level*, which is either sea level or a local base level.

A *graded stream* is one with a delicate balance between its transporting capacity and its available load.

Lateral erosion widens a valley after the stream has become graded.

A valley is lengthened by both *headward erosion* and sediment deposition at the mouth.

Stream terraces can form by erosion of rock benches or by dissection of thick valley deposits during downcutting.

Incised meanders form as (1) river meanders are cut vertically downward following uplift or (2) lateral erosion and downcutting proceed simultaneously.

Terms to Remember

abrasion 362
alluvial fan 373
bar 365
base level 379
bed load 364
braided stream 368
delta 371
dendritic pattern 359
discharge 361
dissolved load 364
distributary 371
divide 358
downcutting 379
drainage basin 358
drainage pattern 358
flood plain 370
graded stream 382
headward erosion 383
hydraulic action 362
hydrologic cycle 356
incised meander 384
lateral erosion 383
meander 368
meander cutoff 368
natural levee 370
oxbow lake 368
point bar 368
pothole 362
radial pattern 359
rectangular pattern 359
saltation 364
sheetwash 357
solution 362
stream 356
stream channel 357
stream gradient 360
stream terrace 383
stream velocity 359
suspended load 364
traction 364
trellis pattern 359
tributary 358

Testing Your Knowledge

Use the questions below to prepare for exams based on this chapter.

1. What factors control a stream's velocity?
2. Describe how bar deposition creates a braided stream.
3. In what part of a large alluvial fan is the sediment the coarsest? Why?
4. What does a trellis drainage pattern tell us about the rocks underneath it?
5. Describe one way that incised meanders form.
6. How does a meander neck cutoff form an oxbow lake?
7. How does a natural levee form?
8. Describe how stream terraces form.
9. Describe three ways in which a river erodes its channel.
10. Name and describe the three main ways in which a stream transports sediment.
11. How does a stream widen its valley?
12. What is base level?
13. The total area drained by a stream and its tributaries is called the
    a. hydrologic cycle b. tributary area
    c. divide d. drainage basin
14. Stream erosion and deposition are controlled primarily by a river's
    a. velocity b. discharge
    c. gradient d. channel shape
    e. channel roughness
15. What is the gradient of a stream that drops 10 vertical metres over a 5-kilometre horizontal distance?
    a. 20 m per km b. 2 m per km
    c. 1 m per km d. 5 m per km
16. What are typical units of discharge?
    a. kilometres per hour b. cubic metres
    c. cubic metres per second d. metres per second
17. Hydraulic action, solution, and abrasion are all examples of stream
    a. erosion b. transportation
    c. deposition
18. Cobbles are more likely to be transported in a stream's
    a. bed load b. suspended load
    c. dissolved load d. all of the above
19. A river's velocity is _____ on the outside of a meander curve compared to the inside.
    a. higher b. equal
    c. lower
20. Sandbars deposited on the inside of meander curves are called
    a. dunes b. point bars
    c. cutbanks d. none of the above
21. Which is not a drainage pattern?
    a. dendritic b. radial
    c. rectangular d. trellis
    e. none of the above
22. The broad strip of land built up by sedimentation on either side of a stream channel is
    a. a flood plain b. a delta
    c. an alluvial fan d. a meander
23. The average time between floods of a given size is
    a. the discharge b. the gradient
    c. the recurrence interval d. the magnitude
24. A platform of sediment formed where a stream flows into standing water is
    a. an alluvial fan b. a delta
    c. a meander d. a flood plain

Exploring Web Resources

www.mcgrawhill.ca/olc/plummer

Go to the Online Learning Centre to review your answers for the Testing Your Knowledge section. This Website also contains additional quizzing, reading, media resources, and some really great animations to further your understanding of streams and floods. Click on the links to go directly to the Websites listed below.

http://atlas.gc.ca

Maps and information on drainage basins and rivers in Canada.

http://water.usgs.gov/

Contains extensive information on water issues throughout the U.S. and many links to USGS data and online publications.

http://water.usgs.gov/public/realtime.html

Contains real-time stream-flow data from USGS gauging stations throughout the U.S.

http://water.usgs.gov/usa/nwis/

Contains historical stream-flow data from USGS gauging stations throughout the U.S.

www.dartmouth.edu/~floods/

The *Dartmouth Flood Observatory* Website contains information on flood detection and satellite images of floods and flood damage from around the world.

http://vcourseware.sonoma.edu/VirtualRiver/Flooding/

California State University, Los Angeles *Virtual River* exercise.

www.msc.ec.gc.ca

Environment Canada (Meteorological Service of Canada) National Water Data Archive for daily, monthly, or instantaneous information for stream flow, water level, and sediment load data for stations across Canada.

Animations

This chapter includes the following animations available on our Online Learning Centre at www.mcgrawhill.ca/olc/plummer.

14.14 Modes of Sediment Transport
14.20 River Meander Development

CHAPTER

15

Groundwater

What is groundwater?
How does groundwater move?
What are aquifers, wells, springs, and streams?
What determines groundwater quality?
What are the causes of groundwater contamination?
How long will groundwater last?
What features can groundwater form?
How do hot springs and geysers form?
How is geothermal energy produced?

How much of Earth's water is found underground? Compared to the oceans, not much. Approximately 0.6 percent of the world's water is groundwater, whereas over 97 percent is ocean water. If we look at fresh water alone, we find that the amount of groundwater is 35 times that of all rivers and lakes in the world. (However, the amount of fresh water stored in glaciers is 3.5 times greater than the amount of groundwater.)

Groundwater is a tremendously important resource. Managing our water resources becomes increasingly difficult as demands increase. Growing cities in arid climates are removing groundwater faster than it can be replenished. Pollution of groundwater by industrial wastes, agricultural pesticides, and other means can render the water unfit for human consumption. Growing population and improvements in lifestyles increasingly impact our water supply.

Groundwater moving through soluble bedrock dissolves the rock to form caves and underground passageways. Collapse of cave roofs can result in the formation of sinkholes, or dolines, such as the one shown here, which lies between Norman Wells and Tulita in the Northwest Territories. *Photo by Nick Eyles*

How water gets underground, where it is stored, how it moves while underground, how we look for it, and, perhaps most important of all, why we need to protect it are the main topics of this chapter.

Also important is how groundwater is related to surface rivers and springs. Groundwater can form distinctive geologic features, such as caves, sinkholes, and petrified wood. It also can appear as hot springs and geysers. Hot groundwater can be used to generate geothermal energy.

Distribution of Water in the Hydrosphere (%)

Oceans	97.2
Glaciers and other ice	2.15
Groundwater	.61
Lakes	
Fresh	.009
Saline	.008
Soil moisture	.005
Atmosphere	.001
Rivers	.0001

WHAT IS GROUNDWATER?

Many communities obtain the water they need from rivers, lakes, or reservoirs, sometimes using aqueducts or canals to bring water from distant surface sources. Another source of water lies directly beneath most towns. This resource is **groundwater,** the water that lies beneath the ground surface, filling the pore space between grains in bodies of sediment and clastic sedimentary rock, and filling cracks and crevices in all types of rock.

Groundwater is a major economic resource, particularly in the dry western areas of Canada and United States where surface water is scarce. Many towns and farms pump great quantities of groundwater from drilled wells. Even cities next to large rivers may pump their water from the ground because groundwater is commonly less contaminated and more economical to use than surface water.

More than 30 percent of the population of Canada (9 million people) relies on groundwater for domestic use. Most of these users (approximately two-thirds) live in rural areas where groundwater pumped from drilled wells supplies towns and farms with an inexpensive and reliable source of water. Other users include smaller municipalities whose primary source of water is groundwater. The amount of groundwater and the predominant use of groundwater in Canada varies significantly by province (figure 15.1). Prince Edward Island is dependent upon groundwater for all its uses, whereas the largest user of groundwater in Alberta, Saskatchewan, and Manitoba is the agricultural industry. Groundwater is used in Ontario, New Brunswick, and the Yukon primarily by municipalities and in Nova Scotia and Newfoundland for rural domestic purposes. Industry is the largest user of groundwater in Quebec, British Columbia, and the Northwest Territories.

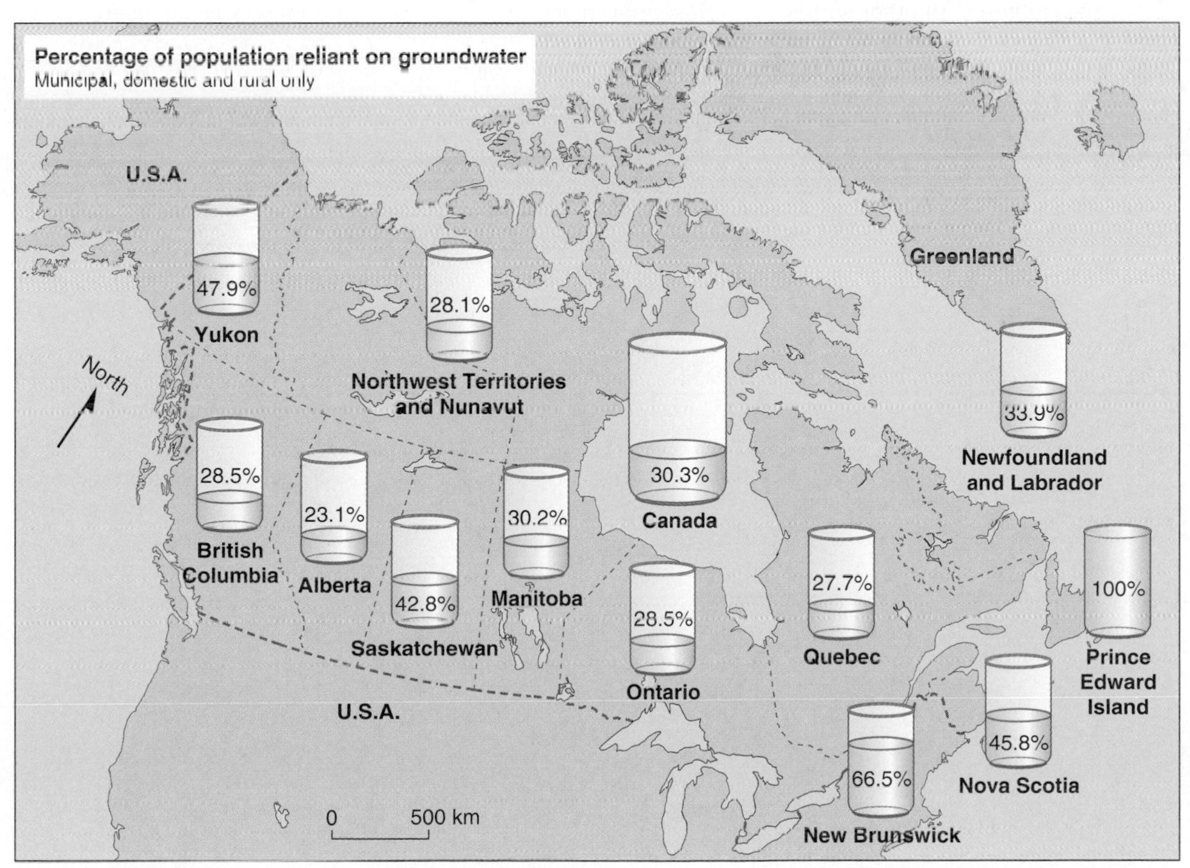

FIGURE 15.1

Percentage of population in each Canadian province reliant on groundwater for domestic use, adapted from Statistics Canada. Environmental Accounts and Statistics Division (special compilation using data from Environment Canada. Municipal Water use Database), and from the Statistics Canada publication, Quarterly estimates of population for Canada, the provinces and the territories, Catalogue 91-001, 2006.

Based on http://www.ec.gc.ca/water/images/nature/grdwtr/a5f6e.htm Environment Canada's Freshwater Website, 2004. Reproduced with the permission of the Minister of Public Works and Government Services, 2007.

The source of groundwater is rain and snow that falls to the ground. A portion of this precipitation percolates down into the ground to become groundwater (see figure 14.1 and the Hydrologic Cycle discussion in chapter 14). How much precipitation soaks into the ground is influenced by climate, land slope, soil and rock type, and vegetation. In general, approximately 15 percent of the total precipitation ends up as groundwater, but that varies locally and regionally from 1 percent to 20 percent.

Porosity and Permeability

Porosity, the percentage of rock or sediment that consists of voids or openings, is a measurement of a rock's ability to hold water. Most rocks can hold some water. Some sedimentary rocks, such as sandstone, conglomerate, and many limestones, tend to have a high porosity and therefore can hold a considerable amount of water. A deposit of loose sand may have a porosity of 30 percent to 50 percent, but this may be reduced to 10 percent to 20 percent by compaction and cementation as the sand lithifies (table 15.1). A sandstone in which pores are nearly filled with cement and fine-grained matrix material may have a porosity of 5 percent or less. Crystalline rocks, such as granite, schist, and some limestones, do not have pores but may hold some water in joints and other openings.

Although most rocks can hold some water, they vary a great deal in their ability to allow water to pass through them. **Permeability** refers to the capacity of a rock to transmit a fluid such as water or petroleum through pores and fractures. In other words, permeability measures the relative ease of water flow and indicates the degree to which openings in a rock interconnect. The distinction between porosity and permeability is important. A rock that holds much water is called *porous;* a rock that allows water to flow easily through it is described as *permeable.* Most sandstones and conglomerates are both porous and permeable. An *impermeable* rock is one that does not allow water to flow through it easily. Unjointed granite and schist are impermeable. Shale can have substantial porosity, but it has low permeability because its pores are too small to permit easy passage of water.

The Water Table

Responding to the pull of gravity, water percolates down into the ground through the soil and through cracks and pores in the rock. Several kilometres down in the crust percolation stops. With increasing depth, sedimentary rock pores tend to be closed by increasing amounts of cement and by the weight of the overlying rock. Moreover, sedimentary rock overlies igneous and metamorphic crystalline basement rock, which usually has very low porosity.

The subsurface zone in which all rock openings are filled with water is called the **saturated zone** (figure 15.2*A*). If a well were drilled downward into this zone, groundwater would fill the lower part of the well. The water level inside the well marks the upper surface of the saturated zone; this surface is the **water table.**

Most rivers and lakes intersect the saturated zone. Rivers and lakes occupy low places on the land surface, and groundwater flows out of the saturated zone into these surface depressions.

TABLE 15.1 Porosity and Permeability of Sediments and Rocks

Sediment	Porosity (%)	Permeability
Gravel	25 to 40	Excellent
Sand (clean)	30 to 50	Good to excellent
Silt	35 to 50	Moderate
Clay	35 to 80	Poor
Glacial till	10 to 20	Poor to moderate
Rock		
Conglomerate	10 to 30	Moderate to excellent
Sandstone		
Well-sorted, little cement	20 to 30	Good to very good
Average	10 to 20	Moderate to good
Poorly sorted, well-cemented	0 to 10	Poor to moderate
Shale	0 to 30	Very poor to poor
Limestone, dolomite	0 to 20	Poor to good
Cavernous limestone	up to 50	Excellent
Crystalline rock		
Unfractured	0 to 5	Very poor
Fractured	5 to 10	Poor
Volcanic rocks	0 to 50	Poor to excellent

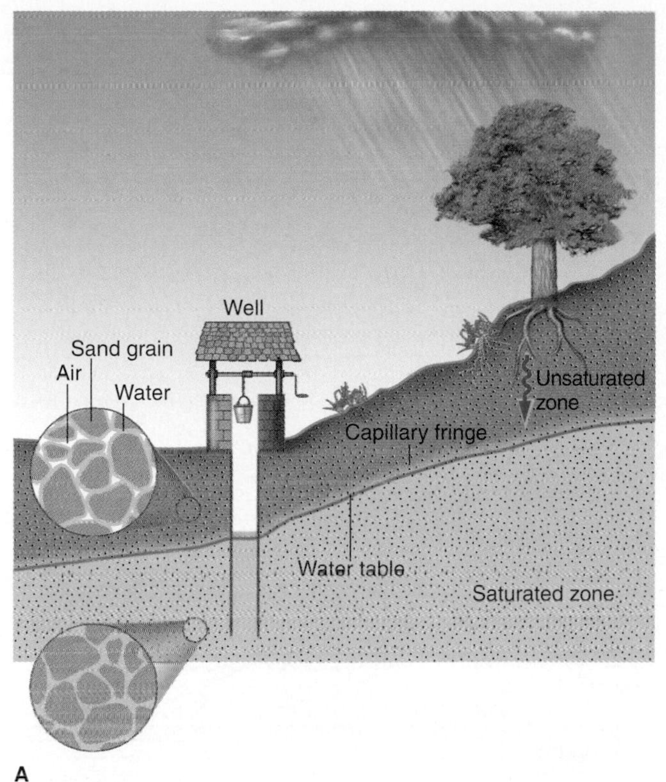

FIGURE 15.2

(*A*) The water table marks the top of the saturated zone, in which water completely fills the rock pore space (inset figure). Above the water table is the unsaturated zone, in which rock openings typically contain both air and water. (*B*) Groundwater fills the abandoned Marmoraton Iron Mine in Ontario, which extends beneath the water table.

Photo by Nick Eyles

The water level at the surface of most lakes and rivers coincides with the water table. Groundwater also flows into mines and quarries cut below the water table (figure 15.2*B*).

Above the water table there is a zone that is generally unsaturated and is referred to as the **unsaturated zone** (figure 15.2*A*). Within the unsaturated zone, surface tension causes water to be held above the water table. The *capillary fringe* is a transition zone with higher moisture content at the base of the vadose zone just above the water table. Some of the water in the capillary fringe has been drawn or wicked upward from the water table (much like water rising up a paper towel if the corner is dipped in water). The capillary fringe is generally less than a metre thick, but may be much thicker in fine-grained sediments and thinner in coarse-grained sediments such as sand and gravel.

Plant roots generally obtain their water from the belt of soil moisture near the top of the unsaturated zone, where fine-grained clay minerals hold water and make it available for plant growth. Most plants "drown" if their roots are covered by water in the saturated zone; plants need both water and air in soil pores to survive. (The water-loving plants of swamps and marshes are an exception.)

A **perched water table** is the top of a body of groundwater separated from the main water table beneath it by a zone that is not saturated (figure 15.3). It may form as groundwater collects above a lens of less permeable shale within a more permeable rock, such as sandstone. If the perched water table intersects the land surface, a line of springs can form along the upper contact of the shale lens. The water perched above a shale lens can provide a limited water supply to a well; it is an unreliable long-term supply.

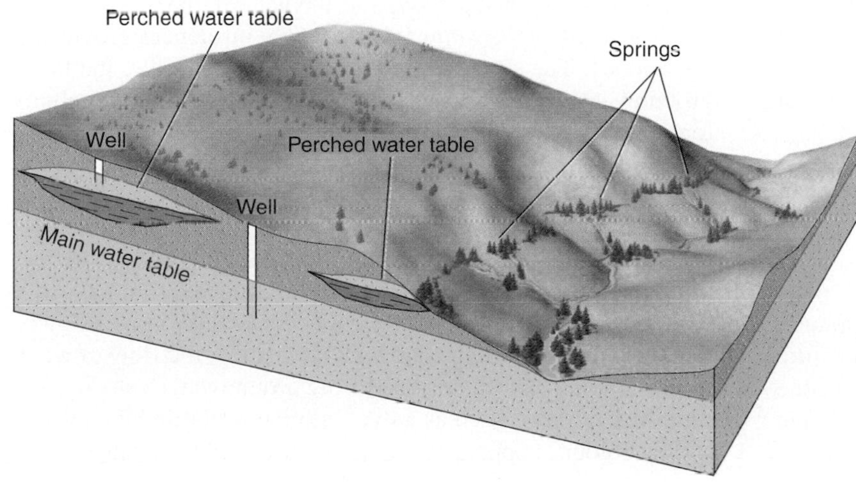

FIGURE 15.3

Perched water tables above lenses of less permeable shale within a large body of sandstone. Downward percolation of water is impeded by the less permeable shale.

IN GREATER DEPTH 15.1

Darcy's Law and Fluid Potential

In 1856, Henry Darcy, a French engineer, found that the velocity at which water moves depends on the *hydraulic head* of the water and on the permeability of the material that the water is moving through.

The hydraulic head of a drop of water is equal to the elevation of the drop plus the water pressure on the drop:

$$\text{Hydraulic head} = \text{elevation} + \text{pressure}$$

In box figure 1*A*, the points A and B are both on the water table, so the pressure at both points is zero (there is no water above points A and B to create pressure). Point A is at a higher elevation than B, so A has a higher hydraulic head than B. The difference in elevation is equal to the difference in head, which is labelled **h.** Water will move from point A to point B (as shown by the dark blue arrow), because water moves from a region of high hydraulic head to a region of low head. The distance the water moves from A to B is labelled **L.** The *hydraulic* gradient is the difference in head between two points divided by the distance between the two points:

$$\text{Hydraulic gradient} = \frac{\text{difference in head}}{\text{distance}} = \frac{\Delta h}{L}$$

In box figure 1*B*, the two points have equal elevation, but the pressure on point C is higher than on point D (there is more water to create pressure above point C than point D). The head is higher at point C than at point D, so the water moves from C to D. In box figure 1*C*, point F has a lower elevation than point G, but F also has a higher pressure than G. The difference in pressure is greater than the difference in elevation, so F has a

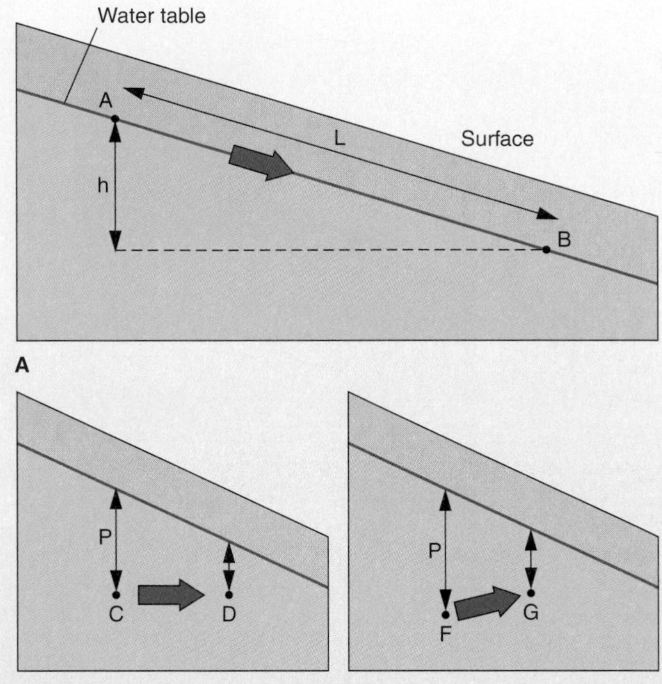

BOX 15.1 ■ FIGURE 1
Groundwater moves in response to hydraulic head (elevation plus pressure). Water movement shown by dark blue arrows. (*A*) Points A and B have the same pressure, but A has a higher elevation; therefore, water moves from A to B. (*B*) Point C has a higher pressure (arrow marked P) than D; therefore, water moves from C to D at the same elevation. (*C*) Pressure also moves water upward from F to G.

Geoscientists can determine the position and shape of the water table across broad regions by mapping water levels encountered in waterwells and boreholes. A variety of software programs are now available to analyze large quantities of waterwell data and to create digital maps of the water table. These maps allow geoscientists to visualize and analyze subsurface water movement, to calculate flow rates, and to predict the position of the water table in areas where no boreholes exist. Such data are essential for exploration of new groundwater resources and for the design of effective groundwater protection measures.

HOW DOES GROUNDWATER MOVE?

Compared to the rapid flow of water in surface streams, most groundwater moves relatively slowly through rock underground. Because it moves in response to differences in water pressure and elevation, water within the upper part of the saturated zone tends to move downward following the slope of the water table (figure 15.4). See box 15.1 for a discussion of Darcy's Law.

The circulation of groundwater in the saturated zone is not confined to a shallow layer beneath the water table. Groundwater may move hundreds of metres vertically downward before rising again to discharge as a spring or to seep into the beds of rivers and lakes at the surface (figure 15.4) due to the combined effects of gravity and the slope of the water table.

The *slope of the water table* strongly influences groundwater velocity. The steeper the slope of the water table, the faster groundwater moves. Water-table slope is controlled largely by topography—the water table roughly parallels the land surface (particularly in humid regions), as you can see in figure 15.4. Even in highly permeable rock, groundwater will not move if the water table is flat.

How fast groundwater flows also depends on the *permeability* of the rock or other materials through which it passes. If rock pores are small and poorly connected, water moves slowly. When openings are large and well connected, the flow of water is more rapid. One way of measuring groundwater velocity is to introduce a tracer, such as a dye, into the water and then watch for the colour to appear in a well or spring some distance away.

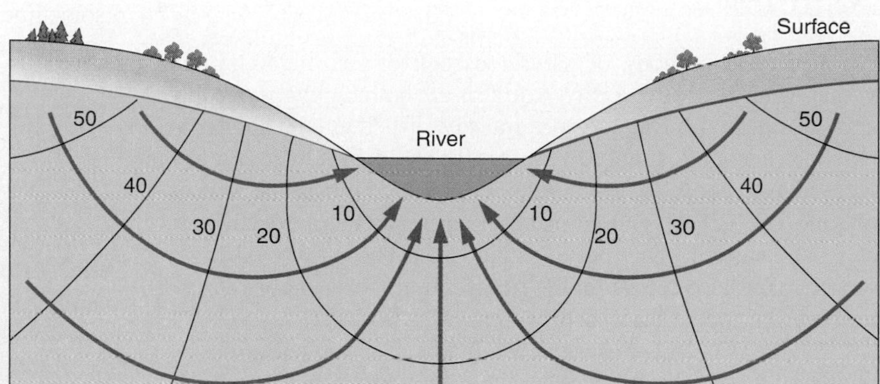

BOX 15.1 ■ FIGURE 2

Dark blue arrows are flow lines, which show direction of groundwater flow. Flow is perpendicular to equipotential lines (black lines with numbers), which show regions of equal hydraulic head. Groundwater generally flows from hilltops toward valleys, emerging from the ground as springs into stream beds and banks, lakes, and swamps.

higher head than G, and water moves from F to G. Note that underground water may move downward, horizontally, or upward in response to differences in head but that it always moves in the direction of the downward slope of the water table above it. One of the first goals of groundwater geoscientists, particularly in groundwater contamination investigations, is to find the slope of the local water table in order to determine the direction (and velocity) of groundwater movement.

The velocity of groundwater flow is controlled by both the permeability of the sediment or rock and the hydraulic gradient. *Darcy's law* states that the velocity equals the permeability multiplied by the hydraulic gradient. This gives the Darcian velocity (or the velocity of water flowing through an open pipe). To determine the actual velocity of groundwater, since groundwater only flows through the openings in sediment or rock, the Darcian velocity must be divided by the porosity.

Groundwater velocity = permeability/porosity × hydraulic gradient

$$V = \frac{K}{n}\frac{\Delta h}{L}$$

(Darcy called **K** the hydraulic conductivity; it is a measure of permeability and is specific to a particular aquifer. The porosity is represented by **n** in the equation.)

Groundwater movement is shown in diagrams in relation to *equipotential lines* (lines of constant hydraulic head). Ground water moves from regions of high head to regions of low head. Box figure 2 illustrates how *flow lines,* which show groundwater movement, cross equipotential lines at right angles as water moves from high to low head.

Such experiments have shown that the velocity of groundwater varies widely, averaging a few centimetres to many metres a day. Nearly impermeable rocks may allow water to move only a few centimetres per *year,* but highly permeable materials, such as unconsolidated gravel or cavernous limestone, may permit flow rates of hundreds or even thousands of metres per day.

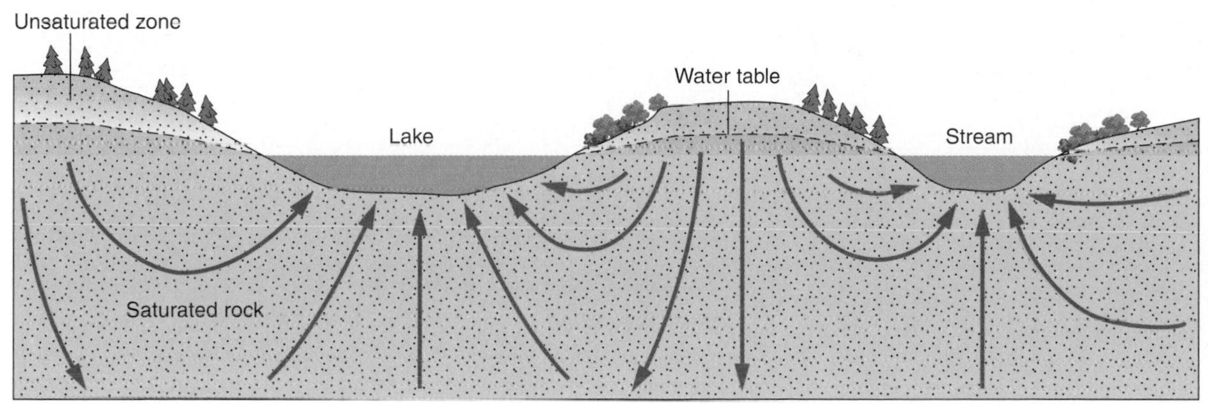

FIGURE 15.4

Movement of groundwater beneath a sloping water table in uniformly permeable rock. Near the surface, groundwater tends to flow parallel to the sloping water table.

WHAT ARE AQUIFERS, WELLS, SPRINGS, AND STREAMS?

Aquifers

An **aquifer** is a body of saturated rock or sediment through which water can move easily. Aquifers are both highly permeable and saturated with water. A well must be drilled into an aquifer to reach an adequate supply of water (figure 15.5). Good aquifers include sandstone, conglomerate, well-jointed limestone, bodies of sand and gravel, and some fragmental or fractured volcanic rocks such as columnar basalt (table 15.1). These favourable geologic materials are sought in "prospecting" for groundwater or looking for good sites to drill water wells (box 15.2).

In Canada, where most of the land surface is underlain by unconsolidated glacial deposits, aquifers are found in permeable sands and gravels and some glacial tills (see table 15.1). Some of the most productive aquifers are located in sand and gravel deposits laid down by glacial rivers that drained the large ice sheets that covered Canada during the Quaternary (see chapter 16). In southern Ontario, aquifers within permeable glaciofluvial deposits supply more than 200,000 people with drinking water. These aquifers are recharged with water that falls as precipitation on the Oak Ridges Moraine, an area to the north of Lake Ontario underlain by thick and extensive Quaternary-age sand and gravel deposits. Some glacial tills can also serve as productive aquifers if they have relatively low clay content or are highly fractured.

Wells drilled in shale or clay beds are not usually very successful because fine-grained deposits, although sometimes quite porous, are relatively impermeable (figure 15.5). Wet mud may have a porosity of 80 percent to 90 percent and even when compacted to form shale may still have a high porosity of 30 percent. Yet the extremely small size of the pores, together with the electrostatic attraction that clay minerals have for water molecules (see chapter 8), prevents water from moving readily through the shale into a well.

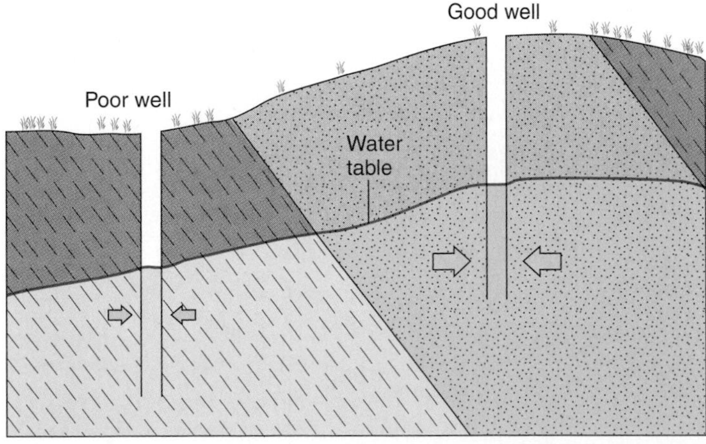

FIGURE 15.5

A well must be installed in an aquifer to obtain water. The saturated part of the highly permeable sandstone is an aquifer, but the less permeable shale is not. Although the shale is saturated, it will not readily transmit water.

Because they are not very porous, crystalline rocks such as granite, gabbro, gneiss, schist, and some types of limestone are not good aquifers. The porosity of such rocks may be 1 percent or less. (Shale and crystalline rocks and clay-rich sediments are sometimes called *aquitards* because they retard the flow of groundwater.) Crystalline rocks that are highly fractured, however, may be porous and permeable enough to provide a fairly dependable water supply to wells (figure 15.6).

Figure 15.7 shows the difference between an **unconfined aquifer,** which has a water table because it is only partly filled with water, and a **confined aquifer,** which is completely filled with water under pressure, and which is usually separated from the surface by a relatively impermeable confining bed, or aquitard, such as shale. An unconfined aquifer is recharged rapidly by precipitation, has a rising and falling water table during wet and dry seasons, and has relatively rapid movement of groundwater through it (figure 15.8). Unconfined aquifers are especially susceptible to contamination by agricultural activities such as intensive animal farming (e.g., Walkerton: see box 15.3). A confined aquifer is recharged slowly through confining shale or clay beds. With very slow movement of groundwater, a confined aquifer may have no response at all to wet and dry seasons. These aquifers generally provide more reliable and less contaminated sources of drinking water than unconfined aquifers.

Wells

A **well** is a deep hole, generally cylindrical, that is dug or drilled into the ground to penetrate an aquifer within the saturated zone (figure 15.5). Usually, water that flows into the well from the saturated rock must be lifted or pumped to the surface. As figure 15.8 shows, a well dug in a valley usually has to go down a shorter distance to reach water than a well dug on a hilltop. During dry seasons the water table falls as water flows out of the saturated zone into springs and rivers. Wells not deep enough to intersect the lowered water table go dry, but the rise of the water table during the next rainy season normally returns water to the dry wells. The addition of new water to the saturated zone is called **recharge.**

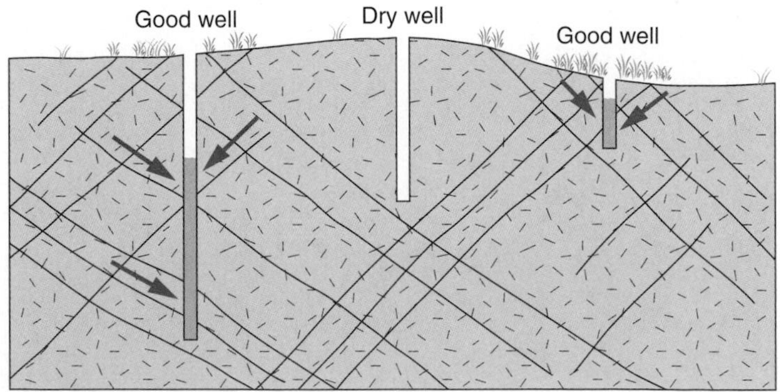

FIGURE 15.6

Wells can obtain some water from fractures in crystalline rock. Wells must intersect fractures to obtain water.

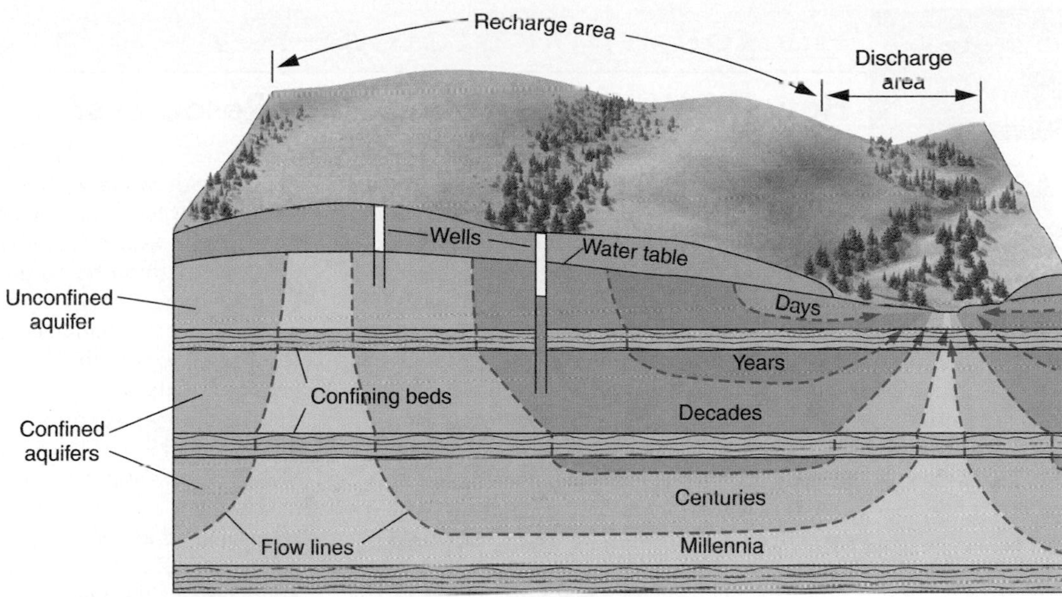

FIGURE 15.7
An unconfined aquifer is exposed to the surface and is only partly filled with water; water in a shallow well will rise to the level of the water table. A confined aquifer is separated from the surface by a confining bed and is completely filled with water under pressure; water in wells rises above the aquifer. Flow lines show direction of groundwater flow. Days, years, decades, centuries, and millennia refer to the time required for groundwater to flow from the recharge area to the discharge area. Water enters aquifers in recharge areas and flows out of aquifers in discharge areas.

When water is pumped from a well, the water table is typically drawn down around the well into a depression shaped like an inverted cone, known as a **cone of depression** (figure 15.9). This local lowering of the water table, called **drawdown,** tends to change the direction of groundwater flow by changing the slope of the water table. In lightly used wells that are not pumped, drawdown does not occur and a cone of depression does not form. In a simple, rural well with a bucket lowered on the end of a rope, water cannot be extracted rapidly enough to significantly lower the water table. A well of this type is shown in figure 15.2*A*.

In unconfined aquifers, water rises in shallow wells to the level of the water table. In confined aquifers, the water is under pressure and rises in wells to a level above the top of the aquifer (figure 15.10). Such a well is called an **artesian well**, and confined aquifers are also called *artesian aquifers*.

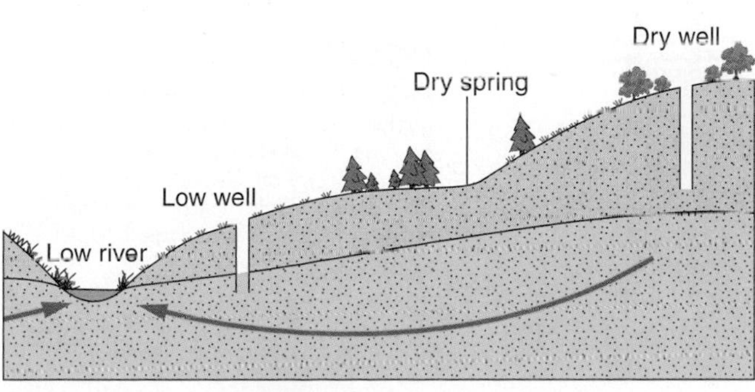

FIGURE 15.8
The water table in an unconfined aquifer rises in wet seasons and falls in dry seasons as water drains out of the saturated zone into rivers. (*A*) Wet season: water table and rivers are high; springs and wells flow readily. (*B*) Dry season: water table and rivers are low; some springs and wells dry up.

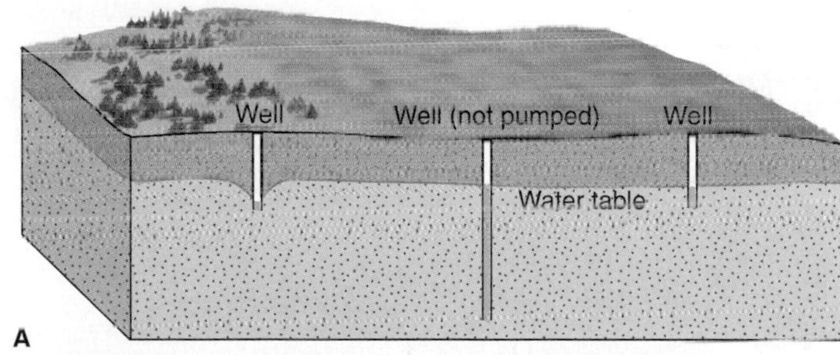

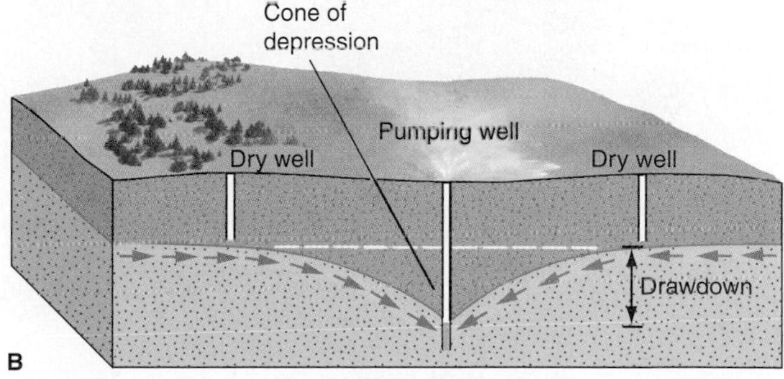

FIGURE 15.9
Pumping well lowers the water table into a cone of depression. If well is heavily pumped, surrounding shallow wells may go dry.

ENVIRONMENTAL GEOLOGY 15.2

How Do We Find Groundwater Resources?

Many wells are drilled or dug without any effort to locate a good aquifer. Many of these wells are successful (especially if only small amounts of water are needed) because most rocks hold some water, which flows into wells that intersect the water table. If a large and dependable supply of water is needed—as for a city water system—specialists in groundwater geology (hydrogeologists) may be called in to locate a promising well site. Hydrogeologists use many methods to locate aquifers. A detailed knowledge of the local rocks is necessary. Therefore, a geologist may map the surface rocks and use electrical, magnetic, and seismic surveys to study subsurface rocks to determine the presence of possible aquifers. Sometimes a small-diameter test well is drilled before the larger, more expensive supply well is sunk. Hydrogeologists look for potentially high-producing aquifers rather than searching for water directly, which would be much more difficult. In some regions, however, the presence of certain plants may be a useful guide to locating water, particularly the depth of the water table.

In areas of Alberta, geologists have used satellite imagery to identify linear zones of fractured bedrock. These zones of fractured rock are more permeable and have greater potential to host productive aquifers than surrounding unfractured rocks. Bedrock channels buried beneath glacial deposits are the targets for groundwater exploration programs in Ontario, Alberta, Saskatchewan, and the Peace River region of B.C., as they contain thick permeable sand and gravel deposits.

Some people search for water by water witching, or *dowsing,* with a divining rod (also sometimes used to search for metals or lost objects). Usually the dowser holds a forked stick horizontally in the hands while walking over an area. The stick is supposed to deflect or twist downward of its own accord when the dowser passes over water. This method has been tried for centuries, the only modification being that a twisted metal rod, often made of a coat hanger, now may be substituted for the stick. Carefully controlled tests conducted by workers in psychic research have shown that water witchers' "success" is equal to or less than pure chance, while geologists' results are superior both to witching and to chance. Records kept on thousands of wells in Australia in the early 1900s show that of wells that were not divined, more than 83 percent produced flows of over 375 litres per hour and 7.4 percent were failures, finding no water at all. Of wells that were divined by water witchers, only about 70 percent produced more than 375 litres per hour, and 14.7 percent were dry. In the early part of the twentieth century the U.S. Geological Survey concluded that any future testing of the results of water witching would be a misuse of public funds.

Despite such findings, many people believe strongly in dowsing. Water witchers themselves devoutly believe they can find water, and in some regions of the United States almost no wells are drilled without a witcher's advice. Dowsers are helped in locating water by the fact that most rocks hold some water, and dowsers often have a longstanding knowledge of a particular region and its potential water resources.

Additional Web Resource

www.agr.gc.ca/pfra/

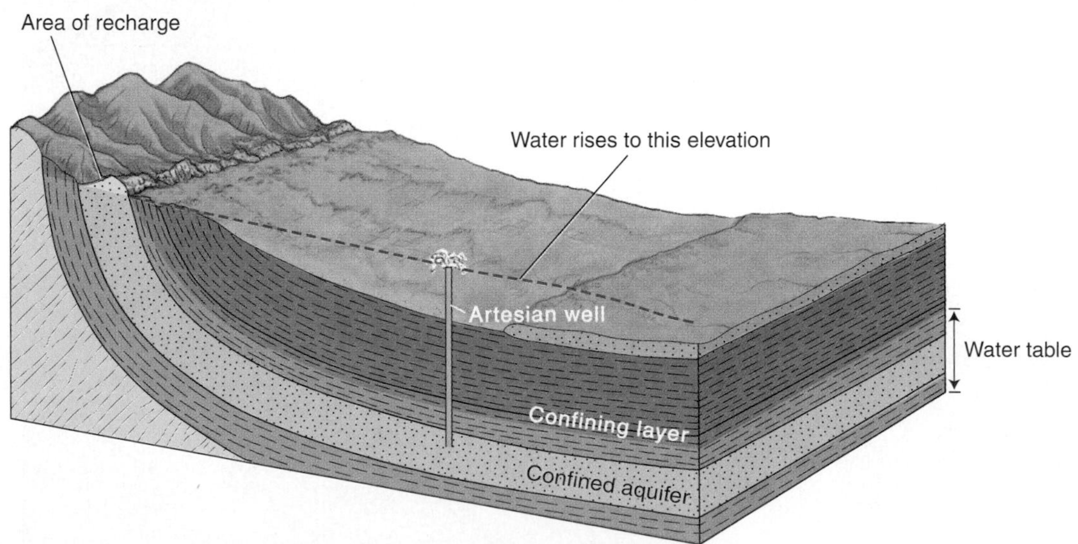

FIGURE 15.10
The Dakota Sandstone in South Dakota is a relatively unusual type of confined aquifer because it is tilted and exposed to the surface by erosion. Water in most wells rose above the land surface when the aquifer was first tapped in the 1800s.

In some artesian wells the water rises above the land surface, producing a flowing well that spouts continuously into the air unless it is capped (figure 15.10). Flowing wells used to occur in South Dakota, when the extensive Dakota Sandstone aquifer was first tapped (figure 15.11), but continued use has lowered the water pressure surface below the ground surface in most parts of the state. Water still rises above the aquifer, but does not reach the land surface.

FIGURE 15.11

Artesian well spouts water above land surface in South Dakota, early 1900s. Heavy use of this aquifer has reduced water pressure so much that spouts do not occur today.

Photo by N. H. Darton, U.S. Geological Survey

Springs and Streams

A **spring** is a place where water flows naturally from rock onto the land surface (figure 15.12). Some springs discharge where the water table intersects the land surface, but they also occur where water flows out from caverns or along fractures, faults, or rock contacts that come to the surface (figure 15.13).

Climate determines the relationship between stream flow and the water table. In rainy regions most streams are **gaining streams;** that is, they receive water from the saturated zone (figure 15.14*A*). The surface of these streams coincides with the water table. Water from the saturated zone flows into the stream through the stream bed and banks that lie below the water table. Because of the added groundwater, the discharge of these streams increases downstream. Where the water table intersects the land surface over a broad area, ponds, lakes, and swamps are found.

In drier climates rivers tend to be **losing streams;** that is, they are losing water to the saturated zone (figure 15.14*B*). The channels of losing streams lie above the water table. The water percolating into the ground beneath a losing stream causes the water table to slope away from the stream. In very dry climates, such as in a desert, a losing stream may be separated or *disconnected* from the underlying saturated zone and a groundwater mound remains beneath the stream even if the stream bed is dry (figure 15.14*C*).

A

B

FIGURE 15.12

(*A*) A large spring issuing from a cavern in limestone, Jasper National Park, Alberta. (*B*) Groundwater issuing as a spring at the contact between permeable and impermeable rocks along the Niagara Escarpment, Ontario.

Photos by Nick Eyles

WHAT DETERMINES GROUNDWATER QUALITY?

Pure water is tasteless and odourless and consists only of hydrogen and oxygen atoms. In nature, water is never found in an entirely pure state as it contains a variety of constituents including inorganic and organic compounds, gases, and microorganisms. The chemistry of the precipitation and recharge water has considerable effect on groundwater chemistry. For example, acid rain, formed downwind of industrial areas (refer to box 8.1), contains high concentrations of sulphur and nitrogen compounds, and in areas close to coastlines precipitation often has high concentrations of sodium chloride.

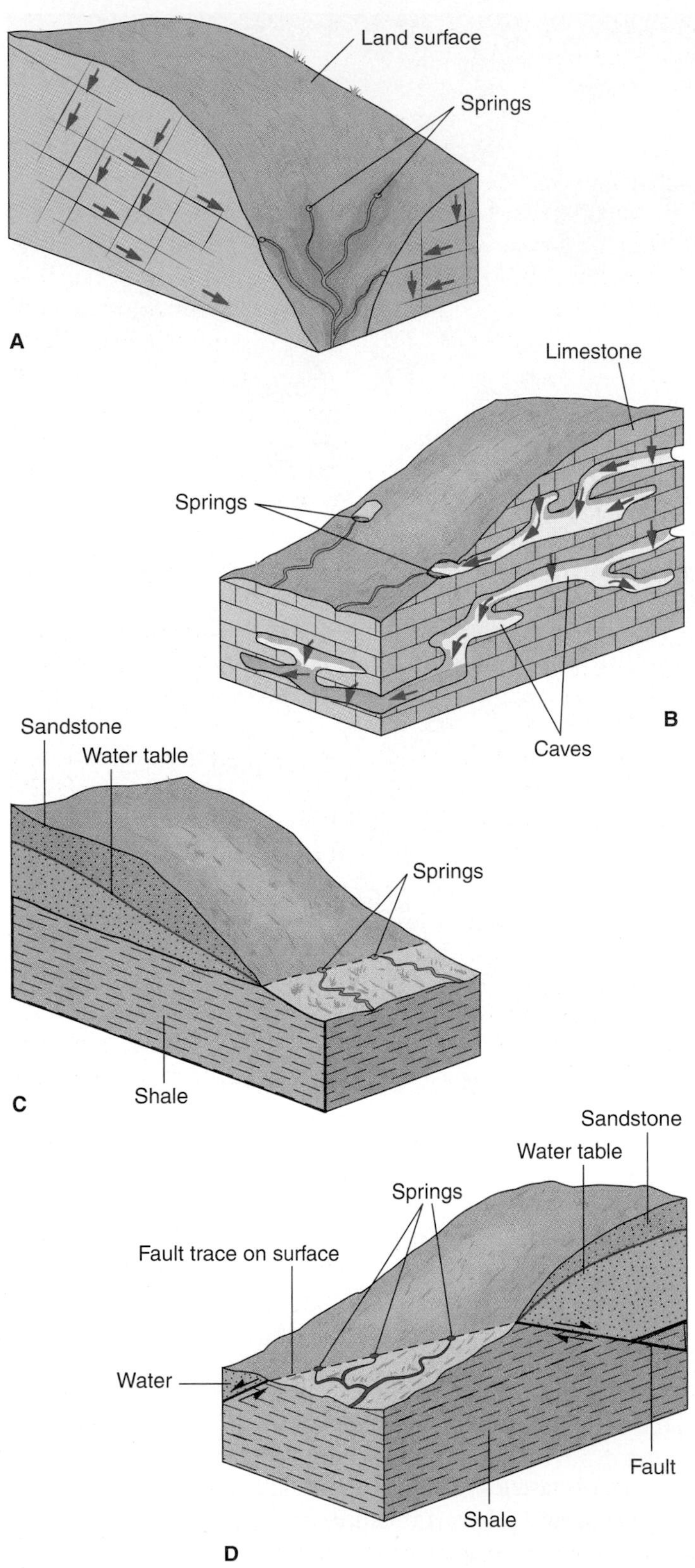

FIGURE 15.13

Springs can form in many ways. (*A*) Water moves along fractures in crystalline rock and forms springs where the fractures intersect the land surface. (*B*) Water enters caves along joints in limestone and exits as springs at the mouths of caves. (*C*) Springs form at the contact between a permeable rock such as sandstone and an underlying less permeable rock such as shale. (*D*) Springs can form along faults when permeable rock has been moved against less permeable rock. Arrows show relative motion along fault.

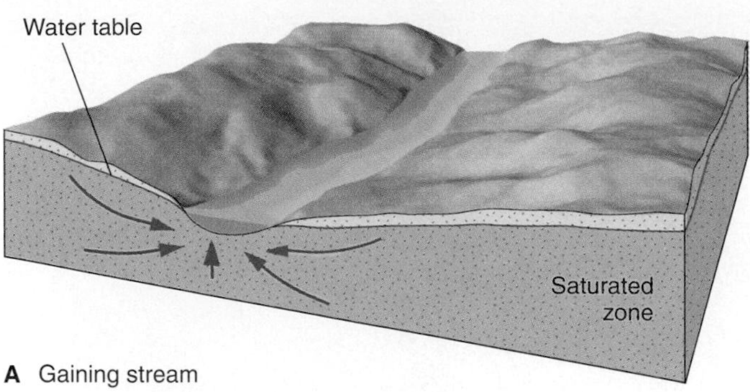

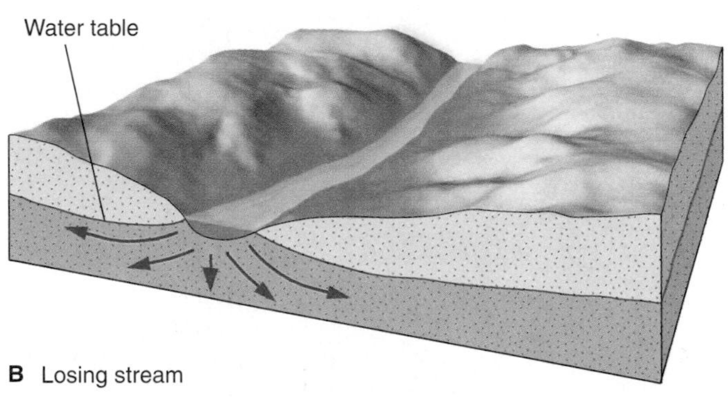

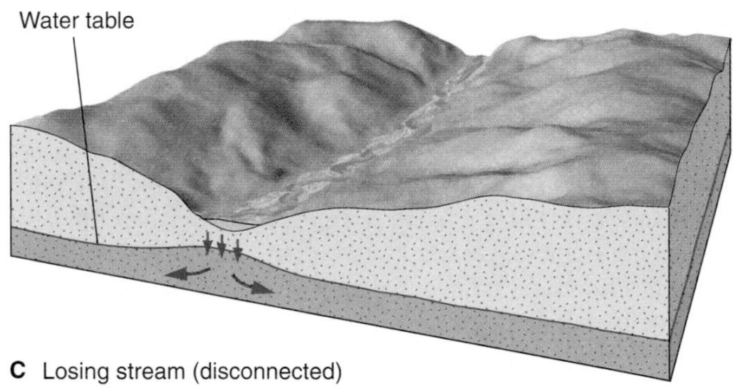

FIGURE 15.14

Gaining and losing streams. (*A*) Stream gaining water from saturated zone. (*B*) Stream losing water through stream bed to saturated zone. (*C*) Water table can be close to the land surface, but disconnected from the surface water beneath a stream bed that intermittently contains water.

The chemistry of groundwater is also affected by the type of rock and soil the water moves through and the rate at which it moves. Groundwater moving through soil dissolves carbon dioxide, creating a weak acid that in turn may dissolve silicate minerals as well as other substances it encounters. This means that groundwater often contains more salts and is harder (contains larger amounts of dissolved calcium) than surface waters in the same location. Slower-moving groundwater generally dissolves more material than faster-moving groundwater, and in some situations substances may be precipitated out from groundwater.

Geoscientists determine groundwater chemistry by measuring the amount of dissolved constituents in the water, which are usually reported in milligrams per litre (mg/L). Groundwater from different locations shows considerable chemical variability. Total dissolved solids in groundwater can range from values of 25 mg/L on the Canadian Shield to 300,000 mg/L in deep saline waters in the Interior Plains.

Groundwater quality is a measure of the suitability of groundwater for a certain purpose. The pH, salinity, and hardness of groundwater can all affect its suitability for different uses, and the presence of contaminants, or undesirable substances, can significantly affect the quality of groundwater as a source of drinking water.

Not all sources of groundwater contamination are human-made. Naturally occurring *minerals within rock and soil* may contain elements such as arsenic, selenium, mercury, and other toxic metals. Circulating groundwater can leach these elements out of the minerals and raise their concentrations to harmful levels within the water. Not all spring water is safe to drink. Like a "bad waterhole" depicted in a Western movie, some springs contain such high levels of toxic elements that the water can sicken or kill humans and animals that drink it. Many desert springs contain such high concentrations of sodium chloride or other salts that their water is undrinkable.

Large amounts of dissolved calcium and magnesium create "hard" waters that cause problems for domestic use. Hard-water areas are associated with deposits of limestone, dolostone, and carbonate rich shale. Although hard water is not usually a health hazard, the abundant calcium and magnesium ions prevent soap from lathering and form scaly deposits inside teakettles and pipes. In some areas, such as locations directly southeast of Ottawa, calcium and magnesium concentrations are so high that water cannot be "softened" by water-softening technology and is not palatable to drink.

Excessively saline water is a problem in some areas of southern Saskatchewan and Alberta. Saline ponds can be recognized by the presence of salt-loving plants and often overlie underground deposits of potash. High levels of arsenic in groundwater are often associated with gold-bearing rocks, most notably in Nova Scotia.

WHAT ARE THE CAUSES OF GROUNDWATER CONTAMINATION?

Groundwater in its natural state tends to be relatively free of contaminants in most areas. Because it is a widely used source of drinking water, contamination of groundwater can be a very serious problem.

Pesticides and *herbicides* (such as DDT and 2,4-D) applied to agricultural crops (figure 15.15*A*) can find their way into groundwater when rain or irrigation water leaches the poisons downward into the soil. *Fertilizers* are also a concern. Nitrate, one of the most widely used fertilizers, is harmful in even small quantities in drinking water. It is estimated that 40 percent of rural wells in Canada are contaminated by nitrates and bacteria to levels exceeding those recommended for drinking water. Unfortunately, over 60 percent of Canadian farmers do not have their well water tested regularly for contaminants (2001 data).

Rain can also leach pollutants from city landfills into groundwater supplies (figure 15.15*B*). Consider for a moment some of the things you threw away last year. A partially empty aerosol can of ant poison? The can will rust through in the landfill, releasing the poison into the ground and into the saturated zone below. A broken thermometer? The toxic mercury may eventually find its way to the groundwater supply. A half-used can of oven cleaner? The dried-out remains of a can of lead-based paint? *Heavy metals* such as mercury, lead, chromium, copper, and cadmium, together with household chemicals and poisons, can all be concentrated in groundwater supplies beneath landfills (figure 15.16).

Liquid and solid wastes from septic tanks, sewage plants, and animal feedlots and slaughterhouses may contain *bacteria, viruses,* and *parasites* that can contaminate groundwater (figure 15.15*C*, box 15.3). Liquid wastes from industries (figure 15.15*D*) and military bases can be highly toxic, containing high concentrations of heavy metals and compounds such as cyanide and PCBs (polychlorinated biphenyls), which are widely used in industry. A degreaser called TCE (trichloroethylene) has been increasingly found to pollute both surface and underground water in numerous regions (box 15.4). Toxic liquid wastes are often held in surface ponds or pumped down deep disposal wells. If the ponds leak, groundwater can become polluted. Deep wells may be safe for liquid waste disposal if they are deep enough, but contamination of drinking water supplies and even surface water has resulted in some localities from improper design of the disposal wells.

Acid mine drainage from coal and metal mines can contaminate both surface and groundwater. It is usually caused by sulphuric acid formed by the oxidation of sulphur in pyrite and other sulphide minerals when they are exposed to air by mining activity. Fish and plants are often killed by the acid waters draining from long-abandoned mines.

Radioactive waste is both an existing and a very serious potential source of groundwater contamination. The shallow burial of *low-level* solid and liquid radioactive wastes from the nuclear power industry has caused contamination of groundwater, particularly as liquid waste containers leak into the saturated zone and as the seasonal rise and fall of the water table at some sites periodically covers the waste with groundwater. The search for a permanent disposal site for solid, *high-level* radioactive waste (now stored temporarily on the surface) is a major national concern for Canada and the United States. The permanent site will be deep underground and must be isolated from groundwater circulation for thousands of years. Salt beds, shale, glassy tuffs, and crystalline rock deep beneath the surface have all been studied, particularly in arid regions where the water table is hundreds of metres below the land surface.

FIGURE 15.15

Some sources of groundwater pollution. (*A*) Pesticides. (*B*) Household garbage. (*C*) Animal waste. (*D*) Industrial toxic waste.
Photo A by Michael Stimmann; Photo B by Frank M. Hanna; Photos C and D from USDA-Soil Conservation Service

Atomic Energy of Canada Ltd (AECL) has investigated the possibility of deep underground storage of nuclear wastes from its CANDU reactors in metamorphic and igneous rocks of the Canadian Shield. Any potential site on the Canadian Shield must be free of joints and other structures that could allow groundwater to flow through storage vaults. An underground research laboratory (URL) was established in Pinawa, Manitoba, consisting of a series of deep tunnels excavated up to 440 m below the ground surface into the Lac du Bonnet granite batholith. Geoscientists at the URL studied the fracture characteristics and thermal and hydrogeochemical properties of the rocks and concluded that they have very low permeability. Groundwater encountered in the rocks at depths of greater than 500 m is saline, reducing, and old, and can be considered as essentially stagnant over a 1-million-year time period, the period of concern for a nuclear waste facility. Even though crystalline rocks of the Canadian Shield may be feasible as hosts for storage sites for high-level nuclear waste, the process of specific site identification is extremely costly, time-consuming, and hindered by community opposition to test programs as well as geological and political uncertainties. A specific nuclear waste disposal site on the Canadian Shield has not yet been selected.

The likely site for disposal of high-level waste in the United States, primarily spent fuel from nuclear reactors, is Yucca Mountain, Nevada, 180 km northwest of Las Vegas. The site would be deep underground in volcanic tuff well above the current (or predicted future) water table, and in a region of very low rainfall. The U.S. Congress, under intense political pressure from other candidate states that did not want the site, essentially chose Nevada in late 1988 by eliminating the funding for the study of all alternative sites. After much controversy over the ultimate safety of the site and objections from Nevada, President George W. Bush approved the Yucca Mountain site in 2002.

Not all groundwater contaminants form plumes within the saturated zone, as shown in figure 15.16. *Gasoline*—which leaks from gas station storage tanks at tens of thousands of locations—is less dense than water and floats upon the water table (figure 15.17). Some liquids such as TCE are heavier than water and sink to the bottom of the saturated zone, perhaps travelling in unpredicted directions upon the surface of an impermeable layer (figure 15.17). Determining the extent and flow direction of groundwater pollution is a lengthy process requiring the drilling of tens, or even hundreds, of costly wells for each contaminated site.

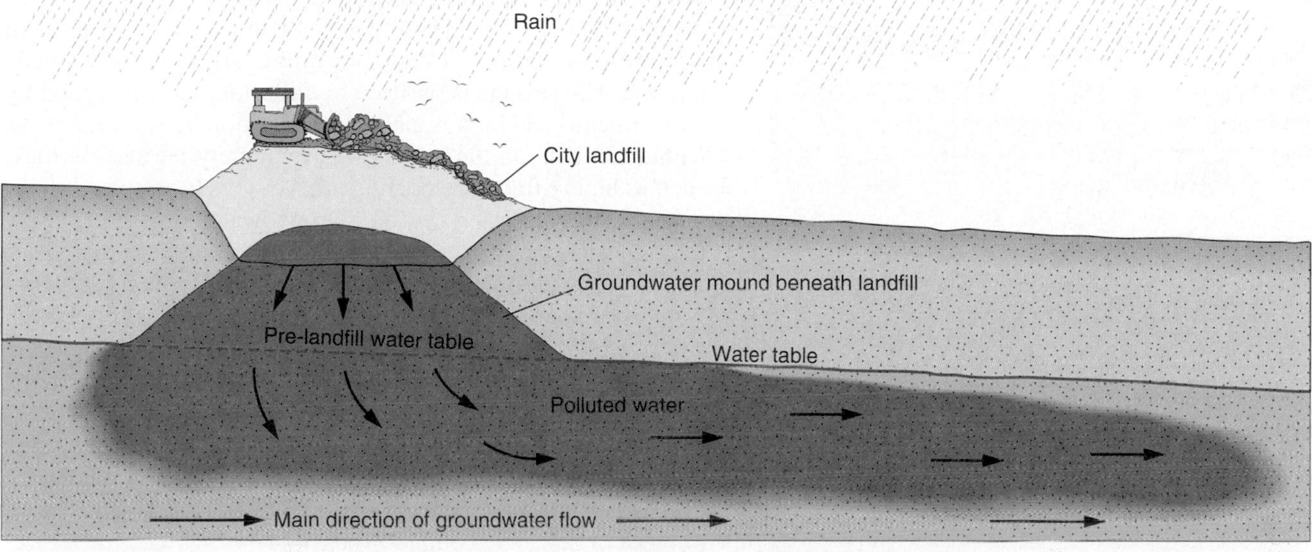

A Cross-section

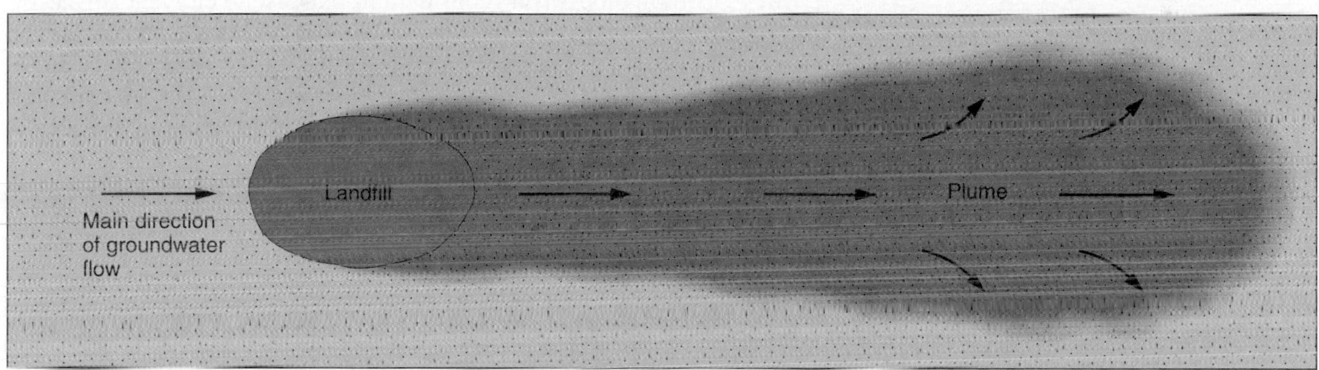

B Map view of contaminant plume. Note how it grows in size with distance from the pollution source.

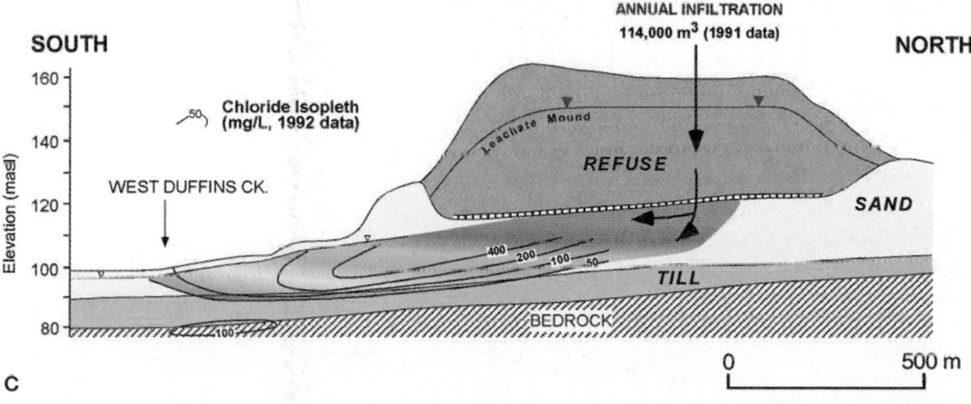

C

FIGURE 15.16

(*A*) and (*B*): Waste piled on the land surface creates a groundwater mound beneath it because the landfill forms a hill, and because the waste material is more porous and permeable than the surrounding soil and rock. Rain leaches pollutants into the saturated zone. A plume of contaminated water will spread out in the direction of groundwater flow. (*C*) Leachate plume generated by contaminants leaking from the closed Brock West Landfill site in Pickering, Ontario. The plume is identified by elevated concentrations of chloride, a common constituent of landfill leachate.

Illustration C Modified from Nick Eyles, 2002

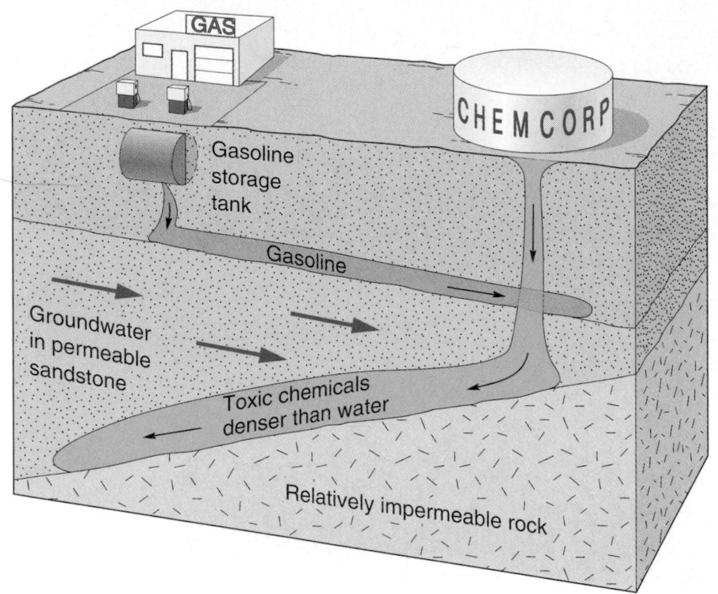

FIGURE 15.17

Not all contaminants move within the saturated zone, as shown in figure 15.16. Gasoline floats on water; many dense chemicals move along impermeable rock surfaces below the saturated zone.

Soil and rock filter some contaminants out of groundwater. This filtering ability depends on the permeability and mineral composition of rock and soil. Under ideal conditions, human sewage can be purified by only 30 to 45 metres of travel through a sandy loam soil (a mixture of clay minerals, sand, and organic humus). The sewage is purified by filtration, ion absorption by clay minerals and humus, and decomposition by soil organisms (figure 15.18). On the other hand, extremely permeable rock, such as highly fractured granite or cavernous limestone, has little purifying effect on sewage. Groundwater flows so rapidly through such rocks that it is not purified even after hundreds of metres of travel (see box 15.3). Some pesticides and toxic chemicals are not purified by passage through rock and soil at all, even soil rich in humus and clay minerals.

Contaminated groundwater is extremely difficult to clean up. Networks of expensive wells may be needed to pump contaminated water out of the ground and replace it with clean water. Because of the slow movement and large volume of groundwater, the cleanup process for a large region can take decades and tens of millions of dollars to complete (box 15.4).

Groundwater contamination can be largely prevented with careful thought and considerable expense. A city landfill can be sited high above the water table and possible flood levels, or located in a region of groundwater discharge rather than recharge. A site can be sealed below by impermeable (and expensive) clay barriers and plastic liners, and sealed off from rainfall by an im-

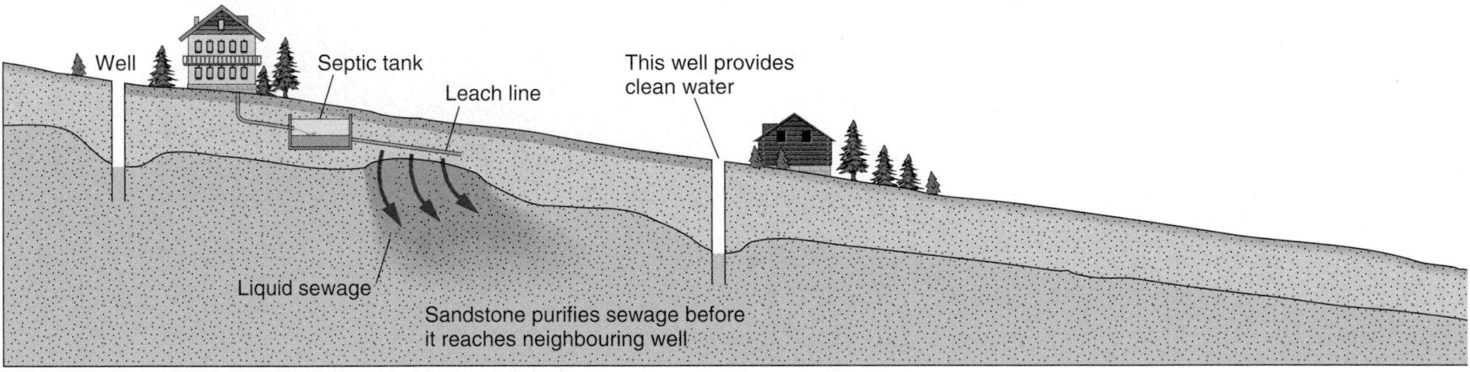

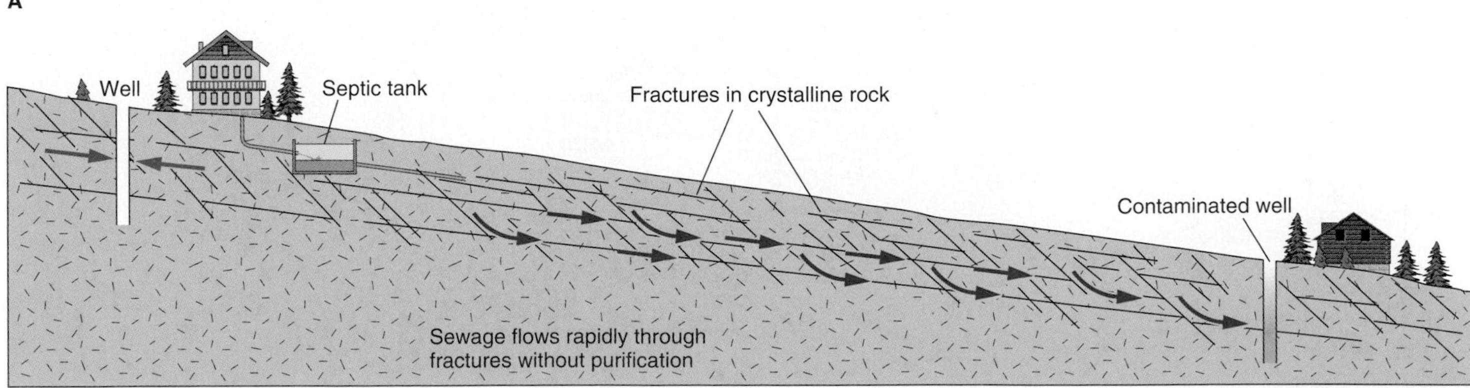

FIGURE 15.18

Rock type and distance control possible sewage contamination of neighbouring wells. (*A*) As little as 30 metres of movement can effectively filter human sewage in sandstone and some other rocks and sediments. (*B*) If the rock has large open fractures, contamination can occur many hundreds of metres away.

ENVIRONMENTAL GEOLOGY 15.3

The Walkerton Water Tragedy

In late May 2000, Canada's most serious case of water contamination occurred when coliform and *E. coli* bacteria in the water consumed by residents of the town of Walkerton, Ontario caused more than 2,300 cases of gastrointestinal illness and 7 deaths. Walkerton, a town of approximately 5,000 people, is dependent on groundwater as its source of drinking water. The water is pumped from several wells comprising a *well field* and is chlorinated at source.

Scientists realized that the source of contamination of Walkerton's water supply had to be close to the wells, as *E. coli* bacteria can survive outside human or animal hosts for only a few days. The most severe contamination was associated with Well #5, which was located close to a farm where manure had recently been spread. Bacteria released from the manure soaked into the ground and contaminated groundwater below the site. Both the groundwater and the bacteria it contained were quickly transported through underlying rocks to the vicinity of Well #5. Groundwater movement in the Walkerton area is particularly rapid, as there is only a thin layer of glacial sediment (less than 2.5 m thick) resting on top of fractured and jointed limestones and dolomites. The unconfined aquifer supplying the town with drinking water is less than 6 m below the ground surface, making it particularly susceptible to contamination from surface sources. Other wells in the area had experienced similar contamination problems in the past, but no protective action by government agencies had been taken.

Important lessons were learned from the Walkerton tragedy. First, a supply of clean, potable groundwater cannot be taken for granted; like any other resource, groundwater should be managed effectively and in a sustainable manner. Second, shallow, unconfined aquifers require protection by careful management of land uses around well fields that involve the restriction of agricultural and industrial practices. Third, to be effective, management plans must be based on a thorough understanding of geological and hydrogeological conditions in the area.

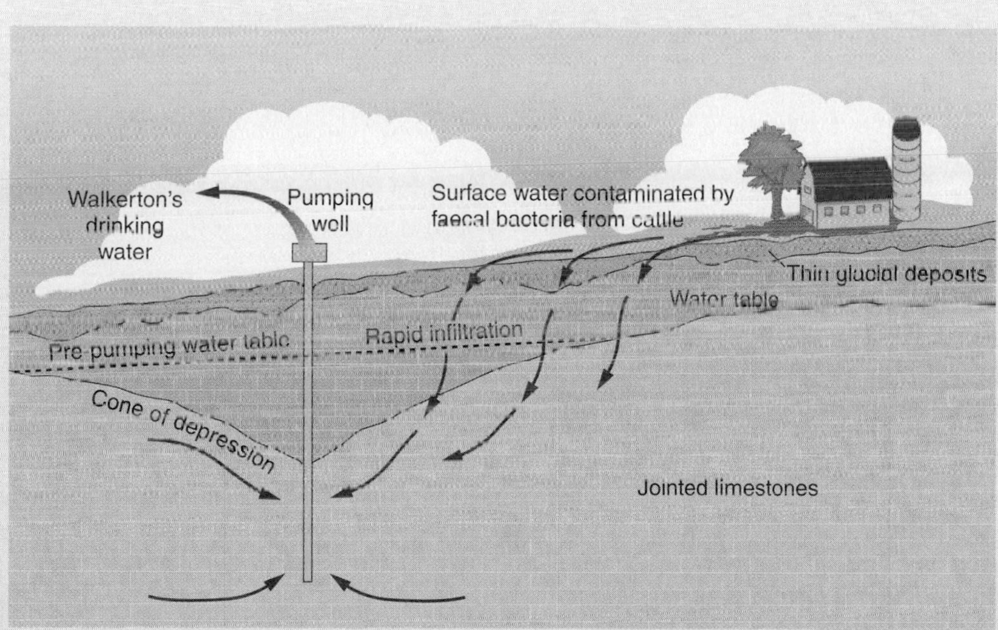

BOX 15.3 ■ FIGURE 1
Rapid infiltration of bacteria-contaminated surface waters into unconfined aquifers in jointed bedrock was partly responsible for the Walkerton tragedy. The illness and loss of life caused by bacteria contamination of drinking water at Walkerton could have been avoided if correct chlorination and monitoring procedures had been followed.

permeable cover. Dikes can prevent surface runoff through or from the site. Although sanitary landfills are expensive, they are much cheaper than groundwater cleanup caused by leakage.

Pumping wells can cause or aggravate groundwater contamination (figure 15.19). Well drawdown can increase the slope of the water table locally, thus increasing the rate of groundwater flow and giving the water less time to be purified underground before it is used (figure 15.19*A*). Drawdown can even reverse the original slope of the water table, perhaps contaminating wells that were pure before pumping began (figure 15.19*B*). Heavily pumped wells near a coast can be contaminated by *saltwater intrusion* (figures 15.19*C* and *D*). Saltwater intrusion is becoming a serious problem as the demand for drinking water increases in rapidly growing coastal communities.

HOW LONG WILL GROUNDWATER LAST?

A local supply of groundwater will last indefinitely if it is withdrawn for use at a rate equal to or less than the rate of recharge to the aquifer. If groundwater is withdrawn faster than it is

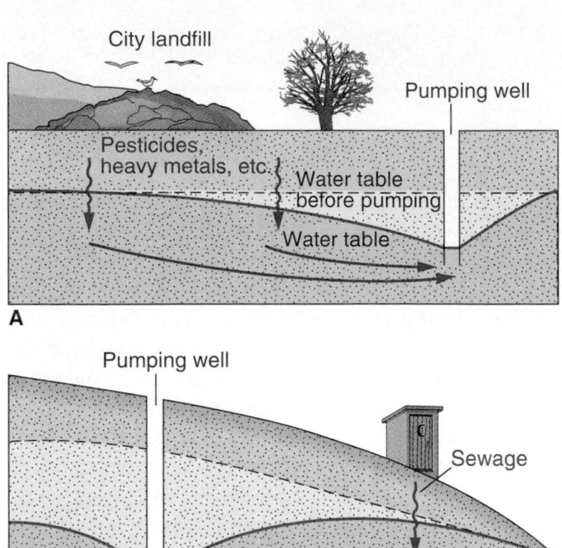

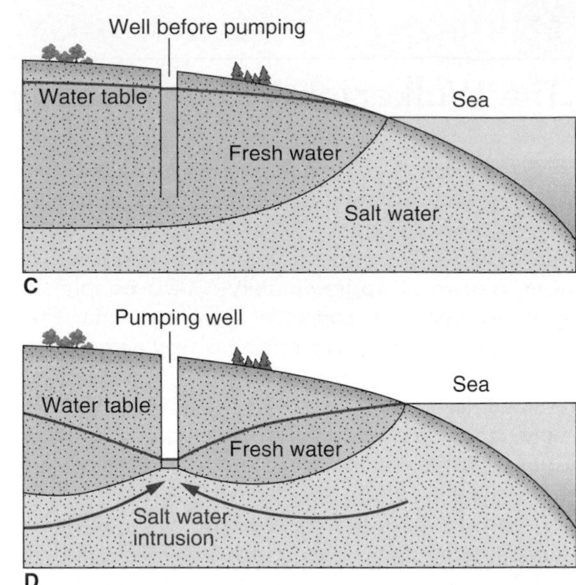

FIGURE 15.19

Groundwater pollution problems caused or aggravated by pumping wells. (*A*) Water table steepens near a landfill, increasing the velocity of groundwater flow and drawing contaminants into a well. (*B*) Water table slope is reversed by pumping, changing direction of the groundwater flow and contaminating the well. (*C*) Well near a coast (before pumping). Fresh water floats on salt water. (*D*) Well in C begins pumping, thinning the fresh water lens and drawing salt water into the well.

being recharged, however, the supply is being reduced and will one day be gone. In parts of Alberta, growing water demands for irrigation, domestic supply, and thermal recovery of bitumen are far exceeding local supplies. In B.C., groundwater levels are declining in areas of intensive urban development such as the Lower Mainland, the Okanagan, and the southeast coast of Vancouver Island. Ontario, Prince Edward Island, and Manitoba have issued moratoriums on new groundwater permits as the sustainable yield of certain aquifers has been reached.

Heavy use of groundwater causes a regional water table to drop. Most of the regional aquifers in Canada are in hydrodynamic equilibrium (recharge equals discharge). However, in parts of western Texas and eastern New Mexico the pumping of groundwater has caused the water table to drop 30 metres over the past few decades. The lowering of the water table means that wells must be deepened and more electricity must be used to pump the water to the surface. Moreover, as water is withdrawn, the ground surface may settle because the water no longer supports the rock and sediment. Mexico City has subsided more than 7 metres and portions of California's Central Valley 9 metres because of extraction of groundwater (figure 15.20). Such *subsidence* can crack building foundations, roads, and pipelines. Overpumping of groundwater also causes compaction and porosity loss in rock and soil, and can permanently ruin good aquifers.

To avoid the problems of falling water tables, subsidence, and compaction, many towns use *artificial recharge* to increase the natural rate of recharge. Natural floodwaters or treated industrial or domestic wastewaters are stored in infiltration ponds on the surface to increase the rate of water percolation into the ground.

FIGURE 15.20

Subsidence of the land surface caused by the extraction of groundwater, near Mendota, San Joaquin Valley, California. Signs on the pole indicate the positions of the land surface in 1925, 1955, and 1977. The land sank 9 metres in 52 years. Since the late 1970s, subsidence decreased to less than a metre due to reduced groundwater pumping and increased use of surface water for irrigation.

Photo by Richard O. Ireland, U.S. Geological Survey

ENVIRONMENTAL GEOLOGY 15.4

The Legacy of Landfilling in Urban Areas

Past land-use practices can have a detrimental effect on groundwater quality and cause widespread soil and water contamination that may cost many millions of dollars to clean up. The legacy of former industrial land uses is reflected in the many abandoned and currently unusable sites in urban areas. The most severely contaminated area in the City of Toronto is the Port Industrial District (box figure 1), which contains more than 2 million tonnes of historic landfill material contaminated by coal ash (generated by thermal generating stations), arsenic, heavy metals, oils, and greases. The Port Industrial District lies in a former embayment on the north shore of Lake Ontario that was infilled after 1912 as a means of remediating a marshland already polluted by sewage and offal from slaughterhouses. Unfortunately, the materials used for infilling the marsh were also contaminated and included garbage, street sweepings, scrap metal, fuel distribution pipes, and fly ash (box figure 2). Contaminated groundwater (leachate) migrating from landfilled areas is now adversely affecting water quality in adjacent water bodies, including Lake Ontario. Similar contaminated materials were used to infill tidal wetlands in New York City (box figure 3); J.F.K. International Airport and La Guardia Airport are both sited on such landfill materials.

The problem of soil and groundwater contamination in the Port Industrial District of Toronto was compounded by the mixed industrial and manufacturing industries it supported. Refining and recycling operations and storage of commodities such as road salt and fuel oils have contributed toward widespread soil and groundwater contamination. Contaminants now found in soil and groundwater include lead, chromium, zinc, cadmium, benzene, xylene, petroleum hydrocarbons, tars, creosotes, polycyclic aromatic hydrocarbons (PAHs), and polychlorinated biphenyls (PCBs) (box figure 4). Remediation plans call for on-site treatment of soil (at costs of up to $500/tonne) and collection, treatment, and monitoring of groundwaters.

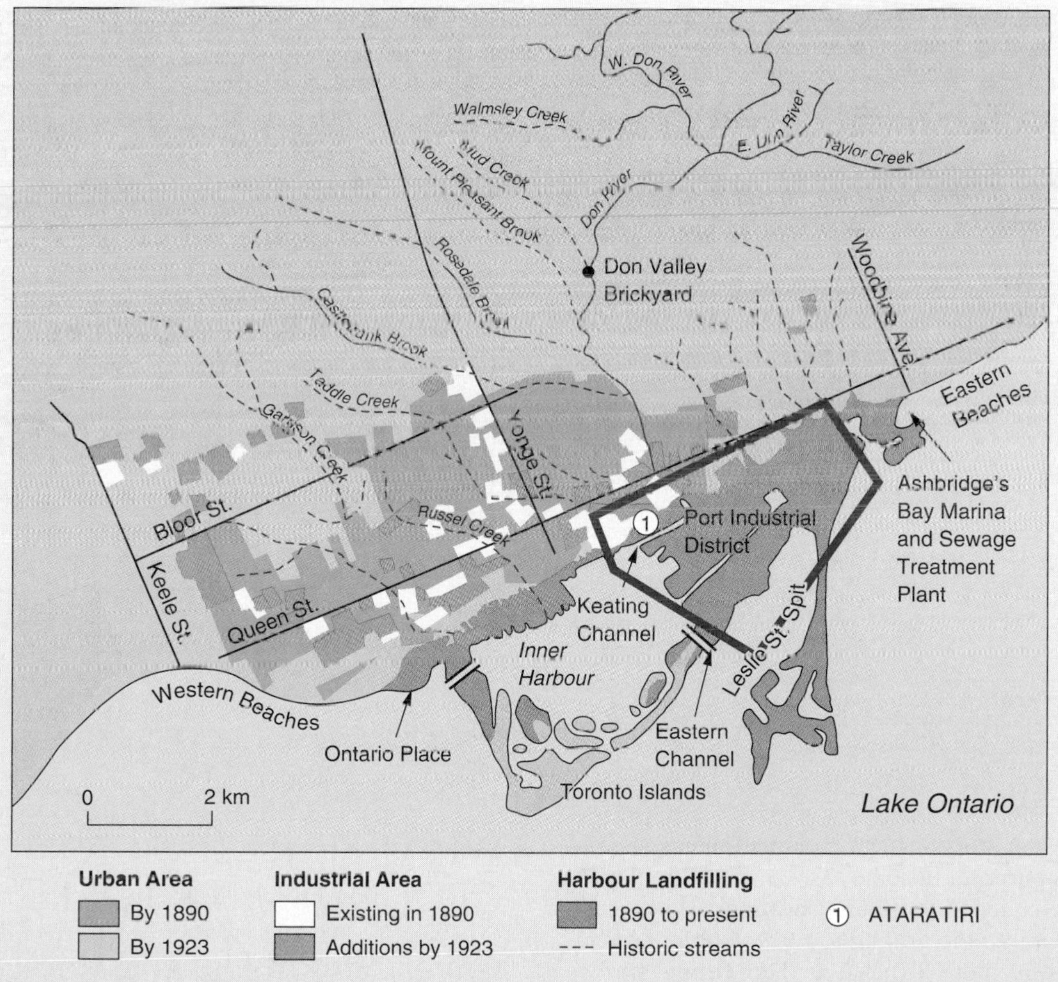

BOX 15.4 ■ FIGURE 1

Extent of landfilling along Toronto harbourfront. The Port Industrial District is underlain by heavily contaminated landfill material.
After Nick Eyles, 1997

BOX 15.4 ■ FIGURE 2

Municipal garbage and street sweeping being used as landfill along the Toronto waterfront, 1923.

Photo PA-84921/Public Archives Canada

In 1988, a large-scale residential development project (called Ataratiri) was designed for part of the Port Industrial District. The 7,000-home project was one of the largest downtown redevelopment projects proposed anywhere in North America. However, the project was cancelled in 1992 due to the prohibitive costs of cleaning up soil and groundwaters and insufficient understanding of the potential health risks posed by contaminants.

Additional Resources

For further information see N. Eyles, 1997. Environmental Geology of a Supercity: The Greater Toronto Area. In *Environmental Geology of Urban Areas*. Ed. N. Eyles. Geological Association of Canada, pp 7–80.

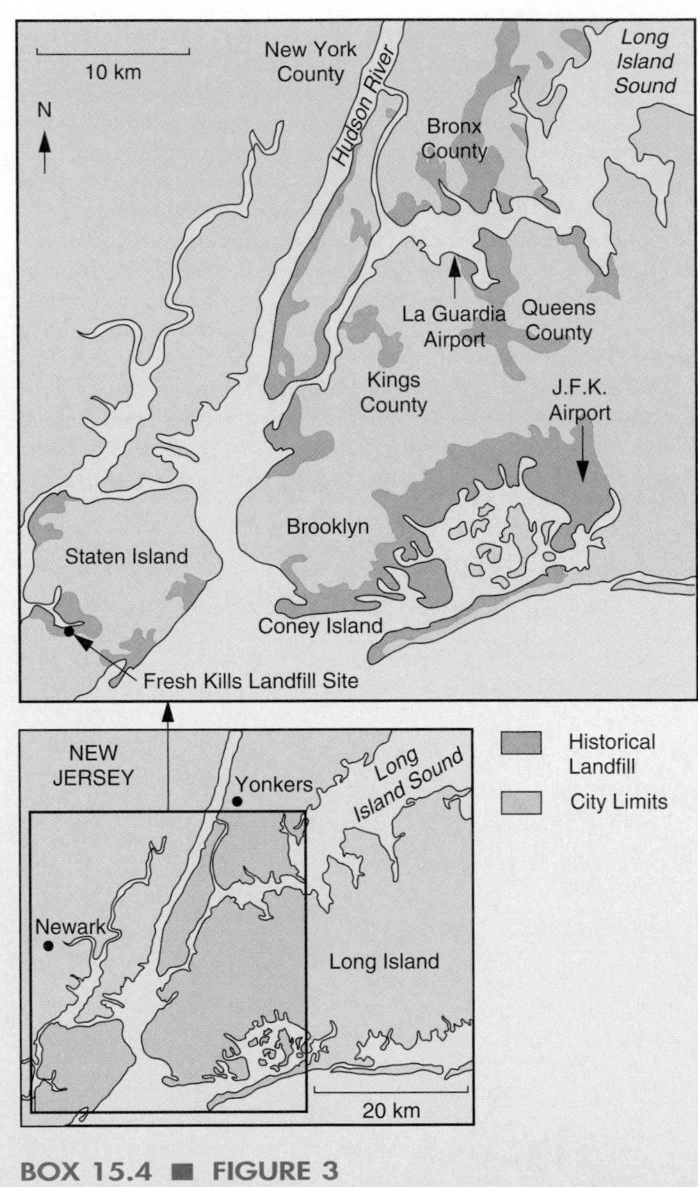

BOX 15.4 ■ FIGURE 3

Historic landfill areas in New York City

Reclaimed, clean water from sewage treatment plants is commonly used for this purpose. In some cases, especially in areas where groundwater is under confined conditions, water is actively pumped down into the ground to replenish the groundwater supply. This is more expensive than filling surface ponds, but it reduces the amount of water lost through evaporation. Groundwater is also extracted and reinjected for aquifer thermal energy systems in several regions of Canada (see box 15.5).

WHAT FEATURES CAN GROUNDWATER FORM?

Caves, Sinkholes, and Karst Topography

Caves (or **caverns**) are naturally formed underground chambers. Most caves develop when slightly acidic groundwater dissolves

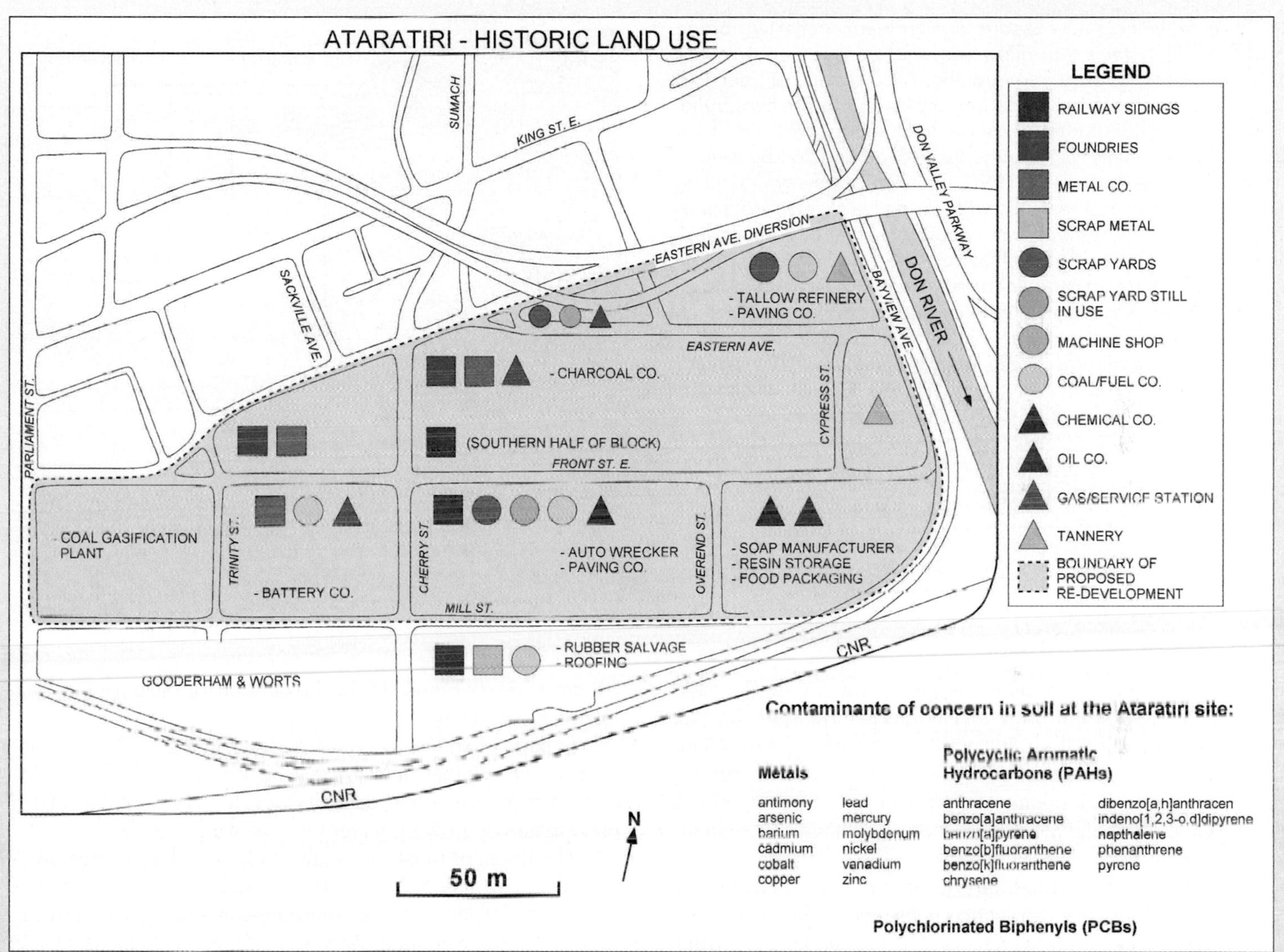

BOX 15.4 ■ FIGURE 4

Legacy of past land uses and principal contaminants at the proposed Ataratiri residential development site, Toronto.

After Nick Eyles, 2002

limestone along joints and bedding planes, opening up cavern systems as calcite is carried away in solution (figure 15.21). Natural groundwater is commonly slightly acidic because of dissolved carbon dioxide (CO_2) from the atmosphere or from soil gases (see chapter 8).

Geoscientists disagree whether limestone caves form above, below, or at the water table. Most caves probably are formed by groundwater circulating below the water table, as shown in figure 15.21. If the water table drops or the land is elevated above the water table, the cave may begin to fill in again by calcite precipitation. The equation below can be read from left to right for calcite solution, and from right to left for the calcite precipitation reaction (see also table 8.1).

Groundwater with a high concentration of calcium (Ca^{++}) and bicarbonate (HCO_3^-) ions may drip slowly from the ceiling of an air-filled cave. As a water drop hangs on the ceiling of the cave, some of the dissolved carbon dioxide (CO_2) may be lost into the cave's atmosphere. The CO_2 loss causes a small

IN GREATER DEPTH 15.5

Aquifer Thermal Energy Storage Systems

Groundwater is now used in aquifer thermal energy storage (ATES) systems that allow seasonal storage and recovery of heat and cold energy in an aquifer. ATES systems extract cold groundwater to cool buildings in summer and to pre-heat buildings in winter (box figure 1). Waste heat energy produced from cooling in the summer is stored in the ground and can be used in the winter for heating. Since 1990, Carleton University in Ottawa has been using groundwater with an ambient temperature of approximately 9°C as a thermal energy source for campus residence buildings. The five-well system used on the campus supplies groundwater at a combined rate of 125 L/sec through simultaneous pumping and reinjection and can be reversed on a seasonal basis. Groundwater circulates in a closed loop and is exposed to minimal environmental contamination. ATES systems also operate at the Pacific Agricultural Research Centre in Agassiz, B.C., the Sussex Hospital in N.B., and the Scarborough Centre in Toronto.

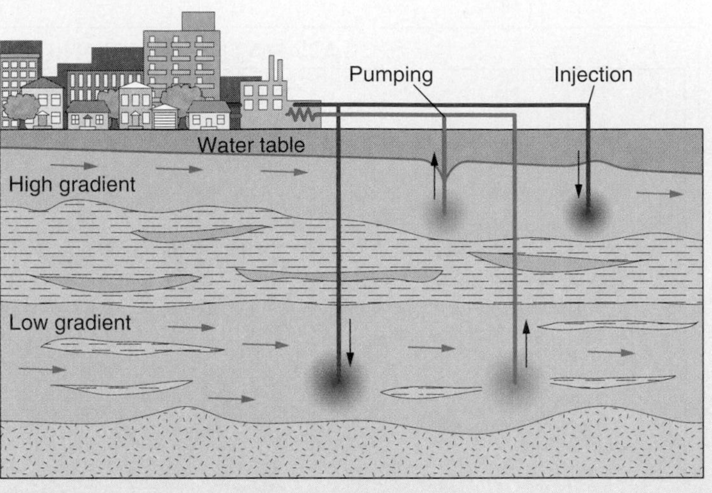

BOX 15.5 ■ FIGURE 1

Aquifer thermal energy storage (ATES) system operating during a cooling season. Cold energy (blue) is extracted from cold storage water wells (stored during a previous heating season). Waste heat from, for example, building air is transferred to the pumped groundwater using a heat exchanger situated in a building. The heated groundwater (red) is then returned to the aquifer. The cold store diminishes in size over the cooling season, while the heat store increases in size. The system is then reversible during the following heating season.

Diagram courtesy of Dr. Diana Allen. Publisher — Simon Fraser University

Additional Web Resources

www.geothermal.ca/utes.html

www.geothermal.ca/carleton.pdf

amount of calcite to precipitate out of the water onto the cave ceiling. When the water drop falls to the cave floor, the impact may cause more CO_2 loss, and another small amount of calcite may precipitate on the cave floor. A falling water drop, therefore, can precipitate small amounts of calcite on both the cave ceiling and the cave floor, and each subsequent drop adds more calcite to the first deposits.

Deposits of calcite (and, rarely, other minerals) built up in caves by dripping water are called *dripstone,* or **speleothems.** **Stalactites** are icicle-like pendants of dripstone hanging from cave ceilings (figure 15.21*B*). They are generally slender and are commonly aligned along cracks in the ceiling, which act as conduits for groundwater. **Stalagmites** are cone-shaped masses of dripstone formed on cave floors, generally directly below stalactites. Splashing water precipitates calcite over a large area on the cave floor, so stalagmites are usually thicker than the stalactites above them. As a stalactite grows downward and a stalagmite grows upward, they may eventually join to form a *column* (figure 15.21*B*). Figure 15.22 shows some of the intriguing features formed in caves.

In parts of some caves, water flows in a thin film over the cave surfaces rather than dripping from the ceiling. Sheetlike or ribbonlike *flowstone* deposits develop from calcite that is precipitated by flowing water on cave walls and floors.

The floors of most caves are covered with sediment, some of which is *residual clay,* the fine-grained particles left behind as insoluble residue when a limestone-containing clay dissolves. (Some limestone contains only about 50 percent calcite.) Other sediment, including most of the coarse-grained material found on cave floors, may be carried into the cave by streams, particularly when surface water drains into a cave system from openings on the land surface.

Solution of limestone underground may produce features that are visible on the surface. Extensive cavern systems can undermine a region so that roofs collapse and form depressions in the land surface above. **Sinkholes** (also known as *dolines*) are

$$\underset{\text{water}}{H_2O} + \underset{\text{carbon dioxide}}{CO_2} + \underset{\text{calcite in limestone}}{CaCO_3} \rightleftarrows \underset{\text{calcium ion}}{Ca^{++}} + \underset{\text{bicarbonate ion}}{2HCO_3^-}$$

→ development of caves (solution)

← development of flowstone and dripstone (precipitation)

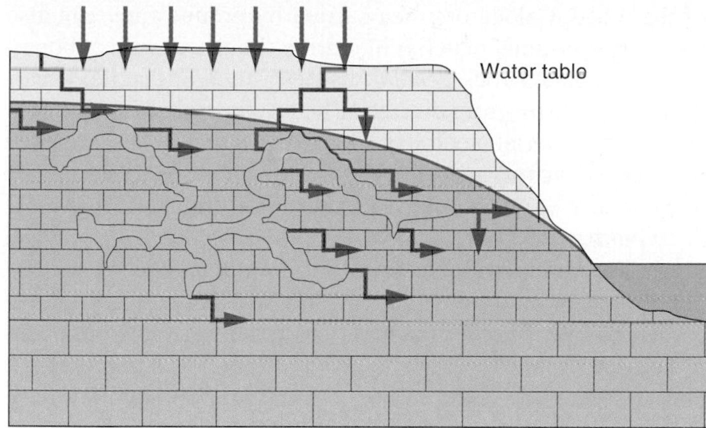

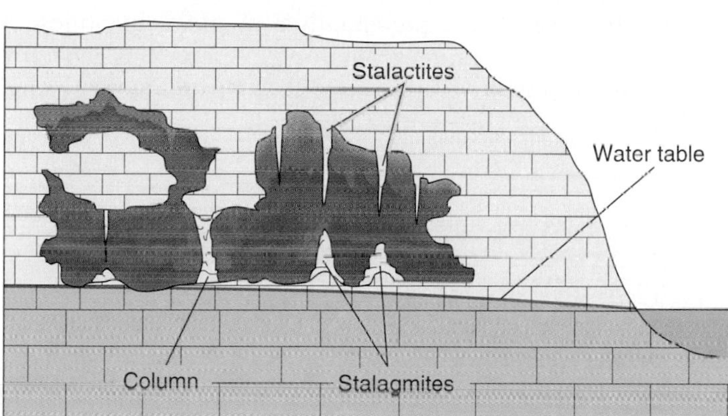

FIGURE 15.21
Solution of limestone to form caves. (*A*) Water moves along fractures and bedding planes in limestone, dissolving it to form caves below the water table. (*B*) Falling water table allows cave system, now greatly enlarged, to fill with air. Calcite precipitation forms stalactites, stalagmites, and columns above the water table.

FIGURE 15.22
Stalactites, stalagmites, and flowstone in Great Onyx Cave, Kentucky.
Photo courtesy of Stanley Fagerlin

closed depressions found on land surfaces underlain by limestone (figure 15.23). They form either by the *collapse* of a cave roof or by *solution* as descending water enlarges a crack in limestone. Limestone regions in Florida, Missouri, Indiana, and Kentucky are heavily dotted with sinkholes. Sinkholes can also form in regions underlain by gypsum or rock salt, which are also soluble in water.

An area with many sinkholes and with cave systems beneath the land surface is said to have **karst topography** (figure 15.24*A*). Karst areas are characterized by a lack of surface streams, although one major river may flow at a level lower than the karst area.

Karst areas are less common in Canada than the U.S., but excellent examples of limestone karst can be found at Castleguard Cave in Banff National Park, Maligne Lake in Jasper National Park, and Nahanni National Park in N.W.T (figure 15.24*B*). Gypsum karst can be found in Windsor, Nova Scotia, and evaporate karst north of Fort McMurray, Alberta.

Streams sometimes disappear down sinkholes to flow through caves beneath the surface. In this specialized instance, a true *underground stream* exists. Such streams are quite rare, however, as most groundwater flows very slowly through pores and cracks in sediment or rock. You may hear people with wells describe the "underground stream" that their well penetrates, but this is almost never the case. Wells tap groundwater in the rock pores and crevices, not underground streams. If a well did tap a true underground river in a karst region, the water would probably be too polluted to drink, especially if it had washed down from the surface into a cavern without being filtered through soil and rock.

Other Effects

Groundwater is important in the preservation of *fossils* such as **petrified wood,** which develops when porous buried wood is either filled in or replaced by inorganic silica carried in by groundwater (figure 15.25). The result is a hard, permanent rock, commonly preserving the growth rings and other details

A

B

C

FIGURE 15.23

(*A*) Sinkholes formed in limestone near Timaru, New Zealand. (*B*) A collapse sinkhole that formed suddenly in Winter Park, Florida, in 1981. (*C*) Large sinkhole (doline) in limestone bedrock, near Norman Wells, NWT.

Photo A © G. R. "Dick" Roberts/Natural Sciences Image Library; Photo B © AP/Wide World Photos; Photo C by Nick Eyles

of the wood. Calcite or silica carried by groundwater can also replace the original material in marine shells and animal bones.

Sedimentary rock *cement,* usually silica or calcite, is carried into place by groundwater. When a considerable amount of cementing material precipitates locally in a rock, a hard rounded mass called a **concretion** develops, typically around an organic nucleus such as a leaf, tooth, or other fossil (figure 15.26*A*).

Geodes are partly hollow, globe-shaped bodies found in volcanic rocks, some limestones, and locally in other rocks. The outer shell is amorphous silica, and well-formed crystals of quartz, calcite, or other minerals project inward toward a central cavity (figure 15.27). The origin of geodes is complex but clearly related to groundwater. Crystals in geodes may have filled original cavities or have replaced fossils or other crystals.

In arid and semi-arid climates, *alkali soil* (solonetz) may develop because of the precipitation of great quantities of sodium salts by evaporating groundwater. Such soil is generally unfit for plant growth. Alkali soil generally forms at the ground surface in low-lying areas. (See chapter 8.)

HOW DO HOT SPRINGS AND GEYSERS FORM?

Hot springs are springs in which the water is warmer than human body temperature. Water can gain heat in two ways while it is underground. First, and more commonly, groundwater may circulate near a magma chamber or a body of cooling igneous rock. In the United States most hot springs are found in the western states, where they are associated with relatively recent volcanism. The hot springs and pools of Yellowstone National Park in Wyoming are of this type.

Groundwater can also gain heat if it circulates unusually deeply in the Earth, perhaps along joints or faults. As discussed in chapter 6, the normal geothermal gradient (the increase in temperature with depth) is 25°C/kilometre. Water circulating to a depth of 2 or 3 kilometres is warmed substantially above normal surface water temperature. The famous springs at Warm Springs, Georgia, at Radium Hot Springs in Kootenay National Park, B.C., and at Banff in the Canadian Rockies have been warmed by deep circulation. Warm water, regardless of its origin, is lighter than cold water and readily rises to the surface.

A **geyser** is a type of hot spring that periodically erupts hot water and steam. The water is generally near boiling (100°C). Eruptions may be caused by a constriction in the underground "plumbing" of a geyser, which prevents the water from rising and cooling. The events thought to lead to a geyser eruption are illustrated in figure 15.28. Water gradually seeps into a partially emptied geyser chamber and heat supplied from below slowly warms the water. Bubbles of water vapour and other gases then begin to form as the temperature of the water rises. The bubbles may clog the constricted part of the chamber until the upward pressure of the bubbles pushes out some of the water above in a gentle surge, thus lowering the pressure on the water in the lower part of the chamber. This drop in pressure causes the chamber

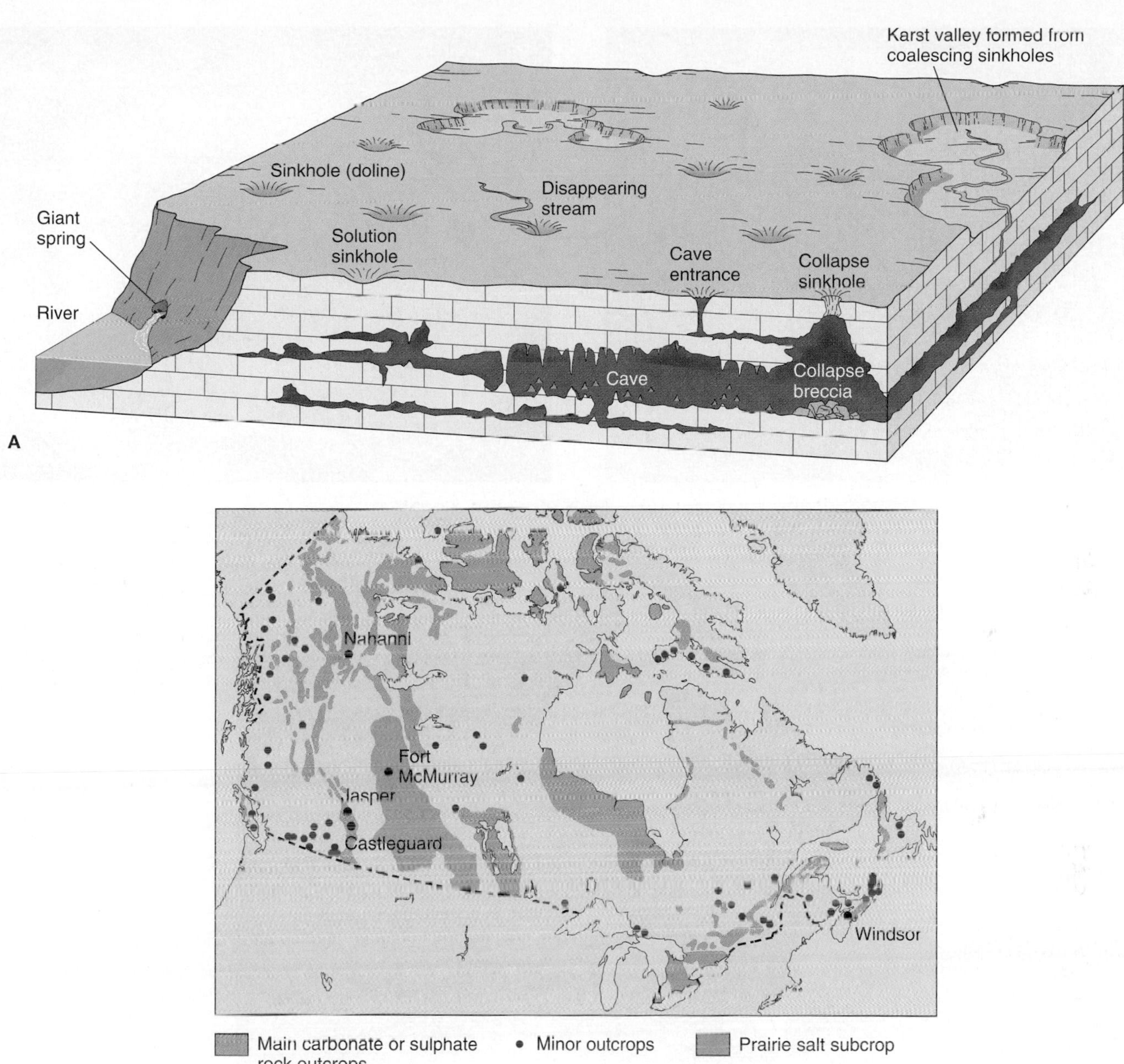

FIGURE 15.24

(*A*) Schematic diagram to show karst topography is marked by underground caves and numerous surface sinkholes (dolines). A major river may cross the region, but small surface streams generally disappear down sinkholes. (*B*) Distribution of karst rocks in Canada. For further information on karst in Canada see Ford, D.C., 1989. Solution process. Fulton, R.J. (ed) Quaternary Geology of Canada and Greenland. Geological Survey of Canada, chapter 9.

Illustration B is after Ford, 1989

water, now very hot, to flash into vapour. The expanding vapour blasts upward out of the chamber, driving hot water with it and condensing into visible steam. The chamber, now nearly empty, begins to fill again and the cycle is repeated. The entire cycle may be quite regular, as it is in Yellowstone's Old Faithful geyser, which averages about 79 minutes between eruptions (though it varies from about 45 to 105 minutes depending on the amount of water left in the chamber after an eruption). Many geysers, however, erupt irregularly, some with weeks or months between eruptions. Geysers are found only in the U.S., Iceland, New Zealand, and Kamchatka.

As hot groundwater comes to the surface and cools, it may precipitate some of its dissolved ions as minerals. *Travertine* is a deposit of *calcite* that often forms around hot springs (figure 15.29), while dissolved *silica* precipitates as *sinter* (called *geyserite* when deposited by a geyser, as shown in figure 15.30). Small travertine deposits are forming at Rabbit Kettle Hot Springs in Nahanni National Park, N.W.T. The composition of the subsurface

FIGURE 15.25

Petrified log in the Painted Desert, Arizona. The log was replaced by silica carried in solution by groundwater. Small amounts of iron and other elements colour the silica in the log.

Photo © Eric & David Hosking/Corbis Media

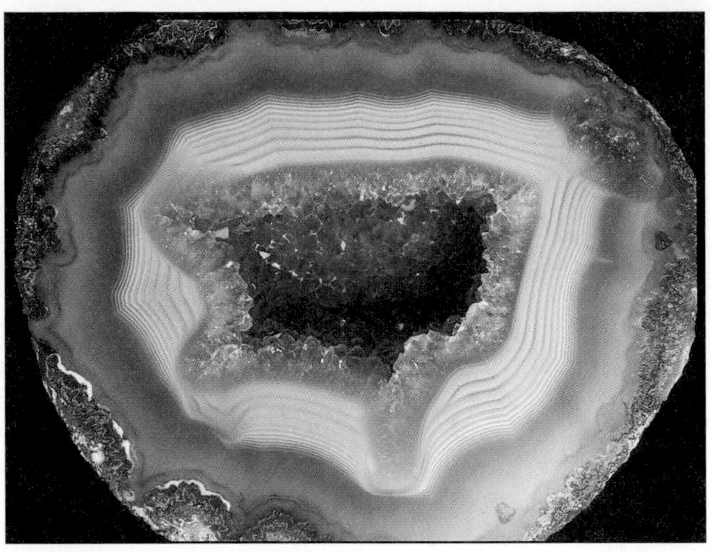

FIGURE 15.27

Concentric layers of amorphous silica are lined with well-formed amethyst (quartz) crystals growing inward toward a central cavity in a geode.

Photo © Martin Land/LANDM/Bruce Coleman

FIGURE 15.26

Large concretion (approximately one metre across) within shales of the Kettle Point Formation, southwestern Ontario. Concretions contain more cement than the surrounding rock and are more resistant to weathering.

Photo by Nick Eyles

rocks generally determines which type of deposit forms, although sinter can indicate higher subsurface temperatures than travertine because silica is harder to dissolve than calcite. Both deposits can be stained by the pigments of bacteria that thrive in the hot water. These thermophyllic bacteria are some of the most primitive of living bacteria and suggest that life may have arisen near hot springs.

A *mudpot* is a special type of hot spring that contains thick, boiling mud. Mudpots are usually marked by a small amount of water and strongly sulphurous gases, which combine to form strongly acidic solutions. The mud probably results from intense chemical weathering of the surrounding rocks by these strong acids (see figure 8.15).

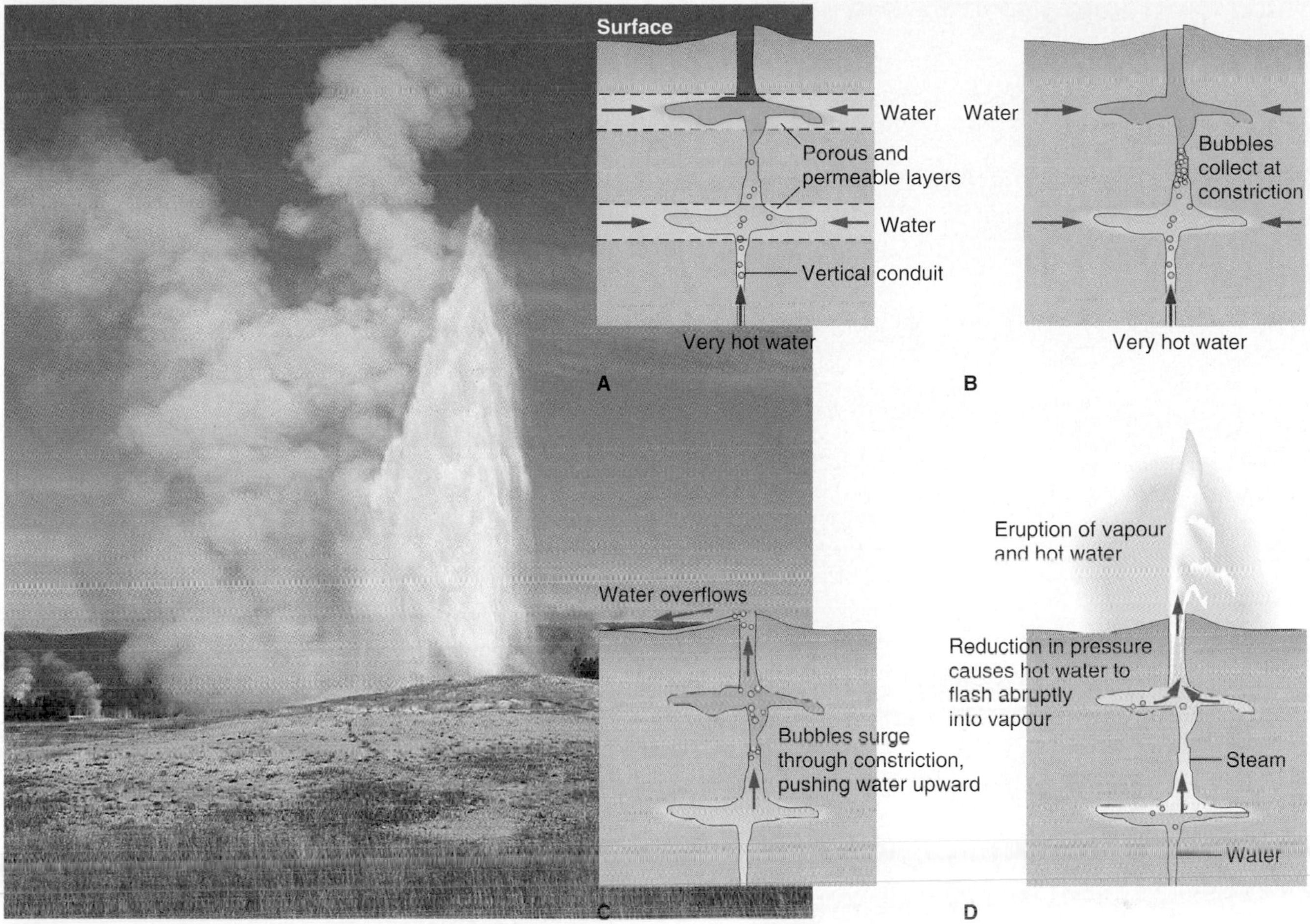

FIGURE 15.28

Eruptive history of a typical geyser in (A) through (D). Photo shows the eruption of Old Faithful geyser in Yellowstone National Park, Wyoming. See text for explanation.
Photo © Hal Beral/Visuals Unlimited

HOW IS GEOTHERMAL ENERGY PRODUCED?

Electricity can be generated by harnessing naturally occurring steam and hot water in areas that are exceptionally hot underground. In such a *geothermal area,* wells can tap steam (or superheated water that can be turned into steam) that is then piped to a powerhouse where it turns a turbine that spins a generator, creating electricity.

Geothermal energy production requires no burning of fuel, so the carbon dioxide emissions of power plants that burn coal, oil, or natural gas are not produced. Although geothermal energy is relatively clean, it has some environmental problems. Workers need protection from toxic hydrogen sulphide gas in the steam, and the hot water commonly contains dissolved ions and metals, such as lead and mercury, that can kill fish and plants if discharged on the surface. Geothermal fluids are often highly corrosive to equipment, and their extraction can cause land subsidence. Pumping the cooled wastewater underground can help reduce subsidence problems.

FIGURE 15.29

Precipitation of calcite in the form of travertine around a hot spring (Mammoth Hot Springs, Yellowstone National Park). Thermophilic bacteria living in the hot water provide the colour.
Photo by Diane Carlson

FIGURE 15.30

Geyserite deposits around the vent of Castle Geyser, Yellowstone National Park.

Photo by David McGeary

FIGURE 15.31

Geothermal power plant at The Geysers, California. Underground steam, piped from wells to the power plant, is discharging from the cooling towers and surrounding wells.

Photo © Roger Ressmeyer/Corbis

Geothermal fields can be depleted. The largest field in the world is at The Geysers in California (figure 15.31), 120 kilometres north of San Francisco. The Geysers field increased its capacity in recent years to 2,000 megawatts of electricity (enough for 2 million people), but production has declined and the field may soon run out of steam.

Mapping of geothermal resources in British Columbia indicates considerable potential for development of a geothermal power plant in the South Meager area north of Vancouver. Exploratory test wells drilled in 2002 recorded temperatures of between 200 and 224 degrees Celsius at relatively shallow depths of 500 to 900 metres. These are the highest temperatures ever reached by any well drilled in Canada at this depth.

Nonelectric uses of geothermal energy include space heating (in Boise, Idaho; Klamath Falls, Oregon; and Reykjavik, the capital of Iceland), as well as paper manufacturing, ore processing, and food preparation.

SUMMARY

About 15 percent of the water that falls on land percolates underground to become *groundwater*. Groundwater fills pores and joints in rock, creating a large reservoir of usable water in most regions.

Porous rocks can hold water. *Permeable* rocks permit water to move through them.

The *water table* is the top surface of the *saturated zone* and is overlain by the *unsaturated zone*.

Local variations in rock permeability may develop a *perched water table* above the main water table.

Groundwater velocity depends on rock permeability and the slope of the water table.

An *aquifer* is porous and permeable and can supply water to wells. A *confined aquifer* holds water under pressure, which can create *artesian wells*.

Gaining streams, springs, and lakes form where the water table intersects the land surface. *Losing streams* contribute to the groundwater in dry regions.

Groundwater can be contaminated by city landfills, agriculture, industry, or sewage disposal. Some pollutants can be filtered out by passage of the water through moderately permeable geologic materials.

A pumped well causes a *cone of depression* that in turn can cause or aggravate groundwater pollution. Near a coast, it can cause *saltwater intrusion*.

Artificial *recharge* can help create a balance between withdrawal and recharge of groundwater supplies, and help prevent subsidence.

Solution of limestone by groundwater forms *caves*, *sinkholes*, and *karst topography*. Calcite precipitating out of groundwater forms *speleothems* such as *stalactites* and *stalagmites* in caves.

Precipitation of material out of solution by groundwater helps form *petrified wood*, other fossils, sedimentary rock cement, *concretions*, *geodes*, and alkali soils.

Geysers and *hot springs* occur in regions of hot groundwater. *Geothermal energy* can be tapped to generate electricity.

Terms to Remember

aquifer 396
artesian well 397
cave (cavern) 408
concretion 412
cone of depression 397
confined (artesian) aquifer 396
drawdown 397
gaining stream 399
geode 412
geyser 412
groundwater 391
groundwater quality 401
hot spring 412
karst topography 411
losing stream 399
perched water table 393
permeability 392
petrified wood 411
porosity 392
recharge 396
saturated zone 392
sinkhole 410
speleothem 410
spring 399
stalactite 410
stalagmite 410
unconfined aquifer 396
unsaturated zone 393
water table 392
well 396

Testing Your Knowledge

Use the questions below to prepare for exams based on this chapter.

1. What conditions are necessary for an artesian well?
2. What distinguishes a geyser from a hot spring? Why does a geyser erupt?
3. What is karst topography? How does it form?
4. What chemical conditions are necessary for caves to develop in limestone? For stalactites to develop in a cave?
5. What causes a perched water table?
6. Describe several ways in which groundwater can become contaminated.
7. Discuss the difference between porosity and permeability.
8. What is the water table? Is it fixed in position?
9. Sketch four different origins for springs.
10. What controls the velocity of groundwater flow?
11. Name several geologic materials that make good aquifers. Define *aquifer*.
12. How does petrified wood form?
13. What happens to the water table near a pumped well?
14. How does a confined aquifer differ from an unconfined aquifer?
15. Porosity is
    a. the percentage of a rock's volume that is openings
    b. the capacity of a rock to transmit a fluid
    c. the ability of a sediment to retard water
    d. none of the above
16. Permeability is
    a. the percentage of a rock's volume that is openings
    b. the capacity of a rock to transmit a fluid
    c. the ability of a sediment to retard water
    d. none of the above
17. The subsurface zone in which all rock openings are filled with water is called the
    a. saturated zone b. water table
    c. vadose zone
18. An aquifer is
    a. a body of saturated rock or sediment through which water can move easily
    b. a body of rock that retards the flow of groundwater
    c. a body of rock that is impermeable
19. Which rock type would make the best aquifer?
    a. shale b. mudstone
    c. sandstone d. all of the above
20. Which of the following determines how quickly groundwater flows?
    a. elevation b. water pressure
    c. permeability d. all of the above
21. Groundwater flows
    a. always downhill
    b. from areas of high hydraulic head to low hydraulic head
    c. from high elevation to low elevation
    d. from high permeability to low permeability
22. The drop in the water table around a pumped well is the
    a. drawdown b. hydraulic head
    c. porosity d. fluid potential

Exploring Web Resources

www.mcgrawhill.ca/olc/plummer

Visit our Online Learning Centre for additional readings, interactive quizzes, and answers to Testing Your Knowledge. Watch animated presentations on artesian wells and geyser eruptions. Explore the other sites listed below by clicking on their direct link found on this Website.

www.ec.gc.ca/water/en/nature/grdwtr/e_gdwtr.htm

Environment Canada Website covering groundwater, aquifers, groundwater use, contamination, and protection.

www.env.gov.bc.ca/wsd/

For information on aquifers, groundwater resources, and water well records in British Columbia.

http://toxics.usgs.gov/toxics/

Various sites and information about cleanup of toxics in surface and groundwater.

http://water.usgs.gov/public/wid/html/bioremed.html

Information about using bioremediation to clean up toxics in the soil, surface, and groundwater.

http://capp.water.usgs.gov/gwa/

Groundwater Atlas of the United States. Good general information about aquifers.

http://water.usgs.gov/

Good general Website that has a lot of links to water topics in the United States from the USGS.

www.caves.org/

Home page of the *National Speleological Society* contains links to Web pages of local interest and access to the NSS bookstore.

Animations

This chapter includes the following animations available on our Online Learning Centre at www.mcgrawhill.ca/olc/plummer.

15.8	Basic Dynamics of Groundwater Movement
15.19 a	Landfill and Cone Depression
15.19 b, c, d	Cone of Depression and Saltwater Intrusion during Groundwater Pumping

CHAPTER

Glaciers, Glaciation, and Permafrost

Why is it important to understand glaciers?
Where do we find glaciers?
How do glaciers form and move?
How do glaciers erode?
What features are produced by glacial erosion?
What features are produced by glacial deposition?
How have glaciers affected us in the past?
What are permafrost and periglacial environments?
How is permafrost affected by climate change?

In the preceding three chapters you have seen how the surface of the land is shaped by mass wasting, running water, and, to some extent, groundwater. Running water is regarded as the erosional agent most responsible for shaping the Earth's land surface. Where glaciers exist, however, they are far more effective agents of erosion, transportation, and deposition. Geologic features characteristic of glaciation are distinctly different from the features formed by running water. Once recognized, they lead one to appreciate the great extent of glaciation during the last 2.5 million years (known as the Quaternary Ice Ages).

Immense ice sheets, covering as much as a third of the Earth's land surface, had a profound effect on the landscape and on our present civilization. The last ice sheet to cover nearly all of Canada (the Laurentide Ice Sheet) had a volume of 33 million cubic kilometres.

Geologists stand at the edge of a large moulin on Athabasca Glacier, Banff National Park, Alberta. Moulins form as meltwater streams enter crevasses on the ice surface and flow towards the glacier base.
Photo by Nick Eyles

Moreover, worldwide climatic changes during the glacial ages distinctively altered landscapes in areas far from the glacial boundaries. For instance, water stored as ice in glaciers came from the oceans, so sea levels were lowered and more land was above sea level.

These episodes of glaciation only ended about 10,000 years ago. Preserved in the rock record, however, is evidence of older glaciations extending back as far as 2.8 billion years ago. The chapter on plate tectonics shows how the record of these ancient glaciations supports the theory of plate tectonics.

To understand how glacial erosion and deposition could have created the features regarded as evidence for past glaciation, it is important to understand how present-day glaciers erode, transport, and deposit material. In other words, the principle of uniformitarianism must be applied to the study of glaciation.

A **glacier** is a large, long-lasting mass of ice, formed on land, that moves downslope under the influence of gravity. It develops as snow is compacted and recrystallized. Glaciers can develop anywhere that, over a period of years, more snow accumulates than melts away or is otherwise lost.

There are two types of *glaciated* terrain on the Earth's surface. **Alpine glaciation** is found in mountainous regions, while **continental glaciation** exists where a large part of a continent (thousands of square kilometres) is covered by glacial ice. In both cases the moving masses of ice profoundly and distinctively change the landscape.

WHY IS IT IMPORTANT TO UNDERSTAND GLACIERS?

The Theory of Glacial Ages

In the early 1800s the hypothesis of past extensive continental glaciation of Europe was proposed. Among the many people who regarded the hypothesis as outrageous was the Swiss naturalist Louis Agassiz. But, after studying the evidence in Switzerland, he changed his mind. In 1837, he published a discourse that eventually led to wide acceptance of the theory. Later, Agassiz travelled widely throughout Europe and North America promoting and extending the theory. Agassiz and his colleagues observed the characteristic erosional and depositional features of present glaciers in the Alps and compared these with similar features found in northern Europe and the British Isles, well beyond the farthest extent of the Alpine glaciers. Based on these observations, Agassiz proposed that very large glaciers had covered most of Europe in the past. Agassiz had to overcome skepticism over past climates being quite different from those of today. At the time, the hypothesis seemed to many geologists to be a violation of the principle of uniformitarianism. Agassiz later came to North America and worked with Canadian and American geologists who had found similar indications of large-scale past glaciation on this continent.

As more evidence accumulated, the hypothesis became accepted as a theory that today is seldom questioned. The **theory of glacial ages** states that at times in the past, colder climates prevailed during which much more of the land surface of Earth was glaciated than at present.

Because the last episode of glaciation was at its peak only about 18,000 years ago, its record has remained largely undestroyed by subsequent erosion and so provides abundant evidence to support the theory. This most recent glacial episode was the last of several glacial ages that alternated with periods of warmer climate (similar to today's climate) around the world.

Our lives and environment today have been profoundly influenced by the effects of past Ice Ages. For example, the Great Lakes, the Finger Lakes, and the thousands of lakes on the Canadian Shield and in states such as Minnesota are the product of deep erosion by past glaciations. The fertile soils of the Canadian Prairies and northern Great Plains of the United States developed on loose debris transported and deposited by glaciers. Some of these glacial deposits contain mineral indicators that have been used to locate rich sources of gold in British Columbia, iron ore in Northern Ontario, and diamonds in the Northwest Territories and Nunavut. Thick deposits of glacial sediment—more than 200 metres thick in some areas of the southern Prairies—store and transmit vast amounts of groundwater. Glacial deposits form the foundations on which Calgary, Edmonton, Winnipeg, Ottawa, and many other Canadian cities and towns are built. The spectacularly scenic areas of the Canadian Rockies and Kluane, Gros Morne, and Auyuittuq National Parks owe their beauty to glacial action, which is responsible for carving them into their present shapes (figure 16.1). Continental glaciation has shaped the landscape of other national parks in Canada including Prince Albert, Riding Mountain, Pukaskwa, La Mauricie, and Kejimkujik.

Before we can understand how continental glaciers deposited sediment across Saskatchewan and formed lakes in Ontario, or how valley glaciers shaped the mountain valleys of Banff and Labrador, we must learn something about present-day glaciers.

FIGURE 16.1
The Alexandra River Valley in Banff National Park, Alberta. The steep valley walls and flat, sediment-covered valley floor are typical of glacially carved valleys.
Photo by Nick Eyles

WHERE DO WE FIND GLACIERS?

Distribution of Glaciers

Glaciers occur both in cold, polar regions, where there is little melting during the summer, and in temperate climates that have heavy snowfall during the winter months. They are found where more snow falls during the cold time of year than can be melted during warm months.

Approximately 2 percent of Canada's land area is currently covered by glaciers. Glaciers are present in British Columbia, Alberta, the Yukon, the Northwest Territories, Nunavut, and the Torngat Mountains of Labrador. Some of the world's largest non-polar glaciers are found along Canada's west coast in B.C. and in the St. Elias Mountains of the Yukon, where precipitation values are extremely high. The Canadian Rockies also host thousands of glaciers and numerous icefields, many of which lie along the Continental Divide and feed major river systems. Whistler and other locations in the Coast Ranges of British Columbia have warmer winters but much higher precipitation at higher elevations than Glacier and Mount Revelstoke National Parks in interior B.C. These parks in turn receive more snowfall than Kootenay, Yoho, Banff, and Jasper National Parks in the Rocky Mountains farther east.

Glaciers are common near the equator in the very high Andes Mountains of South America, due to the colder temperatures at higher altitudes. Glaciers have also developed near the equator in New Guinea and on Mount Kilimanjaro in Africa. However, recent climate change is causing very rapid ablation of these ice masses. At present rates of melting, the "snows of Kilimanjaro" will be completely gone in less than 20 years.

Glaciation is most extensive in polar regions, where little melting takes place at any time of year. The cold, polar regions of Canada's eastern Arctic contain some of the largest and oldest glaciers in North America. On Baffin Island, the 100,000-year-old Barnes Ice Cap and Penny Ice Cap are the only Canadian remnants of the gigantic Laurentide Ice Sheet that covered northern North America about 20,000 years ago. At present about one tenth of the land surface on Earth is covered by glaciers (compared with about one-third during the peak of the glacial ages). Approximately 85 percent of the present-day glacier ice is on the Antarctic continent, covering an area larger than the combined areas of western Europe and the United States; 10 percent is in Greenland. All the remaining glaciers of the world amount to only about 5 percent of the world's freshwater ice. This means that Antarctica is in fact storing most of the Earth's fresh water in the form of ice (see box 16.1). Some have suggested that ice from the Antarctic, towed as icebergs, could be brought to areas of dry climate to alleviate water shortages. It is worth noting that if all of Antarctica's ice were to melt, sea level around the world would rise more than 60 metres. This would flood the world's coastal cities and significantly decrease the land surface available for human habitation.

Types of Glaciers

A simple criterion—whether or not a glacier is restricted to a valley—is the basis for classifying glaciers by form. A **valley glacier** is a glacier that is confined to a valley and flows from a higher to a lower elevation. Like streams, small valley glaciers may be tributaries to a larger trunk system. Valley glaciers are prevalent in areas of alpine glaciation. As might be expected, most glaciers in Canada and the United States, being in mountains, are of the valley type (figure 16.2).

FIGURE 16.2
Valley glacier on the flanks of Mount Logan, Canada's highest mountain.
Photo by C. C. Plummer

In contrast, an **ice sheet** is a mass of ice that is not restricted to a valley but covers a large area of land (more than 50,000 square kilometres). Ice sheets are associated with continental glaciation. Only two places on Earth now have ice sheets, Greenland and Antarctica. A similar but smaller body is called an **ice cap.** Ice caps (and valley glaciers as well) are found in a few mountain highlands (e.g., the Columbia Icefields of Alberta and B.C.), on Baffin Island (Barnes and Penny Ice Caps), on Ellesmere Island (Agassiz Ice Cap), on other Canadian Arctic islands, and on Iceland, Spitzbergen, and Novaya Zemlya. An ice cap or ice sheet flows downward and outward from a central high point, as figure 16.3 shows.

HOW DO GLACIERS FORM AND MOVE?

Formation and Growth of Glaciers

Snow converts to glacier ice in somewhat the same way that sediment turns into a sedimentary rock and then into metamorphic rock; figure 16.4 shows the process. A snowfall can be compared to sediment settling out of water. A new snowfall may be in the form of light "powder snow," which consists mostly of air trapped between many six-pointed snowflakes. In a short time the snowflakes settle by compaction under their own weight and much of the air between them is driven out. Meanwhile, the sharp points of the snowflakes are destroyed as flakes reconsolidate into granules. In warmer climates, partial thawing and refreezing

ENVIRONMENTAL GEOLOGY 16.1

Glaciers as a Water Resource

While Canada has much fresh water its population is clustered in relatively small urban regions, placing exorbitant demands on rivers and lakes in those areas. More than 80 percent of Canada's population lives in cities. Most Canadians (85 percent) live along the southern border with the United States and are concentrated in a few watersheds. With high per capita consumption, it is not surprising that water usage is beginning to reach critical levels in some areas.

Watersheds of the western Prairies are fed substantially by snowmelt and by glaciers along the eastern Rocky Mountains such as the Saskatchewan and Athabasca glaciers that drain the Columbia Icefield. Glaciers are giant reservoirs of water for these watersheds and provide a valuable supply of water, particularly in the summer months when glacier melt rates are high. Unfortunately, these glaciers have lost 25 percent of their mass since 1890, in response to climate warming. This loss continues at an accelerated rate. The 1990s were the warmest on record in the Rockies, with average January temperatures some 2.1°C warmer than the 1960–1999 average. Current summer flows in the glacier-fed Prairie rivers are as much as 84 percent lower than those of the early 20th century; the worst affected is the South Saskatchewan River. This area will be much drier in the future with continuing climate warming. Alberta is the province most vulnerable to water shortages. The Athabasca River feeds the water needed for oil sands extraction in northern Alberta. Each barrel of oil extracted from oil sands requires between three and six barrels of water (see chapter 12). The primary source of this water is the glaciers of the Canadian Rockies. A glacier's state of health is dependent on how much snow and ice is added each winter compared with that lost by melting each summer (referred to as *mass balance*). Since 1970, there have been only seven years with positive mass balances on glaciers along the eastern slopes of the Rockies. It is not impossible that the glaciers of the Rockies (and their water supply to the western Prairie provinces) will be gone in 100 years time.

Further Reading

Schindler, D. W. and Donahue, W. F. 2006. An impending water crisis in Canada's western Prairie Provinces. Proceedings U.S. National Academy of Sciences. 7 pages.
www.pnas.org/cgi/doi/10.1073/pnas.0601568103

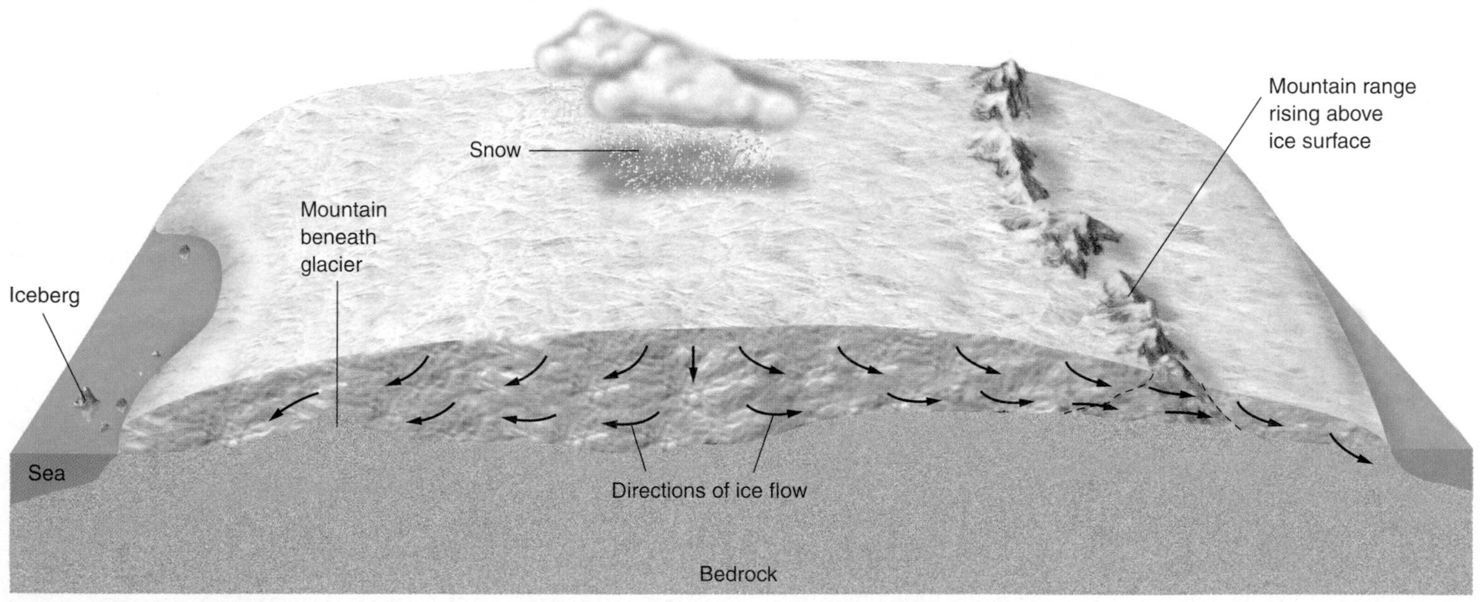

FIGURE 16.3

Diagrammatic cross-section of an ice sheet. Vertical scale is highly exaggerated.

results in coarse granules—the "corn snow" of spring skiing. In colder climates where little or no melting takes place, the snowflakes will recrystallize into fine granules. After the granular snow is buried by a new snowpack, usually during the following winter, the granules are compacted and weakly "cemented" together by ice (figure 16.4*B*). The compacted mass of granular snow, transitional between snow and glacier ice, is called *firn*. Firn is analogous to a sedimentary rock such as sandstone.

Through the years, the firn becomes more deeply buried as more snow accumulates. More air is expelled, the remaining pore space is greatly reduced, and granules forced together recrystallize into the tight, interlocking mosaic of *glacier ice* (figure 16.4*C*).

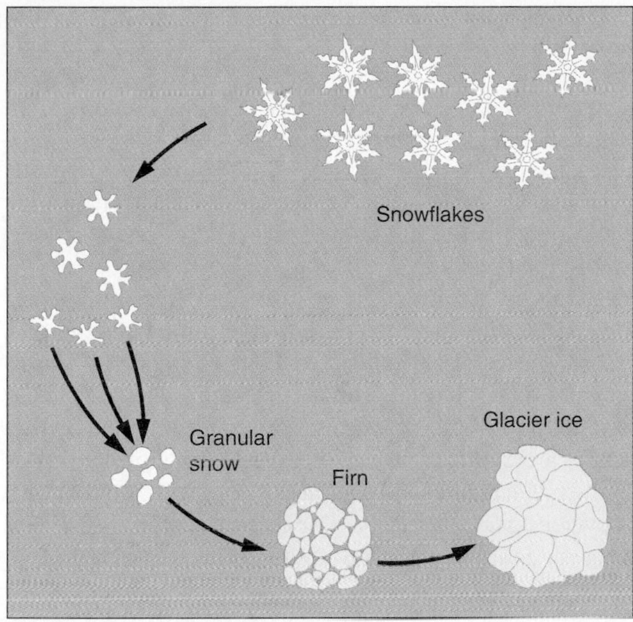

A

B

C

FIGURE 16.4

(*A*) Conversion of snow to glacier ice. (*B*) Crevasses in the accumulation area of the Columbia Icefield, Banff National Park, expose annual snow layers between 1 and 3 m thick. (*C*) Thin slice of an ice core from a glacier. The ice is between sheets of polarizing filters. In polarized light, the colours of individual ice grains vary depending on their crystallographic orientation. Without the polarized light, the ice would be transparent and colourless. Many of the ice grains shown are more than a centimetre in length.

Photo B by Nick Fyles; Photo C by C. C. Plummer

The recrystallization process involves little or no melting and is comparable to metamorphism. Glacier ice is texturally similar to the metamorphic rock quartzite.

Under the influence of gravity, glacier ice moves downward and is eventually **ablated,** or lost. For glaciers in all but the coldest parts of the world, ablation is due mostly to melting, although some ice evaporates directly into the atmosphere. If a moving glacier reaches a body of water, blocks of ice break off (or *calve*) and float free as **icebergs** (figure 16.5). In most of the Antarctic, ablation takes place largely through calving of icebergs and direct evaporation. Only along the coast does melting take place, and there for only a few weeks of the year.

FIGURE 16.5

Icebergs at Cape Bonavista, Newfoundland.

Photo by K. Bruce Lane Photography — www.lanephotography.com

Glacial Budgets

If, over a period of time, the amount of snow a glacier gains is greater than the amount of ice and water it loses, the glacier's budget is *positive* and it expands. If the opposite occurs, the glacier decreases in volume and is said to have a *negative budget.* Glaciers with positive budgets push outward and downward at their edges; they are called **advancing glaciers.** Those with negative budgets grow smaller and their edges melt back; they are **receding glaciers.** Bear in mind that the glacial ice moves downvalley, as shown in figure 16.6, whether the glacier is advancing or receding. In a receding glacier, however, the rate of melt of ice at the terminus exceeds the rate of ice flow, and the terminus melts back. If the amount of snow retained by the glacier equals the amount of ice and water lost, the glacier has a *balanced budget* and is neither advancing nor receding.

The upper part of a glacier, called the **zone of accumulation,** is the part of the glacier with a perennial snow cover (figure 16.6). The lower part is the **zone of ablation,** where ice is lost, or ablated, by melting, evaporation, and calving.

The boundary between these two altitudinal zones of a glacier is an irregular line called the **equilibrium line,** which marks the highest point at which the glacier's winter snow cover is lost during a melt season (figure 16.7).

The equilibrium line may shift up or down from year to year, depending on whether there has been more accumulation or more ablation. Its location therefore indicates whether a glacier has a positive or negative budget. An equilibrium line migrating upglacier over a period of years is a sign of a negative budget, whereas an equilibrium line migrating downglacier indicates that the glacier has a positive budget. If an equilibrium line remains essentially in the same place year after year, the glacier has a balanced budget.

The **terminus,** the lower edge of a glacier, moves farther downvalley when a valley glacier has a positive budget. In a receding glacier the terminus melts back upvalley. Because most glaciers move slowly, migration of the terminus tends to lag several years behind a change in the budget.

An ice sheet with a positive budget increases in volume, advancing its outer margins. If the expanding ice sheet extends into the ocean, an increasing number of icebergs break off and float away in the open sea.

Advancing or receding glaciers are significant and sensitive indicators of climatic change; however, an advancing glacier does not necessarily indicate that the climate is getting colder. It may mean that the climate is getting wetter, or that more precipitation is falling during the winter months, or that the summers are cloudier. It is estimated that a worldwide decrease in the mean annual temperature of about 5°C could bring about a new ice age. Conversely, global climate warming of just a few degrees could significantly reduce the world's glaciers (see boxes 16.2 and 16.3).

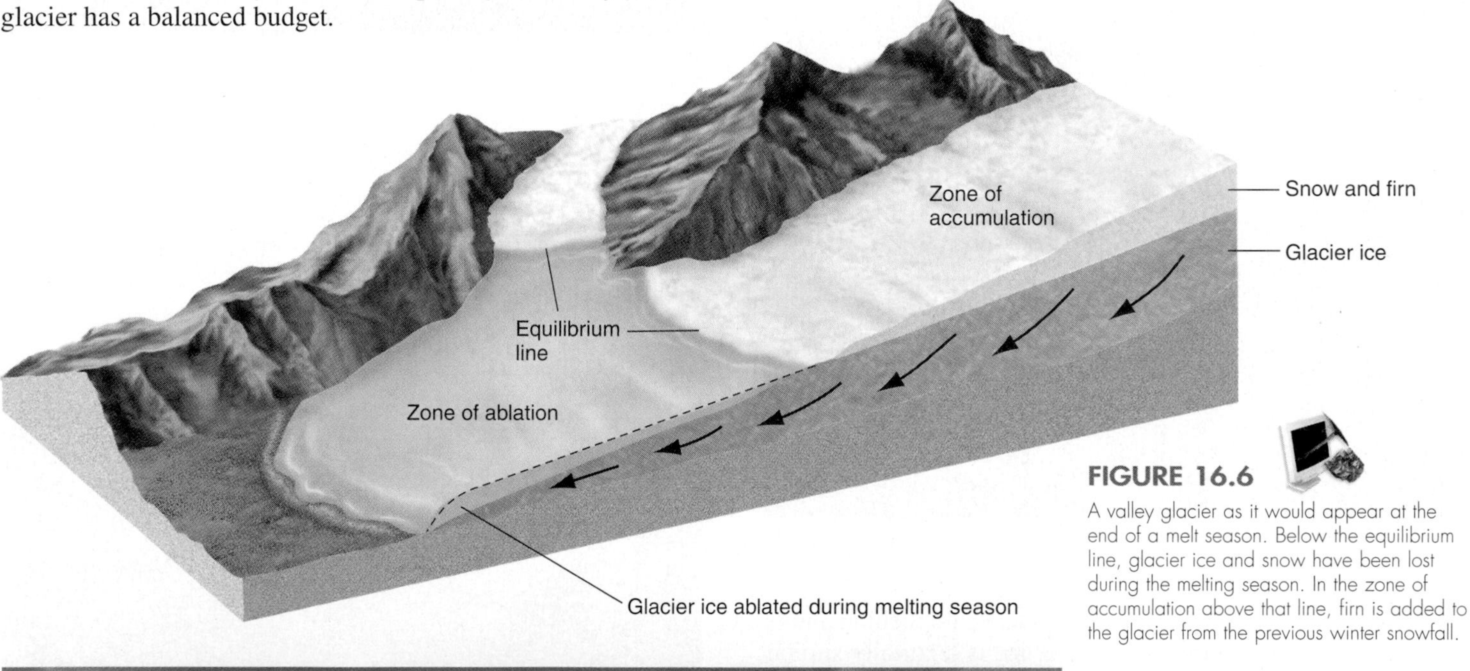

FIGURE 16.6
A valley glacier as it would appear at the end of a melt season. Below the equilibrium line, glacier ice and snow have been lost during the melting season. In the zone of accumulation above that line, firn is added to the glacier from the previous winter snowfall.

FIGURE 16.7
View looking down Saskatchewan Glacier, Banff National Park, near the end of 2006 summer ablation season. The irregular boundary between the snow-covered glacier surface in the foreground and darker glacier ice in the distance marks the approximate position of the equilibrium line.
Photo by Nick Eyles

GEOMATICS 16.2

Observing Glaciers from Space

Polar ice masses such as those in the Antarctic are likely to show relatively rapid response to global climate change, and it is important to monitor their behaviour over time. Considering the remote location and huge size of the Antarctic continent, satellite remote sensing is the only means to provide the type of coverage required for this task. Antarctic ice sheets and glaciers were examined in 1997 and 2000 during the Antarctic Mapping Mission, a joint project between the Canadian Space Agency and NASA, using synthetic aperture radar (SAR) images obtained from the Canadian RADARSAT-1 satellite. The satellite orbits at an elevation of approximately 800 kilometres. Separate images taken at different times can be precisely aligned, and any surfaces that have changed or moved in the time interval can be measured using a technique called *radar interferometry*. This technique allows very precise measurement of glacier movement. Box figure 1 shows the movement of the Lambert Glacier in Antarctica as determined from SAR images taken 24 days apart during the 2000 mission. The different colours on the image represent varying amounts of motion: yellow represents areas of no movement; green represents low flow velocities of between 100 and 300 m per year, typical of small tributary glaciers; and blue represents faster velocities of between 400 and 800 m per year. As the glacier extends out across the Amery Ice Shelf the ice spreads out and thins, increasing velocities to 100 to 1,200 m per year (red areas). These images can be compared with similar images obtained in the future to determine overall change in glacier behaviour over time.

BOX 16.2 ■ FIGURE 1

Ice velocities for the Lambert Glacier measured using RADARSAT-1 synthetic aperture radar (SAR) images taken 24 days apart during the 2000 Antarctic Mapping Mission.

Image courtesy Canadian Space Agency/NASA/Ohio State University, Jet Propulsion Laboratory, Alaska SAR facility

Other types of satellite images are also being used in an international project to map the extent of Earth's glaciers and their changes over time. Advanced Spaceborne Thermal Emission and Reflection Radiometer (ASTER) images are very high resolution and allow even small features to be distinguished on glacier surfaces (box figure 2). Analysis of changes in these features over time can be used to determine the characteristics of ice flow and behaviour. Monitoring such changes is critical to evaluating the effects of global climate warming on ice masses and, ultimately, sea-level change.

A

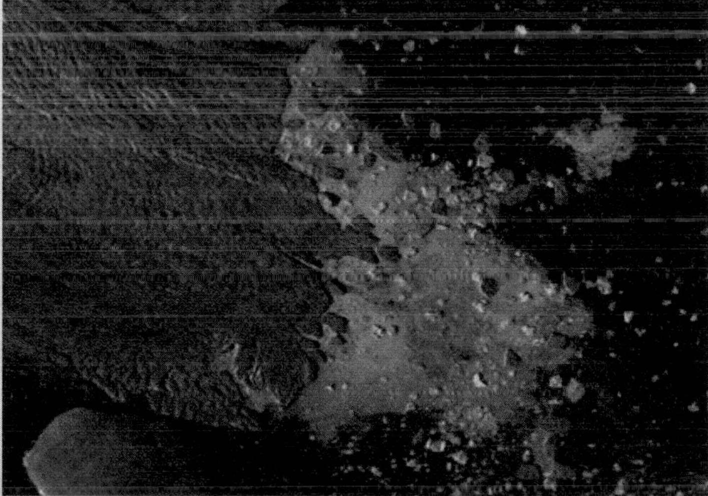

B

BOX 16.2 ■ FIGURE 2

ASTER images of Dobbin Bay on Ellesmere Island.

(A) Dobbin Bay is at the bottom of this image, taken on July 31, 2000. (B) Detail of the floating edge of the Eugenie Glacier's ice tongue shows surface cracks and extensive calving of icebergs in Dobbin Bay.

Image courtesy University of Alberta and NASA/GSFC/MITI/ERSDAC/JAROS, and U.S./Japan ASTER Science Team. From http://visibleearth.nasa.gov/view_rec.php?id=1828

Movement of Valley Glaciers

Valley glaciers move downslope under the influence of gravity, the rate being variable, generally ranging from less than a few millimetres a day to 15 metres a day. The upper part of a glacier—where the volume of ice is greater and slopes tend to be steeper—generally moves faster than ice farther down or on gentler slopes. In this way ice from the higher altitudes keeps replenishing ice lost in the zone of ablation.

Glaciers in temperate climates—where the temperature of the glacier is at or near the melting point for ice—tend to move faster than those in colder regions—where the ice temperature stays well below freezing.

Velocity also varies within the glacier itself (figure 16.8). The central portion of a valley glacier moves faster than the sides (as water does in a stream), and the surface moves faster than the base. How ice moves within a valley glacier has been demonstrated by studies in which holes are drilled through the glacier ice and flexible pipes inserted. Changes in the shape and position of the pipes are measured periodically. The results of these studies are shown diagrammatically in figure 16.8.

Note in the diagram that the base of the pipe has moved downglacier. This indicates **basal sliding,** which is the sliding of the glacier as a single body over the underlying rock. A thin film of meltwater that develops along the base from the pressure of the overlying glacier facilitates basal sliding. Think of a large bar of wet soap sliding down an inclined board.

Note that the lower portion of the pipe is bent in a downglacier direction. The bent pipe indicates **plastic flow** of ice, movement that occurs within the glacier due to the plastic or "deformable" nature of the ice itself. Visualize two neighbouring grains of ice within the glacier, one over the other. Both are moving—carried along by the ice below them; however, the higher of our two ice grains slides over its underlying neighbour a bit farther. The reason the pipe is bent more sharply near the base of the glacier is that pressure from overlying ice results in greater flowage with increasing depth. Deep in the glacier, ice grains are sliding past their underlying neighbours farther than similar ice grains higher up where the pipe is less bent. We should point out that a glacier flows not only because ice grains slide past one another but also because ice grains deform and recrystallize.

In the **rigid zone,** or upper part of the glacier, the pipe has been moved downglacier; however, it has remained unbent. The ice nearer the top apparently rides along passively on the plastically moving ice closer to the base. In the rigid zone grains of ice do not move relative to their neighbours.

Crevasses

Along its length, a valley glacier moves at different rates in response to changes in the steepness of the underlying rock. Typically, a valley glacier rides over a series of rock steps. Where the glacier passes over a steep part of the valley floor, it moves faster. The upper rigid zone of ice, however, cannot stretch to move as rapidly as the underlying plastic-flowing ice. Being brittle, the ice of the rigid zone is broken by the tensional forces; open fissures, or **crevasses,** develop (figure 16.9). Crevasses also form along the margins of glaciers in places where the path is curved, as shown in part of figure 16.9. This is because ice (like water) flows faster toward the outside of the curve. For glaciers in temperate climates, a crevasse should be no deeper than about 40 metres, the usual thickness of the rigid zone. If you are falling down a crevasse, it may be of some consolation that, as you are hurtling to death or injury, you realize on the way down that you will not fall more than 40 metres.

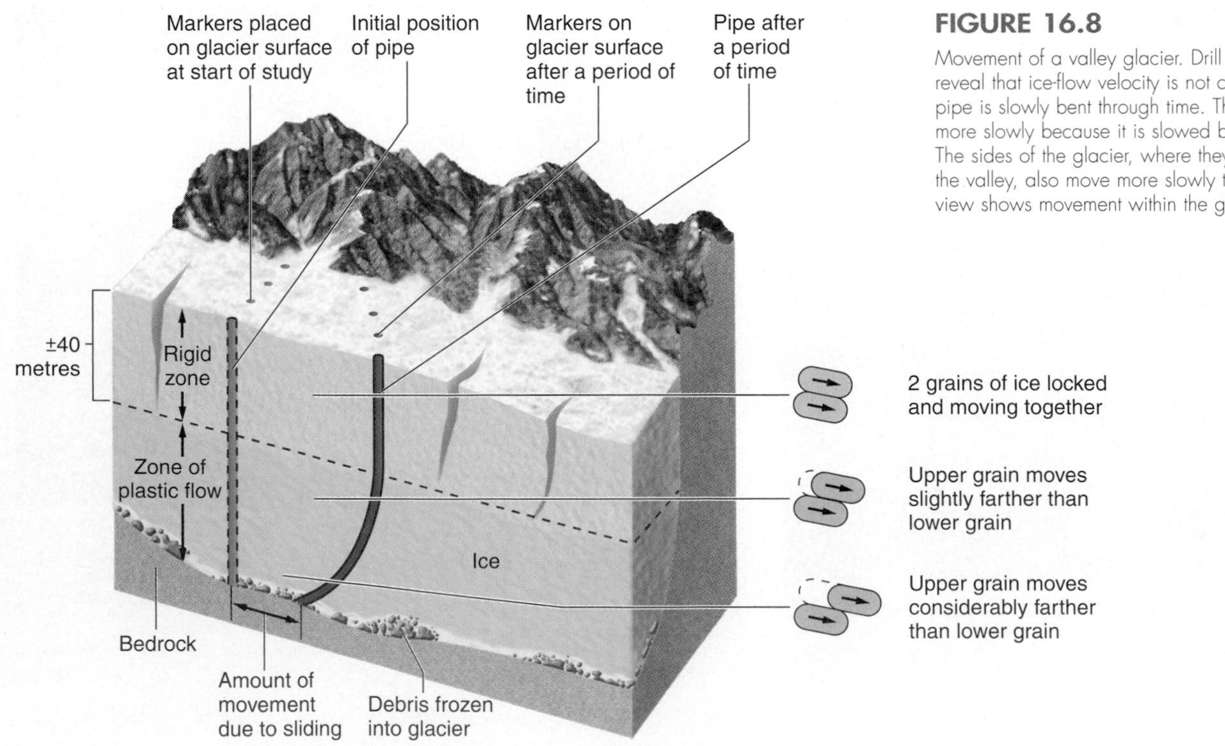

FIGURE 16.8

Movement of a valley glacier. Drill holes drilled through glaciers reveal that ice-flow velocity is not constant with depth. A drill pipe is slowly bent through time. The base of the glacier moves more slowly because it is slowed by friction against bedrock. The sides of the glacier, where they abut the bedrock walls of the valley, also move more slowly than its centre. Cross-sectional view shows movement within the glacier.

IN GREATER DEPTH 16.3

Drilling Through Ice Sheets for a Record of the Past

Glaciers preserve records of precipitation, air temperatures, atmospheric dust, volcanic ash, carbon dioxide, and other atmospheric gases. When snow becomes converted to glacier ice, some of the air that was mixed with the snowflakes becomes bubbles trapped in the glacier ice. By analyzing the air in these bubbles, we are analyzing the air that prevailed when an ancient ice layer formed. Drilling into glaciers and retrieving ice cores allows scientists to sample environmental conditions at the time of ancient snowfalls. A cylindrical core of ice is extracted from a hollow drill after it has penetrated a glacier. The layers in an ice core represent the different layers of snow that converted to glacier ice (box figure 1). Each layer, when analyzed, can reveal information about conditions of the atmosphere at the time the snow accumulated and turned into ice.

As of 2004, the most ambitious drilling project in Antarctica had extracted the longest-ever ice core, representing a record going back 740,000 years. Initial findings confirm that the Earth experienced eight major glaciations (ice ages) during this time.The international team hopes to extract a million years of core before reaching the base of the ice sheet. The next-longest core was drilled at Vostok (described in box 16.4) in Antarctica in the 1990s.

The Vostok core reached a depth of over 3 kilometres and yielded information that may help reconstruct a climate and atmospheric history of the past 420,000 years. Graphs derived from the research project are shown in box figure 2. The team determined temperature by studying hydrogen isotope variation within the ice layers. Methane and carbon dioxide are greenhouse gases. Note how the greenhouse gases correlate with the temperature variations. Recent work demonstrates that temperature changes occur slightly *earlier* than corresponding changes in greenhouse gases. This suggests, in fact, that changing carbon dioxide content of the Earth's atmosphere is not the prime driver, rather a follower of climate change. Also note the five periods when the temperature was warmest. These are the *interglacial* periods during which the North American and European ice sheets disappeared. Two of the five warm periods are emphasized for comparison—the Holocene interglacial epoch (which began about 12,000 years ago and is ongoing) and the previous interglacial (the Sangamon).

Additional Resources

For more on ice sheet drilling, go to the Online Learning Centre at

- www.mcgrawhill.ca/olc/plummer.

BOX 16.3 ■ FIGURE 1

An ice core being examined in a cold laboratory.

Photo by Mark Twickler, University of New Hampshire/National Oceanic and Atmospheric Administration Paleoclimatology Program/Department of Commerce

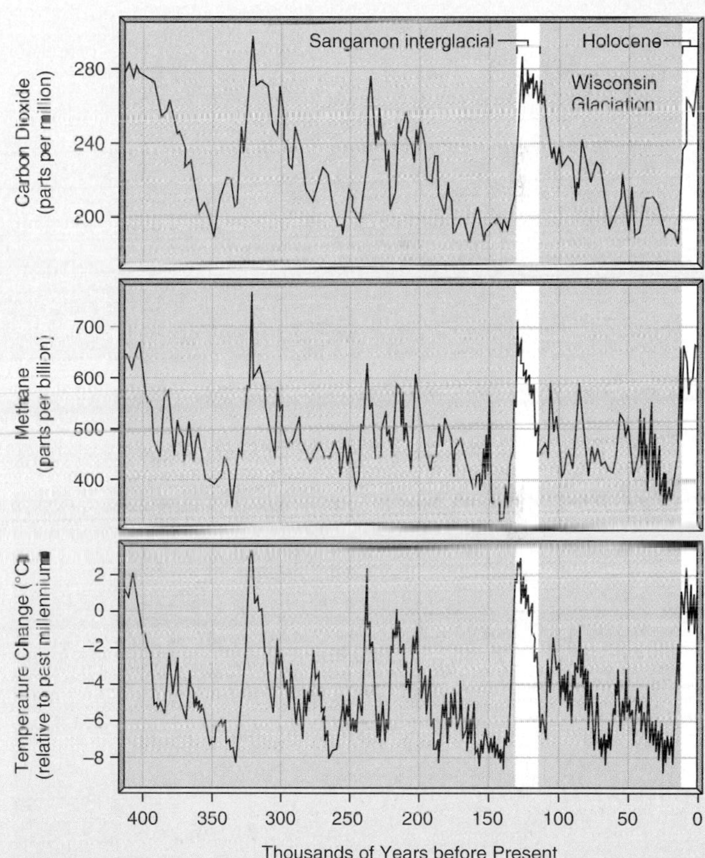

BOX 16.3 ■ FIGURE 2

Temperature, carbon dioxide, and methane content of air at Vostok on the East Antarctic Ice Sheet for the last 420,000 years. See text for explanation. Also note the rapid rising of temperature at the beginning of interglacial periods.

Crevasses

A

B

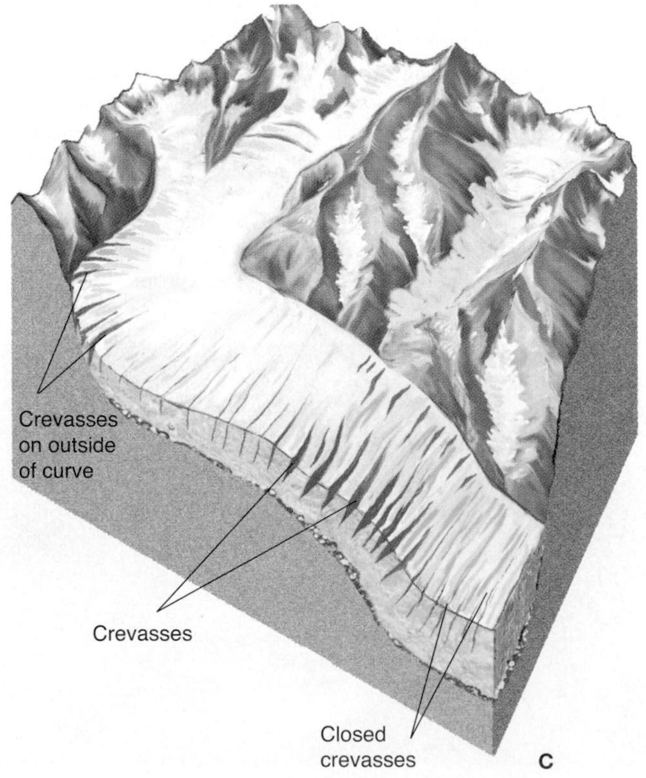

FIGURE 16.9

(*A*) Crevasses on a glacier, looking down from Mount Logan, Yukon Territory. (*B*) Crevasses on the surface of Bering Glacier, Alaska. (*C*) Crevasses form when ice is stretched either as it flows over bedrock "highs" or around curves.

Photo A by C. C. Plummer; Photo B by Nick Eyles

After the ice has passed over a steep portion of its course it slows down, and compressive forces close the crevasses.

Movement of Ice Sheets

An ice sheet or ice cap moves like a valley glacier except that it moves outward from a central high area toward the edges of the glacier (as shown in figure 16.3).

Glaciological research in Antarctica has determined how ice sheets grow and move. Antarctica has two ice sheets; the West Antarctic Ice Sheet is separated by the Transantarctic Mountains from the much larger East Antarctic Ice Sheet (figure 16.10). The two ice sheets join in the low areas between mountain ranges. Both are almost completely within a zone of accumulation because so little melting takes place (ablation is largely by calving of icebergs) and because occasional snowfalls nourish their high central parts. The ice sheets mostly overlie interior lowlands, but also completely bury some mountain ranges. Much of the base of the West Antarctic Ice Sheet is on bedrock that is below sea level. At least one active volcano underlies the West Antarctic Ice Sheet (resulting in a depression in the ice sheet). Where mountain ranges are higher than the ice sheet, the ice flows through as valley glaciers.

At the South Pole (figures 16.10 and 16.11)—neither the thickest part nor the centre of the East Antarctic Ice Sheet—the ice is 2,700 metres thick. The thickest part of the East Antarctic Ice Sheet is 4,776 metres.

Most of the movement of the East Antarctic Ice Sheet is by means of plastic flow. It has been thought that most of the ice sheet is frozen to the underlying rocks and basal sliding takes place only locally. But, the recent discovery of a giant lake and other lakes beneath the thickest part of the Antarctic Ice Sheet (box 16.4) indicates that liquid water at its base is more widespread and basal sliding might be more important than previously thought.

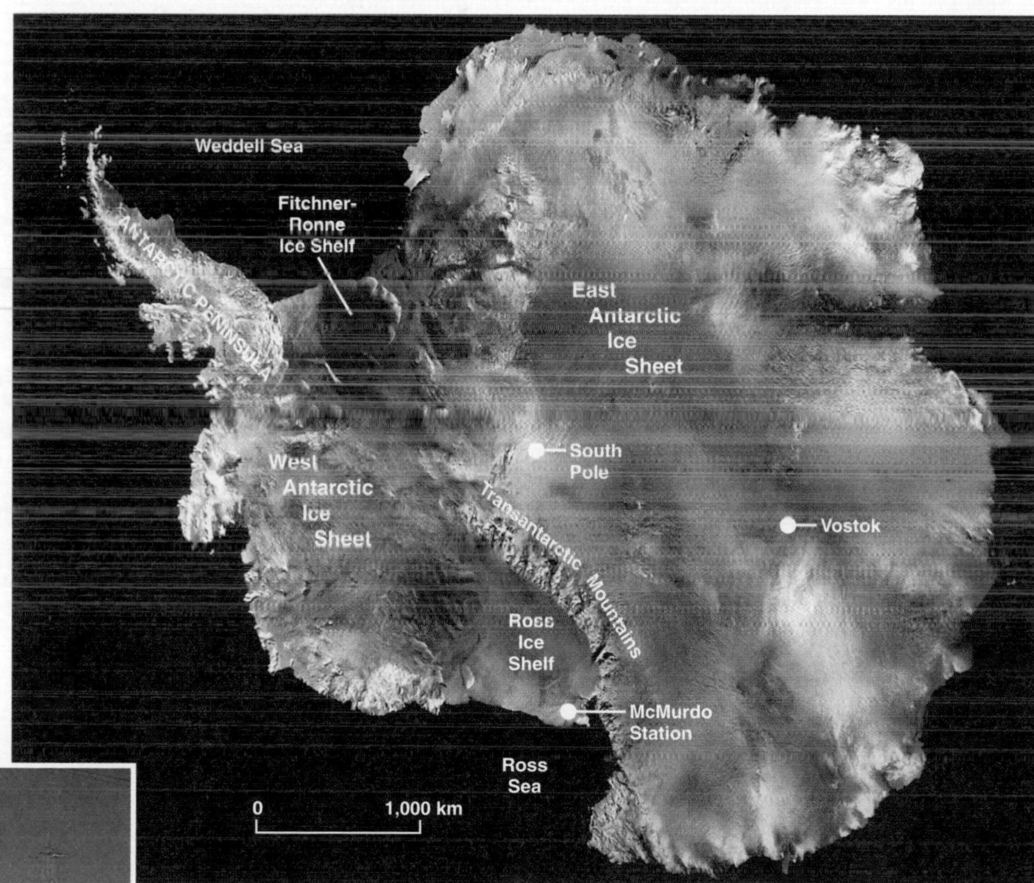

FIGURE 16.10

The Antarctic continent and its ice sheets. Vostok is at the highest part of the East Antarctic Ice Sheet. (False colouring is used to show variations among snow, ice, blue ice, and exposed rock.)

Photo by U.S. Geological Survey/NASA

FIGURE 16.11

The South Pole. Actually, the true South Pole is several kilometres from here. The moving ice sheet has carried the striped pole away from the site of the true South Pole, where the pole was erected in 1956.

Photo by C. C. Plummer

ENVIRONMENTAL GEOLOGY 16.4

Water Beneath Glaciers: Floods, Giant Lakes, and Galloping Glaciers

A Galloping Glacier

Glacial motion is often used as metaphor for slowness ("The trial proceeded at a glacial pace"). But, some glaciers will *surge*—that is, move very rapidly for short periods following years of barely moving at all. The most extensively documented surge (or "galloping glacier") was that of Alaska's Bering Glacier in 1993–94. The Bering Glacier is the largest glacier in continental North America and it surges on a 20- to 30-year cycle. After its previous surge in 1967, its terminus retreated 10 kilometres. In August 1993, the latest surge began. Ice travelled at velocities up to 100 metres per day for short periods of time and sustained velocities of 35 metres per day over a period of several months. The terminus advanced 9 kilometres by the time the surge ended in November 1994. When glaciers surge, the previously slow-moving, lower part of a glacier breaks into a chaotic mass of blocks (box figure 1). Surges are usually attributed to a buildup of water beneath part of a glacier, floating it above its bed. In July 1994, a large flood of water burst from Bering Glacier's terminus, carrying with it blocks of ice up to 25 metres across.

A Flood

Glacial outburst floods (*jokulhlaups*) (box figure 2) are not always associated with surges. In October 1996, a volcano erupted beneath the Vatnajökull Ice Cap in Iceland. The glacier, which is up to 500 metres thick, covers one-tenth of Iceland. Emergency teams prepared for the flood that geologists predicted would follow the eruption. The expected flood took place early in November with a peak flow of 45,000 cubic metres per second! The flood lasted only a few hours; however, it caused between 10 and 15 million dollars worth of damage. Three major bridges were destroyed or damaged, and 10 kilometres of roads were washed away. Because people had been kept away from the expected flood path there were no casualties.

A Giant Lake

One of the world's largest lakes was only recently discovered. But don't expect to take a dip or go windsurfing on it. It lies below the thickest part of the East Antarctic Ice Sheet and is named after the Russian research station Vostok, which is 4,000 metres above the lake at the coldest and most remote part of Antarctica. Lake Vostok was discovered in the 1970s through ice-penetrating radar; however, its extent was unknown until 1996, when satellite-borne radar revealed that the lake is 200 km long and 50 km wide—about the size of Lake Ontario. At its deepest, it is 510 metres, placing it among the ten deepest lakes in the world. Recently, more, but smaller, lakes beneath the East Antarctic Ice Sheet have been discovered.

The lake has been sealed off from the rest of the world for around a million years and it is likely that it contains organisms, such as microbes, dating back to that time. These organisms (and their genes) would not have been affected by modern pollution or nuclear-bomb fallout. By coincidence, the world's deepest ice hole (more than 3 kilometres) was being drilled from Vostok Station above the lake when the size of Lake Vostok was being determined. The ice core from this hole should add to the findings from the Greenland drilling projects (see box 16.3) and provide an even greater picture of the Earth's climate during the ice ages. When the hole was completed in 1997, drilling was halted short of reaching the lake due to fear of contaminating it and harming whatever living organisms might be in the very old water. Study of the lake and its organisms is curtailed until a future generation of scientists can devise means of sampling the waters without altering its ecosystem.

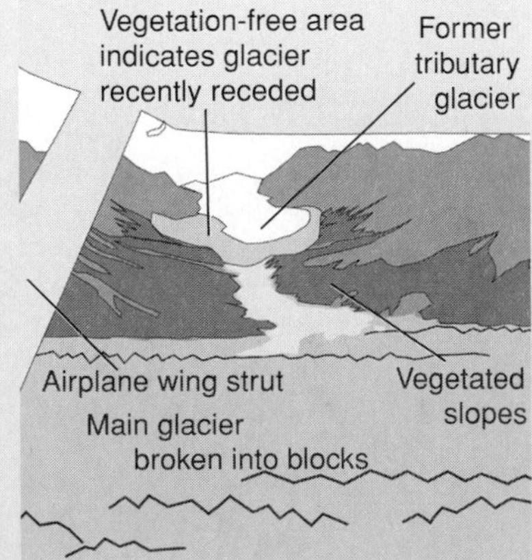

GEOSCIENTIST'S VIEW

BOX 16.4 ■ FIGURE 1

Part of a glacier after a surge (lower part of photo). The debris-covered ice has been broken up into a chaotic mass of blocks. In the background is a small glacier that has retreated up its valley. Photo taken near the Canada–Alaska border.
Photo by C. C. Plummer

A

B

BOX 16.4 ■ FIGURE 2

(A) Glacier outburst flood (*jokulhlaup*) issuing from the Vatnajökull Ice Cap in Iceland in 1996. (B) The flood was triggered by eruption of a volcano beneath the ice. Similar subglacial volcanoes (*tuya*) existed in Northern B.C. in the past (see figure 7.21B); these volcanoes typically have flat, plantar tops.

Photos by Magnus Tumi Gudmundsson, University of Iceland

HOW DO GLACIERS ERODE?

Wherever basal sliding takes place, the rock beneath the glacier is abraded and modified. As meltwater works into cracks in bedrock and refreezes, pieces of the rock are broken loose and frozen into the base of the moving glacier, a process known as *plucking*. This occurs mostly on the down-ice side of bedrock irregularities where stresses on the rock are low. While being dragged along by the moving ice, the rock within the glacier grinds away at the underlying rock, a process called *abrasion* (figure 16.12). Abrasion is most active on the up-ice side of bedrock irregularities where pressures are high. The thicker the glacier, the more pressure on the rocks and the more effective the grinding and crushing.

Pebbles and boulders that are dragged along are *faceted;* that is, given a flat surface by abrasion. Bedrock underlying a glacier is *polished* by fine particles and *striated* (scratched) by sharp-edged, larger particles. Striations and grooves on bedrock indicate the direction of ice movement (figure 16.13).

The grinding of rock across rock produces a powder called **rock flour.** Rock flour is composed largely of very fine (silt- and clay-sized) particles of unaltered minerals (pulverized from chemically unweathered bedrock). When *meltwater* washes rock flour from a glacier, the streams draining the glacier appear milky and lakes into which glacial meltwater flows often appear a milky green colour.

Not all glacier-associated erosion is caused directly by glaciers. Mass wasting takes place on steep slopes created by downcutting glaciers. Frost wedging breaks up bedrock ridges and cliffs above a glacier, causing frequent rockfalls. Snow avalanches bring down loose rocks onto the glacier surface, where they ride on top of the ice. Debris may also fall into crevasses to be transported within or at the base of a glacier, as shown in figure 16.20.

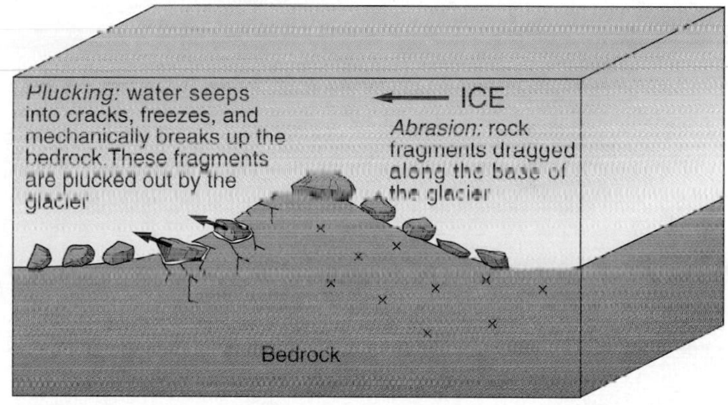

FIGURE 16.12

Plucking and abrasion beneath a glacier.

WHAT FEATURES ARE PRODUCED BY GLACIAL EROSION?

Erosional Landscapes Associated with Alpine Glaciation

We are in debt to glaciers for the rugged and spectacular scenery of high mountain ranges. Figure 16.14 shows how glaciation has radically changed a previously unglaciated mountainous region. The striking and unique features associated with mountain glaciation are due to the erosional effects of glaciers as well as frost wedging on exposed rock.

A

B

FIGURE 16.13

(*A*) Striated bedrock surface in front of Athabasca Glacier, Banff National Park. (*B*) Striated Paleozoic bedrock surface near Ottawa.

Photo A by C.H. Eyles; Photo B by Nick Eyles

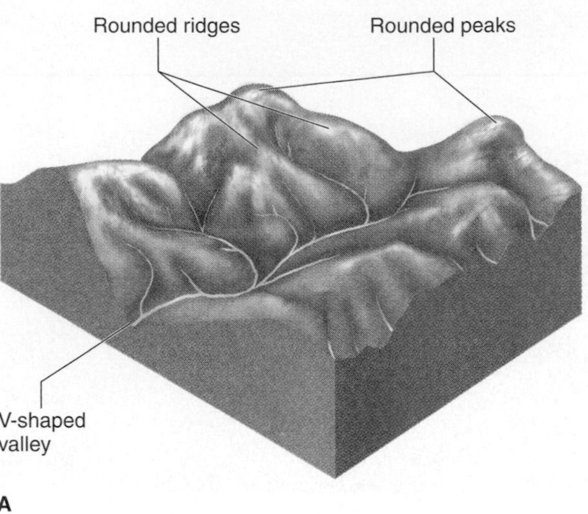

A

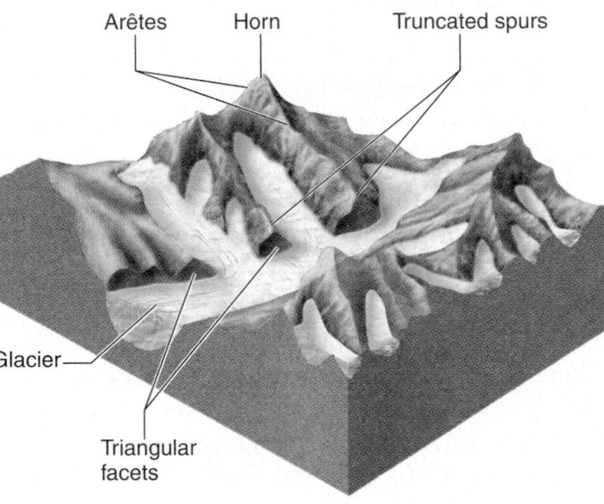

B

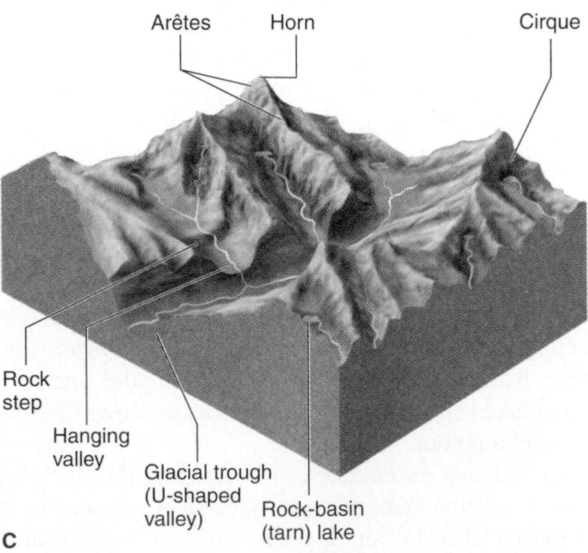

C

FIGURE 16.14

(*A*) A stream-carved mountain landscape before glaciation. (*B*) The same area during glaciation. Ridges and peaks become sharper due to frost wedging. (*C*) The same area after glaciation.

Glacial Troughs

Glacially carved valleys are easy to recognize. **Glacial troughs** have steep bedrock walls and a relatively flat floor (figure 16.15). (These troughs are also called U-shaped valleys.) The flat floor of a glacial trough is not usually formed by bedrock but is underlain by glacial and fluvial sediments that overlie a very irregular bedrock surface. As glaciers recede upvalley, previously deposited sediment collapses along the steep valley walls and alluvial fans build out from the valley walls to smooth the profile into a U shape. It is important to note that glacial erosion does not produce a flat-floored valley. The flat floor is produced by subsequent deposition of glacial and fluvial sediment.

Valley glaciers carve deep troughs along valleys previously occupied by rivers. These valleys often follow structural weakness in bedrock (such as faults). Glacial erosion of ridges that formerly extended into the valley creates **truncated spurs** (figure 16.14*B*).

FIGURE 16.15

Glaciated valley in Gros Morne National Park, Newfoundland. The steep walled "U shaped" valley has been flooded and is accumulating sediment to create a flat valley floor.

Photo by Nick Eyles

FIGURE 16.16

Takkakaw Falls, Yoho National Park, flow over the edge of a hanging valley.

From CP/Larry MacDougal

The amount of eroding a glacier does is dependent on a number of factors including ice velocity, ice mass, the nature and amount of basal debris it carries, and the type of bedrock or sediment over which it flows. Different amounts of erosion occur along the length of a valley glacier as tributary glaciers join (figure 16.14*B*), adding more mass to the trunk glacier (and hence increasing velocity and erosion of the valley floor) and as different rock types are overridden by the ice. Large trunk glaciers carve deeper valleys than the smaller tributary glaciers that join them. After the glaciers retreat, these tributary valleys remain as **hanging valleys** high above the main valley (figure 16.16).

In some parts of glacial troughs, the irregular bedrock floor may be exposed (rather than covered by sediment) and bedrock depressions infilled by lakes. These lakes are called **tarns** (also known as **rock-basin lakes**) and often occur as a series, reminiscent of a string of prayer beads, called *paternoster lakes*. Pinto Lake and Lake Louise in Banff National Park are good examples of tarn lakes (figure 16.17). Lake O'Hara and associated lakes in B.C.'s Yoho National Park are paternoster lakes.

Areas where bedrock is relatively resistant to erosion form *roches moutonnées* (figure 16.18), streamlined bedrock knobs elongated parallel to glacier flow. (The term roches mountonnées was used to describe an assemblage of rounded knobs in the Alps that resembled grazing sheep; in French, *rouche* is rock and *moutonnée* means fleecy or curled.) Roches moutonnées have a gently sloping upglacier end eroded by abrasion, and a blunt "chopped off" downglacier end eroded by plucking processes (refer to figure 16.12).

FIGURE 16.17

Pinto Lake in Banff National Park is an example of a tarn lake.

Photo by Nick Eyles

Cirques, Horns, and Arêtes

A **cirque** is a steep-sided, half-bowl-shaped recess carved into a mountain at the head of a valley carved by a glacier (figure 16.19). In this unique—often spectacular—topographic feature, a large percentage of the snow accumulates that eventually converts to glacier ice and spills over the threshold as the valley glacier starts its downward course.

A cirque is not entirely carved by the glacier itself but is also shaped by the weathering and erosion of the rock walls above the surface of the ice. Frost wedging and avalanches break up the rock and steepen the slopes above the glacier. Broken rock tumbles onto the valley glacier and becomes part of its load, and some rock may fall into a crevasse (*bergschrund*) that develops where the glacier is pulling away from the cirque wall (figure 16.20).

The headward erosional processes that enlarge a cirque also help create the sharp peaks and ridges characteristic of glaciated mountain ranges. A **horn** is the sharp peak that remains after cirques have cut back into a mountain on several sides (figure 16.21).

Frost wedging works on the rock exposed above the glacier, steepening and cutting back the side walls of the valley. Sharp ridges called **arêtes** separate adjacent glacially carved valleys (figure 16.22).

FIGURE 16.18
Roche moutonnée formed by subglacial abrasion and plucking processes, Ontario.
Photo by Nick Eyles

FIGURE 16.19
A cirque occupied by a small glacier in the Canadian Rocky Mountains. The glacier was much larger during the Quaternary ice ages.
Photo by C. C. Plummer

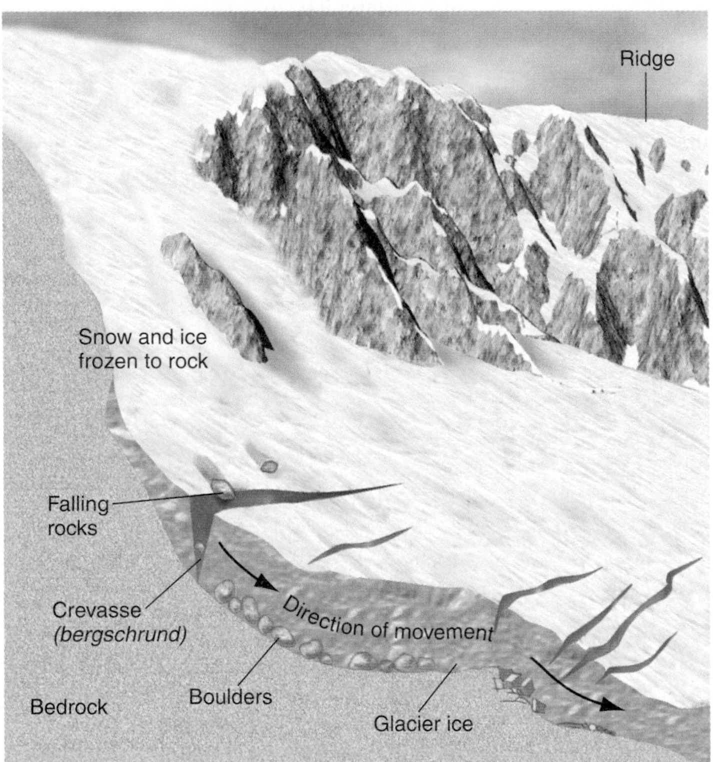

FIGURE 16.20
Cutaway view of a cirque.

FIGURE 16.21
Mount Assiniboine, a horn in the Canadian Rockies.
Photo by Photodisc Blue

FIGURE 16.22
An arete on Mount Logan, Yukon Territory
Photo by C. C. Plummer

Erosional Landscapes Associated with Continental Glaciation

In contrast to the rugged and angular nature of glaciated mountains, an ice sheet tends to produce smoothed, streamlined erosional topography. The rock underneath an ice sheet is eroded in much the same way as the rock beneath a valley glacier.

However, ice sheets tend to scour and smooth the pre-existing landscape rather than deeply dissecting it. It is estimated that, on average, only 50 metres of material was eroded from the Canadian Shield during all of the Quaternary glaciations combined (see chapter 20). This contrasts with much larger amounts of material excavated from glacial troughs (typically several hundred metres or more).

The landscape produced by continental ice sheet erosion is one of smoothed streamlined bedrock knobs (roches moutonnées), and striated and grooved bedrock surfaces (figure 16.23). The orientation of grooves and striations indicates the direction of movement of a former ice sheet.

FIGURE 16.23
Glacially scoured terrain of the Canadian Shield, Great Slave Lake, N.W.T.
Photo © Jim Wark — Airphoto

Ice sheets can be up to 4,000 metres thick and in places may completely blanket the terrain, although they will flow around larger topographic features. The glacially scoured summits of the Appalachian Mountains of Quebec, New Brunswick, and Cape Breton Island show the effects of glacial erosion on the terrain.

WHAT FEATURES ARE PRODUCED BY GLACIAL DEPOSITION?

The rock fragments scraped and plucked from the underlying bedrock and carried along at the base of the ice make up most of the load carried by an ice sheet, but only part of a valley glacier's load. Much of a valley glacier's load comes from avalanches and rock falls onto the ice surface and from rocks eroded from the valley walls.

Some of the rock fragments carried by glaciers are angular, as the pieces have not been tumbled around enough for the edges and corners to be rounded. Distinctive abraded and faceted "bullet-shaped" boulders are produced when debris is dragged along in the base of the glacier (figure 16.24). The debris is unsorted, and clay-sized to boulder-sized particles are mixed together. The unsorted and unlayered rock debris carried and subsequently deposited by a glacier is called **till.**

Glaciers are capable of carrying virtually any size of rock fragment, even boulders as large as a house. An **erratic** is an ice-transported boulder that has not been derived from underlying bedrock but has been transported from elsewhere. If its bedrock source can be found, the erratic indicates the direction of movement of the glacier that carried it. The Foothills Erratics Train of Alberta includes the Okotoks erratic south of Calgary (see chapter 20), which weighs more than 22,000 metric tonnes (about 30,000 grizzly bears).

A

B

FIGURE 16.24

(*A*) Shaped and faceted "bullet-shaped" boulder deposited in front of Saskatchewan Glacier, Banff National Park. The boulder was shaped as it was transported at the base of the glacier. (*B*) Poorly sorted till resting on bedrock, Northern England.

Photos by C. H. Eyles

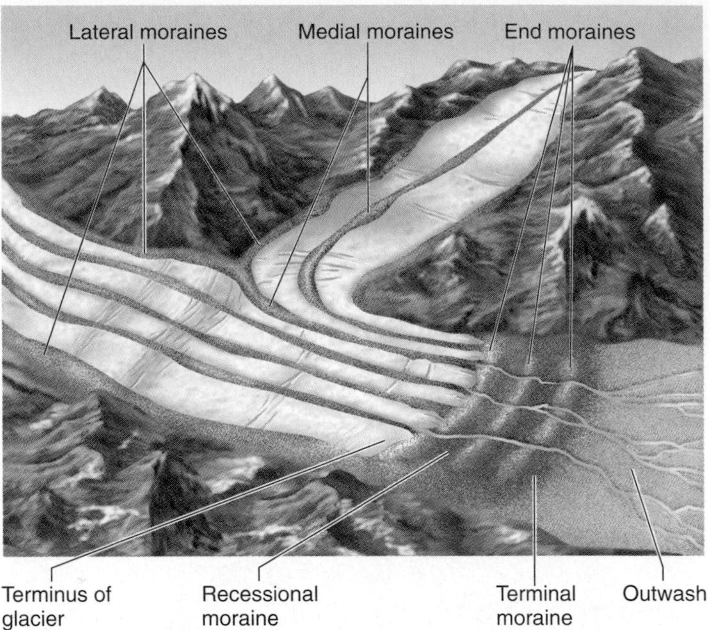

FIGURE 16.25

Moraines associated with valley glaciers.

FIGURE 16.26

Medial and lateral moraines on valley glaciers, Yukon Territory. Ice is flowing toward viewer and to lower right.

Photo by C. C. Plummer

Moraines

When sediment occurs as a body of unsorted and unlayered debris either on a glacier or left behind by a glacier, the body can form one of several types of **moraines. Lateral moraines** are elongate, low mounds of debris that form along the sides of a valley glacier (figures 16.25, 16.26, and 16.27). Rockfall debris from the steep cliffs that border valley glaciers accumulates along the edges of the ice to form lateral moraines.

Where tributary glaciers come together, the adjacent lateral moraines join and are carried downglacier as a single long ridge of debris known as a **medial moraine.** In a large-trunk glacier that has formed from many tributaries, the numerous medial moraines give the glacier the appearance from the air of a multilane highway (figures 16.25 and 16.26).

An actively flowing glacier brings debris to its terminus. If the terminus remains stationary for a few years or advances, a distinct **end moraine,** a ridge of glacial debris, piles up along the front edge of the ice. Valley glaciers build end moraines that are crescent-shaped or sometimes horseshoe-shaped (figures 16.25 and 16.27*B*). The end moraine of an ice sheet takes a similar lobate form, but is much longer and more irregular than that of a valley glacier (figure 16.28).

Geoscientists distinguish two special kinds of end moraines. A *terminal moraine* is the end moraine marking the farthest advance of a glacier. A *recessional moraine* is an end moraine built while the terminus of a receding glacier remains temporarily stationary. A single receding glacier can build several recessional moraines (as shown in figures 16.25, 16.27*C*, and 16.28).

As ice melts, rock debris that has been carried at the base of a glacier is deposited to form a **till plain,** a fairly thin, extensive layer or blanket of till (figure 16.28). Very large areas of Canada that were previously glaciated now have the gently rolling surface characteristic of till plains.

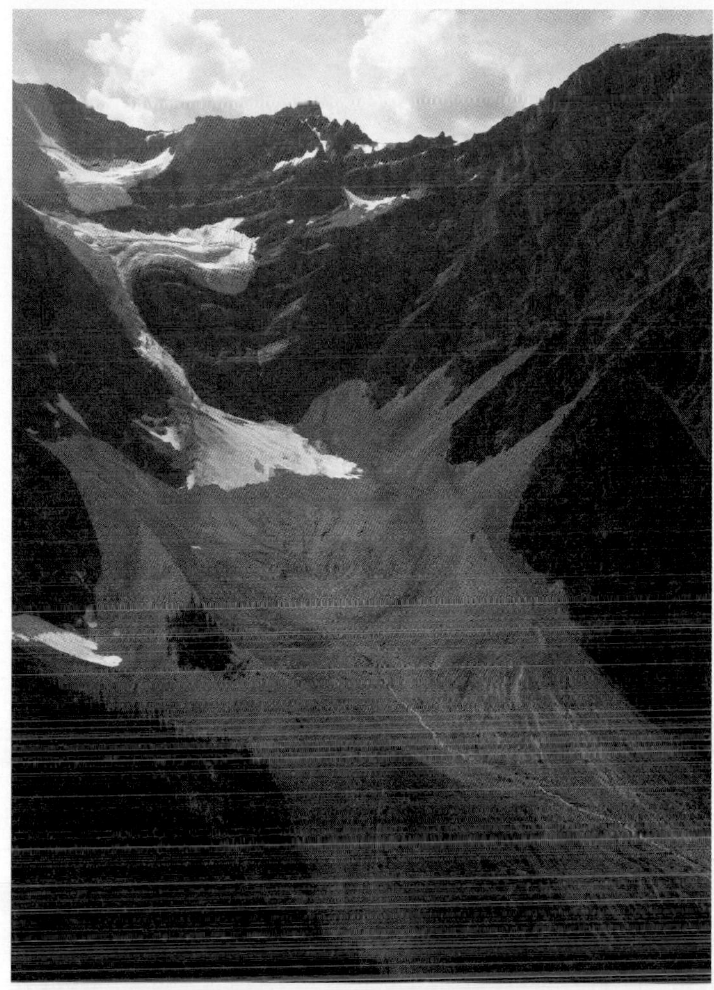
A

B

C

FIGURE 16.27

(*A*) Sharp-crested lateral moraines mark the limits of advance of the Crowfoot Glacier about 11,000 years ago; Banff National Park. (*B*) Arcuate end moraines left behind by retreating glacier on Mount Lawson, B.C. (*C*) Series of end moraine ridges (two are arrowed) mark former positions of the terminus of Athabasca Glacier as it retreats up valley.

Photos A and C by Nick Eyles.
Photo B © Jim Wark — Airphoto

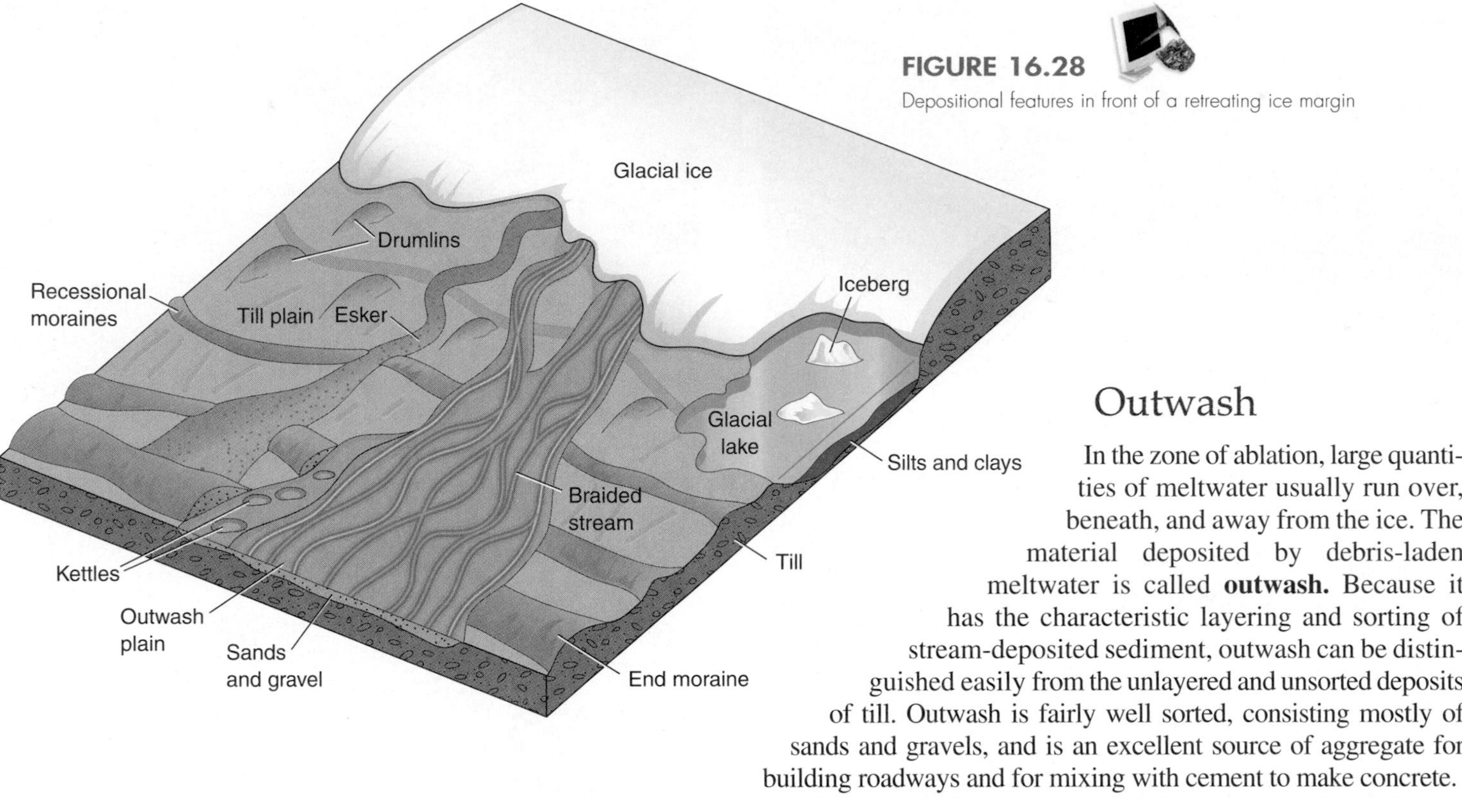

FIGURE 16.28

Depositional features in front of a retreating ice margin

FIGURE 16.29

Drumlin near Morley, Bow Valley, Alberta.
Photo by Nick Eyles

In some areas of past continental glaciation there are bodies of till shaped into streamlined hills called **drumlins** (figures 16.28 and 16.29). A drumlin is shaped like an inverted spoon aligned parallel to the direction of ice movement of the former glacier. The steeper side of a drumlin points upglacier and its gentler end points in the downglacier direction. Because we cannot observe drumlins forming beneath present ice sheets there is uncertainty regarding the process involved in the shaping of till into these streamlined hills.

Outwash

In the zone of ablation, large quantities of meltwater usually run over, beneath, and away from the ice. The material deposited by debris-laden meltwater is called **outwash.** Because it has the characteristic layering and sorting of stream-deposited sediment, outwash can be distinguished easily from the unlayered and unsorted deposits of till. Outwash is fairly well sorted, consisting mostly of sands and gravels, and is an excellent source of aggregate for building roadways and for mixing with cement to make concrete.

An outwash feature of unusual shape associated with former ice sheets and some very large valley glaciers is an **esker,** a long, sinuous ridge of water-deposited sediment (figures 16.28 and 16.30). Eskers can be up to 10 metres high and are formed of gravels and sands, which may be cross-bedded and relatively well-sorted. Eskers are deposited in tunnels within or under glaciers, where meltwater loaded with sediment flows under and out of the ice (figure 16.30*A*).

As meltwater builds thick deposits of outwash alongside and in front of a retreating glacier, blocks of stagnant ice may be surrounded and buried by sediment. When the ice block finally melts (sometimes years later), a depression called a **kettle** forms (figures 16.28 and 16.31*A*). Many of the small scenic lakes (sloughs) in Manitoba, Saskatchewan, and Alberta are kettle lakes. A **kame** is a low mound or irregular ridge of outwash deposits that were originally deposited along the margins or in front of a glacier. When the ice melted the outwash deposits were left as isolated hills (figure 16.31*B*).

The streams that drain glaciers tend to be very heavily loaded with sediment, particularly during the melt season. As they come off the glacial ice and spread out over the outwash deposits, the streams form a braided pattern (see chapter 14). If meltwater streams are unconfined by valley walls they form laterally extensive outwash plains or sandur (plural *sandar*) plains.

The large amount of rock flour (silt) that these streams carry in suspension settles out in quieter waters. In dry seasons or drought, the water may dry up and the silt deposits may be picked up by the wind and carried long distances. Some of the best agricultural soil in the United States has been formed by rock flour that has been redeposited by wind. Such fine-grained, wind-blown deposits of silt are called *loess* (see chapter 17).

A

B

C

FIGURE 16.30

(A) Tunnel underneath a glacier in which sands and gravels have been deposited by a meltwater stream. *(B)* Six-metre-high sinuous esker ridge in Iceland. *(C)* Sinuous, tree-covered esker crosses a till plain in northwestern Manitoba.

Photos A and B by Nick Eyles; Photo C by D. A. Rahm, Courtesy of Rahm Memorial Collection, Western Washington University

A

D

FIGURE 16.31

(A) Large kettle formed by melting of an ice block buried by outwash deposits; Saskatchewan Glacier, Banff National Park. *(B)* Outwash sands and gravels exposed in a kame near Peterborough, Ontario.

Photo A by C. H. Eyles; Photo B by Nick Eyles

Glacial Lakes and Varves

Lakes often occupy depressions carved by glacial erosion but can also form behind dams built by glacial deposition. Commonly a lake forms between a retreating glacier and an end moraine.

In the still water of the lake, clay and silt settle on the bottom in two thin layers—one light-coloured, one dark—that are characteristic of glacial lakes. Two layers of sediment representing one year's deposition in a lake are called a **varve** (figure 16.32). The light-coloured layer consists of slightly coarser sediment (sand or silt) deposited during the warmer part of the year when the nearby glacier is melting and sediment is transported to the lake. This relatively coarse-grained material is rapidly deposited either by currents flowing along the floor of the lake, or from suspension. The dark layer is finer-grained sediment (clay) that slowly settles out of the water during the winter when the lake surface is frozen and the supply of coarser sediment stops due to lack of meltwater input. Each varve couplet represents a year's deposit, and varves are used to estimate how long a glacial lake existed.

FIGURE 16.32
Varves from a former glacial lake. Each pair of light and dark layers represents a year's deposition.
Photo by Nick Eyles

Varved glacial lake deposits formed extensively over Canada as the ice retreated at the end of the last glaciation. The "clay belt" of northern Ontario and Quebec is underlain by varved deposits formed in Glacial Lake Barlow-Ojibway, and similar deposits are associated with former Glacial Lake Peace (northwest Alberta and northeast B.C.) and Glacial Lake McConnell (ancestral Great Slave and Great Bear Lakes).

HOW HAVE GLACIERS AFFECTED US IN THE PAST?

As the glacial theory gained general acceptance during the latter part of the nineteenth century, it became clear that most of Canada as well as much of northern Europe and the northern United States had been covered by great ice sheets during the so-called Ice Age. It also became evident that even areas not covered by ice had been affected because of the changes in climate and the redistribution of large amounts of water.

We now know that the last of the great North American ice sheets melted away from Canada less than 10,000 years ago. In many places, however, till from that ice sheet overlies older tills, deposited by earlier glaciations. The older till is distinguishable from the newer till because the older till was deeply weathered during times of warmer climate between glacial episodes.

The Ice Ages

Geoscientists can reconstruct with considerable accuracy the last episode of extensive glaciation, which covered large parts of North America and Europe and was at its peak about 18,000 years ago. There has not been enough time for weathering and erosion to significantly alter the effects of glaciation. Less evidence is preserved for each successively older glacial episode, because (1) weathering and erosion occurred during warm interglacial periods, and (2) later ice sheets and valley glaciers overrode and obliterated many of the features of earlier glaciation. However, by piecing together the evidence, geoscientists understand that earlier glaciers covered approximately the same region as the more recent ones.

Worldwide climate changes necessary for northern continental glaciation probably began at least 3 million years ago. Antarctica has been glaciated for at least 38 million years (see box 16.5 for some interesting "rocks" found on Antarctic glaciers). The Earth has undergone episodic changes in climate during the Quaternary (approximately the last 2 million years). Actually, the climate changes necessary for a glacial age to occur are not so great as one might imagine. During the height of a glacial age, the worldwide average of annual temperatures was probably only about 5°C cooler than at present. Some of the intervening interglacial periods were probably slightly warmer worldwide than present-day average annual temperatures. For more on causes of ice ages, see box 16.6.

Direct Effects of Past Glaciation in North America

Moving ice abraded vast areas of northern and eastern Canada during the growth of the North American ice sheets (figure 16.33). Most of the sediment was scraped off the Canadian Shield, and the underlying crystalline bedrock was scoured. Many thousands of future lake basins were gouged out of the bedrock (see chapter 20).

Erratics from the northeastern Canadian Shield are found in western Prince Edward Island; iron-bearing boulders from western Labrador were transported to southern Ontario, Illinois, and New Jersey; and erratics from the western shore of Hudson Bay have been transported to Kansas. The distribution of these erratics, together with the orientation of striations and grooves in bedrock, has been used to reconstruct the nature and flow patterns of ice sheets that covered North America during the last glaciation.

The largest ice sheets, located to the west (Keewatin), east (Quebec-Labrador), and north (Innuitian) of Hudson Bay, flowed radially outward and coalesced to form the Laurentide Ice Sheet (figure 16.33*A*). Another ice sheet, the Cordilleran Ice Sheet, flowed radially from central B.C. to central Yukon and Washington State. Smaller ice sheets and ice caps grew in Atlantic Canada and in northern Nunavut, and valley glaciers formed in the western Cordillera, the Torngat Mountains of Labrador, and the ranges of Baffin and Ellesmere Islands. Eventually, all of Canada and the northern part of the U.S., with the exceptions of the northern Yukon, isolated plateaus in southern Alberta and Saskatchewan (notably the Cypress Hills), central Alaska, and the highest mountain summits, were covered by ice.

The exposed, eroded surface of the Canadian Shield contrasts markedly with the Great Plains and Great Lakes–St. Lawrence Valley, where vast amounts of till and glaciolacustrine sediments were deposited. Most of the sediment was deposited on till plains, which along with glacial lake deposits have partially weathered to yield excellent soil for agriculture. Large end moraines that

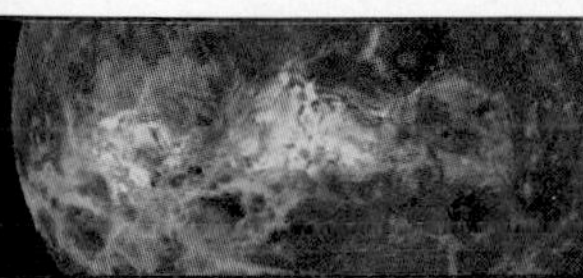

ASTROGEOLOGY 16.5

Mars on a Glacier

Meteorites are extraterrestrial rocks—fragments of material from space that have managed to penetrate Earth's atmosphere and land on Earth's surface. They are of interest not only to astronomers but also to geoscientists, for they help us date Earth (chapter 8) and give us clues to what Earth's interior is like (see chapter 17) because many of the meteorites are thought to represent fragments of destroyed minor planets. Meteorites are rarely found; they usually do not look very different from Earth's rocks with which they are mixed.

The international Antarctic meteorite program has recovered 30,000 specimens during the last three decades. This far exceeds the total collected elsewhere in the past two centuries. Over a thousand meteorites have been collected from one small area where the ice sheet abuts against the Transantarctic Mountains. The reason for this heavy concentration is that meteorites landing on the surface of the ice over a vast area have been incorporated into the glacier and transported to where ablation takes place. The process is illustrated in box figure 1.

A few of the meteorites are especially intriguing. Some almost certainly are rocks from the moon, while several others apparently came from Mars. Their chemistry and physical properties match what we would expect of a Martian rock. But how could a rock escape Mars and travel to Earth? Scientists suggest that a meteorite hit Mars with such force that fragments of that planet were launched into space. Eventually, some of the fragments reached Earth.

In 1996, researchers announced that they found what could be signs of former life on Mars in one of the meteorites collected 12 years earlier in Antarctica. The evidence included carbon-containing molecules that might have been produced by living organisms as well as microscopic blobs that could be fossil alien bacteria. But there are alternate explanations for each line of evidence, and a hot debate has ensued among scientists with opposing viewpoints.

Additional Resource

Antarctic Meteorite Program

- www-curator.jsc.nasa.gov/antmet/index.cfm

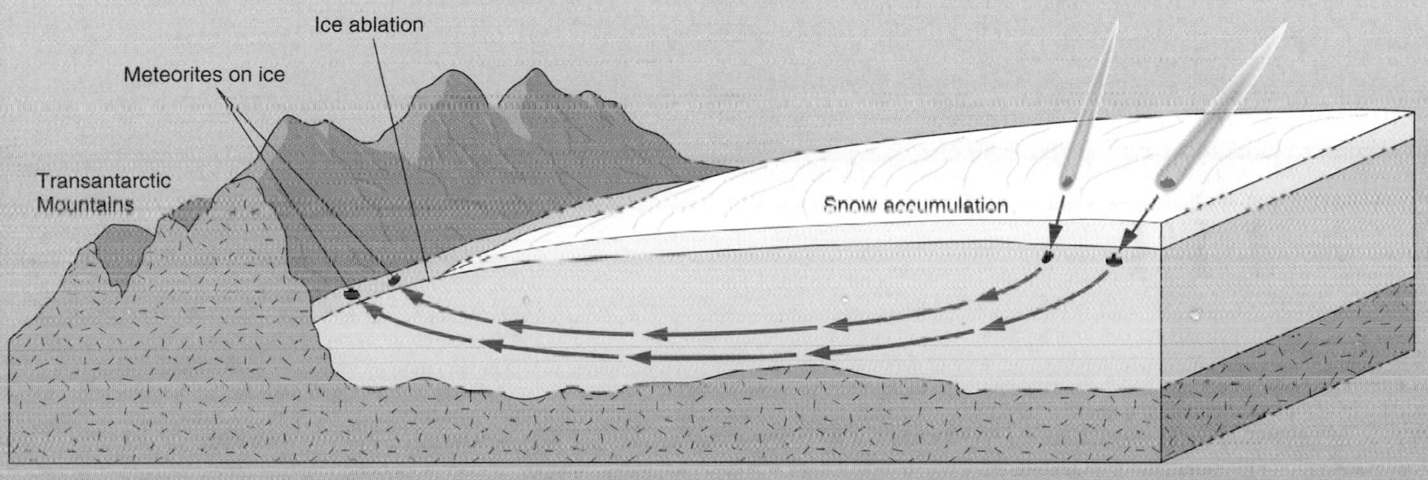

BOX 16.5 ■ FIGURE 1

Diagram showing the way in which meteorites are concentrated in a narrow zone of wastage along the Transantarctic Mountains. Two meteorites are shown as well as the paths they would have taken from the time they hit the ice sheet until they reached the zone of ablation. The vertical scale is greatly exaggerated.

Source: Antarctic Journal of the United States

formed along the southern margin of the Laurentide Ice margin are found at Sable Island (Nova Scotia), Long Island (New York), and Nantucket and Cape Cod (Massachusetts). Numerous drumlins are preserved in southern Ontario and Nova Scotia (see chapter 20), and the downtown areas of Halifax and Boston are built on the flanks of drumlins.

Glaciers have tremendous capacity for forming lakes through both erosion and deposition. Large regions of Canada were covered by ice marginal lakes. Glacial Lake Agassiz covered much of southern Manitoba, as well as parts of Saskatchewan, northwestern Ontario, Minnesota, and North Dakota (figure 16.34). The Red River flows through the clays deposited in Glacial Lake Agassiz, eventually reaching Lake Winnipeg, the last remnant of the great lake.

The Finger Lakes in New York (figure 16.35) lie in long, north–south glacially modified valleys that are dammed by recessional moraines at their southern ends. The Great Lakes are, at least in part, a legacy of continental glaciation. Former stream valleys were widened by the ice sheet eroding weak layers of sedimentary rock into the present lake basins (see chapter 20).

Alpine glaciation was much more extensive throughout the world during the glacial ages than it is now. For example, small

IN GREATER DEPTH 16.6

Causes of Ice Ages

Go to the book's Online Learning Centre at www.mcgrawhill.ca/olc/plummer for a more in-depth presentation of the summary below.

The question of what caused the ice ages has not been completely answered since the theory of glacial ages was accepted more than a century ago. Only in the last few decades have climatologists thought they were beginning to provide acceptable answers.

The primary control on the Quaternary glacial and interglacial episodes seems to be variations in the Earth's orbit and inclination to the sun. The amount of heat from solar radiation received by any particular portion of the Earth is related to the angle of the incoming sun's rays and, to a lesser degree, the distance to the sun. The angle of the Earth's poles relative to the plane of the Earth's orbit about the sun also changes periodically. Variations in orbital relationships and "wobble" of Earth's axis are largely responsible for glacial and interglacial episodes. These provide variations in incoming solar radiation cycles of 21,000, 41,000, and 100,000 years, as calculated by Milutin Milankovitch, a Serbian mathematician, in 1921. Proof of Milankovitch's cycles came from cores of deep sediment taken by oceanographic research ships. Deep-sea sediment provides a fairly precise record of climatic variations over the past few hundred thousand years. The cycles of cooling and warming determined from the marine sediments closely match the times predicted by Milankovitch.

However, the theory fails to explain the absence of glaciation over most of geologic time. Thus, one or more of the other mechanisms listed below (and described in the Website) may have contributed to climate change resulting in ice ages.

- **Changes in the atmosphere.** These changes include the amount of carbon dioxide in the atmosphere. Carbon dioxide has a "greenhouse effect," whereby the more of the gas in the atmosphere, the warmer the global climate. Large volcanic eruptions are known to lower temperature worldwide by placing SO_2 gas and fine dust in the high atmosphere. A series of large, volcanic eruptions might help trigger an ice age.
- **Changes in the positions of continents.** Plate tectonics movement of continents closer to the poles increases the likelihood of glaciation. Movement of northern-hemisphere continents closer to the North Pole has placed land masses in a position more favourable for glaciation.
- **Changes in circulation of sea water.** Land masses block the worldwide free circulation of sea water, affecting which oceans are warmer than others.
- **Changes in land elevation created by tectonics.** Where plates collide or are torn apart, tectonic uplift is a common result. This causes regional cooling and if other requirements are met, such as a sufficient supply of moisture, or latitude, then ice covers can form.
- **Changes in cosmic ray flux (CRF).** As the solar system moves in and out of the spiral arms of the galaxy, the cosmic ray flux changes. Cosmic ray flux is correlated with the Earth's low-level cloud cover and may significantly influence global climate.

glaciers in the Rocky Mountains that now barely extend beyond their cirques were then valley glaciers 10, 50, or 100 kilometres in length. Alpine glaciers from Jasper and Banff extended eastward to join with Laurentide ice east of the foothills of the Rockies (as evidenced by the Foothills Erratic Train; see figure 20.29*A*); Yellowhead Pass and Kicking Horse Pass were filled by thick glaciers. Furthermore, cirques and other features typical of valley glaciers can be found in regions that at present have no glaciers, such as the Gaspé Peninsula and Cape Breton Highlands National Park.

Indirect Effects of Past Glaciation

As the last continental ice sheet wasted away, what effects did the tremendous volume of meltwater have on North American rivers? Rivers that now contain only a trickle of water were huge in the glacial ages. Other river courses were blocked by the ice sheet or clogged with morainal debris. Large dry stream channels have been found that were preglacial tributaries to the Mississippi and other river systems.

Pluvial Lakes

During the glacial ages the climate in North America, even beyond the glaciated parts, was more humid than it is now. Most of the currently arid regions of the western United States had moderate rainfall, as traces or remnants of numerous lakes indicate. These **pluvial lakes** (formed in a period of abundant rainfall) once existed in Utah, Nevada, and eastern California. Some may have been fed by meltwater from mountain glaciers, but most were simply the result of a wetter climate.

Great Salt Lake in Utah is but a small remnant of a much larger body of fresh water called Lake Bonneville, which, at its maximum size, was nearly as large as Lake Michigan is today. Even Death Valley in California—now the driest and hottest place in the United States—was occupied by a deep lake during the Quaternary. The salt flats that were left when this lake dried include rare boron salts that were mined during the pioneer days of the American West.

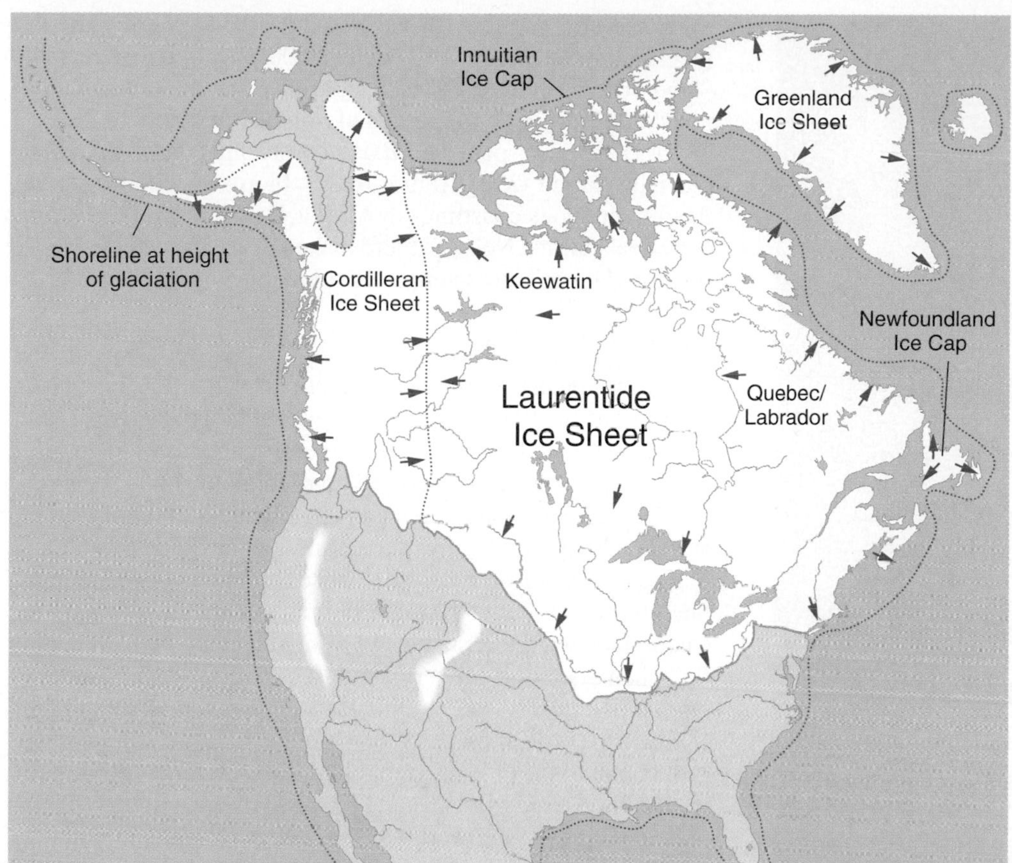

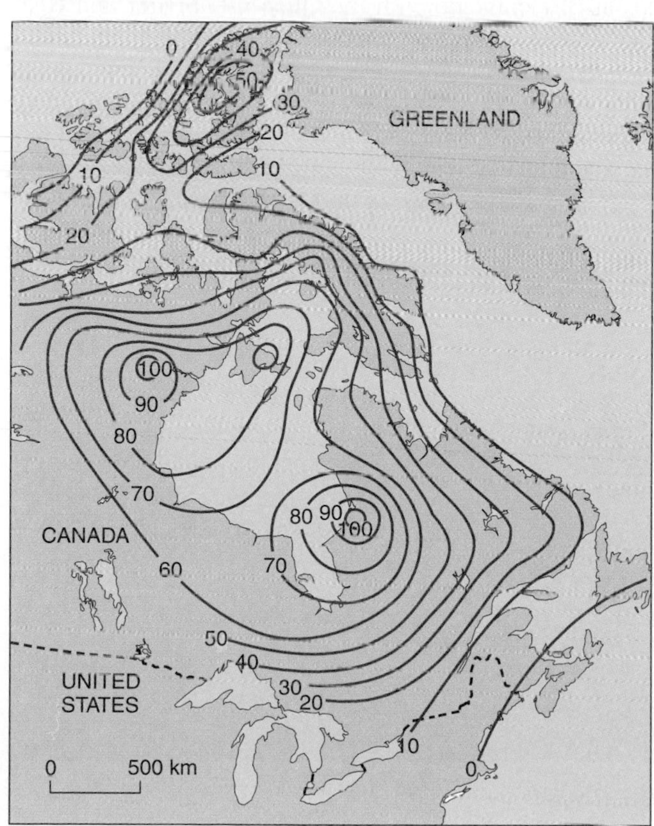

FIGURE 16.33

(A) Extent of maximum ice coverage in North America during the Quaternary Era. Arrows show approximate directions of ice movement. *(B)* Amount of uplift (in metres) of the crust in eastern Canada over the past 6,000 years. Uplift still continues but at a decreasing rate. Large areas of the west and east coasts of Canada are slowly being submerged as the rate of sea level rise outpaces the rate of crustal rebound.

Figure A after C.S. Denny, U.S. Geological Survey, National Atlas of the United States; Figure B from Murphy & Nance 1999. Earth Science Today, Pacific Grove, CA: Brooks/Cole-Wadsworth, 429

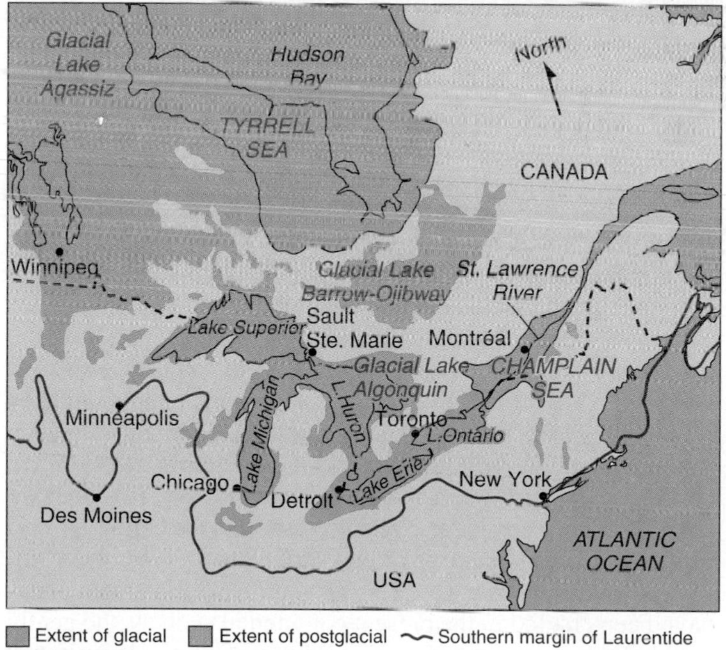

FIGURE 16.34

Large glacial lakes formed around the margin of the Laurentide Ice Sheet as it gradually withdrew from North America after 20,000 years ago. Marine waters also flooded the depressed area around Hudson Bay (Tyrrell Sea) and the St. Lawrence Valley (Champlain Sea).

After Nick Eyles, 2002

FIGURE 16.35

Satellite image of Finger Lakes in New York. Part of Lake Ontario is at the top.
Photo © Advanced Satellite Productions, Inc.

Lowering of Sea Level

All of the water for the great glaciers had to come from somewhere. The water was removed from the oceans, such that sea level worldwide was lower than it is today—at least 130 metres lower, according to scientific estimates.

Recall that if today's ice sheets were to melt, sea level worldwide would rise by more than 60 metres and shorelines would be considerably farther inland. It's important to realize that our present shorelines are not fixed and are very much controlled by climate changes. We should also realize that we are still in a cooler than usual (relative to most of the Earth's history) time, perhaps the lingering effects of the last ice age.

What is the evidence for lower sea level? Stream channels have been charted in the present continental shelves, the gently inclined, now-submerged edges of the continents (described in chapter 18). These submerged channels are continuations of today's major rivers and had to have been above sea level for stream erosion to take place. Bones and teeth from now-extinct mammoths and mastodons have been dredged up from the Atlantic continental shelf, indicating that these relatives of elephants roamed over what must have been dry land at the time. Archaeological evidence also suggests that lowered sea levels during the last glacial episode (about 18,000 to 20,000 years ago) created a "land bridge" between Alaska and Siberia and allowed human migration from Asia to North and South America.

A **fjord** (also spelled *fiord*) is a coastal inlet that is a drowned glacially carved valley (figure 16.36). Fjords are common along the mountainous coastlines of Alaska, British Columbia, Chile, New Zealand, and Norway. Fjords are evidence that valleys eroded by past glaciers were later partly submerged by rising sea level.

Crustal Rebound

The weight of an ice sheet several thousand metres thick depresses the crust of the Earth much as the weight of a person depresses a mattress. A land surface bearing the weight of a continental ice sheet may be depressed several hundred metres.

Crustal depression allowed marine waters to flood the land surface along the St. Lawrence Valley as far inland as Ottawa during deglaciation of eastern Canada, forming the Champlain Sea (figure 16.34). Glaciomarine clays deposited within the Champlain Sea are unstable and form quickclays, which readily fail when disturbed or overloaded (see chapter 13). Marine coverage of the land surface around Hudson Bay was also extensive, forming the Tyrrell Sea.

The amount of crustal rebound that has taken place since ice finally melted from Canada 6,000 years ago (figure 16.33*B*) indicates areas of the crust that experienced the greatest amount of depression. This in turn indicates areas of greatest ice thickness. Note that areas of maximum rebound to the east and west of Hudson Bay coincide with the Quebec/Labrador and Keewatin ice centres that fed the Laurentide Ice Sheet.

Once the glacier is gone, the land begins to rebound slowly to its previous height (see chapter 4 and figures 4.12 and 4.13). Uplifted and tilted shorelines along lakes are an indication of this process. The Great Lakes region is still rebounding as the crust slowly adjusts to the removal of the last ice sheet.

FIGURE 16.36

Pangnirtung Fjord, Cumberland Sound, Baffin Island.
Photo by AVIVA — Paul Nopper.

Evidence of Older Glaciations

Throughout most of geologic time, the climate has been warmer and more uniform than it is today. We think that the late Cenozoic Era is unusual because of the periodic fluctuations of climate and the widespread glaciations. However, glacial ages are not restricted to the late Cenozoic.

Evidence of older glaciation comes from rocks called diamictites. Unsorted rock particles, including angular, striated, and faceted boulders, have been consolidated into a sedimentary rock. In some places, diamictite layers overlie surfaces of older rock that have been polished and striated. Diamictites that originally formed as tills are called **tillites.**

The oldest glacial rocks on the planet are 2.8 billion years old and are found in Southern Africa. One of the oldest glaciations for which we have excellent evidence appears to have taken place around 2.3 billion years ago in what is now Ontario (figure 16.37*A*). Glacial deposits of Neoproterozoic age (between 900 and 550 million years old) are found on every continent on Earth and have prompted some scientists to propose that the Earth was completely frozen, from equator to poles, at that time. However, the length of time over which Neoproterozoic glaciation occurred and the problems of accurately dating these ancient deposits makes it difficult to prove a fully ice-covered Earth (see box 16.7). Key deposits of this age occur in the Mackenzie Mountains of northern British Columbia and the Yukon (figure 16.37*B*).

A

B

FIGURE 16.37

(*A*) Very poorly sorted diamictite of the Gowganda Formation, Ontario. This rock is composed of mixtures of boulders and finer-grain sizes. Some of the larger boulders have been scratched (striated) by ice and record glaciation in what is now northern Ontario 2.3 billion years ago. (*B*) Outcrops of the Rapitan Group in northern British Columbia. These rocks are interpreted to record cold climate conditions 750 million years ago during the Neoproterozoic.

Photo A by Nick Eyles; Photo B by Noel James

Around 450 million years ago, the area that now forms the West African Shield lay close to the south pole. Late Ordovician glacial deposits are now exposed in parts of the Sahara Desert and give a dramatic reminder of the extreme climate changes that may be caused by continental drift. Late Paleozoic glacial deposits (formed between 350 and 250 million years ago) are found on all of the southern continents (South Africa, Australia, Antarctica, India, South America) and demonstrate that these continents were once joined (see chapter 2). The ancient supercontinent of Gondwana slowly migrated across the south pole during the Late Paleozoic, allowing the deposition of extensive glacial deposits (refer to figure 2.4) before it broke up and the continents dispersed.

WHAT ARE PERMAFROST AND PERIGLACIAL ENVIRONMENTS?

Another way that ice can affect geologic processes and human activities on our planet is through the development of *permafrost*. More than 50 percent of the land area of Canada and most of the state of Alaska is underlain by some form of permafrost. **Permafrost** is ground that has been frozen for two or more years. Frozen ground can be present everywhere below the ground surface as a *continuous* zone, or it may be *discontinuous*, where areas of frozen ground are separated by unfrozen areas. In general, the colder areas of northern and Arctic Canada and Alaska are underlain by continuous permafrost, and the more southern areas are affected by discontinuous permafrost (figure 16.38*A*). Alpine permafrost exists in mountainous areas such as the Rockies, where cold temperatures result from high altitudes.

Permafrost Thickness

The thickness of permafrost beneath the ground is determined by the balance between heat loss from the Earth's surface and heat gained from geothermal sources inside the Earth. Temperatures increase by approximately 1°C with every 30- to 60-metre increase in depth within the Earth's crust—this is called the *geothermal gradient*. The base of the permafrost layer occurs where the temperature increase due to the geothermal gradient offsets the freezing temperature penetrating from the surface downward. Thus, the permafrost base lies several hundred metres below the ground surface in extremely cold high-latitude regions but may be only a few metres below the surface in the south (figure 16.38*B*). Only the uppermost surface layer (called the *active layer*) of the permafrost thaws each summer.

IN GREATER DEPTH 16.7

Earth as a Snowball? — Unravelling the Mysteries of Ancient Glaciations

Glacial deposits formed during the Neoproterozoic (between 900 and 540 million years ago) can be found on every continent. Key deposits of this age are beautifully exposed in Canada across the Mackenzie Mountains of northern British Columbia and the Yukon (refer to figure 16.37*B*). The global distribution of Neoproterozoic glacial deposits led a number of geoscientists to suggest that the Earth was completely covered by ice from equator to poles during this time, separating the atmosphere and oceans and bringing the hydrologic cycle to a complete halt. The notion of such global-scale glaciation is the foundation of the "snowball Earth" hypothesis. Eventual melting of the "snowball" is thought to have been caused by buildup of greenhouse gases in the atmosphere, creating brutally hot interglacial conditions. The snowball Earth hypothesis proposes that such extreme climatic events were the trigger behind the early Cambrian diversification of life forms (called the Cambrian Explosion) so notably recorded in Canada in the Burgess Shale (see chapter 20).

Although the snowball Earth hypothesis has attracted many followers, there are problems in matching the theory with geologic evidence. First, snowball Earth conditions require glacial deposits to form simultaneously on the planet, but very few Neoproterozoic deposits can be accurately dated or correlated from one region to another. Second, detailed descriptions and interpretations of the origin of many ancient "tillites" have demonstrated that they did not form directly beneath glacier ice but in marine environments supplied with sediment by glacial meltwater in an active hydrologic system. Icebergs, which can travel long distances from their source areas, contributed glacially shaped and striated boulders to the poorly sorted sediment accumulating on the sea floor. The distribution of such "glaciomarine" deposits cannot be used to directly infer the former extent and position of ice sheets, making reconstruction of Neoproterozoic ice cover extremely difficult. A third problem lies in trying to reconstruct the geographic position of continents in the Neoproterozoic. Given that continents have changed their positions many times in Earth's history, geoscientists rely on paleomagnetic studies of the rocks to recreate their former geographic positioning. There is some evidence to suggest that parts of Australia were cold and lay at low latitudes during the Neoproterozoic, but paleomagnetic data from other areas has been shown to be the result of later *overprinting*, a common problem in old rocks with a long history of burial and tectonics. This makes it difficult to prove that Neoproterozoic glacial rocks were actually deposited simultaneously in a range of paleolatitudes extending from equator to poles.

A key piece of evidence used to support the snowball Earth hypothesis is the dramatic change in carbon isotope ratios during proposed global glaciation events, particularly in the ratios of ^{12}C and ^{13}C. Life forms selectively use the smaller ^{12}C, resulting in higher relative amounts of ^{13}C available to accumulate in sediment. When carbon isotopes are measured in sedimentary rocks, increases in the ratio of ^{12}C to ^{13}C imply reduced biomass. ^{12}Carbon isotope spikes identified in Neoproterozoic deposits are therefore explained in terms of the biosphere being placed "on hold" during episodes of great cold. However, many other factors control carbon isotope ratios, and many similar shifts have been recorded well before and after the Neoproterozoic that are not associated with any known glacial rocks. In addition, it cannot be demonstrated that the shifts in carbon isotope ratios occurred at the same time around the globe.

The debate over the snowball Earth hypothesis continues, although existing dating control on glacial rocks does not indicate synchronous Neoproterozoic glaciations—nor do sedimentological data indicate extremely cold conditions or halting of the hydrologic cycle. Emerging tectonic data suggest that these ancient glacial deposits formed at times of the breakup and dispersal of one or more supercontinents. Uplift prior to continental breakup cooled the newly formed rift margins and allowed glaciers to develop and transport large amounts of sediment from rifted margins into newly formed marine basins. It therefore appears that ancient glaciations can be explained in uniformitarian terms involving comparisons of modern and ancient processes within a plate tectonic framework. There is still much to learn about conditions that existed on Earth during the Neoproterozoic—the snowball Earth hypothesis has certainly played a key role in awakening interest in ancient glaciations during an episode in Earth's history when life forms were rapidly changing.

Ice in Permafrost

The amount and type of ice present in permafrost can vary a great deal. Ice either fills pore spaces between sediment grains (*pore ice*) or grows into large ice bodies that take thousands of years to grow (*ground ice*) and often take the form of downward tapering ice wedges (figure 16.39*A*, *B*). **Ice wedges** form only when the mean annual temperature is below –3°C. When ice wedges melt, they may be infilled with wind-blown or washed-in sediment and preserved as *ice-wedge casts* (figure 16.39*C*). Ice-wedge casts are sometimes used to indicate the former distribution of permafrost under colder climate conditions.

In areas where abundant water is available for freezing, such as lowland areas or deltas, the growth of ground ice raises overlying sediments into conical hills, or **pingos** (figure 16.39*D*). These features form distinctive landmarks on the Mackenzie Delta of the Northwest Territories.

Patterned Ground

The repeated freezing and thawing of the ground surface (active layer) above permafrost causes sorting of coarser materials and the formation of a variety of types of patterned ground. Stone polygons (figure 16.40) and stone stripes are often found on the ground surface over permafrost and have also been identified from photos of the surface of Mars.

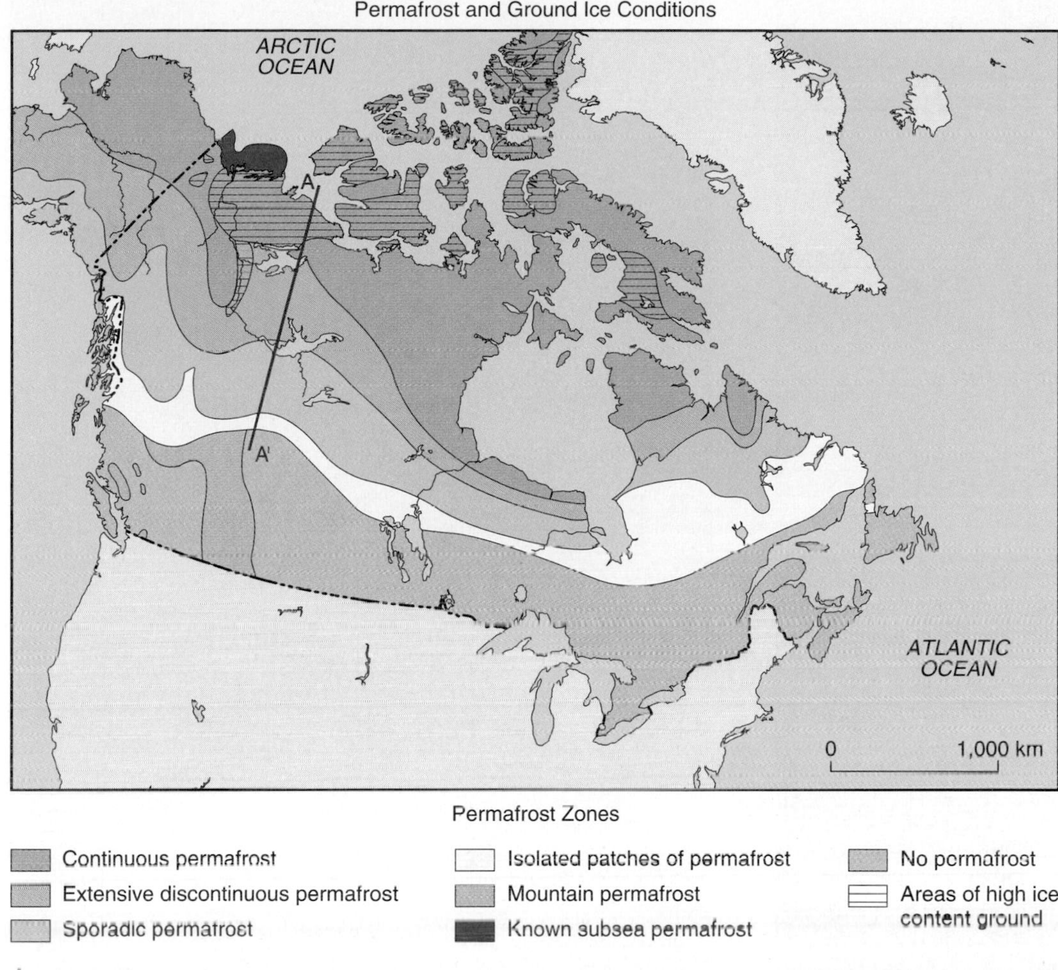

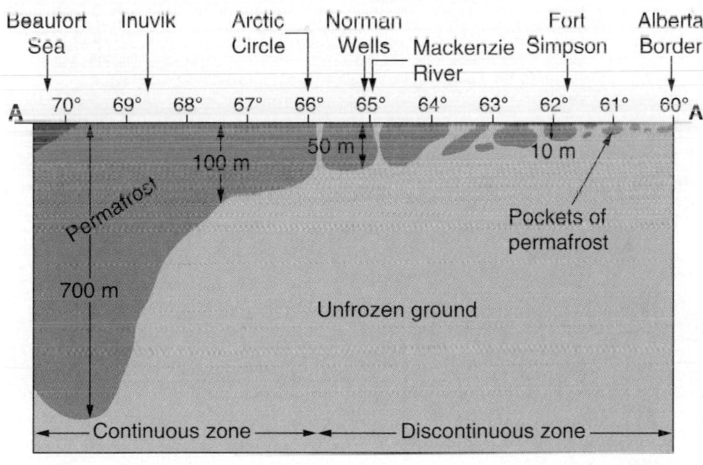

FIGURE 16.38

(*A*) Distribution of permafrost in North America. (*B*) North–south transect across permafrost zone showing decrease in thickness of permafrost toward the south and change from continuous to discontinuous permafrost. (See figure 16.38*A* for location of transect.)

Thermokarst

Melting of the ice within permafrost causes the overlying ground surface to subside, as water occupies approximately 9 percent less volume than ice. The processes associated with subsidence and collapse caused by thawing of permafrost are termed **thermokarst.** Thawing of permafrost may be caused by physical disturbance of the frozen ground and exposure of the ice to warm summer temperatures. This often occurs where rivers or lakes (*thaw lakes*) form on the ground surface in summer (figure 16.41*A*). Thermokarst may also be caused by the release of heat from buildings into the ground (figure 16.41*B*), or by absorption of heat along a road bed (figure 16.41*C*). Most of the problems associated with driving the Alaska Highway are related to disturbances caused by thermokarst. For more on the problems of urban permafrost, see box 16.8.

The 1,250-km-long Alaska pipeline crosses terrain underlain by permafrost (figure 16.42). This posed enormous problems for engineers, because the hot oil moved in the pipeline could thaw the underlying permafrost and cause collapse and rupture of the pipeline. Most of the pipeline is therefore constructed above ground to avoid contact with the frozen ground and is supported on "stilts" equipped with heat exchangers to radiate heat away from the ground. In places, refrigeration equipment in the ground protects against melting.

A

B

C

D

FIGURE 16.39

(*A*) Ground ice (arrowed) exposed in a mine site in central Alaska. (*B*) Ice wedge in sediments in area of continuous permafrost. Ice pick is one metre long. (*C*) Ice-wedge cast in layered Quaternary sediments in eastern England. (*D*) Pingos (ice-cored hills) near Tuktoyaktuk, N.W.T. Each of the pingos is approximately 40 metres high.

Photos A and C by Nick Eyles; Photo B by Geological Survey of Canada, Terrain Science Image. http://gsc.nrcan.gc.ca/beaufort/images/ground_ice4.jpg. Reproduced with the permission of the Minister of Public Works and Government Services Canada, 2007 and Courtesy of Natural Resources Canada, Geological Survey of Canada. Photo D by Greg Brooks. GSC Photo 2002-704. Reproduced with the permission of the Minister of Public Works and Government Services Canada, 2005 and Courtesy of Natural Resources Canada, Geological Survey of Canada.

IN GREATER DEPTH 16.8

Permafrost in Urban Areas

There are many important engineering problems related to urban development in areas underlain by permafrost. In particular, development has to avoid disturbance and thawing of frozen ground. This can be accomplished by constructing roads or buildings on a pad of gravel or fill to insulate the permafrost from any heat released or absorbed by the structure. Structures may also be placed on piles above the ground to allow free movement of cold air and the dissipation of any heat released by the structure. Insulating matting, heat exchangers, and refrigeration units may also be used beneath buildings (box figure 1). The construction of bridges in frozen ground is a particular problem as pilings used for support may be subject to heaving or lifting as water freezes in the uppermost active layer each year. In most cases, bridges are built with deep pilings but alternative structures, such as large-diameter culverts, can also be used.

The provision of municipal services such as water supply and sewage removal is also a problem in urban areas underlain by permafrost. In Inuvik, a town built on the Mackenzie delta, N.W.T., services are placed in above-ground insulated aluminum boxes called *utilidors*, which link individual buildings to a central distribution system (box figure 2*A*). The cost of utilidor systems is high and is warranted only in larger communities. An alternative system is used in Dawson City, Yukon, where services are placed in underground trenches filled with coarse gravel (box figure 2*B*). The supply of water to communities in permafrost regions is a severe limitation to the growth of large urban areas. As wells are not productive in areas of deep permafrost, water must be obtained from surface lakes that do not freeze to the bottom in water. In several communities water is trucked from lakes by water tanker every few days. These lakes are susceptible to contamination and must be protected and monitored to maintain a potable water supply

Source: *French, "Living On Ice: Problems of Urban Development in Canada's North." In N. Eyles (Ed). 1997. Environmental Geology of Urban Areas. Geological Association of Canada, pp. 81–91.*

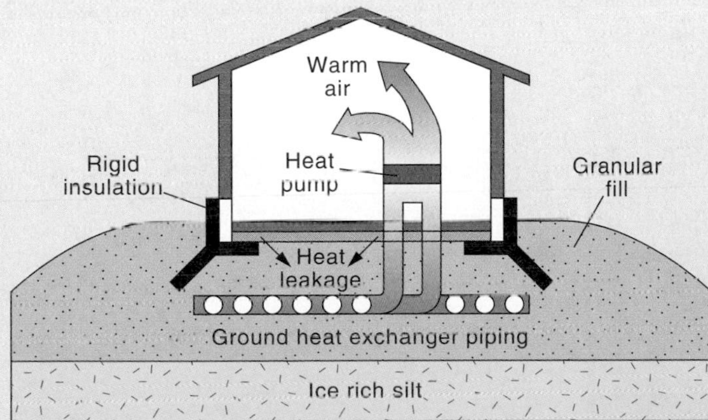

BOX 16.8 ■ FIGURE 1

Building designed to prevent melting of permafrost. Building rests on a thick gravel pad and incorporates heat-pump chilled foundations.

After French, 1997

A

B

BOX 16.8 ■ FIGURE 2

(*A*) Utilidors link each building in Inuvik and contain services such as water and sewage. (*B*) Municipal services in trenches excavated to depths of more than two metres and backfilled with coarse gravel.

Photos from French in Eyles, Environmental Geology of Urban Areas

FIGURE 16.40

Patterned ground in Spitzbergen. Coarse debris forms the outside margins of the polygons; fine sediment is concentrated in the centres.

Photo by Nick Eyles

A

B

C

FIGURE 16.41

(*A*) Bank collapse caused by thawing of permafrost along the banks of a stream in Spitzbergen. (*B*) Building affected by thermokarst (ground subsidence) in Dawson City, Yukon. (*C*) Irregular road surface caused by thawing of underlying permafrost.

Photos by Nick Eyles

FIGURE 16.42

The Alaska pipeline crosses over permafrost in central Alaska. Note the heat-radiating "fins" on top of support structures used to raise the pipeline above the ground.

HOW IS PERMAFROST AFFECTED BY CLIMATE CHANGE?

The effects of climate change are being felt particularly strongly in Canada's north, where much of the ground is permanently frozen (permafrost). A significant amount of this ground will likely thaw as a result of climate change, especially in areas close to the southern boundary of the permafrost zone (figure 16.38). Communities developing in Canada's far north will have to deal with the consequences of frozen ground being disturbed either by climate change or by construction activity.

The Mackenzie Valley of the Northwest Territories has experienced the greatest amount of warming in Canada over the past 100 years (almost 2°C), with 1998 being the warmest year on record in the far north, some 3 to 5 degrees above the long-term norm. Climate change has reduced by 50 percent the number of days the ground remains frozen, and increased temperatures have also caused the depth of summer thawing of permafrost to increase significantly (figure 16.43). In 1970 the ground remained frozen for about 210 days, allowing heavy equipment to move over the delicate surface of the permafrost; the season now starts much later and is only 100 days long. Permafrost below existing buildings and bridges is also at risk of thawing. Future construction activities in permafrozen terranes may have to use artificial refrigeration techniques (such as thermosyphons) to maintain stable conditions beneath built structures.

Land subsidence caused by ice melt allows the development of extensive thaw lakes (figure 16.44*A*) and causes extensive landslides (figure 16.44*B*), creating problems for road and pipeline construction and river traffic. Large areas of Canada's Arctic Coastal lowlands are above sea level only because of the presence of thick ice at depth. If this melts, many coastal communities will be threatened. Climate change also promotes forest fires, which destroy the insulating cover of organic material over frozen ground and enhance thawing. Changing climatic conditions also impact the construction of winter ice roads necessary to supply remote communities and mine sites (figure 16.45) and the maintenance of airstrips and some short-line rail operations, such as the OmniTRAX line to the Port of Churchill in Manitoba. Paved runways are particularly vulnerable to permafrost changes, as they readily absorb solar energy.

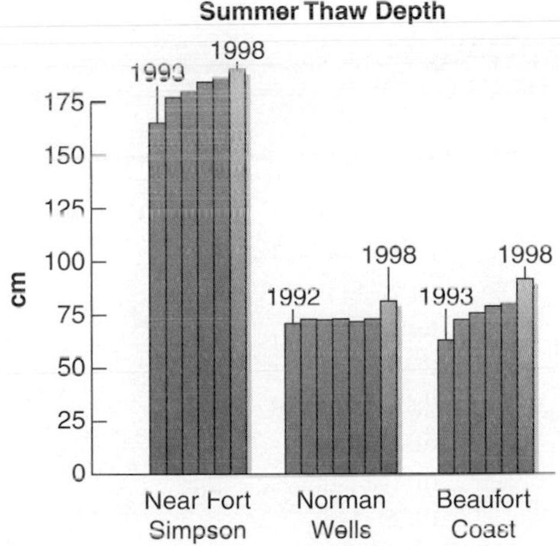

FIGURE 16.43

Thawing depth of permafrost over the period 1993 to 1998.

From Nixon, F.M., Geological Survey of Canada, pers. comm.; Wolfe, S.A., Kotler, E. and Nixon, F.M. 2000. Recent warming impacts in the Mackenzie Delta, Northwest Territories, and northern Yukon Territory coastal areas; Geological Survey of Canada, Current Research 2000-B1, 9 p. Reproduced with the permission of the Minister of Public Works and Government Services Canada, 2005 and Courtesy of Natural Resources Canada, Geological Survey of Canada.

A

B

FIGURE 16.44

(*A*) Thaw lakes in the Mackenzie Valley, Northwest Territories, created by the melt of underlying permafrost. (*B*) Landslides along the banks of the Mackenzie River triggered by recent thawing of permafrost.

Photos by Nick Eyles

FIGURE 16.45

Climate variability impacts the construction of winter ice roads necessary to supply remote communities and mine sites.

Photo courtesy of Diavik Diamond Mines Inc.

SUMMARY

A *glacier* is a large, perennial mass of ice that forms on land and moves under the influence of gravity. A glacier can form wherever more snow accumulates than is lost. *Ice sheets* and *valley glaciers* are the two most important types of glaciers. Glaciers move downward from where the most snow accumulates toward where the most ice is wasted.

A glacier moves both by basal sliding and by internal flow. The upper portion of a glacier tends to remain rigid and is carried along by the ice moving beneath it.

Glaciers advance and recede in response to changes in climate. A receding glacier has a *negative budget* and an advancing one has a *positive budget*. A glacier's budget for the year can be determined by noting the relative position of the *equilibrium line*.

Snow recrystallizes into firn, which eventually becomes converted to glacier ice. Glacier ice is lost (or ablated) by melting, by breaking off as icebergs, and by direct evaporation of the ice into the air.

A glacier erodes by plucking and the grinding action of the rock it carries. The grinding produces rock flour and faceted and polished rock fragments. Bedrock over which a glacier moves is generally polished, striated, and grooved.

A mountain area showing the erosional effects of alpine glaciation possesses relatively straight valleys with U-shaped cross-profiles. A glacial valley often has a *cirque* at its head and descends as a series of rock steps. Small *tarn lakes* are commonly found along the steps and in cirques.

A *hanging valley* indicates that a smaller tributary joined the main glacier. A *horn* is a peak between several cirques. *Arêtes* usually separate adjacent glacial valleys.

A glacier deposits unsorted rock debris or till, which contrasts sharply with the sorted and layered deposits of glacial outwash. Till forms *till plains*, *drumlins*, and various types of moraines.

Fine silt and clay may settle as *varves* in a lake in front of a glacier, each pair of layers representing a year's accumulation.

Multiple till deposits and other glacial features indicate several major episodes of glaciation during the late Cenozoic Era. During each of these episodes, large ice sheets covered most of northern Europe and northern North America, and glaciation in mountain areas of the world was much more extensive than at present. At the peak of glaciation about a third of the Earth's land surface was glaciated (in contrast to the 10 percent of the land surface currently under glaciers). Warmer climates prevailed during interglacial episodes.

The glacial ages also affected regions never covered by ice. Because of wetter climate in the past, large lakes formed in now-arid regions of the United States. Sea level was considerably lower.

Glacial ages also occurred in the more distant geologic past, as indicated by late Paleozoic and Precambrian tillites.

Permafrost is ground that has remained frozen for two or more years and can reach thicknesses of up to several hundred metres in high-latitude regions. *Ground ice*, *ice wedges*, *pingos*, and *patterned ground* are found in areas underlain by permafrost. *Thermokarst* is the subsidence and collapse of the ground surface caused by thawing of permafrost. Dealing with permafrost is a major issue for urban development in Canada's north.

Terms to Remember

ablation (wastage) 423
advancing glacier 423
alpine glaciation 420
arête 434
basal sliding 426
cirque 434
continental glaciation 420
crevasse 426
drumlin 438
end moraine 436
equilibrium line 424
erratic 435
esker 438
fjord 444
glacial trough 433
glacier 420
hanging valley 433
horn 434
iceberg 423
ice cap 421
ice sheet 421
ice kame 438
ice wedge 446
kane 438
kettle 438
lateral moraine 436
medial moraine 436
moraine 436
outwash 438
permafrost 445
pingo 446
plastic flow 426
pluvial lake 442
receding glacier 423
rigid zone 426
rock-basin lake (tarn) 433
rock flour 431
tarns 433
terminus 424
theory of glacial ages 420
thermokarst 447
till 435
till plain 436
tillite 445
truncated spur 433
valley glacier 421
varve 439
zone of ablation 424
zone of accumulation 424

Testing Your Knowledge

Use the questions below to prepare for exams based on this chapter.

1. How do erosional landscapes formed beneath glaciers differ from those that developed in rock exposed above glaciers?
2. How do features caused by stream erosion differ from features caused by glacial erosion?
3. How does material deposited by glaciers differ from material deposited by streams?
4. Why is the North Pole not glaciated?
5. How do arêtes, cirques, and horns form?
6. How does the glacial budget control the migration of the snow line?
7. How do recessional moraines differ from terminal moraines?
8. Alpine glaciation
   a. is found in mountainous regions
   b. exists where a large part of a continent is covered by glacial ice
   c. is a type of glacier
   d. none of the above
9. Continental glaciation
   a. is found in mountainous regions
   b. exists where a large part of a continent is covered by glacial ice
   c. is a glacier found in the subtropics of continents
   d. none of the above
10. At present about ______ % of the land surface of the Earth is covered by glaciers.
   a. 1/2 b. 1
   c. 2 d. 10
   e. 33 f. 50
11. Which is not a type of glacier?
   a. valley glacier b. ice sheet
   c. ice cap d. sea ice
12. The boundary between the zone of accumulation and the zone of ablation of a glacier is called the
   a. firn b. equilibrium line
   c. ablation zone d. moraine
13. Recently geologists have been drilling through ice sheets for clues about
   a. ancient mammals b. astronomical events
   c. extinctions d. past climates
14. Glacial troughs are usually ______ shaped.
   a. V b. U
   c. Y d. all of the above
15. Which is not a type of moraine?
   a. medial b. end
   c. terminal d. recessional
   e. ground f. esker
16. The last episode of extensive glaciation in North America was at its peak about ______ years ago.
   a. 2,000 b. 5,000
   c. 10,000 d. 18,000
17. How fast does the central part of a valley glacier move compared to the sides of the glacier?
   a. faster b. slower
   c. at the same rate
18. What are thought to be the causes of climate change that lead to ice ages?
19. An ice-cored hill is called a
   a. varve b. talus
   c. esker d. pingo
20. The oldest glacial rocks on Earth are:
   a. 28,000 years old b. 28 million years old
   c. 280 million years old d. 2.8 billion years old
21. Draw a N–S cross-section across Canada to show the distribution of permafrost.
22. Why is climate change a problem in northern regions of Canada underlain by permafrost?

Exploring Web Resources

www.mcgrawhill.ca/olc/plummer
Visit our Online Learning Centre for some great animations of glacier development and movement. Check your answers for the Testing Your Knowledge section and try some additional interactive quizzing to further your understanding. This site also provides you with direct links to the sites listed below.

http://nsidc.org/sotc/permafrost.html
National Snow and Ice Data Center. State of the Cryosphere: Permafrost, Insights from a New Northern Hemisphere Map.

http://atlas.gc.ca/site/english/maps/environment/land/permafrost
National Resources Canada's *Atlas of Canada* illustration of permafrost in Canada.

www.geology.gov.yk.ca/publications/summaries/permafrost.html
A discussion of permafrost, Yukon Geological Survey. The site includes a good reference list.

http://gsc.nrcan.gc.ca/permafrost/index_e.php
The Geological Survey of Canada's Website for permafrost research.

http://dir.yahoo.com/Science/Earth_Sciences/Geology_and_Geophysics/Glaciology/

Glaciers and Glaciology—list of sites. This site provides links and descriptions of numerous icy Websites.

www.glacier.rice.edu/

Glacier. Explore Antarctica on Rice University's site. Go to "Ice." There are many topics you can go to for information that expands upon that covered in this book. Examples are "How Do Glaciers Move," "How Do Glaciers Change the Land," "What Causes Ice Ages."

www.crevassezone.org/

Glacier velocity and surface elevation research on the Juneau Icefield, Alaska. Go to "Photo Gallery" to view photos of glacial features and other aspects of the project.

www.museum.state.il.us/exhibits/ice_ages/

Ice Ages. Illinois State Museum's virtual ice ages exhibit. The site features a tape clip showing the retreat of glaciers during the last ice age. You can download the video clip by going to:
www.museum.state.il.us/exhibits/ice_ages/laurentide_deglaciation.html

www-nsidc.colorado.edu/NSIDC/EDUCATION/

National Snow and Ice Data Center's Education Resources Site. General information on snow and ice. You can link to pages on glaciers, avalanches, icebergs.

www.canadiangeographic.ca/mountains/glaciers.asp

A *Canadian Geographic* article on the glaciers of Canada.

Animations

This chapter includes the following animations available on our Online Learning Centre at www.mcgrawhill.ca/olc/plummer.

16.3 Cross section of an Ice Sheet
16.6 Dynamics of Glacial Advance and Retreat
16.9 Crevasse Formation in Glaciers
16.28 Formation of Glacial Features by Deposition at a Wasting Ice Front

CHAPTER

Deserts and Wind Action

Where do deserts form?
What are the characteristics of deserts?
How does wind erode and transport sediment?
What features result from wind deposition?

In chapters 13 through 16 you have seen how the land is sculptured by mass wasting; streams and floods; groundwater; and glaciers, glaciation, and permafrost. Here we discuss the fifth agent of erosion and deposition: wind. Deserts and wind action are discussed together because of the wind's particular effectiveness in dry regions. But wind erosion and deposition can be very significant in cold and semi-arid regions as well. Canada has large desert areas in its arctic regions, and wind is an important agent of erosion and deposition in semi-arid areas such as the Prairies.

The word *desert* may suggest shifting sand dunes. Although moviemakers usually film sand dunes to represent deserts, only small portions of most deserts are covered with dunes. Actually, a **desert** is any region with low rainfall. A region is usually classified as a desert if it has a dry or *arid climate* with less than 25 centimetres of rain per year. Few plants can tolerate low rainfall, so most deserts look barren (see figure 17.1 and box 17.1).

Active sand dunes of the Great Sand Hills, southwestern Saskatchewan. Stabalized dunes can be seen in the background.
Photo by Stephen Wolfe

FIGURE 17.1
A scene from the Mojave Desert in southern California showing widely spaced plants that have adapted to less than 25 centimetres of rain per year.
Photo by Diane Carlson

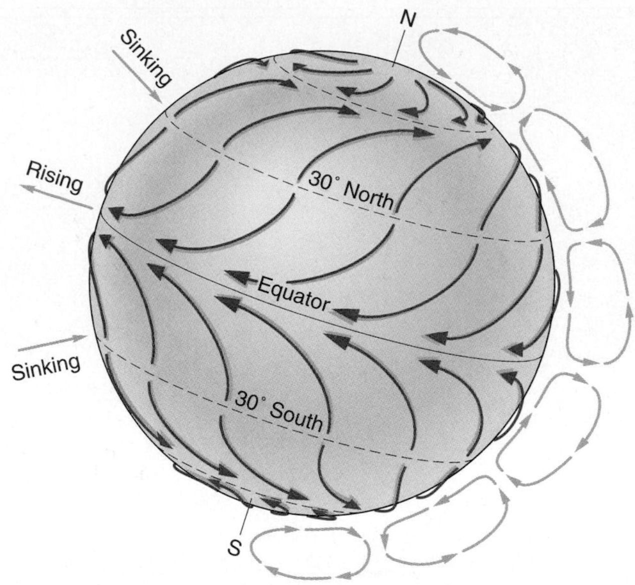

FIGURE 17.2
Global air circulation. Red arrows show surface winds. Blue arrows show vertical circulation of air. Air sinks at 30°N and 30°S latitude and at the poles.

Some specialized types of plants, however, grow well in desert climates despite the dryness. These plants are generally salt-tolerant, and they have extensive root systems to conserve water, so they often are widely spaced (figure 17.1). The leaves are usually very small, minimizing water loss by transpiration; they may even drop off the plants between rainstorms. During much of the year many desert plants look like dead, dry sticks. When rain does fall on the desert the plants become green, and many will bloom.

WHERE DO DESERTS FORM?

The location of most deserts is related to descending air. The global pattern of air circulation is shown in simplified form in figure 17.2. The equator receives the sun's heat more directly than the rest of the Earth. Air warms and rises at the equator, then moves both northward and southward to sink near 30° North latitude and 30° South latitude. The world's best-known deserts lie in a belt 10–15° wide centred on 30° North and South latitude (figure 17.3*A*).

Air sinking down through the atmosphere is compressed by the weight of the air above it. As air compresses it warms up and as it warms it is able to hold more water vapour. Evaporation of water from the land surface into the warm, dry air is so great under belts of sinking air that moisture seldom falls back to Earth in the form of rain. The two belts at 30° North and South latitude characteristically have clear skies, much sunshine, little rain, and high evaporation.

Not all deserts are hot. The cold, descending air near the North and South Poles (figure 17.2) creates *polar deserts* that have an arid climate along with a snow or ice cover. The entire continent of Antarctica is a desert, as are most of Greenland and the northernmost parts of Canada, Alaska, and Siberia. Most of the Canadian Arctic Island region is classified as polar desert, as it receives less than 20 centimetres of precipitation per year (figure 17.3*B*).

In contrast to the belts of descending air centred on 30° and at the poles, the equator is marked by rising air masses that expand and cool as they rise. In cooling, the air loses its moisture, causing cloudy skies and heavy precipitation. Thus a belt of high rainfall at the equator separates the two major belts of deserts.

Not all deserts lie near the 30° latitude belts. Some of the world's deserts are the result of the **rain shadow** effect of mountain ranges (figure 17.4). As moist air is forced up to pass over a mountain range it expands and cools, losing moisture as it rises. The dry air coming down the other side of the mountain compresses and warms, bringing high evaporation with little or no rainfall to the downwind side of the range. This dry region downwind of mountains is the *rain-shadow zone.* The desert regions of the south Okanagan Valley of British Columbia lie in the rain-shadow zone of the coastal mountain ranges (see figure 17.4). Parts of the southwestern United States desert in Nevada and northern Arizona are largely the result of the rain-shadow effect of the Sierra Nevada range in eastern California.

Great distance from the ocean is another factor that can create deserts, since most rainfall comes from water evaporated from the sea. The dry climate of the large arid regions in China, well north of 30° North latitude, is due to their location in a continental interior and to the rain-shadow effect of mountains such as the Himalayas.

Deserts also tend to develop on tropical coasts next to *cold ocean currents.* Cold currents run along the western edges of continents, cooling the air above them. The cold marine air warms up as it moves over land, causing high evaporation and little rain on the coasts. This effect is particularly pronounced on the Pacific coast of South America and the Atlantic coast of Africa, and to a lesser extent in Western Australia.

ENVIRONMENTAL GEOLOGY 17.1

Expanding Deserts

Many geologists and geographers use a two-part definition of *desert.* A desert must have less than 25 centimetres of rain per year or must be so devoid of vegetation that few people can live there. Many dry regions have supported marginally successful agriculture and moderate human populations in the past but are being degraded into barren deserts today by overgrazing, overpopulation, and water diversions. The expansion of barren deserts into once-populated regions is called *desertification.*

Limited numbers of people can exist in dry regions through careful agricultural practices that protect water sources and limit grazing of sparse vegetation. Overuse of the land by livestock and humans, however, can strip it bare and make it uninhabitable. The large desert in northern Africa, shown in figure 17.3*A*, is the Sahara Desert, and the semiarid region (25 to 50 centimetres of rain per year) to the south of it is the Sahel. In the early 1960s, a series of abnormally wet years encouraged farmers in the Sahel to expand their herds and grazing lands.

A severe drought throughout the 1970s and 1980s caused devastation of the plant life of the region as starving livestock searched desperately for food, and humans gathered the last remaining sticks for firewood (box figure 1). Vast areas that were once covered with trees and sparse grass became totally barren, and an acute famine began, killing more than 100,000 people. The desert expanded southward, advancing in some places as much as 50 kilometres per year. The denuded soil in many regions became susceptible to wind erosion, leading to choking dust storms and new, advancing dune fields (some even migrating into cities).

Some of the same problems afflicted the midwestern United States and Canada in the 1930s, as intense land cultivation coupled with a prolonged drought produced the barren Dust Bowl during the time of the Great Depression. Renewed rains and improved soil-conservation practices have reversed the trend in the United States, but the area is still vulnerable to a future drought, and the possibility of future dust storms in prairie provinces and states is very real.

Drought accelerates desertification but is not necessary for it to occur. Overloading the land with livestock and humans can strip marginal regions of vegetation even in wet years.

Diversion of rivers for agricultural use can also cause desertification. Such is the case in the Turkestan Desert where two rivers feeding the Aral Sea (figure 17.3*A*) were diverted to provide irrigation water for agriculture. The Aral Sea, once the fourth-largest inland water body in the world, has decreased in size by nearly half since 1960. Fishing boats are now marooned in a sea of sand as the shoreline has migrated tens of kilometres with the sea continuing to shrink (box figure 2).

Additional Resource

http://pubs.usgs.gov/gip/deserts/desertification/

Good overview of desertification around the world. Includes photographs.

BOX 17.1 ■ FIGURE 1
Desertification in Africa after a period of drought has left inhabitants desperately searching for water.
Photo © Walt Anderson/Visuals Unlimited

BOX 17.1 ■ FIGURE 2
Fishing boats marooned in a sea of sand as the Aral Sea decreases in size due to diversion of water for irrigation.
Photo © David Turnley/Corbis Images

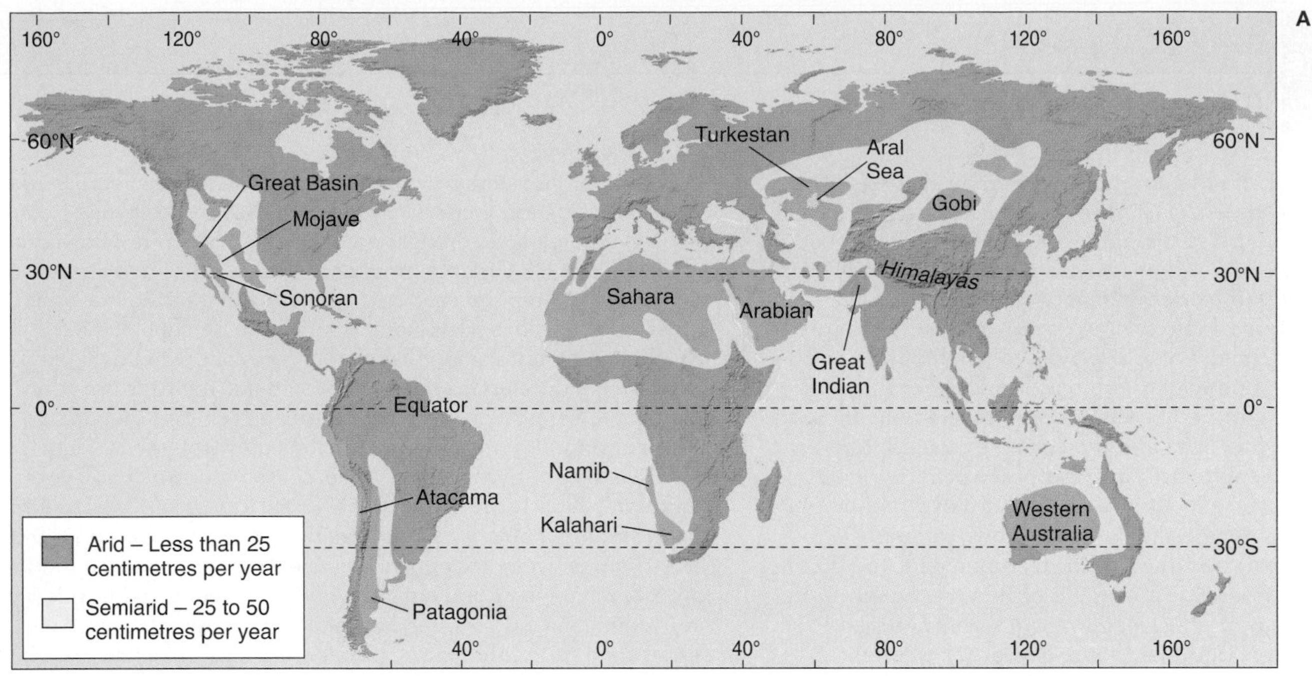

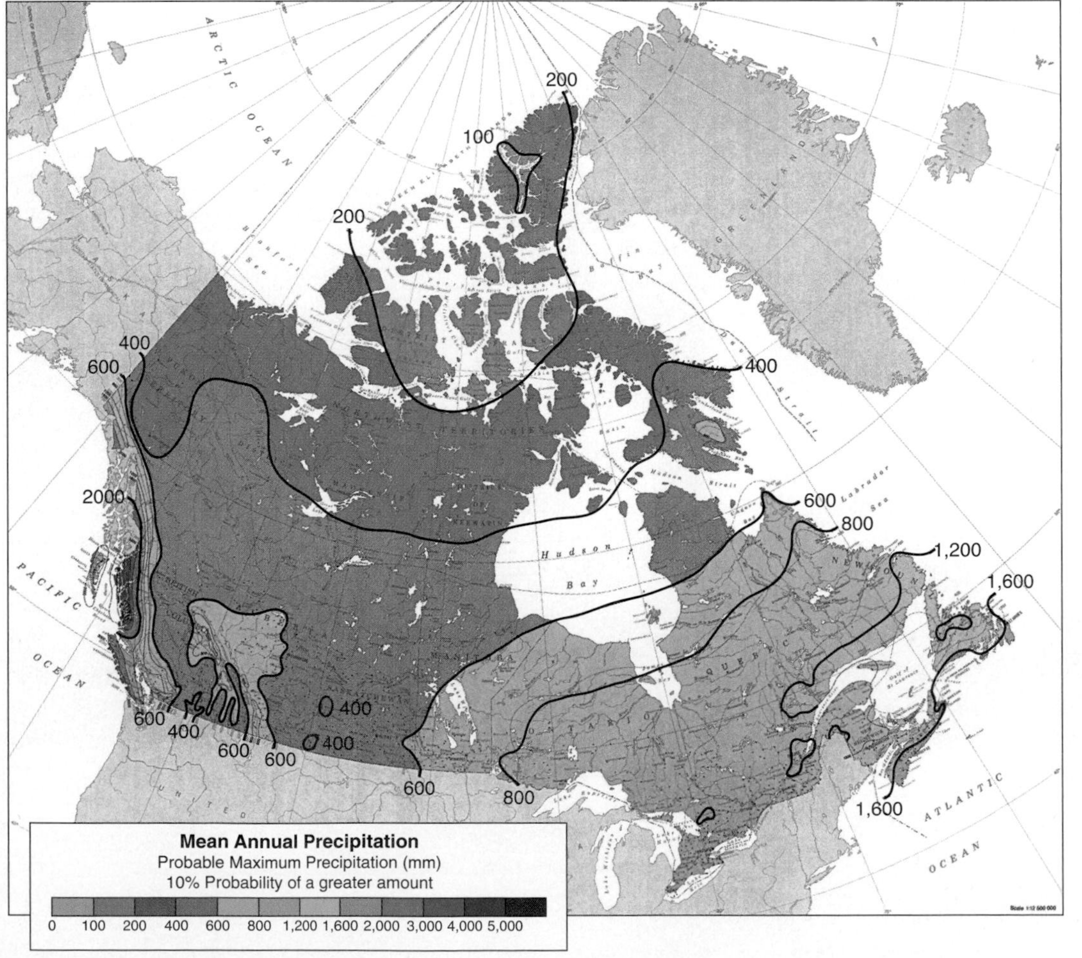

FIGURE 17.3

(A) World distribution of nonpolar deserts. Most deserts lie in two bands near 30°N and 30°S.

Map adapted from U.S. Department of Agriculture.

(B) Mean annual precipitation map of Canada. Large areas of Canada's Arctic Islands are deserts that receive less than 200 mm annual precipitation.

Original map data provided by The Atlas of Canada http://atlas.gc.ca/© 2006. Produced under license from her Majesty the Queen in Right of Canada, with permission of Natural Resources Canada.

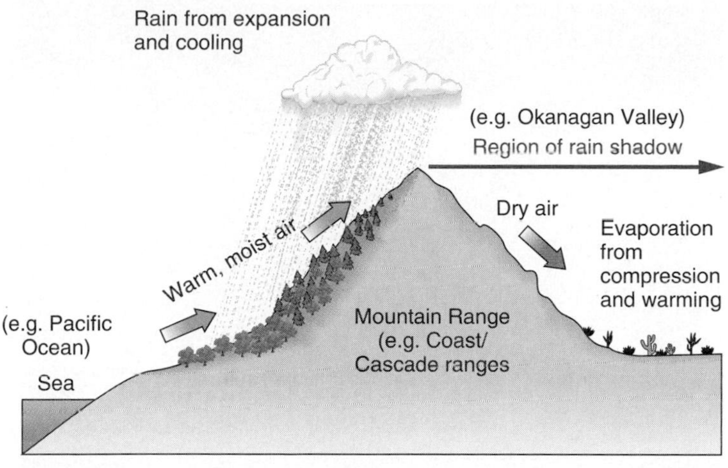

FIGURE 17.4

Rain shadow causes deserts on the downwind side of mountain ranges. Prevailing winds are from left to right.

WHAT ARE THE CHARACTERISTICS OF DESERTS?

Because of their low rainfall, deserts have characteristic drainage and topography that differ from those of humid regions. Desert streams usually flow intermittently. Water runs over the surface after storms, but during most of the year stream beds are dry. As a result, most deserts *lack through flowing streams*. The Colorado River in the southwestern United States and the Nile River in Egypt are notable exceptions. Both are fed by heavy rainfall in distant mountains. The runoff is great enough to sustain stream flow across dry regions with high evaporation.

Many desert regions have *internal drainage;* the streams drain toward landlocked basins instead of toward the sea. The surface of an enclosed basin acts as a local base level. Because each basin is generally filled to a different level than the neighbouring basins, desert erosion may be controlled by many different *local base levels*. As a basin fills with sediment its surface rises, leading to a *rising base level*—a rare situation in humid (wet) regions.

The limited rainfall that does occur in deserts often comes from violent thunderstorms, with a high volume of rain falling in a very short time. Desert thunderstorms may dump more than 13 centimetres of rain in one hour. Such a large amount of rain cannot soak readily into the sun-baked caliche soil, so the water runs rapidly over the land surface, particularly where vegetation is sparse. This high runoff can create sudden local floods of high discharge and short duration called **flash floods**. Flash floods are more common in arid regions than in humid regions. They can turn normally dry stream beds into raging torrents for a short time after a thunderstorm. Because soil particles are not held in place by plant roots, these occasional floods can effectively erode the land surface in a desert region. As a result, desert streams normally are very heavily laden with sediment. Flash floods can easily erode enough sediment to become *mudflows* (see chapter 13).

Desert stream channels are distinctive in appearance because of the great erosive power of flash floods and the intermittent nature of streamflow. Most stream channels are normally dry and covered with sand and gravel that is moved only during occasional flash floods. Rapid downcutting by sediment-laden floodwaters tends to produce narrow canyons with vertical walls and flat, gravel-strewn floors (figure 17.5*A*). Such channels are often called *arroyos* or *dry washes*.

Newcomers to deserts sometimes get into serious trouble in desert canyons in rainy weather. Imagine for a moment that you have camped on the canyon floor in figure 17.5*A* to get out of the strong desert winds. Later that night a towering thunderhead cloud forms, and heavy rain falls on the mountains several kilometres upstream from you. Although no rain has fallen *on* you, you are awakened several minutes later by a distant roar. The roar grows louder until a 3-metre "wall" of water rounds a bend in the canyon, heading straight for you at the speed of a galloping horse. Boulders, brush, and tree trunks are being swept along in this raging flash flood. The walls of the canyon are too steep to climb. Several hikers died during the summer of 1997 when such a wall of water roared down a side canyon in Grand Canyon National Park in Arizona. In August 2004, two people died in Death Valley National Park in California when their vehicle was carried away by a flash flood that washed out most of the roads in the park. Stay out of desert canyons if there is any sign of rain; sleeping in such canyons is particularly dangerous!

The resistance of some rocks to weathering and erosion is partly controlled by climate. In a humid (wet) climate limestone dissolves easily, forming low places on the Earth's surface. In a desert climate the lack of water makes limestone resistant, so it stands up as ridges and cliffs in the desert just as sandstone and conglomerate do. Lava flows and most igneous and metamorphic rock are also resistant. Shale is the least resistant rock in a desert, so it usually erodes more deeply than other rock types and forms gentler slopes or badland topography (figure 17.5*B*).

Although intersecting joints form angular blocks of rock in all climates, desert topography characteristically looks more angular than the gently rounded hills and valleys of a humid region. This may be due indirectly to the low rainfall in deserts. Shortage of water slows chemical weathering processes to the point where few minerals break down to form fine-grained clay minerals. Soils are coarse and rocky, with few chemically weathered products. Plants, which help bind soil into a cohesive layer in humid climates, are rare in deserts, and so desert soils are easily eroded by wind and rainstorms. Downhill creep of thick, fine-grained soil is partly responsible for softening the appearance of jointed topography in humid climates. With thin, rocky soil and slow rates of creep, desert topography remains steep and angular.

Polar Deserts

Much of northern Canada, including the Canadian Arctic Islands, has an arid climate with average annual precipitation values of less than 250 mm (figure 17.3*B*) and mean temperatures in the warmest months of less than 10°C. These areas are considered

A

B

FIGURE 17.5

Desert features. (*A*) Desert stream channel showing dry, gravel-covered floor and steep, vertical sides (cut in sandstone and conglomerate) in Death Valley, California. (*B*) Badland topography (steep V-shaped gullies, sharp ridges) eroded into shale, coal, and sandstone near Drumheller, Alberta. Areas underlain by shale erode more easily and form more gently sloping surfaces.

Photo B by S.Dickie

to be **polar deserts** and are characterized by low-relief bedrock or gravel plains underlain by permanently frozen ground (figure 17.6*A*). Cold temperatures, low precipitation, high winds, and poorly developed soils in polar deserts allow growth of only sparse, dwarf vegetation.

Polar desert conditions affect almost 5 million square kilometres of the Earth's surface, mostly in northern regions of Canada, Siberia, and on the Antarctic continent. In polar desert regions of Canada mean summer temperatures range from 0.5°C in the north to 4.5°C in the south, and winter temperatures range from −30°C to −20°C. Snow usually covers the ground for 10 months of the year. Some of the driest areas in Canada are found in polar desert areas of Ellesmere Island. Quttinirpaaq National Park is a polar desert covering an area of almost 38,000 km^2 on northern Ellesmere Island and includes the Hazen Plateau and Mount Barbeau, the highest mountain in eastern North America at 2,616 m.

The polar desert landscape has many similarities to landscapes found in the American southwest and to those observed on Mars (see online chapter 21). In fact, NASA has a field-based research program based in polar desert regions of Nunavut's Devon Island, as this area has many similar geological and biological features to those on Mars and serves as an excellent terrestrial analogue for scientific investigations and astronaut training (figure 17.6*B*).

Desert Landforms and Features

Not all landforms and features found in deserts are unique to deserts. Many of these features are found in other climatic zones but are particularly well exposed in deserts because of thin soil and sparse vegetation. The type and structure of underlying bedrock has a significant influence on the types of land forms that develop in arid regions.

In areas underlain by flat-lying beds of sedimentary rock, resistant rock layers such as sandstone, limestone, and lava flows form **plateaus**—broad, flat-topped areas elevated above the surrounding land and bounded, at least in part, by cliffs (figure 17.7). The western parts of the Canadian Arctic Islands are underlain by flat-lying Paleozoic and Mesozoic sedimentary

FIGURE 17.6

Polar desert. (*A*) Arid, frost-shattered landscape of Devon Island, Nunavut. (*B*) NASA scientist using polar desert landscape of Haughton Crater, Devon Island, as an analogue for geological conditions of Mars.

Photo A and B from Mars Institute/Haughton Mars

rocks that form extensive plateaus deeply dissected by glacially eroded troughs (figure 17.7*A*). In the southwestern United States, the Colorado Plateau is also underlain by sedimentary rocks that are well exposed in the Grand Canyon and Monument Valley, Arizona (figure 17.7*B*). Remnants of resistant rock layers may be left behind, forming flat-topped mesas or narrow buttes (figure 17.7*C*). A **mesa** is a broad, flat-topped hill bounded by cliffs and capped with a resistant rock layer. A **butte** is a narrow hill of resistant rock with a flat top and very steep sides. Most buttes form by continued erosion of mesas.

The Colorado Plateau is also marked by peculiar steplike folds called *monoclines.* Erosion of monoclines (and other folds) leaves resistant rock layers protruding above the surface as ridges (figure 17.8). A steeply tilted resistant layer erodes to form a *hogback,* a sharp ridge that has steep slopes. A gently tilted resistant layer forms a *cuesta,* with one steep side and one gently sloping side.

Mountain ranges in arid and semi-arid regions are subject to intense erosion caused by heavy rainfall from occasional thunderstorms or spring melt. Rock debris from the mountains, picked up by flash floods and mudflows, is deposited at the base of steep mountain slopes in the form of alluvial fans (figure 17.9). Alluvial fans (described in chapter 14) build up where stream channels abruptly widen as they flow out of narrow canyons onto the open valley floors, causing a decrease in velocity and rapid deposition of sediment.

If no outlet drains the valley, runoff water may collect and form a **playa lake** on the valley floor. Playa lakes are usually very shallow and temporary, lasting for only a few days after a rainstorm. After the lake evaporates, a thin layer of fine mud may be left on the valley floor. The mud dries in the sun, forming a **playa,** a very flat surface underlain by hard, mud-cracked clay (figure 17.10). If the runoff contained a large amount of dissolved salt or if seeping groundwater brings salt to the surface, the flat playa surface may be underlain by a bright white layer of dried salt instead of mud, as on the Bonneville Salt Flats in Utah (see figure 9.26). Playas are common on the floors of faulted valleys and basins of the Basin and Range Province of the southwestern United States.

Continued deposition near the base of the mountains may form a **bajada,** a broad, gently sloping depositional surface formed by the coalescing of individual alluvial fans (figure 17.11). Erosion of the mountain can eventually form a **pediment,** which is a gently sloping surface, commonly covered with a veneer of gravel, cut into the solid rock of the mountain (figure 17.11). A pediment develops uphill from a bajada as the mountain front retreats. It can be difficult to distinguish a pediment from the surface of the bajada downhill, because both have the same slope and gravel cover. The pediment, however, is an erosional surface, usually underlain by solid rock, while the bajada surface is depositional and may be underlain by hundreds of metres of sediment.

A

FIGURE 17.7

Characteristic landforms of arid regions underlain by flat-lying sedimentary rocks. (*A*) Flat-topped plateau in sedimentary rocks behind the McGill High Arctic Research Station on Axel Heiberg Island, Nunavut. (*B*) Mesas and buttes in Monument Valley, Arizona, an area of eroded, horizontal, sedimentary rocks. (*C*) Erosional retreat of a cliff at the edge of a plateau can leave behind mesas and buttes as erosional remnants of the plateau.

Photo A © Lyle G. Whyte; Photo B by David McGeary

B

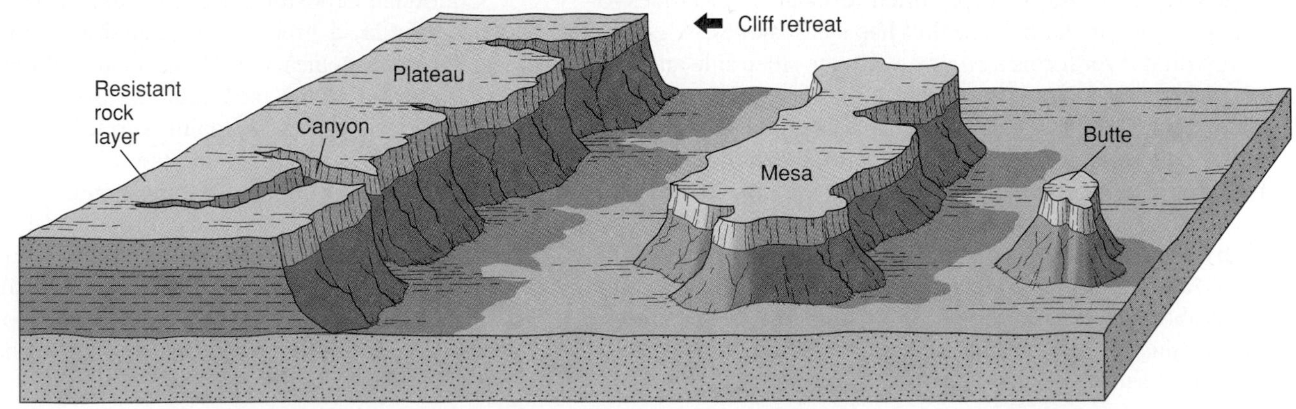

C

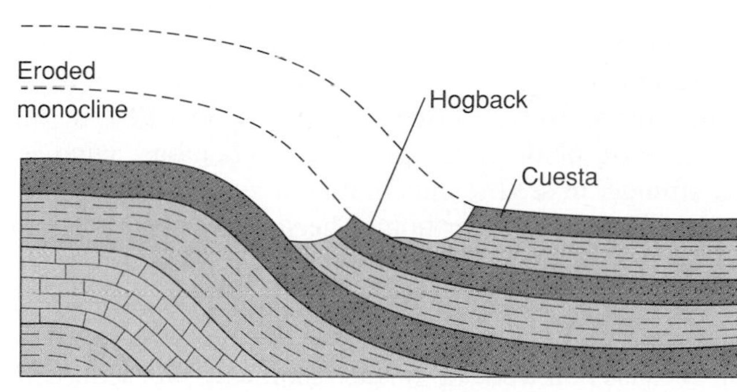

A

B

FIGURE 17.8

(*A*) Steplike monocline folds often erode so that resistant rock layers form hogbacks and cuestas (these features are not unique to deserts).
(*B*) Monocline near Mexican Hat, Utah. Rocks at the top and bottom are horizontal and rocks in the centre are steeply tilted and step down toward the right.
Photo © Marli Miller/Visuals Unlimited

A

B

FIGURE 17.9

(*A*) Alluvial fan formed at the base of a narrow canyon in Quttinirpaaq National Park, Ellesmere Island. (*B*) Basin and Range topography in Death Valley, California. In the distance the fault-bounded Panamint Mountains rise more than 3 kilometres above Death Valley. Giant alluvial fans at the base of the mountains show a braided stream pattern. Fine-grained sediments and salt deposits underlie the playa in the foreground.
Photo A by AIVA — Paul Nopper

FIGURE 17.10

Mud-cracked playa surface.

Photo © Bill Ross/Westlight/Corbis

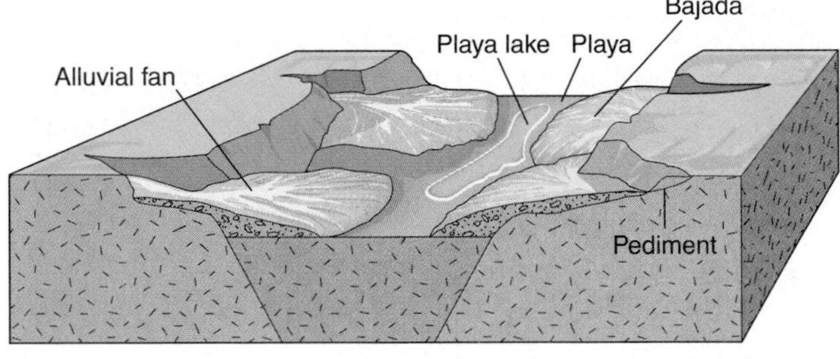

FIGURE 17.11

Landforms developed on the margins and floors of desert basins and valleys.

HOW DOES WIND ERODE AND TRANSPORT SEDIMENT?

Wind Action

Wind can be an important agent of erosion and deposition in any climate, as long as sediment particles are loose and dry. Wind differs from running water in two important ways. Because air is less dense than water, wind can erode only fine sediment—sand, silt, and clay. But wind is not confined to channels as running water is, so wind can have a widespread effect over vast areas.

In general, the faster the wind blows, the more sediment it can move. Wind velocity is determined by differences in air pressure caused by differences in air temperature. As air warms and cools it changes density, and these density changes create air pressure differences that cause wind. Wet climates and cloud cover help buffer changes in air temperature, but in dry climates daily temperature changes can be extreme. In a desert, the temperature may range from 10°C at night to more than 40°C in the daytime. Because of these temperature fluctuations, wind is generally stronger in deserts than in humid regions, commonly exceeding 100 kilometres per hour. The scarcity of vegetation in deserts to slow wind velocity by friction increases the effectiveness of desert winds. Although strong winds are also associated with rainstorms and hurricanes, these winds seldom erode sediment because rain wets the surface sediment. Wet sediment is heavy and cohesive and will not be blown away. Strong winds in the desert, however, blow over loose, dry sediment, so wind is an effective erosional agent in dry climates.

Wind Erosion and Transportation

Thick, choking *dust storms* are one example of wind action (figure 17.12). "Dust Bowl" conditions in the 1930s in the agricultural prairie states lasted for several years (the Dirty Thirties) due to droughts and poor soil-conservation practices (for more on drought conditions in the Canadian Prairies, see box 17.2). Loose silt and clay are easily picked up from barren dry soil, such as in a cultivated field. Wind erosion is even greater if the soil is disturbed by animals or vehicles. Silt and clay can remain suspended in turbulent air for a long time, so a strong wind may carry a dust cloud hundreds of metres upward and hundreds of kilometres horizontally. Dust storms of the 1930s frequently blacked out the midday sun, fertile soil was lost over vast regions ruining many farms, and streets and rivers downwind were filled with thick dust deposits.

Wind-blown sediment is sometimes picked up on land and carried out to sea (figure 17.13). Particles from the Sahara Desert in Africa have been collected from the air over the islands of the West Indies after having been carried across the Atlantic Ocean. A substantial amount of the fine-grained sediment that settles to the sea floor is wind-transported sediment. Ships 800 kilometres offshore have reported dustfalls a few millimetres thick covering their decks.

Volcanic ash can be carried by wind for very great distances. An explosive volcanic eruption can blast ash more than 15 kilometres upward into the air. Such ash may be caught in the high-altitude *jet streams,* narrow belts of strong winds with velocities sometimes greater than 300 kilometres per hour. Following the 1980 eruption of Mount St. Helens in western Washington, a visible ash layer to the east blanketed parts of Washington, Idaho, and Montana. At high altitudes, St. Helens ash could be detected blowing over New York and out over the Atlantic Ocean, 5,000 kilometres from the volcano.

Because sand grains are heavier than silt and clay, sand moves close to the ground in the leaping pattern called *saltation* (as does some sediment in streams). High-speed winds can cause

A

B

FIGURE 17.12

(*A*) Prolonged drought in the years 1929 to 1937, combined with poor agricultural practices, caused extensive wind erosion of dry soils on the Canadian Prairies. (*B*) A wall of dust approaches a town in Kansas in October, 1935. Because of the intensity and duration of the storms in the 1930s, parts of the Great Plains became known as the "Dust Bowl."

Photo A 3742 appears courtesy of the Provincial Archives of Alberta; Photo B by National Oceanic and Atmospheric Administration/Department of Commerce

FIGURE 17.13

Satellite image of a dust storm from the Sahara Desert blowing off the coast of Africa northwestward out into the Atlantic Ocean on February 26, 2000. The thick plume of dust is about the size of Spain, and dust particles from this storm were blown all the way to the west side of the Atlantic. Such storms are fairly common in the Sahara Desert and are the world's greatest supplier of dust

Photo by NASA/Goddard Space Flight Center, The SeaWiFS Project and ORBIMAGE, Scientific Visualization Studio

sandstorms, clouds of sand moving rapidly near the land surface. The moving sand in such a storm can sandblast smooth surfaces on hard rock and scour the windshields and paint of automobiles. Because of the weight of the sand grains, however, sand rarely rises more than 1 metre above a flat land surface, even under extremely strong winds. Therefore, most of the sandblasting action of wind occurs close to the ground (figure 17.14). Telephone poles in regions of wind-driven sand often are severely abraded near the ground. To prevent such abrasion, desert residents pile stones or wrap sheet metal around the base of the poles.

Wind seldom moves particles larger than sand grains, but wind blown sand may sculpt isolated pebbles, cobbles, or boulders into **ventifacts**—rocks with flat, wind-abraded surfaces (figure 17.15). If the wind direction shifts, or if the stone is turned, more than one flat face may develop on the ventifact.

Deflation

The removal of clay, silt, and sand particles from the land surface by wind is called **deflation.** If the sediment at the land surface is made up only of fine particles, the erosion of these particles by the wind can lower the land surface substantially. A **blowout** is a depression on the land surface caused by wind erosion (figure 17.16). A *pillar,* or erosional remnant of the former land, may be left at the centre of a blowout (figure 17.16).

Blowouts are common in the Great Plains states (figure 17.16). One in Wyoming measures 5 by 15 kilometres and is 45 metres deep. The enormous Qattara Depression in northwestern Egypt, more than 250 kilometres long and more than 100 metres *below* sea level, has been attributed to wind deflation. Deflation can continue to deepen a blowout in fine-grained sediment until it reaches wet, cohesive sediment at the water table.

ENVIRONMENTAL GEOLOGY 17.2

Drought on the Canadian Prairies

Drought conditions, characterized by critically low water supply as a result of below-normal precipitation, occur one in every three years on average on the Canadian Prairies. Historical accounts of droughts on the Prairies extend back to the 1800s, with the most severe drought conditions occurring during the 1930s (popularly known as the Dust Bowl or the Dirty Thirties); the most recent major drought occurred in 2001–02. The area of the Prairies most vulnerable to drought is referred to as Palliser's Triangle (named after explorer Captain John Palliser), a dry area lying within the points of Cartwright, Manitoba; Lloydminster, Saskatchewan; and Calgary, Alberta (box figure 1). This area receives annual precipitation amounts of less than 400 mm.

Between 1929 and 1937, drought conditions affected more than 7.3 million hectares of arable land (about 25% of Canada's total arable land) in the Prairie region. Severe wind erosion of the cultivated and exposed, sandy soil resulted in loss of soil, organic matter, and nutrients (see figure 17.12*A*). The productivity of the land declined rapidly, and many farms were abandoned. However, many lessons were learned from this harsh experience, and government agencies such as the Prairie Farm Rehabilitation Administration (PFRA) were established to help farmers develop and implement more effective farming strategies. Drought alleviation programs were also developed, and financial and technical assistance was provided to build water storage reservoirs including dugouts and small dams.

The severe and extensive drought of 2001–02, which caused more than $3.6 billion in losses in Canadian agricultural production, was a timely reminder that rural water supply is a persistent problem on the Prairies (box figure 2). This most recent drought raised a number of important questions regarding the sustainability of water supplies and the potential impact of future climate warming. Water resources on the Prairies are vulnerable to fluctuations in supply caused by changes in the rain/snow mix, the timing of precipitation events, melting of glacial source areas, increased severe weather events, and increased rates of evaporation. Degradation of water quality through human activities such as surface and groundwater contamination, or irrigation causing salinization, also impacts water supply. Demands on water resources are also changing, with increasing population pressures and increased need for irrigation waters. The threat of drought is ever-present on the Canadian Prairies and demands thorough understanding and diligent protection of invaluable surface and groundwater resources.

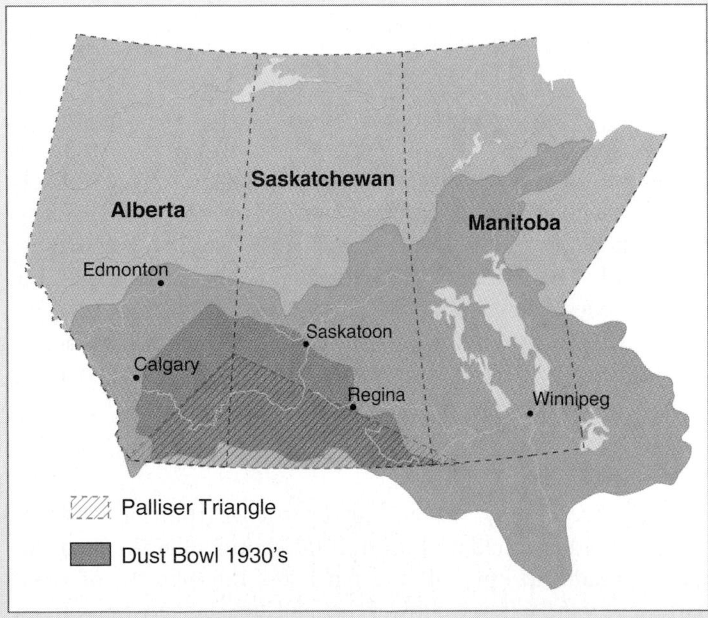

BOX 17.2 ■ FIGURE 1

Palliser's Triangle is one of the driest regions in the Canadian Prairies and was severely affected by drought during the "Dust Bowl" conditions of the 1930s.

"Palliser's Triangle Map of Precipitation compared to Historical Distribution" Agriculture and Agri-Food Canada. Reproduced with permission of the Minister of Public Works and Government Services Canada, 2007.

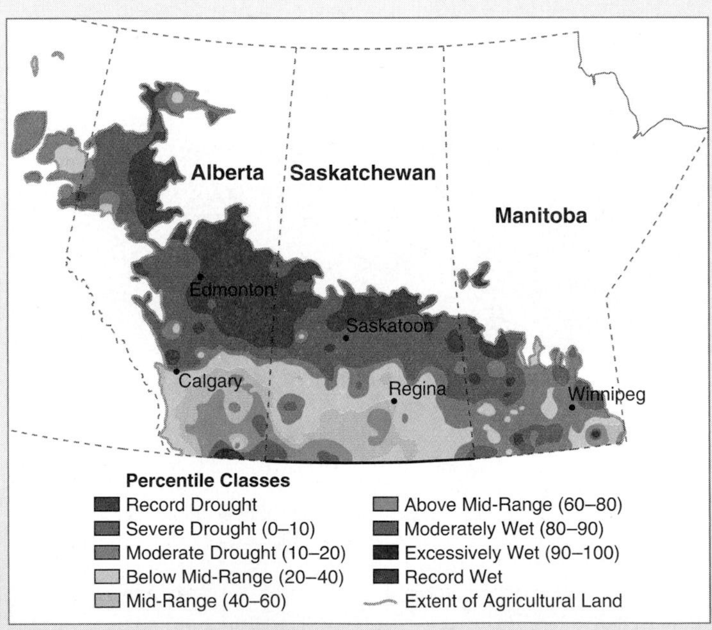

BOX 17.2 ■ FIGURE 2

Extent of the 2001–02 drought in the Prairie region of Canada.

Reproduced with permission of the Minister of Public Works and Government Services Canada, 2007.

A

B

FIGURE 17.14

(A) Wind erosion near the ground has sandblasted the lower metre of this basalt outcrop, Death Valley, California. Hammer for scale. (B) Power pole with its base wrapped in an abrasion-resistant material to minimize wind erosion.

Photo A by David McGeary; Photo B courtesy Paul Bauer

FIGURE 17.15

Ventifacts eroded by blowing sand, Death Valley, California. Most show two flat sides joined by a sharp ridge. Attached sand shows original depth of burial.

A

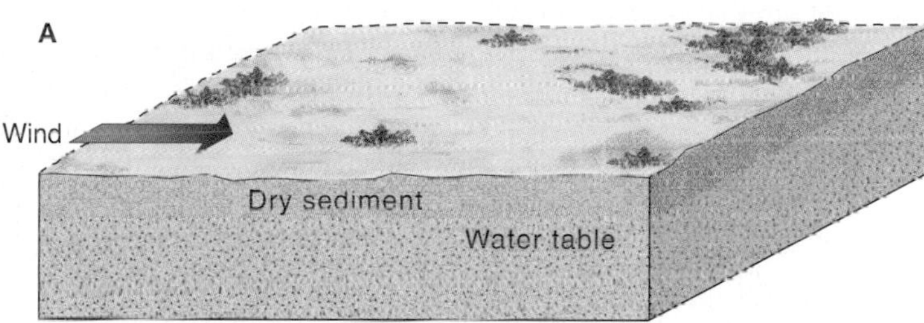

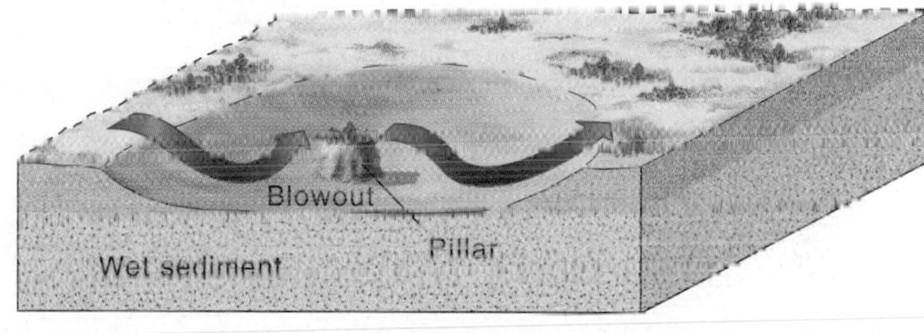

B

FIGURE 17.16

(A) Deflation by wind erosion can form a blowout in loose, dry sediment. Deflation stops at the water table. A pillar, or erosional remnant, may be found in the centre of a blowout. (B) Large blowout near Harrison, Nebraska. Pillar top is the original level of land before wind erosion lowered the land surface by more than 3 metres. The pillar is the erosional remnant at the centre of the blowout.

Photo by N. H. Darton, U.S. Geological Survey

WHAT FEATURES RESULT FROM WIND DEPOSITION?

Loess

Loess is a deposit of wind-blown silt and clay composed of unweathered, angular grains of quartz, feldspar, and other minerals weakly cemented by calcite. Loess has a high porosity, typically near 60 percent. Deposits of loess may blanket hills and valleys downwind of a source of fine sediment, such as a desert or a region of glacial outwash.

China has extensive loess deposits (figure 17.17), more than 100 metres thick in places. Wind from the Gobi Desert carried the silt and clay that formed these deposits. Loess is easy to dig into and has the peculiar ability to stand as a vertical cliff without slumping (figure 17.18), perhaps because of its cement or perhaps because the fine, angular, sediment grains interlock with one another. For centuries the Chinese have dug cavelike homes in loess cliffs. When a large earthquake shook China in 1920, however, many of these cliffs collapsed, burying alive about 100,000 people.

During the glacial ages of the Quaternary, the rivers that drained the margins of glaciers and ice sheets transported and deposited vast amounts of glacial outwash (see chapter 16). Later, winds eroded silt and clay (originally glacial rock flour) from the flood plains of these rivers and blanketed large areas of the U.S. Midwest, Alaska, and the Yukon with a cover of loess (refer to box 20.4). Soils that have developed from the loess in the Midwest are usually fertile and productive.

Sand Dunes

Sand dunes are mounds of loose sand grains heaped up by the wind. Dunes are most likely to develop in areas with strong winds that generally blow in the same direction. Patches of dunes are scattered throughout the southwestern United States desert and also occur in interior regions of Canada. Sand dune areas cover approximately 26,000 km^2 (or 0.27 percent) of the land area of Canada. More extensive dune fields occur on some of the other deserts of the world, such as the Sahara Desert of Africa, which contains vast *sand seas.* Dunes are also commonly found just landward of beaches (figure 17.19), where sand is blown inland. Beach dunes are common along the

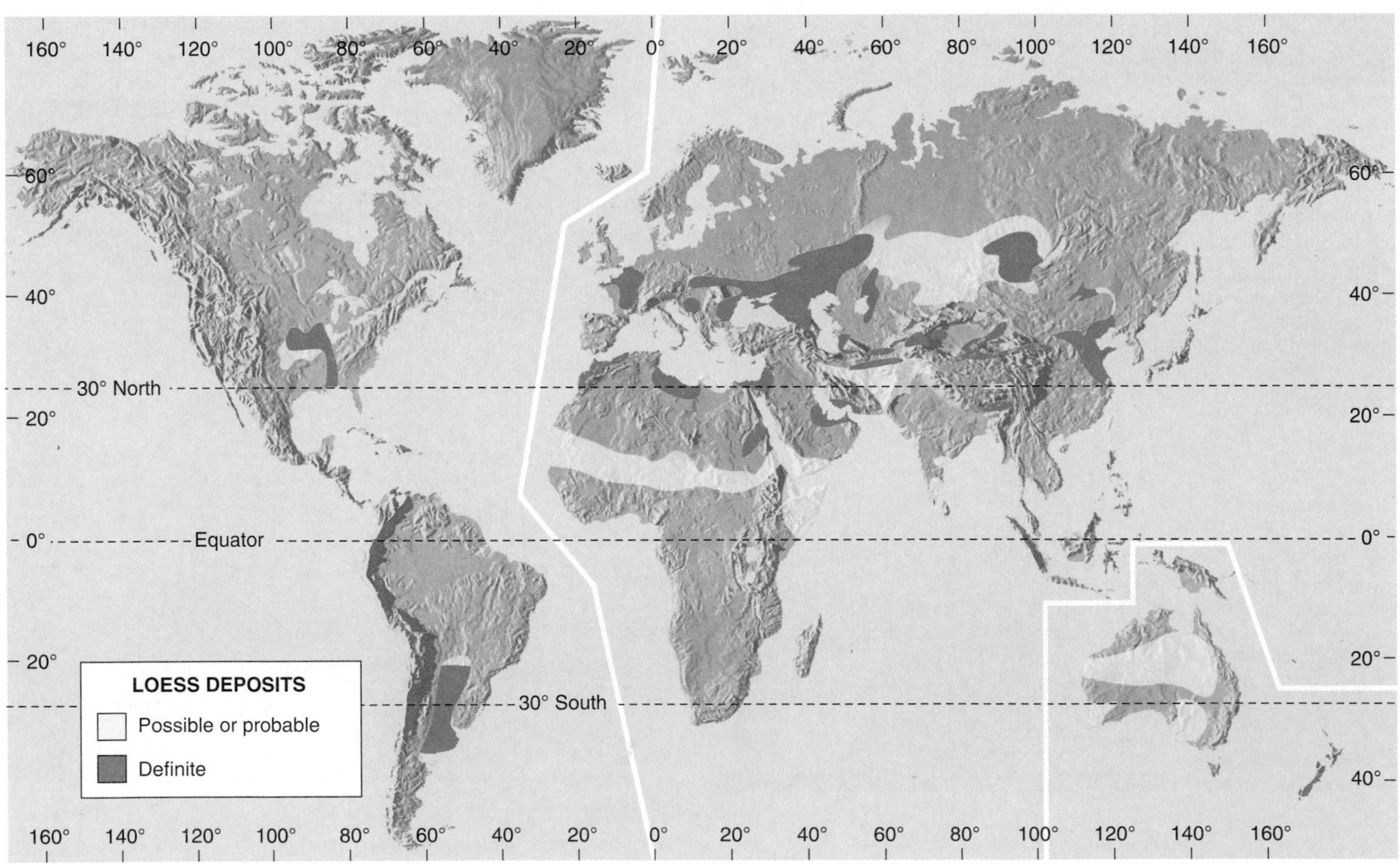

FIGURE 17.17

Major loess-covered areas in the world.

Data from map from U.S. Department of Agriculture; Smalley, I. J., Loess Lithology and Genesis, fig. 1, p. 768, Halsted Press, 1975

FIGURE 17.18

Home built into a steep cliff of loess in central China.
Photo © Stephen C. Porter

FIGURE 17.19

Coastal foredunes on the Queen Charlotte Islands (Haida Gwaii), British Columbia, Canada
Photo by Ian J. Walker

shores of the Great Lakes and along both coasts of North America. Braided rivers (see chapter 14) can also be sources of sand for dune fields. The most northerly region of active sand dunes in the world is on the southern shore of Lake Athabasca in Saskatchewan, where dunes up to 30 metres high and hundreds of metres long are found (box 17.3). The largest sand dune area in the southern prairies is in the Great Sand Hills region of Saskatchewan. Extensive dune fields, with similar features to those on Earth, have also been identified on Mars (see box 17.4).

The mineral composition of the sand grains in sand dunes depends on both the character of the original sand source and the intensity of chemical weathering in the region. Many dunes, particularly those near beaches in humid regions, are composed largely of quartz grains because quartz is so resistant to chemical weathering. Inland dunes, such as the Great Sand Dunes National Monument in Colorado, often contain unstable feldspar and rock fragments in addition to quartz. Some dunes are formed mostly of carbonate grains, particularly those near tropical beaches. At White Sands, New Mexico, dunes are made of gypsum grains, eroded by wind from playa lake beds.

Sand grains found in dunes are commonly well-sorted and well-rounded because wind is very selective as it moves sediment. Fine-grained silt and clay are carried much farther than sand, and grains coarser than sand are left behind when sand moves. The result is a dune made solely of sand grains, commonly all very nearly the same size. The prevalence of well-rounded grains in many dunes also may be due to selective sorting by the wind. Rounded grains roll more easily than angular grains, and so the wind may remove only the rounded grains from a source to form dunes. Wind will often selectively roll oolitic grains from a carbonate beach of mixed oolitic and skeletal grains.

Most sand dunes are asymmetric in cross-section, with a gentle slope facing the wind and a steeper slope on the downwind side. The steep downwind slope of a dune is called the **slip face** (figure 17.20). It forms from loose, cascading sand that generally keeps the slope at the *angle of repose,* which is about 34 degrees for loose, dry sand. Sand is blown up the gentle slope and over the top of the dune. Sand grains fall like snow onto the slip face when they encounter the calm air on the downwind side of the dune. Loose sand settling on the top of the slip face may become oversteepened and slide as a small avalanche down the slip face. These processes form high-angle cross-bedding within the dune. When found in sandstone, such cross-bedding strongly suggests deposition as a dune (refer to figure 9.29).

In passing over a dune, the wind erodes sand from the gentle upwind slope and deposits it downwind on the slip face. As a result, the entire dune moves slowly in a downwind direction. The rate of dune motion is much slower than the speed of the wind, of course, because only a thin layer of sand on the surface of the dune moves at any one time. The dune may move only 10 to 15 metres per year. Over many years, however, the movement of dunes can be significant, a fact not always appreciated by people who build homes close to moving sand dunes.

IN GREATER DEPTH 17.3

Moving Sands in Canada

Canada has more than 120 sand dune fields in Alberta, Saskatchewan, and Manitoba, and sand dunes can be found in many coastal regions of the Atlantic and Pacific seaboards and around lake shorelines. The largest active sand surface in Canada, and the world's most northerly major sand dune field, lies in northwestern Saskatchewan. The Athabasca Sand Dunes Provincial Park forms a 100-km series of dune fields on the south shore of Lake Athabasca (box figure 1). Sand forming the dunes is derived from glacial erosion of soft sandstones of the Athabasca Basin, which were redeposited as drumlins, eskers, and other glacial landforms. In the area south of Lake Athabasca the sands hold insufficient moisture to allow the growth of stabilizing vegetation, and wind reworks the sand into large dunes—some of which are more than 30 m high (box figure 2).

Other significant Canadian sand dune fields include the 2,000-km^2 Great Sand Hills region of Saskatchewan, located near the centre of the arid Palliser Triangle (see box 17.2). This is the largest dune area in the southern Prairies and was characterized by actively moving sand dunes in the late 1800s. These areas are now mostly stabilized with few active areas (box figure 3), in part as a result of careful land management practices. The Middle Sand Hills of Alberta is a 400-km^2 sand dune field that lies north of Medicine Hat, on the west side of the South Saskatchewan River. These sand dunes were actively migrating toward the river in 1937 but have since become vegetated and are now stabilized. In Manitoba, the Spirit Sand Hills consist of sand dunes stabilized by forest vegetation with few bare sandy areas. Historical accounts suggest that these dunes were also more active in the past. Scientists are now concerned that increased aridity in the Canadian Prairies caused by drought or climate warming could result in increased sand dune activity and the loss of agricultural land.

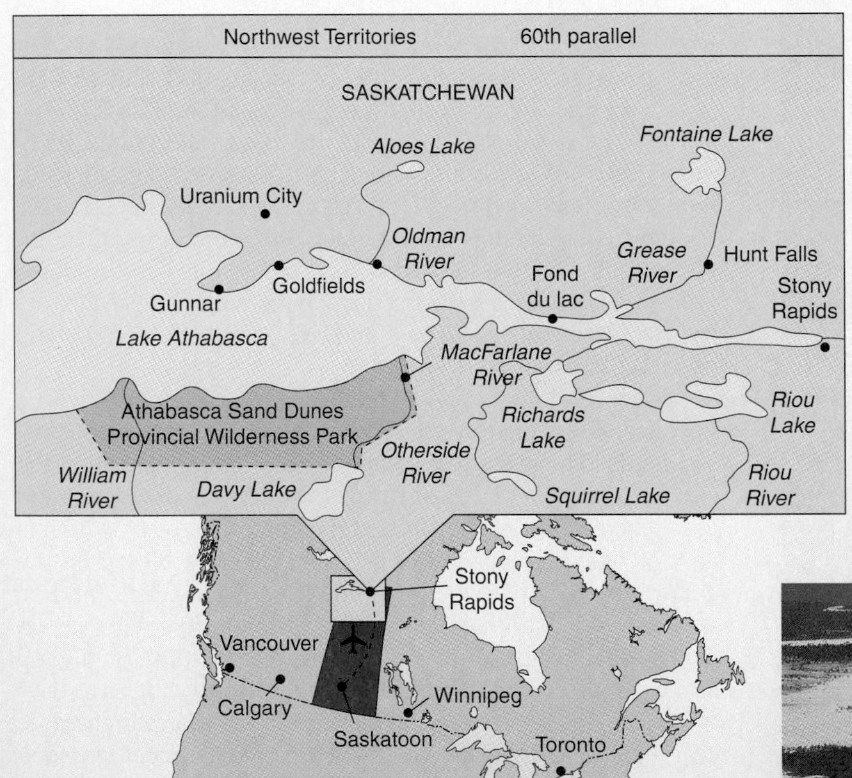

BOX 17.3 ■ FIGURE 1
Location of Athabasca Sand Dunes Provincial Park.
Map from http://www.athabascalake.com/ecoexped/map/html

BOX 17.3 ■ FIGURE 2
Sand dunes along the margin of the William River.
Photo © Robin Karpan

BOX 17.3 ■ FIGURE 3
Active sand dunes in Cantara Bay, Athabasca Sand Dunes, Saskatchewan.
Photo © Robin Karpan

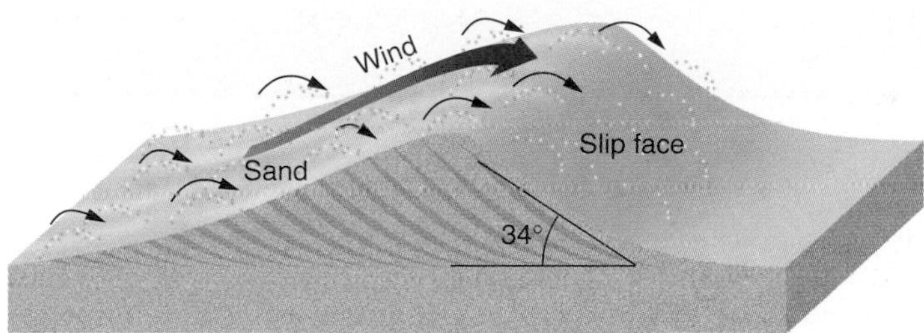

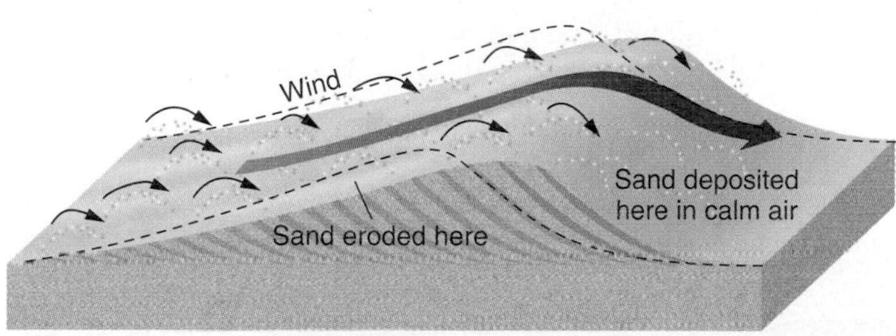

A

B

FIGURE 17.20

(A) A sand dune forms with a gentle upwind slope and a steeper slip face on the downwind side. Sand eroded from the upwind side of the dune is deposited on the slip face, forming cross-beds. Movement of sand causes the dune to move slowly downwind. (B) Strong desert winds (100 km per hour) blowing to the right remove sand from the gently sloping upwind side of this dune. The sand settles onto the steep slip face on the right.
Photo B by David McGeary

If a dune becomes overgrown with grass or other vegetation, movement stops. The Great Sand Hills of Saskatchewan are large dunes that are now stabilized by vegetation. Other inactive dune systems can be found in the Grande Prairie region of Alberta, the Churchill River Valley in Labrador, and in the Chalk River–Petawawa area of Ontario (see also box 17.3). The migration of many beach dunes toward beach homes and roads has been stopped by planting a cover of beach grass over the dunes. Dune-buggy tires can uproot and kill the grass, however, and start the dunes moving again. Prolonged drought or climate warming resulting in increased aridity may also cause renewed sand dune activity.

FIGURE 17.21

Wind ripples on sand surface, Monument Valley, Utah.
Photo by © Doug Sherman

Sand moving over a dune surface typically forms *wind ripples*—small, low ridges of sand produced by saltation of the grains (figure 17.21). The ripples are similar to those formed in sediment by a water current (see chapter 9). Because sand moves perpendicularly to the long dimension of the ripples, a rippled sand surface indicates the direction of sand movement.

Types of Dunes

As figure 17.22 shows, dunes tend to develop certain characteristic shapes, depending on (1) the wind's velocity and direction (that is, whether constant or shifting); (2) the sand supply available; and (3) how the vegetation cover, if any, is distributed.

Where the sand supply is limited, a type of dune called a **barchan** generally develops. The barchan is crescent-shaped with a steep slip face on the inward or concave side. The horns on a barchan dune point in the downwind direction (figure 17.22*A*). Barchan dunes are usually separated from one another and move across a barren surface (figure 17.23). If more sand is available, the wind may develop a **transverse dune,** a relatively straight, elongate dune oriented perpendicular to the wind direction (figures 17.22*B* and 17.24).

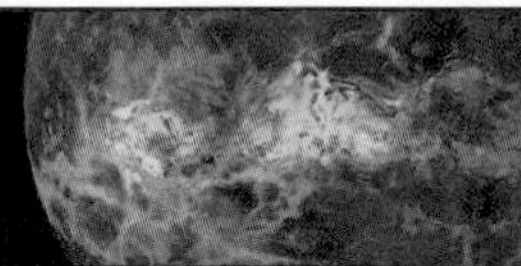

ASTROGEOLOGY 17.4

Wind Action on Mars

Mars has an atmosphere only 1/200th as dense as Earth's but with very strong winds that have been recorded at more than 200 kilometres per hour. The sides of Olympus Mons have been obscured by dust to a height of 15 kilometres, a height made possible by the low gravity on Mars. Although dust storms occur throughout the year on Mars, the greatest number and the largest global dust events (those that cover the entire planet) occur during the southern spring and summer, when the southern polar cap of frozen carbon dioxide begins to sublimate. The difference in air temperature between the polar cap and the warmer surrounding landscape creates large pressure differences that produce high winds and trigger isolated dust storms. In June 2001, the Mars Global Surveyor spacecraft and Hubble Space Telescope recorded a sequence of dust storms that began along the retreating margin of the southern polar cap and near the Hellas impact crater (box figure 1). The individual storms intensified and moved north of the equator in only five days. This was the beginning of one of the largest global dust events observed on Mars in decades, and, for the first time, scientists were able to see how the development and progression of regional dust storms resulted in the entire planet being obscured by dust.

BOX 17.4 ■ FIGURE 1

Storm watch images of Mars from the Hubble Space Telescope show isolated dust storms on June 26, 2001, in the Hellas basin (lower right edge of Mars) and on the northern polar cap. By the end of July, the entire planet was clouded by dust that obscured its surface for several months.

Photo courtesy of NASA, James Bell (Cornell Univ.), Michael Wolff (Space Science Inst.), and the Hubble Heritage Team (STScI/AURA)

BOX 17.4 ■ FIGURE 2

False-colour image of a dune field in the Endurance crater taken by the Mars Exploration Rover Opportunity. The "blue" tint is caused by the presence of hematite-containing spherules ("blueberries") that accumulate on the flat surface between the dunes. Dunes in the foreground are about 1 metre in height.

Photo by NASA/JPL/Cornell

Images from the recent Mars Rover missions show that the windswept surface of the planet contains features similar to those found on Earth. Barchan, star, transverse, and longitudinal sand dunes are prevalent, particularly on the floors of impact craters. The 2004 Mars Rover, Opportunity, recorded images of a dune field in Endurance crater and also discovered the presence of hematite-containing spherules ("blueberries") that accumulate on the flat surfaces (box figure 2). What appear to be *yardangs,* wind-eroded round and elliptical knobs, have been observed in the Medussae Fossae region of Mars. The robotic "geoscientist" Sojourner, launched from the Pathfinder mission, recorded detailed images of rocks with smooth yet pitted surfaces that are similar to wind-scoured ventifacts on Earth (box figure 3).

Additional Resources

For more information and additional images of dust storms and wind features on Mars, visit the NASA Mars Exploration and Planetary Photojournal websites:

- http://mars.jpl.nasa.gov/gallery/duststorms/index.html
- http://mars.jpl.nasa.gov/gallery/sanddunes/index.html
- http://photojournal.jpl.nasa.gov/

BOX 17.4 ■ FIGURE 3
Close-up of the rock named "Moe" from the Pathfinder landing site that shows a smooth but pitted surface similar to wind-abraded rocks, or ventifacts, on Earth; rock is 1 metre in diameter.
Photo by JPL/NASA

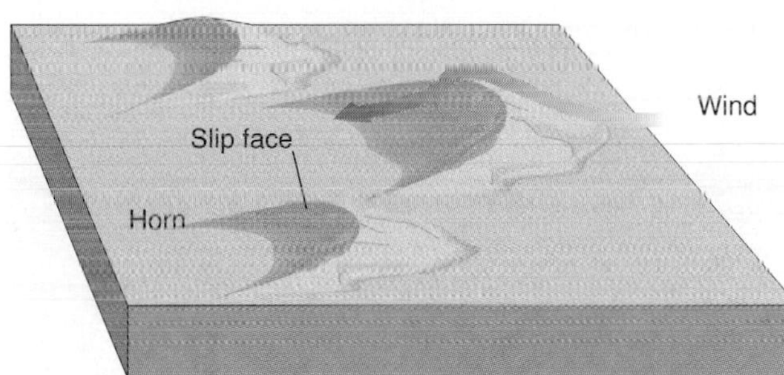

A Barchans

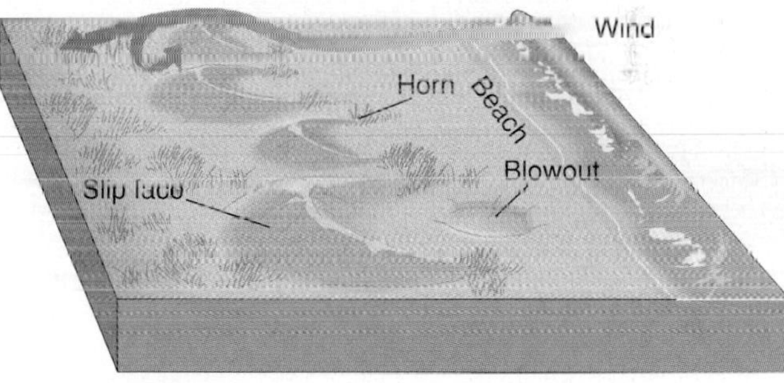

C Parabolic dunes

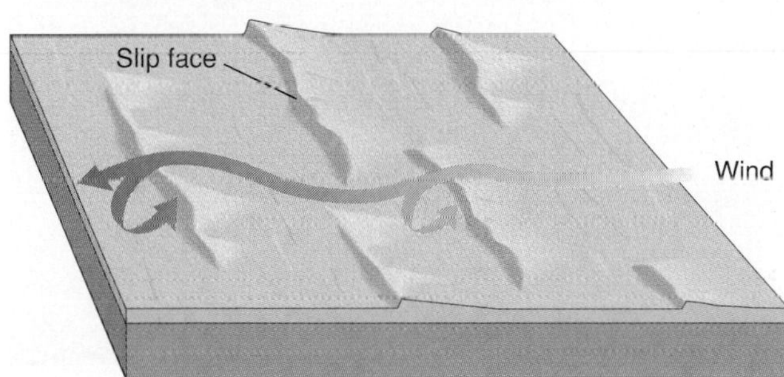

B Transverse dunes

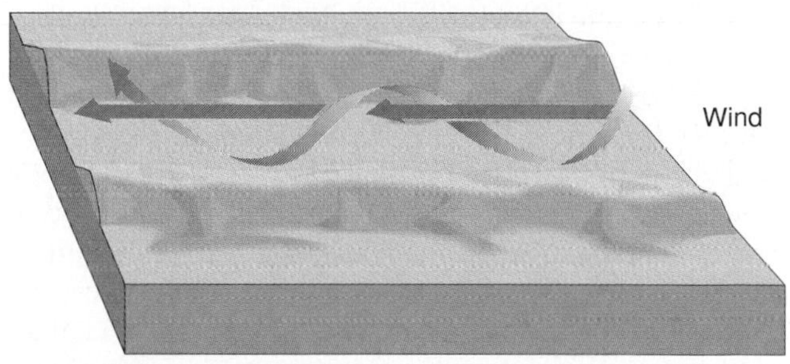

D Longitudinal dunes (seifs)

FIGURE 17.22
Types of sand dunes.

FIGURE 17.23

Barchan dune formed in the Skeleton Coast National Park in Namibia. Prevailing wind blows from left to right and carries a limited supply of sand.

Photo © Gerry Ellis/Minden Pictures

FIGURE 17.24

Transverse dunes in the Great Sand Dunes National Monument, Colorado. Wind blows from left to right.

Photo © Adriel Heisey Photography

A **parabolic dune** is somewhat similar in shape to a barchan dune, except that it is deeply curved and is convex in the downwind direction. The horns point upwind and are commonly anchored by vegetation (figure 17.22*C*). The parabolic dune requires abundant sand and commonly forms around a blowout. Because they require abundant sand and strong winds, parabolic dunes are typically found inland from an ocean beach (figure 17.25). Almost all of Canada's dune areas are dominated by parabolic dunes. All three of these dune shapes develop in areas having steady wind direction, and all three have steep slip faces on the downwind side.

One of the largest types of dunes is the **longitudinal dune** or *seif* (figure 17.22*D*), which is a symmetrical ridge of sand that forms parallel to the prevailing wind direction. Longitudinal dunes occur in long, parallel ridges that are exceptionally straight and regularly spaced. They are typically separated by barren ground or desert pavement. Longitudinal dunes in the Sahara Desert (figure 17.26) are as high as 200 metres and more than 120 kilometres in length. Numerous hypotheses have been proposed to explain the development of longitudinal dunes, but none can adequately explain their spectacular size and regular spacing. It appears that crosswinds are important in piling up sand, which adds to the height of longitudinal dunes, whereas the more constant prevailing wind direction redistributes the sand down the length of the dunes. Smoke-bomb experiments to analyze airflow have shown that the wind spirals down the intervening troughs between longitudinal dunes and may control the regularity of their spacing.

Not all dunes can be classified by an easily recognizable shape. Many of them are quite irregular. To read about two intriguing features of many deserts, see box 17.5. To learn what can be found under desert sand, see box 17.6.

FIGURE 17.25

Parabolic dunes near Pismo Beach, central California. Wind blows from left to right. The ocean and a sand beach are just to the left of the photo.

Photo by Frank M. Hanna

FIGURE 17.26

Longitudinal dunes in the Sahara Desert, Algeria. Photo from Gemini spacecraft at an altitude of about 100 kilometres.

Photo by NASA

IN GREATER DEPTH 17.5

Desert Pavement and Desert Varnish

Two intriguing features can be seen in many deserts, particularly on the surface of old alluvial fans no longer receiving new sediment.

Desert pavement is a thin, surface layer of closely packed pebbles (box figure 1). The pebbles were once thought to be lag deposits, left behind as strong winds blew away all the fine grains of a rocky soil. The pebbles are now thought to be brought to the surface by cycles of wetting and drying, which cause the soil to swell and shrink as water is absorbed and lost by soil particles. Swelling soil lifts pebbles slightly; drying soil cracks, and fine grains fall into the cracks. In this way pebbles move up, while fine grains move down. The surface layer of pebbles protects the land from wind erosion and deflation. When the desert pavement is disturbed (as in the 1991 and 2003 Gulf Wars), dust storms and new sand dunes may result.

Many rocks on the surface of deserts are darkened by a chemical coating known as *desert varnish*. Although the interior of the rocks may be light coloured, a hard, often shiny, coating of dark iron and manganese oxides and clay minerals can build up on the rock surface over long periods of time (box figure 2). These paper-thin coatings can be used to obtain a numerical age of the exposed desert surface by measuring cosmogenic helium-3 isotopes preserved in the desert varnish.

Although no one is quite certain how this coating develops, it seems to be added to the rocks from the outside, for even white quartzite pebbles with no internal source of iron, manganese, or clay minerals can develop desert varnish. One hypothesis is that the clay is windblown, perhaps sticking to rocks dampened by dew. A film of clay on a rock may draw iron and manganese-containing solutions upward from the soil by capillary action, and the presence of the clay minerals may help deposit the dark manganese oxide that cements the clay to the rock. Another more recent hypothesis is that the oxide is deposited biologically by manganese-oxidizing bacteria. Regardless of how the varnish forms, the longer a rock is exposed on a desert land surface, the darker it becomes.

BOX 17.5 ■ FIGURE 1

Desert pavement on an old alluvial fan surface in Death Valley, California. The surface pebbles are closely packed; fine sand underlies the pebbles.

Photo by Diane Carlson

BOX 17.5 ■ FIGURE 2

Petroglyphs carved on this rock cut through the dark desert varnish to show the lighter colour of the interior of the rock, Valley of Fire, Nevada.

Photo by J. Freeberg, U.S. Geological Survey

SUMMARY

Deserts are located in regions where less than 25 cm of rain falls in a year. Such regions are found primarily in belts of descending air at 30° North and South latitude and in polar regions. Arid regions also may be due to the *rain shadow* of a mountain range, great distance from the sea, and proximity to a cold ocean current. Descending air forms cold deserts at the poles.

Desert landscapes differ from those of humid regions of lacking through-flowing streams and in having internal drainage and many local, rising base levels. *Flash floods* caused by desert thunderstorms are effective agents of erosion despite the low rainfall. Limestone is resistant in deserts. Thin soil and slow rates of creep may give desert topography an angular look.

Desert landforms are determined primarily by rock structure. Flat-lying sedimentary rocks of the Western Canadian Arctic and the Colorado Plateau are sculptured into cliffs, *plateaus*, *mesas*, and *buttes*. Landforms developed on the margins and floors of desert basins and valleys include *alluvial fans*, *bajadas*, *playas*, and *pediments*.

Although wind erosion can be intense in regions of low moisture, streams are usually more effective than wind in sculpturing landscapes, regardless of climate.

Fine-grained sediment can be carried long distances by wind, even across entire continents and oceans.

Sand moves by *saltation* close to the ground, occasionally carving *ventifacts*.

Wind can *deflate* a region, creating a *blowout* in fine sediment.

Sand dunes move slowly downwind as sand is removed from the gentle upwind slope and deposited on the steeper *slip face* downwind.

Dunes are classified as *barchans*, *transverse dunes*, *parabolic dunes*, and *longitudinal dunes*, but many dunes do not resemble these types. Dune type depends on wind strength and direction, sand supply, and vegetation.

Terms to Remember

bajada 461
barchan 471
blowout 465
butte 461
deflation 465
desert 455
flash flood 459
loess 468
longitudinal dune (seif) 474
mesa 461
parabolic dune 474
pediment 461
plateau 460
playa 461
playa lake 461
polar deserts 460
rain shadow 456
sand dune 468
slip face 469
transverse dune 471
ventifact 465

Testing Your Knowledge

Use the questions below to prepare for exams based on this chapter.

1. Why do parts of the Canadian Arctic have an arid climate?
2. Sketch a cross-section of an idealized dune, labelling the slip face and indicating the wind direction. Why does the dune move?
3. Describe the characteristics of polar desert areas in Canada.
4. How does a flash flood in a dry region differ from most floods in a humid region?
5. Give two reasons why wind is a more effective agent of erosion in a desert than in a humid region.
6. Name four types of sand dunes and describe the conditions under which each forms.
7. The defining characteristic of a desert is
   a. shifting sand dunes
   b. high temperatures
   c. low rainfall
   d. all of the above
   e. none of the above
8. What are the effects of prolonged droughts on the Canadian prairies?
9. Draw a sketch to show why the Okanagan Valley receives little precipitation.
10. Which is characteristic of deserts?
    a. internal drainage
    b. limited rainfall
    c. flash floods
    d. slow chemical weathering
    e. all of the above
11. The major difference between a mesa and a butte is one of
    a. shape
    b. elevation
    c. rock type
    d. size

12. A very flat surface underlain by a dry lake bed of hard, mud-cracked clay is called a
    a. ventifact
    b. plateau
    c. playa
    d. none of the above

13. Rocks with flat, wind-abraded surfaces are called
    a. ventifacts
    b. pediments
    c. bajadas
    d. none of the above

14. The removal of clay, silt, and sand particles from the land surface by wind is called
    a. deflation
    b. depletion
    c. deposition
    d. abrasion

15. Which is not a type of dune?
    a. barchan
    b. transverse
    c. parabolic
    d. longitudinal
    e. all of the above are dunes

16. A broad ramp of sediment formed at the base of mountains when alluvial fans merge is
    a. a playa
    b. a bajada
    c. a pediment
    d. an arroyo

17. A surface layer of closely packed pebbles is called
    a. desert varnish
    b. deflation
    c. a blowout

Exploring Web Resources

www.mcgrawhill.ca/olc/plummer

Visit our Online Learning Centre to check your answers for the Testing Your Knowledge section and click on the direct links for great sites on deserts and sand dunes.

http://pubs.usgs.gov/gip/deserts/contents/

Online version of *Deserts: Geology and Resources* by A. S. Walker provides a good overview of deserts, processes, and mineral resources.

www.sc.gov.sk.ca

For information on Athabasca Sand Dunes Provincial Park.

www.gov.mb.ca/conservation/parks/popular_parks/spruce_woods/index.html

For information on Spruce Woods Provincial Park.

http://gsc.nrcan.gc.ca/climate/sanddune/index_e.php

Excellent source of information on sand dune and climate change studies in the Prairie provinces.

CHAPTER

18

Coasts and Oceans

How do waves form?
What happens when waves reach the shore?
How do beaches develop?
Why does sand move along a shoreline?
Why are there many different types of coasts?
What features can be found on the ocean floor?
What kinds of sediment accumulate on the ocean floor?

The previous five chapters have dealt with the sculpturing of the land by mass wasting; streams and floods; groundwater; glaciers, glaciation, and permafrost; and deserts and wind. Water waves are another agent of erosion, transportation, and deposition of sediment. Along the shores of oceans and lakes, waves break against the land, building it up in some places and tearing it down in others.

Ordinary ocean waves (as opposed to tsunamis) are created by wind. Waves moving across an ocean transfer the energy derived from wind to processes that operate along shorelines. This energy is used to a large extent in eroding and transporting sediment along the shoreline. Understanding how waves travel and move sediment can help you see how easily the balance of supply, transportation, and deposition of beach sediment can be disturbed. Such disturbances can be natural or human-made, and the changes that result often destroy beachfront homes and block harbours with sand.

Rocky headlands and steep cliffs formed along the Gulf of St. Lawrence shoreline, Gaspe, Quebec. *Photo by Jim Wark — Airphoto*

Beaches have been called "rivers of sand" because breaking waves, as they sort and transport sediment, tend to move sand parallel to the shoreline. Sediment is also moved offshore into the deeper ocean basins where it may be deposited on the continental shelf, rise, or abyssal plain. In this chapter we look at how beaches are formed and also examine the influence of wave action on such coastal features as sea cliffs, barrier islands, and terraces. We will also learn about features of the ocean floor and the processes that lead to sediment transport and deposition in deeper-water environments.

If you spend a week at the shore during the summer, you may not notice any great change in the appearance of the beach while you are there. Even if you spend the whole summer at the seaside, nothing much seems to happen to the beach during those months. Tides rise and fall every day and waves strike the shore, but the sand that you walk on one day looks very much like the sand that you walked on the previous day. The shape of the beach does not appear to change, nor does the sand seem to move very much.

On most beaches, however, the sand *is* moving, in some cases quite rapidly. The beach looks the same from day to day only because new sand is being supplied at about the same rate that old sand is being removed.

Where is the sand going? Some sand is carried out to deep water. Some is piled up and stored high on the beach. But on most shores much more of the sand moves along parallel to the beach in relatively shallow water. In this way, loose sand grains travel hundreds of metres per day along some coasts, especially those subject to strong waves.

On some beaches, sand is being removed faster than it is being replenished. When this happens, beaches become narrower and less attractive for swimming. Where erosion is severe, buildings close to the beach can be undermined and destroyed by waves as the beach disappears (figure 18.1). The sand moved from the beach may be redeposited in inconvenient places, such as across the mouth of a harbour, where it must be dredged out periodically. Because moving sand can create many problems for people in coastal towns and cities, it is important to understand something of how and why the sand moves.

FIGURE 18.1

Beach houses were built too close to the ocean and destroyed by 2004 Hurricane Ivan on Cape San Blas along the coast of Pensacola, Florida.

Photo © Associated Press, St. Petersburg Times/Wide World Photos

Canada has the longest coastline of any country in the world. The total length of Canada's coastline is more than 202,000 km and includes mainland coasts and offshore islands that fringe three oceans. Hence, it is extremely important to understand the long-term processes that affect coastlines and oceans, in particular those influenced by gradual sea-level change.

HOW DO WAVES FORM?

The energy that moves sand along a beach comes from the wind-driven water waves that break upon the shore. As wind blows over the surface of an ocean or a lake, some of the wind's energy is transferred to the water surface, forming the waves that move through the water. The height of waves (and their length and speed) are controlled by the wind speed, the length of time that the wind blows, and the distance that the wind blows over the water (*fetch*). The largest waves form where high winds blow over a long expanse of open water for an extended period of time.

Wave shapes can vary. Short, choppy *seas* in and near a storm create a confused sea surface, often with considerable white foam as strong winds blow the tops off of waves. Long, rolling *swells* form a regular series of similar-sized waves on shores that may be thousands of kilometres from the storms that generated the waves. (Summer surfing waves in southern California can be generated by large storms north of Australia in the southern-hemisphere winter.) When waves break against the shore as *surf,* a large portion of their energy is spent moving sand along the beach.

The height of waves is the key factor in determining wave energy. **Wave height** is the vertical distance between the **crest,** which is the high point of a wave, and the **trough,** which is the low point (figure 18.2). In the open ocean, normal waves have heights of about 0.3 to 5 metres, although during violent storms, including hurricanes, waves can be more than 15 metres high. The highest wind wave ever measured was 34 metres by the anxious crew of a ship in the north Pacific in 1933. (The highest tsunami ever measured, caused by a submarine earthquake rather than wind, was 85 metres, in the Ryukyu island chain south of Japan in 1971; see chapter 3.)

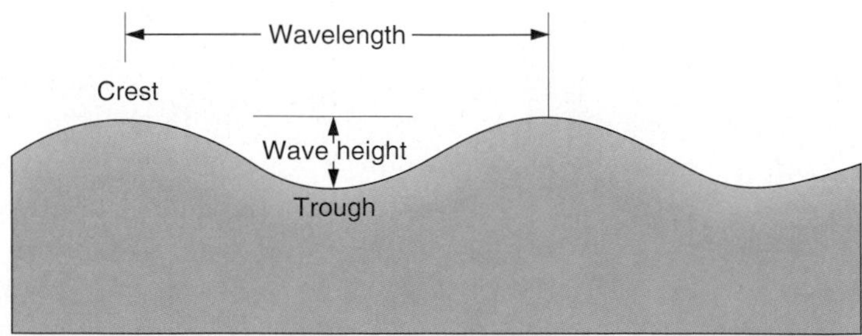

FIGURE 18.2

Wave height is the vertical distance between the wave crest and the wave trough. Wavelength is the horizontal distance between two crests.

Wavelength is the horizontal distance between two wave crests (or two troughs). Most ocean wind waves are between 40 and 400 metres in length, and move at speeds of 25 to 90 kilometres per hour in deep water.

The movement of water in a wave is like the movement of wheat in a field when wind blows across it. You can see the ripple caused by wind blowing across a wheat field, but the wheat does not pile up at the end of the field. Each stalk of wheat bends over when the wind strikes it and then returns to its original position. A particle of water moves in an *orbit,* a nearly circular path, as the wave passes (figure 18.3); the particle returns to its original position after the wave has passed. In deep water, when a wave moves across the water surface, energy moves with the wave; but the water, like the wheat, does not advance with the wave.

At the surface, the diameter of the orbital path of a water particle is equal to the height of the wave (figure 18.3). Below the surface, the orbits decrease in size until the motion is essentially gone at a depth equal to half the wavelength. This is why a submarine can cruise in deep, calm water beneath surface ships that are being tossed by the orbital motion of large waves.

Surf

As waves move from deep water to shallow water near shore, they begin to be affected by the ocean bottom. A wave first begins to "feel bottom" at the level of lowest orbital motion—that is, when the depth to the bottom equals half the wavelength. For example, a wave 150 metres long will begin to be influenced by the bottom at a water depth of 75 metres.

In shallow water the presence of the bottom interferes with the circular orbits, which flatten into ovals (figure 18.4). The waves slow down and their length decreases. Meanwhile, the sloping bottom wedges the moving water upward, increasing the wave height. Because the height is increasing while the length is decreasing, the waves become steeper and steeper until they break. A **breaker** is a wave that has become so steep that the crest of the wave topples forward, moving faster than the main body of the wave. The breaker then advances as a turbulent, often foamy, mass. Breakers collectively are called **surf.** Water in the surf zone has lost its orbital motion and moves back and forth, alternating between onshore and offshore flow.

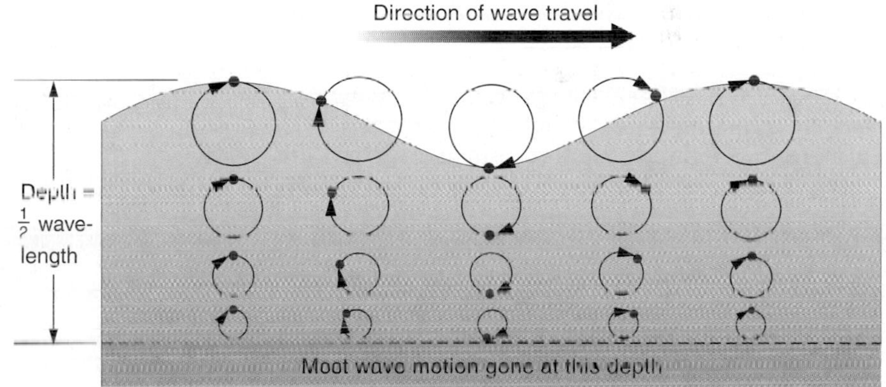

FIGURE 18.3

Orbital motion of water in waves dies out with depth. At the surface, the diameter of the orbits is equal to the wave height.

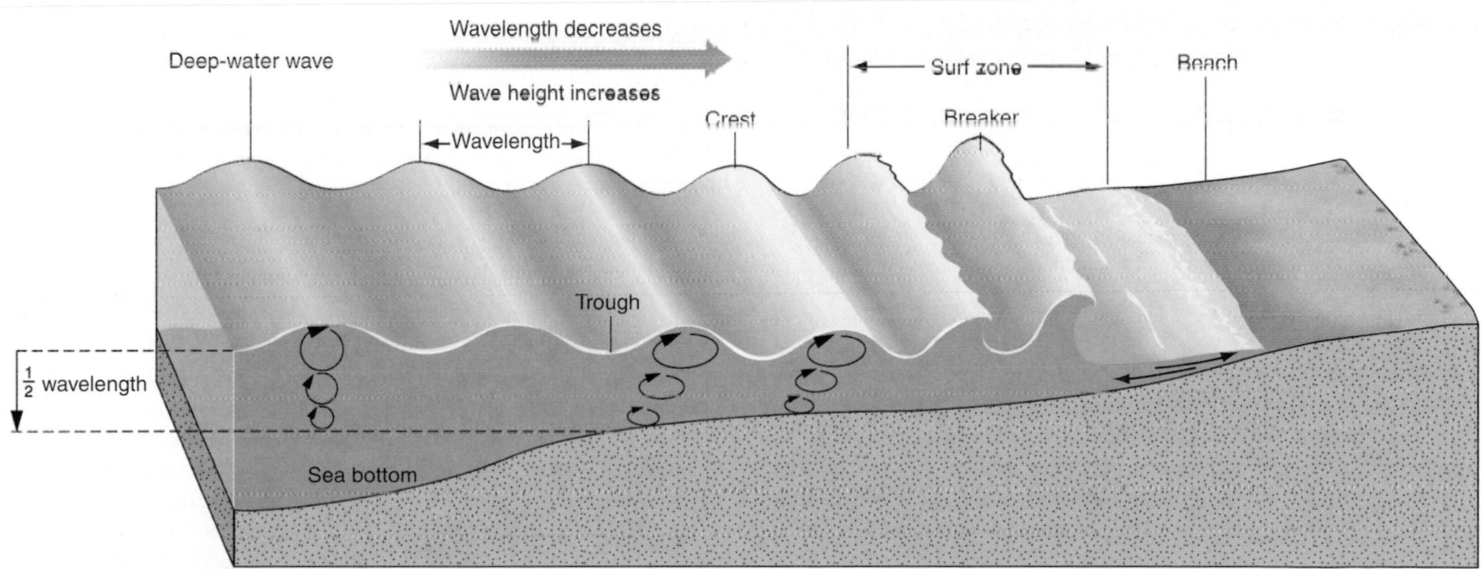

FIGURE 18.4

As a deep-water wave approaches shore, it begins to "feel" the sea bottom and slow down. Circular water orbits flatten and the wave peaks and breaks. In the foamy surf zone, water moves back and forth rather than in orbits.

WHAT HAPPENS WHEN WAVES REACH THE SHORE?

Wave Refraction

Most waves do not come straight into shore. A wave crest usually arrives at an angle to the shoreline (figure 18.5*A*). One end of the wave breaks first, and then the rest breaks progressively along the shore.

This angled approach of a wave toward shore can change the direction of wave travel. One end of the wave reaches shallow water first. This end of the wave "feels bottom" and slows down while the rest of the wave continues at its deep-water speed (figure 18.5*B*). As more and more of the wave comes into contact with the bottom, more of the wave slows down. As the wave slows progressively along its length, the wave crest changes direction and becomes more nearly parallel to the shoreline. This bending of waves is called **wave refraction** (figure 18.5*B*). Wave refraction causes energy to be focused on headlands where waves are higher than in adjacent bays (see figure 18.15 later in this chapter).

Longshore Currents

Although most wave crests become nearly parallel to shore as they are refracted, waves do not generally strike *exactly* parallel to shore. Even after refraction, a small angle remains between the wave crest and the shoreline. As a result, the water in the wave is pushed both *up* the beach toward land and *along* the beach parallel to shore.

Each wave that arrives at an angle to the shore pushes more water parallel to the shoreline. Eventually a moving mass of water called a **longshore current** develops parallel to the shoreline (figure 18.6). The width of the longshore current is about equal to the width of the surf zone. The seaward edge of the current is the outer edge of the surf zone, where waves are just beginning to break; the landward edge is the shoreline. A longshore current can be very strong, particularly when the waves are large. Such a current can carry swimmers hundreds of metres parallel to shore before they are aware that they are being swept along. It is these longshore currents that transport most of the beach sand parallel to shore.

A

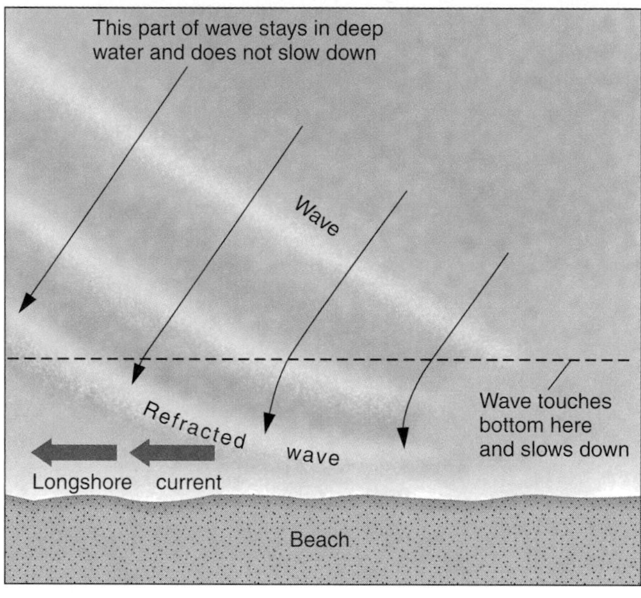

B

FIGURE 18.5

(*A*) These waves are arriving at an angle to the shoreline. They break progressively along the shore, from the upper right of the photo to the lower left. (*B*) Wave refraction changes the wave direction, bending the wave so it becomes more parallel to shore. The angled approach of waves to shore sets up a longshore current parallel to the shoreline.

Photo by David McGeary

Rip Currents

Rip currents are narrow currents that flow straight out to sea in the surf zone, returning water seaward that breaking waves have pushed ashore (figure 18.6). Rip currents travel at the water surface and die out with depth. They pulsate in strength, flowing most rapidly just after a set of large waves has carried a large amount of water onto shore. Rip currents can be important transporters of sediment, as they carry fine-grained sediment out of the surf zone into deep water.

As a single wave comes toward shore, its height varies from place to place. Rip currents tend to develop locally where wave height is low. Rip currents that are fixed in position are apt to be found over channels on the sea floor, because depressions on the bottom reduce wave height. Complex wave interactions can also lower wave height, and rip currents that form because of wave interactions tend to shift position along the shore. Such shifting rip currents are usually spaced at regular intervals along the beach.

Rip currents are fed by water within the surf zone. They flow rapidly out through the surf zone and then die out quickly. Where waves are nearly parallel to a shoreline, longshore feeder currents of equal strength develop in the surf zone on either side of a rip current (figure 18.6*A*). Where waves strike the shore at an angle and set up a strong, unidirectional, longshore current, a rip current is fed from one side by the longshore current, which increases in strength as it nears the rip current (figure 18.6*B*). Rip

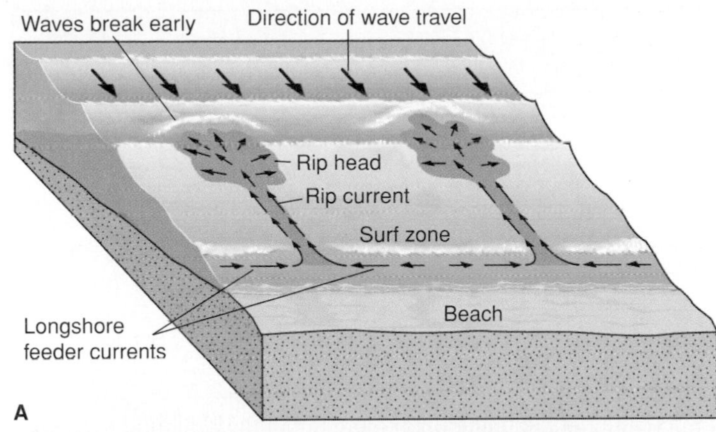

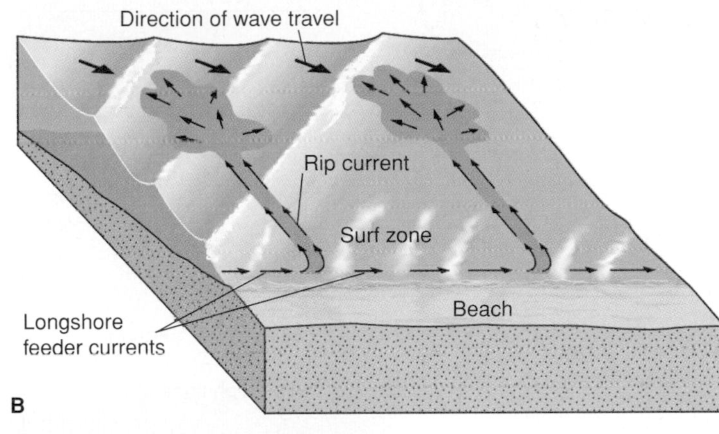

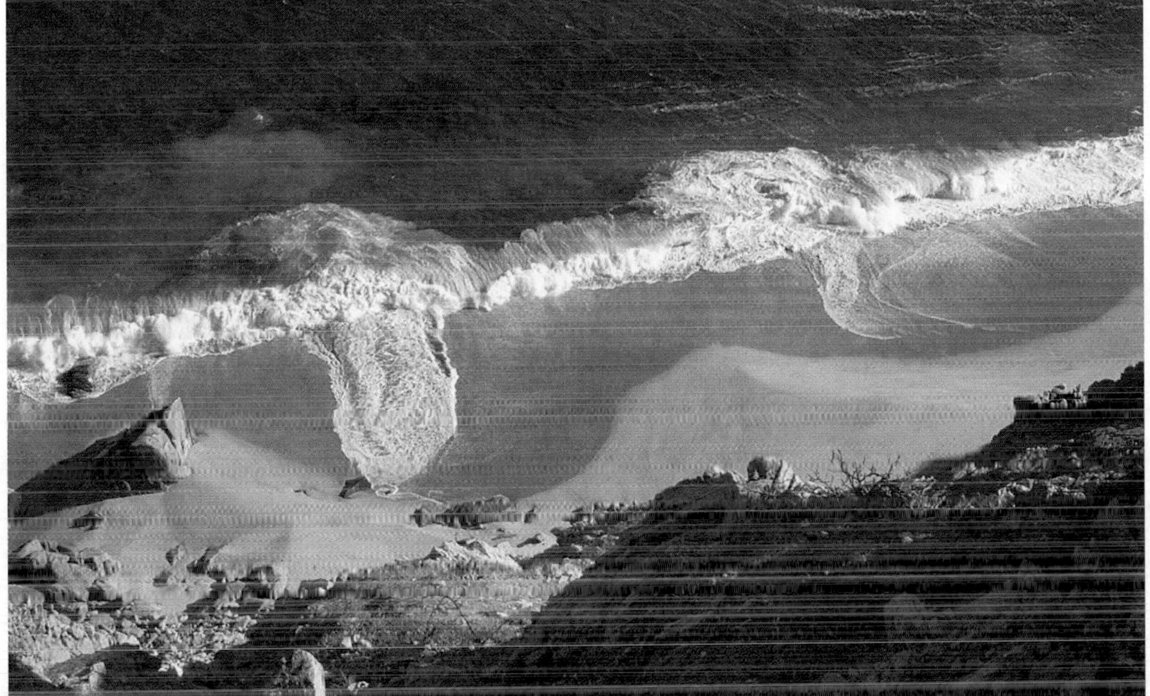

FIGURE 18.6

Rip currents and their feeder currents can develop regardless of the angle of approach of waves. (*A*) Waves approach parallel to shore; feeder currents on both sides of rip currents. (*B*) Waves approach at an angle to the shore; feeder current on only one side of rip current. (*C*) Rip currents carry dirty water and foam seaward; they can cause incoming waves to break early.

Photo © Sanford Berry/ Visuals Unlimited

currents are also found alongside points of land and engineered structures such as jetties and piers, which can deflect longshore currents seaward.

You can easily learn to spot rip currents at a beach. Look for discoloration in the water where sediment is being picked up in the surf zone and moved seaward (figure 18.6*C*). Another sign is incoming waves breaking early within a rip current as they meet the opposing flow. The diffuse heads of rip currents outside the surf zone may be marked at the edge with foam lines. Even on very calm days rips can often be identified by subtle changes in the water surface, such as a different pattern of water ripples or light reflection off the water.

Getting caught in a rip current and being carried out to sea can panic an inexperienced swimmer—even though the trip will stop some distance beyond the surf zone as the rip dies out. A swimmer frightened by being carried away from land and into breaking waves can grow exhausted fighting the current to get back to shore. The thing to remember is that rip currents are narrow. Therefore, you can get out of a rip easily by swimming *parallel* to the beach instead of struggling against the current.

Surfers, on the other hand, often look for rip currents and paddle intentionally into them to get a quick ride out into the high breakers.

Tides

Once or twice daily, the sea level at any point on the Earth's surface rises and falls as a **tide.** Tides are caused by a **tide-generating force** created by the gravitational attraction of the moon and the sun. The largest tide-generating force comes from the moon, which lies closer to the Earth and is almost double that of the sun's influence. The tide-generating force creates two bulges in the Earth's oceans, one on the side of the Earth closest to the moon (the sublunar bulge), and the

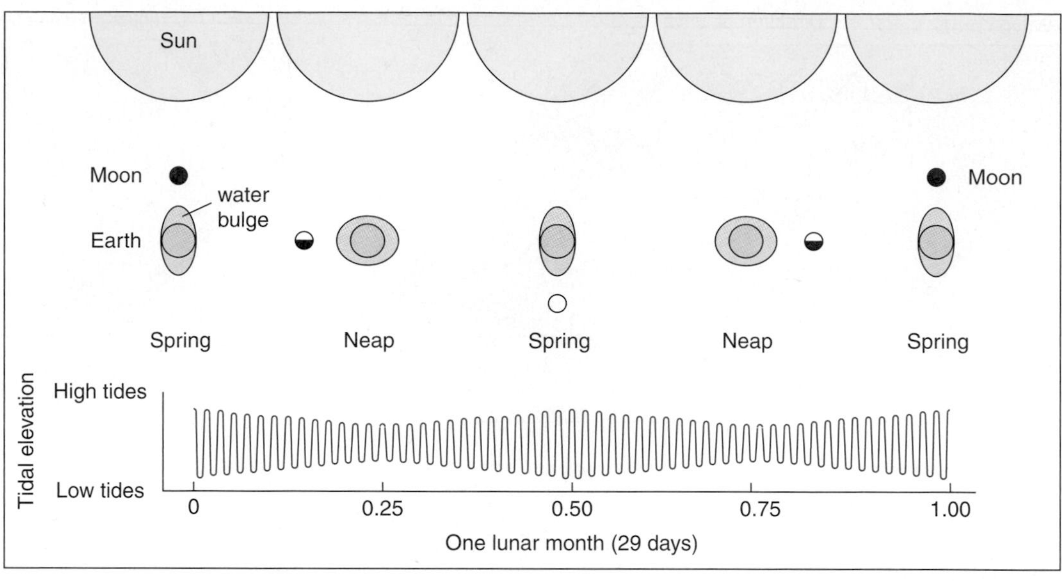

FIGURE 18.7

The tidal driving forces in a lunar cycle.

After Trenhaile, Geomorphology: A Canadian Perspective. *Toronto: Oxford University Press, 1998*

other on the opposite side of the Earth (figure 18.7). As the Earth rotates, these bulges travel around the Earth and cause water levels to rise (*flood tide*) and fall (*ebb tide*). The *tidal range* is the difference in sea level between high tide and low tide. In the open ocean, tidal ranges are less than 1 metre, but this can increase dramatically toward the coast. Some of the highest tidal ranges in the world are experienced in the Bay of Fundy between New Brunswick and Nova Scotia (see box 18.1), where sea level can fluctuate by more than 16 metres (figure 18.8). An incoming tide can form a visible "wave" of water, a *tidal bore*, which can reach speeds of up to 35 km/hr.

Tidal range at any one location also varies according to the interaction of the sun and the moon on a 29-day cycle. Maximum tide-generating forces occur twice a month when the Earth, sun, and moon are aligned. The tides generated at these times have a high range and are called *spring tides*. *Neap tides* occur when the forces of the sun and moon counteract each other.

Most regions experience a *semidiurnal* (twice a day) tidal cycle, with high tides occurring every 12.42 hours. High tides arrive approximately 50 minutes later each day due to the difference in time it takes for the Earth to fully rotate on its axis and the time required for the moon to orbit the Earth.

Tides can cause the shoreline to move considerable distances, especially where the coastline has low relief. The area exposed between high and low tides is called the *intertidal zone*. Extensive *tidal flats* may be exposed at low tide.

HOW DO BEACHES DEVELOP?

A **beach** is a strip of sediment (usually sand or gravel) that extends from the low-water line inland to a cliff or a zone of permanent vegetation. Waves break on beaches, and rising and falling tides may regularly change the amount of beach sediment that is exposed above water (figure 18.9*A*).

The steepest part of a beach is the **beach face,** which is the section exposed to wave action, particularly at high tide. Offshore from the beach face there is usually a **marine terrace,** a broad, gently sloping platform that may be exposed at low tide if the shore has significant tidal action. Marine terraces may be *wave-built* terraces constructed of sediment carried away from the shore by waves, or they may be *wave-cut* rock benches or platforms, perhaps thinly covered with a layer of sediment.

The upper part of the beach, landward of the usual high-water line, is the **berm,** a wave-deposited sediment platform that is flat or slopes slightly landward (figure 18.9*B*). It is usually dry, being covered by waves only during severe storms.

Beach sediment is usually sand, typically quartz-rich because of quartz's resistance to chemical weathering. Heavy metallic minerals ("black sands") can also be concentrated on some beaches as less dense minerals such as quartz and feldspar are carried away by waves or wind (titanium-bearing sands are mined on some beaches in Florida and Australia). Tropical beaches may be made of bioclastic carbonate grains from offshore corals, algae, and shells. Some Hawaiian beaches are made of sand-sized fragments of basalt. Gravel beaches are found on coasts attacked by the high energy of large waves (*shingle* is a regional name for disk-shaped gravel) and are common in Atlantic Canada, British Columbia, and the Arctic. Gravel beaches have a steeper face slope than sand beaches.

In seasonal climates, beaches often go through a summer–winter cycle (figure 18.10). This is due to the greater frequency of storms with strong winds during the winter months, which tend to produce high waves with short wavelengths. These high-energy waves tend to crash onshore and erode sand from the beach face and narrow the berm. Offshore, in less turbulent water, the sand settles to the bottom and builds an underwater sandbar (parallel to the beach) that serves as a "storage facility" for the next summer's sand supply. The following summer, or during calmer weather, lower-energy waves with long wavelengths break over the sandbar

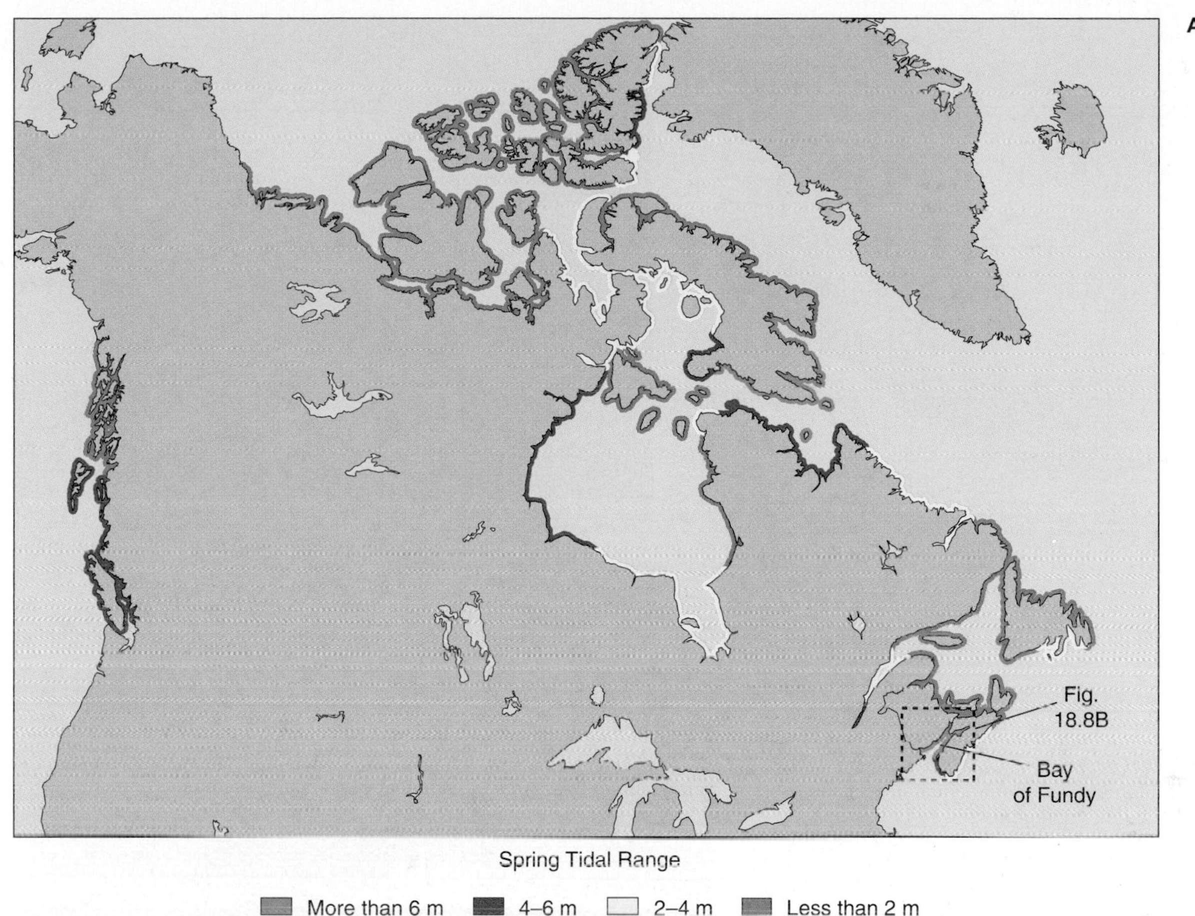

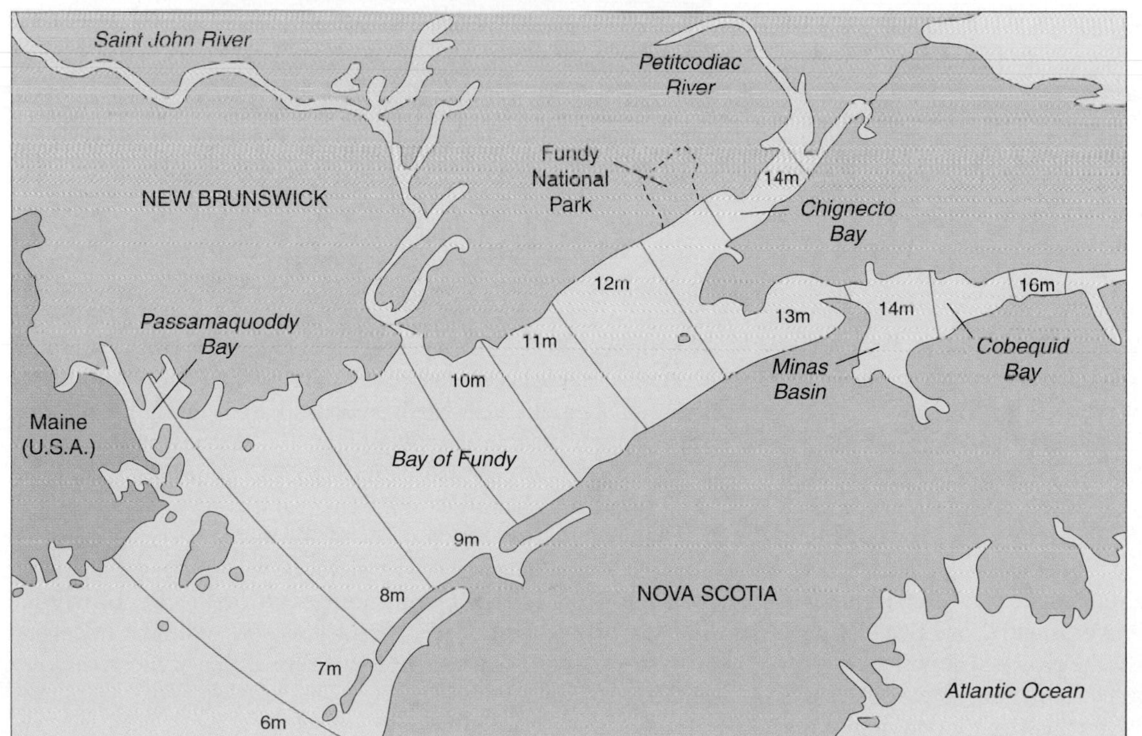

FIGURE 18.8

(*A*) Tidal ranges in Canada.
(*B*) Tidal range in the Bay of Fundy The tidal range increases toward the head of the Bay and reaches more than 16 m east of the Minas Basin.

Figure A after Trenhaile, 1998; Figure B from http://www.bayoffundy.com/mariner/images/tidalmapbig.jpg

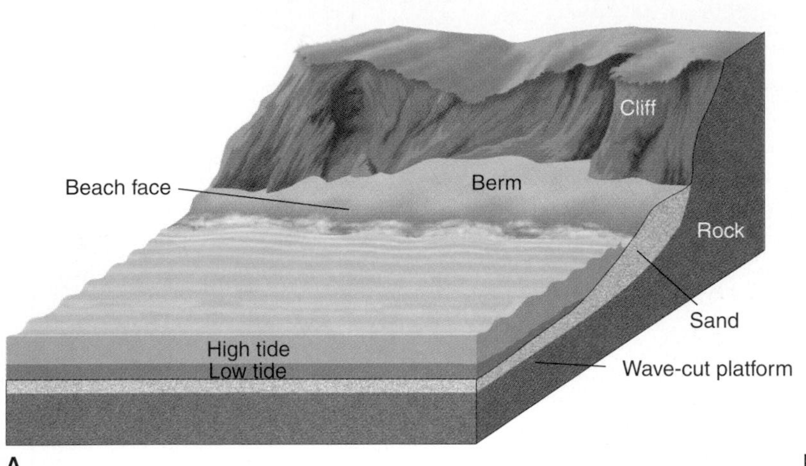

FIGURE 18.9
(*A*) Parts of a beach. (*B*) Beach face and narrow berm (on the far right) at Joggins, Nova Scotia. Parallel ridges on the beach face were formed by waves during previous high tides.
Photo by C.H. Eyles

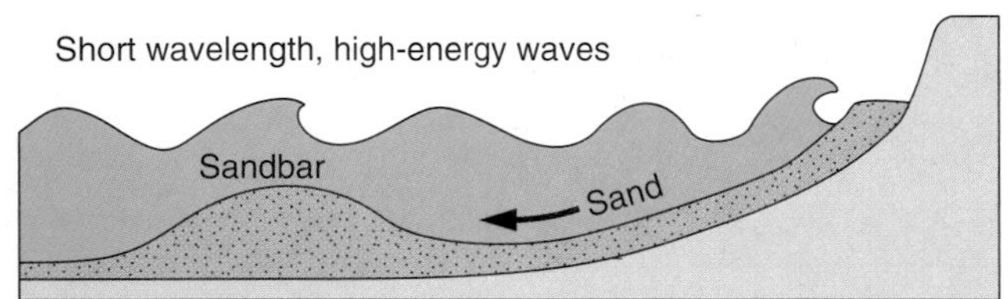

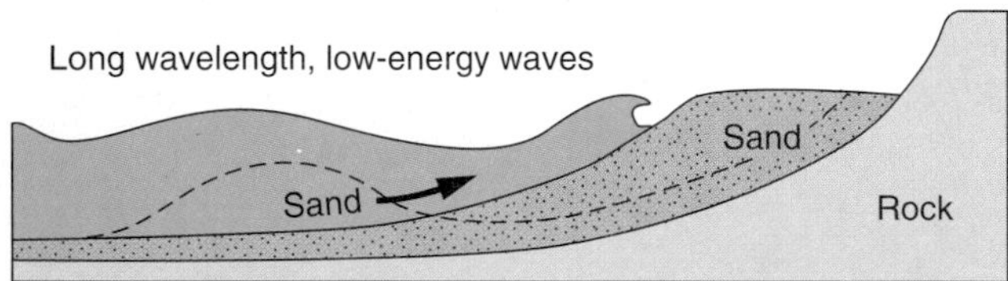

FIGURE 18.10
Seasonal influences on beach development. (*A*) Narrow winter beach. Waves may break once on the winter sandbar, then re-form and break again on the beach face. (*B*) Wide summer beach. (*C*) Modern ridge of sand and gravel (sea-ice rampart) pushed up against fossil rampart by winter pack ice. Tullett Point, western Prince Patrick Island, N.W.T.
Photo GSC Photo Number 2002-341. Reproduced with the permission of the Minister of Public Works and Government Services Canada, 2007 and Courtesy of Natural Resources Canada, and Geological Survey of Canada.

and gradually push the sand back onto the beach face to widen the berm. Each season the beach changes in shape until it comes into equilibrium with the prevailing wave type.

Many winter beaches can be dangerous because of high waves and narrowed beaches. Several beaches along the Pacific coast of North America are nearly free of accidents in the summer, when they are heavily used, but are regularly marked by drownings in the winter as beachwalkers are swept off narrower beaches out to sea by large storm waves. Such rogue waves are a particular hazard on the west coast of Vancouver Island (Long Beach, Pacific Rim National Park), where tourists are drowned almost every year.

Coastal features are affected by ice in many areas of Canada. Sea ice blown onshore by winds can push up mounds and ridges of sediment to form ramparts (figure 18.10*C*). These ice-push features are most common where moving pack ice is present for a large part of the year, such as coastal areas along the western side of the Canadian Arctic Islands and Parry Channel. Boulders and cobbles can also be pressed into underlying sediment by grounded slabs of ice in the intertidal zone to form boulder pavements and *boulder barricades*. Boulder barricades can be found along Hudson and Ungava Bays, along the coast of Labrador, on Baffin Island, and even in the St. Lawrence Estuary. Movement of ice floes across soft sediment in coastal bays and estuaries can also produce linear *ice-scour structures* and depressions.

GEOMATICS 18.1

Mapping Coastal Sediment Types

Coastal environments are dynamic, and monitoring their characteristics and changes over time is essential for sustainable management of coastal environments that are under increasing pressure from climate change, urban development, tourism, pollution, and erosion. Remote sensing of coastline characteristics using synthetic aperture radar (SAR) data obtained from aircraft or spacecraft can be used to map shorelines and identify substrate sediment types. These data can be used to create maps showing progressive shoreline change over time and environmental sensitivity or habitat maps. Polarimetric SAR measures a larger number of parameters than single-channel radar and can be used to identify and map specific coastal features and sediment types. Box figure 1 is a polarimetric SAR image of a gently sloping intertidal flat in the Minas Basin of the Bay of Fundy (see figure 18.8 for location map). Each of the colours shown can be interpreted to represent a different substrate type, as different sediment types will absorb or reflect different amounts of energy emitted by the radar system. This type of remotely sensed image provides a rapid and accurate means of mapping sediment types across the tidal flat, and may be used to identify sensitive areas that are especially threatened by environmental change.

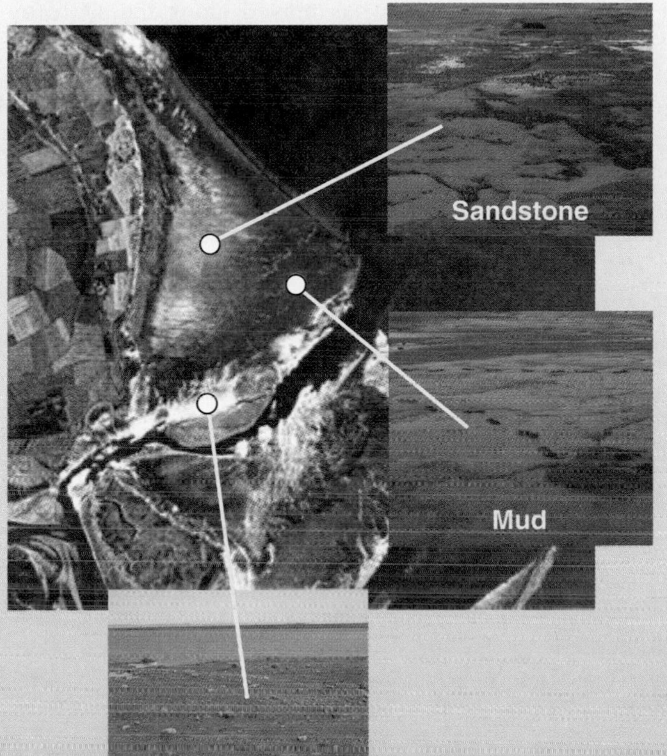

BOX 18.1 ■ FIGURE 1

Multi-polarization SAR image of an inter-tidal flat in the Minas Basin of the Bay of Fundy, Nova Scotia. Green colours represent areas in which sandstone bedrock is exposed. Blue represents areas underlain by mud and white areas are characterised by gravel and boulder deposits. The backscatter contrasts in the dry land area and tidal plain were enhanced independently to allow distinction of the different substrate types shown in this image.

Images from Applications Potential of RADARSAT-2 Supplement One; van der Sanden, J.J.; Thomas, S.J. 2004. Reproduced with the permission of the Minister of Public Works and Government Services Canada, 2007 and Courtesy of Natural Resources Canada, and Canada Centre for Remote Sensing

WHY DOES SAND MOVE ALONG A SHORELINE?

Longshore drift is the movement of sediment parallel to shore when waves strike the shoreline at an angle. Figure 18.11 shows the two ways in which this movement of sediment (usually sand) occurs. Some longshore drift takes place directly on the beach face when waves wash up on land. A wave washing up on the beach at an angle tends to wash sand along at the same angle. After the wave has washed up as far as it can go, the water returns to the sea by running down the beach face by the shortest possible route; that is, straight downhill to the shoreline, not back along the oblique route it came up. (Wave run-up is known as *swash*, the return as *backwash*.) The net effect of this motion is to move the sand in a series of arcs along the beach face.

Much more sand is moved by longshore transport in the surf zone, where waves are breaking into foam. The turbulence of the breakers erodes sand from the sea bottom and keeps it suspended. Even a weak longshore current can move the suspended sand parallel to the shoreline. The sand in the longshore current moves in the same direction as the sand drift on the beach face (figure 18.11).

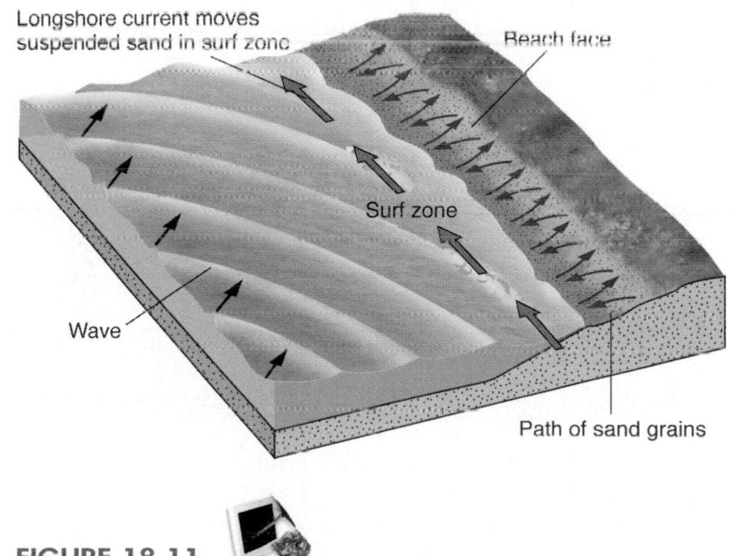

FIGURE 18.11

Longshore drift of sand on the beach face and by a longshore current within the surf zone.

Vast amounts of sand can be moved by longshore transport. The U.S. Army Corps of Engineers estimates that 436,000 cubic metres of sand per year are moved northward by waves at Sandy Hook, New Jersey, and 1,000,000 cubic metres of sand per year are moved southward at Santa Monica, California.

Eventually the sand that has moved along the shore by these processes is deposited. Sediment may build up off a point of land to form a **spit,** a fingerlike ridge of sediment that extends out into open water (figures 18.12*A,B*). A **baymouth bar,** a ridge of sediment that cuts a bay off from the open ocean or lake, is formed by sediment migrating across what was earlier an open bay (figures 18.12*A,C*). Off the western coast of North America, a considerable amount of drifting sand is carried into the heads of underwater canyons, where the sediments slide down into deep, quiet water.

A striking, but rare, feature formed by longshore drift is a **tombolo,** a bar of sediment connecting a former island to the mainland. As shown in figure 18.13, waves are refracted around an island in such a way that they tend to converge behind the island. The waves sweep sand along the mainland (and from the island) and deposit it at this zone of convergence, forming a bar that grows outward from the mainland and eventually connects to the island.

Human Interference with Sand Drift

Several engineered features can interrupt the flow of sand along a beach (figure 18.14). *Jetties,* for example, are rock walls designed to protect the entrance of a harbour from sediment deposition and storm waves. Usually built in pairs, they protrude above the

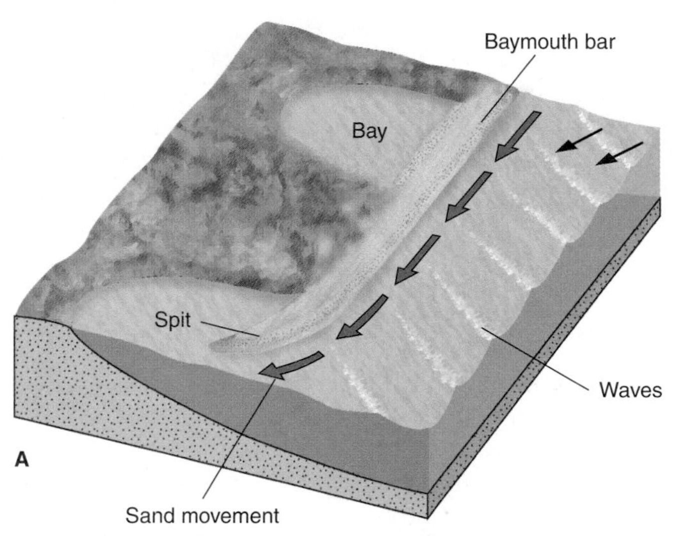

FIGURE 18.12

(*A*) Longshore drift of sand can form spits and baymouth bars. (*B*) Curved spit near Victoria, B.C. (*C*) The Burlington Bar separates Hamilton Harbour (right side of image) from the open waters of Lake Ontario (left side of image).

Photo B by Diane Carlson; Photo C from AIVA — Paul Nopper

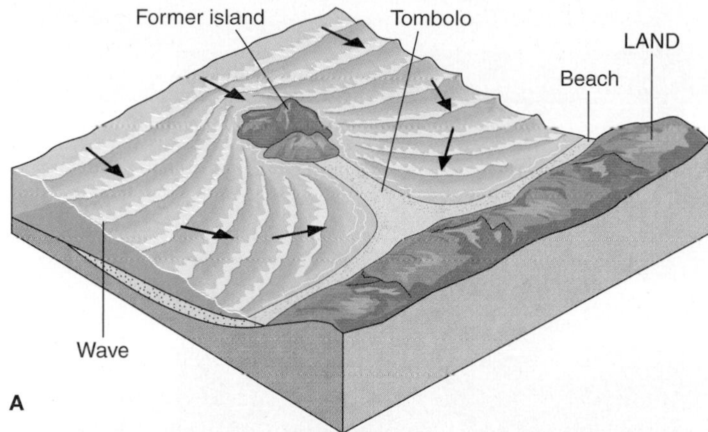

FIGURE 18.13

(*A*) Formation of a tombolo. Wave refraction around an island interrupts the longshore current and creates a sandbar that connects the island with the mainland. (*B*) Presqu'ile Provincial Park formed as a tombolo in Lake Ontario. Note also the extensive baymouth bars connecting headlands in the right-hand side of the image. *Photo from Google Earth*

surface of the water. Figure 18.14*A* shows how sand piles up against one jetty while the beach next to the other, deprived of a sand supply, erodes back into the shore.

Groins are sometimes built in an attempt to protect beaches that are losing sand from longshore drifting. These short walls are built perpendicular to shore to trap moving sand and widen a beach (figure 18.14*B*).

Sand deposition also occurs when a stretch of shore is protected from wave action by a *breakwater,* an offshore structure built to absorb the force of large breaking waves and provide quiet water near shore. When the city of Santa Monica in California built a rock breakwater parallel to the shore to create a protected small-boat anchorage, the lessening of wave action on the shore behind the breakwater allowed sand to build up there (figure 18.14*C*), threatening eventually to fill in the anchorage. The city had to buy a dredge to remove the sand from the protected area and re-deposit it farther along the shore where the waves could resume moving sediment.

A beach attempts to come into equilibrium with the waves that strike it. The type and amount of sediment, the position of the sediment, and especially the movement of the sediment adjust to the incoming wave energy. Whenever human activity interferes with sand drift or wave action, the beach responds by changing its configuration, usually through erosion or deposition in a nearby part of the beach.

Sources of Sand on Beaches

Some beach sand comes from the erosion of local rock, such as points of land or cliffs nearby. On a few beaches replenishment comes from sand stored outside the surf zone in the deeper water offshore. Bioclastic carbonate beaches are formed from the remains of marine organisms offshore. But the greater part of the sand on most beaches comes from river sediment brought down to the ocean. Waves pick up this sediment and move it along the beach by longshore drift.

What happens to a beach if all the rivers contributing sand to it are dammed? Although damming a river may be desirable for many reasons (flood control, power generation, water supply, recreation), when a river is dammed its sediment load no longer reaches the sea (see box 14.1). The sand that supplied the beach in the past now comes to rest in the quiet waters of the reservoir behind the dam. Longshore drift, however, continues to remove sand from beaches even though little new sand is being supplied, and the result is a net loss of sand from beaches. Beaches without a sand supply eventually disappear. To prevent this, some coastal communities have set up expensive programs of building pipelines or draining reservoirs and trucking the trapped sand down to the beaches.

WHY ARE THERE MANY DIFFERENT TYPES OF COASTS?

A beach is just a small part of the **coast,** which is all the land near the sea, including the beach and a strip of land inland from it. Coasts can be rocky, mountainous, and cliffed, as in Newfoundland, Nova Scotia, and British Columbia; or they can be broad, gently sloping plains, as along the Gulf of St. Lawrence, the Hudson Bay coasts of Ontario and Manitoba, and the Beaufort Sea. Wave erosion and deposition can greatly modify coasts from their original shapes. Many coasts have been drowned during the past 15,000 years by the rise in sea level caused by the melting of Quaternary glaciers (see chapter 16). Other coasts have been lifted up by tectonic forces at a rate greater than the rise in sea level so that sea-floor features are now exposed on dry land.

Erosional Coasts

A great many steep, rocky coasts have been visibly changed by wave erosion. Soluble rocks such as limestone dissolve as waves wash against them, and more durable rocks such as granite are

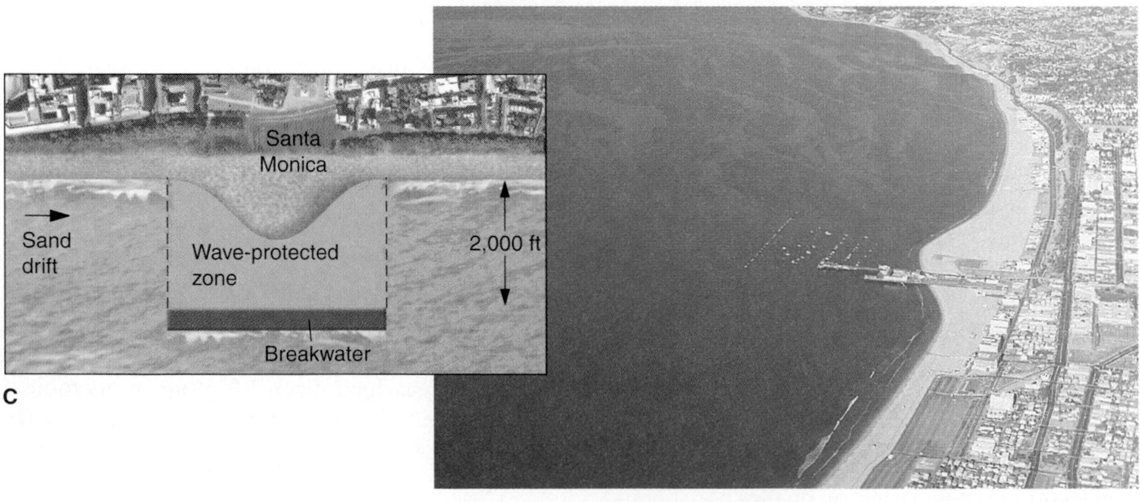

FIGURE 18.14

Sand piles up against obstructions and in areas deprived of wave energy. (*A*) Jetties at Manasquan Inlet, New Jersey. Sand drift to the right has piled sand against the left jetty and removed sand near the right jetty. (*B*) Groins at Ocean City, New Jersey. Sand drift is to the right. (*C*) Breakwater at Santa Monica, California, has caused deposition of sand in the wave-protected zone.

Photos A and B by S. Jeffress Williams: Photo C © John S. Shelton

fractured by the enormous pressures caused by waves slamming into rock (wave impact pressures have been measured as high as 60 metric tonnes per square metre).

An irregular coast with bays separated by rocky **headlands** (points of land) can be gradually straightened by wave action. Because wave refraction bends waves approaching such a coast until they are nearly parallel to shore, most of the waves' energy is concentrated on the headlands, while the bays receive smaller, diverging waves (figure 18.15). Rocky cliffs form from wave erosion on the headlands. The eroded material is deposited in the quieter water of nearby bays, forming broad beaches. **Coastal straightening** of an irregular shore gradually takes place through wave erosion of headlands and wave deposition in bays.

Wave erosion of headlands produces **sea cliffs** or lakeshore bluffs, steep slopes that retreat inland by mass wasting as wave erosion undercuts them (figure 18.16). At the base of sea cliffs are sometimes found *sea caves,* cavities eroded by wave action along zones of weakness in the cliff rock. As headlands on irregular coasts are eroded landward, sea cliffs enlarge until the entire coast is marked by a retreating cliff. On some exposed coasts the rate of cliff retreat can be quite rapid, particularly if the rock is weakly consolidated or if the cliffs are made of unlithified sediment. Some sea cliffs north of San Diego, California and at Cape Cod National Seashore in Massachusetts are retreating at an average rate of 1 metre per year. Because sea-cliff erosion in weak rock is often in the form of large, infrequent slumps (see chapter 13), some portions of these coasts may retreat 10 to 30 metres in a single storm. Average clifftop retreat rates of over 4 m/yr have been recorded along sections of the Nova Scotia coastline underlain by unlithified Quaternary-age sediments. Some of these cliffs have "ocean-view" homes and hotels at their very edges (see box 18.2). Sea cliffs in hard, durable rock such as granite and schist retreat much more slowly.

In Canada, coastal erosion is enhanced by frost action, where water-saturated rocks are subject to weathering caused by repeated freeze–thaw cycles. Such weathering processes are particularly important in the formation of sea cliffs, arches, and stacks along the coasts of eastern Canada and along the shorelines of the Great Lakes, where much of the erosion occurs above the level of wave action (figure 18.17).

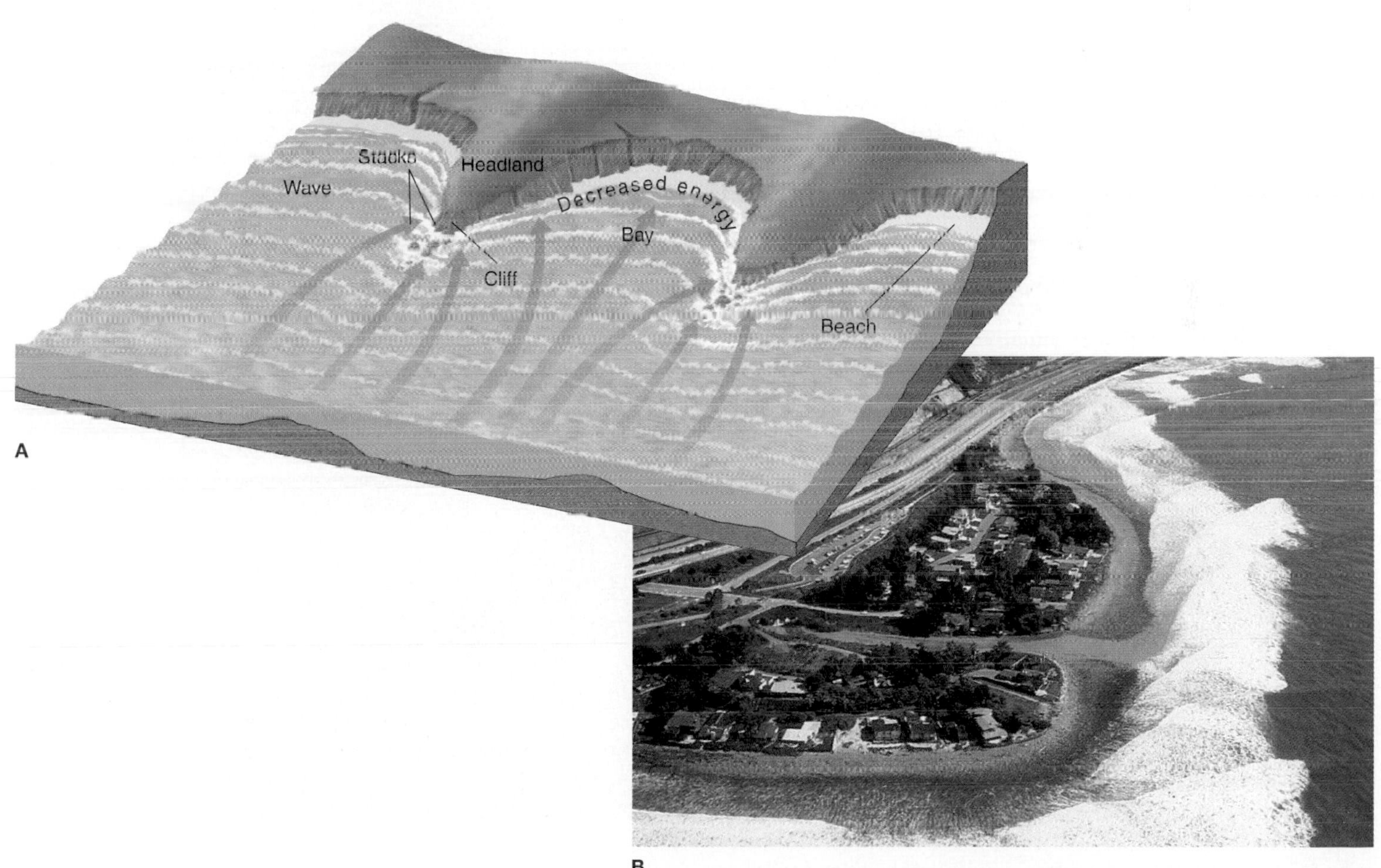

FIGURE 18.15

(*A*) Wave refraction on an irregular coast. Arrows show transport direction of wave energy, concentrated on headlands, spread out in bays. (*B*) Wave refraction around a point of land, Rincon Point, California. Note waves at right centre have been bent 90 degrees from their original direction (parallel to the bottom of the photo).

Photo by Frank M. Hanna

FIGURE 18.16

Retreating wave-cut cliff, north of Bodega Bay, Sonoma County, California. A concrete seawall has been built at the base of the cliff to slow wave erosion and help protect the cliff-edge homes. Note fragments of wave-destroyed structures near the seawall. Seawalls usually increase the erosion of sand beaches.

Photo by David McGeary

A

B

FIGURE 18.17

(*A*) Sea arch and sea stacks at Hopewell Cape, New Brunswick. (*B*) Stacks ("flowerpots") eroded from dolostone exposed along the Niagara Escarpment, Georgian Bay, Lake Huron.

Photo A by Panoramic Images/Getty; Photo B by S. Fletcher

Seawalls may be constructed along the base of retreating cliffs to prevent wave erosion (figure 18.16). Seawalls of giant pieces of broken stone (rip-rap) or concrete tetrahedrons are designed to absorb wave energy rather than allow it to erode cliff rock. Vertical or concave seawalls of concrete, such as the one in Galveston, Texas (see box 18.3 figure 3), are designed to reflect wave energy seaward rather than allow it to impact the shore. Some of the reflected energy, however, is focused at the base of the seawall, which eventually undermines it and causes the seawall to collapse. Reflection of waves from a seawall also increases the amount of wave energy just offshore, often increasing the amount of sand erosion offshore. Thus, a seawall designed to protect a sea cliff (and the buildings at its edge) may in some cases destroy a sand beach at the base of the cliff. Seawalls are difficult and expensive to build and maintain, and they may destroy beaches, but political pressure to build more of them will increase as the sea level rises in the future.

Wave erosion produces other distinctive features in association with sea cliffs. A **wave-cut platform** (or *terrace*) is a horizontal bench of rock formed beneath the surf zone as a coast retreats by wave erosion (figure 18.18). The platform widens as the sea cliffs retreat. The depth of water above a wave-cut platform is generally 6 metres or less, coinciding with the depth at which turbulent breakers actively erode the sea bottom. **Stacks** are erosional remnants of headlands left behind as the coast retreats inland (figure 18.17 and 18.18). They form small, rocky islands off retreating coasts, often directly off headlands (figure 18.15). **Arches** (or *sea arches*) are bridges of rock left above openings eroded in headlands or stacks by waves. The openings are eroded in spots where the rock is weaker than normal, perhaps because of closely spaced fractures (figure 18.17*A*).

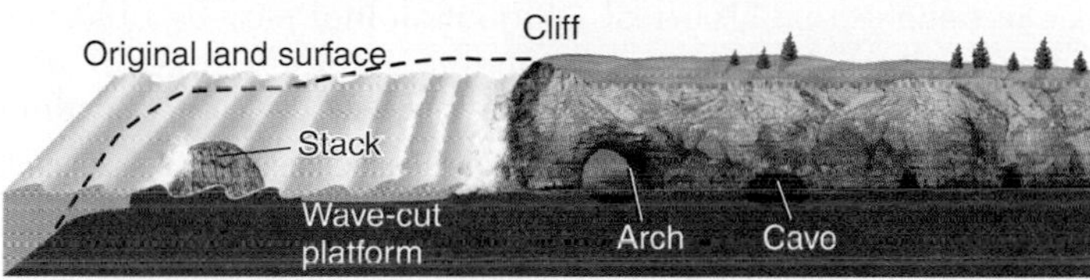

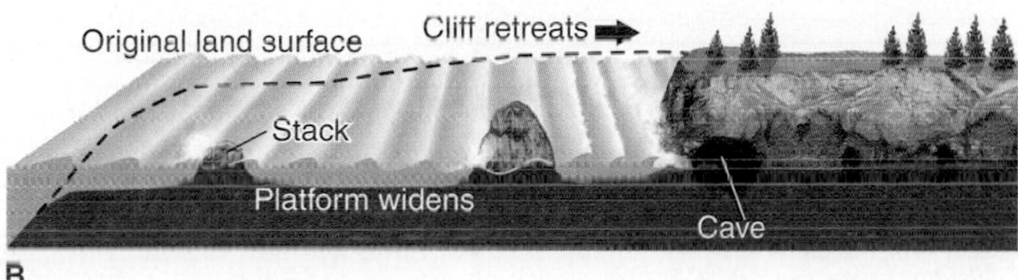

FIGURE 18.18

(*A*) A wave-cut platform (the wide, horizontal bench of dark rock at the base of the cliffs) is exposed at low tide, La Jolla, California. (*B*) A wave-cut platform widens as a cliff retreats.

Photo by David McGeary

Depositional Coasts

Many coasts are gently sloping plains and show few effects of wave erosion. Such coasts are found along most of the Atlantic Ocean and Gulf of Mexico shores of North America. These coasts are primarily shaped by sediment deposition, particularly by longshore drift of sand.

Coasts such as these are often marked by **barrier islands**—ridges of sand that parallel the shoreline and extend above sea level (figure 18.19). These barrier islands may have formed from sand eroded by waves from deeper water offshore, or they may be greatly elongated sand spits formed by longshore drift. The slowly rising sea level associated with the melting of Quaternary glaciers may have been a factor in their development. A protected lagoon separates barrier islands from the mainland. Because the lagoon is protected from waves, it provides a quiet waterway for boats. A series of such lagoons stretches almost continuously from New York to Florida, and many also exist along the Gulf Coast, forming an important route for barge traffic. Prince Edward Island National Park and Kouchibouguac National Park (New Brunswick) are also located on barrier islands. As tides rise and fall, strong tidal currents may wash in and out of gaps between barrier islands, distributing sand in submerged *tidal deltas* both landward and seaward of the gaps.

FIGURE 18.19

(*A*) A barrier island on a gently sloping coast. A lagoon separates the barrier island from the mainland, and tidal currents flowing in and out of gaps in the barrier island deposit sediment as submerged tidal deltas. (*B*) A barrier island near Pensacola, Florida: open ocean to right, lagoon to left, and mainland Florida on far left. Light-coloured lobes of sand within lagoon were eroded from the barrier island by hurricane waves, which washed entirely across the island and into the lagoon.

Photo by Frank M. Hanna

ENVIRONMENTAL GEOLOGY 18.2

Communities and Coastal Erosion

Coastal erosion is a severe problem in many communities located along the shores of lakes or the ocean. In areas where the coastline is rocky retreat may be barely noticeable, but in areas where the coast is composed of loose sands, gravels, and mud, erosion rates may be high. In Atlantic Canada, erosion rates of up to 12 metres per year have been recorded, although most coastlines retreat at rates of less than 1 metre per year. Several islands have disappeared from the Atlantic coastline during historic times due to wave erosion, and shoreline structures such as roads, docks, homes, and lighthouses are being increasingly affected (box figure 1). Unconsolidated cliffs of sand and silt along the Pacific coast close to Vancouver were eroding at average rates of 60 centimetres per year before gravel was placed along the shoreline to slow erosion rates in 1982.

Coastal erosion is also a severe problem along lake shorelines, particularly along the Great Lakes. The erosion of spectacular cliffs of unconsolidated sand and silt along the Scarborough Bluffs near Toronto at average rates of up to 1 m per year threatened clifftop communities during the 1980s (box figure 2*A*). This prompted the construction of extensive toe protection structures along the foot of the cliffs (box figure 2*B*).

Erosion along Arctic coasts is directly affected by the amount of sea ice that develops each year. Most Arctic coastlines are protected from wave action by the development of shore ice. However, in years when little sea ice develops, shoreline erosion can cause severe damage. The community of Tuktoyaktuk on the coast of the Beaufort Sea experienced considerable shoreline erosion (box figure 3) in recent years when ice development was limited. Shoreline protection structures are currently being built to protect this coastline, which is particularly vulnerable to the effects of climate change.

Strategies for coastal zone management in the Canadian Arctic benefit from historical information that may be obtained from analysis of successive satellite images. Landsat and RADARSAT images of the Canadian Beaufort Sea coastline obtained over the period 1973 to 1999 have been used to create a composite image showing areas particularly susceptible to coastal erosion (box figure 4).

A

B

BOX 18.2 ■ FIGURE 1

(A) Massive erosion of sidewalks, curbs, and pavement in Corner Brook, Nfld.
(B) Parts of an apartment building falls into the Sable River in Jonquiere, 1996.
Photo A CP/Corner Brook Western Star/Cory Hurley; photo B CP/Jacques Boissinot

A

B

BOX 18.2 ■ FIGURE 2

(A) Communities in Scarborough, Ontario threatened by erosion of coastal bluffs in the 1980s. *(B)* Toe berm constructed along the base of the Scarborough Bluffs to reduce wave erosion. The indentations along the berm are designed as fish habitats.

Photos by Nick Eyles

BOX 18.2 ■ FIGURE 3

Cliff erosion along the Beaufort Sea Coast has accelerated in recent years due to reduced sea ice development.

Photo by Steve Solomon, Geological Survey of Canada Atlantic

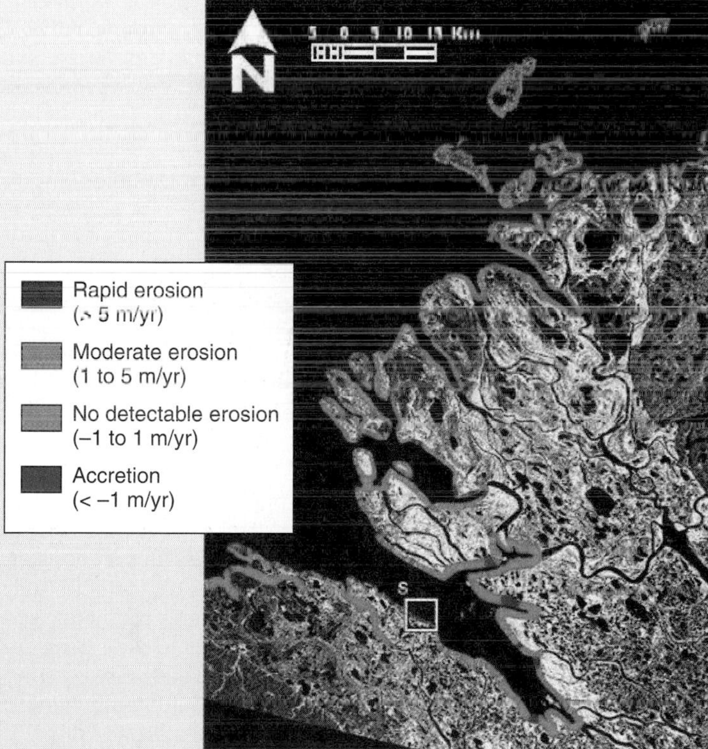

BOX 18.2 ■ FIGURE 4

Map showing coastal erosion classes along part of the Beaufort Sea coastline. A RADARSAT image acquired in 1999 was chosen as the background.

RADARSAT image of Beaufort Sea Coastline 1999 From http://ccrs.nrcan.gc.ca/radar/marine/beaufort_e.php Reproduced with the permission of the Minister of Public Works and Government Services Canada, 2005 and Courtesy of Natural Resources Canada, Geological Survey of Canada

ENVIRONMENTAL GEOLOGY 18.3

The Effects of Rising Sea Level

Long Term

Sea level has risen about 130 metres in the past 15,000 years as the Quaternary glaciers melted, adding water to the oceans. If all the glacial ice on Earth were to melt, sea level would rise 70 metres, drowning many coastal areas.

On low-lying coasts a rise in sea level would cause ocean waves (particularly storm waves) to extend much farther inland than before, flooding the land (box figure 1*A*). Barrier islands migrate landward. In the United States from New Jersey south to Florida, and along the Gulf of Mexico to southern Texas, the coastal land is very flat. A small rise in sea level can send sea water many kilometres inland.

On steep coasts a rise in sea level can accelerate the erosion of coastal cliffs, destroying oceanfront property (box figure 1*B*).

Some scientists predict that global sea level will rise in the next century as a result of global warming by the greenhouse effect. Burning of coal and oil increased the amount of carbon dioxide in the atmosphere by more than 10 percent in the twentieth century. The carbon dioxide and other gases may trap more of the sun's energy on Earth, warming the air and the sea. Warm air would accelerate the melting of glaciers on Greenland and Antarctica. Warm sea water expands. Together, these two effects could raise sea level.

Predictions on the amount of sea level rise by the year 2100 vary widely. Early predictions of a rise of 1 to 2 metres have recently been revised downward to .3 to .6 metre as computer models were refined (by considering cloud cover, for example). A rise of even 1 metre on a low-lying coast could destroy thousands of beachfront houses and hotels.

Global warming is an emotional and controversial subject. The recent rise in atmospheric carbon dioxide is indisputable, but its relationship to global climate change is debatable. The debate over the causes of global warming does not change the fact that glaciers are now melting (as they have for 15,000 years) and sea level is now rising. Low-lying coastal areas of Atlantic Canada are particularly sensitive to the effects of sea-level rise as the land surface in the area is slowly subsiding. In Halifax, sea level is rising at a rate of 36 cm per century due to the combined effects of land subsidence and global sea-level rise.

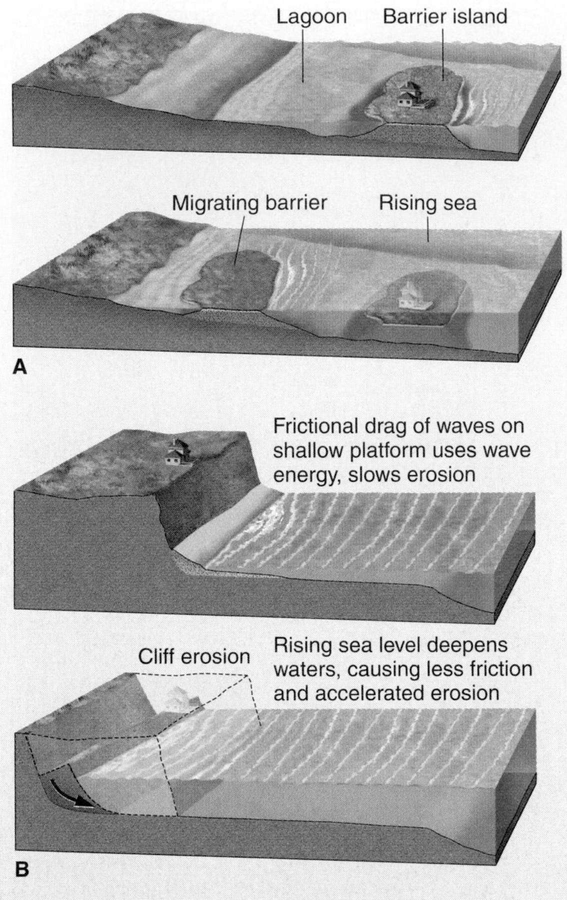

BOX 18.3 ■ FIGURE 1
Rising sea level can cause erosion of both gentle (*A*) and steep (*B*) coasts, leading to destruction of buildings.

Short Term

Hurricanes, which are common along the Atlantic and Gulf coasts, tend to cause short-term rises in sea level called *storm surges.* Hurricanes consist of strong winds rotating counterclockwise (in the northern hemisphere) around a region of very low air pressure. They may be 500 kilometres in diameter and contain winds up to 300 kilometres per hour. The low air pressure in the centre of a hurricane allows the sea surface to rise forming a broad dome, and this rise in sea level is accentuated by strong onshore winds, which pile water against shore (box figure 2). Storm surges can easily raise sea level 5 metres, and are most devastating at high tide. A storm surge 8 metres high struck Galveston, Texas in 1900, completely covering the barrier island that Galveston is built upon. High storm waves on top of the high sea level destroyed countless buildings, and 6,000 people died, many by drowning. Hurricane-tracking programs cut the death tolls of storm surges today by providing advance warning to coastal communities. In September 1989, hurricane Hugo hit South Carolina with 220-kilometre winds and a 5-metre storm surge north of Charleston, causing $10 billion in damage. Because more than 500,000 people were evacuated along the low-lying coast, however, the death toll was only 29.

In August 2005 Hurricane Katrina made landfall along the Central Gulf Coast as a category 4 storm and caused extensive shoreline damage and flooding in New Orleans as the storm surge breached the levee system that protects the city. Over 1,300 lives were lost, 1 million people were evacuated, and damage is estimated at between $100 and $200 billion—making Katrina the most expensive natural disaster in U.S. history.

Large storms also commonly batter the low-lying coastal areas of Atlantic Canada, and many communities have invested in costly coastal protection measures. The west coast of Canada has an average of three storms per year with hurricane-force winds, but storm surges are less significant than on the east coast due to the steep coastal topography.

Careful planning for the use of coastal land in your lifetime will be necessary to lessen destruction caused by storm surges and long-term sea-level rise. Developed land at the ocean's edge may face a grim choice in the future—abandonment of existing buildings or expensive "armouring" of the coast with structures such as seawalls in hopes of protecting the buildings from wave damage (box figure 3).

Further Reading

Williams, S. J., K. Dodd, and K. K. Gohn. 1990. Coasts in crisis. U.S. Geological Survey Circular 1075. Online version can be found at http://pubs.usgs.gov/circular/c1075/.

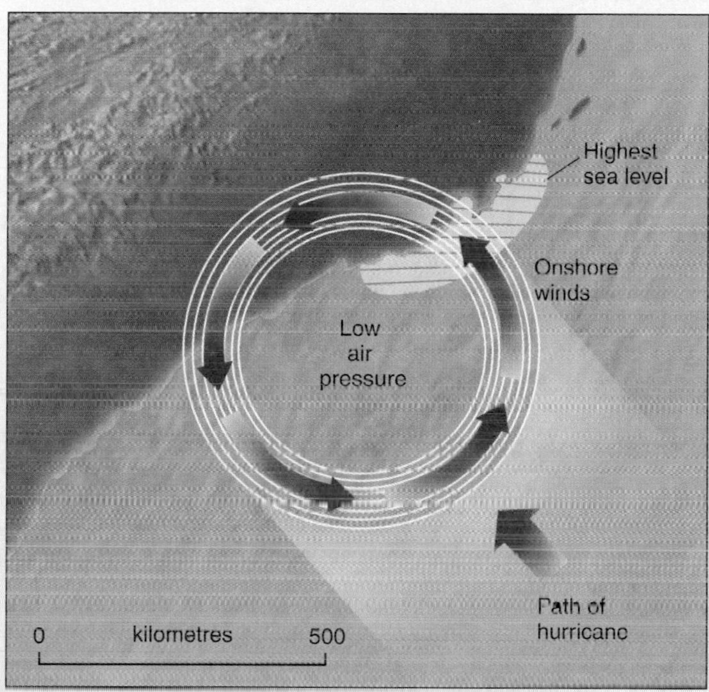

BOX 18.3 ■ FIGURE 2

Strong onshore winds in a hurricane pile water against the shore, forming a storm surge (high sea level) that may cause severe flooding on a low-lying coast. This is particularly a problem where irregular-shaped bays and estuaries (i.e. Lake Pontchartrain) funnel the water onshore. Damage is worse at high tide.

A

B

BOX 18.3 ■ FIGURE 3

(A) Satellite image of Hurricane Katrina taken on August 28, 2005 as it grew to a monster Category 5 hurricane in the Gulf of Mexico with sustained winds up to 233 kilometres per hour. (B) Damage caused by Hurricane Katrina in Gulfport, Mississippi where splintered wood from houses completely destroyed by the high winds and storm surge washed inland along with shipping containers, boats, and recreational vehicles

Photo A by NOAA; Photo B © Paul J. Richards/AFP/Getty Images

Some barrier islands along the Atlantic and Gulf Coasts are densely populated. Atlantic City (New Jersey), Ocean City (Maryland), Miami Beach (Florida), and Galveston (Texas) are examples of cities built largely on barrier islands. In some of these cities, houses, luxury hotels, and condominiums are clustered near the edge of the sea; many are built upon the loose sand of the island (figure 18.20). These developed areas are vulnerable to late-summer hurricanes that sooner or later bring huge storm waves onto these coasts, eroding the sand and undermining the building foundations at the water's edge. The barrier island systems found along the Mississippi and Louisiana Gulf coasts have been extensively damaged in recent hurricane seasons; Hurricane Katrina in August 2005 almost destroyed the Chandeleur Islands, which lie approximately 100 km east of New Orleans (figure 18.21).

Nonmarine deposition may also shape a coast. Rapid sedimentation in *deltas* by rivers can build a coast seaward (see chapter 14). *Glacial deposition* can form shoreline features. Several islands off the New England coast were glacially deposited; Long Island, New York formed from the deposition of a recessional end moraine.

FIGURE 18.20
Hotels built upon the loose sand of a barrier island, Miami Beach, Florida.
Photo © Werner Bertsch/Bruce Coleman

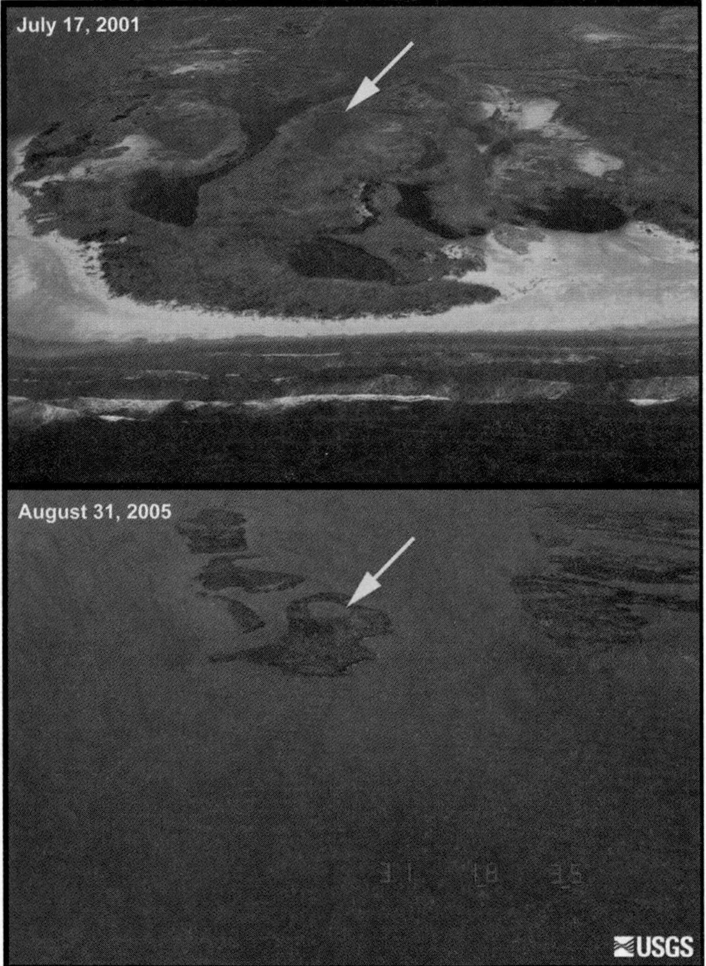

FIGURE 18.21
Part of the Chandeleur islands before (upper image—July 2001) and after (lower image—August 2005) extensive damage caused by recent hurricanes including Hurricane Katrina, which occurred two days before the lower photo was taken. Few recognizable landforms were left on the island chain following the hurricanes.
Photo from USGS: http://coastal.er.usgs.gov/hurricanes/katrina/photo-comparisons/chandeleur.html

Drowned Coasts

Drowned (or *submergent*) coasts are common because sea level has been rising worldwide for the past 15,000 years. During the glacial ages of the Quaternary, sea level was 130 metres below its present level. The shallow sea floor near the continents was then dry land and rivers flowed across it, cutting valleys. As the great ice sheets melted, sea level began to rise, drowning the river valleys. These drowned river mouths, called **estuaries,** mark many coasts today (figure 18.22). They extend inland as long arms of the sea. Fresh water from rivers mixes with the sea water to make most estuaries brackish. The quiet, protected environment of estuaries makes them very rich in marine life, particularly the larval forms of numerous species. Unfortunately, cities and factories built on many estuaries to take advantage of quiet harbours are severely polluting the water and the sediment of the estuaries. The poor circulation that characterizes most estuaries hinders the flushing away of this pollution, and estuary shellfish are sometimes not safe to eat as a result. Halifax Harbour, Nova Scotia is an estuary that has become severely polluted by sewage discharged into the harbour from numerous points.

Drowned coasts may be marked by **fjords,** glacially cut valleys flooded by rising sea level (figure 18.22*B*). They form in the same way as estuaries, except they were cut by glacial ice rather than rivers during low-sea-level stands (see chapter 16).

Uplifted Coasts

Uplifted (or *emergent*) coasts have been elevated by deep-seated tectonic forces. The land has risen faster than sea level, so parts of the old sea floor are now dry land.

Marine terraces form just offshore from the beach face, as described earlier in this chapter. These terraces can be wave-cut platforms caused by erosion of rock associated with cliff retreat, or they can be wave-built terraces caused by deposition of sediment. If the shore is elevated by tectonic uplift, these flat surfaces will become visible as *uplifted marine terraces* (figure 18.23). They formed below the ocean surface but are visible now because of uplift. The tectonically unstable Pacific coast of North America has many areas marked by uplifted terraces, along with the erosional coast features described earlier.

Coasts Shaped by Organisms

The growth of coral and algal *reefs* offshore can shape the character of a coast. The reefs act as a barrier to strong waves, protecting the shoreline from most wave erosion. Carbonate sediments blanket the sea floor on both sides of a reef and usually form a carbonate sand beach on land (see figure 9.18). Southernmost Florida has a coast of this type.

Branching *mangrove roots* dominate many parts of the southeastern United States coast. The roots dam wave and current action, creating a quiet environment that provides a haven for the larval forms of many marine organisms, and may trap fine-grained sediment. Mangroves also deposit layers of organic peat on low-lying coasts.

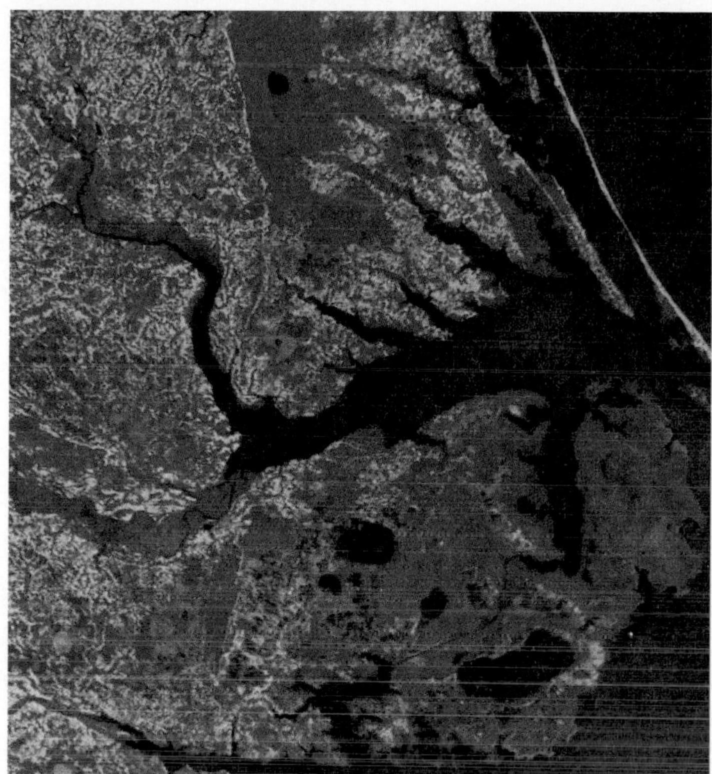

A

B

FIGURE 18.22

(A) Landsat satellite photo of estuaries Albemarle and Pamlico Sounds, North Carolina. Barrier islands are visible in upper right. Infrared image shows vegetation as red. (B) Steep-sided glacial trough flooded by ocean waters to form a fjord, Keel Bay, eastern Baffin Island.

Photo A by NASA; Photo B by GSC Photo 2002–236. Keel Bay, eastern Baffin Island. Photographer: Douglas Hodgson. Reproduced with the permission of the Minister of Public Works and Government Services Canada, 2007 and Courtesy of Natural Resources Canada, Geological Survey of Canada.

FIGURE 18.23

Uplifted marine terrace, northern California. The flat land surface at the top of the sea cliff was eroded by wave action, then raised above sea level by tectonic uplift. The rock knob on the terrace was once a stack.

Photo by David McGeary

WHAT FEATURES CAN BE FOUND ON THE OCEAN FLOOR?

Beyond the coastline bordering continents and islands, the ocean floor generally slopes gently seaward. Along continental margins this gently sloping area of the floor is called the **continental shelf.** The continental shelf passes seaward into the more steeply sloping **continental slope** that leads down to the deep ocean floor (figure 18.24). The topography of the ocean floor depends to a large extent on the plate tectonics setting in which it is found (see chapter 2).

A **passive continental margin,** such as that found on the eastern coast of North America, is not tectonically active and usually lies at the margin of an extending ocean (e.g., the Atlantic). The ocean floor in these passive-margin settings is characterized by a broad continental shelf, a continental slope, and a continental rise that passes into the flat **abyssal plain** (figure 18.24). Modern continental shelves are underlain by continental crust and, even though they are covered by sea water today, are truly part of the continents (figure 18.25).

Active continental margins, mainly found around the Pacific Rim, are associated with earthquakes and volcanoes. In these active-margin settings, the continental shelf is often narrow and the continental slope extends much deeper, to form one wall of an **oceanic trench** (figure 18.24). Abyssal plains are seldom found off active margins. The deep ocean floor seaward of trenches is hilly and irregular, unlike the flat and featureless abyssal plains. Encircling the globe are a series of **mid-oceanic ridges,** usually (but not always) near the centre of an ocean. Conical seamounts rise above the sea floor in some regions (figure 18.24).

Continental Shelves

A continental shelf is the shallow submarine platform at the edge of a continent and inclines very gently seaward, generally at an angle of 0.1 degrees. Continental shelves vary in width. On the Pacific coast of North America the shelf is only a few kilometres wide, but off Newfoundland in the Atlantic Ocean it is about 500 kilometres wide. Portions of the shelves in the Arctic Ocean off Siberia and northern Europe are even wider. Water depth over a continental shelf tends to increase regularly away from land, with the outer edge of the shelf being about 100 to 200 metres below sea level.

The continental shelves of the world are usually covered with sediment, and most of this is derived from land. Rivers transport sediment to the oceans, where it is then moved by currents generated by waves and tides. In relatively shallow water near to shore, sand-sized sediment can be transported by currents and waves producing a zone of sand deposition in the shallow areas of a shelf. Some of this sand may be moved father offshore during storms when waves and currents are particularly powerful. The outer parts of continental shelves are usually blanketed by mud, as only fine-grained sediment is able to be carried by the more slowly moving currents that operate in deeper water. However, many outer continental-shelf areas are covered by coarse-grained sediment that was deposited near to the shore when sea levels were much lower. The growth and decay of glaciers on the continents during the past 2 million years caused sea level to rise and fall many times, by up to 200 metres. Coarse-grained sediments on the outer continental shelves are considered to be "relic" deposits that are not related to modern depositional processes.

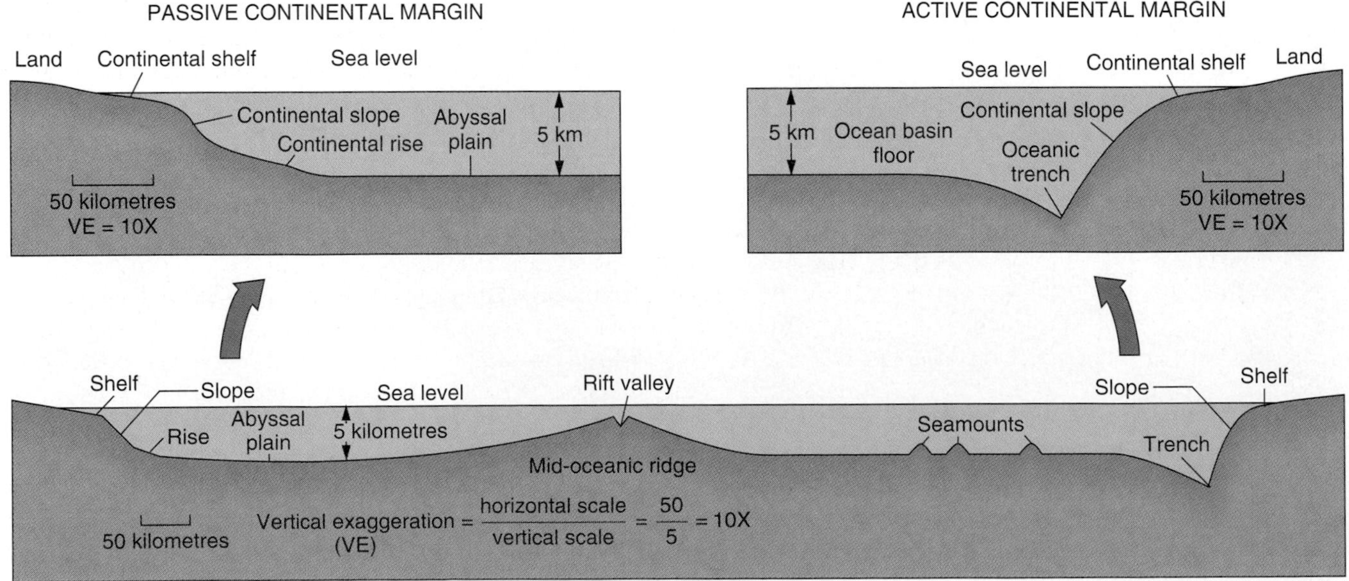

FIGURE 18.24

Profiles of sea floor topography. The vertical scales differ from the horizontal scales, causing vertical exaggeration, which makes slopes appear steeper than they really are. The bars for the horizontal scale are 50 kilometres long, while the same distance vertically represents only 5 kilometres, so the drawings have a vertical exaggeration of 10.

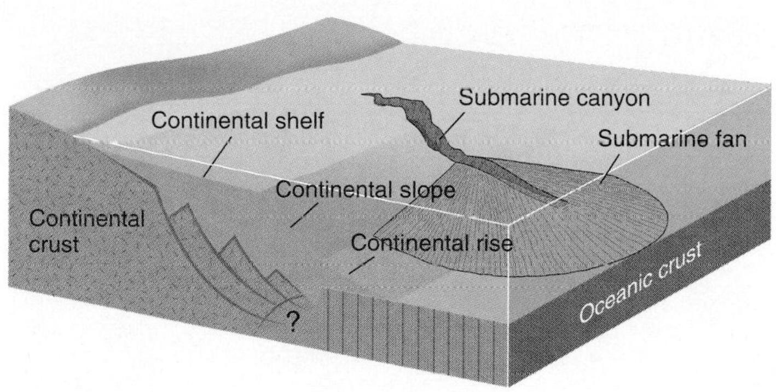

FIGURE 18.25

The continental shelf lies upon continental crust, and the continental rise lies upon oceanic crust. The complex transition from continental crust to oceanic crust lies under the continental slope.

Some areas of the continental shelf are not supplied with sediment from rivers; the water is clear, with little suspended material. Such clear, sediment-free waters allow the development of **reefs** in warm, shallow areas where sunlight can penetrate to the ocean floor. Reefs are wave-resistant ridges of coral, algae, and other calcareous organisms that stand above the surrounding sea floor, which is often covered by calcareous sediment derived from the reef (see figure 9.19). Three important types are *fringing reefs, barrier reefs*, and *atolls* (figure 18.26).

Fringing reefs are flat, tablelike reefs attached directly to shore. The seaward edge is marked by a steep slope leading down into deeper water. Many of the reefs bordering the Hawaiian Islands are of this type.

Barrier reefs parallel the shore but are separated from it by wide, deep lagoons. This type of reef is shown in figure 9.18. The lagoon has relatively quiet water because the reef shelters it by absorbing the energy of large, breaking waves. A barrier reef lies about 8 kilometres offshore of the Florida Keys, a string of islands south of Miami. On a much grander scale is the Great Barrier Reef off northeastern Australia. It extends for about 2,000 kilometres along the coast, and its seaward edge lies up to 250 kilometres from shore. Another long barrier reef lies along the eastern coast of the Yucatan Peninsula in Central America, and others surround many islands in the South Pacific.

Atolls are circular reefs that rim lagoons. They are surrounded by deep water. Small islands of calcareous sand may be built by waves at places along the reef ring. The diameter of atolls varies from 1 to more than 100 kilometres. Numerous atolls dot the South Pacific. The Bikini and Eniwetok atolls were used through 1958 for the testing of nuclear weapons by the United States.

Following the 4-year cruise of the HMS *Beagle* in the 1830s, Charles Darwin proposed that these three types of reefs are related to one another by subsidence of a central volcanic island, as shown in figure 18.26. A fringing reef initially becomes established near the island's shore. As the volcano slowly subsides because of tectonic lowering of the sea floor, the reef becomes a barrier reef, because the corals and algae grow rapidly upward, maintaining the reef's position near sea level. Less and less of the island remains above sea level, but the reef grows upward into shallow, sunlit water, maintaining its original size and shape. Finally the volcano disappears completely below sea level, and the reef becomes a circular atoll. Drilling through atolls in the 1950s showed that these reefs were built on deeply buried volcanic cores, thus confirming Darwin's hypothesis of 120 years before.

Continental Slopes

A continental slope is a relatively steep slope that extends from a depth of 100 to 200 metres at the edge of the continental shelf down to oceanic depths of up to 5 km (figure 18.24). The average angle of slope for a continental slope is 4 to 5 degrees, although locally some parts are much steeper.

Because the continental slopes are more difficult to study than the continental shelves, less is known about them. The greater depth of water and the locally steep inclines on the continental slopes hinder rock dredging and drilling and make the results of seismic refraction and reflection harder to interpret. However, it is likely that the thick continental crust, found under the land and continental shelves, grades into thin oceanic crust below the continental slope (figure 18.25). Sedimentation on continental slopes is dominated by deposition of land-derived sediment by turbidity currents (see below).

Submarine Canyons

Submarine canyons are V-shaped valleys that run across continental shelves and down continental slopes (figure 18.25). On narrow continental shelves, such as those off the Pacific coast of North America, the heads of submarine canyons may be so close to shore that they lie within the surf zone (figure 18.27). On wide shelves, such as those off the Atlantic coast of North America, canyon heads usually begin near the outer edge of the continental shelf tens of kilometres from shore. Great fan-shaped deposits of sediment called **submarine fans** are found at the base of many submarine canyons (figure 18.25). Submarine fans are made up of land-derived sediment that has moved down the submarine canyons. Along continental margins that are cut by submarine canyons, many coalescing fans may build up at the base of the continental slope.

Submarine canyons are erosional features, but how rock and sediment are removed from the steep-walled canyons is controversial. Erosional agents probably vary in relative importance from canyon to canyon. Divers have filmed *down-canyon movement of sand* in slow, glacierlike flow and in more rapid sand falls. This sand movement, which has been observed to cause erosion of rock, is particularly common in Pacific-coast canyons, which collect great quantities of sand from longshore drift (figure 18.27). *Bottom currents* have been measured moving up and down the canyons in a pattern of regularly alternating flow, in some cases apparently caused by ocean tides. The origin of these currents is not well understood, but they often move fast enough to erode and transport sediment. *River erosion* may have helped to cut the upper part of canyons when the drop in sea level during Quaternary glaciations left canyon heads above the water, such as the extension of the Hudson River into the Atlantic.

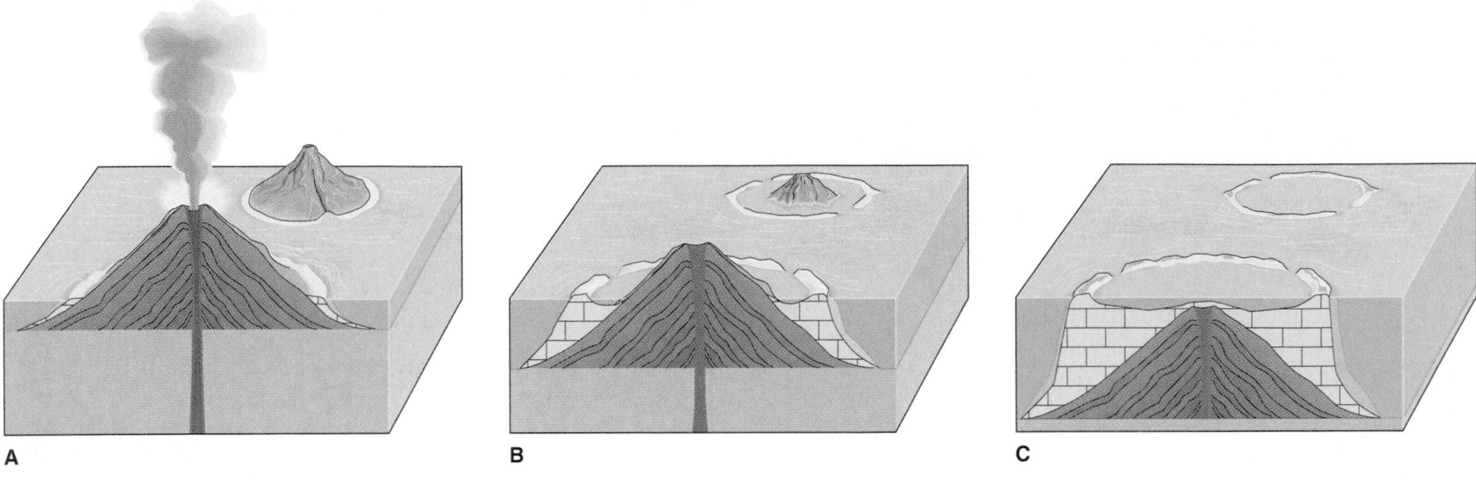

D

FIGURE 18.26

Types of coral-algal reefs. (*A*) Fringing reefs are attached directly to the island. (*B*) Barrier reefs are separated from the island by a lagoon. (*C*) Atolls are circular reefs with central lagoons. Charles Darwin proposed that the sequence of fringing, barrier, and atoll reefs form by the progressive subsidence of a central volcano, accompanied by the rapid upward growth of corals and algae. (*D*) Reefs on the island of Mooréa in the Society Islands, south-central Pacific (Tahiti in the background). Living corals appear brown. Fringing reef next to shoreline, barrier reef at breaker line; light blue lagoon between reefs is covered with carbonate sand. The island is an extinct volcano, heavily eroded by stream action.

Photo © David Hiser/Stone/Getty Images

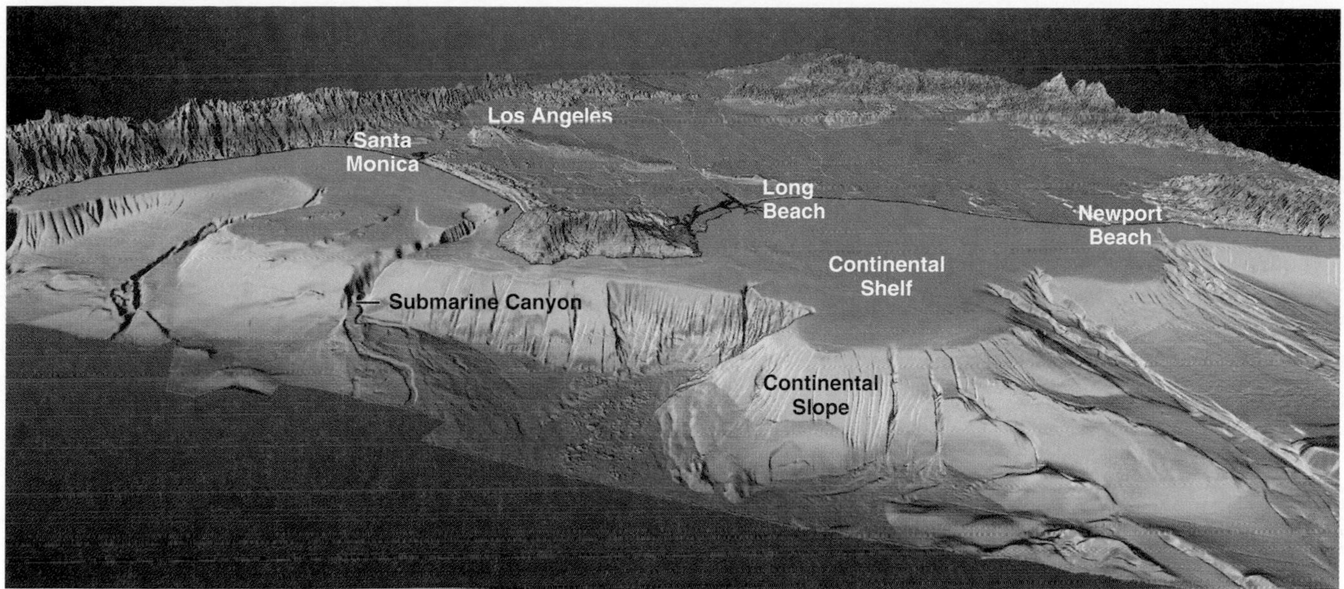

FIGURE 18.27

Continental shelf and slope off the coast of southern California are cut by submarine canyons that channel sediments to the deep ocean. Vertical exaggeration is approximately 6X. To view other high-resolution multibeam images of the coastline, visit http://walrus.wr.usgs.gov/pacmaps/la-pers1.html.
Image by Jim Gardner, U.S. Geological Survey

Many (but not all) submarine canyons are found off land canyons or rivers, which tends to support the view that river erosion helped shape them. It is unlikely, however, that the deeper parts of submarine canyons were ever exposed as dry land.

Turbidity Currents

In addition to the canyon-cutting processes just described, turbidity currents probably play the major role in canyon erosion and are responsible for transporting and depositing sediment on continental slopes. **Turbidity currents** are great masses of sediment-laden water that are pulled downhill by gravity. The sediment-laden water is heavier than clear water, so the turbidity current flows down the continental slope until it comes to rest on the flat abyssal plain at the base of the slope (see figure 9.14). Turbidity currents are thought to be generated by underwater earthquakes and landslides, strong surface storms, and floods of sediment-laden rivers discharging directly into the sea on coasts with a narrow shelf. Although large turbidity currents have not been directly observed in the sea, small turbidity currents can be made and studied in the laboratory.

Indirect evidence also indicates that turbidity currents occur in the sea. The best evidence comes from the breaking of submarine cables that carry telephone and telegraph messages across the ocean floor. Figure 18.28 shows a downhill sequence of cable breaks that followed a 1929 earthquake in the Grand Banks region off Newfoundland. This sequence of cable breaks has been interpreted to be the result of an earthquake-caused turbidity current flowing rapidly down the continental slope.

If cable breaks are caused by turbidity currents, they give good evidence of the currents' dramatic size, speed, and energy. Breaks in Grand Banks cables continued for more than 13 hours after the 1929 earthquake, the last of the series occurring more than 700 kilometres from the epicentre. The velocity of the flow that caused the breaks has been calculated to be from 15 to 60 kilometres per hour. Sections of cable more than 100 kilometres long were broken off and carried away, both ends of a missing section being broken simultaneously. Attempts to find broken cable sections were fruitless, and it is assumed that they were buried by sediment. The Grand Banks turbidity current is estimated to have moved a volume of more than 175 km^3 of sediment, and its deposits now cover a large part of the Sohm Abyssal Plain.

The Grand Banks 1929 cable breaks are not unique. Cables crossing submarine canyons are broken frequently, particularly after river floods and earthquakes. In the submarine canyons off the Congo (Zaire) River of Africa and off the Magdalena River in Colombia, for example, cables break every few years.

Additional indirect evidence for the existence of turbidity currents comes from the graded bedding and shallow-water fossils in the sediments that make up the continental rise and the abyssal plains.

The Continental Rise

Along the base of many parts of the continental slope lies the **continental rise,** a wedge of sediment that extends from the lower part of the continental slope to the deep sea floor. The continental rise, which slopes at about 0.5 degrees, more gently than the continental slope, typically ends in a flat abyssal plain at a depth of about 5 kilometres. The rise rests upon oceanic crust (figure 18.25).

Sediments appear to be deposited on the continental rise in two ways—by turbidity currents flowing *down* the continental slope and by *contour currents* flowing *along* the continental slope.

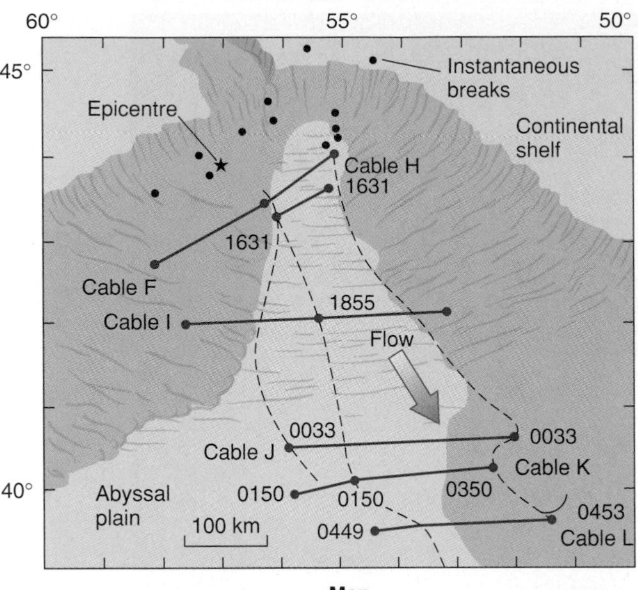

A

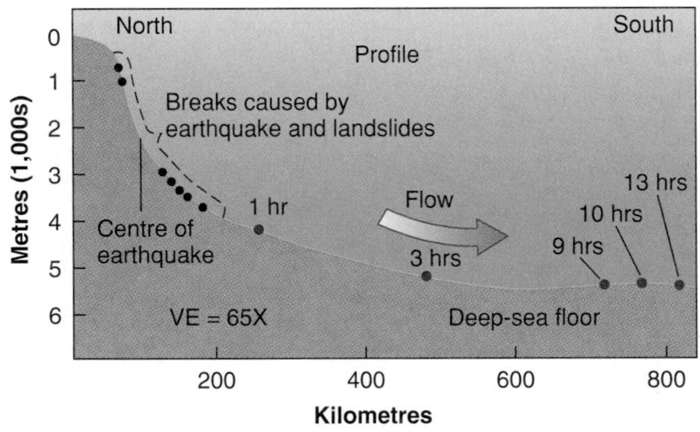

B

FIGURE 18.28

Submarine cable breaks following the Grand Banks earthquake of 1929. (A) Map view of the cable breaks. Black dots near the epicentre show locations of cable breaks that were simultaneous with the earthquake (cables not shown for these breaks). Coloured dots show cable breaks that followed the earthquake, with the time of each break shown (on the 24-hour clock). Segments of cable more than 100 kilometres long were broken simultaneously at both ends and then carried away. Dashes show sea-floor channels that probably concentrated the flow of a turbidity current, increasing its velocity. (B) Profile showing the time elapsed between the quake and each cable break.

A From H. W. Menard, 1964, Marine Geology of the Pacific, *copyright McGraw-Hill, Inc. B from B. C. Heezen and M. Ewing, 1952,* American Journal of Science.

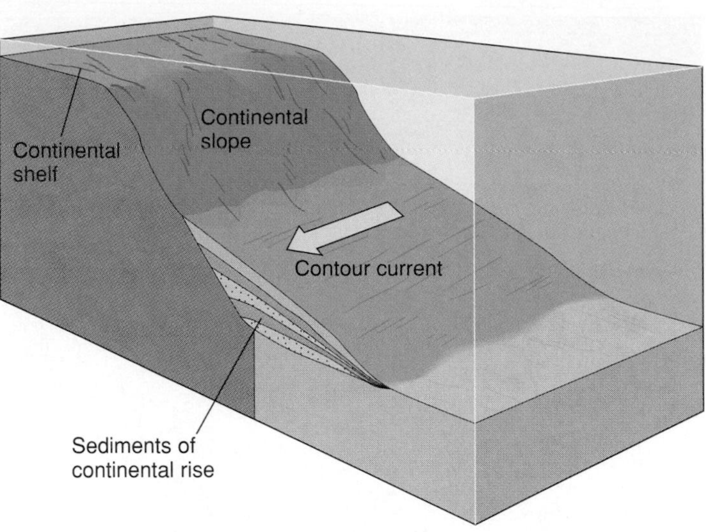

FIGURE 18.29

A contour current flowing along the continental margin shapes the continental rise by depositing fine sediment. Coarse layers within the continental rise were deposited by turbidity currents flowing down the continental slope.

Cores of sediment recovered from most parts of the continental rise show layers of fine sand or coarse silt interbedded with layers of fine-grained mud. The mineral grains and fossils of the coarser layers indicate that the sand and silt came from the shallow continental shelf. Some transporting agent must have carried these sediments from shallow water to deep water. The coarse layers also exhibit graded bedding, which indicates that they settled out of suspension according to size and weight; therefore, the transporting agent for these sediments was most likely turbidity currents. The continental rise in these locations probably formed as turbidity currents deposited abyssal fans at the base of a continental slope.

Sediments in some other parts of the continental rise, however, are uniformly fine-grained and show no graded bedding. This sediment appears to have been deposited by the regular ocean currents that flow along the sea bottom rather than by the intermittent turbidity currents that occasionally flow downslope.

A **contour current** is a bottom current that flows parallel to the slopes of the continental margin—*along* the contour rather than *down* the slope (figure 18.29). Such a current runs south along the continental margin of North America in the Atlantic Ocean. Flowing at the relatively slow speed of a few centimetres per second, this current carries a small amount of fine sediment from north to south. The current is thickest along the continental slope and gets progressively thinner seaward. The thick landward part of the current carries and deposits the most sediment. The thinner seaward edge of the current deposits less sediment. As a result, the deposit of sediment beneath the current is wedge-shaped, becoming thinner away from land. Similar contour currents apparently shape parts of the continental rise off other continents as well.

Abyssal Plains

Abyssal plains are very flat regions usually found at the base of the continental rise. Seismic profiling has shown that abyssal plains are formed of horizontal layers of sediment. The gradual deposition of sediment buried an older, more rugged topography that can be seen on seismic profiler records as a rock basement beneath the sediment layers (see figure 2.11). Samples of abyssal plain sediment show that much of it is derived from land. Graded bedding within sediment layers suggests deposition by turbidity currents.

Abyssal plains are the flattest features on Earth. They generally have slopes less than 1:1,000 (less than 1 metre of vertical drop for every 1,000 metres of horizontal distance) and some have slopes of only 1:10,000. Most abyssal plains are 5 kilometres deep below the ocean surface.

Not all parts of the deep-ocean basin floor consist of abyssal plains. The deep floor is normally very rugged, broken by faults into hills and depressions and dotted with volcanic seamounts. Abyssal plains form only where turbidity currents can carry in enough sediment to bury and obscure this rugged relief. If the sediment is not available or if the bottom-hugging turbidity currents are stopped by a barrier such as an oceanic trench, then abyssal plains cannot develop.

WHAT KINDS OF SEDIMENT ACCUMULATE ON THE OCEAN FLOOR?

The basaltic crust of the sea floor is covered in many places with layers of sediment. This sediment is either *terrigenous,* derived from land, or *pelagic,* settling slowly through seawater.

Terrigenous sediment is land-derived sediment that has found its way to the sea floor. The sediment that makes up the continental rise and the abyssal plains is mostly terrigenous and apparently has been deposited by turbidity currents or similar processes. Once terrigenous sediment has found its way down the continental slope, contour currents may distribute it along the continental rise. On active continental margins, oceanic trenches may act as traps for terrigenous sediment and prevent it from spreading out onto the deep sea floor beyond the trenches.

Pelagic sediment is sediment that settles slowly through the ocean water. It is made up of fine-grained clay and the skeletons of microscopic organisms (figure 18.30). Fine-grained pelagic clay is found almost everywhere on the sea floor, although in some places it is masked by other types of sediments that accumulate rapidly. The clay is mostly derived from land; part of it may be volcanic ash. This sediment is carried out to sea primarily by wind, although rivers and ocean currents also help to distribute it.

Microscopic shells and skeletons of plants and animals also settle slowly to the sea floor when marine organisms of the surface waters die. In some parts of the sea, such as the polar and equatorial regions, great concentrations of these shells have built unusually thick pelagic deposits.

The constant slow rain of pelagic clay and shells occurs in all parts of the sea. Although the rate of accumulation varies from place to place, pelagic sediment should be expected on all parts of the sea floor. For a discussion of resources that are being extracted from the sea, see box 18.4.

FIGURE 18.30

Photograph (taken through a scanning electron microscope) of pelagic sediment from the floor of the Pacific Ocean. The sediment is made up of microscopic skeletons of single-celled marine organisms (large objects are foraminifera; smaller, sieve-like ones are radiolaria about 0.05 mm in diameter).

Photo © Dr. Richard Kessel & Dr. Gene Shih/Visuals Unlimited

ENVIRONMENTAL GEOLOGY 18.4

Geologic Riches in the Sea

Many resources are currently being extracted from the sea floor and from sea water, and in some instances there is a great potential for increased extraction.

Offshore oil and gas are the most valuable resources now being taken from the sea. More than one-sixth of the United States' oil production (and more than one-quarter of world production) comes from drilling platforms set up on the continental shelf (box figure 1). Oil and gas have been found within deeper parts of the sea floor, such as the continental slope and continental rise. Producing oil from these deeper regions is much more costly than present production; oil spills from wells in deep water would be especially hard to control.

Other important resources are dredged from the sea floor. *Phosphorite* can be recovered from shallow shelves and banks and used for fertilizers. *Gold, diamonds,* and heavy *black sands* (which are black because they contain metal-bearing minerals) are being separated from the surface sands and gravels of some continental shelves by specially designed ships.

Manganese nodules (box figure 2) cover many parts of the deep sea floor, notably in the central Pacific. These black, potato-sized lumps contain approximately 25 percent manganese, 15 percent iron, and up to 2 percent nickel and 2 percent copper, along with smaller amounts of cobalt. Although there are international legal problems concerning who owns them, larger industrial countries such as the United States and Canada may mine them, particularly for copper, nickel, and cobalt. The United States is also interested in manganese; it imports 95 percent of its manganese, which is critical to producing some types of steel.

Metallic brines and sediments, first discovered in the Red Sea, are deposited by submarine hot springs active at the rift valley of the mid-oceanic ridge crest. The Red Sea sediments contain more than 1 percent copper and more than 3 percent zinc, together with impressive amounts of silver, gold, and lead (worth U.S.$25 billion, according to a 1983 estimate). Because of their great value, the sediments will probably be mined even though they are at great depth. Deposits similar to those of the Red Sea—although not of such great economic potential—have been found on several other parts of the ridge.

A few substances can be extracted from the salts dissolved in sea water. Approximately two-thirds of the world's production of *magnesium* is obtained from sea water, and in many regions *sodium chloride* (table salt) is obtained by solar evaporation of sea water.

BOX 18.4 ■ FIGURE 1

Offshore oil drilling platform Hibernia is located 320 kilometres off the coast of Newfoundland and began producing oil in 1997. As many as 50 different wells can be drilled from a single platform.

Photo courtesy Chevron Canada Resources

BOX 18.4 ■ FIGURE 2

Dense concentrations of manganese nodules on the floor of the abyssal plain (depth 5,350 metres) in the northeast Atlantic Ocean.

Photo © Institute of Oceanographic Sciences/NERC/Photo Researchers, Inc.

SUMMARY

Wind blowing over the sea surface forms waves, which transfer some of the wind's energy to shorelines. Orbital water motion extends to a depth equal to half the wavelength.

As a wave moves into shallow water, the ocean bottom flattens the orbital motion and causes the wave to slow and peak up, eventually forming a *breaker* whose crest topples forward. The turbulence of *surf* is an important agent of sediment erosion and transportation.

Wave refraction bends wave crests and makes them more parallel to shore. Few waves actually become parallel to the shore, and so *longshore currents* develop in the surf zone. *Rip currents* carry water seaward from the surf zone.

A beach consists of a *berm*, *beach face*, and *marine terrace*. Summer beaches have a wide berm and a smooth offshore profile. Winter beaches are narrow, with offshore bars.

Longshore drift of sand is caused by the waves hitting the beach face at an angle and also by longshore currents.

Deposition of sand that is drifting along the shore can form *spits* and *baymouth bars*. Drifting sand may also be deposited against jetties or groins or inside breakwaters.

Rivers supply most sand to beaches, although local erosion may also contribute sediment. If the river supply of sand is cut off by dams, the beaches gradually disappear.

Coasts may be erosional or depositional, drowned or uplifted, or shaped by organisms such as corals and mangroves.

Coastal straightening by waves is caused by headland erosion and by deposition within bays.

A coast retreating under wave erosion can be marked by *sea cliffs*, a *wave-cut platform*, *stacks*, and *arches*.

Waves can form *barrier islands* off gently sloping coasts. River and glacial deposition can also shape coasts.

Drowned coasts are marked by *estuaries* and *fjords*. *Uplifted marine terraces* characterize coasts that have risen faster than the recent rise in sea level.

The *continental shelf* and the steeper *continental slope* lie under water along the edges of continents. They are separated by a change in slope angle at a depth of 100 to 200 metres. Corals and algae living in warm, shallow water construct *fringing reefs*, *barrier reefs*, and *atolls*.

Submarine canyons are cut into the continental slope and outer continental shelf by a combination of turbidity currents, sand flow and fall, bottom currents, and river erosion during times of lower sea level. Graded bedding and cable breaks suggest the existence of *turbidity currents* in the ocean. *Submarine fans* form as sediment collects at the base of submarine canyons.

The *continental rise* and *abyssal plains* may form from sediment deposited by turbidity currents. The continental rise may also form from sediment deposited by *contour currents* at the base of the continental slope.

Terrigenous sediment is composed of land-derived sediment deposited near land by turbidity currents and other processes. *Pelagic sediment* is made up of wind-blown dust and microscopic skeletons that settle slowly to the sea floor.

The crest of the mid-oceanic ridge lacks *pelagic sediment*.

Terms to Remember

abyssal plain 500, 504
active continental margin 500
arch (sea arch) 492
atoll 501
barrier island 493
barrier reef 501
baymouth bar 488
beach 484
beach face 484
berm 484
breaker 481
coast 489
coastal straightening 491
continental rise 503
continental shelf 500
continental slope 500
contour current 504
crest (of wave) 480
estuary 498
fjord 498
fringing reef 501
headland 491
longshore current 482
longshore drift 487
marine terrace 484
mid-oceanic ridge 500
oceanic trench 500
passive continental margin 500
pelagic sediment 505
reef 501
rip current 482
sea cliff 491
spit 488
stack 492
submarine canyon 501
submarine fan 501
surf 481
terrigenous sediment 505
tide 483
tide-generating force 483
tombolo 488
trough (of wave) 480
turbidity current 503
wave-cut platform 492
wave height 480
wavelength 481
wave refraction 482

Testing Your Knowledge

Use the questions below to prepare for exams based on this chapter.

1. Show in a sketch how longshore drift of sand can form a baymouth bar.
2. In a sketch, show how and why sand moves along a beach face when waves approach a beach at an angle.
3. How are summer beaches different from winter beaches? Discuss the reasons for these differences.
4. What would happen to the beaches of most coasts if all the rivers flowing to the sea were dammed? Why?
5. What does the presence of an estuary imply about the recent geologic history of a region?
6. Describe how waves can straighten an irregular coastline.
7. Describe the transition of deep-water waves into surf.
8. Show in a sketch the refraction of waves approaching a straight coast at an angle. Explain why refraction occurs.
9. What is a longshore current? Why does it occur?
10. What is a rip current? Why does it occur? How do you get out of a rip current?
11. The path a water particle makes as a wave passes in deep water is best described as
    a. elliptical b. orbital
    c. spherical d. linear
12. The easiest method of escaping a rip current is to
    a. swim toward the shore
    b. swim parallel to the shore
    c. swim away from the shore
13. Why is beach sediment typically quartz-rich sand?
    a. other minerals are not deposited on beaches
    b. quartz is the only mineral that can be sand-sized
    c. quartz is resistant to chemical weathering
    d. none of the above
14. Longshore drift is
    a. the movement of sediment parallel to shore when waves strike the shoreline at an angle
    b. a type of rip current
    c. a type of tide
    d. the movement of waves
15. Which structure would interfere with longshore drift?
    a. jetties b. groins
    c. breakwaters d. all of the above
16. What is the most common source of sand on beaches?
    a. sand from river sediment brought down to the ocean
    b. land next to the beach
    c. offshore sediments
17. Which would characterize an erosional coast?
    a. headlands b. sea cliffs c. stacks
    d. arches e. all of the above
18. Which would characterize a depositional coast?
    a. headlands b. sea cliffs c. stacks
    d. arches e. barrier islands
19. A glacial valley drowned by rising sea level is
    a. a fjord b. an estuary
    c. a tombolo d. a headland
20. The surf zone is
    a. the region in which waves break
    b. water less than one half-wavelength in depth
    c. where the longshore current flows
    d. all of the above
21. The storm surge of a hurricane is
    a. the highest winds
    b. the tallest waves
    c. the dome of high water in the centre of the hurricane
    d. the area of high pressure within the storm
22. Sketch an active continental margin and a passive continental margin, labelling all their parts. Show approximate depths.
23. Which is true of the continental shelf?
    a. it is a shallow submarine platform at the edge of a continent
    b. it inclines very gently seaward
    c. it can vary in width
    d. all of the above
24. Sketch a cross-section of a fringing reef, a barrier reef, and an atoll.
25. The average angle of slope for a continental slope is
    a. 1–2 degrees b. 3–4 degrees
    c. 4–5 degrees d. greater than 10 degrees
26. What is a submarine canyon? How do submarine canyons form?
27. What is a turbidity current? What is the evidence that turbidity currents occur on the sea floor?
28. Describe two different origins for the continental rise.
29. Discuss the appearance, structure, and origin of abyssal plains.
30. Great masses of sediment-laden water that are pulled downhill by gravity are called
    a. contour currents b. bottom currents
    c. turbidity currents d. traction currents
31. Describe the two main types of sea-floor sediment.
32. Reefs parallel to the shore but separated from it by wide, deep lagoons are called
    a. fringing reefs b. barrier reefs
    c. atolls d. lagoonal reefs
33. Pelagic sediment could be composed of
    a. fine-grained clay
    b. skeletons of microscopic organisms
    c. volcanic ash
    d. all of the above
34. What part of the continental margin marks the true edge of the continent?
    a. continental shelf b. continental slope
    c. continental rise d. abyssal plain

Exploring Web Resources

www.mcgrawhill.ca/olc/plummer
Visit our Online Learning Centre to check your answers for the Testing Your Knowledge section and watch some great animations created to better your understanding of waves, beaches, coastal shift, reefs, and atolls. Click on the direct links to go to the other Websites listed below.

http://marine.usgs.gov/
Web page for the *Coastal and Marine Geology Program* of the U.S. Geological Survey contains information about numerous geologic studies of U.S. coastal areas.

http://pubs.usgs.gov/circ/c1075/
Online version of *Coasts in Crisis.* U.S. Geological Survey Circular 1075.

www.bayoffundy.com
Information about the Bay of Fundy and tides.

http://woodshole.er.usgs.gov/
Web page for the U.S. Geological Survey *Woods Hole Field Center* for coastal and marine research contains information and data from ongoing scientific projects.

www-ccs.ucsd.edu/
Web page for the *Center for Coastal Studies at the Scripps Institute of Oceanography* provides information about their research and access to data collected from various coastal studies projects.

www.esdim.noaa.gov/ocean_page.html
Oceans Web page from *National Oceanic and Atmospheric Administration (NOAA)* provides numerous links to oceanography research projects and data.

http://seawifs.gsfc.nasa.gov/ocean_planet.html
Smithsonian Ocean Planet Exhibit, containing many links to sea-floor topics such as hydrothermal vents and microbiology of deep-sea vents.

www.noaa.gov
National Oceanographic and Atmospheric Administration (NOAA) has many links to numerous sites relating to the oceans.

www.ngdc.noaa.gov/mgg/mggd.html
NOAA Marine Geology and Geophysics Division Website contains world sea-floor maps and information on ocean-drilling data and samples.

www-odp.tamu.edu/
Ocean Drilling Program (ODP) Website contains information on the nature and history of the ocean crust and provides data from research projects.

www.whoi.edu/VideoGallery/
Woods Hole Oceanographic Institution contains video clips of black smokers, exotic organisms at mid-oceanic ridges, and oceanographic research using vessels and submersibles.

www.pmel.noaa.gov/vents/
NOAA Vents Program Website provides photos, video clips, data, and research program activities about the investigation of submarine volcanoes and hydrothermal venting around the world.

Animations

This chapter includes the following animations available on our Online Learning Centre at www.mcgrawhill.ca/olc/plummer.

18.10 Seasonal Cycle of a Beach
18.11 Wave Refraction and Longshore Movement of Sand and Water

CHAPTER

19

Time and Geology

What is uniformitarianism?
How can the sequence of past geological events be determined?
How can rock units be traced from one area to another?
How do we use relative dating to understand geological time?
How can we determine the absolute age of rocks?
What is radiocarbon dating?
When can isotopic dating techniques be used?
How old is the Earth?

The immensity of geologic time is hard for humans to perceive. It is unusual for someone to live a hundred years, but a person would have to live 10,000 times that long to observe a geologic process that takes a million years. In this chapter we try to help you develop a sense of the vast amounts of time over which geologic processes have been at work.

Geoscientists working in the field or with maps or illustrations in a laboratory are concerned with relative time—unravelling the sequence in which geologic events occurred. For instance, a geoscientist working in southern Ontario can determine that glacial sediments found near the ground surface are younger than the layered sedimentary rocks they cover and that the deformed igneous and metamorphic rocks underlying them, that form part of the North American craton, are significantly older. But this tells us nothing about how long ago any of the rocks formed. To determine how many years ago rocks formed, we need the specialized techniques of radioactive isotope dating. Through isotopic dating we have been able to determine that the rocks forming the North American craton range in age between 4 billion years and 1 billion years old.

Deformed gneiss more than 1 billion years old is exposed on islands in Georgian Bay, Ontario. The gneiss forms part of the Canadian Shield, the ancient core of the North American continent. *Photo by C. H. Eyles*

This chapter explains how to apply several basic principles to decipher a sequence of events responsible for geologic features. These principles can be applied to many aspects of geology—as, for example, in understanding geologic structures (chapter 11). Understanding the complex history of continental building (chapter 20) also requires knowing the techniques for determining relative ages of rocks.

Determining age relationships between geographically widely separated rock units is necessary for understanding the geologic history of a region, a continent, or the whole Earth. Substantiation of the plate tectonics theory depends on intercontinental correlation of rock units and geologic events, piecing together evidence that the continents were once one great body.

Widespread use of fossils led to the recognition that certain strata recurred in the same order over large areas. This helped geologists identify stratigraphic successions and eventually allowed the creation of the standard geologic time scale. Originally based on relative age relationships, the subdivisions of the standard geologic time scale have now been assigned numerical ages in thousands, millions, and billions of years through isotopic dating. Think of the geologic time scale as a sort of calendar to which events and rock units can be referred. Its major subdivisions are referred to elsewhere in this book.

WHAT IS UNIFORMITARIANISM?

Until the 1800s, almost all people living in Western culture accepted the church concept of Earth being only a few thousand years old. On the other hand, Chinese and Hindu cultures believed the age of Earth was vast beyond comprehension—more in line with what has now been determined scientifically. In the Christendom of the seventeenth and eighteenth centuries, formation of all rocks and other geologic events were placed into a biblical chronology. This required that features we observe in rocks and landscapes were created supernaturally and catastrophically. The sedimentary rocks with marine fossils (clams, fish, etc.) that we find in mountains thousands of metres above sea level were believed to have been deposited by a worldwide flood (Noah's flood) that inundated all of Earth, including its highest mountains, in a matter of days. Because no known physical laws could account for such events, they were attributed to divine intervention. In the eighteenth century, however, James Hutton, a Scotsman who is regarded as the father of modern geology, realized that geologic features could be explained through present-day processes (figure 19.1). He recognized that our mountains are not permanent, but have been carved into their present shapes and will be worn down by the slow agents of erosion now working on them. He realized that the great thicknesses of sedimentary rock we find on the continents are products of sediment removed from land and deposited as mud and sand in seas. The time required for these processes to take place had to be incredibly long. Hutton upset conventional thinking (the world was believed to be fewer than 6,000 years old) by writing in 1788, "We find no sign of a beginning—no prospect for an end." Hutton's writings received scant notice until his ideas were given widespread attention by Charles Lyell (figure 19.1) in a landmark book, *Principles of Geology*. Lyell referred to Hutton's concept that geologic processes operating at present are the same processes that operated in the past as the principle of **uniformitarianism.** The principle is stated more succinctly as "The present is the key to the past."

A

B

FIGURE 19.1

(*A*) James Hutton (left) and Sir Charles Lyell (right) recognized that landscapes had evolved over many millions of years by the processes that we still see operating today: "The present is the key to the past." (*B*) Siccar Point, Scotland, where near-vertical Silurian-age rocks (centre right) are overlain by much younger Upper Devonian/Lower Carboniferous rocks (upper left). Hutton recognized that contact between the two rock types represents a very long period of geologic time.

Photo A (left) from Scottish National Portrait Gallery (PG 2686); Photo B from Copyright © C.E. Ford 2004

The term uniformitarianism is a bit unfortunate, because it suggests that changes take place at a uniform *rate.* Hutton recognized that sudden, violent events, such as a major, short-lived volcanic eruption, also influence the Earth's history. In some countries *actualism* is used in place of uniformitarianism. The term actualism comes closer to conveying Hutton's principle that the same processes and natural laws that operated in the past are those we can actually observe or infer from observation as operating at present. It is based on the assumption, central to the sciences, that physical laws are independent of time and location. Under present usage, uniformitarianism has the same meaning as actualism for most geologists.

We now realize that geology involves time periods much greater than a few thousand years. But how long? For instance, were rocks of the Canadian Shield formed closer to 10,000 or 100,000 or 1,000,000 or 1,000,000,000 years ago? What geologists needed was some "clock" that began working when rocks formed. Such a clock was found when radioactivity was discovered. Dating based on radioactivity (discussed later in this chapter) allows us to determine a rock's **numerical age** (also known as *absolute age*)—age given in years or some other unit of time. Geologists working in the field or in a laboratory with maps, cross-sections, and photographs are more concerned with **relative time,** the *sequence* in which events took place, than with the number of years involved.

These statements show the difference between absolute age and relative time: "Canada came into Confederation after the signing of the Magna Carta but before the Second World War." This statement gives the time of an event (Confederation) relative to other events. But in terms of numerical age we could say: "Confederation took place about one and a half centuries ago." Note that a numerical age does not have to be an *exact* age, merely age given in units of time. Because most geologic problems are more concerned with the sequence of events, we discuss relative time first.

HOW CAN THE SEQUENCE OF PAST GEOLOGICAL EVENTS BE DETERMINED?

Relative Time

The geology of an area may seem, at first glance, to be hopelessly complex. A nongeologist might think it impossible to decipher the sequence of events that created such a geologic pattern; however, a geologist has learned to approach seemingly formidable problems by breaking them down to a number of simple problems. As an example, the geology of the Grand Canyon, shown diagrammatically later in this chapter in figure 19.17, can be analyzed in four parts: (1) horizontal layers of rock; (2) inclined layers; (3) rock underlying the inclined layers (plutonic and metamorphic rock); and (4) the canyon itself, carved into these rocks.

Principles Used to Determine Relative Age

Most of the individual parts of the larger problem are solved by applying several simple principles while studying the exposed rock. In this way the sequence of events or the relative time involved can be determined. Contacts are particularly useful for deciphering the geologic history of an area. (**Contacts,** as described in previous chapters, are the surfaces separating two different rock types or rocks of different ages.) To explain various principles, we will use a fictitious place that bears some resemblance to the Grand Canyon. We will call this place, represented by the block diagram of figure 19.2, Cold Canyon. The formation names are also fictitious. (*Formations,* as described in chapter 9, are bodies of rock of considerable thickness with recognizable characteristics that make each distinguishable from adjacent rock units. They are named after local geographic features, such as towns or landmarks.) Note the contacts between the tilted formations, the horizontal formations, the granite, and the dike. What sequence of events might be responsible for the geology of Cold Canyon? (You might briefly study the block diagram and see how much of the geologic history of the area you can decipher before reading further.)

Our interpretations are based mainly on layered rock (sedimentary or volcanic). The subdiscipline of geology that uses interrelationships between layered rock (mostly) or sediment to interpret the history of an area or region is known as *stratigraphy.* Stratigraphy uses four principles to determine the geologic history of a locality or a region. These are the principles of (1) original horizontality, (2) superposition, (3) lateral continuity, and (4) cross-cutting relationships. These principles will be used in interpreting figure 19.2.

Original Horizontality

The principle of **original horizontality** states that beds of sediment deposited in water formed as horizontal or nearly horizontal layers (as described in chapter 9). (All the layers in figure 19.2 represent sedimentary rock originally deposited in a marine environment.)

Note in figure 19.2 that the Thompson River Formation and overlying rock units (Okanagan Formation, Penticton Formation, and Osoyoos Lake Limestone) are horizontal. Evidently their original horizontal attitude has not changed since they were deposited. However, the Kamloops, Revelstoke, Abbotsford, and Copper Creek Formations must have been tilted after they were deposited as horizontal layers. By applying the principle of original horizontality, we have determined that a geologic event—tilting of bedrock—occurred after the Copper Creek, Abbotsford, Revelstoke, and Kamloops Formations were deposited on a sea floor. We can also see that the tilting event did not affect the Thompson River and overlying formations. (A reasonable conclusion is that tilting was accompanied by uplift and erosion, all before renewed deposition of younger sediment.)

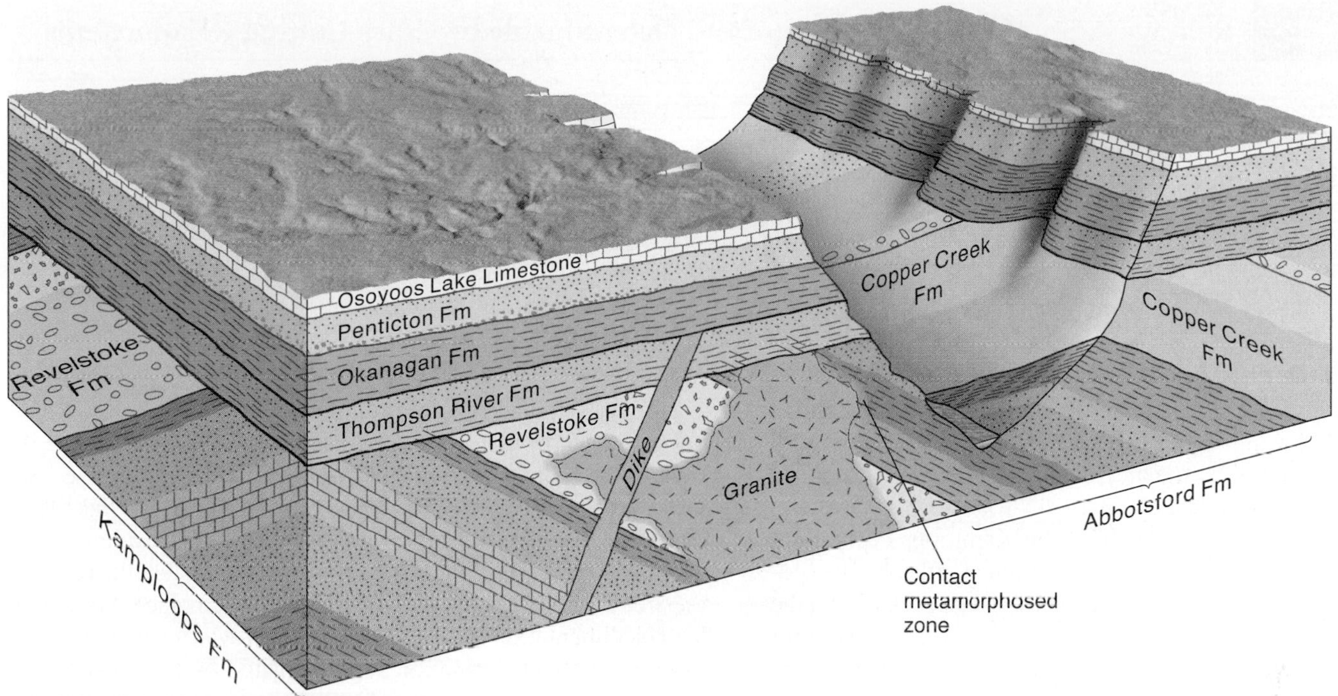

FIGURE 19.2

Block diagram representing the Cold Canyon area.

Superposition

The principle of **superposition** states that within a sequence of undisturbed sedimentary or volcanic rocks the layers get younger going from bottom to top.

Obviously, if sedimentary rock is formed by sediment settling onto the sea floor, then the first (or bottom) layer must be there before the next layer can be deposited on top of it. The principle of superposition also applies to layers formed by multiple lava flows, where one lava flow is superposed on a previously solidified flow.

Applying the principle of superposition, we can determine that the Osoyoos Lake Limestone is the youngest layer of sedimentary rock in the Cold Canyon area. The Penticton Formation is the next oldest formation, and the Thompson River Formation is the oldest of the still horizontal sedimentary rock units. Similarly, we assume that the inclined layers were originally horizontal (by the first principle). By mentally restoring them to their horizontal position (or "untilting" them), we can see that the youngest formation of the sequence is the Copper Creek Formation and that the Abbotsford, Revelstoke, and Kamloops Formations are progressively older.

Lateral Continuity

The principle of **lateral continuity** states that an original sedimentary layer extends laterally until it tapers or thins at its edges. This is what we expect at the edges of a depositional environment, or where one type of sediment interfingers laterally with another type of sediment as environments change. In figure 19.2 the bottom bed of the Penticton Formation tapers, as we would expect from this principle. We are not seeing any other layers taper, either because we are not seeing their full extent within the diagram or because they have been truncated (cut off abruptly) due to later events.

Cross-cutting Relationships

The fourth principle can be applied to determine the remaining age relationships at Cold Canyon. The principle of **cross-cutting relationships** states that a disrupted pattern is older than the cause of disruption. A layer cake (the pattern) has to be baked (established) before it can be sliced (the disruption).

To apply this principle, look for disruptions in patterns of rock. Note that the valley in figure 19.2 is carved into the horizontal rocks as well as into the underlying tilted rocks. The sedimentary beds on either side of the valley appear to have been sliced off, or *truncated,* by the valley. (The principle of lateral continuity tells us that sedimentary beds normally become thinner toward the edges rather than stop abruptly.) So the event that caused the valley must have come after the sedimentation responsible for deposition of the Osoyoos Lake Limestone and underlying formations. That is, the valley is younger than these layers. We can apply the principle of cross-cutting relationships to contacts elsewhere in figure 19.2 with the results shown in table 19.1.

Table 19.1 Relative Ages of Features in Figure 19.2 Determinable by Cross-Cutting Relationships

Feature	Is Younger Than	But Older Than
Valley (canyon)	Osoyoos Lake Limestone	
Okanagan Fm.	Dike	Penticton Fm.
Dike	Thompson River Fm.	Okanagan Fm.
Thompson River Fm.	Copper Creek Fm. and granite	Dike
Granite	Abbotsford Fm.	Thompson River Fm.

We can now describe the geological history of the Cold Canyon area represented in figure 19.2 on the basis of what we have learned through applying the principles. Figures 19.3 through 19.12 show how the area changed over time, progressing from oldest to youngest events.

By *superposition,* we know that the Kamloops Formation, the lowermost rock unit in the tilted sequence, must be the oldest of the sedimentary rocks as well as the oldest rock unit in the diagram. From the principle of *original horizontality,* we infer that these layers must have been tilted after they formed. Figure 19.3 shows initial sedimentation of the Kamloops Formation taking place. If the entire depositional basin were shown, the layer would be tapered at its edges, according to the principle of *lateral continuity.*

Superposition indicates that the Revelstoke Formation was deposited on top of the Kamloops Formation. Deposition of the Abbotsford and Copper Creek Formations followed in turn (figure 19.4).

The truncation of bedding in the Kamloops, Revelstoke, and Abbotsford Formations by the granite tells us that the granite intruded sometime after the Abbotsford Formation was formed (this is an *intrusive contact*). Although figure 19.5 shows that the granite was emplaced before tilting of the layered rock, we cannot determine from looking at figure 19.2 whether the granite intruded the sedimentary rocks before or after tilting. We can, however, determine through *cross-cutting relationships* that tilting and intrusion of the granite occurred before deposition of the Thompson River Formation. Figure 19.6 shows that the rocks in the area have been tilted and erosion has taken place. Sometime later, sedimentation was renewed and the lowermost layer of the Thompson River Formation was deposited on the erosion surface, as shown in figure 19.7. Contacts representing buried erosion surfaces such as these are called *unconformities* and are discussed in more detail later in this chapter.

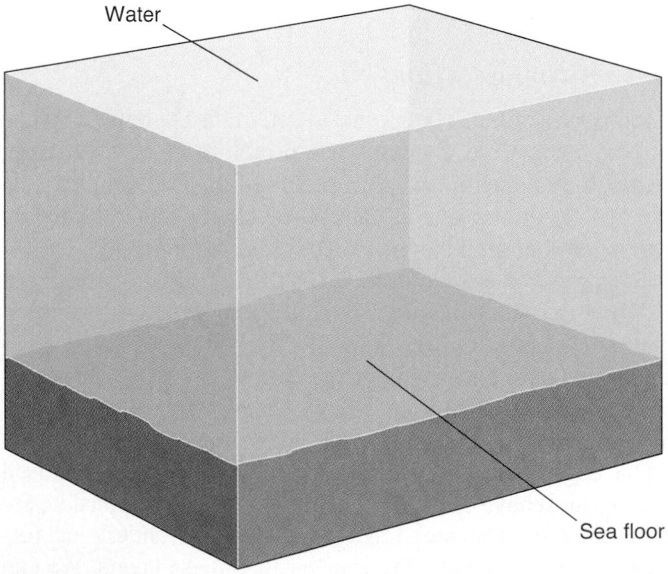

FIGURE 19.3

The area during deposition of the initial sedimentary layer of the Kamloops Formation.

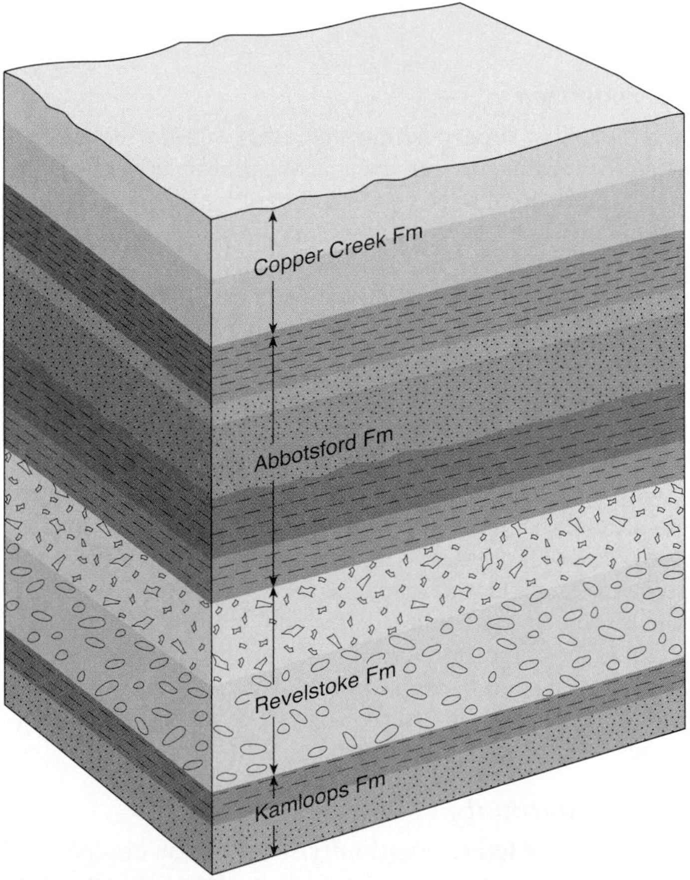

FIGURE 19.4

The area before intrusion of the granite.

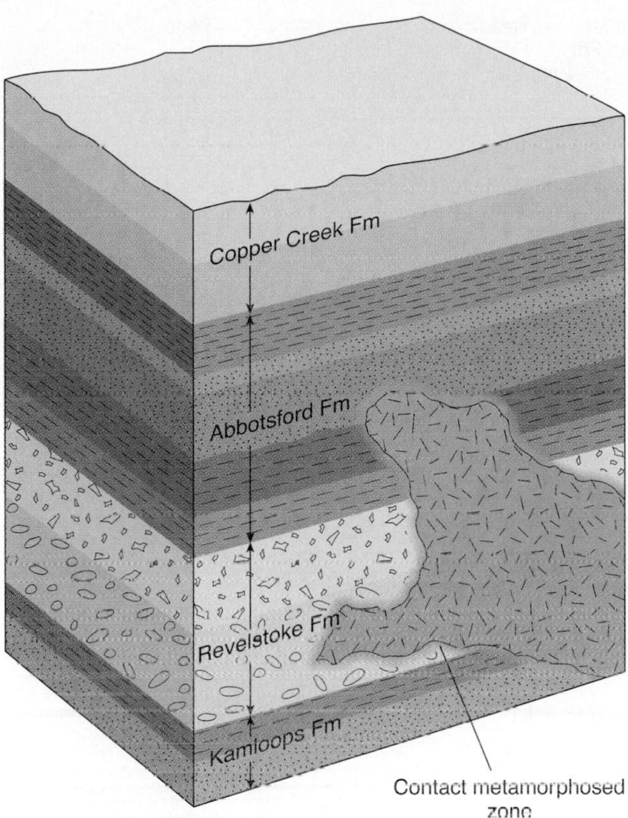

FIGURE 19.5

The area before layers were tilted.

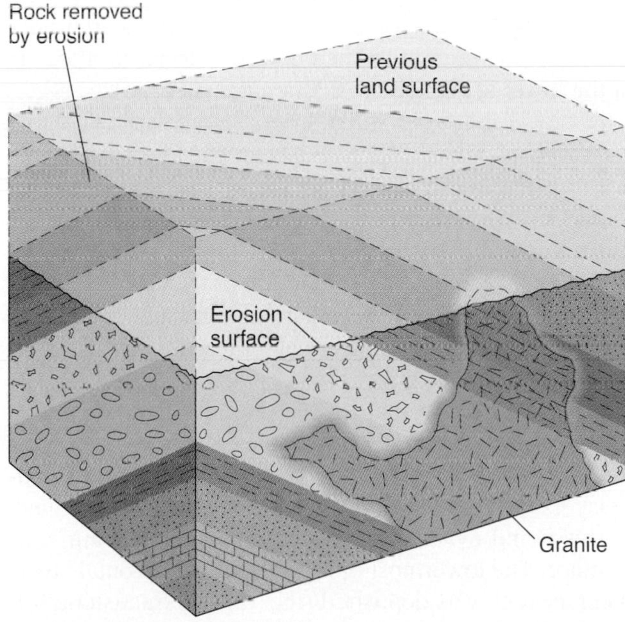

FIGURE 19.6

The area before deposition of the Thompson River Formation. Dashed lines show rock probably lost through erosion.

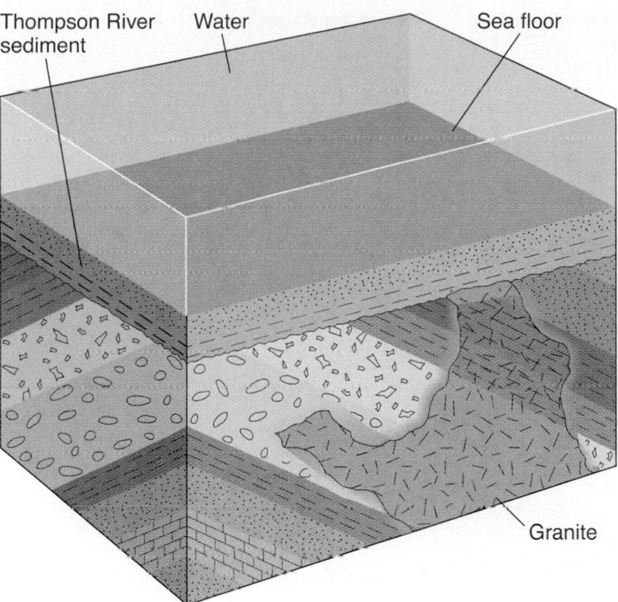

FIGURE 19.7

The area at the time the Thompson River Formation was being deposited.

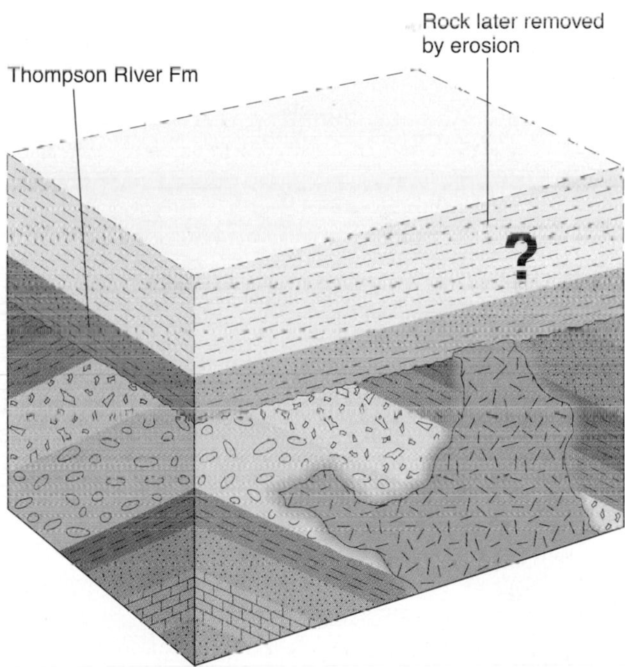

FIGURE 19.8

Area before intrusion of dike. Thickness of layers above the Thompson River Formation is indeterminable.

After the Thompson River Formation was deposited, an unknown additional thickness of sedimentary layers was deposited, as shown in figure 19.8. This can be determined through application of cross-cutting relationships. The dike is truncated by the Okanagan Formation; therefore, it must have extended into some rocks that are no longer present, such as shown in figure 19.9. Figure 19.10 shows the area after the erosion that truncated the dike took place.

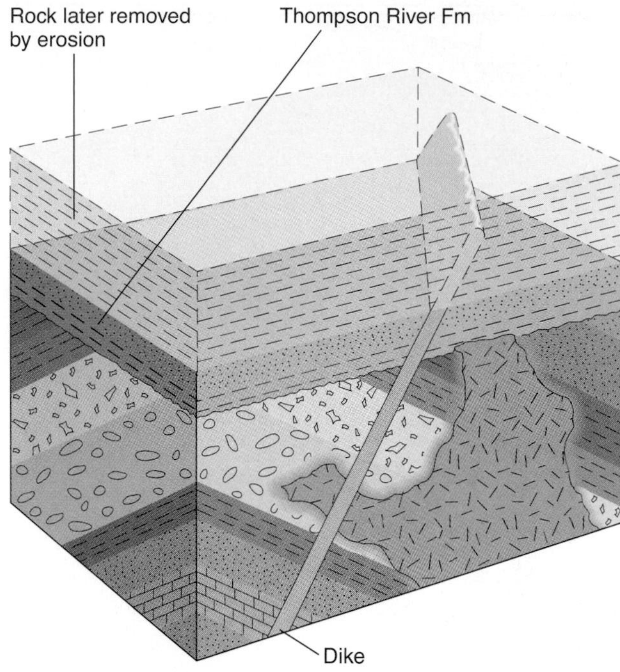

FIGURE 19.9
Dike intruded into the Thompson River Formation and preexisting, overlying layers of indeterminate thickness.

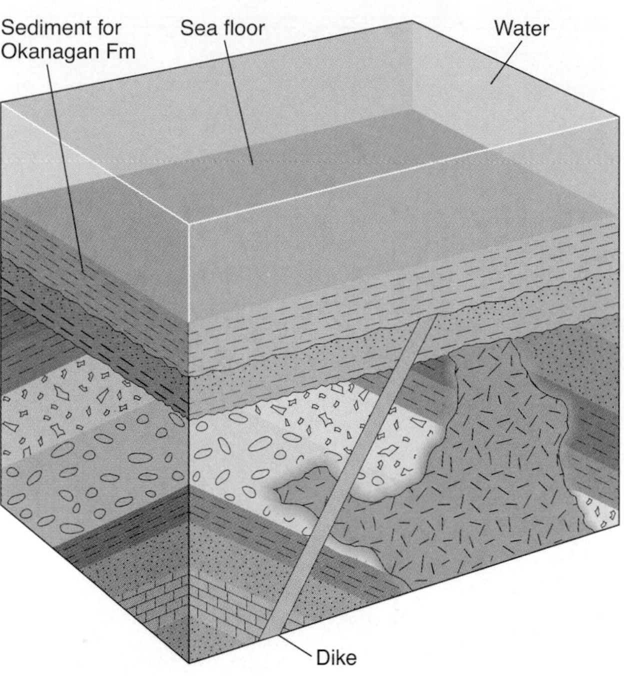

FIGURE 19.11
Sediment being deposited that will become part of the Okanagan Formation.

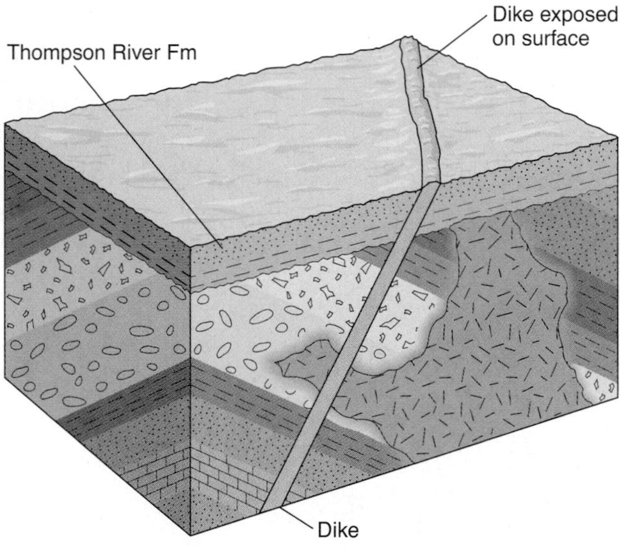

FIGURE 19.10
The area after rock overlying the Thompson River Formation, along with part of the dike, was removed by erosion.

Once again, sedimentation took place as the lowermost layer of the Okanagan Formation blanketed the erosion surface (figure 19.11). Sedimentation continued until the uppermost layer (top of the Osoyoos Lake Limestone) was deposited. At some later time, the area was raised above sea level and the stream began to carve the canyon (figure 19.12). Because the valley sides truncated the youngest layers of rock, we can determine from figure 19.2 that the last event was the carving of the valley.

Note that there are limits on how precisely we can determine the relative age of the granite body. It definitely intruded *before* the Thompson River Formation was deposited and *after* the Abbotsford Formation was deposited. As no contacts can be observed between the Copper Creek Formation and the granite, we cannot say whether the granite is younger or older than the Copper Creek Formation. Nor, as mentioned earlier, can we determine whether the granite formed before, during, or after the tilting of the lower sequence of sedimentary rocks.

We must also point out some special circumstances under which these principles do not apply. For example, intense deformation by tectonic forces can overturn or disrupt beds so much that the principle of superposition cannot be used (see chapter 11). A geoscientist must avoid being dogmatic in applying principles.

Now, if you take a look at the diagram of the Grand Canyon (see figure 19.17), you should be able to determine the sequence of events. The sequence (going from older to younger) is as follows. Regional metamorphism took place resulting in the Vishnu Schist of the lower part of the Grand Canyon. Erosion followed and levelled the land surface. Sedimentation followed, resulting in the Grand Canyon Series rocks. These sedimentary layers were subsequently tilted (they were also faulted, although this is not evident in the diagram). Once again, erosion took place. The lowermost of the currently horizontal layers of sedimentary rock was deposited (the Tapeats Sandstone followed by the Bright Angel Shale). Subsequently, each of the layers progressively higher up the sequence formed. Finally, the stream (the Colorado River) eroded its way through the rock, carving the Grand Canyon.

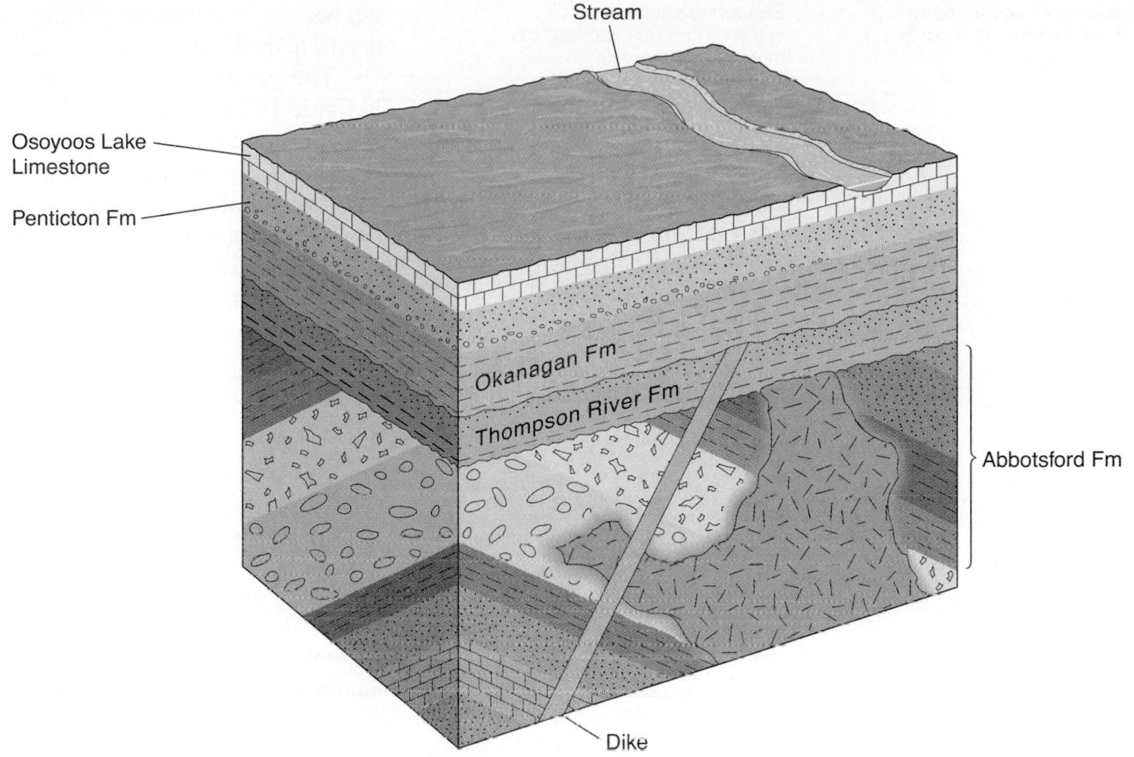

FIGURE 19.12

Same area as shown in figure 19.2 but before valley was carved into rock.

Other Time Relationships

Other characteristics of geology can be applied to help determine relative ages (figure 19.13). The portion of the Abbotsford Formation (the tilted layers) immediately adjacent to the granite body has been *contact metamorphosed* (or "baked"). This indicates that the Abbotsford Formation had to be there before intrusion of the hot, granitic magma. The base of the Thompson River Formation in contact with the granite would not be contact metamorphosed, because it was deposited after the granite had cooled (and was exposed by erosion).

The principle of **inclusion** states that fragments included in a host rock are older than the host rock. In figure 19.13, the granite contains inclusions (xenoliths) of the tilted sedimentary rock. Therefore, the granite is younger than the tilted rock. The rock overlying the granite has granite pebbles in it. Therefore, the granite is older than the horizontal sedimentary rock.

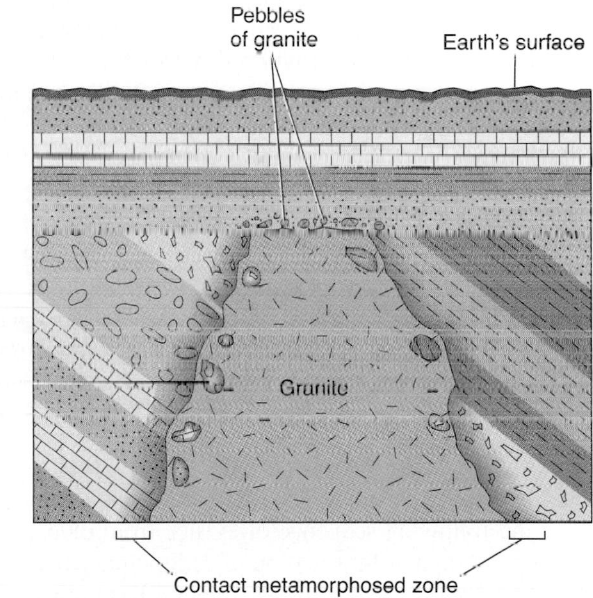

FIGURE 19.13

Age relationships indicated by contact metamorphism, inclusions (xenoliths) in granite, and pebbles of granite.

Unconformities

In this and earlier chapters, we noted the importance of *contacts* for deciphering the geologic history of an area. In chapters 6 and 9 we described intrusive contacts and sedimentary contacts. Faults (described in chapter 11) are a third type of contact. The final important type of contact is an *unconformity.* Each type of contact has a very different implication about what took place in the geologic past.

An **unconformity** is a surface (or contact) that represents a *gap in the geologic record,* with the rock unit immediately above the contact being considerably younger than the rock beneath (figure 19.1*B*). Most unconformities are buried erosion surfaces. Unconformities are classified into three types—disconformities, angular unconformities, and nonconformities—with each type having important implications for the geologic history of the area in which it occurs.

Disconformities

In a **disconformity,** the contact representing missing rock strata separates beds that are parallel to one another. Probably what

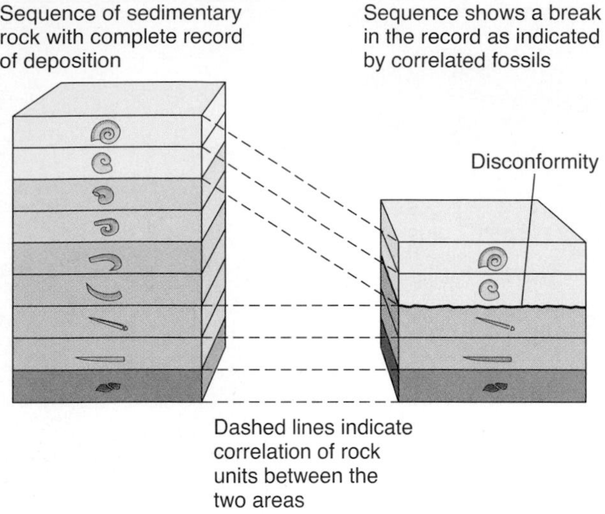

FIGURE 19.14

Schematic representation of a disconformity. The disconformity is in the block on the right.

has happened is that older rocks were eroded away parallel to the bedding plane; renewed deposition later buried the erosion surface (figure 19.14).

Because it often appears to be just another sedimentary contact (or bedding plane) in a sequence of sedimentary rock, a disconformity is the hardest type of unconformity to detect in the field. Rarely, a telltale weathered zone is preserved immediately below a disconformity. Usually the disconformity can be detected only by studying fossils from the beds in a sequence of sedimentary rocks. If certain fossil beds are absent, indicating that a portion of geologic time is missing from the sedimentary record, it can be inferred that a disconformity is present in the sequence. Although it is most likely that some rock layers are missing because erosion followed deposition, in some instances neither erosion nor deposition took place for a significant amount of geologic time.

Angular Unconformities

An **angular unconformity** is a contact in which younger strata overlie an erosion surface on tilted or folded layered rock. It implies the following sequence of events, from oldest to youngest: (1) deposition and lithification of sedimentary rock (or solidification of successive lava flows if the rock is volcanic); (2) uplift accompanied by folding or tilting of the layers; (3) erosion; (4) renewed deposition (usually preceded by subsidence) on top of the erosion surface (figure 19.15). Figures 19.1*B*, 19.2, and 19.13 also show angular unconformities, but with simple tilting rather than folding of the older beds.

Nonconformities

A **nonconformity** is a contact in which an erosion surface on plutonic or metamorphic rock has been covered by younger sedimentary or volcanic rock (figure 19.16). A nonconformity generally indicates deep or long-continued erosion before subsequent burial, because metamorphic or plutonic rocks form at considerable depths in the Earth's crust.

The geologic history implied by a nonconformity, shown in figure 19.16, is (1) crystallization of igneous or metamorphic rock at depth; (2) erosion of at least several kilometres of overlying rock (the great amount of erosion further implies considerable uplift of this portion of the Earth's crust); (3) deposition of new sediment, which eventually becomes sedimentary rock, on the ancient erosion surface. Figures 19.2 and 19.13 also show nonconformities; however, these represent erosion to a relatively shallow depth as the rocks intruded by the pluton have not been regionally metamorphosed, as was the case for those in figure 19.16.

HOW CAN ROCK UNITS BE TRACED FROM ONE AREA TO ANOTHER?

In geology, **correlation** usually means determining time equivalency of rock units. Rock units may be correlated within a region, a continent, and even between continents. Various methods of correlation are described below along with examples of how the principles we described earlier in this chapter are used to determine whether rocks in one area are older or younger than rocks in another area.

Physical Continuity

Finding **physical continuity**—that is, being able to trace physically the course of a rock unit—is one way to correlate rocks between two different places. In figure 19.17, the prominent white layer of cliff-forming rock, the Coconino Sandstone, exposed along the upper part of the Grand Canyon can be seen all the way across the photograph. You can physically follow this unit for several tens of kilometres, thus verifying that, wherever it is exposed in the Grand Canyon, it is the same rock unit.

(Incidentally, the Coconino Sandstone is also shown on figure 9.36.) The Grand Canyon is an ideal location for correlating rock units by physical continuity. However, it is not possible to follow this rock unit from the Grand Canyon into another region because it is not continuously exposed. We usually must use other methods to correlate rock units between regions.

Similarity of Rock Types

Under some circumstances, correlation between two regions can be made by assuming that similar rock types in two regions formed at the same time. This method must be used with extreme caution, especially if the rocks being correlated are common ones.

To show why correlation by similarity of rock type does not always work, we can try to correlate the white, cliff-forming Coconino Sandstone in the Grand Canyon (shown in figure 19.17) with a rock unit of similar appearance in Zion National Park about 100 kilometres away. Both units are white sandstone.

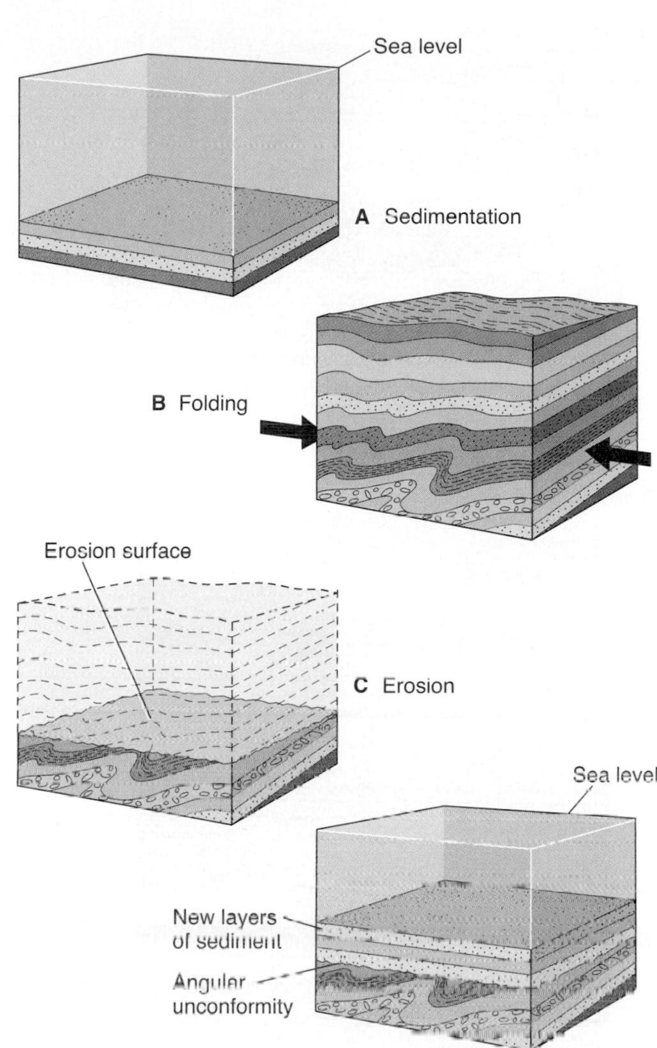

FIGURE 19.15

A particular sequence of events (*A–D*) producing an angular unconformity. Marine-deposited sediments are uplifted and folded (probably during plate tectonic convergence). Erosion removes the upper layers. The area drops below sea level (or sea level rises) and renewed sedimentation takes place. (An angular unconformity can also involve terrestrial sedimentation.) (*E*) is an angular unconformity at Cody, Wyoming.

Photo by C. C. Plummer

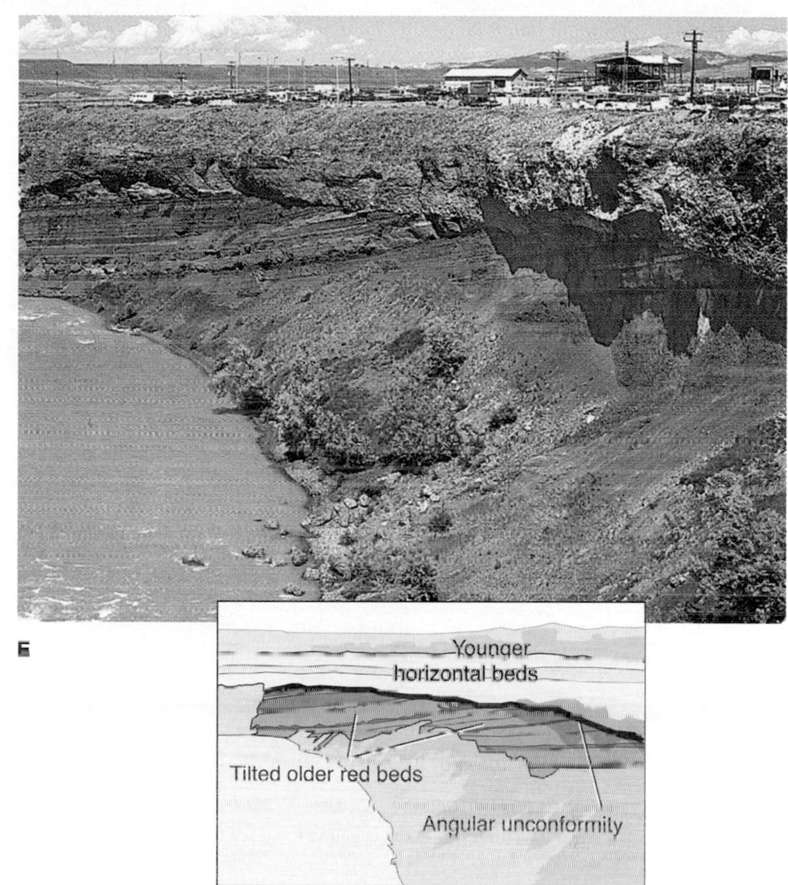

Cross-bedding indicates that both were once a series of sand dunes. It is tempting to correlate them and conclude that both formed at the same time. But if you were to drive or walk from the rim of the Grand Canyon (where the Coconino Sandstone is *below* you), you would get to Zion by ascending a series of layers of sedimentary rock stacked on one another. In other words, you would be getting into progressively younger rock, as shown diagrammatically in figure 19.17. In short, you have shown through *superposition* that the sandstone in Zion (called the Navajo Sandstone) is younger than the Coconino Sandstone.

Correlation by similarity of rock types is more reliable if a very unusual sequence of rocks is involved. If you find in one area a layer of green shale on top of a red sandstone that, in turn, overlies basalt of a former lava flow, and then find the same sequence in another area, you probably would be correct in concluding that the two sequences formed at essentially the same time.

When the hypothesis of continental drift was first proposed (see chapter 2), important evidence was provided by correlating a succession of rocks (figure 19.18) consisting of glacially deposited sedimentary rock (*tillites*, described in chapter 16 on glaciation), overlain by continental sandstones, shales, and coal beds. These strata are in turn overlain by basalt flows. The succession is found in parts of South America, Australia, Africa, Antarctica, and India. It is very unlikely that an identical succession of rocks could have formed on each of the continents if they were widely separated, as they are at present. Therefore, the continents on which the succession is found are likely to have been part of a single super-continent on which the rocks were deposited. Fossils found in these rocks further strengthened the correlation.

In some regions, a *key bed,* a very distinctive layer, can be used to correlate rocks over great distances. An example is a layer of volcanic ash produced from a very large eruption and distributed over a significant portion of a continent.

Correlation by Fossils

Fossils are common in sedimentary rock, and their presence is important for correlation. Plants and animals that lived at the

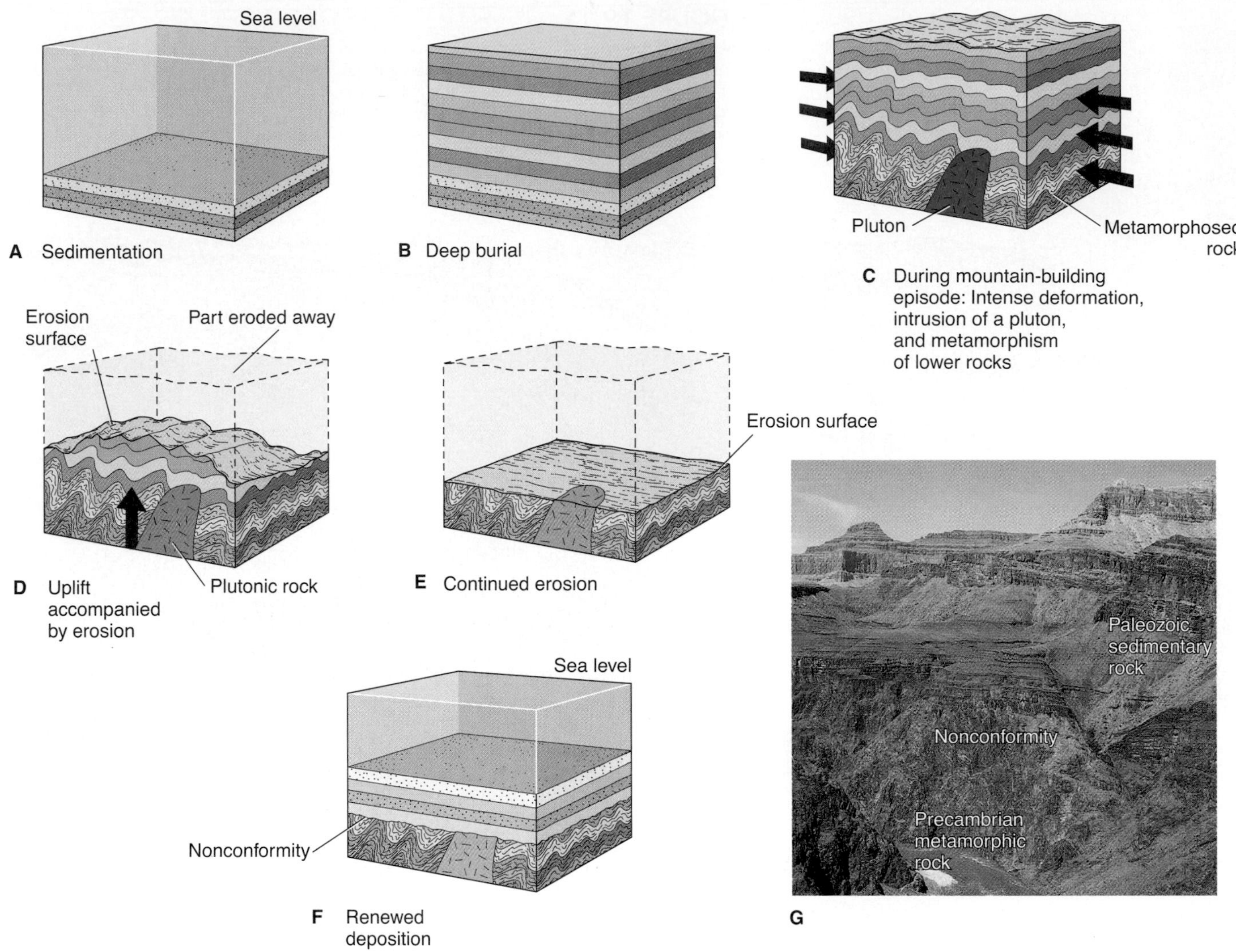

FIGURE 19.16

(A–F) Sequence of events implied by a nonconformity underlain by metamorphic and plutonic rock. (G) A nonconformity in Grand Canyon, Arizona. Paleozoic sedimentary rocks (the Grand Canyon series) overlie vertically foliated Precambrian metamorphic rocks (Vishnu Schist).
Photo by C. C. Plummer

time the rock formed were buried by sediment, and their fossil remains are preserved in sedimentary rock. Most of the fossil species found in rock layers are now extinct—99.9 percent of all species that ever lived are extinct. (The concept of *species* for fossils is similar to that in biology.)

In a thick sequence of sedimentary rock layers, the fossils nearer the bottom (that is, in the older rock) are more unlike today's plants and animals than are those near the top. As early as the end of the eighteenth century, naturalists realized that the fossil remains of creatures of a series of "former worlds" were preserved in Earth's sedimentary rock layers. In the early nineteenth century, a self-educated English surveyor named William Smith realized that different sedimentary layers are characterized by distinctive fossil species and that *fossil species succeed one another through the layers in a predictable order.* Smith's discovery of this principle of **faunal succession** allowed rock layers in different places to be correlated based on their fossils. We now understand that faunal succession works because there is an evolutionary history to life on Earth. Species evolve, exist for a time, and go extinct. Because the same species never evolves twice (extinction is forever), any period of time in Earth's history can be identified by the species that lived at that time. *Paleontologists,* specialists in the study of fossils, have patiently and meticulously over the years identified many thousands of species of fossils and determined the time sequence in which they existed. Therefore, sedimentary rock layers anywhere in the world can be assigned to their correct place in geologic history by identifying the fossils they contain.

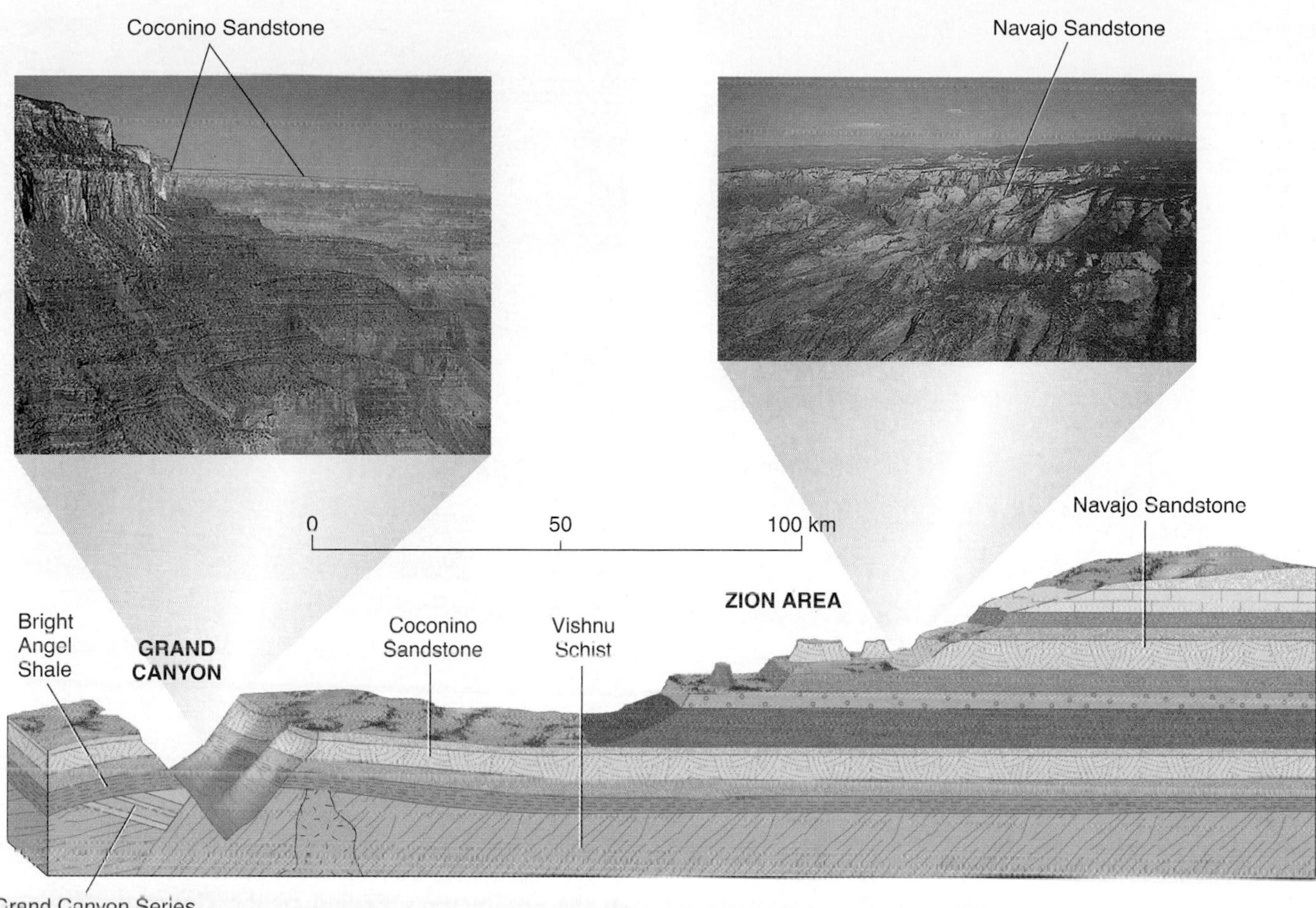

FIGURE 19.17

Schematic cross-section through part of the Colorado Plateau showing the relationship of the Coconino Sandstone, the white cliff-forming unit in the left photo, in Grand Canyon, to the Navajo Sandstone, white unit in the right photo, at Zion National Park.
Photos by C. C. Plummer

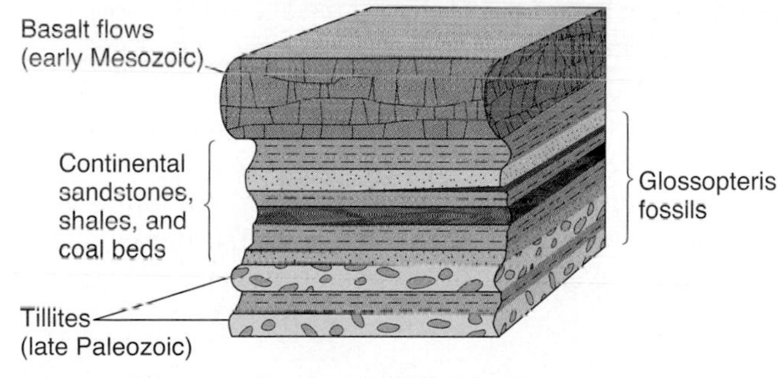

FIGURE 19.18

Rock successions similar to this are found in India, Africa, South America, Australia, and Antarctica. The rocks in each of these localities contain the fossil plant *Glossopteris*.

Ideally, a geologist hopes to find an **index fossil,** a fossil from a very short-lived, geographically widespread species known to exist during a specific period of geologic time. A single index fossil allows the geologist to correlate the rock in which it is found with all other rock layers in the world containing that fossil.

Many fossils are of little use in time determination because the species thrived during too large a portion of geologic time. Sharks, for instance, have been in the oceans for a long time, so discovering a shark's tooth in a rock is not very helpful in determining the rock's relative age.

A geologist is likely to find a **fossil assemblage,** several different fossil species in a rock layer. A fossil assemblage is generally more useful for dating rocks than a single fossil is, because the sediment must have been deposited at a time when all the species represented existed (figure 19.19).

Some fossils are restricted in geographic occurrence, representing organisms adapted to special environments. But, many former organisms apparently lived over most of the Earth, and fossil assemblages from these may be used for worldwide correlation. Fossils in the lowermost horizontal layers of the Grand Canyon are comparable to ones collected in Wales, Great Britain, and many other places in the world (the trilobites in figure 19.20 are an example). We can therefore correlate these rock units and say they formed during the same general span of geologic time.

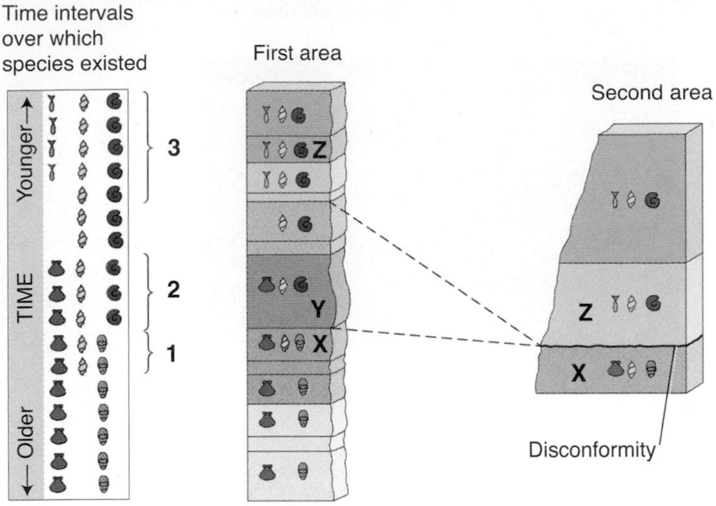

FIGURE 19.19

The use of fossil assemblages for determining relative ages. Rock X contains three fossils. Therefore it must have formed during time interval 1. Rock Y contains three fossils. Therefore, it must have formed during time interval 2. Rock Z contains three fossils. Therefore, it must have formed during time interval 3. In the second area, fossils of time interval 2 are missing. Therefore, the surface between X and Z is a disconformity.

FIGURE 19.20

Early Cambrian trilobite found in the Burgess Shale at the Walcott Quarry, Mount Wapta, B.C.

Photo by M. Radomski

HOW DO WE USE RELATIVE DATING TO UNDERSTAND GEOLOGICAL TIME?

The Standard Geologic Time Scale

Geoscientists can use fossils in rock to refer the age of the rock to the **standard geologic time scale,** a worldwide relative time scale. Based on fossil assemblages, the geologic time scale subdivides geologic time. On the basis of fossils found, a geoscientist can say, for instance, that the rocks of the lower portion of horizontal layers in the Grand Canyon formed during the *Cambrian Period.* This implicitly correlates these rocks with certain rocks in Wales (in fact, the period takes its name from *Cambria,* the Latin name for Wales) and elsewhere in the world where similar fossils occur.

The geologic time scale, shown in a somewhat abbreviated form in table 19.2, has had tremendous significance as a unifying concept in the physical and biological sciences. The working out of the evolutionary chronology by successive generations of geologists and other scientists has been a remarkable human achievement. The geologic time scale, representing an extensive fossil record, consists of three **eras,** which are subdivided into **periods,** which are, in turn, divided into **epochs.** (Remember that this is a relative time scale.)

Precambrian denotes the vast amount of time that preceded the Paleozoic Era (which begins with the Cambrian Period). The **Paleozoic Era** (meaning "old life") began with the appearance of complex life (trilobites, for example), as indicated by fossils. Rocks older than Paleozoic contain few fossils. This is because creatures with shells or other hard parts, which are easily preserved as fossils, did not evolve until the beginning of the Paleozoic.

The **Mesozoic Era** (meaning "middle life") followed the Paleozoic. On land, dinosaurs became the dominant animals of the Mesozoic. We live in the **Recent** (or **Holocene**) **Epoch** of the **Quaternary Period** of the **Cenozoic Era** (meaning "new life"). The Quaternary also includes the most recent ice ages, which were part of the **Pleistocene Epoch.**

It is noteworthy that the fossil record indicates that mass extinctions, in which a large number of species have become extinct, have occurred a number of times in the geologic past. The two greatest mass extinctions define the boundaries between the three eras (see boxes 19.1 and 19.2).

Fossils have been used to determine ages of the horizontal rocks in the Grand Canyon. All are Paleozoic. The lowermost horizontal formations (figure 19.17) are Cambrian, above which are Mississippian, Pennsylvanian, and Permian rock units. By referring to the geologic time scale (table 19.2), we can see that Ordovician, Silurian, and Devonian rocks are not represented. Thus an unconformity (buried erosion surface) is present within the horizontally layered rocks of Grand Canyon.

HOW CAN WE DETERMINE THE ABSOLUTE AGE OF ROCKS?

Counting annual growth rings in a tree trunk will tell you how old a tree is. Similarly, layers of sediment deposited annually in glacial lakes can be counted to determine how long those lakes existed (*varves,* as these deposits are called, are explained in chapter 16). But only since the few decades following the discovery of radioactivity in 1896 have scientists been able to determine numerical ages of rock units. We have subsequently been able to as-

TABLE 19.2 Geologic Time Scale

Era	Period		Epoch
Cenozoic	*Quaternary*		Holocene (Recent)
			Pleistocene
		Neogene	Pilocene
			Miocene
	***Tertiary*	Paleogene	Oligocene
			Eocene
			Paleocene
Mesozoic	*Cretaceous*		
	Jurassic		
	Triassic		
Paleozoic	*Permian*		
	Pennsylvanian	Carboniferous*	
	Mississippian		
	Devonian		
	Silurian		
	Ordovician		
	Cambrian		
Precambrian Time			

*Outside of North America, Carboniferous Period is used rather than Pennsylvanian and Mississippian.

**In 2003, the International Commission on Stratigraphy recommended dropping Tertiary and replacing it with Paleogene and Neogene (shown in red, along with their boundaries). It will probably take the larger geologic community a long time, if ever, to adapt to the change.

sign numerical values to the geologic time scale and determine how many years ago the various eras, periods, and epochs began and ended. We can now state that the Cenozoic Era began some 65 million years ago, the Mesozoic Era started about 250 million years ago, and the Precambrian ended (or the Paleozoic began) about 545 million years ago. The Precambrian includes most of geologic time, because the age of the Earth is commonly regarded as about 4.5 to 4.6 billion years.

The oldest rocks found on the Earth are from northwestern Canada and have been dated at 4.03 billion years old. In 2001, the oldest known mineral was dated at 4.4 billion years old, which is much older than the oldest rock dated so far. The mineral, a zircon crystal from Australia, was likely originally in a granite. Scientists who have studied this mineral think that its chemical makeup indicates that the granite formed from a magma that had a component of melted sedimentary rock. This would indicate that seas existed much earlier than geologists had previously thought possible.

Isotopic Dating

Radioactivity provides a "clock" that begins working when radioactive elements are sealed into newly crystallized minerals. The rates at which radioactive elements decay can be measured and duplicated in many different laboratories. Therefore, if we can determine the ratio of a particular radioactive element and its decay products in a mineral, we can calculate how long ago that mineral crystallized.

Determining the age of a rock through its radioactive elements is known as **isotopic dating** (previously, and somewhat inaccurately, called *radiometric dating*). Geoscientists who specialize in this important field are known as *geochronologists.*

Isotopes and Radioactive Decay

As discussed in chapter 5, every atom of a given element possesses the same number of protons in its nucleus. The number of neutrons, however, need not be the same in all atoms of the same element. The **isotopes** of a given element have different numbers of neutrons, but the same number of protons.

Uranium, for example, commonly occurs as two isotopes, uranium-238 (^{238}U) and uranium-235 (^{235}U). The former has 238 protons and neutrons in its nucleus, whereas the latter has 235. (^{238}U is, by far, the most abundant of naturally occurring uranium isotopes. Only 0.72 percent of uranium is ^{235}U; however, this is the isotope used for nuclear weapons and power generators.) For both isotopes, 92 (the atomic number of uranium) nuclear particles must be protons and the rest neutrons.

Radioactive decay is the spontaneous nuclear change of isotopes with unstable nuclei. Energy is produced with radioactive decay. Emissions from radioactive elements can be detected by a Geiger counter or similar device, and, in high concentrations, can damage or kill humans (see box 19.3).

Nuclei of radioactive isotopes change primarily in three ways (figure 19.21). An *alpha* (α) *emission* is the ejection of two protons and two neutrons from a nucleus. When an alpha emission takes place the atomic number of the atom is reduced by two and its atomic mass number is reduced by four. After an alpha emission, ^{238}U becomes ^{234}Th (thorium), which has an atomic number of 90. The original isotope (^{238}U) is referred to as the *parent isotope.* The new isotope (^{234}Th) is the *daughter product.*

Beta (β) *emissions* involve the release of an electron from a nucleus. To understand this, we need to explain that electrons, which have virtually no mass and are usually in orbit around the nucleus, are also in the nucleus as part of a neutron. A neutron is a proton with an electron inside of it, thus it is electrically neutral. If an electron is emitted from a neutron during radioactive decay, the neutron becomes a proton and the atom's atomic number is increased by one. For example, when ^{234}Th undergoes a beta emission, it becomes ^{234}Pa, an element with an atomic number of 91. Note that the atomic mass number has not changed. This is because the weight of an electron is negligible.

The third mode of change is *electron capture,* whereby a proton in the nucleus captures an orbiting electron. The proton becomes a neutron. The atom becomes a different element having an atomic number one less than its parent isotope.

IN GREATER DEPTH 19.1

Highlights of the Evolution of Life through Time

The history of the biosphere is preserved in the fossil record. Through fossils, we can determine their place in the evolution of plants and animals as well as get clues as to how extinct creatures lived. The oldest readily identifiable fossils found are prokaryotes—microscopic, single-celled organisms that lack a nucleus. These date back to around 3.5 billion years (b.y.) ago, so life on Earth is at least that old. It is likely that even more primitive organisms date back further in time but are not preserved in the fossil record. Fossils of much more complex, single-celled organisms that contained a nucleus (eukaryotes) are found in rocks as old as 1.4 b.y. These are the earliest living creatures to have reproduced sexually. Colonies of unicellular organisms likely evolved into multicellular organisms. Multicellular algae fossils date back at least a billion years.

Imprints of larger multicellular creatures appear in rocks of late Precambrian age, about 700 to 550 million years ago (m.y.). These resemble jellyfish and worms. The earliest true animals are soft-bodied organisms (the Ediacaran biota) that lived between 565 and 543 m.y. and are now found as fossils in the Mistaken Point Formation of Newfoundland (box figure 1; see box 20.1).

Sedimentary rocks from the Paleozoic, Mesozoic, and Cenozoic Eras have abundant fossils. Large numbers of fossils appeared early in the Cambrian Period. Trilobites (see figure 19.20) evolved into many species and were particularly abundant during the Cambrian. Trilobites were arthropods that crawled on muddy sea floors and are the oldest fossils with eyes. They became less significant later in the Paleozoic, and finally, all trilobites became extinct by the end of the Paleozoic.

BOX 19.1 ■ FIGURE 1

Imprint of one of the earliest multi-cellular animals known (between 635 and 543 m.y. old) from the Mistaken Point Formation, Newfoundland. The animal may have been an early type of jellyfish.

Photo Image: 36052: With permission of the Royal Ontario Museum © ROM.

The most primitive fishes, the first vertebrates, date back to late in the Cambrian (box figure 2). Fishes similar to currently living species (including sharks) flourished during the Devonian (named after Devonshire, England). The Devonian is often called the "age of fishes." Amphibians evolved from air-breathing fishes late in the Devonian. These were the first land vertebrates. However, invertebrate land animals date back to the latest Cambrian, and land plants first appeared in the Ordovician. Reptiles and early ancestors of mammals evolved from amphibians in Pennsylvanian time or perhaps earlier.

The Paleozoic ended with the greatest mass extinction ever to occur on Earth. More than 95 percent of species that existed died out.

During the Mesozoic, new creatures evolved to occupy ecological domains left vacant by extinct creatures. Dinosaurs and mammals evolved from the animal species that survived the great extinction. Dinosaurs became the dominant group of land animals (box figure 3). Birds likely evolved from dinosaurs in the Mesozoic. Large, now extinct, marine reptiles lived in Mesozoic seas. Ichtyosaurs, for example, were up to 20 metres long, had dolphinlike bodies, and were probably fast swimmers. Flying reptiles, pterosaurs, some of which had wingspans of almost 10 metres, soared through the air.

The Cretaceous Period (and Mesozoic Era) ended with the second-largest mass extinction (around 75 percent of species were wiped out).

The Cenozoic is often called the age of mammals. Mammals, which were small, insignificant creatures during the Mesozoic, evolved into the many groups of mammals (whales, bats, canines, cats, elephants, primates, and so forth) that occupy Earth at present. Many species of mammals evolved and became extinct throughout the Cenozoic. Hominids (modern humans and our extinct ancestors) have a fossil record dating back 6 m.y. and likely evolved from a now-extinct ancestor common to hominids, chimpanzees, and other apes.

BOX 19.1 ■ FIGURE 2

Primitive Mesozoic-age fish covered with thick bony scales. Found at Wapiti Lake, B.C.

Photo Image: 46204: With permission of the Royal Ontario Museum © ROM.

BOX 19.1 ■ FIGURE 3
Upper Cretaceous Hadrosaur skeleton found at the Red Deer River, Alberta.
Photo Image: 00787: With permission of the Royal Ontario Museum © ROM.

We tend to think of mammals' evolution as being the great success story (because we are mammals); mammals, however, pale in comparison to insects. Insects have been around far longer than mammals and now account for an estimated 1 million species.

Additional Resources

University of California Museum of Paleontology
Find the fossils mentioned here.
www.ucmp.berkeley.edu/

The Paleontology Portal
Another site to find out about fossils. You can search by type of creature, by time, or by location.
www.paleoportal.org/

Royal Tyrell Museum, Drumheller
Go on a virtual tour of some of the exhibition halls that feature dinosaurs of western Canada.
www.tyrellmuseum.com

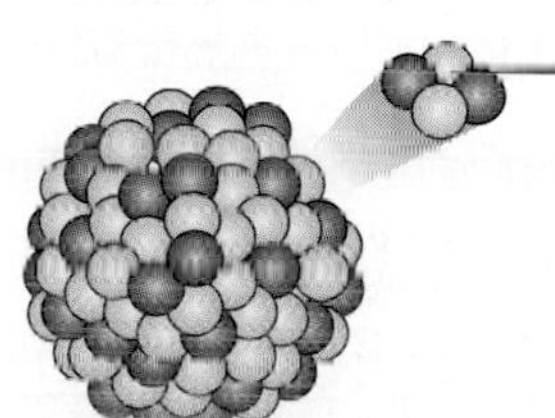

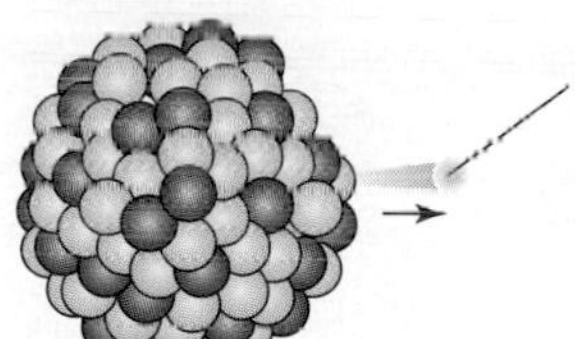

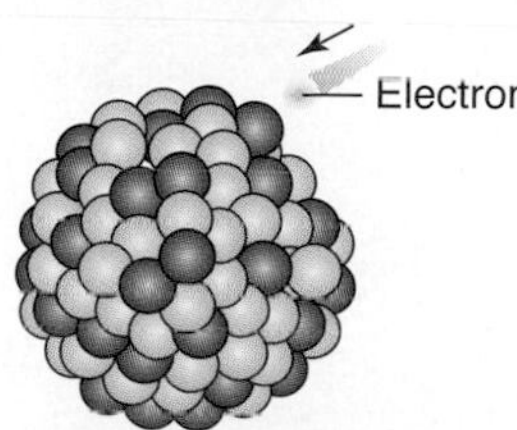

FIGURE 19.21
Three modes of radioactive decay.

An example of this is the potassium-argon system, discussed below, in which ^{40}K becomes ^{40}Ar. The parent isotope, potassium, has an atomic number of 19 and the atomic number of argon, the daughter product, is 18.

Figure 19.22 shows how ^{238}U decays to ^{206}U (lead-206) in a series of alpha and beta emissions. The important point is not the intermediate steps but the starting and ending isotopes. In the process, ^{238}U loses 10 protons, so that the daughter product has an atomic number of 82 (which is lead) and a total of 32 protons and neutrons, so the new atomic mass number is 206. ^{206}Pb can be produced only by the decay of ^{238}U.

To understand how isotopic dating works, it is important to recognize that if a large number of atoms of a given radioactive isotope are present in a rock or mineral, the *proportion* (or percentage) of those atoms that will radioactively decay over a given time span is constant, for example, if you have 100,000 atoms of isotope X and over a period of a million years, a quarter of those atoms (25,000) radioactively decay (the proportion would be 1 in 4). You would have the same proportion of 1 in 4 if you started out with 300,000 atoms: after a million years, 75,000 of the atoms would have decayed. The proportional number of atoms that periodically decay is apparently unaffected by chemical reactions or by the high pressures and high temperatures of the Earth's interior.

The rate of proportional decay for isotopes is expressed as **half-life,** the time it takes for a given amount of a radioactive isotope to be reduced by one-half. (The other half disintegrates into daughter products and energy.) The half-lives of some isotopes created in nuclear reactors are in fractions of a second.

IN GREATER DEPTH 19.2

Demise of the Dinosaurs—Was It Extraterrestrial?

Dinosaurs dominated the continents during the Mesozoic Era. Now they prey on the imaginations of children of all ages and are featured in media ranging from movies to cartoons. It's hard to accept that beings as powerful and varied as dinosaurs existed and were wiped out. But the fossil record is clear—when the Mesozoic came to a close, dinosaurs became extinct. Not a single of the numerous dinosaur species survived into the Cenozoic Era. Not only did the dinosaurs go, but about 75 percent of all plant and animal species, marine as well as terrestrial, were extinguished. This was one of Earth's "great dyings"—an even "greater dying" was when the Paleozoic Era ended with the extinction of over 95 percent of all species. Most major extinctions have been gradual, and scientists usually have attributed them to climate changes.

A couple decades ago, geologist Walter Alvarez, his father, physicist Luis Alvarez, and two other scientists proposed a hypothesis that the dinosaur extinction was caused by the impact of an asteroid (box figure 1). This was based on the chemical analysis of a thin layer of clay marking the boundary between the Mesozoic and Cenozoic Eras (usually referred to as the K-T boundary—it separates the Cretaceous [K] and Tertiary [T] Periods). The K-T boundary clay was found to have about 30 times the amount of the rare element iridium as is normal for crustal rocks. Iridium is relatively abundant in meteorites and other extraterrestrial objects such as comets, and the scientists suggested that the iridium was brought in by an extraterrestrial body.

A doomsday scenario is visualized in which an asteroid 10 kilometres in diameter struck Earth (box figure 2). The asteroid would have blazed through the atmosphere at astonishing speed and, likely, impacted at sea. Part of the ocean would have been vapourized and a crater created on the ocean floor. There would have been an earthquake much larger than any ever felt by humans. Several-hundred-metre-high waves would crisscross the oceans, devastating life anywhere near shorelines. The lower atmosphere would have become intolerably hot, at least for a short period of time. The atmosphere worldwide would have been altered and the climate cooled because of the increased blockage of sunlight by dust particles suspended in the upper atmosphere.

For a while, the hypothesis was hotly debated. Other scientists hypothesized that the extinctions were caused by exceptionally large volcanic activity. Further evidence supporting the asteroid hypothesis accumulated. K-T layers throughout the world were found to have grains of quartz that had been subjected to shock metamorphism (see box 7.1). Microscopic spheres of glass that formed when rock melted from the impact and droplets were thrown high into the air were also found in the K-T layers. Sediment that appeared to have been deposited by giant sea waves was found in various locations.

The asteroid hypothesis advocates predicted that a large meteorite crater should be found someplace on Earth that could be dated as having formed around 65 million years ago, when the Mesozoic ended.

In 1990, the first evidence for the "smoking gun" crater was found. The now-confirmed crater is over 200 kilometres in diameter and centred along the coast of Mexico's Yucatan peninsula at a place called Chicxulub. The crater at Chicxulub, now buried beneath younger sedimentary rock, is the right size to have been formed by a 10-kilometre asteroid.

The existence of the crater was confirmed by geologists going over Mexican oil company records compiled during drilling for oil at Yucatan and finding breccias of the right age buried in the Chicxulub area. Breccias, due to meteorite impact, are common at known meteorite craters. The evidence for an asteroid impact is overwhelming. However, the impact may not be entirely to blame for dinosaur extinction.

BOX 19.2 ■ FIGURE 1

Was the extinction of the dinosaurs caused by an asteroid impact?

Artwork by Joe Tucciarone

BOX 19.2 ■ FIGURE 2

Impact of a large asteroid would cause vaporization of ocean waters, creation of a crater, huge ocean waves, and dramatic climate change as dust particles blocked sunlight.

© Brand X pictures/Punchstock

As the Mesozoic was ending, huge lava floods were taking place in India over a 2- to 3-million-year period and building a basalt plateau larger than the Columbia Plateau described in chapter 4. Gas emissions very likely had an effect on the atmosphere and worldwide climate. It is noteworthy that Earth's largest mass extinction at the end of the Paleozoic has often been blamed on even larger lava floods that occurred in Siberia. However, recently found evidence strongly indicates that a major asteroid impact also occurred at that time. (Recently, geologists reported that they found evidence for an appropriate crater buried beneath sedimentary rocks off the coast of Australia. Other geologists remain skeptical.) It seems to many geologists that having the two largest mass extinctions associated with an impact and lava floods is more than coincidence. Some have suggested that the lava floods were somehow triggered by the asteroid impacts, even though they were in different parts of the world.

We will probably never know exactly how big a role either the asteroid or the lava floods played in the unfortunate extinction of dinosaurs. But "unfortunate" is from the perspective of dinosaurs, not humans. The only mammals in the Cretaceous were inconsequential, rat-sized creatures. They survived the K-T extinction and, with dinosaurs no longer dominating the land, evolved into the many mammal species that populate Earth today, including humans.

Additional Resource

Walking with Dinosaurs

Visit *Tyrannosaurus rex* and other famous dinosaurs and read more about the K-T extinction.

www.bbc.co.uk/dinosaurs/

Naturally occurring isotopes that we find in rocks have very long half-lives (table 19.3). ^{40}K has a half-life of 1.3 billion years. If you began with one milligram of ^{40}K, 1.3 billion years later one-half milligram of ^{40}K would remain. After another 1.3 billion years, there would be one-fourth of a milligram, and after another half-life only one-eighth of a milligram. (Note that two half-lives do not equal a whole life.)

To determine the age of a rock by using ^{40}K, the amount of ^{40}K in that rock must first be determined by chemical analysis. The amount of ^{40}Ar (the daughter product) must also be determined and then used to calculate how much ^{40}K was present when the rock formed. By knowing how much ^{40}K was originally present in the rock and how much is still there, we can calculate the age of the rock on the basis of its half-life mathematically (see box 19.4). The graph in figure 19.23*A* demonstrates the mathematical relationship between a radioactively decaying isotope and time.

WHAT IS RADIOCARBON DATING?

Because of its short half-life of 5,730 years, radiocarbon dating is useful in dating materials and events accurately back to only about 40,000 years—about seven half-lives. (However, new techniques allow some scientists to push the limit to nearly double that time.) The technique is most useful in archaeological dating and for very young geologic events (recent, or Holocene, volcanic and glacial features, for instance). It is also used to date historical artifacts. For instance, the Dead Sea Scrolls, the oldest of the surviving biblical manuscripts, were radiocarbon dated and their ages ranged from the third century B.C. to 68 A.D. These ages are consistent with estimates previously made by archaeologists and other scholars.

Radiocarbon dating is fundamentally different from the parent–daughter systems described previously in that ^{14}C is being created continuously in the atmosphere (figure 19.23*B*). Carbon (atomic number 6) is in the air as part of CO_2. It is mostly the stable isotope ^{12}C. However, ^{14}C is created in the atmosphere when cosmic radiation bombards nitrogen (N), atomic number 7. A neutron strikes and is captured by an ^{14}N atom. A proton is expelled from the nucleus and the atom becomes ^{14}C. The nucleus of the newly created carbon atom is unstable and will, sooner or later, through a beta emission, revert to ^{14}N. The electron is emitted from the atom as radiation. The rate of production of ^{14}C approximately balances the rate at which ^{14}C reverts to ^{14}N so that the level of ^{14}C remains essentially constant in the atmosphere.

Living matter incorporates ^{12}C and ^{14}C into its tissues (figure 19.23*B*); the ratios of ^{12}C and ^{14}C in the new tissues are usually the same as in the atmosphere. On dying, the plant or animal ceases to build new tissue. The ^{14}C disintegrates radioactively at the fixed rate of its half-life (5,730 years). The radioactive emissions per gram of carbon from the plant or animal remains are determined. The greater the radioactivity, the younger the sample. By comparing the radioactivity per gram to previously determined standards, we can indirectly determine the ratio of ^{12}C to radioactive ^{14}C in organic remains, and we can determine the time elapsed since the death of the organism.

Cosmogenic Isotope Dating

During the past couple decades, another dating technique has been added to geoscientists' numerical age determination arsenal. *Cosmogenic isotope dating,* or *surface exposure dating,* uses the effects of constant bombardment by neutron radiation coming from deep space (cosmogenic) of material at Earth's surface.

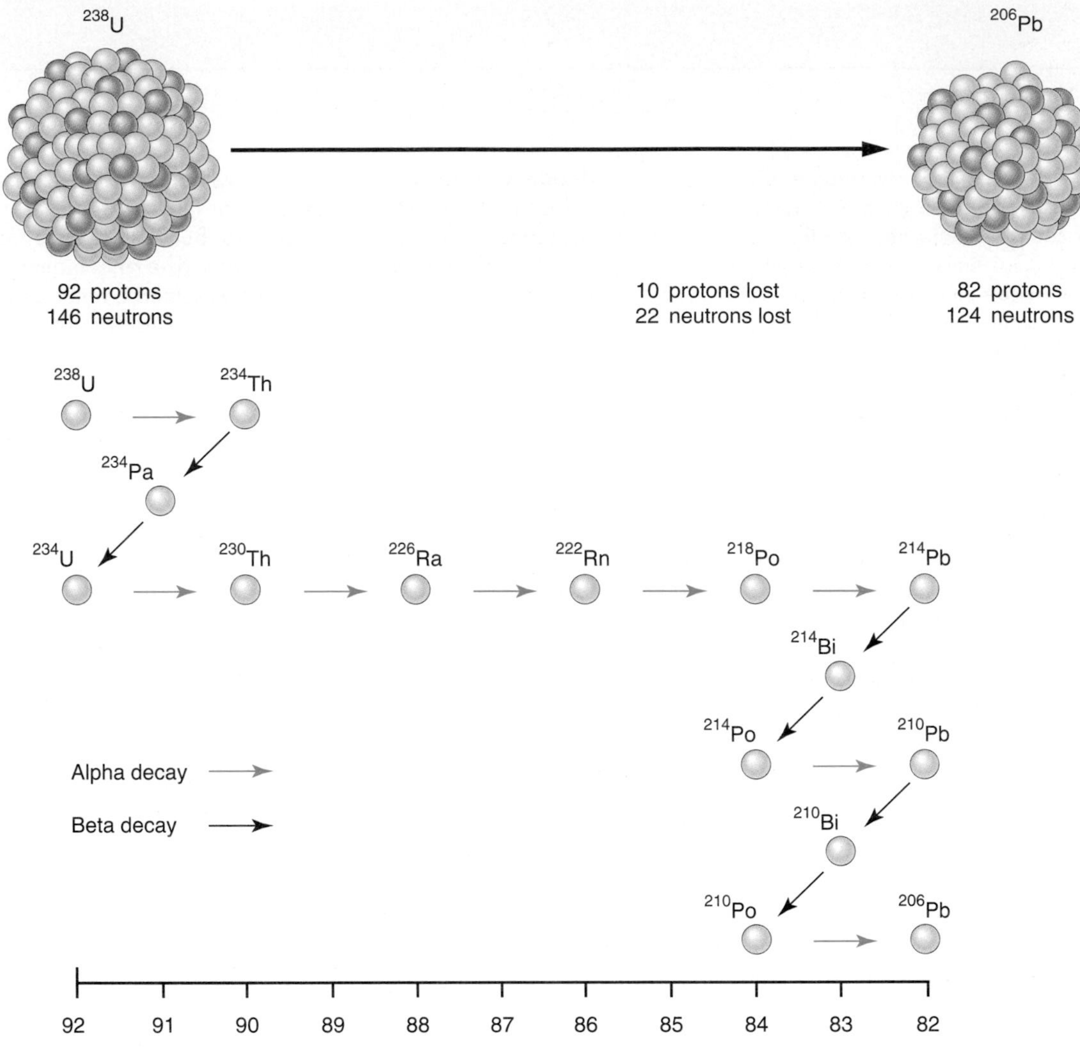

FIGURE 19.22

Uranium 238 decays to lead 206. The different intermediate steps in the process are shown below the models of the nuclei of ^{238}U and ^{206}Pb. Refer to appendix C or the periodic table of elements in appendix D for names of the elements shown.

TABLE 19.3 Radioactive Istopes Commonly Used for Determining Ages of Earth's Materials

Parent Isotope	Half-Life	Daughter Product	Effective Dating Range (years)
K-40 ^{40}K	1.3 billion years	^{40}Ar	100,000–4.6 billion
U-238 ^{238}U	4.5 billion years	^{206}Pb	10 million–4.6 billion
U-235 ^{235}U	713 million years	^{207}Pb	10 million–4.6 billion
Th-232 ^{232}Th	14.1 billion years	^{208}Pb	10 million–4.6 billion
Rb-87 ^{87}Rb	49 billion years	^{87}Sr	10 million–4.6 billion
C-14 ^{14}C	5,730 years	^{14}N	100–40,000

ENVIRONMENTAL GEOLOGY 19.3

Radon, a Radioactive Health Hazard

Radon is an odourless, colourless gas. Every time you breathe outdoors you inhale a harmless, minute amount of radon. If the concentration of radon that you breathe in a building is too high, however, you could, over time, develop lung cancer. It is one of the intermediate daughter products in the radioactive disintegration of ^{238}U to ^{206}Pb. It has a half-life of only 3.8 days.

Concentrations of radon are highest in areas where the bedrock is granite, gneiss, limestone, black shale, or phosphate-rich rock—rocks in which uranium is relatively abundant. Concentrations are also high where glacial deposits are made of fragments of these rocks. Even in these areas, radon levels are harmless in open, freely circulating air. Radon may dissolve in groundwater or build up to high concentrations in confined air spaces such as caves or basements (box figure 1).

The U.S. Environmental Protection Agency (EPA) regards 5 million American homes to have unacceptable radon levels in the air. Scientists outside of the EPA have concluded that the standards the EPA is using are too stringent. They think that a more reasonably defined danger level means that only 50,000 homes have radon concentrations that pose a danger to their occupants. Health Canada recommends that the guidelines for exposure to radon gas should be 800 becquerels per cubic metre as the annual average concentration in a normal living area. Less than one tenth of one percent of all homes in Canada have radon levels that exceed the recommended guidelines.

Radon was first recognized in the 1950s as a health hazard in uranium mines, where the gas would collect in poorly ventilated air spaces. Radon lodges in the respiratory system of an individual and, as it deteriorates into daughter products, the subatomic particles given off cause damage to lung tissue. Three-quarters of the uranium miners studied were smokers. Thus, it is difficult to determine the extent to which smoking or radon induced lung cancer. (All studies show, however, that smoking and exposure to high radon levels are more likely to cause lung cancer than either alone.)

Interpolating the high rates of cancer incidence from the uranium miners to the population exposed to the very much lower radium levels in homes, as the EPA has done, is scientifically questionable.

What should you do if you are living in a high-radon area? First, have your house checked to see what the radon level is. Then read up on what acceptable standards should be. In most buildings with a high radon level, the gas seeps in from the underlying soil through the building's foundation. If a building's windows are kept open and fresh air circulates freely, radon concentrations cannot build up. But, houses are often kept sealed for air conditioning during the summer and heating during the winter. Air-circulation patterns are such that a slight vacuum sucks the gases from the underlying soil into the house. Thus, radon concentrations might build up to dangerous levels.

The problem may be solved in several ways (aside from leaving windows open winter and summer). Basements can be made airtight so that gases cannot be sucked into the house from the soil. Air-circulation patterns can be altered so that gases are not sucked in from underlying soil, or they are mixed with sufficient fresh, outside air.

If you are purchasing a new house, it would be a good idea to have it tested for radon before buying, particularly if the house is in an area of high-uranium bedrock or soil.

Related Web Resource

Radon in earth, air, and water

http://sedwww.cr.usgs.gov/radon/radonhome.html

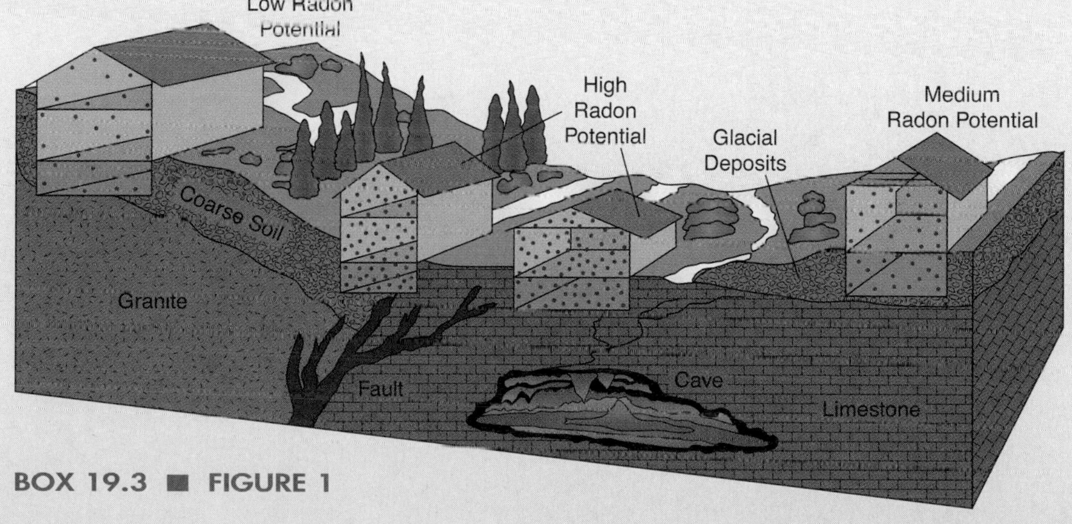

BOX 19.3 ■ FIGURE 1

IN GREATER DEPTH 19.4

Calculating the Age of a Rock

(NOTE: This box is not intended for the mathematically challenged.)

The relationship between time and radioactive decay of an isotope is expressed by the following equation (which is used to plot curves such as shown in figure 19.23).

$$N = N_0 e^{-\lambda t}$$

N is the number of atoms of the isotope at time t, the time elapsed. N_0 is the number of atoms of that isotope present when the "clock" was set. The mathematical constant e has a value of 2.718. λ is a decay constant—a proportionality constant that relates the rate of decay of an isotope to the number of atoms of that isotope remaining.

The relationship between λ and the half-life (t_{hl}) is

$$\lambda = \frac{\ln 2}{t_{hl}} = \frac{0.693}{t_{hl}}$$

Replacing λ in the first equation and converting that equation to natural logarithmic (to the base e) form, we get

$$t = \frac{t_{hl}}{.693} \ln \frac{N}{N_0}$$

N/N_0 is the ratio of parent atoms at present to the original number of parent atoms.

As an example, we will calculate the age of a mineral using ^{235}U decaying to ^{207}Pb. Table 19.3 indicates that the half-life is 713 million years. A laboratory determines that, at present, there are 440,000 atoms of ^{235}U and that the amount of ^{207}Pb indicates that when the mineral crystallized, there were 1,200,000 atoms of ^{235}U. (We assume that there was no ^{207}Pb in the mineral at the time the mineral crystallized.) Plugging these values into the formula, we get

$$t = \frac{713{,}000{,}000}{.693} \ln \frac{440{,}000}{1{,}200{,}000}$$

Solving this gives us 1,032,038,250 years. Rounded off, we can say the mineral formed 1.032 billion years ago.

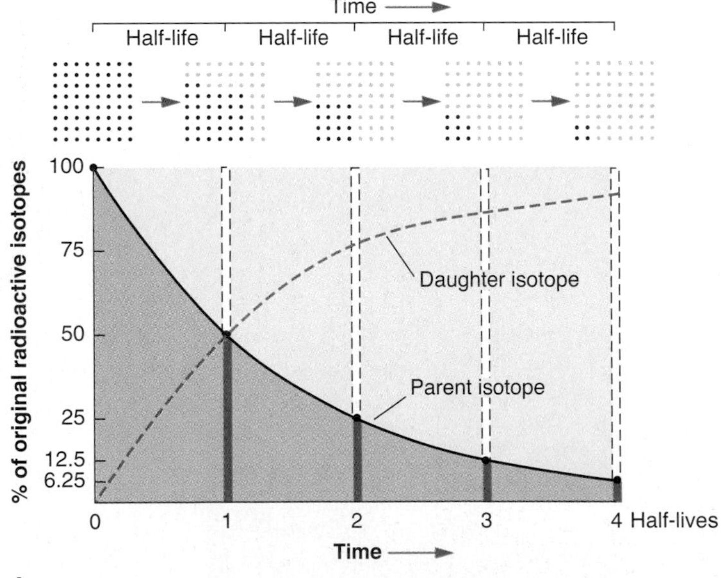

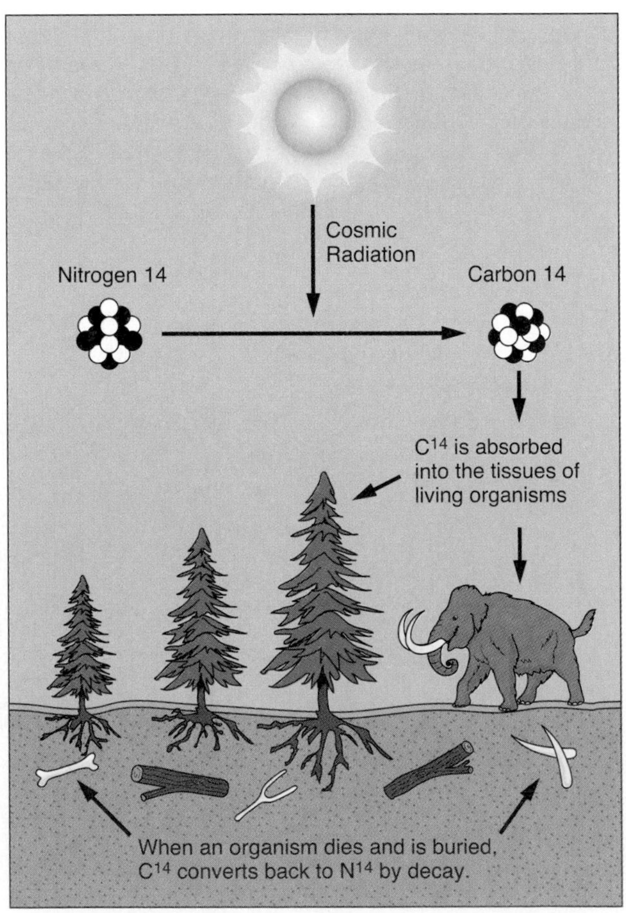

FIGURE 19.23

(A) The curve used to determine the age of a rock by comparing the percentage of radioactive isotope remaining in time to the original amount. Darker bars show the amount left after each half-life. Dashed bars show the amount disintegrated into daughter product and lost nuclear particles. The numbers of dots in the squares above the graph are proportional to the numbers of atoms. (B) Principles of radiocarbon dating.

The high-energy particles hit atoms in minerals and alter their nuclei. For instance, when the atoms in quartz are hit, oxygen is converted to beryllium-10 (^{10}Be) and silicon is changed to aluminum-26 (^{26}Al). The concentrations of these isotopes increase at a constant rate once a rock surface is exposed to the atmosphere because the influx of cosmogenic radiation is uniform over time. The length of time a rock surface has been exposed can be calculated by knowing the rate of increase of a cosmogenic isotope and determining the amount of that isotope in a mineral at a rock's surface.

One application of cosmogenic dating has been to determine how long ago boulders were deposited by advancing glaciers during the geologically recent ice ages (see chapter 16). However, dates obtained are minimum ages, because snow that covered the boulders for part of the year reduced their exposure to cosmogenic radiation.

WHEN CAN ISOTOPIC DATING TECHNIQUES BE USED?

When we are dating a rock, we are usually attempting to determine how long ago that rock formed. But exactly what is being dated depends on the type of rock and the isotopes analyzed. For a metamorphic rock, we are likely to be dating a time during the millions of years of the cooling of that rock rather than the peak of high temperature during metamorphism. Some techniques determine isotopic ratios for a whole rock, while others use single minerals within a rock. Usually, an isotopic date determines how long ago the rock or mineral became a closed system; that is, how long ago it was sealed off so that neither parent nor daughter isotopes could enter or leave the mineral or rock. Different isotopic pairs have different closure temperatures; when a rock cools below that temperature, the system is closed and the "clock" starts. For instance, the ^{40}K ^{40}Ar isotopic pair has closure temperatures ranging from 150°C to 550°C, depending on the mineral. (Ar is a gas and gets trapped in different crystal structures at different temperatures.)

Generally, the best dates are obtained from igneous rocks. For a lava flow, which cools and solidifies rapidly, the age determined gives us the time of eruption and solidification. On the other hand, plutonic rocks, which may take more than a million years to solidify, will not necessarily yield the time of intrusion, but the time at which a mineral cooled below the closure temperature. Dating metamorphic rocks usually means determining the time of cooling following the end of a metamorphic episode. Sedimentary rocks are difficult to date reliably.

For an isotopic age determination to be accurate, several conditions must be met. To ensure that the isotopic system has remained closed, the rock collected must show no signs of weathering or hydrothermal alteration. Second, one should be able to infer there were no daughter isotopes in the system at the time of closure or make corrections for probable amounts of daughter isotopes present before the "clock" was set. Third, there must be sufficient parent and daughter atoms to be measurable by the instrument (a mass spectrometer) being used. And, of course, technicians and geochronologists must be highly skilled at working sophisticated equipment and collecting and processing rock specimens.

Whenever possible, geochronologists will use more than one isotope pair for a rock. The two U-Pb systems (table 19.3) can usually be used together and provide an internal cross-check on the age determinate. Using the K-Ar or another system would provide more confidence in our age determination.

How Reliable Is Isotopic Dating?

Half-lives of radioactive isotopes, whether short-lived, such as used in medicine, or long-lived, such as used in isotopic dating, have been found not to vary beyond statistical expectations. The half-life of each of the isotopes we use for dating rocks has not changed with physical conditions or chemical activity, nor could the rates have been different in the distant past. It would violate laws of physics for decay rates (half-lives) to have been different in the past. Moreover, when several isotopic dating systems are painstakingly done on a single ancient igneous rock, the same age is obtained, or we understand the reason for differences in ages, it confirms that the decay constants for each system are indeed constant.

Comparing isotopic ages with relative age relationships confirms the reliability of isotopic dating. For instance, a dike that crosscuts rocks containing Cenozoic fossils gives us a relatively young isotopic age (less than 65 million years old), whereas a pluton truncated by overlying sedimentary rocks with Paleozoic fossils yields a relatively old age (greater than 250 million years). Many thousands of similar determinations have confirmed the reliability of radiometric dating.

Combining Relative and Numerical Ages

Radiometric dating can provide numerical time brackets for events whose relative ages are known. Figure 19.24 adds radiometric dates for each of the two igneous bodies in the fictitious Cold Canyon area of figure 19.2. The date obtained for the granite is 540 million years B.P. (before present), while the dike formed 78 million years ago. We can now state that the Abbotsford Formation and older tilted layers formed before 540 million years ago (though we cannot say how much older they are). We still do not know whether the Copper Creek Formation is older or younger than the granite because of the lack of cross-cutting relationships. The Thompson River Formation's age is bracketed by the age of the granite and the age of the dike. That is, it is between 540 and 78 million years old. The Okanagan City and overlying formations are younger than 78 million years old; how much younger we cannot say.

Isotopic dates from volcanic ash layers or lava flows interlayered between fossiliferous sedimentary rocks have been used to assign absolute ages to the geologic time scale (figure 19.25). Isotopic dating has also allowed us to extend the time scale back into the Precambrian. There is, of course, a margin of uncertainty in each of the given dates. The beginning of the Paleozoic, for instance, was regarded until recently to be 570 million years ago, but

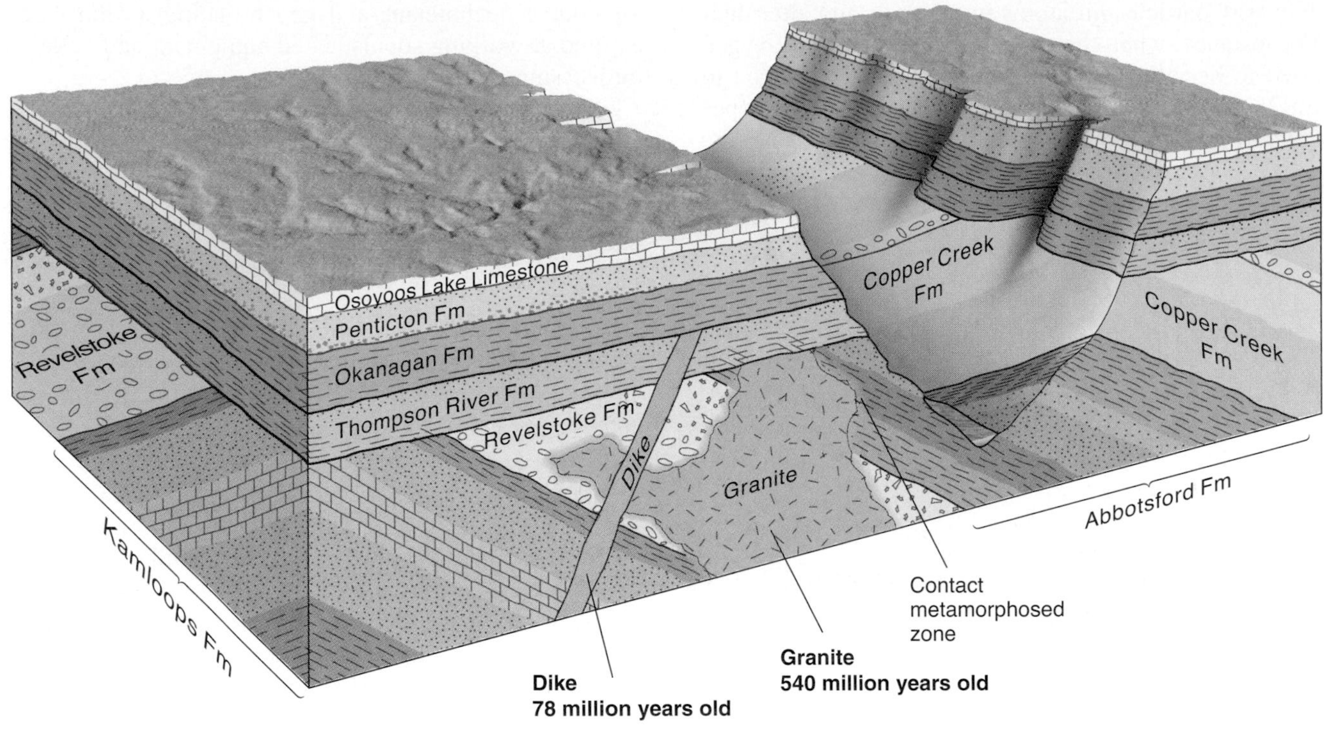

FIGURE 19.24

The Cold Canyon area as shown in figure 19.2 but with isotopic dates for igneous rocks indicated.

with an uncertainty of ±30 million years. Recent work has fixed the age as 544± 1 million years. There are inherent limitations on the dating techniques as well as problems in finding the ideal rock for dating. For instance, if you wanted to obtain the date for the end of the Paleozoic Era and the beginning of the Mesozoic Era, the ideal rock would be found where there is no break in deposition of sediments between the two eras, as indicated by fossils in the rocks. But the difficulties in dating sedimentary rock mean you would be unlikely to date such rocks. Therefore, you would need to date volcanic rocks interlayered with sedimentary rocks found as close as possible to the transitional sedimentary strata. Alternatively, isotopically dated intrusions, such as dikes, whose cross-cutting relationships indicate that the age of intrusion is close to that of the transitional sedimentary layers, could be used to approximate the absolute age of the transition.

Isotopic dating has shown that the Precambrian took up most of geologic time (87 percent). Obviously, the Precambrian needed to be subdivided. The three major subdivisions of the Precambrian are the **Prearchean** (or **Hadean**), the **Archean,** and the **Proterozoic** (Greek for beginning life). Each is regarded as an **eon,** the largest unit of geological time. A fourth, and youngest, eon is the **Phanerozoic** (Greek for visible life). The Phanerozoic eon is all of geologic time with an abundant fossil record; in other words, it is made up of the three eras that followed the Precambrian.

HOW OLD IS THE EARTH?

In 1625 Archbishop James Ussher determined that the Earth was created in the year 4004 B.C. His age determination was made by counting back generations in the Bible. This would make the Earth 6,000 years old at present. That very young age was largely taken for granted by Western countries. By contrast, Hindus at the time regarded the Earth as very old. According to an ancient Hindu calendar, the year A.D. 2000 would be year 1,972,949,101.

With the popularization of uniformitarianism in the early 1800s, Earth scientists began to realize that the Earth must be very old—at least in the hundreds of millions of years. They were dealt a setback by the famous English physicist Lord Kelvin. Kelvin, in 1866, calculated from the rate at which Earth loses heat that the Earth must have been entirely molten between 20 and 100 million years ago. He later refined his estimate to between 20 and 40 million years. He was rather arrogant in scoffing at Earth scientists who hypothesized that uniformitarianism indicated a much older age for Earth. The discovery of radioactivity in 1896 invalidated Kelvin's claim because it provided a heat source that he had not known about. When radioactive elements decay, heat is given off and that heat is added to the heat already in the Earth. The amount of radioactive heat given off at present approximates the heat the Earth is losing. So, for all practical purposes, Earth is not getting cooler.

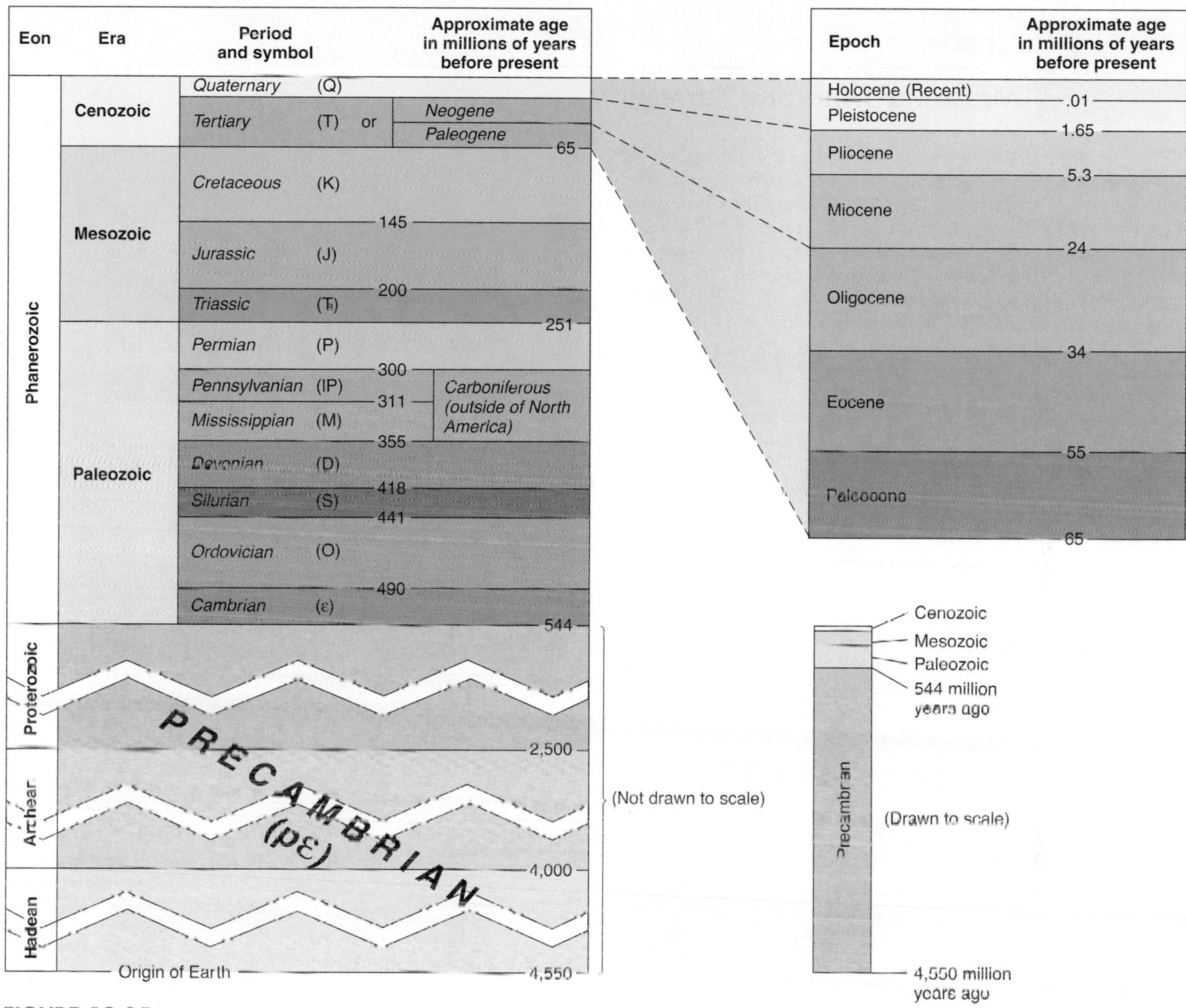

FIGURE 19.25

The geologic time scale. The small diagram to the right shows the Precambrian and the three eras at the same scale. Note that the Precambrian accounts for almost 90 percent of geologic time.

After A. V. Okulitch, 1999, Geological Survey of Canada, Open File 3040

The discovery of radioactivity also provided the means to determine how old the Earth is (see box 19.5). In 1905, the first crude isotopic dates were done and indicated an age of 2 billion years. But since then, we have dated rocks on Earth that are twice that age.

Earth is now regarded as between 4.5 and 4.6 billion years old—much older than the oldest rock found. Because erosion and tectonic activity have recycled the original material at Earth's surface, we cannot determine Earth's age from its rocks. The age determination comes primarily from dates obtained from meteorites and lunar rocks. Most meteorites are regarded as fragments of material that did not coalesce into a planet. The oldest dates obtained from meteorites and lunar rocks are in the 4.5 to 4.6 billion-year range. It is highly likely that the planets and other bodies of the solar system, including Earth, formed at approximately the same time.

Comprehending Geologic Time

The vastness of geologic time (sometimes called deep time) is difficult for us to comprehend. One way of visualizing deep time is to imagine the CN Tower in Toronto, the world's tallest freestanding structure at 553.33 m tall, representing the 4.5 billion years of the Earth's history (figure 19.26). Each metre of the tower represents approximately 8 million years.

GEOMATICS 19.5

Mapping Time and Terrain

GIS (geographic information systems) and digital mapping technology have been used to produce an integrated geologic and shaded relief map of North America (box figure 1). The map has been compiled by geoscientists from the U.S. Geological Survey, the Geological Survey of Canada, and Mexico's Consejo Recursos de Minerales, and shows patterns of rock age and distribution draped across a shaded relief map of the North American continent. The website that displays the map (http://nationalatlas.gov/articles/geology/a_timeterrain.html) includes a "zoom in" feature that allows close-up visualization of a selected area. There is also a "features" option that provides more detailed information on several important geologic features (such as the Manicouagan Crater in Quebec) and allows users to view the distribution of separate rock types (igneous, sedimentary, plutonic, metamorphic) or rocks of different ages through an interactive legend (box figure 2). The composite map can be used to reconstruct the complex geologic history that formed our continent, including episodes of continental collision and breakup, mountain-building, surficial erosion, volcanism, and glaciation.

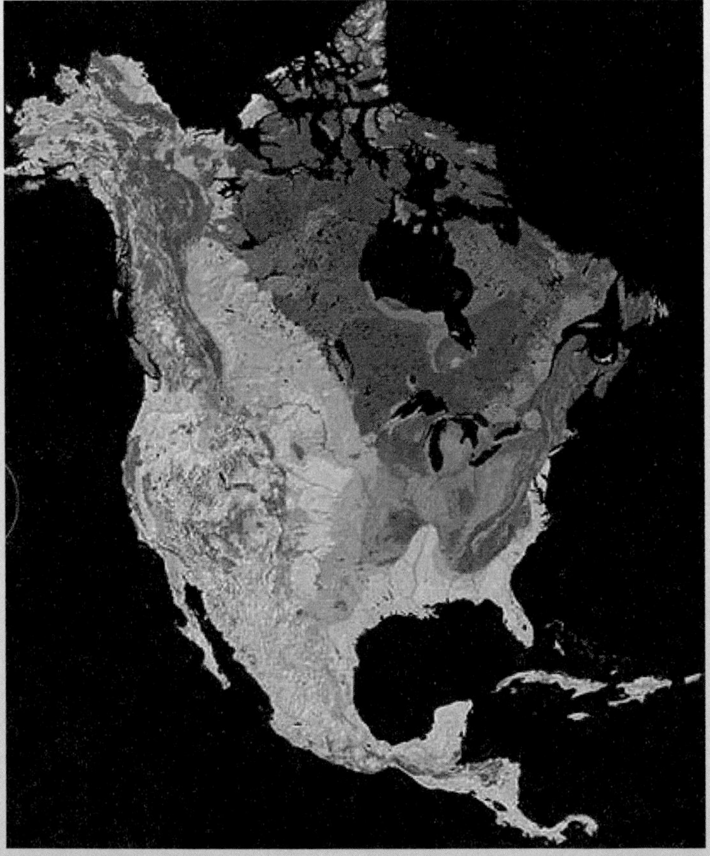

BOX 19.5 ■ FIGURE 1

The North America Tapestry of Time and Terrain map.

U.S. Geological Survey. Geologic Investigations Series I-2781. The North American Tapestry of Time and Terrain By Kate E. Barton, David G. Howell, and José F. Vigil

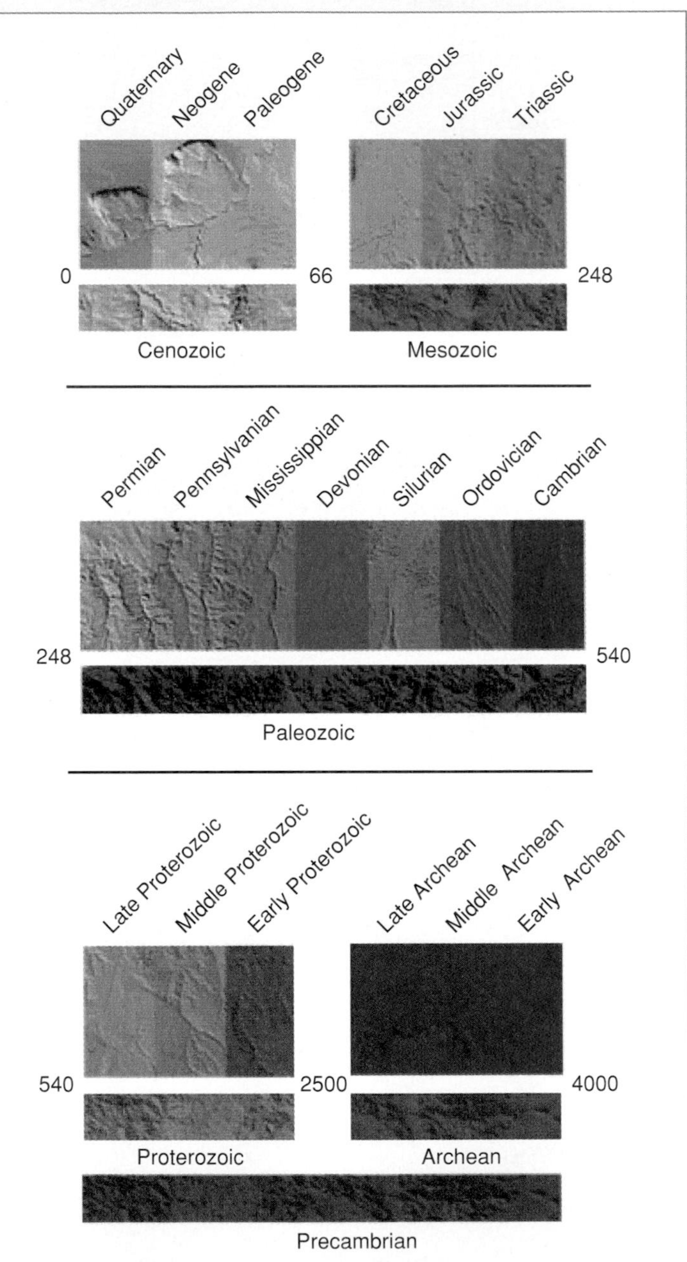

BOX 19.5 ■ FIGURE 2

Legend for the composite geological and shaded relief map of North America shown in box figure 1. Clicking on any one of the coloured boxes in the legend displayed on the website listed below will link to a map showing the distribution of those rock units in North America.

From National Atlas of the United States® and The National Atlas of the United States of America® are registered trademarks of the United States Department of the Interior

The evolution of humans on Earth occurred in the upper 50 cm of the highest antenna on the tower, and time since the last great ice sheets melted across northern North America is represented by the uppermost millimetre.

Another way to get a sense of geologic time is to compare it to a motion picture. A movie is projected at a rate of 32 frames per second; that is, each image is flashed on the screen for only 1/32 of a second, giving the illusion of continuous motion. But suppose that each frame represented 100 years. If you lived 100 years, one frame would represent your whole lifetime.

If we were able to show the movie on a standard projector, each 100 years would flash by in 1/32 of a second. It would take only 1/16 of a second to go back to a time when the first four provinces of Canada came into Confederation. The 2,000-year-old Christian era would be on screen for 3/4 of a second. A section showing all time back to the last major ice age would only be less than seven seconds long. However, you would have to sit through almost six hours of film to view a scene at the close of the Mesozoic Era (perhaps you would see the last dinosaur die). And to give a complete record from the beginning of the Paleozoic Era, this epic film would have to run continuously for two days. You would have to spend more than two weeks (16 days) in the theatre, without even a popcorn break between reels, to see a movie entitled "The Complete Story of Earth, from Its Birth to Modern Civilization."

Thinking of our lives as taking less than a frame of such a movie can be very humbling. From the perspective of being stuck in that one last frame, geologists would like to know what the whole movie is like—or, at least, get a synopsis of the most dramatic parts of the film.

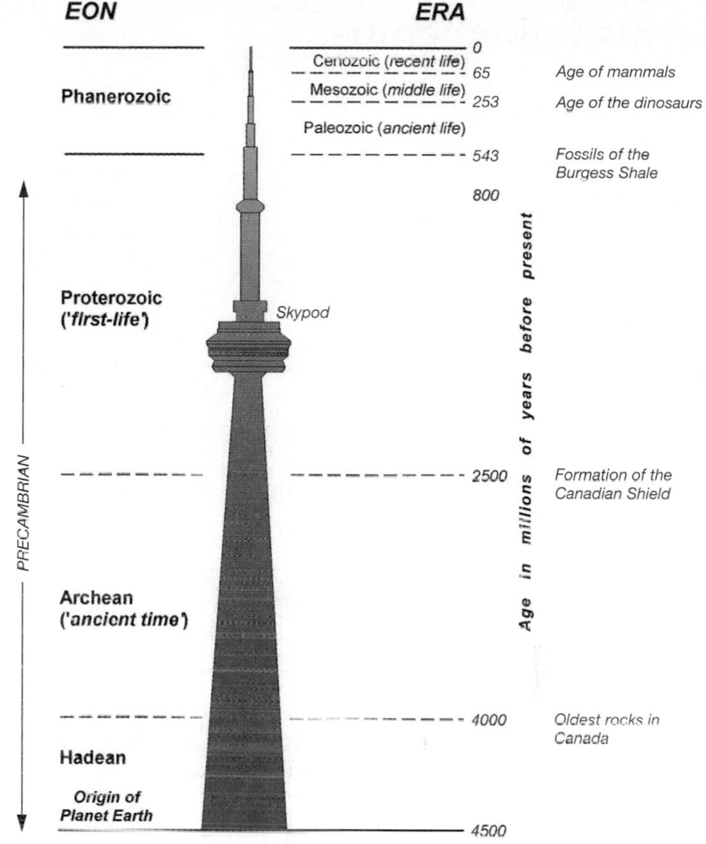

FIGURE 19.26

Geologic time represented by the CN Tower in Toronto.

SUMMARY

The principle of *uniformitarianism*, a fundamental concept of geology, states that the present is the key to the past.

Relative time, or the sequence in which geologic events occur in an area, can be determined by applying the principles of *original horizontality*, *superposition*, *lateral continuity*, and *cross-cutting relationships*.

Unconformities are buried erosion surfaces that help geologists determine the relative sequence of events in the geologic past. Beds above and below a *disconformity* are parallel, generally indicating less intense activity in the Earth's crust. An *angular unconformity* implies that folding or tilting of rocks took place before or around the time of erosion. A *nonconformity* implies deep erosion because metamorphic or plutonic rocks have been exposed and subsequently buried by younger rock.

Rocks can be correlated by determining the physical continuity of rocks between the two areas (generally this works only for a short distance). A less useful means of correlation is similarity of rock types (which must be used cautiously).

Fossils are used for worldwide correlation of rocks. Sedimentary rocks are assigned to the various subdivisions of the *geologic time scale* on the basis of fossils they contain, which are arranged according to the principle of *faunal succession*.

Numerical age—how many years ago a geologic event took place—is generally obtained by using *isotopic dating techniques*. Isotopic dating is accomplished by determining the ratio of the amount of a radioactive isotope currently in a rock or mineral being dated to the amount originally present. The time it takes for a given amount of an isotope to decay to half that amount is the *half-life* for that isotope. Numerical ages have been determined for the subdivisions of the geologic time scale.

Terms to Remember

Testing Your Knowledge

Use the questions below to prepare for exams based on this chapter.

1. Why is it desirable to find an index fossil in a rock layer? In the absence of index fossils, why is it desirable to find several fossils in a rock unit to determine relative age?

2. Suppose you had a radioactive isotope *X* whose half-life in disintegrating to daughter product *Y* is 120,000 years. By calculating how much it took to make the present amount of *Y*, you determine that, originally, the rock contained 8 grams of isotope *X*. At present only 1/4 gram of *X* is in the rock. How many half-lives have gone by? How old is the rock?

3. By applying the various principles, draw a cross-section of an area in which the following sequence of events occurred. The relative time relationship for all events should be clear from your single cross-section that shows what the geology looks like at present.
   a. Metamorphism took place during the Archean. During later Precambrian time, uplift and erosion reduced the area to a plane.
   b. Three layers of marine sedimentary rock were deposited on the plain during Ordovician through Devonian time.
   c. Although sedimentation may have taken place during the Mississippian through Permian, there are currently no sedimentary rocks of that age in the area.
   d. A vertical dike intruded all rocks that existed here during the Permian.
   e. A layer of sandstone was deposited during the Triassic.
   f. All of the rocks were tilted 45° during the early Cretaceous. This was followed by erosion to a planar surface.
   g. The area dropped below sea level, and two layers of Tertiary sedimentary rock were deposited on the erosion surface.
   h. Uplift and erosion during the Quaternary resulted in a slightly hilly surface.
   i. Following erosion, a vertical dike fed a small volcano.

4. Name as many types of contacts (e.g., intrusive contact) as you can.

5. Using information from box 19.4, calculate the age of a feldspar. At present, there are 1.2 million atoms of ^{40}K. The amount of ^{40}Ar in the mineral indicates that originally there were 1.9 million ^{40}K atoms in the rock. Use a half-life of 1.3 billion years. (Hint: The answer is 862 million years.)

6. "Geological processes operating at present are the same processes that have operated in the past" is the principle of
   a. correlation
   b. catastrophism
   c. uniformitarianism
   d. none of the above

7. "Within a sequence of undisturbed sedimentary rocks, the layers get younger going from bottom to top" is the principle of
   a. original horizontality
   b. superposition
   c. cross-cutting
   d. none of the above

8. If rock A cuts across rock B, then rock A is ______ rock B.
   a. younger than
   b. the same age as
   c. older than

9. Which is a method of correlation?
   a. physical continuity
   b. similarity of rock types
   c. fossils
   d. all of the above

10. Eras are subdivided into
    a. periods
    b. eons
    c. ages

11. Periods are subdivided into
    a. eras
    b. epochs
    c. ages

12. Which division of geologic time was the longest?
    a. Precambrian
    b. Paleozoic
    c. Mesozoic
    d. Cenozoic

13. Which is a useful radioactive decay scheme?
    a. $^{238}U/^{206}Pb$
    b. $^{235}U/^{207}Pb$
    c. $^{40}K/^{40}Ar$
    d. $^{87}Rb/^{87}Sr$
    e. all of the above

14. C-14 dating can be used on all of the following except
    a. wood
    b. shell
    c. the Dead Sea Scrolls
    d. granite
    e. bone

15. Concentrations of radon are highest in areas where the bedrock is
    a. granite
    b. gneiss
    c. limestone
    d. black shale
    e. phosphate-rich rock
    f. all of the above

16. Which is not a type of unconformity?
    a. disconformity
    b. angular unconformity
    c. nonconformity
    d. triconformity

17. A geoscientist could use the principle of inclusion to determine the relative age of
    a. fossils
    b. metamorphism
    c. shale layers
    d. xenoliths

18. The oldest abundant fossils of complex multicellular life date from the
    a. Precambrian
    b. Paleozoic
    c. Mesozoic
    d. Cenozoic

19. A contact between parallel sedimentary rock that records missing geologic time is
    a. a disconformity
    b. an angular unconformity
    c. a nonconformity

Exploring Web Resources

www.mcgrawhill.ca/olc/plummer

Visit the Online Learning Centre to learn more about geological time. Review your answers to Testing Your Knowledge and try some of the other quizzes. Explore the additional readings, media resources, and direct links to the other sites listed below.

www.ucmp.berkeley.edu/exhibit/index.php

Paleontology without Walls. University of California Museum of Paleontology virtual exhibit. Click on Geologic Time.

http://vearthquake.calstatela.edu/VirtualDating/

Virtual Dating. This site provides an excellent, interactive way of learning how isotopic dating works. You can change data presented and watch graphs and other illustrations change accordingly. Quizzes help you understand the material.

www.talkorigins.org/origins/faqs/faq-youngearth.html

The Age of the Earth. Topics include isotopic dating, the geologic time scale, and changing views of the age of the Earth. The site also explores the creation/evolution controversy.

http://asa.calvin.edu/ASA/resources/Wiens.html

Radiometric Dating: a Christian Perspective. At this Website you can get a very thorough knowledge of isotopic dating, how it works, and how it has been used to determine the age of the Earth and other events. The author addresses concerns of people who feel that an old Earth is incompatible with their religious beliefs.

http://depts.washington.edu/cosmolab

Information regarding the use of cosmogenic isotopes for dating.

CHAPTER

20

Geological History of Canada

What are the main geological "building blocks" of North America?
How did the North American continent evolve?
How were the Atlantic provinces added to Canada?
Where did British Columbia come from?
Where can we find dinosaurs in Canada?
How did the Canadian Rockies form?
When did the ice sheets develop?
Why is it important to understand the geological history of Canada?

This chapter uses much of what has been discussed about geological processes earlier in the book to examine the geological history of our own country. Knowledge of Canada's geology is far from complete and is always being updated by new information—such as that produced by Lithoprobe, for example (see chapter 4). Canadians have prospered economically by finding and using natural resources such as copper, gold, oil, coal, iron, and nickel; the recent and highly publicized discovery of diamonds in the Northwest Territories continues this trend. Canada's economic well being is dependent on geological resources and the finding and use of new mineral and energy deposits, such as oil and gas, requires new generations of geoscientists armed with better knowledge of the varying geological conditions across Canada.

Poorly sorted Quaternary-age sediments exposed in the Hoodoos at Banff. These deposits were formed as debris flows were released onto the floor of the Bow Valley during glacier retreat. *Photo by Nick Eyles*

Access to and exploitation of these resources also requires knowledge of modern physical processes operating on the sea floors off Canada's coasts, in the permafrost terrains of the north, and in the earthquake- and landslide-prone mountain valleys of the west. At the same time, as urban populations grow so does society's dependence on the geosciences. The finding and protection of groundwater resources; the safe disposal of wastes; the design of foundations for buildings, roads, and other infrastructure; the location of sufficient quantities of construction materials such as sand and gravel; and the assessment of earthquake risk all require a detailed understanding of Canadian geology.

CANADA: A YOUNG NATION, BUT AN OLD COUNTRY

As a nation, Canada has been in existence since 1867. In that year, the Act of Confederation (also known as the British North America Act) brought together the provinces of Ontario, New Brunswick, Nova Scotia, and Quebec to create a larger and more powerful political entity. The country was welded together as an economic unit by a transcontinental railway, the completion of which was celebrated by the driving of the "last spike" in 1885 (figure 20.1). Later, other provinces saw the wisdom of joining this larger political and economic group; the last to join was Newfoundland in 1949.

In the very same fashion, but operating over a vastly longer time period, the North American continent was assembled by plate tectonics processes that brought together many smaller land masses. When fused and locked together, they have created the geological mosaic of the present-day continent. The process of continental building has not been a simple one and has taken more than 4 billion years to accomplish (figure 20.2).

FIGURE 20.1

Driving the last spike on the Canadian Pacific Railroad in November 1885 at Craigellachie in British Columbia. The building of the railway across the Canadian Shield resulted in exploration of the country's geology and the discovery of mineral resources such as the rich copper and nickel ores at Sudbury.

Public Archives of Canada

Construction of North America began at least 4,000 million years ago with the formation of the **Acasta Gneiss** of the Northwest Territories, which now forms part of the Slave Province of the Canadian Shield. The construction process has not been a simple one, as the continental mass we now call North America has itself been part of much larger land masses (called "supercontinents") that have since broken up. What remains today is a geological mosaic of fragments of the many former land masses that were fused together by plate tectonics processes. The building of North America was essentially complete by 65 million years ago, although the modern landscape is the result of geologically recent glaciations that have occurred in the last 2.5 million years. The last ice sheet left the southern portions of the country only 12,000 years ago, and finally melted in Labrador 6,000 years ago. Remnants of this vast ice sheet still survive on Baffin Island today as the Penny and Barnes Ice Caps (see chapter 16).

WHAT ARE THE MAIN GEOLOGICAL "BUILDING BLOCKS" OF NORTH AMERICA?

A Geologic Jigsaw

When one looks at a geological map of North America (such as the one inside the front cover of this book), the phrase "jigsaw puzzle" might come to mind. The northern part of the continent, in Canada, is underlain by the exposed part of the ancient core or **craton** of North America; this exposed part is called the *Canadian Shield* and consists predominantly of very old, Archean and Proterozoic rocks (figure 20.3*A*, *B*). These rocks range in age from 4 billion to approximately 1 billion years old and are largely devoid of fossils.

The craton is composed of a complex assemblage of several distinct geologic *provinces*. **Geologic provinces** are broad regions of similar rocks, usually covering many thousands of square kilometres, with characteristics that differ significantly from rock types present in adjacent areas. Individual geologic provinces have been subdivided into smaller units called *subprovinces*, which are fault-bounded units containing similar rock types, structures, and mineral deposits. Both geologic provinces and subprovinces were identified by geologists in the nineteenth century on the basis of broad-scale field mapping and form the basis of the earliest geologic maps of Canada.

Recent understanding of plate tectonics processes has clarified the origin of provinces and subprovinces. Provinces and subprovinces are now widely recognized to be **terranes.** Terranes are discrete fragments of oceanic or continental material that have been added to a craton at an active margin by accretion. (Note the different spelling and meaning from *terrain*, a physiographic term referring to topography.)

FIGURE 20.2A

Eon	Era	Period		Epoch
Phanerozoic	Cenozoic	Quaternary (Q)		Recent or Holocene
				Pleistocene
		Tertiary (T)	Neogene	Pliocene
				Miocene
			Paleogene	Oligocene
				Eocene
				Paleocene
	Mesozoic	Cretaceous (K)		
		Jurassic (JR)		
		Triassic (TR)		
	Paleozoic	Permian (P)		
		Carboniferous (C)	Pennsylvanian (P)	
			Mississippian (M)	
		Devonian (D)		
		Silurian (S)		
		Ordovician (O)		
		Cambrian (€)		
Precambrian	Proterozoic Eon			
	Archean Eon			
	Hadean Eon			

m.y.	
	Present-day interglacial
0.01	End of last glaciation
	Erosion and sculpting of Great Lake basins, fiords and mountains by glaciers
1.7	
	Beginning of continental glaciation in Canada
5.3	First glaciers in Alaska
24	
33.5	
	First glaciers in Antarctica as globe cools but warm-loving forests still flourish in Canadian arctic
54.8	
	Alberta's Tar Sands form
	Dinosaurs wiped out by giant meteorite strike
65	Greenland breaks free of North America
	Creation of Rocky Mountain fold belt
145	British Columbia's copper deposits form
	Orogenies in Western Canada, **British Columbia built by accretion of terranes**
200	
	Pangea breaks up–early Atlantic Ocean forms and ancestral North America breaks free
251	
	Alleghanian Orogeny completes the building of the Appalachian Mountain fold and thrust belt
300	
311	Formation of **Pangea**–coal deposits of Maritime provinces formed
355	Alberta's oil-bearing limestones form Saskatchewan's salt deposits accumulate
418	**Acadian Orogeny** in eastern Canada marks arrival of Avalonia
441	**Maritime Canada begins to be built by accretion of terranes during Taconic Orogeny**
490	Iapetus Ocean begins to close–several orogenies in eastern Canada
	Cambrian explosion of life begins at 540 Ma; Burgess Shale and Ediacaran fauna
544	**Rodinia** breaks up–Iapetus Ocean forms at 620 Ma Assembly of **Rodinia**–Grenville Orogeny at 1000 Ma completing the formation of the North American craton Sudbury Structure formed by giant meteorite strike at 1,850 Ma **Nena** forms
2,500	Gowganda glaciation at c. 2,400 Ma
	Arctica forms–first clear evidence of ancient bacterial life
4,000	Acasta gneiss of Northwest Territories formed–oldest Canadian rocks
4,600	FORMATION OF PLANET EARTH

A

B

FIGURE 20.2

(A) Time scale for the building of Canada's geology. Numbers are in millions of years. (B) Cambrian-Ordovician boundary (arrowed) exposed at Green Point, Gros Morne National Park, Newfoundland.

Photo by Nick Eyles

These likely originated as small continents and remnants of ocean-floor crust, each with its own complex geological history, and were welded together by plate tectonics processes to form the North American craton. Most of the North American craton was assembled between 1 and 4 billion years ago.

The full geographic extent of the craton is not immediately apparent from a map of the geology of North America as its outermost margins are buried by layers of younger "cover rocks" that reach thicknesses of more than 10 km (figure 20.3*A*). The North American craton is the largest in the world and extends from the Atlantic seaboard, west beneath the Rocky Mountains, and south as far as Texas. Greenland also consists of a detached mass of the same craton that broke off from North America during the opening of the northern North Atlantic Ocean some 80 million years ago.

The North American Craton versus the Canadian Shield

It is important to be aware of the difference between the **North American craton** and the **Canadian Shield.** The former refers to a large, continent-sized block of distinct geology making up the basement of much of North America (and Greenland). The Shield is the exposed part of the craton, and consists of a gently undulating surface that rises inconspicuously, almost like an arch, in its centre. The term "shield" was introduced by the Austrian geologist Eduard Suess around 1912. This was in reference to its dome-like form resembling a warrior's shield when placed flat on the ground.

The Canadian Shield is a large landform called a **peneplain** (derived from the Latin meaning *almost* a plain), which is a surface of low relief and great areal extent and age. Erosion and bevelling of the ancient rocks of the craton had created this peneplain by about 800 million years ago. The outer, gently sloping margins of the Shield are buried below younger sedimentary rocks, and the ancient peneplain surface now forms an unconformity between the craton below and younger rocks above. This unconformity can be seen in the sidewalls of the Grand Canyon in Arizona, where it separates metamorphic rocks of the craton from overlying Paleozoic cover rocks (see figures 19.16 and 19.17). This is the very same surface that is exposed many hundreds of kilometres to the north as the Canadian Shield (figure 20.3*A*).

The cover rocks that bury the outer margins of the North American craton are fossiliferous sedimentary strata of Paleozoic and Mesozoic age (figure 20.3*C*). These sedimentary rocks were deposited when the outer margins of the craton were depressed and flooded by shallow seas. This occurred primarily during mountain-building episodes (orogenies) that resulted from the collision of other land masses with the craton. During these events, the outer margins of the craton were depressed by the great weight of large mountain belts (including volcanoes) and their thick piles of sediment. Many other continents on Earth show the same basic anatomy of a central, ancient *craton* created by the fusion of many separate geological provinces (microcontinents), buried around its margins by younger sedimentary **cover strata.**

The Geologic Jigsaw of the North American Craton

This question has fascinated geologists since the middle of the nineteenth century—however, only since 1970, with the advent of plate tectonics, has an effective explanation been available. Early geologists recognized that, in general, rocks got younger as one travelled away from the geographic interior of North America to its coasts. Also, rocks that were uniform in character over enormous distances changed abruptly across very sharp boundaries. As early as 1870, several geologists were using the term "province" to describe the many distinct blocks of geology they discovered within the Canadian Shield. Famous Canadian geologist Sir William Logan was the first to write of "geological provinces" in the early 1860s; the term is still in use to describe areas of the Shield with distinctive geological characteristics (e.g., the Superior Province, shown in figure 20.4).

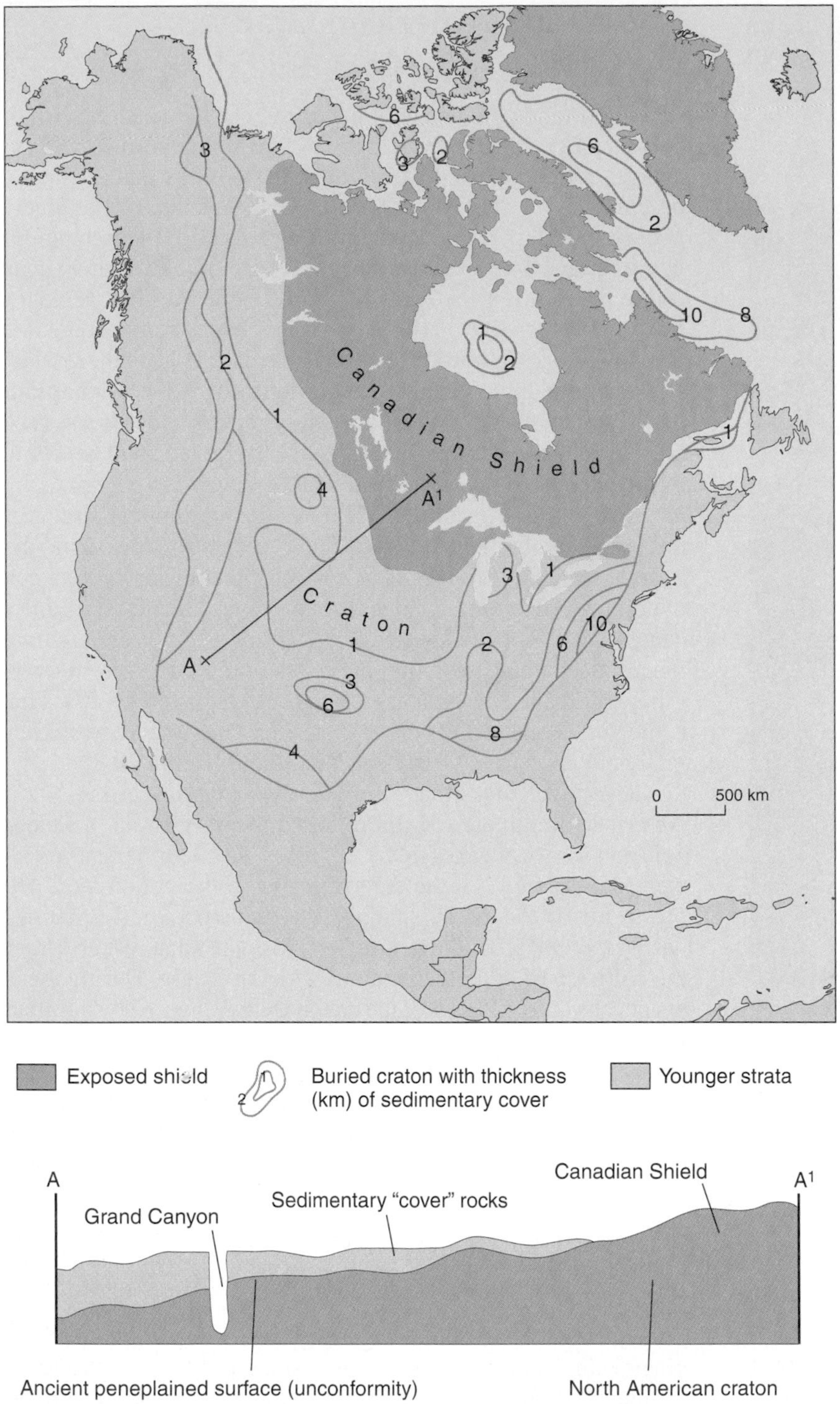

Canadian Shield

Craton

A

A¹

0 500 km

Exposed shield

Buried craton with thickness (km) of sedimentary cover

Younger strata

Canadian Shield

Sedimentary "cover" rocks

Grand Canyon

Ancient peneplained surface (unconformity)

North American craton

A

B

C

FIGURE 20.3

(*A*) A simplified version of the geological map of North America (such as that shown on the inside of the front cover of this book). The Canadian Shield is the exposed portion of the North American craton in the geographic centre of North America; the craton is covered by much younger sedimentary rocks ("cover rocks") on its margins. Schematic cross-section A-A' shows that the ancient peneplained surface of the craton, which forms an unconformity where buried by younger strata, also forms the exposed surface of the Canadian Shield. (*B*) The low-relief surface of the Canadian Shield. (*C*) Paleozoic "cover rocks" exposed along the Niagara Gorge, Ontario. The North American craton lies 600 m below.

Figure A by Nick Eyles; Photos B and C by Nick Eyles

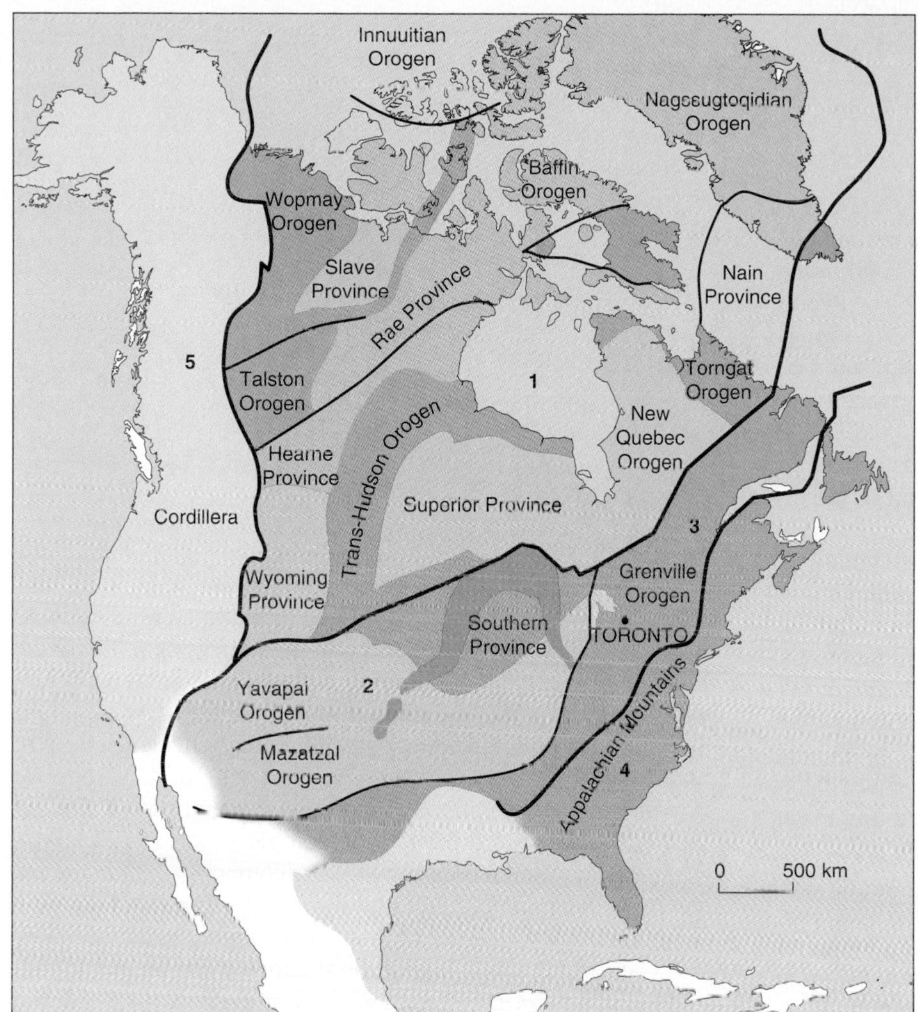

FIGURE 20.4

The growth of North America showing the five principal building blocks of which the continent is composed and when each block was added. The boundary between areas 3 and 4 in eastern Canada is known today as Logan's Line, after its discovery by Canada's most famous geologist, Sir William Logan, along the St. Lawrence River near Quebec City in 1863.

1. The original North American continent, **Arctica**, which started to form about 2.5 billion years ago from smaller continents and was completed by about 1.9 billion years ago when old Archean cratons (e.g. Slave, Nain provinces) were welded together by the Trans-Hudson Orogen and others.
2. Added to the North American continent during the formation of **Nena** about 1.8 billion years ago after the Penokean Orogeny.
3. Added during the formation of **Rodinia** about 1.3 billion years ago during the Grenville Orogeny.
4. Added during the formation of **Pangea** about 600 million–300 million years ago.
5. Added after the breakup of **Pangea** about 250 million years ago

Early settlers of Canada took a keen interest in geology and soon opened mines; some of the earliest gold mines were in Nova Scotia. Discoveries of coal soon followed, such as at Joggins, Nova Scotia, but these were essentially random finds by prospectors. Systematic study of the geology of Canada by professional geologists began with the newly formed Geological Survey of Canada (GSC) in the early 1840s, at a time when few topographic maps or roads existed. An impetus behind the formation of the GSC was the need to better understand the country's geological resources, mostly coal. Its first director was Sir William Logan, who sent Alexander Murray to work in Newfoundland and sponsored William Dawson to study the geology of Nova Scotia. Logan's magnificent map of the Geology of Canada (actually parts of Manitoba, southern Ontario, and Quebec) was published in 1869 and was a major achievement. The combination of a large, barely explored country and complex geology offered major challenges, both physical and mental. Canadian geologist A.C. Lawson declared in 1913 that "no one geologist, not for that matter two, can hope to become familiar with the details of more than a very small proportion of the entire field. The work to be done is appalling in its magnitude."

Even today, much of the Canadian Shield remains to be mapped in detail, but the basic geological survey started in the 1840s was essentially completed with the aid of broad-scale helicopter and other airborne surveys in the 1970s. At that time, emerging revelations about the workings of plate tectonics by geologists studying the sea floor provided the key to the origin of the Canadian Shield. The plate tectonics revolution identified that the Earth's outer surface was made of lithospheric plates that are always in the process of being newly built, destroyed, or welded together to form larger plates (see chapter 2). It is now realized that these processes have been operating on the Earth in some form over the past 4 billion years and are responsible for bringing together the various geological provinces of Canada. Processes operating in the modern world are thus being used to explain the evolution of the ancient North American craton, a working principle known as *uniformitarianism* (see chapter 19). In turn, the model of plate tectonics is also being tested against the ancient geological record preserved in the Canadian Shield.

HOW DID THE NORTH AMERICAN CONTINENT EVOLVE?

Stages in the Evolution of the North American Continent

Figure 20.4 shows the essential building blocks now recognized by geologists within the North American continent. Within the oldest part of the continent (Area 1 on figure 20.4), several geological provinces (such as Rae, Superior, and Wyoming) are rimmed by intensely deformed rocks that form ancient *orogens* (such as Torngat and Wopmay). As a broad generalization, geologic provinces can be regarded as the remains of individual continents that collided (see box 20.1). The **orogens** consist of crushed and deformed rocks that represent the remains of mountain belts or volcanic arcs formed during collision.

The earliest part of Canada's geological history is not well known and therefore is subject to much debate. Much remains to be discovered. Some geologists suggest that the growth of the North American continent can be broken down into five different stages. Each stage is characterized by a major plate tectonics event, when ancestral North America either collided with or ripped apart from other land masses. This process of repeated continental aggradation and breakup is known as the Wilson Cycle (see chapter 2), and has resulted in the development of **supercontinents** at certain times in Earth's history. Because of this cyclic process of continental aggradation and dispersal, each of the present-day continents (now widely separated on the globe) has a broadly similar geologic history. Hence, it is possible to work on the geology of Australia or South America, for example, and recognize the same tectonic events as those that affected parts of North America.

It is important to remember that our understanding of the early stages of Canada's geological history is preliminary at best. Some of the stages of development are speculative and their timing uncertain. However, one model of the five stages of continental growth and their timing is discussed below.

Stage 1—Arctica: North America in the Archean

The formation of the central part of the North American continent (Area 1 on figure 20.4) spans the entire Archean Era (4–2.5 billion years ago). Some of the oldest rocks so far dated on Earth are found in the Slave Province (Acasta Gneiss of the Northwest Territories, see chapter 10) and are thought to be between 3.96 and 4.05 billion years old. These rocks formed part of an ancient continent that some have named **Arctica** (figure 20.5). Speculative evidence suggests that Arctica had begun to form by 2.5 billion years ago, but its final assembly was not completed until after 2 billion years ago. The continent included much of present-day Siberia.

The Slave Province is today the focus of much mineral-exploration activity, especially for diamonds that occur in kimberlite pipes penetrating the ancient shield rocks (see box 4.4). Similarly, the Superior Province, which also formed part of Arctica, is of special importance to Canadians because of its great mineral wealth. Approximately 450 significant mineral deposits have been discovered to date within the Superior Province. More than 50 percent of Canada's entire annual gold production and 30 percent of its zinc, copper, and silver are mined from the Superior Province. Much undoubtedly remains to be discovered.

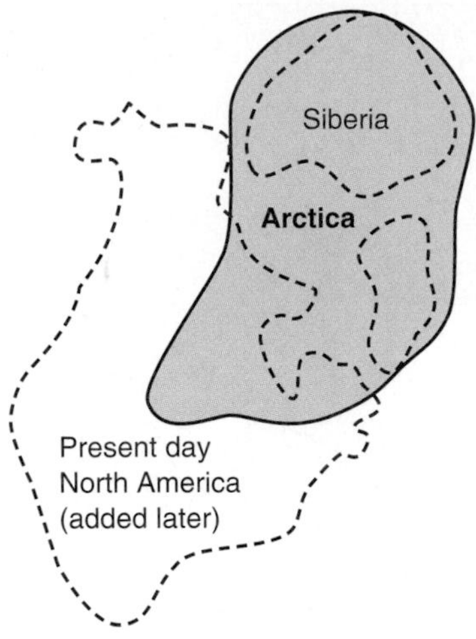

FIGURE 20.5
The continent Arctica about 2 billion years ago (Area 1 on figure 20.4).

Detailed mapping of the ancient rocks of the Superior Province reveals smaller **subprovinces** composed of distinct rock types that record particular geologic events (figure 20.6). For example, *plutonic subprovinces* are composed of granites and record the intrusion of giant plutons into the Superior Province. *Greenstone subprovinces* consist of metamorphosed sea-floor volcanic rocks, such as **basalt**, originally formed on the floor of ancient Archean oceans (figure 20.7). These subprovinces contain thick banded iron deposits that form the major source of the world's iron and record the first large-scale production of oxygen by single-celled cyanobacteria on Earth. The *metasedimentary subprovinces* consist of deep-sea Archean ocean sediments. Together, the various subprovinces within the Superior Province record clear evidence of the operation of plate tectonics activity in the ancient past, much as we see it today.

The southern continental margin of Arctica was the site of a major glaciation (the **Gowganda glaciation**). This glaciation is one of the oldest recorded on Earth and is dated at around 2.4 billion years ago. The glacial deposits of the Gowganda Formation (figure 16.37*A*) form part of the Huronian Supergroup of Ontario, famous for the uranium deposits found near Elliot Lake.

Stage 2—Nena and Rodinia: North America in the Proterozoic

The next events in the development of the North American continent span the entire Proterozoic Era (from about 2.5 billion to 570 million years ago). The southern part of the Canadian Shield, including the Southern Province and the Yavapai and Mazatzal Orogenies (Area 2 on figure 20.4), had been added to Arctica by about 1.9 billion years ago or shortly thereafter, to form a larger land mass some geologists call **Nena.** The final

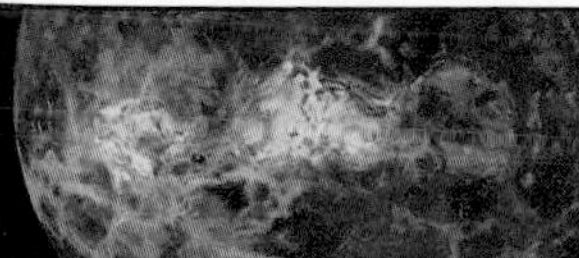

ASTROGEOLOGY 20.1

The Sudbury Impact Structure: Collision of an Ancient Meteorite

The huge 60-kilometre-long and 30-kilometre-wide crater preserved at Sudbury was originally circular and formed by a colossal meteorite impact around 1.8 billion years ago (box figure 1). Recent models suggest impact by a meteorite about 10 kilometres in diameter that broke through the Earth's crust melting vast volumes of melt rocks. This catastrophic event is also recorded in the Sudbury region by so-called **shatter cones** (box figure 2) and breccia and glass called *pseudotachylite* (or breccia; box figure 3), which are typical of meteorite impact structures worldwide. The impact crater was squeezed into its present oval shape during the *Penokean Orogeny*, shortly after it formed.

The Sudbury structure is famous for its rich nickel, copper, and platinum ores (box figure 4). These ores record the upwelling of mantle magmas as a result of the penetration of the meteorite through the crust to a depth of 35 km. The ore consists of massive sulphides, typically in the form of the minerals pentlandite, pyrrhotite, pyrite, and chalcopyrite, with about 2 percent nickel and platinum derived from the original meteorite and melted mantle rocks.

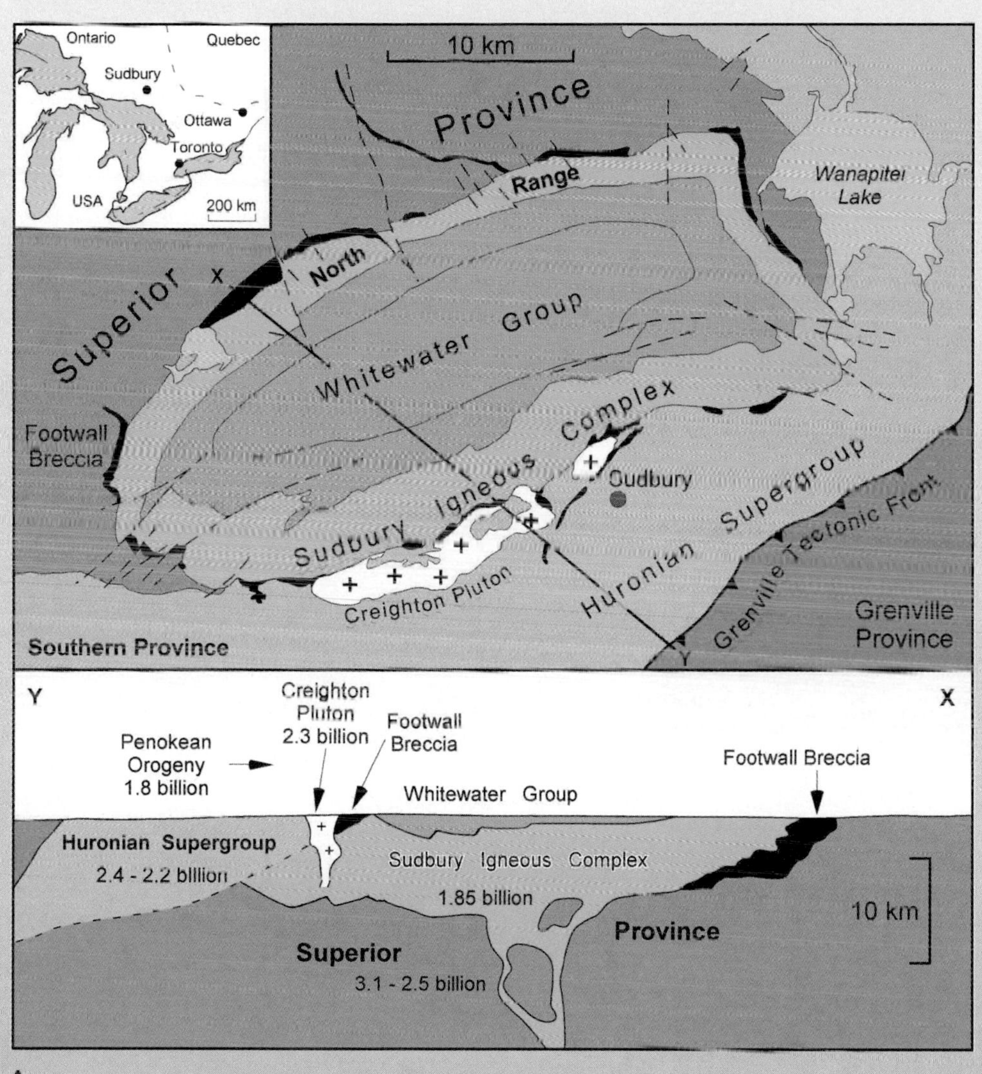

A

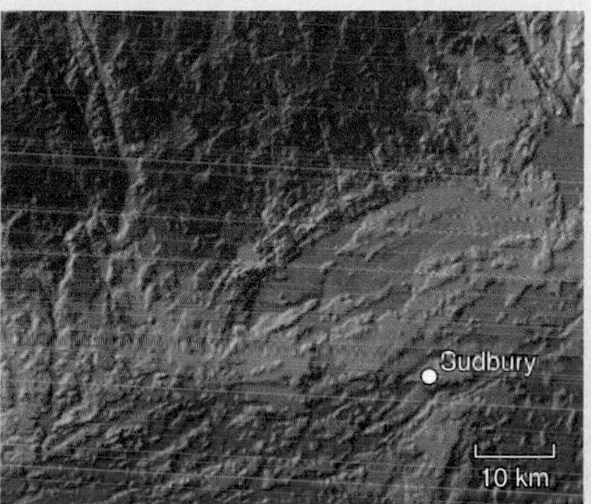

B

BOX 20.1 ■ FIGURE 1

Sudbury meteorite impact structure.

(A) Geological map and schematic cross-section.

(B) Digital elevation model showing the oval-shaped impact crater as an area of low topography (green and blue colours) to the north of Sudbury.

Photo B from www.unb.ca/passc/ImpactDatabase/images.html

BOX 20.1 ■ FIGURE 2

V-shaped shatter cones form when rocks are struck violently during a meteorite impact event. Shatter cones occur widely around the Sudbury area, especially along the access road to the Laurentian University campus.

Photo by Nick Eyles

BOX 20.1 ■ FIGURE 3

Breccia formed by the disintegration and mixing of rock when hit by a large meteorite and broken into angular fragments. This breccia is common in the Sudbury area.

Photo by Nick Eyles

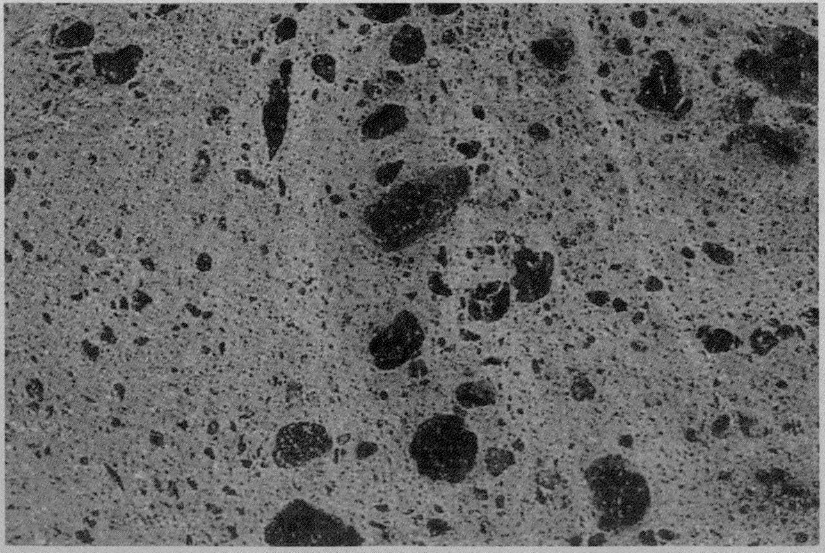

BOX 20.1 ■ FIGURE 4

Close-up image of fine-grained copper sulphide ore with small fragments of dark-coloured country rock. Width of view is 10 cm.

Photo: GSC ESS photo library #1995-225C. Reproduced with the permission of the Minister of Public Works and Government Services Canada, 2007 and Courtesy of Natural Resources Canada, and Geological Survey of Canada.

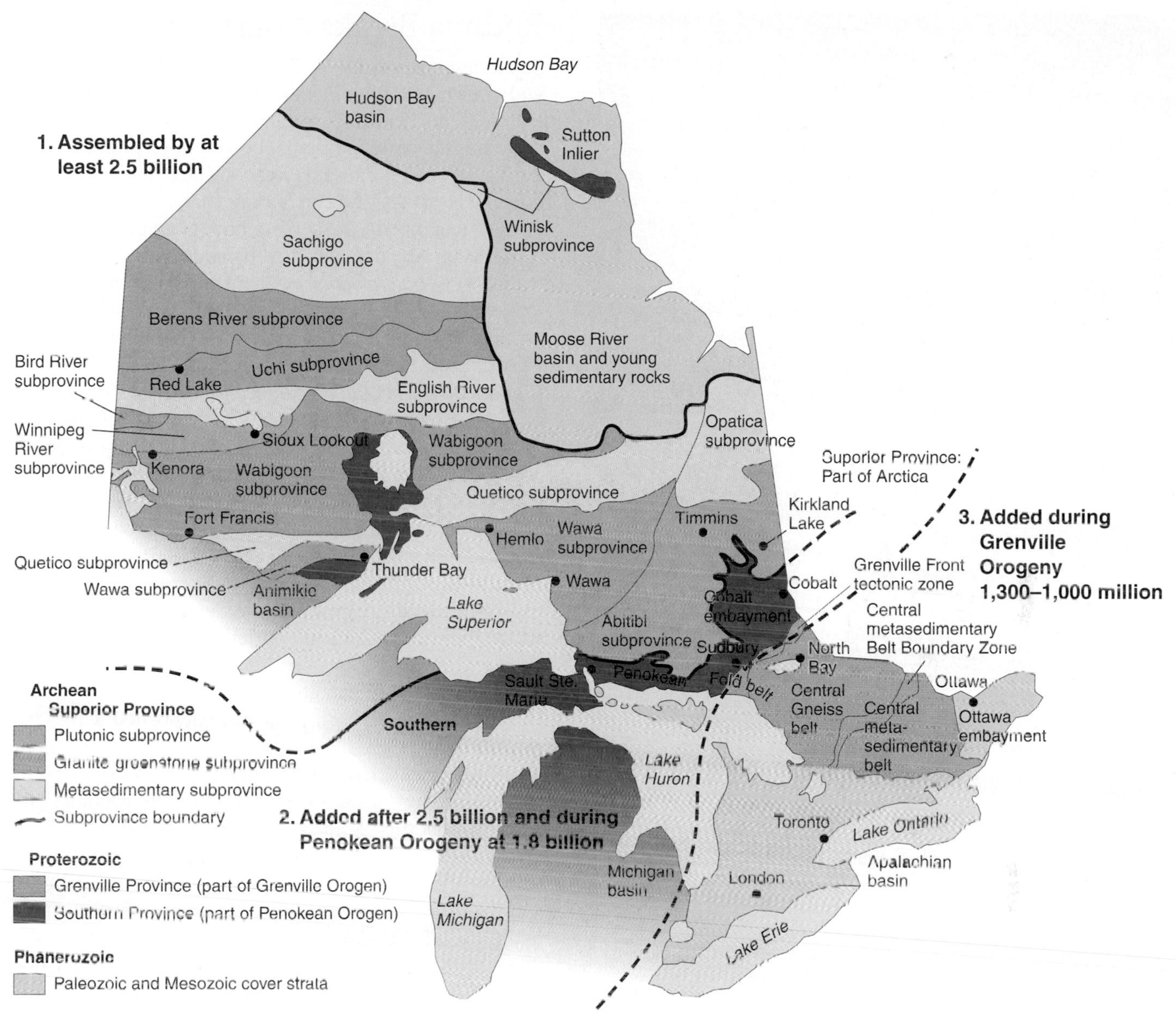

FIGURE 20.6

Geological map of Ontario showing structure within the Superior Province (1), younger Southern (2) and Grenville (3) provinces (areas 2 and 3 on Figure 20.4). *Fitzhenry and Whiteside Limited*

assembly of Nena is recorded in Ontario by the *Penokean Orogeny*. This created a major Himalayan-type mountain range near what is now the northern end of Lake Huron (figure 20.6). The range was destroyed by erosion, but its deep roots remain as the Penokean Fold Belt along the northern border of Lake Huron in Ontario.

Stage 3—The Grenville Orogeny and Formation of Rodinia

The **Grenville Orogeny** was the result of the long-lived collision between ancestral South and North America between 1.3 and 1 billion years ago (figure 20.8*A*). Much of what is now eastern North America from southern Greenland to Northern Mexico was formed at this time by accretion of smaller land masses (Area 3 on figure 20.4). As a result of the collision, a huge supercontinent called **Rodinia** was created. Rocks accreted and deformed during the orogeny underlie much of southern Ontario and Quebec, extending through the Maritimes and into Newfoundland.

The Grenville orogenic belt (which can be called either the Grenville Province or the Grenville Orogen) is dominated by beautifully banded gneisses, highly metamorphosed sediments, and igneous rocks (figure 20.9). These once formed the deep roots of an ancient mountain range that some call the Grenville Mountains (figure 20.10). By 800 million years ago the forces of erosion had reduced the mountains to a peneplain. This surface has, in fact, changed little over the ensuing 800 million years and is represented

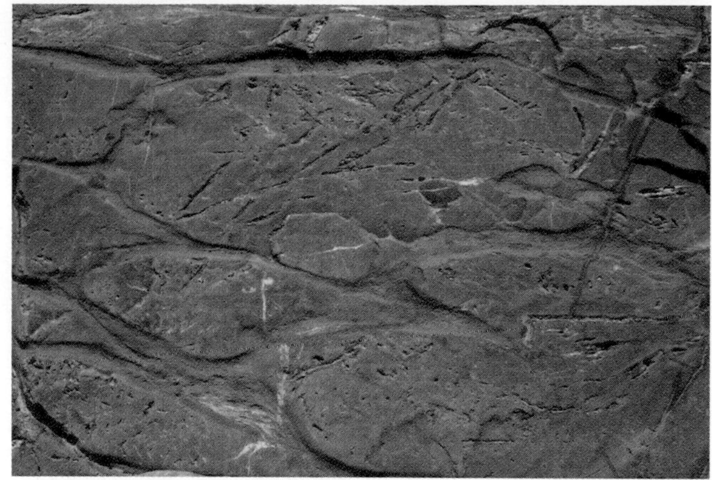

A

B

FIGURE 20.7

(*A*) Ancient pillow basalts formed when magma erupted and cooled underwater. Enormous thicknesses of pillow basalts make up the greenstone belts of the Slave and Superior provinces. Hot water circulating through the sea floor at the time of their formation scavenged and concentrated metals as "hydrothermal" deposits now mined for gold and other minerals. (*B*) The Giant Mine near Yellowknife in the Northwest Territories extracted gold from pillow basalts between 1947 and 1999. The process left arsenic-rich waste on the edge of Great Slave Lake, which is now being cleaned up.

Photos by Nick Eyles

today by the low-relief landscape of the exposed Canadian Shield. The Canadian Shield forms one of the most extensive and ancient landforms in the world and is remarkable because geologists have little detailed knowledge of how such a landscape formed.

Detailed mapping shows that the Grenville Province is, in fact, made up of many smaller terranes. These represent the highly deformed remnants of island arcs, microcontinents, and pieces of ocean floor that were not subducted and destroyed during the orogeny but were scraped off and accreted onto the eastern edge of Rodinia. The latest phase of the stage was the intrusion of many granite plutons and dikes that welded together the various terranes (refer to figure 20.15 later in this chapter).

Rodinia Breaks Apart

Supercontinents such as Rodinia are inherently unstable and with time inevitably self-destruct. This is because their large extent prevents heat from escaping from the Earth's interior and promotes the buildup of giant convecting "plumes" in the mantle. These plumes cause the land surface to rise in the form of a dome. Eventually, the convecting forces below are sufficiently great to tear, or rift, the supercontinent apart.

This process of continental breakup commences with the formation of a *triple junction*, which consists of interlinked *grabens* that eventually grow and widen into a new ocean basin. Initial breakup of Rodinia started around 750 million years ago and finished at about 570 million years ago, during the earliest Cambrian (figure 20.8*B*). The first break occurred along the western margin of North America when Antarctica and Australia broke away to form an ancestral Pacific Ocean (called the Panthalassic Ocean). Millions of years later, the last tear occurred along the eastern margin of North America and saw Europe and Africa drift off to form an ancestral Atlantic Ocean called the **Iapetus Ocean** (figure 20.8*B*). The process of continental rifting by the formation of interlinked grabens has left a legacy of failed grabens (called *aulacogens*) that are now preserved deep within North America, where they are infilled and buried by younger cover strata (see chapter 2).

The Iapetus Ocean closely resembled the modern Atlantic Ocean, and coastal sediments were deposited across extensive shallow continental shelves. These areas of warm water provided well-lit, nutrient-rich habitats for an enormous variety of marine organisms. These include corals, molluscs, brachiopods, trilobites, and echinoderms. Some paleontologists have speculated that the breakup of Rodinia and the opening up of a wealth of new habitats around the margins of the newly formed Iapetus Ocean was the major stimulus to the proliferation of organisms evident in the so-called **Cambrian Explosion** of early lifeforms such as those of the Burgess Shale and the Mistaken Point Formation (see boxes 1.11 and 20.2).

Stage 4—Pangea: North America in the Late Paleozoic and Some of the Mesozoic

The Iapetus Ocean was not long-lived and by about 480 million years ago was beginning to close. A land mass called Baltica (consisting of much of modern Europe) once more approached the eastern seaboard of North America (then called *Laurentia*, figure 20.8*C*) and eventually collided to form *Laurasia*. Collision of the two land masses caused a major orogenic event, the *Taconic Orogeny*. (The equivalent event in Europe is known as the Caledonian Orogeny and formed mountains in Wales, Scotland, and Scandinavia.) An arc of volcanic islands rose high above the active subduction zone off eastern North America and huge andesitic volcanoes, much like those found in today's Andes Mountains or in Japan and the Philippines, spewed volcanic ash into the continental interior (figure 20.11). Today, these ash beds provide important *marker horizons* within the thick sedimentary strata that

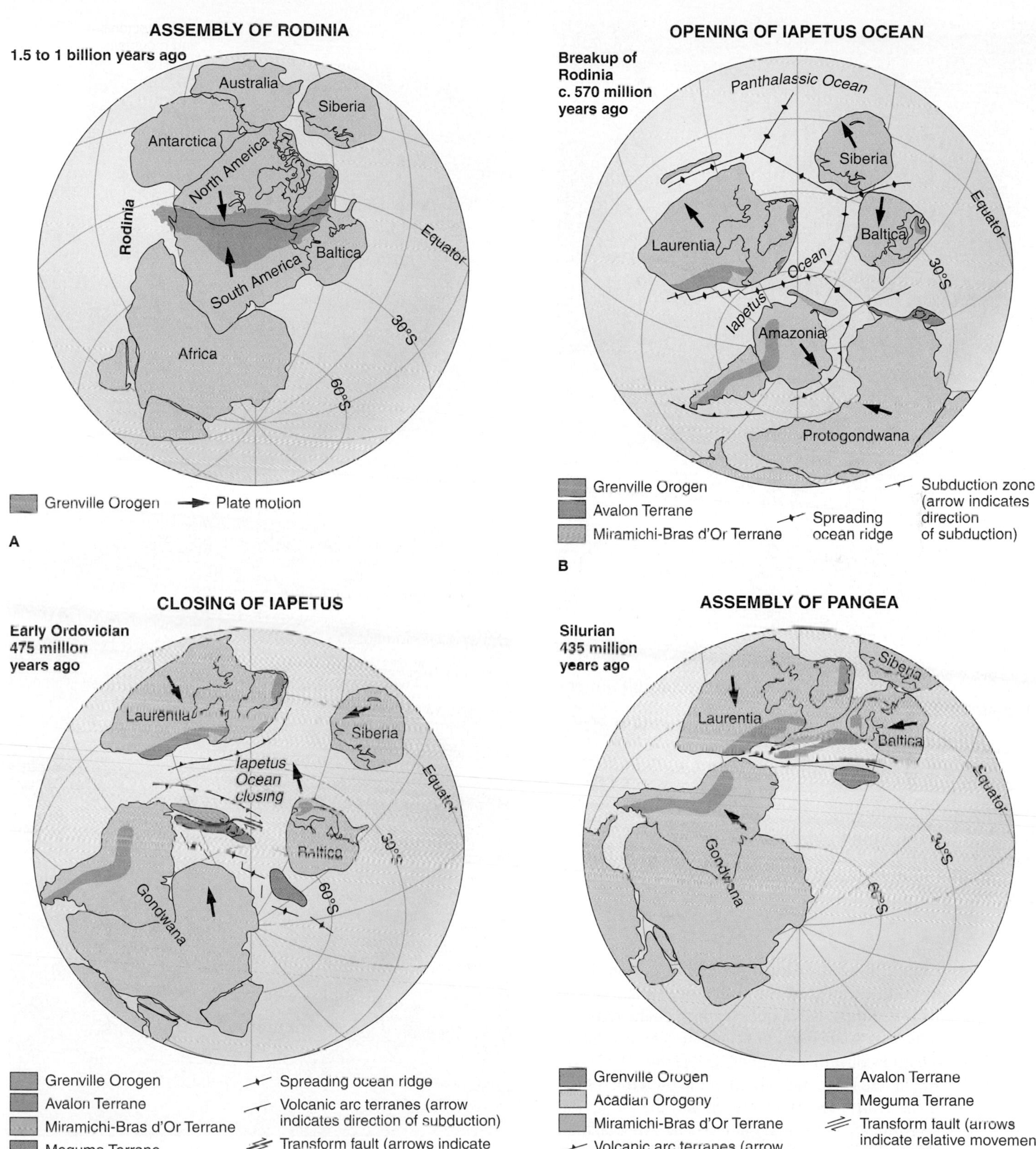

FIGURE 20.8

The building of eastern Canada. (*A*) Collision of North and South America caused the Grenville Orogeny and created the supercontinent Rodinia. (*B*) The breakup of Rodinia and development of the ancestral Atlantic Ocean, the Iapetus Ocean. (*C*/*D*) Closure of the Iapetus Ocean. Exotic far-travelled terranes were added (accreted) to eastern North America during the Ordovician (*C*) and Silurian (*D*). Parts of the floor of the Iapetus Ocean were shoved on land and are now preserved as ophiolites in western Newfoundland (see figure 20.13). The same process of terrane accretion built much of the province of British Columbia during the Mesozoic era (see figure 20.20).

FIGURE 20.9

Banded gneisses near Parry Sound, Ontario were produced at great depths below colliding land masses during the Grenville Orogeny between 1.3 and 1 billion years ago. These rocks are now exposed on the Canadian Shield as a result of slow uplift and erosion.

Photo by Nick Eyles

cover the outer margins of the North American craton. Each bed, now composed of weathered ash called *bentonite*, can be "fingerprinted" by reference to its unique chemical characteristics, providing geologists with an invaluable means of correlating strata across large areas of mid-continent.

Closure of the Iapetus Ocean resulted in the buckling of the outer margins of the North American craton and the formation of downwarped sedimentary basins, such as the Appalachian Basin, in the continent's interior (figure 20.11). These are called *intracratonic basins*, as they form on the craton and are underlain by continental crust rather than oceanic crust. Slow subsidence of the sea floor and the continual addition of new sediment allowed great thicknesses of sedimentary strata to accumulate in these basins over time. More than 10 km of Paleozoic sediments are preserved in parts of the Appalachian Basin (figure 20.3).

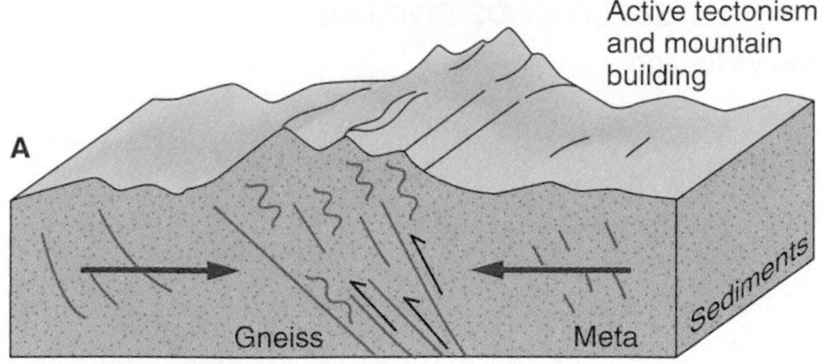

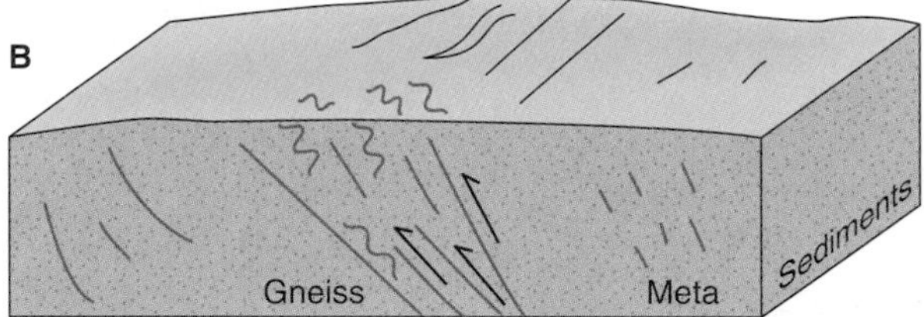

C

FIGURE 20.10

(A,B) Mountains formed during the Grenville Orogeny were eroded to form an almost flat peneplain by 800 million years ago. (C) The Canadian Shield forms an 800-million-year-old, low-relief landform that has barely been modified by geologically recent processes such as glaciation. Taltson River, Alberta.

Photo C © Jim Wark — Airphoto

IN GREATER DEPTH 20.2

The Oldest Animals in the World

The Cambrian explosion of new species, including those with backbones (chordates), is recorded by the enigmatic fossils of the Burgess Shale, found in Yoho National Park in British Columbia (see chapter 1). The world-famous Burgess Shale is exposed on the slopes of Mount Wapta (box figure 1). Mount Wapta forms a part of the Cathedral Escarpment, which formed the edge of an ancient carbonate reef. Some 505 million years ago, this reef overlooked deep water along the margin of western North America and the fine-grained muds that accumulated along its base formed the Burgess Shale. Strange-looking fossils were discovered in the Burgess Shale in 1909 by Charles Walcott—the site of the richest fossil finds is named Walcott's Quarry and is now a UNESCO World Heritage Site (see box 1.1). Slightly older but similar fossil groups have been found in China (the Chengjiang deposits), indicating that Burgess-type organisms were more widespread than previously thought. Fossils even older than those in the Burgess shales, some of the oldest definitive multicellular animals known to science—are also Canadian. These were formed by soft-bodied organisms and as a group are called the *Ediacara fauna*. Their ages range from about 565 to 543 million years; the oldest yet known anywhere in the world occur in Newfoundland in the Mistaken Point Formation (box figure 2). A very distinctive Ediacaran fossil is *Charnia masoni*, a large frond-like organism, almost 2 metres in length, resembling the shape of a modern-day lichen but that lived on the sea floor. Other Ediacaran organisms (called the "Twitya discs") are found in strata deposited at the same time as those of the Mistaken Point Formation but that outcrop in the Mackenzie Mountains of the Northwest Territories on the opposing paleo-Panthalassic coastline of North America.

BOX 20.2 — FIGURE 2
Outcrops of the Mistaken Point Formation, Avalon Peninsula, Newfoundland. This site is now protected as a UNESCO Biosphere preserve. Strata consist of layers of mudstone in which delicate fossils are preserved.
Photo by Nick Eyles

BOX 20.2 — FIGURE 1
Mount Wapta (peak on left) and the Walcott Quarry (circled) in Yoho National Park, British Columbia.
Photo by Nick Eyles

BOX 20.2 — FIGURE 3
Frond-like organisms belonging to the Ediacaran fauna, found in deep-sea rocks at Mistaken Point, Newfoundland.
Photo by Nick Eyles

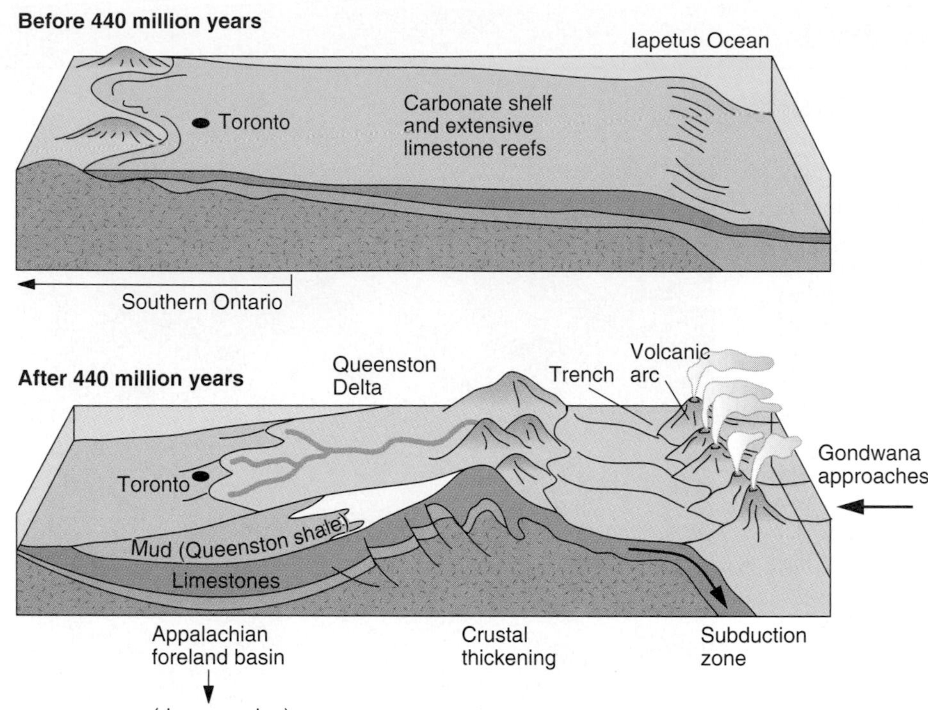

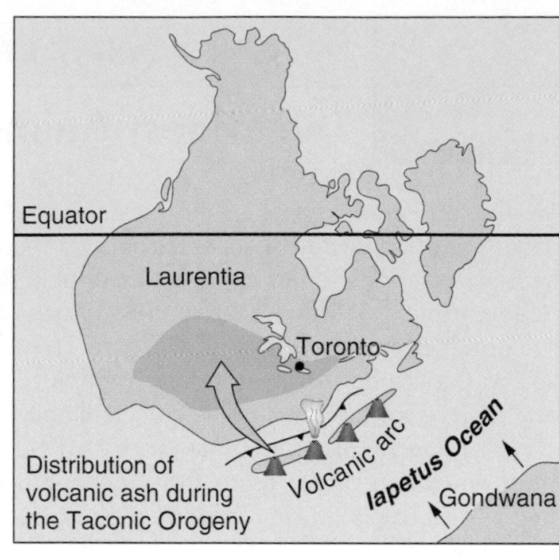

FIGURE 20.11

The Taconic Orogeny. Destruction of the Iapetus Ocean and its floor resulted in crustal thickening along the edge of eastern North America. Associated downwarping at the edge of the shield created shallow seas in which fossiliferous sedimentary strata such as shales, sandstones, and limestones were deposited.

Enormous deltas, such as the Queenston Delta, drained the north slopes of the Taconic Mountains and spread fossiliferous sediment inland, across the outer margins of the North American craton (figure 20.12). At times when little sediment was being transported to the basins, extensive tropical reefs hosting a great wealth of marine organisms flourished in the warm, clear, shallow waters (figure 20.13). Thick limestone deposits are the remains of these reefs and now cover much of Southern Ontario, outcropping spectacularly along the Niagara Escarpment (figure 20.3*C*). The reefs were episodically killed off by influxes of muddy and sandy sediment that deposited shales and sandstones. Canada's first oil strike in 1858 was at Petrolia in southwestern Ontario, where oil is found at shallow depths. It is derived from the breakdown of organic matter in early Paleozoic shales of Devonian age and migrated upward to the ground surface as oil seeps. The commercial pumping of oil at Petrolia heralded the start of the world's oil and gas industry and predated the major discovery at Titusville, Pennsylvania in 1859.

In eastern Canada, the closure of the Iapetus Ocean (figure 20.8*C*,*D*) resulted in portions of the entire ocean floor, including parts of spreading centres and their gabbroic and basaltic rocks, being thrust up as *ophiolites* (see chapter 2). These ophiolites are now preserved high and dry in southern Quebec and along the western coast of Newfoundland (figure 20.14). Geologists can walk over these areas and see first-hand the structure and composition of mid-ocean spreading centres and determine how oceanic crust is produced. Study of these ancient rocks provided important ground truth in support of the emerging plate tectonics ideas of the 1970s. By studying the ancient record much has been learned of the operation of modern processes.

FIGURE 20.12

Brightly coloured, thinly bedded shales and sandstones of the deltaic Queenston Shale, Bronte Creek, Ontario.

Photo by Nick Eyles

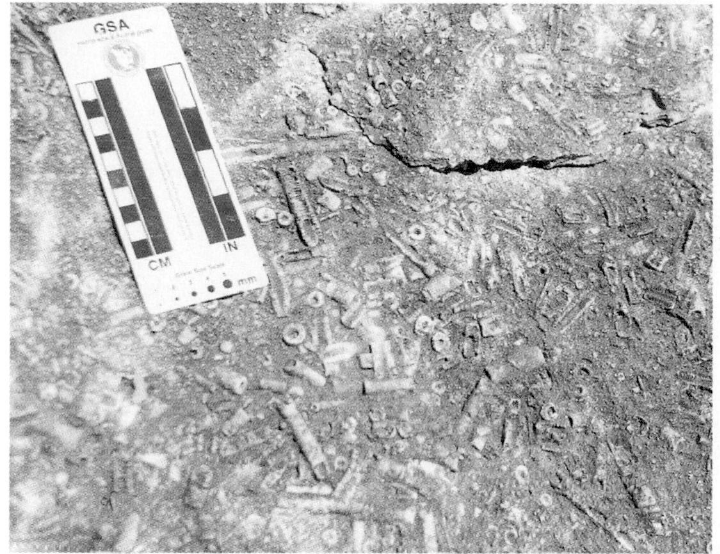

A

B

C

FIGURE 20.13

Fossiliferous Paleozoic limestones. (*A*) Broken stems of sea lilies (crinoids). (*B*) Cephalopod fossil. (*C*) Burrows preserved in limestones.

Photos by Nick Eyles

FIGURE 20.14

Pillow lavas, Betts Cove, Newfoundland. Section is approximately 10 metres high. These result from the rapid cooling of basaltic lava erupted on the ancient floor of the Iapetus Ocean. They form part of a larger slab of oceanic crust and sediment (called an ophiolite) that was shoved onto the North American continent when the Iapetus Ocean closed (figure 20.16).

HOW WERE THE ATLANTIC PROVINCES ADDED TO CANADA?

The Taconic Orogeny marks the beginning of the construction of another supercontinent, called *Pangea*. Pangea was completed in the late Carboniferous when Laurasia docked with Gondwana, which consisted of most of the so-called "southern" continents of Australia, South America, India, and Antarctica. This docking event occurred in several stages and involved an initial Devonian-age collision known as the *Acadian Orogeny* and a later Carboniferous collision called the *Alleghenian Orogeny*, which were responsible for the buckling and thrusting of previously deposited sediments. The present-day *Appalachian Mountains* (figure 20.15) are the eroded remnants of these orogenic events. Huge folds and thrust sheets in the Appalachians were created in much the same way as the Canadian Rockies were to be developed much later (see below).

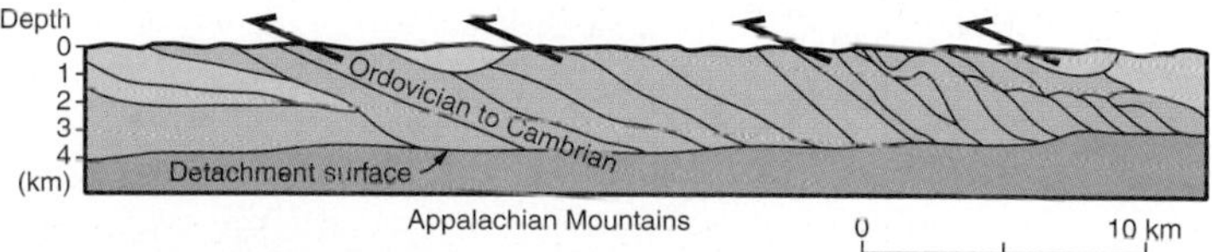

FIGURE 20.15

Geologic cross-section through the modern-day Appalachian Mountains of eastern North America showing folds and thrusts produced during the Acadian and Alleghanian Orogeny, which marked the formation of the supercontinent Pangea.

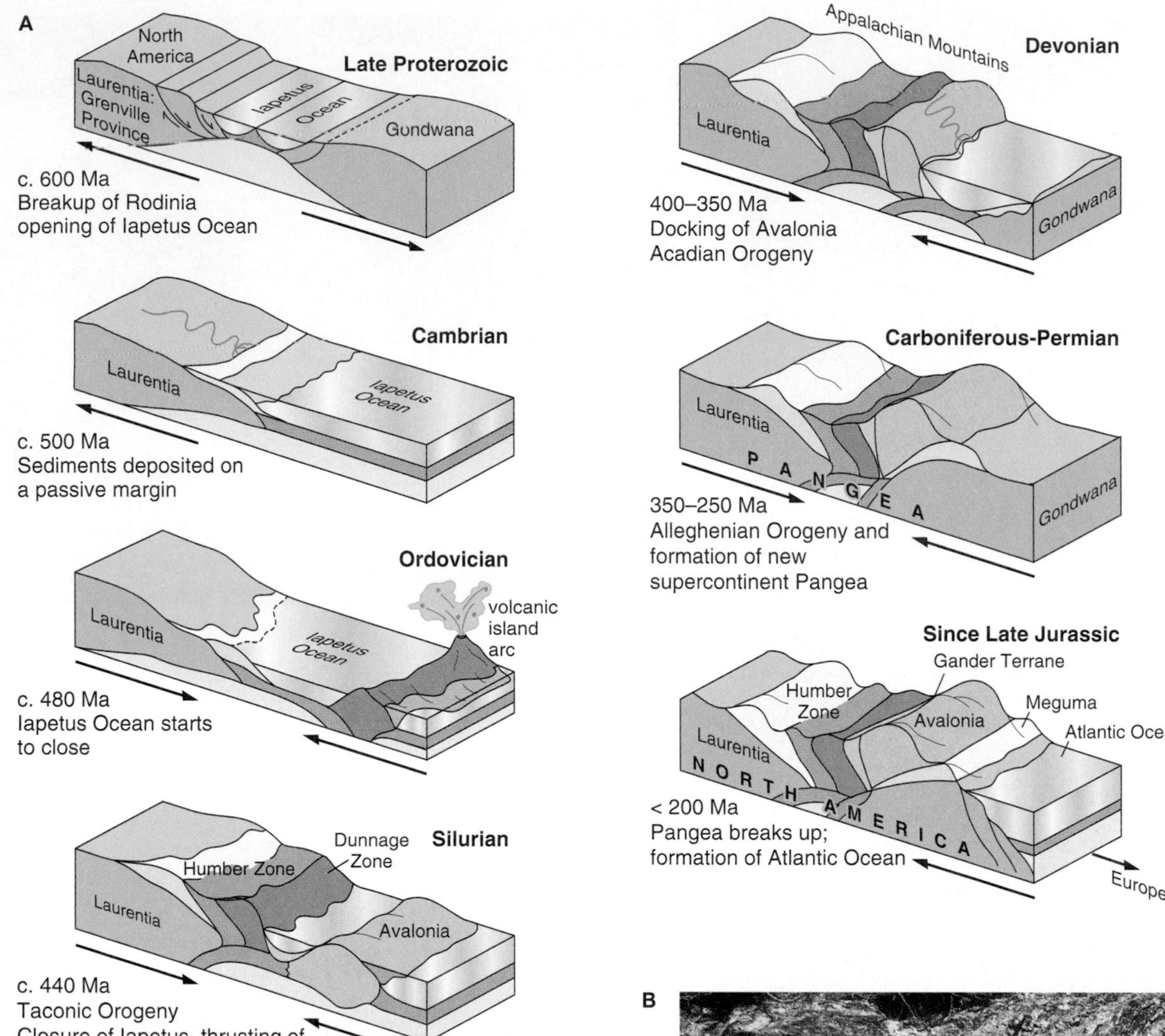

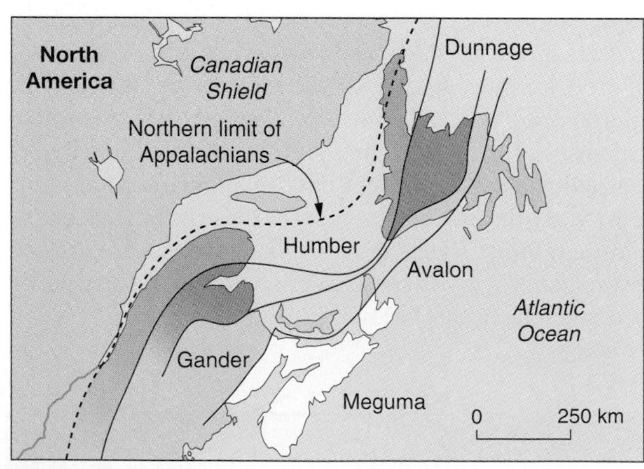

FIGURE 20.16

Formation of Maritime Canada. (*A*) Opening and subsequent closure of the Iapetus Ocean and formation of the Atlantic provinces in the Paleozoic. The sequence of ocean closure, subduction, and accretion of terranes also occurred in western Canada during the Mesozoic (see figure 20.20). Ocean opening and closing is referred to as a Wilson Cycle. The Grenville province records ocean closure during the Proterozoic between 1,300 and 1,000 million years ago. In eastern Canada, the opening of the Iapetus Ocean around 600 million years ago began a new cycle, and the opening of the Atlantic Ocean 200 million years ago marks the beginning of another cycle. (*B*) Folded green and red marine shales at Black Point, west coast of Newfoundland, record closure of the Iapetus Ocean about 400 million years ago.

Photo B by Nick Eyles

In eastern Canada, several large geological blocks (terranes) were added to early North America as micro-continents were thrust and accreted onto the continent during the Taconic and Acadian/Alleghanian Orogenies (figure 20.16). This is how much of what are now the modern-day Atlantic provinces came into being. The most famous of these accreted terranes are the Miramichi-Bras d'Or, Meguma, and Avalon terranes that now make up much of New Brunswick, Cape Breton Island, Nova Scotia, and Newfoundland (figure 20.8*D*). These originated on the far side of the Iapetus Ocean during the Cambrian (figure 20.8*B*,*C*) and now form broad west–east trending geologic belts that lie parallel to the modern Atlantic coastline.

During the late Carboniferous, the Alleghanian Orogeny also produced the extensive **Maritimes Basin** in eastern Canada. This basin was composed of many smaller interconnected basins separated by active faults and lay within the highlands of the Appalachian mountain belt (figure 20.17). The Maritimes Basin extends from central Nova Scotia and southern New Brunswick east into Newfoundland. Huge alluvial fans fed sediment from volcanoes into deep lakes in which microscopic algae thrived. Fine-grained, organic-rich sediments accumulating in these basins would eventually produce oil shales, such as those of southern New Brunswick.

Cordilleran mountain belt
Canadian Shield
Williston Basin
Appalachian Foreland Basin
Michigan Basin
Cincinnati Arch
Maritimes Basin
Illinois Basin
Nashville Dome
Appalachian mountain belt
Permian Basin
Ouachita mountain belt
0 500 km

A

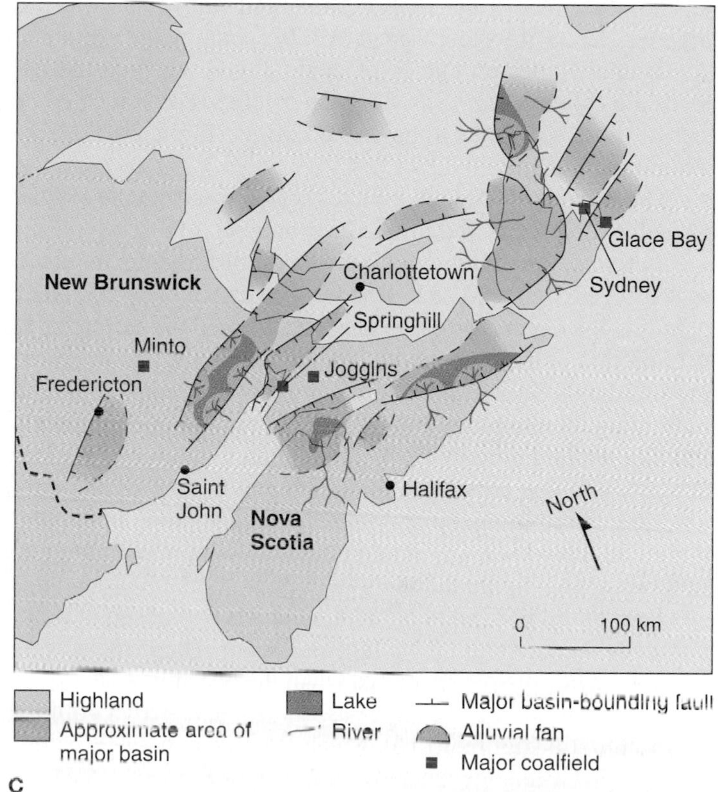

C

B

D

FIGURE 20.17

(*A*) Paleozoic orogenies along the eastern margin of North America created several large sedimentary basins separated by basement highs (arches) (e.g., Cincinnati Arch) in the underlying craton. (*B*) Goderich Salt Mine, Ontario. Salt deposits are a prominent component of the Michigan Basin reflecting strong evaporation when mid-continent North America lay across the equator 400 million years ago. (*C*) Maritimes Basin of eastern Canada and major coalfields. (*D*) Famous fossil cliffs at Joggins, Nova Scotia. Lithified tree trunks preserved within these coal-bearing deposits provided the first clues about the origin of coal in the mid-nineteenth century.

C after Atlantic Geoscience Society; Photos B and D by Nick Eyles

By the late Carboniferous (around 325 million years ago), extensive swamps formed in the Maritime provinces, which at that time straddled the equator. Thick peats accumulated in these swamps and eventually compressed and hardened to form coal. The coals are associated with fluvial and shallow marine deposits that record the repeated rise and fall of sea level as large ice sheets grew and waned over the south polar regions of Gondwana. Repeated cycles of coal, fluvial, and marine deposits are called *cyclothems*. The rich coalfields of Nova Scotia, such as those at Sydney, Joggins, and Minto (figure 20.17*C, D*), formed at this time.

The clearest evidence for plate collisions during the closure of the Iapetus Ocean and for the later breakup of Pangea occurs in Newfoundland. The island is made up of three distinct geological zones (figure 20.18*A*). The western part of Newfoundland, including the Great Northern Peninsula, formed the eastern edge of Laurentia at the time of the Iapetus Ocean. In this zone, the Bay of Islands Ophiolite Complex consists of oceanic crust and underlying mantle rocks thrust up and over the margin of North America as the Iapetus Ocean closed (figure 20.16*A*). The contact between the crust and the mantle is known as the Mohorovicic discontinuity (or Moho for short), and an ancient Moho is well exposed on the slopes of Table Mountain in western Newfoundland (figure 20.18*B*).

The central portion of Newfoundland is composed of rocks that made up the floor of the Iapetus Ocean. These rocks were crumpled against North America when the ocean was destroyed during the formation of Pangea as Eurasia and Africa collided with Laurentia (figure 20.16). Much later, as explained below, the Atlantic Ocean opened and a fragment of Eurasia and Africa was left attached to North America to form the eastern third of the island. These rocks can now be found on the Avalon Peninsula (figure 20.18*A*). Matching these rocks with their counterparts on the other side of the Atlantic Ocean provided key evidence to indicate that the modern Atlantic Ocean is a relatively young geological feature resulting from the breakup of Pangea.

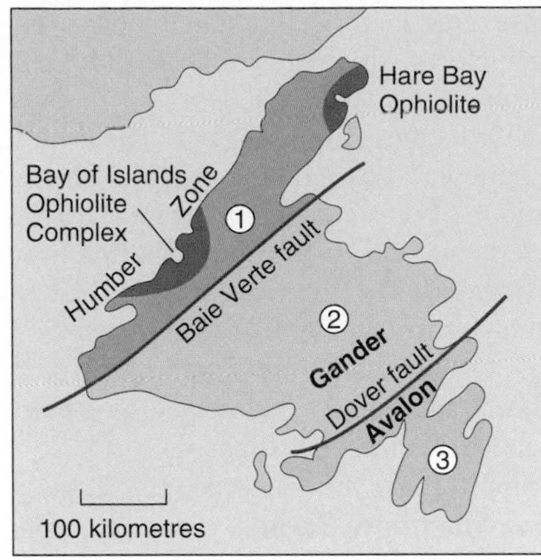

Three Zones of Newfoundland

① Continental margin of ancient North American Plate. Now comprises the Humber Zone.

② Island-arc volcanic rocks and Iapetus Ocean sediments consisting of Dunnage and Gander terranes.

③ Rock left behind when Eurasian/African Plate wrenched loose from Pangea (Avalon terrane)

A

B

FIGURE 20.18

(*A*) Geology of Newfoundland as established in the 1970s. Areas in black consist of sedimentary, volcanic, and ophiolitic rock and record thrusting of portions of the Iapetus Ocean floor over the ancient margin of North America as the ocean closed during the Taconic Orogeny. Tracing this structure westward into the Atlantic Provinces showed that eastern Canada is made up of accreted terranes (figure 20.16*A*). (*B*) Table Mountain, western Newfoundland; the contact between crust and mantle rocks in the Bay of Islands Ophiolite Complex is shown (dashed line).

Photo B by Nick Eyles

WHERE DID BRITISH COLUMBIA COME FROM?

Stage 5—Canada in the Mesozoic: Pangea Breaks Apart and British Columbia Is Swept Up

Some 200 million years ago, Pangea was beginning to break up, just as Rodinia had done 500 million years earlier. By the middle part of the Jurassic, around 165 million years ago, the northern part of Pangea (Laurasia; see figure 20.19) was beginning to detach from its southern part (Gondwana), forming the narrow equatorial *Tethys Ocean*.

Slowly North America began to move away from Africa, and in the late Jurassic the present-day Atlantic Ocean was born (figure 20.19). This process involved the formation of aulacogens, such as the Ottawa–Bonnechere Graben and the St. Lawrence Rift (see box 2.4). Eastern Canada eventually became a passive continental margin as North America began to track westward. In turn, the former passive margin of western Canada became an active margin characterized by subduction processes. At this time, the area now called British Columbia is thought to have existed as chains of volcanic islands, hot spots, ocean-floor rocks, and microcontinents far offshore in the ancestral Pacific Ocean. These land masses were eventually swept up and accreted onto western North America as the continent drifted westward to form the mountain belts of British Columbia, Alberta, the Yukon, and the Northwest Territories (collectively referred to as the **Canadian Cordillera**). The migration of the land masses that now make up the Cordillera is recorded by

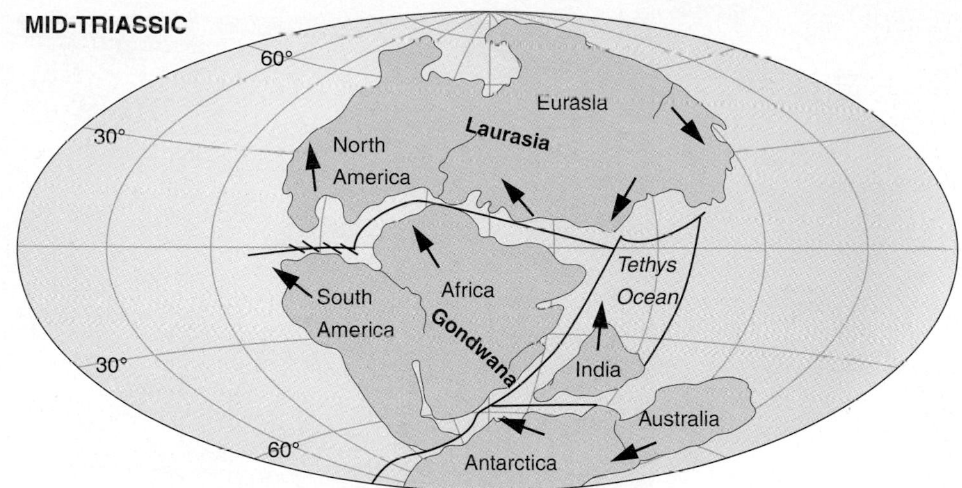

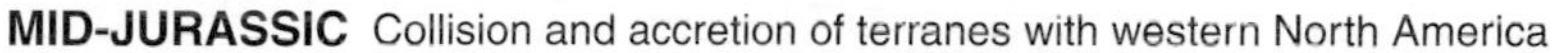

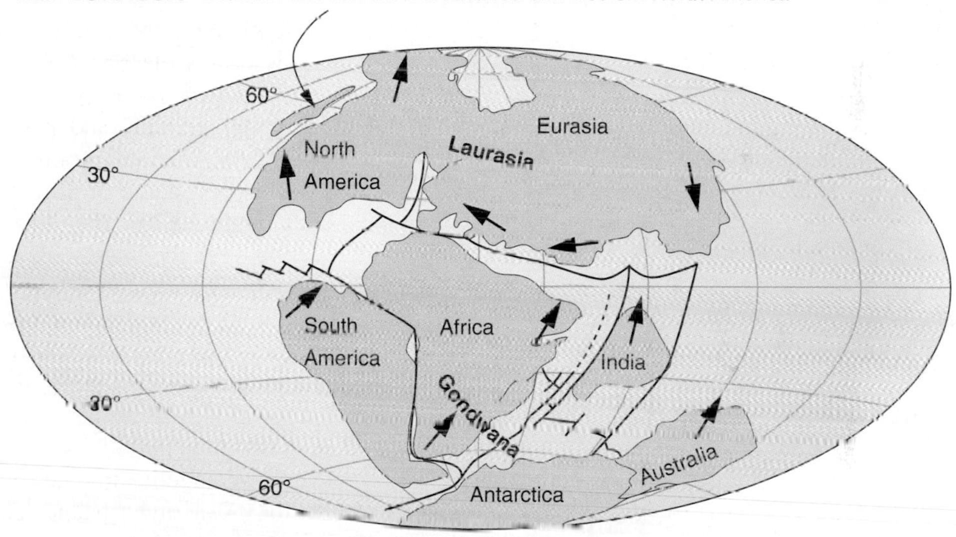

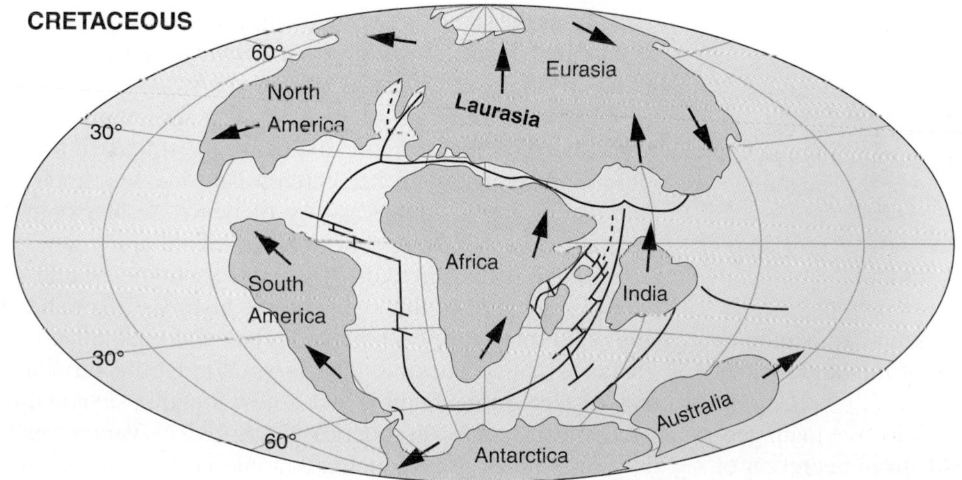

FIGURE 20.19

(A) Global geography approximately 220 million years ago showing Laurasia and Gondwana separated by the narrow Tethys Ocean. (B) Tethys Ocean begins to close and Atlantic Ocean opens as southern Gondwana breaks up (150 million years ago). (C) The western movement of North America in the Cretaceous swept up several microcontinents and volcanic arcs to form what is now the province of British Columbia (see figure 20.20). Continental movements at this time radically changed ocean circulation patterns and global climates, contributing to global cooling (see figure 20.25).

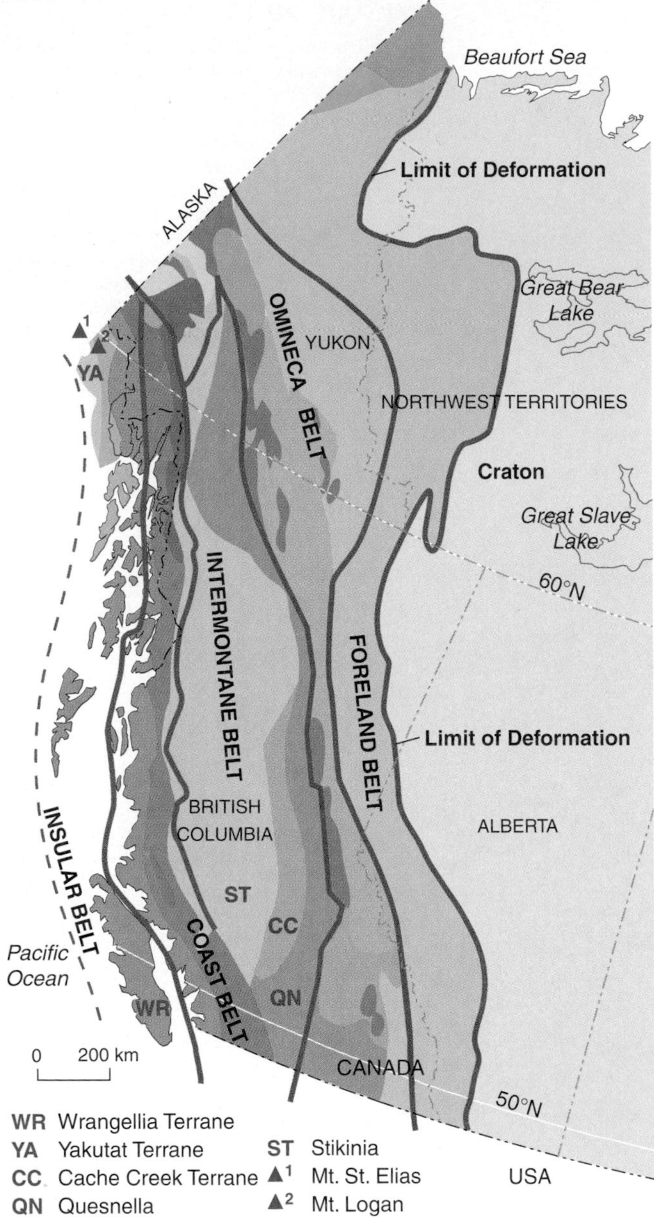

FIGURE 20.20

(*A*) Major geological belts and terranes of western Canada (in different colours: selected terranes are identified on figure). (*B*) Highly deformed rocks exposed near Cache Creek, southern B.C., contain Permian-age fossils common in Asia. Discovered in 1971, they provided the first clues that much of British Columbia originated elsewhere and was added to western North America as far-travelled blocks of crust called terranes.

Figure A from www.uni-mainz.de/~mezger/Text/Cordilleran-Belts.jpg; Photo B by Nick Eyles

fossils in sedimentary strata and paleomagnetic data stored in igneous rocks that allow the former source of the land masses to be determined.

The Canadian Cordillera is subdivided into five main geological belts (figure 20.20*A*). The belts result from accretion of far-travelled terranes, which has occurred almost continuously since the mid-Jurassic. The oldest collisional event has historically been referred to as the *Columbian Orogeny*, but this term is slowly disappearing from usage. During this early event the so-called *Intermontane Belt* (consisting of Stikinia, Cache Creek terrane, and part of Quesnellia) docked with western North America, around 185 million years ago. Far to the east, this first phase of thrusting affected rocks of the Foreland Belt (figure 20.20). The Omineca belt marks the boundary or *suture* between the Intermontane and Foreland belts and is a zone of intensely metamorphosed rocks intruded by plutons.

A second major orogenic event, poorly dated, occurred between 150 and 100 million years ago. This event has been called the *Laramide Orogeny*, but geologists now think the orogeny was not a single event and the term is almost abandoned. This time period saw the accretion of the *Insular Belt* consisting of Wrangellia and the Alexander terranes that underlie modern-day Vancouver Island. This collision triggered further eastward thrusting of the rocks that now make up the Rocky Mountains, when resistant Cambrian limestones were thrust over soft Mesozoic shales to form the magnificent panorama of the Front Ranges around Banff. The timing of this second phase cannot be pinned exactly, as the suture between the Insular and Intermontane belts is obscured by the *Coast Plutonic Complex* where the rocks range in age from about 150 to 50 million years.

The many different geologic terranes that now comprise the Canadian Cordillera (figure 20.20) are the squeezed and highly deformed remnants of the accreted land masses. Each terrane has undergone a similar history of being pushed onto North America and then being smeared northward along major intracontinental strike-slip faults. The presence of these faults greatly complicates unravelling the geologic history of British Columbia as terranes have been moved relative to each other over big distances since docking. The large Wrangellia terrane (WR) was broken during faulting and is now found in several different areas of western North America (figure 20.21). Vancouver Island is a large block of Wrangellia that was added to western North America after 200 million years ago. Much of Wrangellia is composed of old oceanic crust, including pillow lavas and

A

B

FIGURE 20.21

(*A*) The Karmutsen Volcanics, west of Campbell River, Vancouver Island, form part of the Wrangellia terrane. (*B*) Highly deformed gneiss exposed on the foreshore below Mile 0 of the Trans-Canada Highway in Victoria, B.C.

Photos by Nick Eyles

FIGURE 20.22

Mount St. Elias (5,540 metres above sea level), on the border between the Yukon and Alaska, is one of several large mountains raised by collision of the Yakutat terrane and the North American continent. Folded strata recordintense compression associated with collision.

Photo by Nick Eyles

FIGURE 20.23

Glaciated mountains of the Coast Belt of British Columbia consist predominantly of eroded granite plutons.

Photo by Nick Eyles

associated volcanic rocks such as those forming the Karmutsen Volcanics west of Campbell River (figure 20.21*A*). Highly deformed metamorphic rocks that represent lower crustal rocks formed at the base of the Wrangellia Terrane are also exposed near Victoria on Vancouver Island (figure 20.21*B*).

Today, the Yakutat Terrane (YA) is in the process of docking against Alaska with its easternmost margin defined by a major strike-slip fault and high mountains. Intense compressive forces between the terrane and North America uplifted the mountain massifs of Mount St. Elias (figure 20.22; 5,540 metres above sea level) and Mount Logan (5,060 metres above sea level), both of which lie on the border between the Yukon and Alaska. Much of the Coast Belt of British Columbia, which lies inboard of the Wrangellia Terrane (figure 20.20*A*) is composed of granite plutons intruded between 170 and 45 million years ago as Wrangellia collided with western North America (figure 20.23). The rounded, dome-like form of the plutons is reflected in the subdued shape of mountains within the Coastal Plutonic Complex of British Columbia, which contrasts with the more sharply etched mountains of the Rockies (refer to figure 20.26 later in this chapter).

The accretionary growth of the Cordillera of western North America in the Mesozoic is very similar in style to the way in which Atlantic Canada was added to North America in the Paleozoic (figure 20.16). Both record the growth of continental crust by collisions of smaller land masses against the North American craton.

During the collisional events that formed the Cordillera, thick successions of strata that had earlier accumulated along the western passive margin of Rodinia were pushed eastward and compressed to form mountains in the interior of British Columbia. Sediment from the eroding mountains was shed eastward into the **Western Interior Sedimentary Basin**. A classic example of a foreland basin (figure 20.24), it extended from modern-day Mexico in the south to the Arctic in the north. Foreland basins develop in response to the weight of thickened thrust belts (on the margins of mountain belts) depressing the surrounding crust (see the examples of the Taconic and Appalachian foreland basins in eastern Canada; figure 20.11).

The Western Interior Sedimentary Basin experienced many changes in water depths allowing alternations of shallow marine sandstones and deeper-water shales. The organic-rich shales, now deeply buried, have acted as source rocks for oil and gas that migrated and collected in porous, shallow-water sandstones, to form large oil and gas reservoirs (see chapter 12). Complex thrust structures created in what is now western Alberta by Cordilleran tectonics far to the west helped to form *hydrocarbon traps*, where oil and gas that would otherwise escape to the Earth's surface became trapped. Western Canada's first major oil discovery took place in 1913 at Turner Valley, south of Calgary. Coal deposits within the Mesozoic strata were discovered near Banff during construction of the Canadian Pacific railway in the 1880s.

In Saskatchewan, rich potash deposits formed as evaporites in the Western Interior Sedimentary Basin. Potassium chloride (the mineral sylvite) is an essential nutrient for plant growth and is a major component of modern fertilizers. Saskatchewan is the world's largest producer of potash.

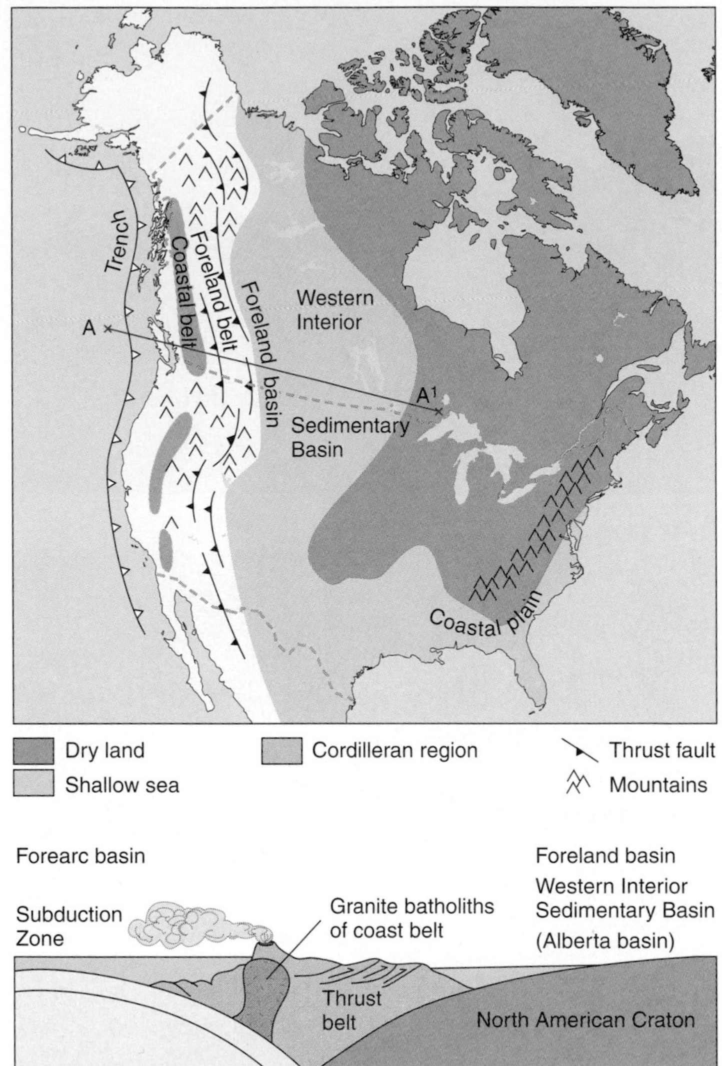

FIGURE 20.24

Paleogeography of western North America during the Late Cretaceous, and schematic cross-section across the thrust belt and Western Interior Sedimentary Basin.

WHERE CAN WE FIND DINOSAURS IN CANADA?

The Western Interior Sedimentary Basin is famous not only for oil and gas but also for its rich fossil remains, principally those of dinosaurs. Dinosaurs roamed subtropical swamps along the shores of the Late Cretaceous seaways. Rapid deposition of sediment in these environments allowed preservation of fossil material, and rich finds of dinosaur fossils are now made in rocks exposed in the many deeply cut valleys of the badlands area of Alberta. More than 150 dinosaur species have been identified in Dinosaur Provincial Park, and the Royal Tyrrell Museum of Paleontology at Drumheller houses many of the major dinosaur discoveries found in the area (figure 20.25). The museum is named after the geologist J.B. Tyrrell, who found the first fossil remains in Alberta in 1884. Important collections of Canadian dinosaurs are also kept at the Canadian Museum of Nature in Ottawa and the Royal Ontario Museum in Toronto (go to http://images.rom.on.ca/public for photos of many specimens).

Dinosaur finds are not restricted to western Canada, and important dinosaur fossils have been found in Nova Scotia. Canada's oldest dinosaur fossils (Triassic age) were found at Burntcoat Head, Nova Scotia, and dinosaur bones and footprints—some made by *Coelophysis*, a small, bipedal dinosaur—were found near Rossway, Nova Scotia. Early Jurassic rocks exposed in the cliffs at Wasson Bluff are particularly rich in dinosaur fossils, including the remains of prosauropods (animals "before the lizard feet") and some of the more agile carnivorous dinosaurs.

FIGURE 20.25

Albertosaurus, one of the large carnivorous dinosaurs found in Dinosaur Provincial Park, Alberta. Late Cretaceous fluvial, deltaic, and shallow marine sediments are widely exposed in the badlands of Alberta and are host to many dinosaur fossils.

Photo by Nick Eyles

Why Did the Dinosaurs Disappear So Suddenly?

The very youngest Late Cretaceous rocks of Alberta also contain the record of a massive meteorite impact that hit the Yucatan Peninsula about 65 million years ago. The impact is marked in Mexico by the 100-km-wide Chicxulub crater, and at the Cretaceous-Tertiary boundary (K-T boundary) by abrupt global changes in fossil types. This meteorite impact event is thought to have contributed to the demise of the dinosaurs as well as 75 percent of all plant species and 90 percent of the plankton species in the oceans.

HOW DID THE CANADIAN ROCKIES FORM?

In addition to adding what is now British Columbia to North America, the Mesozoic and Cenozoic Orogenies of western North America helped create one of Canada's most spectacular landscapes, the Canadian Rocky Mountains. The Canadian Rockies are formed of layered sedimentary rocks that were displaced eastward by about 150 kilometres as a result of compression caused by collisional events along the western margin of North America. The layered sedimentary rocks detached from underlying granites of the craton and were folded and thrust over one another (figure 20.26). Hard sedimentary rocks, such as limestones that had formed on the margins of Rodinia, were moved as intact slabs over much younger and softer Mesozoic rocks along major thrust faults such as the Rundle and McConnell thrusts. The thrusting process was lubricated by water-rich shale within the Mesozoic rocks. Harder limestones form slab-sided mountains in the Front and Main ranges of the Rockies, such as Mount Rundle near Banff (figure 20.26*B*,*C*). Farther east, in the Foothills, thrusts are composed entirely of the softer and more easily eroded Mesozoic rocks. It is important to bear in mind, however, that the spectacular mountain landscapes we now see in the Canadian Rockies, where deep wide valleys are juxtaposed with steep sided peaks, developed in only the recent geologic past. The mountains of the Canadian Rockies were formed as a result of deep glacial erosion and mass wasting during the glaciations of the past 2.5 million years. By cutting deep valleys, glaciers can create even bigger mountains. The load on the underlying mantle is reduced, promoting additional uplift of mountains. The Rockies contain old rocks, but they are relatively young mountains.

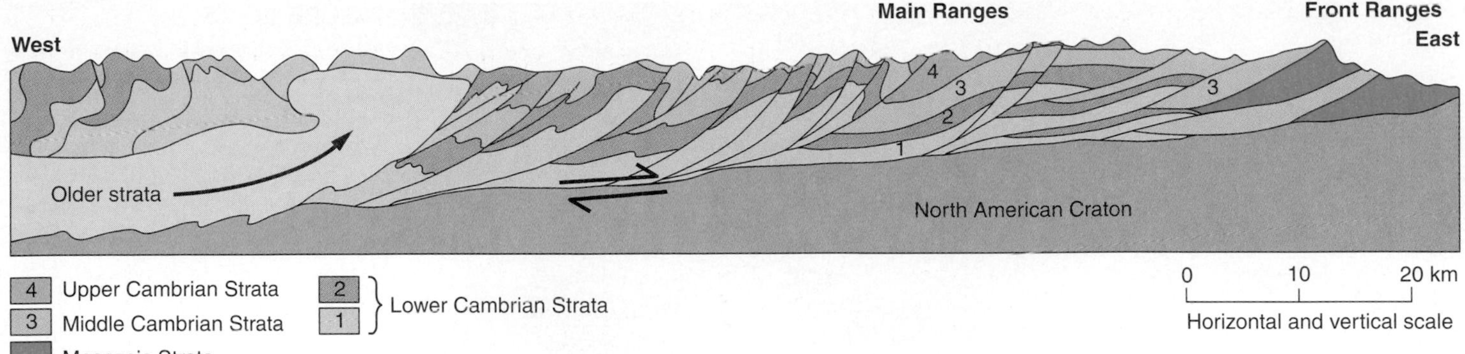

A

B

C

FIGURE 20.26

(A) Geologic structure of the Canadian Rocky Mountains. These structures were formed as a result of eastward thrusting during the Mesozoic Orogeny. (B) Mountain landforms of the Rockies such as these around Banff are relatively young and formed as a result of deep glacial erosion in the last 2.5 million years. Note the steeply dipping rock strata in Mount Rundle (left of centre). The strata were deformed by large thrusts during the Laramide Orogeny (150–100 million years ago), when much of British Columbia was added to western North America. Resistant limestones form the mountains, whereas the valleys have been eroded into softer shales. (C) Castle Mountain in the Main Ranges of the Rocky Mountains consists of Cambrian limestones that were thrust eastward as large flat slabs and were gently warped into broad synclines. Glacial erosion has produced a different mountain form to that seen in the Front Ranges.

Photos by Nick Eyles

WHEN DID THE ICE SHEETS DEVELOP?

Canada in the Cenozoic: The Arctic Cools Down

The latest (and unfinished) stage in Canada's ongoing geological evolution did not involve major tectonic events but climate. Climate changes have alternated from full glacial conditions when ice sheets covered most of Canada to interglacial conditions such as the present day. Under interglacial conditions, limited ice masses survive only in the Canadian Arctic, for example on Baffin Island (figure 20.27) or at high elevations in the Rockies. Erosion by ice flowing from the centre of ice sheets to their outer margins profoundly altered Canada's landscapes (see chapter 16).

Global climates during the Mesozoic are considered to have been relatively warm and equable. Paleoclimatic evidence obtained from the sediments that accumulate on the floors of oceans (and thus retain a near-continuous record of global temperatures) indicate that global climates began to cool down around 60 million years ago. As late as 45 million years ago, in the Eocene, warm-loving redwoods and broadleaved trees still flourished in the high Arctic of Canada. The Buchanan Lake Formation of Axel Heiberg Island (almost 80 degrees North latitude) contains 25 layers of mummified (not mineralized) tree stumps and logs and the remains of some of the animals that lived in the forest. Trees were well adapted to living three months in total darkness.

FIGURE 20.27

Buchan Gulf, Baffin Island, seen from Baffin Bay. Deep fjords have been eroded by glaciers draining Baffin Island ice caps.

Photo: GSC Photo 2002–239. Photo by Douglas Hodgson. Reproduced with the permission of the Minister of Public Works and Government Services Canada, 2005 and Courtesy of Natural Resources Canada, Geological Survey of Canada.

Formation of the Antarctic Ice Sheet at about 40 million years ago marked the beginning of more pronounced global cooling (figure 20.28). This cooling phase was to eventually result in the formation of mountain glaciers at high elevations in the northern hemisphere at about 10 million years ago and full-fledged continental ice sheets in North America and Europe at about 2.5 million. Key contributors to this global climate change were radical changes in ocean and atmospheric circulation caused by the closure of the Tethys Ocean as India slammed into Asia (figure 20.19), and the uplift of the enormous Himalayan mountain system. At about 2.5 million years ago, the isthmus of Panama that joins North, Central, and South America formed as a volcanic arc by subduction. Prior to this event water (and heat) could be exchanged from one ocean to another, but development of the isthmus separated the Pacific from the Atlantic, and changed the configuration of ocean currents. Evidence suggests that the Arctic Ocean and Canada's northern coastline first began to freeze over at this time. Arctic Canada became a cold place characterized by widespread permafrost.

Canada in the Quaternary: Ice Sheets Come and Go

Over the past 2 million years or so, Canada has been repeatedly covered by enormous ice sheets. The Cordilleran Ice Sheet grew in western Canada and covered the Rockies and much of British Columbia. Only the highest mountain peaks protruded through the ice (as **nunataks**). Another, much larger ice sheet (the *Laurentide Ice Sheet*) extended from the Prairies through the High Arctic to Labrador and reached down into the United States (see figure 16.33). Newfoundland had its own ice cap, yet other areas—such as the Yukon—were severely cold and ice-free.

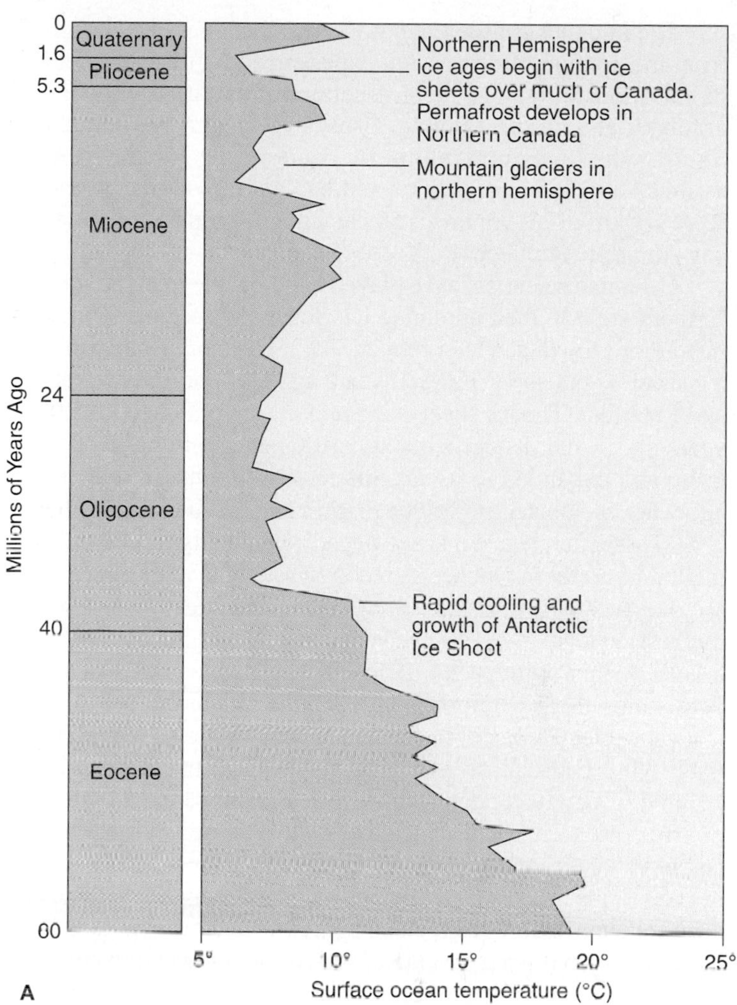

FIGURE 20.28

(*A*) Global climate changes over the past 60 million years. (*B*) Glacially eroded trough incised into high plateau, Gros Morne National Park, Newfoundland. Glacial erosion has unloaded the crust promoting uplift of the plateau, part of the Canadian Shield (figure 20.3).

Photo by Nick Eyles

The thickness of the Laurentide Ice Sheet has been inferred from the amount of depression experienced by the crust below its enormous weight; by reconstructions of its surface profile, or gradient; and by the amount of sea-level lowering created by its growth. At its maximum, the Laurentide Ice Sheet was as much as 3 kilometres thick and had a total volume of about $33 \times 10^6 km^3$. This volume is 50 percent greater than the present-day Antarctic Ice Sheet.

Canadian scientists have played a major role in mapping the former extent of the Laurentide Ice Sheet and reconstructing its history of growth and later retreat. The last major advance of the ice sheet began about 100,000 years ago, over northern Quebec and Labrador. The ice sheet oscillated in size until about 25,000 years ago, when it reached southern Quebec and parts of Ontario, and expanded to its maximum size around 18,000 years ago when its lobate, finger-like margin reached into the northern U.S. Thereafter, the ice sheet began to thin; its last remnants melted in northern Quebec as recently as 6,000 years ago.

Ice flows from central areas of an ice sheet (where it is thickest) to its margins at speeds of up to several hundred metres a year. Armed with rocky debris frozen into its base, or with debris moving as a deforming layer below its base, flowing ice is a very effective agent of erosion and deposition. By abrading underlying bedrock or sediment the ice creates an enormous volume of newly created sediment that is deposited either as a poorly sorted sediment called *till*, or that is transported under and beyond the ice sheet by meltwaters and eventually deposited in lakes or the ocean. Great thicknesses of glaciomarine sediments accumulate offshore from glaciated regions. The wide range of landforms left behind by glacial activity is reviewed in chapter 16 and is only briefly summarized here.

Canadian Landscapes Produced by Glacial Erosion

The most extensive landscape created by large-scale, areal erosion by glaciers and ice sheets is the Canadian Shield, comprising smoothed and eroded bedrock and numerous lakes. The high plateau of Labrador, Newfoundland, and Baffin Island, into which narrow, steep-sided troughs are cut (figure 20.28), forms a classic Canadian landscape. The plateau represents the uplifted margins of the Canadian Shield, which were raised when the Atlantic Ocean opened during the Mesozoic. Rifting allowed the broken edges of the Shield to gently rise in elevation (geologists call this **passive margin uplift**). Fjords are cut by fast-flowing streams of ice descending from the plateau, and their floors have been eroded many hundreds of metres below sea level. The rugged fjord-indented coastline of British Columbia is another example of a classic glaciated landscape.

In western Canada, glacial erosion created the final striking form of the Rocky Mountains, consisting of high angular mountain tops with frost-shattered summits and narrow, glacially excavated troughs (figure 20.26*B*). In many places, the slopes of glaciated valleys are so steep they are unstable and prone to large landslides. Landsliding is also common along the unstable banks of rivers that cross permafrozen ground in Canada's northern regions (figure 20.29).

Thousands of glacially eroded rock basins across Canada trap huge volumes of freshwater and form lakes, such as the Great Lakes (see box 20.5). These freshwater reservoirs, the largest in the world, are of international strategic importance given the growing demand for clean drinking water by expanding urban areas in Canada and the U.S.

FIGURE 20.29

Landslide on the banks of the Mackenzie River, Northwest Territories. Increased landsliding in the area is the result of recent climate change and the melt of deeply frozen ground (permafrost) allowing sediments to slide downslope.

Photo by Nick Eyles

IN GREATER DEPTH 20.3

Tors and Gold: Relics of Early Cenozoic Warmth?

There is some evidence to indicate that during the warmth of the early Cenozoic—around 40 million years ago—the Canadian Shield was exposed to deep chemical weathering that allowed the production of a thick clayey layer called *saprolite*. Saprolite is widespread in the modern tropics and sometimes reaches thicknesses of 500 metres. Unweathered portions of the parent rock survive as *core stones*. As the upper, near-surface parts of the saprolite are eroded with time, core stones are exposed on the surface as *tors*. It is thought by some that many of the rounded boulders of gneiss and granite that are now scattered throughout the tills of southern Canada started off as core stones on the deeply weathered portions of the Canadian Shield only to be moved south by Quaternary ice sheets. Some tors still survive in the Arctic in areas where ice was thin and cold, and glacial erosion was ineffective (box figure 1).

The famous alluvial gold deposits along the Fraser River and in the Cariboo district of British Columbia, and also at Dawson City in the Yukon (site of the famous 1898 Gold Rush), owe something to deep weathering during the early Cenozoic. Gold that was finely dispersed in bedrock was chemically concentrated during weathering leaving large nuggets at the base of the saprolite. When the climate cooled and weathered debris was reworked by rivers and glaciers, the heavy gold nuggets were concentrated in rich placer deposits (box figure 2).

BOX 20.3 ■ FIGURE 1
Marble tors several metres high formed by weathering on the Melville Peninsula, Nunavut.
GSC Photo 2002-480 by Lynda Dredge. Reproduced with the permission of the Minister of Public Works and Government Services Canada, 2005 and Courtesy of Natural Resources Canada, Geological Survey of Canada.

BOX 20.3 ■ FIGURE 2
Placer gold panned from alluvial gold in the Cariboo district of B.C.
Photo by Nick Eyles

Canadian Landscape Features Produced by Glacial Deposition

The extensive deposits of glacial sediment left across northern North America allowed deep soils to develop in the last 10,000 years, creating productive agricultural land. Huge volumes of groundwater are also stored in thick glacial sediments that cover much of southern Canada, forming economically important *aquifers*.

A classic Canadian landscape created by glacial deposition is that of the *drumlin field*. Drumlin fields typically contain many hundreds of elongate hills, some of which are many tens of kilometres in length, up to 100 metres high, and 250 metres wide, while others are much narrower (the narrowest are called *flutes*). Famous drumlin fields in Canada occur in the Peterborough area of southern Ontario and around Yarmouth in Nova Scotia. The magnificent drumlins of the Livingstone Lake drumlin field in northern Alberta are illustrated in figure 20.30.

The well-ordered, streamlined form of drumlin fields contrasts with the haphazard look of glacial deposits forming hummocky moraine. This landscape type covers many thousands of square kilometres of the prairies of central and western Canada (figure 20.31). High-standing hummocks are randomly interspersed with hollows that are often filled with water. This landform may have originated as glacial sediment carried on the surface of stagnant ice (ice that is no longer moving), and was slowly deposited as underlying ice wasted away. Alternately, areas of hummocky moraine may have developed when stagnant ice pressed into soft till below, creating a dimpled surface.

In the mid nineteenth century, as Canadian geologists sought to test Louis Agassiz's newly introduced hypothesis of continental glaciation, many were struck by the presence of "findling boulders," also called **erratics.** These are boulders that have been transported long distances (sometimes almost half a continent) away from their source areas by glacial ice. Some of the best-known erratics include the Foothills Erratics Train of Alberta and the Bleasdell Boulder of southern Ontario (figure 20.32). The former records the collapse of a nunatak near present-day Jasper and the transport of landslide debris on the ice surface. The debris was left scattered out on the prairie as the ice thinned and retreated.

IN GREATER DEPTH 20.4

Cold Winds and Ancient Soils

While the rest of Canada lay under ice as much as 3 kilometres thick, parts of the Yukon Territory (and Alaska) were ice-free. It was simply too dry and cold for glacial ice to build up in these areas. Alaska and the Yukon experienced periglacial conditions (see chapter 16) and large parts of the landscape became buried under deposits of windblown silt and sand (called *loess*), which in places reach considerable thicknesses (box figure 1). During interglacial epsiodes, when the ice sheets largely disappeared from the rest of Canada, soils formed on the surface of these windblown deposits. The soils were subsequently buried when glaciation was renewed and are now preserved within the thick successions of loess. Ancient soils often contain abundant plant and animal fossils and preserve an excellent record of past climatic conditions.

BOX 20.4 ■ FIGURE 1

Thick deposits of windblown silt (loess) have been terraced in this road cut in Alaska.
Photo by Nick Eyles

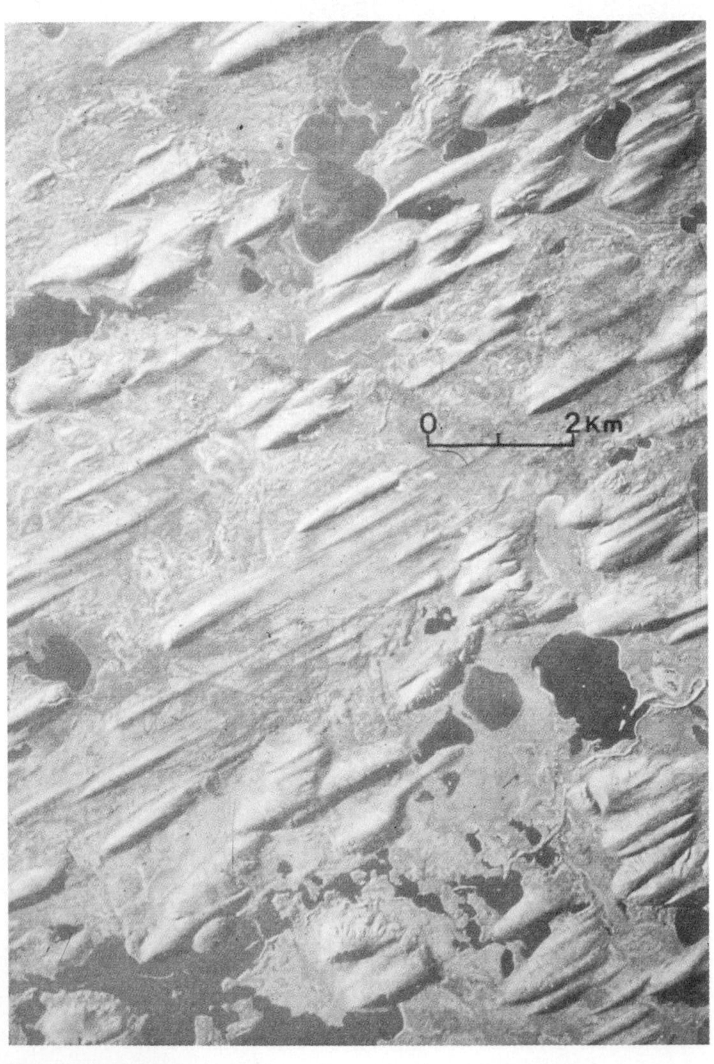

FIGURE 20.30

Drumlins in the Livingston Lake drumlin field in northern Alberta. Ice flows from lower left to upper right of photo. Note the rounded, upglacier ends of the drumlins.

FIGURE 20.31

Glaciated landscape consisting of hummocky moraine, central Alberta.
Photo by Nick Eyles

As big as the Okotoks erratics are, they are dwarfed by the enormous slabs of rock (called **megablocks**) that were moved *under* the Laurentide Ice Sheet where it flowed over softer sedimentary rocks in Alberta and Saskatchewan. Some are 100 metres thick and extend over 1,000 square kilometres. They are composed of relatively soft shale, have no surface expression in the landscape, and were discovered only by drilling. Large ice-thrust ridges (figure 20.33) forming broad arcuate belts on the land surface are closely associated with hummocky moraine in central Alberta and Saskatchewan and have a similar origin. These features are examples of *glaciotectonic* deformations and illustrate the ability of ice sheets to quarry and deeply disturb the landscape over which they flow.

IN GREATER DEPTH 20.5

How Did the Great Lakes Form?

Another expression of deep glacial erosion is the Great Lakes of mid-continent North America. The modern lakes fill deep bedrock basins whose floors reach well below sea level. Evidence suggests that the lakes are the result of glacial erosion and overdeepening of a much older mid-continent river system (box figures 1 and 2). Portions of the old river valleys, now plugged and buried by glacial sediment, survive in places. As big as they are, the modern lakes are smaller remnants of much larger lakes that were dammed when the margin of the Laurentide Ice Sheet blocked lake outlets. The biggest of these lakes is called *Glacial Lake Agassiz* (after Louis Agassiz; see chapter 16). Silts and clays deposited in Glacial Lake Agassiz form a huge flat plain that extends over much of Saskatchewan, Manitoba, and Northern Ontario (see figure 16.34). Sediments deposited in much larger precursors of Lake Ontario are exposed in the 100-metre-high Scarborough Bluffs to the east of Toronto (box figure 3). The five Great Lakes (Erie, Huron, Michigan, Ontario, and Superior) form the largest body of freshwater in the world. Lake Superior is the largest of the Great Lakes (and is in fact the largest lake in the world) with a surface area of 82,000 km^2, of which about one-third is in Canada. The Great Lakes watershed covers 766,000 km^2 and is home to 25 percent of Canadians, 10 percent of Americans. The lakes extend 1,200 kilometres from west to east and are unique in that they are interconnected to form a continuous body of freshwater. The St. Lawrence Seaway–Great Lakes Waterway was completed in 1959 and is the world's longest inland waterway, stretching 3,790 kilometres from the mouth of the St. Lawrence River near the Atlantic Ocean to the head of Lake Superior.

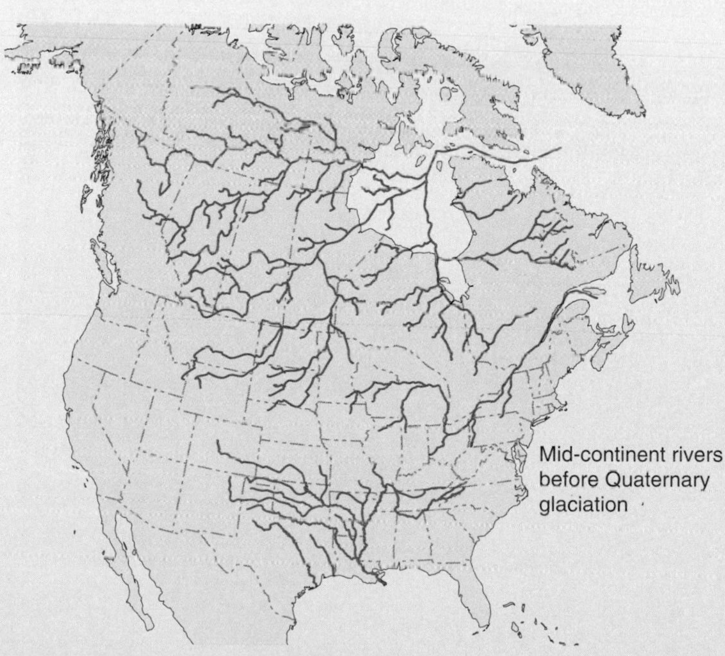

BOX 20.5 ■ FIGURE 1

Pre-glacial river system of mid-continent North America, around 5 million years ago.

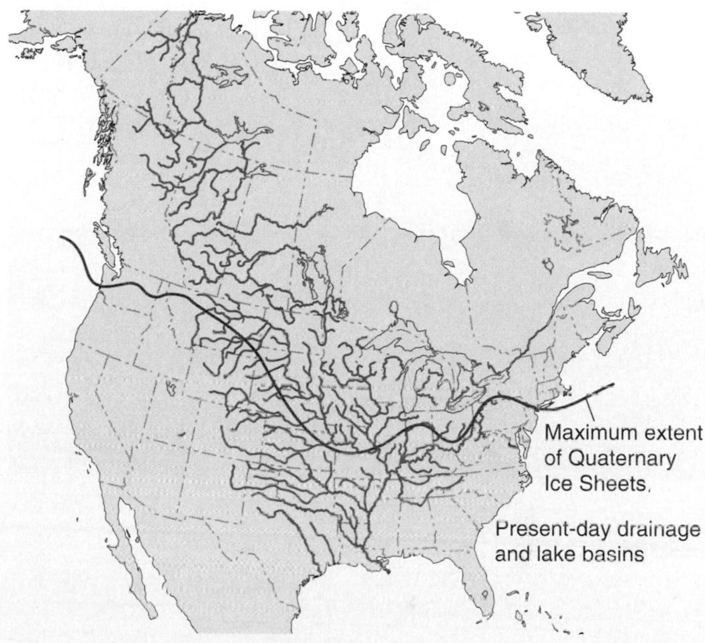

BOX 20.5 ■ FIGURE 2

Deep rock basins that now form the Great Lakes were cut by ice sheets only 2.5 million years ago. Note very different drainage patterns in Canada after glaciation.

BOX 20.5 ■ FIGURE 3

The Scarborough Bluffs east of Toronto expose a world-famous succession of glacial lake sediments that contain a lengthy record of past lake-level and climate changes.

Photo by Nick Eyles

A

B

FIGURE 20.32

Glacial erratics. (*A*) The Big Rock at Okotoks, Alberta forms part of the Foothills Erratic Train, a series of erratic sandstone blocks transported from their source area near Jasper. (*B*) The Bleasdell Boulder weighs more than 33,000 tonnes and is composed of metamorphosed sedimentary and volcanic rocks derived from the Canadian Shield; near Trenton, Ontario.

Photos by Nick Eyles

A

B

FIGURE 20.33

Three-kilometre-wide belt of ice thrust ridges, Alberta. (*A*) Arcuate ice thrust ridges associated with hummocky moraine (upper left of photo). (*B*) Steeply dipping ice-thrust Cretaceous bedrock at Stettler, southern Alberta.

Photos by Nick Eyles

FIGURE 20.34

The Lemieux Landslide at South Nation River valley, Ontario. This was a retrogressive landslide that occurred on June 20, 1993.

Lemieux Landslide, Ont. June 1993. Photograph by Greg Brooks. Reproduced with the permission of the Minister of Public Works and Government Services Canada, 2007 and Courtesy of Natural Resources Canada, Geological Survey of Canada.

The soft glaciomarine clays of eastern Canada underlie large areas of the St. Lawrence Valley and are prone to failure as quickclay slides (chapter 13). These clays were deposited on the floor of the Champlain Sea as the Laurentide Ice Sheet was retreating northward about 11,000 years ago (see figure 16.34). Glaciomarine clays are unstable because when deposited they contain salts in their structure derived from sea water. These salts are slowly leached out of the clays by percolating groundwater, weakening the clay structure and leaving them prone to failure if they are disturbed by earthquakes or construction activity. These quickclays give rise to spectacular landslides and mud flows (figure 20.34).

Canada's Geology: A Project in Progress?

Canada's geological evolution continues as the entire country moves westward at velocities of up to 6 centimetres a year (caused by the drift of the North American plate) and surface processes dissect and change its surface. Major earthquakes and intermittent volcanic activity along the west coast testify to ongoing subduction of the Juan de Fuca plate under North America. Other western earthquakes are the expression of continued displacement along the major strike-slip faults separating the various terranes of British Columbia (figure 20.20). The *intracratonic earthquakes* that occur in central and eastern Canada record displacement along buried aulacogens (e.g., the Ottawa-Bonnechere Graben, see box 2.4) and ancient terrane boundaries in the craton. These structures are said to be "reactivated" by within-plate stresses as North America slowly drifts westward.

In addition, although scarcely noticeable to people, large areas of the country continue to slowly rise in elevation recording unloading of the crust (*isostatic rebound*) following the retreat of the ice sheets. More obvious expressions of ongoing geological activity are landslides, rockfalls, and cliff retreat recording the continual processes of mass movement, erosion, and sediment transport.

The major Canadian rivers continue to shape the landscape, eroding gorges at Niagara and depositing vast amounts of sediment to form large deltas on which we depend for resources, agriculture, and urban land. The Fraser River delta in B.C. has grown more than 25 kilometres seaward in the past 10,000 years and now hosts large urban areas and industrial developments, as well as valuable agricultural land and salmon spawning grounds (figure 20.35). Future growth and stability of the delta is now in question as dredging of sand to maintain shipping channels limits sand deposition and parts of the delta front are slowly being eroded. Landslides have occurred on the submarine slope of the Fraser River delta in the past and any future failure could cause damage to the coal port, ferry terminal, or submarine electric cables.

Looking to the future, if climate change continues as it has in the past, Canada will cool down and the next Laurentide Ice Sheet will start to form over northern Quebec and Labrador in about 5,000 years. However, long before then, Canadians might be adapting to a much warmer, but possibly more variable, climate resulting from anthropogenic increases to greenhouse gas concentrations in the atmosphere. The possibility of this global warming raises many questions for Canadians, such as what the future of the Great Lakes and Arctic Ocean are in a warmer climate, and how we can safeguard coastal infrastructure and communities from erosion and flooding as sea level continues to rise.

IN GREATER DEPTH 20.6

4 Billion Years of Canadian Geology in Summary

In chapter 2 we showed how the formation and breakup of supercontinents formed a natural repetitive cycle in Earth history; these cycles, repeated many times over the last 4 billion years, are called Wilson Cycles in honour of J. Tuzo Wilson. Canada's complex geology can be simplified using the concept of Wilson Cycles.

Canada's highly varied geology records at least two complete cycles in the last 1,000 million years (box figure 1). The first ended with the formation of *Rodinia* and the addition of the Grenville Belt to eastern North America. Much of present-day central and southern Ontario, Quebec, and Labrador were added at this time.

The breakup of Rodinia started in the west at 750 Ma and somewhat later in the east at about 600 Ma. North America was then surrounded by passive margins along the Paleo-Pacific and Iapetus Oceans. Marine waters and sediments flooded the low-relief interior (the craton), resulting in extensive deposits of richly fossiliferous limestone and shales. Those deposited about 540 to 500 million years ago record the earliest diversification of complex organisms (Burgess Shale etc.).

About 440 million years ago, small crustal blocks (terranes) began to loom into view along the east coast as the Iapetus Ocean began to close. These eventually collided with eastern North America in a series of orogenic events and form the basement of the present-day Maritime Provinces of eastern Canada. Many terranes had an African origin. During ocean closure, old Iapetus Ocean crust was shoved onto North America as an ophiolite complex. These events mark the beginning of the formation of *Pangea*, a second supercontinent. Large basins (e.g., Appalachian and Michigan basins) formed in the interior of the craton pushed down by crustal thickening along its eastern maritime rim. Great thicknesses of sedimentary rocks, including coal, accumulated in these basins derived from the erosion of the Appalachian Mountains and others formed by orogenic activity.

Inevitably, Pangea broke up, forming the early Atlantic Ocean about 200 million years ago. North America began to be pushed westward, colliding with small crustal blocks that were gathering in the Pacific Ocean. Their eventual collision added 500 km

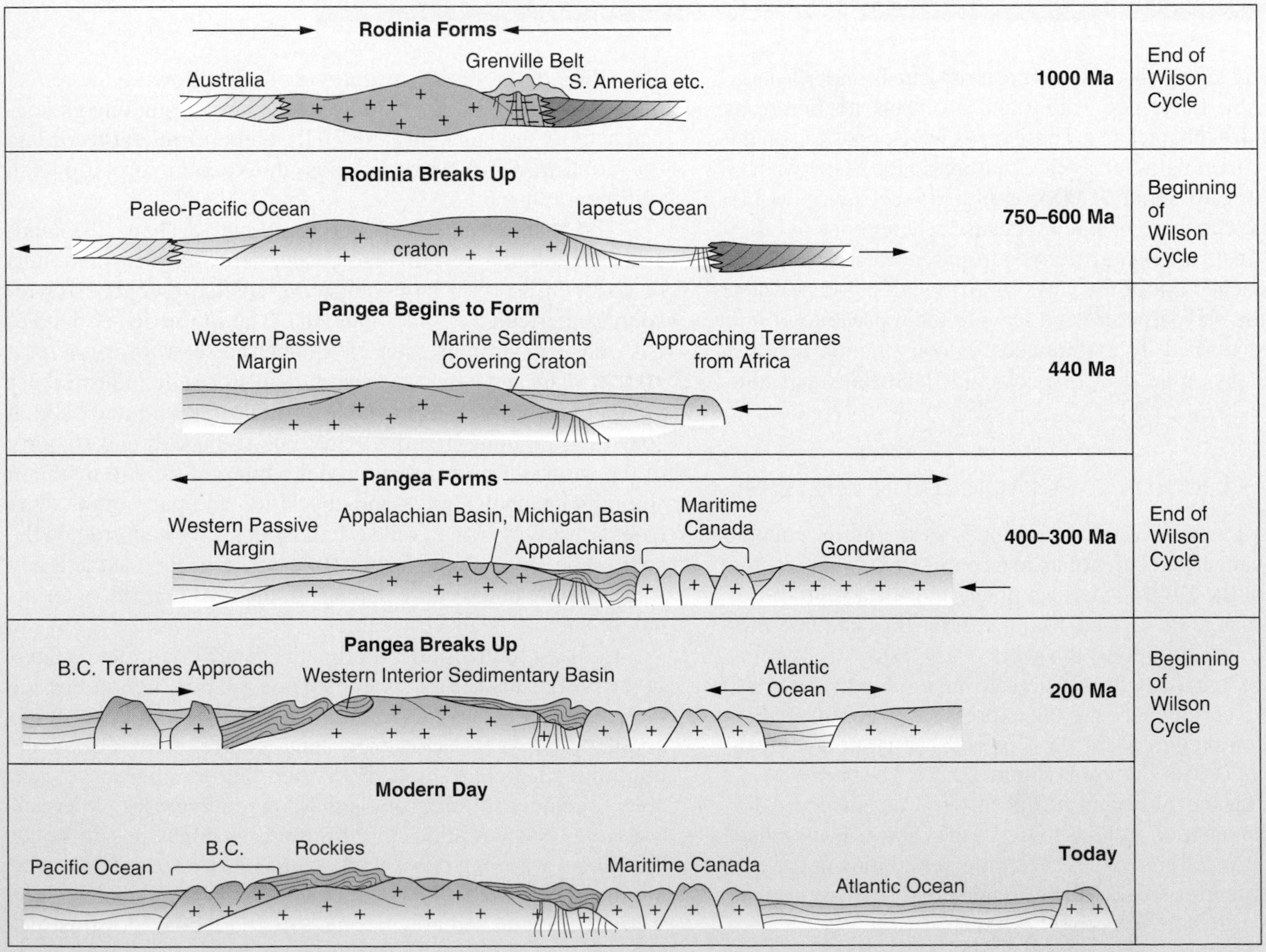

BOX 20.6 ■ FIGURE 1

to North America, forming what is now British Columbia, the Yukon, and much of Alaska. Older marine rocks from the earlier passive margin were shoved eastward and when sculpted by glaciation in the last 2.5 million years produced the Rocky Mountains. The thickened crust produced the Western Interior Sedimentary Basin in what is now Alberta. Dinosaurs languished along its coast until extirpated 65 million years ago. Organic matter deposited and buried in the basin was slowly cooked into oil and gas.

Climate cooling of the last 2.5 million years shaped the landscape, cutting the Great Lakes (among many others), carving deep fiords along the coasts, and leaving thick sediments in which rich soils could develop.

The craton of North America records a series of much older Wilson Cycle extending back to almost 4 billion years beginning with the Acasta Gneiss. The old oceanic crust and volcanic rocks trapped during many collisions of terranes as the craton slowly grew in size are rich in minerals.

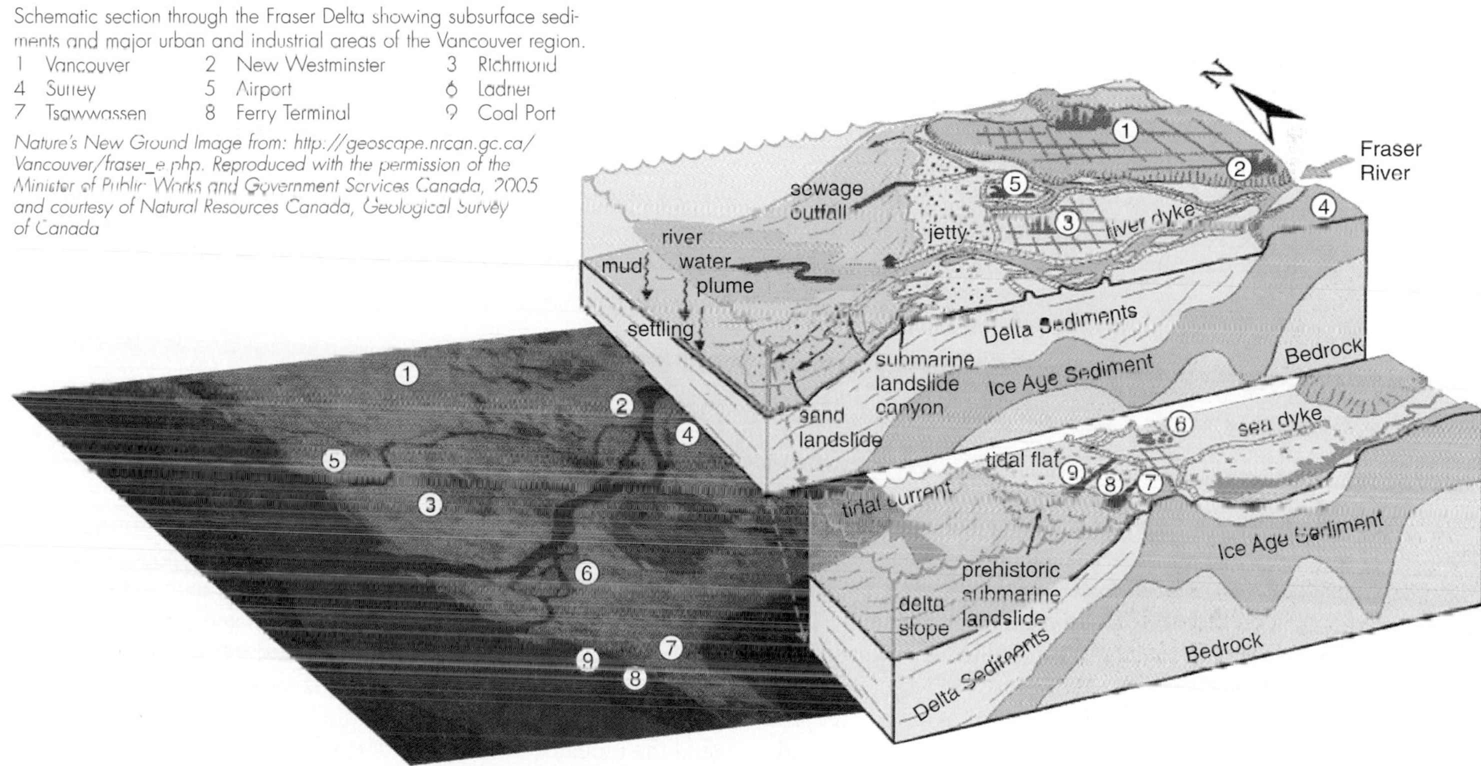

FIGURE 20.35

Schematic section through the Fraser Delta showing subsurface sediments and major urban and industrial areas of the Vancouver region.

1 Vancouver	2 New Westminster	3 Richmond
4 Surrey	5 Airport	6 Ladner
7 Tsawwassen	8 Ferry Terminal	9 Coal Port

Nature's New Ground Image from: http://geoscape.nrcan.gc.ca/Vancouver/fraser_e.php. Reproduced with the permission of the Minister of Public Works and Government Services Canada, 2005 and courtesy of Natural Resources Canada, Geological Survey of Canada

WHY IS IT IMPORTANT TO UNDERSTAND THE GEOLOGICAL HISTORY OF CANADA?

Geoscientists carry out a wide range of professional activities where a detailed understanding of Canada's geological history is a fundamental prerequisite. They increasingly work in consulting companies as part of interdisciplinary teams alongside engineers, biologists, and planners. A major part of their work is to conduct environmental assessments designed to minimize the impact of new development and associated land-use changes.

Canada has become one of the world's most urbanized countries, with 80 percent of its population living in urban areas. While this would suggest that we are no longer as reliant on our natural geological environment as we once were, the opposite is true as we try to build cities that are sustainable environmentally. As cities expand, the hazards posed by earthquakes, landslides, and floods increase as a result. Urban expansion and development commonly requires reuse of *brownfield sites* (sites previously used for commercial, industrial, or residential purposes) to spare precious agricultural land. This demands improved ways of mapping contaminants in the subsurface and removing them. The issue of drinking water is now firmly on the Canadian agenda, especially after the tragic events in Walkerton, Ontario in 2001 when drinking water tainted by animal waste killed seven people (see box 15.3). Protection of the quality and quantity of ground and surface waters for future uses, including possible export of Canada's water to the United States, requires detailed understanding of watersheds, their geology, and their functions. Water is the resource of the future, with impending shortages in the U.S. that will place pressure on Canada to export it. The western Prairie Provinces of Canada lie in the lee of the Rocky Mountains, and surface waters derive from snow and ice melt in the mountains (see box 16.1). This source is threatened by climate change. Groundwater is at risk from urbanization in central Canada and agriculture in the western provinces, and this creates a need for effective groundwater protection schemes in many areas (figure 20.36). No longer is groundwater "out of sight, out of mind."

Much new mineral wealth remains to be discovered on the Canadian Shield using new geological techniques and concepts. The recent development of the Canadian diamond mining industry is a classic example of new wealth created as a consequence of the work of geoscientists. Building on experience gained in the Atlantic provinces, renewed oil and gas exploration both in the Arctic frontier and along the British Columbia coast face physical challenges that can be met only with a proper understanding of geological and environmental conditions. The search for our own origins and the history of other organisms and past climates is also dependent on our ability to interpret the geologic record. All of these activities create a need for continuing research and investigation by new generations of geoscience students, armed with new questions and new techniques. What we learn in Canada can also be applied globally.

FIGURE 20.36

Road sign identifying groundwater protection zone.

Photo by Nick Eyles

SUMMARY

Canada's lengthy geologic history began more than 4 billion years ago, when the North American craton began to form. Many separate land masses were fused together by plate tectonics processes over several billion years to create a *craton* consisting of geologically distinct *provinces*. Deformed rocks of the craton identify areas where land masses were sutured together, and are termed *orogens*. These ancient rocks of North America record repeated episodes of supercontinent formation and breakup and are exposed on the *Canadian Shield*.

The outermost margins of the North American craton have been buried by younger *cover rocks*. These sedimentary rocks were deposited in extensive shallow seas (*intracratonic* and *foreland* basins) that covered parts of the craton when it was depressed by the weight of developing mountain ranges. Most of Atlantic Canada originated as parts of other continents that were accreted onto the eastern seaboard of North America through repeated opening and closing of ancestral forms of the Atlantic Ocean. Western Canada also formed by the accretion of many former land masses (*terranes*) onto North America as it slowly drifted westward after the opening of the modern Atlantic Ocean. The Canadian Rockies record compression caused by these collisional events on the western margin. Canadian sedimentary rocks contain a rich fossil record including some of the earliest animal fossils (*Ediacaran fauna*) and numerous species of dinosaur that roamed the swamps around inland seas.

The most recent events in Canada's geologic history involve climate changes that saw the gradual cooling of warm climates, and the repeated growth and decay of enormous ice sheets that covered almost all of Canada. Many of the landscapes most familiar to Canadians are the product of glacial processes active during the past 2 million years. *Glacial erosion* is responsible for the formation of steep-walled glacial troughs and fjords in mountainous areas, streamlined landforms of the Canadian Shield, and excavation of the Great Slave, Great Bear, and Great Lakes basins. *Glacial depositional features* include drumlin fields, hummocky moraine, and extensive plains underlain by till and outwash that host productive aquifers. Canada's geological history continues to evolve as landslides alter the form of mountains and valleys, quickclays fail, deltas develop, and coastal erosion continues.

Terms to Remember

Acasta Gneiss 539
Arctica 543, 544
basalt 544
Cambrian Explosion 548
Canadian Cordillera 556
Canadian Shield 541
cover strata 541
craton 539
erratic 565
geologic province 539
Gowganda glaciation 544
Grenville Orogeny 547
Iapetus Ocean 548
Maritimes Basin 555
megablock 566
Nena 543, 544
North American craton 541
nunatak 563
orogen 544
Pangea 543
passive margin uplift 564
peneplain 541
Rodinia 543, 547
shatter cones 545
subprovinces 544
supercontinent 544
terrane 539
Western Interior Sedimentary Basin 560

Testing Your Knowledge

Use the questions below to prepare for exams based on this chapter.

1. How did construction of the Canadian Pacific Railroad aid in geologic exploration?
2. Where would you find the oldest rocks in Canada?
3. Why are parts of the North American craton buried by cover rocks?
4. Explain uniformitarianism. Do you think this working principle is always valid?
5. What are thought to be the five stages of evolution of the North American craton?
6. What are the three main types of subprovinces within the Superior Province?
7. What happened in the Grenville Orogeny?
8. Where in Canada would you find the oldest fossil animals?
9. How did the Appalachian Basin form? What kinds of sediment accumulated in the basin and what is the economic significance of these deposits?
10. What happened in eastern North America during the collision of Laurasia and Gondwana?
11. How did the major coal deposits in the Maritimes form? What are cyclothems?
12. Describe the three geological parts of Newfoundland. Why do rocks in Newfoundland provide clues as to the age of the Atlantic Ocean?
13. How does geology and rock type dictate the shape of mountains in western Canada?
14. Explain the formation of the Canadian Cordillera.
15. What is a foreland basin?
16. The Canadian Rockies are young landforms made of old rocks. Explain this statement.
17. Where in Canada would you find fossils indicating warm climatic conditions in the Tertiary?
18. Why are quickclays a problem in eastern Canada?
19. How did the Great Lakes form?
20. What is a Wilson Cycle? How many such cycles are recorded by Canada's geology?

Exploring Web Resources

gsc.nrcan.gc.ca
Geological Survey of Canada Website with links to other geological sites of interest.

http://geoscape.nrcan.gc.ca
Geoscape Canada: Geology and geoscience issues for Canadian cities, including Vancouver.

www.gov.ns.ca/natr/meb/field/start.htm
Virtual field trip of the landscapes of Nova Scotia.

www.tyrrellmuseum.com
Royal Tyrrell Museum, Drumheller Website. Virtual tour, information about local sites.

APPENDIX A

IDENTIFICATION OF MINERALS

Each mineral is identified by a unique set of physical or chemical properties. To determine some of these properties requires specialized equipment and techniques. Most common minerals, however, can be distinguished from one another by tests involving simple observations. Cleavage is an especially useful property. If cleavage is present, you should determine the number of cleavage directions, estimate the angles between cleavage directions, and note the quality of each direction of cleavage. Other easily performed tests and observations check for hardness (abbreviated H), lustre, and colour, and determining crystal form (if present). A simple chemical test can be made using dilute hydrochloric acid to see if the mineral effervesces.

The identification tables included here can be used to identify the most common minerals (the rock-forming minerals) and some of the most common ore minerals. For identifying less common minerals, refer to one of the books or Websites on mineralogy listed at the end of chapter 5. Mineral identification takes practice, and you will probably want to verify your mineral identifications with a geology instructor.

Because the common rock-forming minerals are the ones you are most likely to encounter, we have included a simple key for identifying them. The key is based on first determining whether or not the mineral is harder than glass and then checking other properties that should lead to identification of the mineral. You should verify your identification by seeing whether other properties of your sample correspond to those listed for the mineral in table A.1.

Ore minerals are usually distinctive enough that a key is unnecessary. To identify an ore mineral, read through table A.2 and determine which set of properties best fits the unknown mineral.

Key for Identifying Common Rock-Forming Minerals

Determine whether a fresh surface of the mineral is harder or softer than glass. If you can scratch the mineral with a knife blade, the mineral is softer than glass.

I. Harder than glass—knife will not scratch mineral. (If softer than glass, go to II.)
   A. Determine if cleavage is present or absent (this may require careful examination). If cleavage is absent proceed to 1.; if cleavage is present, proceed to B.
      1. Vitreous lustre
         a. Olive green or brown—*olivine*
         b. Reddish brown or in equidimensional crystals with twelve or more faces—*garnet*
         c. Usually light-coloured or clear—*quartz*
      2. Metallic lustre
         a. Bright yellow—*pyrite*
      3. Greasy or waxy lustre
         a. Mottled green and black—*serpentine*
   B. Cleavage present. Determine the number of directions of cleavage in an individual grain or crystal.
      1. Two directions, good, at or near 90°—*feldspar*
         a. If striations are visible on cleavage surfaces—*plagioclase*
         b. If pink or salmon-coloured—*potassium feldspar* (or *orthoclase*)
         c. If white or light grey without striations, it could be either type of feldspar
      2. Two directions, fair, at 90°
         a. Dark green to black—*pyroxene* (usually augite)
      3. Two directions, excellent, not near 90°
         a. Dark green to black—*amphibole* (usually hornblende)

II. Softer than glass—knife scratches mineral
   A. No cleavage detectable
      1. Earthy lustre, in masses too fine to distinguish individual grains—*clay group* (for instance, *kaolinite*)
   B. Cleavage present
      1. One direction
         a. Perfect cleavage in flexible sheets—*mica:*
            Clear or white—*muscovite mica*
            Black or dark brown—*biotite mica*
      2. Three directions
         a. All three perfect and at 90° to each other (cubic cleavage)—*halite*
         b. All three perfect and not near 90° to each other:
            If effervesces in dilute acid—*calcite*
            If effervesces in dilute acid only after being pulverized—*dolomite*

TABLE A.1 Diagnostic Properties of the Common Rock-Forming Minerals Mineral groups are in uppercase letters.

Name (mineral groups shown in capitals)	Chemical Composition	Chemical Group	Diagnostic Properties	Other Properties
AMPHIBOLE (A mineral group in which *hornblende* is the most common member.)	$XSi_8O_{22}(OH)_2$ (*X* is a combination of Ca, Na, Fe, Mg, Al)	Chain silicate	2 good cleavage directions at 60° (120°) to each other.	H = 5–6 (barely scratches glass). Hornblende is dark green to black; tends to form in needle-like or elongate crystals; vitreous lustre.
Augite (see Pyroxene)				
Biotite (see Mica)				
Calcite	$CaCO_3$	Carbonate	3 excellent cleavage directions, *not* at right angles (they define a rhombohedron). H = 3. Effervesces vigorously in weak acid.	Usually white, grey, or colourless; vitreous lustre. Clear crystals show double refraction.
CLAY MINERALS (*Kaolinite* is a common example of this large mineral group.)	Compositions include $XSi_4O_{10}(OH)_8$ (*X* is Al, Mg, Fe, Ca, Na, K)	Sheet silicate	Generally microscopic crystals. Masses of clay minerals are softer than fingernail. Earthy lustre. Clay-like smell when damp.	Seen as a chemical weathering product of feldspars and most other silicate minerals. A constituent of most soils.
Dolomite	$CaMg(CO_3)_2$	Carbonate	Identical to calcite (rhombohedral cleavage, H = 3) except effervesces in weak acid only when pulverized.	Usually white, grey, or colourless. Vitreous lustre.
FELDSPAR (Most common group of minerals.)	Framework	Framework silicates	H = 6 (scratch glass) 2 good cleavage directions at about 90° to each other	Vitreous lustre but surface may be weathered to clay, giving an earthy lustre. Perfect crystal, shaped like an elongated box.
The group includes:				
Potassium feldspar (orthoclase)	$KAlSi_3O_8$		White, pink, or salmon-coloured.	Never has striations on cleavage surfaces.
Plagioclase (sodium and calcium feldspar)	Mixture of: $CaAl_2Si_2O_8$ and $NaAlSi_3O_8$		White, light to dark grey, rarely other colours. *May* have striations on cleavage surfaces.	Calcium-rich varieties generally a darker grey and may show an iridescent play of colours.
GARNET	$XAl_2Si_3O_{12}$ (*X* is a combination of Ca, Mg, Fe, Al, Mn)	Isolated silicate	No cleavage. Usually reddish brown. Tends to occur in perfect equidimensional crystals, usually 12 sided. H = 7.	Rarely yellow, green, or black. Usually found in metamorphic rocks. Vitreous lustre.
Gypsum	$CaSO_4{\cdot}2H_2O$	Sulphate	H = 2. 1 good and 2 perfect cleavage directions. Vitreous or silky lustre.	Clear, white, or pastel colours. Flexible cleavage fragments.

TABLE A.1 Diagnostic Properties of the Common Rock-Forming Minerals
Mineral groups are in uppercase letters. *(continued)*

Name (mineral groups shown in capitals)	Chemical Composition	Chemical Group	Diagnostic Properties	Other Properties
Halite	NaCl	Halide	3 excellent cleavage directions at 90° to each other (cubic). H = 2 1/2. Salty taste. Soluble in water.	Usually clear or white.
Hematite (*see* Ore mineral table)				
Hornblende (*see* Amphibole)				
Kaolinite (*see* Clay)				
MICA	$K(X)(AlSi_3O_{10})(OH)_2$	Sheet silicate	1 perfect cleavage direction (splits easily into flexible sheets).	H = 2–3. Vitreous lustre.
The group includes:				
Biolite	(*X* is Mg, Fe, and Al)		Black or dark brown.	
Muscovite	(*X* is Al)		White or transparent.	
Olivine	X_2SiO_4 (*X* is Fe, Mg)	Isolated silicate	No cleavage. Generally olive green or brown. H = 6–7 (scratches glass). Vitreous lustre.	Usually as small grains in mafic or ultramafic igneous rocks.
Orthoclase (*see* Feldspar)				
Plagioclase (*see* Feldspar)				
Pyrite ("fool's gold")	FeS_2	Sulphide	H = 6 (scratches glass). Bright, yellow, metallic lustre. Black streak.	Commonly occurs as perfect crystals: cubes or crystals with five-sided faces. Weathers to brown.
PYROXENE (A mineral group; *Augite* is most common member.)	$XSiO_3$ (*X* is Fe, Mg, Al, Ca)	Chain silicate	2 fair cleavage directions at 90° to each other.	H = 6. Augite is dark green to black. Vitreous lustre; usually stubby crystals.
Quartz	SiO_2	Framework silicate	H = 7. No cleavage. Vitreous lustre. Does not weather to clay.	Almost any colour but commonly white or clear. Good crystals have six-sided "column" with complex "pyramid" on top.
Serpentine	$Mg_6Si_4O_{10}(OH)_8$	Sheet silicate	Hardness variable but softer than glass. Mottled green and black. Greasy lustre. Fractures along smooth curved surfaces.	Sometimes fibrous (asbestos).

TABLE A.2 Diagnostic Properties of the Most Common Ore Minerals

Name	Chemical Composition	Diagnostic Properties	Other Properties
Azurite	$Cu_3(CO_3)_2(OH)_2$	Azure blue; effervesces in weak acid.	H = 3–4.
Bauxite	$Al_2O_3 \cdot nH_2O$	Earthy lustre. A variety of clay. Generally pea-sized spheres included in a fine-grained mass.	
Bornite	Cu_3FeS_4	Metallic lustre, tarnishes to iridescent purple colour.	Grey streak; H = 3 (softer than glass).
Chalcopyrite	$CuFeS_2$	Metallic lustre, brass-yellow. Softer than glass.	Black streak.
Cinnabar	HgS	Scarlet red, bright red streak.	Softer than glass. Generally an earthy lustre.
Galena	PbS	Metallic lustre, grey; 3 directions of cleavage at 90° (cubic). High specific gravity.	Softer than glass; grey streak.
Gold	Au	Metallic lustre, yellow. H = 3 (softer than glass, can be pounded into thin sheets, easily deformed).	Yellow streak; high specific gravity.
Halite	NaCl	Salty taste; 3 cleavage directions at 90° (cubic)	Clear or white; easily soluble in water.
Hematite	Fe_2O_3	Red-brown streak.	Either in earthy reddish masses or in metallic, silver-coloured flakes or crystals.
Limonite	$Fe_2O_3 \cdot nH_2O$	Earthy lustre; yellow-brown streak.	Yellow to brown colour; softer than glass.
Magnetite	Fe_3O_4	Metallic lustre, black; magnetic.	Harder than glass; black streak.
Malachite	$Cu_2(CO_3)(OH)_2$	Bright-green colour and streak.	Softer than glass; effervesces in weak acid.
Sphalerite	ZnS	Brown to yellow colour; 6 directions of cleavage.	Lustrelike resin; yellow or cream-coloured streak; softer than glass.
Talc	$Mg_3Si_4O_{10}(OH)_2$	White, grey, or green; softer than fingernail (H = 1).	Greasy feel.

APPENDIX B

IDENTIFICATION OF ROCKS

Igneous Rocks

Igneous rocks are classified on the basis of texture and composition. For some rocks, texture alone suffices for naming the rock. For most igneous rocks, composition as well as texture must be taken into account. Ideally, the mineral content of the rock should be used to determine composition; but for fine-grained igneous rocks, accurate identification of minerals may require a polarizing microscope or other special equipment. In the absence of such equipment, we rely on the colour of fine-grained rocks and assume the colour is indicative of the minerals present.

To identify a common igneous rock, use either table 6.1 or follow the key given below.

Key for Identifying Common Igneous Rocks

I. What is the texture of the rock?

- A. Is it glassy (a very vitreous lustre)? If so, it is *obsidian,* regardless of its chemical composition. Obsidian exhibits a pronounced conchoidal fracture.
- B. Does it have a frothy appearance? If so, it is *pumice.* Pumice is light in weight and feels abrasive (it probably will float on water).
- C. Does it have angular fragments of rock embedded in a volcanic-derived matrix? If so, it is a *volcanic breccia.* If the precise nature of the rock fragments and matrix can be identified, modifiers may be used; for instance, the rock may be an *andesite* breccia or a *rhyolite* breccia.
- D. Is the rock composed of interlocking, very coarse-grained minerals? (The minerals should be more than 1 centimetre across.) If so, the rock is a *pegmatite.* Most pegmatites are mineralogically equivalent to granite, with feldspars and quartz being the predominant minerals.
- E. Is the rock entirely coarse-grained? (That is, does it have an interlocking crystalline texture in which nearly all grains are more than 1 mm across?) If so, go to part II of this key.
- F. Is the rock *entirely* fine-grained? (Are grains less than 1 mm across or too fine to distinguish with the naked eye?) If so, go to part III of this key.
- G. Is the matrix fine-grained with some coarse-grained minerals visible in the rock? If so, go to part III and add the adjective *porphyritic* to the name of the rock.

II. Igneous rocks composed of interlocking coarse-grained minerals.

- A. Is quartz present? If so, the rock is a *granite.*
  Confirmation: Granite should be composed predominantly of feldspar—generally white, light grey, or pink (indicating high amounts of potassium or sodium in the feldspar). Rarely are there more than 20% ferromagnesian minerals in a granite.
- B. Are quartz and feldspar absent? If so, the rock should be composed entirely of ferromagnesian minerals and is *ultramafic.*
  Confirmation: Identify the minerals as being olivine or pyroxene (or less commonly, amphibole or biotite).
- C. Does the rock have less than 50% feldspar and no quartz? If so, the rock should be a *gabbro.*
  Confirmation: Most of the rock should be ferromagnesian minerals. Plagioclase can be medium or dark grey. There would be no pink feldspars.
- D. Is the rock composed of 30% to 60% feldspar (and no quartz)? If so, the rock is a *diorite.*
  Confirmation: Feldspar (plagioclase) is usually white to medium grey but never pink.

III. Igneous rocks that are fine-grained.

- A. Can quartz be identified in the rock? If so, the rock is a *rhyolite.*
- B. If the rock is too fine-grained for you to determine whether quartz is present but is white, light grey, pink, or pale green, the rock is most likely a *rhyolite.*
- C. Is the rock composed predominantly of ferromagnesian minerals? If so, the rock is *basalt.*
- D. If the rock is too fine-grained to identify ferromagnesian minerals but is black or dark grey, the rock is probably a *basalt.*
  1. Does the rock have rounded holes in it? If so, it is a *vesicular basalt* or *scoria.*
- E. Is the rock composed of roughly equal amounts of white or grey feldspar and ferromagnesian minerals (but no quartz)? If so, the rock is an *andesite.*
  Confirmation: Most andesite is porphyritic, with numerous identifiable crystals of white or light grey feldspar and lesser amounts of hornblende crystals within the darker, fine-grained matrix. Andesite is usually medium to dark grey or green.

Sedimentary Rocks

The following key shows how sedimentary rocks are classified on the basis of texture and composition. The descriptions of the rocks in the main body of the text provide additional information, such as common rock colours. *Equipment* needed for identification of sedimentary rocks includes a bottle of dilute hydrochloric acid, a hand lens or magnifying glass, a millimetre scale, a glass plate for hardness tests, and a pocketknife or rock hammer.

Begin by testing the rock for carbonate minerals by applying a small amount of dilute hydrochloric acid to the surface of the rock.

1. The rock does not effervesce (fizz) in acid, or effervesces weakly, but when powdered by a knife or hammer, the powder effervesces strongly. If so, the rock is *dolomite.*
2. The rock does not effervesce at all, even when powdered, or effervesces only in some places, such as the cement between grains. Go to part I of this key.
3. The rock effervesces strongly. The rock is *limestone.* Go to part II of this key to determine limestone type.

I. With a hand lens or magnifying glass, determine if the rock has a clastic texture (grains cemented together) or a crystalline texture (visible, interlocking crystals).

- A. If clastic:
  1. Most grains are more than 2 mm in diameter.
     - a. Angular grains *sedimentary breccia.*
     - b. Rounded grains—*conglomerate.*
  2. Most grains are between 1/16 and 2 mm in diameter. Rock feels gritty to the fingers. *Sandstone.*
     - a. More than 90% of the grains are quartz—*quartz sandstone.*
     - b. More than 25% of the grains are feldspar—*arkose.*
     - c. More than 25% of the grains are fine-grained rock fragments, such as shale, slate, and basalt—*lithic sandstone.*
     - d. More than 15% of the rock is fine-grained matrix—*graywacke*
  3. Rock is fine-grained (grains less than 1/16 mm in diameter). Feels smooth to fingers.
     - a. Grains visible with a hand lens—*siltstone.*
     - b. Grains too small to see, even with a hand lens.
       1. Rock is laminated, fissile—*shale.*
       2. Rock is unlayered, blocky—*mudstone.*
- B. If crystalline:
  1. Crystals fine to coarse, hardness of 2—*rock gypsum.*
  2. Coarse crystals that dissolve in water—*rock salt.*
- C. Hard to determine if clastic or crystalline:
  1. Very fine-grained, smooth to touch, conchoidal fracture, hardness of 6 (scratches glass), nonporous—*chert* (*flint* if dark)
  2. Very fine-grained, smooth to touch, breaks into flat chips—*shale.*
  3. Black or dark brown, readily broken, soils fingers—*coal.*

II. *Limestone* may be clastic or crystalline, fine- or coarse-grained, and may or may not contain visible fossils. Usually grey, tan, buff, or white. Some distinctive varieties are:

- A. *Bioclastic limestone*—clastic texture, grains are whole or broken fossils. Two relatively rare varieties are:
  1. *Coquina*—very coarse, recognizable shells, much open pore space.
  2. *Chalk*—very fine-grained, white or tan, soft and powdery.
- B. *Oolitic limestone*—grains are small spheres (less than 2 mm in diameter), all about the same size.
- C. *Travertine*—coarsely crystalline, no pore space, often contains different-coloured layers (bands).

Metamorphic Rocks

The characteristics of a metamorphic rock are largely governed by (1) the composition of the parent rock and (2) the particular combination of temperature, confining pressure, and directed pressure. These factors cause different textures in rocks formed under different sets of conditions. For this reason, texture is usually the main basis for naming a metamorphic rock. Determining the composition (e.g., mineral content) is necessary for naming some rocks (e.g., *quartzite*), but for others the minerals present are used as adjectives to describe the rock completely (e.g., *biotite* schist).

Metamorphic rocks are identified by determining first whether the rock has a *foliated* or *nonfoliated* texture.

Key for Identifying Metamorphic Rocks

I. If the rock is *nonfoliated,* then it is identified on the basis of its mineral content.

- A. Does the rock consist of mostly quartz? If so, the rock is a *quartzite.* A quartzite has a mosaic texture of interlocking grains of quartz and will easily scratch glass.
- B. Is the rock composed of interlocking coarse grains of calcite? If so, it is *marble.* (The individual grains should exhibit rhombohedral cleavage; the rock is softer than glass.)
- C. Is the rock a dense, dark mass of grains mostly too fine to identify with the naked eye? If so, it probably is a *hornfels.* A hornfels may have a few larger crystals of uncommon minerals enclosed in the fine grained mass.

II. If the rock is *foliated,* determine the type of foliation and then, if possible, identify the minerals present.

- A. Is the rock very fine-grained and does it split into sheetlike slabs? If so, it is *slate.* Most slate is composed of extremely fine-grained sheet silicate minerals, and the rock has an earthy lustre.
- B. Does the rock have a silky sheen but otherwise appear similar to slate? If so, it is a *phyllite.*
- C. Is the rock composed mostly of visible grains of platy or needle like minerals that are approximately parallel to one another? If so, the rock is a *schist.* If the rock is composed mainly of mica, it is a *mica schist.* If it also contains garnet, it is called a *garnet mica schist.* If hornblende is the predominant mineral, the rock is a *hornblende schist.* If talc prevails, it is a *talc schist* (sometimes called soapstone).
- D. Are dark and light minerals found in separate lenses or layers? If so, the rock is a *gneiss.* The light layers are composed of feldspars and perhaps quartz, whereas the darker layers commonly are formed of biotite, amphibole, or pyroxene. A gneiss may appear similar to granite or diorite but can be distinguished from the igneous rocks by the foliation.

APPENDIX C

THE ELEMENTS MOST SIGNIFICANT TO GEOLOGY

TABLE C.1

Atomic Number	Name	Symbol	Atomic Weight	Some Usual Charge of Ions	Atomic Number	Name	Symbol	Atomic Weight	Some Usual Charge of Ions
1	Hydrogen	H	1.0	+1	29	Copper	Cu	63.5	+2
2	Helium	He	4.0	0 inert	30	Zinc	Zn	65.4	+2
3	Lithium	Li	6.9	+1	33	Arsenic	As	74.9	+3
4	Beryllium	Be	9.0	+2	35	Bromine	Br	79.9	—
5	Boron	B	10.8	+3	37	Rubidium	Rb	85.5	+1
6	Carbon	C	12.0	+4	38	Strontium	Sr	87.3	+2
7	Nitrogen	N	14.0	+5	40	Zirconium	Zr	91.2	—
8	Oxygen	O	16.0	−2	42	Molybdenum	Mo	95.9	+4
9	Fluorine	F	19.0	−1	47	Silver	Ag	107.9	+1
10	Neon	Ne	20.2	0 inert	48	Cadmium	Cd	112.4	—
11	Sodium	Na	23.0	+1	50	Tin	Sn	118.7	+4
12	Magnesium	Mg	24.3	+2	51	Antimony	Sb	121.8	+3
13	Aluminum	Al	27.0	+3	52	Tellurium	Te	127.6	—
14	Silicon	Si	28.1	+4	55	Cesium	Cs	132.9	—
15	Phosphorus	P	31.0	+5	56	Barium	Ba	137.4	+2
16	Sulphur	S	32.1	−2	60	Neodymium	Nd	144	+3
17	Chlorine	Cl	35.5	−1	62	Samarium	Sm	150	+3
18	Argon	Ar	39.9	0 inert	74	Tungsten	W	183.9	—
19	Potassium	K	39.1	+1	78	Platinum	Pt	195.2	—
20	Calcium	Ca	40.1	+2	79	Gold	Au	197.0	—
22	Titanium	Ti	47.9	+4	80	Mercury	Hg	200.6	+2
23	Vanadium	V	50.9		82	Lead	Pb	207.2	+2
24	Chromium	Cr	52.0		83	Bismuth	Bi	209.0	—
25	Manganese	Mn	54.9	+4, +3	86	Radon	Rn	222	0 inert
26	Iron	Fe	55.8	+2, +3	88	Radium	Ra	226.1	
27	Cobalt	Co	58.9		90	Thorium	Th	232.1	—
28	Nickel	Ni	58.7	+2	92	Uranium	U	238.1	—
					94	Plutonium	Pu	239.0	—

APPENDIX D

PERIODIC TABLE OF ELEMENTS

6	Atomic number
C	Symbol
12.01	Atomic weight

Legend: Representative elements; Noble gases; Transition metals; Metalloids; Lanthanides; Actinides

Transition metals: groups 3–12

1 Group IA	2 IIA	3 IIIB	4 IVB	5 VB	6 VIB	7 VIIB	8 VIIIB	9 VIIIB	10 VIIIB	11 IB	12 IIB	13 IIIA	14 IVA	15 VA	16 VIA	17 VIIA	18 VIIIA
1 H 1.008																	2 He 4.00
3 Li 6.94	4 Be 9.01											5 B 10.81	6 C 12.01	7 N 14.01	8 O 16.00	9 F 19.00	10 Ne 20.18
11 Na 22.99	12 Mg 24.31											13 Al 26.98	14 Si 28.09	15 P 30.97	16 S 32.06	17 Cl 35.45	18 Ar 39.95
19 K 39.10	20 Ca 40.08	21 Sc 44.96	22 Ti 47.90	23 V 50.94	24 Cr 52.00	25 Mn 54.94	26 Fe 55.85	27 Co 58.93	28 Ni 58.69	29 Cu 63.54	30 Zn 65.37	31 Ga 69.72	32 Ge 72.61	33 As 74.92	34 Se 78.96	35 Br 79.91	36 Kr 83.80
37 Rb 85.47	38 Sr 87.62	39 Y 88.91	40 Zr 91.22	41 Nb 92.91	42 Mo 95.94	43 Tc (98)	44 Ru 101.07	45 Rh 102.90	46 Pd 106.42	47 Ag 107.87	48 Cd 112.41	49 In 114.82	50 Sn 118.69	51 Sb 121.75	52 Te 127.60	53 I 126.90	54 Xe 131.29
55 Cs 132.91	56 Ba 137.34	57 La 138.91	72 Hf 178.49	73 Ta 180.95	74 W 183.85	75 Re 186.21	76 Os 190.2	77 Ir 192.22	78 Pt 195.08	79 Au 196.97	80 Hg 200.59	81 Tl 204.37	82 Pb 207.19	83 Bi 208.98	84 Po (209)	85 At (210)	86 Rn (222)
87 Fr (223)	88 Ra 226.03	89 Ac 227.03	104 Rf (261)	105 Ha (262)	106 Sg (263)	107 Ns (262)	108 Hs (265)	109 Mt (266)									

58 Ce 140.12	59 Pr 140.91	60 Nd 144.24	61 Pm (145)	62 Sm 150.36	63 Eu 151.96	64 Gd 157.25	65 Tb 158.92	66 Dy 162.50	67 Ho 164.93	68 Er 167.26	69 Tm 168.93	70 Yb 173.04	71 Lu 174.97
90 Th 232.04	91 Pa 231.04	92 U 238.03	93 Np (237)	94 Pu (244)	95 Am (243)	96 Cm (247)	97 Bk (247)	98 Cf (251)	99 Es (252)	100 Fm (257)	101 Md (258)	102 No (259)	103 Lr (260)

Elements with an atomic number greater than 92 are not naturally occurring.

APPENDIX E

SELECTED CONVERSION FACTORS

TABLE E.1

	English Unit	Conversion Factor	Metric Unit	Conversion Factor	English Unit
Length and Distance	inch (in)	2.54	centimetres (cm)	0.4	inch (in)
	foot (ft)	0.3048	metre (m)	3.28	feet (ft)
	inch (in)	0.026	metre (m)	39.4	inches (in)
	mile, statute (mi)	1.61	kilometres (km)	0.62	mile (mi)
Area	square inch (in^2)	6.45	square centimetres (cm^2)	0.16	square inch (in^2)
	square foot (ft^2)	0.093	square metre (m^2)	10.8	square feet (ft^2)
	square mile (mi^2)	2.59	square kilometres (km^2)	0.39	square mile (mi^2)
	acre	0.4	hectare	2.47	acres
Volume	cubic inch (in^3)	16.4	cubic centimetres (cm^3)	0.06	cubic inch (in^3)
	cubic yard (yd^3)	0.76	cubic metre (m^3)	1.3	cubic yards (yd^3)
	cubic foot (ft^3)	0.0283	cubic metre (m^3)	35.5	cubic feet (ft^3)
	quart (qt)	0.95	litre	1.06	quarts (qt)
Weight	ounce (oz)	28.3	grams (g)	0.04	ounce (oz)
	pound (lb)	0.45	kilogram (kg)	2.2	pounds (lb)
	ton, short (2,000 lb)	907	kilograms (kg)	0.001	ton, short
	ton, short	0.91	tonne, metric	1.1	ton, short
Temp.	degrees Fahrenheit (°F)	−32° × 5/9	degrees Celsius (°C) (centigrade)	× 1.8 + 32°	degrees Fahrenheit (°F)

APPENDIX F

ROCK SYMBOLS

Shown below are the rock symbols used in the text. In general, these symbols are used by all geologists, although they sometimes are modified slightly.

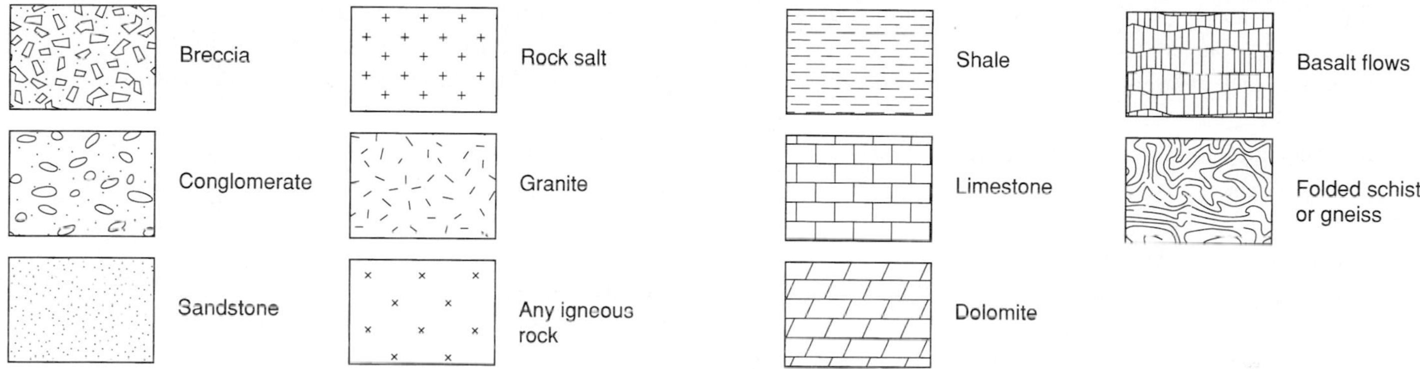

APPENDIX G

COMMONLY USED PREFIXES, SUFFIXES, AND ROOTS

abyss deep (Greek)
alluvium deposited by flowing water (Latin)
anti- opposite (Greek)
archea (archaeo)- ancient (Greek)
astheno- weak, lack of strength (Greek)
ceno recent (Greek)
circum- about, around, round about (Latin)
clast broken (Greek)
-cline tilted, gradient (Greek)
de- lower, reduce, take away (Latin)
dis- separation, opposite of (Latin)
ex- out of, away from (Greek)
feld field (Swedish, German)
folium leaf (Latin)
geo- Earth (Greek)
glomero- cluster (Latin)
hydro- water (Greek)
iso- equal (Greek)
-lith stone or rock (Greek)
meso- middle (Greek)
meta- change (Greek)
-morph form, shape (Greek)
paleo- ancient (Greek)
ped- foot (Latin)
pelagic pertaining to the ocean (Greek)
petro- stone or rock (Greek)
phanero- visible, evident (Greek)
pheno- large, conspicuous ("to show" in Greek)
pluto- deep-seated (from Roman god of the underworld or infernal regions)
pre- before, in front (Latin)
proto- first, primary, primitive (Greek)
pyro- fire (Greek)
spar crystalline material (German)
-sphere ball (Greek)
stria small groove, streak, band (Latin)
sub- under, less than (Latin)
super- above, more than, in addition to (Latin)
syn- together, at the same time (Latin, Greek)
tecto- means building or constructing in Greek and Latin; in geology it means movement or structures caused by internal forces.
terra, terre pertaining to the Earth (Latin)
thermal pertaining to heat (Greek)
trans- over, beyond, through, across (Latin)
xeno- strange, foreign (Greek)
zoo, zoic- animal (Greek)

GLOSSARY

A

ablation The loss of glacial ice or snow by melting, evaporation, or breaking off into icebergs; also called *wastage*.

abrasion The grinding away of rock by friction and impact during transportation.

abyssal plain Very flat sediment-covered region of the deep-sea floor, usually at the base of the continental rise.

Acasta Gneiss Formed 4,000 million years ago in the Northwest Territories, now part of the Slave Province of the Canadian Shield.

accretion The process of building large bodies of matter through collisions and gravitational attraction.

active continental margin A margin consisting of a continental shelf, a continental slope, and an oceanic trench.

advancing glacier Glacier with a positive budget, so that accumulation results in the lower edges being pushed outward and downward.

aftershock Small earthquake that follows a main shock.

A horizon The top layer of soil, characterized by the downward movement of water; also called *zone of leaching*.

alluvial fan Large fan-shaped pile of sediment that usually forms where a stream's velocity decreases as it emerges from a narrow canyon onto a flat plain at the foot of a mountain range.

alpine glaciation Glaciation of a mountainous area.

andesite Fine-grained igneous rock of intermediate composition. Up to half of the rock is plagioclase feldspar with the rest being ferromagnesian minerals.

angle of dip A vertical angle measured downward from the horizontal plane to an inclined plane.

angular unconformity An unconformity in which younger strata overlie an erosion surface on tilted or folded layered rock.

anomaly A deviation from average readings of magnetic strength.

anticline An arched fold in which the rock layers usually dip away from the axis of the fold.

aquifer A body of saturated rock or sediment through which water can move readily.

arch (sea arch) Bridge of rock left above an opening eroded in a headland by waves.

Archean Eon The oldest eon of the Earth's history.

Arctica The original North American continent.

arête A sharp ridge that separates adjacent glacially carved valleys.

artesian well A well in which water rises above the aquifer.

aseismic ridge Submarine ridge with which no earthquakes are associated.

asthenosphere A region of the Earth's outer shell beneath the lithosphere. The asthenosphere is of indeterminate thickness and behaves plastically.

atoll A circular reef surrounding a deeper lagoon.

atom Smallest possible particle of an element that retains the properties of that element.

atomic mass number The total number of neutrons and protons in an atom.

atomic number The total number of protons in an atom.

avalanche The fastest type of mass movement by flow.

axial plane A plane containing all of the hinge lines of a fold.

B

bajada A broad, gently sloping, depositional surface formed at the base of a mountain range in a dry region by the coalescing of individual alluvial fans.

bar A ridge of sediment, usually sand or gravel, that has been deposited in the middle or along the banks of a stream by a decrease in stream velocity.

barchan A crescent-shaped dune with the horns of the crescent pointing downwind.

barrier island Ridge of sand parallelling the shoreline and extending above sea level.

barrier reef A reef separated from the shoreline by the deeper water of a lagoon.

basal sliding Movement in which the entire glacier slides along as a single body on its base over the underlying rock.

basalt A fine-grained, mafic, igneous rock composed predominantly of ferromagnesian minerals and with lesser amounts of calcium-rich plagioclase feldspar.

base level A theoretical downward limit for stream erosion of the Earth's surface.

batholith A large discordant pluton with an outcropping area greater than 100 square kilometres.

baymouth bar A ridge of sediment that cuts a bay off from the ocean.

beach Strip of sediment, usually sand but sometimes pebbles, boulders, or mud, that extends from the low-water line inland to a cliff or zone of permanent vegetation.

beach face The section of the beach exposed to wave action.

bedding An arrangement of layers or beds of rock.

bedding plane A nearly flat surface separating two beds of sedimentary rock.

bed load Heavy or large sediment particles in a stream that travel near or on the stream bed.

Benioff zone Distinct earthquake zone that begins at an oceanic trench and slopes landward and downward into Earth at an angle of about 30° to 60°.

berm Platform of wave-deposited sediment that is flat or slopes slightly landward.

B horizon A soil layer characterized by the accumulation of material leached downward from the A horizon above; also called *zone of accumulation*.

block Large angular pyroclast.

blowout A depression on the land surface caused by wind erosion.

body wave Seismic wave that travels through the Earth's interior.

bomb Large spindle- or lens-shaped pyroclast.

bonding Attachment of an atom to one or more adjacent atoms.

Bowen's reaction series The sequence in which minerals crystallize from a cooling basaltic magma.

braided stream A stream that flows in a network of many interconnected rivulets around numerous bars.

breaker A wave that has become so steep that the crest of the wave topples forward, moving faster than the main body of the wave.

brittle strain Cracking or rupturing of a body under stress.

butte A narrow pinnacle of resistant rock with a flat top and very steep sides.

C

caldera A volcanic depression much larger than the original crater.

Cambrian Explosion Proliferation of early lifeforms such as those of the Burgess Shale and the Mistaken Point Formations.

Canadian Cordillera Land masses swept up and accreted onto western North America.

Canadian Shield The exposed part of the North American craton; consists of a gently undulating surface that rises inconspicuously in its centre.

cave (cavern) Naturally formed underground chamber.

cement The solid material that precipitates in the pore space of sediments, binding the grains together to form solid rock.

cementation The chemical precipitation of material in the spaces between sediment grains, binding the grains together into a hard rock.

Cenozoic Era The most recent of the eras; followed the Mesozoic Era.

chain silicate structure Silicate structure in which two of each tetrahedron's oxygen ions are shared with adjacent tetrahedrons, resulting in a chain of tetrahedrons.

chemical sedimentary rock A rock composed of material precipitated directly from solution.

chemical weathering The decomposition of rock resulting from exposure to water and atmospheric gases.

chert A hard, compact, fine-grained sedimentary rock formed almost entirely of silica.

chill zone In an intrusion, the finer-grained rock adjacent to a contact with country rock.

C horizon A soil layer composed of incompletely weathered parent material.

circum-Pacific belt Major belt around the edge of the Pacific Ocean on which most composite volcanoes are located and where many earthquakes occur.

cirque A steep-sided, amphitheatre-like hollow carved into a mountain at the head of a glacial valley.

clastic sedimentary rock A sedimentary rock composed of fragments of pre-existing rock.

clastic texture An arrangement of rock fragments bound into a rigid network by cement.

clay Sediment composed of particles with diameter less than 1/256 mm.

clay mineral A hydrous aluminum-silicate that occurs as a platy grain of microscopic size with a sheet silicate structure.

cleavage The ability of a mineral to break along preferred planes.

coal A sedimentary rock formed from the consolidation of plant material. It is rich in carbon, usually black, and burns readily.

coal bed methane Water and natural gas trapped in the interior of coal.

coarse-grained rock Rock in which most of the grains are larger than 1 millimetre (igneous) or 2 millimetres (sedimentary).

coast The land near the sea, including the beach and a strip of land inland from the beach.

coastal straightening The gradual straightening of an irregular shoreline by wave erosion of headlands and wave deposition in bays.

columnar structure Volcanic rock in parallel, usually vertical columns, mostly six-sided; also called *columnar jointing*.

compaction A loss in overall volume and pore space of a rock as the particles are packed closer together by the weight of overlying material.

composite volcano (stratovolcano) A volcano constructed of alternating layers of pyroclastics and rock solidified from lava flows.

compressive stress A stress due to a force pushing together on a body.

concretion Hard, rounded mass that develops when a considerable amount of cementing material precipitates locally in a rock, often around an organic nucleus.

cone of depression A depression of the water table formed around a well when water is pumped out; it is shaped like an inverted cone.

confined aquifer (artesian aquifer) An aquifer completely filled with pressurized water and separated from the land surface by a relatively impermeable confining bed, such as shale.

confining pressure Pressure applied equally on all surfaces of a body; also called *geostatic* or *lithostatic pressure*.

conglomerate A coarse-grained sedimentary rock (grains coarser than 2 mm) formed by the cementation of rounded gravel.

contact Boundary surface between two different rock types or ages of rocks.

contact (thermal) metamorphism Metamorphism under conditions in which high temperature is the dominant factor.

continental drift A concept suggesting that continents move over the Earth's surface.

continental glaciation The covering of a large region of a continent by a sheet of glacial ice.

continental rise A wedge of sediment that extends from the lower part of the continental slope to the deep-sea floor.

continental shelf A submarine platform at the edge of a continent, inclined very gently seaward generally at an angle of less than 1°.

continental slope A relatively steep slope extending from a depth of 100 to 200 metres at the edge of the continental shelf down to oceanic depths.

contour current A bottom current that flows parallel to the slopes of the continental margin (along the contour rather than down the slope).

convection (convection current) A very slow circulation of a substance driven by differences in temperature and density within that substance.

convergent plate boundary A boundary between two plates that are moving toward each other.

core The central zone of the Earth.

correlation In geology, correlation usually means determining time equivalency of rock units. Rock units may be correlated within a region, a continent, and even between continents.

country rock Any rock that was older than and intruded by an igneous body.

covalent bonding Bonding due to the sharing of electrons by adjacent atoms.

cover strata Younger sedimentary strata that bury the more ancient craton around its margins.

crater (of a volcano) A basinlike depression over a vent at the summit of a volcanic cone.

craton Portion of a continent that has been structurally stable for a prolonged period of time.

creep Very slow, continuous downslope movement of soil or debris.

crest (of wave) The high point of a wave.

crevasse Open fissure in a glacier.

cross-bedding An arrangement of relatively thin layers of rock inclined at an angle to the more nearly horizontal bedding planes of the larger rock unit.

cross-cutting relationship A principle or law stating that a disrupted pattern is older than the cause of disruption.

cross-section *See* geologic cross-section.

crude oil A liquid mixture of naturally occurring hydrocarbons.

crust The outer layer of rock, forming a thin skin over the Earth's surface.

crustal rebound The rise of the Earth's crust after the removal of glacial ice.

crystal form Arrangement of various faces on a crystal in a definite geometric relationship to one another.

crystalline Describing a substance in which the atoms are arranged in a regular, repeating, orderly pattern.

crystalline texture An arrangement of interlocking crystals.

crystallization Crystal development and growth.

crystal settling The process whereby the minerals that crystallize at a high temperature in a cooling magma move downward in the magma chamber because they are denser than the magma.

Curie point The temperature below which a material becomes magnetized.

D

decompression melting Occurs when a body of hot mantle rock moves upward and the pressure is reduced to the extent that the melting point drops to the temperature of the body.

debris Any unconsolidated material at the Earth's surface.

debris fall A free-falling mass of debris.

debris flow Mass wasting in which motion is taking place throughout the moving mass (flow). The common varieties are earthflow, mudflow, and debris avalanche.

debris slide Rapid movement of debris as a coherent mass.

deflation The removal of clay, silt, and sand particles from the land surface by wind.

delta A body of sediment deposited at the mouth of a river when the river velocity decreases as it flows into a standing body of water.

dendritic pattern Drainage pattern of a river and its tributaries, which resembles the branches of a tree or veins in a leaf.

density Weight per given volume of a substance.

deposition The settling or coming to rest of transported material.

depth of focus Distance between the focus and the epicentre of an earthquake.

desert A region with low precipitation (usually defined as less than 25 cm per year).

diapir Bodies of rock (e.g., rock salt) or magma that ascend within the Earth's interior because they are less dense than the surrounding rock.

differential stress When pressures on a body are not of equal strength in all directions.

differential weathering Varying rates of weathering resulting from some rocks in an area being more resistant to weathering than others.

differentiation Separation of different ingredients from an originally homogeneous mixture.

dike A tabular, discordant intrusive structure.

diorite Coarse-grained igneous rock of intermediate composition. Up to half of the rock is plagioclase feldspar and the rest is ferromagnesian minerals.

dip *See* angle of dip, direction of dip.

dip-slip fault A fault in which movement is parallel to the dip of the fault surface.

directed pressure *See* differential stress.

direction of dip The compass direction in which the angle of dip is measured.

discharge In a stream, the volume of water that flows past a given point in a unit of time.

disconformity A surface that represents missing rock strata but beds above and below that surface are parallel to one another.

dissolved load The portion of the total sediment load in a stream that is carried in solution.

distributary Small shifting river channel that carries water away from the main river channel and distributes it over a delta's surface.

divergent plate boundary Boundary separating two plates moving away from each other.

divide Line dividing one drainage basin from another.

dolomite A sedimentary rock composed mostly of the mineral dolomite.

downcutting A valley-deepening process caused by erosion of a stream bed.

drainage basin Total area drained by a stream and its tributaries.

drainage pattern The arrangement in map view of a river and its tributaries.

drawdown The lowering of the water table near a pumped well.

drumlin A long, streamlined hill made of till.

ductile Capable of being moulded and bent under stress.

E

Earth system A small part of the larger solar system that has its own component parts, or subsystems: the atmosphere, the hydrosphere, the biosphere, and the geosphere.

earthflow Slow-to-rapid mass wasting in which debris moves downslope as a very viscous fluid.

earthquake A trembling or shaking of the ground caused by the sudden release of energy stored in the rocks beneath the surface.

earthy lustre A lustre giving a substance the appearance of unglazed pottery.

elastic The ability of a deformed body to recover its original shape after the reduction or removal of stress.

elastic limit The maximum amount of stress that can be applied to a body before it deforms in a permanent way by bending or breaking.

elastic rebound theory The sudden release of progressively stored strain in rocks results in movement along a fault.

electron A single, negative electric charge that contributes virtually no mass to an atom.

element A substance that cannot be broken down to other substances by ordinary chemical methods. Each atom of an element possesses the same number of protons.

end moraine A ridge of till piled up along the front edge of a glacier.

environment of deposition The location in which deposition occurs, usually marked by characteristic physical, chemical, or biological conditions.

eon The largest unit of geological time.

epicentre The point on the Earth's surface directly above the focus of an earthquake.

epoch Each period of the standard geologic time scale is divided into epochs (e.g., Pleistocene Epoch of the Quaternary Period).

equilibrium line An irregular line marking the highest level to which the winter snow cover on a glacier is lost during a melt season. (Also called *snow line*.)

era Major subdivision of the standard geologic time scale (e.g., Mesozoic Era).

erosion The physical removal of rock by an agent such as running water, glacial ice, or wind.

erratic An ice-transported boulder that does not derive from bedrock near its present site.

esker A long, sinuous ridge of sediment deposited by glacial meltwater.

estuary Drowned river mouth.

evaporite Rock that forms from crystals precipitating during evaporation of water.

exfoliation The stripping of concentric rock slabs from the outer surface of a rock mass.

exfoliation dome A large, rounded landform developed in a massive rock, such as granite, by the process of exfoliation.

exploration geologist Geoscientists who work for exploration companies looking for gold, silver, or diamonds.

extrusive rock Any igneous rock that forms at the Earth's surface, whether it solidifies directly from a lava flow or is pyroclastic.

F

fall The situation in mass wasting that occurs when material free-falls or bounces down a cliff.

fault A fracture in bedrock along which movement has taken place.

faunal succession A principle or law stating that fossil species succeed one another in a definite and recognizable order; in general, fossils in progressively older rock show increasingly greater differences from species living at present.

feldspars Group of most common minerals of the Earth's crust. All feldspars contain silicon, aluminum, and oxygen and may contain potassium, calcium, and sodium.

ferromagnesian mineral Iron/magnesium-bearing mineral, such as augite, hornblende, olivine, or biotite.

fine-grained rock A rock in which most of the mineral grains are less than 1 millimetre across (igneous) or less than 1/16 mm (sedimentary).

fjord A coastal inlet that is a glacially carved valley, the base of which is submerged.

flank eruption An eruption in which lava erupts out of a vent on the side of a volcano.

flash flood Flood of very high discharge and short duration; sudden and local in extent.

flood plain A broad strip of land built up by sedimentation on either side of a stream channel.

flow A type of movement that implies that a descending mass is moving downslope as a viscous fluid.

focus The point within the Earth from which seismic waves originate in an earthquake.

fold Bend in layered bedrock.

foliation Parallel alignment of textural and structural features of a rock.

footwall The underlying surface of an inclined fault plane.

forearc basin Offshore accumulations of enormous quantities of sediment transported by streams.

foreland basin Inland seas in which shallow marine sediments accumulate.

formation A body of rock of considerable thickness that has a recognizable unity or similarity making it distinguishable from adjacent rock units. Usually composed of one bed or several beds of sedimentary rock, although the term is also applied to units of metamorphic and igneous rock. A convenient unit for mapping, describing, or interpreting the geology of a region.

fossil Traces of plants or animals preserved in rock.

fossil assemblage Various different species of fossils in a rock.

fracture The way a substance breaks where not controlled by cleavage.

framework silicate structure Crystal structure in which all four oxygen ions of a silica tetrahedron are shared by adjacent ions.

fringing reef A reef attached directly to shore. (*See* barrier reef.)

frost action Mechanical weathering of rock by freezing water.

frost heaving The lifting of rock or soil by the expansion of freezing water.

frost wedging A type of frost action in which the expansion of freezing water pries a rock apart.

G

gabbro A mafic, coarse-grained igneous rock composed predominantly of ferromagnesian minerals and with lesser amounts of calcium-rich plagioclase feldspar.

gaining stream A stream that receives water from the zone of saturation.

gas hydrates An unusual mixture of ice and gas in which methane is trapped in ice crystals.

gelifluction The flow of water-saturated debris over impermeable material.

geode Partly hollow, globelike body found in limestone or other cavernous rock.

geologic cross-section A representation of a portion of the Earth in a vertical plane.

geologic map A map representing the geology of a given area.

geologic province Nineteenth-century term for major geologic subdivisions of the North American craton.

geologic resources Valuable materials of geologic origin that can be extracted from the Earth.

geologic time scale A worldwide relative scale of geologic time divisions.

geology The scientific study of the planet Earth.

geomatician A geoscientist who collects, organizes, analyzes, and creates images from spatial and geographic data available in digital form.

geophysics The application of physical laws and principles to a study of the Earth.

geoscientist A professional who deals with environmental problems from the finding and managing of drinking water to managing radioactive waste.

geothermal gradient Rate of temperature increase associated with increasing depth beneath the surface of the Earth (normally about 25°C/km).

geyser A type of hot spring that periodically erupts hot water and steam.

glacial trough Landforms that, in cross-profile, have steep bedrock walls and a relatively flat floor.

glacier A large, long-lasting mass of ice, formed on land by the compaction and recrystallization of snow, which moves because of its own weight.

glassy (or vitreous) lustre A lustre that gives a substance a glazed, porcelainlike appearance.

gneiss A metamorphic rock composed of light and dark layers or lenses.

gneissic The texture of a metamorphic rock in which minerals are separated into light and dark layers or lenses.

Gowganda glaciation A major glaciation at the southern continental margin of Arctica.

graded bed A single bed with coarse grains at the bottom of the bed and progressively finer grains toward the top of the bed.

graded stream A stream that exhibits a delicate balance between its transporting capacity and the sediment load available to it.

granite A felsic, coarse-grained, intrusive igneous rock containing quartz and composed mostly of potassium- and sodium-rich feldspars.

gravel Rounded particles coarser than 2 mm in diameter.

gravity meter An instrument that measures the gravitational attraction between the Earth and a mass within the instrument.

Grenville Orogeny The result of a long-lived collision between ancestral South and North America between 1.3 and 1 billion years ago.

greywacke A sandstone with more than 15 percent fine-grained matrix between the sand grains.

groundwater The water that lies beneath the ground surface, filling the cracks, crevices, and pore space of rocks.

groundwater quality A measure of the suitability of groundwater for a certain purpose, determined by its pH, salinity, and hardness.

guyot Flat-topped seamount.

H

half-life The time it takes for a given amount of a radioactive isotope to be reduced by one-half.

hanging valley A smaller valley that terminates abruptly high above a main valley.

hanging wall The overlying surface of an inclined fault plane.

hardness The relative ease or difficulty with which a smooth surface of a mineral can be scratched; commonly measured by Mohs' scale.

headland Point of land along a coast.

headward erosion The lengthening of a valley in an uphill direction above its original source by gullying, mass wasting, and sheet erosion.

heat flow Gradual loss of heat (per unit of surface area) from the Earth's interior out into space.

heavy crude Dense, viscous petroleum that flows slowly or not at all.

hematite A type of iron oxide that has a brick-red colour when powdered; Fe_2O_3.

hinge line Line about which a fold appears to be hinged. Line of maximum curvature of a folded surface.

horn A sharp peak formed where cirques cut back into a mountain on several sides.

hornfels A fine-grained, unfoliated metamorphic rock.

hot spring Spring with a water temperature warmer than human body temperature.

hydraulic action The ability of water to pick up and move rock and sediment.

hydrologic cycle The movement of water and water vapour from the sea to the atmosphere, to the land, and back to the sea and atmosphere again.

hydrothermal rock Rock deposited by precipitation of ions from solution in hot water.

hypothesis A tentative theory.

I

Iapetus Ocean The ancestral Atlantic Ocean.

iceberg Block of glacier-derived ice floating in water.

ice cap A glacier covering a relatively small area of land but not restricted to a valley.

ice kame A low mound or irregular ridge of outwash deposits that were originally deposited along the margins or in front of a glacier.

ice sheet A glacier covering a large area (more than 50,000 square kilometres) of land.

ice wedge Landform found in areas underlain by permafrost.

igneous rock A rock formed or apparently formed from solidification of magma.

incised meander A meander that retains its sinuous curves as it cuts vertically downward below the level at which it originally formed.

inclusion A fragment of rock that is distinct from the body of igneous rock in which it is enclosed.

index fossil A fossil from a very short-lived species known to have existed during a specific period of geologic time.

intensity A measure of an earthquake's size by its effect on people and buildings.

intermediate rock Rock with a chemical content between felsic and mafic compositions.

intrusion (intrusive structure) A body of intrusive rock classified on the basis of size, shape, and relationship to surrounding rocks.

intrusive rock Rock that appears to have crystallized from magma emplaced in surrounding rock.

ion An electrically charged atom or group of atoms.

ionic bonding Bonding due to the attraction between positively charged ions and negatively charged ions.

island arc A curved line of islands.

isoclinal fold A fold in which the limbs are parallel to one another.

isolated silicate structure Silicate minerals that are structured so that none of the oxygen atoms are shared by silica tetrahedrons.

isostasy The balance or equilibrium between adjacent blocks of crust resting on a plastic mantle.

isostatic adjustment Concept of vertical movement of sections of the Earth's crust to achieve balance or equilibrium.

isotherm A line along which the temperature of rock (or other material) is the same.

isotopes Atoms (of the same element) that have different numbers of neutrons but the same number of protons.

isotopic dating Determining the age of a rock or mineral through its radioactive elements and decay products (previously and somewhat inaccurately called *radiometric* or *radioactive* dating).

J

joint A fracture or crack in bedrock along which essentially no displacement has occurred.

joint set Joints oriented in one direction approximately parallel to one another.

Jovian planets Planets that have low densities: Jupiter, Saturn, Uranus, and Neptune.

K

karst topography An area with many sinkholes and a cave system beneath the land surface and usually lacking a surface stream.

kettle A depression caused by the melting of a stagnant block of ice that was surrounded by sediment.

L

landslide The general term for a slowly to very rapidly descending mass of rock or debris.

lateral continuity Principle that states that an original sedimentary layer extends laterally until it tapers or thins at its edges.

lateral erosion Erosion and undercutting of stream banks caused by a stream swinging from side to side across its valley floor.

lateral moraine A low ridgelike pile of till along the side of a glacier.

laterite Highly leached soil that forms in regions of tropical climate with high temperatures and very abundant rainfall.

lava Magma on the Earth's surface.

lava flow In Hawaii, lava that extrudes out of fissures in the ground.

left-lateral fault A strike-slip fault in which the block seen across the fault appears displaced to the left.

LFH Designation for the uppermost layer of soil that consists of non-decomposed and highly decomposed organic material; L (leaf), F (needle), H (humus).

limb Portion of a fold shared by an anticline and a syncline.

limestone A sedimentary rock composed mostly of calcite.

limonite A type of iron oxide that is yellowish-brown when powdered; $Fe_2O_3 \cdot nH_2O$.

lithification The consolidation of sediment into sedimentary rock.

lithosphere The rigid outer shell of the Earth, 70 to 125 or more kilometres thick.

lithospheric plates Large pieces of the Earth's surface that move over the asthenosphere and interact with each other at plate boundaries.

loam Soil containing approximately equal amounts of sand, silt, and clay.

loess A fine-grained deposit of wind-blown dust.

longitudinal dune (seif) Large, symmetrical ridge of sand parallel to the wind direction.

longshore current A moving mass of water that develops parallel to a shoreline.

longshore drift Movement of sediment parallel to shore when waves strike a shoreline at an angle.

losing stream Stream that loses water to the zone of saturation.

Love waves A type of surface seismic wave that causes the ground to move side to side in a horizontal plane perpendicular to the direction the wave is travelling.

lustre The quality and intensity of light reflected from the surface of a mineral.

M

mafic rock Silica-deficient igneous rock with a relatively high content of magnesium, iron, and calcium.

magma Molten rock, usually mostly silica. The liquid may contain dissolved gases as well as some solid minerals.

magmatic arc A line of batholiths or volcanoes. Generally the line, as seen from above, is curved.

magnetic field Region of magnetic force that surrounds the Earth.

magnetic polarity time scale Records the pattern of magnetic reversals over time.

magnetic pole An area where the strength of the magnetic field is greatest and where the magnetic lines of force appear to leave or enter the Earth.

magnetic reversal A change in the Earth's magnetic field between normal polarity and reversed polarity. In normal polarity the north magnetic pole, where magnetic lines of force enter the Earth, lies near the geographic North Pole. In reversed polarity the south magnetic pole, where lines of force leave the Earth, lies near the geographic North Pole (the magnetic poles have exchanged positions).

magnetite An iron oxide that is attracted to a magnet.

magnetotellurics A new geophysical approach being used in remote regions of the Canadian Arctic to investigate and map structures within the underlying crust and mantle.

magnetometer An instrument that measures the strength of the Earth's magnetic field.

magnitude A measure of the energy released during an earthquake.

mantle A thick shell of rock that separates the Earth's crust above from the core below.

mantle plume Narrow column of hot mantle rock that rises and spreads radially outward.

marble A coarse-grained rock composed of interlocking calcite (or dolomite) crystals.

marine terrace A broad, gently sloping platform that may be exposed at low tide.

Maritimes Basin A series of interconnected mountains and basins that existed in the Maritime provinces from the late Devonian to the Permian.

mass wasting (or mass movement) Movement, caused by gravity, in which bedrock, rock debris, or soil moves downslope in bulk.

matrix Fine-grained material found in the pore space between larger sediment grains.

meander A pronounced sinuous curve along a stream's course.

meander cutoff A new, shorter channel across the narrow neck of a meander.

mechanical weathering The physical disintegration of rock into smaller pieces.

medial moraine A single long ridge of till on a glacier, formed by adjacent lateral moraines joining and being carried downglacier.

Mediterranean-Himalayan belt (Mediterranean belt) A major concentration of earthquakes and composite volcanoes that runs through the Mediterranean Sea, crosses the Mideast and the Himalaya, and passes through the East Indies.

megablock Large mass of rock moved by a glacier.

mesa A broad, flat-topped hill bounded by cliffs and capped with a resistant rock layer.

Mesozoic Era The era that followed the Paleozoic Era and preceded the Cenozoic Era.

metallic lustre Lustre giving a substance the appearance of being made of metal.

metamorphic rock A rock produced by metamorphism.

metamorphism The transformation of pre-existing rock into texturally or mineralogically distinct new rock as a result of high temperature, high pressure, or both, but without the rock melting in the process.

metasomatism Metamorphism coupled with the introduction of ions from an external source.

mid-oceanic ridge A giant mountain range that lies under the ocean and extends around the world.

migmatite Mixed igneous and metamorphic rock.

mineral A naturally occurring, crystalline solid that has a specific chemical composition.

modified Mercalli scale Scale expressing intensities of earthquakes (judged on amount of damage done) in Roman numerals ranging from I to XII.

Mohorovičić discontinuity The boundary separating the crust from the mantle beneath it (also called *Moho*).

Mohs' hardness scale Scale on which ten minerals are designated as standards of hardness.

moment magnitude An earthquake magnitude calculated from the strength of the rock, surface area of the fault rupture, and the amount of rock displacement along the fault.

moraine A body of till either being carried on a glacier or left behind after a glacier has receded.

mud crack Polygonal crack formed in very fine-grained sediment as it dries.

mudflow A flowing mixture of debris and water, usually moving down a channel.

muscovite Transparent or white mica that lacks iron and magnesium.

N

natural gas A gaseous mixture of naturally occurring hydrocarbons.

natural levee Low ridges of flood-deposited sediment formed on either side of a stream channel, which thin away from the channel.

nebula A large volume of interstellar gas and dust.

negative gravity anomaly Less than normal gravitational attraction.

negative magnetic anomaly Less than average strength of the Earth's magnetic field.

Nena A supercontinent that existed between 1.9 and 1.3 billion years ago. Acronym for Northern Europe and North America.

neutron A subatomic particle that contributes mass to an atom and is electrically neutral.

nonconformity An unconformity in which an erosion surface on plutonic or metamorphic rock has been covered by younger sedimentary or volcanic rock.

nonmetallic lustre Lustre that gives a substance the appearance of being made of something other than metal (e.g., glassy).

nonrenewable resource A resource that forms at extremely slow rates compared to its rate of consumption.

normal fault A fault in which the hanging-wall block moved down relative to the footwall block.

normal polarity A period (such as the present) when magnetic lines of force flow from the south pole to the north pole and compass needles point to the north.

North American craton The ancient continent-sized block of distinct geology that forms the basement of much of North America and Greenland.

nucleus Protons and neutrons form the nucleus of an atom. Although the nucleus occupies an extremely tiny fraction of the volume of the entire atom, practically all the mass of the atom is concentrated in the nucleus.

numerical age Age given in years or some other unit of time.

nunatak An isolated bedrock peak completely surrounded by glacial ice.

O

oblique-slip fault A fault with both strike-slip and dip-slip components.

obsidian Volcanic glass.

oceanic trench A narrow, deep trough parallel to the edge of a continent or an island arc.

O horizon Dark-coloured soil layer that is rich in organic material and forms just below surface vegetation.

oil *See* crude oil.

oil field An area underlain by one or more oil pools.

oil pool Underground accumulation of oil.

oil sand Asphalt-cemented sand deposit.

oil shale Shale with a high content of organic matter from which oil may be extracted by distillation.

oil trap A set of conditions that hold petroleum in a reservoir rock and prevent its escape by migration.

open fold A fold with gently dipping limbs.

open-pit mine Mine in which ore is exposed at the surface in a large excavation.

ophiolite A distinctive rock sequence found in many mountain ranges on continents.

ore Naturally occurring material that can be profitably mined.

ore mineral A mineral of commercial value.

organic sedimentary rock Rock composed mostly of the remains of plants and animals.

original horizontality The deposition of most water-laid sediment in horizontal or near-horizontal layers that are essentially parallel to the Earth's surface.

orogeny An episode of intense deformation of the rocks in a region, generally accompanied by metamorphism and plutonic activity.

outgassing Occurs when water and gaseous elements are released during volcanic eruptions.

outwash Material deposited by debris-laden meltwater from a glacier.

overturned fold A fold in which both limbs dip in the same direction.

oxbow lake A crescent-shaped lake occupying the abandoned channel of a stream meander that is isolated from the present channel by a meander cutoff and sedimentation.

P

paleomagnetism A study of ancient magnetic fields.

Paleozoic Era The era that followed the Precambrian and began with the appearance of complex life, as indicated by fossils.

Pangea A supercontinent that broke apart 200 million years ago to form the present continents.

parabolic dune A deeply curved dune in a region of abundant sand. The horns point upwind and are often anchored by vegetation.

parent rock Original rock before being metamorphosed.

partial melting Melting of the components of a rock with the lowest melting temperatures.

passive continental margin A margin that includes a continental shelf, continental slope, and continental rise that generally extends down to an abyssal plain at a depth of about 5 kilometres.

passive margin uplift Uplift of the margin of a continent caused by rifting.

pediment A gently sloping erosional surface cut into the solid rock of a mountain range in a dry region; usually covered with a thin veneer of gravel.

pelagic sediment Sediment made up of fine-grained clay and the skeletons of microscopic organisms that settle slowly down through the ocean water.

peneplain A nearly flat erosional surface presumably produced as mass wasting, sheet erosion, and stream erosion reduce a region almost to base level.

perched water table A water table separated from the main water table beneath it by a zone that is not saturated.

period Each era of the standard geologic time scale is subdivided into periods (e.g., the Cretaceous Period).

peridotite A plutonic rock composed mostly of olivine with little or no feldspar.

permafrost Ground that remains permanently frozen for many years.

permeability The capacity of a rock to transmit a fluid such as water or petroleum.

petrified wood A material that forms as the organic matter of buried wood is either filled in or replaced by inorganic silica carried in by groundwater.

petroleum Crude oil and natural gas. (Some geologists use petroleum as a synonym for oil.)

Phanerozoic Eon Segment of geologic time. Includes all time following the Precambrian.

phenocryst Any of the large crystals in porphyritic igneous rock.

phyllite A metamorphic rock in which clay minerals have recrystallized into microscopic micas, giving the rock a silky sheen.

physical continuity Being able to physically follow a rock unit between two places.

pillow structure Rocks, generally basalt, formed in pillow-shaped masses fitting closely together; caused by underwater lava flows.

pingo Landform found in areas underlain by permafrost

planetismals Smaller, irregularly shaped planets.

plastic Capable of being moulded and bent under stress.

plastic flow Movement within a glacier in which the ice is not fractured.

plate A large, mobile slab of rock making up part of the Earth's surface.

plateau Broad, flat-topped area elevated above the surrounding land and bounded, at least in part, by cliffs.

plateau basalts Layers of basalt flows that have built up to great thicknesses.

plate tectonics A theory that the Earth's surface is divided into a few large, thick plates that are slowly moving and changing in size. Intense geologic activity occurs at the plate boundaries.

playa A very flat surface underlain by hard, mud-cracked clay.

playa lake A shallow temporary lake (following a rainstorm) on a flat valley floor in a dry region.

Pleistocene Epoch An epoch of the Quaternary Period characterized by several glacial ages.

plunging fold A fold in which the hinge line (or axis) is not horizontal.

pluton An igneous body that crystallized deep underground.

plutonic rock Igneous rock formed at great depth.

pluvial lake A lake formed during an earlier time of abundant rainfall.

point bar A stream bar (*see* definition) deposited on the inside of a curve in the stream, where the water velocity is low.

polarity *See* magnetic reversal.

polar desert Areas characterized by low-relief bedrock or gravel plains underlain by permanently frozen ground.

polar wandering An apparent movement of the Earth's poles.

pore space The total amount of space taken up by openings between sediment grains.

porosity The percentage of a rock's volume that is taken up by openings.

porphyritic rock An igneous rock in which large crystals are enclosed in a matrix (or ground mass) of much finer-grained minerals or obsidian.

positive gravity anomaly High gravity values measured over an area underlain by denser rocks than those of the surrounding region (e.g., a granite pluton).

positive magnetic anomaly Greater than average strength of the Earth's magnetic field.

pothole Depression eroded into the hard rock of a stream bed by the abrasive action of the stream's sediment load.

Precambrian The vast amount of time that preceded the Paleozoic Era.

Prearchean (Hadean) Eon Along with the Archean and the Proterozoic, one of the three subdivisions of the Precambrian.

pressure release A significant type of mechanical weathering that causes rocks to crack when overburden is removed.

prokaryotes Microorganisms that are the earliest life forms preserved in the geologic record.

Proterozoic Eon Segment of Precambrian time.

proton A subatomic particle that contributes mass and a single positive electrical charge to an atom.

pumice A frothy volcanic glass.

P wave A compressional wave (seismic wave) in which rock vibrates parallel to the direction of wave propagation.

P-wave shadow zone The region on the Earth's surface, 103° to 142° away from an earthquake epicentre, in which P waves from the earthquake are absent.

pyroclast Fragment of rock formed by volcanic explosion.

pyroclastic cone A volcano constructed of loose rock fragments ejected from a central vent.

pyroclastic flow Turbulent mixture of pyroclastics and gases flowing down the flank of a volcano.

pyroxene group Mineral group, all members of which are single-chain silicates.

Q

quartzite A rock composed of sand-sized grains of quartz that have been welded together during metamorphism.

Quaternary Period The youngest geologic period; includes the present time.

quickclay Clays that are originally deposited in saline waters; the salt forms an important part of the clay structure.

R

radial pattern A drainage pattern in which streams diverge outward like spokes of a wheel.

radioactive decay The spontaneous nuclear disintegration of certain isotopes.

rain shadow A region on the downwind side of mountains that has little or no rain because of the loss of moisture on the upwind side of the mountains.

Rayleigh waves A type of surface seismic wave that behaves like a rolling ocean wave and causes the ground to move in an elliptical path.

receding glacier A glacier with a negative budget, which causes the glacier to grow smaller as its edges melt back.

Recent (Holocene) Epoch The present epoch of the Quaternary Period.

recharge The addition of new water to an aquifer or to the zone of saturation.

recrystallization The development of new crystals in a rock, often of the same composition as the original grains.

rectangular pattern A drainage pattern in which tributaries of a river change direction and join one another at right angles.

recumbent fold A fold overturned to such an extent that the limbs are essentially horizontal.

reef A resistant ridge of calcium carbonate formed on the sea floor by corals and coralline algae.

regional metamorphism Metamorphism that takes place at considerable depth underground.

regression The fall of sea level; as water depths reduce, the shoreline moves basinward.

relative time The sequence in which events took place (not measured in time units).

relief The vertical distance between points on the Earth's surface.

reserves The discovered deposits of a geologic material that are economically and legally feasible to recover under present circumstances.

reservoir rock A rock that is sufficiently porous and permeable to store and transmit petroleum.

residual soil Soil that develops directly from weathering of the rock below.

resources The total amount of a geologic material in all its deposits, discovered and undiscovered (*see* reserves).

reverse fault A fault in which the hanging-wall block moved up relative to the footwall block.

reversed polarity Periods when the lines of magnetic force run from the south pole to the north pole and compass needles point to the south.

rhyolite A fine-grained, felsic, igneous rock made up mostly of feldspar and quartz.

Richter scale A numerical scale of earthquake magnitudes.

rift valley A tensional valley bounded by normal faults. Rift valleys are found at diverging plate boundaries on continents and along the crest of the mid-oceanic ridge.

right-lateral fault A strike-slip fault in which the block seen across the fault appears displaced to the right.

rigid zone Upper part of a glacier in which there is no plastic flow.

rip current Narrow currents that flow straight out to sea in the surf zone, returning water seaward that has been pushed ashore by breaking waves.

ripple mark Any of the small ridges formed on sediment surfaces exposed to moving wind or water. The ridges form perpendicularly to the motion.

rock Naturally formed, consolidated material composed of grains of one or more minerals. (There are a few exceptions to this definition.)

rock avalanche A very rapidly moving, turbulent mass of broken-up bedrock.

rock-basin lake A lake occupying a depression caused by glacial erosion of bedrock.

rock cycle A theoretical concept relating tectonism, erosion, and various rock-forming processes to the common rock types.

rockfall Rock falling freely or bouncing down a cliff.

rock flour A powder of fine fragments of rock produced by glacial abrasion.

rockslide Rapid sliding of a mass of bedrock along an inclined surface of weakness.

Rodinia A supercontinent that formed about 1,000 million years ago.

rotational slide In mass wasting, movement along a curved surface in which the upper part moves vertically downward while the lower part moves outward. Also called a *slump*.

rounding The grinding away of sharp edges and corners of rock fragments during transportation.

S

saltation A mode of transport that carries sediment downcurrent in a series of short leaps or bounces.

sand Sediment composed of particles with a diameter between 1/16 mm and 2 mm.

sand dune A mound of loose sand grains heaped up by the wind.

sandstone A medium-grained sedimentary rock (grains between 1/16 mm and 2 mm) formed by the cementation of sand grains.

saturated zone A subsurface zone in which all rock openings are filled with water.

scale The relationship between distance on a map and the distance on the terrain being represented by that map.

schist A metamorphic rock characterized by coarse-grained minerals oriented approximately parallel.

schistose The texture of a rock in which visible platy or needle-shaped minerals have grown essentially parallel to each other under the influence of directed pressure.

scientific method A means of gaining knowledge through objective procedures.

sea cliff Steep slope that retreats inland by mass wasting as wave erosion undercuts it.

sea-floor spreading The concept that the ocean floor is moving away from the mid-oceanic ridge and across the deep ocean basin, to disappear beneath continents and island arcs.

seamount Conical mountain rising 1,000 metres or more above the sea floor.

sediment Loose, solid particles that can originate by (1) weathering and erosion of pre-existing rocks, (2) chemical precipitation from solution, usually in water, and (3) secretion by organisms.

sedimentary breccia A coarse-grained sedimentary rock (grains coarser than 2 mm) formed by the cementation of angular rubble.

sedimentary rock Rock that has formed from (1) lithification of any type of sediment, (2) precipitation from solution, or (3) consolidation of the remains of plants or animals.

sedimentary structure A feature found within sedimentary rocks, usually formed during or shortly after deposition of the sediment and before lithification.

seismic reflection The return of part of the energy of seismic waves to the Earth's surface after the waves bounce off a rock boundary.

seismic refraction The bending of seismic waves as they pass from one material to another.

seismic sea wave *See* tsunami.

seismic wave A wave of energy produced by an earthquake.

seismogram Paper record of Earth vibration.

seismograph A seismometer with a recording device that produces a permanent record of Earth motion.

shale A fine-grained sedimentary rock (grains finer than 1/16 mm in diameter) formed by the cementation of silt and clay (mud). Shale has thin layers (laminations) and an ability to split (fissility) into small chips.

shatter cone V-shaped cones that form when rocks are struck violently during a meteorite impact event.

shearing Movement in which parts of a body slide relative to one another and parallel to the forces being exerted.

shear strength In mass wasting, the resistance to movement or deformation of material.

shear stress Stress due to forces that tend to cause movement or strain parallel to the direction of the forces.

sheet joints Cracks that develop parallel to the outer surface of a large mass of expanding rock, as pressure is released during unloading.

sheet silicate structure Crystal structure in which each silica tetrahedron shares three oxygen ions.

sheetwash Water flowing down a slope in a layer.

shield volcano Broad, gently sloping cone constructed of solidified lava flows.

sidescan sonar Measures the intensity of sound reflected from the ocean floor and provides detailed images and information about sediments and bedforms on the sea floor.

silica A term used for oxygen plus silicon.

silicate A substance that contains silica as part of its chemical formula.

silicic (felsic) rock or magma Silica-rich igneous rock or magma with a relatively high content of potassium and sodium.

silicon–oxygen tetrahedron Four-sided, pyramidal object that visually represents the four oxygen atoms surrounding a silicon atom; the basic building block of silicate minerals. Also called a *silica tetrahedron* or a *silicon tetrahedron*.

sill A tabular intrusive structure concordant with the country rock.

silt Sediment composed of particles with a diameter of 1/256 to 1/16 mm.

sinkhole A closed depression found on land surfaces underlain by limestone.

slate A fine-grained rock that splits easily along flat, parallel planes.

slaty Describing a rock that splits easily along nearly flat and parallel planes.

slaty cleavage The ability of a rock to break along closely spaced parallel planes.

slide In mass wasting, movement of a relatively coherent descending mass along one or more well-defined surfaces.

slip face The steep, downwind slope of a dune; formed from loose, cascading sand that generally keeps the slope at the angle of repose (about 34°).

snow line *See* equilibrium line.

soil A layer of weathered, unconsolidated material on top of bedrock; often also defined as containing organic matter and being capable of supporting plant growth.

soil horizon Any of the layers of soil that are distinguishable by characteristic physical or chemical properties.

solution Usually slow but effective process of weathering and erosion in which rocks are dissolved by water.

sorting Process of selection and separation of sediment grains according to their grain size (or grain shape or specific gravity).

source area The locality that eroded to provide sediment to form a sedimentary rock.

source rock A rock containing organic matter that is converted to petroleum by burial and other postdepositional changes.

specific gravity The ratio of the mass of a substance to the mass of an equal volume of water, determined at a specified temperature.

speleothem Dripstone deposit of calcite that precipitate from dripping water in caves.

spheroidally weathered boulder Boulder that has been rounded by weathering from an initial blocky shape.

spit A fingerlike ridge of sediment attached to land but extending out into open water.

spring A place where water flows naturally out of rock onto the land surface.

stack A small rock island that is an erosional remnant of a headland left behind as a wave-eroded coast retreats inland.

stalactite Iciclelike pendant of dripstone formed on cave ceilings.

stalagmite Cone-shaped mass of dripstone formed on cave floors, generally directly below a stalactite.

standard geologic time scale A worldwide relative scale of geologic time divisions.

stock A small discordant pluton with an outcropping area of less than 100 square kilometres.

strain Change in size (volume) or shape of a body (or rock unit) in response to stress.

streak Colour of a pulverized substance; a useful property for mineral identification.

stream A moving body of water, confined in a channel and running downhill under the influence of gravity.

stream channel A long, narrow depression, shaped and more or less filled by a stream.

stream gradient Downhill slope of a stream's bed or the water surface, if the stream is very large.

stream terrace Steplike landform found above a stream and its flood plain.

stream velocity The speed at which water in a stream travels.

stress A force acting on a body, or rock unit, that tends to change the size or shape of that body, or rock unit. Force per unit area within a body.

striations (1) On minerals, extremely straight, parallel lines; (2) Glacial—straight scratches in rock caused by abrasion by a moving glacier.

strike The compass direction of a line formed by the intersection of an inclined plane (such as a bedding plane) with a horizontal plane.

strike-slip fault A fault in which movement is parallel to the strike of the fault surface.

strip mine A mine in which the valuable material is exposed at the surface by removing a strip of overburden.

stromatolites Mound-like organic structures formed by the trapping of sediment particles by prokaryotic bacteria. These are some of the oldest fossils on Earth.

structural basin A structure in which the beds dip toward a central point.

structural dome A structure in which beds dip away from a central point.

structural geology The branch of geology concerned with the internal structure of bedrock and the shapes, arrangement, and interrelationships of rock units.

subduction The sliding of the sea floor beneath a continent or island arc.

submarine canyon V-shaped valleys that run across the continental shelf and down the continental slope.

submarine fan Fan-shaped deposits of sediment found at the base of many submarine canyons.

subprovinces Fault-bounded subdivisions of a geological province that contain similar rock types, structures, and mineral deposits.

supercontinent A giant conglomeration of all the continents on Earth.

superposition A principle or law stating that within a sequence of undisturbed sedimentary rocks, the oldest layers are on the bottom, the youngest on the top.

surf Breaking waves.

surface wave A seismic wave that travels on the Earth's surface.

suspended load Sediment in a stream that is light enough in weight to remain lifted indefinitely above the bottom by water turbulence.

suture zone Contact zone separating large blocks of continental crust characterized by highly deformed rocks recording former continental collisions.

S wave A seismic wave propagated by a shearing motion, which causes rock to vibrate perpendicular to the direction of wave propagation.

S-wave shadow zone The region on the Earth's surface (at any distance more than 103° from an earthquake epicentre) in which S waves from the earthquake are absent.

syncline A fold in which the layered rock usually dips toward an axis.

T

talus An accumulation of broken rock at the base of a cliff.

tarn *See* rock-basin lake.

tensional stress A stress due to a force pulling away on a body.

terminus The lower edge of a glacier.

terrane (tectonostratigraphic terrane) A region in which the geology is markedly different from that in adjoining regions.

terrestrial planets Planets that are small, dense, and rocky: Mercury, Venus, Earth, and Mars.

terrigenous sediment Land-derived sediment that has found its way to the sea floor.

texture A rock's appearance with respect to the size, shape, and arrangement of its grains or other constituents.

theory of glacial ages At times in the past, colder climates prevailed during which significantly more of the land surface of the Earth was glaciated than at present.

thermokarst The subsidence and collapse of the ground surface caused by thawing of permafrost.

thrust fault A reverse fault in which the dip of the fault plane is at a low angle to horizontal.

tide Occurs when the sea level rises and falls.

tide-generating force The gravitational attraction of the moon and the sun.

till Unsorted and unlayered rock debris carried by a glacier.

till plain A thin, extensive layer or blanket of till, formed as ice melts and rock debris is deposited at the base of a glacier.

tillite Lithified till.

tombolo A bar of marine sediment connecting a former island or stack to the mainland.

traction Movement by rolling, sliding, or dragging of sediment fragments along a stream bottom.

transform fault The portion of a fracture zone between two offset segments of a mid-oceanic ridge crest.

transform plate boundary Boundary between two plates that are sliding past each other.

transgression The rise of sea level; when water depths increase the shoreline moves landward.

translational slide In mass wasting, movement of a descending mass along a plane approximately parallel to the slope of the surface.

transportation The movement of eroded particles by agents such as rivers, waves, glaciers, or wind.

transported soil Soil not formed from the local rock but from parent material brought in from some other region and deposited, usually by running water, wind, or glacial ice.

transverse dune A relatively straight, elongate dune oriented perpendicular to the wind.

travel-time curve A plot of seismic-wave arrival times against distance.

trellis pattern A drainage pattern consisting of parallel main streams with short tributaries meeting them at right angles.

tributary Small stream flowing into a large stream, adding water to the large stream.

trough (of wave) The low point of a wave.

truncated spur Triangular facet where the lower end of a ridge has been eroded by glacial ice.

tsunami Huge ocean wave produced by displacement of the sea floor; also called *seismic sea wave*.

tuff A rock formed from fine-grained pyroclastic particles (ash and dust).

turbidity current A flowing mass of sediment-laden water that is heavier than clear water and therefore flows downslope along the bottom of the sea or a lake.

U

ultramafic rock Rock composed entirely or almost entirely of ferromagnesian minerals.

unconfined aquifer A partially filled aquifer exposed to the land surface and marked by a rising and falling water table.

unconformity A surface that represents a break in the geologic record, with the rock unit immediately above it being considerably younger than the rock beneath.

uniformitarianism Principle that geologic processes operating at present are the same processes that operated in the past. The principle is stated more succinctly as "The present is the key to the past." *See* actualism.

unsaturated zone A zone above the water table that is generally unsaturated.

V

vadose zone A subsurface zone in which rock openings are generally unsaturated and filled partly with air and partly with water; above the saturated zone.

valley glacier A glacier confined to a valley. The ice flows from a higher to a lower elevation.

varve Two thin layers of sediment, one dark and the other light in colour, representing one year's deposition in a lake.

vein Narrow, root-like structure in a rock filled with minerals or other rock types.

vent The opening in the Earth's surface through which a volcanic eruption takes place.

ventifact Boulder, cobble, or pebble with flat surfaces caused by the abrasion of wind-blown sand.

vesicle A cavity in volcanic rock caused by gas in a lava.

viscosity Resistance to flow.

volcanic breccia Rock formed from large pieces of volcanic rock (cinders, blocks, bombs).

volcanic dome A steep-sided, dome- or spine-shaped mass of volcanic rock formed from viscous lava that solidifies in or immediately above a volcanic vent.

volcanic neck An intrusive structure that apparently represents magma that solidified within the throat of a volcano.

volcanism Volcanic activity, including the eruption of lava and rock fragments and gas explosions.

volcano A hill or mountain constructed by the extrusion of lava or rock fragments from a vent.

W

water table The upper surface of the zone of saturation.

wave-cut platform A horizontal bench of rock formed beneath the surf zone as a coast retreats because of wave erosion.

wave height The vertical distance between the crest (the high point of a wave) and the trough (the low point)

wavelength The horizontal distance between two wave crests (or two troughs).

wave refraction Change in direction of waves due to slowing as they enter shallow water.

weathering The group of processes that change rock at or near the Earth's surface.

well A hole, generally cylindrical and usually walled or lined with pipe, that is dug or drilled into the ground to penetrate an aquifer below the zone of saturation.

X

xenolith Fragment of rock distinct from the igneous rock in which it is enclosed.

Z

zone of ablation That portion of a glacier in which ice is lost.

zone of accumulation (1) That portion of a glacier with a perennial snow cover; (2) *See* B horizon (a soil layer).

zone of leaching The lower part of the A horizon, characterized by the downward movement of water.

INDEX

D

H

I

J

K

N

O

P

Q

R

T